Ernst Bamberg, Wilhelm Schoner
Editors

# The Sodium Pump

Structure Mechanism, Hormonal
Control and its Role in Disease

Prof. Dr.
ERNST BAMBERG
Max-Planck-Institut
für Biophysik
Kennedyallee 70
60596 Frankfurt

Prof. Dr.
WILHELM SCHONER
Institut für Biochemie
Universität Gießen
Frankfurter Str. 100
35392 Gießen

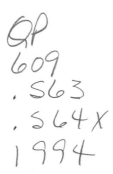

Die Deutsche Bibliothek – CIP-Einheitsaufnahme

The **sodium pump** : structure mechanism, hormonal control and
its role in disease / Ernst Bamberg ; Wilhelm Schoner (ed.). –
Darmstadt : Steinkopff ; New York : Springer, 1994
  ISBN 3-7985-0961-1 (Steinkopff) Gb.
  ISBN 0-387-91466-8 (Springer) Gb.
NE: Bamberg, Ernst [Hrsg.]

Copyright © 1994 by Dietrich Steinkopff Verlag GmbH & Co. KG, Darmstadt
Medical Editor: Jens Fabry – English Editor: James C. Willis – Production: Heinz J. Schäfer

Printed in Germany

Druck und Bindung: Druckerei Zechner, Speyer
Printed on acid-free paper

# Preface

The sodium of animal cell membranes converts the chemical energy obtained from the hydrolysis of adenosine 5′-triphosphate into a movement of the cations $Na^+$ and $K^+$ against an electrochemical gradient. The gradient is used subsequently as an energy source to drive the uptake of metabolic substrates in polar epithelial cells and to use it for purposes of communications in excitable cells. The biological importance of the sodium pump is evident from the fact that between 20–70% of the cell's metabolic energy is consumed for the pumping process. Moreover, the sodium pump is an important biological system involved in regulatory processes like the maintenance of the cells' and organism's water metabolism. It is therefore understandable that special cellular demands are handled better by special isoforms of the sodium pump, that the expression of the sodium pump and their isoforms is regulated by hormones as is the activity of the sodium pump via hormone-regulated protein kinases. Additionally, the sodium pump itself seems to be a receptor for a putative new group of hormones, the endogenous digitalis-like substances, which still have to be defined in most cases in their structure. This group of substances has its chemically well known counterpart in steroids from plant and toad origin which are generally known as "cardiac glycosides". They are in medical use since at least 200 years in medicine in the treatment of heart diseases.

Since the first description of the sodium pump's enzymological equivalent, $Na^+/K^+$-ATPase, the study of this membrane protein has attracted a still increasing number of scientists because the sodium pump is used as a model system to understand the molecular events of the conversion of chemical energy into the vectorial process of transport. Many attempts have been made to get insight into the structure of the pump, the mechanism of assembly of the subunits and the sorting of the proteins involved into the proper location within the cytoskeleton, the organisation of the gene structure and its regulation of expression. Medical aspects, moreover, like the intentions to understand the role of the pump in sodium, potassium and water metabolism as well as the activity regulation by natriuretic homones and cardiocative steroids and the therapeutic difficulty in handling these hormones and drugs in disease give still new impacts to the field.

To facilitate the exchange of information, seven international conferences have been held on the subject, the first in 1973 in New York, sponsored by the New York Academy of Sciences; the second in 1978 in connection with Aarhus University, Denmark. Since then it has been a tradition to hold a conference every third year: in 1981 at Yale University, New Haven, Connecticut; in 1984 in Cambridge University, England; in 1987 in Fuglsø again in connection with Aarhus University, Denmark; in 1990 at the Marine Biological Laboratory in Woods Hole, Massachusetts as a meeting of the Society of General Physiologists. The 7th International Conference on THE SODIUM PUMP in 1993 was held from

September 5–11 in Todtmoos, Germany. This conference with almost 250 participants from all over the world was organized as the 105th Conference of the Gesellschaft für Biologische Chemie and was supported financially primarily by the Deutsche Forschungsgemeinschaft, Bonn–Bad Godesberg and the Max-Planck-Gesellschaft zur Förderung der Wissenschaften, München. The generous support by the Deutsche Forschungsgemeinschaft helped a great number of scientists from eastern european countries and the former Soviet Union to attend and to contribute to the conference. The financial support by funds of the Oswalt Stiftung des Instituts für physikalische Grundlagen der Medizin, the Gesellschaft für Biologische Chemie, the Giessener Hochschulgesellschaft and the pharmaceutical companies Astra Hässle AB, Mölndal/Sweden, Bayer AG, Wuppertal/Germany, Knoll AG, Ludwigshafen/Germany greatly smoothened problems of organization. On behalf of the organizers, we thank all who took care of the practical arrangements, and those who helped us to run the conference.

The proceedings of all conferences on THE SODIUM PUMP have developed to be a major source of information and reference for people interested in reviews on all aspects of the sodium pump as well for the researcher in the field on detailed questions. With this knowledge in mind, the editors tried to assemble scientists in Todtmoos to review the status of the knowledge in their special field of scientific interest. As will be evident, major progress has been made in the last 3 years in understanding of a number of issues of general importance. The reviews on almost all fields of the topic are commented and supported by extended abstracts of the scientific contributions presented in poster form during the whole week of the conference. They attracted a continuous stimulus for exchange of ideas and further research. The editors hope that reading of reviews and short reports will stimulate further progress in our understanding of structure, mechanism and hormonal control as well as of biological and medical significance of the sodium pump of the mammalian cell.

Ernst Bamberg
Wilhelm Schoner

# Contents

## Gene organization, analysis of the sodium pump by molecular biology

## Cell Biology

## Structure and reaction mechanisms of the sodium pump

XII

## Electrogenicity and voltage dependence

## Spectroscopy as a tool in Na$^+$/K$^+$-ATPase

# Hormonal regulation of the sodium pump

# Endogenous digitalis-like substances

# Role of P-type ATPases in disease

# Transcription factors regulating the Na$^+$/K$^+$-ATPase genes

K. Kawakami, Y. Suzuki-Yagawa, Y, Watanabe, K. Ikeda, K. Nagano

Department of Biology, Jichi Medical School, Yakushiji, Minamikawachi, Kawachi, Tochigi, 329-04 Japan

## Introduction

Na$^+$/K$^+$-ATPase is composed of two subunits named α and β. Each subunit has three isoforms of α1, α2 and α3 for α, β1, β2 and β3 for β (4,11,18,25). Expression of each isoform gene is regulated tissue specifically and developmentally (5,20,21). Each isoform has its unique affinities for ligands including Na$^+$, K$^+$ and ATP and for an inhibitor ouabain (6,9,10). It is believed that any of the α isoforms could be assembled with any of the β isoforms, suggesting that the gene expression of each isoform gene is involved in the production of functional difference of the assembled enzyme. Each isoform gene has been isolated from various species and characterized (12,15,17,26,27, 28,30,32).

To understand the molecular mechanism of gene regulation, it is essential to identify *cis*-acting elements and *trans*-acting factors of each gene and analyze how they regulate the expression of the gene. We have initiated a systematic analysis of *cis*-elements in three isoform genes of α1, α2 and β2 with use of several cultured cell lines and astrocyte primary cultured cells (13,14,29) . Some of the factors binding to the elements were identified by gel retardation analyses and their binding regions were identified by DNase I footprinting and methylation interference analyses.

### *Cis*-elements and *trans*-acting factors of isoform genes

Housekeeping α1 subunit gene is composed of *cis*-elements to which multiple factors bind. The α1 subunit gene is expressed virtually in all tissues, although its expression level is different depending on tissues and developmental stages (5,21). To understand the mechanism how the gene is expressed in a wide variety of tissues, we analyzed the *cis*-elements by transient transfection assay using five cell lines (MDCK, HepG2, L6, 3Y1 and B103) from different tissue origins. The 5' flanking sequence of the gene was fused to the reporter luciferase gene. Analysis of 5' sequential deletion mutations revealed that the region between -102 to -61 was a positive regulatory element common to all the cell types examined. The region between -61 to -49 containing the Sp1 consensus sequence also functions as a positive regulatory element. The region between -102 to -61 enhanced the luciferase activity 5 to 10 fold depending on cell types. We named the region ARE (*Atp1a1* Regulatory Element). We could not find out any known consensus elements except the ATF binding sequence (GTGACGT) from -71 to -65 to which bacterially expressed ATF1 (33) can bind. Linker substitution mutations possessing *Bgl*II linkers at positions -73 to -64 (LS1), at -87 to -78 (LS2) and at -98 to -89 (LS3) in ARE did not fully impair the positive regulatory function, suggesting that there are multiple positive and/or negative regulatory elements in the region. To identify transacting factors binding to ARE, we prepared nuclear extracts from MDCK and B103 cells. By gel retardation analysis using the DNA fragment from -102 to -58 as a probe,

1

three complexes (C1, C2 and C3) were formed in the MDCK nuclear extract, while two complexes (C1 and C2) were formed in B103 nuclear extract. C1 and C2 were observed in all cell types examined (B103, MDCK, HeLa, Namalva, L6 and 3T3), while C3 complex was observed only in restricted cell types (MDCK, HeLa, L6 and 3T3).

The binding regions of C1, C2 and C3 were analyzed by gel retardation competition assays. LS1 fragment competed with the formation of all three complexes in both extracts. LS2 fragment competed with none of the complexes. LS3 fragment competed with the C1 and C2 both in MDCK and B103 nuclear extracts but had only a marginal effect on the formation of the C3. These findings were confirmed by DNase I footprinting and methylation interference experiments. The summary is shown in Fig. 1. The protected regions were identical between C1 and C2. They were from -88 to -69 for upper strand and from -90 to -71 for lower strand. Methylation of eight guanine bases (indicated by closed triangles in Fig. 1) interfered formation of the C1 and C2 complexes. The protected region of C3 was from -88 to -79 for upper strand and -90 to -71 for lower strand. Methylation of eight guanine bases (four of them different in position in C1 and C2; indicated by closed triangles in Fig. 1) interfered with C3 complex formation. These results clearly indicated that the binding regions of C1 and C2 overlap with that of C3 but are distinct from it. By gel retardation competition assays, the sequence required for formation of the C1 and C2 complexes on ARE was revealed to be GGTTGCNNNGG. This sequence is also found in several other genes involved in energy metabolism including mouse mitochondrial malate dehydrogenase (24) and rat pyruvate kinase (3), suggesting that the sequence might be involved in the common regulation among energy metabolizing enzyme genes.

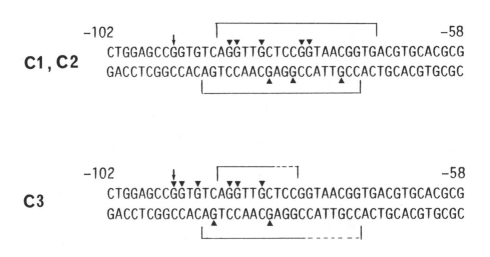

**Figure 1.** Summary of DNase I footprinting and methylation interference analyses of MDCK nuclear extract with α1 subunit gene promoter probe (from -102 to -58). The protected regions are indicated. Hypersensitive sites are shown by arrows. The G residues that interfered with are marked by closed triangles.

C1, C2 and C3 are altered not only with cell types but also with cell cycle.

To examine whether the expression of the α1 subunit gene is modulated during cell cycle, we performed northern blot analysis using total RNA from the synchronized cells with rat $Na^+/K^+$-ATPase α1 subunit cDNA as a probe. mRNA in Go state was repressed compared to the level of growing cells, while after serum stimulation, the mRNA levels were gradually increased and remained constant until 30 h. This raised the possibility that the change of ARE binding factors correlate with the change in transcription level of the α1 subunit gene, although we have not excluded the possibility that the stability of the mRNA changed.

In looking for the binding factors to ARE during cell cycle, we prepared nuclear extracts from BALB/c-3T3 cells of various growth stages. Cells were synchronized by serum starvation and serum stimulation. In growing cells, the major gel retardation complexes identified with the ARE probe were similar to those observed in MDCK nuclear extract (compare C1, C2 and C3 in Fig. 2, lanes 1 and 5). The properties of complexes were similar to those from MDCK nuclear extract on the basis of specific competition by three linker substitution mutations (Fig. 2, lanes 1-4 and 5-8). When cells entered into Go state, the relative quantity of C1 decreased and the relative mobility of C3 increased to become C3* (compare lanes 5 and 9). The specific competition with three linker substitution mutations was identical to that of growing cells (lanes 10-12). After 2h of

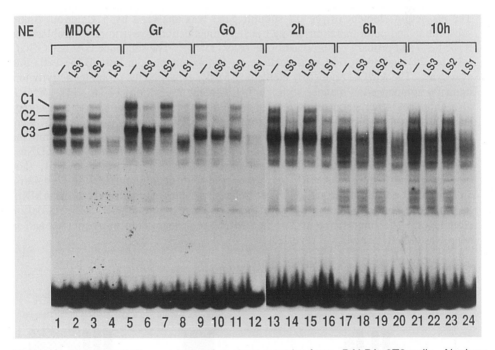

**Figure 2.** Gel retardation analysis of nuclear extracts from BALB/c-3T3 cells. Nuclear extracts from MDCK cells (lanes 1-4), BALB/c-3T3 growing cells (lanes 5-8), cells of Go state (lanes 9-12), 2h after serum stimulation (lanes 13-16), 6h after stimulation (lanes 17-20) and 10h after stimulation (lanes 21-24) were analyzed with ARE probe. Competitors added are shown above the lanes. The positions of the complexes C1, C2 and C3 are indicated.

serum stimulation, the competition pattern of the gel retardation complexes was similar to that of Go (lanes 13-16). After 6h, C1 disappeared and C3* became to be competed with by LS3 (lanes 17-20). Furthermore, several faster migrating complexes that are competed by LS1 were observed. The results are essentially the same as those after 10h (lanes 21-24). These results imply that different sets of transacting factors become involved in the transcription of the gene in late G1 and S phases.

_Cis_-element of α2 subunit gene in muscle cell is composed of Sp1 binding elements and E box (8). To analyze the _cis_-acting elements of the gene for the expression in muscle, we searched for _cis_-acting elements of the α2 subunit gene using rat L6 myoblast cells by transient transfection assays. By 5' deletion mutation analysis, the region between the positions -175 to -108 was identified as a positive regulatory region. In the region, the distal E box (-144 to -139) acts as a negative regulatory element and the Sp1 consensus sequence (-123 to -118) and the GGGAGG sequence (-114 to -109) act as positive regulatory elements. Gel retardation analysis revealed that the binding factors are an E-box binding protein and Sp1. T4 DNA polymerase footprinting analysis revealed that there are three overlapping Sp1 binding sites in the region and that Sp1 binds to one of the three sites in a mutually exclusive manner (see details; Ikeda, K. et al. in this volume.)

Astrocyte specific β2 subunit gene is regulated by AMRE to which Sp1 binds.
The _cis_-acting element of the β2 subunit gene was analyzed by 5' deletion, linker substitution and point mutations using transient transfection assays in rat neuroblastoma B103 cells and primary cultured astrocyte cells (13,14). The sequence GAGGCGGGG (-87 to -79) was identified as a positive regulatory element in both B103 cells and astrocytes. The AMOG regulatory element (AMRE) enhanced the promoter activity in a mutually compensating manner with the Sp1 element at position -147 to -142. AMRE also enhances other gene promoters such as myelin basic protein gene and herpes simplex virus thymidine kinase gene promoters. Binding factors to the AMRE were identified as Sp1 from the following observations using nuclear extracts from the astrocytes and B103 cells. The interaction of the factors with AMRE analyzed by DNase I footprinting and methylation interference analyses was similar to that of Sp1. The binding of the factors to AMRE was competed with an oligonucleotide containing authentic Sp1 consensus sequence. The bacterially expressed Sp1 binds to the AMRE. Sp1-specific antibody interfered with the formation of the AMRE gel retardation complexes. The importance of Sp1 in AMOG gene regulation in astrocytes and in neuroblastoma B103 cells was revealed.

**Transcription factors of α1 subunit gene**

**Figure 3.** Diagram of the predicted AREB6 protein. Six $C_2$-$H_2$ fingers are represented by filled boxes and one $C_2$-HC finger is represented by shaded box. Glutamic acid-rich region and Ser/Thr rich region are also indicated.

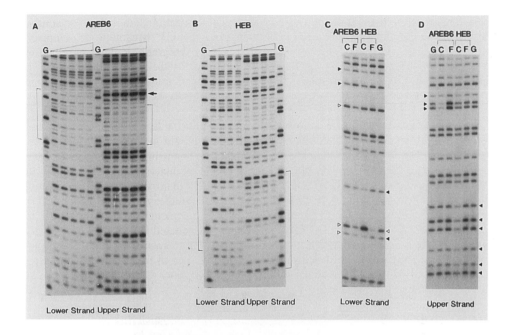

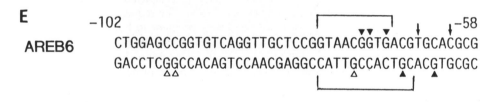

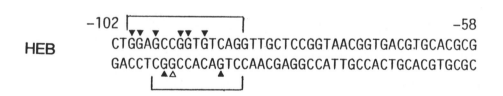

**Figure 4.** DNase I footprinting and methylation interference analyses of the bacterially expressed AREB6 and HEB protein. (A) and (B): DNase I footprint of AREB6 and HEB. Increasing amount of the purified GST-AREB6 fusion protein (A) or GST-HEB fusion protein (B) was added. Protected regions are shown. Hypersensitive sites are shown by arrows. Maxam-Gilbert G residues are in the adjacent lanes. (C) and (D): Methylation interference analysis. The G residues that are interfered with are marked by close triangles and those that are enhanced are marked by open triangles. (E): Summary of DNase I footprinting and methylation interference experiments.

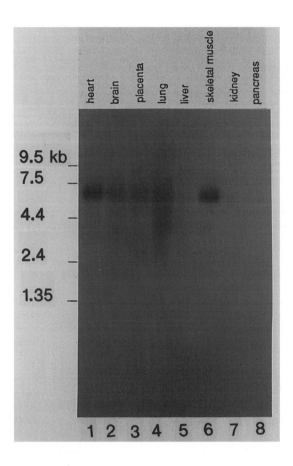

**Figure 5.** Northern blot analysis of AREB6 mRNA in various human tissues. PolyA$^+$ RNA from human heart (lane 1), brain (lane 2), placenta (lane 3), lung (lane 4), liver (lane 5), skeletal muscle (lane 6), kidney (lane 7) and pancreas (lane 8) were applied. Positions of RNA size markers are shown.

To identify the binding proteins to ARE of the $\alpha$1 subunit gene, we screened HeLa cell expression library with a DNA fragment containing ARE sequence as a probe. We obtained two cDNA clones encoding specific DNA binding proteins to ARE. One of the clones was HEB (7) and the other was AREB6, an extended clone of Nil-2-a, which was originally isolated as a negative regulator of IL-2 gene (31). The nucleotide sequence of AREB6 contains one long open reading frame encoding 1124 amino acids. The deduced amino acid sequence revealed that it contains seven putative zinc-finger motifs clustered as four and three. They are $C_2$-$H_2$ type zinc fingers (19) except the fourth finger that is a $C_2$-HC type (16). A glutamic acid-rich region is found at the C-terminal portion, and a Ser/Thr rich region is observed between the two zinc fingers. Both of the regions might act as transactivation domains. A diagram of AREB6 protein is shown in Fig. 3.

The binding region of the AREB6 and HEB proteins to the ARE were analyzed by DNase I footprinting and methylation interference analyses using GST-fusion proteins that were purified by a Glutathion Sepharose column. The results are summarized in Fig 4. The residues from -100 to -87 of the upper strand and -96 to -87

6

of the lower strand were protected from digestion by GST-HEB fusion protein. The residues from -77 to -68 of the upper strand and from -77 to -65 of the lower strand were protected from digestion by GST-AREB6 fusion protein. Methylation of eight guanine residues interfered (indicated by closed triangles in Fig. 4) and one guanine enhanced (indicated by open triangle) the formation of the gel retardation complex with GST-HEB fusion protein, while five guanines interfered (closed triangles) and three guanines enhanced (open triangles) the formation of the complex with GST-AREB6 fusion protein. The protected region from DNase I and interfered G residues are distinct from those obtained by C1, C2 and C3 observed in MDCK and B103 cell nuclear extracts (compare Fig. 2) (29). These results strongly suggest that the isolated cDNA clones code for proteins distinct from factors in C1, C2 and C3.

The tissue distribution of AREB6 mRNA was analyzed by northern blot analysis of polyA$^+$ RNA from human tissues. mRNAs were abundant in heart and skeletal muscle, moderately abundant in brain, placenta and lung, while scarcely detectable in liver, kidney and pancreas (Fig. 5).

To examine whether AREB6 and HEB modulate the $\alpha 1$ subunit gene expression, we performed co-transfection experiments using three cell lines. 3 or 6 $\mu$g of the AREB6 expression plasmid enhanced the promoter activity more than 2 fold in MDCK cells while 9 $\mu$g gave no effect. In 3Y1 cells, AREB6 repressed the promoter activity in a dose dependent manner. In COS7 cells, no effect of AREB6 was observed. HEB expression plasmid enhanced the expression about 5 fold in COS7 cells, while repression of 23% in MDCK cells and moderate repression of 37 % in 3Y1 cells were observed. Results are summarized in Table I. These results suggest that AREB6 and HEB can enhance or repress the $\alpha 1$ subunit gene expression depending on cell types and on the quantity of AREB6 and HEB.

Table I. Effect of HEB and AREB6 on the *Atp1a1* gene expression

| Reporter gene (9µg) | HEB ($\mu$g) | AREB6 ($\mu$g) | Relative Luciferase Act.[a] | | |
|---|---|---|---|---|---|
| | | | COS7 | MDCK | 3Y1 |
| pA1U-102LF | 0 | | 100 | 100 | 100 |
| | 3 | | 103±31 | 239±58 | 97±2 |
| | 6 | | 117±9 | 206±47 | 82±1 |
| | 9 | | 124±7 | 84±29 | 34±7 |
| | | 3 | 224±7 | 56±26 | 78±6 |
| | | 6 | 357±7 | 37±43 | 48±2 |
| | | 9 | 509±7 | 23±10 | 37±2 |

[a]9µg of pCDM8 (expression vector) was co-transfected and relative luciferase activity was normalized as 100. ND: Not determined. Averages of two to six independent transfections and standard deviations are shown.

## Conclusion

The involvement of the transcription factor Sp1 is a common feature among the genes of $\alpha 1$, $\alpha 2$ and $\beta 2$. It has been thought that Sp1 is ubiquitously expressed in all mammalian cell nuclei and involved in the transcription of genes expressed in all cell types (1). Recently, it is suggested that Sp1 is likely to play an important regulatory role in cellular process during development and differentiation (22,23) and in the determination of tissue specificity (2) besides its general role in the transcription of housekeeping genes.

At least 6 distinct transacting factors (C1, C2 and C3 factors, ATF1, AREB6 and HEB) bind to ARE of the α1 subunit gene and the binding regions of some of the factors overlap each other. Sp1 binding sequence exists just downstream of ARE. C1 and C2 factors, Sp1 and ATF1 are ubiquitous, while C3 factor, AREB6 and HEB are tissue specific (7). This suggests that the housekeeping α1 subunit gene is regulated by different sets of transcription factors depending on cell types and that complex interactions of the factors are involved in the regulation (Fig. 6). Cell cycle analysis further supports the notion that different repertoires of the factors are involved in the regulation of the gene in different phases of the cell cycle. *In vitro* transcription system using purified upstream transcription factors and basic transcription factors will reveal their respective involvement and mode of action in the housekeeping α1 subunit gene expression.

**A conceptual scheme of ARE (ATP1A1 regulatory element) and its binding factors.**

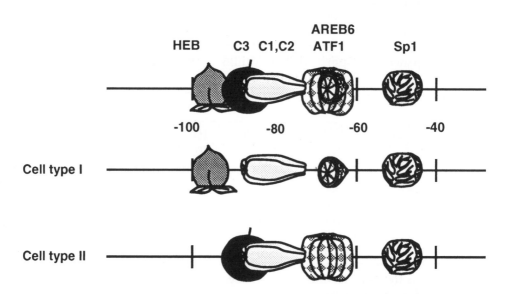

**Figure 6.** A conceptual scheme of ARE (*Atp1a1* regulatory element) and its binding factors. Transcription factors binding to the ARE region were symbolized as fruits and vegetables. Ubiquitous factors are shown as pepper, lemon and cabbage, which are available throughout the year. Tissue specific factors are indicated as peach, tomato and pumpkin, which are usually seasonal.

*Acknowledgments*

We thank Drs. Ken Yanagisawa and Shin-ichi Tominaga for preparing synchronized BALB/c-3T3 cells and RNA from the cells and Yasutaka Hirayama for sequencing AREB6 cDNA clone. This work was supported by grants from Ministry of Education, Science and Culture of Japan.

**References**

1.  Briggs MR, Kadonaga JT, Bell SP, Tjian R (1986) Purification and biochemical characterization of the promoter-specific transcription factor, Sp1. Science 234:47-52
2.  Chen H-M, Pahl HL, Scheibe RJ, Zhang D-E, Tenen DG (1993) The Sp1 transcription factor binds the CD11b promoter specifically in myeloid cells in vivo and is essential for myeloid-specific promoter activity. J Bio Chem 268:8230-8239
3.  Cognet M, Lone YC, Vaulont S, Kahn A, Marie J (1987) Structure of the rat L-type pyruvate kinase gene. J Mol Bio 196:11-25
4.  Good PJ, Richter K, Dawid IB (1990) A nervous system-specific isotype of the $\beta$ subunit of $Na^+,K^+$-ATPase expressed during early development of *Xenopus laevis*. Proc Natl Acad Sci USA 87:9088-9092
5.  Herrera VLM, Emanuel JR, Ruiz-Opazo N, Levenson R, Nadal-Ginard B (1987) Three differentially expressed Na,K-ATPase $\alpha$ subunit isoforms: Structural and functional implications. J Cell Bio 105:1855-1865
6.  Horisberger J-D, Jaunin P, Good PJ, Rossier BC, Geering K (1991) Coexpression of $\alpha_1$ with putative $\beta_3$ subunits results in functional $Na^+/K^+$ pumps in *Xenopus* oocytes. Proc Natl Acad Sci USA 88:8397-8400
7.  Hu J-S, Olson EN, Kingston RE (1992) HEB, a helix-loop-helix protein related to E2A and ITF2 that can modulate the DNA-binding ability of myogenic regulatory factors. Mol Cell Bio 12:1031-1042
8.  Ikeda K, Nagano K, Kawakami K (1993) Anomalous interaction of Sp1 and specific binding of an E-box binding protein with the regulatory elements of the Na,K-ATPase $\alpha2$ subunit gene promoter. Eur J Biochem in press:
9.  Jaisser F, Canessa CM, Horisberger J-D, Rossier BC (1992) Primary sequence and functional expression of a novel ouabain-resistant Na,K-ATPase. J Bio Chem 267:16895-16903
10. Jewell EA, Lingrel JB (1991) Comparison of the substrate dependence properties of the rat Na,K-ATPase $\alpha1$, $\alpha2$, and $\alpha3$ isoforms expressed in HeLa cells. J Bio Chem 266:16925-16930
11. Kawakami K, Nojima H, Ohta T, Nagano K (1986) Molecular cloning and sequence analysis of human Na,K-ATPase $\beta$-subunit. Nucl Acids Res 14:2833-2844
12. Kawakami K, Okamoto H, Yagawa Y, Nagano K (1990) Regulation of $Na^+,K^+$-ATPases I. Cloning and analysis of the 5'-flanking region of the rat *NKAB2* gene encoding the $\beta2$ subunit. Gene 91:271-274
13. Kawakami K, Suzuki-Yagawa Y, Watanabe Y, Nagano K (1992) Identification and characterization of the *cis*-elements regulating the rat AMOG (Adhesion Molecule on Glia)/Na,K-ATPase $\beta2$ subunit gene. J Biochem 111:515-522
14. Kawakami K, Watanabe Y, Araki M, Nagano K (1993) Sp1 binds to the Adhesion Molecule on Glia regulatory element that functions as a positive transcription regulatory element in astrocytes. J Neurosci Res 35:138-146

15. Kawakami K, Yagawa Y, Nagano K (1990) Regulation of $Na^+,K^+$-ATPases I. Cloning and analysis of the 5'-flanking region of the rat *NKAA2* gene encoding the $\alpha2$ subunit. Gene 91:267-270

16. Kim JA, Hudson LD (1992) Novel member of the zinc finger superfamily: A $C_2$-HC finger that recognizes a glia-specific gene. Mol Cell Bio 12:5632-5639

17. Lane LK, Shull MM, Whitmer KR, Lingrel JB (1989) Characterization of two genes for the human Na,K-ATPase $\beta$ subunit. Genomics 5:445-453

18. Martin-Vasallo P, Dackowski W, Emanuel JR, Levenson R (1989) Identification of a putative isoform of the Na,K-ATPase $\beta$ subunit. J Bio Chem 264:4613-4618

19. Miller JA, McLachlan AD, Klug A (1985) Repetitive zinc-binding domains in the protein transcription factor TFIIIA from xenopus oocytes. EMBO J 4:1609-1614

20. Orlowski J, Lingrel JB (1988) Differential expression of the Na,K-ATPase $\alpha1$ and $\alpha2$ subunit genes in a murine myogenic cell line. J Bio Chem 263:17817-17821

21. Orlowski J, Lingrel JB (1988) Tissue-specific and developmental regulation of rat Na,K-ATPase catalytic $\alpha$ isoform and $\beta$ subunit mRNAs. J Bio Chem 263:10436-10442

22. Robidoux S, Gosselin P, Harvey M, Leclerc S, Guerin SL (1992) Transcription of the mouse secretary protease inhibitor p12 gene is activated by the developmentally regulated positive transcription factor Sp1. Mol Cell Biol 12:3796-3806

23. Saffer JD, Jackson SP, Annarella MB (1991) Developmental expression of Sp1 in the mouse. Mol Cell Biol 11:2189-2199

24. Setoyama C, Ding S, Choudhury K, Joh T, Takeshima H, Tsuzuki T, Shimada K (1990) Regulatory regions of the mitochondrial and cytosolic isoenzyme genes participating in the malate-aspartate shuttle. J Bio Chem 265:1293-1299

25. Shull GE, Greeb J, Lingrel JB (1986) Molecular cloning of three distinct forms of the $Na^+, K^+$-ATPase $\alpha$-subunit from rat brain. Biochemistry 25:8125-8132

26. Shull MM, Pugh DG, Lingrel JB (1989) Characterization of the human Na,K-ATPase $\alpha2$ gene and identification of intragenic restriction fragment length polymorphisms. J Bio Chem 264:17532-17543

27. Shull MM, Pugh DG, Lingrel JB (1990) The human Na,K-ATPase $\alpha1$ gene: Characterization of the 5'-flanking region and identification of a restriction fragment length polymorphism. Genomics 8:451-460

28. Shyjan AW, Canfiele VA, Levenson R (1991) Evolution of the Na,K- and H,K-ATPase $\beta$ subunit gene family: Structure of the murine Na,K-ATPase $\beta$ 2 subunit gene. Genomics 11:435-442

29. Suzuki-Yagawa Y, Kawakami K, Nagano K (1992) Housekeeping Na,K-ATPase $\alpha1$ subunit gene promoter is composed of multiple *cis* elements to which common and cell type-specific factors bind. Mol Cell Bio 12:4046-4055

30. Takeyasu K, Hamrick M, Barnstein AM, Fambrough DM (1993) Structural analysis and expression of a chromosomal gene encoding an avian $Na^+/K^+$-ATPase $\beta1$-subunit. Biochim Biopys Acta 1172:212-216

31. Williams TM, Moolten D, Burlein J, Romano J, Bhaerman R, Godillot A, Mellon M, Raucher III FJ, Kant JA (1991) Identification of a zinc finger proteinthat inhibits IL-2 gene expression. Science 254:1791-1794

32. Yagawa Y, Kawakami K, Nagano K (1990) Cloning and analysis of the 5'-flanking region of rat $Na^+/K^+$-ATPase $\alpha$ 1 subunit gene. Biochim Biophys Acta 1049:286-292

33. Yoshimura T, Fujisawa J, Yoshida M (1990) Multiple cDNA clones encoding nuclear proteins that bind to the tax-dependent enhancer of HTLV-1: all contain a leucine zipper structure and basic amino acid domain. EMBO J 9:2537-2542

# Expression of Functional $Na^+/K^+$-ATPase in Yeast

Robert A. Farley[†≠], Kurt A. Eakle[†], Georgios Scheiner-Bobis[†], and Kena Wang[†]

[†]Department of Physiology & Biophysics, and [≠]Department of Biochemistry, University of Southern California School of Medicine, 2025 Zonal Avenue, Los Angeles, CA 90033

## Introduction

Investigations of the structure and mechanism of the $Na^+/K^+$-ATPase (sodium pump) have been greatly facilitated by the determination of the amino acid sequences of the $\alpha$ and $\beta$ subunits of the enzyme using recombinant DNA methodologies (13,18-20). The availability of cloned DNAs for different isoforms of each subunit has made it possible for investigators to transfect these DNAs into cells, either in their unmodified forms or in forms containing specific mutations, for subsequent analysis of the structural basis of pump activity. Several heterologous expression systems for $Na^+/K^+$-ATPase have been examined by different investigators, and certain experimental advantages may be attributed to each one of these. Because all animal cells contain endogenous $Na^+/K^+$-ATPase molecules that might contribute to the measured activity in the heterologous expression system, however, the choice of the most appropriate expression cell is not obvious. We had previously observed that the ATPase activity measured in membranes from the yeast *Saccharomyces cerevisiae* was not inhibited by ouabain and that no high-affinity binding sites for [$^3$H]ouabain could be detected in yeast membranes. These observations indicated that yeast cells do not contain endogenous $Na^+/K^+$-ATPase. Therefore, in order to avoid potential problems that may arise from the heterologous expression of the $Na^+/K^+$-ATPase in the presence of endogenous pump subunits, an expression system for $Na^+/K^+$-ATPase was developed using the yeast *Saccharomyces cerevisiae*.

There are several potential advantages for the expression of $Na^+/K^+$-ATPase in yeast. As indicated above, the most significant advantage is the absence of endogenous sodium pump molecules in the yeast cell. The activity of the pumps expressed in yeast can be analyzed in the absence of any background activity. An additional advantage for the expression of sodium pumps in yeast is the availability of a variety of yeast cloning vectors containing different auxotrophic markers and different promoters. Yeast strains characterized by different phenotypes are also available, and yeast genetics can be used to manipulate the yeast genome in ways that are favorable for certain experiments. This has proven to be useful on several occassions. After transfection of plasmid DNA, yeast cells are stably transformed and cells containing the expression plasmids can easily be selected by growth on nutrient-deficient media. The cost of maintaining yeast cultures is also very low. During the development of the expression system for $Na^+/K^+$-ATPase in yeast, it also became apparent that there are certain problems that are associated with the

11

expression of Na$^+$/K$^+$-ATPase in this organism. These problems will be discussed at the end of this chapter.

**Experimental Procedures**

*Plasmid construction and cell growth* - The construction of plasmids for the expression of Na$^+$/K$^+$-ATPase subunits in yeast, transformation of yeast cells, cell growth, and preparation of samples for analysis of Na$^+$/K$^+$-ATPase activities have been described previously (2,10,17). Two expression strategies have been used. In one (10), a large plasmid (pCGY1406$\alpha\beta$) with cDNA encoding the $\alpha$1 subunit of sheep Na$^+$/K$^+$-ATPase (20) and the $\beta$1 subunit of dog Na$^+$/K$^+$-ATPase was used to constitutively express both subunits in the yeast cell at the same time. In more recent experiments (2,3), separate plasmids have been used to direct the synthesis of the $\alpha$ subunit and $\beta$ subunit polypeptides. In these experiments, the $\alpha$ subunit is constitutively expressed under the control of the yeast phosphoglycerate kinase (PGK) promoter on a plasmid derived from YEp1PT (8), whereas the $\beta$ subunit expression is under the control of the inducible yeast GAL1 promoter on plasmid pG1T (2). This promoter is active when cells are grown using galactose as the carbon source, and is rapidly repressed when cells are grown in the presence of glucose.

*Assays of Na$^+$/K$^+$-ATPase enzymatic activities* - Procedures have previously been described for the measurement in yeast of the ouabain-sensitive hydrolysis of ATP and *p*-nitrophenyl phosphate (10,21), high-affinity binding of [$^3$H]ouabain (3,10), and $^{86}$Rb uptake (17). It is important to note that because of the relatively low yield of heterologously expressed Na$^+$/K$^+$-ATPase molecules in yeast, it has been necessary to extract yeast microsomal membranes with SDS (11) in order to measure ouabain-sensitive ATPase or nitrophenylphosphatase activities. Unlike kidney membranes, extraction of yeast membranes with SDS does not purify the sodium pumps to homogeneity, but only enriches the membranes for this protein. Including inhibitors of other ATPases in the assay buffer, such as NaN$_3$, increases the percentage of total ATPase activity in these membranes that is sensitive to ouabain. In order to measure $^{86}$Rb uptake into yeast cells (17) it has been necessary to use a yeast strain in which the high-affinity potassium transporter, encoded by the TRK1 gene, has been inactivated (6).

**Results**

*Expression of functional Na$^+$/K$^+$-ATPase in yeast* - Immunoprecipitation of [$^{35}$S]methionine-labeled yeast cells after transfection with plasmids encoding Na$^+$/K$^+$-ATPase subunits demonstrated that each polypeptide was synthesized by the yeast (10). The $\beta$ subunit is glycosylated, although no evidence for glycosylation beyond core glycosylation is apparent. Pulse-chase experiments demonstrated that all three potential extracellular glycosylation sites on the $\beta$ subunit contain carbohydrate.

Functional pump activity was inferred on the basis of several criteria, including the appearance of ouabain-sensitive ATPase and nitrophenylphosphatase activities in yeast membranes (10). Because yeast cells have other proteins that are capable of hydrolyzing these substrates, the inhibition of these activities in yeast membranes by ouabain is always less than 100%. The binding of $[^3H]$ouabain to yeast membranes can also be used as an indicator of functional sodium pumps since ouabain binds to a phosphoenzyme intermediate that is formed during the catalytic cycle of the enzyme (22). All of the high-affinity binding of $[^3H]$ouabain to yeast membranes can be attributed to the heterologously expressed enzyme, since yeast lack an endogenous $Na^+/K^+$-ATPase. Both ouabain binding and ouabain-sensitive enzymatic activities have been detected only in yeast cells expressing both $\alpha$ and $\beta$ subunits of the sodium pump. Neither of these characteristics were detected in cells expressing either $\alpha$ alone or $\beta$ alone. Thus, it was concluded that the functional sodium pump requires both of these subunits. Ouabain binding measurements indicated that the $\alpha\beta$ complex is translocated to the surface in the yeast cell, and that there are approximately 200-500 functional sodium pumps per cell.

The two-plasmid expression system in yeast has also been used to investigate whether a third subunit is required for sodium pump function (17). A third subunit was suggested after a small proteolipid was labeled in kidney membranes with photoaffinity derivatives of cardiotonic steroids (4,5), and the molecular and immunological characteristics of a small candidate $\gamma$ subunit have recently been described (14,15). Yeast cells were transfected with plasmid pCGY1406$\alpha\beta$ and also a plasmid encoding the rat $\gamma$ subunit (pA-RG1). Ouabain-sensitive activities were compared in samples from these cells and from cells transfected with only pCGY1406$\alpha\beta$, but no differences in the activities of $Na^+/K^+$-ATPase could be measured in the two samples. Conclusions regarding the significance of the $\gamma$ subunit in $Na^+/K^+$-ATPase function based on these results, however, must be interpreted with caution. There are several possible explanations for the absence of an effect of $\gamma$ in these experiments, such as a regulatory role for $\gamma$ that does not occur in yeast cells. Nevertheless, because sodium pump activities could be measured in cells or membranes containing only $\alpha$ and $\beta$ subunits, these results support the previous conclusion that the $\alpha$ and $\beta$ subunits are both necessary and sufficient for the performance of pump functions.

**Table I: Characteristics of Na,K-ATPase Molecules Expressed in *S. cerevisiae***

| Parameter | Value | Parameter | Value |
|-----------|-------|-----------|-------|
| turnover number | 3000-8000/min | $K_{0.5}$ (Na$^+$) | 1 - 2.5 mM |
| $K_D$ (ouabain) | 5 - 20 nM | $K_{0.5}$ (K$^+$) | 1 - 2 mM |
| $B_{max}$ (ouabain) | 3 - 10 pmol/mg | $K_{0.5}$ (ATP) | 0.13 $\mu$M |
| ATPase/pNPPase | 5.2 - 7.8 | $K_{0.5}$ (P$_i$) | 16 - 46 $\mu$M |

The characteristics of heterologously expressed sodium pumps that have been measured in yeast cells are the same as the characteristics of the pumps purified from mammalian kidney membranes. It has also been observed that functional pumps can be synthesized from α and β subunits that are derived from different species. Table I summarizes the properties of the enzyme assembled in the yeast membranes from sheep α1 and dog β1 subunits.

The apparent affinities ($K_{0.5}$) of the pump for different ligands were determined from the ligand concentration-dependence of [$^3$H]ouabain binding (Na$^+$, ATP, and Pi) or from the potassium concentration-dependence of the antagonism of [$^3$H]ouabain binding. The $K_{0.5}$ values for Na$^+$, ATP, and Pi are in good agreement with the $K_M$ values for these ligands for phosphoenzyme formation in purified Na$^+$/K$^+$-ATPase. These results indicate that the rate-limiting step in the [$^3$H]ouabain binding assay is likely to be phosphoenzyme formation.

*Site-Directed Mutagenesis* - On the basis of chemical modification experiments, K480 was suggested to be essential for Na$^+$/K$^+$-ATPase activity (7). Expression of Na$^+$/K$^+$-ATPase in yeast was used to test this hypothesis by examining the effects on sodium pump activity of the replacement of K480 with other amino acids (21). The mutant pumps were expressed at levels similar to the wild type pumps, and they bound ouabain with $K_D$ values that were the same as those of the normal pumps. This observation indicates that the mutations do not introduce large-scale structural changes in the protein. Because all of the mutants retained the ability to hydrolyze ATP, it was concluded that K480 is not essential for enzymatic activity. The replacement of lysine with glutamic acid, however, significantly reduced the apparent affinity of the pumps for both ATP and phosphate. The $K_{0.5}$ value for ATP was increased from 1.1 μM to 17.8 μM, and the $K_{0.5}$ value for phosphate was increased from 16 μM to 74 μM by this change. Replacement of lysine with either alanine or arginine did not significantly affect the apparent affinity of the enzyme for either of these ligands. From these data, it was concluded that although the side chain of K480 does not participate in the hydrolysis reaction, it is likely that K480 is located within the ATP binding site of the protein where it may interact with the phosphate groups of the nucleotide. A similar conclusion regarding the interaction of K480 with the phosphate moiety of ATP was made by Hinz and Kirley on the basis of chemical modification of Na$^+$/K$^+$-ATPase by adenosine diphosphopyridoxal (7).

*Assembly of hybrid pumps* - The two-plasmid expression strategy described above has been used to express several combinations of sodium pump α and β isoforms in yeast. In addition to the Na$^+$/K$^+$-ATPase β subunits, the gastric H$^+$/K$^+$-ATPase β subunit was shown to assemble in yeast with the Na$^+$/K$^+$-ATPase α1 and α3 subunits into complexes that bind [$^3$H]ouabain in either a Na$^+$ and ATP-dependent reaction or in a Mg$^{2+}$ and P$_i$-dependent reaction (3). The amino acid sequence of the H$^+$/K$^+$-ATPase β subunit is 30-

35% identical to the sequences of $Na^+/K^+$-ATPase $\beta 1$ and $\beta 2$ subunits. This similarity is sufficient to enable the $H^+/K^+$-ATPase $\beta$ subunit to assemble into hybrid pumps with the $\alpha 1$ and $\alpha 3$ subunits of the sodium pump, however, in the hybrid pumps the $K_{0.5}$ for potassium antagonism of ouabain binding is shifted to higher concentrations. This observation suggests that the structure of the $\beta$ subunit can influence the conformation of the binding sites on the pump for potassium. Assembly of the $H^+/K^+$-ATPase $\beta$ subunit with $Na^+/K^+$-ATPase $\alpha$ subunit has also been observed by others using a different expression system (9,16).

In order to identify the regions of the $\beta$ subunit that influence pump affinity for $K^+$, chimeric molecules were made in which the extracellular regions of the $H^+/K^+$-ATPase $\beta$ subunit and the $Na^+/K^+$-ATPase $\beta$ subunit were interchanged. The chimeric $\beta$ subunits are called NH$\beta 1$, in which the cytoplasmic and transmembrane regions of the polypeptide are from the rat $Na^+/K^+$-ATPase $\beta 1$ subunit and the extracellular region is from the rat $H^+/K^+$-ATPase $\beta$ subunit, and HN$\beta 1$, in which the cytoplasmic and transmembrane regions are from the $H^+/K^+$-ATPase $\beta$ subunit and the extracellular region is from the $Na^+/K^+$-ATPase $\beta$ subunit. The chimeric $\beta$ subunits were expressed in yeast together with the $Na^+/K^+$-ATPase $\alpha 1$ or $\alpha 3$ subunits, and the antagonism of $[^3H]$ouabain binding by $K^+$ was measured. Results obtained in a typical experiment using 10 mM KCl are summarized in Table II as percent of maximum $[^3H]$ouabain binding measured in the absence of KCl.

### Table II: Binding of $[^3H]$Ouabain to $Na^+/K^+$-ATPase in Yeast Membranes in the presence of 10 mM KCl

| Subunits Expressed | Percent Maximum Binding |
|---|---|
| $\alpha 1 + \beta 1$ | 18% |
| $\alpha 1 + HK\beta$ | 45% |
| $\alpha 1 + NH\beta 1$ | 42% |
| $\alpha 3 + \beta 1$ | 24% |
| $\alpha 3 + HK\beta$ | 61% |
| $\alpha 3 + NH\beta 1$ | 66% |

These results demonstrate that the difference in the apparent affinity for potassium of the pumps assembled with $Na^+/K^+$-ATPase $\beta$ subunits or $H^+/K^+$-ATPase $\beta$ subunits is due primarily to differences in the structures of the extracellular regions of these polypeptides. If the binding sites for $Na^+$ and $K^+$ are located within the $\alpha$ subunit of the protein (1,12), then these results indicate that the structure of the $\beta$ subunit indirectly influences the conformation of the ion binding sites on the $\alpha$ subunit through protein-protein interactions. An alternative interpretation that can not be ruled out, however, is that amino acids of the $\beta$ subunit may also contribute directly to the structure of the ion binding sites. A more complete characterization of the interaction of the hybrid pumps with potassium has been described by Eakle et al. (submitted for publication).

In many assays of $Na^+/K^+$-ATPase function, a competition has been observed between $Na^+$ and $K^+$ ions. Since the structure of the β subunit can influence the interaction of the pumps with $K^+$, the heterologous expression of $Na^+/K^+$-ATPase in yeast was used to investigate the possibility that the interaction of the pump with $Na^+$ may also be affected by the structure of the β subunit. The binding of [$^3$H]ouabain to pumps expressed in yeast membranes was measured in an ATP-dependent reaction that normally also requires $Na^+$. Preliminary results of these investigations are shown in Figure 1. In the absence of added $Na^+$, both α1/NHβ1 and α3/NHβ1 bind more [$^3$H]ouabain than either α1/β1 or α3/β1.

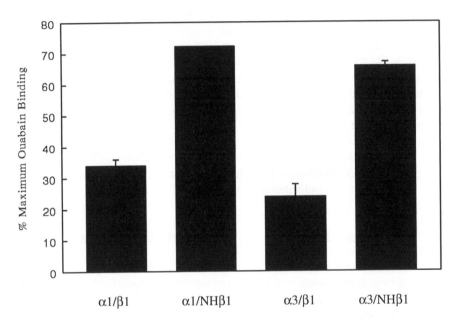

Figure 1: Ouabain binding to $Na^+/K^+$-ATPase assembled with either β1 or NHβ1 subunits. [$^3$H]ouabain binding to yeast membranes was measured in the presence of ATP and $MgCl_2$, and either 10 mM NaCl or in the absence of added NaCl. The amount of ouabain bound in the absence of NaCl is expressed for each pump complex as the percent of maximum binding in the presence of 10 mM NaCl.

Control experiments showed that removing ATP from the binding reaction reduces the [$^3$H]ouabain binding to background levels, indicating that binding is dependent on phosphoenzyme formation. Other controls indicate that only a small part of the phosphoenzyme formed in the absence of added $Na^+$ can be attributed to phosphorylation of the pumps by inorganic phosphate. Measurement of the sodium content by flame photometry also indicated that the concentration of sodium was less than 1 mM in the samples without additional added NaCl. Although the presence of lower concentrations of sodium in the samples has not been excluded, the increased binding of ouabain to $\alpha1/NH\beta1$ and $\alpha3/NH\beta1$ in the absence of added $Na^+$ suggests that another ion may substitute for $Na^+$ in the phosphorylation reaction of the pumps containing chimeric $\beta$ subunits. Other explanations are possible; nevertheless, these results indicate that the structure of the $\beta$ subunit can also influence the interaction of the pumps with sodium ions as well as with potassium ions.

## Limitations of $Na^+/K^+$-ATPase Expression in Yeast

There are certain limitations that have been observed to accompany the expression of $Na^+/K^+$-ATPase in yeast. The most significant limitation is the low yield of the expressed enzyme. There may be several reasons for the low yield, and an exhaustive examination of these possible factors has not been done in this laboratory. We have noticed that differences in yeast strains and promoters can influence yield to some extent. It is not possible to do protein chemical experiments on the enzyme expressed in yeast at the currently-observed levels, but in this regard the yeast expression system does not differ from the other expression systems for $Na^+/K^+$-ATPase currently in use. Another potential disadvantage of the expression of $Na^+/K^+$-ATPase in yeast is the uncertainty regarding cellular regulation of the enzyme in an environment in which it is not normally found. Thus, although it is possible that the $\gamma$ subunit participates in the regulation of $Na^+/K^+$-ATPase activity in animal cells, it was not possible to demonstrate this in yeast.

## Conclusions

The heterologous expression of $Na^+/K^+$-ATPase in higher eukaryotes is complicated by the assembly of heterologously-expressed subunits with endogenous subunits, as well as the possiblilty that heterologous expression may alter the expression of the endogenous subunits. The ability to express different $\alpha$ and $\beta$ subunits in yeast in the absence of endogenous polypeptides provides investigators with a powerful tool with which to examine the structural basis of $Na^+/K^+$-ATPase function. The expression of $Na^+/K^+$-ATPase in yeast has been used to demonstrate that the functional enzyme complex requires both $\alpha$ and $\beta$ subunits. Yeast cells will synthesize and assemble functional sodium pumps from different combinations of $\alpha$ and $\beta$ subunits, and will transport the complexes to the cell surface. The characteristics of sodium pumps expressed in yeast are identical to those of pumps located in animal cell membranes. Site-directed mutagenesis, and chimeric $\beta$ subunit molecules have been used together with the yeast expression system to demonstrate that K480 is not essential for ATP hydrolysis, although it is probably located within the ATP binding site of the protein, and that the $\beta$ subunit can influence the interaction of the ion pump with the transported cations.

**Acknowledgements**

We thank all of the members of this laboratory who have contributed to the development of the yeast expression system over the past several years. In addition, we thank the investigators outside of this laboratory who have generously provided to us materials developed in their laboratories. This work was supported by the US Public Health Service grants GM28673, NSF grant DMB-8919336, and American Heart Association Greater Los Angeles Affiliate grants 809-IG, 926-F1, and 983-F1.

**References**

1.  Capasso, J. M., S. Hoving, D. M. Tal, R. Goldshleger, and S. J. D. Karlish. Extensive digestion of Na,K-ATPase by specific and nonspecific proteases with preservation of cation occlusion sites. *J. Biol. Chem.* 267: 1150-1158, 1992.

2.  Eakle, K. A., B. Horowitz, K. S. Kim, R. Levenson, and R. A. Farley. Expression and assembly of different $\alpha$- and $\beta$-subunit isoforms of Na,K-ATPase in yeast. In: *The Sodium Pump: Recent Developments*, edited by J. H. Kaplan and P. DeWeer. New York: The Rockefeller University Press, 1991, p. 125-130.

3.  Eakle, K. A., K. S. Kim, M. A. Kabalin, and R. A. Farley. High-affinity ouabain binding by yeast cells expressing Na,K-ATPase $\alpha$ subunits and the gastric H,K-ATPase $\beta$ subunit. *Proc. Natl. Acad. Sci. USA* 89: 2834-2838, 1992.

4.  Forbush, B. , J. Kaplan, and J. Hoffman. Characterization of a new photoaffinity derivative of ouabain: labelling of the large polypeptide and of a proteolipid component of Na,K-ATPase. *Biochemistry* 17: 36671978.

5.  Forbush, B. , III. Cardiotonic steroid binding to Na,K-ATPase. *Current Topics in Membranes and Transport* 19: 167-201, 1983.

6.  Gaber, R. F., C. A. Styles, and G. R. Fink. TRK1 encodes a plasma membrane protein required for high-affinity potassium transport in Saccharomyces cerevisiae. *Mol. Cell. Biol.* 8: 2848-2859, 1988.

7.  Hinz, H. R. and T. L. Kirley. Lysine 480 is an essential residue in the putative ATP site of lamb kidney Na,K-ATPase. Identification of the pyridoxal 5'-diphospho-5'-adenosine and pyridoxal phosphate reactive residue. *J. Biol. Chem.* 265: 10260-10265, 1990.

8.  Hitzeman, R. A., D. W. Leung, L. J. Perry, W. J. Kohr, H. L. Levine, and D. V. Goeddel. Secretion of human interferons by yeast. *Science* 219: 620-625, 1983.

9. Horisberger, J. , P. Jaunin, M. A. Reuben, L. S. Lasater, D. C. Chow, J. G. Forte, G. Sachs, B. C. Rossier, and K. Geering. The H,K-ATPase β subunit can act as a surrogate for the β subunit of Na,K-pumps. *J. Biol. Chem.* 266: 19131-19134, 1991.

10. Horowitz, B. , K. A. Eakle, G. Scheiner-Bobis, G. R. Randolph, C. Y. Chen, R. A. Hitzeman, and R. A. Farley. Synthesis and assembly of functional mammalian Na,K-ATPase in yeast. *J. Biol. Chem.* 265: 4189-4194, 1990.

11. Jorgensen, P. L. Purification and characterization of Na,K-ATPase. III. Purification from the outer medulla of mammalian kidney after selective removal of membrane components by SDS. *Biochim. Biophys. Acta* 356: 36-52, 1974.

12. Karlish, S. J. D., R. Goldshleger, and W. D. Stein. A 19 kDa C-terminal tryptic fragment of the α chain of Na,K-ATPase is essential for occlusion and transport of cations. *Proc. Natl. Acad. Sci. USA* 87: 4566-4570, 1990.

13. Kawakami, G. K. and S. Numa. Primary structure of the alpha-subunit of Torpedo californica Na,K-ATPase deduced from cDNA sequence. *Nature* 316: 733-736, 1985.

14. Mercer, R. W., D. Biemesderfer, D. P. Bliss,Jr., J. H. Collins, and B. Forbush,III. Molecular cloning and immunological characterization of the γ polypeptide, a small protein associated with the Na,K-ATPase. *J. Cell Biol.* 121: 579-586, 1993.

15. Mercer, R. W., D. Biemsderfer, D. P. Bliss,Jr., J. H. Collins, and B. Forbush,III. Molecular cloning and immunological characterization of the g subunit of the Na,K-ATPase. In: *The Sodium Pump: Recent Developments*, edited by J. H. Kaplan and P. DeWeer. New York: The Rockefeller University Press, 1991, p. 37-42.

16. Noguchi, S. , M. Maeda, M. Futai, and M. Kawamura. Assembly of a hybrid from the α subunit of Na,K-ATPase and the β subunit of H,K-ATPase. *Biochem. Biophys. Res. Comm.* 182: 659-666, 1992.

17. Scheiner-Bobis, G. , K. A. Eakle, K. S. Kim, and R. A. Farley. Expression of DNA for alpha, beta, and gamma polypeptides of Na, K-ATPase in yeast. Does gamma have a function in Na,K-ATPase activity?. In: *The Sodium Pump: Structure, Mechanism, and Regulation*, edited by J. H. Kaplan and P. DeWeer. New York: Rockefeller University Press, 1991,

18. Shull, G. E., J. Greeb, and J. B. Lingrel. Molecular cloning of three distinct forms of the Na,K-ATPase alpha-subunit from rat brain. *Biochemistry* 25: 8125-8132, 1986.

19. Shull, G. E., L. K. Lane, and J. B. Lingrel. Amino-acid sequence of the beta-subunit of the Na,K-ATPase deduced from a cDNA. *Nature* 321: 429-431, 1986.

20. Shull, G. E., A. Schwartz, and J. B. Lingrel. Amino-acid sequence of the catalytic subunit of the Na,K-ATPase deduced from a complementary DNA. *Nature* 316: 691-695, 1985.

21. Wang, K. and R. A. Farley. Lysine 480 is not an essential residue for ATP binding or hydrolysis by Na,K-ATPase. *J. Biol. Chem.* 267: 3577-3580, 1992.

22. Yoda, S. and A. Yoda. Phosphorylated intermediates of Na,K-ATPase proteoliposomes controlled by bilayer cholesterol. Interaction with cardiac steroid. *J. Biol. Chem.* 262: 103-109, 1987.

# Expression of Functional Na$^+$/K$^+$-ATPase in Insect Cells Using Baculovirus

R.W. Mercer, G. Blanco, A.W. De Tomaso, J.C. Koster, and Z.J. Xie

Department of Pharmacology, Medical College of Ohio, Toledo, Ohio 43699 USA and Department of Cell Biology and Physiology, Washington University School of Medicine, St. Louis, Missouri 63110 USA

## Introduction

Expression systems for the Na$^+$/K$^+$-ATPase have been hindered because most commonly used cell lines have high endogenous levels of the Na pump. In contrast, some insect cells have little or no Na$^+$/K$^+$-ATPase activity. Recently, the insect baculovirus, *Autographa californica* nuclear polyhedrosis virus (AcNPV) has been adapted for the overproduction of recombinant proteins in insect cells. In this system, a foreign gene replaces the nonessential AcNPV polyhedrin gene. The foreign gene is then under control of the polyhedrin promoter, a promoter that is extremely active late in the infective cycle. Using the baculovirus expression system a wide variety of exogenous genes have been produced in insect cells, most commonly a cell line (*Sf*-9) from the ovary of the fall armyworm *Spodoptera frugiperda*. A major advantage of this system is that the insect cells perform most of the post-translational modifications found in mammalian cells. Infected insect cells will accurately complete the glycosylation, signal peptide cleavage, proteolytic processing, phosphorylation, palmitylation and myristylation dictated by the primary structure of the polypeptide. In addition, insect cells will successfully express and assemble multimeric proteins and rigorously sort recombinant proteins to their proper cellular domains (20). Using this system, we have produced functional Na$^+$/K$^+$-ATPase activity in insect cells (4,8). Ouabain-sensitive ATPase activity, isolation of a sodium-dependent potassium- and ouabain-sensitive ATP phosphorylated intermediate, isolation of a ouabain-stimulated P$_i$ phosphorylated intermediate, and ouabain-sensitive $^{86}$Rb uptake, all demonstrate that the baculovirus-induced Na$^+$/K$^+$-ATPase is identical to the native enzyme. In addition, because each subunit can be produced separately, we have obtained some unexpected and surprising results concerning the expression of the individual subunits.

## Localization of Na$^+$/K$^+$-ATPase α and ß subunits in infected *Sf*-9 cells

The cellular localization of the independently expressed α and ß subunits in infected cells was analyzed using confocal microscopy (Fig. 1). To clear the intracellular organelles of proteins in biosynthetic transit, cells were preincubated for 1 hr in cycloheximide, an inhibitor of protein synthesis. As shown, the α1 subunit is localized primarily to the plasma membrane. In addition, the ß1 subunit is also delivered to the plasma membrane in the absence of the α-subunit. Significant intracellular labeling of α1 or ß1 polypeptides in singly infected *Sf*-9 cells preincubated in cycloheximide was not observed. In addition, the uninfected *Sf*-9 cell shows little reactivity to the Na$^+$/K$^+$-

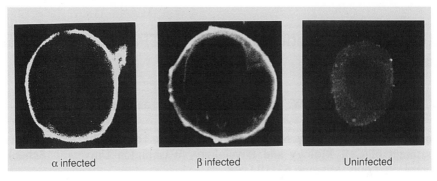

| α infected | β infected | Uninfected |

Figure 1. Localization of Na$^+$/K$^+$-ATPase α and ß subunit subunits in infected *Sf*-9 cells. Procedures are described before (4,8).

ATPase antiserum. Similar results are seen with the α2, α3 and ß2 polypeptides (4; unpublished results). Consequently, it appears that the rat Na$^+$/K$^+$-ATPase α and ß subunits can be delivered to the plasma membrane in the absence of the other subunit.

Immunoprecipitation of the α-subunit and purification on 5%-20% sucrose gradients by velocity sedimentation indicates that the subunit is not associated with any other protein (8). Membrane preparations were solubilized in 4% CHAPS and subjected to sucrose gradient velocity sedimentation. Fractions were collected and concentrated by TCA precipitation, separated by SDS-PAGE, and immunoblotted. Fig. 2A and 2B demonstrate that these gradients can resolve the unassociated α1 subunit from the assembled α1ß1 complex. When the α1 subunit is expressed alone, the α1 polypeptides sediment to a peak at approximately 5S. When the α1 subunit is coexpressed with an excess of ß1 (multiplicity of infection [MOI] ß1 > MOI α1), the α1 polypeptides sediment to a different peak of approximately 9.5S. As shown, this peak coincides with the native α1 polypeptides from rat kidney microsomes (2B), and also purified rat and dog kidney enzyme (not shown). Fig. 2B and 2C confirm that these two peaks correspond to unassociated α1 polypeptides and α1ß1 assembled complexes. Cells were infected to give an excess of α1 subunit over the ß1 subunit (MOI α1 > MOI ß1); membrane preparations were solubilized and resolved. Under these conditions the α1 polypeptides resolved into the same two 5S and 9.5S peaks (Fig. 2B). A portion of these infected cells were metabolically labeled for 4 hours with $^{35}$S-methionine, chased for 2 hours, solubilized, and resolved on separate gradients. The peak fractions were immunoprecipitated with an α specific monoclonal (mAb). As shown in Fig. 2C, the α1 polypeptides from the 5S peak are not associated with any other proteins, while in the 9.5S fraction the ß1 subunit coimmunoprecipitated with the α1 subunit. Thus the 5S fraction represents the unassociated α1 subunit, while the 9.5S fraction corresponds to the α1ß1 assembled complex.

The ability of the unassociated α1 polypeptides to resolve into a single peak suggests that the unassociated α1 subunit is in a stable conformation and is not forming a nonspecific, denatured aggregate. This is analogous to the situation found in *Xenopus* oocytes, where preexisting α-subunits in the endoplasmic reticulum (ER) are able to associate with newly synthesized ß subunits (1). While it is clear that the unassociated α-subunit in the infected *Sf*-9 cells exits the ER to be delivered to the plasma membrane, it is not known if this subunit corresponds to an assembly intermediate. In addition, we have found allow amount of endogenous insect Na$^+$/K$^+$-ATPase in the

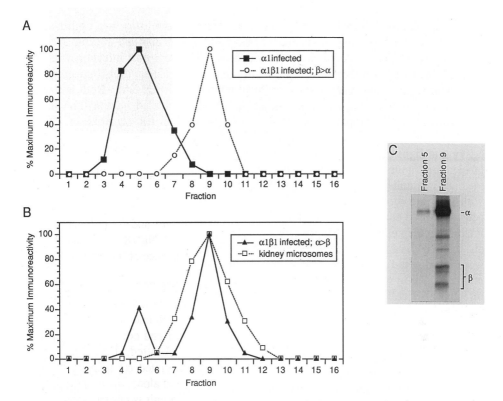

Figure 2. Sucrose gradient velocity sedimentation analysis of the α1 subunit. Solubilized membranes sedimented on 5%-20% sucrose gradients were fractionated, concentrated by TCA precipitation, resolved and quantitated as described (8). Apparent sedimentation coefficients used in the text were interpolated from standards analyzed on a separate gradient. *A*, α1 infected alone (-■-) and α1ß1 coinfection where MOI ß1 > MOI α1 (-O-). *B*, α1ß1 coinfection where MOI α1 > MOI ß1 (-▲-) and native α1 in rat kidney microsomes (-□-). *C*, Immunoprecipitations of peak fractions from α1ß1 coinfection where MOI α1 > MOI ß1 (B;-▲-) for unassociated α1 (left, fraction 5) and assembled α1ß1 (right, fraction 9).

uninfected *Sf*-9 cells. However, there is no increase in ouabain-sensitive ATPase activity when each subunit is expressed alone, suggesting that no functional hybrid enzymes are formed. The ability to resolve the α1 polypeptides from the α1ß1 assembled complexes also suggests that the α1 subunit is not associated with the endogenous *Sf*-9 cell ß-subunit.

The independent targeting of the α and ß subunits to the plasma membrane is in contrast to other studies that suggest the assembly and intracellular transport of the $Na^+/K^+$-ATPase is analogous to several other multisubunit proteins in that only correctly assembled complexes exit the ER and are properly delivered to their final destination (reviewed in 11). Two possible explanations of this phenomenon are that this invertebrate cell line does not "edit" the assembly of multisubunit proteins, or that it loses this ability when infected. To test these possibilities we obtained the baculoviruses corresponding to the murine heavy and light chains and infected *Sf*-9 cells with each virus alone, or coinfected cells with both. Three days post-infection cells were preincubated in cycloheximide for 1 hr and intracellular distribution of the immunoglobulin chains analyzed by confocal microscopy (Fig. 3A). Both heavy and light chains show strong staining around the swollen infected nucleus and diffuse patterns within the cytoplasm, reminiscent of mammalian ER distribution. In two other experiments, cells were either infected individually or coinfected with both the heavy and light chain subunits. Three days post-infection, cells were metabolically labeled for 2 hours with $^{35}$S-methionine, then chased for an additional 4 hours. The cell medium was collected and the cells isolated and lysed; both were immunoprecipitated with goat anti-mouse immunoglobulin magnetic beads. As shown in Fig. 3B, when cells expressed each immunoglobulin separately, the majority of the polypeptides were retained in the cells. In the singly infected cells approximately 97% of the total heavy chain and 84% of the light chain polypeptides were retained intracellularly. When cells were coinfected with the heavy and light chains, over 99% of the labeled immunoglobulin was secreted. These results clearly demonstrate that the baculovirus-infected *Sf*-9 cell has the ability to recognize and retain unassembled proteins and properly process assembled complexes.

When single subunits of the T-cell receptor (7), or the acetylcholine receptor (6) are expressed in cells that lack the endogenous protein, the subunits are retained in the ER and degraded. In contrast, the baculovirus-infected *Sf*-9 cell targets both the Na$^+$/K$^+$-ATPase α and ß subunits to the plasma membrane independently of each other. The absence of significant intracellular staining, strongly suggests that the high levels of expression are not saturating an *Sf*-9 ER retention mechanism, but that the subunits are targeted to the plasma membrane in this cell. Therefore the infected *Sf*-9 cell may simulate the native system; that formation of rodent αß complexes is not required for exit from the endoplasmic reticulum. This premise allows αß subunit assembly outside the ER, and allows for individual subunits to be targeted alone, phenomena that have been previously described (12,16).

In contrast, results from other experimental systems suggest that α and ß subunit association is required for exit from the ER and subsequent targeting to the plasma membrane. It appears in *Xenopus* oocytes (1) that the α and ß subunits must be associated before the complex can exit the ER and the α-subunit can acquire cation specific conformational changes thought to be a prerequisite for functional maturation. In addition, when avian α or ß polypeptides were separately transfected into mouse Ltk⁻ cells, it was observed using immunofluorescent microscopy that the ß subunit was localized to the plasma membrane when transfected alone, presumably associated with the endogenous murine α-subunit (24). However, when the avian α-subunit was expressed the majority of the α-subunit was found intracellularly, presumably unassociated with the endogenous murine ß-subunit, and thus unable to exit the ER (23). In mouse cells transfected with both subunits, the majority of immunofluorescent staining was at the plasma membrane (10). However, in neither of these studies is there any direct evidence of ER retention of unassociated α polypeptides. The infected *Sf*-9 cells ability to retain improperly assembled proteins (4,8), yet target unassembled

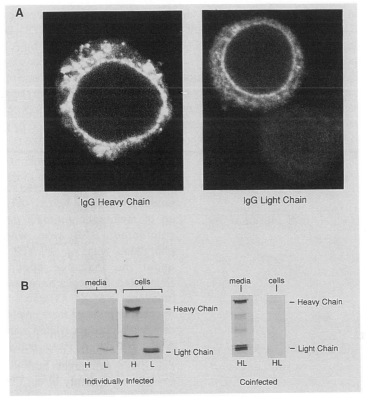

Figure 3. Expression of murine IgG heavy and light chains in *Sf*-9 cells. *Sf*-9 cells were singly or coinfected with two baculoviruses for the IgG heavy and light chains. 72 hours post-infection the infected cells were analyzed for distribution of the expressed subunits. *A*, After a 2 hour incubation in 100 µg/ml cycloheximide cells were fixed and analyzed by confocal microscopy. *B*, Cells were metabolically labeled for 2 hours and after an additional 4 hour chase both the media and cell lysates were immuno-precipitated. Panel 1 shows the results when each subunit is expressed alone. Panel 2 shows a coinfection with both subunits. When each subunit is expressed separately most of the IgG polypeptides are retained intracellularly. In contrast, when both chains are simultaneously expressed the polypeptides are secreted into the medium.

Na$^+$/K$^+$-ATPase $\alpha$ and ß subunits to the plasma membrane is an intriguing result, but it remains to be shown if this is representative of Na$^+$/K$^+$-ATPase assembly in mammalian cells.

### Activity of the independent $\alpha$ subunit

Although all enzymatic properties of the Na$^+$/K$^+$-ATPase have been ascribed to the $\alpha$-subunit it has never been shown if the subunit has activity separate from the ß-subunit. Recombinant baculoviruses containing the cDNAs coding for the rodent Na$^+$/K$^+$-ATPase $\alpha$1, $\alpha$2, $\alpha$3 and ß1 subunits were used to infect *Sf*-9 cells. As described before, the infected cells produce only the individual Na$^+$/K$^+$-ATPase polypeptides when infected.

As shown in Fig. 4A, activity of the virally produced αß polypeptides can be demonstrated by the ability of the α-subunit to be phosphorylated by ATP. During the reaction cycle, the α-subunit of the $Na^+/K^+$-ATPase is phosphorylated by ATP at a ß-aspartyl carboxyl group (asp-371) (2,22). This aspartate group is conserved in all $Na^+/K^+$-ATPase α-subunits. Phosphorylation by ATP requires the presence of both Mg and Na ions; in the presence of $K^+$ the phosphate is rapidly released from the subunit as $P_i$. As shown, the $Na^+/K^+$-ATPase α-subunit from the αß coinfected Sf-9 cells can be phosphorylated by ATP. This $Na^+$-dependent phosphorylation is inhibited by ouabain if ouabain is bound to the enzyme before exposure to ATP. In addition, the α-subunit is completely dephosphorylated in the presence of $K^+$. Because the reaction cycle of the

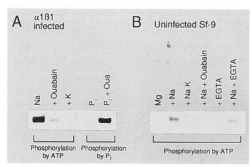

Figure 4. A, α1ß1 infected Sf-9 cells produce functional $Na^+/K^+$-ATPase. 30 µg of membrane proteins from α1ß1 coinfected Sf-9 cells were phosphorylated with $\gamma[^{32}P]$-ATP or $[^{32}P_i]$ under the indicated conditions. B, Uninfected Sf-9 cells have low levels of functional $Na^+/K^+$-ATPase. 100 µg of membrane proteins from uninfected Sf-9 cells were phosphorylated with $\gamma[^{32}P]$-ATP. $Na^+/K^+$-ATPase activity of the uninfected Sf-9 cells is approximately 20-100 fold less than the α1ß1 coinfected cells.

$Na^+/K^+$-ATPase is reversible, the α-subunit can also be phosphorylated directly with $P_i$. The phosphorylation of the α-subunit by $P_i$ is stimulated by the presence of ouabain. Although it is more difficult to detect, the $Na^+/K^+$-ATPase from the uninfected Sf-9 cells has an identical pattern of phosphorylation (Fig. 4B). The insect $Na^+/K^+$-ATPase α-subunit is phosphorylated only in the presence of $Na^+$; EGTA does not affect the $Na^+$-dependent phosphorylation and $Mg^{2+}$ alone cannot support phosphorylation.

A completely different type of α-subunit phosphorylation is seen when the α-subunit is expressed in the absence of the ß-subunit (Fig. 5). As shown, phosphorylation of the α-subunit can occur with only $Mg^{2+}$ present. This $Mg^{2+}$-dependent phosphorylation is not inhibited by ouabain, however, 150 mM $Tris^+$, $K^+$ or $Na^+$ greatly reduce phosphorylation. The lack of inhibition by ouabain is consistent with the finding that $^3H$-ouabain does not bind to the unassociated α-subunit (15; unpublished results). The ß-independent phosphorylation of the α-subunit is not restricted to the α1 isoform as both α2 and α3 exhibit identical $Mg^{2+}$-dependent phosphorylation (data not shown). Increasing concentrations of either $Na^+$, $K^+$ or $Tris^+$ inhibit α-subunit phosphorylation (data not shown). The inhibition by these ions appears to be caused by increased ionic strength as choline, a cation not thought to influence $Na^+/K^+$-ATPase enzymatic activity, also inhibits the independent α-subunit phosphorylation. Moreover, the inhibition of phosphorylation is not influenced by chloride ions, as substitution of

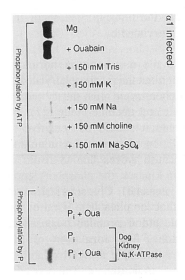

Figure 5. The independent α-subunit has catalytic activity. 30 μg of membrane proteins from α1 infected *Sf*-9 cells were phosphorylated with γ[$^{32}$P]-ATP or [$^{32}$P$_i$] as described.

chloride with sulphate has no affect on the inhibition. Taken together these results suggest that the phosphorylation of the unassociated α-subunit is sensitive to the ionic strength of the medium. At higher ionic strengths the phosphorylation of the independent α-subunit is inhibited. However, even at physiological ionic strength, the unassociated α-subunit has catalytic activity. The unassociated α-subunit can only be phosphorylated by ATP; the subunit is not phosphorylated by Pi either in the presence or absence of ouabain.

In addition to increasing ionic strength, 1 mM EGTA reduces the $Mg^{2+}$-dependent phosphorylation of the α-subunit (not shown). The inhibition of α-subunit phosphorylation by EGTA suggests that the phosphorylation may be dependent upon $Ca^{2+}$. However, if the phosphorylation of the α-subunit in the $Mg^{2+}$ medium is a result of contaminating Ca ions, then addition of $Ca^{2+}$ to the $Mg^{2+}$-EGTA reaction should eliminate the inhibition. The inhibition of α-subunit phosphorylation by EGTA is not influenced by $Ca^{2+}$. The addition of 1 mM $Ca^{2+}$ to the $Mg^{2+}$-EGTA medium did not alter the inhibition by 1 mM EGTA. At these concentrations of $Ca^{2+}$ and EGTA, the free $Ca^{2+}$ concentration is approximately 5 μM (21), higher than would be expected in the $Mg^{2+}$ medium. Consequently the inhibition by EGTA appears to be mediated through a factor other than $Ca^{2+}$. EGTA may inhibit activity by chelating another metal ion that is essential for the phosphorylation of the α-subunit, however, it is possible that EGTA inhibits activity in a manner other than its metal-binding abilities. In either event, EGTA is useful in defining the activity of the unassociated α-subunit as an EGTA-sensitive, ATPase activity. This activity can be identified in membranes from α infected cells as an EGTA-sensitive $Mg^{2+}$-dependent ATPase activity that is not present in the membranes from uninfected *Sf*-9 cells. In addition, the $Mg^{2+}$-dependent ATPase of the α infected cells is reduced by ≈40% under conditions of high ionic strength (130 mM choline). The $Mg^{2+}$-dependent ATPase activity remaining at the higher ionic strength is completely inhibited by EGTA. These results are in agreement with the phosphorylation studies and demonstrate that the phosphorylation of the unassociated α-subunit represents the ATPase activity of the enzyme.

As previously mentioned, during the catalytic cycle the α-subunit is normally phosphorylated at an aspartate residue. If the independent α-subunit is phosphorylated at an aspartate residue, then the characteristics of the phosphoenzyme must satisfy certain criteria. The phosphointermediate should be acid-stable and alkaline-labile (22). In addition, because the intermediate is an acylphosphate it should be sensitive to hydroxylamine. This is in contrast to the phosphorylation of serine, threonine and tyrosine, which results in an esterphosphate that is resistant to hydroxylamine treatment (9,13,19). Also, unlike most serine or threonine protein kinases, the phosphorylation of the α-subunit should be sensitive to inhibition by vanadate. Characterization of independent α-subunit phosphorylation demonstrated that the phosphate bond of the unassociated α-subunit phosphointermediate is acid-stable and alkaline-labile consistent with phosphorylation at the normal aspartyl residue. The formation of the phosphointermediate is also inhibited by vanadate and is sensitive to hydroxylamine, compatible with it being an acylphosphate and not an esterphosphate. These results provide strong evidence that the independent α-subunit is phosphorylated at the normal aspartate residue. Taken together this work provides strong evidence that the α-subunit, in the absence of the ß subunit, can be phosphorylated and mediate an ATPase activity. It is not clear if the enzyme is transporting ions or if this activity is physiologically significant.

**Assembly at the plasma membrane into functional enzyme**

As shown above, both α and ß subunits of the $Na^+/K^+$-ATPase are targeted to the plasma membrane when individually expressed in *Sf*-9 cells. In contrast, the infected *Sf*-9 cell retains murine light and heavy chains when expressed alone and only secretes functional immunoglobulin when both chains are simultaneously expressed (8). Infected *Sf*-9 cells also retain the $α_1$ subunit of the GABA receptor intracellularly and only direct it to the plasma membrane when coexpressed with the $ß_1$ subunit (3). Although heteromeric proteins appear to require assembly for delivery, homomeric plasma membrane proteins are delivered directly to the plasma membrane. Therefore the baculovirus-infected *Sf*-9 cell, like mammalian cells, has the ability to recognize and retain unassembled polypeptides, suggesting that the delivery of the unassociated $Na^+/K^+$-ATPase α and ß subunits is physiologically relevant. These results imply that α and ß assembly may occur, not only in the ER, but at the plasma membrane as well. This type of assembly for the $Na^+/K^+$-ATPase has been recently suggested in insulin sensitive tissues (16). To test if the α and ß subunits can assemble at the plasma membrane we decided to take advantage of a unique property of the infected *Sf*-9 cells. The 64K envelope glycoprotein of baculovirus mediates pH-dependent membrane fusion (18). This fusion activity can be used to fuse infected cells into large syncytia. Infected cells growing in suspension are allowed to settle at a high density onto a culture plate. Fusion is initiated by a 10 minute shift to pH 5.1; as shown in Fig. 6 this pH shift results in a large syncytia of fused cells. Under these conditions approximately 50-60% of the infected cells fuse into syncytia. This cell fusion appears to be limited to the plasma membrane. When *Sf*-9 cells are infected separately with murine heavy and light

28

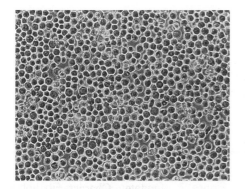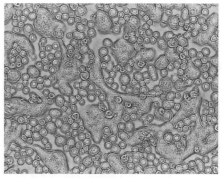

Figure 6. Unfused (left) and fused (right) infected *Sf*-9 cells.

chain immunoglobulins and fused, confocal microscopy demonstrates that the separate, cytoplasmic chains to not associate. To determine if the α and ß subunits can assemble at the plasma membrane, cells grown in suspension were infected with either the α1 or ß1 baculovirus. After 72 hours the α and ß infected cells were plated together and allowed to attach for two hours; the cells were then metabolically labeled with $^{35}$S-methionine, chased for 2 hours and treated with cycloheximide. This protocol of labeling and treatment with cycloheximide prior to fusion assures that once fusion is initiated only pre-fusion polypeptides are detected and allowed to associate. To start fusion, the medium was shifted to pH 5.1 for ten minutes. The cells were then incubated for 1 hour, solubilized and the α-subunit immunoprecipitated with an α specific mAb. Immunoprecipitation of the α subunit from fused α and ß cells results in the coprecipitation of the ß-subunit. This association is not seen in cells treated identically, except not receiving the pH shift. These results demonstrate that the α and ß subunits at the plasma membrane can associate and assemble into an enzyme complex.

To determine if the fused α and ß subunits are active, the ouabain-sensitive K$^+$ uptake of the fused cells was assayed. As shown in Fig. 7 only cells that had been individually infected with α or ß baculoviruses and subsequently fused exhibited an increase in ouabain-sensitive uptake. Fusions between cells expressing only one subunit showed no such increase in activity. Also there was no increase in activity in the mixture of α and ß infected cells that have not been fused. These results supply strong evidence that the α and ß subunits at the plasma membrane can assemble into functional Na$^+$/K$^+$-ATPase enzyme. It is clear that under normal conditions the majority of α and ß assembly occurs in the ER, however, these subunits can assemble into functional Na$^+$/K$^+$-ATPase at the plasma membrane. The exact functional significance of the plasma membrane assembly of the Na$^+$/K$^+$-ATPase must await further study. In addition, it will have to be determined if the α and ß subunits at the plasma membrane can dissociate and re-associate with other subunits to form functional enzyme. This type of assembly has been suggested to explain the insulin-responsive assembly of α2ß1 complexes (16). Thus if different combinations of the subunits have diverse functional characteristics then the regulated dissociation and re-association of the different isoforms may be important in enzyme regulation. The molecular mechanisms leading to this type of assembly remains to be investigated.

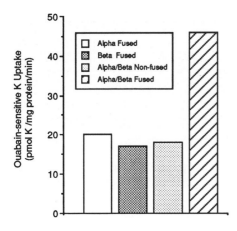

Figure 7. Ouabain-sensitive Rb uptake into fused, infected *Sf*-9 cells.

## Expression and characterization of Na$^+$/K$^+$-ATPase isoforms

Recombinant baculoviruses containing the cDNAs for the rodent $\alpha$ and $\beta$ Na$^+$/K$^+$-ATPase isoforms were used to infect *Sf*-9 insect cells. Coinfections combining the different $\alpha$ isoforms with $\beta$1 and $\beta$2 were performed. After 72 hrs post-infection, *Sf*-9 cells expressed high amounts of the corresponding Na$^+$/K$^+$-ATPase subunits, as determined by immunoblot analysis using Na$^+$/K$^+$-ATPase isoform specific antibodies. As we previously reported (4), in the case of cells coinfected with $\alpha$1, $\alpha$2 and $\alpha$3 in combination with $\beta$1, catalytically active complexes were obtained. To determine if and which $\alpha$ isoforms can associate with $\beta$2, metabolic labeling and immunoprecipitation experiments were performed. Immunoprecipitations demonstrate that all three $\alpha$ isoforms are able to assemble with the $\beta$2 subunit. Furthermore, all the $\alpha$ and $\beta$ combinations are catalytically active; Na$^+$/K$^+$-ATPase activity assays performed on whole cells, showed that *Sf*-9 cells infected with $\alpha$1$\beta$2, $\alpha$2$\beta$2 and $\alpha$3$\beta$2 displayed activities 5 to 14 times higher than the uninfected cells or the cells singly infected with $\beta$2 (5).

## Characterization of the H$^+$/K$^+$-ATPase and H$^+$/K$^+$- and Na$^+$/K$^+$-ATPase chimeric $\alpha$-subunits

To investigate the cellular distribution of the H$^+$/K$^+$-ATPase $\alpha$ and $\beta$ subunits in *Sf*-9 cells, infected cells were analyzed using confocal microscopy. While the H$^+$/K$^+$-ATPase $\beta$-subunit in *Sf*-9 cells is localized both intracellularly and at the surface, the H$^+$/K$^+$-ATPase $\alpha$-subunit is exclusively found intracellularly (17). When coexpressed with the $\beta$-subunit some of the $\alpha$-subunit is delivered to the plasma membrane, however, the majority still remains intracellular. These results suggest that the H$^+$/K$^+$-ATPase $\alpha$-subunit may contain signals that allow for its intracellular localization. In contrast, SREHP (serine rich entamoeba histolytic protein) and the anion transporter exhibit strong plasma membrane labeling demonstrating that the infected *Sf*-9 cell is able to recognize and target proteins from diverse sources. These results suggest that the H$^+$/K$^+$-ATPase $\alpha$-subunit contains a signal(s) that targets it to intracellular compartments.

To examine the domains responsible for function and sorting of the countertransporting P-type ATPases a series of six chimeras were constructed between the $\alpha$-subunits of the

$Na^+/K^+$-ATPase and the $H^+/K^+$-ATPase (17). Using the cDNAs for the $Na^+/K^+$-ATPase and $H^+/K^+$-ATPase $\alpha$ subunits, the regions corresponding to the $N$-terminal transmembrane region, the cytoplasmic domain, and the $C$-terminal regions were interchanged to create fusion polypeptides. Using baculovirus these chimeras were coexpressed with the $Na^+/K^+$- or $H^+/K^+$- ß subunits in $Sf$-9 cells. All recombinant polypeptides were shown to associate with the membrane fraction of the infected $Sf$-9 cells. All chimeras examined, regardless of the $C$-terminal domain expressed, were capable of stable association with both the $Na^+/K^+$- ß and $H^+/K^+$- ß subunits in detergent extracts. Coimmunoprecipitation of the polypeptides serves as an important operational definition of the assembled state since all immune precipitates are subjected to several stringent washes prior to separation from the antibody. Generally only proteins that form stable subunit associations will copurify with one another and the antibody. This also suggests that the chimeric polypeptides are correctly folded and properly inserted into the membrane. As observed in $Xenopus$ oocytes, the $Na^+/K^+$-ATPase $\alpha$-subunit assembles with the expressed $H^+/K^+$- ß subunit and this interaction is also resistent to detergent treatment (14). Presently, we are characterizing the enzymatic activity of the native and chimeric proteins under various reaction conditions in order to establish ion requirements, ß subunit selectivity, and inhibitor specificity.

*Acknowledgments*

This work was supported by NIH Grants HL 36573, DK 45181 and GM 39746.

**References**

1. Ackermann U and Geering K (1990) Mutual dependence of Na,K-ATPase alpha- and beta- subunits for correct posttranslational processing and intracellular transport. Febs Lett 269:105–108
2. Bastide F, Meissner G, Fleischer S, and Post RL (1973) Similarity of the active site of phosphorylation of the adenosine triphosphatase for transport of sodium and potassium ions in kidney to that for transport of calcium ions in the sarcoplasmic reticulum of muscle. J Bio Chem 248: 8385-8391
3. Birnir B, Tierney ML, Howitt SM, Box GB and Gage PW (1992) A combination of $\alpha$1 and ß1 subunits is required for formation of detectable GABA-activated chloride channels in Sf9 cells. Proc R Soc Lond B 250:307-312
4. Blanco G, Xie ZJ and Mercer RW (1993) Functional expression of the $\alpha$2 and $\alpha$3 isoforms of the Na,K-ATPase in baculovirus-infected insect cells. Proc Natl Acad Sci USA 90:1824-1828
5. Blanco G, Xie ZJ and Mercer RW (1994) Function expression of $Na^+/K^+$-ATPase $\alpha$ and ß isofroms. This volume
6. Blount, P and Merlie JP (1990) Mutational analysis of muscle nicotinic acetylcholine receptor subunit assembly. J Cell Bio 111:2613–2622
7. Bonifacino JS, Suzuki CK, Lippincott–Schwartz J, Weissman AM and Klausner RD (1989) Pre-Golgi degradation of newly synthesized T-cell antigen receptor chains: intrinsic sensitivity and the role of subunit assembly. J Cell Bio 109:73–83
8. De Tomaso AW, Xie ZJ, Liu G and Mercer RW (1993) Expression, targeting and assembly of functional Na,K-ATPase polypeptides in baculovirus-infected insect cells.

J Bio Chem 268:1470-11478

9.    Duclos B, Marcandier S, and Cozzone AJ (1991) Chemical properties and separation of phosphoamino acids by thin-layer chromatography and/or electrophoresis. Methods Enzymol 201: 10-21

10.   Fambrough, DM, Wolitzky BA, Taormino JP, Tamkun MM, Takeyasu K, Somerville, D, Renaud KJ, Lemas MV, Lebovitz RM, Kone BC, Hamrick M, Rome J, Inman EM, and Barnstein A (1991) A cell biologist's perspective on sites of Na,K-ATPase regulation. Soc Gen Physiol Ser 46:17–30

11.   Geering K (1991) Posttranslational modifications and intracellular transport of sodium pumps: importance of subunit assembly. In: De Weer P, Kaplan JH (eds) The Sodium Pump. Rockefeller University Press, New York, pp 32-43

12.   Gottardi, CJ and Caplan MJ (1993) An ion-transporting ATPase encodes multiple apical localization signals. J Cell Bio 121:283-293

13.   Hokin LE, Sastry PS, Galsworthy PR, and Yoda A (1965) Evidence that a phosphorylated intermediate in a brain transport adenosine triphosphatase is an acyl phosphate. Proc Natl Acad Sci USA 54:177-184

14.   Horisberger JD, Jaunin P, Reubens MA, Lasater LS, Chow DC, Forte JG, Sachs G,Rossier BC, and Geering K (1991) The $H^+/K^+$-ATPase ß-subunit can act as a surrogate for the ß-Subunit of $Na^+/K^+$-pumps. J Bio Chem 266:19131-19134

15.   Horowitz B, Eakle KA, Scheiner–Bobis G, Randolph GR, Chen CY, Hitzeman RA, and Farley RA (1990) Synthesis and assembly of functional mammalian Na,K-ATPase in yeast. J Bio Chem 265:4189–4192

16.   Hundal HS, Marette A, Mitsumoto Y, Ramlal T, Blostein R and Klip A (1992) Insulin induces translocation of the $\alpha2$ and ß2 subunits of the Na/K-ATPase from intracellular compartments to the plasma membrane in mammalian skeletal muscle. J Bio Chem 267:5040-5043

17.   Koster JC, Reuben MA, Sachs G, Mercer RW (1994) Expression of $Na^+/K^+$-ATPase and $H^+/K^+$-ATPase fusion proteins in *Sf*-9 insect cells. This volume

18.   Leikina E, Onaran HO and Zimmerberg J (1992) Acidic pH induces fusion of cells infected with baculovirus to form syncytia. FEBS lett 304:221-224

19.   Lipmann F, and Tuttle LC (1945) A specific micromethod for the determination of acyl phosphates. J Bio Chem 159:21-28

20.   Luckow, VA (1990) Cloning and expression of heterologous genes in insect cells with baculovirus vectors. In: Ho C, Prokop A and Bajpai R (eds) Recombinant DNA Technology and Applications. McGraw-Hill, New York

21.   Martell AE, and Smith RM (1977) Critical Stability Constants, vol.3, Other Organic Ligands. Plenum Press, New York

22.   Post, RL and Kume S (1973) Evidence for an aspartyl phosphate residue at the active site of sodium and potassium ion transport adenosine triphosphatase. J Bio Chem 248: 6993-7000

23.   Takeyasu K, Tamkun MM, Renaud KJ and Fambrough DM (1988) Ouabain-sensitive $(Na^++K^+)$-ATPase activity expressed in mouse L cells by transfection with DNA encoding the alpha-subunit of an avian sodium pump. J Bio Chem 263:4347–4354

24.   Takeyasu K, Tamkun MM, Siegel NR and Fambrough DM (1987) Expression of hybrid $(Na^++K^+)$-ATPase molecules after transfection of mouse Ltk⁻ cells with DNA encoding the beta-subunit of an avian brain sodium pump. J Bio Chem 262:10733–10740

# Sp1 and an E-Box Binding Protein Regulate Na$^+$/K$^+$-ATPase α2-Subunit Gene

K. Ikeda, K. Nagano, K. Kawakami
Department of Biology, Jichi Medical School, Yakushiji, Minamikawachi, Kawachi, Tochigi, 329-04 Japan

## Introduction

Na$^+$/K$^+$-ATPase α2 subunit gene exhibits tissue-specific and developmental patterns of expression in the rat. For example, it is expressed specifically in muscle and neural tissues. The mRNA level of the α2 subunit gene in muscle tissue increases dramatically during rat myogenesis (3). In this study, we report promoter analysis of rat Na,K-ATPase α2 subunit gene *(Atp1a2)*. To identify *cis* elements and *trans*-acting factors, we used L6 rat myoblast cells as a model for skeletal muscle.

## Identification of regulatory elements in the 5'-flanking region of *Atp1a2*.

For mapping *cis* elements responsible for the *Atp1a2* promoter activity in L6 cells, we constructed a series of 5' sequential deletion mutations of *Atp1a2*-luciferase chimeric constructs and tested for their promoter activity by transient transfection assays. Analysis of 5' deletion mutations revealed that the region between -175 and -108 was a positive regulatory element (10-fold). In this region, there are three known consensus sequences for transcription regulation: two E boxes (CANNTG) and one Sp1 binding consensus sequence (Fig. 1A). Site-specific mutation analysis of these elements suggested that the distal E box acts as a negative regulatory element and that the Sp1 binding consensus and a GGGAGG sequences act as positive regulatory elements, together yielding the full activity level of pA2-175/LΔ5'.

## Identification of the *trans*-acting factors binding to the *Atp1a2* promoter.

To identify DNA-binding transcription regulatory factors to the elements, gel retardation assays using nuclear extract from L6 cells were performed. Using the DNA fragment from -175 to -99 of *Atp1a2* as a probe, four specific retarded complexes (C1, C2, C3 and C4) were observed. The observation that the formation of C1, C2 and C4 was interfered with by human Sp1 antibody, indicates that the binding factors in C1, C2 and C4 are Sp1. The formation of C3 was specifically competed with by the E-box oligonucleotide (5'-GATCCCCCCAA<u>CACCTG</u>CTGCCTGA-3') (1) , but was not competed with by the mutant E-box oligonucleotide (5'-GATCCCCCCAA<u>CACGGT</u>CTGCCTGA-3') . These results suggest that the formation of C3 depends on the sequence CANNTG (E box). The formation of C3 was not disrupted with myogenin antibody.

## Interaction of the binding factors with *Atp1a2*.

33

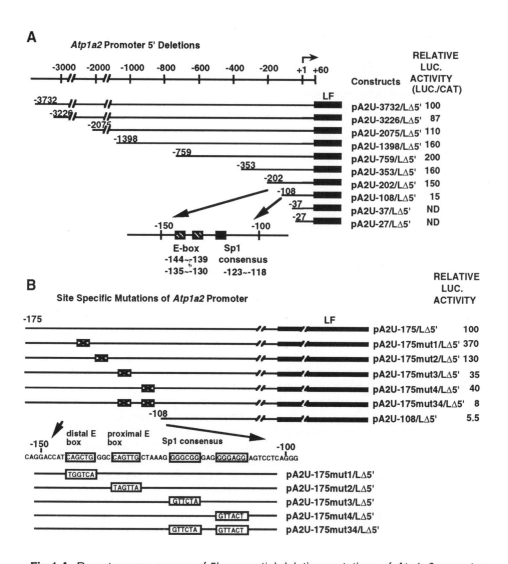

**Fig.1.A.** Reporter gene assays of 5' sequential deletion mutations of *Atp1a2* promoter in L6 cells. The terminal nucleotide positions of the deletions are indicated. 6.7 µg of each plasmid was cotransfected with 0.4 µg of pEFBOS-CAT as an internal control into L6 cells. RELATIVE LUC. ACTIVITY was normalized with respect to CAT activity in the same cell lysate and was presented as a percentage of the luciferase activities obtained in pA2-3732/LΔ5'. The result represents a typical data in four separate experiments, each done as duplicate. The lower line shows the positions of two E boxes (CANNTG) and the Sp1 consensus sequence.

**B.** Reporter gene assays of site specific mutations of *Atp1a2* promoter in L6 cells. The sequence between -153 and -98 of *Atp1a2* and the mutated nucleotides introduced are shown in the lower panel. RELATIVE LUC. ACTIVITY was presented as a percentage of the luciferase activity obtained in pA2-175/LΔ5', which was set at 100.

In order to identify protein-bound regions and guanine residue contacts of C1 and C2, we performed DNase I footprinting and methylation interference experiments. Results are summarized in Fig. 2A. GGCGGGAGGGGAGGAGTCCTCA (-122 to -101) on the upper strand and CCTCC (-114 to -118) on the lower strand were protected from DNaseI digestion. Methylation of guanine residues (marked with closed triangles) between -122 and -104, including the Sp1 consensus and the GGGAGG sequences, interfered with the complex formation. The recognition sequence for the binding factor of C3 was determined by methylation interference experiment (Fig. 2B). Methylation of

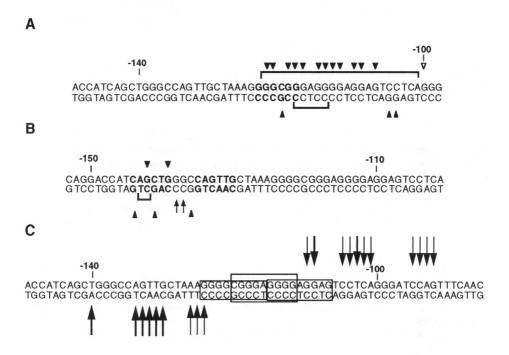

**Fig.2.A.** DNase I footprinting and methylation interference analyses of C1 and C2. The regions of protection in DNase I footprinting analysis are indicated by solid lines. Guanine positions that interfered with the binding in methylation interference analysis are indicated by closed triangles. A position of overrepresented guanine in the retarded probe is marked with an open triangle. The Sp1 consensus GGGCGG is indicated by bold letters.
**B.** DNase I footprinting and methylation interference analyses of C3. The region of protection in DNase I footprinting analysis is indicated by a solid line. Positions of enhanced guanines in the retarded probe are marked with arrows. Guanine positions that interfered with the binding in methylation interference analysis are indicated by closed triangles. Two E boxes are indicated by bold letters.
**C.** T4 DNA polymerase footprinting of Sp1 with *Atp1a2*. Thick arrows denote strong stops and thin arrows denote weak stops. Solid boxes are three Sp1 binding elements identified.

two of the guanines on the upper and three guanines on the lower strands interfered with the complex formation (marked with closed triangles). The contact points of the binding factor in C3 corresponded to the distal E box. In the DNase I footprinting experiment, the protected region (-142 to -143) and hypersensitive sites (-137 and -138) were observed around the distal E box weakly but reproducibly. Because the central NN sequence and the flanking sequences of the E box are known to be important for specific binding of E protein, this E box binding protein might not bind to the proximal E box.

To clarify the molecular interactions of Sp1 with elements of *Atp1a2*, T4 DNA polymerase footprinting was performed using purified human Sp1. The results indicated that there are three binding sites for Sp1 within the region centered around positions -109, -103 and -93 of the upper strand and three stops centered around positions -125, -132 and -140 of the lower strand. Results are summarized in Fig. 2C. Taken together with the result of gel retardation assay that one Sp1 binds to the probe, we conclude that one Sp1 binds to one of the three elements, which overlap with each other, in a mutually exclusive manner. The molecular interaction pattern of Sp1 with *Atp1a2* we observed in this experiment is interpreted as a mixture of three independent bindings. Binding of each Sp1 may result in kinetic synergism in transcription initiation (2).

*Acknowledgements*

We thank Dr. Suresh Subramani for pSV0A/LΔ5', Dr. Atsuko Fujisawa-Sehara for pSV2CAT, Dr. Shigekazu Nagata for pEF-BOSCAT, Dr. Yo-ichi Nabeshima for discussion and Mrs. Kuniko Takase for technical assistance. We also thank Dr. Akira Ishihama for discussion and critical reading of the manuscript.
This work was supported by grants from the Ministry of Education, Science and Culture of Japan.

# References

1.  Davis RL, Cheng P-F, Lassar AB, Weintraub H (1990) The MyoD DNA binding domain contains a recognition code for muscle-specific gene activation. Cell 60: 733-746
2.  Herschlag D, Johnson FB (1993) Synergism in transcriptional activation: a kinetic view. Genes Dev 7: 173-179
3.  Orlowski J, Lingrel JB (1988) Tissue-specific and developmental regulation of rat Na,K-ATPase catalytic α isoform and β subunit mRNAs. J. Bio Chem 263: 10436-10442

# Cloning and Analysis of the 5'-Flanking Region of the Rabbit Na$^+$/K$^+$-ATPase α2 Subunit Gene

Trine Kjærsig, Peter L. Jørgensen and Li-Mei Meng

Biomembrane Research Centre, August Krogh Institute, Copenhagen University, DK-2100, Denmark.

## Introduction

The Na$^+$/K$^+$-ATPase consists of one α and one β subunit. In mammals, there are at least three isoforms of the α subunit. The α1 subunit is expressed in most tissues, while the expression of α2 and α3 subunits is highly tissue-specific. The α2 gene is expressed in brain, heart and muscle. To investigate the regulation in muscle, we isolated and sequenced a 5'-flanking region of the rabbit Na$^+$/K$^+$-ATPase α2 subunit gene. Here we report the data on sequencing and characterization of the promoter.

## Nucleotide Sequence analysis

A lambda clone, λLMM52.3, hybridizing to a *Bst*E II fragment (nucleotides 56 to 877) of the rat Na$^+$/K$^+$-ATPase α2 subunit cDNA (9), was isolated from a rabbit liver genomic DNA library (7). The nucleotide sequence of this clone was determined. Comparing the sequence of λLMM52.3 with those of the rat α subunit genes of Na$^+$/K$^+$-ATPase (9) revealed that λLMM52.3 carried 10 kb of rabbit genomic DNA, which contained a part of the α2 subunit gene including approx. 1.4 kb 5'-flanking region, introns 1 to 5, exons 1 to 5 and a part of exon 6 (Fig.1).

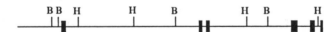

**Figure 1**. Restriction map of λLMM52.3. Exons are indicatied by solid bars. E, *Eco*RI; H, *Hind* III.

Comparison of the nucleotide sequence of the 5'-flanking region of the rabbit α2 gene with those of the rat and the human α2 genes (4,10) revealed high similarity at three regions (Fig.2). The overall homology between the three species at regions A, B, and C were 80%, 65% and 75%, respectively. Sequence elements which have been shown to be involved in the control of expression of various muscle specific genes were found in region C. These include the E-box consensus sequence, CGNNTG (2), and the GArC core sequence, G(A/T)$_{4,5}$TCT (8). Both motifs were perfectly conserved between the three species. The E-box is recognized by the MyoD family of muscle specific transcription factors (2), and the GArC-box is presented in several muscle specific enhancers (8). In region B, no sequence similarity was found to known regulatory sequence elements. In region A, a Sp 1 consensus sequence (GGGCGG) and a TATA box like sequence, TATTTAAA, are perfectly conserved between the three species. A number of other noteworthy sequence elements are also found in this region. i) A GGGGAGGAG sequence adjacent to the Sp1 binding site. The GGGGAGGAG

37

**REGION A**

```
                                    Sp1          PAL                    GTTT(C/G)
Human  -236  GGCCAGTGGCTAAA GGGGCGGG A GGGGAGGAGT CCTCAGGGATCCT GTTTCAACAAAC-GTT
Rabbit -231  GGCCAGTTGCTAAA GGGGCGGG A GGGGAGGAGC CCTCGGGGATCCT GTTTCAACAAACTGTT
Rat    -241  GGCCAGTTGCTAAA GGGGCGGG A GGGGAGGAGT CCTCAGGGATCCA GTTTCAACAAAC-GTT
             ****** ****** [********] * [********* ]**** ******* [********** *** 

                                  GA-rich                    TATA-like
Human  -175  TCTTTC G-G AGGAGGGGAAGGCGGGGGAGAGGGGGAGAAGGA CC TATTTAAA GCTACCCTGT
Rabbit -169  TCTTTC CCG AGAAGGGGAAGGCGGGGGCGAGGGGGAGAGGGA CC TATTTAAA GCTACCCAGT
Rat    -180  TCTTTC CCC AGAAGGGGAAGGCGGGAGTGAGGGGGAGAGGGA CC TATTTAAA GCTACCCTGT
             ******] [** ************** * ********** ***] ** [********] ******* **

                 Inr-like
Human  -115  TGCTTTGG CTTTCTCTGTTGCC--- -AGGGTCTCCGACTGTCCCAGACGGGCTGGTGTGGGCTTG
Rabbit -108  GGCTTGGG CTGTCTCTGTTGCCTGC CAGGGTCTCCGGCTGCCCCAGACGGGCCGGTGTGGGCTTG
Rat    -119  TGCTCTGG CTCTCTCTGCTGTCTGC CAGGGTCTCCAGCTGCCCCAGACAGGC-GGTGTGGTCTTG
             ***  ** [** ****** ** *  ] ********* *** ******* *** ******* ****

                                                                    +1
Human   -54  GGATCCTCCTGGTGACCTCTCCCGC-TAAGGTCCCTCAGCCACTCTGCCCCAAG ATG GGCCGTGG
Rabbit  -44  G-ACCCTCT--GCAGCCT--------AGGTCCCTCAGCCACTCTGCCCCAAG ATG GGTCGCGG
Rat     -55  GGATCCTCCTGGTGACCTTTCCAGCCTAGGTCCCCTCAGCCACTCTGCCCCAAG ATG GGTCGTGG
             * * ****  * ***        * *  ******************** [***] ** ** ** **
```

**REGION B**

```
Human  -1501  AGACCTGCCCCA-CTCCGCTGGCTTCCCCTCACTCCATCTCTGTGCCCAGAGAAACAGGGGTACTCC
Rabbit -1272  AGACCTGTCCCAGCACGGCTGGCTTCCCTGCACCCCATCTCCGTGCCCAGAGAAATGGGGACACTTC
Rat    -1696  AGACCAGCCCCTGCTCAGCGGGTATCCACTCACTCCATCTCTGTGCCTAGAGAAATGGGGACGCTAC
              ***** * *** * * ** **  *** *** ******* ***** ******* ***   ** *

Human  -1435  AGGGGAATCCTGTCCTCTGACTGTTCCCAGTAGAAGGCGGGAGCAGAATTTTTTCT-TTTCTCTCTG
Rabbit -1205  AGAAAGATCCTGTCCTCCGACTGCTCTGGGCAGAAGCCGGG-GCAGGATTTTTTCT-TCTCTGTCTG
Rat    -1629  AGAAGAACCTTGTCCTCTAACTGCTCCTGGTAGATGGGGGCAGAAGTCTTTCTTCTCTCTGTCCCCA
              **     * * ******* **** **  *  * *** *  ** * **  *** **** * * *

Human  -1389  -CCC-CTTCTG
Rabbit -1160  -CCCTCTTCTG
Rat    -1584  ACCCCCTTCTG
              *** ******
```

**REGION C**

```
                          GArC                                              E-box
Human  -1737  CCTGTAAT GGAAAA-TCT CAACTGCAG--CCGGAGGGGGGCCAGTTATTTAGGGAATT CAGATG
Rabbit -1483  TCTGTAAT GGAAAA-TCT CAACAGCAG-GCCCGGATGGGTGGGTTATTTAGGGAATT CAGATG
Rat    -1973  CCTGTAAT GGAAAAATCT CAGCGGCAGAGCCGGAGGTGGCCAGTTATTTAGGGAATT CAGATG
              ******* [****** ***] ** * **** ** *    ** *************** [******]

Human  -1677  TTAGCCTTGTGCCCTGAGAGTGGATTTGCTAGAGGAGAGGCTGATAATCCCTCTGCAGGGAG
Rabbit -1422  CGAGACCTGTGCCCTGAGAGTGGATTTGCTGGAGGAGAGGTTGATAGTCCCTCTGCAGGGAG
Rat    -1910  TTAGCCCGGTGCTCTGAGAGTAGATTTCCTGAACAGGAGGTCCATAATCCCTCTACAGGGAG
              ** ***** ******** ***** ** * *    **** *** ********* *******
```

**Figure 2.** Alignment of nucleotide sequences of the 5'-flanking region of $\alpha_2$ gene of Na$^+$/K$^+$-ATPase. The alignment was carried out by the Clustal V program. The noteworthy sequence elements are indicated. Asterisks indicate identical nucleotides. The first nucleotide of the translation initiation codon (ATG) was numbered as +1.

sequence has recently been shown to be a binding site for a nuclear factor, PAL (3). The Sp1-PAL motif has been found in the promoter region of genes encoding neurofilaments. ii) A 33 bp purine-rich sequence was found just upstream of the TATA-like sequence. Purine-rich sequences have been observed at similar positions in many genes, and their regulatory effects on transcription have been studied to some extent (5). iii) A TC-rich stretch of nucleotides surrounding the transcription starts, contained TCTCTG and TG(T/C)C motifs which have been found in the initiator elements of various genes (1,6). iv) Three copies of a GTTT(C/G) sequence are found at similar position in all three spieces (Fig.2).

## Promoter activity of the 5'-flanking region

In order to see whether the 5'-flanking region of rabbit α2 gene was able to drive transcription, the 1.45 kb 5'-terminal DNA fragment (nucleotides -1485 to -43 from translation initiation) of the insert of λLMM52.3 was cloned into the polylinker of pCAT-basic (Promega Corp. Madison). The resulting construct, pTNK68 (Fig.3), was co-transfected with pSVβGal (Promega Corp.) into both non-myogenic cells, NIH3T3 (fibroblast), and myogenic cells, C2C12. pCAT-control (Promega Corp.), which contains the SV40 promoter and enhancer driving the *cat* gene, and pCAT-basic (Promega Corp.), which is promoter- and enhancerless, were included as a positive control and a negative control, respectively. The results showed that the 5'-flanking region contained a C2C12 cell specific promoter/enhancer (Fig.3).

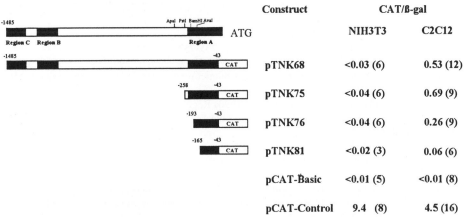

| Construct | CAT/ß-gal | |
|---|---|---|
| | NIH3T3 | C2C12 |
| pTNK68 | <0.03 (6) | 0.53 (12) |
| pTNK75 | <0.04 (6) | 0.69 (9) |
| pTNK76 | <0.04 (6) | 0.26 (9) |
| pTNK81 | <0.02 (3) | 0.06 (6) |
| pCAT-Basic | <0.01 (5) | <0.01 (8) |
| pCAT-Control | 9.4 (8) | 4.5 (16) |

**Figure 3.** Expression of CAT in non-myogenic and myogenic cells under the control of 5'-flanking sequences of rabbit *a*2 subunit gene of Na$^+$/K$^+$-ATPase. The top line shows the location of the three conserved regions (solid bars). Restriction sites used in 5'-sequential deletions are indicated. The cells were maintained in Dulbecco's modified medium (Gibco BRL) supplemented with 10% fetal calf serum (Imperial) in a $CO_2$ incubator (5% $CO_2$) at 37°C. Each 15 μg of the plasmid was cotransfected with 10 μg of pSVβGal into cells (2-3 x10$^5$ cells in 6-cm dishes) by the method of calcium phosphate co-precipitation. The cells were then harvested and lysed 48 h later. CAT activity was assayed by the CAT-ELISA system (Boehringer Mannheim). β-gal activity was measured in each cell extract to compensate for the variations in transfection efficiency. Enzyme activities are expressed as OD/μg protein/h. The figures in parentheses indicates the number of transfections.

## Identification of the core promoter sequences

In order to identify the minimal DNA segment required for the C2C12 specific promoter activity, 5'-deletion clones of pTNK68 were made (Fig.3). The constructs were transfected into NIH3T3 and C2C12 cells. The results showed that a C2C12 cell specific basal promoter aktivity was located between nucleotides -165 to -43 from translation initiation (Fig.3), wich contained the 33 bp purine-rich sequence and the initiator like sequence. The three copies of the GTTT(C/G) sequence aktivated the basal promoter by 4 fold. DNA sequence between nucleotides -258 to -194, which contained the Sp1 and the PAL binding sites activated the basal promoter activity by additional 2.5 fold. Work is in progress to identify the transcription factors involved in the regulation of the muscle specific core promoter.

## Acknowledgments

We are thankful to Dr. Jerry B. Lingrel for providing a rat $Na^+/K^+$-ATPase $\alpha2$ cDNA, to Dorte Meinertz for the excellent technique assistance. This work was supported by the grants from the Biomembrane Research Centre and Danish medical Science Research Council.

## References

1.  Du H, Roy AL, Roeder RG (1993) Human transcription factor USF stimulates transcription through the initiator elements of the HIV-1 and the Ad-ML promoters. EMBO J. 12: 501-511.
2.  Edmondson DG, Olson EN (1993) Helix-loop-helix proteins as regulators of muscle-specific transcription. J. Biol. Chem. 268: 755-758.
3.  Elder GA, Liang Z, Lee N, Friedrich Jr VL, Lazzarini RA (1992) Novel DNA binding proteins participate in the regulation of human neurofilament H gene expression. Mol. Brain Res.15:85-98
4.  Kawakami K, Yagawa Y, Nagano K (1990) Regulation of $Na^+,K^+$-ATPases. I. Cloning and analysis of the 5'-flanking region of the rat *NKAA2* gene encoding the $\alpha2$ subunit. Gene 91: 267-270
5.  Kennedy GC, Rutter WJ (1992) Pur-1, a zinc-finger protein that binds to purine-rich sequences, transactivates an insulin promoter in heterologous cells. Proc. Natl. Acad. Sci. USA 89: 11498-11502.
6.  Lenormand JL, Guillier M, Leibovitch SA (1993) Identification of a cis-acting element responsible for muscle specific expression of the c-mos protooncogene. Nucleic Acids Res. 21: 695-702.
7.  Lücking K, Jørgensen PL, Meng LM (1993) Proc. VII Int. Conf. on the Sodium Pump. Steinkopff Verlag, Darmstadt, Germany. submitted.
8.  Mably JD, Sole MJ, Liew CC (1993) Characterization of the GArC motif - A novel cis-acting element of the human cardiac myosin heavy chain genes. J. Biol. Chem. 268: 476-482.
9.  Shull GE, Greeb J, Lingrel JB (1986) Molecular Cloning of three distinct forms of the $Na^+,K^+$-ATPase $\alpha$-subunit from rat brain. Biochemistry 25: 8125-8132.
10. Shull MM, Pugh DG, and Lingrel JB (1989) Characterization of the human Na,K-ATPase $\alpha2$ gene and identification of intragenic restriction fragment length polymorphisms. J. Biol. Chem. 264: 17532-17543

# Positive and Negative Transcriptional Elements in the Chicken Na+/K+-ATPase α1-Subunit Gene

Hui-Ying Yu and Kunio Takeyasu

Department of Medical Biochemistry and Biotechnology Center, The Ohio State University, Columbus, Ohio 43210 USA

## Introduction

Coordinate gene regulation may be achieved by the action of trans-acting factors with multiple DNA-binding sites within a given gene. These DNA-protein interactions can occur in several different ways. Some act in a relatively constitutive manner (28), while others are tissue specific (25,17) and are subjected to regulation by environmental signals such as metal ions (8), serum (20), mitogens (5), and phorbol esters (1). Although much is known about cis elements which interact with trans factors and stimulate transcription, there are some cis elements that are responsible for negative regulation (2,7,14,15,16). Recently two DNA-binding proteins which repress transcription have been cloned and characterized. A Drosophila homeodomain protein (eve) has been shown to repress transcription of Ultrabithorax gene (4). A protein GCF that binds to G/C-rich sequences has been found to repress transcription in the promoters of several genes (12). Such negative factors may interact with the same or overlapping DNA sequences that bind activating transcription factors including Sp1. In this report, we characterize the positive and negative regulatory mechanisms involved in a chicken house keeping gene encoding the Na+/K+-ATPase α1-subunit.

## Property of the Core Promoter

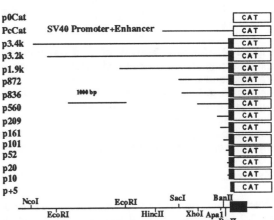

**Figure 1.** Reporter gene constructs of 5'-sequential deletion mutants. p3.2kbCAT, p1.9kbCAT, p836bpCAT, p209-bpCAT were constructed by exonuclease III deletion. p872-bpCAT, p560bpCAT, p161bp-CAT, p101bpCAT and p52bp-CAT were constructed by restriction enzyme digestion. P20bpCAT, P10bpCAT, P+5bp-CAT were constructed by polymerase chain reaction (PCR).

Cloning and basic characteristics of the chicken Na+/K+-ATPase α1-subunit gene (~25 kilobase long) have been described (24), the characteristics of which are also found in mammalian Na+/K+-ATPase α1-subunit genes (13,23,27) and in other house keeping genes (18,21); e g., lack of an obvious TATA-box and presence of multiple G/C-rich domains and multiple cap sites. A set of expression plasmids consisting of the structural gene of chloramphenicol acetyl-transferase (CAT) and various lengths of the 5'-flanking region (-3.4 kilobases to +5 bases from the cap site) of the chicken α1-subunit gene were constructed (Figure 1), and their promoter activities were measured in primary tissue cultures of chicken skeletal muscle, myocardiac muscle and fibroblasts (Figure 2a,b). The results show that the first 101 nucleotides of 5'-flanking DNA, which do not contain TATA box and CAT box but have 5 Sp1 sites, is capable of directing the highest level of transcription in both muscles and fibroblasts.

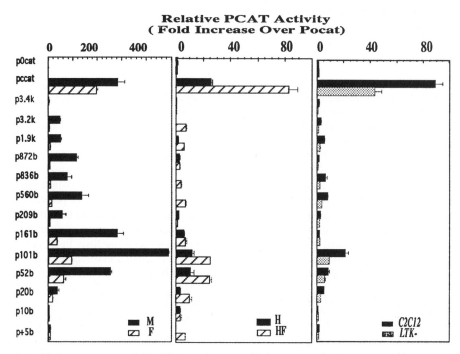

**Figure 2.** Promoter activity of of 5'-deletion mutants in different cell types. Four different primary tissue-cultured cells from chick embryo (3,19,26) were subjected to transfection by calcium phosphate precipitation (9). SV40 promoter with enhancer was used as a positive control and RSV drived β-galactosidase as an internal control. The reporter gene assay (22) demonstrates the existence of a common regulatory mechanism of the Na+/K+-ATPase α1-subunit gene. The data also suggest that the α1-subunit gene is subjected to multiple positive and negative regulation. H: cardiac myocyte; HF: cardiac fibrobrast; M: skeletal muscle; MF: skeletal fibroblast.

### Existence of Negative Regulatory Elements

When 56 bp 5'-upstream sequence was added (p161), the promoter activity was decreased by 2-3 fold. Introduction of additional 48 bp 5'-flanking region to this construct (p209) further repressed the promoter activity by 3-5 fold in all cell-types (Figure 2a,b). This suggests that the core-promoter activity of the α1-subunit gene appears to be modulated in both cell-types by the sequence between -101 and -208. This potential region for cis-acting inhibitory elements is designated as the first negative regulatory region (NRR1). This expression pattern was observed in other species (Figure 2c). The promoter activities of the deletion constructs detected in C2C12 and LTK‾cells from mouse indicated that the chicken α1-subunit gene could be regulated in a similar manner in mammalian species, although the overall activity was lower.

NRR1 was further characterized by gel retardation assays using two different oligonucleotides, Oligo-1 (-208 to -161) and Oligo-2 (-161 to -115) (Figure 3). Upon incubation with myocardiac fibroblast nuclear extract, the both oligonucleotide probes formed two major retarded bands which specifically disappear in the presence of unlabeled excess probe. In addition, the formation of Oligo-2/protein complexes were competed out by Oligo-1 (Figure 3), and vice versa (data not shown). The two 5'-regions of the α1-subunit gene represented by the two oligonuleotide probes possibly interact with the transacting factors that recognize the common sequences conserved (repeated) in these two probes.

**Figure 3.** Gel retardation analysis of the 5'-region (-115 to -208) of the chicken Na+/K+-ATPase α1-subunit gene. $^{32}$P-labeled Oligo-1 and -2 were incubated with nuclear extracts from myocardiac fibroblasts (6), and DNA/protein complexes were separated by 5% polyacrylamide gel electrophoresis (10). Unlabeled oligonucleotide was used as a specific competitor, and a polylinker sequence of pBluescript II was used as unspecific competitor. Two major DNA/protein complexes (c1 and c2) were identified. Oligo-2/protein complexes were eliminated by excess Oligo-1, suggesting multiple DNA-binding proteins are common to two separate regions of the gene.

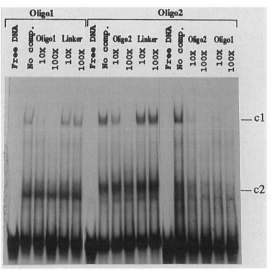

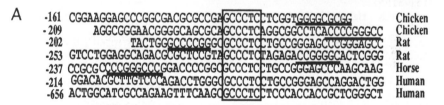

A
```
-161 CGGAAGGAGCCCGGCGACGCGCCGA GCCCTC TCGGTGGGGCGCGG      Chicken
-209     AGGCGGGAACGGGGCAGCGCA GCCCTC AGGCGGCCTCACCCCGGGCC   Chicken
-202           TACTGGGCCCCGGGC GCCCTC CTGCCGGGAGCCCGGGAGCC    Rat
-253 GTCCTGGAGGCAGACGCGCTCCGTA GCCCTC TAGAGACCGGGGCACTCGGG   Rat
-237 CCGCGCCCCGGGCCCGGACCCCGGC GCCCTC CTGCCGGGAGCCCAAGCAAG   Horse
-214 GGACACGCTTGTCCC AGACCTGGGC GCCCTC CTGCCGGGAGCCAGGACTGG   Human
-656 ACTGGCATCGCCAGAAGTTTCAAGC GCCCTC CTCCCACCACCGCTCGGGCT   Human
```

B
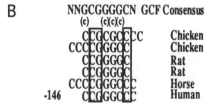

**Figure 4.** Sequences shared by all the known Na+/K+-ATPase α1-subunit genes. Positions of the 5'-most nucleotides are indicated with respect to the transcription start site. A: NRE1 is boxed and NRE 2 is underlined. B: Alignment of NRE2 (GCF consensus sequence).

A comparison of the 5'-flanking sequences between α1-subunit genes cloned from different species and classes revealed two consensus sequences, GCCCTC (NRE1) and (G/C)CG(G/C)(G/C)(G/C)C (NRE2) (Figure 4). These sequences were found in the two oligonucleotide probes. NRE2 is homologous to the well-characterized silencer element, the GCF consensus sequence NN(G/C)CG(G/C)(G/C)(G/C)CN, and NRE1 is a new element, to which no consensus sequence has been reported. These two elements have also been founded in rat horse and humman, suggesting that these elements might be involved in the regulation of the Na$^+$/K$^+$-ATPase $\alpha$1-subunit genes in general.

## Cell-type Specific Regulatory Sites

Although the first 209 nucleotides of 5'-flanking sequence are not rigidly restricted in terms of cell-type specificity (Figure 2), the further 351 nucleotides exhibited cell-type specificity; 3 fold increase in muscle cells but only 1.5 fold in fibroblast compared to the activity of p209. This suggests that this sequence, -560 to -209 may direct cell-type specific regulation (Yu and Takeyasu, in preparation). This region possibly contains additional sites for activating and inhibitory elements.

### References
1. Auwerx, J.H., Chait, A., Deeb, S.S. (1989) Proc. Natl. Acad. Sci. USA 86: 1133-1137.
2. Baniahmad, A., Muller, M., Steiner, C., Renkawitz, R. (1987) EMBO J. 6: 2297-2307.
3. Barry,W.H., Biedert, S., Miura, D.S., Smith, T.W. (1981) Circ. Res. 49: 141-149.
4. Biggin, M. D., Tjian, R. (1989) Cell 58: 433-440.
5. Bohnlein, E., Lowenthal, J. W., Siekevitz, M., Ballard, D.W., Franza, B. R., Greene, W. C. (1988) Cell 53: 827-836.
6. Dignam, J. D., Lebovitz, R. M., Roeder, R. G. (1983) Nucleic Acids Res. 11: 1475-1489.
7. Farrell, F. X., Sax, M., Zehner, Z.Z. (1990) Mol. Cell. Bio. 10: 2349-2358.
8. Furst, P., Hu, S., Hackett, R., Hamber, D. (1988) Cell 55: 705-717.
9. Gorman, C.M., Moffat, L.F., Howard, B.H. (1982) Mol. Cell. Biol. 2: 1044-1051.
10. Hai, T., Lin, F., Coukos, W.J., Green, M.R. (1989) Genes Dev. 3: 2083-2090.
11. Herbomel, P., Bourachot, B., Yaniv, M. (1984) Cell 39: 653-.
12. Kageyama, R., Pastan, I. (1989) Cell. 59: 815-825.
13. Kano, I., Nagai, F., Satoh, K., Ushiyama, K., Nakao, T., Kano, K. (1989) FEBS Lett. 250: 91-98.
14. Laimonis, L., Holmgren-Konig, M., Khoury, G. (1986) Proc. Natl. Acad. Sci. USA 83: 3151-3155.
15. Larsen, P.R., Harney, J. W., Moore, D.D. (1986) Proc. Natl. Acad. Sci. USA 83: 8283-8287.
16. Lin, Z., Edenberg, H.J., Carr, L.G. (1993) J. Biol. Chem. 268:10260-10267.
17. Maeda, M., Oshima, K., Tamura, S., Kaya, S., Mahmood, S., Reuben, M., Lasater, L., Sachs, G., Futai, M. (1991) J. Biol. Chem. 266: 21584-21588.
18. Miller, J.H. (1972) In "Experiments in molecular genetics". Cold Spring Harbor Labotatory. N.Y.
19. Orlowski, J., Lingrel, J.B. (1990) J. Biol . Chem. 265: 3462-3470
20. Rittling, S.R., Coutinho, L., Amram, T., Kolbe, M. (1989) Nucleic Acids Res. 17: 1619-1633.
21. Scriver, C. R., Claw, L.L. (1980) Annu. Rev. Genet. 14: 179-202.
22. Seed. B., Sheen, J.Y. (1988) Gene (Amst) 67: 271-277.
23. Shull, M.M., Pugh, D.G., Lingrel, J.B. (1990). Genomics 6: 451-460.
24. Tamura, S., Oshiman, K., Nishi, T., Mori, M., Maeda, M., Futai, M. (1992) FEBS Lett. 298: 137-141.
25. Takeyasu, K., Mizushima, A., Barnstein, A., Hamrick, M., Fambrough, D. M. (1990) The Sodium Pump: Recent Developments. 46:11-17
26. Wolitzky, B.A., Fambrough, D.M. (1986) J. Biol. Chem. 261: 9990-9999.
27. Yagawa, Y., Kawakami, K., Nagano, K. (1990) Biochim. Biophys. Acta. 1049: 286-292.
28. Yagawa, Y., Kawakami, K., Nagano, K. (1992) Mol. Cell. Biol. 12: 4046-4055.

# Transcriptional Regulation of Na$^+$/K$^+$-ATPase $\alpha$1 Gene Expression by 8-Bromo-Cyclic AMP in Renal Cells

M. Ahmad, L. Olliff, N. Weisberg, and R. M. Medford

Department of Medicine, Division of Cardiology, Emory University School of Medicine, Atlanta, GA 30322 USA

## Introduction

Several factors such as ouabain and cAMP inhibit Na$^+$/K$^+$-ATPase (NAKA) activity and ion transport function. Interestingly, inhibition of NAKA activity with ouabain results in an increase in the number of expressed NAKA functional units (12). The molecular mechanisms mediating this apparently compensatory response to NAKA inhibition are unknown. A consequence of NAKA inhibition is an increase in intracellular Na$^+$ among other ionic alterations. It is thus of particular interest that intracellullar Na$^+$ and/or Ca$^{2+}$ induce the transcription of the NAKA $\alpha$1 catalytic subunit gene. A linkage between NAKA activity and transport function and the intracellular ionic milieu may underly the molecular mechanism regulating NAKA gene expression in response to inihbitors of NAKA activity and function.

cAMP inhibits NAKA activity in renal and several other cell types (1,6,13,14). Whethter and to what degree cAMP regulates NAKA gene expression at the transcriptional or posttranscriptinal levels is not well understood. Structural analysis of the NAKA $\alpha$1 promoter suggests that cAMP regulated transcriptional proteins may play a role in NAKA $\alpha$1 gene expression. The NAKA $\alpha$1 gene promoter (9,15)contains two elements proximal to the transcriptional initiation site, CRE(-69) and CRE(-87), that are homologous to cyclic AMP/Ca$^{2+}$ response element (CRE) (3,7,8,10). CRE(-69) has a critical core motif (TGACG) which is present in the promoter of several genes that are inducible by cAMP while CRE(-87) has half intact site(TGAC). CRE binding (CREB) protein is activated by a phosphorylation event through two different types of kinases: cyclic AMP dependent protein kinase A and calcium/calmodulin dependent kinases(CaM kinases). This raises the possiblity that through ATF/CREB $\alpha$1 gene transcription may be modulated both by cAMP and an ionic signal. In this study, we establish that cAMP regulates the transcriptional activity of the NAKA $\alpha$1 promoter through ATF/CREB prtoeins in a feline kidney cell line (CRFK).

## 8-bromo-cAMP induces transcriptional activity of NAKA $\alpha$1 promoter

The NAKA $\alpha$1 promoter (coordinates -481 to +197) was attached to pBLCAT3, a promoterless pUC18 based chloramphenicol acetyl transferase (CAT) gene vector, (pA1PCAT10) and was transiently transfected into CRFK cells. The CAT assays were performed after 24 hour stimulation of the cells with 8-bromo-cAMP. The CAT activity is presented as percentage of chloramphenicol converted to acetyl chloramphenicol (% conversion). As shown in Figure 1, 8-bromo-cAMP induced CAT activity driven by the NAKA $\alpha$1 promoter three-fold in CRFK cells. This correlated well with the approximately 2-fold induction of $\alpha$1 mRNA accumulation in CRFK cells (data not shown). Under the same conditions 8-bromo-cAMP was unable to induce the CAT activity of reporter genes CAT13(NAKA $\alpha$1 promoter attached to pBLCAT3

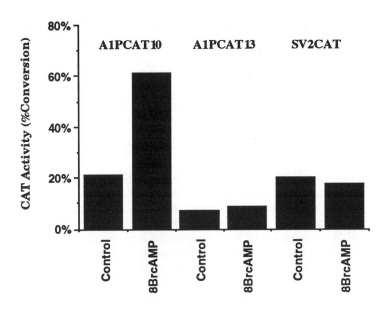

Figure 1. 8-bromo-cAMP induces the transcriptional activity of the NAKA α1 Promoter in CRFK cells.
pAlPCAT10 (10μg) is transiently transfected into CRFK cells using a standard calcium phosphate procedure. 5μg of a Rous sarcoma virus promoter/enhancer driven β-galactosidase reporter gene (pRSVβ-gal) is cotransfected as a control to correct for any differences in transfection efficiencies. The CAT activity is determined after stimulating the cells with 8-bromo-cAMP(1 mM) for 24 hours. The acetylated and unacetylated forms of the chloramphenicol are separated by thinlayer chromatography and the percentage of chloromphenicol converted to acetylated form is determined (% conversion). The CAT activity of pAlPCAT10 and pAlPCAT13 and pSV2CAT in the absence and presence of 8-bromo-cAMP is presented as a bar chart.

in reverse orientation) and pSV2CAT (SV2, simian virus promoter enhancer, driven CAT reporter gene). These results demonstrate that cAMP induces transcriptional activation of the NAKA α1 promoter in CRFK cells.

## 8-bromo-cAMP induces transcriptional Activity of CRE-containing PUC-1 Core driven heterologous promoter

CRE(-69) and CRE(-87) are located in a highly conserved region of the NAKA α1 promoter termed PUC-1 core (-104 to -57) (Figure 2A)(2). To determine whether this region could confer cAMP responsiveness to a heterologous promoter, PUC1 was inserted into pC"(an enhancerless SV40 promoter driven CAT reporter gene)(pB1CAT) and was transiently transfected into CRFK cells. The CAT activity was determined after treating the cell with 8-bromo-cAMP for 24 hours. As shown in Figure 2B, 8-bromo-cAMP induces the basal activity of pB1CAT at least 3 fold. Under the same conditions 8-bromo-cAMP was unable to significantly induce pC"CAT (lacking PUC-1 core) compared with pB1CAT.

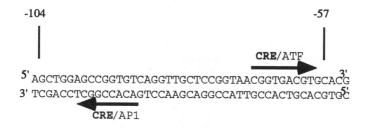

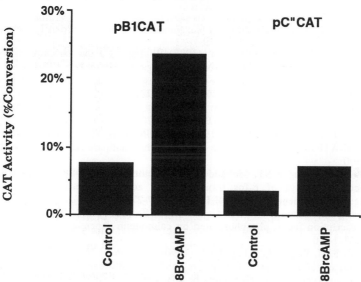

Figure 2. 8-bromo-cAMP induces the transcriptional activity of the PUC-1 core driven heterologous promoter in renal cells.
(A) Structure of the PUC-1 core of NAKA $\alpha 1$ promoter. (B) pB1CAT (10µg) is transiently transfected into CRFK cells using a standard calcium phosphate procedure. 5 µg of pRSVβ-gal is contransfected as a control to correct for transection efficiencies. The CAT activity is determined after stimulating the cells with 8-bromo-cAMP. The acetylated and unacetylated forms of chloramphenicol were separated by thinlayer chromatography and the percentage of chloramphenicol converted to acetylated form was calculated(% conversion).

These result suggest that PUC-1 core of the NAKA $\alpha 1$ promoter contains functional CRE(s) which are responsible, at least in part, for the transactivation of NAKA $\alpha 1$ promoter by cAMP. This transcriptional mechanism is likely responsible, at least in part, for the induction of $\alpha 1$ mRNA accumulation by cAMP in CRFK cells and may underly the observed induction of NAKA units in other cell types such as neurons exposed to forskolin (11). Whether these same elements also mediate a transcriptional effect that couples elevated intracellular $Na^+$ and free $Ca^{2+}$ levels with NAKA inhibition (4,5), needs to be determined. Since CRE has been shown to mediate the effect of both elevated cAMP and intracellular free $Ca^{2+}$ (3,7,8,10)it is likely that the different pathways that increase the intracellular cAMP and/or free $Ca^{2+}$ levels also utilize the same element(s), CRE(-69) and/or CRE(-87), for the transactivation of NAKA $\alpha 1$ gene.

47

## Acknowledgment

This work was supported in part by NIH grant RO1 HL45933 to RMM. RMM is an Established Investigator of the American Heart Association.

## References

1. Lingham RB, Sen AK. (1982) Regulation of rat brain (Na+ +K+)-ATPase activitybycyclic AMP. Biochim Biophys Acta 688: 475-485
2. Medford RM, Ahmad M, Olliff L Weisburg NK (1991) Transcriptional regulation of $Na^+$, $K^+$-ATPase $\alpha$-1 gene expression: Evidence for oncogene- and hormone mediated transcriptional factor interactions. In: Bonvalet J, Farman N, Lombes M Rafestin-Oblin M (ed.), Aldosterone: Fundamental Aspects. John Libbey, Abbaye de Fontevraud (France), pp. 131-139
3. Montminy MR, Bilezikjian LM. (1987) Binding of a nuclear protein to the cyclic AMP response element of the somatostatin gene. Nature 328: 175-178
4. Rayson BM. (1991) [Ca2+]i regulates transcription rate of the Na+/K(+)-ATPase alpha 1 subunit. J Biol Chem 266: 21335-21338
5. Rayson BM. (1993) Calcium: a mediator of the cellular response to chronic Na+/K(+)-ATPase inhibition. J Biol Chem 268: 8851-8854
6. Satoh T, Cohen HT, Katz AI. (1992) Intracellular signaling in the regulation of renal Na-K-ATPase. I. Role of cyclic AMP and phospholipase A2. J Clin Invest 89: 1496-1500
7. Sheng M, Dougan ST, McFadden G, Greenberg ME. (1988) Calcium and growth factor pathways of c-fos transcriptional activation require distinct upstream regulatory sequences. Mol Cell Biol 8: 2787-2796
8. Sheng M, Thompson MA, Greenberg ME. (1991) CREB: a Ca(2+)-regulated transcription factor phosphorylated by calmodulin-dependent kinases. Science 252: 1427-1430
9. Shull MM, Pugh DG, Lingrel JB. (1990) The human Na, K-ATPase alpha 1 gene: characterization of the 5'-flanking region and identification of a restriction fragment length polymorphism. Genomics 6: 451-460
10. Silver BJ, Bokar JA, Virgin JB, Vallen EA, Milsted A, Nilson JH. (1987) Cyclic AMP regulation of the human glycoprotein hormone alpha-subunit gene is mediated by an 18-base-pair element. Proc Natl Acad Sci USA 84: 2198-2202
11. Swan AC, Steketee J. (1989) Forskolin infusion in vivo increases oubain binding in brainn. Brain Res 476: 351-353
12. Taormino JP, Fambrough DM. (1990) Pre-translational regulation of the (Na+ + K+)-ATPase in response to demand for ion transport in cultured chicken skeletal muscle. J Biol Chem 265: 4116-4123
13. Tria E, Luly P, Tomasi V, Trevisani A, Barnabei O. (1974) Modulation by cyclic AMP in vitro of liver plasma membrane (Na+-K+)-ATPase and protein kinases. Biochim Biophys Acta 343: 297-306.
14. Tung P, Pai G, Johnson DG, Punzalan R, Levin SR. (1990) Relationships between adenylate cyclase and Na+, K+-ATPase in rat pancreatic islets. J Biol Chem 265: 3936-3939
15. Yagawa Y, Kawakami K, Nagano K. (1990) Cloning and analysis of the 5'-flanking region of rat Na+/K(+)-ATPase alpha 1 subunit gene. Biochim Biophys Acta 1049: 286-292

# Cloning of the Dog Na$^+$/K$^+$-ATPase α1-Subunit

Z.J. Xie, H. Li, Q. Liu, Y. Wang, A. Askari and R.W. Mercer

Department of Pharmacology, Medical College of Ohio, Toledo, Ohio 43699 USA and Department of Cell Biology and Physiology, Washington University School of Medicine, St. Louis, Missouri 63110 USA

## Introduction

Many of the kinetic, proteolytic and structural studies on the Na$^+$/K$^+$-ATPase have been conducted on the enzyme isolated from the dog kidney. While the cDNA for the dog ß1-subunit is available (1), the cDNA coding for the α-subunit has not been isolated. The availability of the dog α-subunit cDNA will make it possible to compare previous studies with results obtained from the expression of the dog kidney subunit. The purpose of this study was to isolate the cDNA coding for the α-subunit from the dog kidney.

## Isolation of dog Na$^+$/K$^+$-ATPase α-subunit cDNA clones

The strategy for the cloning of the dog Na$^+$/K$^+$-ATPase α1-subunit is based on the procedure of RT-PCR amplification (6). 1 μg of poly (A$^+$) mRNA from dog kidney was reverse transcribed using random hexamers (New England Biolabs, Beverly, MA). The reaction mixture was diluted to 1 ml with TE buffer (10 mM Tris-Cl, pH 8.0, 1 mM EDTA) and stored at 4° C. 1 μl of the diluted reaction mixture was used in a 100 μl PCR reaction consisting of 1 pmol of each primer, 10 mM Tris-Cl, pH 8.3, 50 mM KCl, 2 mM MgCl$_2$, 200 μM each dNTP, and 2 units of *Taq* DNA polymerase (Perkin-Elmer, Norwalk, CT). Two sets of PCR primers were used to amplify the 5' and 3' ends of the cDNA. For the 5' end of the mRNA the primer was based on the 5'-untranslated region of the mammalian sequences (5,8,9) and consisted of: 5'-AAGCTAGCCGCCATGGGGAAGGGGG. The corresponding 3' primer was based on the sequence from a partial dog α1-subunit cDNA (kindly provided by R. Farley): 5'-ATCTCCATCACGGAGCCACAGC. The set of primers designed to amplify the 3' end of the cDNA were based on the mammalian sequences and consisted of: 5'-TTGCTGGTCTTTGTAACAGG and 5'-TCCACAACGTGCAGGACACG. According to the available Na$^+$/K$^+$-ATPase α1-subunit cDNA sequences, there is a unique *Afl* II restriction site in the middle of the mammalian coding regions, and this restriction site is also present in the dog kidney partial cDNA sequence. The two sets of PCR primers were therefore designed to incorporate this unique *Afl* II site into the amplified segments. Cycles were as follows: 94°, 5 min; 75° C 5 min (*Taq* polymerase was added during this period of time); 94°C 1 min, followed by 25 cycles consisting of 94° C, 20 s, 56° C, 30 s, and 72° C 30 s. The PCR products were electrophoresed on a 1% agarose gel. Both 5' end and 3' end DNA fragments were excised, and eluted separately. The eluted DNA fragments were subcloned into pBluescript SK (Strategene) previously digested with *EcoR* V, and linked with a single ddT using DNA terminal transferase (3).

49

The T3 sense subclones for both 5' end and 3' end cDNA clones were selected by restriction enzyme digestion, and the subclones were sequenced to confirm that the clones were the Na+/K+-ATPase α1-subunit. Both of the 5' end and 3' end clones were then digested with *EcoR* I and *Afl* II, and ligated to obtain the full length dog kidney α1-subunit cDNA designated pBS SK dog α.

## Sequencing and sequence comparison of the cloned dog kidney Na$^+$/K$^+$-ATPase α1-Subunit cDNA

Both strands of the dog α1 cDNA were sequenced using the method of Sanger et al. (7). The deduced amino acid sequence from the cloned cDNA shows the identical *N*-terminal amino acid sequence of the dog kidney enzyme (GRNKYEPAA?SE) as reported with the exception of third amino acid N (2). It also has the identical *N*-terminal sequence of 83 kD chymotrysin fragment (4). Overall this clone shows 90% identity at the nucleic acid and 97% identity at amino acid levels with the α1-subunit cDNA of sheep kidney (8). Interestingly there is a deletion of both lysine and serine residues at the *N*-terminal of the peptides in this dog kidney clone which was found in both sheep and pig kidney α-subunit cDNAs, comparing to the rat α-subunit cDNA (5,8,9). The dog kidney α cDNA has the identical ouabain-sensitive H1-H2 ectodomain as the sheep cDNA (QAATEEEPQNDN). It also has the conserved sequences for both phosphorylation and FITC binding.

## *In vitro* translation of the cloned dog kidney Na$^+$/K$^+$-ATPase α1-Subunit cDNA

pBS SK dog α was translated *in vitro* using a TNT coupled reticulocyte lysate system (Promega, Madison, WI). Proteins labeled with $^{35}$S-methionine were analyzed by SDS-PAGE (7.5% gel). Using this system a 100 kD protein was translated from the pBS SK dog α (not shown). In summary these data indicate that the cloned cDNA is the dog kidney Na+/K+-ATPase α1-subunit.

```
AAGCTAGCCGCCATGGGGAAGGGGGTCGGACGTGATAAATATGAACCTGCGGCCGTTTCAGAGCATGGCGACAAAAAAAAGGCCAAGAAA        90
              MetGlyLysGlyValGlyArgAspLysTyrGluProAlaAlaValSerGluHisGlyAspLysLysLysAlaLysLys

GAAAGGGATATGGATGAACTGAAGAAAGAAGTTTCTATGGATGACCATAAACTGAGCCTTGATGAACTTCATCGCAAATATGGAACAGAC       180
GluArgAspMetAspGluLeuLysLysGluValSerMetAspAspHisLysLeuSerLeuAspGluLeuHisArgLysTyrGlyThrAsp

CTGAGTCGAGGCCTAACAACCGCTCGAGCTGCTGAGATCCTGGCTCGAGATGGTCCCAATGCCCTCACCCCACCTCCCACCACTCCCGAA       270
LeuSerArgGlyLeuThrThrAlaArgAlaAlaGluIleLeuAlaArgAspGlyProAsnAlaLeuThrProProProThrThrProGlu

TGGGTCAAGTTCTGTCGGCAGCTGTTTGGAGGTTTCTCGATGTTACTGTGGATTGGAGCGATTCTTTGTTTCTTAGCTTATGGCATCCAA       360
TrpValLysPheCysArgGlnLeuPheGlyGlyPheSerMetLeuLeuTrpIleGlyAlaIleLeuCysPheLeuAlaTyrGlyIleGln

GCTGCCACGGAAGAGGAACCTCAAAATGATAATCTATATCTTGGTGTGGTACTATCAGCTGTTGTCATCATTACTGGCTGTTTCTCCTAC       450
AlaAlaThrGluGluGluProGlnAsnAspAsnLeuTyrLeuGlyValValLeuSerAlaValValIleIleThrGlyCysPheSerTyr

TATCAAGAAGCTAAAAGTTCAAAGATCATGGAATCCTTCAAAAACATGGTTCCTCAGCAAGCACTTGTGATTCGAAATGGTGAAAAAATG       540
TyrGlnGluAlaLysSerSerLysIleMetGluSerPheLysAsnMetValProGlnGlnAlaLeuValIleArgAsnGlyGluLysMet

AGCATCAACGCAGAGGAGGTTGTAATCGGGGACTTGGTGGAAGTGAAAGGAGGAGACCGAATCCCTGCTGATCTCGAATCATATCTGCC        630
SerIleAsnAlaGluGluValValIleGlyAspLeuValGluValLysGlyGlyAspArgIleProAlaAspLeuArgIleIleSerAla

AACGGCTGCAAGGTGGATAACTCCTCGCTCACTGGTGAATCAGAGCCCCAGACTAGATCTCCAGACTTCACAAATGAAAACCCCCTGGAA       720
AsnGlyCysLysValAspAsnSerSerLeuThrGlyGluSerGluProGlnThrArgSerProAspPheThrAsnGluAsnProLeuGlu

ACGAGGAATATTGCCTTCTTCTCAACCAACTGCGTGAAAGGTACTGCGCGCGGCATTGTTGTATACACTGGGGATCGCACTGTCATGGGA       810
ThrArgAsnIleAlaPhePheSerThrAsnCysValLysGlyThrAlaArgGlyIleValValTyrThrGlyAspArgThrValMetGly

AGAATTGCCACACTTGCTTCTGGGCTGGAAGGAGGCCAGACTCCCATCGCTGCAGAAATTGAACATTTTATCCATATCATCACGGGTGTG       900
ArgIleAlaThrLeuAlaSerGlyLeuGluGlyGlyGlnThrProIleAlaAlaGluIleGluHisPheIleHisIleIleThrGlyVal
```

50

```
GCCGTGTTTCTGGGCGTGTCCTTCTTTATCCTTTCTTTGATCCTTGAGTATACCTGGCTTGAGGCCGTCATCTTTCTCATCGGAATCATC    990
AlaValPheLeuGlyValSerPhePheIleLeuSerLeuIleLeuGluTyrThrTrpLeuGluAlaValIlePheLeuIleGlyIleIle

GTAGCCAATGTGCCAGAAGGTCTGCTGGCCACTGTCACGGTATGTCTGACCCTCACTGCCAAACGCATGGCAAGGAAAAACTGCTTAGTG   1080
ValAlaAsnValProGluGlyLeuLeuAlaThrValThrValCysLeuThrLeuThrAlaLysArgMetAlaArgLysAsnCysLeuVal

AAGAACTTAGAAGCTGTGGAGACCTTGGGATCCACGTCCACCATCTGCTCAGATAAAACTGGAACTCTGACTCAGAACCGGATGACCGTT   1170
LysAsnLeuGluAlaValGluThrLeuGlySerThrSerThrIleCysSerAspLysThrGlyThrLeuThrGlnAsnArgMetThrVal

GCCCACATGTGGTTCGACAATCAAATCCACGAAGCCGACACGACAGAGAATCAGAGTGGTGTCTCGTTCGATAAGAGTTCAGCCACCTGG   1260
AlaHisMetTrpPheAspAsnGlnIleHisGluAlaAspThrThrGluAsnGlnSerGlyValSerPheAspLysSerSerAlaThrTrp

CTCGCTCTGTCCAGAATTGCGGGTCTTTGTAACAGGGCGGTGTTTCAGGCTAACCAGGAAAACCTGCCTATCCTTAAGCGGGCAGTTGCA   1350
LeuAlaLeuSerArgIleAlaGlyLeuCysAsnArgAlaValPheGlnAlaAsnGlnGluAsnLeuProIleLeuLysArgAlaValAla

GGTGACGCCTCAGAGTCCGCACTCTTAAAATGCATCGAGCTGTGCTGTGGTTCTGTGAAGGAGATGAGAGATCGATATGCCAAGATTGTG   1440
GlyAspAlaSerGluSerAlaLeuLeuLysCysIleGluLeuCysCysGlySerValLysGluMetArgAspArgTyrAlaLysIleVal

GAGATCCCCTTCAACTCTACCAACAAGTACCAGCTGTCCATCCACAAGAACCCTAACACGTCTGAACCCCGACACCTGTTGGTGATGAAG   1530
GluIleProPheAsnSerThrAsnLysTyrGlnLeuSerIleHisLysAsnProAsnThrSerGluProArgHisLeuLeuValMetLys

GGTGCTCCGGAAAGGATCCTGGACCGCTGCAGCTCCATCCTCCTGCACGGCAAGGAGCAGCCTCTGGATGAGGAGCTGAAGGATGCCCTT   1620
GlyAlaProGluArgIleLeuAspArgCysSerSerIleLeuLeuHisGlyLysGluGlnProLeuAspGluGluLeuLysAspAlaLeu

CAGAATGCCTACCTGGAGCTGGGCGGCCTCGGAGAGCGAGTGCTGGGTTTCCGCCACCTTTTCCTGCCAGATGAACAGTTTCCTGAAGGG   1710
GlnAsnAlaTyrLeuGluLeuGlyGlyLeuGlyGluArgValLeuGlyPheArgHisLeuPheLeuProAspGluGlnPheProGluGly

TTCCAGTTTGACACTGATGATGTGAATTTCCCTGTTGAGAACCTCTGCTTTGTGGGCTTCATCTCCATGATTGGCCCTCCACGGGCTGCT   1800
PheGlnPheAspThrAspAspValAsnPheProValGluAsnLeuCysPheValGlyPheIleSerMetIleGlyProProArgAlaAla

GTTCCTGACGCTGTGGGCAAATGCCGAGGTGCTGGAATTAAGGTCATTATGGTCACTGGAGACCATCCGATCACAGCCAAAGCCATTGCC   1890
ValProAspAlaValGlyLysCysArgGlyAlaGlyIleLysValIleMetValThrGlyAspHisProIleThrAlaLysAlaIleAla

AAAGGTGCAGGCATCATCTCAGAAGGCAATGAGACCGTGGAAGACATTGCTGCCCGTCTCAACATCCCAGTGAGGCAGGTGAACCCCAGA   1980
LysGlyAlaGlyIleIleSerGluGlyAsnGluThrValGluAspIleAlaAlaArgLeuAsnIleProValArgGlnValAsnProArg

GACGCCAAGGCCTGTGTGGTACATGGAAGCGACCTGAAAGACATGACCTCCGAGCAGCTGGATGGCATTCTGAAGTACCACACGGAGATC   2070
AspAlaLysAlaCysValValHisGlySerAspLeuLysAspMetThrSerGluGlnLeuAspGlyIleLeuLysTyrHisThrGluIle

GTGTTTGCCAGGACCTCTCCTCAGCAGAAGCTTATCATTGTGGAAGGCTGCCAGAGACAGGGGGCTATTGTGGCTGTAACTGGTGATGGC   2160
ValPheAlaArgThrSerProGlnGlnLysLeuIleIleValGluGlyCysGlnArgGlnGlyAlaIleValAlaValThrGlyAspGly

GTGAATGACTCTCCAGCCTTGAAGAAGGCAGACATTGGGGGTTGCCATGGGGATTGTTGGCTCCGACGCGTCTAAGCAAGCTGCCGACATG   2250
ValAsnAspSerProAlaLeuLysLysAlaAspIleGlyValAlaMetGlyIleValGlySerAspAlaSerLysGlnAlaAlaAspMet

ATTCTTCTAGATGACAACTTTGCGTCAATTGTGACTGGAGTAGAAGAAGGTCGTCTGATCTTTGATAACTTGAAGAAATCCATTGCCTAC   2340
IleLeuLeuAspAspAsnPheAlaSerIleValThrGlyValGluGluGlyArgLeuIlePheAspAsnLeuLysLysSerIleAlaTyr

ACCCTGACCAGTAACATTCCAGAGATCACCCCCTTCCTGATATTTATTATTGCAAACATTCCACTACCACTGGGGACTGTCACCATCCTC   2430
ThrLeuThrSerAsnIleProGluIleThrProPheLeuIlePheIleIleAlaAsnIleProLeuProLeuGlyThrValThrIleLeu

TGCATTGACTTGGGCACAGACATGGTCCCTGCTATCTCCCTGGCTTATGAGCAAGCTGAGAGCGACATCATGAAGAGACAGCCCAGAAAC   2520
CysIleAspLeuGlyThrAspMetValProAlaIleSerLeuAlaTyrGluGlnAlaGluSerAspIleMetLysArgGlnProArgAsn

CCCAAAACGGACAAGCTTGTGAATGAGCGGCTGATCAGCATGGCCTATGGACAGATCGGTATGATCCAGGCCTTGGGGGGGCTTCTTCACT   2610
ProLysThrAspLysLeuValAsnGluArgLeuIleSerMetAlaTyrGlyIleGlnIleGlyMetIleGlnAlaLeuGlyGlyPhePheThr

TACTTTGTGATTCTGGCTGAGAACGGCTTCCTCCCGACCCACCTGCTGGGGCTCCGAGTGGACTGGGATGACCGCTGGATCAACGACGTG   2700
TyrPheValIleLeuAlaGluAsnGlyPheLeuProThrHisLeuLeuGlyLeuArgValAspTrpAspAspArgTrpIleAsnAspVal

GAGGACAGCTATGGGCAGCAGTGGACCTACGAACAGAGGAAAATCGTGGAGTTCACTTGCCACACGGCCTTCTTCGTCAGTATCGTGGTG   2790
GluAspSerTyrGlyGlnGlnTrpThrTyrGluGlnArgLysIleValGluPheThrCysHisThrAlaPhePheValSerIleValVal

GTGCAGTGGGCTGACTTGGTCATCGTAAGACCAGGAGGAACTCAGTCTTCCAGCAGGGGATGAAGAACAAGATCCTAATATTTGGCCTC   2880
ValGlnTrpAlaAspLeuValIleCysLysThrArgArgAsnSerValPheGlnGlnGlyMetLysAsnLysIleLeuIlePheGlyLeu

TTTGAAGAGACGGCCCTTGCTGCTTTCCTGTCCTACTGCCCTGGAATGGGTGTTGCCCTGAGGATGTACCCCCTCAAACCTACCTGGTGG   2970
PheGluGluThrAlaLeuAlaAlaPheLeuSerTyrCysProGlyMetGlyValAlaLeuArgMetTyrProLeuLysProThrTrpTrp

TTCTGTGCCTTTCCCTACTCTCTTCTCATCTTCGTGTACGACGAAGTCCGAAAACTCATCATCAGGCGACGCCCTGGCGGCTGGGTAGAG   3060
PheCysAlaPheProTyrSerLeuLeuIlePheValTyrAspGluValArgLysLeuIleIleArgArgArgProGlyGlyTrpValGlu

AAAGAAACCTACTACTAGACCCCCGTGTCCTGCACGTTGTGAATCAAGCTT                                          3111
LysGluThrTyrTyr
```

Figure 1. Nucleotide and deduced amino acid sequences of the dog kidney Na$^+$/K$^+$-ATPase $\alpha$1-subunit.

*Acknowledgments*

We thank Dr. R. Farley for providing the partial cDNA sequence for the dog $\alpha$1-subunit. This work was supported by National Institutes of Health Grant HL 36573.

51

# References

1. Brown TA, Horowitz B, Miller RP, McDonough AA, and Farley RA (1987) Molecular cloning and sequence analysis of the (Na+, K+)-ATPase ß subunit from dog kidney. Biochim Biophys Acta 912:244-253

2. Cantley LC (1981) Structure and mechanism of the (Na,K)-ATPase. Curr Topics Bioenerg 11: 201-237

3. Holton TA and Graham MW (1990) A simple and efficient method for direct cloning of PCR products using ddT-tailed vectors. Nucleic Acids Res 19: 1156.

4. Huang WH, Ganjeizadeh M, Wang Y, Chiu IN, and Askari A (1990) Autoregulation of the phosphointermiate of Na+/K+-ATPase by the amino-terminal domain of the α-subunit. Biochim Biophys Acta 1030: 65-72

5. Ovchinnikov YA, Modyanov NN, Broude NE, Petrukhin KE, Grishin AV, Arzamazova NM, Aldanova NA, Monastyskaya GS, and Sverdlov ED (1986) Pig kidney Na+, K+-ATPase primary structure and spatial organization. FEBS Lett 201: 237-245

6. Saiki RK, Scharf F, Faloona F, Mullis KB, Horn GT, Ehrlich HA, and Arnheim N (1985) Enzymatic amplification of ß-globin genomic sequences and restriction site analysis for diagnosis of sickle cell anemia. Science 230: 1350-1354

7. Sanger F, Nicklen S, and Coulson AR (1977) DNA sequencing with chain terminating inhibitors. Proc Nalt Acad Sci USA 74:5463-5467

8. Shull GE, Greeb J, and Lingrel JB (1986) Molecular cloning of three distinct forms of the Na+,K+-ATPase α-subunit from rat brain. Biochemistry 25: 8125-8132

9. Shull GE, Schwartz A, and Lingrel JB (1985) Amino-acid sequence of the catalytic subunit of the (Na+-K+)ATPase deduced from a complementary DNA. Nature 316: 691-695

# Cloning and Characterization of a New Member of the Family of Na⁺/K⁺-ATPase Genes

**Karl Lücking, Peter L. Jørgensen and Li-Mei Meng**

Biomembrane Research Centre, August Krogh Institute, Copenhagen University, DK-2100, Denmark.

## Introduction

The Na⁺/K⁺-ATPase consists of one $\alpha$ and one $\beta$ subunit. In mammals, three isoforms of the $\alpha$ subunit, and two isoforms of the $\beta$ subunit have been identified by molecular genetic and immunological techniques. Two additional genes ($\alpha$C and $\alpha$D) with high similarity to the $\alpha$ genes have also been cloned and partly sequenced. The functional status of these two genes remains to be determined. Here we report the data on cloning and characterization of a new Na⁺/K⁺-ATPase $\alpha$ subunit related gene from rabbit.

## Cloning and sequencing

A rabbit liver genomic DNA library constructed in EMBL3 Sp6/T7 was purchased from Clontech (Palo Alto, CA). The cDNA encoding rat Na⁺/K⁺-ATPase $\alpha$1 subunit (6), a *Bst*E II fragment (nucleotides 56 to 877) of the rat Na⁺/K⁺-ATPase $\alpha$2 subunit cDNA (6), and a *Pst* I fragment (nucleotides 56 to 1685) of the rat $\alpha$3 cDNA (6) were labelled with biotin using a nick translation kit (Tropix, Bedford, MA). Through screening the library with the biotin labelled probes at low stringency and analyzing the positive clones with restriction endonuclease, three different positive clones, λLMM31.1, λLMM52.2 and λLMM52.3 were isolated. The insert from each clone was subcloned into pBluescript II SK (Stratagene, San Diego, CA), and the nucleotide sequence of a portion of each clone was determined by the dideoxy chain termination method. Comparison of the sequences of the rabbit clones with those of the rat Na⁺/K⁺-ATPase $\alpha$ subunit isoform genes (6), showed that λLMM31.1 and λLMM52.3 contained a part of the $\alpha$1 and $\alpha$2 genes, respectively. Lambda LMM52.2 differed

**Figure 1.** Restriction map of ΛLMM52.2. The exons are indicated by solid bars. B, *Bam*H I; H, *Hin*d III. The clone was isolated from the rabbit genomic DNA library by plaque lift. The rat cDNAs were labelled with biotin. Hybridization was performed at 65°C overnight in 20 ml buffer containing 5% Dextran Sulphate, 1 mM EDTA (pH 8.0), 0.2% Heparin, 1 M NaCl, 0.5% Polyvinylpyrrolidone, 4% SDS, 50 mM Tris-Cl (pH 7.5). Following the hybridization, the filters were washed two times for 5 min in 2 x SSC, 1% SDS, and then two times for 20 min in 1 x SSC, 1% SDS. All washes were performed at 65°C. The positive clones were visualized by Chemiluminescence (Tropix, Bedford, MA).

53

from λLMM31.1 and λLMM52.3 (3) both by restriction fragment mapping and by DNA sequencing. The restriction map of this clone is shown in Fig. 1. The nucleotide sequences of all coding regions and a major part of the introns were determined. The results showed that λLMM52.2 contained sequences corresponding to the exons 6 to 8 of the α genes of Na$^+$/K$^+$-ATPase (Fig.2). However, these sequences were different from those of the α1 and α2 genes of the Na$^+$/K$^+$-ATPase of rabbit (Meng LM, unpublished). The nucleotide sequence of rabbit α3 gene was not available, but comparison of the derived amino acid sequences showed that the amino acid sequence of λLMM52.2 was 82% identical to those of α3 subunit of human and rat. In the same region, there is 100% homology with respect to amino acid sequences between human and rat α3 subunit. Even the more distant related chicken α3 sequence showed 97% identity to that of human. Thus, we can conclude that λLMM52.2 is different from α3. The derived amino acid sequence of LMM52.2 was also compared with that of αD (5) and that of H$^+$/K$^+$-ATPase (2,4). The results revealed that λLMM52.2 is less related to these genes than to the α genes of Na$^+$/K$^+$-ATPase (Fig.3). On the basis of

```
CGAGCTCTGGTCATCCGAGGTGGGGAGAAGATGCAGATTCTTGTAAAGGATGTGGTGTTGGGAGATCTGGTAGAGTGAAAGGGGGAG
                               ‾‾‾‾‾‾‾‾‾‾‾‾‾‾‾‾‾‾
                               Oligo 1

ACCGAATCCCTGCCGACCTGCGGCTCATCTCTTCACAAGGATGTAAG♦GTGGACAACTCGTCCCTGACCGGGGAGTCGGAGCCCCAG
TCCCGCTCTCCTGACTTCACCCACGAGAATCCCCTGGAGACCCGCAACATCTGCTTCTTCTCCACCAATTGTGTAGAAG♦GGACAGC
CCGGGGCATTGTGATTGCTACAGGAGACTCCACGGTGATGGGCAGGATCGCCTCACTGACATCAGGCCTGGCAGTTGGCCAGACCCC
CATAGCTGCCGAGATTGAACACTTCATCCGTCTGATCACCATGATGGCCCTCTTCTTTGGTGTCACTTTTTTCGGGCTCTCACTCAT
                                                     TACTACCGGGAGAAGAAACC
                                                     ‾‾‾‾‾‾‾‾‾‾‾‾‾‾‾‾‾‾‾
                                                     Oligo 2

CCTGGGTTATGGCTGGCTGGAGGCTGTCATTTTCCTTATTGGCATCATTGTGGCCAATGTGCCTGAGGGGCTATTGGCCACTGTCAC
```

**Figure 2.** Nucleotide sequences of exons 6, 7 and 8 of λLMM52.2. Exon boundaries are marked with the symbol ♦. Location of oligos used for reverse transcription and PCR are underlined.

```
LMM52.2               RALVIRGGEKMQILVKDVVLGDLVEVKGGDRIPADLRLISSQGCKVDNSSLTGESEPQ 100%
Alfa 1                Q-----N----S-NAEE--V----------------I--AN--------------- 82 %
Alfa 2                Q-----E------NAEE--V----------V-----I----H--------------- 85 %
Alfa 3                Q-----E-----VNAEE--V-----I-----V-----I--AH--------------- 81 %
Alfa D                Q-----DS--KT-PSEQL-V--I-------Q----I-VL-----R------------ 64 %
Gastric H, K-ATPase   Q-T---D-D-F--NADQL-V-----M-----V---I-ILAA---------------- 63 %
Colon H, K-ATPase     Q-----DA--KV-SAEQL-V--V---I----Q----I--VF---------------- 64 %

LMM52.2               SRSPDFTHENPLETRNICFFSTNCVEGTARGIVIATGDSTVMGRIASLTSGLAVGQTP
Alfa 1                T------N---------A---------------VY---R-------T-A---EG----
Alfa 2                T---E----------------------------R-------T-A---E--R--
Alfa 3                T----C--D--------T-------------V-V----R-------T-A---E--K--
Alfa D                P--SE--------K----Y--T-L----VT-M--N---R-II-H----A--VGNEK--
Gastric H, K-ATPase   T---EC---S-------A----M-L----Q-L-VN---R-II------A--VENEK--
Colon H, K-ATPase     A--TE--------K--G-Y--T-L----T----N---R-II------A--VGSEK--

LMM52.2               IAAEIEHFIRLITMMALFFGVTFFGLSLILGYGWLEAVIFLIGIIVANVPEGLLATVT
Alfa 1                ---------HI--GV-V-L--S--I-----E-T-------------------------
Alfa 2                --M------Q---GV-V-L--S--V-------S------------------------
Alfa 3                --I------Q---GV-V-L--S--I-------T------------------------
Alfa D                --I-----VHIVAGV-VSI-IL--IIAVS-K-QV-DSI-------------------
Gastric H, K-ATPase   --I-----VDI-AGL-IL--A---IVAMCI--TF-R-MV-FMA-V---Y---------
Colon H, K-ATPase     --I-----VHIVAGV-VSIDII--ITAVCMK-YV-D-I----S--------------
```

**Figure 3.** Alignment of amino acid sequences of exons 6 to 8. Overall homology to λLMM52.2 is indicated.

this analysis we suggest that LMM52.2 represents a new member of the family of Na$^+$/K$^+$-ATPase genes.

## Detection of transcription product of λLMM52.2

To determine whether λLMM52.2 is transcriptionally competent, cytoplasmic RNA from various rabbit tissues was prepared (1). cDNA was synthesized by Superscript II (Gibco BRL) and by using a λLMM52.2 specific oligonucleotide (oligo 2, Fig.2) as the primer. The products were amplified by Vent$_R$ DNA polymerase (New England Biolabs) on a thermal cycler (Hybaid) with oligo 1 and oligo 2 as primers (Fig.2). The PCR products were then electrophoresised on an 1% agarose gel. The results showed that the λLMM52.2 transcript existed in the cytoplasmic RNA of rabbit brain, eye, heart, lung and liver (Fig.4, data for liver not shown), whereas no λLMM52.2 transcript was detected in RNA from colon, kidney, muscle, ovary, pancreas and tongue (data not shown). The PCR product was confirmed by restriction digestion and DNA sequencing to be identical to the exon sequences of λLMM52.2. These results demonstrate that λLMM52.2 is expressed at least at the level of mRNA, and expression of this gene is restricted to some tissues. Work is in progress to clone and sequence the rest of cDNA of λLMM52.2.

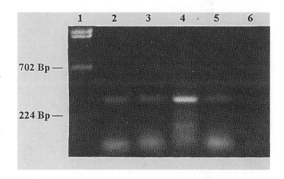

**Figure 4.** Detection of λLMM52.2 transcript in different rabbit tissues. Reverse transcription and PCR were performed as described in text. Product yield is not comparable between the samples due to different number of PCR cycles.
Lanes: 1: *BstE*II digested Lambda. DNA. 2: Brain; 3: Eye; 4: Heart; 5: Lung; 6: Negative control.

## Acknowledgments

We are thankful to Dr. Jerry B. Lingrel for providing rat Na$^+$/K$^+$-ATPase α subunits cDNA, to Dr. N. N. Modyanov for the αD amino acid sequence, and to Dorte Meinertz for the excellent technique assistance. This work was supported by the grants from the Biomembrane Research Centre and Danish Medical Science Research Council.

## References

1.  Clemens MJ (1984) Purification of Eukaryotic Messenger RNA. In "Transcription and Translation, a practical approach". Eds. Hames BD, Higgins SJ. IRL PRESS: pp215.
2.  Crowson MS, Shull GE (1992) Isolation and characterization of a cDNA encoding the putative distal colon H$^+$,K$^+$-ATPase. J. Biol. Chem. 267: 13740-13748.
3.  Kjærsig T, Jørgensen PL, Meng LM (1993) Proc. VII Int. Conf. on the Sodium Pump. Steinkopff Verlag, Darmstadt, Germany. submitted.
4.  Maeda, M., Oshiman, K.I., Tamura, S., Futai, M. (1990) Human gastric H$^+$,K$^+$-ATPase gene. Similarity to Na$^+$/K$^+$-ATPase genes in exon/intron organization but difference in

control region. J. Biol. Chem. 265: 9027-9032.

5. Modyanov NN, Petrukhin KE, Sverdlov VE, Grishin AV, Orlova MY, Kostina MB, Makarevich OI, Broude NE, Monastyrskaya GS, Sverdlov ED (1991) The family of human Na,K-ATPase genes - ATP1AL1 gene is transcriptionally competent and probably encodes the related ion transport ATPase. FEBS 278: 91-94.

6. Shull GE, Greeb J, Lingrel JB (1986) Molecular Cloning of three distinct forms of the $Na^+,K^+$-ATPase $\alpha$-subunit from rat brain. Biochemistry 25: 8125-8132.

# The Cloning of Homologous Na$^+$/K$^+$-ATPase Genes from New Species

C.P. Cutler, I.L. Sanders, G.A. Luke, N. Hazon, G. Cramb

School of Biological and Medical Sciences, University of St. Andrews, St. Andrews, Fife, U.K. KY16 9TS

## Introduction

The investigation of new models for the study of the mechanisms of regulation of ion transport in new species such the European eel (*Anguilla anguilla*), requires the use of gene probes to enable the measurement of mRNA levels using nucleic hybridisation techniques. One possibility, for proteins such as Na$^+$/K$^+$-ATPase, is to use genes already available from other species for this purpose. However, the cross reactivity of these genes, with diverse species such as the eel is likely to be low, especially in the case of the Na$^+$/K$^+$-ATPase β (β1) subunit gene. Consequently, the only alternative is to clone the genes concerned from the model organism. Normal cloning procedures requiring the screening of up to 10$^7$ colonies are both expensive and time consuming. However, using the recently developed PCR (polymerase chain reaction) technology (Perkin Elmer/Cetus), fragments of genes such as Na$^+$/K$^+$-ATPase, (where regions of amino acid homology are known to exist between species) can be quickly and easily cloned, and used as gene probes. This poster illustrates the methods used for the cloning of the Na$^+$/K$^+$-ATPase α and β genes from the European eel. These methods are however, easily adapted to any gene where known regions of amino acid or nucleotide homology exist.

## Methods

### Primer design

Initially, regions of high amino acid (or nucleotide) homology were identified. For the α subunit, these were coincidentally located in the same region of the amino acid chain as the fluorescein isothiocyanate (FITC) binding region and the 5-(p-fluorosulphonyl)-benzoyl-adenosine (FSBA) reactive site (2). For the β subunit, a region in the 3' untranslated portion of β subunit genes (3' homologous box) known to have high homology between species (8), as well as a region located at approximately amino acid 68-76 of the sequence were selected. These regions of amino acids were reverse translated to yield all the possible combinations of nucleotides which could code for them. Sense and antisense PCR primers were then made from these nucleotide sequence using inosine/cytidine wobbles where the nucleotide sequence was uncertain (5).

### PCR Techniques

RNA was prepared from the gills of 3 eels which had been adapted to seawater for 4 days. Messenger RNA (mRNA) was extracted from this using an oligo dT cellulose column (6). Using 5 µg of mRNA 1st strand cDNA was manufactured with Superscript II reverse transcriptase (Gibco BRL). PCR was performed using 5% of the total cDNA produced. The optimal primer concentration was determined empirically to be 5µM, using 200µM dNTPs, 0.5 units of Taq DNA polymerase (Perkin Elmer/Cetus) and standard enzyme buffer (1.5mM $MgCl_2$), in a reaction volume used of 25 µl. The reaction was cycled through 94°C, a variable annealing temperature, and 72°C for 1 minute each for 30 cycles. The optimal annealing temperature was determined empirically for each primer pair, and was normally in the range 44-60°C.

Cloning and Sequencing

PCR reaction products were electrophoresed through low melting point agarose (Gibco BRL). DNA bands were excised, purified using Magic DNA clean up kit (Promega) and then cloned into a plasmid using a TA Cloning kit (Invitrogen). Transformed bacterial colonies were extracted for plasmid DNA and, inserted DNA fragments were excised by restriction enzyme digestion. Plasmids containing DNA inserts were sequenced using Sequenase T7 DNA polymerase (USB).

Using the sequence of the known DNA fragments, primers were designed to extended the known sequence on both sides of the original fragment (see Fig. 1). One primer was made to an unknown homologous region of the gene as before and a second primer was made from within the previously identified sequence. In this way fragments were produced to extend the known sequence to near both ends of the gene ($Na^+/K^+$-ATPase α gene). The 3' end of the $Na^+/K^+$-ATPase α gene was cloned using a technique which is essentially the same as outlined in 3' RACE (rapid amplification of cDNA ends; Gibco BRL) kit. An antisense primer was used which contained a poly dT region which bound to the poly A tail of the mRNA. The 5' end was cloned using a 5' RACE kit (Gibco BRL). This used 2 internal antisense primers (1 for reverse transcriptase and 1 for PCR) and an external anchor sense primer. The cDNA was tailed with poly dC at the 5' end, and PCR was performed using the anchor primer which contained a poly dG tract.

The initial fragment of the $Na^+/K^+$-ATPase β subunit gene produced had an internal 86 nucleotide deletion (see Fig. 1). This region was re-amplified using exact sequence primers from either side of the deletion. The PCR fragment produced was sequenced directly without cloning. The 5' end of the β subunit gene was cloned and sequenced in the same fashion as the α subunit gene.

## Results and Discussion

The eel $Na^+/K^+$-ATPase α subunit gene has an open reading frame of 1022 amino acids with 112 untranslated nucleotides at the 5' end and 126 untranslated nucleotides at the 3' end (data not shown). When considering sequence data generated by PCR, the fidelity of the enzyme (Taq DNA polymerase) should be taken into account, as errors are occasionally introduced into the sequence. Consequently, a small number of amino acids will be incorrect.

A gene alignment with the sequences of other vertebrate $Na^+/K^+$-ATPase α or α1 subunit genes (data not shown) illustrates that the eel gene represents an α1 type isoform. Whether or not other α isoforms exist in fish species is at present unknown. The gene alignment also illustrates that, as expected, the eel $Na^+/K^+$-ATPase α subunit gene shares the highest homology (89.4%) with that of the White Sucker another Teleost fish (7). It is slightly surprising perhaps that the homology shared with the ray (*Torpedo californica;* 85.6%:(3)) an elasmobranch fish is lower than that shared with the clawed toad (*Xenopus laevis* ; 88.1%:(9)), the giant toad (*Bufo marinus*; 87.2%:(1)) or indeed other mammalian or avian species.

a,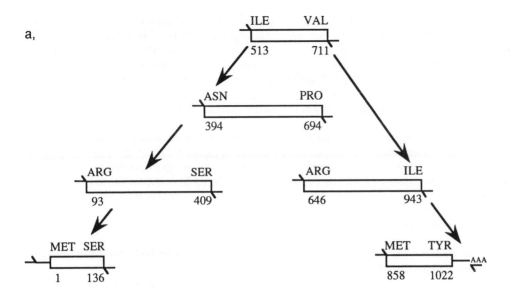

b,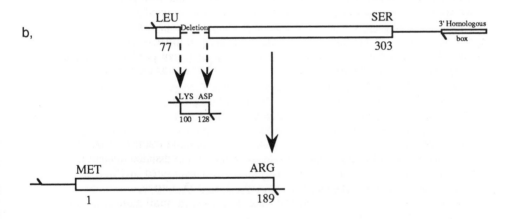

**Figure 1.** Schematic diagram of the PCR cloning of the Na⁺/K⁺-ATPase α (a) and β (b) genes from the European eel (*Anguilla anguilla*). Where open boxes represent PCR fragments, numbers and their associated letters represent amino acids and their position in the gene sequence, and ➘ represents the position of PCR primers.

end (data not shown). Similarly to the α subunit gene, an alignment of the β subunit gene with the sequences of other vertebrate Na$^+$/K$^+$-ATPase β or β1 subunit genes illustrates that the eel gene represents a β1 type isoform. Whether or not other β isoforms exist in fish species is at present unknown. Also similarly to the α subunit gene, the eel Na$^+$/K$^+$-ATPase β subunit gene shares the highest homology with that of the clawed toad (*Xenopus laevis;* 61.4%:(9)), the giant toad (*Bufo marinus*; 58.9%:(1)) and other mammalian or avian species, rather than with the ray (*Torpedo californica;* 56.2%:(4)).

The cloning of genes such as those of the Na$^+$/K$^+$-ATPase α and β subunits illustrates that gene probes from new species can be produced rapidly. This enables new models of ion transport to be further evaluated by quantitative nucleic acid hybridisation, as well as allowing antibodies to synthetic peptides to be produced from the amino acid sequences generated.

*Acknowledgements*

These studies were supported by both the Wellcome Trust and the Natural Environment Research Council (NERC).

## References

1. Jaisser F, Canessa CM, Horisberger JD, Rossier BC (1992) Primary Sequence and Functional Expression of a Novel Ouabain-resistant Na,K-ATPase. J Biol Chem 267: 16895-16903
2. Jorgensen PL, (1988) Overview: Structural Basis for Coupling of E$_1$-E$_2$ Transitions in αβ-Units of Renal Na,K-Translocation. Prog Clin Biol Res 268A: 16-38
3. Kawakami K, Noguchi S, Noda M, Takahashi H, Ohta T, Kawamura M, Nojima H, Nagano K, Hirose T, Inayama S, Hayashida H, Miyata T, Numa S (1985) Primary Structure of the α-Subunit of *Torpedo californica* (Na$^+$+K$^+$)-ATPase Deduced from the cDNA Sequence. Nature 316: 733-736
4. Noguchi S, Noda M, Takahashi H, Kawakami K, Ohta T, Nagano K, Hirose T, Inayama S, Kawamura M, Numa S (1986) Primary Sructure of the β-Subunit of *Torpedo californica* (Na$^+$+K$^+$)-ATPase Deduced from the cDNA Sequence. FEBS Letts196: 315-320
5. Ohtsuka E, Matsuki S, Ikehara M, Takahashi Y, Matsubara K (1985) An Alternative Approach to Deoxyoligonucleotides as Hybridization Probes by Insertion of Deoxyinosine at Ambiguous Codon Positions. J Biol Chem 260: 2605-2608
6. Sambrook J, Fritsch EF, Maniatis T (1989) Molecular Cloning: A Laboratory Manual. 2nd edition, Cold Spring Harbour Laboratory Press, New York, USA
7. Schoenrock C, Morley SD, Okawara Y, Lederis K, Richter D (1991) Sodium and Potassium ATPase of the Teleost fish *Catstomus commersoni* : Sequence, Protein Structure and Evolutionary Conservation of the α Subunit. Biol Chem Hoppe-Seyler 372: 16895-16903
8. Takyasu K, Tamkun MM, Siegel NR, Fambrough DM (1987) Expression of Hybrid (Na$^+$+K$^+$)-ATPase Molecules after Transfection of Mouse Ltk- Cells with DNA Encoding the β-Subunit of an Avian Brain Sodium Pump. J Biol Chem 262: 10733-10740
9. Verry F, Kairouz P, Schaerer E, Fuentes P, Geering K, Rossier BC, Kraehenbuhl JP (1989) Primary Sequence of *Xenopus laevis* Na$^+$-K$^+$-ATPase and its localization in A6 Kidney Cells. Am J Physiol 256: F1034-F1043

# Functional expression of the gastric H,K-ATPase in *Xenopus* oocytes

P. M. Mathews, D. Claeys°, F. Jaisser, K. Geering, J.-P. Kraehenbuhl°, and B. C. Rossier

Institut de Pharmacologie et de Toxicologie, Université de Lausanne, CH 1005 Lausanne, Switzerland and °Institut Suisse de Recherche Experimental sur le Cancer et Institut de Biochimie, Université de Lausanne, CH 1066 Epalinges, Switzerland.

## Introduction

The gastric parietal cell H,K-ATPase mediates the secretion of acid by an energy dependent, electroneutral exchange of intracellular $H^+$ against extracellular $K^+$. Among the different known P-type ATPases, the $\alpha$-subunits of the vertebrate H,K-ATPases and Na,K-ATPases share the greatest sequence identity (~65 % amino acid identity). Like the Na,K-ATPase, the functional complex of the H,K-ATPase is an $\alpha/\beta$ heterodimer. Although the $\beta$-subunits of the Na,K-ATPase and the H,K-ATPase are distinct proteins, it has recently been shown that the $\beta$-subunit of the H,K-ATPase can substitute for the $\beta$-subunit of the Na,K-ATPase (1). Despite these similarities, these two pumps are functionally distinct with characteristic sensitivities to inhibitors. For example, the Na,K-ATPase from most species is sensitive to inhibition by cardiac glycosides such as ouabain whereas the gastric H,K-ATPase is not. The converse case is also true: the gastric H,K-ATPase is sensitive to both the $K^+$ competitive inhibitor SCH 28080 and to the covalent inhibitor omeprazole while the Na,K-ATPase is resistant.

While the $\alpha$- and $\beta$-subunits of the gastric H,K-ATPase have been cloned from a number of mammalian species (4, 5, 6), functional analysis of this pump has predominantly relied on protein isolated from *in vivo* source such as mammalian gastric vesicles (for example 2, 7; reviewed in 6). As any detailed analysis of structure-function relationships of the H,K-ATPase would be greatly facilitated by the expression of functional pump in a heterologous system, we have expressed gastric H,K-ATPase in *Xenopus* oocytes. In the oocyte, both mouse and *Xenopus* $\alpha$-subunit protein are expressed, can assemble with $\beta$-subunit, and form functional pumps in the plasma membrane.

## $\alpha_{H,K}$ protein expression in oocytes

We have isolated from mouse and *Xenopus* cDNAs encoding the $\alpha$-subunit of the gastric H,K-ATPase ($\alpha_{H,K}$); the full sequence of these two cDNAs will be reported elsewhere (manuscript in preparation). The deduced amino acid sequence for the mouse $\alpha_{H,K}$ has 99% identity with that of rat (6) whereas the mouse and *Xenopus* proteins share 84% sequence identity. For expression experiments in *Xenopus* oocytes, we have co-expressed the mouse and *Xenopus* $\alpha_{H,K}$-subunits along with the gastric $\beta_{H,K}$-subunit from rabbit (5). Previous work from our group has demonstrated that the rabbit $\beta_{H,K}$-subunit is expressed in the plasma membrane of oocytes following cRNA injection, and that it will assemble with the $\alpha_1$-subunit of *Xenopus* to form functional Na,K-pumps (1).

That the mouse and *Xenopus* $\alpha_{H,K}$-subunits are expressed in cRNA injected oocytes is demonstrated in Fig. 1. Following metabolic labelling and a 24 hour chase, oocytes were extracted in non-denaturing conditions that preserve the $\alpha/\beta$ heterodimer interaction and a mAb raised against rabbit $\beta_{H,K}$ was used to immunoprecipitate the complex (3). Lane 1 shows the immunoprecipitation pattern on SDS-PAGE from oocytes injected with $\beta_{H,K}$ cRNA alone. The mature $\beta_{H,K}$ protein migrates as a diffuse band with an apparent molecular weight of ~60 to 80 kDa. In addition to immunoprecipitating the exogenously expressed $\beta_{H,K}$, the endogenously expressed Na,K-ATPase $\alpha_1$-subunit of the oocyte is co-precipitated, giving a band of weak intensity at ~95 kDa. When oocytes are injected with both $\alpha_{H,K}$ and $\beta_{H,K}$ cRNA (lanes 2 and 3), the $\beta_{H,K}$ and the co-precipitated $\alpha_{H,K}$ proteins are apparent. Mouse and *Xenopus* $\alpha_{H,K}$ proteins expressed in oocytes typically migrate as a doublet of ~92 and 95 kDA. The $\beta_{H,K}$ protein migrates somewhat more rapidly when $\alpha_{H,K}$ is co-expressed. Presently, we do not have an explanation for either of these observations. As the complex carbohydrate form of the $\beta_{H,K}$-subunit was detected in this pulse-chase experiment, we can conclude that the $\alpha_{H,K}/\beta_{H,K}$ complex has been transported from the ER along the secretory pathway at least through the Golgi apparatus.

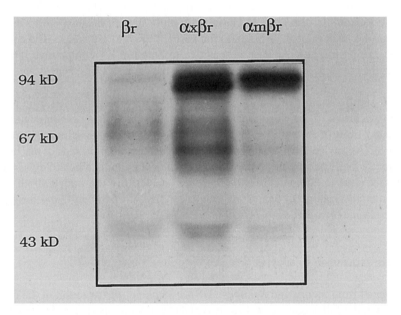

**Figure 1.** Co-precipitation of the H,K-ATPase $\alpha$-subunit with the $\beta$-subunit from cRNA injected oocytes. Immunoprecipitation from oocytes injected with rabbit $\beta_{H,K}$ cRNA alone (2 ng/oocyte; lane 1); *Xenopus* $\alpha_{H,K}$ (10 ng/oocyte) and rabbit $\beta_{H,K}$ cRNAs (lane 2); and mouse $\alpha_{H,K}$ (10 ng/oocyte) and rabbit $\beta_{H,K}$ cRNAs (lane 3). Protocol as described in the text.

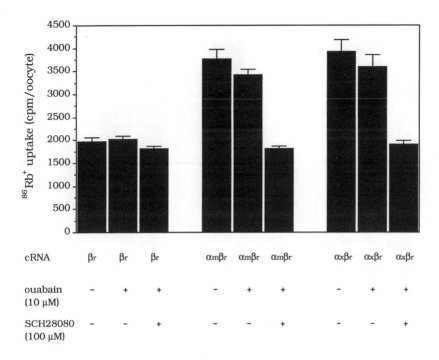

**Figure 2.** Functional expression of gastric H,K-ATPase in oocytes. Oocytes were injected either with rabbit $\beta_{H,K}$ cRNA alone or together with cRNA encoding mouse or *Xenopus* $\alpha_{H,K}$ as in Fig. 1. Three days later oocytes were pre-treated for 10 min. with 10 $\mu$M ouabain. Oocytes were then incubated in 0.5 mM K$^+$ and $^{86}$Rb$^+$ for 12 min. with the indicated inhibitors, washed, and the $^{86}$Rb$^+$ uptake in single oocytes determined. Data are expressed as the mean $\pm$ SE (n = 10).

### Functional H,K-ATPase in oocytes

Since the $\beta$-subunit of the H,K-ATPase was found to be terminally glycosylated and associated with the $\alpha$-subunit, we examined whether the subunits were assembled in the plasma membrane of the oocyte as a functional pump. To detect H,K-ATPase activity in the plasma membrane, we have used a $^{86}$Rb$^+$ uptake assay (1). Oocytes injected with $\beta_{H,K}$ cRNA alone were used to determine background levels of $^{86}$Rb$^+$ incorporation after pre-inhibiting the endogenous oocyte Na,K-ATPase with 10 $\mu$M ouabain (Fig. 2). This background incorporation of $^{86}$Rb$^+$ was not found to be sensitive either to further ouabain inhibition or SCH 28080. When either mouse or *Xenopus* $\alpha_{H,K}$ cRNA was co-injected with $\beta_{H,K}$ cRNA, the incorporation of $^{86}$Rb$^+$ was approximately doubled when compared to background. Ouabain, at a concentration that completely inhibits the endogenous *Xenopus* Na,K-ATPase (10 $\mu$M), did not significantly affect this activity. 100 $\mu$M SCH 28080, however, completely abolished the

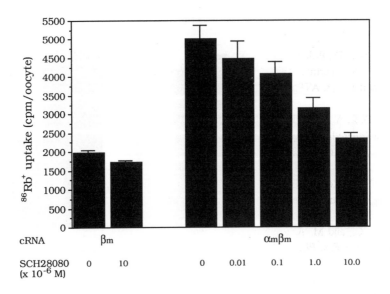

**Figure 3.** Dose-dependent inhibition of $^{86}$Rb$^+$ uptake by SCH 28080. Oocytes were injected with mouse $\beta_{H,K}$ cRNA alone or together with mouse $\alpha_{H,K}$ cRNA and assayed, as in Fig. 2. Data are expressed as the mean $\pm$ SE (n = 9-12). SCH 28080 concentrations as indicated.

$\alpha_{H,K}$ dependent $^{86}$Rb$^+$ uptake. Interestingly, similar levels of activity were seen with the expression of the $\alpha_{H,K}$ from either species when co-expressed with the rabbit $\beta_{H,K}$.

The dose-dependence inhibition of $^{86}$Rb$^+$ uptake by SCH 28080 in oocytes expresssing mouse $\alpha_{H,K}/\beta_{H,K}$ heterodimers is shown in Fig. 3. From this experiment, the apparent Ki was estimated to be in the sub-$\mu$M range, in good agreement with previous determinations on gastric membrane preparations (2, 7; reviewed in 5). Our data suggest that the expressed $\alpha_{H,K}/\beta_{H,K}$ heterodimer behaves in oocytes as does the native enzyme. This expression system should allow us to further study the transport properties of the H,K-ATPase in relation to those already characterized for the Na,K-ATPase.

**Acknowledgements**

We would like to thank P. Mangeat for the anti-rabbit $\beta_{H,K}$ monoclonal antibody and G. Sachs for the rabbit $\beta_{H,K}$ cDNA. This work was supported by the Swiss National Fund for Scientific Research (grant # 31-33598.92 to BCR and # 32-30011.92 to J.-P.K.) and a post-doctoral fellowship from the Etat de Vaud to PMM.

64

# References

1.  Horisberger, J.-D., P. Jaunin, M. A. Reuben, L. S. Lasater, D. C. Chow, J. G. Forte, G. Sachs, B. C. Rossier, K. Geering. 1991. The H,K-ATPase β-subunit can act as a surrogate for the β-subunit of the Na,K-ATPase. *J. Biol. Chem.* 266:19131-19134.

2.  Keeling, D. J., A. G. Taylor, and C. Schudt. 1989. The binding of a $K^+$competitive ligand, 2-methyl,8-(phenylmethoxy)imidizo(1,2-a)pyridine 3-acetonitrile, to the gastric ($H^+ + K^+$)-ATPase. *J. Biol. Chem.* 264:5545-5551.

3.  Mercier, F., H. Reggio, G. Devilliers, D. Bataille, and P. Mangeat. 1989. A marker of acid-secreting membrane movement in rat parietal cell. *Biol. Cell.* 67:7-20.

4.  Rabon, E. C., and M. A. Reuben. 1990. The mechanism and structure of the gastric H,K-ATPase. *Annu. Rev. Physiol.* 52:321-344.

5.  Reuben, M. A., L. S. Lasater, and G. Sachs. 1990. Characterization of a β-subunit of the gastric $H^+/K^+$-transporting ATPase. *Proc. Natl. Acad. Sci U.S. A.* 87:6767-6771.

6.  Shull, G. E., and J. B. Lingrel. 1986. Molecular cloning of the rat stomach ($H^+ + K^+$)-ATPase. 1986. *J. Biol. Chem.* 261:16788-16791.

7.  Wallmark, B., C. Briving, J. Fryklund, K. Munson, R. Jackson, J. Mendlein, E. Rabon, and G. Sachs. 1987. Inhibition of gastric $H^+,K^+$-ATPase and acid secretion by SCH 28080, a substituted pyridyl(1,2a)imidazole. *J. Biol. Chem.* 262:2077-2084.

# Over-expression of the peptide containing the ATP-binding site in Na/K-ATPase α-subunit

T. Ohta, M. Kuroda, M. Yoshii, and H. Hayashi

Institute of Basic Medical Sciences, University of Tsukuba,
Tsukuba, Ibaraki, 305 JAPAN

## Introduction

The deduced primary structure of Na/K-ATPase from the cloned nucleotide sequence revealed that the structure of ATP-hydrolysing active site was highly conserved among P-type ATPases (1, 2, 4). However, the minute conformation and molecular mechanism of ion transport have not been made clear yet. X-ray crystallography may be an efficient method to reveal the conformation of the protein, but such a hydrophobic membrane-protein has been unable to be purifed without using detergents. We made an assumption that out-of-membrane domain of Na/K-ATPase α-subunit facing to hydrophilic side with catalytic activity could be crystallized like regular soluble proteins (3). We report here that the hydrophilic, catalytic domain (50kDa) of α-subunit has been over-expressed in *E.coli* and purified without using any detergent, and discuss on the characterization for the peptide.

### Construction of the fusion plasmid, Na/K-ATPase catalytic domain gene on pMAL expression vector

The catalytic domain of Na/K-ATPase protruding to the cytoplasm is located at the region between two transmembrane regions; M4 and M5. The DNA fragment (1.3kbp) corresponding to the amino acid sequences from Gly[368] to Asp[814] (*Torpedo* α-subunit) was isolated from the cDNA clone by digesting with *Bam*HI and *Tth*III I. This fragment was cloned into a fusion protein expression vector, pMAL-c2, at the sites of *Bam*HI and *Hind*III (Fig.1). The fusion protein was cleaved into the maltose-binding protein

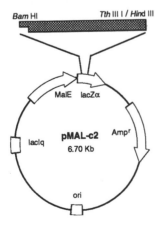

**Figure 1.** Construction of the fusion plasmid (pMAL expression vector and Na/K-ATPase active site gene). The loci of the associated genes in the pMAL plasmid are shown and the inserted Na,K-ATPase gene fragment is by a solid bar.

(MBP) and the Na/K-ATPase by the protease, factor Xa, which recognized 4 amino acid sequence of Ile-Glu-Gly-Arg, leaving four vector-derived amino acid residues to the Na,K-ATPase peptide.

## Over-expression of the MBP/Na,K-ATPase fusion protein

The fusion plasmid was transformed into the *E. coli* strain PR745. IPTG was added to the cloned cell culture (A600 of 0.5) to give a final concentration of 0.5mM. After 2 hours of induction, the fusion protein was remarkably over-expressed in *E. coli* and after 4 hours the amount of the protein reached to a maximum. The fusion protein, (pMAL/AS7), a molecular mass of ~90kDa was identified by Western blotting with antiserum of canine Na/K-ATPase α-subunit (Fig. 2).

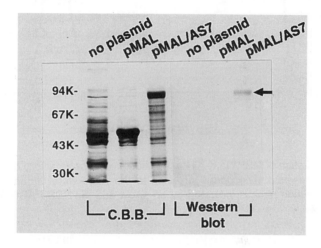

**Figure 2.** SDS-PAGE and Western Blotting of the fusion protein (pMAL/AS7). The protein identfied by the immunodetection with antiserum of Na/K-ATPase α-subunit was shown by an arrow.

The expressed protein formed 'inclusion body' in the cytoplasm and was precipitated by centrifugation at 1,000 x g for 10 min. The 162mg of the protein was recovered from one liter of the culture. It was solubilized in 4M urea and purified with an affinity chromatography through an amylose-resin. The bound protein was eluted by 10mM maltose. The peak of eluted protein fraction and the corresponding SDS-PAGE pattern are shown in Fig. 3. The purified fusion protein was recovered about 40% and, was digested with factor Xa at 4°C for 2days. Then the Na/K-ATPase catalytic domain peptide in the flow-through fractions was separated from the fusion protein by the chromatography through the second amylose-resin after cleavage with factor Xa.

However it was difficult to recover the catalytic domain peptide in large amount, because the fator Xa cleaved non-specifically the fusion protein into small peptides.

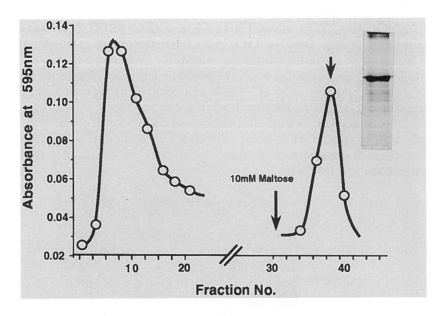

**Figure 3.** Purification of MBP/Na,K-ATPase fusion protein by an affinity chromatography through amylose-resin. The bound proteins to the resin were eluted out by 10mM maltase. The elution peak is indicated by an arrow and the SDS-PAGE pattern of the fraction is shown.

### Characterization of the over-expressed protein

To characterize the properties of the expressed protein, a CD (circular dichroism) was measured, and then the ATP-binding activity to the purified fusion protein was examined by using an ATP derivative, TNP-ATP (5). The resulting CD spectrum showed a typical spectrum of α-helix with a ratio of 8% (data not shown). A preliminary result showed it had ATP-binding activity but needed to be confirmed. These results strongly suggest that the over-expressed Na/K-ATPase peptide should retain the native structure in *E. coli*.

*Acknowledgements*

We thank Prof. Masasuke Yoshida and Dr. Hidero Muneyuki for technical support on the CD measurement and the ATP-binding assay. These studies were supported by the grant from the Ministry of Education, Science and Culture of Japan.

# References

1. Kawakami K., Noguchi S., Noda M., Takahashi H., Ohta T., Kawamura M., Nojima H., Nagano K., Hirose T., Inayama S., Hayashida H., Miyata T., and Numa S. (1985) Primary structure of the α-subunit of *Torpedo californica* (Na,K)-ATPase deduced from cDNA sequence. Nature 316: 733-736
2. MacLennan D. H., Brandl C. J., Korczak B., and Green N. M. (1985) Amino acidsequence of a Ca+Mg-dependent ATPase from rabbit muscle sarcoplasmic reticulum, deduced from its complementary DNA sequence. Nature 316: 696-700
3. Miki K., Kaida S., Kasai N., Iyanagi T., Kobayashi K., and Hayashi K. (1987) Crystallization and preliminary X-ray crystallographic study of NADH-cytochrome$b_5$ reductase from pig liver microsomes. J. Biol. Chem. 262: 11801-11802
4. Shull G. E. and Lingrel J. B. (1986) Molecular cloning of the rat stomach (H+K)-ATPase. J. Biol. Chem. 261: 16788-16791
5. Watanabe T, and Inesi G. (1982) The use of 2',3'-*O*-(2,4,6-trinitrophenyl)adenisine 5'-triphosphate for studies of nucleotide interaction with sarcoplasmic reticulum vesicles. J. Biol. Chem. 257: 11510-11516

# Expression of Na$^+$,K$^+$-ATPase in *Saccharomyces cerevisiae*

Per Amstrup Pedersen, Vibeke Foersom and Peter Leth Jørgensen
Biomembrane Research Center, August Krogh Institute University of Copenhagen
Universitetsparken 13, 2100 Copenhagen OE, Denmark.

## INTRODUCTION.

The lack of an efficient expression system that allows direct analysis of ligand binding and enzymatic intermediates of the Na$^+$,K$^+$ pump constitutes a bottleneck for studying the mechanisms for Na$^+$,K$^+$-transport across the membranes.

The ideal expression system should have the following characteristics; the host cell should not have any endogenous Na$^+$,K$^+$-ATPase acticity, modern molecular biology tools should be applicable to the host as should classical genetic screening and selection techniques. It should be possible to grow cells in large volumes in inexpensive media and to obtain a high expression level. The yeast *Saccharomyces cerevisiae* theoretically fulfills all these requirements and is therefore a potential alternative to mammalian cells for the heterogenous expression of Na$^+$,K$^+$-ATPase. Yeast cells have the capacity to express the $\alpha$ and $\beta$ subunits of the Na$^+$,K$^+$-ATPase in a functional form in the plasma membrane, but the expression level is rather low (1). The aim of this study has been to investigate whether the expression level of the Na$^+$,K$^+$-pump can be improved in yeast by the use of a plasmid with a controlable copynumber.

## RESULTS.

Our expression system is based on an episomal yeast expression vector, pEMBLyex4 (2). pEMBLyex4 is an *Escherichia coli/Saccharomyces cerevisiae* shuttle vector. pEMBLyex4 contains a CYC-GAL10 hybrid promoter, a URA3 gene used as a selection marker, a LEU2 allele (*leu2-d*) which complements an auxotrophy poorly, and the 2$\mu$ origin of replication. The copynumber of pEMBLyex4 is dependent on the growth medium. Growth in the presence of leucine results in a copynumber around 20 per haploid chromosome while growth in the absence of leucine increases the plasmid copynumber to approximately 200 per haploid chromosome. Genes to be expressed in yeast are cloned into the polylinker separating the promotor and terminator regions of pEMBLyex4.

We wanted to investigate whether an increase in plasmid copynumber could increase the expression level of the Na$^+$,K$^+$-pump. We chose to express the pig Na$^+$,K$^+$-ATPase in pEMBLyex4. The pig $\alpha$1 and $\beta$1 cDNA clones were a kind gift from Dr. Kostya Petrukhin (Columbia University, New York). cDNA from the pig $\alpha$ and $\beta$ subunits and several heavily engineered versions of the pEMBLyex4 plasmid were used to construct the plasmid pPAP1466 (Fig.1). Expression of the $\alpha$1 and $\beta$1 subunit cDNAs in pPAP1466 is controlled by the CYC-GAL10 galactose regulated promoter.

Yeast strain BJ5457 was electro transformed with pPAP1466 resulting in the strain BJ5457(pPAP1466) designated PAP1478. Microsomal membranes were isolated from galactose induced PAP1478.

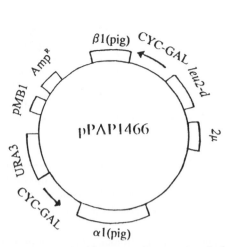

**Fig.1:** Construction of the yeast $\alpha1$, $\beta1$ expression vector derived from pEMBLyex4. Abbreviations used: $\alpha1$(pig), the $\alpha1$ subunit from pig Na$^+$,K$^+$-ATPase; $\beta1$, the $\beta1$ subunit from pig Na$^+$,K$^+$-ATPase; CYC-GAL, a hybrid CYC-GAL10 promotor; $2\mu$, the $2\mu$ origin of replication; URA3, the orotidine-5'-P-decarboxylase gene; leu2-d, a poorly expressed allele of the $\beta$-isopropylmalate dehydrogenase gene; ORI, the pMB1 origin of replication; Amp$^R$, a $\beta$-lactamase gene; T, the PMA1 terminator region.

Fig. 2 shows a Western blot of these crude microsomal membranes using an anti-$\alpha$-antibody (anti-KETYY-antibody) kindly provided by Jack Kyte and an anti-$\beta1$-antibody provided by Robert Levenson. The anti-KETYY-antibody recognizes the five C-terminal residues of all sequenced $\alpha$ subunits while the anti-$\beta1$-antibody is raised against amino acid residues 152-320 of the rat $\beta1$ subunit.

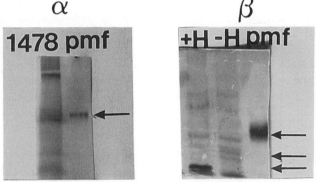

**Fig.2:** A Western blot of SDS-PAGE separated crude microsomal membranes isolated from PAP1478 and SDS-PAGE separated pig microsomal fraction (PMF) using the anti-KETYY serum and the anti-$\beta$-serum. The positions of the $\alpha$ and $\beta$ subunits are marked with arrows. Abbreviations: $\alpha$, the blot was developed with the anti-$\alpha$-serum: $\beta$, the blot was developed with the anti-$\beta$-serum: $+/-$ H indicates pretreatment or not of the crude membranes with Endoglycosidase H prior to SDS-PAGE.

Table 1 shows the results of equilibrium $^3$H-ouabain binding to crude microsomal membranes isolated from PAP1478. No high affinity saturable $^3$H-ouabain binding was detected in crude microsomal membranes isolated from BJ5457 transformed with pPAP1401 (expressing only the $\alpha1$ subunit) or pPAP1446 (expressing only the $\beta1$ subunit)(data not shown).

71

| [3-ouabain] | ouabain sites |
|---|---|
| 10nM | 5.9 pmol/mg |
| infinite | 20 pmol/mg |

**Table 1:** The number of $^3$H-ouabain binding sites present in a crude microsomal fraction isolated from galactose induced PAP1478. A $K_D$ value of 21nM was determined from Scatchard analysis and used in calculation of the binding capacity.

The microsomal membranes were fractionated on a continuous sucrose gradient to investigate the intracellular distribution of ouabain binding sites in yeast cells. The Na$^+$,K$^+$-ATPase was found primarily to be located in the plasma membrane by the following two criteria. The ouabain binding shows a maximum at the position in the sucrose gradient corresponding to the density of the yeast plasma membrane and the Na$^+$,K$^+$-ATPase cofractionates with the yeast plasma membrane as judged by Western blotting using an anti-H$^+$-ATPase serum and the anti-$\alpha$1-serum. The H$^+$-ATPase is a marker enzyme for the yeast plasma membrane. Figure 3 shows the distribution of high affinity ouabain binding sites in a continuous sucrose gradient and Figure 4 shows a Western blot of the same fractions using an anti-H$^+$-ATPase serum and an anti-$\alpha$1-serum.

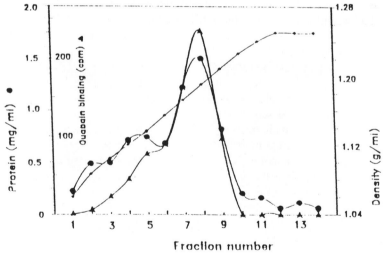

**Fig.3:** Distribution of high affinity $^3$H-ouabain binding sites and protein in a continuous sucrose gradient of microsomal membranes prepared from galactose induced PAP1478.

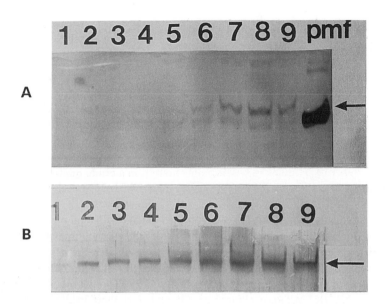

**Fig.4:** Distribution of the $Na^+,K^+$-ATPase $\alpha 1$ and the yeast $H^+$-ATPase in a continuous sucrose gradient of microsomal membranes prepared from galactose induced PAP1478. Proteins were identified using an anti-$\alpha$-serum (A) and an anti-$H^+$-ATPase serum (B), respectively. The positions of the $\alpha 1$ subunit and the $H^+$-ATPase are indicated with arrows.

## CONCLUSION.

We have investigated the ability of *Saccharomyces cerevisiae* to synthesize the pig $Na^+,K^+$-pump using a plasmid with a controllable copynumber. The $Na^+,K^+$-pump was shown primarily to be located in the plasma membrane (Fig.3 and Fig.4). The $\beta 1$-subunit was shown to be glycosylated and most of the carbohydrate was found to exist in an Endoglycosidase H sensitive form (Fig.2).

From equilibrium ouabain binding the number of pumps were estimated to be 20 pmol/mg protein in a crude microsomal membrane preparation (Table 1). This capacity is comparable to the amount of $Na^+,K^+$-ATPase expressed in purified membranes from transfected mammalian cells, while the amount present in purified membranes from Baculovirus infected insect cells is ten fold lower.

The data presented here show that the amount of $Na^+,K^+$-pumps produced in *S. cerevisiae* can be increased by increasing the plasmid copynumber. We are presently devellopping a purification scheme for the $Na^+,K^+$-pumps produced in yeast.

**REFERENCES.**
1.Horowitz et al. 1990. Synthesis and assembly of functional $Na^+,K^+$-ATPase in yeast. J. Biol. Chem. 265 : 4189 - 4194

2.Cesareni, G. and Murray, J.A.H. Plasmid vectors carrying the replication origin of filamentous single stranded phages. In: J.K.Setlow (ed) Genetic engineering, principles and methods. Vol 9. Plenum Press, New York.

# Regulation of Na$^+$/K$^+$-ATPase α1 and ß1 Expression in Epithelial Cells

K.K. Grindstaff, G. Blanco and R.W. Mercer

Department of Cell Biology and Physiology, Washington University School of Medicine, St. Louis, MO 63110 USA

## Introduction

In nearly all epithelia the Na$^+$/K$^+$-ATPase is confined to the basolateral membrane where it establishes the transepithelial Na$^+$ gradient that drives the vectoral transport of solutes and ions across the epithelium. In some epithelial diseases such as polycystic kidney disease and ischemia the cell surface polarity of the Na$^+$/K$^+$-ATPase can be disrupted resulting in a loss of epithelial function and integrity (4,7). Though other investigators have examined Na$^+$/K$^+$-ATPase expression (3) and α subunit localization (1,2) in polarized cells, there is seldom reference as to the expression and cellular localization of the specific subunit isoforms. We have found by RNA and immunoblot analysis of Madin Darby canine kidney (MDCK) and Caco-2 (human colon carcinoma) cells that both cell lines express only α1 and ß1 isoforms (data not shown). Likewise in agreement with what other investigators have seen in MDCK cells using non-isoform specific antibodies (2), the α1 and ß1 subunits are primarily localized to the lateral plasma membrane in both cell lines (data not shown). With this in mind, we wished to express the individual Na$^+$/K$^+$-ATPase subunits in polarized epithelial cells to elucidate the role of each isoform in directing cellular sorting of the enzyme and in turn to identify possible structural features of each subunit that may participate in determining the localization.

## Results

We have expressed the Na$^+$/K$^+$-ATPase α1 and ß1 subunits in several mammalian cell lines using the vaccinia virus T7 RNA polymerase expression system. Other researchers have demonstrated the utility of vaccinia recombinants to investigate the polarized targeting of exogenously expressed polypeptides in epithelial cultures (6). For all constructs, the cDNAs were subcloned into the vaccinia shuttle vector (pTM3) at the initiation site (*Nco*1), resulting in the identical start site without extraneous 5' untranslated (UT) sequences. By coinfection with a recombinant vaccinia virus that contains the T7 RNA polymerase gene under control of a vaccinia early-late promoter and a second recombinant virus that contains the target genes downstream of the T7 promoter, high levels of expression can be obtained early in infection (5). Infection of cells derived from African green monkey kidney (BSC 40 and CV 1), mouse connective tissue (L929), or human amnion (Wish cells) with viral recombinants containing the α1 and ß1 isoforms of the Na$^+$/K$^+$-ATPase, the glucose transporters, GLUT1 and GLUT4, or the capsid protein of the Sindbis virus all result in the production of the appropriate

74

protein products as assayed on immunoblots. However, several epithelial-like cell lines, MDCK, Caco-2, Madin Darby bovine kidney (MDBK), and LLC-PK1 (pig kidney), fail to synthesize vaccinia directed α1 and ß1 polypeptides, yet they express the other polypeptides at levels comparable to those seen in the non-epithelial cell lines (Table 1). This suggests that the expression of both α1 and ß1 is specifically regulated in the epithelial cell lines investigated. Results from pulse chase experiments indicate that the lack of α1 and ß1 expression as assayed by immunoblotting is not due to rapid post-translation degradation of the proteins (data not shown).

**Table 1.** Survey of vaccinia recombinant expression in mammalian cell lines

| Cell Line (Morphology) | T7 RNA Polymerase | Sindbis Capsid Protein | α1 Polypeptide | ß1 Polypeptide |
|---|---|---|---|---|
| BSC 40 (fibroblast-like) | yes | yes | yes | yes |
| CV 1 (fibroblast-like) | yes | yes | yes | yes |
| L929 (fibroblast-like) | yes | yes | yes | yes |
| WISH (fibroblast-like) | yes | yes | yes | yes |
| MDCK (epithelial-like) | yes | yes | no | no |
| MDCK type II (epithelial-like) | yes | yes | no | no |
| Caco-2 (epithelial-like) | yes | yes | no | no |
| MDBK (epithelial-like) | yes | yes | no | no |
| LLC-PK1 (epithelial-like) | yes | yes | no | no |

Results are based on immunoblot analysis of protein isolated from cells infected at an multiplicity of infection (moi) of 1 and 10 for each virus (VTF 7.3; VVpTM3 SS; VVpTM3 α1; VVpTM3 ß1). BSC 40 and MDCK type II (kindly provoided by Dr. W.J. Nelson) cells were also tested for their ability to express recombinant GLUT1 and GLUT4. Both cells lines expressed comparable levels of GLUT1 and GLUT4 as directed by the vaccinia recombinants, VVpTM3 GLUT1 VVpTM3 GLUT4.

Analysis of total RNA isolated from MDCK cells infected with increasing multiplicities of infection (moi) of recombinant α1, ß1, and Sindbis capsid virus demonstrates that all three mRNAs are produced at significantly higher levels than in uninfected cells (Figure 1). While these MDCK cells express Sindbis capsid protein at an moi of 1, they fail to synthesize α1 and ß1 polypetide over endogenous background (Figure 2), even though the message for both at an moi of 25 is well over that seen for the Sindbis capsid protein, even at an moi of 1 and 5. Likewise construction of α1 recombinants either lacking the 3' untranslated region (VVpTM3 α1-Δ3'end) or possessing the 3' untranslated region of the Sindbis capsid or GLUT4 cDNAs (VVpTM3 α1-SinS, VVpTM3 α1-

GLUT4, respectively) has no affect on α1 expression in either cell type (Figure 3). While this regulation of α1 and ß1 expression may be partially due to transcription, our data suggests that the regulation is primarily at the level of translation since sufficient levels of α1 and ß1 message are produced in recombinant infected cells. Additionally, the regulation of α1 does not appear to be due to elements within the untranslated regions of the message. We are currently investigating whether a similar regulation of expression exists with the other Na$^+$/K$^+$-ATPase isoforms.

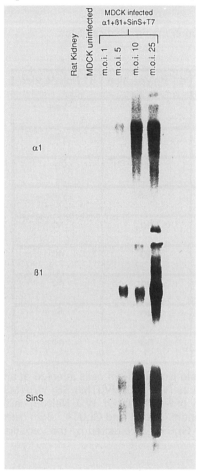

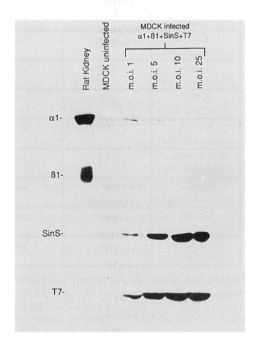

**Figure 1.** Northern analysis of recombinant induced mRNA synthesis with increasing moi in MDCK cells. Total RNA was isolated from polarized MDCK cells infected with vaccinia recombinants (VVpTM3 α1; VVpTM3 ß1; VVpTM3 SinS; VTF T7) at an moi of 1, 5, 10, and 25. 5 µg/lane of total RNA was separated on formaldehyde gels and screened with probes corresponding to the specific messages.

**Figure 2.** Immunoblot analysis of recombinant induced protein synthesis with increasing moi in MDCK cells. Whole cell homogenates were isolated from polarized MDCK cells infected with vaccinia recombinants (VVpTM3 α1; VVpTM3 ß1; VVpTM3 SinS; VTF T7) at an moi of 1, 5, 10, and 25. 25 µg/lane of protein was separated by SDS-PAGE (7.5%) and analyzed by immunoblotting.

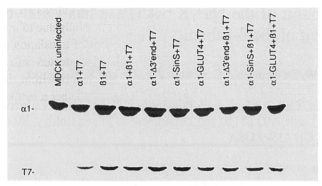

Figure 3: Expression of α1-3' UT recombinants in MDCK cells. Whole cell homogenates were isolated from polarized MDCK cells infected at an moi of 10 with each of the following vaccinia recombinants: VVpTM3 α1; VVpTM3 ß1; VVpTM3 α1-Δ3'end; VVpTM3 α1-SinS; VVpTM3 α1-GLUT4; VTF T7. 25μg/lane of protein was separated by SDS-PAGE (7.5%) and analyzed by immunoblotting. When expressed in BSC 40 cells, all chimeras produced polypeptides of the appropriate size (data not shown).

*Acknowledgments*

We thank Dr. Charles Rice for assistance with the vaccinia system. This work was supported by National Institutes of Health Grant DK 45181.

## References

1. Caplan MJ, Anderson HC, Palade GE, Jamieson JD (1986) Intracellular sorting and polarized cell surface delivery of (Na,K)ATPase, an endogenous component of MDCK cell basolateral plasma membranes. Cell 46:623-631

2. Hammerton RW, Krzeminski KA, Mays RW, Ryan TA, Wollner DA, Nelson WJ (1991) Mechanism for regulating cell surface distribution of Na,K-ATPase in polarized cells. Science 254:847-850

3. Mircheff AK, Bowen JS, Yiu SC, McDonough AA (1992) Synthesis and translocation of Na⁺-K⁺-ATPase α-and ß-subunits to plasma membrane in MDCK cells. Am J Physiol 262:C470-483

4. Molitoris BA, Hoilien CA, Dahl R, Ahnen DJ, Wilson PD, Kim J (1988) Characterization of ischemia-induced loss of epithelial polarity. J Membrane Biol 106:233-242

5. Elroy-Stein O, Fuerst TR, Moss B (1989) Cap-independent translation of mRNA conferred by encephalomyocarditis virus 5' sequence improves the preformance of the vaccinia virus/bacteriophage T7 hybrid expression system. Proc Natl Acad Sci 86:6126-6130

6. Stephens EB, Compans RW, Earl P, Moss B (1986) Surface expression of viral glycoproteins is polarized in epithelial cells infected with recombinant vaccinia viral vectors. EMBO J 5:237-245

7. Wilson PD, Sherwood AC, Palla K, Du J, Watson R, Norman JT (1991) Reversed polarity of Na⁺-K⁺-ATPase: mislocation to apical plasma membranes in polycystic kidney disease epithelia. Am J Physiol 260:F420-430

# The α-Subunit of the Na+/K+-ATPase has Catalytic Activity Independent of the ß-subunit

G. Blanco, A.W. De Tomaso, J.C. Koster, Z.J. Xie and R.W. Mercer

Department of Cell Biology and Physiology, Washington University School of Medicine, St. Louis, MO 63110 and Department of Pharmacology, Medical College of Ohio, Toledo, OH 43699 USA

## Introduction

A functional Na+/K+-ATPase enzyme consists of two polypeptides: a large catalytic α-subunit and a smaller glycosylated ß-subunit. Although all catalytic properties of the Na+/K+-ATPase are associated with the α-subunit, it has never been determined if this subunit has activity independent of the ß-subunit. Using the baculovirus expression system we have previously demonstrated that when expressed alone, the rodent α-subunit exists in a stable conformation unassociated with other proteins in the infected Sf-9 cell (3). Here we show that the α-subunit has catalytic activity independent of the ß-subunit. This ATPase activity is dependent on $Mg^{2+}$, does not require $Na^+$ or $K^+$, and is not inhibited by ouabain. However, the independent α-subunit ATPase activity is inhibited by EGTA, vanadate and increasing ionic strength. The inhibition of the ATPase activity by EGTA is not abolished by the addition of $Ca^{2+}$ suggesting that EGTA inhibits activity in a manner other than its ion-binding abilities. Phosphorylated intermediates of the independent α-subunit were acid-stable and alkaline-labile, as well as sensitive to hydroxylamine, strongly suggesting the presence of an acylphosphate and thus phosphorylation at an aspartate residue. The physiological role, if any, of this independent α-subunit activity is unknown.

## Results

Recombinant baculoviruses containing the cDNAs coding for the rodent Na+/K+-ATPase α1, α2, and α3 subunits were used to infect Sf-9 cells. The infected cells produce only the individual Na+/K+-ATPase polypeptides when infected; when coinfected with α and ß baculoviruses, functional Na+/K+-ATPase activity is expressed (2,3). Activity of the virally produced αß polypeptides can be demonstrated by the ability of the α-subunit to be phosphorylated by ATP. During the reaction cycle, the α-subunit of the Na+/K+-ATPase is phosphorylated by ATP at a ß-aspartyl carboxyl group (asp-371) (1,7). This aspartate group is conserved in all Na+/K+-ATPase α-subunits. Phosphorylation by ATP requires both Mg and Na ions; in the presence of $K^+$ the phosphate is released from the subunit as $P_i$. The baculovirus induced Na+/K+-ATPase α-subunit from the αß coinfected Sf-9 cells can be phosphorylated by ATP. This $Na^+$-dependent phosphorylation is inhibited by ouabain if ouabain is bound to the enzyme before exposure to ATP. In addition, the α-subunit is completely dephosphorylated in the presence of $K^+$. The α-subunit can also be directly phosphorylated with $P_i$. Because ouabain binds with high affinity to the phosphorylated intermediate, the phosphorylation of the α-subunit by $P_i$ is stimulated by the presence of ouabain.

Although it is more difficult to detect, the $Na^+/K^+$-ATPase from the uninfected *Sf-9* cells has an identical pattern of phosphorylation. The insect $Na^+/K^+$-ATPase α-subunit is phosphorylated only in the presence of $Na^+$; EGTA does not affect the $Na^+$-dependent phosphorylation and $Mg^{2+}$ alone cannot support phosphorylation.

A completely different type of α-subunit phosphorylation is seen when the α-subunit is expressed in the absence of the ß-subunit (Fig. 1). As shown, phosphorylation of the α-subunit can occur with only $Mg^{2+}$ present. This $Mg^{2+}$-dependent phosphorylation is not inhibited by ouabain, however, 150 mM $Tris^+$, $K^+$ or $Na^+$ greatly reduce phosphorylation. The ß-independent phosphorylation of the α-subunit is not restricted to the α1 isoform as both α2 and α3 exhibit identical $Mg^{2+}$-dependent phosphorylation.

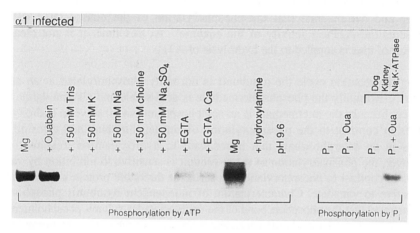

Figure 1. The independent α-subunit can be phosphorylated by ATP. 30 µg of membrane proteins from α1 infected *Sf-9* cells were phosphorylated with γ[$^{32}$P]-ATP or [$^{32}$P$_i$].

Increasing concentrations from 5 to 150 mM of either $Na^+$ or $K^+$ inhibit α-subunit phosphorylation. However, choline, a cation not thought to influence $Na^+/K^+$-ATPase enzymatic activity, also inhibits the independent α-subunit phosphorylation. Therefore, although $Na^+$, $K^+$ and $Tris^+$ can influence the phosphorylation of the native enzyme it appears that these ions do not directly affect the phosphorylation of the independent α-subunit. Moreover, the inhibition of phosphorylation is not influenced by chloride ions, as substitution of chloride with sulphate has no affect on the inhibition. Taken together these results suggest that the phosphorylation of the unassociated α-subunit is sensitive to the ionic strength of the medium. At higher ionic strengths the phosphorylation of the independent α-subunit is inhibited. However, even at physiological ionic strength, the unassociated α-subunit has catalytic activity. The unassociated α-subunit can only be phosphorylated by ATP; the subunit is not phosphorylated by P$_i$ either in the presence or absence of ouabain.

In addition to increasing ionic strength, 1 mM EGTA reduces the $Mg^{2+}$-dependent phosphorylation of the α-subunit. The inhibition of α-subunit phosphorylation by EGTA suggests that the phosphorylation may be dependent upon $Ca^{2+}$. However, the addition of $Ca^{2+}$ to the $Mg^{2+}$-EGTA reaction medium does not influence the inhibition of the α-subunit phosphorylation. The addition of 1 mM $Ca^{2+}$ to the $Mg^{2+}$-EGTA medium does not alter the inhibition by 1 mM EGTA. Consequently the inhibition by EGTA appears to be mediated through a factor other than $Ca^{2+}$. EGTA may inhibit activity by chelating another metal ion that is essential for the phosphorylation of the α-subunit, however, it is possible that EGTA inhibits activity in a manner other than its metal-binding abilities. In either event, EGTA is useful in defining the activity of the unassociated α-subunit as an EGTA-sensitive, ATPase activity. This activity, which is not present in the membranes from uninfected Sf-9 cells, can be measured in membranes from α infected cells. In addition, the $Mg^{2+}$-dependent ATPase of the α infected cells is reduced by ≈40% under conditions of high ionic strength (130 mM choline). The $Mg^{2+}$-dependent ATPase activity remaining at the higher ionic strength is completely inhibited by EGTA. These results are in agreement with the phosphorylation studies and demonstrate that the phosphorylation of the unassociated α-subunit represents the ATPase activity of the enzyme. At this time, it is not clear if the transport of ions is coupled to the hydrolysis of ATP.

During the reaction cycle the α-subunit is normally phosphorylated at an aspartate residue. Normally the phosphointermediate is acid-stable and alkaline-labile (5). In addition, because the intermediate is an acylphosphate it is sensitive to hydroxylamine. This is in contrast to the phosphorylation of serine, threonine and tyrosine, which results in an esterphosphate that is resistant to hydroxylamine treatment (4,5,6). Moreover, the phosphorylation of the α-subunit is sensitive to inhibition by vanadate. This is in contrast to phosphorylation by serine or threonine protein kinases which are insensitive to vanadate. Characterization of independent α-subunit phosphorylation demonstrated that the phosphate bond of the unassociated α-subunit phosphointermediate is acid-stable and alkaline-labile consistent with phosphorylation at the normal aspartyl residue. Compatible with phosphorylation at the normal aspartate residue the formation of the phosphointermediate is also inhibited by vanadate and is sensitive to hydroxylamine. Thus it appears that the independent α-subunit is phosphorylated at the normal aspartate residue.

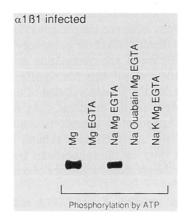

Figure 2. Independent α-subunit and αβ phosphorylation of membrane proteins from α1β1 infected Sf-9 cells.

To test if the EGTA-sensitive phosphorylation of the α-subunit occurs in the presence of the ß-subunit, *Sf*-9 cells were infected with both α and ß viruses. The MOI of the α baculovirus was greater than that of the ß baculovirus. Under these circumstances both EGTA-sensitive and $Na^+$-sensitive phosphorylation patterns are seen. As shown in Fig. 2, in the presence of $Mg^{2+}$ the α-subunit is phosphorylated and this phosphorylation is sensitive to EGTA. Presumably this activity is mediated by the unassociated α-subunit. When $Na^+$ is added to the $Mg^{2+}$-EGTA medium, α-subunit phosphorylation is again seen. This $Na^+$-dependent α-subunit phosphorylation is inhibited by ouabain and $K^+$ and is likely mediated by the αß enzyme complex. These results suggest that when the α-subunit is expressed in excess of the ß-subunit the unassociated α-subunits exhibit an EGTA-sensitive activity. However, we cannot exclude the possibility that the activities observed are a result of individual cells expressing only the α-subunit. This work provides strong evidence that the α-subunit, in the absence of the ß subunit, can be phosphorylated and mediate an ATPase activity. At this time it is not clear if the enzyme is transporting ions or if this activity is physiologically significant.

*Acknowledgements*

This work was supported by National Institute of Health grants (GM 39746 and HL 36573).

**References**

1.     Bastide F, Meissner G, Fleischer S, and Post RL (1973) Similarity of the active site of phosphorylation of the adenosine triphosphatase for transport of sodium and potassium ions in kidney to that for transport of calcium ions in the sarcoplasmic reticulum of muscle. J Bio Chem 248: 8385-8391

2.     Blanco G, Xie ZJ and Mercer RW (1993) Functional expression of the α2 and α3 isoforms of the Na,K-ATPase in baculovirus infected insect cells. Proc Natl Acad Sci USA 90: 1824-1828

3.     De Tomaso AW, Xie ZJ, Liu G and Mercer RW (1993) Expression, targeting and assembly of functional Na,K-ATPase polypeptides in baculovirus-infected insect cells. J Bio Chem 268:1470-11478

4.     Duclos B, Marcandier S, and Cozzone AJ (1991) Chemical properties and separation of phosphoamino acids by thin-layer chromatography and/or electrophoresis. Methods Enzymol 201: 10-21

5.     Hokin LE, Sastry PS, Galsworthy PR, and Yoda A (1965) Evidence that a phosphorylated intermediate in a brain transport adenosine triphosphatase is an acyl phosphate. Proc Natl Acad Sci USA 54:177-184

6.     Lipmann F, and Tuttle LC (1945) A specific micromethod for the determination of acyl phosphates. J Bio Chem 159:21-28

7.     Post, RL and Kume S (1973) Evidence for an aspartyl phosphate residue at the active site of sodium and potassium ion transport adenosine triphosphatase. J Bio Chem 248: 6993-7000

# Functional Expression of Na+/K+-ATPase α and ß Isoforms

G. Blanco, Z.J. Xie and R.W. Mercer

Department of Cell Biology and Physiology, Washington University School of Medicine, St. Louis, MO 63110 and Department of Pharmacology, Medical College of Ohio, Toledo, OH 43699 USA

## Introduction

Both the catalytic (α) and glycosylated (ß) subunits of the Na+/K+-ATPase are encoded by a family of genes. Three Na+/K+-ATPase α isoforms (α1, α2 and α3) and two ß (ß1 and ß2) have been identified in mammals (3). The differential properties of the α and ß polypeptides in tissue expression, development and hormonal stimulation, suggest that each isoform might be playing a specific physiological role. The study of the properties of the Na+/K+-ATPase isoforms has been complicated by their coexistence in different tissues and because of the difficulty that their isolation represents. To gain insight into the functional characteristics of the Na+/K+-ATPase isoforms, we have successfully used the baculovirus expression system to express the rodent α1, α2 and α3 Na+/K+-ATPase isoforms, separately and in combination with ß1 in Sf-9 insect cells (1). These cells have low levels of endogenous Na+/K+-ATPase and are able to express high amounts of the baculovirus induced Na+/K+-ATPase polypeptides. The coexpression of α1, α2 and α3 with ß1 resulted in the production of catalytically competent Na+/K+-ATPase molecules as determined by ouabain-sensitive ATPase activity, ouabain and K-sensitive ATP phosphorylation and [$^3$H]ouabain binding. α2ß1 and α3ß1 isozymes display a high sensitivity to ouabain, with α3ß1 the more sensitive. At present, the different α and ß isoform combinations that result in active Na+/K+-ATPase enzyme have not been completely determined. Here we show coexpression of the α1, α2 and α3 isoforms in combination with the ß2 subunit. All α isoforms are able to associate with the ß2 subunit in virally infected insect cells and the Na+/K+-ATPase isozymes produced are catalytically active.

## Results

Recombinant baculoviruses containing the cDNAs for the rodent α and ß Na+/K+-ATPase isoforms were prepared according to standard procedures (2) and used to infect Sf-9 insect cells. Coinfections combining the different α isoforms with ß1 and ß2 were performed. After 72 hrs post-infection, Sf-9 cells expressed high amounts of the corresponding Na+/K+-ATPase subunits, as determined by immunoblot analysis using Na+/K+-ATPase isoform specific antibodies (data not shown). As we previously reported (1), in the case of cells coinfected with α1, α2 and α3 in combination with ß1, catalytically active complexes were obtained. To determine if and which α isoforms can associate with ß2, metabolic labeling and immunoprecipitation experiments were performed.

82

As shown in Fig. 1, in insect cells coinfected with the corresponding viruses, all three α isoforms are able to assemble with the ß2 subunit. Furthermore, all the α and ß combinations are catalytically active; Na$^+$/K$^+$-ATPase activity assays performed on whole cells (4), showed that *Sf*-9 cells infected with α1ß2, α2ß2 and α3ß2 displayed activities 5 to 14 times higher than the uninfected cells or the cells singly infected with ß2.

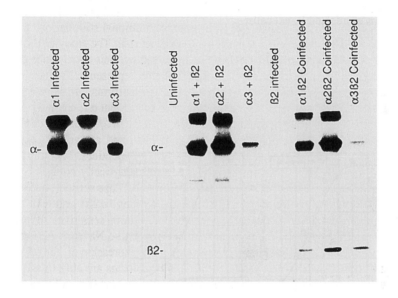

**Figure 1.** Uninfected and 48 hrs infected *Sf*-9 cells were incubated for 1 hr in a methionine free medium and pulse labeled for 20 min with $^{35}$S-methionine. Cells were lysed with 1% CHAPS in 150 mM NaCl, 25 mM Hepes pH 7.4; the insoluble material was pelleted in a microfuge (10 min at 15,000 x g). Immunoprecipitation of the labeled proteins was done using a monoclonal antibody that recognizes all α isoforms of the Na$^+$/K$^+$-ATPase (5α, gift of Dr. D. Fambrough) and magnetic beads coated with a goat anti-mouse antibody. Precipitated proteins were then separated by SDS/PAGE and subjected to fluorography. Analysis was performed on cells infected with α1, α2, α3 and ß2 viruses, cells coinfected with α1ß2, α2ß2 and α3ß2, and on a mixture of cells separately infected with α isoforms and ß2 (α1 + ß2, α2 + ß2 and α3 + ß2).

An interesting property, that may influence the physiological roles of the Na$^+$/K$^+$-ATPase isoforms, is their sensitivity to cardiotonic steroids (3). In a previous paper, we determined the ouabain sensitivity of the α1β1, α2β1 and α3β1 Na$^+$/K$^+$-ATPase isozymes (1). When compared to α1β1 from rat kidney, α2β1 and α3β1 display a high sensitivity to ouabain, α3β1 being more sensitive. Here, we studied the apparent ouabain affinities of the α1β2, α2β2 and α3β2 isozymes, by performing dose response curves for the ouabain inhibition of Na$^+$/K$^+$-ATPase activity of cells coinfected with the different α isoforms and β2 (Fig. 2). Table 1 depicts the $K_{0.5}$ obtained for these isozymes. When coexpressed with β2, α2 and α3 show a higher sensitivity to the cardiotonic steroid than α1. This kinetic behavior agrees with that displayed by the isozymes that contain β1; however, the association with β2 produces a slight change in the ouabain affinity pattern displayed by each isoform. This apparent difference in ouabain sensitivity still remains to be investigated. Currently, we are studying other properties of the Na$^+$/K$^+$-ATPase isozymes resulting from the different α and β isoform associations in infected insect cells.

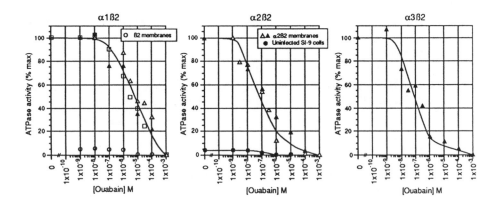

**Figure 2.** Dose response curves for the ouabain inhibition of α1β2, α2β2 and α3β2 isozymes of the Na$^+$/K$^+$-ATPase. ATPase activity was determined in whole cell preparations from uninfected cells or after 72 hrs infection with α1β2, α2β2, α3β2 and β2. Cells were washed with 10 mM Tris-HCl, EGTA (pH 7.4) and permeabilized by pretreatment with 0.01 mg alamethicin/mg total protein for 10 min at 25° C. Aliquots containing 0.1-0.2 mg of protein were added to the reaction mixture (130 mM NaCl, 20 mM KCl, 3 mM MgCl$_2$, 0.2 mM EGTA, 5 mM NaN$_3$, 30 mM Tris-HCl pH 7.4, 3 mM ATP) and the indicated ouabain concentrations. Incubation was at 37° C for 30 min. ATPase was determined by the release of $^{32}$Pi from [γ$^{32}$P]-ATP. Specific Na$^+$/K$^+$-ATPase activity was the Na$^+$/K$^+$-dependent hydrolysis inhibitable by 1 mM ouabain. Symbols are the average of triplicate determinations performed on different infections.

**Table 1.** Apparent ouabain affinities ($K_{0.5}$ values) of the different Na+/K+-ATPase isozymes

| Isoform | $K_{0.5}$ (nM) | |
|---|---|---|
| Uninfected | 80 | (65 to 99) |
| $\alpha 1 \beta 1$ (rat kidney) | $91 \times 10^3$ | (75 to 110) |
| $\alpha 2 \beta 1$ | 64 | (56 to 74) |
| $\alpha 3 \beta 1$ | 23 | (20 to 27) |
| $\alpha 1 \beta 2$ | $7 \times 10^3$ | (4 to 11) |
| $\alpha 2 \beta 2$ | 119 | (76 to 188) |
| $\alpha 3 \beta 2$ | 104 | (51 to 216) |

Constants for the interaction of ouabain with different Na+/K+-ATPase isozymes were obtained from [³H] ouabain binding studies (uninfected, $\alpha 2 \beta 1$ and $\alpha 3 \beta 1$) and from ouabain inhibition of Na+/K+-ATPase activity (rat kidney $\alpha 1 \beta 1$, $\alpha 1 \beta 2$, $\alpha 2 \beta 2$ and $\alpha 3 \beta 2$). Calculations were done by nonlinear regression analysis of the data (Marquart-Levenberg algorithm) from one to three different experiments performed in triplicate. Numbers in parenthesis are the 95% confidence intervals.

*Acknowledgements*

We thank Drs. K Sweadner and D Fambrough for providing antibodies. This work was supported by National Institute of Health grants ( GM 39746 and HL 36573).

**References**

1.  Blanco G, Xie ZJ and Mercer RW (1993) Functional expression of the α2 and α3 isoforms of the Na,K-ATPase in baculovirus infected insect cells. Proc Natl Acad Sci USA 90: 1824-1828
2.  O'Reilly DR, Miller LK and Luckow VA (1992) in Baculovirus expression vectors. A laboratory manual. W.H. Freeman Company ed.
3.  Sweadner, KJ (1989) Isozymes of the Na,K-ATPase. Biochim Biophys Acta 988: 185-220
4.  Xie ZJ, Wang Y, Ganjeizadeh M, McGee R and Askari A (1989) Determination of total (Na+K)-ATPase activity of isolated or cultured cells. Anal Biochem 183: 215-219

# Ca$^{2+}$/Calmodulin-Sensitive Na$^+$/K$^+$-ATPase Activity Expressed in Chimeric Molecules between the Plasma Membrane Ca$^{2+}$-ATPase and the Na$^+$/K$^+$-ATPase α-Subunit

Toshiaki Ishii and Kunio Takeyasu

Department of Medical Biochemistry and Biotechnology Center, The Ohio State University, Columbus, Ohio 43210 USA

## Introduction

The plasma membrane (PM) Ca$^{2+}$-ATPase belongs to a family of ion pumps (P-type ATPase) which undergoes a series of conformational changes including a phosphorylated intermediate during the course of ATP-catalysis and cation transport (10). It differs from the sarcoplasmic reticulum (SR) Ca$^{2+}$-ATPase in its subcellular localization, and consists of a single-type catalytic subunit in contrast to the multi-subunit H$^+$/K$^+$- and Na$^+$/K$^+$-ATPases. An additional distinct feature of the PM Ca$^{2+}$-ATPase is that its function is regulated by Ca$^{2+}$/calmodulin.

A striking difference between the primary structures of PM Ca$^{2+}$-ATPases and other cation-transporting ATPases resides at the carboxy terminus; the PM Ca$^{2+}$-ATPase carries extra ~150 amino acids at its carboxy terminus in which two possible binding domains of calmodulin can be found (14). Proteolytic cleavage of this region of the PM Ca$^{2+}$-ATPase results in a loss of calmodulin-sensitivity and a full activation of the Ca$^{2+}$-ATPase activity in the absence of calmodulin (8). Furthermore, synthetic peptides that copy the calmodulin binding domains of the PM Ca$^{2+}$-ATPase inhibit the activity of this truncated ATPase (4). These results have led to the proposal of the existence of an auto-inhibitory domains within the PM Ca$^{2+}$-ATPase molecules; the calmodulin-binding domains at the carboxy terminus interact with not-yet-identified domains within the ATPase and repress the activity in the absence of calmodulin. The binding of calmodulin to these domains would remove the auto-inhibition, and thus, stimulate the PM Ca$^{2+}$-ATPase activity. A series of attempts have been made towards identification of possible regions within the PM Ca$^{2+}$-ATPase that might interact with the "calmodulin-binding domains (CBD)" (1,5). However, concrete answers for this question have not yet been obtained, although some possible candidate fragments have been forwarded.

In this report (15), we have constructed and expressed a chimeric ATPase in which the carboxy terminal region (Arg1043 - Leu1198 including CBD) of the rat brain PM Ca$^{2+}$-ATPase II was connected to the carboxy terminus of the ouabain sensitive chicken Na$^+$/K$^+$-ATPase α1 subunit. This chimeric molecule has retained ouabain-sensitive ATPase activity which can be further stimulated by Ca$^{2+}$ and calmodulin in a dose-dependent manner. These results suggest that the Na$^+$/K$^+$-ATPase α1-subunit contains the necessary molecular device(s) except CBD for the regulation by Ca$^{2+}$/calmodulin.

## Methods

Construction of chimeric cDNA:    For the construction of chimera NNN-CBD (Figure 1), oligonucleotide directed mutagenesis using polymerase chain reaction (PCR) was employed to create unique endonuclease recognition sites, Xho I and Sal I, at the encoding position of Pro1011 in the chicken Na$^+$/K$^+$-ATPase α1-subunit and Arg1043 in the rat brain PM Ca$^{2+}$-ATPase II, respectively. The nucleotide sequence encoding Met1 to Pro1011 of the Na$^+$/K$^+$-ATPase α1-subunit and Arg1043 to Leu1198 of the PM Ca$^{2+}$-ATPase II was synthesized by PCR, and ligated using Xho I/Sal I sites as a chimeric junction. Nucleotide sequence of PCR fragments was determined by the standard DNA sequencing technique (12). When mutated sequences were found, they were replaced with the wild-type sequences using appropriate restriction enzyme sites available near the mutated regions. The final construction of chimeric cDNA were ligated into one of the polylinker sites in a mammalian expression vector, pRC/CMV.

Other methods:  Detailed descriptions for cell culture, DNA transfection, crude membrane preparation, [$^3$H]ouabain binding, and ATPase assay are provided elsewhere (see Takeyasu et al. in this volume).

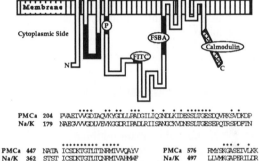

**Figure 1.** Schematic model of the chimera NNN~CBD. Homologous regions between the PM Ca$^{2+}$-ATPase II and the Na$^+$/K$^+$-ATPase α1-sub-unit were shaded.

```
PMCa 204  PVAEIVVGDIAQVKYGDLLPADGILIQGNDLKIDESSLITGESDQVRKSVDKDP
Na/K 179  NABGVVVGDLVEVKGGDRIPADLRIISANGCKVDNSSLITGESEPQIRSPDFIN

PMCa 447  NATA ICSDKTGTLTTNRMTVVQAYV        PMCa 576  RMYSKGASETVLKK
Na/K 362  STST ICSDKTGTLTQNRMTVAHMWF        Na/K 497  LLVMKGAPERILDR

PMCa 656  VVGIEDPVRPEVPEAIRKCQRAGITVRMVTGDNINIARAIAIKCGI IHPGEDFL
Na/K 595  LISMIDPPRAAVPDAVGKCRSAGIKVIMVTGDHPITAKAIAKGVGI ISDGNETV

PMCa 767  ROVV AVTGDGTNDGPALKKADVGFAMGIAGTDVAKEASDI ILTDDNFSSIVKGAVMWGRNVY
Na/K 707  GAIV AVTGDGVNDSPALKKADIGVAMGIAGSDVSKQAADMILLDDNFASIVTGVEBGRLIF
```

## Results and Discussion

<u>Ouabain-binding property of the chimeric molecule suggest that carboxy terminal modification of the Na$^+$/K$^+$-ATPase α-subunit affects the affinity for ouabain.</u>

The search for ouabain-binding domains has led to the conclusion that they are localized within the amino terminal region rather than the carboxy terminal region (11). As an extension of this line of evidence, we have previously demonstrated that the N-terminal 200 amino acids of the Na$^+$/K$^+$-ATPase are sufficient to bind ouabain and exert inhibitory function even after incorporation into the corresponding region of the SR Ca$^{2+}$-ATPase (7). On the other hand, another line of evidence has indicated that the β-subunit of the Na$^+$/K$^+$-ATPase can modulate the characteristics of ATPase functions including affinities for K$^+$ and ouabain (3). We have previously demonstrated that the carboxy-terminal 160 amino acids of the α-subunit are responsible for the assembly with the β-subunit (9). Thus, there seems to be an interaction between the carboxy- and amino-terminal regions via a long range conformational changes.

Figure 2 shows the properties of [$^3$H]ouabain-binding in three cell lines (Ltk (control for endogenous mouse enzymes), NNN (expressing chicken wild type Na$^+$/K$^+$-ATPase α1 subunit) and NNN~CBD (expressing the chimeric molecules)). NNN exhibited a high-affinity ouabain-binding with a Kd value of ~350 nM inconsistent with previous reports (7). In contrast, NNN~CBD showed much lower affinity for ouabain than those seen in NNN, although this affinity is still higher than that of endogenous mouse enzymes.

These results directly demonstrate that the affinity for ouabain can be altered by a modification of the carboxy terminal region, probably through a long range interference. Immunoprecipitation using a monoclonal IgG specific to the chicken β-subunit has confirmed that the chimeric molecule, NNN~CBD actually assembles with the β-subunit. Therefore, the apparent lower affinity for ouabain observed in NNN~CBD is not simply due to a loss of interaction with the β-subunit, although an allosteric effect via a subtle change in subunit assembly caused by this modification cannot be excluded.

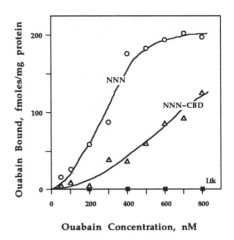

**Figure 2.** Ouabain-binding to NNN~CBD

Ca$^{2+}$/calmodulin-dependent, ouabain-inhibitable Na$^+$/K$^+$-ATPase activity

Under normal condition without calmodulin and Ca$^{2+}$, NNN showed ouabain-inhibitable Na$^+$/K$^+$-ATPase activity (chicken-type with an IC$_{50}$ value of ~1μM) over the background of endogenous mouse Na$^+$/K$^+$-ATPase activity (mouse-type with an IC$_{50}$ value of ~100μM) (7). In the same condition, NNN~CBD did not show any detectable Na$^+$/K$^+$-ATPase activity over the mouse background, but, in the presence of calmodulin and Ca$^{2+}$, NNN~CBD showed detectable ouabain-inhibitable Na$^+$/K$^+$-ATPase activity over the mouse activity, although the IC$_{50}$ value for ouabain was larger than in NNN, being consistent with [$^3$H]ouabain-binding data (Figure 2). The concentration of calmodulin required for the Na$^+$/K$^+$-ATPase activity of NNN-CBD (EC$_{50}$ ~400nM) is very similar to that for

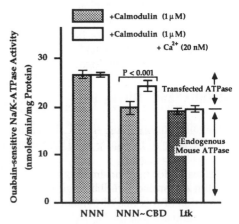

**Figure 3.** Activation of NNN~CBD by Ca$^{2+}$/calmodulin (n=4)

the PM Ca$^{2+}$-ATPase (6). In Figure 3, effects of 100 nM free Ca$^{2+}$ on the Na$^+$/K$^+$-ATPase activities of Ltk, NNN and NNN~CBD were monitored in the presence of a fixed concentration of calmodulin (1μM).

The effect of Ca$^{2+}$ on the calmodulin-stimulated Na$^+$/K$^+$-ATPase of NNN~CBD was biphasic; the Na$^+$/K$^+$-ATPase activity in NNN~CBD was stimulated at lower free Ca$^{2+}$ concentrations (up to ~100nM), and then inhibited at higher free Ca$^{2+}$ concentrations. By contrast, the Na$^+$/K$^+$-ATPase activities of NNN and Ltk were not stimulated, and, instead, subjected to a sole inhibition by Ca$^{2+}$ ions in the same concentration range of free Ca$^{2+}$ ion (20nM~1μM). Thus, the newly detected Na$^+$/K$^+$-ATPase activity of NNN~CBD is Ca$^{2+}$/calmodulin-dependent.

## Possible domains of the Na⁺/K⁺-ATPase α-subunit that interact with the calmodulin binding regions of the PM Ca²⁺-ATPase.

$Ca^{2+}$/calmodulin had dual effects (activation and inhibition) on the chimeric ATPase, while exhibiting a single inhibitory effect on the wild-type Na⁺/K⁺-ATPase (Figures 3). In crude systems such as cell homogenates used in the present study, the activity of the Na⁺/K⁺-ATPase can be regulated by protein kinases including PKA (13) and PKC (2) via phosphorylation at positions other than Asp395. This could partly explain the inhibitory effect of $Ca^{2+}$/calmodulin on both the wild-type and chimeric Na⁺/K⁺-ATPase. On the other hand, the stimulatory effect of $Ca^{2+}$/calmodulin was seen only in the chimeric ATPase, NNN~CBD, indicating a direct regulation of the Na⁺/K⁺-ATPase via the CBD incorporated in the carboxy-terminus of the α1-subunit. The calmodulin binding regions possibly interact with the α1-subunit via structurally homologous domains.

Figure 1 compares the homologous domains between the rat brain PM $Ca^{2+}$-ATPase ll and the chicken Na⁺/K⁺-ATPase α1-subunit. Five homologous domains were identified as candidates for the sites that might interact with the calmodulin binding domain. One is located between M2 and M3, and others are located in the catalytic domain between M4 and M5. According to the previous reports using a photoactivatable derivative of the synthetic calmodulin binding regions (1,5), the CBD interacts with two sites; Ile206-Val271 and Cys537-Thr544. The calmodulin-binding domain seems to fold over extensive protruding regions including the large catalytic domain between M4 and M5 and the short intracellular loop between M2 and M3. It is also considered that acceptor sites and effective sits for the calmodulin binding domain might be different (1). It is of interest that the amino acid residues 206-271 which are located between M2 and M3 is one of the highly homologous domain between the PM $Ca^{2+}$- and the Na⁺/K⁺-ATPase.

## References

1. Carafoli, E. (1992) J. Biol. Chem. 267: 2115-2118.
2. Chibalin, A.V., Vasilets, L.A., Hennekes, H., Pralong, D., Geering, K. (1992) J. Biol. Chem. 267:22378-22384.
3. Eakle, K.A., Kim, K.S., Kabalin, M.A., Farley, R.A. (1992) Proc. Natl. Acad. Sci. USA 89: 2834-2838.
4. Enyedi, A., Vorherr, T., James, P., McCormick, D.J., Filoteo, A.G., Carafoli, E., Penniston, J.T. (1989) J. Biol. Chem. 264: 12313-12321.
5. Falchetto, R., Vorherr , T., Brunner, J., Carafoli, E. (1991) J. Biol. Chem. 266: 2930-2936.
6. Filoteo, A.G., Enyedi, A., Penniston, J.T. (1992) J. Biol. Chem. 267: 11800-11805.
7. Ishii, T., Takeyasu, K. (1993) Proc. Natl. Acad. Sci. USA in press.
8. James, P., Vorherr, T., Krebs, J., Morelli, A., Castello, G., McCormick, D.J., Penniston, J.T., Flora, A.D., Carafoli, E. (1989) J. Biol. Chem. 264: 8289-8296.
9. Lemas, M.V., Takeyasu, K., Fambrough, D.M. (1992) J. Biol. Chem. 267: 20987-20991.
10. Pedersen, P.L., Carafoli, E. (1987) Trends Biochem. Sci. 12: 146-150.
11. Price, E.M., Lingrel, J.B. (1988) Biochemistry 27: 8400-8408.
12. Sanger, F., Nicklen, S., Coulson, A.R. (1977) Proc. Natl. Acad. Sci. USA 74: 5463-5467.
13. Satoh, T., Cohen, H.T., Katz, A.I. (1992) J. Clin. Invest. 89: 1496-1500.
14. Vorherr, T., James, P., Krebs, J., Enyedi, A., McCormic, D.J., Penniston, J.T., Carafoli, E. (1990) Biochemistry 29: 355-365.
15. We thank Dr. G.K. Shull for a gift of PM $Ca^{2+}$-ATPase cDNAs. KT is an Established Investigator of the AHA. This work was supported by a grant from NIH (GM 44373).

# Expression of Na+/K+-ATPase and H+/K+-ATPase fusion proteins in *Sf*-9 insect cells

J.C. Koster, M.A. Reuben, G. Sachs and R.W. Mercer

Department of Cell Biology and Physiology, Washington University School of Medicine, St. Louis, Missouri 63110 USA and Department of Medicine, University of California at Los Angeles, School of Medicine, Los Angeles, California 90024 USA

## Introduction

The baculovirus-directed expression system has been utilized extensively to express a variety of soluble and transmembrane polypeptides in infected *Sf*-9 insect cells. In most instances the expressed polypeptides receive the proper post-translational modifications and assume a functionally active conformation even when multimerization is required. Recently, reconstitution of ouabain-inhibitable Na+/K+-ATPase activity has been demonstrated in *Sf*-9 insect cells coexpressing baculovirus induced $\alpha 1$ and $\beta 1$-subunit isoforms (1). Infections with recombinant baculovirus expressing the $\alpha$ or $\beta$-subunit alone showed no appreciable level of ouabain-sensitive Na+/K+-ATPase activity although both subunits were delivered to the plasma membrane independent of one another. Additional kinetic studies of uninfected insect cells demonstrated negligible levels of endogenous Na+/K+-ATPase activity and an absence of SCH 28080-inhibitable H+/K+-ATPase activity (data not shown). Thus, cultured *Sf*-9 insect cells represent an attractive model system for functional studies in which recombinant $\alpha$-subunits can be expressed in combination with specific $\beta$-subunit isoforms and assayed for altered enzymatic activity under controlled reaction conditions. At present, the exact polypeptide regions and residues that comprise the ligand binding domains of the countertransporting P-type ATPases are unresolved. Through this current study a greater understanding of the regions of the $\alpha$-subunit involved in cation sensitivity (Na+ vs. H+), inhibitor binding (ouabain vs. SCH 28080), and $\beta$-subunit interaction can be ascertained.

## Results

Currently, the complete identity of the residues within the countertransporting ATPases that confer ion responsiveness, $\beta$-subunit interaction, and inhibitor binding are unresolved. To elucidate the functional domains of the countertransporting P-type ATPases a series of six chimeras were constructed between the $\alpha$-subunits of the Na+/K+-ATPase and the homologous gastric H+/K+-ATPase ($\approx 63\%$ amino acid identity). Using the cDNAs, the regions corresponding to the *N*-terminal transmembrane region, the cytoplasmic domain, and the *C*-terminal regions were interchanged to create fusion Na+/K+- and H+/K+-ATPase polypeptides (Fig.1). Chimeras were coexpressed with the Na+/K+ or H+/K+ $\beta$-subunits in *Sf*-9 insect cells using the baculovirus expression system. Using this system recombinant polypeptides were shown to associate with the membrane fraction. All chimeras examined, regardless of the *C*-terminal domain expressed, were capable of stable and specific association with both the Na+/K+ β2 and H+/K+ β-subunits in detergent extracts (Fig.3). This suggests that the chimeric polypeptides were correctly folded and properly inserted into the membrane. As observed in both *Xenopus* oocytes (3) and yeast cells (2), the Na+/K+-ATPase α-subunit assembled with the expressed H+/K+ β-subunit and this interaction was also resistent to detergent treatment. Interestingly, immunofluorescent localization demonstrated that the HHN chimera, similar to the native H+/K+ α-subunit, is targeted intracellularly in infected *Sf*-9 cells (Fig 4). Conversely, the NNH chimera is localized primarily at the cell surface in a similar distribution to the native Na+/K+-ATPase α1-subunit.

Presently, we are characterizing the enzymatic activity of the native and chimeric proteins under various reaction conditions in order to establish ion requirements, β–subunit selectivity, and inhibitor specificity.

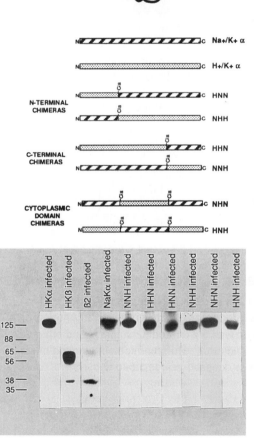

**Figure 1.** Construction of chimeric α-subunits. Full length α-subunit cDNAs from the rat Na+/K+-ATPase (α1) and the rabbit H+/K+-ATPase were used to generate the series of chimeric constructs depicted. Unique *Cla* I restriction sites were generated at fusion junctions to effectively divide native α-subunits into distinct *N*-terminal, cytoplasmic, and *C*-terminal domains in accordance with the 8 transmembrane model for α-subunit topology. Chimeras are designated based on the position of the structural domain: N for Na+/K+-ATPase derived, H for H+/K+-ATPase derived.

**Figure 2.** Expression of Na+/K+ : H+/K+-ATPase fusion polypeptides in infected Sf-9 insect cells. 25 µg of protein from infected *Sf*-9 membrane preparations were separated on SDS-PAGE (7.5%) and analyzed by immunoblotting. Specific monoclonal and polyclonal antibodies with known epitopes in the Na+/K+-ATPase and H+/K+-ATPase α-subunits were used to confirm structural organization of chimeric polypeptides.

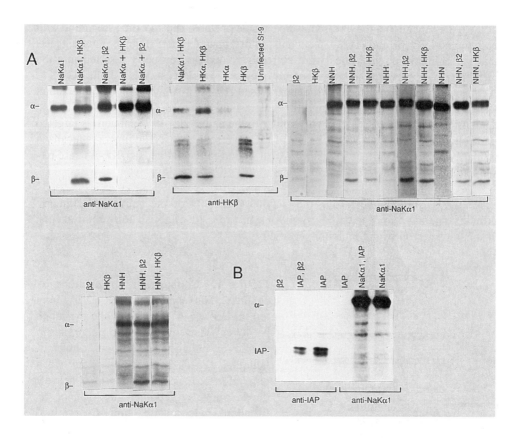

**Figure 3. A** Stable association of chimeric and native α-subunits with the β-subunits in immunoprecipitates from dually infected *Sf*-9 cells. 48h infected cells were metabolically labeled with $^{35}$S[Met], lysed in RIPA buffer (1% Triton, 0.1% SDS, HBS) and centrifuged 13,000 x g for 10 min. Supernatent was incubated overnight at 4°C with the specific monoclonal antibody as indicated. Immunoprecipitates were washed three times with RIPA buffer prior to separation on SDS-PAGE (7.5%). In control lanes, detergent lysate from cells monoinfected with the α1 and β-subunit were mixed at equal volume prior to immunoprecipitation. We are currently assessing the ability of the chimeras to associate with the β1-subunit. **B.** Coexpression of the unrelated integrin associated protein (IAP) with either Na$^+$/K$^+$-ATPase α1 or β2 subunits does not lead to stable association in detergent lysates. The integrin associated protein, an unrelated 45 kD transmembrane protein, was coexpressed with the α1 and β2-subunits and protein from solubilized *Sf*-9 cells was immunoprecipitated with the specific monoclonal antibodies indicated. Immunoprecipitates were processed as described above.

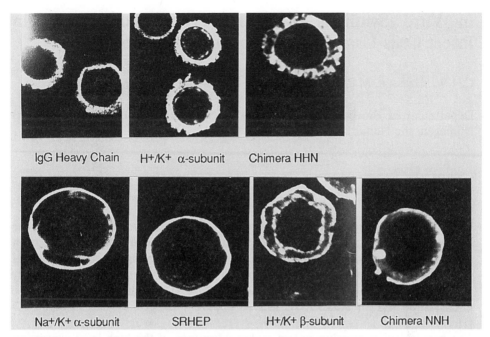

| IgG Heavy Chain | H+/K+ α-subunit | Chimera HHN |
| Na+/K+ α-subunit | SRHEP | H+/K+ β-subunit | Chimera NNH |

**Figure 4.** Immunofluorescent localization of baculovirus-expressed polypeptides in monoinfected *Sf*-9 cells. At 48h post infection cells were treated for 2h with cyclohexamide (100 μg/ml), fixed in 4% paraformaldehyde and subsequently incubated with rhodamine conjugated goat anti-mouse antibody for fluorescent detection by confocal microscopy. **1.** Murine IgG Heavy Chain Subunit, the IgG heavy chain when expressed without the light chain has previously been shown to be retained by the infected cell and to give a distinctive intracellular distribution characteristic of the H2 homodimer (1) **2.** H+/K+-ATPase α-subunit **3.** Chimera HHN **4.** Na+/K+-ATPase α1-subunit **5.** Serine-Rich *Entamoeba histolytic* Protein (SRHEP), a plasma membrane associated protein of entamoeba **6.** H+/K+-ATPas β-subunit **7.** Chimera NNH

*Acknowledgments*

We thank Dr. John Forte for providing the monoclonal antibody against the H+/K+ β-subunit. This work was supported by NIH Grant HL 36573 and GM 39746.

**References**

1.  DeTomaso AW, Xie ZJ, Liu G, and Mercer RW (1993) Expression, targeting, and assembly of functional Na,K-ATPase in baculovirus-infected insect cells. J Bio Chem 268: 1470-1478
2.  Eakle KA, Kim KS, Kabalin MA, and Farley RA (1992) High-affinity ouabain binding by yeast cells expressing Na+,K+-ATPase α-subunits and gastric H+,K+-ATPase ß-subunit. Proc Natl Acad Sci USA 89:2834-2838
3.  Horisberger JD, Jaunin P, Reubens MA, Lasater LS, Chow DC, Forte JG, Sachs G,Rossier BC, and Geering K (1991) The H,K-ATPase ß-subunit can act as a surrogate for the ß-Subunit of Na,K-pumps. J Bio Chem 266: 19131-19134

# In Vitro Synthesis of H$^+$/K$^+$-ATPase Polypeptides in Insect Cells Using the Baculovirus Expression System

C.H.W. Klaassen, M.P. De Moel, H.G.P. Swarts and J.J.H.H.M. De Pont

Department of Biochemistry, University of Nijmegen, PO Box 9101, 6500 HB Nijmegen, the Netherlands

## Introduction

H$^+$/K$^+$-ATPase is the enzyme responsible for gastric acid secretion catalysing the active exchange of intracellular H$^+$ for luminal K$^+$ ions and generating a proton gradient of more than 6 units across the apical membrane of gastric parietal cells. The enzyme consists of two subunits, a catalytic $\alpha$-subunit of approximately 114 kD and a heavily glycosylated ß-subunit of 34 kD which due to its extensive glycosylation has on SDS-PAGE gels an apparent molecular mass of 60-80 kD [1].

The baculovirus expression system makes advantage of the high level expression of certain viral proteins not essential for viral replication in a cell line from the fall armyworm *Spodoptera frugiperda* (Sf9 cells) [2]. Two such proteins are polyhedrin and p10 protein. Both proteins can reach expression levels of more than 30% in Sf9 cells, dependent on the stage of infection. Replacing polyhedrin or p10 coding sequences by homologous recombination between wild-type viral DNA and a transfer vector DNA has led to the production of recombinant viruses expressing large amounts of recombinant proteins. Many cytosolic proteins as well as complex membrane proteins, including Na$^+$/K$^+$-ATPase [3], have successfully been expressed using this system.

We report here the *in vitro* synthesis of H$^+$/K$^+$-ATPase polypeptides using the baculovirus expression system. By constructing a single recombinant baculovirus expressing both H$^+$/K$^+$-ATPase subunits we have been able to synthesize *in vitro* a functional H$^+$/K$^+$-ATPase [4].

## Methods

Using standard molecular biology techniques, both H$^+$/K$^+$-ATPase subunits were subcloned separately into baculovirus expression vectors pAcDZ1 [5] and pAcAS3 [6] containing the E. coli lacZ coding sequences for convenient screening for recombinant viruses in plaque assays. Next, these transfer vectors were used for recombination into wild-type AcNPV DNA to produce recombinant viruses expressing either H$^+$/K$^+$-ATPase $\alpha$-subunit or ß-subunit from either the polyhedrin or the p10 promoter, respectively. Using two successive recombination events at the polyhedrin and p10 locus, respectively, we produced recombinant viruses expressing both H$^+$/K$^+$-ATPase subunits (for a more detailed describtion of the production of these viruses see [4]). A total of 12 different recombinant viruses expressing either H$^+$/K$^+$-ATPase subunits or control viruses have been produced (Table 1).

**Table 1**.

| no. | viral code | polyhedrin locus | p10 locus |
|-----|-----------|-----------------|-----------|
| 1 | DZ1 | ß-gal | p10 |
| 2 | DZα | α-subunit, ß-gal | p10 |
| 3 | DZß | ß-subunit, ß-gal | p10 |
| 4 | AS3 | polyhedrin | ß-gal |
| 5 | ASα | polyhedrin | α-subunit, ß-gal |
| 6 | ASß | polyhedrin | ß-subunit, ß-gal |
| 7 | DLZα | α-subunit | p10 |
| 8 | DLZαAS3 | α-subunit | ß-gal |
| 9 | DLZαASß | α-subunit | ß-subunit, ß-gal |
| 10 | DLZß | ß-subunit | p10 |
| 11 | DLZßAS3 | ß-subunit | ß-gal |
| 12 | DLZßASα | ß-subunit | α-subunit, ß-gal |

Overview of recombinant baculoviruses produced. behind each viral code, H$^+$/K$^+$-ATPase subunit(s) expressed in the polyhedrin locus or p10 locus are indicated, ß-gal = ß-galactosidase.

**Results.**

Sf9 cells infected with these recombinant baculoviruses express the corresponding H$^+$/K$^+$-ATPase subunit(s) as demonstrated by western blot (figure 1). Expression of the α-subunit leads to formation of an immunoreactive protein, with an apparent molecular weight of 95 kD, which is exactly the same size as that of native H$^+$/K$^+$-ATPase purified from pig gastric mucosa [7]. This protein band is recognized by several different antisera to the α-subunit (monoclonal as well as polyclonal antibodies).

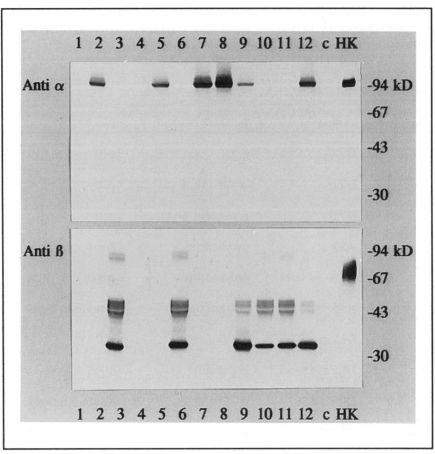

**Figure 1.**
Western blot of Sf9 cells infected with recombinant baculoviruses. Numbers correspond to viral numbers listed in Table 1; c = not infected Sf9 cells; HK = rat H⁺/K⁺-ATPase

Expression of the ß-subunit also leads to production of an immunoreactive protein which appears both as a single protein band of 34 kD and partly as a smear of approximately 40-50 kD. This smear can be explained by different glycosylation patterns of the protein produced in the insect cells. The protein band at 34 kD could be identified as the naked core protein of the ß-subunit. Indeed when infected cells were incubated in the presence of 5 $\mu$g/ml tunicamycin, an inhibitor of the N-glycosylation process, the smear identified by the anti-ß-subunit antisera diasppeared, leading to an increase of the immunoreactive protein band at 34 kD. Although the *in vitro* produced ß-subunit in the absence of tunicamycin is glycosylated, glycosylation patterns were not as extensive as in the native enzyme which appeared on SDS-PAGE as a protein with an apparent molecular mass of 60-80 kD, while the glycosylated ß-subunit as produced by Sf9 cells had a molecular mass between 40 and 50 kD.

96

We next studied ATP-dependent phosphoryation capacities of purified Sf9 membranes and found that only membranes obtained from cells infected with a recombinant virus expressing both $H^+/K^+$-ATPase subunits (i.e. Table 1, no. 9) exhibit a $K^+$ and SCH 28080 sensitive phosphorylation capacity similar to native $H^+/K^+$-ATPase from pig gastric mucosa [4]. We therefore conclude that both subunits are essential for the phosphorylation capacity of $H^+/K^+$-ATPase.

*Acknowledgements.*

This work was sponsored by the Netherlands foundation for scientific research under grant no. 900-522-086.

**References.**

1. Klaassen CHW and De Pont JJHHM (1994) Gastric H/K-ATPase. Cell Physiol Biochem 4: 114-134
2. Vlak JM, Keus RJA (1990) Baculovirus expression vector system for production of viral vaccines. In: Viral Vaccines. Wiley-Liss, New York, pp 91-128
3. De Tomaso AW, Xie ZJ, Liu GQ, Mercer RW (1993) Expression, targeting, and assembly of functional Na,K- ATPase polypeptides in baculovirus-infected insect cells. J Biol Chem 268: 1470-1478
4. Klaassen CHW, Van Uem TJF, De Moel MP, De Caluwe GLJ, Swarts HGP, De Pont JJHHM (1993) Functional expression of gastric H,K-ATPase using the baculovirus expression system. FEBS Lett 329: 277-282
5. Zuidema D, Schouten A, Usmany M, Maule AJ, Belsham GJ, Roosien J, Klinge-Roode EC, Van Lent JWM, Vlak JM (1990) Expression of cauliflower mosaic virus gene I in insect cells using a novel. Polyhedrin-based baculovirus expression vector. J Gen Virol 71: 2201-2209
6. Vlak JM, Schouten A, Usmany M, Belsham GJ, Klinge-Roode EC, Maule AJ, Van Lent JWM, Zuidema D (1990) Expression of cauliflower mosaic virus gene I using a baculovirus vector based upon the P10 gene and a novel selection method. Virology 178: 312-320
7. Swarts HGP, Van Uem TJF, Hoving S, Fransen JAM, De Pont JJHHM (1991) Effect of free fatty acids and detergents on H,K-ATPase - The steady-state ATP phosphorylation level and the orientation of the enzyme in membrane preparations. Biochim Biophys Acta 1070: 283-292

# Functional domains of sarcoplasmic reticulum Ca$^{2+}$-ATPase studied by site-directed mutagenesis

J.P. Andersen, B. Vilsen

Institute of Physiology, University of Aarhus,
DK-8000 Aarhus C, Denmark

## Introduction

The availability of the cDNA clones of P-type ATPases has opened up the possibility for introducing defined point mutations in the proteins by in vitro oligo-nucleotide-directed mutagenesis of the cDNA. This approach can be used to search for residues involved in ligand binding or in critical conformational transitions, provided a suitable system is available for functional expression of the mutant cDNA. In addition to the requirement for a high expression level of the exogenous mutant cDNA, the absence of a significant contribution from endogenous pumps is demanded. The sarcoplasmic reticulum Ca$^{2+}$-ATPase cDNA has been succesfully expressed in COS-1 cells, at levels more than 100-fold higher than the expression level of the endogenous COS-1 cell Ca$^{2+}$-pump. This has permitted the functional analysis of more than 200 different Ca$^{2+}$-ATPase mutants (2-11,16-17,19-22,24). The general strategy has been to assay first for the ATP-driven active Ca$^{2+}$ uptake and ATPase activity in the isolated microsomal fraction containing the mutant enzyme, and thereafter for phosphorylation from ATP and P$_i$, its Ca$^{2+}$ and substrate dependence, as well as the dephosphorylation kinetics in the presence and absence of ADP. Recently, it has proved possible to measure Ca$^{2+}$-occlusion in the mutants (24).

The mutagenesis work with the Ca$^{2+}$-ATPase has led to the functional classification of the amino acid substitutions indicated by the various symbols in Fig. 1. A relatively small number of residues (between 20 and 30) have been found to be essential, in the sense that substitution with any other residue led to a block of Ca$^{2+}$ transport activity. Many more residues could be replaced without significant functional changes. The latter mutations are not the least important, since they serve as a background high-lighting the interesting mutations that give rise to important functional changes.

The mutants are analysed under the assumption that the minimum functional unit in ATP hydrolysis and Ca$^{2+}$ transport is a single Ca$^{2+}$-ATPase peptide chain which binds two calcium ions (1). This hypothesis has been supported by a number of reproducible data from many independent laboratories. Functional changes induced by amino acid alterations are thus ascribed to prevention of conformational changes or protein-ligand interactions, but not to prevention of dimeric peptide-peptide interactions.

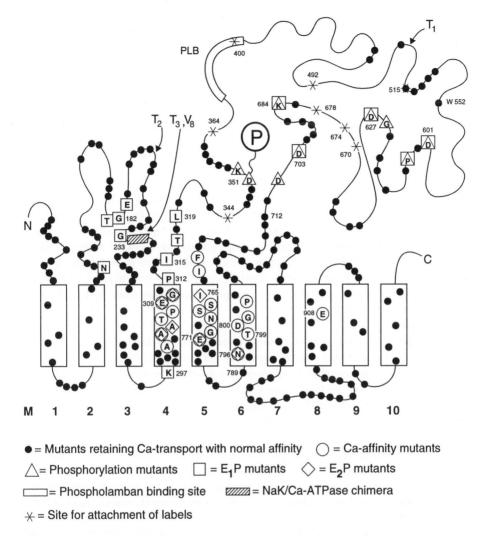

**Figure 1.** Functional classification of amino acid substitutions in the sarcoplasmic reticulum $Ca^{2+}$-ATPase. Based on data in (2-11,16-17,19-22,24) and unpublished work. Ten transmembrane helices (M1-M10) are shown, as suggested in (15). The segments linking M1-M5 to the cytplasmic domains are usually denoted S1-S5 ("stalk helices").

## 1. The "Ca²⁺-site" mutants

The mutagenesis approach has led to major progress towards the identification of amino acid residues involved in formation of the ion binding/occlusion sites. In Fig. 1 the open circles symbolize residues whose substitution has resulted in mutants with reduced $Ca^{2+}$ affinity relative to the wild-type enzyme, as observed in assays for $Ca^{2+}$-activated $Ca^{2+}$ transport and/or phosphorylation. These residues, all of which are located in, or close to, the putative transmembrane segments M4, M5, M6, and M8, constitute a maximum estimate of the peptide domain sphere involved in $Ca^{2+}$ binding.

Six residues with oxygen-containing side chains, $Glu^{309}$, $Glu^{771}$, $Asn^{796}$, $Thr^{799}$, $Asp^{800}$, and $Glu^{908}$ were originally pinpointed as likely constituents of a $Ca^{2+}$ binding pocket (6,9). This proposal was based on two sets of findings with mutants in which the negative charge or the oxygen atom(s) had been removed from the side chain of one of the residues: (a) the mutants (which were completely deficient in $Ca^{2+}$ transport) did not phosphorylate from ATP in the presence of $Ca^{2+}$, and (b) they phosphorylated from $P_i$ even in the presence of $Ca^{2+}$. On the contrary, in the wild-type enzyme $Ca^{2+}$ binding at the high-affinity transport sites activates phosphorylation from ATP and inhibits phosphorylation from $P_i$.

The alternatives for interpretation of these phosphorylation data are:
(I)   One or both $Ca^{2+}$ sites were disrupted by the mutation.
(II)  The $E_1$-$E_2$ conformational equilibrium was displaced in favor of $E_2$.
(III) The catalytic site or its coupling with the $Ca^{2+}$-binding sites was altered. In particular, the ability of bound $Ca^{2+}$ to confer the enzyme its ability to receive the γ-phosphoryl group from ATP might have been defective.

Studies were undertaken with the purpose of further resolving this issue. One of the questions we have addressed is whether an extension of the range of $Ca^{2+}$ concentrations would allow observation of phosphorylation from ATP in the mutants with alterations to the above mentioned six residues with oxygen-containing side chains. In the previous work (6,9), the $Ca^{2+}$ concentration was 100 $\mu$M, i.e. 100-fold that required for half-saturation of phosphorylation from ATP in the wild-type enzyme. We carried out ATP phosphorylation experiments at $Ca^{2+}$ concentrations ranging up to 12.5 mM. It was found that mutants $Glu^{309}$→Gln, $Glu^{309}$→Lys, $Asn^{796}$→Ala, and $Asp^{800}$→Asn were unable to phosphorylate from ATP, even at this high $Ca^{2+}$ concentration, whereas mutants $Glu^{309}$→Ala, $Glu^{771}$→Gln, $Thr^{799}$→Ala, and $Glu^{908}$→Ala became able to phosphorylate from ATP, when the $Ca^{2+}$ concentration was in the millimolar range (Table 1). Hence, for the latter group of mutants possibility III could be discarded. Since the binding of two calcium ions is required to activate phosphorylation from ATP, it may be concluded that in the phosphorylating mutants either $Ca^{2+}$ site was able to function.

The second question that we addressed is whether the affinity of one or of both $Ca^{2+}$ sites was reduced in the above mentioned mutants. Since there is evidence that in the wild-type enzyme the two calcium ions bind through a stepwise mechanism, in which the binding of the first calcium ion is sufficient to prevent phosphorylation from $P_i$ (13), the previous observation (6,9) of maximum phosphorylation from $P_i$ in the presence of 100 $\mu$M $Ca^{2+}$ would suggest that the amino acid substitution always hit the site binding the first calcium ion in the sequence or, alternatively, that possibility II mentioned above is the correct explanation of the results. By carrying out $Ca^{2+}$ titrations of phosphorylation from $P_i$ under neutral pH conditions, under which the $E_1$-$E_2$ equilibrium of free enzyme is less in favor of $E_2$ than under the acid pH conditions used in (6), we have now been able to show that $P_i$-phosphorylation can be prevented by $Ca^{2+}$ in all of the above mentioned mutants, provided the $Ca^{2+}$ concentration is sufficiently high. Interestingly, the apparent affinities for $Ca^{2+}$ determined by this titration differed widely among the mutants (Table 1). For the $Glu^{771} \rightarrow Gln$, $Thr^{799} \rightarrow Ala$, and $Glu^{908} \rightarrow Ala$ mutants, the apparent $Ca^{2+}$ affinity for the $Ca^{2+}$-inhibition of $P_i$-phosphorylation was in the millimolar range, close to the apparent $Ca^{2+}$ affinity determined by titration of $Ca^{2+}$-activation of phosphorylation from ATP. This suggests that these mutations reduced the affinity of the particular site binding the first calcium ion in the sequence or, alternatively, that possibility II is correct. In mutants $Glu^{309} \rightarrow Gln$ and $Asn^{796} \rightarrow Ala$, the apparent affinity for $Ca^{2+}$ determined in the $P_i$-phosphorylation assay was similar to that of the wild-type $Ca^{2+}$-ATPase. Hence, in these mutants at least the site binding the first calcium ion must have been intact. Since these mutants were unable to phosphorylate from ATP at $Ca^{2+}$ concentrations, which prevented $P_i$-phosphorylation completely, the data obtained with mutants $Glu^{309} \rightarrow Gln$ and $Asn^{796} \rightarrow Ala$ are not well explained by possibility II. Either, the mutations $Glu^{309} \rightarrow Gln$ and $Asn^{796} \rightarrow Ala$ rendered the second $Ca^{2+}$ binding step in the sequence defective, or possibility III was operative for these mutants. The mutant $Asp^{800} \rightarrow Asn$ displayed intermediate affinity for $Ca^{2+}$ in the $P_i$-phosphorylation assay, but did not phosphorylate from ATP. Therefore, this mutation may have hit both $Ca^{2+}$ sites in a common $Ca^{2+}$ binding pocket, or possibility III was operative.

The perspective that some of these mutants might be able to bind both calcium ions, and that it was the signal transduction between the $Ca^{2+}$-binding domain and the phosphorylation site which was defective, led us to search for alternative assays for the intactness of the $Ca^{2+}$ binding sites. It proved possible to apply the previously developed procedure for studies of CrATP-induced $Ca^{2+}$ occlusion in non-phosphorylated enzyme (1,23). The use of CrATP to stabilize the $Ca^{2+}$-occluded enzyme form was combined with molecular sieve HPLC of detergent-solubilized protein. With this technique, chromatopgraphic separation of the $Ca^{2+}$-ATPase from other $Ca^{2+}$-binding proteins was accomplished simultaneously with the measurement of the occluded $^{45}Ca^{2+}$ remaining associated with the separated protein. In contrast to equilibrium binding measurements of bound $Ca^{2+}$, which require large amounts of pure enzyme, the occlusion assay could be carried out with the expressed mutant and wild-type $Ca^{2+}$-ATPases in the crude COS-1 cell microsomal fraction (24). None of the mutants $Glu^{309} \rightarrow Gln$, $Glu^{771} \rightarrow Gln$, $Asn^{796} \rightarrow Ala$, and $Asp^{800} \rightarrow Asn$ were able to occlude $Ca^{2+}$ at $Ca^{2+}$ concentrations up to 1 mM (the upper limit permitting accurate measurements). Since the occlusion induced with CrATP occurs without phosphorylation of the enzyme (23,24), it now seems to be excluded that the inability of mutants $Glu^{309} \rightarrow Gln$, $Asn^{796} \rightarrow Ala$, and $Asp^{800} \rightarrow Asn$ to undergo $Ca^{2+}$-activated phosphorylation from ATP was caused exclusively by a defective catalysis of the transfer of the $\gamma$-phosphoryl group of ATP. Step(s) preceding

**Table 1.** Phosphorylation analysis of mutants with alterations to residues in M4, M5, M6, and M8

| Mutant | $K_{0.5}$ for $Ca^{2+}$-activation of phosphorylation from ATP | $K_{0.5}$ for $Ca^{2+}$-inhibition of phosphorylation from $P_i$ | Half-life for dephosphorylation of $E_2P$ |
|---|---|---|---|
| | $\mu M$ | | s |
| Wild type | 0.6 | 2-5 | <2 |
| Glu[309]→Gln | n.p.[a] | 5 | ~15 |
| Glu[309]→Ala | ~2,500 | ~100 | ~7 |
| Glu[309]→Lys | n.p. | ~10 | ~15 |
| Glu[771]→Gln | 500 | 500 | >20 |
| Asn[796]→Ala[b] | n.p. | 6 | ~5 |
| Thr[799]→Ala[b] | 3,035 | 3,230 | <2 |
| Asp[800]→Asn | n.p. | 260 | <2 |
| Glu[908]→Ala[b] | 440 | 230 | <2 |
| Ala[305]→Val | >3 | | ~7 |
| Ala[306]→Val | 0.3 | | ~7 |
| Gly[310]→Val | ~20 | ~20 | ~20 |
| Phe[760]→Gly[b] | 210 | 5 | <2 |
| Ile[765]→Gly[b] | 0.3 | | ~7 |

[a]n.p., no phosphorylation observed at the $Ca^{2+}$ concentrations tested (up to 12.5 mM)
[b]Andersen JP, Vilsen B previously unpublished, the measurements were carried out at pH 7.0 as described in (2). All other data were from (2,5,7,22).

the transfer of the $\gamma$-phosphoryl group af ATP must have been defective in these mutants. Since the evidence from the $P_i$-phosphorylation assay discussed above suggested that one of the two high-affinity $Ca^{2+}$ binding sites was left intact in the mutants Glu[309]→Gln and Asn[796]→Ala, it might have been expected that these mutants would be able to occlude one calcium ion in the presence of CrATP. The finding, that neither of the two calcium ions became occluded in the Glu[309]→Gln and Asn[796]→Ala mutants can be accounted for in terms of a model for $Ca^{2+}$ binding, in which $Ca^{2+}$

occlusion occurs as a result of the closure of a common binding pocket for the two ions (1,23). The defective $Ca^{2+}$ occlusion might result either from inability to bind the second calcium ion or from inability to close a "flickering gate" to the binding pocket (12).

In addition to the residues mentioned above, a few others with oxygen-containing side chains, $Ser^{766}$, $Ser^{767}$, and $Asn^{768}$ located in M5, have been implicated in $Ca^{2+}$ binding by mutagenesis studies (Fig. 1), but the effect on $Ca^{2+}$ sensitivity of replacing these residues with Ala was much less dramatic (2-5 fold reduction of apparent $Ca^{2+}$ affinity with preserved transport capability) than the effect observed upon replacement of the critical six residues (9). The $Ca^{2+}$ affinity was also found to be reduced after mutation of two prolines ($Pro^{308}$ and $Pro^{803}$) and three glycines ($Gly^{310}$, $Gly^{770}$ and $Gly^{801}$) located next to the critical carboxylic acid residues in M4, M5, and M6 (5,19). The extent to which alterations to the glycines affected the apparent $Ca^{2+}$ affinity depended on the size of the side chain used for replacement, the larger the side chain the more reduction of $Ca^{2+}$ affinity. The mutants $Gly^{310} \rightarrow Val$ and $Gly^{801} \rightarrow Val$ were unable to occlude $Ca^{2+}$ (24). The importance of the proline and glycine residues for $Ca^{2+}$ binding can be attributed to the unique properties of these amino acids. Prolines have the ability to break $\alpha$-helical structures and may create kinks and flexible loops for ion binding to juxtaposed carboxylate groups and exposed backbone carbonyls. Since a single hydrogen atom constitutes the only side chain of glycine, this amino acid is helpful in exposing backbone carbonyls and in providing flexibility, so the backbone can wrap around the cation. Parallel mutagenesis studies of the $Na^{+}/K^{+}$-ATPase have shown that the prolines present in this enzyme at positions homologous to those of $Pro^{308}$, $Pro^{803}$, and $Gly^{770}$ in the $Ca^{2+}$-ATPase may play a role in the binding of $Na^{+}$ and $K^{+}$ (18, and Vilsen B this volume).

We have also found that the apparent $Ca^{2+}$ affinity of the $Ca^{2+}$-ATPase detected in the phosphorylation assay with ATP was reduced more than 100-fold, when the phenylalanine $Phe^{760}$ at the top of M5 was substituted by glycine. On the other hand, the $Ca^{2+}$ inhibition of phosphorylation from $P_i$ occurred with normal apparent affinity in this mutant (Table 1). This apparent discrepancy between the $Ca^{2+}$ affinities determined in the two types of phosphorylation assay makes the $Phe^{760} \rightarrow Gly$ mutant resemble the mutants $Glu^{309} \rightarrow Gln$ and $Asn^{796} \rightarrow Ala$ discussed above, in the sense that the second, but not the first, $Ca^{2+}$ binding step in the sequence was rendered defective. Some of the hydrophobic residues in M4 have also been shown to be of importance for the $Ca^{2+}$ affinity (7). This is consistent with the hypothesis that the cation specificity is determined by steric factors imposed by the packing of hydrophobic side chains and not by the oxygen-containing side chains in the membrane, which generally are highly conserved among P-type ATPases with different cation specificity.

## 2. Phosphorylation-negative mutants

Mutagenesis analysis has shown that several of the residues located in the central cytoplasmic domain are crucial to formation of the phosphoenzyme intermediate (triangles in Fig. 1). Not unexpectedly, the phosphorylated aspartic acid residue, $Asp^{351}$, itself, is one of these residues. Generally, residues belonging to this category are highly conserved within the P-type ATPase family, and some have been implicated in ATP binding by structure prediction and affinity labeling with nucleotide-related reagents. The adenosine-binding fold has not been localized. So far, the studies of mutants have failed to pinpoint residues with a specific role in binding of the adenosine part of the ATP molecule. Hence, phosphorylation was lost in the mutants symbolized by triangles, irrespective of whether ATP or $P_i$ was used as substrate. The defective phosphorylation with either substrate points to a role for the mutated residue in interaction with the phosphoryl group or with the catalytic $Mg^{2+}$, or to a role in local conformational changes associated with such interaction.

Some of the residues whose substitution has led to phosphorylation-negative mutants can be replaced with certain amino acids without the loss of phosphorylation (Fig. 1, residues double labeled with triangles and squares). One example is $Lys^{684}$ located in the highly conserved "hinge-domain". If this residue is replaced with Ala, His, or Gln, phosphorylation is defective, but replacement with Arg, retaining the positive charge, allows phosphorylation from ATP, but not phosphorylation from $P_i$ (20). Moreover the conversion of $E_1P$ to $E_2P$ was blocked in the $Lys^{684} \rightarrow Arg$ mutant, so that this mutant resembled the $E_1P$-type mutants described below. It is possible that the positive charge is required for stabilization of a transition state in the phosphoryl transfer reaction with ATP, and that the bulky and highly polar guanidinium group of arginine interfered with the active site closure thought to be associated with the $E_2P$ state (20).

Another interesting residue in the hinge domain is $Asp^{703}$. Phosphorylation was retained in mutants $Asp^{703} \rightarrow Asn$ and $Asp^{703} \rightarrow Glu$ (11), whereas mutant $Asp^{703} \rightarrow Ala$ was unable to phosphorylate from either ATP or $P_i$ (20). It has sometimes been suggested that $Asp^{703}$, which is highly conserved within the P-type ATPase family, might bind the catalytic $Mg^{2+}$. However, we demonstrated that the $Asp^{703} \rightarrow Ala$ mutant is able to form a $Ca^{2+}$-occluded complex with CrATP (24). Therefore, it seems that $Asp^{703}$ is not required for binding the cation in its complex with ATP.

## 3. $E_1P$-type mutants

Given the proposed locations of the catalytic site in the cytoplasmic region and the $Ca^{2+}$ binding/occlusion sites in the membrane domain, it is clear that the coupling between ATP hydrolysis and ion translocation requires long-distance communication through conformational changes. Residues critical to the energy interconversion have been identified by kinetic studies of the phosphoenzyme formed in mutant $Ca^{2+}$-ATPases.

The squares in Fig. 1 indicate amino acid substitutions that have led to a block of the transition of the phosphoenzyme intermediate from the ADP-sensitive $E_1P$ form to the ADP-insensitive $E_2P$ form with resulting accumulation of $E_1P$. The first residue to be implicated in this transition on the basis of mutagenesis studies was $Pro^{312}$ in the upper part of M4 (19). Replacement of $Pro^{312}$ with Ala resulted in a complete block of $Ca^{2+}$ uptake, although the ability to phosphorylate from ATP in a normal $Ca^{2+}$-dependent manner was preserved. When EGTA was added to chelate free $Ca^{2+}$ and thereby terminate new formation of phosphoenzyme, there was no decay of the $E_1P$ phosphoenzyme intermediate in the Pro[312]→Ala mutant, whereas the wild-type phosphoenzyme intermediate decayed rapidly through conversion of $E_1P$ to $E_2P$ and hydrolysis of $E_2P$ (19). The phosphoenzyme formed from ATP in the Pro[312]→Ala mutant was not only stable, it also retained the ability to transfer the phosphoryl group back to ADP for several minutes after the addition of EGTA in the presence of ionophore. Since the reaction with ADP to reverse enzyme phosphorylation normally requires the presence of $Ca^{2+}$ bound to the enzyme, it could be deduced that in the mutant phosphoenzyme $Ca^{2+}$ remained bound in an occluded state inaccessible to chelation. The dephosphorylation of the $E_2P$ intermediate formed in the backdoor reaction with $P_i$ took place at a normal rate in the Pro[312]→Ala mutant. Hence, the reason for the stability of the phosphoenzyme formed from ATP must have been a block of the $E_1P→E_2P$ conformational change. Replacement of Pro[312] in $Ca^{2+}$-ATPase by a leucine, which is the residue present at the homologous position in $Na^+/K^+$-ATPase, also produced an enzyme with defective $E_1P→E_2P$ transition. In this case, however, the inhibition was less severe than with alanine. There was also some indication that the replacement of Pro[312] with either Ala or Leu led to higher apparent $Ca^{2+}$ affinity of the $Ca^{2+}$-ATPase, i.e. to a displacement of the equilibrium of the dephosphoenzyme in favor of $E_1$ (19). Parallel mutagenesis studies of $Na^+/K^+$-ATPase have demonstrated that substitution of the leucine (Leu[332]) homologous to Pro[312] in $Ca^{2+}$-ATPase with Ala leads to a displacement of the conformational equilibrium of $Na^+/K^+$-ATPase in favor of $E_1$ (18) and to a reduced turnover number (Vilsen B this volume). The crucial role of Pro[312] in the conformational change may be connected with its ability to kink the M4 helix at the top and thereby introduce a defect in helix packing in the transmembrane sector. This might constitute an important aspect of the mechanism leading to channel opening and release of the occluded ions at the luminal surface. It is possible that to some extent the bulky side chain of the leucine present in $Na^+/K^+$-ATPase can mimic the effect of the kink dictated by proline.

Mutations located in the stalk segment S4 linking M4 with the catalytic site have also been found to slow the $E_1P→E_2P$ transition (Fig. 1). Either of the two mutations giving rise to the most conspicuous block of $E_1P→E_2P$ transition introduced a charged residue in place of a hydrophobic residue (Ile[315]→Arg and Leu[319]→Arg). If an α-helical structure is assumed for S4, these mutations would destroy the amphipatic character of the helix. By contrast, Ala[320], although juxtaposed with Leu[319] in the primary sequence, would be located on the polar surface of the S4 helix, and we found little functional consequence of replacement of this residue with Arg (21). In addition, the substitution of the highly conserved glycine Gly[233], located in the C-terminal part of the smaller cytoplasmic loop near S3, with either Glu, Val, or Arg blocked the $E_1P→E_2P$ transition (4). The replacements of Gly[233], Ile[315], and Leu[319] led not only to mutant enzymes that were unable to form $E_2P$ in the forward direction from $E_1P$, but also the backdoor phosphorylation with inorganic phosphate was defective, contrary to the situation with the Pro[312]→Ala mutant, in which $P_i$-phosphorylation was unaffected. On the basis of this finding it is suggested that while the proline residue at the top of M4 participates in the transmission of signals between the catalytic site and the cation sites, the residues near S3 and in S4 might be involved in the $E_1P→E_2P$ transition of the catalytic site itself.

## 4. A vanadate-insensitive chimeric mutant

Recently, we substituted the 7-residues segment LysIleArgAspGlnMetAla240 located in the stalk segment S3, right after Gly[233], with the corresponding $Na^+/K^+$-ATPase segment ArgIleAlaThrLeuAlaSer. This chimera behaved as a new phenotypic variant of $Ca^{2+}$-ATPase, which was insensitive to inhibition by vanadate (3). Analysis of the phosphoenzyme intermediates showed that in the chimeric mutant $E_2P$ accumulated at steady state under buffer conditions at which the $E_1P$ form accumulated in the wild type, and the rate of dephosphorylation of $E_2P$ was found to be reduced in the mutant. As there was no significant difference between the chimeric mutant and the wild-type $Ca^{2+}$-ATPase with respect to the apparent affinity for $P_i$ measured under equilibrium conditions, it may be concluded that in the chimeric mutant the decrease in the "off" rate for the phosphoryl group must have been accompanied by a similar reduction of the "on" rate (3). In combination, these observations suggest that the transition state in the hydrolysis of the aspartyl phosphate bond in the $E_2P$ intermediate was destabilized in the chimera. There was no change in the apparent affinity for ATP in the chimera. The vanadate insensitivity may thus be explained by the resemblance of vanadate to the pentacoordinated transition state of the phosphoryl group. It is tempting to speculate that in the 3-dimensional structure of the $E_2P$ state the 7-residues segment substituted in the chimeric mutant might be located close to the catalytic cleft.

## 5. $E_2P$-type mutants

Our studies of the phosphoenzyme intermediates of mutants have revealed a number of mutants which resemble the chimeric mutant described above in the sense that the rate of dephosphorylation of the $E_2P$ intermediate was reduced, but which in contrast to the chimera were characterized by an increased equilibrium binding affinity for $P_i$ (mutations indicated by diamonds in Fig. 1, see also Table 1). This type of behaviour, suggesting a stabilization of the $E_2P$ form rather than a destabilization of the transition state, was first observed in mutants with alterations to Gly[310], the residue adjacent to the critical glutamic acid residue Glu[309] in M4 (5). By reexamination of the "$Ca^{2+}$-site mutants" we found that mutants Glu[309]→Gln, Glu[309]→Lys, Glu[771]→Gln, and Asn[796]→Ala displayed a reduced rate of $E_2P$ dephosphorylation, whereas mutants Thr[799]→Ala, Asp[800]→Asn, and Glu[908]→Ala dephosphorylated at normal rates (Table 1). Contrary to what we first believed (5), there is no obligatory linkage of the stabilization of $E_2P$ to a defective binding of $Ca^{2+}$ at the cytoplasmic high-affinity sites. Mutants Ala[306]→Val and Ile[765]→Gly displayed a normal high affinity for $Ca^{2+}$ in the phosphorylation assay with ATP, although the rate of dephosphorylation of $E_2P$ was reduced in these mutants (Table 1).

The amino acid residues whose substitutions have led to $E_2P$-type mutants are located in the putative transmembrane segments M4, M5, and the lower part of M6 (Asn[796]), and include the acidic residues Glu[309] and Glu[771]. Since recent evidence shows that protons are countertransported by the $Ca^{2+}$-ATPase (14) in a way similar to the $K^+$-transport by $Na^+/K^+$-ATPase, it is tempting to speculate that the roles for Glu[309], Glu[771] and Asn[796] in the dephosphorylation of $E_2P$ might be associated with the donation of

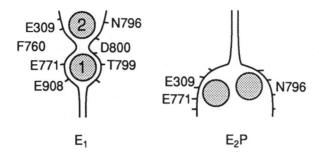

**Figure 2.** Working hypothesis for the assignment of residues to $Ca^{2+}$ and $H^+$ transport/occlusion sites.

oxygens to form binding/occlusion site(s) or an access channel for the $H^+$ (possibly $H_3O^+$) to be countertransported. By analogy with the $Na^+/K^+$-ATPase, it is conceivable that in a $H^+$-countertransport mechanism for the $Ca^{2+}$-ATPase the hydrolysis of $E_2P$ needs triggering by binding of the countertransported protons. The residues with hydrophobic side chains $Ala^{305}$, $Ala^{306}$, and $Ile^{765}$ might participate more indirectly in formation of the binding/occlusion site(s), or in the conformational change(s) mediating long-range signal transmission between the proton binding site(s) and the phosphorylation site.

## Conclusion

The left part of Fig. 2 shows a working hypothesis for the structure of the high affinity $Ca^{2+}$ binding pocket based on the functional assignment of the residues to two consecutive $Ca^{2+}$ binding steps discussed in Section 1. The residues whose substitution led to reduced apparent affinity in the binding of the first calcium ion that interferes with phosphorylation from $P_i$ are assigned to Site 1, whereas the residues whose substitution prevented phosphorylation from ATP as well as $Ca^{2+}$ occlusion, but not the $Ca^{2+}$ inhibition of phosphorylation from $P_i$, are assigned to Site 2.

The right part of Fig. 1 shows our proposal for the involvement of $Glu^{309}$, $Glu^{771}$, and $Asn^{796}$ in $H^+$-countertransport, based on the reduced rate of dephosphorylation of $E_2P$ observed for mutants with substitutions of these residues.

The demonstration of the importance of M4 for the $E_1P$-$E_2P$ transition, in conjunction with the data indicating that M4 and M5 are directly or indirectly involved in cation binding, as well as in dephosphorylation of $E_2P$, suggests that M4 and M5 play pivotal roles in the long-range communication between the catalytic site and the ion binding domain. This can be seen as a natural consequence of the direct physical link between these two transmembrane segments and the central cytoplasmic domain. $E_1P$-$E_2P$ conformational changes originating in the catalytic site may be transmitted through the "stalk" to M4. Conversely, conformational changes elicited at the ion binding sites upon ion binding or dissociation may be transmitted from M4 and M5 back through the "stalk" to the catalytic center to alter its reactivity.

*Acknowledgements*

We thank Janne Petersen, Jytte Jørgensen, Karin Kracht and Lene Jacobsen for excellent technical assistance. We would also like to express our gratitude to Dr. D.H. MacLennan, in whose laboratory our mutagenesis work was initiated, and to Dr. N.M. Green for continuing discussion. This research was supported by grants from the Danish Biomembrane Research Centre, the Danish Medical Research Council, and the NOVO-Nordisk Foundation.

## References

1.  Andersen JP (1989) Monomer-oligomer equlibrium of sarcoplasmic reticulum Ca-ATPase and the role of subunit interaction in the $Ca^{2+}$ pump mechanism. Biochim Biophys Acta 988:47-72
2.  Andersen JP, Vilsen B (1992) Functional consequences of alterations to $Glu^{309}$, $Glu^{771}$, and $Asp^{800}$ in the $Ca^{2+}$ ATPase of sarcoplasmic reticulum. J Biol Chem 267:19383-19387
3.  Andersen JP, Vilsen B (1993) Functional consequences of substitution of the 7-residues segment LysIleArgAspGlnMetAla240 located in the stalk helix S3 of the $Ca^{2+}$-ATPase of sarcoplasmic reticulum. Biochemistry 32: issue 38
4.  Andersen JP, Vilsen B, Leberer E, MacLennan DH (1989) Functional consequences of mutations in the $\beta$-strand sector of the $Ca^{2+}$-ATPase of sarcoplasmic reticulum. J Biol Chem 264:21018-21023
5.  Andersen JP, Vilsen B, MacLennan DH (1992) Functional consequences of alterations to $Gly^{310}$, $Gly^{770}$, and $Gly^{801}$ located in the transmembrane domain of the $Ca^{2+}$-ATPase of sarcoplasmic reticulum. J Biol Chem 267:2767-2774
6.  Clarke DM, Loo TW, Inesi G, MacLennan DH (1989) Location of high affinity $Ca^{2+}$ binding sites within the predicted transmembrane domain of the sarcoplasmic reticulum $Ca^{2+}$-ATPase. Nature (London) 339:476-478
7.  Clarke DM, Loo TW, Rice WJ, Andersen JP, Vilsen B, MacLennan DH (1993) Functional consequences of alterations to hydrophobic amino acids located in the $M_4$ transmembrane sector of the $Ca^{2+}$-ATPase of sarcoplasmic reticulum. J Biol Chem 268:18359-18364
8.  Clarke DM, Maruyama K, Loo TW, Leberer E, Inesi G, MacLennan DH (1989) Functional consequences of glutamate, aspartate, glutamine, and asparagine mutations in the stalk sector of the $Ca^{2+}$-ATPase of sarcoplasmic reticulum. J Biol Chem 264:11246-11251

9.      Clarke DM, Loo TW, MacLennan DH (1990) Functional consequences of alterations to polar amino acids located in the transmembrane domain of the $Ca^{2+}$-ATPase of sarcoplasmic reticulum. J Biol Chem 265:6262-6267

10.    Clarke DM, Loo TW, MacLennan DH (1990) Functional consequences of mutations of conserved amino acids in the $\beta$-strand domain of the of sarcoplasmic reticulum. J Biol Chem 265:14088-14092

11.    Clarke DM, Loo TW, MacLennan DH (1990) Functional consequences of alterations to amino acids located in the nucleotide binding domain of the $Ca^{2+}$-ATPase of sarcoplasmic reticulum. J Biol Chem 265:22223-22227

12.    Forbush III B (1987) Rapid release of $^{42}K$ or $^{86}Rb$ from two distinct transport sites on the Na,K-pump in the presence of $P_i$ or Vanadate. J Biol Chem 262:11116-11127

13.    Fujimori T, Jencks WP (1992) Binding of two $Sr^{2+}$ ions changes the chemical specificities for phosphorylation of the sarcoplasmic reticulum calcium ATPase through a stepwise mechanism. J Biol Chem 267:18475-18487

14.    Levy D, Seigneuret M, Bluzat A, Rigaud JL (1990) Evidence for proton countertransport by the sarcoplasmic reticulum $Ca^{2+}$-ATPase during calcium transport in reconstituted proteoliposomes with low ionic permeability. J Biol Chem 265:19524-19534

15.    MacLennan DH, Brandl CJ, Korczak B, Green NM (1985) Amino-acid sequence of a $Ca^{2+}$ + $Mg^{2+}$-dependent ATPase from rabbit muscle sarcoplasmic reticulum, deduced from its complementary DNA sequence. Nature (London) 316:696-700

16.    Maruyama K, MacLennan DH (1988) Mutation of aspartic acid-351, lysine-352, and lysine-515 alters the $Ca^{2+}$ transport activity of the $Ca^{2+}$-ATPase expressed in COS-1 cells. Proc Natl Acad Sci USA 85:3314-3318

17.    Maruyama K, Clarke DM, Fujii J, Inesi G, Loo TW, MacLennan DH (1989) Functional consequences of alterations to amino acids located in the catalytic center (isoleucine 348 to threonine 357) and nucleotide-binding domain of the $Ca^{2+}$-ATPase of sarcoplasmic reticulum. J Biol Chem 264:13038-13042

18.    Vilsen B (1992) Functional consequences of alterations to $Pro^{328}$ and $Leu^{332}$ located in the 4th transmembrane segment of the $\alpha$-subunit of the rat kidney $Na^+,K^+$-ATPase. FEBS Lett 314:301-307

19.    Vilsen B, Andersen JP, Clarke DM, MacLennan DH (1989) Functional consequences of proline mutations in the cytoplasmic and transmembrane sectors of the $Ca^{2+}$-ATPase of sarcoplasmic reticulum. J Biol Chem 264:21024-21030

20.    Vilsen B, Andersen JP, MacLennan DH (1991) Functional consequences of alterations to amino acids located in the hinge domain of the $Ca^{2+}$-ATPase of sarcoplasmic reticulum. J Biol Chem 266:16157-16164

21.    Vilsen B, Andersen JP, MacLennan DH (1991) Functional consequences of alterations to hydrophobic amino acids located at the $M_4S_4$ boundary of the $Ca^{2+}$-ATPase of sarcoplasmic reticulum. J Biol Chem 266:18839-18845

22.    Vilsen B, Andersen JP (1992) Mutational analysis of the role of $Glu^{309}$ in the sarcoplasmic reticulum $Ca^{2+}$-ATPase of frog skeletal muscle. FEBS Lett 306:247-250

23.    Vilsen B, Andersen JP (1992) Interdependence of $Ca^{2+}$ occlusion sites in the unphosphorylated sarcoplasmic reticulum $Ca^{2+}$-ATPase complex with CrATP. J Biol Chem 267:3539-3550

24.    Vilsen B, Andersen JP (1992) CrATP-induced $Ca^{2+}$ occlusion in mutants of the $Ca^{2+}$-ATPase of sarcoplasmic reticulum. J Biol Chem 267:25739-25743

# Conservation Patterns Define Interactions Between Transmembrane Helices of P-type Ion Pumps

N.M. Green

The National Institute for Medical Research, The Ridgeway, Mill Hill, London NW7 1AA.

## Introduction

For several years we have been analysing the sequences of P-type ion pumps in order to obtain structural information to provide a basis for understanding the mechanism of their action. There are now over one hundred sequences in the data banks, about half of which are for sodium or calcium pumps. From a study of the location of conserved and variable regions, combined with secondary structure prediction and the use of information from chemical modification and site directed mutagenesis, we have put forward a model for the $Ca^{2+}$-pump (1), which, although it cannot be claimed to be definitive, is consistent with almost all of the extensive experimental results.

At the same time, in collaboration with Stokes and Toyoshima, we have been extending the earlier work of Taylor et al. (2), in an attempt to obtain direct structural information from crystals. In spite of efforts in many laboratories no crystals suitable for X-ray crystallography have been obtained, so we have been restricted by the small size to electron microscopy, using the electron crystallographic techniques devised by Henderson and Unwin applied to frozen hydrated specimens. Thin crystals (1-2μ) in diameter have yielded a projection map at 6Å resolution (3). More recently, long helical tubes were obtained which have given a three-dimensional structure at 14Å resolution (4). It may be possible to obtain a higher resolution structure by tilting the crystals in the microscope. Work is in progress in this direction but it is a long term project. In the absence of better crystals, our best hope for some immediate progress is to try and integrate information from the sequence comparisons with the existing low resolution model and it is this approach which will be considered here.

Our current model for the $Ca^{2+}$-pump is shown in Fig. 1. It is consistent in terms of mass distribution with the 14Å, 3D, structure, which is not surprising since information from the latter was used to aid the modelling. Before discussing our current work, something should be said about the validity of extending this model to the sodium pump. The functional parallels between the two pumps have long been recognized from both kinetic and structural studies. At the same time the marked differences in ion stoichiometry and specificity, as well as the requirement of the sodium pump for an extra β-subunit have led to some reservations about their similarity. The recent demonstration of proton countertransport by the $Ca^{2+}$-pump, providing an analogy to the role of $K^+$ in the sodium pump, should remove some of these (6,7).

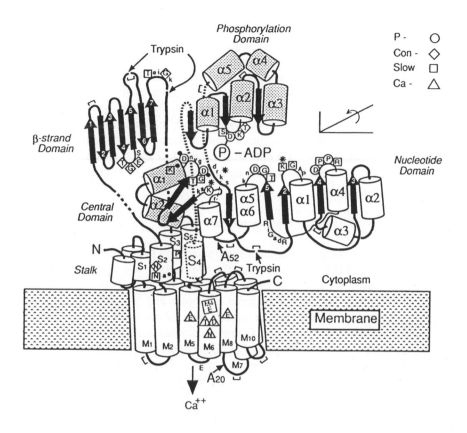

Figure 1: Model for P-type ion pumps based on the $Ca^{2+}$-pump.

Two cytoplasmic regions (antiparallel β and parallel β) are linked to transmembrane helices by amphipathic helical segments, which form the narrow stalk seen in the 3D structure. An aspartic acid residue on a bend in the first parallel β sheet domain is phosphorylated by ATP bound in a cleft between it and the second parallel β domain. Conformational changes accompanying phosphorylation are transmitted to two $Ca^{2+}$-binding sites, which have been located by site mutagenesis in $M_4$, $M_5$, $M_6$ and $M_8$ (8). The $Ca^{2+}$ ions become occluded and after further changes can escape on the lumenal side. A number of sites of functional importance have been identified by affinity labelling (*) and then further characterized by site directed mutagenesis (5). Five types of mutant could be distinguished: (1) Upper case letters, no effect; (2) ☐, slow transport; (3) ◇, no transport, normal phosphorylation (con-); (4) △, no $Ca^{2+}$ control of phosphorylation (Ca-) (5) ○, no transport, no phosphorylation, (P- ). Many sites which have been changed in the transmembrane region are not shown here, but can be found in Fig. 4. This figure is reproduced from Ref. 1, with permission of Acta Physiol. Scand. Green and Stokes (1992).

At the sequence level there is only 30% overall identity between pumps from different families, but in the 300 residues forming the core of the cytoplasmic domain this rises to 40% and this includes all the main functionally characterized regions, except the transmembrane segments. Here the similarity is less convincing (20% identity). Although the overall hydrophobicity patterns are the same, detailed matching of several of the hydrophobic segments ($M_1$, $M_2$, $M_3$, $M_7$, $M_9$, $M_{10}$) is difficult, relying on unequivocal matching of nearby extramembraneous residues and the assumption that no gaps occurred between these 'anchor' segments and the membrane. Conservation is good only in segments 4, 5 and 6 which also contain several conserved polar residues, shown by mutagenesis (8) to be critical for cation activated phosphorylation.

**Analysis of variability in the transmembrane region.**

To provide a more critical evaluation of the similarity in the transmembrane region we have made use of the observation that the hydrophobic residues which face the lipid are much more variable than those which face the core. This was first noted by Blanck and Oesterhelt (9) for bacterial rhodopsin and halorhodopsin. It was generalized by Rees et al (10) and has recently been substantiated in great detail by Baldwin (11) in a study of 7-helix, G-protein coupled receptors. The validity of this approach rests on the assumption that all members of a family have the same structure and that any variation in sequence indicates a solvent (or lipid) exposed site and that it is not correlated with any change in structure. The best way to ensure this is to compare sequences only within highly conserved groups (>90% identity). Unfortunately, there are only 20 members in each of the $Ca^{2+}$ and $Na^+$ families, so that the data would be too sparse to define all the variable sites. In contrast there are about 200 sequences known for the family of 7-helix receptors and a representative selection of these were grouped into 49 small families, within which the sequence identity was greater than 90% (10). Given our small database, we have included sequences down to 60% identity within each family in the TM segments, checking to ensure that when the more remote variants were added to the family they did not disturb the pattern of relationships.

Two families, each of nine sequences were chosen for the analysis (Fig. 2) omitting those which showed only small variations from other members of the group. The $Ca^{2+}$ family was the more diverse (mean pairwise identity over the whole sequence = 77%), since it included three pumps from protozoa, whereas Hydra was the most primitive organism with a defined sodium pump. The mean pairwise identity of the $Na^+$ family was 83%, though to increase the variability of the highly conserved $M_4$, $M_5$, and $M_6$, we sometimes included the sequences of the gastric and the distal colon $H^+/K^+$ pumps. With one exception, discussed below, the sequences of the transmembrane segments were aligned as shown earlier (12) and each of the ten segments was allocated 30 positions. Usually, three residues on each side of the central 20 residue hydrophobic core were included and two blank spaces at each end were left to accommodate extra residues in the longer hydrophobic segments.

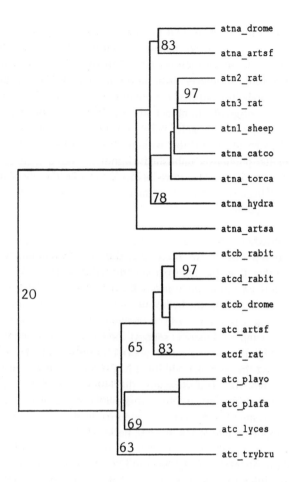

Figure 2: Evolutionary tree based on transmembrane segments of selected Na$^+$- and Ca$^{2+}$- pumps.

The segments of sequence used consisted of 272 residues which included the ten transmembrane segments with two or three flanking residues for each. They were aligned as described elsewhere (12) and the tree was derived by the 'pileup' program (Wisconsin GCG). The sequence names are those used in the Swissprot data base or have been adapted to that style. They include shark (catco), torpedo (torca), tomato (lyces), brine shrimp (artsa and artsf), plasmodium (plafa and playo) and trypanosomes (trybru). The numbers give the % identity between sequences which separate at that branch of the tree.

113

The 'plotsimilarity' program of the Wisconsin GCG package was used to display the variation along the sequence as a mean pairwise identity at each position. Comparison of these plots for the seven most variable helices of the two families (Fig. 3) shows that, disregarding end effects, minima occur close to twice in every seven residues ($\alpha$-helical periodicity) and that the patterns for the two families are in phase. This supports the original alignment and suggests that the variable hydrophobic residues define a face of the helix with structural significance, which can plausibly be assumed to be its interaction with the lipid solvent. The only significant deviation from the previously published alignment was a shift of the $Na^+$ pump sequence by one residue in $M_7$, introducing a gap into the connecting loop between $M_6$ and $M_7$. This also improved the matching of individual residues with those in $M_7$ of the $Ca^{2+}$-pump.

It is noteworthy that the periodicity is well preserved in helices $M_8$ and $M_{10}$ since in some models of the pump these have been assigned to extramembraneous regions. The observation that it is the hydrophobic rather than the polar sites which vary, strongly supports an intramembraneous location for $M_8$ and $M_{10}$. If these helices were outside the membrane the hydrophobic sites should be conserved. Another outstanding question concerning the transmembrane helices remains unanswered. In the sodium pump field there is support for a single transmembrane helix in place of $M_5$ and $M_6$, leaving the rather polar N-terminus of $M_5$ in the cytoplasm. If this were correct then the periodicity of $M_5$ should continue through the 6-residue luminal loop into $M_6$. Unfortunately, the N-terminus of $M_6$ is too well conserved for any clear answer to emerge from this evidence. However, in the $H^+/K^+$-pump the extracytoplasmic location of the omeprazole site places the 5-6 loop outside the cell and supports separate $M_5$ and $M_6$ helices (13).

We have not shown the similarity plots for $M_4$, $M_5$, and $M_6$ because apart from the C-terminus of $M_5$ they are highly (>80%) conserved and show no clear periodicity. Their orientation is however defined by the conserved polar sites which have been shown by mutagenesis to be part of the $Ca^{2+}$-binding sites and/or the gating mechanism (8). We have therefore displayed the information on helical wheels (Fig. 4), in which variable sites defined by the minima of the similarity plots are encircled, whereas triangles, diamonds and squares indicate the location of various mutants. Some wheels are also shown for the less conserved helices to show the excellent definition of the variable regions. In five of the transmembrane helices the boundaries between variable and conserved sites is clearly defined and only one or two residues are found in the 'wrong' half of the circumference. We have also determined the boundaries using a larger number (nine) of small homogeneous families, including proton pumps and plasma membrane $Ca^{2+}$-pumps, along the lines suggested by Baldwin (11), with similar results. Taken in conjunction with the previous work on the bacterial rhodopsin and 7-helix receptor familes, this gives confidence that the analysis can be used to define the probable number of neighbours for each helix. In $M_4$, $M_5$ and $M_6$ the variable sites are fewer and less well clustered, while in $M_7$ and $M_8$ the variability changes significantly from one end of the helix to the other, particularly in the $Ca^{2+}$-pumps where variation at the

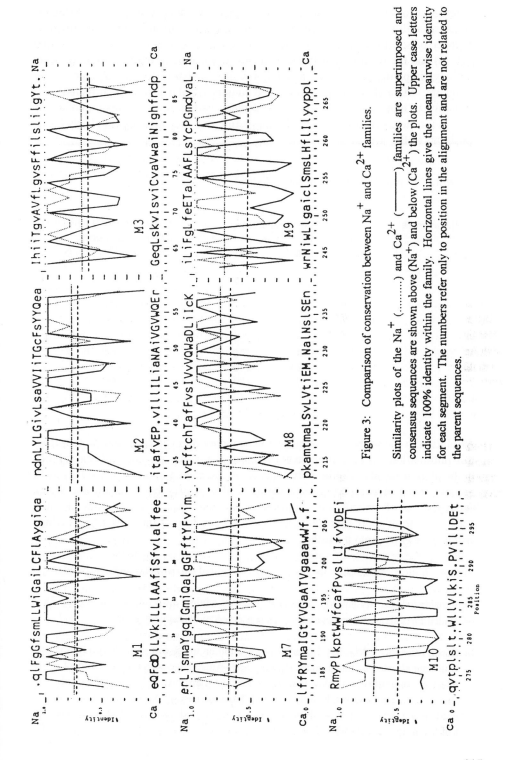

Figure 3: Comparison of conservation between Na$^+$ and Ca$^{2+}$ families.

Similarity plots of the Na$^+$ (........) and Ca$^{2+}$ (————) families are superimposed and consensus sequences are shown above (Na$^+$) and below (Ca$^{2+}$) the plots. Upper case letters indicate 100% identity within the family. Horizontal lines give the mean pairwise identity for each segment. The numbers refer only to position in the alignment and are not related to the parent sequences.

115

C-terminus of $M_7$ and N-terminus of $M_8$ is notably high. The lack of information from variability in $M_4$, $M_5$ and $M_6$ is compensated by information from the large number of site mutants in these helices (Fig. 4). $M_4$ has been almost saturated (14,15) and the results show that only a few sites can be changed without effect on activity, implying that it has at least four and probably five neighbours in the structure. $M_5$ and $M_6$ have well defined conserved polar faces which include sites sensitive to mutagenesis and there is not much variation on the hydrophobic face except for the C-terminal half of $M_5$, suggesting that they also occupy fairly buried positions in the structure.

TABLE 1. Angular exposure of helices to lipid

| TM segment | Lipid Exposure (°) | | Neighbouring helices (n) | |
|---|---|---|---|---|
| | $Ca^{2+}$ | Na+ | $Ca^{2+}$ | $Na^+$ |
| 1 | 120 | 120 | 4 | 4 |
| 2 | 160 | 160 | 3-4 | 3-4 |
| 3 | 200 | 220 | 2-3 | 2-3 |
| 4 | 60 | 60 | 5 | 5 |
| 5 | 100 | 120 | 4-5 | 4 |
| 6 | 140 | 140 | 4 | 4 |
| 7 | 180 | 240 | 3 | 2 |
| 8 | 100 | 180 | 4-5 | 3 |
| 9 | 240 | 220 | 2 | 2 |
| 10 | 240 | 200 | 2 | 2-3 |

A solution to the problem of defining the boundary between lipid and protein from this information has been proposed (16), but this analysis has not yet been applied to the present results. Preliminary conclusions are shown in Table I, but these will probably need revision. Further analysis is also in progress to determine which arrangements of helices are consistent with these results, taking into account constraints from short connecting loops and the location of the stalk helices which will bring $M_2$-$M_5$ into a cluster.

Figure 4: Helical wheels show the boundary between lipid exposed and internal residues.

(a) Variable helices, $M_1$ and $M_2$. Residues found at minima of the identity plots (Fig. 3). Radial lines show the boundary between conserved and variable regions.

(b) Conserved helices, $M_4$, $M_5$, $M_6$ and $M_8$. In addition to the few variable residues, sites which are affected by mutation are shown . Sites which have been mutated without effect are underlined; □, slow transport; ◇ , no transport, $E_1P \rightarrow E_2P$ blocked (con-); ⟨ , no transport, hydrolysis of $E_2P$ blocked (Pase-); Δ , no $Ca^{2+}$ control of phosphorylation    (Ca-); ∇, High $K_m$ for $Ca^{2+}$. ⟨ and ∇   are types of mutant additional to those shown in Fig. 2.

116

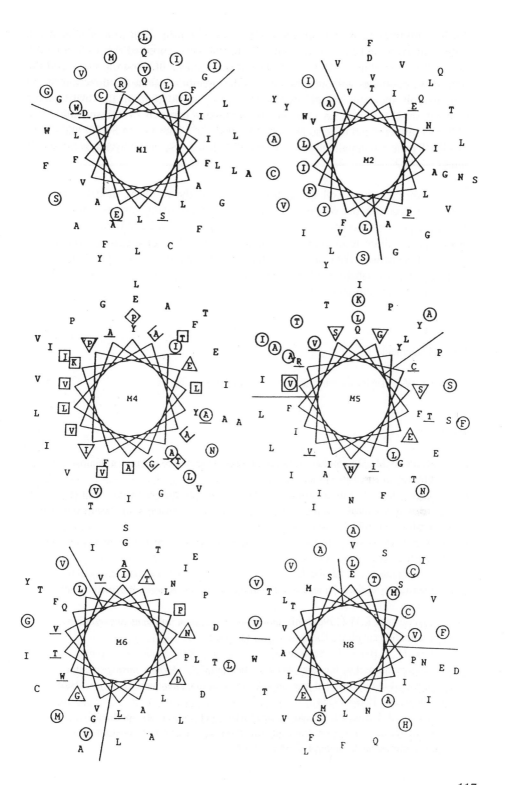

117

Further information which can be used can be obtained from a similar study of conservation in the helices of the stalk. Here it is to be expected that the conserved sites will be buried and that the variable and polar sites will be on the surface. If the stalk helix continues without a break into the membrane then the direction of the conservation moment should remain the same, while that of the hydrophobic moment should reverse on crossing the boundary between cytoplasm and lipid. This was in fact found to be so for both $Na^+$ and $Ca^{2+}$ families in $S_2$, $S_3$ and $S_5$. In $S_1$ and $S_4$ the amphipathicity was less well defined, $S_4$ being hydrophobic on all sides.

It is not yet clear how well we shall be able to fit the resulting model to the low resolution structure, but we think that this approach of using various constraints to define the positional relationships of transmembrane helices and then matching the model to a medium resolution structure will be valuable for a field in which high resolution structural data are still very hard to acquire. A related approach, supplemented by labelling of hydrophobic residues in the lipid phase, is being applied to the sodium pump (17).

Acknowledgement

I thank Dr. W.R. Taylor for much useful advice and discussion.

References

1.    Green, N.M., D.L. Stokes (1992) Structural modelling of P-type ion pumps. Acta Physiol. Scand. 146:59-68.
2.    Taylor, K.A., L. Dux, A. Martonosi (1986) Three-dimensional reconstruction of negatively stained crystals at the $Ca^{2+}$-ATPase from muscle sarcoplasmic reticulum. J. Mol. Biol. 187:417-427.
3.    Stokes, D.L., N.M. Green (1990) Structure of $Ca^{2+}$-ATPase: Electron microscopy of frozen-hydrated crystals at 6A resolution in projection. J. Mol. Biol. 213:529-538.
4.    Toyoshima, C., H. Sasabe, D.L. Stokes (1992) Three-dimensional cryo-electron microscopy of the calcium ion pump in the sarcoplasmic reticulum membrane. Nature 360:469-471.
5.    MacLennan, D.H. (1990) Molecular tools to elucidate problems in excitation contraction coupling. Biophys. J. 58:1355-1365.
6.    Levy D., A Bluzat, M. Seigneuret, J.L. Rigaud. (1990). Evidence for proton countertransport by the sarcoplasmic reticulum $Ca^{2+}$-ATPase during calcium transport in reconstituted proteoliposomes with low ionic permeability. J. Biol. Chem. 265:19524-19534.
7.    Yu, X., S. Carrol, J-L Rigaud, G. Inesi (1993) $H^+$ countertransport and electrogenicity of the sarcoplasmic reticulum $Ca^{2+}$ pump in reconstituted proteoliposomes. Biophys. J. 64:1232-1242.

8. Clarke, D.M., T.W. Loo, G. Inesi, D.H. MacLennan (1989) Location of high affinity $Ca^{2+}$-binding sites within the predicted transmembrane domain of the sarcoplasmic reticulum $Ca^{2+}$-ATPase. Nature 339:476-478.

9. Blanck, A. and D. Oesterhelt (1987) The halo-opsin gene. II. Sequence, primary structure of halorhodopsin and comparison with bacteriorhodopsin. EMBO J. 6:265-273.

10. Rees, D.C., L. De Antonio, D. Eisenberg (1989) Hydrophobic organization of membrane proteins. Science, 245:510-513.

11. Baldwin J.M. (1993) The probable arrangement of the helices in G protein-coupled receptors. EMBO J. 12:1693-1703.

12. Green, N.M. (1989) ATP-driven cation pumps: alignment of sequences. Biochem. Soc. Trans. 17:972-974.

13. Munson, K.B., C. Gutierrez, V.N. Balaji, K. Rammarayan, G. Sachs. (1991) Identification of an extracytoplasmic region of $H^+$, $K^+$-ATPase labelled by a $K^+$-competitive photoaffinity inhibitor. J. Biol. Chem. 266:18976-18988.

14. Vilsen, B., J.P. Andersen, D.H. MacLennan (1991b) Functional consequences of alterations by hydrophobic amino acids located at the $M_4S_4$ boundary of the $Ca^{2+}$-ATPase of sarcoplasmic reticulum. J. Biol. Chem. 266:18839-18845.

15. Clarke, D.M. , T.W. Loo, W.J. Rice, J.P. Andersen, B. Vilsen and D.H. MacLennan (1993) Functional consequences of alterations to hydrophobic amino acids located in the $M_4$ transmembrane sector of the $Ca^{2+}$-ATPase of Sarcoplasmic reticulum. J. Biol. Chem. 268:183259-18364..

16. Taylor, W.R., D.T. Jones, and N.M. Green (1993) A method for alpha-helical integral membrane protein fold prediction. Proteins. (in the Press).

17. Modyanov, N.M., N.M. Vladimirova, D.I. Gulyaev, R.G. Efremov (1992) Architecture of the sodium pump molecule. Ann. N.Y. Acad. Sci. 671:134-146.

# Structural Studies of Ca$^{2+}$-ATPase by Cryoelectron Microscopy

D.L. Stokes[$], N.M Green[£], C. Toyoshima[¥]

[$]Dept. of Molecular Physiology and Biological Physics, Univ. of Virginia Health
  Sciences Center, Charlottesville, VA 22908;
[£]National Inst. for Medical Research, The Ridgeway, London NW7 1AA;
[¥]Frontier Research Program, RIKEN, Wako, Saitama 351-01 Japan

The Ca$^{2+}$-ATPase from sarcoplasmic reticulum (SR) is exceedingly well characterized
in terms of kinetics, ligand binding and amino acid sequence. For example, particular
events in the reaction cycle have been assigned to specific amino acid residues by site-
directed mutagenesis and characterization of ligand binding. Various spectroscopies
have documented conformational changes and measured distances between labelled
amino acids. Typical of most membrane proteins, however, structural information has
lagged far behind these functional studies. This is not for lack of effort as three
different crystal forms have been studied by electron microscopy (11-13), oriented
pellets have been studied by x-ray diffraction (2), and countless, unreported
crystallization trials have failed to produce crystals suitable for x-ray crystallographic
analysis. However, we are currently studying two crystal forms by cryoelectron
microscopy and report here on results pertaining both to the conformational state of
molecules within each crystal lattice and to their physical structures.

## Different Crystal Forms

Several crystal forms of Ca$^{2+}$-ATPase from rabbit SR have been discovered, all of them
by Dux and Martonosi. The first two crystals forms were induced in the native SR
membrane by incubation in solutions containing either vanadate and EGTA (11) or
lanthanides (13). In both cases upon crystallization, the spherical vesicles from SR
elongate into tubes (Fig. 1) that are ~600-800 Å in diameter and can grow to be 10 μm
long if large vesicles are used. These tubes have high membrane curvature along one
direction (circumference of the tube) and virtually no curvature normal to this direction
(long axis of tube). This morphology is a consequence of simultaneous intermolecular
bonding between the small luminal domains (inside the tubes) and between the much
larger cytoplasmic domains (outside the tubes, see Fig. 5 for relative sizes). Simply
stated, molecules are pushed farther apart on the outer surface and pulled together on
the inner surface thus generating strong curvature. More specifically, molecules wrap
around the tubes and uncontrained contacts are made between neighboring ribbons
along the long axis of tubes. Thus, molecules are helically arranged around the surface
of the tube and this helicity can be used for three-dimensional reconstruction when the
cylindrical shape of the tubes are preserved by rapid freezing. In general, tubes usually

loose their helicity during negative staining due to flattening on the support film during drying.

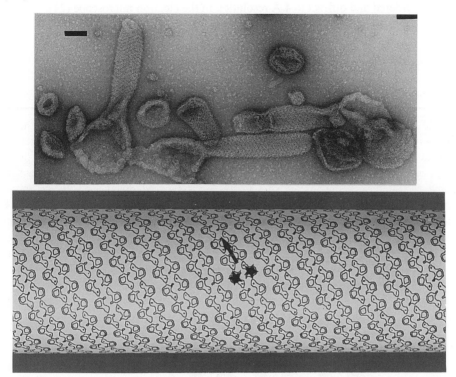

**Figure 1.** An electron micrograph showing 3 short tubes in negative stain and a representation of their helical symmetry. The tubes have grown out of larger vesicles, the residual of which can be seen at one end of the each tube and which appears to be devoid of protein. In solution, the tubes are cylindrical, but after drying in negative stain they have flattened onto the carbon support film. Smaller vesicles of SR are also present and, although they are unable to form tubes, some of them show evidence of array formation. In the diagram below, each pear-shape represents one $Ca^{2+}$-ATPase molecule. The two molecules marked with a ★ are related by two-fold symmetry; they are linked into a dimer ribbon (arrow), which spirals around the vesicle inducing high curvature in one direction, thus generating a tube. Scale bar = 1000 Å

A very different crystal form was grown in detergent at high $Ca^{2+}$ concentrations (12), which unlike the tubes is a three-dimensional crystal with space group C2 (22,25). These are unlike x-ray quality crystals of photosynthetic reaction center due to the bilayer structure that dominates their morphology (15). In fact, these crystals are best characterized as ordered stacks of bilayers and unlike the reaction center contain significant amounts of lipid (estimated at ~25:1 molar ratio to $Ca^{2+}$-ATPase, ref. 22). The detergent ($C_{12}E_8$) serves to randomize the orientation of $Ca^{2+}$-ATPase molecules in the bilayer and the presence of equal numbers of cytoplasmic domains on both sides

121

of the bilayer eliminates the need for curvature. Thus, under optimal growth conditions, these crystals are large (1-5 μm), flat plates containing 3-10 bilayers. They are well ordered and diffract to 4 Å resolution in the electron microscope (21).

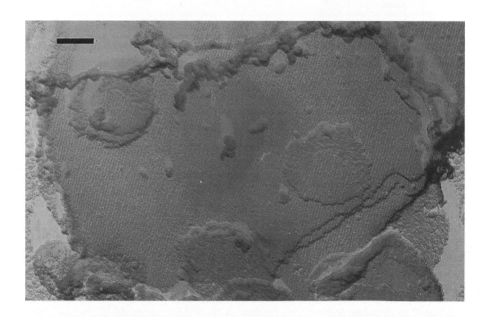

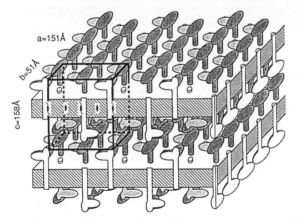

**Figure 2.** A thin, three-dimensional crystal together with a packing diagram for this crystal form. The image is of a freeze-dried, unidirectionally shadowed crystal and steps in crystal thickness are visible at the lower, right edge of the crystal. Thus, it is evident that most of this particular crystal is 3 or 4 unit cells thick. The packing diagram shows that molecules protrude from both surfaces of the bilayer,which is composed of lipid, detergent and protein. Stacking of layers is mediated by contacts between the top-most surfaces of cytoplasmic heads.

## Growth Conditions Suggest Different Conformational States

The growth conditions for these three crystal forms suggests their stabilization of different conformational states of $Ca^{2+}$-ATPase. This is not surprising as the original rationale for using ligands such as vanadate and lanthanides was presumably to lock molecules into a single, well defined conformation that could then be induced to crystallize. A very simple scheme for the reaction cycle of $Ca^{2+}$-ATPase is shown below, where $Ca^{2+}$ is being pumped from the outside of SR to the inside.

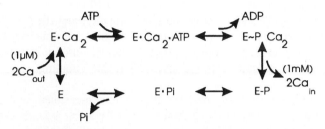

Based on the idea that vanadate acts as a transition state analogue for phosphate, the vanadate-induced tubes were initially thought to contain molecules in the E-P conformation (11). Based on the binding of lanthanides to $Ca^{2+}$ sites, lanthanide-induced tubes were assumed to be composed of the $E \cdot Ca_2$ conformation (13). The conformation in thin, three-dimensional crystals was less apparent because $Ca^{2+}$ requirements for crystallization were rather high (> 0.2 mM and optimally 20 mM) and could thus be producing conformations with either low or high $Ca^{2+}$ affinity (12). Subsequent study by ourselves and others has resolved these issues to a large extent.

*Vanadate-Induced Tubes.* Soon after the discovery of the vanadate-induced tubes, it was noted that vanadate formed a series of oligomers in solution and that decavanadate was the specific oligomer that promoted crystallization (17,27). The spectroscopic effects of decavanadate on $Ca^{2+}$-ATPase were distinct from those of monovanadate, suggesting that the conformational effects of the two oligomers on non-crystalline $Ca^{2+}$-ATPase were different (9). Tubes are disrupted by $Ca^{2+}$, but differing sensitivities were observed depending on whether monovanadate (sensitive to μM $Ca^{2+}$) or decavanadate (sensitive to mM $Ca^{2+}$) was used (28); nevertheless, this sensitivity suggested that the Ca-free state was required for crystal formation: either E or E-P. The presence of nucleotide had no effect on crystallization (3,10). The distinction between E and E-P seemed to be made by thapsigargin, which was shown both to trap $Ca^{2+}$-ATPase in the Ca-free, E conformation and to promote vanadate-induced tube formation even in the presence of 0.25 mM $Ca^{2+}$ (19).

We have extended these observations and conclude that vanadate-induced tubes form from either of the Ca-free conformations E, E·ATP, or E-P. First, we observed tube formation as a function of $Ca^{2+}$ concentration under several different conditions. From native SR, tubes form only at < 0.3 μM $Ca^{2+}$, which correlates well with the affinity of transport sites measured by both ATPase activity and $Ca^{2+}$ uptake. However, we discovered that preformed crystals (formed by incubating SR in solutions containing EGTA and decavanadate overnight at 4°C) were stable to $Ca^{2+}$ concentrations up to 1 mM, suggesting that breakup employs a different mechanism than prevention. This difference may explain contradictory results previously observed for decavanadate vs.

monovanadate with rabbit SR because decavanadate promotes more rapid tube formation. Tubes from scallop SR are stable up to 3 $\mu$M Ca$^{2+}$, well above the measured affinity of transport sites (5), but these also should be considered as preformed crystals since they exist *in situ* (14). When we included thapsigargin, we found that tube formation was insensitive to Ca$^{2+}$ concentration up to 10 mM. Addition of AMP-PCP or labelling with Cr-ATP had no effect on tube formation and the Ca$^{2+}$-sensitivity of Cr-ATP labelled tubes was identical to that of native SR. We conclude therefore, that binding of nucleotide to Ca$^{2+}$-ATPase in the Ca-free, E conformation does not greatly alter its conformation, at least not in the same way as Ca$^{2+}$ binding.

*Three-dimensional crystals.* In contrast, thin, three-dimensional crystals require at least 0.1 mM Ca$^{2+}$ and any of these conformationally active ligands prevents crystal formation: thapsigargin, AMP-PCP, Cr-ATP. Inhibition of ATPase activity by Cr-ATP lasts 1-2 weeks, after which time activity begins to return due to hydrolysis of ligands between Cr$^{3+}$ and the active site. We found that crystals eventually do form, but only after a large proportion of molecules have lost Cr-ATP. This observation confirms that a specific conformational effect of Cr-ATP on Ca$^{2+}$-ATPase prevents crystallization. Thus, we conclude that the E·Ca$_2$ conformation is required for formation of three-dimensional crystals and that significant conformational changes in Ca$^{2+}$-ATPase occur either upon nucleotide binding or upon Ca$^{2+}$ removal thereby preventing crystallization. Such conformational changes are well documented by other spectroscopic techniques, such as tryptophan fluorescence for binding of Ca$^{2+}$ and low-angle x-ray diffraction for ATP binding (reviewed in ref. 1). The concentration of Ca$^{2+}$ required for crystallization is rather high (100 $\mu$M) compared to the affinity of transport sites (<1 $\mu$M) suggesting that additional, low affinity Ca$^{2+}$ sites must also be occupied for crystallization.

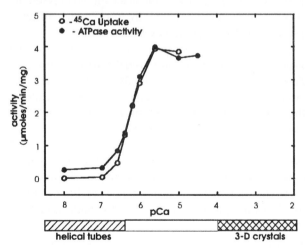

**Figure 3.** Correlation of Ca$^{2+}$-ATPase activity with formation both of helical tubes and thin, three-dimensional crystals. Both Ca$^{2+}$ uptake and ATPase activity are plotted against Ca$^{2+}$ concentration. At the bottom are shown Ca$^{2+}$ concentrations at which helical tubes (diagonal hatch) and three-dimensional crystals (cross hatch) form.

## Structure Analysis of Thin, Three-dimensional Crystals

After their initial discovery and characterization by Martonosi and colleagues (12,18,25), we improved growth conditions for thin, three-dimensional crystals by purification of $Ca^{2+}$-ATPase and by careful regulation of the lipid-to-detergent ratio (22). Thereafter, rapid freezing and imaging of crystals in the frozen-hydrated state showed that they were very well ordered in the plane of the bilayer (a-b crystal plane) with electron diffraction to ~4 Å resolution. A projection map was determined at 6 Å resolution, which is difficult to interpret due to extensve overlap of molecules in this densely packed crystal form (21). This density is roughly double that in vanadate-induced tubes from sarcoplasmic reticulum because, unlike this native membranes, molecules protrude from both sides of the bilayer in the detergent-induced, three-dimensional crystals (Fig. 2). Nevertheless, a tentative arrangement of 10 transmembrane helices was proposed to account for the intramembranous features of this map; this arrangement should now be reevaluated based on our recent reconstruction from vanadate-induced tubes, which is described in the next section.

Since this published work, we have continued our work on these three-dimensional crystals, aiming towards a reconstruction from this crystal form. In particular, we have observed frozen-hydrated crystals parallel to the membrane plane and the corresponding diffraction patterns show that these crystals are well ordered normal to the bilayers (i.e., along the c axis); this is a prerequisite to successful analysis of these crystals. Also, the previously proposed orientation of molecules within the a-b plane has recently been verified by observing replicas of freeze-dried, shadowed crystals (shown in Fig. 2). These replicas represent the surface structure of the crystals and thus show only the cytoplasmic domains of $Ca^{2+}$-ATPase from the top layer, whereas previous images were from negatively stained crystals and therefore represent the projection of density from throughout the crystal. After Fourier averaging of these shadowed crystals, we are able to directly observe the orientation of the asymmetric cytoplasmic domain in the crystal lattice (Fig. 4). These observations confirm the original proposals for the orientation of molecules in the unit cell. Finally, we have obtained higher resolution electron diffraction both from untilted crystals and from tilted crystals by improving crystallization conditions and methods of preparation for electron microscopy. Most importantly, the large crystal aggregates that are normally obtained can now be dispersed to provide isolated crytals on microscope grids. Such isolation substantially improves flatness, because crystals within aggregates were usually inclined, bent, and curved due to the underlying crystals. Flatness is critical for the crystal to diffract as a single, coherent domain, thus allowing the collection of high resolution information. Also, the crystallization conditions have been slightly refined to generate larger crystals - 2-5 µm vs. 1-2 µm previously - which substantially improves the quality of electron diffraction patterns. Given these improvements in our specimen preparation methods, we routinely obtain electron diffraction to better than 4 Å resolution from untilted specimens, and to ~8 Å from specimens tilted to 45°. This data is being processed and will ultimately contribute to a three-dimensional data set for reconstruction.

**Figure 4.** Density map for thin, three dimensional crystals that have been freeze-dried and unidirectionally shadowed. This map represents an image of the surface of crystals and can therefore be compared with the top surface in the diagram in Fig. 2. The orientation of the triangular densities in this map confirms the proposed orientation of pear-shaped molecules in this diagram. The map was obtained by averaging 3 images of crystals that had been shadowed from the same direction (from the right in this map). Unit cell axes (a and b) are indicated; positive density is represented by solid contours and negative density by dotted contours

### Structure analysis of Vanadate-Induced Tubes

Vanadate and EGTA were discovered to induce array formation in the SR membrane by Dux and Martonosi (11), following the example of Skriver et al. (20) in their crystallization of $Na^+/K^+$-ATPase. The crystal packing is apparently identical to that observed naturally in scallop muscle under rigor conditions (4). Three-dimensional reconstructions were determined both from vanadate-induced arrays in rabbit SR and from the naturally occuring arrays in scallop SR after negative staining at 25-30 Å resolution (6,23). Once again, we found that rapid freezing was a better preservative of structure and also that large vesicles of SR produced very long, helical tubes (also produced from scallop SR). Unlike a previous report of a single frozen-hydrated tube (24), we were able to separate tubes into three symmetry classes and employing helical reconstruction methods we obtained three independent, averaged structures (26). The molecular features seen in the three reconstructions were entirely consistent and the best reconstruction, representing an average of four tubes, had a resolution of 14 Å. From this reconstruction we could clearly distinguish parts of the molecule in the cytoplasm, membrane, and lumen and could identify domains within each of these

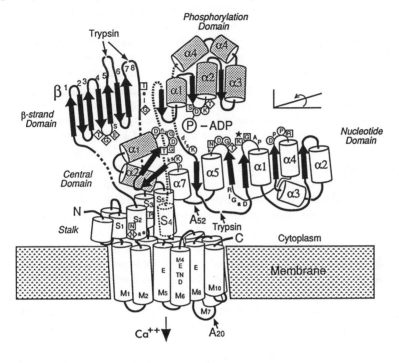

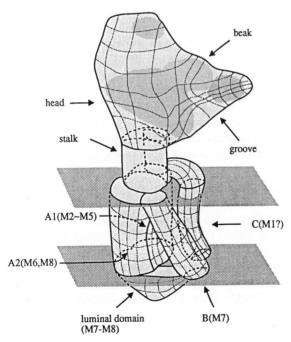

**Figure 5**. Comparison of structure predictions (N.M. Green, this volume) with the structure obtained by helical reconstruction (26). Tentative assignments for 8 of the predicted 10 transmembrane helices are shown; the other two helices may be accomodated by the reconstruction but have not yet been assigned.

127

parts. As expected, the luminal part is quite small (~5% of the total mass) and appears as a single domain. The cytoplasmic part is considerably more complicated and, overall, resembles the head and neck of a bird. The neck, also referred to as the stalk, divides into 2 unequal segments ~15 Å above the bilayer and leaves a large cavity in the back of the stalk. The bird's beak contains another striking cavity that opens down towards the bilayer surface. Either of these two cavities would accomodate ATP. However, various studies have identified a wide range of residues interacting with ATP from three different domains and the cavity in the beak would more easily explain these results. Furthermore, structure predictions suggest a penta-helical stalk and these five stalk helices would therefore be expected to line the cavity in the stalk; there is little know interaction between these helices and ATP. There is alot of mass at the top of the cytoplasmic part, ~50 Å above the membrane surface, and a discrete ball off density is observed to interact with four neighboring molecules at this level, thus stabilizing the crystal lattice. Decavanadate has been shown to be the oligomeric species that induces crystallization and the size and location of this density suggests that it arises from decavanadate. Specifically, this density ball is the right size (12-14 Å) and lies right on the two-fold symmetry axis, well separated from surrounding protein densities.

The membrane part of the structure is characterized by three distinct segments, denoted A, B, and C. Segment A is the largest and lies directly below the stalk; B and C are under, but not directly attached to, the beak. Segment A contributes to the 100-Å long column of density that runs the full length of the molecule; it divides into two pieces near the luminal surface of the membrane, thus defining A1 and A2. Based on structure predictions, we suggest that segment A contains transmembrane helices M2-M6 and M8, with M2-M5 being in the part directly under the stalk and M6 and M8 being to one side. It has been suggested on the basis of Ca-specific effects of site-directed mutagenesis (8) that the Ca-binding site is formed by a cluster of residues in M4,5,6,8; according to our current assignments, this group of transmembrane helices would lie within segment A at the edge of the stalk. Segment B is strongly inclined relative to the membrane surface and we have suggested M7 as a plausible candidate for this density, given its run of 28 consecutive hydrophobic residues. The small luminal domain rejoins B with A, which we have assigned to the loop between M7 and M8. This assignment is based on structure predictions of only one long loop on the luminal surface and on the demonstrated antigenicity of this loop from the luminal side of the membrane (7,16). Segment C is curved and is continuous with a lobe coming off the stalk at the cytoplasmic side of the membrane. We tentatively suggest that the N-terminal helices lie within this lobe and that segment C contain M1. This assignment is based on the observation that the stalk is not large enough to accomodate 5 helices, as was previously predicted, and that the first stalk helix is distinctly non-helical in the related structure of $Na^+/K^+$-ATPase. We have not assigned the locations of M9 and M10, but they might well exist within the structure. The boundary of the molecule within the membrane is poorly determined due to the low contrast and uncertainties as to the exact cutoff density for protein; a small adjustment to this cutoff will change the volume within the membrane considerably. Thus the question of the number of transmembrane helices accomodated by the model requires further consideration.

128

*Acknowledgements*

We thank Cindy Klevickis and Jean-Jacques Lacapere for preparation of Cr-ATP, Dan Czajkowsky and Margaretta Allietta for their work with freeze-dried crystals. This work was supported in part by NIH grant AR40997 to D.L.S.

# References

1. Bigelow DJ, Inesi G (1992) Contributions of chemical derivatization and spectroscopic studies to the characterization of the $Ca^{2+}$ transport ATPase of sarcoplasmic reticulum. Biochim. Biophys. Acta 1113:323-338
2. Blasie JK, Pascolini D, Asturias F, Herbette LG, Pierce D, Scarpa A (1990) Large-scale structural changes in the sarcoplasmic reticulum ATPase appear essential for calcium transport. Biophys. J. 58:687-693
3. Buhle EL, Knox BE, Serpersu E, Aebi U (1983) The Structure of the $Ca^{2+}$ ATPase as Revealed by Electron Microscopy and Image Processing of Ordering Arrays. J. Ultrastr. Res. 85:186-203
4. Castellani L, Hardwicke PM (1983) Crystalline structure of sarcoplasmic reticulum from scallop. J. Cell. Biol. 97:557-561
5. Castellani L, Hardwicke PM, Franzini-Armstrong C (1989) Effect of $Ca^{2+}$ on the dimeric structure of scallop sarcoplasmic reticulum. J. Cell. Biol. 108:511-520
6. Castellani L, Hardwicke PM, Vibert P (1985) Dimer ribbons in the three-dimensional structure of sarcoplasmic reticulum. J. Mol. Biol. 185:579-594
7. Clarke DM, Loo TW, MacLennan DH (1990) The Epitope for Monoclonal Antibody A20 (Amino Acids 870-890) is Located on the Luminal Surface of the $Ca^{2+}$-ATPase of Sarcoplasmic Reticulum. J. Biol. Chem. 265:17405-17408
8. Clarke DM, Loo TW, Inesi G, MacLennan DH (1989) Location of high affinity $Ca^{2+}$-binding sites within the predicted transmembrane domain of the sarcoplasmic reticulum $Ca^{2+}$-ATPase. Nature (Lond.) 339:476-478
9. Coan C, Scales DJ, Murphy AJ (1986) Oligovanadate binding to sarcoplasmic reticulum ATPase. Evidence for substrate analogue behavior. J. Biol. Chem. 261:10394-10403
10. Dux L, Martonosi A (1983) The regulation of ATPase-ATPase interactions in sarcoplasmic reticulum membrane. I. The effects of $Ca^{2+}$, ATP, and inorganic phosphate. J. Biol. Chem. 258:11896-11902
11. Dux L, Martonosi A (1983) Two-dimensional arrays of proteins in sarcoplasmic reticulum and purified $Ca^{2+}$-ATPase vesicles treated with vanadate. J. Biol. Chem. 258:2599-2603
12. Dux L, Pikula S, Mullner N, Martonosi A (1987) Crystallization of $Ca^{2+}$-ATPase in detergent-solubilized sarcoplasmic reticulum. J. Biol. Chem. 262:6439-6442
13. Dux L, Taylor KA, Ting-Beall HP, Martonosi A (1985) Crystallization of the $Ca^{2+}$-ATPase of sarcoplasmic reticulum by calcium and lanthanide ions. J. Biol. Chem. 260:11730-11743
14. Franzini-Armstrong C, Ferguson DG, Castellani L, Kenney L (1986) The density and disposition of Ca-ATPase in in situ and isolated sarcoplasmic reticulum. Ann. N. Y. Acad. Sci. 483:44-56
15. Kuhlbrandt W (1988) Three-dimensional crystallization of membrane proteins. Quart. Rev. Biophys. 21:429-477

16. Matthews I, Sharma RP, Lee AG, East JM (1990) Transmembranous Organization of $(Ca^{2+}-Mg^{2+})$-ATPase from Sarcoplasmic Reticulum: Evidence for Lumenal Location of Residues 877-888. J. Biol. Chem. 265:18737-18740

17. Maurer A, Fleischer S (1984) Decavanadate is responsible for vanadate-induced two-dimensional crystals in sarcoplasmic reticulum. J. Bioenerg. Biomembr. 16:491-505

18. Pikula S, Mullner N, Dux L, Martonosi A (1988) Stabilization and crystallization of $Ca^{2+}$-ATPase in detergent-solubilized sarcoplasmic reticulum. J. Biol. Chem. 263:5277-5286

19. Sagara Y, Wade JB, Inesi G (1992) A Conformational Mechanism for Formation of a Dead-end Complex by the Sarcoplasmic Reticulum ATPase with Thapsigargin. J. Biol. Chem. 267:1286-1292

20. Skriver E, Maunsbach AB, Jorgensen PL (1981) Formation of Two-Dimensional Crystals in Pure Membrane-Bound $Na^+/K^+$-ATPase. FEBS Lett. 131:219-222

21. Stokes DL, Green NM (1990) Structure of CaATPase: Electron Microscopy of Frozen-Hydrated Crystals at 6 Å Resolution in Projection. J. Mol. Biol. 213:529-538

22. Stokes DL, Green NM (1990) Three-dimensional crystals of Ca-ATPase from sarcoplasmic reticulum: symmetry and molecular packing. Biophys. J. 57:1-14

23. Taylor KA, Dux L, Martonosi A (1986) Three-dimensional reconstruction of negatively stained crystals of the $Ca^{++}$-ATPase from muscle sarcoplasmic reticulum. J. Mol. Biol. 187:417-427

24. Taylor KA, Ho MH, Martonosi A (1986) Image analysis of the $Ca^{2+}$-ATPase from sarcoplasmic reticulum. Ann. N. Y. Acad. Sci. 483:31-43

25. Taylor KA, Mullner N, Pikula S, Dux L, Peracchia C, Varga S, Martonosi A (1988) Electron microscope observations on $Ca^{2+}$-ATPase microcrystals in detergent-solubilized sarcoplasmic reticulum. J. Biol. Chem. 263:5287-5294

26. Toyoshima C, Sasabe H, Stokes DL (1993) Three-dimensional cryo-electron microscopy of the calcium ion pump in the sarcoplasmic reticulum membrane. Nature 362:469-471

27. Varga S, Csermely P, Martonosi A (1985) The binding of vanadium (V) oligoanions to sarcoplasmic reticulum. Eur. J. Biochem. 148:119-126

28. Varga S, Csermely P, Mullner N, Dux L, Martonosi A (1987) Effect of chemical modification on the crystallization of $Ca^{2+}$-ATPase in sarcoplasmic reticulum. Biochim. Biophys. Acta 896:187-195

# The Topology of Sarcoplasmic Reticulum Ca$^{2+}$-ATPase and Na$^+$,K$^+$-ATPase in the M5/M6 Region

J.V. Møller, B. Juul, Y.-J. Lee, M. le Maire, P. Champeil

Institute of Biophysics, Ole Worms Allé 185, University of Aarhus, DK-8000 Aarhus C, Denmark and Département de Biologie Cellulaire et Moléculaire, Centre d'Etudes de Saclay, F-91191 Gif-sur-Yvette, France.

While there is general agreement concerning the existence of 4 hydrophobic membrane traverses in the N-terminal part of P-type ATPases, the exact topology of the C-terminal, membraneous domain is still a matter of dispute. For sarcoplasmic reticulum (SR) Ca$^{2+}$-ATPase 3 pairs of transmembrane helices (M5-M10) were proposed, leading to a 10-helical model for this ATPase, cf. Fig. 1. For Na$^+$,K$^+$-ATPase fewer membrane traverses were considered probable, leading to 7 helical (8) or 8 helical (3) models. In the 7 helical model, regions corresponding to M6, M8, and M10 were placed outside the lipid membrane, based on protein-chemical and immunochemical evidence. However, since there is strong evidence for cytosolic exposure of both the N-terminus and C-terminus, 8 helical models are now favored for Na$^+$,K$^+$-ATPase, but exactly how the transmembrane segments would be positioned in the membrane remains undefined. Concerning the different models proposed for Ca$^{2+}$-ATPase and Na$^+$,K$^+$-ATPase it also needs to be asked, if it is plausible that their topology should be different, considering the fact that the hydropathic profiles of the two enzymes are strikingly similar.

In the present communication we report data which were designed to provide experimental evidence on the existence of M5 and M6 in Ca$^{2+}$-ATPase and Na$^+$,K$^+$-

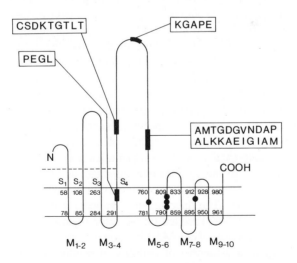

Fig. 1. 10 helical model for SR Ca$^{2+}$-ATPase (1), with sequence motifs characteristic of P-type ATPases and the intramembraneous position of C-terminal putative Ca$^{2+}$-binding amino acid residues (●, Ref. 2)

131

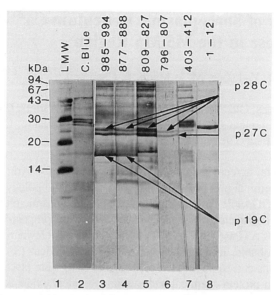

Fig. 2. Immunoblot obtained after treating SR vesicles at pH 6.5 with proteinase K. Polypeptides were separated by SDS gel electrophoresis on 16% acrylamide tricine gels and blotted onto PVDF membranes. Lanes 3-8 show the reactivity of the digest using antipeptide antibodies specific for the sequence indicated above each lane and visualized by antibodies conjugated to peroxidase.

ATPase. The data were obtained on intact SR vesicles, with cytosolic side exposed, and tight pig kidney microsomes, with extracellular side exposed, using a combined approach of proteolytic digestion and sequence specific antibodies. Fig. 2 shows that after treatment of SR vesicles with proteinase K two bands are formed with apparent molecular masses below 30 kDa, which react with antibodies against p985-994 (the C-terminus), p877-888, p809-827, and, less strongly, p796-807. By N-terminal sequencing of the corresponding region of the Coomassie Blue stained lane of the blot, the following sequence was obtained for p27C: VEEGRAIN--, which together with the immunological evidence identifies this as p747-994, while N-terminal sequencing data on p28C confirms that this is a mixture of slightly larger variants of p27C, starting at D737 or V734.

In addition to p27C and p28C, a distinct band with an apparent molecular mass below 20 kDa (p19C) is present which reacts with antibody against the C terminus and p877-888. The N-terminal sequence of p19C is determined to be DPPRSKEP-- which, together with its immunoreactive properties, identifies it as p818-994. During proteinase K treatment the integrity of the vesicles remains intact as indicated by the retention of undegraded proteins (calsequestrin, glycoprotein) in the pellet after ultracentrifugal sedimentation (not shown). The data therefore indicate that the 740-750 and 815-820 hydrophilic regions of Ca$^{2+}$-ATPase are on the same (cytosolic) side of the membrane, consistent with the existence of an M5-M6 loop in the intervening region, and the presence in this region of a number of amino acid residues that probably are directly involved in cation translocation (2). In addition to p19C, another band with a slightly higher molecular mass and a strong immunoreactivity against the p809-827 antibody is observed. This presumably corresponds to a cut closer to M6, but the N-terminal

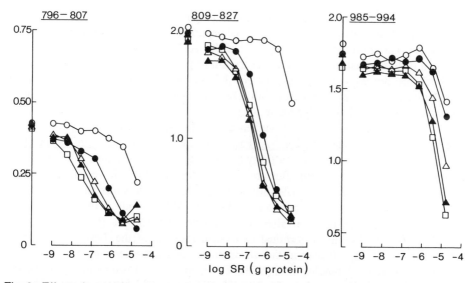

Fig. 3. Effect of proteinase K treatment on the reactivity of sarcoplasmic reticulum vesicles with C-terminal antibodies. SR vesicles were treated with proteinase K for 0 min (O), 12.5 min (●), 25 min (Δ), 50 min (▲), and 75 min (□). Different amounts of SR vesicles (0-15 μg protein) were incubated with antibody against either p796-807, or p809-827, or p985-994 in Elisa wells coated with SDS/carbonate denatured SR protein and blocked with 3 % BSA. Antigen/antibody complex on the plastic was quantitated by a peroxidase assay.

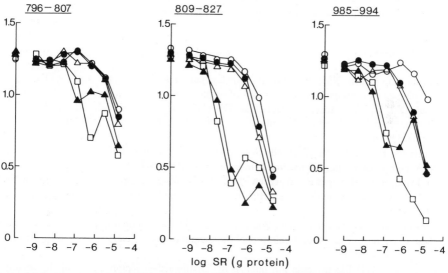

Fig. 4. Effect of non-solubilizing and solubilizing concentrations of $C_{12}E_8$ on the immuno-reactivity of sarcoplasmic reticulum $Ca^{2+}$-ATPase. Vesicles, pretreated with $C_{12}E_8$, were subjected to by competitive Elisa as described in Fig. 3. The various curves refer to the following concentrations of free $C_{12}E_8$: O, 0 mg ; ●, 0.0175 mg/ml (non-solubilizing); Δ, 0.030 mg/ml (non-solubilizing); ▲, 0.050 mg/ml (just solubilizing); □, 0.10 mg/ml (fully solubilizing).

133

sequence of this peptide has not been obtained as yet.

p19C is analogous to the C-terminal peptide fragment described by Karlish and coworkers (5) after proteolytic cleavage of $Na^+,K^+$-ATPase under certain conditions. By treatment of tight renal microsomes with trypsin or proteinase K we observed that only the β-subunit, not the α-polypeptide of $Na^+,K^+$-ATPase was susceptible to proteolytic degradation (not shown), consistent with a recent report of a cytosolic exposure of the N-terminus of the 19 kDa polypeptide (4).

We have also used the sequence specific antibodies in competitive Elisa experiments with undenatured $Ca^{2+}$-ATPase. As can be seen from Fig. 3 intact SR vesicles are characterized by a low affinity for the C-terminal antibodies. A large increase in antibody affinity is observed after proteinase K treatment of SR vesicles (Fig. 3). It is of interest to note that the p796-807-antibody, directed towards the putative M6-segment, reacts to almost the same extent as the p809-827-antibody. This is also the case for the intact polypeptide chain after solubilization with $Ca_{12}E_8$ (Fig.4). This could indicate that several of the C-terminal aminoacid residues of the 796-807 epitope are indeed exposed, being situated at the membrane/cytosolic border region. This would place the putative $Ca^{2+}$-binding sites (N796, T799, and D800) towards the cytosolic side and, assuming a similar topology for $Na^+,K^+$-ATPase, would be consonant with electrophysiological evidence for a long access channel for release of bound cation towards the extracellular aspect of this ATPase (6). With the monoclonal antibody $VG_2$ against $Na^+,K^+$-ATPase, reported to have a similar epitope location as our p796-807 and p809-827 antibodies, Mohraz et al. (7) also obtained evidence for exposure of this region, but at the extracellular side. For these data to conform with the present results would imply a different location for the $VG_2$ epitope, perhaps corresponding to the M5/M6 turn, which is earlier in the sequence than previously considered (8).

# References

1. Brandl CJ, Green NM, Korczak B, MacLennan DH (1986) Two $Ca^{2+}$-ATPase genes: homologies and mechanistic implications of deduced amino acid sequences. Cell 44: 597-607.
2. Clark DM, Loo TW, Inesi G, MacLennan DH (1989) Location of high affinity $Ca^{2+}$-binding sites within the predicted transmembrane domain of the sarcoplasmic reticulum $Ca^{2+}$-ATPase. Nature 339: 476-478.
3. Jørgensen PL, Andersen JP (1988) Structural basis for $E_1$-$E_2$ conformational transitions in Na,K-pump and Ca-pump proteins. J. Membr. Biol. 103: 95-120
4. Karlish SDJ, Goldshleger R, Jørgensen PL (1993) Location of $Asn^{831}$ of the α-chain of Na/K-ATPase at the cytoplasmic surface. Implications for topological models. J. Biol. Chem. 268: 3471-3478.
5. Karlish SDJ, Goldshleger R, Stein WD (1990) A 19 kDa C-terminal tryptic fragment of the α-chain of Na,K-ATPase is essential for occlusion and transport of cations. Proc. Natl. Acad. Sci. USA 87: 4566-4570.
6. Läuger P (1991) Electrogenic ion pumps. In "Distinguished lecture series of the society of general physiologist, Vol. 5, pp168-225, Sinauer Associates, Mass, USA.
7. Mohraz M, Arystarkhova E, Sweadner KJ (1993) Membrane topology of the α-subunit of Na,K-ATPase. Proceedings from this meeting.
8. Ovchinnikov YA (1987) Probing the folding of membrane proteins. TIBS 12: 434-438.

# Changes of molecular structure and interaction in the catalytic cycle of sarcoplasmic reticulum $Ca^{2+}$- ATPase.

## Infrared difference spectra of $Ca^{2+}$ binding, nucleotide binding, phosphorylation and phosphoenzyme conversion

Andreas Barth, Holger Georg, Werner Kreutz, Werner Mäntele

Institut für Biophysik und Strahlenbiologie der Universität Freiburg, Albertstr. 23, D-79104 Freiburg

**Introduction:** Infrared difference spectroscopy was used to get more insight into the transport mechanism of sarcoplasmic reticulum $Ca^{2+}$-ATPase. Several partial reactions were induced directly in the infrared cuvette via photolytic release of ligands or substrates from photolabile precursors [1,2]. From the spectra before and after the release a difference spectrum can be calculated that reflects only those changes of infrared absorption which are associated with the catalytic reactions and the triggering photolysis reaction. The accuracy of this method is extremely high, since it avoids possible inaccuracies when spectra of two samples prepared in different enzyme states are compared.

The following reaction steps of the catalytic cycle were investigated (see scheme 1): Step 1, $Ca^{2+}$ binding to the $Ca^{2+}$ free form of the ATPase; step 2, formation of the ADP sensitive phosphoenzyme $Ca_2E_1$-P from $Ca_2E_1$ [1,2]; step 3, phosphoenzyme conversion and $Ca^{2+}$ release; step 2+3, formation of the ADP insensitive phosphoenzyme $E_2$-P; step 4, nucleotide binding [2] and step 5, the effects of $\gamma$-phosphate binding and phosphorylation.

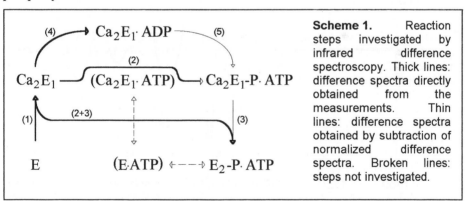

**Scheme 1.** Reaction steps investigated by infrared difference spectroscopy. Thick lines: difference spectra directly obtained from the measurements. Thin lines: difference spectra obtained by subtraction of normalized difference spectra. Broken lines: steps not investigated.

**Methods:** Infrared difference spectra were recorded in the region from 1800 to 900 $cm^{-1}$ in $H_2O$ and $D_2O$. Infrared difference signals due to the photolysis reaction were subtracted. Thus, the obtained spectra are predominantly composed of the absorbance changes of ATPase and nucleotide. The difference spectra were normalized to a uniform protein content in order to compare spectra from different partial reactions. In the case of the $Ca^{2+}$ binding reaction intrinsic fluorescence was measured simultaneously to the infrared absorbance.

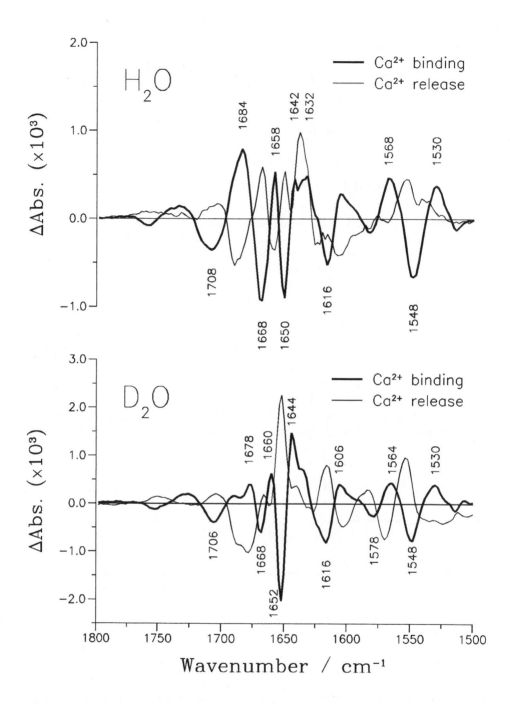

**Fig. 1.** Infrared difference spectra in $H_2O$ and $D_2O$ of $Ca^{2+}$ binding to the $Ca^{2+}$ free ATPase (thick lines, step 1 in scheme 1) and of $Ca_2E_1$-P to $E_2$-P phosphoenzyme conversion and $Ca^{2+}$ release (thin lines, step 3 in scheme 1). The numbers give the band positions of the thick line spectra.

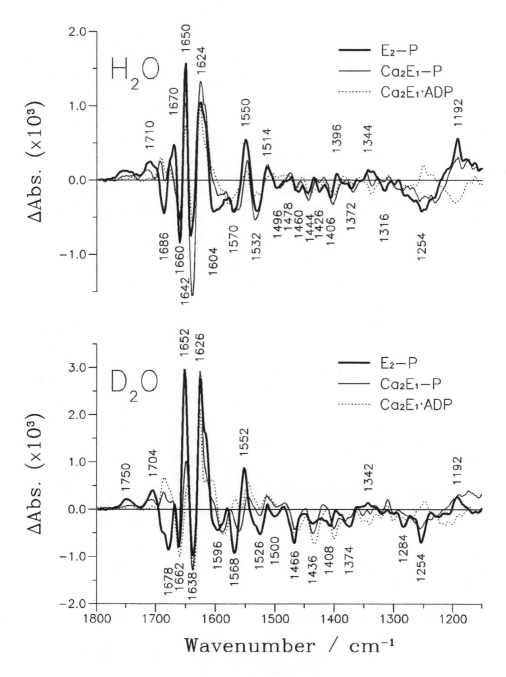

**Fig 2.** Infrared difference spectra in $H_2O$ and $D_2O$ of ADP binding (dotted lines, step 4 in scheme 1), formation of the ADP-sensitive phosphoenzyme (thin lines, step 2 in scheme 1), and formation of the ADP-insensitive phosphoenzyme (thick lines, steps 2+3 in scheme 1). The numbers give the band positions of the thick line spectra.

137

**Results:** All reaction steps caused distinct changes of the vibrational infrared spectrum which were characteristic for each reaction step but comparable for all steps in the number and magnitude of the changes. Most pronounced were absorbance changes in the amide I spectral region sensitive to protein secondary structure. However, they were small - less than 1 % of the total protein absorbance - indicating that the reaction steps are associated with small and local conformational changes of the polypeptide backbone instead of affecting large protein domains. No absorbance change was observed in the region of the strongest lipid headgroup absorbance ($1736 \text{ cm}^{-1}$), suggesting that the lipid structure is not affected by the reaction steps investigated.

The infrared difference spectra of $Ca^{2+}$ binding (see Fig. 1, step 1 in scheme 1) gave evidence for the participation of carboxyl groups from Asp or Glu in the binding reaction, either becoming deprotonated or stronger hydrogen bonded (bands above $1700 \text{ cm}^{-1}$). A pair of difference signals (1548/1568 or $1564 \text{ cm}^{-1}$) can be attributed to the release of $Ca^{2+}$ from carboxylate groups, since they were also observed in model compound spectra of $Ca^{2+}$ binding to carboxylate groups. The difference bands of the $Ca^{2+}$ binding reaction could be correlated to the $Ca^{2+}$ induced rise of intrinsic ATPase fluorescence supporting their attribution to structural changes of the ATPase.

Nucleotide (ADP) binding (see Fig. 2, step 4 in scheme 1) induces conformational changes in the ATPase polypeptide backbone with $\alpha$-helical structures ($1650 \text{ cm}^{-1}$) and presumably $\beta$-sheet (1624, 1675 to $1700 \text{ cm}^{-1}$) or turn structures (1675 to $1700 \text{ cm}^{-1}$) involved. As far as the nucleotide is concerned, dehydration of the $\alpha$-phosphate is supported by our spectra (1170 to $1250 \text{ cm}^{-1}$), whereas there is no hint for a change of hydrogen bonding to the adenine $NH_2$ group.

Phosphorylation (see Fig. 2, step 2 in scheme 1) is accompanied by the appearance of a keto group that we tentatively assign to the phosphorylated residue Asp[351] ($1714 \text{ cm}^{-1}$). Upon phosphorylation most of the bands due to nucleotide binding remain, indicating nucleotide binding to the phosphoenzyme.

Phosphoenzyme conversion and $Ca^{2+}$ release (see Fig. 1, step 3 in scheme 1) produced a difference spectrum that was almost completely the reflected image of the $Ca^{2+}$ binding difference spectrum. Therefore, structural changes of $Ca^{2+}$-ATPase induced by $Ca^{2+}$ binding are reversed by phosphoenzyme conversion and $Ca^{2+}$ release.

**Acknowledgements:** We like to thank Prof. W. Hasselbach (MPI Heidelberg) for the gift of $Ca^{2+}$-ATPase and Prof. R. Goody (MPI Dortmund) for providing caged ADP.

**References**

(1) Barth A, Kreutz W and Mäntele M (1990) Molecular changes in the sarcoplasmic reticulum calcium ATPase during catalytic activity. FEBS Lett 277: 147-150.
(2) Barth A, Mäntele W and Kreutz W (1991) Infrared spectroscopic signals arising from ligand binding and conformational changes in the catalytic cycle of sarcoplasmic reticulum calcium ATPase. Biochim Biophys acta 1057: 115-123.

# $Ca^{2+}$ binding to sarcoplasmic reticulum $Ca^{2+}$-ATPase in the absence of chelators

Otto Hansen, Jørgen Jensen

Institute of Physiology, Biomembrane Research Centre, Aarhus University, DK-8000 Aarhus C, Denmark.

## Introduction

The $Ca^{2+}$ affinity of the sarcoplasmic reticulum $Ca^{2+}$-ATPase is in the submicromolar range. Since buffer solutions will contain $Ca^{2+}$ above this level, $Ca^{2+}$ concentrations in binding studies have to be controlled by chelators such as EGTA. Obviously the accuracy of the calculated free concentrations of $Ca^{2+}$ is crucial for determination of the characteristic features of $Ca^{2+}$ binding to the enzyme. The relationship between calculated $Ca^{2+}$ concentrations and binding in e.g. Hill plots is usually interpreted as positive cooperativity in binding. This interpretation seems compatible with 2 $Ca^{2+}$ sites per phosphorylation unit of $Ca^{2+}$-ATPase encompassing only half of the 110-kDa peptides present in purified $Ca^{2+}$-ATPase preparations. Apparently no role is assigned to half of the protein (for refs., see 1). Artefactually high Hill numbers have been noticed even in soluble monomers (1) and recently it was emphasized that EGTA should be used as $Ca^{2+}$ chelator with caution (2).

Since we have experimental evidence for non-cooperativity in $Sr^{2+}$ binding to $Ca^{2+}$-ATPase and for high-affinity ADP and ATP binding to every 110-kDa peptide present (3) an alternative approach in $Ca^{2+}$ binding was attempted in this communication.

## Materials and Methods

$Ca^{2+}$-ATPase was prepared from rabbit fast-twitch skeletal muscle according to de Meis and Hasselbach (4) and gently extracted with low concentrations of deoxycholate as described (3). The specific activity of 3 batches of enzyme was around 30 $\mu mol \cdot mg^{-1} \cdot min^{-1}$ at 37 °C. From $CaCl_2, 2H_2O$ (Merck) and $^{45}Ca^{2+}$ (Danish Atomic Energy Commission, Isotope Lab. Risø) a stock solution of 1 mM $(^{45}Ca)Ca^{2+}$ and a stock solution containing 25 mM TES (pH 7.5, 0-2 °C) and enzyme in sucrose/Hepes buffer were prepared. The two solutions were mixed and redistilled $H_2O$ added so as to give final concentrations of 20 mM TES, 30 mM sucrose, 0.5 mM Hepes, 1.39 mg enzyme per ml and varying (so far unknown) concentrations of $Ca^{2+}$. Radioactivity was determined on aliquots. After incubation at 0-2 °C for 15 min separation of free and bound $Ca^{2+}$ was carried out by centrifugation for 1 hour at 100,000·g. From the radioactivity and the concentration of free $Ca^{2+}$ determined by atomic absorption spectrophotometry (Philips PU 9200; Pye Unicam, Cambridge, UK) the specific activity of $(^{45}Ca)Ca^{2+}$ of the supernatant was known. Assuming that the specific activity of the incubation medium was the same as that of the supernatant, the total concentration of $Ca^{2+}$ during incubation was calculated. The difference between the two concentrations makes bound $Ca^{2+}$.

## Results and Discussion

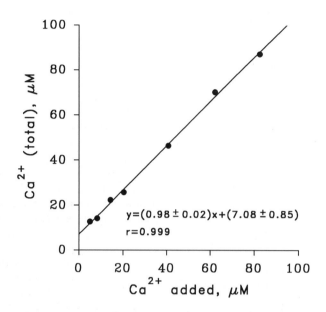

Figure 1. Linear regression of $Ca^{2+}$ data in binding experiments. Added $Ca^{2+}$ is plotted against the total concentration of $Ca^{2+}$ during binding calculated from the radioactivity of the medium and the specific activity of the supernatant.

The total concentration of $Ca^{2+}$ during enzyme incubation was calculated from the radioactivity of the medium and the specific activity of the supernatant after enzyme precipitation. The precision of the method is illustrated in Fig. 1. A plot of added $Ca^{2+}$ versus calculated total calcium fits nicely a straight line with a slope not significant from 1. The ordinate intercept would indicate a contribution of $7.08 \pm 0.85$ $\mu$M $Ca^{2+}$ from buffer solution and the specific enzyme batch but different intercepts were obtained with different enzyme batches.

The difference between the calculated total concentration of $Ca^{2+}$ during enzyme incubation and that one measured in the supernatant after enzyme precipitation represents bound $Ca^{2+}$. Part of the enzyme-bound $Ca^{2+}$ takes place at specific sites, another part is unspecific contamination. $Ca^{2+}$ binding was carried out with 3 $Ca^{2+}$-ATPase batches. Binding data were analyzed according to a 2 component model

$$y = \frac{a \cdot x}{b+x} + \frac{c \cdot x}{d+x}$$

140

in which x and y represent free and bound $Ca^{2+}$, a and c the low-affinity (unspecific) and high-affinity (specific) component, and b and d their respective dissociation constants.

The resolution of $Ca^{2+}$-binding data into 2 components was justified by a nearly 500-fold difference in their apparent affinities. In Fig. 2 is shown specific $Ca^{2+}$ binding of the 3 enzyme batches. The method is restricted for determination of binding in a relatively narrow range of bound/free values and not at very low concentrations of free $Ca^{2+}$. With this reservation in mind it is seen that data are compatible with 10.5 nmoles of $Ca^{2+}$ bound per mg protein and with one homogeneous population of $Ca^{2+}$ receptors with an apparent dissociation constant of 0.5 $\mu$M.

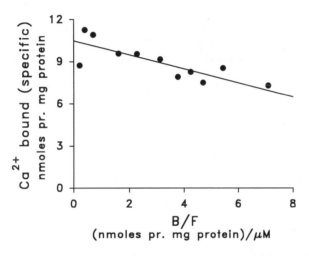

Figure 2. A Scatchard-type plot of specifically bound $Ca^{2+}$ versus bound/free (B/F) $Ca^{2+}$ obtained with 3 different $Ca^{2+}$-ATPase batches.

The results obtained with $Ca^{2+}$ are in line with our previous experience with the $Ca^{2+}$ congener $(^{85}Sr)Sr^{2+}$ used under exactly the same conditions in an attempt at avoiding chelators (3). $Sr^{2+}$ has a lower affinity for $Ca^{2+}$-ATPase and moreover, the apparent affinity is reduced by contaminating $Ca^{2+}$ whereas the $Sr^{2+}$-binding capacity is unchanged. In strontium binding the specific $(^{85}Sr)Sr^{2+}$ receptors were monitored by subtraction of the binding taking place after thermal inactivation of the enzyme. Net $Sr^{2+}$ binding was compatible with one homogeneous population of receptors with a capacity of around 10 nmoles per mg protein and with an apparent dissociation constant of 9.5 $\mu$M.

It was recently discovered by us that the nucleotide binding capacity corresponded with the number of 110-kDa peptides present. With pure, but otherwise native, $Ca^{2+}$-ATPase from sarcoplasmic reticulum (similar to that of the present study) ATP and ADP binding of relatively high affinity took place in the absence of $Mg^{2+}$. Binding capacities of 8.4-

8.5 nmoles per mg protein were obtained (3). On this background and in the absence of cooperativity in $Sr^{2+}$ binding we propose a stoichiometry of $Ca^{2+}$ receptors to nucleotide receptors to 110-kDa peptides as 1:1:1 and a monomeric structure of $Ca^{2+}$-ATPase. The apparent cooperativity noticed in Hill plots of $Ca^{2+}$-binding data obtained with $Ca^{2+}$ chelators may be an artefact. The interpretation of cooperativity is intimately connected with a steady-state phosphorylation of roughly half of the peptides present indicating 2 $Ca^{2+}$ ions bound per active unit (for refs., see 1). Considerations on the structure of $Ca^{2+}$-ATPase based on the nucleotide binding capacity obtained from equilibrium binding studies are to us better substantiated.

Nucleotide and $Sr^{2+}/Ca^{2+}$ binding data are thus compatible with receptors on every 110-kDa peptide present and a monomeric structure of $Ca^{2+}$-ATPase. This is a striking difference to another, closely related P-type pump, the $Na^{+}/K^{+}$-ATPase. For the sodium pump it is generally accepted that there is an equal number of sites for high-affinity nucleotide-, ouabain-, and vanadate-binding as well as phosphorylation (for refs., see 5). Maximum nucleotide and ouabain binding in the range 3.2-3.6 nmoles per mg protein hardly exceeds half of the expected maximum value favouring a dimeric structure of $Na^{+}/K^{+}$-ATPase, $(\alpha\beta)_2$ (6,7).


## Acknowledgements

Thanks are due to Edith Møller, Bente Mortensen and Toke Nørby for expert technical assistance.

## References

1. Andersen, JP (1989) Monomer-oligomer equilibrium of sarcoplasmic reticulum Ca-ATPase and the role of subunit interaction in the $Ca^{2+}$ pump mechanism. Biochim Biophys Acta 988:47-72
2. Forge V, Mintz E, Guillain F (1993) $Ca^{2+}$ binding to sarcoplasmic reticulum ATPase revisited. I. Mechanism of affinity and cooperativity modulation by $H^{+}$ and $Mg^{2+}$. J Biol Chem 268:10953-10960
3. Jensen J, Hansen O (1993) Binding of nucleotide and strontium to purified sarcoplasmic reticulum $Ca^{2+}$-ATPase. Submitted
4. de Meis L, Hasselbach W (1971) Acetyl phosphate as substrate for $Ca^{2+}$ uptake in skeletal muscle microsomes. J Biol Chem 246:4759-4763
5. Hansen O, Jensen J, Nørby JG, Ottolenghi P (1979) A new proposal regarding the subunit composition of $(Na^{+}+K^{+})$ATPase. Nature 280:410-412
6. Jensen J (1992) Heterogeneity of pig kidney Na,K-ATPase as indicated by ADP- and ouabain-binding stoichiometry. Biochim Biophys Acta 1110:81-87
7. Hansen O (1992) Heterogeneity of Na,K-ATPase from kidney. Acta Physiol Scand 146:229-234

# Electrogenicity and Countertransport Properties of Reconstituted Sarcoplasmic Reticulum $Ca^{2+}$-ATPase.

Flemming Cornelius and Jesper V. Møller
Institute of Biophysics, University of Aarhus, DENMARK

## INTRODUCTION

The extent to which active $Ca^{2+}$-transport is accompanied by an obligatory exchange with positive charge or, basically, is a purely electrogenic process has not been satisfactorily clarified. We report here some recent observations concerning the role of $Ca^{2+}/H^+$ exchange in active $Ca^{2+}$-transport by reconstituted sarcoplasmic reticulum $Ca^{2+}$-ATPase.

As evidence against a significant role of countertransport we previously reported that $Ca^{2+}$-transport in tight vesicles, prepared by cholate reconstitution at a high lipid to protein ratio, is accompanied by the development of an appreciable transmembrane potential of $\Delta V \approx 90$ mV, inside positive (1). The electric net current flow ($I$) accompanying $Ca^{2+}$-transport was calculated according to the equation:

$$I = C_m A_m (dV/dt)_o /e$$

where $A_m$ and $C_m$ is the surface area and capacitance ($1\mu F/cm^2$) of the lipid bilayer, $(dV/dt)_o$ is the initial rate of development of the membrane potential, and $e$ is the electrostatic charge ($1,6 \times 10^{-19}C$). These calculations indicated that the electrogenic net current can be accounted almost quantitatively by the initial rate of $Ca^{2+}$-transport, implying that $Ca^{2+}$-transport mainly is an electrogenic process, unaccompanied by other ions.

However, by the use of a different reconstitution procedure, employing β-octylglucoside and a different medium ($SO_4^{2-}$ instead of $Cl^-$, absence of $Ca^{2+}$-precipitating anion (phosphate) inside the vesicles), *Levy et al.* (4) found that $Ca^{2+}$ accumulation could be increased by addition of the weak acid FCCP, facilitating proton translocation. This was attributed to collapse of a pH-gradient (vesicle alkalinization) which developed in the absence of FCCP, and which was monitored with the aid of pyranine trapped inside the vesicles. In combination with other evidence for $H^+$-ejection these studies demonstrated the existence of a $Ca^{2+}$-$H^+$ counter transport mechanism. It was later found by *Yü et al.* (5) that a trans-membrane potential developed concomitant with the pH-gradient. Thus, the electrogenic nature of $Ca^{2+}$-transport was also observed under these conditions, and a stoichiometry of $2H^+$: $2Ca^{2+}$ was suggested.

A problem inherent in the previous approaches is that estimates of stoichiometric ratios are based on several assumptions, and that it is not easy to sort out effects, arising from the development of membrane potential from those that arise from the presence of pH gradients *per se*. For instance, it is not clear to what extent a pH-gradient is the result of (i) $Ca^{2+}/H^+$ countertransport, or (ii) is passively imposed by the membrane potential.We have therefore reinvestigated the role of $H^+$ as counterions during the active uptake of $Ca^{2+}$.

## PROCEDURES

Cholate reconstituted liposomes of dioleoylphosphatidylcholine were treated with addition of $\beta$-octylglucoside to achieve incomplete (state 2) solubilization (3) in a medium usually containing 10 mM Pipes (pH 6.5), 100 mM $Na_2SO_4$, 10 mM $Li_2SO_4$, 1 mM $Mg^{2+}$, and 300 $\mu$M pyranine. Then $C_{12}E_8$-solubilized $Ca^{2+}$-ATPase was added at a weight ratio lipid-to-protein of 100:1. $\beta$-Octylglucoside was removed with BioBeads to produce tight and large ($\sim$ 160 nm diameter) vesicles with incorporated $Ca^{2+}$ATPase. After reconstitution, the preparations were centrifuged through Penefsky columns to remove extraneous pyranine and to be able to vary the pH of the outer medium, relative to that of the intravesicular medium.

Changes in intravesicular pH were followed spectrofluorometrically in the presence of pyranine on a Spex fluorometer by excitation at 440 nm and emission > 515 nm. Changes in membrane potential were followed by changes in oxonol fluorescence (1) after excitation at 580 nm and emission >660 nm. Energized $Ca^{2+}$-uptake by reconstituted vesicles, incubated with 0.1 mM $Ca^{2+}$ + $^{45}Ca^{2+}$ and 1 mM Mg ATP, was measured by rapid filtration through Biorex columns (2).

## RESULTS

During ATP-energized $Ca^{2+}$-uptake it was easy to demonstrate that alkalinization of vesicle contents took place. Interestingly, alkalinization at an even higher initial rate was observed after addition of the $Li^+$-ionophore AS701 to proteoliposomes, containing an equal concentration of $Li^+$ inside and outside (**Fig 1 A**). In separate experiments with oxonol this set-up was found to short-circuit the membrane potential otherwise generated by the $Ca^{2+}$-pump, (**Fig. 1 B**). These data indicate that the pH-gradient does not arise secondary to the development of a transmembrane potential: $H^+$-ejection from vesicles can therefore be ascribed to $Ca^{2+}/H^+$ exchange via the $Ca^{2+}$-pump. However, the $H^+$: $Ca^{2+}$ stoichiometry, calculated from the initial rates of $Ca^{2+}$-transport and $\Delta$pH development, is low amounting to $\sim$ 0.3-0.6 $H^+$: 2 $Ca^{2+}$ at $pH_o$ 7.0. This stoichiometry is decreased even further to around $\sim$ 0.1 $H^+$: 2 $Ca^{2+}$ at pH 7.3.

We have also examined the effect of CCCP (a weak acid which like FCCP acts as a $H^+$-conducting ionophore), alone and in combination with AS701, on $Ca^{2+}$-uptake and $H^+$-equilibration. The effect of CCCP is qualitatively similar to that of AS701: $\Delta$pH rises steeper than in the absence of ionophore, and steady state alkalinization of vesicle interior is only moderately decreased (cf. **Table I**), indicating that $Ca^{2+}/H^+$ exchange overshadows the ionophore effect on passive proton conduction. Combined addition of CCCP and AS701 results in a substantially larger increase of passive proton conductance by $H^+/Li^+$ exchange, and reduces steady state $\Delta$pH, but the $H^+/Ca^{2+}$ exchange rate *via* the pump is increased to $2H^+$:$2Ca^{2+}$ per ATP split under these conditions (**Table I**).

Addition of ionophores (AS701, monensin) or CCCP produces an identical and moderate stimulation of $Ca^{2+}$-uptake (**Fig. 2**). This was found to correlate with dissipation of the membrane potential, but to be independent of the pH gradient.

144

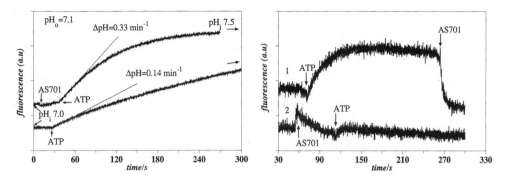

**Fig. 1** Left panel shows alkalization inside proteoliposomes generated by ATP activation of reconstituted $Ca^{2+}$-ATPase as detected by pyranine fluorescence . The experiments were performed in the absence or presence of the $Li^+$-ionophore AS701, which effectively clamps the membrane potential to zero due to the equal $Li^+$ concentration inside and outside the proteoliposomes. pH of the medium ($pH_o$) was 7.1. Calibration of the fluorescence response with pH was performed by titration with NaOH in the presence of monensin, which equilibrates internal pH with medium pH. The right panel depicts the presence (curve 1) or absence (curve 2) of a transmembrane potential during such conditions (± AS701) as measured by the oxonol VI fluorescence. The maximum transmembrane potential developed corresponds to about 60 mV, inside positive.

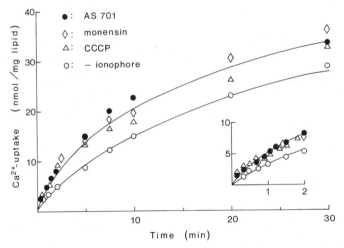

**Fig. 2.** The effect of ionophores and CCCP on ATP-dependent $Ca^{2+}$-uptake by reconstituted $Ca^{2+}$-ATPase.

145

**Table I.** Calculated steady-state pH-gradient and stoichiometries of $H^+:Ca^{2+}$ during various conditions of ionophores.

| $pH_o$ | ionophore | ΔpH (steady state) | Exchange Ratio ($H^+$: $Ca^{2+}$) |
|---|---|---|---|
| 7.1 | 0 | 0.60 | 0.25 |
| 7.1 | AS701 | 0.38 | 0.29 |
| 6.9 | CCCP | 0.35 | 0.50 |
| 6.9 | CCCP + AS701 | 0.15 | 1.0 |

## CONCLUSIONS

(1) ΔpH changes develop independently of the membrane potential and do not seem to have much, or any effect on $Ca^{2+}$-transport (**Figs. 1, 2**).

(2) The $H^+/Ca^{2+}$ stoichiometry is low and variable, but stimulated by $H^+$-conducting ionophore (**Table I**)

(3) Dissipation of membrane potential leads to an increase of $Ca^{2+}$-transport (**Fig. 2**)

(4) It is suggested that during enzyme turnover a distinct proton conducting pathway opens. This is different from the $Ca^{2+}$-channel and possibly located at the protein-lipid interface at a position where transfer can be facilitated by "$H^+$-ionophores". In the absence of ionophores, the $Ca^{2+}$-pump exchanges $Ca^{2+}$ for $H^+$ in a non-obligatory fashion, in which $Ca^{2+}$-transport only occasionally is coupled to proton countertransport.

## REFERENCES

1. Cornelius, F and Møller, JV (1991) FEBS lett. 284:46-50
2. Cornelius, F and Skou, JC (1985) Biochim. Biophys. Acta 818:211-221
3. Kragh-Hansen, U, le Maire, M, Nöel, J-P, Gulik-Krzywicki, T, and Møller, JV (1993) Biochemistry 32:1648-1656
4. Levy, D, Seigneuret, M, Bluzat, A, and Rigaud, J-L (1990) J. Biol. Chem. 265:19524-19534
5. Yü, X, Carroll,S, Rigaud, J-L, and Inesi, G (1993) Biophys. J. 64:1232-1242

# Endo- Exocytotic Sorting of $Na^+/K^+$-Pumps in Developing Oocytes and Embryos of *Xenopus laevis*

G. Schmalzing

Max-Planck-Institute of Biophysics, Department of Cell Physiology, Heinrich-Hoffmann-Str. 7, D-60528 Frankfurt/Main, Germany

## Introduction

Amphibian oocytes are physiologically arrested at the $G_2/M$ border of the first meiotic prophase until progesterone triggers the resumption of the meiotic cell cycle. Within a few hours, immature oocytes are transformed by a process designated meiotic maturation into mature oocytes ready to be shed and fertilized (see Ref. 9 for review). In parallel to the well known nuclear changes, a profound reorganization of the plasma membrane takes place that is reflected by a marked reduction of the surface area (8), a near complete membrane depolarization, and a downregulation of virtually all transmembrane transport systems (see Ref. 15 for review). As a consequence of these changes, the plasma membrane of mature oocytes has an extremely low ionic permeability, in contrast to the plasma membrane of immature oocytes, which exhibits high permeability for ions.

Following fertilization, the early embryo enters the rapid cleavage period that comprises twelve rapid and synchronous cell divisions. Since almost no transcription can be detected during the rapid cleavage period (12), the early developmental program depends completely on the recruitment of maternal constituents such as proteins and nucleic acids that are synthesized and stored during oogenesis. When the rapid cleavage period ends after about 7h, as much as 4,000 cells have been formed. The embryo then consists of an epithelium surrounding a cavity (the blastocoel) and is called a blastula. A cleavage cavity that gives rise to the formation of the blastocoel is present for the first time at the 4-cell stage and is sealed off from the outside medium by tight junctions, which form between the apical regions of contacting cells. This enables the embryo to create and maintain an inner ionic milieu with a composition similar to that of the extracellular milieu of the adult animal. Since injection of ouabain into the intercellular space of *Xenopus* embryos prevents blastocoel development, it has been proposed that the cells lining the blastocoel possess $Na^+/K^+$-pumps that vectorially extrude $Na^+$ into the cavity. Water follows passively to produce an isotonic solution, thus forming the blastocoel fluid (23).

A vectorial transport of $Na^+$ requires an asymmetrical distribution of $Na^+/K^+$-pumps. Mechanisms that have been described to account for a polarized distribution of membrane proteins include sorting of newly synthesized proteins in the Golgi complex and vectorial delivery to the appropriate membrane domain, selective removal of proteins from one of the membrane domains, and interaction with the membrane cytoskeletal proteins fodrin and ankyrin (11). Here I show that a polarized distribution of $Na^+/K^+$-pumps between the apical and basolateral domains of early embryonic cells results from (i) a complete removal of $Na^+/K^+$-pumps from the plasma membrane of the oocyte (the later apical domain) just before fertilization, and (ii) the vectorial delivery of the endocytosed $Na^+/K^+$-pumps to the new plasma membrane (the basolateral domain) formed during the cell divisions.

147

### Immature *Xenopus* oocytes have two pools of Na⁺/K⁺-pumps

Functional Na⁺/K⁺-pumps of fully-grown *Xenopus* oocytes consist of the α1 subunit isoform (26) and the ß3 subunit isoform (4,13). Fig. 1 illustrates that two pools of functional Na⁺/K⁺-pumps can be distinguished in fully-grown *Xenopus* oocytes: (i) a surface pool of Na⁺/K⁺-pumps that is readily accessible for [³H]ouabain from the external side of the plasma membrane; (ii) an intracellular pool of Na⁺/K⁺-pumps that becomes accessible for [³H]ouabain after permeabilization of the intracellular membranes by detergents (20). In the normal oocyte, the intracellular Na⁺/K⁺-pumps are continuously exchanged for surface expressed pumps by endocytosis and recycling to the cell surface in a way that the number of surface Na⁺/K⁺-pumps remains virtually constant (18). The rapid cycling of Na⁺/K⁺-pumps between intracellular membranes and the plasma membrane may reflect the high endocytotic activity of the *Xenopus* oocyte.

Beside these pools of complete Na⁺/K⁺-pumps, fully-grown oocytes contain a pool of immature α1 subunits. These α1 subunits are retained in the endoplasmic reticulum in an apparently nonfunctional form due to lack of enough endogenous ß subunits for pairing to Na⁺/K⁺-pump holoenzymes (3). Increasing the synthesis of ß subunits by sole injection of cRNA coding for an exogenous ß1 or ß2 subunit, for instance of the mouse, increases the number of functional Na⁺/K⁺-pumps in oocytes (18,21) at the expense of nonfunctional α1 subunits.

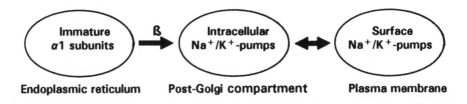

**Figure 1.** Na⁺/K⁺-pump pools of fully-grown *Xenopus* oocytes. During oogenesis, oocytes accumulate a large number of Na⁺/K⁺-pumps (cf. Fig. 2), which reside in both intracellular membranes and in the plasma membrane. In addition, *Xenopus* oocytes contain in the ER structurally and functionally immature α1 subunits that lack a ß subunit. Since α1 subunits need to assemble to a complete α/ß Na⁺/K⁺-pump complex to exit the ER and to commence the transport Na⁺ and K⁺ (see Ref. 17 for review), it follows that the number of newly formed Na⁺/K⁺-pumps is limited by the supply of ß subunits.

Fully-grown oocytes of Dumont stages V and stage VI (1) synthesize significant amounts of α1 subunits, but the synthesis of endogenous ß subunits is barley detectable (7). This suggests that a major fraction of the large number of Na⁺/K⁺-pumps found in these cells might have been formed already at earlier oogonial stages. Oocytes grow over a period of weeks to several months due to the uptake of vitellogenin from the blood by receptor-mediated endocytosis. According to competitive PCR measurements, the mRNA coding for the α1 subunit of the *Xenopus* Na⁺/K⁺-pump attains a maximum concentration already in young oocytes (stage III) and then declines (S. Friehs and G. Schmalzing, unpublished results). Fig. 2 illustrates that the number of surface Na⁺/K⁺-pumps becomes maximal already in stage IV oocytes and then remains constant despite the further marked increase in the cell surface area. By contrast, the intracellular pool of $Na^+/K^+$-pumps does not appear before stage III, but then increases continuously as long as the oocytes grow (Fig. 2). Since $Na^+/K^+$-pumps are synthesized at a very low rate in fully-grown oocytes (7), the $Na^+/K^+$-pumps synthesized at earlier stages seem to be preserved during development. The appearance of the intracellular $Na^+/K^+$-pumps parallels the appearance of endocytotic activity in these cells, since neither stage I nor II oocytes can incorporate vitellogenin (1).

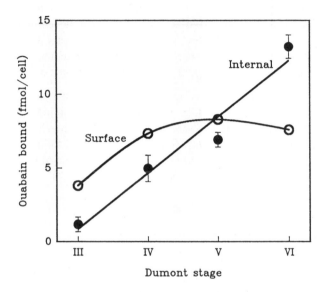

**Figure 2.** Formation of an intracellular pool of $Na^+/K^+$-pumps during development of *Xenopus* oocytes. During oogenesis, oocytes accumulate a large number of $Na^+/K^+$-pumps, which are located in intracellular membranes and in the plasma membrane. The number of intracellular $Na^+/K^+$-pumps (●) was calculated by subtracting the number of surface binding sites for [³H]ouabain (○) from the total number of binding sites, which was assessed on detergent-permeabilized oocytes exactly as described (20). All data points represent average values (±SEM) from 9-11 oocytes.

### Internalization of all surface $Na^+/K^+$-pumps during meiotic maturation

Meiotic maturation of amphibian oocytes has been found to be associated with a complete inhibition of ouabain-sensitive $Rb^+$ at the time around the appearance of the maturation spot (14,27). In the *Xenopus* oocyte, other transport activities such as alanine uptake and chloride uptake have been shown to be also reduced, however, by about 50% only, whereas active $Rb^+$ uptake declines to virtually zero (14).

Electron microscopy (cf. Figs. 5a and 5c) and measurements of the electrical membrane capacitance had shown that the cell surface area of oocytes decreases by about 70% during meiotic maturation (8). This suggested that a pronounced endocytotic event accompanies maturation and prompted us to investigate whether the downregulation of the $Na^+/K^+$-pump activity may also result from internalization of the pump molecules.

Indeed, as illustrated in Fig. 3, all the Na+/K+-pump molecules that can no longer be detected at the cell surface after completion of maturation can be recovered in the interior of cells made completely permeable for [³H]ouabain by detergents (16). Changes in the subcellular distribution of the Na+/K+-pumps could also be demonstrated by sucrose gradient fractionation, which showed that maturation induced a shift of Na+/K+-pumps from the plasma membrane fraction to membranes with lower buoyant density (16).

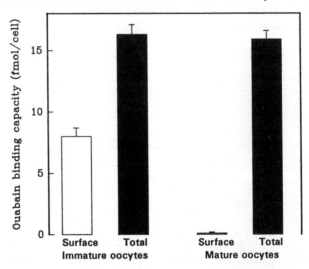

**Figure 3.** Na+/K+-pumps that disappear from the plasma membrane during meiotic maturation can be recovered in the cell interior. Oocytes matured *in vitro* and immature control oocytes were assayed for [³H]ouabain binding either intact (*open columns*) or after permeabilization with detergent (*filled columns*) under conditions that rendered all intracellular Na+/K+-pumps accessible for [³H]ouabain (16). All data points represent average values (±SEM) from 9-11 oocytes.

The idea that Na+/K+-pumps disappear from the plasma membrane by endocytosis was further supported by immunocytochemistry, as illustrated in Fig. 4. In sections through immature oocytes, the plasma membrane is intensely labelled, consistent with the high number of surface Na+/K+-pumps of these cells (*left panels* in Fig. 4). The less intense immunofluorescence of the plasma membrane at the animal hemisphere (Fig. 4c) than at the vegetal hemisphere (Fig. 4a) indicates that some polarized distribution of Na+/K+-pumps does exist. By contrast, in sections from meiotically mature oocytes the plasma membrane remains unstained both at the vegetal pole (Fig. 4b) and at the animal pole (Fig. 4d). At the vegetal pole, at least, islets of α1 subunits are visible underneath the plasma membrane that seem to represent internalized Na+/K+-pumps.

The endocytosed Na+/K+-pumps could be phosphorylated by back-door phosphorylation with inorganic [³²P]phosphate (16). To incorporate phosphate, the enzyme must be capable of adopting the $E_2P$ configuration, giving support for the view that the endocytosed Na+/K+-pumps are functional (16). Exogenous Na+/K+-pumps synthesized from injected cRNAs coding for the *Torpedo* α1 subunit and the mouse ß1 subunit are also internalized during maturation. Immunoprecipitation of these exogenous $Na^+/K^+$-pumps under native conditions by a monoclonal antibody to the ß1 subunit, followed by SDS-polyacrylamide gel electrophoresis and fluorography showed that the internalized $Na^+/K^+$-pumps are not subjected to degradation, but rather maintained in a structurally unimpaired form during the lifetime of the meiotically mature oocytes (G. Schmalzing and S. Kröner, unpublished results). Taken collectively, it is apparent from these results that oocytes contain a large intracellular stockpile of functional $Na^+/K^+$-pumps when they are ready for fertilization after completion of meiotic maturation.

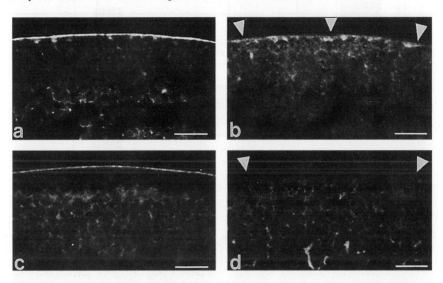

**Figure 4** Immunocytochemical evidence for a translocation of $Na^+/K^+$-pumps from the surface to the cell interior during meiotic maturation (13). Paraffin sections taken along the animal-vegetal axis were incubated with a polyclonal antibody to the $a1$ subunit of the *Xenopus* $Na^+/K^+$-pump, followed by biotinylated goat anti-rabbit F(ab)", and rhodamine-conjugated streptavidin. *Left panels*, immature oocyte; *right panels*, meiotically mature oocyte. *a,b*, vegetal pole; *c,d*, animal pole. *Bars* 25 $\mu$m. *Arrows*, location of the plasma membrane.

Activation of protein kinase C by phorbol esters mimics part of the effects induced by progesterone (22,25). Phorbol 12-myristate 13-acetate, for instance, greatly stimulates the endocytotic activity of oocytes and causes a rapid disappearance of all microvilli, as shown in Fig. 5*b*. Parallel measurements of the membrane area from electron micrographs and the ouabain binding capacity show that each reduction of the surface area is associated with a proportional decrease in the number of surface $Na^+/K^+$-pump. However, in oocytes treated with phorbol esters the number of $Na^+/K^+$-pumps never declines below a certain threshold, which is given by the minimum surface area of an oocyte that has lost all its microvilli and resembles a smooth ball (Fig. 5*b*). We attribute this incomplete removal of surface $Na^+/K^+$-pumps to a nonselective type of endocytosis. In contrast, during meiotic maturation all the $Na^+/K^+$-pumps disappear from the plasma membrane indicating that $Na^+/K^+$-pumps are removed by an additional process, designated selective

151

endocytosis in Fig. 5c. It is possible that the complete removal of Na⁺/K⁺-pumps from the plasma membrane during meiotic maturation results from a combination of both selective and nonselective endocytosis, as indicated in Fig. 5c.

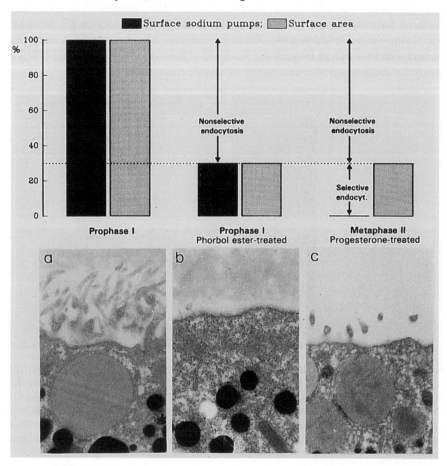

**Figure 5** Na⁺/K⁺-pumps are internalized by selective and nonselective endocytosis. Immature oocytes (*a, prophase I* ), oocytes treated for 30 min with 50 nM phorbol myristate acetate (*b*), and oocytes matured *in vitro* by progesterone (*c, metaphase II* ) were fixed with glutaraldehyde and processed for electron microscopy as described (25). The length of the plasma membrane was measured on a series of electron micrographs to assess the surface area (*upper panels*). In addition, the cells were assayed for surface binding of [³H]ouabain. The number of the surface Na⁺/K⁺-pumps (*filled columns*) and the surface area (*hatched columns*) under the various conditions are given in percent of those of nontreated, immature control oocytes.

## Na⁺/K⁺-pumps reenter the plasma membrane upon elevation of cytosolic free Ca²⁺

Fertilization is known to be associated with a wave of exocytosis named cortical reaction that by discharge of the contents of the cortical granules gives rise to the fertilization envelope. The cortical reaction can be induced parthenogenetically, i.e. without sperm, by calcium ionophores (24) or phorbol esters. To examine whether the intracellular $Na^+/K^+$-pumps can in principle be triggered to reenter the plasma membrane, we induced meiotically mature oocytes to accomplish the cortical reaction by exposing them to A23187, a calcium ionophore. Indeed, as illustrated in Fig. 6, a certain fraction of the intracellular $Na^+/K^+$-pumps was reinserted into the plasma membrane during the treatment with the calcium ionophore.

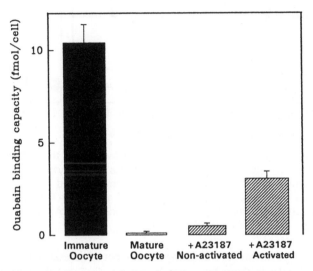

**Figure 6** $Na^+/K^+$-pumps reappear in the plasma membrane upon parthenogenetic activation of oocytes (16). Immature oocytes, meiotically mature oocytes, and mature oocytes incubated at 10 $\mu$M A23187 for 20 min were assayed for surface binding of [$^3$H]ouabain. Activation was scored by the contraction of the pigment at the animal pole. Mature oocytes exposed to A23187 that did not accomplish the cortical reaction were also analysed. All data points represent average values ($\pm$ SEM) from 9-11 oocytes.

The question whether *Xenopus* oocytes possess a regulated pathway for the recruitment and plasma membrane insertion of intracellular $Na^+/K^+$-pumps was addressed further by using permeabilized oocytes to manipulate the internal milieu. Digitonin-permeabilized secretory cells are known to maintain their secretory function and release granule contents such as neurotransmitters, hormones and enzymes when challenged with $Ca^{2+}$. As shown in Fig. 7a, micromolar free $Ca^{2+}$ concentrations induced a marked increase in the number of [$^3$H]ouabain binding sites (19). The half-maximal increase was observed at 0.4 $\mu$M $Ca^{2+}$, a concentration known to induce exocytosis in a variety of permeabilized secretory cells. In addition to $Ca^{2+}$, MgATP is required for compound exocytosis from most permeabilized cell systems. As apparent from Fig. 7b, $Ca^{2+}$ completely failed to induce an increase in the number of [$^3$H]ouabain binding site in the presence of $Mg^{2+}$ when ATP was absent. Likewise, the $Ca^{2+}$ effect was abolished in the presence of ATP when $Mg^{2+}$ was omitted (Fig. 7c).

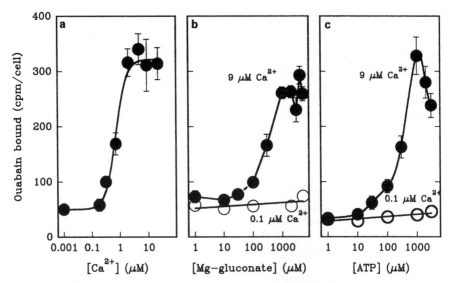

**Figure 7** Ca$^{2+}$- and MgATP-dependent pathway for the plasma membrane insertion of Na$^+$/K$^+$-pumps (16). Meiotically mature oocytes were permeabilized at 10 $\mu$M digitonin and then challenged with the indicated concentrations of free Ca$^{2+}$ (adjusted by CA/EGTA-buffers,) and *a*, 1 mM MgATP; *b*, 1 mM ATP and Mg$^{2+}$ as indicated; *c*, 1 mM Mg$^{2+}$ and ATP as indicated. All data points represent average values ($\pm$ SEM) from 9-11 oocytes.

Altogether, the requirements for increasing the number of accessible ouabain binding sites are remarkably similar to those for secretory exocytosis. Since the intracellular Na$^+$/K$^+$-pumps are not directly accessible for ouabain from the cytosol (20), it is intriguing to assume that the Ca$^{2+}$- and MgATP-stimulated increase in the number of accessible ouabain bindings sites reflects fusion of exocytotic vesicles containing Na$^+$/K$^+$-pumps with the plasma membrane. The possible orientation of Na$^+$/K$^+$-pumps in internal and surface membranes is shown schematically in Fig. 8. After fusion with the cell surface, the luminal side of the membranes faces out, such that the Na$^+$/K$^+$-pumps assume their plasma membrane orientation, in which the ouabain binding site is on the external side.

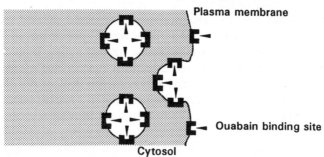

**Figure 8** Orientation of Na$^+$/K$^+$-pumps in internal and surface membranes. The ouabain binding domains of intracellular Na$^+$/K$^+$-pumps face to the luminal side of intracellular membranes. Fusion of the exocytotic vesicles with the plasma membrane causes the Na$^+$/K$^+$-pumps to assume their plasma membrane orientation.

154

## Recruitment of maternal Na$^+$/K$^+$-pumps after fertilization

Oscillations in intracellular free Ca$^{2+}$ corresponding in frequency to cell division have been measured in the cleaving *Xenopus* embryo (5). If we assume that the main function of the Ca$^{2+}$-regulated pathway (cf. Fig. 7) is the delivery of Na$^+$/K$^+$-pumps to the growing surface of the early *Xenopus* embryo, it might be speculated that each Ca$^{2+}$ transient is associated with the recruitment of a certain fraction of the intracellular Na$^+$/K$^+$-pumps, depending on the duration and size of the Ca$^{2+}$ rise (19). It is possible that the Na$^+$/K$^+$-pumps are inserted simultaneously with other maternal membrane proteins, for instance K$^+$ channels, from entities that serve as a general source of constituents needed for the assembly of plasma membranes (10).

To trace the fate of the maternal Na$^+$/K$^+$-pumps after fertilization, we took advantage of the sidedness of [$^3$H]ouabain binding. Already at the 2-cell stage, *Xenopus* embryos possess tight junctions at the outermost edge of the intercellular spaces, which act as seals between the external medium and the intercellular fluid. Since ouabain cannot pass through tight junctions or membranes, the ouabain binding capacity of intact embryos is a measure of the number of Na$^+$/K$^+$-pumps of the membrane domain that forms the outer surface of the embryo. We find that intact embryos up to the blastula stage completely fail to bind ouabain (Ref. 20; P. Kuhl-Mrozek and G. Schmalzing, unpublished results), indicating that Na$^+$/K$^+$-pumps once removed from the plasma membrane during maturation do not re-enter the outer cell surface of the embryo.

Since the adhesion of the embryonic cells is Ca$^{2+}$-dependent, they dissociate from each other when the embryo is reared in a Ca$^{2+}$- and Mg$^{2+}$-free medium. The dissociated cells are confined to approximately normal position by the vitelline layer, but ouabain has now access to all the intercellular spaces of the embryo. Hence, the ouabain binding capacity of dissociated embryos is a measure for the number of Na$^+$/K$^+$-pumps of all plasma membranes of the embryo. We find that the ouabain binding capacity of dissociated embryos increases at each cell division (cf. Fig. 9; P. Kuhl-Mrozek and G. Schmalzing, unpublished results). Comparison of these data with those obtained with intact embryos indicates that Na$^+$/K$^+$-pumps appear only in the intercellular plasma membranes, which are formed by the cell divisions, but not in the plasma membrane domains constituting the outer surface of the embryo (cf. Fig. 10). The appearance of Na$^+$/K$^+$-pumps in the newly formed plasma membranes fits to the fact that these membranes are constructed entirely from internal sources. The preexisting surfaces of the mature oocyte, which is also called the primary oocyte membrane and forms the outer cell surface of the embryo, does not contribute membrane to the intercellular plasma membranes. A restriction of the Na$^+$/K$^+$-pumps to the newly formed plasma membranes could also be shown by immunostaining of paraffin sections through embryos of various stages by a polyclonal antibody to the *Xenopus* α1 subunit (Ref. 20; P. Kuhl-Mrozek, K. Geering, and G. Schmalzing, unpublished results).

To assess the total number of Na$^+$/K$^+$-pumps of *Xenopus* embryos, the [$^3$H]ouabain binding capacity was measured in permeabilized embryos to expose the ouabain binding sites of all Na$^+$/K$^+$-pumps including the intracellular ones (cf. Fig. 8). As shown in Fig. 9, the total number of Na$^+$/K$^+$-pumps present in the plasma membrane plus the intracellular membranes stays constant from the immature oocyte up to the morula stage (*dotted line* in Fig. 9). Since the morula stage is attained within 4h after fertilization and the half life of

155

Na⁺/K⁺-pumps is well above 2 days in oocytes, it is clear that most if not all the $Na^+/K^+$-pumps up to the morula stage are of maternal origin.

Beyond the morula stage, we observe a net increase in the number of $Na^+/K^+$-pumps, which depends on protein synthesis (20). Since the mRNA coding for the α1 subunit is not translated to a significant extent in early embryos (6), it has been postulated that maternal mRNA coding for a $Na^+/K^+$-pump ß subunit becomes translationally competent soon after fertilization (6). Newly synthesized ß subunits could then combine with preformed but unassembled α1 subunits to generate functional α1/ß $Na^+/K^+$-pump complexes, similar to exogenous ß subunits when introduced into ooctyes by cRNA injection (3,18,21). At the blastula stage, nearly all $Na^+/K^+$-pumps reside in plasma membranes (Fig. 9).

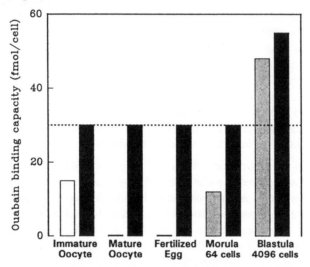

**Figure 9** Resorting of Na⁺/K⁺-pumps from the cell surface to the cell interior and back to intercellular plasma membranes. Oocytes and embryos were assayed for [³H]ouabain binding under the conditions detailed in the text to determine the number Na⁺/K⁺-pumps present on the cell surface (open columns, immature oocyte to fertilized egg), intercellular plasma membranes of embryos (hatched columns), and in all membranes including the intracellular ones (filled columns).

### Summary

The endo- exocytotic resorting of $Na^+/K^+$-pumps that takes place during early development is summarized schematically in Fig. 10. During oogenesis, *Xenopus* oocytes accumulate a large stockpile of $Na^+/K^+$-pumps, which reside in the plasma membrane and in intracellular membranes. During meiotic maturation that normally occurs a few hours before fertilization, all surface $Na^+/K^+$-pumps are internalized. Beginning with the first cleavage, the internalized $Na^+/K^+$-pumps are reinserted into the newly formed intercellular plasma membranes, but do not appear in the membrane domain that forms the outer surface of the embryo (cf. Figs. 9 and 10). Therefore, the final consequence of this endo- exocytotic cycle is a polarized distribution of $Na^+/K^+$-pumps, which resembles the apical-basolateral polarity of epithelia if the outer plasma membrane of the embryo is considered the apical domain and the intercellular plasma membrane the basolateral domain. This asymmetric $Na^+/K^+$-pump distribution provides a morphological basis for a vectorial transport of $Na^+$ that leads to the formation of the blastocoel. Immunostaining of embryos showed further that $Na^+/K^+$-pumps are segregated preferentially to the cells of the animal hemisphere of early embryos lining the blastocoel cavity (2). This additional polarized distribution of $Na^+/K^+$-pumps along the animal-vegetal axis of the embryo explains the fact that the blastocoel is formed in the animal hemisphere (2).

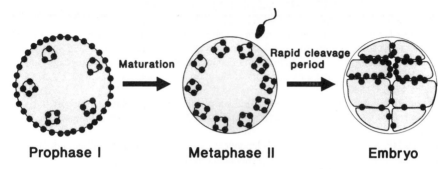

**Prophase I**          **Metaphase II**          **Embryo**

**Figure 10** Polarized distribution of $Na^+/K^+$-pumps as a consequence of endo- and exocytosis. Immature ( *prophase I* ) oocytes contain $Na^+/K^+$-pumps both in intracellular membranes and in the plasma membrane. Since the surface $Na^+/K^+$-pumps are internalized during meiotic maturation, mature oocytes ( *metaphase II* ) contain a large stockpile of intracellular $Na^+/K^+$-pumps. After fertilization, the stored $Na^+/K^+$-pumps are reinserted into the newly formed intercellular plasma membranes, giving rise to a apical-basolateral type of cell polarity.

*Acknowledgement*

I thank Dr. Masaru (University of Occupational and Environmental Health, Kitakyushu, Japan) for the gift of the cDNA encoding the α1 subunit of the $Na^+/K^+$-pump of *Torpedo californica*. The work in the laboratory of the author was supported by the Deutsche Forschungsgemeinschaft (Sonderforschungsbereich 169, project A8).

**References**

1. Dumont JN (1972) Oogenesis in *Xenopus laevis* (Daudin). I. Stages of oocyte development in laboratory maintained animals. J Morphol 136: 153-180
2. Friehs S and Schmalzing G (1993) Segregation of maternal $Na^+/K^+$-pump α1 and ß3 mRNAs to animal blastomeres of early *Xenopus* embryos, this volume
3. Geering K, Theulaz I, Verrey F, Häuptle MT, Rossier BC (1989) A role for the ß-subunit in the expression of functional $Na^+$-$K^+$-ATPase in *Xenopus* oocytes. Am J Physiol 257: C851-C858
4. Good PJ, Richter K, Dawid IB (1990) A nervous system-specific isotype of the ß subunit of $Na^+$,$K^+$-ATPase expressed during early development of *Xenopus laevis*. Proc Natl Acad Sci USA 87: 9088-9092
5. Grandin N, Charbonneau M (1991) Intracellular free calcium oscillates during cell division of *Xenopus* embryos. J Cell Biol 112: 711-718
6. Han Y, Pralong-Zamofing D, Ackermann U, Geering K (1991) Modulation of Na,K-ATPase expression during early development of *Xenopus laevis*. Dev Biol 145: 174-181
7. Jaunin P, Horisberger JD, Richter K, Good PJ, Rossier BC, Geering K (1992) Processing,

157

intracellular transport, and functional expression of endogenous and exogenous α-ß$_3$ Na,K-ATPase complexes in *Xenopus* oocytes. J Biol Chem 267: 577-585

8. Kado RT, Marcher K, Ozon R (1981) Electrical membrane properties of the *Xenopus laevis* oocyte during progesterone-induced meiotic maturation. Dev Biol 84: 471-476

9. Maller JL, Krebs EG (1980) Regulation of oocyte maturation. Curr Top Cell Regul 16: 271-311

10. Müller AHJ, Gawantka V, Ding XY, Hausen P (1993) Maturation-induced internalization of ß$_1$-integrin by *Xenopus* oocytes and formation of the maternal integrin pool. Mechan Develop 42: 77-88

11. Nelson WJ, Veshnock PJ (1987) Ankyrin binding to (Na$^+$+K$^+$) ATPase and implications for the organization of membrane domains in polarized cells. Nature 328: 533-535

12. Newport J, Kirschner M (1982) A major developmental transition in early *Xenopus* embryos. I. Characterization and timing of cellular changes at the midblastula stage. Cell 30: 675-686

13. Pralong-Zamofing D, Yi QH, Schmalzing G, Good P, Geering K (1992) Regulation of α1-ß3 Na$^+$/K$^+$-ATPase isozyme during meiotic maturation of *Xenopus laevis* oocytes. Am J Physiol 262: C1520-C1530

14. Richter HP, Jung V, Passow H (1984) Regulatory changes of membrane transport and ouabain binding during progesterone-induced maturation of *Xenopus* oocytes. J Membr Biol 79: 203-210

15. Richter HP, Schwarz W (1991) Na$^+$-dependent carrier transport in amphibian oocytes and its regulation during development. Comp Biochem Physiol 10: 86-103

16. Schmalzing G, Eckard P, Kröner S, Passow H (1990) Downregulation of surface sodium pumps by endocytosis during meiotic maturation of *Xenopus laevis* oocytes. Am J Physiol 258: C 179-184

17. Schmalzing G, Gloor S (1994) Na$^+$/K$^+$-pump beta-subunits: structure and functions. Cell Physiol Biochem 4: 96-114

18. Schmalzing G, Gloor S, Omay H, Kröner S, Appelhans H, Schwarz W (1991) Up-regulation of sodium pump activity in *Xenopus laevis* oocytes by expression of heterologous ß1 subunits of the sodium pump. Biochem J 279: 329-336

19. Schmalzing G, Kröner S (1990) Micromolar free calcium exposes ouabain binding sites in digitonin-permeabilized *Xenopus laevis* oocytes. Biochem J 269: 757-766

20. Schmalzing G, Kröner S, Passow H (1989) Evidence for intracellular sodium pumps in permeabilized *Xenopus laevis* oocytes. Biochem J 260: 395-399

21. Schmalzing G, Kröner S, Schachner M, Gloor S (1992) The adhesion molecule on glia (AMOG/ß2) and α1 subunits assemble to functional sodium pumps in *Xenopus* oocytes. J Biol Chem 267: 20212-21216

22. Schmalzing G, Mädefessel K, Haase W and Geering K (1991) Evidence for protein kinase C-induced internalization of sodium pumps in *Xenopus laevis* oocytes. In: De Weer P Kaplan JH (eds) The Sodium Pump: Recent Developments. Rockefeller University Press, New York, pp 465-470

23. Slack C, Warner AE (1973) Intracellular and intercellular potentials in the early amphibian embryo. J Physiol Lond 232: 313-330

24. Steinhardt RA, Epel D, Carroll Jun EJ, Yanagimachi R (1974) Is calcium ionophore a universal activator for unfertilized eggs? Nature 252: 41-43

25. Vasilets LA, Schmalzing G, Mädefessel K, Haase W, Schwarz W (1990) Activation of protein kinase C by phorbol ester induces down-regulation of the Na$^+$/K$^+$-ATPase in oocytes of *Xenopus laevis*. J Membr Biol 118: 131-142

26. Verrey F, Kairouz P, Schaerer E, Fuentes P, Geering K, Rossier BC, Kraehenbuhl J-P (1989) Primary sequence of *Xenopus laevis* Na$^+$-K$^+$-ATPase and its localization in A6 kidney cells. Am J Physiol 256: F1034-F1043

27. Weinstein SP, Kostellow AB, Ziegler DH, Morrill GA (1982) Progesterone-induced downregulation of an electrogenic Na$^+$,K$^+$-ATPase during the first meiotic division in amphibian oocytes. J Membr Biol 69: 41-48

# Determinants of Cell Surface Distributions of $Na^+/K^+$-ATPase in Polarized Epithelial Cells

J. A. Marrs, W. J. Nelson

Department of Molecular and Cellular Physiology, Beckman Center for Molecular and Genetic Medicine, Stanford University School of Medicine, Stanford, CA 94305-5426, USA

## Introduction

While there has been emphasis in recent years on understanding the protein structure, catalytic function and assembly of $Na^+/K^+$-ATPase (23), relatively less attention has been given to the mechanisms involved in regulating restricted distributions of $Na^+/K^+$-ATPase on the plasma membrane of cells. The function of $Na^+/K^+$-ATPase is to maintain plasma membrane potential, cell volume and, in general, cellular homeostasis; this requires that functional protein is expressed at the cell surface. However, in transporting epithelia, localization of $Na^+/K^+$-ATPase to the cell surface alone is not sufficient for cellular function (1). These cells play critical roles in regulating the ionic composition of biological compartments that are separated by the epithelium. In the kidney, for example, the renal tubule comprises a closed monolayer of cells that regulate the vectorial transport (reabsorption) of ions and solutes from the lumen of the tubule to the blood supply. In contrast, in the choroid plexus the epithelium controls vectorial transport (secretion) from the blood supply to the ventricular space to form the cerebral-spinal fluid. Opposite directions of ion transport across transporting epithelium is determined by the distribution of ion channels, exchanges and transporters between domains of the plasma membrane that face different biological compartments; these domains are defined as apical (facing the lumen or closed compartment) and basal-lateral (facing the blood supply) [see Figure 1]. Defining the mechanisms involved in regulating the distributions of these proteins is central to our understanding of the function of this important cell type in normal and disease states.

## Regulation of $Na^+/K^+$-ATPase in a Reabsorptive Transporting Epithelium

In the proximal tubule, the epithelium regulates the reabsorption of $Na^+$ from the ultrafiltrate to the blood supply. Due to the presence of cation-selective tight junctions at the apical contact site, ions do not pass through the paracellular space between cells. Instead, ions must be transported across the epithelium. $Na^+$ enters the cell down its electrochemical gradient through ion channels and exchangers localized in the apical membrane. $Na^+$ is then pumped out of the cell by $Na^+/K^+$-ATPase which is restricted to the basal-lateral membrane (1).

159

To investigate mechanisms involved in restricting the distribution of $Na^+/K^+$-ATPase to the basal-lateral membrane, we have used an established cell line derived from kidney distal tubule epithelium termed Madin-Darby canine kidney (MDCK) (18). These cells have many of the structural and functional characteristics of the cell of origin, including the formation of tight junctions and a polarized distribution of proteins to structurally and functionally distinct apical and basal-lateral membrane domains. MDCK cells can be grown to confluency on permeable filter supports, thereby allowing biochemical access to both either the apical and basal-lateral membrane domains (18). These feature are important in quantitative studies to examine the delivery and steady state distribution of proteins in different membrane domains (Figure 2).

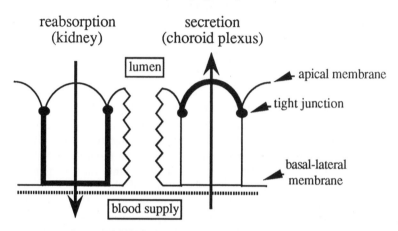

**Figure 1.** Transporting epithelia form a barrier between two biological compartments (lumen, blood supply) and regulate the ionic composition of those compartments by vectorial transport of ions and solutes across the epithelium. Depending on the distribution of ion transporters between the apical (lumen) and basal-lateral (blood supply) domains of the plasma membrane, the epithelium is reabsorptive (eg. kidney) or secretory (eg. choroid plexus).

To determine how cells develop a restricted cell surface distribution of $Na^+/K^+$-ATPase, we established confluent monolayers of MDCK cells on filters, induced cell-cell contact and analyzed the development of cell surface polarity Cell surface distribution of Na/K-ATPase was determined qualitatively by laser-scanning confocal microscopy, and quantitatively by cell surface biotinylation, immunoprecipitation of $Na^+/K^+$-ATPase subunits with specific antibodies from cell extracts, followed by detection of biotinylated Na/K-ATPase using [125I]-streptavidin after transfer of immunoprecipitated proteins from an SDS-gel to nitrocellulose (5). Our results showed that the generation of cell surface polarity of the $Na^+/K^+$-ATPase after induction of cell-cell contact is slow, requiring ~72hrs by which time >90% of the Na/K-ATPase is located on the basal-lateral membrane. During the development of polarity, we found that $Na^+/K^+$-ATPase is transiently expressed on the apical membrane, as also demonstrated in our preliminary studies of Na/K-ATPase distributions in developing renal tubules *in vivo* (see below) (5).

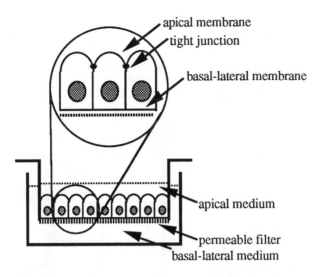

**Figure 2.** Distributions of apical and basal-lateral membrane domains in polarized epithelial cells growing on permeable filters in a plastic petri dish insert. The formation of tight junctions at the apex of the lateral membrane seals the confluent monolayer of cells. Thereby, the medium bathing the basal-lateral surface of the cells is separate from that bathing the apical cell surface. Biotinylated cross-linking reagents can be added from one side of the monolayer resulting in selective labelling of protein on only one membrane domain.

We investigated the delivery of newly-synthesized $Na^+/K^+$-ATPase to the cell surface during the development of cell surface polarity. Cells on pairs of filters were metabolically labelled, biotinylated on the apical or basal-lateral membrane, and $Na^+/K^+$-ATPase subunits were immunoprecipitated with specific antibodies from cell extracts. The immunoprecipitates were dissociated in SDS or low pH, and biotinylated proteins recovered subsequently by precipitation with avidin-agarose; proteins were detected by SDS-PAGE and fluorography. Results showed clearly that $Na^+/K^+$-ATPase subunits are not targeted specifically to the basal-lateral membrane at times up to 6-7 days after the generation of cell surface polarity at steady state; in parallel experiments, we showed that other marker proteins are vectorially delivered to the basal-lateral membrane, indicating that $Na^+/K^+$-ATPase result is not an artifact of the cells and techniques used (5).

How is steady state polarity of Na/K-ATPase achieved when newly-synthesized protein was being delivered to both the apical and basal-lateral membranes? In a pulse-chase experiment together with detection of protein on the membrane by biotinylated crosslinking as described above, we determined the half-life of newly-synthesized $Na^+/K^+$-ATPase on the apical and basal-lateral membranes, respectively. We found that $Na^+/K^+$-ATPase delivered to the basal-lateral membrane became metabolically stabilized ($t_{0.5}$ >36hrs), while $Na^+/K^+$-ATPase delivered to the apical membrane was rapidly removed ($t_{0.5}$ <2hrs). Thus, cell surface polarity of $Na^+/K^+$-ATPase is regulated by differences in the stability of the protein in the apical and basal-lateral membranes (Figure 3) (5).

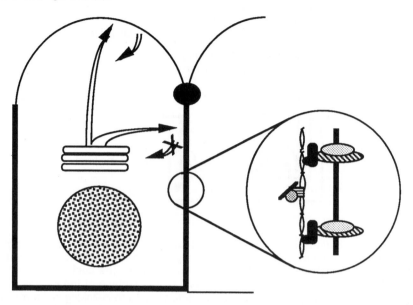

**Figure 3.** In MDCK cells, newly-synthesized $Na^+/K^+$-ATPase is delivered from the Golgi complex to both apical and basal-lateral membrane domains, although the absolute ratio varies between cell lines (see text). $Na^+/K^+$-ATPase delivered to the basal-lateral membrane is retained through binding to the membrane-cytoskeleton (inset) and accumulates (thick line). Based upon our understanding of the erythrocyte, the membrane-cytoskeleton comprises a fodrin lattice interlinked by ternary complexes of protein 4.1, adducin, and short actin oligomers. The lattice is linked through ankyrin, which binds to fodrin, to membrane proteins (eg. $Na^+/K^+$-ATPase). In contrast, $Na^+/K^+$-ATPase delivered to the apical membrane is rapidly internalized and degraded, resulting in the accumulation of little or no $Na^+/K^+$-ATPase (thin line).

Studies by another group (4) showed that a different clone of MDCK cells exhibited a greater degree of sorting and delivery of $Na^+/K^+$-ATPase directly to the basal-lateral membrane. We obtained a low passage number of this MDCK cell clone (strain II). We have compared the sorting of $Na^+/K^+$-ATPase in these two clones of MDCK cells. Results have revealed that Na/K-ATPase is preferentially delivered to the basal-lateral membrane of strain II cells, and approximately equally to both the apical and basal-lateral membranes of our cell clone (strain II/8). However, in strain II cells, ~10 - 20% of newly-synthesized $Na^+/K^+$-ATPase is also delivered to the apical membrane domain, although we do not detect $Na^+/K^+$-ATPase in the apical membrane domain at steady state indicating that the distribution of the protein is also refined by differential degradation of $Na^+/K^+$-ATPase delivered to the apical membrane, as described previously in the strain II/8 cells (see above). It is important to note that the delivery and steady state distribution of other apical and basal-lateral membrane proteins is similar in both clones of MDCK cells, indicating that there is not a general difference in the sorting pathways between these cells (Mays, van Meer, Nelson, in preparation).

162

At present, the differences in the regulation of $Na^+/K^+$-ATPase sorting prior to delivery to the plasma membrane is poorly understood. However, analysis of glycosphingolipid sorting and distribution in these different clones of MDCK cells indicates an inverse correlation between the sorting of glycosphingolipids to the apical membrane domains and delivery of $Na^+/K^+$-ATPase to the basal-lateral membrane. Preliminary studies indicate that glycosphingolipids are poorly sorted in the strain II/8 clone of MDCK cells compared to strain II cells; the apical : basal-lateral ratio of glycosphingolipids is ~1:1 and 2:1, respectively (Mays, van Meer, Nelson, in preparation). These results may have implications for the sorting of basal-lateral membrane proteins such as $Na^+/K^+$-ATPase. It has been proposed that glycosphingolipid sorting is important in the sorting of apical proteins in the TGN through the co-clustering of specific proteins in glycosphingolipid patches (12, 14, 18); this has been partially confirmed through the analysis of the sorting of proteins linked to the plasma membrane via a glycophosphoinositol (GPI-) moeity. In addition, it was proposed that these glycosphingolipid/protein clusters would exclude other proteins which would then be diverted into the basal-lateral membrane pathway. It is possible that in cells that sort glycosphingolipids, $Na^+/K^+$-ATPase is excluded from the apical pathway and, hence, is diverted into the basal-lateral pathway (strain II cells); in the absence of glycosphingolipid sorting (strain II/8 cells) $Na^+/K^+$-ATPase may enter both pathways (Mays, van Meer, Nelson, in preparation).

In both clones of MDCK cells the steady state distribution of $Na^+/K^+$-ATPase is exquisitely restricted to the basal-lateral membrane domain, although newly-synthesized protein is delivered also to the apical membrane. The differences in turnover of $Na^+/K^+$-ATPase in the apical and basal-lateral membrane domains may reflect association of these populations of $Na^+/K^+$-ATPase with the membrane-cytoskeleton (see Figure 3). Previous studies have established that components of the membrane-cytoskeleton, ankyrin and fodrin, co-localize in MDCK cells and renal epithelium with $Na^+/K^+$-ATPase (6, 11, 15). In addition, *in vitro* studies with purified membrane-bound $Na^+/K^+$-ATPase, ankyrin ad fodrin have shown that these proteins bind in a specific order and with high affinity (11, 16). In this hierarchy of protein interactions, $Na^+/K^+$-ATPase binds to ankyrin ($K_D$ ~$10^{-9}$M), and fodrin heterotetramers($[\alpha\beta]_2$) bind to ankyrin. In the erythrocyte, where interactions between another transmembrane ion exchanger ($Cl^-/HCO_3^-$-exchanger, Band 3) and ankyrin-spectrin (fodrin) has been well studied, spectrin molecules are interlinked by a complex of protein that include protein 4.1 and 4.9, adducin, actin oligomers, and tropomyosin (2). This ternary complex generates a 2-dimensional protein matrix that underlies the cytoplasmic surface of the plasma membrane. The formation of this next step in the hierarchical assembly of the membrane-cytoskeleton is poorly understood in epithelial cells, although assembly (see below) of the membrane-cytoskeleton coincides with increased insolubility in non-ionic detergents and decreased turnover of the proteins (15, 17), and co-localization of protein 4.1 and adducin with the membrane-cytoskeleton (2), both of which would be consistent with assembly of a macromolecular protein complex at the plasma membrane (see Figure 3). We will discuss in the last section how assembly of the membrane-cytoskeleton may be restricted to the basal-lateral membrane.

# Regulation of Na$^+$/K$^+$-ATPase in a Secretory Transporting Epithelium

The choroid plexus epithelium plays a central role in the secretion of cerebrospinal fluid by vectorial transport of ions and solutes from the blood supply to the brain ventricles. In the choroid plexus epithelium Na$^+$/K$^+$-ATPase is restricted to the apical membrane domain (3, 7-9, 19, 22, 24). We have shown by immunofluorescence microscopy that Na$^+$/K$^+$-ATPase colocalized on the apical plasma membrane of chick and rat choroid plexus with fodrin, and ankyrin in chick choroid plexus (7). Several observations indicate that in the choroid plexus fodrin, ankyrin and Na$^+$/K$^+$-ATPase may comprise a multi-protein complex (7). First, Na$^+$/K$^+$-ATPase is relatively resistant to extraction from the choroid plexus in buffers containing Triton X-100. The proportion of Na$^+$/K$^+$-ATPase that remained insoluble under these extraction conditions was similar to those of ankyrin and fodrin; these data are consistent with Na$^+$/K$^+$-ATPase being associated with the Triton X-100 insoluble cytoskeleton. Second, analysis of extracted protein complexes by sucrose density gradients revealed that Na$^+$/K$^+$-ATPase cosedimented with fodrin and ankyrin in a peak at ~10.5 S. Third, the cosedimenting proteins were further separated by electrophoresis in polyacrylamide gels under nondenaturing conditions. The electrophoretic mobility of these protein was found to be closely similar to that of the protein complex extracted from MDCK cells after the same series of separations in sucrose gradient and non-denaturing polyacrylamide gels. Analysis of protein distributions in the non-denaturing gel revealed that the slower migrating of the two protein bands reacted with antibodies specific for the α- and β-subunits of Na$^+$/K$^+$-ATPase, fodrin and ankyrin.

These results indicate strongly that the cosedimentation of these proteins in the sucrose gradient reflected the presence of a high molecular weight complex of fodrin, ankyrin and Na$^+$/K$^+$-ATPase. The faster migrating of the two protein bands reacted with ankyrin and fodrin antibodies, but not with antibodies specific for either the α- and β-subunit of Na$^+$/K$^+$-ATPase. Previous studies of membrane-cytoskeletal protein complexes isolated from MDCK cell extracts or reconstituted *in vitro* from purified

proteins (see above) have characterized these two protein bands in non-denaturing gels as containing fodrin tetramers bound to ankyrin (faster migrating complex), and fodrin tetramers and ankyrin bound to integral membrane proteins (slower migrating complex). By analogy, we suggest that the slower migrating protein complex isolated from choroid plexus contains fodrin tetramers and ankyrin complexed with Na$^+$/K$^+$-ATPase (7).

## Induction of Membrane-Cytoskeleton Assembly in Polarized Epithelial Cells

Together, these studies of cells with opposite vectorial transport functions and distributions of Na$^+$/K$^+$-ATPase show a common complex between Na$^+$/K$^+$-ATPase and proteins of the membrane-cytoskeleton. It is significant that the same subunits of Na$^+$/K$^+$-ATPase (α1, β1) are expressed in both choroid plexus and renal epithelium (7). This indicates that the generation of different Na$^+$/K$^+$-ATPase distributions is not simply due to unique apical or basal-lateral sorting signals within the protein, and implies that the sorting machinery must accommodate plasticity in the delivery of $Na^+/K^+$-ATPase to either membrane domain. In both cell types, assembly of the membrane-cytoskeleton may retain $Na^+/K^+$-ATPase in the membrane resulting in its accumulation. But, how is assembly of the membrane-cytoskeleton in these different membrane domains regulated?

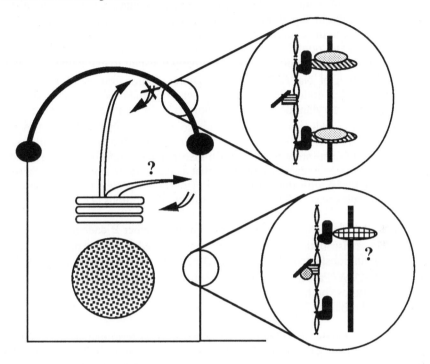

**Figure 4.** In choroid plexus cells, $Na^+/K^+$-ATPase is localized exclusively to the apical membrane. The membrane-cytoskeleton is predominantly localized at the apical membrane, but also at the lateral membrane; at the apical membrane, $Na^+/K^+$-ATPase is complexed with ankyrin and fodrin. At present, it is unclear whether newly-synthesized $Na^+/K^+$-ATPase is delivered directly to the apical membrane and retained, or a portion is also delivered to the basal-lateral membrane where it is internalized and degraded (see text).

In MDCK cells, assembly of the membrane-cytoskeleton and accumulation of $Na^+/K^+$-ATPase to the (basal-) lateral membrane domain is induced by E-cadherin-mediated cell-cell adhesion (10, 15, 17, 18). Direct evidence for this inductive role of E-cadherin was obtained by analyzing protein distributions in confluent monolayers of fibroblasts transfected with E-cadherin, or with E-cadherin that contained a deletion of the cytoplasmic domain that included the binding site for cytoskeletal proteins. Immunofluorescence showed that $Na^+/K^+$-ATPase was specifically localized to regions of cell-cell contact in the cells containing E-cadherin, but not in cells expressing the mutant E-cadherin that had a truncated cytoplasmic domain. Significantly, the distribution of fodrin paralleled that of $Na^+/K^+$-ATPase in both cell types. Since $Na^+/K^+$-ATPase and E-cadherin have been shown to be associated with the membrane-cytoskeleton (ankyrin and fodrin) in MDCK cells, we suggest that E-cadherin initiates assembly of the membrane-cytoskeleton at regions of cell-cell contact, and $Na^+/K^+$-ATPase becomes stabilized within this complex (10).

We did not detect E-cadherin expression in the choroid plexus (7). However, our analysis of the complement of cadherins, and a previous study, showed that a different member of the cadherin superfamily, termed B-cadherin (13, 21), is the predominant cadherin expressed in the choroid plexus epithelium (7,13). We localized cadherins to the lateral membrane domain, and showed that it colocalized with a subset of the membrane-cytoskeleton. Cadherins were relatively resistant to extraction by Triton X-100, similar to that of $Na^+/K^+$-ATPase. Direct demonstration of a membrane-cytoskeletal complex containing B-cadherin was obtained by sucrose gradient and nondenaturing polyacrylamide gel electrophoresis. In sucrose gradients, a ~10.5 S complex containing cadherins was detected, also similar to that of $Na^+/K^+$-ATPase. Cadherin reactivity in nondenaturing gels was complex. The high molecular weight complex that contained ankyrin and fodrin also reacted with antiserum against this cadherin. The additional reactivity suggests that a considerable amount of cadherin was either not complexed with these membrane-cytoskeletal proteins, or the membrane-cytoskeletal complex was labile under these extraction conditions (7).

We used the fibroblast-expression assay to examine whether B-cadherin also regulates $Na^+/K^+$-ATPase distribution (7). B-cadherin was detected at regions of cell-cell contact in the B-cadherin transfected fibroblasts. In contrast to the E-cadherin transfected cells, however, $Na^+/K^+$-ATPase and fodrin did not colocalize with B-cadherin, even in densely seeded cell cultures; both proteins were diffusely distributed similar to that seen in untransfected fibroblasts (7).

The steady state levels of cadherin expression were determined in the transfected fibroblasts to examine whether differences in E- and B-cadherin induction of $Na^+/K^+$-ATPase redistribution were due to lower levels of B-cadherin expression (7). Immunoblots using antiserum against the C-terminal cytoplasmic domain of E-cadherin showed that B-cadherin expression was 85% of the E-cadherin expression in the transfected fibroblast lines, and that no cadherin was detected in the untransfected fibroblasts. In addition, we detected similar levels of catenins coimmunoprecipitated with E- and B-cadherin from transfected fibroblasts. These results show that the different functional property of E- and B-cadherin to induce $Na^+/K^+$-ATPase redistribution is not due to differences in either levels of expression of cadherins, or affinity of the two cadherins for catenins, and therefore may represent an intrinsic difference in the properties of each cadherin.

That cadherins and $Na^+/K^+$-ATPase are each complexed with the membrane-cytoskeleton, but have different membrane domain distributions in the choroid plexus epithelium is strikingly different from the situation in absorptive epithelia where all these proteins colocalize. To interpret these results with regard to our current views on how epithelial cells generate polarized distributions of $Na^+/K^+$-ATPase (see above), we propose two possible mechanisms for generating apically localized $Na^+/K^+$-ATPase in the choroid plexus epithelium (7).

First, Na$^+$/K$^+$-ATPase may be exclusively delivered from the *trans* Golgi network to the apical plasma membrane domain. In this way, the cell would bypass the cadherin/membrane-cytoskeletal complex which might stabilize Na$^+$/K$^+$-ATPase at the lateral plasma membrane. Na$^+$/K$^+$-ATPase delivered to the apical plasma membrane may accumulate at this location due to assembly with the membrane-cytoskeleton. At present, we have no insight into the intracellular trafficking of proteins in the choroid plexus; in the absence of an established cell line or methods for primary culture of these cells, this is difficult to resolve. However, the fact that B-cadherin is localized to the (basal-) lateral membrane in all cell types where it is expressed, including absorptive epithelia (kidney and intestine) and secretory epithelia (choroid plexus) (13), indicates that polarized distributions of proteins are not simply reversed in the choroid plexus relative to absorptive epithelia. Rather, the change in polarity seems to be restricted to a subset of proteins that includes Na$^+$/K$^+$-ATPase and perhaps other ion transporters, and components of the membrane-cytoskeleton.

Second, Na$^+$/K$^+$-ATPase may be delivered to both cell surface domains, but due to differences between the cadherin-associated membrane-cytoskeletal proteins of absorptive and secretory epithelial cells, Na$^+$/K$^+$-ATPase is not stabilized in the lateral membrane and is internalized. This is supported by the fact that in fibroblasts expressing B-cadherin, neither Na$^+$/K$^+$-ATPase or fodrin are induced to accumulate at sites of cell-cell contact (7); in the presence of E-cadherin both of these proteins accumulate at contact sites in these cells, a process which is dependent on the presence of the complete E-cadherin cytoplasmic domain (10). These results indicate that, despite having 88% amino acid identity in their cytoplasmic domain, B- and E-cadherin differ in their respective abilities to nucleate membrane-cytoskeleton assembly and to induce reorganization of Na$^+$/K$^+$-ATPase. Considering that the sequence differences between B- and E-cadherin are distributed throughout the cytoplasmic domain, it is remarkable that these two cadherins have different abilities to produce Na$^+$/K$^+$-ATPase polarity. Further studies are underway to identify specific domains of these cadherins and the protein-protein interactions required to generate these different functions.

## Synopsis

Our studies of Na$^+$/K$^+$-ATPase trafficking and localization in different polarized epithelia cells have begun to uncover mechanisms involved in generating specific distributions of the protein in reabsorptive and secretory epithelia. Significantly, we have found that a common factor is a complex between Na$^+$/K$^+$-ATPase and components of the membrane-cytoskeleton. We propose that one function of this complex is to specifically retain Na$^+$/K$^+$-ATPase in the membrane; in the absence of membrane-cytoskeletal attachment, Na$^+$/K$^+$-ATPase appears to be rapidly removed from the membrane. Na$^+$/K$^+$-ATPase and membrane-cytoskeleton assembly at specific sites on the membrane appear to be specifically induced by E-cadherin mediated cell-cell contacts; a closely related cadherin, B-cadherin, does not appear to share this activity with E-cadherin. Na$^+$/K$^+$-ATPase may also be delivered directly from the Golgi complex to specific domains of the plasma membrane, where it interacts with the membrane-cytoskeleton. It is not clear whether Na$^+$/K$^+$-ATPase contains an endogenous sorting signal(s), or whether sorting in the Golgi complex is determined by exclusion of the protein from domains enriched in glycosphingolipids. Further studies of the cell biology of Na$^+$/K$^+$-ATPase are likely to complement the ongoing analyses of the expression and function of this important protein.

### Acknowledgements

This work was supported by grants from the NIH (GM35527) and American Cancer Society to W. J. Nelson. J. A. Marrs is the recipient of a postdoctoral fellowship from the NIH, and W. J. Nelson is an Established Investigator of the American Heart Association.

### References

1.  Almers W, Stirling C (1984) Distribution of transport proteins over animal cell membranes. J. Membr. Biol. 77:169-186
2.  Bennett V (1990) Spectrin-based membrane skeleton: A multipotential adaptor between plasma membrane and cytoplasm. Physiological Reviews 70:1029-1065
3.  Ernst SA, Palacios II JR, Siegel GJ (1986) Immunocytochemical localization of Na$^+$/K$^+$-ATPase catalytic polypeptide in mouse choroid plexus. J. Histochem. Cytochem. 34:189-195
4.  Gottardi CJ, Caplan MJ (1993) Delivery of Na$^+$/K$^+$-ATPase in polarized epithelial cells. Science . 260:552-554
5.  Hammerton RW, Krzeminski KA, Mays RW, Ryan TA, Wollner DA, Nelson WJ (1991) Mechanism for regulating cell surface distribution of Na$^+$/K$^+$-ATPase in polarized epithelial cells. Science 254:847-850
6.  Koob R, Zimmerman M, Schoner W, Drenkhahn D (1987) Co-localization and co-precipitation of ankyrin and Na$^+$/K$^+$-ATPase in kidney epithelial cells. Eur. J. Cell Biol. 45:230-237
7.  Marrs JA, Napolitano EW, Murphy-Erdosh C, Mays RW, Reichardt L, Nelson WJ (1993) Distinguishing roles of the membrane-cytoskeleton and cadherin mediated cell-cell adhesion in generating different Na$^+$/K$^+$-ATPase distributions in polarized epithelia. J. Cell Biol. 123:149-164.
8.  Masuzawa T, Ohta T, Kawakami K, Sato F (1985) Immunocytochemical localization of Na$^+$/K$^+$-ATPase in the canine choroid plexus. Brain 108:625-646
9.  Masuzawa T, Ohta T, Kawamura M, Nakahara N, Sato, F (1984) Immunohistochemical localization of Na$^+$/K$^+$-ATPase in the choroid plexus. Brain Res. 302:357-362
10. McNeill H, Ozawa M, Kemler R, Nelson WJ (1990) Novel function of the cell adhesion molecule uvomorulin as an inducer of cell surface polarity. Cell 62:309-316
11. Morrow JS, Cianci CD, Ardito T, Mann AS, Kashgarian M (1989) Ankyrin links fodrin to the alpha subunit of Na$^+$/K$^+$-ATPase in Madin-Darby canine kidney cells and in intact renal tubule cells. J. Cell Biol. 108:455-465
12. Mostov K, Apodaca G, Aroeti B, Okamoto C (1992) Plasma membrane protein sorting in polarized epithelial cells. J. Cell Biol. 116:577-583
13. Napolitano EW, Venstrom K, Wheeler EF, Reichardt LF (1991) Molecular cloning and characterization of B-cadherin, a novel chicken cadherin. J Cell Biol 113:893-905
14. Nelson WJ (1992) Regulation of cell surface polarity from bacteria to mammals. Science 258:948-955
15. Nelson WJ, Veshnock PJ (1986) Dynamics of membrane-skeleton (fodrin) organization during development of polarity in Madin-Darby canine kidney epithelial cells. J. Cell Biol. 103:1751-1765

16. Nelson WJ, Veshnock PJ (1987) Ankyrin binding to $Na^+/K^+$-ATPase and implications for the organization of membrane domains in polarized cells. Nature 328:533-536

17. Nelson WJ, Veshnock PJ (1987) Modulation of fodrin (membrane skeleton) stability by cell-cell contact in Madin-Darby canine kidney epithelial cells. J. Cell Biol. 104:1527-1537

18. Rodriguez-Boulan E, Powell SK (1992) Polarity of epithelial and neuronal cells. Ann. Rev. Cell Biol. 8:395-427

19. Seigel GJ, Holm C, Schreiber JH, Desmond T, Ernst SA (1984) Purification of mouse brain $Na^+/K^+$-ATPase catalytic unit, characterization of antiserum, and immunocytochemical localization in cerebellum, choroid plexus, and kidney. J. Histochem. Cytochem. 32:1309-1318

20. Siemers K, Wilson R, Mays RW, Ryan TA, Wollner DA, Nelson WJ (1993) Delivery of $Na^+/K^+$-ATPase in polarized epithelial cells. Science 260:554-556

21. Sorkin BC, Gallin WJ, Edelman GM, Cunningham BA (1991) Genes for two calcium-dependent cell adhesion molecules have similar structures and are arranged in tandem in the chicken genome. Proc. Natl. Acad. Sci. USA 88:11545-11549

22. Spector R, Johanson CE (1989) The mammalian choroid plexus. Sci. Amer. 131:68-74.

23. Sweadner KJ (1989) Isoenzymes of the $Na^+/K^+$-ATPase. Biochem. Biophys. Acta 988:185-220

24. Wright EM (1972) Mechanisms of ion transport across the choroid plexus. J. Physiol. 226:545-571

# Physiologic Relevance of the α2 Isoform of Na,K-ATPase in Muscle and Heart

A.A. McDonough, K.K. Azuma, C.B. Hensley, and C. E. Magyar

Department of Physiology and Biophysics, University of Southern California School of Medicine, Los Angeles, CA 90033 USA

## Introduction

The existence of isoforms of Na,K-ATPase suggests the potential for isoform specific function, distribution or regulation. Under the distribution category is the potential for tissue specific expression as well as subcellular compartment specific isoform expression. Under the regulation category is the potential for isoform specific acute regulation of existing pumps as well as chronic regulation of the pool size of isoforms. The picture is further complicated by the fact that there are isoforms of not only the α catalytic subunit but also the ß glycoprotein subunit that make up the heterodimeric sodium pump. Skeletal muscles express the fairly ubiquitous α1 and ß1 subunits as well as the more restricted α2 and ß2 subunits. Cardiac muscles express these subunits (ß2 only reported at mRNA to our knowledge) as well as α3 (24). Thus, there is the potential for four and six different heterodimers to be expressed in skeletal and cardiac muscle, respectively, and the potential for each having a unique function, pattern of subcellular distribution and/or regulation, as summarized in the compartmental model in Figure 1.

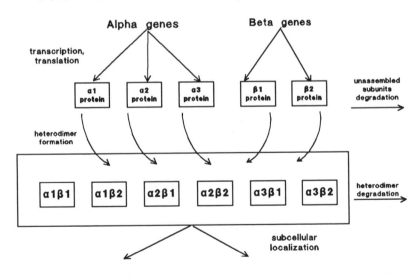

Figure 1. Determinants of sodium pump isoform expression in a tissue.

This review focuses on distribution and regulation of the Na,K-ATPase α2 isoform. At the level of function, studies by Jewell and Lingrel (19) suggest that α1 and α2 have very similar enzymatic properties when expressed in cultured HeLa cells. We summarize our findings on isoform specific distribution and regulation in skeletal and cardiac muscle, as well as our findings on which isoform of ß is regulated with α2. In addition, we speculate about the physiologic consequences of regulation of α2 vs α1 in heart and muscle.

### Regulation of skeletal muscle Na,K-ATPase α and ß isoform expression

Skeletal muscle is the largest organ in the body and possesses significant Na,K-ATPase activity. In the rat, expression of the α2 catalytic isoform mRNA is four fold higher than α1 mRNA (23). While it is difficult to appraise the relative levels of the two isoforms at the protein level in rat because of the ouabain resistance of the α1 isoform, immunologic and ouabain binding studies suggest that α2 isoform may predominate over α1 (26). Our own work comparing α1 and α2 expression at the protein level in skeletal muscle relative to that in brain by immunoblot concluded that the two isoforms were expressed at a similar ratio to that in found in brain, albeit both at 40 fold lower concentrations (2). Clausen and colleagues have conducted extensive studies in skeletal muscle on regulation of sodium pump activity and ouabain binding (5,6). They have established that there is long term regulation of both activity and ouabain binding by $K^+$ deficiency, thyroid hormone status, exercise training, and development. Relevant to this review, they determined that $K^+$ deficiency and hypothyroidism decreased both Na,K-ATPase activity and ouabain binding in skeletal muscle, and hyperthyroidism increased these properties of skeletal muscle. We tested the hypothesis that long term regulation of sodium pump concentration in skeletal muscle by hypokalemia and thyroid status was a function of isoform specific regulation of expression of α and ß.

**Hypokalemia:** To test the hypothesis that the decrease in skeletal muscle Na,K-ATPase activity and ouabain binding reflected an isoform specific decrease in expression, we induced hypokalemia in rats by 2 week $K^+$ deficient diet (1). In hypokalemic skeletal muscle α2 levels fell 82%, with no change in α1 or ß1 protein levels (Figure 2). More recently, we established that ß2 protein expression is decreased significantly and coordinately with the fall in α2 (Figure 3). Our results demonstrate that the expression of α2ß2, and not α1ß1, is specifically depressed in hypokalemia (signals controlling the changes are unidentified). Na,K-ATPase enzymatic activity decreased 40% in hypokalemic muscle indicating that α2ß2 contributes significantly to this activity. These results provide an elegantly simple molecular explanation for the selective loss of $K^+$ from skeletal muscle which serves to buffer the fall in plasma $K^+$. Intracellular $K^+$ in skeletal muscle presumably falls because $K^+$ uptake via the pump is depressed when α2 expression is depressed. This effect is greatest in skeletal muscle where α2 is the major isoform and far less in cardiac muscle where α2 makes up only 10-25% of the pumps. These finding provide a physiologic rationale for the existence of α2ß2 isoforms of Na,K-ATPase in potassium homeostasis (21).

171

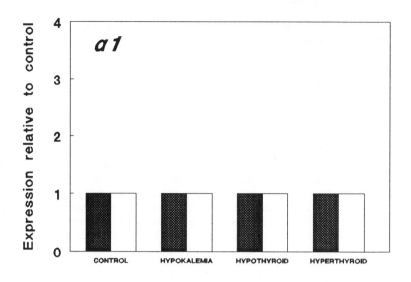

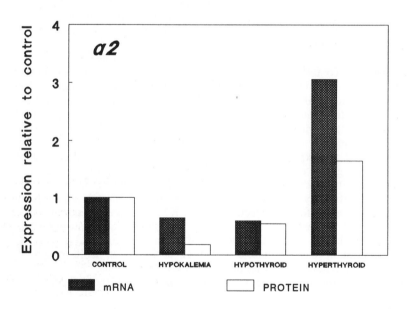

Figure 2.   Summary of the regulation of skeletal muscle Na,K-ATPase α subunit isoform expression at mRNA and protein levels in the rat.   Results of studies in hypokalemic, hypothyroid, and hyperthyroid rats, expressed relative to a constant amount of total RNA or homogenate protein, are normalized to the values found in control rats, defined as = 1.   Upper panel: alpha 1 expression.   Lower panel: alpha 2 expression.

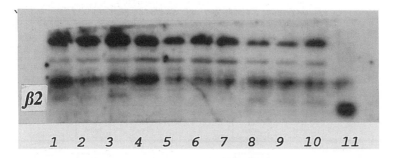

Figure 3. Regulation of ß2 expression in hypokalemic skeletal muscle. Immunoblot analysis of ß2 protein in crude membrane fractions from control and hypokalemic skeletal muscle. Lanes 1-3: 100 ug membrane protein from 3 different control rats. Lanes 4-7: 100 ug membrane protein from 4 different hypokalemic rats. Lanes 8-10: 50 ug membrane protein from the same samples resolved in lanes 1-3. Lane 11: brain homogenate. The characterization of ß2 in skeletal muscle has been previously described (2). While ß2 is barely detected in the hypokalemic samples, a comparison of these samples with the 100 $\mu$g and 50 $\mu$g control samples indicates that the decrease in ß2 in hypokalemia is more than 50%.

**Thyroid hormone:** In order to determine the pattern of thyroid hormone (T3) regulation of Na,K-ATPase isoforms expressed in rat skeletal muscle, three states were studied: euthyroid, hypothyroid, and hyperthyroid (hypothyroids injected with 1 $\mu$g T3/g body weight for up to 16 days) (2). As summarized for $\alpha$ isoforms in Figure 2, there was no change in $\alpha$1 or ß1 mRNA or protein levels in hypo- or hyperthyroid states. In contrast, $\alpha$2 was highly regulated: decreasing to half of euthyroid in the hypothyroids, and increasing 1.5 fold in the hyperthyroids. ß2 mRNA and protein were regulated coordinately in direction and magnitude with regulation of $\alpha$2 (2). These finding suggest that the T3 responsive enzyme in skeletal muscle is the $\alpha$2ß2 heterodimer.

Thus, we conclude that regulation of Na,K-ATPase expression in skeletal muscle is, indeed, isoform specific: regulation of expression of the $\alpha$2ß2 heterodimer, and not the $\alpha$1ß1 heterodimer, occurs in hypokalemic, hypothyroid, and hyperthyroid states, as illustrated in the compartmental diagram in Figure 4. The fact that $\alpha$2 mRNA and protein levels are both altered in these conditions suggests that the regulation is largely pre-translational.

In contrast to the pattern of long term regulation of $\alpha$2ß2 abundance in skeletal muscle illustrateed in Figure 4, Hundal et al (17) have reported that acute insulin treatment of skeletal muscle induces translocation of $\alpha$2 and ß1 subunits from different intracellular sources to the plasma membrane, suggesting that the insulin sensitive heterodimer is $\alpha$2ß1.

### Regional distribution of sodium pump isoforms in heart

The issue of regional distribution of sodium pump isoforms in the heart is an important one because the sodium pump is the receptor for cardiac glycosides and because of the differential sensitivity of isoforms to inhibition by cardiac glycosides.

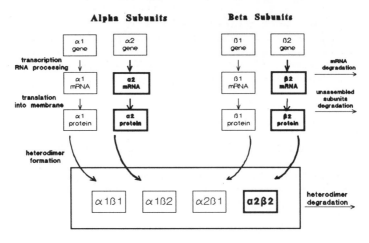

Figure 4. Isoform specific regulation of Na,K-ATPase subunits in skeletal muscle.

In the rat heart the fairly ubiquitous α1 isoform is considered a "low affinity isoform" with a Kd of $10^{-5}$M, while the Kds of the "high affinity" α2 and α3 isoforms are in the range of $10^{-7}$M (3, 26). In the human heart, two ouabain Kds have been measured (7): $2 \times 10^{-9}$M and $2 \times 10^{-8}$M. Since therapeutic levels of digoxin, between $1\text{-}2 \times 10^{-9}$M (20), would be predicted to occupy 50% of the higher affinity isoform and only 5% of the lower affinity isoform in the human heart, it has been postulated, but not demonstrated, that the therapeutic effects of digoxin are a function of occupancy of the higher affinity isoform, the isoform assignment of which has not been determined. Another suggestion that must be considered is that the effects of cardiac glycosides on cardiac contractility are mediated through a sympathoinhibitory effect of cardiac glycosides in heart failure patients, presumably mediated by high affinity isoforms of Na,K-ATPase but independent of their expression in cardiomyocytes (10).

It is only during the past two decades that a proposal for the cellular mechanism of action of cardiac glycosides emerged (20): cardiac glycosides bind specifically to sodium pumps located in myocardial plasma membranes inhibiting transport and enzymatic activity. An increase in intracellular $Na^+$ reduces the driving force for Na/Ca exchanger activity which leads to an increase in cellular stores of calcium and increased contractile force (4, 9, 12). The Na/Ca exchanger is expressed primarily in the T-tubular system of the myocyte, closest to the site of calcium release (11). We tested the hypothesis that there was spatial coupling of one of the sodium pump isoforms, which would control local intracellular $Na^+$ concentration in this same region as the Na/Ca exchanger. At the cellular level, we have examined the distribution of α1, α2 and ß1 isoforms in adult rat ventricular myocytes by immunofluorescence and find that α1, α2 and ß1 are all expressed in both the T-tubular system and the peripheral sarcolemma. Additionally, the pattern is not significantly altered in rats treated with thyroid hormone which elevates Na,K-ATPase expression (see below and Figure 5). This pattern contrasts to the restricted

174

distribution of the Na/Ca exchanger, thus, we do not have evidence for spatial coupling of one isoform vs the other to the Na/Ca exchanger.

In the rat heart Zahler *et al.*(27) applied *in situ* hybridization to determine the pattern of expression of $\alpha$ isoform mRNAs and found that the $\alpha2$ and $\alpha3$ levels were expressed primarily in the conduction system. We examined regional distribution of $\alpha$ and ß isoforms in rat heart by Western and Northern analysis. We found, in contrast to the findings of Zahler *et al*, that $\alpha2$ and $\alpha3$ mRNA and $\alpha2$ protein ($\alpha3$ protein not detected) were expressed at equivalent levels in septum (which is enriched in conduction system) and left ventricle. This result does agree with the findings of Zahler *et al* in human heart where they found equivalent amounts of $\alpha2$ mRNA expressed in septum and left ventricle (28).

### Regulation of cardiac muscle Na,K-ATPase $\alpha$ and ß isoform expression

In the rat cardiac ventricle $\alpha2$ type sodium pumps represent at most 25% of the total $\alpha$ expressed, and $\alpha1$ type pumps the remaining 75%, $\alpha3$ when detected, only represents a minimal percentage. In contrast to skeletal muscle, ß2 protein has not been detected in the heart (25), although its mRNA is present and is regulated with $\alpha2$ in some conditions (see below). We speculate that the ß2 protein is there but very difficult to detect with the available antibody probes. Since the enzymatic characteristics of $\alpha1$ and $\alpha2$ are nearly indistinguishable (19), the physiologic significance of the regulation of $\alpha2$ type pumps remains to be established. We speculate that significant decreases in $\alpha2$ expression will lead to small changes in intracellular $Na^+$, which will decrease Na/Ca exchanger activity significantly because of the very steep dependence of this transporter on intracellular sodium ($[Na^+]i$) (9). In support of this notion, Grupp *et al* (12) have shown that $\mu M$ concentrations of ouabain, which would inhibit $\alpha2$ but not $\alpha1$ in rat heart, produce concentration dependent increases in $[Na^+]i$ and twitch tension in the rat heart. Similar studies in C2C12 cell myotubes in which $\alpha2$ type pumps contribute to 16% of total Na,K-ATPase activity also demonstrate that ouabain inhibition of $\alpha2$, but not $\alpha1$ activity increases $[Na^+]i$ 15% (15). In summary, we postulate that a although a specific decrease in expression of $\alpha2$ type pumps in heart will cause only a small change in $[Na^+]i$ (since the change is limited by the low percentage of $\alpha2$ in the heart cell), the change will decrease Na/Ca exchanger activity because of the steep dependence of the exchanger on $[Na^+]i$.

**Hypokalemia:** Hypokalemia is a common problem in the heart failure patient. The renin-angiotensin-aldosterone system and the sympathetic nervous system are activated as cardiac performance declines, and both systems contribute to depletion of potassium (20). It has been established that aldosterone increases expression of cardiac $\alpha1$ (18). Along with cardiac glycosides to improve left ventricular function, the patient is often placed on a diuretic which exacerbates the hypokalemia, and a potassium supplement to counteract the hypokalemia. Nonetheless, control of plasma potassium continues to be a very serious problem because it predisposes the patient to not only ventricular arrhythmias, but also digitalis toxicity. Hypokalemia increases digoxin binding to myocardial cells, illustrating that the inotropic effects of cardiac glycosides depend to a great extent on the concentration of extracellular potassium (20). In addition, there is isoform specific regulation of Na,K-ATPase in hypokalemia. Specifically, we have shown that in rat models of hypokalemia $\alpha2$ isoform expression is depressed in heart by more than half (Figure 5) while $\alpha1$ and ß1 levels are unaltered (we have not yet detected ß2 protein in heart).

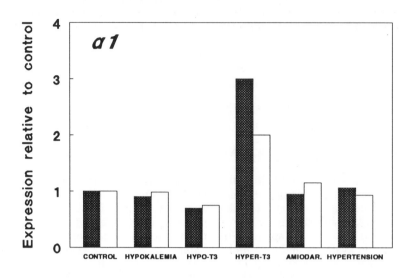

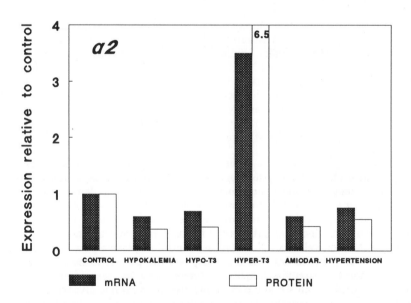

Figure 5. Summary of the regulation of cardiac muscle Na,K-ATPase α subunit isoform expression at mRNA and protein levels in the rat. Results of studies in hypokalemic, hypothyroid, hyperthyroid, amiodarone treated and hypertensive rats, expressed relative to a constant amount of total RNA or homogenate protein, are normalized to the values found in control rats, defined as = 1. Upper panel: alpha 1 expression. Lower panel: alpha 2 expression.

As illustrated in Figure 6, we suggest that digitalis toxicity associated with hypokalemia is due to a combination of effects including: 1) decreased plasma $K^+$ which increases affinity of digitalis for the pump, 2) decreased expression of skeletal muscle α2 which normally serves as a reservoir and buffer compartment for digitalis, and 3) a decrease in heart α2 expression which compounds the effects of digitalis inhibition.

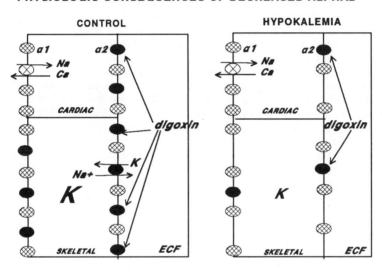

Figure 6. Physiologic consequences of decreased α2 expression associated with hypokalemia. The black ovals represent α2 type pumps and the hatched ovals represent α1. The boxes labelled cardiac and skeletal refer to the intracellular compartments of these tissues, ECF is the extracellular fluid. The α2/α1 ratio is much higher in skeletal than cardiac muscle in the control. Hypokalemia specifically depresses α2 type pumps in both tissues.

**Thyroid hormone:** Thyroid hormone regulates cardiac function at many levels including heart rate, velocity of contraction and metabolic rate. The increase in metabolic rate is ouabain inhibitable, thus, attributed to increased Na,K-ATPase activity. In order to assess the isoform specificity of the increase, we measured Na,K-ATPase α1, α2 and ß1 mRNA and protein expression in both whole ventricle (16) and cardiomyocytes (13) isolated from euthyroid, hypothyroid and hyperthyroid (hypothyroids injected daily with 1 μg T3/gm body weight) rats (13). As summarized for α subunits in Figure 5, α1, α2, and ß1 subunit protein levels are decreased in hypothyroid heart to 0.55, 0.42 and 0.57 of euthyroid, predicting the 0.53 fold decrease in Na,K-ATPase activity (13). In hyperthyroids, both mRNA and protein levels of all three subunits increased significantly; the change in the protein levels of

$\alpha1$ and $\alpha2$ and $\beta1$ were 2, 6.5, and 1.4 fold, respectively. Na,K-ATPase activity changed coordinately with $\beta1$ abundance suggesting that $\beta1$ was assembling with both $\alpha1$ and $\alpha2$, and may limit the upregulation response (as reviewed previously, 22). We conclude from these studies that regulation of both $\alpha1$ and $\alpha2$ type pumps contribute to the changes in ouabain sensitive metabolic rate associated with hypo- and hyperthyroidism. However, the contribution of $\alpha2$ becomes more far more significant in the transition from hypo- to hyperthyroidism since $\alpha2$ protein levels increase 12 fold and $\alpha1$ only 2 fold during the transition.

**Amiodarone:** Studies have suggested that chronic treatment with amiodarone, a class III antiarrhythmic may have effects on the myocardium that resemble those of hypothyroidism. Since hypothyroidism is associated with decreased expression of Na,K-ATPase $\alpha1$, $\alpha2$ and $\beta1$ subunits in heart and $\alpha2$ and $\beta2$ subunits in skeletal muscle, we aimed to test the hypothesis that chronic *in vivo* administration of amiodarone would result in similar changes (14). Rats were treated with 40ug/gm body weight amiodarone. After 6 wk of treatment, cardiac $\alpha1$ and $\beta1$ levels were at control levels and $\alpha2$ was depressed to 0.4 of control. At the mRNA level $\alpha2$ and $\beta2$ decreased to 0.6 fold of control (Figure 5). There was no effect of amiodarone on skeletal muscle Na,K-ATPase expression. We conclude that the effects of amiodarone on myocardial Na,K-ATPase expression are only similar to those of hypothyroidism in regards to the decrease in $\alpha2$ (and perhaps $\beta2$) expression. The lack of effect on $\alpha1$ and $\beta1$ expression in heart at 6 weeks, and lack of effect on skeletal muscle Na,K-ATPase expression, all seen in hypothyroidism, indicates that amiodarone's actions are isoform and tissue specific.

**Hypertension:** We have studied two models of hypertension: the hyperinsulinemic fructose fed rat model and the Goldblatt two kidney-1 clip model. In both models, Na,K-ATPase $\alpha1$ and $\beta1$ expression were unaffected while $\alpha2$ mRNA and protein levels decreased to about 0.5 of control (Figure 5). As discussed in the previous section, this specific decrease in $\alpha2$ may result in a slight rise in intracellular $Na^+$ which will decrease Na/Ca exchanger activity, increase cellular $Ca^{+2}$ and contractility. Whether this response contributes to hypertension or is an epiphenomenon associated with cardiac hypertrophy remains to be established.

In regards to the treatment of heart failure, hypothyroidism, hypokalemia, and amiodarone treatment are all associated with decreased $\alpha2$ expression as well as increased susceptibility to the toxic effects of digitalis (20). While this increased susceptibility has been traditionally attributed to increased affinity of the glycoside for the pump (in hypokalemia) or decreased renal clearance (amiodarone), we postulate that a component of this increased sensitivity to toxicity is due to decreased expression of the high affinity isoform, which is the biochemical equivalent of cardiac glycoside inhibition of this isoform. Whether these expression patterns are relevant to digitalis therapy vs. toxicity warrants further study.

*Acknowledgements*

These studies were supported by NIH grants DK34316, HL44404, and HL39295, and fellowship support to CBH and KKA from the NIH and the American Heart Association Greater Los Angeles Affiliate.

178

# References

1. Azuma, K.K., Hensley, C.B., Putnam, D.S., and McDonough, A.A. (1991) Hypokalemia decreases Na,K-ATPase $\alpha2$ but not $\alpha1$ isoform abundance in heart, muscle and brain. Am. J. Physiol.: Cell Physiol. 260: C958-C964
2. Azuma, K.K., Hensley, C.B., Tang, M.J., McDonough, A.A. (1993) Thyroid hormone specifically regulates skeletal muscle Na,K-ATPase $\alpha2$ and $\beta2$ isoforms. Am. J. Physiol. 265:, in press.
3. Berger, H.J., Werdan, K., and Erdmann E. (1988) Regulation of cardiac glycoside receptors with different affinities for cardiac glycosides in cultured rat heart cells. In: The $Na^+$, $K^+$-pump, Part B: Celluar Aspects. (Alan R.Liss, Inc., 1988), p. 345-352
4. Bers, D.M., Lederer, W.J., and Berlin, J.R. (1990) Intracellular Ca transients in rat cardiac myocytes: role of Na-Ca exchange in excitation-contraction coupling. Am J Physiol 258: C944-C954
5. Clausen, T. (1990) Significance of Na,K-ATPase pump regulation in skeletal muscle. NIPS 5:148-151
6. Clausen, T. and Everts, M.E. (1989) Regulation of the Na,K-pump in skeletal muscle. Kidney Int'l 35:1-13
7. De Pover, A, and Godfraind, T. (1979) Interaction of ouabain with $(Na^+ + K^+)$ATPase from human heart and from guinea-pig heart. Biochem Pharmac 28: 3051-3056
8. Desilets, M. and Baumgarten, C.M. (1986) Isoproterenol directly stimulates the Na-K pump in isolated cardiac myocytes. Am. J. Physiol. 251, H218-H225
9. Eisner, D.A. (1990) The Wellcome Prize Lecture: Intracellular Sodium in Cardiac Muscle--Effects on Contraction. Experimental Physiology 75, 437-457
10. Ferguson, D.W., Berg. W.J., Sanders, J.S., Roach P.J., Kempf, J.S., and Kienzle, M.G. (1989) Sympathoinhibitory Responses to digitalis glycosides in heart failure patients. Circulation 80: 65-77
11. Frank, J.S., Mottino, G., Reid, D., Molday, R.S. and K.D. Philipson. (1992) Distribution of the $Na^+$-$Ca^{+2}$ exchange protein in mammalian cardiac myocytes: An immunofluorescence and Immunocolloidal gold-labeling study. J. Cell Biol. 117: 337-345
12. Grupp, I., Im, WK, Lee, Co.O, Lee, SW, Pecker, M.S., and Schwartz, A. (1985) Relation of sodium pump inhibition to positive inotropy at low concentrations of ouabain in rat heart muscle. J Physiol 360: 149-160
13. Hensley, C.B., Tang, M.J., Azuma, K., K. and McDonough, A.A. (1992) Thyroid hormone induction of rat myocardial Na,K-ATPase: Relationship between $\alpha1$, $\alpha2$ and $\beta1$ mRNA and protein levels at steady state. Am. J. Physiol. 262: C484-492
14. Hensley, C.B., Bersohn, M.M., Sarma, J.S.M. Singh, B.N. and A.A. McDonough. Amiodarone decreases Na,K-ATPase $\alpha2$ and $\beta2$ expression specifically in cardiac ventricle. J. Cell Mol Cardiol. in press.
15. Higham, S.C., Melikian, J., Karin, N.J., Ismail-Beigi, and Pressley, T.A. Na,K-ATPase expression in C2C12 cells during myogenesis: minimal contribution of $\alpha2$ isoform to Na,K transport. J. Memb. Biol. 131:129-136
16. Horowitz, B. Hensley, C.B., Quintero, M., Azuma, K.K., Putnam, D. and McDonough, A.A. (1990) Differential regulation of Na,K-ATPase $\alpha1$, $\alpha2$ and $\beta$ subunit mRNA levels by thyroid hormone. J. Biol. Chem. 265:14308-14314
17. Hundal, H.S., Marette, A., Mitsumoto, Y., Ramlal, T., Blostein, R., and Klip, A. (1992) Insulin indices translocation of the $\alpha2$ and $\beta1$ subunits of the Na,K-ATPase from intracellular compartments to the plasma membrane in mammalian skeletal muscle J. Biol. Chem. 267:5040-5043
18. Ikeda, U., Hyman, R., Smith, T.W., and Medford R.M. (1991) Aldosterone-mediated regulation of Na, K-ATPase gene expression in adult and neonatal rat cardiocytes. J Biol Chem 266: 12058-12066
19. Jewell, E.A. and Lingrel, J.B. (1991) Comparison of the substrate dependence properties of the rat Na, K-ATPase a1, a2, and a3 isoforms expressed in HeLa cells. J. Biol. Chem. 266, 16925-16930
20. Marcus, F.I., L. H. Opie and E.H. Sonnenblick. (1991) Digitalis and Other Inotropes. In: Drugs and the Heart, W.B. Saunders, Philadelphia
21. McDonough, A.A., Azuma, K.K., Lescale-Matys, L., Tang, M.J., Nakhoul, F., Hensley, C.B. and Komatsu, Y. (1992) Physiologic Rationale for Multiple Sodium Pump Isoforms -

Differential Regulation of α1 vs α2 by Ionic Stimuli. In Ion Motive ATPases. Ann. New York Acad. Sci: 671:156-169

22. McDonough, A.A., Geering, K., and Farley, R.A. (1990) The sodium pump needs its b subunit. FASEB J. 4, 1598-1605

23. Orlowski, J. and Lingrel, J.B. (1988) Tissue-specific and developmental regulation of rat Na,K-ATPase catalytic a isoform and b subunit mRNAs. J. Biol. Chem. 263, 10436-442

24. Shamraj, O.L., Melvin, D., and Lingrel, J.B. (1991) Expression of Na, K-ATPase isoforms in human heart. Biochem Biophys Res Comm 179: 1434-1440

25. Shyjan, A.W., Gottardi, C., and Levenson, R. (1990) The Na,K-ATPase β2 subunit is expressed in rat brain and copurifies with Na,K-ATPase activity. J.Biol.Chem. 265:5166-5169

26. Sweadner, K.J. (1989) Isozymes of the Na/K-ATPase. Biochim. Biophys. Acta 988, 185-220, 1989.

27. Zahler, R., Brines, M., Kashgarian, M., Benz, E.J., Jr., and Gilmore-Hebert, M. (1992). The cardiac conduction system in the rat expresses the a2 and a3 isoforms of the Na,K-ATPase. Proc. Natl. Acad. Sci. USA 89, 99-103

28. Zahler, R., Gilmore-Hebert, M., Baldwin, J.C., Franco, K., and Benz, E.J. (1993) Expression of α isoforms of the Na,K-ATPase in human heart Biochim. Biophys. Acta 1149: 189-194

180

# $Na^+/K^+$-ATPase of Neurons and Glia

Victor A. Canfield, Richard Cameron, Robert Levenson

Department of Cell Biology and Section of Neurobiology, Yale University School of Medicine, P.O. Box 3333, New Haven, CT 06510 USA

## Introduction

The application of molecular genetics techniques to the study of the $Na^+/K^+$-ATPase has provided insight into the evolutionary history and molecular heterogeneity of this important enzyme. It is now known that in mammalian and avian species, there are three isoforms of the $\alpha$ subunit (9,24,33) and two isoforms of the $\beta$ subunit (16,19). Each $\alpha$ and $\beta$ subunit is encoded by a separate gene mapping to a discrete chromosomal location in both the mouse (12,15) and human (36) genomes.

The challenge presented by the heterogeneity of $Na^+/K^+$-ATPase subunits and the complexity of $\alpha/\beta$ subunit interaction if formidable, however. Several lines of evidence suggest that each of the three $\alpha$ subunit isoforms may be capable of association with either of the two $\beta$ subunit isoforms to form functional holoenzyme. The current picture of $Na^+/K^+$-ATPase therefore suggests the potential for six structurally (and functionally) distinct isoenzymes. However, the very existence of multiple isoenzymes has made it inherently difficult to study the enzymatic parameters and biochemical properties of individual $Na^+/K^+$-ATPase isoenzymes, primarily because purification of each of the individual isoenzymes has not been achieved.

The potential extent of $Na^+/K^+$-ATPase isoenzyme heterogeneity is best exemplified by the pattern of $\alpha$ and $\beta$ subunit isoform expression in the mammalian central nervous system. Polypeptide products of each of the three $\alpha$ and two $\beta$ subunit genes have been detected in rat brain (25,28). An additional level of complexity is provided by the observation that more than one isoenzyme may be expressed within many neural cell types (3,18). It seems reasonable to wonder, therefore, why there are so many $Na^+/K^+$-ATPase isoenzymes, and what specialized functions these isoenzymes may play in neural function.

$Na^+/K^+$-ATPase plays an important role in mediating the electrical activity of the brain. In neurons, the activity of the enzyme maintains the $Na^+$ and $K^+$ gradients that are essential for nerve-impulse generation (32). Uptake of neurotransmitters and efflux of calcium are also coupled to the activity of the enzyme (10). In astrocytes, $Na^+/K^+$-ATPase plays a key role in potassium uptake during periods of neuronal cell activity (5,29,30). Recently, the astroglial cell isoenzyme has also been proposed to have a function other than ion transport, namely in mediating selective neuron-astrocyte interactions in the developing central nervous system (1,6).

In view of the diverse roles played by $Na^+/K^+$-ATPase in the central nervous system, it seems reasonable to suppose that individual $Na^+/K^+$-ATPase isoenzymes may not be functionally redundant, but rather have evolved in response to differing physiological demands. Knowledge of the distribution of Na,K-ATPase in specific neural cell types may therefore provide important clues to the normal physiological function of distinct Na$^+$/K$^+$ATPase isoenzymes. To address this issue, we have analyzed the expression pattern of Na$^+$/K$^+$-ATPase $\alpha$ and $\beta$ subunit isoforms within the rodent and primate central nervous system. Our results indicate that both neurons and astrocytes possess multiple, yet distinct Na$^+$/K$^+$-ATPase isoenzymes.

## Na,K-ATPase Expression in Astrocytes

To evaluate expression of Na$^+$/K$^+$-ATPase isoforms in astrocytes, primary mixed cultures of glial cells were prepared from the cerebral cortex of P0 rat pups as described by McCarthy and DeVellis (17). After 9 days in culture, O-2A progenitor cells and oligodendrocytes were removed from the monolayer. The predominant cell class (>95%) present in these glial cell cultures comprised cells with an epithelial-like morphology and an antigen phenotype of GFAP$^+$, vimentin$^+$, RC1$^+$ and Rat-401$^+$, a morphology and antigen composition consistent with type-1 astrocytes (4,21).

Immunoblots of membrane fractions prepared from cortical type-1 astrocytes were probed with antisera specific for the rat Na$^+$/K$^+$-ATPase $\alpha$1, $\alpha$2, $\alpha$3, $\beta$1, and $\beta$2 subunits (25,28). Adult rat kidney and brain microsomes were run in control lanes. The results are shown in Fig. 1. As previously described (27), $\alpha$1 antiserum was reactive with a polypeptide of ~100 kDa in kidney and brain, while $\alpha$2 and $\alpha$3 antiserum reacted with polypeptides of ~105 kDa in brain only. In astrocyte membranes, $\alpha$1 and $\alpha$2 subunits were detected at similar levels. In contrast, $\alpha$3 subunits were undetectable in membrane fractions obtained from these cells (Fig. 1A). The expression profile of $\beta$ subunits in astrocyte membranes is shown in Fig. 1B. Antiserum specific for the rat $\beta$1 subunit was reactive with $\beta$1 subunit polypeptides in kidney. However, no $\beta$1 subunits were detectable in membrane fractions from type-1 astrocytes. On immunoblots probed with $\beta$2-specific antiserum, a 48 kDa polypeptide was present in brain and a 50 kDa polypeptide was detected in astrocytes. These results indicate that cerebral cortical type-1 astrocytes specifically express Na$^+$/K$^+$-ATPase $\alpha$1, $\alpha$2, and $\beta$2 subunits.

The expression pattern of $\alpha$ and $\beta$ subunits in primary cultures of type-1 astrocytes suggests the potential formation of isoenzymes containing $\alpha$1/$\beta$2 and $\alpha$2/$\beta$2 subunit combinations in this cell type (reviewed in Fig. 3). The possibility that type-1 astrocytes possess two distinct Na$^+$/K$^+$ isoenzymes is supported by the observation that these cells contain both ouabain-resistant (contributed by the $\alpha$1 subunit) and ouabain-sensitive (contributed by the $\alpha$2 subunit) components of Na$^+$/K$^+$-ATPase activity (3). Previous *in situ* hybridization studies indicate that transcripts encoding the rat $\beta$2 subunit are restricted exclusively to glia (34), while the immunoblotting results reported here show relatively high levels of $\beta$2 subunit polypeptide expression in cerebral cortical type-1 astrocytes. In developing cerebellum, $\beta$2 subunits have been localized to processes of Bergmann glial cells, and to astrocytes located in the internal granular cell layer subsequent to the completion of granule cell migration (1). The localization of $\beta$2 subunits to both cerebral cortical layers and cerebellar astrocytes indicates that $\beta$2 subunit expression is specific to astrocytes, but more than likely is not restricted to specific regional astrocyte subpopulations.

182

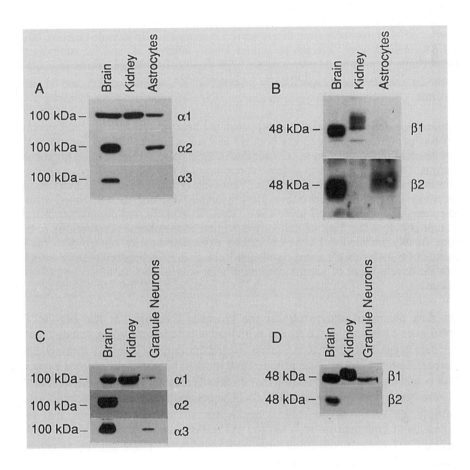

**Figure 1.** Expression of Na+/K+-ATPase α and β subunits in cerebellar granule neurons and cortical astrocytes. Microsomal membrane fractions were prepared from cerebellar granule neurons (P8 rat pups) and type-1 cortical astrocytes (P0) rats. Solubilized membrane proteins were size-fractionated by SDS-PAGE, transferred to nitrocellulose filters, and probed with α or β subunit isoform-specific antibodies. **A** and **B.** Cortical astrocytes. **C** and **D.** Cerebellar granule neurons. The positions of molecular mass markers are indicated at the left.

## Na⁺/K⁺-ATPase Expression in Neurons

To assay expression of Na,K-ATPase in neurons, cerebellar granule neurons were isolated from the cerebellum of P8 rat pups by discontinuous Percoll gradient centrifugation essentially as described by Hatten (7). Primary hippocampal neurons were prepared from hippocampi of E18 rat embryos as previously described (2,3). As shown in Fig. 1 (C and D), the expression pattern of $\alpha$ and $\beta$ subunits in granule neurons differed markedly from that observed in astrocytes. In granule neurons, $\alpha 1$, $\alpha 3$, and $\beta 1$ subunits were detected at similar levels, whereas $\alpha 2$ and $\beta 2$ subunits were undetectable in membrane preparations from these cells. These results indicate that cerebellar granule neurons specifically express Na⁺/K⁺-ATPase $\alpha 1$, $\alpha 3$, and $\beta 1$ subunit isoforms.

Immunoblot analysis revealed that hippocampal neurons (comprising both pyramidal and non-pyramidal cells) express both $\alpha 1$ and $\alpha 3$ subunits (Fig. 2). In addition, we also detected the expression of $\alpha 2$ subunits in membrane preparations of hippocampal neurons. We were unable to detect expression of $\beta 1$ or $\beta 2$ subunit polypeptides on immunoblots prepared from these cells. However, *in situ* hybridization histochemistry has shown that mRNA encoding the $\beta 1$ subunit isoform is associated with the hippocampal pyramidal cell layer (34). Thus, it is likely that expression of the $\beta 1$ subunit is below the limit of antibody detection in membrane preparations obtained from freshly isolated cells which lacked extensive neurite outgrowth. In fact, immunoblot analysis indicates a significant increase in $\beta 1$ subunit abundance associated with the development of neurite outgrowth after several days of hippocampal neuron culture (3).

The data presented above suggest the potential formation of Na⁺/K⁺-ATPase isoenzymes consisting of $\alpha 1/\beta 1$ and $\alpha 3/\beta 1$ subunit combinations in granule neurons. Our data suggest that Na⁺/K⁺-ATPase expression is more complex in hippocampal neurons. Immunoblot analysis revealed that purified hippocampal neurons express all three $\alpha$ subunit isoforms, indicating that these cells may contain three distinct Na⁺/K⁺-ATPase isoenzymes (reviewed in Fig. 3). The possibility that neurons actually possess multiple Na⁺/K⁺-ATPase isoenzymes is supported by experiments that reveal expression of both ouabain-resistant ($\alpha 1$ subunit-containing isoenzyme) and ouabain-sensitive ($\alpha 3$ subunit-containing isoenzyme) Na⁺/K⁺-ATPase activity in cerebellar granule neurons obtained from rat brain (3).

## Immunofluorescence Analysis of $\alpha 3$ Subunit Expression

Previous *in situ* hybridization studies (23,24), and the immunoblotting results presented here, suggest a neuron-specific distribution for the Na⁺/K⁺-ATPase $\alpha 3$ subunit. To further address this issue, we used indirect immunofluorescence microscopy to localize expression of this $\alpha$ subunit isoform in both non-neural periphal tissue and primate brain. As an example, Fig. 4A shows the staining pattern obtained with the a3-specific antiserum on sections of rat skeletal muscle. Immunostaining is restricted exclusively to innervating nerve bundles. Although immunostaining is also detected at the nerve ending, immunostaining is completely absent from the muscle fiber as well as the muscle end plate. These results indicate that $\alpha 3$ subunits are specifically expressed in peripheral nerve and not in skeletal muscle fibers *per se*.

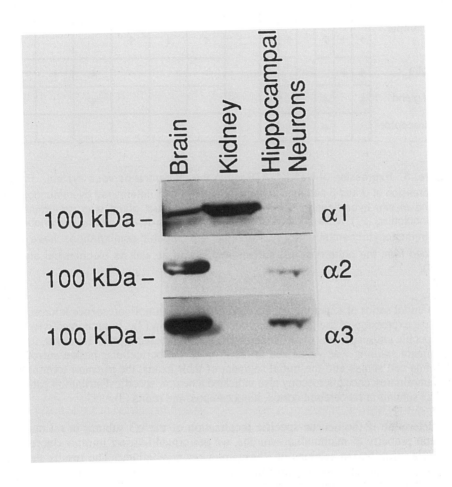

**Figure 2.** Expression of Na$^+$/K$^+$-ATPase $\alpha$ subunits in hippocampal neurons. Microsomal membrane fractions were prepared from primary cultures of hippocampal neurons (E18 rat embryos). Solubilized membrane proteins were fractionated by SDS-PAGE, transferred to a nitrocellulose filter, and probed with antibodies specific for the $\alpha$1, $\alpha$2, and $\alpha$3 subunits. The positions of the molecular mass markers are shown at the left.

| | subunits | | | | | isoenzymes | | | | | |
|---|---|---|---|---|---|---|---|---|---|---|---|
| | α1 | α2 | α3 | β1 | β2 | α1β1 | α2β1 | α3β1 | α1β2 | α2β2 | α3β2 |
| cerebellar granule neurons | + | | + | + | | + | | + | | | |
| hippocampal neurons | + | + | + | + | | + | + | + | | | |
| astroglia | + | + | | | + | | | | + | + | |
| pineal gland | + | | + | | + | | | | + | | + |
| photoreceptor | | | + | | + | | | | | | + |

**Figure. 3.** Expression of Na$^+$,K$^+$-ATPase in the rat central nervous system. Pattern of expression of α and β subunits in neurons and glia was determined by immunoblotting (3) and *in situ* hybridization (34). Expression in pineal gland was determined by immunoblotting (27), while photoreceptor expression was determined by immunoblotting and immunocytochemistry (22). Potential α/β subunit combinations have been deduced from the pattern of α/β subunit expression as well as biochemical analysis (3,27).

In an initial series of experiments, we used indirect immunofluorescence microscopy to localize expression of the α3 subunit isoform in rat brain (3). Immunoreactivity of the α3-specific antiserum was observed throughout all cerebellar layers. However, the most prominant staining was visualized at the level of the pericellular basket surrounding Purkinje cell somae and the initial segment of their axons, the *pinceau* terminals (3). Immunofluorescence microscopy also indicated a neuron-specific distribution pattern for the α3 subunit in rat cerebral cortex, hippocampus, and retina (3).

To determine if the neuron-specific localization of the α3 subunit in rat may be a general property of mammalian neurons, we performed indirect immunofluorescence analysis using frozen sections of Rhesus monkey cerebellum. The results shown in Fig. 4 (B-D) indicate that the distribution of the α3 subunit in primate cerebellum is remarkably similar to that observed in rat. As previously observed in rat cerebellum (3), the staining pattern produced by α3 antiserum in primate cerebellum is associated with the pericellular basket surrounding Purkinje cell somae, the dendritic processes of Purkinje cells (Fig. 4B), and with initial segments of basket cell axons, the *pinceau* terminals (Fig. 4C). In the internal granule cell layer, immunolabeled profiles of granule cell bodies are outlined sharply against a background of immunolabeled neuropil (Fig. 4D).

The immunoblot and immunolocalization results described above provide strong evidence for the neuron-specific distribution of the Na$^+$/K$^+$-ATPase α3 isoform. In rat, α3 subunits have been detected in projection neurons such as cerebellar Purkinje cells and hippocampal and cerebral cortical pyramidal neurons, as well as interneurons such as cerebellar granule, stellate, and basket cells (3). These results indicate that α3 subunit

186

expression is not restricted to functionally distinct neuronal cell populations. However, the pattern of $\alpha 3$ subunit distribution in monkey brain is virtually identical to that observed in rat. This observation suggests a common physiological function for the $\alpha 3$ subunit in both rodents and primates.

## Functional Significance of $Na^+/K^+$-ATPase Isoenzyme Diversity

A critical gap in our understanding of $Na^+/K^+$-ATPase physiology concerns the significance of isoenzyme diversity. In view of the apparent heterogeneity in $\alpha/\beta$ subunit interactions, the functional characterization of individual $Na^+/K^+$-ATPase isoenzymes becomes an essential prerequisite for understanding the significance of isoenzyme diversity. The identification of cell types in the central nervous system expressing limited combinations of $\alpha$ and $\beta$ subunits provides a framework for addressing this important issue. For example, biochemical characterization of the kinetic properties of the $Na^+/K^+$-ATPase $\alpha 3/\beta 2$-containing isoenzyme of pineal gland revealed that this ATPase isoenzyme exhibited a higher apparent affinity for $Na^+$ than the kidney $\alpha 1/\beta 1$-containing isoenzyme (27). In contrast, an $\alpha 3/\beta 1$ isoenzyme expressed in transfected HeLa cells showed a lower apparent affinity for $Na^+$ than an $\alpha 1/\beta 1$-containing isoenzyme (11). These results suggest the possibility that the $Na^+$ affinity of the $Na^+/K^+$-ATPase may be governed, in part, by association with a specific $\beta$ subunit isoform. By examining the substrate requirements of $Na^+/K^+$-ATPases expressed in neurons and glia, it may be possible to derive a much clearer picture of the functional relationships among the six $Na^+/K^+$-ATPase isoenzymes.

Although the biochemical properties of the neuronal $\alpha 3$ isoenzyme have not as yet been characterized, localization of $\alpha 3$ subunit expression within cerebellar granule neurons provides a strategy for addressing this issue. As discussed above, $Na^+/K^+$-ATPase isoenzymes in these cells are likely to consist of $\alpha 1/\beta 1$ and $\alpha 3/\beta 1$ subunit combinations. By exploiting the difference in ouabain sensitivity between the rat $\alpha 1$ and $\alpha 3$ subunit isoforms, it should now be possible to analyze the biochemical properties of the granule neuron $\alpha 3$ subunit-containing isoenzyme. It will clearly be of interest to compare the substrate requirements of the neuronal $\alpha 3/\beta 1$ subunit isoenzyme with the pineal gland isoenzyme that is composed of $\alpha 3$ and $\beta 2$ subunits. It is possible that these two isoenzymes may exhibit very similar substrate requirements. Alternatively, $\alpha 3/\beta 1$ and $\alpha 3/\beta 2$ isoenzymes may exhibit marked differences in their affinity for $Na^+$ or $K^+$. A result of the latter type would suggest that the $\beta$ subunit contributes to the $Na^+$ or $K^+$ affinity of the $\alpha 3$ subunit.

The localization of expression of multiple $Na^+/K^+$-ATPase isoenzymes to individual neurons brings up an additional question of interest. Why is more than one isoenzyme produced in a single cell type? It is possible that coexpression of two or more isoenzymes within a single cell is designed to insure that inactivation of one isoenzyme does not result in cell death. The difference in $Na^+$ activation between the $\alpha 1$ and $\alpha 3$ isoenzymes (27), however, suggests an alternative hypothesis. The answer could well be that the isoenzymes perform distinct, though possibly overlapping functions, within a single neuron. For example, it is known that intracellular $Na^+$ concentrations within neurons varies considerably in response to electrical activity (13,32). Expression of two distinct $Na^+/K^+$-ATPases, one of which is activated at low intracellular $Na^+$, and one

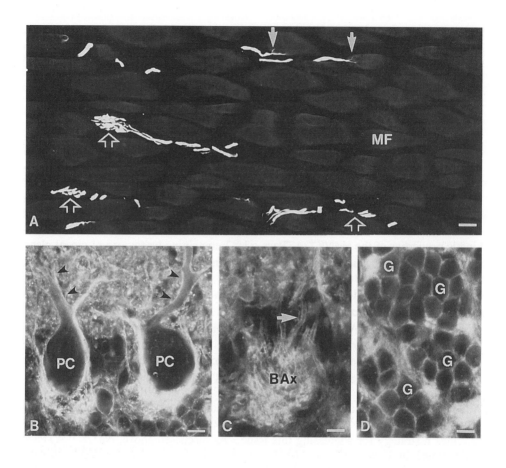

**Figure 4.** Immunofluorescent staining of Na⁺/K⁺-ATPase α3 subunits in rat skeletal muscle and monkey cerebellum. **A.** Frozen, 10 µM thick section of rat skeletal muscle. Open arrows, innervating nerve bundles; solid arrows, nerve teminals; MF, muscle fiber. Scale bar=25 µM. **B-D.** Frozen, 8 µM thick, sections of monkey cerebellar Purkinje cell layer (**B,C**) and internal granule cell layer (**D**). PC, Purkinje cell; BAx, basket cell axons; G, granule neuron. Arrows in **C** outline the hollowed profiles of basket cell axons overlying a Purkinje cell body. Scale bars=7µM

which is activated at high intracellular $Na^+$, suggests that the isoenzymes could act separately or in tandem to buffer intracellular $Na^+$ levels over a broad concentration range. Alternatively, it is possible that the two isoenzymes are distributed in different locations on the cell surface. Differences in subcellular localization would permit differential activation of the isoenzymes in specific membrane microdomains.

In contrast to neurons, which represent a specific expression site of the $\alpha3$ subunit, expression of the $\beta2$ subunit appears to be astroglia cell-specific. The increased extracellular potassium generated during neuronal activity is buffered and redistributed into plasma and cerebrospinal fluid by an uptake system that is expressed in glial cells (20,29,35). Capacity for enhanced potassium uptake may be mediated by a glial cell $Na^+/K^+$-ATPase that is more efficient at higher extracellular $K^+$ concentrations than the neuronal isoenzymes. In an initial series of experiments, Henn et al. (8) demonstated that the glial $Na^+/K^+$-ATPase was more active at high $K^+$ levels than was neuronal $Na^+/K^+$-ATPase. In contrast to the kidney, axolemmal, and pineal gland isoenzymes, all of which are active at $K^+$ concentrations of 1-2 mM, glial $Na^+/K^+$-ATPase appeared to be most active at $K^+$ concentrations in the 5-10 mM range. Taken together, these observations are consistent with the view, first suggested by Sweadner (31), that the glial cell $Na^+/K^+$-ATPase may function as a $K^+$ rather than a $Na^+$ pump. By taking advantage of the difference in ouabain sensitivity between the rat $\alpha1$ and $\alpha2$ isoforms, it should now be possible to characterize the enzymatic properties of the $\alpha1/\beta2$ and $\alpha2/\beta2$-containing isoenzymes in purified glial cell preparations.

The localization of $Na^+/K^+$-ATPase $\alpha$ and $\beta$ subunit isoform expression in the central nervous system is relevant to questions of developmental neurobiological interest. Because expression of $\alpha3$ subunits is restricted to neurons, and $\beta2$ subunits to glia, these $Na^+/K^+$-ATPase isoforms may serve as useful markers for studying neuronal and glial cell lineages. Finally, it has not escaped our notice that the availability of genomic DNA clones containing the promoter regions of the $\alpha3$ and $\beta2$ (26) genes should permit the targeted expression of heterologous genes in neurons and glia.

*Acknowledgements*

These studies were supported by grants to R. Levenson from the NIH (HL-39263 and GM-49023) and to P. Rakic (NS-14841 and NS-22807).

# References

1. Antonicek H, Pershon E, Schachner M (1987) Biochemical and funcional characterization of a novel neuron-glia adhesion molecule that is involed in neuronal migration. J Cell Biol 104: 1587-1595
2. Banker GA, Cowan WM (1977) Rat hippocampal neurons in dispersed cell culture. Brain Res 126: 379-425
3. Cameron R, Klein L, Shyjan AW, Rakic P, Levenson R (1993) Neurons and astroglia express distinct subsets of Na,K-ATPase $\alpha$ and $\beta$ subunits. Mol Brain Res 21: 333-343
4. Cameron RS, Rakic P (1991) Glial cell lineage in the cerebral cortex: a review and synthesis. Glia 4: 124-137
5. Futamachi KJ, Pedley TA (1976) Glial cells and extracellular potassium: their relationship in mammalian cortex. Brain Res 109: 311-322

6.   Gloor S, Antonicek H, Sweadner KJ, Franck R, Moos M, Schachner M (1990) The adhesion molecule on glia (AMOG) is a homologue of the β subunit of the Na,K-ATPase. J Cell Biol 110: 165-174

7.   Hatten ME (1985) Neuronal regulation of astroglial morphology and proliferation. J Cell Biol 100: 384-386

8.   Henn FA, Jaljamae H, Hamberger A (1972) Glial cell function: active control of extracellular $K^+$ concentration. Brain Res 43: 437-443

9.   Herrera VL, Emanuel JR, Ruiz-Opazo N, Levenson R, Nadal-Ginard B (1987) Three differentially expressed Na,K-ATPase α subunit isoforms: structural and functional implications. J Cell Biol 105: 1055-1065

10.  Iverson LL, Kelly JS (1975) Uptake and metabolism of gamma-aminobutyric acid by neurones and glial cells. Biochem Pharmacol 24: 933-938

11.  Jewell EA, Lingrel JB (1991) Comparison of the substrate dependence properties of the rat Na,K-ATPase α1, α2 and α3 isoforms expressed in HeLa cells. J Biol Chem 266: 16925-16930

12.  Kent RB, Fallows D, Geissler E, Glaser T, Emanuel JR, Lalley PA, Levenson R, Housman DE (1987) Genes encoding α and β subunits of the Na,K-ATPase are located on three different chromosomes in the mouse. Proc Natl Acad Sci USA 84: 5369-5372

13.  Lasser-Ross N, Ross WN (1992) Imaging voltage and synaptically activated sodium transients in cerebellar Purkinje cells. Proc R Soc Lond 247: 35-39

14.  Lemas V, Rome J, Taormino J, Takeyasu K, Fambrough DM (1991) Analysis of isoform specific regions within the α-and β-subunits of the avian Na,K-ATPase. In: Kaplan JH, DeWeer P (eds) The Sodium Pump: Recent Developments. Rockefeller University Press, N.Y. pp 117-123

15.  Malo D, Schurr E, Levenson R, Gros P (1990) Assignment of the Na,K-ATPase β2 subunit gene (Atpb-2) to mouse chromosome 11. Genomics 6: 697-699

16.  Martin-Vasallo P, Dackowski W, Emanuel JR, Levenson R (1989) Identification of a putative isoform of the Na,K-ATPase β subunit: primary structure and tissue-specific expression. J Biol Chem 264: 4613-4618

17.  McCarthy JD, Devellis J (1980) Preparation of separate astroglial and oligodendroglial cell cultures from rat cerebral tissue. J Cell Biol 85: 890-902

18.  McGrail KM, Phillips JM, Sweadner KJ (1991) Immunofluorescent localization of three Na,K-ATPase isoenzymes in the rat central nervous system: both neurons and glia can express more than one Na,K-ATPase. J Neurosci 11: 381-391

19.  Mercer RW, Schneider JW, Savitz A, Emanuel JR, Benz EJ Jr, Levenson R (1986) Rat brain Na,K-ATPase β chain gene: primary structure, tissue-specific expression, and amplification in ouabain-resistant HeLa $C^+$ cells. Mol Cell Biol 6: 3884-3890

20.  Orkand RK, Nicholls JG, Küffler SW (1966) Effect of nerve impulses on the membrane potential of glial cells in the central nervous system of amphibia. J Neurophysiol 29: 788-806

21.  Raff MC (1989) Glial cell diversification in the rat optic nerve. Science 243: 1450-1455

22.  Schneider BG, Shyjan AW, Levenson R (1991) Co-localization and polarized distribution of Na,K-ATPase α3 and β2 subunits in photoreceptor cells. J Histochem Cytochem 39: 507-517

23.  Schnieder JW, Mercer RW, Gilmore-Hebert M, Utset MF, Lai C, Greene A, Benz EJ Jr (1988) Tissue specificity, localization in brain, and cell-free translation of mRNA encoding the A3 isoform of Na,K-ATPase. Proc Natl Acad Sci USA 85: 284-288

24.  Shull GE, Greeb J, Lingrel JB (1986) Molecular cloning of three distinct forms of the Na,K-ATPase α subunit from rat brain. Biochemistry 25: 8125-8132

25.  Shyjan AW, Levenson R (1989) Antisera specific for the α1, α2, α3 and β subunits of the Na,K-ATPase: differential expression of α and β subunits in rat tissue membranes. Biochemistry 28: 4531-4535

190

26.  Shyjan AW, Canfield VA, Levenson R (1991) Evolution of the Na,K- and H,K-ATPase β subunit gene family: structure of the murine Na,K-ATPase β2 subunit gene. Genomics 11: 435-442

27.  Shyjan AW, Cena V, Klein DC, Levenson R (1990) Differential expression and enzymatic properties of the Na,K-ATPase α3 isoenzyme in rat pineal glands. Proc Natl Acad Sci USA 87: 1178-1182

28.  Shyjan AW, Gottardi C, Levenson R (1990) The Na,K-ATPase β2 subunit is expressed in rat brain and copurifies with Na,K-ATPase activity. J Biol Chem 265: 5166-5169

29.  Somjen GG (1979) Extracellular potassium in the mammalian central nervous system. Annu Rev Physiol 41: 159-177

30.  Somjen GG (1987) Functions of glial cells in the cerebral cortex. In: Jones EG, Peters A (eds) Cerebral Cortex: Further Aspects of Cortical Function, Including Hippocampus. Plenum Press, N.Y. pp 1-39

31.  Sweadner KJ (1989) Isozymes of the $Na^+/K^+$-ATPase. Biochim Biophys Acta 988: 185-220

32.  Thomas RC (1972) Electrogenic sodium pump in nerve and muscle cells. Physiol Rev 52: 563-594

33.  Takeyasu K, Lemas V, Fambrough DM (1990) Stability of $Na^+/K^+$-ATPase α-subunit isoforms in evolution. Am J Physiol 259: C619-C630

34.  Watts AG, Sanchez-Watts G, Emanuel JR, Levenson R (1991) Cell-specific expression of mRNAs encoding Na,K-ATPase α- and β-subunit isoforms within the rat central nervous system. Proc Natl Acad Sci USA 88: 7425-7429

35.  Wright EM (1972) Mechanisms of ion transport across the choroid plexus. J Physiol 226: 545-571

36.  Yang-Feng TL, Schneider JW, Lindgren V, Shull MM, Benz EJ Jr, Lingrel LB, Francke U (1988) Chromosomal localization of human $Na^+,K^+$-ATPase alpha and beta subunit genes. Genomics 2: 128-138

# Expression of a Truncated α Subunit of the $Na^+,K^+$-ATPase (α-1-T) in Smooth Muscle Cells

Russell M. Medford[1], Thomas A. Pressley[2], Robert W. Mercer[3] and Julius C. Allen[4]

[1]Department of Medicine, Division of Cardiology, Emory University School of Medicine, Atlanta, GA 30322, [2]Department of Physiology and Cell Biology, University of Texas Medical School, Houston, Texas 77030, [3]Department of Cell Biology and Physiology, Washington University School of Medicine, St. Louis, MO 63110, [4] Departments of Medicine and Physiology, Baylor College of Medicine, Houston, TX 77030

## Introduction

The $Na^+,K^+$-ATPase (NAKA) is an essential transmembrane enzyme complex that couples ATP hydrolysis with the transport of $Na^+$ and $K^+$ ions across the plasma cell membrane against their respective electrochemical gradients. The $Na^+/K^+$-, $Ca^{2+}$ and $H^+/K^+$- active ion transport pumps of eukaryotic cells share significant similarities in tertiary structure, a modest degree of amino acid sequence identity , and similar enzymatic mechanisms that classify them as $E_1$-$E_2$ or P-type class of transport ATPases (12). The 100kDa α: 35kDa β complex constituting the Na,K-ATPase is encoded by at least three α isoform genes and two β subunit genes. These isoforms are characterized by high degrees of sequence homology and functional identity but with distinct tissue-specific, developmental and hormonal patterns of regulation.

In contrast to the kidney, brain and other "classical" tissues, relatively little is known of the biochemistry of NAKA in vascular smooth muscle (VSM). VSM membrane preparations used to study the Na,K-ATPase are characterized by a large residual Mg-ATPase activity ranging from 30-70% of total ATPase activity. Multiple purification schemes have been unsuccessful in separating this component from the Na,K-ATPase activity. The basis for this inability to further purify Na,K-ATPase activity is unclear but may be due to tissue-specific differences in the structure of the Na,K-ATPase expressed in VSM, its association with cell membrane components, tissue-specific expression of Mg-ATPase's, or a combination of all three. The protein(s) responsible for this Mg-ATPase activity are also unknown. This has considerably hampered careful biochemical and functional studies of the VSM Na,K-ATPase. Biochemical data on NAKA of VSM is lacking due to its low abundance and possible differences in membrane construction thus affecting its purification. The low enzyme activities and content of VSM NAKA (3 8) make it difficult to assess any functional role of the enzyme besides that of cation regulation. Yet, the regulation of Na+ pump content and activity in VSM is clearly important in the modulation of intracellular Ca++, and hence contractility as well as other functions, through ionic interactions with the Na+/H+ and Na+/Ca++ exchange systems (8 13). Indeed, NAKA has been hypothesized to be involved in the pathogenesis of hypertension (10) in part through its effects on vascular smooth muscle reactivity (6) although a genetic linkage in experimental animal models has not yet been established (15). Nevertheless, in response to ionic signals associated with an increase in intracellular Na+, second messengers such as cAMP, adrenergic receptor signals ,or mechanical signals such as pressure, VSM NAKA may utilize functional and regulatory mechanisms not operative in other tissues (1 10).

As an essential prelude to understanding the role of the Na pump in VSM, we have made two recent and intriguing discoveries regarding the composition and structure of the VSM NAKA    First, in Allen et al., 1991 (2),  we performed Northern filter hybridization and  polymerase chain reaction (PCR) analysis of canine saphenous vein smooth muscle cells and detected only very small amounts of α-1 "like" mRNA, and undetectable α-2 and α-3 mRNA. As the same cells contain relatively abundant levels of β subunit mRNA, β- is in fact in vast excess over α-mRNA. Consistent with this, polyclonal antisera raised against the canine kidney NAKA holoenzyme readily detects the β-subunit by Western blot analysis, but not the expected ~100 kDa α subunit, from the same vascular smooth muscle (data not shown). It is generally accepted that in order for functional pumps to exist in the sarcolemma, a one-to-one ratio of α-to-β complex is required. These observations might explain the extremely low number of Na pump sites contained per unit vascular tissue and some of its unique properties regarding mitogen-induced increases in Na+ influx and accumulation (4) and in its complex contractile-relaxation control system.

The nature and potential complexity of Na+ pump structure and function in vascular smooth muscle was further developed in a series of studies linking the structure and processing of the α-1 subunit gene with the apparent "absence " of a vascular smooth muscle α subunit mRNA or protein. In Medford et al., 1991 (14), we presented the first evidence of the expression and putative structure of a truncated α-1 (α-1-T) protein expressed in vascular smooth muscle. The major findings of our work 1) vascular smooth muscle utilizes alternative RNA processing of the α-1 gene to express a structurally distinct ~65,000 MW isoform, α-1-T (truncated), 2) analysis of both its mRNA and protein structure reveals that α-1-T encodes only the first 554 amino acids of α-1 to Gly554, then terminates with an intron encoded 27 residue peptide as illustrated in Figure 1

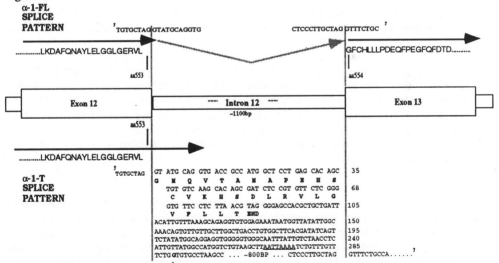

**Figure 1**: Rat α-1-T is Derived from α-1 by Alternative RNA Processing (adapted from (14))

and , 3) by Western blot analysis, α-1-T represents an evolutionarily conserved truncated NAKA isoform expressed in vascular smooth muscle.This is also the first demonstration that alternative RNA processing is a mechanism for the generation of significant protein isoform diversity of the Na+,K+-ATPase multigene family. Figure 2 illustrates the structure of α-1-T (adapted from (14)).

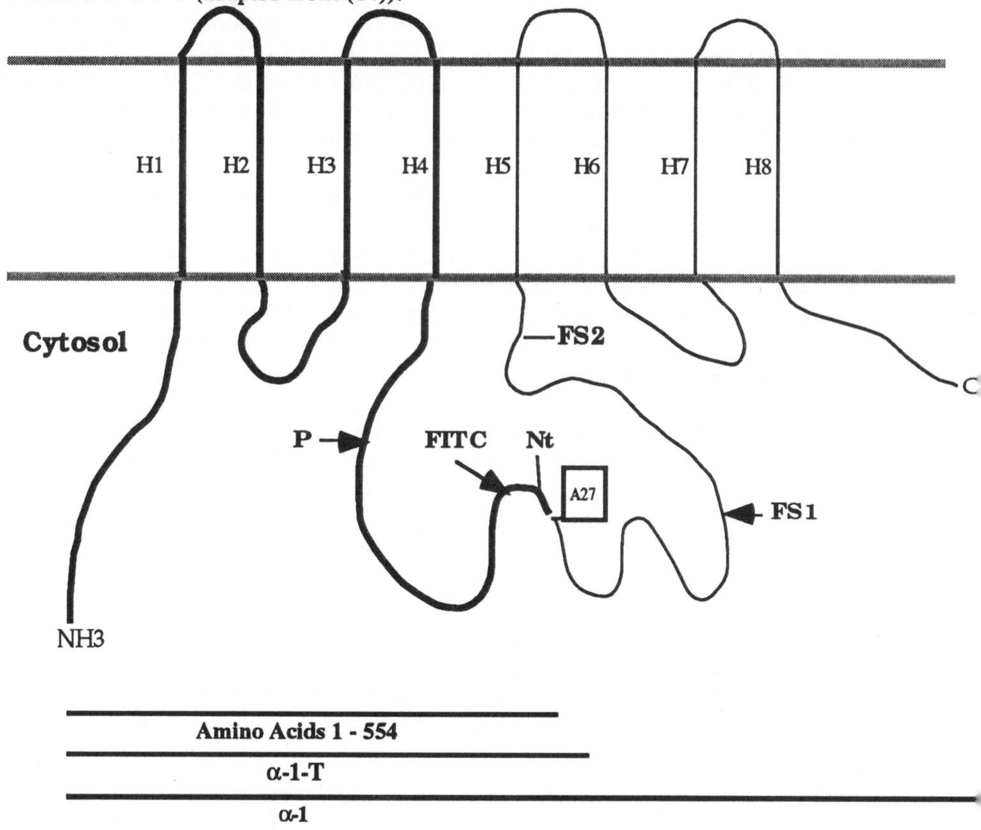

Figure 2: Comparison of the Predicted Structures of α-1 and α-1-T (adapted from (14)).By hydropathy analysis, the bold line represents the predicted shared amino acid sequences between α-1 and α-1-T. The thin line represents the remaining predicted structure of α-1. Selected regions : P-phosphorylation domain, Nt-nucleotide binding region,  FITC binding region, FS1 and FS2- FSBA binding regions, A27-the α-1-T carboxy terminal peptide.

Whether α-1-T alone, or even in some association with the β subunit, binds and transports Na$^+$ and K$^+$ ions appears more problematic. Analogous to studies of the Ca$^{++}$ binding and transport sites in the related Ca$^{++}$-pump (12), elements essential to Na$^+$ and K$^+$ mediated enzyme activity (J. Lingrel, these Proceedings) and β-subunit binding (D. Fambrough, these Proceedings) appear to be located in the carboxyl half of the α subunit, beyond Gly$^{554}$. As α-1-T does not contain this region, one hypothetical model predicts that α-1-T may function as an ATP binding and hydrolysis enzyme, but not exhibit, or exhibit altered, Na$^+$ or K$^+$ dependent enzyme or transport activity. In addition, it may not necessarily be localized to the plasma membrane given its lack of a β-subunit interaction region.  Alternatively, α-1-T may associate as a dimer and thus reprise essential elements of the carboxy region and reprise NAKA enzyme and transport activity.

A central question concerning α-1-T is its functional role. In this report, we have begun to address this critical issue by analysis of the physiochemical and membrane associative properties of α-1-T in both vascular smooth muscle and recombinant expression cDNA systems.

## Results

To provide some insights into the potential functional significace of both the alternative RNA splicing mechanism and intron encoded DNA sequence of α-1-T, we have begun a study of the evolutionary conservation of intron 12 through PCR analysis of genomic DNA. As shown in Figure 3, we have used rat primers to establish that the PCR fragments of predicted size can be generated using human genomic DNA. This suggests that in addition to rat and dog, the human α-1 gene may possess a similar nucleotide structure at the exon12-intron12 boundary. By extrapolation, this would suggest that α-1-T might be expressed in human tissue analogous to rat and dog.

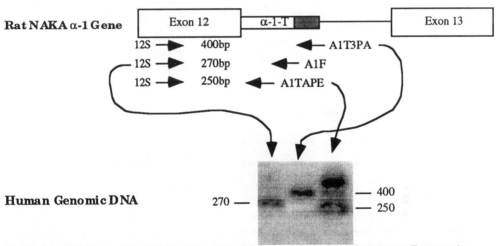

**Figure 3**. Evidence for Structural and Sequence Homology between Rat and a Human Form of α-1-T. 1ug of human genomic DNA was subjected to 30 cycles of PCR amplification (52oC, 42oC, 72oC cycle) using oligonucleotide probes commonto the rat and human α-1 cDNA (12S) and probes complementary tothe rat α-1-T carboxy tail (A1TAPE, A1F) and 3' untranslated region (A1T3PA). PCR products were size fractionated on 1% agarose, transferred to nitrocellulose, and hybridized to a full length rat α-1 cDNA probe with final wash stringency of 55oC, 2XSSC.

We have begun to address the issue of how α-1-T is associated with the cell membrane in vascular smooth muscle tissue compared with another α-1-T expressing tissue, the kidney. Our finding of α-1-T expression in both vascular and renal tissue provides us with an experimental model to compare the membrane organization of α-1-T in tissue with little or no α subunit expression (vascular smooth muscle) compared with tissue expressing large amounts of α subunit (renal). We hypothesize that α-1-T and α-1 may share similar assembly mechanisms (ie β subunit association) defining their final incorporation into the plasma membrane. In addition, α-1-T may exhibit ATPase properties distinct from α. Indeed, we have preliminary data supporting this notion. Microsomal membranes were isolated from carotid smooth muscle, canine renal cortex. The membranes were either treated (+) or not treated (-) with SDS in the 3:1 ratio described by Jorgensen (11) to purify Na,K-ATPase activity from the kidney. Following re-centrifugation and pelleting of the microsomes, 50ug of microsomal membrane protein from both treatments were analyzed by Western blot analysis with anti-α subunit specific antibody for the presence of α and α-1-T isoforms. As shown in Figure 4, both the 100kDa α and 65kDa α-1-T proteins are found in untreated renal microsomes (-SDS) (lane 3). However, only the 100 kDa α subunit is retained in the renal microsomal membrane following SDS treatment (+SDS) (lanes 4). In dramatic contrast, the same analysis of microsomal membranes prepared from canine carotid artery smooth muscle demonstrated that α-1-T is retained fully in the microsomal membrane following SDS treatment and that varying amounts of SDS detergent treatment affected neither the proportion of Na,K-ATPase activity (data not shown) in the fraction or the amount of associated α-1-T (lanes 1 and 2). This preliminary data supports our initial hypothesis by demonstrating a significant difference in membrane organization between the α and α-1-T isoforms between the kidney and vascular smooth muscle. Additionally, this may explain why α-1-T has not been previously identified following Na,K-ATPase purification from the kidney.

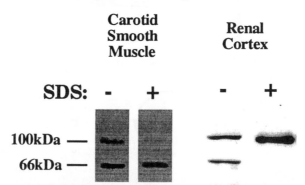

**Figure 4** - Differential Membrane Association of α-1 and α-1-T Na,K-ATPase Isoforms in Vascular Smooth Muscle Compared to Renal Microsomal Membrane. Following membrane isolation, protein was size fractionated on SDS-PAGE, transferred to filter paper, and exposed to anti-α subunit specific antibody. A horseradish peroxidase tagged secondary antibody was used for visualization by chemiluminesence.

To complement our studies in vascular smooth muscle and renal membranes, we have utilized a baculovirus expression vector to begin a further characterization of the membrane associative properties of α-1-T in the insect SF9 cell line. Using protocols described by Blanco et al., 1993 (5) and DeTomaso et al, 1993 (7), recombinant baculoviruses containing the cDNA for the α-1-T isoform of the rat Na,K-ATPase was prepared and used to infect SF-9 cells, an insect cell line derived from the ovary of the fall armyworm Spodoptera frugiperda. By using this system, α-1-T subunits that were antigenically and electrophoretically indistinguishable from native α-1-T were produced. Using an 35S-methionine pulse labeling technique and immunoprecipitation with an α-specific antibody directed against the amino terminus (and thus common to both α-1 and α-1-T), we demonstrate in Figure 4, that co-transfection of both the α-1-T and β2 subunit expression vectors yield and α-1-T:β2 membrane complex (lane 5). This result is similar to that observed after co-transfection with full length α-1 and β2 (lane 3). Membrane isolation and detergent dissociation studies demonstrate that both α-1-T and α-1 form similarly stable membrane complexes with β2 (data not shown). This suggests that α-1-T encodes sufficient information to associate with a beta subunit in an apparantly analagous manner to that of full-length α-1.

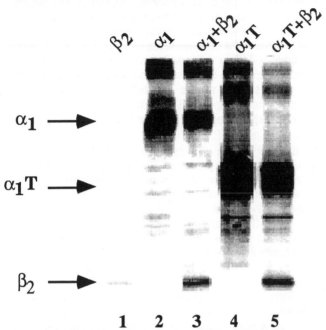

**Figure 5.** Association of α-1 and α-1-T subunits with co-expressed beta subunits in the baculovirus/SF9 expression system. Transfection with one or a combination of expression vectors are as described above each lane. In all lanes, $^{35}$S labeled protein was immunoprecipitated with an α-subunit specific antibody directed against the amino terminus, size fractionated on SDS-PAGE and visualized by autoradiography.

## Discussion

A central question concerning α-1-T is its functional role. While experimental data is not yet available to directly answer whether and through what mechanism α-1-T functions in smooth muscle physiology or ion transport, significant progress has been made in defining aspects of α-1-T expression and structure that will eventually lead to an understanding of its function. Our studies of α-1-T expression strongly suggest tissue-specific and developmental roles for α-1-T and argue against an "aberrant" truncation event. In addition to our previously published studies of rat and dog, evolutionary conservation of the "intron 12" sequence encoding α-1-T in human would suggest a role for selective pressure to maintain an essential functional or structural role. This would not be expected from a noncoding or functionally insignificant intron sequence. α-1 and α-1-T appear to exhibit differential associations with the cell membrane that are complementary between vascular smooth muscle and kidney. This is consistent with the notion that kidney and smooth muscle utilize different posttranslational processing mechanisms to integrate the α-1 and α-1-T subunits into the cell membrane. That α-1-T encodes sufficient information to associate with the cell membrane and associate with the β-subunit is strongly suggested by our co-expression studies using the baculovirus/SF9 cell line.

Based on the criteria of detergent dissociation, the α-1-T protein appears to be a more integral part of the VSM membrane relative to renal membranes. This tissue-specific pattern of membrane association suggests a possible correlation with the expression of α-1. It is intriguing to speculate that in those tissues that express little or no α-1, such as VSM, α-1-T stably associates with a membrane component to form an active ion transport complex. The presence of a ouabain-sensitive active Na,K trasnporter in VSM in the absence of detectable α-1, 2 or 3 supports this hypothesis. In those tissues that express siginficant amounts of α-1, such as the kidney, competition for post-translational assembly factors, such as the β subunit(s), may prevent α-1-T from forming stable membrane complexes. Thus, α-1-T may not exhibit classic, transmembrane active ion transport properties in these tissues.

It is important to note that our studies only establish that α-1-T associates in a tissue-specific manner with a microsomal membrane fraction. Whether this represents the plasma membrane, or another cellular membrane compartment, will require more precise techniques than tissue homogenization and simple centrifugation. In addition, there is no definitive evidence that the α-1-T of the kidney and α-1-T from VSM are identical. Our data suggests that since two different antibodies to two different epitopes react strongly the 66kDa α-1-T, that these proteins, if not identical, are very similar. It is our hypothesis that they are identical and result from alternative splicing of the α-1 gene.

Both the vascular smooth muscle membrane studies and baculovirus expression studies suggest that α-1-T may form a strong association with both the cell membrane and a β-subunit. This would suggest that other regions of the a subunit, other than the COOH region identified by D. Fambrough (these Proceedings), may play a role in β subunit interaction. Alternatively, another model concerns the association of α-1-T with the β-subunit through a third protein component X. In this model, protein X would be a transmembrane protein containing the domains for cation binding and transport that are presumably missing in α-1-T. Theoretically, a functional, three-component $Na^+/K^+$ ion-transport complex could consist of α-1-T transducing the energy from ATP hydrolysis through non-covalent interactions to produce the conformational changes in protein X needed for transport. The β-subunit then would interact with α-1-T via the X protein possibly in a role similar to that of the classical α subunits. This three-component model runs counter to the observation that in all the known eukaryotic $E_1$-$E_2$ or P-type ion transporters ($Ca^+$ pumps, $H^+/K^+$ pump, $Na^+$ pumps) energy transduction is dependent on covalent bonds between energy-transducing and cation binding-transport regions located on the same polypeptide (12)). However, in support of our model, a three-component member of the same P-type family of transport ATPases does function as the $K^+$-ATPase of Escherichia coli (9). Commonly referred to as Kdp, it consists of three polypeptides: a 72 kDa KdpB protein that shares significant sequence identities to the ATP binding and phosphorylation domains, as well as overall structure of the amino half of the the large P-type subunits (such as NAKA α ), a 21 kDA KdpC subunit that resembles that of the NAKA β subunit, and a 59 kDA KdpA subunit that has no known homologue but is highly hydrophobic, binds K+ and is involved in cation transmembrane movement.

198

The precise role of α-1-T in cellular function remains to be established. Our studies demonstrate that α-1-T is a membrane protein thus fulfilling an important criteria for its putative role in transmembrane, active ion transport function. In addition, α-1-T appears to associate with β subunits in co-expression sytems. Tissue-specific differences in this membrane association and potentially β subunit interactions also suggest that α-1-T may exhibit tissue-specific functional characteristics.

**Acknowledgments.** This work was supported in part by grants from the National Heart, Lung and Blood Institute of the NIH to RMM, RWM, TAP(HL39846) and JCA(HL24585). RMM is an Established Investigator of the American Heart Association.

## References
1. Allen J, Navran S  Jemelka S (1987). Functional role of Na pump modulation in vascular smoooth muscle. In: JA Bevan et al. (ed.), Vascular Neuroeffector Mechanisms. IRL Press, Washington, DC, pp. 217-224
2. Allen JC, Medford RM, Zhao X  Pressley TA. (1991) Disproportionate alpha- and beta-mRNA sodium pump subunit content in canine vascular smooth muscle. Circ Res 69: 39-44
3. Allen JC, Navran SS  Kahn AM. (1986) Na+-K+-ATPase in vascular smooth muscle. Am J Physiol 250:C536-C540
4. Allen JC, Navran SS, Seidel CL, Dennison DK, Amann JM  Jemelka SK. (1989) Intracellular Na+ regulation of Na+ pump sites in cultured vascular smooth muscle cells. Am J Physiol 256: C786-C792
5. Blanco G, Xie Z  Mercer R. (1993) Functional expression of the alpha2 and alpha3 isoforms of the Na,K-ATPase in baculovirus infected insect cells. Proc Natl Acad Sci USA 90: 1824-8
6. Bohr DF, Dominiczak AF  Webb RC. (1991) Pathophysiology of the vasculature in hypertension. Hypertension
7. DeTomaso A, Xie Z, Liu G  Mercer R. (1993) Expression, targeting and assembly of functional Na,K-ATPase polypeptides in baculovirus-infected insect cells. J Biol Chem 268: 1470-1478
8. Eggermont J, Vrolix M, Raeymaekers L, Wuytack F  Casteels R. (1986) Circ Res 62: 266-278
9. Epstein W, Walderhaug MO, Polarek JW, Hesse JE, Dorus E  Daniel JM. (1990) The bacterial Kdp K(+)-ATPase and its relation to other transport ATPases, such as the Na+/K(+)- and Ca2(+)-ATPases in higher organisms. Philos Trans R Soc Lond [Biol] 326: 479-86
10. Herrera VL, Chobanian AV  Ruiz Opazo N. (1988) Isoform-specific modulation of Na+, K+-ATPase alpha-subunit gene expression in hypertension. Science 241: 221-3
11. Jorgensen P. (1974) Isolation of Sodium-Potassium ATPase. Meth Enzymol 32B: 277-290
12. Jorgensen PL  Andersen JP. (1988) Structural basis for E1-E2 conformational transitions in Na,K-pump and Ca-pump proteins. J Membr Biol 103: 95-120
13. Kahn AM, Allen JC  Shelat H. (1988) Na+-Ca2+ exchange in sarcolemmal vesicles from bovine superior mesenteric artery. Am J Physiol 254: C441-449
14. Medford RM, Hyman R, Ahmad M, Allen JC, Pressley TA, Allen PD  Nadal GB. (1991) Vascular smooth muscle expresses a truncated Na+, K(+)-ATPase alphα-1 subunit isoform. J Biol Chem 266: 18308-12
15. Rapp JP  Dene H. (1990) Failure of alleles at the Na+, K(+)-ATPase alpha 1 locus to cosegregate with blood pressure in Dahl rats. J Hypertens 8: 457-62

# The C-Terminus of Na+,K+-ATPase β-Subunit Contains a Hydrophobic Assembly Domain.

A.T. Beggah, P. Beguin, P. Jaunin, *M.C. Peitsch, K.Geering.

Institut de Pharmacologie et de Toxicologie de l'Université, CH-1005 Lausanne and
*Institut de Biochimie de l'Université, CH-1066 Epalinges, Switzerland

## Introduction

Assembly of catalytic α-subunits of Na+,K+-ATPase with β-subunits is a prerequisite for the structural and functional maturation of newly synthesized α-subunit and its intracellular transport from the ER to the plasma membrane (2). Little is known on the structural domains in α- and β-subunits that are involved in subunit assembly. Recently, Renaud et al (7) have reported that cytoplasmic and transmembrane deletion mutants of β-subunits are all able to assemble with α-subunits, provided that they are capable to insert into ER membranes. Together with the observation that deletions of 11 amino acids from the C-terminal ectodomain abolishes assembly (6), these results suggest that the C-terminal domain of β-subunits contain important structural information for subunit assembly.

In this study, we have constructed cDNAs encoding β3-subunits of Xenopus Na+,K+-ATPase (3) that contain point mutations in the last 10 amino acids of the C-Terminus (267 GRVTFKVKITE277) and have analyzed the assembly competence of these mutants with Xenopus α-subunits (8).

## Results and Discussion

Sequence comparison between different β-isoforms reveals a certain degree of diversity among the 10 most C-terminal amino acids. Significantly, however, positions of some positively charged and of hydrophobic amino acids are conserved in all β-subunits. When the C-terminal amino acids are aligned to form a β-strand like structure, hydrophilic and hydrophobic amino acids are exposed on opposite sides of the polypeptide, creating two putative assembly domains. To test the respective importance of the hydrophilic charged and the hydrophobic amino acids in the assembly with α-subunits, single and multiple point mutations were introduced into the cDNA of the Xenopus β3 -subunit, according to the PCR method of Nelson and Long (5). The mutants were expressed in Xenopus oocytes from cRNA and their ability to assemble with α-subunits was analyzed by following the stabilization and cellular accumulation of co-expressed α-subunits (1) and by measuring expression of functional α-β complexes at the cell surface (4).

Table I: Effects of mutations in the C-terminus of Xenopus β3-subunits on assembly with α-subunits and cell surface expression of Na$^+$,K$^+$-ATPase

|   |   |   | (I) assembly efficiency | (II) ouabain binding | (III) Rb uptake |
|---|---|---|---|---|---|
| A. | 1 | β3 wt | +++ | +++ | +++ |
|   | 2 | d267 (Δ 10aa) | - | - | - |
|   | 3 | R268Q | +++ | nd | nd |
|   | 4 | R268E | +++ | nd | nd |
|   | 5 | K272Q | +++ | nd | nd |
|   | 6 | K272E | +++ | nd | nd |
|   | 7 | R268Q/K272Q | +++ | +++ | +++ |
|   | 8 | R268E/K272E | ++ | +++ | +++ |
|   | 9 | R268Q/K272E | +++ | nd | nd |
| B. | 1 | β3 wt | +++ | | |
|   | 2 | V269N | +++ | | |
|   | 3 | F271N | + | | |
|   | 4 | V273N | ++ | | |
|   | 5 | I275N | +++ | | |
|   | 6 | V269N/F271N | - | | |
|   | 7 | V269A/F271A | ++ | | |
|   | 8 | V273A/I275A | ++ | | |
|   | 9 | V269A/F271A/ V273A/I275A | - | | |

Mutants affected in hydrophilic (positively charged) (A) or in hydrophobic (B) amino acids of the C-terminus were tested for their assembly efficiency with α-subunits (I) and for their ability to form functional Na$^+$,K$^+$ pumps at the cell surface by measuring ouabain binding to intact oocytes (II) or Rb uptake (III) as described (4).

Table IA summarizes the results that we obtained with the mutants affected in the hydrophilic charged amino acids R268 and K272 on the assembly efficiency with α-subunits. In contrast to the deletion mutant (d267) that lacks 10 amino acids of the C-terminus and was not able to associate with α-subunits, all single or double point mutants produced stable α-β complexes with a similar efficiency than wild type β3-subunits. Consistently, the number and activity of cell surface expressed Na$^+$,K$^+$-pumps as assessed by ouabain binding and Rb flux measurements was increased to a similar extent in oocytes expressing α-wild type β complexes or α-mutant β complexes (Table IA). Thus, these data indicate that the two highly conserved positively charged amino acids in the C-terminus of β-subunits are not directly involved in the interaction with α-subunits and are not responsible for the lack of association observed with the deletion mutant d267.

The results were quite different with the mutants affected in the hydrophobic amino acids (Table IB). Single mutants in which V269, F271, V273 or I275 were changed to the polar uncharged asparagine (N) showed a similar (V269N, V273N and I275N) or a decreased (F271N) efficiency of assembly compared to β3 wild type and the double mutant V269N/F271N was no longer able to assemble with α-subunits. Finally, replacement of 2 hydrophobic residues by the less hydrophobic alanine resulted in mutants with a reduced assembly efficiency (V269A/F271A and V273A/I275A) and replacement of all 4 hydrophobic residues with alanine (V269A/F271A/V273A/I275A) abolished the ability to assemble with α-subunits (Table IB). Together these data support the hypothesis that the hydrophobic domain in the most C-terminal part of the β-subunit of $Na^+,K^+$-ATPase participates in the interaction with α-subunits.

## Acknowledgements

The studies were supported by the Swiss National Fund for Scientific Research (Grant No 31-26241-89 and 31-33676-92).

## References

1.  Ackermann U, Geering K (1990) Mutual dependence of Na,K-ATPase α-subunits and β-subunits for correct posttranslational processing and intracellular transport. FEBS Lett 269: 105-108
2.  Geering K (1991) The functional role of the β-subunit in the maturation and intracellular transport of Na,K-ATPase. FEBS Lett 285: 189-1992
3.  Good P J, Richter K, Dawid IB (1990) A nervous system-specific isotype of the β-subunit of $Na^+,K^+$-ATPase expressed during early development of Xenopus-laevis. Proc Natl Acad Sci USA 87: 9088-9092
4.  Jaunin P, Horisberger JD, Richter K, Good PJ, Rossier BC, Geering K (1992) Processing, intracellular transport and functional expression of endogenous and exogenous α-β3 Na,K-ATPase complexes in Xenopus oocytes. J Biol Chem 267: 577-585
5.  Nelson RM, Long GL (1989) A general method of site-specific mutagenesis using a modification of the Thermus aquaticus polymerase chain reaction. Anal Biochem 180: 147-151
6.  Renaud KJ, Fambrough DM (1991) Molecular analysis of Na,K-ATPase subunit assembly. In: Kaplan JH, De Weer P (eds) The Sodium Pump:Recent Developments, Vol. 46, The Rockefeller University Press, New York, pp 25-29
7.  Renaud K J, Inman EM Fambrough DM (1991) Cytoplasmic and transmembrane domain deletions of Na,K-ATPase β-subunit. Effects on subunit assembly and intracellular transport. J Biol Chem 266: 20491-20497
8.  Verrey F, Kairouz P, Schaerer E, Fuentes P, Geering K, Rossier BC, Kraehenbuhl JP (1989) Primary sequence of Xenopus laevis $Na^+,K^+$-ATPase and its localization in A6 kidney cells. Am J Physiol 256: F1034-F1043

# Synthesis and secretion of the extracellular domain of the β1 subunit isoform of NA+/K+-ATPase into the culture medium of CHO cells

S. Gloor, K. Nasse

Neurobiology, Federal Institute of Technology, CH-8093, Switzerland

## Introduction

The membrane topology of the β subunit of Na+/K+-ATPase suggests that the protein has one single membrane spanning segment about 35 amino-acids behind the N-terminus, followed by a huge extracellular domain. Therefore, a recombinant β1 subunit protein lacking the intracellular N-terminal amino-acids as well as the computer predicted transmembrane domain should be directed to the secretory pathway if provided with the appropriate signal peptide sequences at the new N-terminal end. We have chosen this approach for two reasons. First, secretion into the culture medium would clearly demonstrate that the β subunit of the Na+/K+-ATPase β subunit is anchored by one single transmembrane segment. Second, since structural analysis of membrane anchored proteins is still a difficult task to solve, the secreted extracellular domain could be a suitable target for such analysis as has been shown for example with the extracellular domain of CD4 (1).

## Results and Discussion

Type-II membrane proteins are inserted into the membrane with the N-terminus facing the cytoplasmic side and they do not have a N-terminal signal peptide sequence which is cleaved during translocation through the membrane of the ER. Instead, the transmembrane domain functions both as a signal peptide sequence, targetting the nascent protein chain to the ER

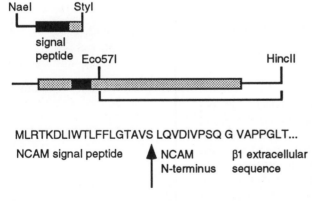

Figure 1. Construction of the chimeric NCAM - β1 subunit clone. The Eco57I (blunt) - HincII fragment of β1 has been ligated to the NCAM signal sequence fragment NaeI -StyI (blunt). The new amino-acid sequence is indicated. The arrow marks the normal signal peptidase cleavage site.

203

and a stop transfer domain (2). To direct the extracellular domain of the β1 subunit to the secretory pathway, we have used a cDNA fragment encoding the signal peptide sequence and the nine first amino-acids of the neural cell adhesion molecule (NCAM), a type-I membrane protein and cloned the cDNA encoding the extracellular domain of β1 (3) behind it (Fig. 1). We have previously been able to show that this approach is feasible when expressing the extracellular domain of the mouse β2 subunit (3). The chimeric construct has been cloned into the eukaryotic expression vector pEE14 (5). Expression of foreighn genes is driven by the human cytomegalovirus promoter and transfected Chinese hamster ovary (CHO) cells are selected by growth in the presence of methionine sulfoximine (5). The culture medium of several stably transfected and cloned CHO cells has been analysed for the presence of the recombinant protein by Western Blot analysis using a polyclonal anti mouse β1 antibody. Fig. 2, lane 1 shows that the recombinant protein migrates as a diffuse smear with an apparent molecular weight of about 35 - 50 kDa.

To analyse the carbohydrates added to the recombinant protein in CHO cells we have made use of endoglycosidase H (Endo H) and N-Glycosidase F (Glyco F). Endo H is able to cleave N-linked oligosaccharides only when they are of high mannose or hybrid type. Fig. 2, lane 2, shows that the recombinant protein is resistant to Endo H cleavage, but can be digested with Glyco F, which removes complex type sugars (lane 3). The apparent molecular weight of the deglycosylated protein is about 28 kDa which corresponds quit well to the calculated molecular weight of 27.9 kDa when the protein is cleaved at the site utilized in the case of native NCAM (Fig. 1, arrow). In contrast to the mouse β2 subunit which is also O-glycosylated in this system (4), the β1 subunit carries only N-linked carbohydrates. It is not clear whether this difference reflects minor structural variations between the two isoforms. Interestingly, the β2 subunit isolated from mouse brain seems to be O-glycosylated also *in vivo*.

The recombinant protein can be immunoaffinity purified from culture supernatants of CHO cells using a mouse β1 specific monoclonal antibody. Fig. 3 shows a sample of purified protein analysed in a Silver gel. Since this antibody exclusively recognizes the native β1 subunit, it is likely that the truncated recombinant protein has a three dimen-

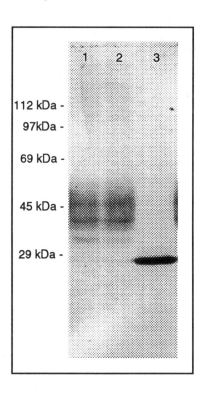

Figure 2. Carbohydrate analysis of the immunoaffinity purified extracellular domain of mouse b1. Lane 1 shows the purified protein probed with a polyclonal anti mouse b1 antibody. Lane 2 shows the protein after treatment with Endo H. Lane 3 shows the protein after complete removal of N-linked carbohydrates with Glyco F.

sional structure which is highly similar to the intact β1 subunit.

Our approach has demonstrated that the extracellular domain of the Na⁺/K⁺-ATPase β1 subunit can be directed to the secretory pathway in eukaryotic cells and isolated from the culture medium in high quality by immunoaffinity purification. Thus, the system could present a convenient way to generate large amounts of this domain for both structural and functional studies.

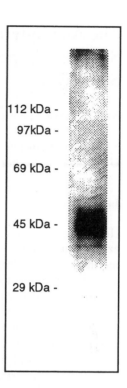

Figure 3. Immunoaffinity purified extracellular domain of mouse β1 analyzed in 0.1 %SDS-10 % PAGE. 100 ng of the purified recombinant protein have been applied and the protein visualized by the Silver stain method.

*Acknowledgements*

We thank Melitta Schachner for financial support.

**References**

1. Davis SJ, Ward HA, Puklavec MJ, Willis AC, Williams AF, Barclay AN (1990) High level expression in Chinese hamster ovary cells of soluble forms of CD4 T lymphocte glycoprotein including glycosylation variants. J Biol Chem 265: 10410-10418.

2. Lipp J, Dobberstein B (1986) Signal recognition particle-dependent membrane insertion of mouse invariant chain: A membrane-spanning protein with a cytoplasmically exposed amino terminus. J Cell Biol 102: 2169-2175.

3. Gloor S (1989) Cloning and nucleotide sequence of the mouse Na,K-ATPase β1 subunit. Nucl Acid Res 17: 10117.

4. Gloor S, Nasse K, Essen LO, Appel (1992) Production and secretion in CHO cells of the extracellular domain of AMOG/β2, a type-II membrane protein. Gene 120: 307-312.

5. Bebbington, C (1991) Expression of antibody genes in nonlymphoid mammalian cells. Methods 2: 136-145.

# The Amino-terminal Transmembrane Region of the $\beta$ Subunit is Important for Na$^+$/K$^+$-ATPase Complex Stability

Kurt A. Eakle $^\diamond$, Shyang-Guang Wang $^\diamond$, and Robert A. Farley $^{\diamond\#}$.

$^\diamond$Department of Physiology and Biophysics and $^\#$Department of Biochemistry, University of Southern California, Los Angeles, CA 90033 USA

## Introduction

Heterologous expression of the $\alpha$ and $\beta$ subunits of Na$^+$/K$^+$-ATPase allows for the expression of altered forms of the Na$^+$/K$^+$-pump which may be useful in determining the structural basis of pump function. Most systems for heterologous expression are limited by the low abundance of protein expression, the difficulty in producing assembled complexes with enzymatic activity, and a background of endogenous Na$^+$/K$^+$ ATPase activity. While the level of heterologous expression of Na$^+$/K$^+$-ATPase in yeast (*Saccharomyces cerevisiae*) is limited, a major advantage of this system is the lack of an endogenous Na$^+$/K$^+$-ATPase enzyme. Thus, activities which are unique to the Na$^+$/K$^+$-ATPase can be assayed with virtually no background. In particular, the ability of Na$^+$/K$^+$-ATPase to bind cardiac glycosides such as ouabain with high affinity has proven to be a reliable and sensitive marker for the assembly of functional enzyme complexes in yeast. Ouabain binding depends on phosphorylation of the enzyme and is sensitive to both Na$^+$ and K$^+$ which are thought to cause conformational shifts in Na$^+$/K$^+$-ATPase structure related to the transport of ions across the cell membrane. (1)

Work in our lab and elsewhere has demonstrated that the $\beta$ subunit for the H$^+$/K$^+$-ATPase (HK$\beta$) is capable of assembling with $\alpha$ subunits of the Na$^+$/K$^+$-ATPase (NK$\alpha$) into hybrid complexes which appear to be functional ion-transport ATPases. (2,3,4) Like Na$^+$/K$^+$-ATPase, co-expression of the HK$\beta$ subunit and (presumably) assembly with the NK$\alpha$ subunit leads to the appearance of a mature trypsin resistant conformation for NK$\alpha$ (4). These hybrid complexes are capable of high-affinity ouabain binding (2) and electrophysiological measurements suggest that they are capable of active ion transport (3). However, data indicate that these hybrid complexes may have a lower affinity for K$^+$ than Na$^+$/K$^+$-ATPase and imply a role for the $\beta$ subunit in defining K$^+$ binding sites involved in ion transport. Similar effects on K$^+$ affinity have been observed for complexes formed from the $\alpha$1 and $\beta$3 isoforms of toad bladder Na$^+$/K$^+$-ATPase (5).

### Hybrid Complexes of HK$\beta$ and NK$\alpha$ are Sensitive to SDS Extraction.

Figure 1 shows the results when microsomal membranes from yeast expressing the $\alpha$1 isoform of NK$\alpha$ in combination with either the $\beta$1 subunit of Na$^+$/K$^+$-ATPase

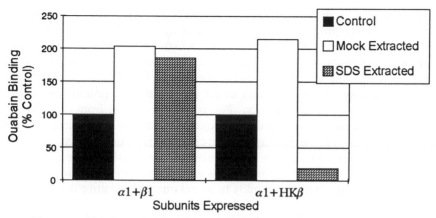

Figure 1. SDS Extraction of $\alpha1+\beta1$ or $\alpha1+HK\beta$ expressed in yeast. [3H]Ouabain binding was assayed as described (2).

or HK$\beta$ are extracted with SDS to a final concentration of 1 mg/ml (protein concentration 3.6 mg/ml, other conditions as described 6,7). The bars represent the amount of [$^3$H]ouabain binding in the SDS extracted membranes normalized to the amount of [$^3$H]ouabain binding in unextracted microsomes (control). [$^3$H]ouabain binding in unextracted microsomes was 1.6 pmol/mg microsomal protein for $\alpha1+\beta1$ and 0.9 pmol/mg protein for $\alpha1+HK\beta$. Complexes of $\alpha1+\beta1$ showed a significant enhancement in the amount of [$^3$H]ouabain binding upon SDS extraction and purification on sucrose gradients. In contrast, complexes of $\alpha1+HK\beta$ showed a marked loss of [$^3$H]ouabain binding under the same conditions. Similar experiments with the $\alpha3$ isoform of NK$\alpha$ combined with either $\beta1$ or HK$\beta$ showed the same effect (data not shown). The open bars in figure 1 represent a mock extraction done at the same time substituting water for the SDS solution and purifying the membranes on sucrose gradients as done with the SDS extracted samples. In this case, both $\alpha1+\beta1$ and $\alpha1+HK\beta$ complexes show significant increases in the amount of [$^3$H]ouabain binding. This argues that the loss of ouabain binding is a specific effect due to treatment with SDS. SDS at sufficiently high concentrations is known to dissociate $\alpha/\beta$ complexes of Na$^+$/K$^+$-ATPase leading to a loss of both enzymatic activity and the ability to bind [$^3$H]ouabain (7). The fact that $\alpha1+HK\beta$ complexes are sensitive to SDS extraction under the same conditions where $\alpha1+\beta1$ complexes are resistant to SDS treatment suggests that the HK$\beta$ subunit forms a less stable complex with $\alpha1$ than the $\beta1$ subunit.

Table 1. ATPase activity of yeast microsomes. ($\mu$mol Pi released/min/mg protein)

| Membranes | - Ouabain | + Ouabain | Na$^+$/K$^+$-ATPase Specific Activity |
|---|---|---|---|
| $\alpha1+\beta1$ Unextracted | 0.1751 | 0.1637 | 0.0114 |
| $\alpha1+\beta1$ SDS Extracted | 0.0471 | 0.0341 | 0.0130 |
| $\alpha1+HK\beta$ Unextracted | 0.1262 | 0.1373 | (-0.0110) |
| $\alpha1+HK\beta$ SDS Extracted | 0.0337 | 0.0331 | 0.0006 |

The activity of yeast microsomes before and after SDS extraction was measured by the coupled spectrophotometric assay as described (6) and the results are shown in Table 1. Unextracted yeast membranes have a high background of ATPase activity which is not ouabain sensitive. This results in a low ratio of ouabain sensitive ATPase activity (due to heterologous expression of $Na^+/K^+$-ATPase) to ouabain insensitive activity. SDS extraction under our conditions primarily reduces the ouabain insensitive ATPase activity. For yeast microsomes from cells expressing the $\alpha1+\beta1$ combination, we can detect a ouabain sensitive ATPase activity that is ~30% of the total ATPase activity. However, SDS extraction of yeast microsomes with the $\alpha3+HK\beta$ combination shows virtually no ouabain sensitive ATPase activity. Thus, the loss of ouabain binding seen upon SDS extraction of $\alpha1+HK\beta$ membranes correlates with an inability to detect ouabain sensitive ATPase activity.

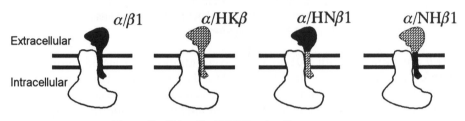

Figure 2. Chimeric $\beta1/HK\beta$ subunits

**Chimeric $\beta1/HK\beta$ subunits show that SDS sensitivity/resistance maps to the amino-terminus of the $\beta$ subunit.**

*In vitro* mutagenesis and cDNA cloning were used to exchange portions of the $\beta1$ and $HK\beta$ cDNAs at a position corresponding to amino acid 77 of the rat $\beta1$ sequence - approximately 15 amino acids past the putative transmembrane region of the $\beta$ subunit. The $NH\beta1$ chimera has the amino-terminus of the $Na^+/K^+$-ATPase $\beta1$ subunit including its transmembrane domain combined with the extracellular domain of the $HK\beta$ subunit. The complementary chimeric construct, $HN\beta1$, has the amino-terminus of the $HK\beta$ subunit combined with the extracellular domain of the $\beta1$ subunit. (The details of the cloning will be published elsewhere.) These constructs are represented pictorially in Figure 2. Co-expression of either chimeric $\beta$ subunit with the $\alpha1$ subunit of $Na^+/K^+$-ATPase results in the appearance of ouabain binding complexes in yeast microsomes. Typically, expression results in the appearance of 0.1-1.0 pmol [$^3$H]ouabain binding/mg microsomal protein. Figure 3 shows the results when these microsomes are extracted with SDS. As before, controls of $\alpha1+\beta1$ and $\alpha1+HK\beta$ show SDS resistance and sensitivity, respectively. When the $NH\beta1$ chimera is combined with $\alpha1$, the resulting complex is resistant to SDS extraction. In contrast, the combination of $\alpha1+HN\beta1$ is sensitive to SDS-extraction. These results indicate that the amino-terminal region of the $\beta$ subunit plays an important role in determining the stability of $Na^+/K^+$-ATPase $\alpha/\beta$ complexes. This suggests that there are strong protein-protein

208

interactions between the transmembrane and/or cytoplasmic domain of the $\beta$ subunit and the $\alpha$ subunits of Na$^+$/K$^+$-ATPase. Further chimeric $\beta$ constructs will help localize the region(s) of $\alpha/\beta$ interaction.

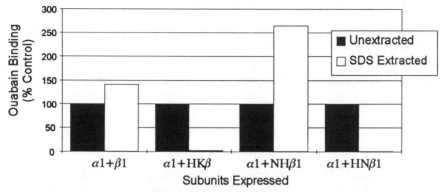

Figure 3. SDS Extraction of $\alpha 1/\beta$ or $\beta$ chimera complexes.

*Acknowledgements*

We would like to thank Ron Hitzeman for providing us with the YEp1PT yeast expression vector and strain 30-4 which was used for yeast expression, Ed Benz for providing the rat $\alpha 3$ and rat $\beta 1$ cDNAs, Jerry Lingrel for providing the sheep $\alpha 1$ cDNA, and Robert Levenson for providing the rat HK$\beta$ cDNA. Supported by AHA-GLAA Initial Investigator Award 983 F1-1 to K.A.E. and US Public Health Service Grants GM28673 and HL39295 and NSF Grant DMB-8919336 to R.A.F.

**References**

1. Yoda S, Yoda A (1986) ADP and K$^+$ sensitive phosphorylated intermediate of Na$^+$/K$^+$-ATPase. J Bio Chem 261:1147-1152
2. Eakle KA, Kim KS, Kabalin MA, Farley RA (1992) High-affinity ouabain binding by yeast cells expressing Na$^+$/K$^+$-ATPase $\alpha$ subunits and the gastric H$^+$/K$^+$-ATPase $\beta$ subunit. Proc Natl Acad Sci USA 89:2834-2838
3. Horisberger JD, Jaunin P, Reuben MA, Lasater LS, Chow DC, Forte JG, Sachs G, Rossier BC, Geering K (1991) The H$^+$/K$^+$-ATPase $\beta$-subunit can act as a surrogate for the $\beta$-subunit of Na$^+$/K$^+$-ATPase. J Bio Chem 266:19131-19134
4. Noguchi S, Maeda M, Futai M Kawamura M (1992) Assembly of a hybrid from the $\alpha$ subunit of Na$^+$/K$^+$-ATPase and the $\beta$ subunit of Na$^+$/K$^+$-ATPase. Biochem Biophys Res Comm 182:659-666
5. Jaisser F, Canessa CM, Horisberger JD, Rossier BC (1992) Primary sequence and functional expression of a novel ouabain resistant Na,K-ATPase. J Bio Chem 267:16895-16903
6. Wang K, Farley RA (1992) Lysine 480 is not an essential residue for ATP binding or hydrolysis by Na$^+$/K$^+$-ATPase. J Bio Chem 267:3577-3580
7. Jorgensen PL (1974) Purification and characterization of Na$^+$/K$^+$-ATPase. Biochim Biophys Acta 356:36-52

# Construction and Expression of Chimeric cDNAs of Na$^+$/K$^+$- and H$^+$/K$^+$-ATPase β-subunits

S. Ueno[1], M. Kusaba[3], K. Takeda[2], F. Izumi[1], and M. Kawamura[2]

[1]Department of Pharmacology and [2]Biology, University of Occupational and Environmntal Health, Kitakyushu, 807 Japan
[3]Department of Biochemical Engineering and Science, Faculty of Computer Science and Systems Engineering, Kyushu Institute of Technology, Iizuka, 820 Japan

## Introduction

Na$^+$/K$^+$-ATPase and H$^+$/K$^+$-ATPase consist of a catalytic α-subunit and a glycosylated β-subunit. Both β-subunits have similar structural features - they span the membrane once, leaving a short N-terminal portion in the cytoplasmic side, and a large portion including C-terminus outside of the cell where there are three highly conserved disulfide bridges(for a review, see ref.1). H$^+$/K$^+$-ATPase β-subunit forms a stable assembly with Na$^+$/K$^+$-ATPase α-subunit, but the resulting αβ complex is not functional, at least in ATPase activity(5). Construction of chimeras between Na$^+$/K$^+$- and H$^+$/K$^+$-ATPase β-subunit, and the expression of the chimeras with Na$^+$/K$^+$-ATPase α-subunit in *Xenopus* oocytes should allow us to determine which regions of Na$^+$/K$^+$-ATPase β-subunit are required for the expression of functional αβ complex.

## Materials and Methods

The cDNA of H$^+$/K$^+$-ATPase β-subunit from pig was kindly provided by Dr. M. Futai (Department of Organic Chemistry and Biochemistry, Institute of Scientific and Industrial Research, Osaka University, Osaka, Japan). It was necessary to modify it for the efficient expression in *Xenopus* oocytes (7) : the plasmid was constructed to carry the 5'- and 3'-flanking regions of *Torpedo californica* Na$^+$/K$^+$-ATPase β-subunit(pSPT β)(4). Then, two new unique restriction sites were created by PCR or site-directed mutagenesis within the coding regions of Na$^+$/K$^+$- and H$^+$/K$^+$-ATPase β-subunits ; one was a *Sna*BI site located within the membrane-spanning domain and the other was a *Eco*T22I site between the second and the third disulfide bridge. After cleavage of the β-subunits with *Sna*BI and *Eco*T22I, we constructed six types of chimeric β-subunit cDNAs by combining the resulting fragments as shown in Fig. 1.

Messenger RNAs for wild-type and chimeric β-subunits as well as Na$^+$/K$^+$-ATPase α-subunit were synthesized *in vitro* by using SP6 RNA polymerase(2,3). Microinjection of mRNAs into *Xenopus* oocytes, identification of the translation products, and assay of ATPase activity of oocytes microsomes and ouabain binding were carried out as described previously(6).

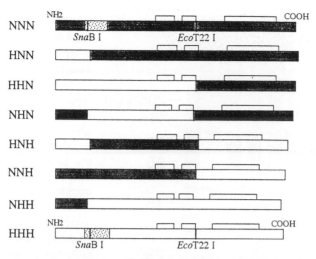

**Figure 1. Constructs of chimeric Na$^+$/K$^+$- and H$^+$/K$^+$-ATPase β-subunits.**
Schematic representations of wild-type β-subunits, NNN(Na$^+$K$^+$-ATPase) and HHH(H$^+$K$^+$-ATPase), and chimeras, HNN, HHN, NNH, NHH, HNH, and NHN are shown. These constructions are set by *three capital letters* which represent the regions from N-terminus to SnaBI site, from SnaBI site to EcoT22I site, and from EcoT22I site to C-terminus of Na$^+$K$^+$-ATPase(N) and H$^+$K$^+$-ATPase(H). Na$^+$K$^+$-ATPase(*open bars*), H$^+$K$^+$-ATPase(*filled bars*), and the membrane-spanning domain(*dotted bars*) are shown in the schemas. The positions of disulfide bridges are also indicated.

**Results and Discussion**

Immunoprecipitation of biosynthetically [$^{14}$C]leucine labeled proteins in oocytes injected with mRNA for the β-subunit demonstrated that not only the wild-type but also chimeric β-subunits were expressed in large amounts in *Xenopus* oocytes(data not shown). We next investigated whether the chimeric β-subunits can assemble or not with Na$^+$/K$^+$-ATPase α-subunit. As shown in Fig. 2 , all chimeric β-subunits were co-immnoprecipitated with anti-α-subunit antiserum, indicating that every chimeric β-subunit can form a complex with the α-subunit. The mobility of the chimeric β-subunits on SDS-PAGE was different due to the difference in number of oligosaccharide chains attached and/or the mass of polypeptides.

To determine whether these αβ complexes were functional or not, we assayed ouabain-sensitive ATPase activity and the ouabain binding capacity of oocytes injected with mRNA for wild-type and chimeric β-subunits together with mRNA for the α-subunit(Table 1). Among the complexes, only the α-HNN complex had a comparable increase in both activities to those of α-NNN complex, i.e. the wild-type. The α-HHN and α-NNH complexes had no increase in both activities, but the α-NHH and α-HHH complexes had a small but significant increase only in ouabain binding.

All chimeric β-subunits that we constructed in this study assembled with Na$^+$/K$^+$-ATPase α-subunit in *Xenopus* oocytes, but the resulting αβ complexes were not functional in ATPase activity, except the α-HNN complex. These data suggest that the sequence

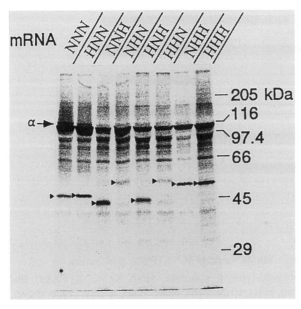

mRNA  NNN HNN NNH NHN HNH HHN NHH HHH

— 205 kDa
,116
97.4
— 66
— 45
— 29

**Figure 2. Assembly of wild-type and chimeric β-subunits with α-subunit in *Xenopus* oocytes.**

Oocytes were injected with wild-type or chimeric β-subunits mRNAs(10ng/oocyte) together with mRNAα(10ng/oocyte), and incubated at 19°C for 3 days in the presence of [¹⁴C]leucine. Triton X-100 extracts of the oocytes were subjected to immnoprecipitation using anti-α-subunit antiserum(6). The immnoprecipitates were subjected to SDS-PAGE, followed by fluorography. The β-subunits are indicated by *arrow heads*.

| mRNAβ injected | ATPase activity ( μmolPi/mg/hr) | | | | Ouabain binding (fmol/oocyte) | |
|---|---|---|---|---|---|---|
| | exp.1 | exp.2 | exp.3 | exp.4 | exp.1 | exp.2 |
| uninjected | 0.3 | 1.7 | 2.5 | 0.6 | 1.3 | 1.0 |
| NNN | 1.9 | 7.8 | 7.2 | 2.2 | – | 7.2 |
| HNN | – | – | 5.0 | 1.5 | 8.1 | 8.0 |
| HHN | – | – | – | 0.6 | 1.1 | 0.9 |
| NNH | – | – | – | 0.5 | 1.5 | 1.0 |
| NHH | – | – | – | 0.5 | 3.8 | 2.0 |
| HHH | 0.3 | 1.8 | 2.4 | 0.4 | 2.9 | 1.9 |

bars : not determined

**Table 1. Ouabain-sensitive ATPase activities and ouabain binding capacities of oocytes co-injected with chimeric β-subunit mRNAs and mRNA α.**

Oocytes were injected with wild-type or chimeric β -subunits mRNA(10ng/oocyte) together with mRNAα(10ng/oocyte), and incubated at 19°C for 3 days. The assay of ouabain-sensitive ATPase activity was performed as described previously(6). For ouabain binding assay, defolliculated oocytes were incubated in modified Barth's medium containing 100nM[³H]ouabain at 19°C for 1 hr (6).

from N-terminus to *Sna*BI site in Na⁺/K⁺-ATPase β-subunit is not an essential requirement for forming the functional enzyme. They also raise the possibility that there exist some sequences in the β -subunit required for functional expression , from the *Sna*BI to *Eco*T22I site, as well as from the *Eco*T22I site to the C-terminus. Construction of several other types of chimeric β -subunits are in progress in our laboratory in order to identify these sequences.

*Acknowledgements*

We thank Dr. M. Futai, Osaka, Japan, for providing the cDNA for the pig H$^+$/K$^+$-ATPase β-subunit, and Ms. Atsuko Sugino for her typing the manuscript.

# References

1. Jørgensen PL, Andersen JP (1988) Structural basis for E1-E2 conformation transitions in Na,K-pump and Ca-pump proteins. J Membr Biol 103: 93-120
2. Konarska MM, Padgett RA, Sharp PA (1984) Recognition of cap structure in splicing in vitro of mRNA precursors. Cell 38: 731-736
3. Melton DA, Krieg PA, Rebagliati MR, Maniatis T, Zinn K, Green MR (1984) Efficient in vitro synthesis of biologically active RNA and RNA hybridization probes from plasmids containing a bacteriophage SP6 promoter. Nucleic Acids Res 12: 7035-7056
4. Noguchi S, Higashi K, Kawamura M (1990) A possible role of the β-subunit of (Na,K)-ATPase in facilitating correct assembly of the α-subunit into the membrane. J Biol Chem 265: 15991-15995
5. Noguchi S, Maeda M, Futai M, Kawamura M (1992) Assembly of a hybrid from the a subunit of Na$^+$/K$^+$-ATPase and the β subunit of H$^+$/K$^+$-ATPase. Biochem Biophys Res Commun 182: 659-666
6. Noguchi S, Mishina M, Kawamura M, Numa S (1987) Expression of functional (Na$^+$+K$^+$)-ATPase from cloned cDNAs. FEBS lett 225: 27-32
7. Noguchi S, Ohta T, Takeda K, Ohtsubo M, Kawamura M (1988) Ouabain sensitivity of a chimeric α subunit (Torpedo / Rat) of the (Na,K)ATPase expressed in Xenopus oocyte. Biochem Biophys Res Commun 155: 1237-1243

213

# Segregation of Maternal Na$^+$/K$^+$-pump α1 and ß3 mRNAs to Animal Blastomeres of Early *Xenopus* Embryos

S. Friehs, G. Schmalzing

Max-Planck-Institute of Biophysics, Department of Cell Physiology, Heinrich-Hoffmann-Str. 7, D-60528 Frankfurt/Main, Germany

## Introduction

During oogenesis *Xenopus* oocytes accumulate a large stockpile of Na$^+$/K$^+$-pump molecules (10) consisting of α1 and ß3 subunits (7). Fully-grown oocytes continue to synthesize significant amounts of α1 subunits, but the formation of new α1/ß3 complexes is strictly limited by a very low level of synthesis of endogenous ß3 subunits (1). The α1 subunits that fail to find a ß3 subunit are maintained in an assembly-competent form (1) and can be rescued to leave the endoplasmic reticulum (ER) by pairing with exogenous ß1 subunits synthesized later from injected cRNAs (9).

Meiotic maturation that prepares fully-grown oocytes for fertilization is associated with a redistribution of all the surface Na$^+$/K$^+$-pump molecules from the plasma membrane to internal membranes (8). Following fertilization, the early embryo proceeds through twelve rapid and synchronous cell divisions in the virtual absence of transcription. The maternal Na$^+$/K$^+$-pumps are not subjected to degradation, but reinserted from internal stores into the newly formed interblastomeric plasma membranes of the early embryo. Besides complete α1/ß3 Na$^+$/K$^+$-pumps, also the maternal pool of α1 subunits that lack a ß subunit (3) as well as α1-, ß1-, and ß3-specific mRNAs (5) are inherited by the embryos. Since specific functions of cells may arise from the localization and subsequent translation of the mRNAs, we have investigated the spatial distribution of the maternal mRNAs encoding the α1/ß3 Na$^+$/K$^+$-pump isozyme during the development of oocytes and early embryos of *Xenopus laevis*.

## Localization of α1 and ß3 mRNAs during oogenesis

Competitive PCR measurements revealed that the mRNA coding for the α1 subunit of the *Xenopus* Na$^+$/K$^+$-pump accumulates rapidly during oocyte development, reaching a maximum concentration already during Dumont stage III, and then declines (not shown). Fig. 1a shows by *in situ* hybridization that the α1 mRNA is evenly spread in high concentrations throughout the cytoplasm of young oocytes (stage I and stage II). Staining for the ß3 subunit-specific mRNA (2) revealed virtually the same uniform distribution (not shown). In later stages of oogenesis, however, both α1 and ß3 subunit mRNAs become concentrated around the nucleus and are in general more abundant in the cytoplasm of the animal than of the vegetal hemisphere (Fig. 1b). The animal hemisphere of stage VI oocytes shows radial band-shaped or network-shaped localizations of the α1 mRNA (Fig. 1d). In the vegetal hemisphere, spots and patches of the α1 mRNA are observed (Fig. 1c).

214

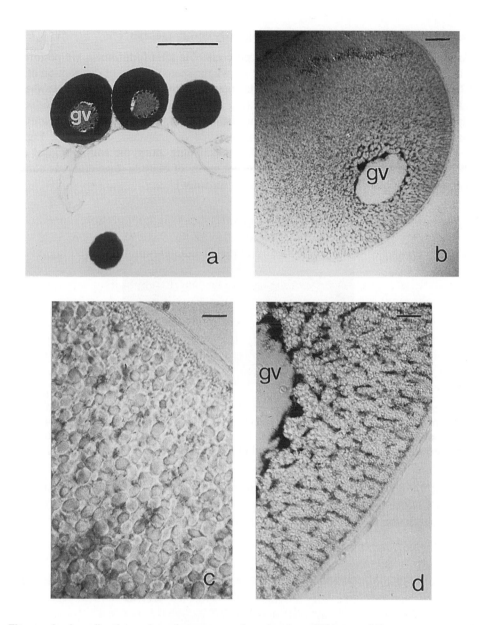

**Figure 1.** Localization of endogenous α1 subunit mRNA at different stages of oogenesis by *in situ* hybridization. Oocytes from an albino line were fixed in formaldehyde/acetic acid and embedded in paraffin. Sections cut at 5 μm were dewaxed, rehydrated, and hybridized with a PCR-generated probe internally labelled with digoxigenin (Boehringer Mannheim) and complementary to the *Xenopus* α1 mRNA. DNA-RNA hybrids were detected with an alkaline phosphatase-conjugated antibody to digoxigenin and the bromochloroindolyl phosphate/nitro blue tetrazolium (BCIP/NBT) substrate. **a**, Early stage oocytes; **b**, stage VI oocyte; **c**, vegetal region of a stage VI oocyte; **d**, animal region of a stage VI oocyte. gv, germinal vesicle. Bars in a and b, 100 μm; bars in c and d, 25 μm.

215

### Localization of α1 and ß3 mRNAs during early embryogenesis

To trace the fate of the Na⁺/K⁺-pump-specific mRNAs after fertilization, we performed whole mount *in situ* hybridizations. Fig. 2 shows the spatial pattern of distribution of the ß3 mRNA in a 4-cell embryo, two morula stages and two blastula stages. Both the α1 mRNA (not shown) and the ß3 mRNA (Fig. 2) are almost exclusively segregated to blastomeres arising from the animal hemisphere. Apparently, the uneven mRNA distribution along the animal-vegetal axis observed on immature oocytes (cf. Fig. 1) becomes more pronounced after fertilization. Sectioning along the animal-vegetal axis shows that the cells surrounding the blastocoel cavity of blastula-stage embryos contain high concentrations of ß3 mRNA (indicated by an arrow).

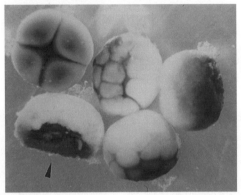

**Figure 2.** Detection of Na⁺/K⁺-pump ß3-subunit mRNA at different stages of early *Xenopus* development. Fixed and rehydrated embryos were treated with proteinase K, then acetylated and refixed, and finally hybridized (4) with a digoxigenin-labelled antisense ß3 cRNA probe. RNA-RNA hybrids were visualized by an antibody to digoxigenin conjugated to alkaline phosphatase and the BCIP/NBT substrate.

Total RNA was extracted separately from animal and vegetal blastomeres of 4-cell embryos. Nothern blots of these RNA preparations showed that the ß3 mRNA is 20 times more abundant in the animal blastomeres than in the vegetal blastomeres (Fig. 3).

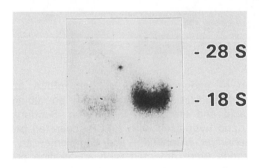

**Figure 3.** Northern blot of total RNA extracted from animal and vegetal blastomeres of 4-cell embryos. Total RNA extracts from animal or vegetal blastomeres were hybridized to an antisense ß3 cDNA probe labelled with ³²P by random priming.

216

Fig. 4 shows that also the α1 subunit polypeptide is concentrated in the animal hemisphere. This suggests that the protein product may also be localized, although we cannot rule out that most of the α1 subunits detected by the antibody are of maternal origin. Since the growth rate of the blastocoel cavity depends on the $Na^+/K^+$ transport rate, the segregation of maternal $Na^+/K^+$-pump polypeptides and mRNAs to animal blastomeres might be a prime cause for the formation of the blastocoel in the animal hemisphere.

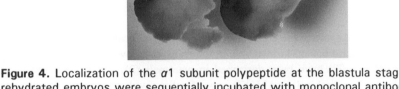

**Figure 4.** Localization of the α1 subunit polypeptide at the blastula stage. Fixed and rehydrated embryos were sequentially incubated with monoclonal antibody 5α to the α1 subunit (6), a peroxidase-conjugated secondary antibody, and diaminobenzidine. The embryo was cut along the animal-vegetal axis to make the blastocoel visible.

*Acknowledgement*

The *Xenopus* ß3 cDNA was kindly provided by Dr. Klaus Richter (University of Salzburg, Austria). We thank Dr. Richter and Günter Lepperdinger for their help with the whole mount *in situ* hybridizations and the hospitality during the stay of one of us (S.F.) in Dr. Richter's laboratory in Salzburg as a recipient of an EMBO short term fellowship.

**References**

1.  Geering K, Theulaz I, Verrey F, Häuptle MT, Rossier BC (1989) A role for the ß-subunit in the expression of functional $Na^+$-$K^+$-ATPase in *Xenopus* oocytes. Am J Physiol 257: C851-C858
2.  Good PJ, Richter K, Dawid IB (1990) A nervous system-specific isotype of the ß subunit of $Na^+,K^+$-ATPase expressed during early development of *Xenopus laevis*. Proc Natl Acad Sci USA 87: 9088-9092
3.  Han Y, Pralong-Zamofing D, Ackermann U, Geering K (1991) Modulation of Na,K-ATPase expression during early development of *Xenopus laevis*. Dev Biol 145: 174-181
4.  Hemmati-Brivanlou A, Frank D, Bolce ME, Brown BD, Sive HL, Harland RM (1990) Localization of specific messenger RNAs in *Xenopus* embryos by whole-mount *in situ* hybridization. Development 110: 325-330
5.  Kairuz P, Corthesy I, Rossier BC (1990) Developmental regulation of Na,K-ATPase isoforms during early development in *Xenopus laevis*. J Gen Physiol 96: 67a
6.  Lebovitz RM, Takeyasu K, Fambrough DM (1989) Molecular characterization and expression of the $(Na^+ + K^+)$-ATPase alpha-subunit in *Drosophila melanogaster*. EMBO J 8: 193-202
7.  Pralong-Zamofing D, Yi QH, Schmalzing G, Good P, Geering K (1992) Regulation of α1-ß3 $Na^+/K^+$-ATPase isozyme during meiotic maturation of *Xenopus laevis* oocytes. Am J Physiol 262: C1520-C1530
8.  Schmalzing G, Eckard P, Kröner S, Passow H (1990) Downregulation of surface sodium pumps by endocytosis during meiotic maturation of *Xenopus laevis* oocytes. Am J Physiol 258: C 179-184
9.  Schmalzing G, Gloor S, Omay H, Kröner S, Appelhans H, Schwarz W (1991) Up-regulation of sodium pump activity in *Xenopus laevis* oocytes by expression of heterologous ß1 subunits of the sodium pump. Biochem J 279: 329-336
10. Schmalzing G, Kröner S, Passow H (1989) Evidence for intracellular sodium pumps in permeabilized *Xenopus laevis* oocytes. Biochem J 260: 395-399

# Isoform Specific Antisera for the Na$^+$/K$^+$-ATPase ß1 and ß2 Subunits: Localization in Fetal Rat Tissues

L.M. González-Martínez, J. Avila Marrero, E. Martí, E. Lecuona, P. Martín-Vasallo

Laboratorio de Biología del Desarrollo (LBD), Departamento de Bioquímica y Biología Molecular. Universidad de La Laguna, 38206 La Laguna, Tenerife, Spain

## Introduction

In the Na$^+$/K$^+$-ATPase system, the alpha subunit has been implicated in the ion pumping process and the ß subunit has no known role in ion transport. Two different isoforms of the ß subunit (ß1 and ß2) have been described in mammals (3). The ß2 isoform is an adhesion molecule on glial cells (AMOG), specifically involved in neuron-astrocyte adhesion (1). The ß2 isoform has no obvious adhesion function in other cell types, and no adhesion function whatsoever has been reported for the ß1 isoform. In order to understand further the localization and functional implications, we have generated a series of isoform-specific antibodies against the human Na$^+$/K$^+$-ATPase ß1 and ß2 isoforms. Polyclonal rabbit antisera were raised against truncated ß-isoform molecules. These proteins were made in *E. coli* with pET expression vectors (5). Immunohistochemical preparations of 15-days-old whole rat embryos and on nerve cell cultures obtained from those, showed an specific staining pattern, particularly enhanced in the apical membrane of ear semicircular ducts and choroid plexus.

## ß isoforms localization in embryonic rat tissues

Figure 1 shows the antisera test for isoform specificity on Western blots of total extracts of bacterial cells expressing both of the ß isoform truncated proteins. Antisera reacted only with its corresponding antigen. These results established that these human ß subunit antisera are specific for their respective isotypes. These anti-human ß subunit antibodies have been tested on immunoblots containing microsomes from human, bovine and rat tissues with identical intensity (data not shown). To further characterize that the native antigen is the Na$^+$/K$^+$-ATPase ß subunit, we performed a series of deglycosylation experiments to show that the bands recognized by these antibodies are glycoproteins that, once deglycosylated, have a core peptide of the predicted molecular weight, as it was previously published by Shyjan, Gottardi and Levenson regarding to the rat Na$^+$/K$^+$-ATPase ß subunit antibodies (4), figure 2, panels A and B.

Neither ß1 nor ß2 mRNA has been detected by Northern blot in adult rat liver (3). The liver of two weeks old rats shows an slightly expression of the ß1 isoform and the same organ of neonatal rats expresses appreciable levels of specific ß2-mRNA (3),

218

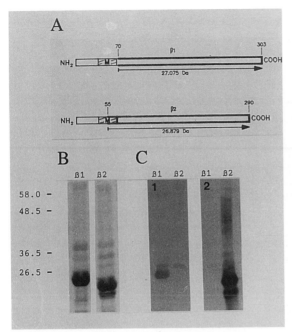

A

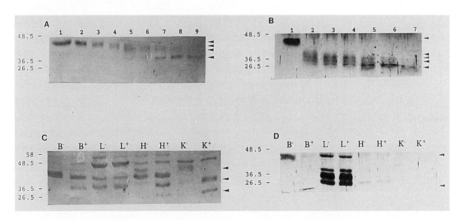

Figure 1. Panel A - Scheme of the Na⁺/K⁺-ATPase ß1 and ß2 subunit isoforms showing their intramembrane domain (M) and (lined with an arrow) the corresponding region to the truncated peptides used as antigen. Panel B - ß1 and ß2 antigens. Coomasie Blue stain of an SDS-containing 12.5 % polyacrylamide gel showing the induced ß1 and ß2 truncated proteins. Panel C - Specificity of antisera raised against ß1 and ß2 truncated proteins. Total bacterial protein after induction with IPTG were transferred to Immobilon filters and probed with sera anti-ß1 (1) and -ß2 (2) truncated proteins.

Figure 2. Na⁺/K⁺-ATPase ß isoforms deglycosylation. Panel A (ß1) and B (ß2). 0.3 Na⁺/K⁺-ATPase units of adult rat brain microsomal fractions were digested with (lanes 1 to 9) 0, 1.56x10⁻³, 3.1x10⁻³, 6.25x10⁻³, 0.0125, 0.025, 0.05, 0.1 and 0.2 of N-glycosidase F units in ß1 deglycosylation experiments and (lanes 1 to 7) 0, 0.0125, 0.025, 0.05, 0.1, 0.2 and 0.4 units in ß2 deglycosylation experiments, for 2 h at 37°C. Protein expression of Na⁺/K⁺-ATPase ß subunit isoforms in rat embryo (E15). Panel C (ß1) and D (ß2).Microsomes containing 0.18 Na⁺/K⁺-ATPase units/lane were fractionated by SDS-PAGE, 10 % polyacrylamide, transferred to Immobilon and probed with specific antisera. Molecular weigth markers (10⁻³) were run in a parallel lane of the same gel and are indicated. B, brain. L, liver. H, heart. K, kidney. + and -, treated or not treated with N-glycosidase F.

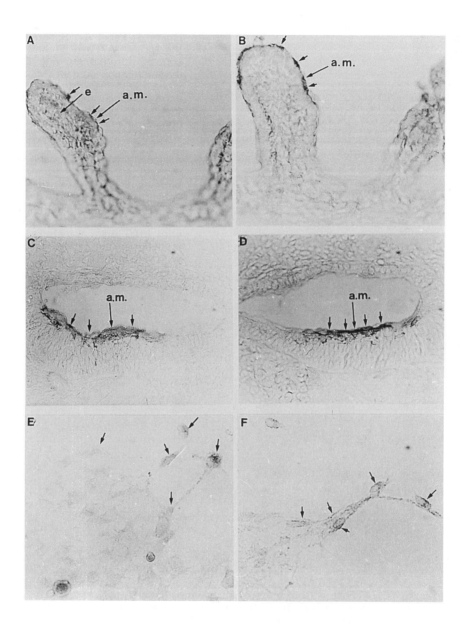

Figure 3. Immunolocalization of Na$^+$/K$^+$-ATPase ß subunit isoforms. Panel A (ß1) and B (ß2). Ten microns cryostat sections through the fourth cerebral ventricle of a rat embryo (E15) showing part of choroid plexus. Immunostaining shows expression of both ß subunit isoforms at the apical membranes (am) of the cuboidal epithelium of the plexus. Panel C (ß1) and D (ß2). Ten microns cryostat sections through the semicircular canals of the inner ear of a rat embryo (E15). Immunostainig for anti-ß1 and anti-ß2 subunits is discretelly related to the apical membranes of the thickened epithelium of the ampulla. Remaining epithelial cells do not exhibit immunostainig. Panel E (ß1) and F (ß2). Rat embryo (E15) hippocampal primary culture cells. Magnification 1700X.

the rat fetal liver expresses both isoforms at close similar levels (data not shown). We wanted to confirm the expression at the protein level in fetal brain, liver, heart and kidney by setting an immunoblot with the same amount of $Na^+/K^+$-ATPase activity in every lane, figure 2, panels C and D. We found ß1 isoform in brain, heart and kidney and ß2 protein only in brain. Neither ß1 nor ß2 protein was detected in liver. Which ß isoform is responsible for that $Na^+/K^+$-ATPase activity still remains to be clarified.

The cell and tissue specific localization of the ß subunit isoforms was studied in para-sagittal slices of E15 rat embryo. Controls of immunoreaction using non-immune rabbit serum are always negative. Surprisingly, among a general shadowing of organs, tissues and cells, the stainig was more intense in inner ear areas and 4th ventricle choroid plexus. Every cuboidal cell expresses either ß1 or ß2 immunostaining at their apical membranes (in contact with the cerebrospinal fluid) figure 3, panels A and B. This localization is in agreement with the cell role in the cerebrospinal fluid production.

In the semicircular canals of the membranous labyrinth of the inner ear, immunohistochemistry for both antisera also reveals a discrete pattern of distribution related to the ampullary crests (the thickened epithelium of the ducts). Ampullary crests contain vestibular hair cells and supporting cells, both kind of cells are integral parts of the collumnar epithelia, which apical membranes are in contact to the endolymphatic fluid. Apical membranes of these cells express ß1 and ß2 isoforms, figure 3, panels C and D. Iwano et al. in a quantitative immunogold localization of the $Na^+/K^+$-ATPase alpha subunit in the rat cochlear duct find a basolateral gold particles precipitation of the interdental and Hensen's cells but a rare o no label in hair cells (2). This is opposite of our pattern of findigs of $Na^+/K^+$-ATPase ß isoforms, not only regarding to the kind of cells but also to de pole of the cell. Taking together both data one could conclude that either, as in fetal liver, the ratio alfa/beta is not equimolecular or there is another ß isoform in mammals which complete the 1:1 alpha/beta molecular ratio.

Aknowledgments: We thank Dr. Teresa Alonso for sequence checking of plasmid constructions and Dr. C. Hdez Calzadilla for help at all times. This work has been supported by grants from Fondo de Investigaciones Sanitarias de la Seguridad Social FIS-93/0831, and Consejería de Educación del Gobierno Autónomo de Canarias 92/08.03.90 (P M-V). LM G-M is a FIS Formación de Personal Investigador Fellow.

# References

1.      Gloor, S., Antonicek, H., Sweadner , K.J., Pagliusi S., Frank R., and Schachner, M. (1990) The adhesion molecule on glia (AMOG) is a homologhe of the beta subunit of the Na,K-ATPase. J. Cell. Biol. 110, 165-174.
2.      Iwano, T., Yamamoto, A., Omori, K., Kawasaki, K., Kumazawa, T., Tashiro, Y. (1990) Quantitative immunogold localization of $Na^+/K^+$-ATPase alpha subunit in the tympanic wall of rat cochlear duct. J. Histochem. Cytochem. 38, 225-232.
3.      Martín Vasallo, P., Dackowski, W., Emanuel, J.R. and Levenson R. (1989) Identification of a putative isoform of the Na,K-ATPase beta subunit: primary structure and tissue, specific expression. J. Biol. Chem. 264, 46131-4618.
4.      Shyjan, A.W., Gottardi, C., Levenson, R. (1990). The Na,K-ATPase ß2 subunit isoform is expressed in rat brain and copurifies with Na,K-ATPase activity. J. Biol. Chem. 265, 5166-5169.
5.      Studier, F.W., Rosenberg, A.H., Dunn, J.J., Dubendorff, J.W. (1990) Use of T7 RNA Polymerase to direct expression of cloned genes. Methods in Enzymology, Goeddel, D.V. ed. vol 185, pp. 60-89. Acad. Press Inc, New York.

# Polyclonal Antibodies to Extramembrane Domains of Na$^+$/K$^+$-ATPase $\alpha$1 and $\alpha$3 Isoforms

M.C. Antonelli♦, M. Costa Lieste♦, A. Saredi♦, E. Malchiodi♠ and W.L. Stahl♣.

♦IQUIFIB, Facultad de Farmacia y Bioquímica,♠Cátedra de Inmunología, Facultad de Farmacia y Bioquímica, Buenos Aires, ARGENTINA.
♣VA Medical Center, Seattle, Washington, USA.

## Introduction

The catalytic $\alpha$ subunit of Na$^+$/K$^+$-ATPase (EC 3.6.1.3.) exists in at least three isoforms and contains the binding site for ouabain at the external side of the membrane. Analysis of the primary sequence of this subunit led to the proposal that it presents 4-5 extracellular domains, some implicated in ouabain binding (4, 5). Polyclonal antibodies raised against synthetic peptides in non-homologous regions of the specific a isoform are useful to delineate sequences of importance in ouabain binding, as well as determining the sidedness of the C-terminus (1,3).

In a first approach, we have generated antibodies to synthetic peptides corresponding to the C-terminus (RPGGWVEKETYY) and N-terminus (GRDKYEPAAVSE) of $\alpha$1 isoform and to the H1-H2 region (QAGTEDDPSGDN) of $\alpha$3 isoform. The peptides were coupled to carrier molecules and injected into rabbits following a standard protocol of immunization. The sera obtained were characterized by ELISA and Western Blots. Immunocytochemistry was performed using fixed rat brain sections and $^3$H-ouabain binding was determined in brain sections using quantitative autoradiography (QAR).

## Results and discussion:

Results obtained show that antisera to C- and N-terminus and H1-H2 of Na$^+$/K$^+$-ATPase $\alpha$ subunit have a titer of 30000, 51200 and 72900 respectively determined by ELISA. Western blots showed a band corresponding to the $\alpha$ subunit that was immunorecognized by the antisera at dilutions up to 1:10000. Immunocytochemical assays on brain sections showed a positive immunoreactivity in cells membranes and capillary endothelia. The distribution of the immunoreactivity of C- and N-terminus antisera seems to be more uniformely distributed than H1-H2 antisera. H1-H2 antibodies appears to be preferentially associated with cell bodies in the cerebral cortex and to pyramidal cells of the CA1/CA3 region of the hippocampus and the granular cells of the dentate gyrus (Fig. 1 ). This distribution resembles mRNA localization of the different isoforms in rat brain (6). $^{3}$H-Ouabain binding was performed in 12 $\mu$m brain cryostat sections incubated with 45 nM $^{3}$H-Ouabain in a $Mg^{2+}/P_i$ medium as previously described (2 ). Ouabain binding was not modified in the presence of the antisera.

These results show that polyclonal antibodies to synthetic peptides homologous to the C-and N-terminal of the $\alpha$1 subunit and H1-H2 region of the $\alpha$3 subunit of $Na^+/K^+$-ATPase can be obtained with high affinity and specificity for the active holoenzyme. The distribution of the $\alpha$1 isoform appears to be more uniformly distributed and probably associated with both neurons and glia. The $\alpha$3 isoform seems to be mainly associated with neuronal cells. The fact that ouabain binding was not altered in the presence of the antisera may confirm the hypothesis that the C-terminal is at the cytoplasmic side of the membrane and that more than one extracellular domain may be involved in the interaction of ouabain with its binding site.

Figure 1. Immunocytochemical localization of $Na^+/K^+$-ATPase $\alpha$ subunit in rat brain hippocampus with the peroxidase technique. A) Preimmune serum. B) C-terminus antiserum. C) N-terminus antiserum. D) H1-H2 antiserum. Sera dilution 1:1000.

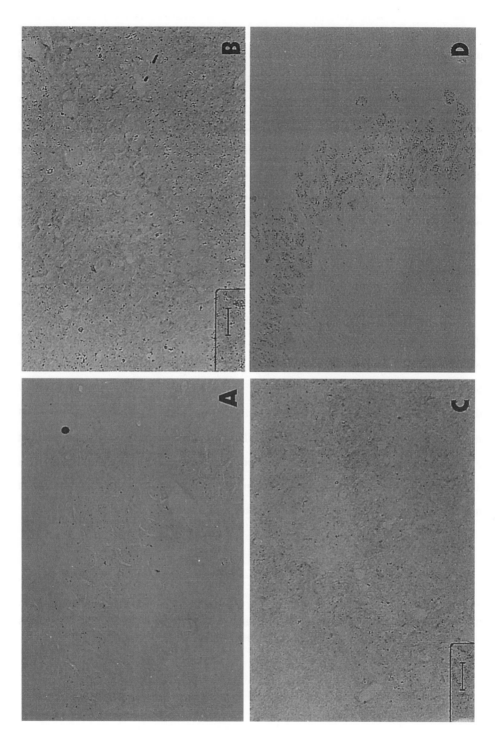

Figure 1.

# References

1. Antolovic R, Bruller H.J, Bunk S, Linder D, Schoner W (1991) Epitope mapping by amino-acid-sequence-specific antibodies reveals that both ends of the $\alpha$ subunit of Na$^+$,K$^+$-ATPase are located on the cytoplasmic side of the membrane. Eur J Biochem 199:195-202

2. Antonelli M.C, Baskin D.G, Garland M, Stahl W.L (1989) Localization and Characterization of Binding Sites with High Affinity for [$^3$H]Ouabain in Cerebral Cortex of Rabbit Brain Using Quantitative Autoradiography. J Neurochem 52:193-200

3. Rowe PM, Link WT, Hazra AK, Pearson PG, Albers RW (1988) Antibodies to synthetic peptides as probes of the structure of the Na$^+$,K$^+$-ATPase. Prog Clin Biol Res 268B:115-120

4. Shull G.E, Greeb J, Lingrel J.B (1986) Molecular Cloning of Three Distinct Forms of the Na$^+$,K$^+$-ATPase $\alpha$-Subunit from Rat Brain. Biochemistry 25:8125-8132

5. Shyjan A.W, Levenson R (1989) Antisera Specific for the $\alpha$1, $\alpha$2, $\alpha$3, and $\beta$ Subunits of the Na$^+$,K$^+$-ATPase: Differential Expression of $\alpha$ and $\beta$ Subunits in Rat Tissue Membranes. Biochemistry 28:4531-4535

6. Stahl W.L, Baskin D.G (1990) Histochemistry of ATPases. J. Histochem. Cytochem .38, 8:1099-1122

225

# Differential Axonal Transport of Individual (Na,K)-ATPase α Subunit Isoforms in Rat Sciatic Nerve

M. Mata, S. Datta, C-F. Jin, D.J. Fink

Department of Neurology and GRECC, VAMC, University of Michigan, Ann Arbor Michigan, 48105 USA

## Introduction

All three isoforms of the α subunit of Na/K-ATPase are found in neurons, with each isoform displaying a distinct regional, cellular, and subcellular distribution. *In situ* hybridization studies with riboprobes specific for α1, α2 and α3 isoforms have demonstrated the presence of mRNAs for each of those isoforms in neurons (1-4). All neurons appear to contain α2 and α3 isoform mRNA, while some neurons also contain α1 mRNA. Glia contain only α1 and α2 isoform mRNAs (5). Electron microscopic immunocytochemical studies have demonstrated the presence of the (Na,K)-ATPase both at the node of Ranvier (6), and in the internodal axolemma of large myelinated fibers (7), though the ultrastructural distribution of individual isoforms along the axolemma of large myelinated fibers has not been unambiguously defined.

We exploited the phenomenon of axonal transport to study the dynamics of isoform distribution within axons of the peripheral nervous system. (Na,K)-ATPase, like other membrane bound proteins, was demonstrated to be carried by rapid axonal transport in sciatic nerve, by the accumulation of ouabain binding sites at a ligature placed on the nerve (8), but that study predated the identification of catalytic subunit isoforms. We used a similar approach with Western blot to determine the identification of individual isoforms at a ligature on the sciatic nerve.

Male Sprague Dawley rats (250-300 gm) were anesthetized with chloral hydrate and the sciatic nerve exposed in the gluteal region. Two ligatures of 4-0 prolene were tied approximately 1 cm apart along the nerve in the exposed region. 24 hr later the animals were sacrificed by decapitation, the nerve removed, and cut into 3 mm segments. Four segments proximal to the first ligature, 3 segments distal to the second ligature, and one 3 mm segment between the ligatures were collected. The tissue was homogenized in 75-100 ml of 5 mM NaCl, 50 mM Tris (pH 7.0), and debris removed by centrifugation at 10,000 RPM for 15 min. A membrane fraction was prepared from the supernatant by centrifugation at 100,000 x g for 1 hr. We have previously shown that all of the immunoreactive (Na,K)-ATPase in sciatic nerve is found in this fraction (7).

The amount of a isoform-specific peptide in each segment was determined using Western blot. 15 mg of membrane fraction protein was separated by 6% SDS-PAGE and transferred to a nitrocellulose membrane (Hybond-ECL, Amersham). The membranes were blocked with 5% dried milk in Tris buffered saline Tween (TBS-T) and incubated overnight with polyclonal antibodies (1:1000) directed specifically against α1, α2 or α3 peptides (Upstate Biotechnologies). The specificities of the antibodies, which were raised against synthetic peptides, has been previously reported (9). Each experiment was performed with 3 different animals for each isoform (because the majority of the membrane protein from an individual segment was applied to the gel for the Western blot), and was repeated twice. The results of all experiments were identical, and representative blots from the ligated nerves are shown below.

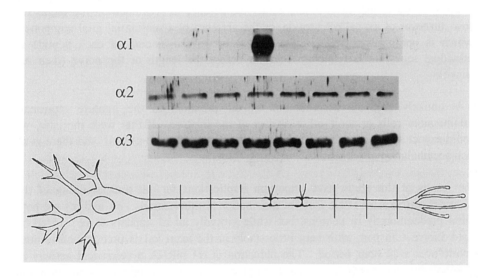

**FIGURE 1:** Western blot of individual (Na,K)-ATPase α isoforms in sequential 3 mm segments of sciatic nerve, 24 hours after application of a ligature about the nerve, as indicated in the diagram. The cell body is indicated schematically, and segments proximal to the ligature are to the left, those distal to the ligature to the right. 15 mg of membrane fraction protein was run in each lane.

227

Accumulation of Na/K-ATPase immunoreactivity at the ligature consisted almost entirely of α1 peptide. The segment immediately proximal to the ligature contained substantially more α1 immunoreactivity than segments more distant from the ligature in either the proximal or distal direction. This data is consistent with the rapid axonal transport of that catalytic subunit isoform. There is no accumulation of the α1 peptide distal to the second ligature, suggesting that there is no significant retrograde axonal transport of that isoform. α1 subunit mRNA is found in motor neurons of spinal cord and in DRG neurons, as well as in Schwann cells of sciatic nerve. However, this result implies that the major portion of α1 peptide in the nerve is axonal, since glial (Na,K)-ATPase would not be transported, and would not accumulate at the ligature.

a2 and α3 isoforms did not show accumulation at the ligature. α3 isoform mRNA is present only in neurons, and is not found in glia. Therefore, the absence of accumulation at the ligature suggests either that α3 is not rapidly transported, which would be unusual for a membrane bound enzyme, or that the amount of α3 that is rapidly transported in a 24 hour period is only a small fraction of the total α3 in the nerve, so that the accumulation is not detectable. α2 isoform is present in both glia and neurons, so we cannot determine whether the absence of accumulation represents the slow turnover of axonal α2 peptide, or the presence of a substantial glial component (which is not transported). In the control nerve, the amount of each peptide in individual segments is relatively constant across the length of the nerve (data not shown).

It is unlikely that the accumulation of α1 proximal to the ligature represents inflammatory cells or some other artifact of the ligature. If that were the case, we would expect to see an accumulation about the second ligature as well, and there is no such accumulation of α1 peptide.

The results of this study have important implications for our understanding of the dynamics of (Na,K)-ATPase isoform distribution within neurons. α1 and α3 are both found predominantly in neurons, but while virtually all a1 appears to be carried by rapid axonal transport, with very little stable in the axon, α3 is predominantly non-mobile over a 24 hour period. The induction of α1 mRNA by electrical activity in hypothalamic neurons (10) has led us to propose that α1 isoform functions to pump $Na^+$ which enters with electrical depolarization. We would propose that α1 is found predominantly at the nerve terminals and at the node of Ranvier, and for that reason the bulk is rapidly transported. α3, which is not transported rapidly, may represent an axolemmal form, present along both the internodal axolemma and the axolemma of unmyelinated fibers in a non-mobile form. Because α2 is found in glia as well, it is not possible to determine whether its stability represents the predominant glial location, or immobility of the axonal fraction of that isoform. Further definition of the distribution of these isoforms awaits the development of antibodies that can be used for electron microscopic immunocytochemical studies.

*Acknowledgment*

Figure 1 is reprinted from *Brain Research* with the permission of the publisher. We thank Mr. Dan Cutler for his excellent assistance in illustration. This work was supported by grants from the Veteran's Administration (MM and DF), the Zyma Foundation (MM), and the NIH (DF).

# References

1.    Filuk PE, Miller MA, Dorsa DM, Stahl WL. Localization of messenger RNA encoding isoforms of the catalytic subunit of Na,K-ATPase in rat brain by in situ hybridization histochemistry. *Neurosci Res Comm.* 1989;**5**:155-162.
2.    Mata M, Siegel GJ, Hieber V, Beaty MW, Fink DJ. Differential distribution of (Na,K)-ATPase alpha isoforms in the peripheral nervous system. *Brain Res.* 1991;**546**:47-54.
3.    Hieber V, Siegel GJ, Fink DJ, Beaty MW, Mata M. Differential distribution of (Na,K)-ATPase alpha isoforms in the central nervous system. *Cell Mol Neurobiol.* 1991;**11**:253-262.
4.    Watts AG, Sanchez-Watts G, Emanuel JR, Levenson R. Cell-specific expression of mRNAs encoding Na,K-ATPase alpha-and beta-subunit isoforms within the rat central nervous system. *Proc Natl Acad Sci USA.* 1991;**88**:7425-7429.
5.    Mata M, Fink DJ, Hieber V, Knapp PE. Isoform-specific expression of (Na,K)-ATPase in glial cells. *J Cell Biol.* 1991;**115**:310a.[Abstract]
6.    Ariyasu RG, Nichol JA, Ellisman MH. Localization of sodium/potassium adenosine triphophatase in multiple cell types of the murine nervous system with antibodies raised against the enzyme from kidney. *J Neurosci.* 1985;**5**:2581-2596.
7.    Mata M, Fink DJ, Ernst SA, Siegel GJ. Immunocytochemical demonstration of (Na,K)-ATPase in internodal axolemma of myelinated fibers. *J Neurochem.* 1991;**57**:184-192.
8.    Lombet A, Laduron P, Mourre C, Hacomet Y, Laxdunski M. Axonal transport of Na+, K+-ATPase indentified as a ouabain binding site in rat sciatic nerve. *Neurosci Lett.* 1986;**64**:177-183.
9.    Shyjan AW, Levenson R. Antisera specific for the alpha1, alpha2, alpha3, and beta subunits of the Na,K-ATPase: Differential expression of alpha and beta subunits in rat tissue membranes. *Biochem.* 1989;**28**:4531-4535.
10.   Mata M, Hieber V, Beaty M, Clevenger M, Fink DJ. Activity-dependent regulation of Na,K-ATPase alpha isoform mRNA expression in vivo. *J Neurochem.* 1992;**59**:622-626.

# Differential Expression of $Na^+/K^+$-ATPase Isoforms in Cultured Rodent Cardiac and Skeletal Muscle Cells

Elena Arystarkhova, Kathleen J.Sweadner

Neurosurgical Research,Massachusetts General Hospital, Boston, MA 02114, and Department of Cellular and Molecular Physiology, Harvard Medical School, Boston, MA 02115, USA

## INTRODUCTION

Cardiac and skeletal muscle are both known to express more than one isoform of the catalytic subunit of $Na^+/K^+$-ATPase (rev.in 6). Moreover, both types of myocytes exhibit changes in α isoform expression as they mature *in vivo*. In the rat, where the difference between isoforms in ouabain affinity is the most pronounced, the α1 isoform (low-affinity receptor) is the major alpha subunit gene product in the heart (at mRNA and protein levels) throughout all developmental stages, whereas expression of α2 and α3 isoforms is more plastic. The alpha 3 isoform is found primarily in fetal and neonatal heart, and it is replaced by the alpha 2 isoform during postnatal maturation of the animal (2). Expression of $Na^+/K^+$-ATPase isoforms in rat skeletal muscle is also complex and plastic in response to various stimuli. Two isoforms of sodium pump with different affinities for ouabain - α1 and α2 - undergo changes in level of expression during development and under some physiological circumstances (1,3).

Here we have investigated whether these changes can be reproduced in culture, and whether the pattern of isoform expression can be hormonally controlled. Two types of muscle cells were analyzed: primary cultures of cardiomyocytes from 1-day old rats and the murine myogenic C2C12 cell line. Isozyme-specific monoclonal antibodies - McK1 or 6F(α1), McB2(α2), and F9G10 (α3),- were used to detect each protein by either Western blot analysis or immunostaining of the cells.

## DEVELOPMENTAL CHANGES IN $Na^+/K^+$-ATPase ISOFORM EXPRESSION IN CULTURE

The murine myogenic C2C12 cell line is a suitable model to follow developmental aspects of $Na^+/K^+$-ATPase isoform regulation since it can be easily transformed from undifferentiated myoblasts to myotubes. We found that only the α1 subunit is expressed in mononucleated blast cells, while both α1 and α2 proteins were detected in fused multinucleated myotubes, correlating with reports on mRNA detection in this cell line (4). Time-course analysis revealed a gradual appearance of α2 during the differentiation period (Fig.1A). Simultaneously, there was a slight decrease in the level of expression of α1 protein. The switch occurred during 4-5 days in medium containing 2% horse serum following 4-5 days of proliferation in medium supplemented with 10% fetal bovine serum. Therefore, this cell line mimics the changes in $Na^+/K^+$-ATPase isoform expression observed in developing rodent skeletal muscle tissue, although more abruptly than during postnatal life of the rat (7).

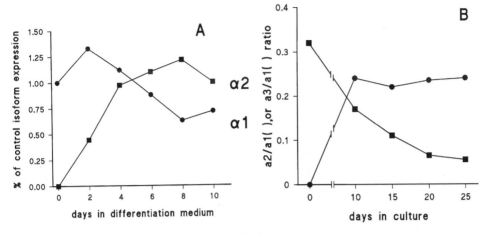

**Figure 1.** Developmental transition of Na$^+$/K$^+$-ATPase isoform expression in cultured skeletal (A) and cardiac (B) muscle cells. Myocyte cultures were maintained in 2% (A) or 10%(B) horse serum for different periods of time. Electrophoretic blots of crude membrane preparations were stained with isoform-specific antibodies and scanned densitometrically as described in (7).

Primary cultures of rat neonatal cardiocytes were established according to procedures used by Orlowski and Lingrel (5). We found that freshly dissociated newborn rat heart cells, as well as myocytes enriched by preplating, expressed α1 and α3 isoforms, while nonmyocardial cells expressed only α1. No α2 was detected either in cardiomyocytes or fibroblasts at this point. To follow the developmental transition of Na$^+$/K$^+$-ATPase isoforms *in vitro*, cardiac cells were grown for 25 days in a medium containing 10% horse serum. Crude membrane fractions of cardiomyocytes taken at different time points were used for Western blot analysis. We found that the level of α1 protein remained more or less constant through all stages of culture maturation, accompanied by an increase of α2 and gradual fall of α3, thus reproducing the pattern of isoform transition characteristic of the period of postnatal life of the rat (2). These data were independently confirmed by immunofluorescent staining of cardiomyocyte cultures of different ages. α2 protein became visible only after 9-10 days in culture in medium supplemented with horse serum. α3 levels gradually declined (data not shown). Densitometric analysis of representative immunoblots (Fig.1B) revealed, however, that the reciprocal switch of α3 to α2 proteins in cultures of cardiac myocytes took place one week earlier than it occurred in the rat heart during postnatal maturation (2). Therefore, we hypothesized that the transition from α3 to α2 isoforms may be defined by humoral factors circulating in serum rather than by a genetic program.

The developmental switch of Na$^+$/K$^+$-ATPase isoform expression *in vivo* coincides temporally with functional innervation of the heart by the sympathetic nervous system, known to undergo major developmental changes in the period between 1 and 3 weeks after birth. The efferent sympathetic pathway becomes functional only after the second postnatal week. Intrigued by this coincidence, we asked whether sympathetic nerve activity may set the timing of cellular maturational events in the heart. To test this hypothesis, cardiac myocytes were grown alone, or in physical contact with sympathetic neurons from superior cervical ganglia of newborn rats in presence of horse serum for different periods of time. However we did not see any significant changes in the pattern of isoform expression between control and co-culture (data not shown). Trophic factors released from sympathetic neurons are thus unlikely to coordinate time-dependent expression of these proteins

We have thus been able to demonstrate *in vitro* that both cardiac myocytes and C2C12 skeletal muscle cells can recapitulate under certain conditions the $Na^+/K^+$-ATPase isoform developmental changes seen *in vivo*. Since serum was required for both cultures, hormonal factors may be important for the control of isoform expression.


HORMONAL CONTROL OF $Na^+/K^+$-ATPase ISOFORM EXPRESSION *IN VITRO*

It is well established that thyroid hormone, T3, up-regulates $Na^+/K^+$-ATPase in a wide variety of tissues including cardiac and skeletal muscle (1,3). Recent experiments *in vivo* indicate that α2ß2 heterodimers are T3-responsive enzymes in rat skeletal muscle, while α1ß1, or α2ß1 are potential targets for thyroid action in rat heart (3). In newborn rat hearts, however, $Na^+/K^+$-ATPase gene expression is not affected by hypothyroid status in conditions that completely block myosin heavy chain gene expression changes (7).

To elucidate a possible role of these hormones as primary physiological modulators responsible for differential expression of $Na^+/K^+$-ATPase isoforms during development, cardiac myocyte cultures were exposed to serum-free chemically defined medium (SFM). *In vitro* studies by Orlowski and Lingrel (5) had demonstrated that in the absence of any hormone, only α1 and traces of α3 mRNA were expressed in cardiomyocytes, while addition of T3 led to significant induction of both α2 and α3 mRNAs without any effect on α1 mRNA. Dexamethasone alone induced α2 mRNA and suppressed α3 mRNA. Our results with isoform protein detection were somewhat different. In primary cultures of cardiomyocytes both α1 and α3 isoforms were detected during time course from 0 to 20 days, but no, or only tiny amounts of α2 protein. Addition of T3 (or T4) to cultures of cardiomyocytes grown in SFM resulted in a rapid and significant induction of α2 (at least 5 times over the control), whereas there was little significant change in levels of expression of α1 and α3 proteins. Dexamethasone added alone to SFM did not induce expression of α2, and, moreover, resulted in a significant decrease in the level of α1. Cells treated for 5 days with T3 followed by dexamethasone+T3 (5 days more) exhibited the same pattern of α1 and α2 proteins as cardiomyocytes treated only with T3 (data not shown). There was no effect of dexamethasone on α3 protein expression.

The use of SFM also led to the supression of α2 protein expression in C2C12 cells, thus implying that α2 biosynthesis is under control of circulating factors, since otherwise the cells appeared phenotypically normal. When cells formed myotubes in SFM, we were

232

not be able to detect any α2 even after 5 days in culture. In contrast to cardiomyocytes, addition of dexamethasone, but not thyroid hormone, led to appearance of α2 protein. However, if a population of well-established myotubes (already expressing α2) was transferred to SFM, the level of this protein was only reduced. Treatment of such cultures with 0.01-1 μM dexamethasone resulted in 1.7+/- 0.3 fold overexpression of α2 protein in comparison with control cultures.

Therefore, although there is a lot of similarity in morphology and biochemistry between these two striated muscle types, the regulation of $Na^+/K^+$-ATPase isoform expression does not follow (at least in *in vitro* systems) a similar paradigm in cardiac and skeletal muscle cells. Such plasticity in hormonal response suggests that transcriptional control by hormone-binding transcription factors, if occurring at all, must be subject to override by other factors. Post-transcriptional and even translational and post-translational regulation of gene expression is possible, as is the regulation of gene transcription by factors that are only indirectly influenced by the presence of T3 and dexamethasone.

## ACKNOWLEDGEMENT

We thank Drs. K.Campbell and D.Fambrough for their generous gifts of monoclonal antibodies F9G10 and 6F. This work was supported by NIH grant HL 36271.

## REFERENCES

1.  Horowitz B., Hensley C.B., Quintero M.,Azuma K.K., Putnam D., McDonough A.A. (1990) Differential regulation of Na,K-ATPase α1, α2, and ß subunit mRNA levels by thyroid hormone. J.Biol.Chem. 266: 14308-14314
2.  Lucchesi P.A., Sweadner K.J. (1991) Postnatal changes in Na,K-ATPase isoform expression in rat cardiac ventricle: conservation of biphasic ouabain affinity. J.Biol.Chem. 266: 9327-9331
3.  Azuma K.K., Hensley C.B., Tang M.-J., McDonough A.A. (1993) Thyroid hormone specifically regulates skeletal muscle Na,K-ATPase α2 and ß2 isoforms. Am.J.Physiol. 265: C680-C687
4.  Orlowski J., Lingrel J.B. (1988) Differential expression of the Na,K-ATPase α1 and α2 subunit genes in a murine myogenic cell line. Induction of the α2 isozyme during myocyte differentiation. J.Biol.Chem. 263:17817-17821
5.  Orlowski J., Lingrel J.B. (1990) Thyroid and glucocorticoid hormones regulate the expression of multiple Na,K-ATPase genes in cultured neonatal rat cardiac myocytes. J.Biol.Chem. 265: 3462-3470
6.  Sweadner K.J. (1989) Isozymes of $Na^+/K^+$-ATPase. Biochim.Biophys.Acta 988: 185-220
7.  Sweadner K.J., McGrail K.M., Khaw B-A. (1992) Discoordinate regulation of isoforms of Na,K-ATPase and myosin heavy chain in hypothyroid postnatal rat heart and skeletal muscle. J.Biol.Chem. 267: 769-773

# Na,K-ATPase Abundance and Activity in HeLa Cells Transfected with Rat α Isoforms

D.S. Putnam, E.A. Jewell[*], L. Lescale-Matys, C.E. Magyar, and A.A. McDonough

Department of Physiology and Biophysics, University of Southern California School of Medicine, Los Angeles, CA 90033, and [*]Department of Molecular Genetics, Biochemistry and Microbiology, University of Cincinnati College of Medicine, Cincinnati, Ohio 45267-0524, USA

## Introduction

Since Na,K-ATPase is a heterodimer of an α catalytic subunit and a ß glycoprotein subunit assembled in a 1:1 stoichiometry, synthesis of αß heterodimers depends on the rate of synthesis of both α and ß subunits. We have previously demonstrated that ß is synthesized in excess over α subunits in two renal epithelial cell lines: LLC-PK$_1$ cells, where ß is synthesized in 3-fold excess over α (4), and MDCK cells where ß is synthesized in 5 fold excess (8). In both cell lines there is a component of degradation of nascent ß, presumably the excess unassembled subunits, during the first 30-60 min after synthesis. There is no detectable degradation of nascent α during this period, and subsequently there is coordinate degradation of α and ß, presumably αß heterodimers, with a t$_{1/2}$ of 10-12 hr (4).

While the kinetics of α and ß synthesis and degradation described above suggest that α is rate limiting for αß assembly in these cells, regulation studies in LLC-PK$_1$ cells as well as in whole animals suggest that synthesis of ß is actually rate limiting. When LLC-PK$_1$ cells are incubated in low K$^+$, ß, but not α, mRNA levels increase and are associated with coordinate accumulation of both α and ß subunits and two fold increase in surface expression of sodium pumps (3,5). In two models of chronic decreased renal proximal tubule sodium transport in the rat, hypothyroidism and high salt diet, Na,K-ATPase activity is decreased (46% and 20% respectively), ß abundance at the protein level is depressed to the same extent as activity, and α subunit abundance is unchanged (1,6). Taken together with the assembly results in unperturbed cells, these findings suggest the hypothesis that ß subunit synthesis is limiting for αß heterodimer formation despite the fact that ß is synthesized in stoichiometric excess. In other words, the ß:α ratio limits αß formation (7). While this hypothesis supports the results seen in hypothyroidism and high salt diet where sodium pump expression is depressed in response to decreased ß expression, it does not explain how increased ß can stimulate accumulation of both α and ß in low K$^+$ treated LLC-PK$_1$ cells. We are left to postulate that either there is a component of nascent α degradation that is too rapid for us to measure, or that α translatability increases either due to, or independent of, increased ß synthesis.

In this report we examine the question of the determinants of αß assembly and accumulation from the perspective of regulation of α rather than ß expression. We have studied a set of HeLa cell lines overexpressing either the ouabain resistant rat α1, α2 or α3 catalytic subunits which were mutated to ouabain resistance. These cells have been described previously (2). We aimed to determine the effect of over-expressing rat α subunit isoforms in these cells on Na,K-ATPase activity (both the

234

ouabain sensitive HeLa and ouabain resistant rat components) and the effects of overexpression on the levels of endogenous HeLa $\alpha1$ and $\beta1$ subunits.

## Methods
The HeLa cells expressing rat $\alpha1$, $\alpha2$ or $\alpha3$ were selected and grown in ouabain containing media as previously described (2), the wild type HeLa cells were grown without ouabain. The methods for RNA and protein extraction and analysis, and Na,K-ATPase activity assays are as previously described in our studies of LLC-PK$_1$ cells (3).

## Na,K-ATPase subunit mRNA levels
It has been established that the transfected cDNAs to rat $\alpha1$ and the mutated forms of $\alpha2$ and $\alpha3$ are actually being expressed as mRNA (2). These mRNAs are translated into functional pumps because survival of the host cells in ouabain selection medium depends on expression of functional ouabain resistant rat sodium pumps. Figure 1 illustrates that the expression levels of $\alpha1$ and $\beta1$ mRNAs in the transfected cell lines are not significantly different from that seen in the wild type cells (note that the $\alpha1$ cDNA will detect both the HeLa and transfected rat $\alpha1$). Equivalent loading in each lane was verified by ethidium bromide staining of the 18 and 28s ribosomal RNA in both the gel and the blot (Figure 1). Thus, there is no evidence for consistent positive or negative feedback from expression of rat sodium pumps in HeLa cells to altered expression of endogenous HeLa $\alpha$ and $\beta$ transcripts.

## Na,K-ATPase subunit protein levels
The challenge of detection of $\alpha$ and $\beta$ subunits in the transfected HeLa cells resides in the discrimination between rat and human isoforms. With rat specific and isoform specific antisera from K. Sweadner (McK1 for $\alpha1$, McB2 for $\alpha2$) and with anti-rat $\alpha3$ from T. Pressley (anti-TED) we detected expression of the transfected isoforms in the HeLa cells (not shown). In the non-transfected wild type cells these antisera did not detect human $\alpha1$. Human $\alpha1$ was detected with anti-chicken $\alpha1$ from D. Fambrough (#5), and human $\beta1$ was detected with an anti-guinea pig holoenzyme antibody made in this lab (#4014). The Fambrough #5 anti-$\alpha1$ antibody also detects rat $\alpha1$ and rat $\alpha3$ subunits expressed in a yeast expression system (data from R.A. Farley laboratory) and this is supported by the results in the transfected cell lines. Cells were not transfected with rat $\beta$ so cross reaction was not an issue. Figure 2A summarizes the immunoblot results obtained when the wild type and transfected cells (each assayed over a range from 15 to 60 $\mu$g/lane) were probed with Fambrough #5 (detection of $\alpha1$ and $\alpha3$), and 4014 ($\beta1$). Expression of $\alpha1$ was depressed, compared to wild type, in the rat $\alpha1$ and rat $\alpha2$ cell lines, this in spite of the fact that antibody #5 detects rat and human $\alpha1$ in the rat $\alpha1$ cell line. The autoradiographic signal obtained with #5 in the rat $\alpha3$ cell line (where both human $\alpha1$ and rat $\alpha3$ will be detected) was similar to that in the wild type. The level of human $\beta1$ expression in the transfected cells was not significantly altered from that seen in the wild type HeLa. These results suggest that there is not excess or spare $\beta$ available that can be stabilized by assembly with the additional $\alpha$ expressed by transfection in these cells.

## Na,K-ATPase enzymatic activity
The immunoassay illustrated in Figure 2A cannot predict whether the total number of $\alpha$ subunits in the cell is increased by transfection because the antibody signal is a function of affinity for the antigen as well as number of antigens. The fact that human $\beta$ subunit abundance remains constant suggests that the total number of pumps does not increase, if all $\beta$s are resident in functional $\alpha\beta$ heterodimers. Since the endogenous

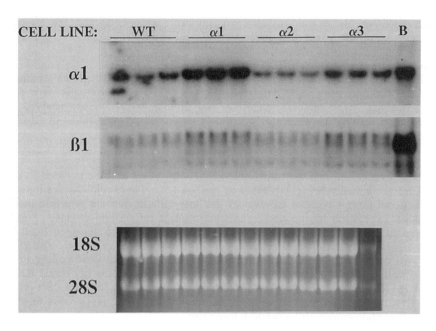

Figure 1. Detection of α1 and ß1 mRNA in 20ug samples of total RNA from wild type (WT) and Hela cell lines and brain (B). Lower panel show ethidium bromide stained gel.

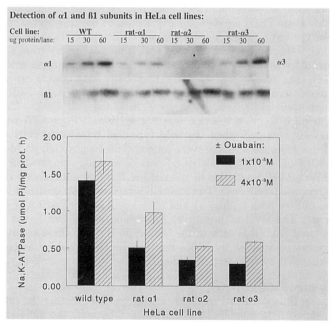

Figure 2. A: Immunoblot analysis of α1 and α3, and ß1 subunit in homogenate protein from WT and transfected HeLa cell lines. B: Na,K-ATPase activity in homogenates from WT and transfected HeLa cell lines.

human Na,K-ATPase is very ouabain sensitive and the transfected rat isoforms all more ouabain resistant, we assessed the activity of human-human vs. rat-human $\alpha$-ß heterodimers. We determined in a dose response assay that the wild type HeLa Na,K-ATPase activity was fully inhibited by $10^{-5}$M ouabain, which did not significantly inhibit the rat type (assessed in rat kidney), and that the rat type was inhibited by 4 x $10^{-3}$M ouabain. Na,K-ATPase activity was assessed in the cell lines with 0, $10^{-5}$M and 4 x $10^{-3}$M ouabain. The results are shown in Figure 2B. Total ouabain sensitive activity of all three of the rat $\alpha$ transfected cells was suppressed 40 to 65% compared to wild type. In other words, overexpression of $\alpha$ subunits in these cells was associated with less, not more, Na,K-ATPase activity. The activity of the endogenous human pumps (inhibited by $10^{-5}$M ouabain) was suppressed 65 - 80% in the transfected cells. The activity of the rat type pumps (the difference between activity measured with 4 x $10^{-3}$M and $10^{-5}$M ouabain) was less than 30% of wild type activity.

The results of this study suggest that during the competition between transfected rat and endogenous human $\alpha$ subunits for assembly with human ß, there is a decrease in the number of active heterodimers formed. The fact that the abundance of ß does not decrease in the cell lines compared to wild type suggests that either there are spare ß not assembled with $\alpha$, $\alpha$ß heterodimers that are not functional, or that the chimeric heterodimers are not as active as the human $\alpha$ß heterodimers. Finally, the results suggest that, like our findings in low $K^+$ incubated LLC-PK$_1$ cells, ß may limit $\alpha$ß heterodimer assembly.

*Acknowledgements*
These studies were supported by NIH grants DK34316 and HL44404, and fellowship support to LLM from the American Heart Association Greater Los Angeles Affiliate.

**References**
1. Horowitz, B. Hensley, C.B., Quintero, M., Azuma, K.K., Putnam, D. and McDonough, A.A. (1990) Differential regulation of Na,K-ATPase $\alpha$1, $\alpha$2 and ß subunit mRNA levels by thyroid hormone. J. Biol. Chem. 265:14308-14314
2. Jewell, E.A. and Lingrel, J.B. (1991) Comparison of the substrate dependence properties of the rat Na, K-ATPase a1, a2, and a3 isoforms expressed in HeLa cells. J. Biol. Chem. 266, 16925-16930
3. Lescale-Matys, L., Hensley, C.B., Crnkovic-Markovic, R., Putnam, D. and McDonough, A. A. (1990) Low $K^+$ increases Na,K-ATPase abundance in LLC-PK1/Cl4 cells by differentially increasing $\beta$ and not $\alpha$ subunit mRNA. J. Biol. Chem. 265:17935-17940
4. Lescale-Matys, L., Putnam, D.S. and McDonough, A.A. (1993) Na,K-ATPase $\alpha$1- and ß1- subunit degradation: evidence for multiple subunit specific rates. Am. J. Physiol. 264:C583-C590.
5. Lescale-Matys, L., Putnam, D.S. and McDonough, A.A. (1993) Surplus sodium pumps: How low $K^+$ incubated LLC-PK$_1$ cells respond to $K^+$ restoration. Am. J. Physiol , in press.
6. McDonough, A.A., Azuma, K.K., Lescale-Matys, L., Tang, M.J., Nakhoul, F., Hensley, C.B. and Komatsu, Y. Physiologic Rationale for Multiple Sodium Pump Isoforms - Differential Regulation of $\alpha$1 vs $\alpha$2 by Ionic Stimuli. Ann. New York Acad. Sci. 671:156-169, 1992.
7. McDonough, A.A., Geering, K., and Farley, R.A. (1990) The sodium pump needs its beta subunit. FASEB J. 4: 1598-1605.
8. Mircheff, A.K., Bowen, J.W., Yiu, S.C., and McDonough, A.A.(1992) Synthesis and translocation of Na,K-ATPase $\alpha$ and $\beta$ subunits to the plasma membrane in MDCK cells. Am.J. Physiol. 262 (Cell Physiol. 31): C470-C483.

# Aldosterone Induces a Delayed Parallel Increase in Total and Cell-Surface $Na^+/K^+$-ATPase in A6 Kidney Cells

J. Beron and F. Verrey

Institute of Physiology, University of Zürich, Winterthurerstrasse 190, CH-8057 Zürich, Switzerland

## Introduction

The reabsorption of $Na^+$ across tight epithelia is mediated by the apical amiloride-sensitive $Na^+$ channel and the basolateral $Na^+/K^+$-ATPase which extrudes $Na^+$ from the cells and provides the driving force for this transport. Adrenal steroids stimulate this transport in A6 kidney cells cultured on filters as in other tight epithelia. This action can be divided in three phases: (a) a lag period, (b) an early response during which an increase in $Na^+$ reabsorption appears to result from a change in the activity of preexisting apical $Na^+$ channels, and (c) a late response which appears to be partially supported by quantitative changes at the level of the transporters (4). It has been recently shown that the early increase in $Na^+$ transport, which precedes the transcriptionally mediated increase in the rate of $Na^+/K^+$-ATPase subunit synthesis, is paralleled by an increase in the rate of ouabain binding to the intact cell monolayers (in K-free buffer). This effect is independent of apical $Na^+$ influx but depends on ongoing transcription and translation (2). Here we test the hypothesis that it is due to an increase in the number of active pump sites which in turn would be mediated by the translocation of $Na^+/K^+$-ATPase molecules to the cell surface (2). Furthermore, we test whether the known induction of $Na^+/K^+$-ATPase subunit synthesis leads to a late increase in the number of total and cell-surface expressed $Na^+$ pumps.

## Lack of early increase in $Na^+/K^+$-ATPase cell-surface expression

Aldosterone ($10^{-6}$M) induced an increase in the initial rate of ouabain binding (2.1-fold, 3 hours after hormone addition). This effect was parallel to but independent of the increase in transepithelial $Na^+$ reabsorption, since it was not inhibited by apical amiloride (Fig. 1). To measure changes in the cell-surface expression of the $Na^+/K^+$-ATPase the basolateral surface proteins were labeled at 4°C using sulfosuccinimidobiotin or enzyme-mediated radioiodination (3,5). After immunoprecipitation with subunit-specific antibodies, cell-surface $Na^+/K^+$-ATPase was detected by blotting with streptavidin or by autoradiography. Since the $\alpha$ subunit was not efficiently labeled by either method, only the surface expression of the ß subunit could be evaluated. Changes in the total pool of $Na^+/K^+$-ATPase were measured by Western blotting. Three hours after aldosterone addition there was no significant increase in total and cell-surface $Na^+/K^+$-ATPase, indicating that the early increase in the initial rate of ouabain binding was not due to the translocation of preexisting pumps to the cell surface (Fig. 2) but to the activation of preexisting ones.

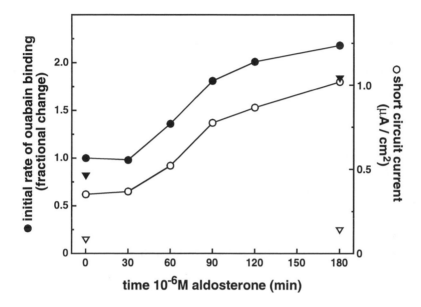

**Figure 1.** Time course of the early effect of aldosterone ($10^{-6}$M) on Na$^+$ transport measured as short circuit current and on the initial rate of ouabain binding measured in K-free buffer. The triangles represent values obtained in the presence of apical amiloride ($10^{-5}$M) added three hours before the measurements.

## Late increase in total and cell-surface Na$^+$/K$^+$-ATPase

Aldosterone ($10^{-6}$M) induced a late increase in total and cell-surface Na$^+$/K$^+$-ATPase ß subunit which was detectable 5 hours after its addition and was 1.7-fold after 20 hours of treatment (2.5-fold after 5d) (Fig. 2). This increase was parallel to that of the total Na$^+$/K$^+$-ATPase pool measured by Western blotting ($\alpha$ and ß subunit), at least at the time resolution of these experiments (Fig. 2). In contrast to the initial rate of ouabain binding which was increased during the early phase of aldosterone action (Fig. 1), the total pump sites measured by saturation ouabain binding showed a delayed increase which was parallel to that of the total Na$^+$/K$^+$-ATPase measured by Western blotting (data not shown). This difference was probably due to the fact that the experimental conditions used for binding ouabain at saturation produced *per se* a maximal activation of the pumps (data not shown), so that this test measured the total number of available pumps, whereas the initial rate of ouabain binding (K-free buffer) measured the state of activation and/or the number of active pumps.

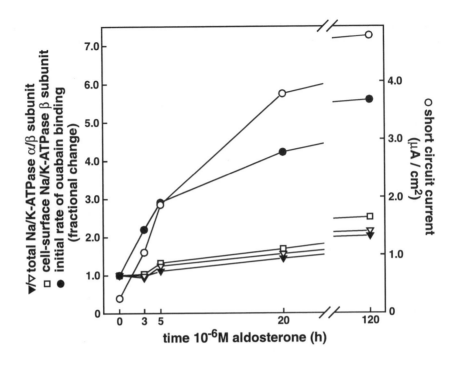

**Figure 2.** Time course of the aldosterone effect on the total and cell-surface Na$^+$/K$^+$-ATPase, on the initial rate of ouabain binding (K-free buffer) and on the Na$^+$ transport. The relative changes in cell-surface Na$^+$/K$^+$-ATPase ß subunit were measured by surface biotinylation followed by immunoprecipitation and blotting with streptavidin. The changes in total subunit pools were measured by Western blotting.

## Discussion

We propose that the early aldosterone induced increase in the initial rate of ouabain binding (in K-free buffer), which is independent of the apical Na$^+$ influx, reflects an *in situ* activation of pumps already present at the basolateral surface, since no increase in the amount of cell-surface Na$^+$/K$^+$-ATPase was observed during that period. In contrast, during the late response to aldosterone (later than three hours after hormone addition) an increase in the amount of cell-surface Na$^+$/K$^+$-ATPase which parallels that of the total Na$^+$/K$^+$-ATPase pool can account for most of the observed increase in the initial rate of ouabain binding. This late increase in the number of pump units is in contrast to a previous report (1) and appears to be the late consequence of an early increase in Na$^+$/K$^+$-ATPase subunit gene transcription (6) which is followed by an accumulation of the specific mRNAs and an increase in subunit synthesis (7). In

240

conclusion, aldosterone ($10^{-6}$M) induces at the basolateral surface of A6 cells an early *in situ* activation of preexisting pumps and a delayed accumulation of new pumps.

*Aknowlegments*

We thank Uschi Bolliger for cell culture work and Christian Gasser for the artwork. This study was supported by grant 31.30132.90 from the Swiss National Science Foundation.

## References

1.  Paccolat MP, Geering K, Gaeggeler HP, Rossier BC (1987) Aldosterone regulation of $Na^+$ transport and $Na^+$-$K^+$-ATPase in A6 cells: role of growth conditions. Am. J. Physiol. 252:C468-C476

2.  Pellanda AM, Gaeggeler HP, Horisberger JD, Rossier BC (1992) Sodium-independent effect of aldosterone on initial rate of ouabain binding in A6 cells. Am. J. Physiol. 262:C899-C906

3.  Sargiacomo M, Lisanti MP, Graeve L, Le Bivic A, Rodriguez- Boulan E (1989) Integral and peripheral protein composition of the apical and the basolateral membrane domains in MDCK cells. J. Membr. Biol. 107:277-286

4.  Verrey F (1990) Regulation of gene expression by aldosterone in tight epithelia. Semin. Neph. 10:410-420

5.  Verrey F, Kairouz P, Schaerer E, Fuentes P, Geering K, Rossier BC, Kraehenbuhl JP (1989) Primary sequence of Xenopus laevis $Na^+$- $K^+$-ATPase and its localization in A6 kidney cells. Am. J. Physiol. 256:F1034-F1043

6.  Verrey F, Kraehenbuhl JP, Rossier BC (1989) Aldosterone induces a rapid increase in the rate of Na,K-ATPase gene transcription in cultured kidney cells. Mol. Endocrinol. 3:1369- 1376

7.  Verrey F, Schaerer E, Zoerkler P, Paccolat MP, Geering K, Kraehenbuhl JP, Rossier BC (1987) Regulation by aldosterone of $Na^+$,$K^+$-ATPase mRNAs, protein synthesis, and sodium transport in cultured kidney cells. J. Cell Biol. 104:1231-1237

241

# Differential aldosterone effects on two epithelial cell pools of [³H]ouabain binding sites

Jesper M. Nielsen and Peter L. Jørgensen

Biomembrane Research Centre, August Krogh Institute, Copenhagen University, 2100 Copenhagen OE, Denmark.

## Introduction

The A6 cell line, derived from the kidney of *Xenopus laevis* has earlier been shown to respond to aldosterone with an increase in the rate of $Na^+$ transport across the epithelium (8). The rate-limiting steps for transcellular $Na^+$ transport are the passive flux of $Na^+$ into the cytoplasm through amiloride-sensitive $Na^+$-channels in the apical membrane, and the active extrusion of $Na^+$ by $Na^+/K^+$-pumps in the basolateral membrane (4). Aldosterone is believed to augment the transcellular $Na^+$ transport by a two-step process. In the early or acute aldosterone-induced response, the molecular activity of existing $Na^+/K^+$-pumps in the basolateral membrane is increased in response to an upregulation of the apical membrane $Na^+$ conductance (1), whereas the late or chronic phase of aldosterone action involves upregulation of the amount of $Na^+/K^+$-pump sites in the basolateral membrane (3,4).

In A6 cells, aldosterone has been shown to modulate the abundance of mRNA and protein of the $Na^+/K^+$-pump subunits as a consequence of altered transcription rates from the $Na^+/K^+$-pump genes (10,11). In accordance with this effect on the biosynthesis of $Na^+/K^+$-pump proteins, incubation with aldosterone has in two separate studies been reported to increase the number of [³H]ouabain binding sites in A6 cells (2,5), although one group found no such effect of aldosterone on the equilibrium [³H]ouabain binding capacity of A6 cells (7). In these earlier studies, [³H]ouabain binding was performed to either homogenates or intact cells by use of different assays, but it was not evaluated if all [³H]ouabain binding sites were detected.

It is important to consider whether [³H]ouabain detects all functional $Na^+/K^+$-pumps in the epithelial cells, and if aldosterone may affect the distribution between intracellular and plasma membrane pools of $Na^+/K^+$-pump sites. In the present work, we therefore determined the capacity of [³H]ouabain binding in dispersed intact A6 cells and in A6 cells where the membranes had been broken by incubation in $NaDodSO_4$, as previously described for *Xenopus* oocytes (9). To examine the effect of aldosterone on the two cellular pools of $Na^+/K^+$-pumps, it was determined if treatment of A6 cells with aldosterone had a differential effect on the apparent and latent number of [³H]ouabain binding sites. The apparent number is determined as the binding capacity of intact cells, and the latent number is calculated as the difference between the total number of sites (measured after incubation with an optimum concentration of $NaDodSO_4$) and the apparent number of sites.

## Methods

*Cell culture* - A6 cells for this study were a gift from Dr B.C. Rossier (Department of Pharmacology, University of Lausanne, Switzerland). The cells were maintained at 28°C in a humidified incubator with 5 % $CO_2$ in air. The growth medium was Dulbecco's Modified Eagle Medium (Gibco, 041-01885) with 20 % water and 10 % fetal calf serum. In addition the medium was supplemented with 2 mM glutamine, 100 U/ml penicillin and 100 $\mu$g/ml streptomycin. Cells for experiments were plated on petri dishes (Nunc, 20.7 $cm^2$) at a density of 4 x $10^4$ cells/$cm^2$, and were fed thrice a week with 5 ml of growth medium. Serum was removed from the cells 24 hours before incubation with aldosterone (Sigma, A 8661) in serum-free growth medium. All experiments were performed 6-9 days after inoculation of cells.

*Dispersion of cells* - Growth medium was removed and the monolayer of cells was washed 4 times with 5 ml of ice-cold dispersion buffer (200 mM sucrose, 1 mM EGTA and 10 mM MOPS-Tris pH 7.2). Thereafter, the intact cells were brought into a single cell suspension by incubation with 0.6 ml of dispersion buffer for 1 hour at 4°C.

*Detergent treatment of cells* - In order to break membranes and demask any latent [³H]ouabain binding sites, dispersed intact cells were treated with NaDodSO₄ (Sigma, L 4390). Equal volumes of cell suspension (2,0 mg protein/ml) and NaDodSO₄ (in dispersion buffer) were mixed gently and incubated at 25°C for 1 hour. Hereafter the mixture was immediately diluted into the [³H]ouabain binding medium.

*[³H]ouabain binding* - Equilibrium binding was determined at 0.1 ml protein/ml in the presence of [³H]ouabain (New England Nuclear, 23 Ci/mmol) in 200 mM sucrose, 1 mM vanadate, 3 mM MgSO₄, 1 mM EGTA and 10 mM MOPS-Tris pH 7.2, in a

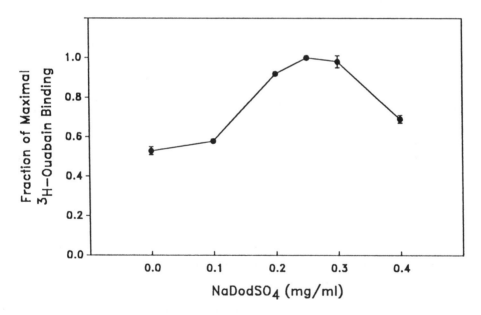

**Figure 1.** Effect of NaDodSO₄ on equilibrium binding of [³H]ouabain to dispersed A6 cells. Cells from three petri dishes were pooled, and aliquots were treated with the indicated concentration of NaDodSO₄ before binding was carried out in the presence of 1,6 x $10^{-7}$ M [³H]ouabain. Each value is the mean ± S.E. of triplicate determinations. After incubation with 0.25 mg/ml NaDodSO₄ the [³H]ouabain binding capacity is increased from 1.83 ± 0.06 to 3.48 ± 0.03 pmol/mg protein.

total volume of 1 ml. For determination of unspecific binding, 1 mM of unlabeled ouabain was added. After 1 hour incubation at 37°C, the samples were cooled on ice for 20 minutes. Unbound [$^3$H]ouabain was removed by vacuum filtration. 0.9 ml aliquots were applied on 0.45 μm Millipore HA mixed cellulose ester filters and washed 5 times with 2 ml of ice-cold washing-buffer (200 mM sucrose, 1 mM vanadate, 3 mM MgSO$_4$, 1 mM EGTA and 10 mM MOPS-Tris pH 7.2). Filters were dissolved in 4 ml of Filter Count™ scintillation fluid (Packard), and radioactivity was determined by liquid scintillation spectrometry.

## Results and Conclusions

The dispersed intact A6 cells have a [$^3$H]ouabain binding capacity of 1.9 ± 0.1 pmol/mg protein (2.3 x 10$^5$ sites/cell) (n = 10). After incubation of the dispersed cells with an optimum concentration of NaDodSO$_4$ (0.25 mg/ml) the capacity for [$^3$H]ouabain binding is increased to 4.0 ± 0.1 pmol/mg protein (4.9 x 10$^5$ sites/cell) (n = 10). The $K_D$ of the equilibrium [$^3$H]ouabain binding is changed from a value of 17-20 nM when binding is carried out to dispersed intact A6 cells, to a value of 3-4 nM when binding

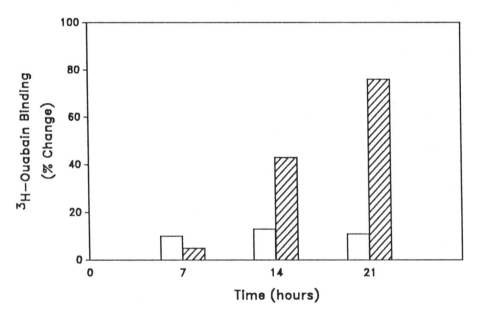

**Figure 2.** Effect of aldosterone on the latent (open bars) and apparent (hatched bars) number of [$^3$H]ouabain binding sites in A6 cells, expressed as the change compaired to controls. Cells on petri dishes were incubated with 100 nM aldosterone for 0 h (n = 10), 7 h (n = 4), 14 h (n = 4) and 21 h (n = 5). The cells from each petri dish were dispersed and divided into two aliquots of which one was treated with 0.25 mg/ml NaDodSO$_4$. Thereafter, equilibrium binding was performed to cells from both aliquots at a concentration of 1.5-1.6 x 10$^{-7}$ M [$^3$H]ouabain. The apparent number of sites represents the binding capacity of intact cells, whereas the latent number of sites is found by subtracting the apparent from the total number of sites found after treatment of cells with an optimum concentration of NaDodSO$_4$, as shown in figure 1.

is performed to A6 cells that have been preincubated in 0.25 mg/ml NaDodSO$_4$. However, from extrapolation of Scatchard plots it is clear that the demasking of a latent pool of sites after incubation in detergent cannot be explained by the difference in apparent affinity for [$^3$H]ouabain (data not shown). Incubation of A6 cells with aldosterone does not appear to alter the affinity of cellular sites for [$^3$H]ouabain.

The present study shows that A6 cells contain two pools of [$^3$H]ouabain binding sites, an apparent and a latent pool. The latent pool of sites is demasked after treatment of the dispersed cells with detergent, and this pool accounts for approximately half the total number of cellular sites. The size of the latent pool is similar to that previously observed in *Xenopus* oocytes (9). The latent pool may consist of Na$^+$/K$^+$-pump molecules in translocation from the site of synthesis in the ER through the Golgi apparatus to the surface membrane. It is also possible that the latent pool comprises Na$^+$/K$^+$-pump sites in storage vesicles that may associate with the plasma membrane upon hormone stimulation.

As shown in figure 2, aldosterone does have a differential effect on the two pools of [$^3$H]ouabain binding sites in A6 cells. After incubation of the cell cultures with aldosterone for 7-21 hours, there is only a small increment in the latent pool of sites (10-13 %), while the number of sites in the apparent pool is significantly increased (43-76 %) after 14-21 hours.

The time course for the action of aldosterone on the two pools of sites in A6 cells suggests that aldosterone does not mediate an increased transfer of existent Na$^+$/K$^+$-pumps from an intracellular storage pool to the plasma membrane. Aldosterone stimulation rather seems to increase the synthesis of new Na$^+$/K$^+$-pumps that subsequently are transported to the surface membrane. The rate of increase in the total number of cellular [$^3$H]ouabain binding sites is slow, in agreement with an upregulation of pump sites through an effect of aldosterone on the rate of biosynthesis of $\alpha\beta$-units. The rate of appearance of new Na$^+$/K$^+$-pump sites is similar to that previously observed in mammalian kidney (6) and surface cells of the colon (12).

# References

1.      Garty H (1986) J Membr biol 90: 193-205
2.      Handler JS, Preston AS, Perkins FM, Matsumura M (1981) Ann N Y Acad Sci 372: 442-54
3.      Jørgensen PL (1972) J Steroid Biochem 3: 181-91
4.      Jørgensen PL (1986) Kidney Int 29: 10-20
5.      Leal T, Crabbé J (1989) J Steroid Biochem 34: 581-4
6.      O'Neil RG, Hayhurst RA (1985) J Membr Biol 85: 169-79
7.      Pellanda AM, Gaeggeler HP, Horisberger JD, Rossier BC (1992) Am J Physiol 262: C899-906
8.      Perkins FM, Handler JS (1981) Am J Physiol 241: C154-9
9.      Schmalzing G, Kröner S, Passow H (1989) Biochem J 260: 395-9
10.     Verrey F, Schaerer E, Zoerkler P, Paccolat MP, Geering K, Kraehenbuhl JP, Rossier BC (1987) J Cell Biol 104: 1231-7
11.     Verrey F, Kraehenbuhl JP, Rossier BC (1989) Mol Endocrinol 3: 1369-76
12.     Wiener H, Nielsen JM, Klaerke DA, Jørgensen PL (1993) J Membr Biol 133: 203-11

# Branchial Na$^+$/K$^+$-ATPase Expression in the European Eel (Anguilla anguilla) following Saltwater Acclimation

G.A. Luke, C.P. Cutler, I.L. Sanders, N. Hazon, G. Cramb

School of Biological and Medical Sciences, University of St. Andrews, St. Andrews, Fife, U.K. KY16 9TS

## Introduction

The physiological adaptation of the euryhaline eel from freshwater (FW) to sea water (SW) environments is accompanied by increases in branchial chloride cell Na$^+$/K$^+$-ATPase activity. Branchial chloride cells are known to have unique morphological characteristics which are associated with their ability to adapt to large alterations in the external ionic environment. In FW, gill chloride cells possess an extensive tubular network which is an extension of the basolateral membrane compartment. The vast majority of Na$^+$/K$^+$-ATPase enzyme units have been shown to be located in the membranes of this tubular network (1). In FW environments, sodium is postulated to be retained by the pumping of this ion (by Na$^+$/K$^+$-ATPase) into the tubular network, where it diffuses down its concentration gradient into the internal milieu. When eels are transferred to a SW environment, the sodium gradient across the epithelium is oriented in the opposite direction, and several characteristic responses to this situation are activated (1). Chloride cells undergo extensive morphological and functional adaptations associated with the expression of Na$^+$/K$^+$-ATPase during the natural migration of these fish from freshwater to seawater (a change in osmolality of around 1000 mOsmol/kg). Proliferation of chloride cells within the gill epithelium occurs in such a way that where once a single chloride cell, with tight junctions between it and its respiratory cell neighbours existed, a group of chloride cells are present. The lateral spaces between these chloride cells are then partially open to the external but not internal milieu [2]. In conjunction with these changes, the apical pole of the basolateral tubular network is extended, and is in contact with the newly opened lateral spaces (2).
Using homologous cDNA probes to eel Na$^+$/K$^+$-ATPase $\alpha$ and $\beta$ subunits we have examined the expression of these two proteins in the branchial epithelium during the period of adaptation from FW to SW environments.

## Methods

Quantification of Na$^+$/K$^+$-ATPase $\alpha$ and $\beta$ subunit mRNAs

Total RNA and poly (A$^+$) mRNA were purified by a methods similar to those used previously (3). Homologous eel cDNAs for both $\alpha$ and $\beta$ subunits of Na$^+$/K$^+$-ATPase were isolated by the polymerase chain reaction (PCR) technique. A synthetic oligonucleotide to rat 18S rRNA was prepared and used as a quantitative marker for the total amount of RNA blotted. The 18S rRNA probes' sequence was homologous between most eukaryote species.
Northern blot hybridisations were employed to establish the optimum conditions for use with the quantitative dot blots. Total RNA samples (0.5 μg) were dot blotted onto filters and hybridised consecutively with probes for Na$^+$/K$^+$-ATPase $\alpha$ and $\beta$ subunits and the 18S rRNA probe. After each hybridisation the blots were washed twice to remove all

traces of radioactivity and then re-hybridised with the next probe. Processed autoradiographs were scanned by densitometry. In all blots a series of radioactive standards (1-500 cpm) were included to assess the linearity of the film response.

Membrane preparation, $Na^+/K^+$-ATPase activity and drinking rates

Eels were killed by decapitation and the gills removed blotted and the mucosal layer homogenised in 10 volumes of buffer containing protease inhibitors. The homogenate was centrifuged at 705g (10 mins). The supernatant was removed and recentrifuged at 31360g (45 min). The pellet was resuspended in buffer containing phenylmethylsulphonyl fluoride (PMSF) and dithiothreitol (DTT). The protein content was determined using a protein assay kit (Biorad). Potassium-stimulated ouabain-sensitive $Na^+/K^+$-ATPase activities were determined by the procedure of Bers (4). The drinking rate was determined in free-swimming fish using a modified version of the isotopic dilution technique (5).

**Results**

$Na^+/K^+$-ATPase expression

Northern blots of eel gill total RNA probed with $Na^+/K^+$-ATPase $\alpha$ and $\beta$ subunit probes hybridised to single mRNA species of approximately 3.8 kb and 2.5 kb respectively (data not shown). Transfer of eels to saltwater initiated a tri-phasic response in branchial $Na^+/K^+$-ATPase activity. Immediately upon transfer, $Na^+/K^+$-ATPase activity rose to a transient maximum 3.5-fold increase within 6 hours and thereafter declined over the following 18 hours to activities found in freshwater (Fig. 1a). This was followed by a second phase where activities gradually increased from 3 days post-transfer, leading to levels at 21 days which matched the 6 hour peak. This second phase was associated with 3-fold and 6-fold increases in $\alpha$ and $\beta$ subunit mRNAs respectively (Fig. 1b,c). The increased levels of enzyme activity and $\alpha$ and $\beta$ subunit mRNA expression were not sustained and by 6 months levels had returned to near those found in freshwater.

Drinking rate

In acute studies, an increase in drinking from a basal FW value of $0.0227 \pm 0.008$ ml/kg/hr to $0.251 \pm 0.065$ ml/kg/hr after one hour SW transfer was observed. This increase was maintained for 90 min after SW transfer, and thereafter started to decline, although the drinking rate after four hours was still significantly higher than the FW basal rate. In the long term studies an additional two phased response of drinking to SW adaptation was found. The level of drinking increased reaching a peak of $0.167 \pm 0.06$ ml/kg/hr two days after transfer. A decline was then observed after 4-5 days but then, the drinking rate increased again stabilising at $0.49 \pm 0.12$ ml/kg/hr between 7-14 day post transfer.

**Discussion**

Results indicate that transfer of eels from FW to SW initiates an immediate and transient drinking response to compensate for loss of body water to the hyperosmotic environment. This drinking response is followed by another transient 3.5-fold increase in

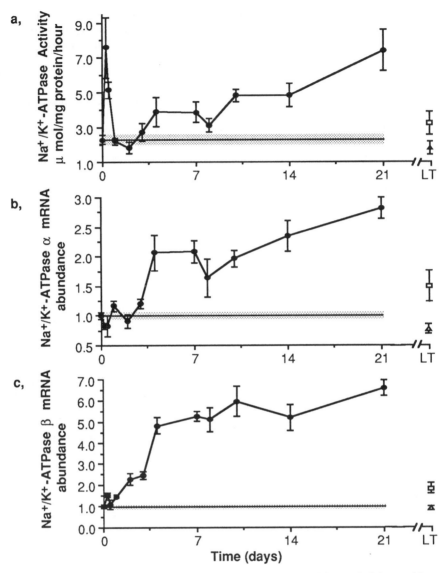

**Figure 1.** Time course of changes in Na$^+$/K$^+$-ATPase activity and alpha and beta subunit mRNA expression in eel gill following transfer from FW to SW environments. The graphs show the changes in maximal Na$^+$/K$^+$-ATPase activity (a), as well as changes in the relative mRNA abundance, compared to 18S rRNA, of the Na$^+$/K$^+$-ATPase alpha (b) and beta (c) subunits during the 21 days of acclimation. The horizontal line with shaded error bar represents the mean and SEM of FW fish measured at various time points during this period (n=18 except for a, where n=17). All other error bars represent standard errors where n=6 except in a, where n=4 on day 2, n=5 on day 4 and n=2 on day 21. Long term SW (□) and FW (△) adapted fish (LT; 6 months) are also shown.

248

branchial $Na^+/K^+$-ATPase activity which presumably compensates for the increased sodium load. However, experiments by Forrest et al (6) found no increase 2 hours after transfer, suggesting that the onset of this acute increase in $Na^+/K^+$-ATPase activity must lie between 2-6 hours after transfer. The mechanism of up-regulation of $Na^+/K^+$-ATPase activity during this time is not known but it is not associated with increased mRNA production.

The second phase of increased $Na^+/K^+$-ATPase activity (up to 3.5-fold) is similar in magnitude and time course to that described by others (7), and is associated with 3-fold and 6-fold increases in the expression of $Na^+/K^+$-ATPase $\alpha$ and $\beta$ subunit mRNAs respectively. This elevated expression of the enzyme parallels the increase in drinking rate found in SW adapted animals. This increase in $Na^+/K^+$-ATPase expression coincides with the reported 3-fold increase in branchial chloride cell numbers, during adaptation (7). The mechanisms controlling $Na^+/K^+$-ATPase activity and $\alpha$ and $\beta$ subunit expression are unknown, although the regulation of this protein may be related in part to the transient increases in plasma cortisol concentration which were also found in these fish (results not shown).

The level of $Na^+/K^+$-ATPase activity and $\alpha$ and $\beta$ subunit mRNA expression was found to have decreased in 6 month-adapted eels. This suggests that after 3 weeks transfer of animals to SW there is a reduced requirement for $Na^+/K^+$-ATPase expression in the branchial epithelium. The reason for this is unclear, but may reflect changes in expression of other branchial epithelial transporting systems or perhaps may result from a reduction in total body water loss and therefore a reduced requirement for drinking.

## Acknowledgements

These studies were supported by both the Natural Environment Research Council (NERC) and the Wellcome Trust.

## References

1. Karnaky KJ,Jr, Kinter LB, Kinter WB, Stirling CE (1976) Teleost chloride cell. II. Autoradiographic localization of gill Na-K-ATPase in killifish. *Fundulus heteroclitus* adapted to low and high salinity environments. J Cell Biol 70: 157-177

2. Sardet C, Pisam M, Meatz J (1979) The surface epithelium of teleostean fish gills: Cellular function and functional adaptations of the chloride cell in relation to salt adaptation. J Cell Biol 80: 96-117

3. Cutler CP, Cramb G, Lamb (1988) Quantitative analysis of sodium pump-specific mRNA from Human endothelial (HeLa) and Canine kidney (MDCK) cell cultures. Prog Clin Biol Res 268B: 59-64

4. Bers DM (1979) Isolation and characterisation of cardiac sarcolemma. Biochim Biophys Acta 555: 131-146

5. Hazon N, Balment RJ, Perrott M, O'Toole LB (1989) The renin angiotensin system and vascular and dipsogenic regulation in elasmobranches. Gen.Comp.Endocr 74: 230-236

6. Forrest JNJr, Cohen, AD, Schon DA, Epstein FH (1973) Na transport and Na-K-ATPase in gills during adaptation to seawater: Effects of cortisol. Am J Physiol 224: 709-713

7. Thomson AJ, Sargent JR (1977) Changes in the level of chloride cells and $(Na^++K^+)$ dependent ATPase in the gills of yellow and silver eels adapting to sea water. J Exp Zool 200: 33-40

# Differentiation of $Na^+/K^+$-ATPase by digoxigenin-labelled lectins on blots. Evidence for organ-typical glycoforms

M. Benallal, B.M. Anner

Laboratory of Experimental Cell Therapeutics, Geneva University Medical Centre, CH-1211 Geneva-4, Switzerland.

## Introduction

Glycan units can exhibit a high degree of structural variations on the same molecule resulting in different glycoforms for a molecule that is essentially carrying out the same function (3,4). Since glycosylation is performed by the host cell (8), it might reflect the original source of a protein. Consequently, the particular physiological, biochemical or pathological conditions in the body at the time the protein is synthesised might result in variable sugar groupings on a same polypeptide. In view of designing cell-specific drug delivery systems, we wished to know if an ubiquitous transmembrane protein, $Na^+,K^+$-ATPase (NKA) could be exploited as an indicator of cell localisation and status. To see whether NKA could be glycosylated differently in distinct organs, highly selective and sensitive digoxigenin-labelled lectins (3) were used in combination with isoform-specific immunoblots to compare kidney and brain enzyme isolated from various rat strains. Lectins are ubiquitous proteins which recognise and bind to sugar groupings of glycoproteins (5) just like antibodies recognise proteinaceous epitopes.

## Lectins differentiate rat brain and kidney NKA

The NKA exists in various isoforms of unknown functional differences encoded by distinct genes: the kidney contains essentially $\alpha 1$ and $\beta 1$ forms whereas the brain contains $\alpha 1$, $\alpha 2$, $\alpha 3$, $\beta 1$ and $\beta 2$ forms (11). With regard to NKA glycosylation, the $\beta$-subunit is a known glycoprotein and the glycosylation pattern of the renal $\beta 1$-form is currently established by chemical analysis (12). The electrophoretic mobility of the $\beta$-subunit on SDS-polyacrylamide gels varies depending on whether the protein is isolated from kidney or brain (10). Results from endoglycosidase F digestion (10) together with restriction endonuclease mapping and DNA sequencing of cDNA libraries from rat kidney and brain tissue (13) suggest that this difference in the apparent molecular weight of the kidney and brain $\beta$-subunit could be due at least in part to differences in their carbohydrate moieties. Glycosylation of the $\alpha 1$ form has been reported recently and shown to be essentially on the cytoplasmic membrane side (6,7) as foreseeable by the intracellular localisation of the N-glycosylation consensus sequences (9). However, comparison between the glycosylation patterns of NKA isolated from different organs has not yet been done.

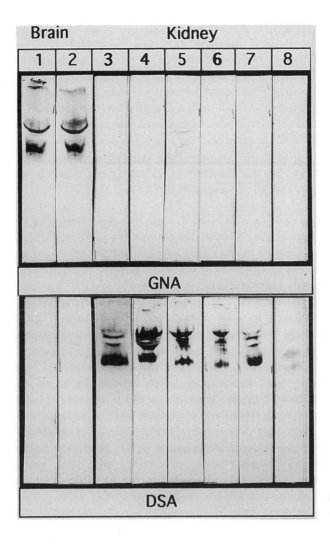

**Figure 1.** GNA and DSA lectin-blots of Na,K-ATPase isolated from brain and kidney of genetically hyper- and normotensive rats and outbred rats: Milan .hypertensive rats (lanes 1,3), Milan normotensive rats (lanes 2,4), spontaneous hypertensive rats (lanes 1,5), normotensive control Wistar Kyoto rats (lane 6), outbred rats (lane 7) and molecular weight standards (lane 8).

251

To look for hypothetical differences in sugar groupings of NKA, the enzymes were isolated by the dodecyl sulphate extraction procedure (1), separated by gel electrophoresis (not shown), electroblotted to nitro-cellulose and incubated for one hour at 20° C with five digoxigenin-labelled lectins chosen for their specific sugar recognition properties. The bound lectins were detected by an alkaline phosphatase labelled antidigoxigenin antibody with a detection limit of about 0.1 $\mu$g. Two of these five lectins turned out to recognise fundamentally distinct sugar groupings on brain and kidney NKA: in brain preparations, the two NKA subunits, $\alpha$ and $\beta$ strongly reacted with *Galanthus Nivalis Agglutinin* (GNA) and not at all with *Datura stramonium Agglutinin* (DSA). On the contrary, in kidney preparations, $\alpha$ and $\beta$ were both identified by DSA and did not react with GNA (Fig. 1).

That the strongly positive reaction of the mannose-recognising GNA with brain NKA was due to specific binding to sugar and not to some non-specific alkaline phosphatase reaction was documented by the totally negative reaction of the lectin DSA applied in the same conditions and previously tested for positive reaction with asialofetuin and fetuin (not shown). The same controls documented specific interaction between DSA and kidney NKA. To control the specificity of the surprising and strong response of both $\alpha$- and $\beta$-subunits to GNA in brain and DSA in kidney, the lectins were pre-treated respectively with the competing sugars $\alpha$-methyl-mannopyrannoside and galactose known to compete with the glycoprotein's sugar residues: the GNA-NKA interaction as well as the DSA-NKA interaction were abolished by the added sugar (data not shown). Taken together, the results indicate that in rat, sugar moieties on NKA vary with tissular origin. While the $\alpha$ and $\beta$ subunits of rat brain NKA both carry terminal mannoses specifically recognised by GNA and typical for high mannose type N-glycans, the sugar moieties on $\alpha$- and $\beta$-subunits of rat kidney react with DSA which recognises the disaccharide galactose $\beta$(1-4) N-acetylglucosamine, typical for complex and hybrid N-glycan structures, and the monosaccharide N-acetylglucosamine. Interestingly, the high mannose form corresponds to a biosynthetic precursor of the complex and hybrid N-glycans (8).

GNA and DSA are able to distinguish rat brain from rat kidney NKA independently of the protein isoforms expressed in these organs. By immunoblotting, $\alpha$1, $\alpha$2, $\alpha$3, $\beta$1 and $\beta$2 were detected in brain, $\alpha$1 and $\beta$1 in kidney (data not shown). Yet, the brain isoforms turned out to be exclusively stained by GNA. Though the brain $\alpha$1 is present at an amount detectable by immunoblotting, it does not react with the very sensitive DSA (2); thus, it can hardly be glycosylated in the same way than the DSA reactive renal $\alpha$1. Organ typical glycosylation might render heterogeneous protein isoforms homogenous with regard to their glycans.

## Conclusion

Using highly selective lectins in combination with isoform specific immunoblots, we discovered organ-specific glycoforms of NKA. Thus, a same polypeptide is diversified by organ-typical sugars and, at the same time, distinct protein isoforms occurring in a same organ become homogenous by host cell glycosylation. The sugars look like the

252

visiting card of NKA, defining its tissular belonging. This finding might be of great biological significance since glycomoieties, owing to their structural diversity and complexity, are at the kernel of the fundamental lock-and key complementary system and participate in cell typing, intercellular recognition and cell adhesion. The recognition properties of lectins could serve as a new tool to detect and characterise NKA glycoforms. In view of our findings, the Na-pump could be classified not only in terms of isoforms on the basis of their amino-acid sequence but also in terms of glycoforms, according to their glycosylation, the former identified by antibodies and the latter by lectins.

*Acknowledgements*

We are grateful to Prof. G. Bianchi for gift of Milan genetically hyper- or normotensive rat strains. Supported by the Swiss National Science Foundation (grant No. 31-25666.88) and the Ernest and Lucie Schmidheiny Foundation.

**References**

1.  Dzhandzhugazyan KN, Jørgensen PL (1985) Asymetric orientation of amino groups in the alpha-subunit and the beta-subunit of $Na^+/K^+$-ATPase in tight right-side-out vesicles of basolateral membranes from outer medulla. Biochem Biophys Acta 817: 165-173

2.  Haselbeck A, Schickaneder E., von der Eltz H, Hösel W (1990) Structural characterization of glycoprotein carbohydrate chains by using digoxigenin-labeled lectins on blots. Anal. Biochem. 191: 25-30

3.  Hosoi M, Kim S, Yamamoto K (1991) Evidence for heterogenity of glycosylation of human renin obtained by using lectins. Clin. Sci. 81: 393-399

4.  Liao J., Heider H, Sun M, Brodbeck U (1992) Different glycosylation in acetylcholinesterases from mammalian brain and erythrocytes. J. Neurochem. 58:1230-1238

5   Lis H, Sharon N (1986) Lectins as molecules and tools. Ann. Rev. Biochem. 55: 35-67

6.  Pedemonte CH, Sachs G., Kaplan JH (1990) An intrinsic membrane glycoprotein with cytosolically oriented N-linked sugars. Proc. Natl. Acad. Sci. USA 87: 9783-9789

7.  Pedemonte CH., Kaplan JH (1992) A monosaccharide is bound to the sodium pump α subunit. Biochemistry 31: 10465-10470

8.  Roth J (1987) Subcellular organization of glycosylation in mammalian cells. Biochim. Biophys. Acta 906: 405-436

9.  Shull GE., Greeb J, Lingrel JB (1986) Molecular cloning of three distinct forms of the $Na^+/K^+$-ATPase α-subunit from rat brain. Biochemistry 25: 8125-8132

10. Sweadner KJ, Gilkeson RC (1985).Two Isozymes of $Na^+/K^+$-ATPase have distinct antigenic determinants. J.Biol. Chem. 260: 9016-9022

11. Sweadner KJ (1989) Isozymes of the $Na^+/K^+$-ATPase. Biochim.Biophys. Acta 988: 185-220

12. Treuheit MJ, Costello CE., Kirley TL Structures of the complex glycans found on the β subunit of $Na^+/K^+$-ATPase. J.Biol. Chem. 268: 13914-13919

13  Young R, Shull GE, Lingrel JB.(1987) Multiple mRNAs from rat kidney and brain encode a single $Na^+/K^+$-ATPase β subunit protein. J.Biol. Chem. 262: 4905-4910

# Structural Requirements for the Interaction between the Na$^+$/K$^+$-ATPase $\alpha$- and ß-Subunits

M.V. Lemas, M. Hamrick, M. Emerick, K. Takeyasu[†], B. Hwang, M. Kostich and D.M. Fambrough

Department of Biology, The Johns Hopkins University, Baltimore, MD, 21218 USA
[†]Biotechnology Center, The Ohio State University, Columbus, OH, 43210 USA

## Introduction

Identification of domains involved in assembly of the Na,K-ATPase $\alpha$- and $\beta$- subunits is a major focus of our lab. These studies can lead to an understanding of the mechanisms controlling subunit assembly and may provide information about the membrane topology of the $\alpha$-subunit and folding properties of the $\beta$-subunit extracellular domains. Our approach has taken advantage of the unique properties of several monoclonal antibodies that recognize the chicken $\beta$1-subunit. Expression of chimeric subunits and/or subunits lacking portions of the polypeptide chains, followed by immune precipitations with these monoclonal antibodies, has allowed us to identify minimal regions within both the $\alpha$- and $\beta$-subunit amino acid sequences that are sufficient for assembly.

Monoclonal antibodies $\beta$24 (mAb-$\beta$24) and $\beta$29 (mAb-$\beta$29) have been used to precipitate Na,K-ATPase $\alpha$-$\beta$ complexes from detergent extracts of transiently or stably transfected cells expressing the chicken $\beta$-subunit (3,10). The immune precipitation of $\alpha$-subunits (or chimeric catalytic subunits) with the $\beta$-subunit serves as an operational definition of the assembled state. mAb-$\beta$24 preferentially recognizes properly folded chicken $\beta$-subunits and only weakly recognizes denatured $\beta$-subunits. The opposite characteristics describe mAb-$\beta$29, which preferentially recognizes denatured chicken $\beta$-subunits and only weakly recognizes native $\beta$-subunits. While the residues that define the mAb-$\beta$24 are not well established, those for the mAb-$\beta$29 are more certain and have been limited to a region (amino acids 127 - 153) that encompasses the most amino-terminal disulfide-bonded domain (Fig. 1). The characteristics of the mAb-$\beta$29, together with data from immune precipitation of $\beta$-subunit deletion mutants using mAb-$\beta$29 (see below), suggest that when the $\beta$-subunit is properly folded, the most amino-terminal disulfide-bonded domain, or some portion of it, is buried within the tertiary structure of the $\beta$-subunit or $\alpha$/$\beta$ complex itself. Thus, a properly folded $\beta$-subunit would preclude mAb-$\beta$29 binding to the epitope. These unique antibody characteristics have allowed us to distinguish the conformational states of several expressed chimeric and deletion mutants of the chicken $\beta$-subunit.

254

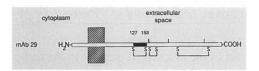

Figure 1. Region (black) of the chicken β1-subunit recognized by mAb-β29.

We have shown that the chicken β1-subunit assembles with each of the chicken α-isoforms (6). Similar immune precipitation experiments were performed with the chicken Na,K-ATPase β2-subunit (7) and the chicken H,K-ATPase β-subunit (referred to as HKβ) (Yu, manuscript submitted). Because no antibodies that recognize chicken β2 or HKβ were available, we tagged the amino termini of the β-subunits (chicken β1, β2, and HKβ) with a 10 amino acid peptide. This peptide sequence, from the human oncogene product of *c-myc*, is recognized by monoclonal antibody 9E10 (mAb-9E10) (2). Because each of the β-subunits contains the same epitope located at the same position, yields from immune precipitations can be compared. Coexpression of each of the chicken α-isoforms (11,12) with each of the "myc-tagged" β-subunits, followed by immune precipitations with mAb-9E10, was performed as previously described (7). All three chicken α-isoforms assemble with each of the chicken Na,K-ATPase β1 and β2 subunits and the H,K-ATPase β-subunit (Fig. 2).

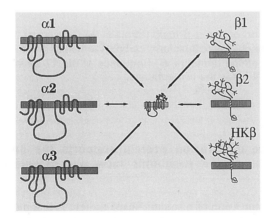

Figure 2. Cartoon representing the ability of each of the chicken α-isoforms to assemble with each of the chicken β-subunits. The three β-subunits are depicted with the *c-myc* epitope tag at their amino termini.

255

## Expression of a functional α-β fusion ATPase.

We have constructed cDNA encoding a "single-subunit" Na,K-ATPase (called α-β) by linking the amino terminus of the chicken β1-subunit to the carboxyl terminus of the chicken α3 subunit through a short peptide (seventeen amino acids) (Fig. 3). One objective in constructing this α-β cDNA is to resolve structural features that determine asymmetric surface expression in polarized cells. We expressed α-β in both transiently transfected COS-1 and in stably transfected mouse L-cells (1). These transfected cells expressed high levels of α-β, much of which appeared at the cell surface when visualized by immunofluorescence microscopy with mAb-β24, and the L-cells expressed high-affinity ouabain-binding sites characteristic of the chicken α3 subunit. The α-β L-cell lines also exhibited ouabain-sensitive Rb-uptake.

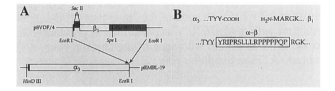

Figure 3. Construction of the α-β fusion protein. A) Cloning strategy used to link the encoding DNA of the chicken α3-isoform carboxyl terminus to the encoding DNA of the chicken β1-isoform amino terminus. B) Sequence of the 17 amino acid peptide linker (boxed residues) that fuses the two subunits.

## Expression of the α-β fusion protein supports the prediction that the carboxyl terminus of the α-subunit faces the cytoplasmic side of the membrane.

One advantage in expression of a subunit-fused protein is that monoclonal antibodies, which recognize an extracellular epitope on chicken β1-subunit, can now be used to detect chicken Na,K-ATPases at the cell surface in which the chicken β-subunit is always found associated with the chicken α3-subunit. Previously, it was impossible to distinguish those β-subunits at the cell surface that associated with chicken α-subunit from those with the endogenous α-subunit in cells cotransfected with both subunits. Linking the two subunits together has essentially created an extracellular epitope for the chicken α-subunit. Detection of chicken Na,K-ATPases can then be accomplished without permeabilizing the cells. Of course, it is important to determine whether the fused α-β remains intact during its expression. In studies from both transiently transfected and stably transfected cells, only a single polypeptide of $M_r$ ~135 kDa is immune precipitated with mAb-β24 from detergent extracts of cells expressing the α-β. The presence of only a single large peptide indicates that the α-β exists as an intact fusion protein and is not cleaved at the linkage site during biosynthesis, or during transport from the ER to the cell surface, or once established at the plasma membrane. Because the mAb-β24 only recognizes an epitope on the native β-subunit in its correct conformation (10) and detects the α-β protein at the surface of unpermeabilized cells, then the β-portion of the α-β protein is correctly oriented in the membrane and its large extracellular domain must be properly folded. Studies from our lab have shown that misfolded β-subunits do not get transported to the plasma membrane and remain in the ER (3,10). Jaunin et. al have shown that subunit assembly is a prerequisite for exit from the ER (4). These data, together with the functional studies of the α-β pump (1) strongly suggest that both the α- and β-portions of the α-β fusion protein are properly folded and assembled into a functional holoenzyme. Since the β-subunit amino terminus is on the cytoplasmic side of the membrane, the carboxyl terminus of the α-subunit must be cytoplasmic as well. This orientation of the α-subunit carboxyl terminus would be consistent with most of the current topological models (see review, 8). It is unlikely that the peptide linker between the carboxyl terminus of the α-subunit and the amino terminus of the β-subunit would traverse the membrane to accommodate an α-subunit with an extracellular carboxyl terminus. This is due to the physiochemical nature of the amino acids. The linker contains a string of prolines and several charged amino acids. Therefore, the carboxyl terminus of the α-subunit of the Na,K-ATPase normally resides on the cytoplasmic side of the membrane.

### Domains of the β-subunit involved in assembly.

Previous studies in our lab have shown that the amino terminus and nearly 80% of the amino acids that define the membrane spanning domain of the chicken β1-subunit are not required for assembly with the α-subunit (10). However, some deletions within the membrane spanning domain of the β-subunit impaired membrane insertion during subunit biosynthesis (5). Thus, it was unclear whether the remainder of the membrane spanning domain was involved in assembly. To determine unambiguously whether the membrane spanning domain is involved in assembly, we constructed cDNA encoding the cytoplasmic and membrane spanning domains of rat dipeptidyl peptidase (DPPIV) fused with the extracellular domain of the chicken β1-subunit (called DPPβ; and the reciprocal chimera was also constructed, βDPP) (Fig. 4). DPPIV is a Type-II membrane protein (9) with the same membrane spanning orientation as the β-subunit. Immune precipitation with mAb-β24 from cells expressing chicken α1-subunit and either DPPβ or βDPP chimeras revealed that the extracellular domain of the β1-subunit is both necessary and sufficient for assembly. Furthermore, immunofluorescence microscopy showed that the DPPβ chimera is transported efficiently to the plasma membrane. However, the yields of α-subunit that coprecipitated with the DPPβ chimeric subunit were much less than those observed with the wild-type chicken β-subunit. We consistently found that the α:DPPβ ratio was approximately 35% of the wild-type α:β ratio. These results suggest that replacement of the cytoplasmic and membrane spanning domains of the β-subunit 1) decreases the efficiency of α- and β- subunit assembly in the ER, and/or 2) decreases the stability of assembled α/β complexes in the intact cell, or during the immune precipitation procedure.

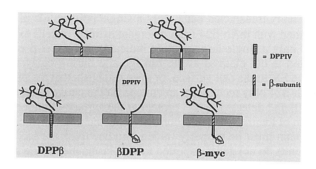

Figure 4. Cartoon representing β1-subunit deletion mutants and chimeras. The cytoplasmic and membrane-spanning domains of the DPPIV and β-subunit are depicted with different patterns (see legend in figure). The amino terminal *c-myc* epitope tag is shown on the βDPP chimera and the full-length chicken β-subunit.

**Regions of the β-subunit extracellular domain required for assembly with the α-subunit.**

To define further the domains of the β-subunit involved in assembly, several carboxyl terminal deletion mutants constructed from the chicken β1-subunit were examined with our immune precipitation assays (Fig. 5A). Deletions of 4, 11, 19, 92, and 146 amino acids from the carboxyl terminus of the β-subunit (called C4, C11, C19, C92, C146 respectively) do not prevent assembly with the α-subunit (3). However, immune precipitation analysis showed that all these deletion mutants exhibited decreased α:β ratios when compared to wild-type subunits. Interestingly, deletions of only a few carboxyl terminal amino acids (C4, C11, or C19), reduces the efficiency of immune precipitation by mAb-β24 and increases the efficiency of immune precipitation by mAb-β29. As mentioned earlier, mAb-β29 recognizes partially denatured β-subunit. Apparently, some amount of structural perturbation of the β-subunit can be tolerated in subunit assembly. Therefore, it appears that these carboxyl terminal amino acids are not required for subunit assembly, but may affect the folding of the extracellular domain. Immunofluorescence microscopy analysis of each of the deletion mutants suggests that assembly alone is not sufficient to guarantee the exit from the ER since none of these mutants was ever observed at the plasma membrane (3).

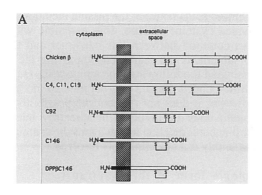

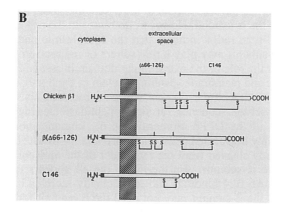

Figure 5. Schematic representations of A) deletion mutants B) Internal deletion mutant. Areas of black at the amino termini of the C92 and C146 identify the *c-myc* epitope tag. The cytoplasmic and membrane-spanning domains, representing DPPIV residues as part of the DPPβC146, are colored black. The approximate location of the three disulfide-bonded domains and the glycosylation sites are also shown.

**The most amino-terminal disulfide-bonded domain of the β-subunit may be involved in assembly with the α-subunit.**

The ability of the α-subunit to assemble with the DPPβ chimera (see earlier) and with C146 deletion mutant suggested that the β-subunit domain critical for assembly must be within the extracellular 96 amino acids from the plasma membrane to the most amino-terminal disulfide-bonded domain. To determine whether these amino acids are sufficient for subunit assembly, DNA encoding the combined features of the DPPβ chimera and C146 deletion mutant, was constructed (Fig. 5B). Immune precipitation analysis revealed that no α-subunit coprecipitated with this DPPβC146 mutant. Perhaps replacement of the C146 mutant with the DPPIV cytoplasmic and membrane spanning domains significantly decreased the efficiency of subunit assembly and/or the stability of α/β complexes, especially during the precipitation assay. Interestingly, expression of an β-subunit mutant with an internal deletion (β(del 66-126)), which

259

removes all but three amino acids between the plasma membrane and the most amino-terminal disulfide-bonded cysteine residue, does assemble with the α-subunit. The extracellular region in common between this construct and the C146, which also assembles with the β-subunit, is the region encompassing the most amino-terminal disulfide-bonded domain and the following nine amino acids (see Fig. 5B). This could suggest that the disulfide-bonded domain and the nine amino acids are sufficient for subunit assembly. However, it does not rule out the possibility that other regions of the β-subunit ectodomain may be involved in assembly interactions with the α-subunit. Clearly, other β-subunit deletion mutants must be examined before the specific domains involved in assembly can be definitively resolved.

### Expression of a soluble form of the β-subunit.

Results from immune precipitation assays using β-subunit chimeras and deletion mutants confirmed that the ectodomain of the β-subunit is both necessary and sufficient for assembly with the α-subunit. However, decreases in the α:β ratios of these mutants compared to wild-type ratios indicate that the cytoplasmic and membrane-spanning domains of the β-subunit may play a role (direct or indirect) in subunit assembly. To investigate further the role of the extracellular domain in assembly and the significance of the cytoplasmic and membrane spanning domain of the β-subunit, a secreted form of the β-subunit (sβ) extracellular domain was constructed. This was accomplished by introducing a consensus sequence for signal peptidase cleavage into the membrane spanning domain of the DPPβ chimera (Fig. 6). Immune precipitation, from both transiently and stably transfected cells, confirmed that the extracellular domain of the β-subunit is efficiently cleaved from the remainder of the molecule and that this sβ can assemble with the α-subunit. Furthermore, the sβ appears to be properly folded (determined by mAb-β24 binding) and is efficiently secreted into the medium. This demonstrates that the β-subunit may not need the α-subunit for proper folding of its extracellular domains. Quantitative analysis of immune precipitation of sβ from stably transfected L-cell lines revealed that the α:sβ ratio was 30% of the α:DPPβ ratio (thus α:sβ ratio is ~10% of wild-type α:β assembly) suggesting that the loss of membrane tethering decreased the efficiency of the β-subunit extracellular domain assembly with the α-subunit and/or stability of assembled α/β complexes. In these L-cell lines, sβ is found at the plasma membrane when visualized by immunofluorescence microscopy with mAb-β24 and is believed to be bound to the mouse α-subunit.

Figure 6. Consensus sequence for signal peptidase cleavage was introduced into the DPPβ chimera and resulted in the expression of a secreted form of the β1-subunit ectodomain.

## Domains of the α-subunit involved in assembly.

Using chimeric cDNA encoding a sarcoplasmic/endoplasmic reticulum Ca-ATPase and regions of the Na,K-ATPase α-subunit, we previously showed that 161 amino acids of the α-subunit carboxyl terminus are sufficient for assembly with the β-subunit (7). This study supports to the view that the two catalytic subunits have the same topology in the membrane. To investigate further the specific domain(s) of the α-subunit involved in assembly, we have constructed several more chimeric cDNAs encoding even smaller regions of the α-subunit. Expression of these chimeric catalytic subunits together with the chicken β-subunit, followed by immune precipitation using mAb-β24, allowed us to seek the minimal α-subunit amino acid sequence that would still mediate subunit assembly. A chimera containing the last 103 amino acids of the carboxyl terminus of the α-subunit, which replaced the corresponding region of the Ca-ATPase, does not assemble with the β-subunit. However, a chimera containing 59 amino acids of the α-subunit (called EC59) (Fig. 7), a region predicted to include the 7th membrane spanning domain and an extracellular domain ending prior to the 8th membrane spanning domain (predictions based on a 10 membrane spanning domain model of the α-subunit), does assemble with the β-subunit (Lemas, manuscript submitted).

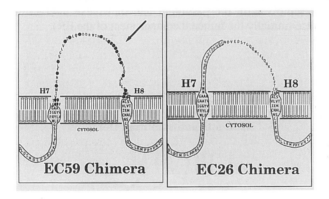

Figure 7. Cartoon representing the EC59 and EC26 proposed ectodomains depicted in the membrane and flanked by membrane-spanning domains H7 and H8 (remainder of the catalytic subunit is composed of Ca-ATPase residues). Capitalized letters outside shaded areas represent Na,K-ATPase α-subunit residues. EC59 shows black circles over amino acid residues found to be identical in both Na,K-ATPase and H,K-ATPase α-subunit sequences. An arrow indicates a cluster of these conserved residues.

This suggested that the minimal regions of the α-subunit sufficient for subunit assembly may be a region predicted to be exposed to the extracellular space. Because the extracellular domain of the β-subunit was found to be both sufficient and necessary for assembly with the α-subunit, it is reasonable to suppose that extracellular regions of the α-subunit are involved in assembly with the β-subunit. Sequence alignments of the amino acid residues in the proposed extracellular domain (flanked by the 7th and 8th membrane spanning domains) revealed a cluster of residues well-conserved among all the known α-subunits of the Na,K-ATPase and H,K-ATPase families (Fig. 7). This cluster is predicted to be located in the extracellular loop just prior to the proposed 8th membrane spanning domain. We wondered whether these amino acid residues were critical for assembly of the α-subunit with the β-subunit. A chimera (called EC26) that replaced the corresponding residues of the Ca-ATPase with 26 amino acids of the α-subunit (which included the cluster of conserved residues) was expressed in cells together with the β-subunit. Immune precipitations showed that the EC26 chimera assembles with the β-subunit and exhibited chimera:β ratios similar to those of the chimera with 161 α-subunit residues. The data clearly demonstrate that these 26 residues of the α-subunit are sufficient for subunit assembly. However, are these residues actually exposed at the external surface of the membrane? To address this question, we expressed the EC26 together with the DPPβ (a chimera composed of the extracellular domains of the β-subunit fused to the cytoplasmic and membrane spanning domains of the DPPIV protein) and followed with immune precipitations using the mAb-β24. If the α-subunit residues of the EC26 chimera are on the same side of the membrane as the extracellular domain of the β-subunit (part of the DPPβ), then these residues would be available for assembly with the DPPβ chimera (Fig. 8). Indeed, immune precipitations have determined that the EC26 chimera can assemble with the DPPβ chimera. Therefore, we conclude that these 26 residues of the α-subunit are sufficient for assembly with the β-subunit and that these interactions occur in the extracellular space (initially the lumenal compartment of the ER).

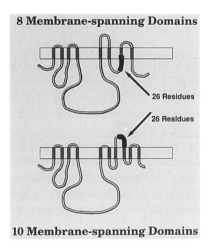

Figure 8. Approximate position of the α-subunit 26 residues (from EC26 chimera) fitted to a model of the Na,K-ATPase α-subunit with 8 or 10 membrane spanning domains.

Acknowledgments:

We would like to thank Misti Ushio, James Naurot, Christine Hatem, and Delores Somerville for all their excellent technical assistance. These studies were supported by grant NS-23241.

References:

1. Emerick M , Fambrough DM (1993) *J. Biol. Chem.*, in press
2. Evan GI, Lewis GK, Ramsay G, Bishop JM (1985) Isolation of Monoclonal Antibodies Specific for Human *c-myc* proto-oncogene Product. *Mol. Cell. Biol.* 5, 3610-3616
3. Hamrick M, Renaud KJ, Fambrough DM (1993) *J. Biol. Chem.*, in press
4. Jaunin P, Horisberger J-D, Richter K, Good PJ, Rossier BC, Geering K (1992) Processing, Intracellular Transport, and Functional Expression of Endogenous and Exogenous α-β3 Na,K-ATPase Complexes in *Xenopus* Oocytes. *J. Biol. Chem.* 267, 577-585
5. Kawakami K, Nagano K (1988) The Transmembrane Segment of the Human Na,K-ATPase β-Subunit Acts as the Membrane Incorporation Signal. *J. Biochem. (Tokyo)* 103, 54-60
6. Kone BC, Takeyasu K, Fambrough DM (1991) Structure-Function Studies of Na/K-ATPase Isozymes. In: *The Sodium Pump: Recent Developments* (Kaplan, JH and DeWeer, P, eds.) pp. 265-269, Rockefeller University Press, New York.
7. Lemas MV, Takeyasu K, Fambrough DM (1992) The Carboxyl-Terminal 161 Amino Acids of the Na,K-ATPase α-subunit are Sufficient for Assembly with the β-subunit. *J. Biol. Chem.* 267, 20987-20991
8. Lingrel JB, Orlowski J, Shull MM, Price EM (1990) Molecular Genetics of Na,K-ATPase. *Prog. in Nucleic Acid Res.* 38, 37-89
9. Ogata S, Misumi Y, Ikehara Y (1989) Primary Structure of Rat Liver Dipeptidyl Peptidase IV Deduced from its cDNA and Identification of the NH2-terminal Signal Sequence as the Membrane-Anchoring Domain. *J. Biol. Chem.* 264, 3596-3601
10. Renaud KJ, Inman EM, Fambrough DM (1991) Cytoplasmic and Transmembrane Domain Deletions of Na,K-ATPase β-Subunit. *J. Biol. Chem.* 266, 20491-20497
11. Takeyasu K, Tamkun MM, Renaud KJ, Fambrough DM (1988) Ouabain-Sensitive (Na+ + K+)-ATPase Activity Expressed in Mouse L Cells by Transfection with DNA Encoding the α-subunit of an Avian Sodium Pump. *J. Biol. Chem.* 263, 4347-4354
12. Takeyasu K, Lemas MV, Fambrough DM (1990) Stability of (Na+ + K+)-ATPase α-subunit Isoforms in Evolution. *Am. J. Physiol.* 259, C619-630

263

# Structure and Function of Ion Pumps Studied by Atomic Force Microscopy and Gene-transfer Experiments Using Chimeric Na+/K+- and Ca2+-ATPases

Kunio Takeyasu, Jose K. Paul, Mehdi Ganjeizadeh, M. Victor Lemas*, Shusheng Wang, Huiying Yu, Toshiyuki Kuwahara, and Toshiaki Ishii

Department of Medical Biochemistry and Biotechnology Center, The Ohio State University, Columbus, Ohio 43210, USA and *Department of Biology, The Johns Hopkins University, Baltimore, Maryland 21218, USA.

Using novel chimeric $Ca^{2+}$-ATPases modified with portions of $Na^+/K^+$-ATPase, and vice versa, we have identified critical domains that govern ion sensitivity, inhibitor specificity, subunit assembly, and activation by $Ca^{2+}$/calmodulin. $Na^+$ and $K^+$ sensors have been identified in nonconserved regions of the $Na^+/K^+$-ATPase sequences; the $Na^+$ sensor, with 69 amino acids, is localized at the amino-terminal region of the $\alpha$ subunit, whereas the $K^+$ sensor, with 161 residues, resides at the carboxy terminus of the same subunit. The assembly domain with the $\beta$ subunit was identified within the C-terminal 161 amino acids of the $Na^+/K^+$-ATPase $\alpha$ subunit (Figure 1). When this region was incorporated into the corresponding region of the sarcoplasmic reticulum (SR) $Ca^{2+}$-ATPase, the heterodimer of the $Na^+/K^+$-ATPase $\beta$ subunit and the chimeric SR $Ca^{2+}$-ATPase was formed (25; for details, see Lemas et al. in this volume). The ouabain-sensitive domain was localized within the N-terminal 200 amino acids of the $Na^+/K^+$-ATPase. These variable amino-terminal 200 amino acids, when incorporated into the corresponding regions of the SR $Ca^{2+}$-ATPase, conferred ouabain sensitivity to the SR $Ca^{2+}$-ATPase (15). Addition of the carboxy-terminal 150 amino acids of the plasma membrane (PM) $Ca^{2+}$-ATPase to the $Na^+/K^+$-ATPase $\alpha$ subunit conferred $Ca^{2+}$/calmodulin sensitivity on the $Na^+/K^+$-ATPase, indicating the existence of potential domains in the $Na^+/K^+$-ATPase that can interact with the carboxy terminus of the PM $Ca^{2+}$-ATPase (for details, see Ishii and Takeyasu in this volume). The successful functional expression of additional chimeric ATPases will allow identification of other critical regions, such as domains responsible for targeting of distinct ATPases to the specific subcellular membranes. This section reviews our recent work on chimeric ATPases. We also address potential abilities of atomic force microscopy (AFM; for details, see Paul et al. in this volume) for analysis of molecular structures of the P-type ATPases.

## A. Ideas behind gene-transfer experiments using chimeric ATPases

The sodium-pump ($Na^+/K^+$-ATPase) is activated by $Na^+$ and $K^+$ ions, not by $Ca^{2+}$ ions, and inhibited by ouabain (13), whereas the sarcoplasmic/endoplasmic reticulum calcium-pump (SERCA-ATPase) requires $Ca^{2+}$ ions for activation, and is blocked by thapsigargin (23,29). The plasma membrane calcium-pump (PM $Ca^{2+}$-ATPase) is unique in terms of its regulation by $Ca^{2+}$/calmodulin (9). Distinct ion pumps have unique amino acid sequences distributed over the major portion of the individual proteins, possibly responsible for ion and inhibitor specificities, whereas some

264

sequences (~30% of total amino acids) are evolutionarily conserved and form a common ATP-catalytic domain in the middle of the enzymes (17).

We expressed chimeric chicken molecules between distinct P-type ATPases by transfection of encoding chimeric cDNAs into mammalian cells. The constructed chimeric chicken cDNAs (see Figure 2) were transfected into mouse L cells that were expressing the chicken $Na^+/K^+$-ATPase β subunit. After selection for G418 resistance, the expression of chimeric ATPases in mouse cells were assured by binding of chicken-specific monoclonal IgG's, 5 and 5D2 (see below, Detailed Methods). This heterologous expression system has been useful for identifying highly expressed exogenous functions over endogenous activities (15,25-27,39,41,42). This type of approach has also been successful in soluble proteins in general (3), and recently been applied for some membrane proteins with multiple transmembrane domains (11,24,43). We have mainly chosen the evolutionarily conserved regions as chimeric junctions. This strategy is based on and dependent upon the functional similarity (ATPase) and the conservation of higher order structures (transmembrane topology). Judging from the properties of the expressed chimeric ATPase molecules, the higher order structures as ion pumps should be reasonably well preserved (15,25-27,39).

**Figure 1.** Working model for structure and function of the $Na^+/K^+$-ATPase that accommodates all the results obtained with SERCA-ATPase as a parental molecule. The model includes ion sensors, ouabain binding (15) and subunit assembly domains (25). The conserved regions used as chimeric junctions (o) are shown. The minimum structural requirements for $Ca^{2+}$- and thapsigargin-sensitive SERCA-ATPase are the segments between Ile163 and Gly354 and between Lys712 and Ser830. The functions of the further amino- and carboxy-

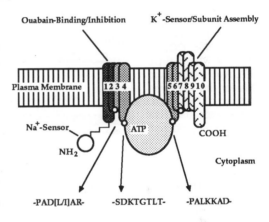

terminal regions of the SERCA ATPase are not identified yet, while the corresponding regions in the $Na^+/K^+$-ATPase are now defined respectively as the $Na^+$- and $K^+$-ion sensors. These implications will be applicable to elucidation of the structure and function of P-type ATPase in general.

Major research efforts have been directed towards identification of specific ion- and inhibitor-binding domains (4,5,8,12,34,35,37,38). Natural or artificial point mutations and chemical modification in transmembrane segments have provided initial evidence for functional domains responsible for inhibitor (4,5,37) and ion binding (8,12,34,38) by these ATPases. Our results obtained by the use of chimeric molecules complement the information obtained in point-mutation and chemical modification studies.

## Functional definition of ATPase activities

Functionality of the chimeric molecules was assessed on the basis of ATPase activities (15); the $Na^+/K^+$-ATPase activity was defined as the ouabain-inhibitable ATP cleavage and the SERCA-ATPase activity as the thapsigargin-inhibitable $Ca^{2+}$-dependent ATP cleavage. Since we use permanent cell lines which stably express exogenous proteins at constant levels, the quantitative aspects of the ATPase activity measured in this report are not variable between experiments. Therefore, all the data on ATPase activity reported here was expressed as actual activity (moles/min/mg protein), and can be compared directly among the data obtained from different cell lines with 10-15% experimental variation.

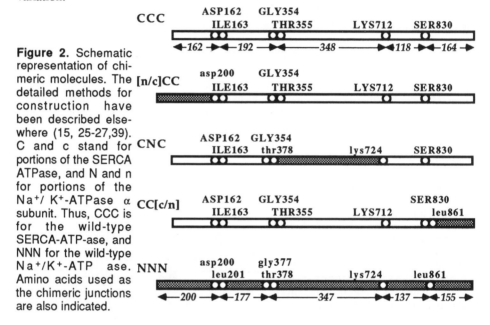

**Figure 2.** Schematic representation of chimeric molecules. The detailed methods for construction have been described elsewhere (15, 25-27,39). C and c stand for portions of the SERCA ATPase, and N and n for portions of the $Na^+/ K^+$-ATPase α subunit. Thus, CCC is for the wild-type SERCA-ATP-ase, and NNN for the wild-type $Na^+/K^+$-ATP ase. Amino acids used as the chimeric junctions are also indicated.

## B. Thapsigargin-binding sites are distinct from the ouabain-binding sites in thapsigargin- and ouabain-sensitive chimeras.

Transfected cells expressing distinct chimeric ATPases (CCC, [n/c]CC, CNC, and CC[c/n] at similar levels did not exhibit any detectable $Na^+/K^+$-ATPase activity, but did exhibit SERCA-ATPase activity at a level ~10, ~3, ~2, and ~5 times higher, respectively, than that intrinsic to mouse Ltk cells at optimum $Ca^{2+}$ concentrations. Some of such data are shown in Figure 3A. Thapsigargin is a tumor-promoting plant toxin which is known to raise the intercellular $Ca^{2+}$ ion concentration (40) by inhibiting the SERCA-ATPase (23,29,36). The specificity of thapsigargin to SERCA-ATPase is a strong advantage in this heterologous expression system. It is highly feasible to examine the SERCA-ATPase function of transfected proteins even if the activity is relatively low, as long as the expressed molecules retain both $Ca^{2+}$- and thapsigargin-sensitivities.

Among these chimeric ATPases, [n/c]CC, in which the amino-terminal 200 amino acids of the $Na^+/K^+$-ATPase α1 subunit were incorporated into the corresponding portion of SERCA1 was actually ouabain-sensitive (Figure 3B). This indicates that [n/c]CC possesses two distinct binding sites; one for ouabain and the other for thapsigargin. Taking into account the thapsigargin-sensitive property of CNC, in which the middle

266

Ca$^{2+}$-ATPase segment (Gly354-Lys712) is replaced with the corresponding domain of the Na$^+$/K$^+$-ATPase, thapsigargin should interact with the Ca$^{2+}$-ATPase domains either Ile163-Gly354 (including M3 and M4) or Lys712-carboxy end (including M5 - M10). Clarke et al. (8) have found that charged amino acids within the transmembrane segments M4, M5, M6 and M8 of the SERCA-ATPase play a key role in Ca$^{2+}$ binding to the SERCA-ATPase. Thapsigargin has been known to compete with Ca$^{2+}$ ions in ATP hydrolysis (36). Therefore, it is reasonable to conclude that the membraneous amino- and carboxy-terminal regions of the SERCA ATPase are critical for thapsigargin-and Ca$^{2+}$-sensitive ATPase activity.

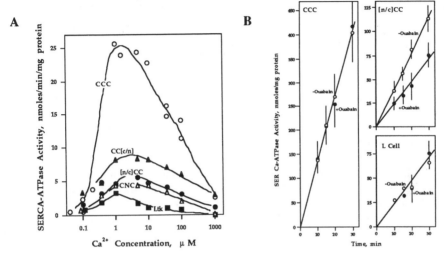

**Figure 3.** A: Thapsigargin-inhibitable and Ca$^{2+}$-sensitive ATPase (SERCA-ATPase) activity requires the transmembrane domains M3, M4, M5, and M6. The existence of SERCA-ATPase activities in CCC, [n/c]CC, CNC, and CC[c/n] in transfected mouse L cells demonstrates the minimum requirement for thapsigargin inhibition and Ca$^{2+}$ activation: the segments between Ile163 and Gly354 and/or between Lys712 and Ser830 of SERCA1. Ca$^{2+}$ ions activate SERCA-ATPase at lower concentrations and inhibit activity at higher concentrations. The SERCA-ATPases in all cell lines tested here were similarly subject to these dual effects of Ca$^{2+}$ ions, and retain thapsigargin- and Ca$^{2+}$-sensitivity. B: Ouabain (4 mM) inhibits the SERCA-ATPase activity of [n/c]CC (15).

## Minimum requirement for SERCA-ATPase activity

[n/c]CC and CC[c/n], in which portions of the membraneous amino or carboxy-terminal region of the SERCA-ATPase were replaced with the corresponding portions of the Na$^+$/K$^+$-ATPase, are also thapsigargin- and Ca$^{2+}$-sensitive ATPases. This set of results further restrict the possible location of the domains responsible for thapsigargin-inhibition to the segments between Ile163 and Gly354 and/or between Lys712 and Ser830, because these domains of SERCA1 are critical for the thapsigargin- and Ca$^{2+}$-sensitive ATPase activity. Thus, SERCA-ATPase activity requires the domains containing transmembrane segments M3, M4, M5, and M6. (For further restriction of the thapsigargin-sensitive domains to the transmembrane segments M3 and M4, see Andersen in this volume.)

## C. Identification of functional domains of the Na⁺/K⁺-ATPase using the SERCA1 as a parental molecule

Several domains critical for the Na⁺/K⁺-ATPase activity were identified by using the SR Ca$^{2+}$-ATPase as a parental molecule, and monitoring the SERCA-ATPase functions modified with the characteristics of the Na⁺/K⁺-ATPase.

**The region between Ala70 and Asp200 of the Na,K-ATPase α1 subunit is sufficient for ouabain binding, but other regions can affect the ouabain-binding property.**

The experiments in Figure 3 demonstrate that the 200 amino-terminal amino acids of the Na⁺/K⁺-ATPase α1 subunit are sufficient to exert ouabain-dependent inhibition even after incorporation into the corresponding portion of the Ca$^{2+}$-ATPase (15). The ouabain-binding ability of the amino-terminal region (Figure 4) was retained after deletion of 69 amino acids from this region, [Δn/c]CC, (manuscript in preparation), indicating that the region between Ala70 and Asp200 of the Na⁺/K⁺-ATPase α1 subunit is sufficient for ouabain binding. However, the affinity of [Δn/c]CC for ouabain was found to be lower than that of the wild-type NNN. The property of ouabain binding can be also changed by a variety of point mutations within the regions other than the amino-terminal 200 amino acids (see Lingrel et al. and Rossier et al. in this volume); e.g., the mutations in the ectodomain between M3 and M4 and between M5 and M6, and the mutations in the transmembrane domains resulted in a decrease of affinity for ouabain. These observations collectively suggest that the minimum requirement for ouabain binding is the region between Leu69 and Asp200, but other regions can affect the property of the ouabain interaction.

**Figure 4.** Identification of ouabain-binding sites within the amino-terminal 200 amino acids of the Na⁺/K⁺-ATPase α1 subunit in an isolated environment; i.e., in the SERCA-ATPase. [³H]ouabain binding to the homogenate of transfected cells were measured in two separate experiments (o,●) (15).

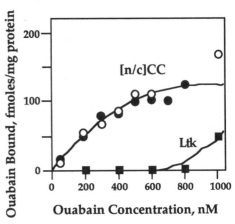

An interesting observation by Blostein et al. (2), apparently inconsistent with our present results, describes ouabain inhibition of ⁸⁶Rb flux in a chimeric molecule consisting of the amino-terminal 1/2 of the H⁺/K⁺-ATPase α subunit and the carboxy-terminal 1/2 of the Na⁺/K⁺-ATPase α subunit. They suggested that ouabain binds to the carboxy-terminal 1/2 of the Na⁺/K⁺-ATPase, because the H⁺/K⁺-ATPase is generally thought to be ouabain-insensitive. However, a recent report by Mathews and Rossier (in this volume) indicates that the Xenopus H⁺/K⁺-ATPase can be inhibited by ouabain as well as SCH28080 (a specific inhibitor for the H$^+$/K$^+$-ATPase) with high affinity. Since the H$^+$/K$^+$- and Na$^+$/K$^+$-ATPases share a high degree of amino-acid conservation (30), it might be reasonable to speculate that the H$^+$/K$^+$-ATPase in general possesses a potential ouabain-binding site, which can mediate ouabain inhibition in some species and/or under favorable conditions. In light of this, a plausible explanation consistent with all the available information is as follows: Site(s) in the amino-terminal domain of the H$^+$/K$^+$-ATPase with a potential ability to bind ouabain can be converted to a true ouabain site(s) by virtue of interaction with proper structural signals from elsewhere in the molecule; e.g., the carboxyterminal 1/2 of the Na$^+$/K$^+$-ATPase α subunit in case of the chimera (2) or that of the native H$^+$/K$^+$-ATPase α subunit in case of certain tissues and/or species (Rossier et al. in this volume).

### Ouabain-binding does not require β subunits of the Na,K-ATPase.

We previously identified the subunit assembly domain within the carboxy-terminal 161 amino acids of the Na$^+$/K$^+$-ATPase α subunit, and demonstrated that CC[c/n] assembles with the Na$^+$/K$^+$-ATPase β subunit (15). Thus, the NNN is expected to assemble with the β subunit, while the CCC, NCC, and [n/c]CC lack the domain required for assembly with the β subunit (25). [$^3$H]ouabain binding to [Δn/c]CC indicates that the short amino-terminal fragment (Leu69 to Asp200) of the Na$^+$/K$^+$-ATPase α1 subunit is sufficient for accepting ouabain with high affinity without the β subunit. This is intriguing with respect to previous reports which demonstrated that the β subunit is required for the α-subunit to acquire the ouabain-sensitive and Na$^+$/K$^+$-dependent ATPase activity (14,31,33). It might be the case that the carboxy-terminal region of the wild-type α subunit could induce a conformational change in remote regions of the α subunit upon assembling with the β subunit, and thus, influence the ouabain-binding ability of the amino-terminal domain. The sensitivity of ouabain-binding to the concentration of K$^+$ ions can be modulated by different combination of β subunits; e.g. either Na$^+$/K$^+$-ATPase β1 subunit or H$^+$/K$^+$-ATPase β subunit (10). A modification of the carboxy-terminal region of the Na$^+$/K$^+$-ATPase α subunit also decreases the affinity for ouabain (see Ishii and Takeyasu in this volume). Thus, there seems to be an allosteric interaction between the carboxy- and amino-terminal regions via long-range conformational changes. Therefore, the ouabain-binding region of NNN, but not of NCC or [n/c]CC, might require the β subunit for its correct conformation. Currently, it has not been clarified how the β subunit directly or indirectly (via conformational changes) affects the binding of ouabain. Nevertheless, our present results suggest that the β subunit is not required for ouabain-binding per se, although it can modulate the binding.

### The region between Met1 and Leu69 can act as a sensor for Na$^+$ ions.

The chimera ([n/c]CC), in which the amino-terminal amino acids (Met1 to Asp162) of the SR Ca$^{2+}$-ATPase were replaced with the corresponding portion of the Na$^+$/K$^+$-ATPase α1 subunit (Met1 to Asp200), retained thapsigargin- and Ca$^{2+}$-sensitive ATPase activity, although the activity was lower than the wild-type SR Ca$^{2+}$-ATPase. This SERCA-ATPase activity of [n/c]CC was drastically stimulated toward the level of wild-type Ca$^{2+}$-ATPase, CCC, with increasing concentrations of Na$^+$ ions in the assay medium, while the SERCA-ATPase activity of CCC, CNC, CC[c/n] or the endogenous mouse Ca$^{2+}$-ATPase was not affected (Figure 5). When the amino-terminal 69 amino acids, Met1 to Leu69, were deleted in [n/c]CC ([Δn/c]CC), the SERCA-ATPase was not activated by Na$^+$ ions, although this deletion chimera, [Δn/c]CC still retains the ouabain-binding ability (manuscript in preparation). These results strongly indicate that the region, Met1 to Leu69, of the Na⁺/K⁺-ATPase α subunit confers Na⁺-ion sensitivity to the SERCA1 and that it plays a role as a sensor for Na⁺ ions, although the actual transport mechanisms of Na⁺ ions are still not clear.

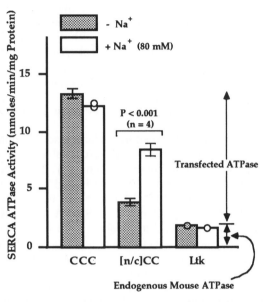

**Figure 5.** Amino-terminus of the Na⁺/K⁺-ATPase acts as a Na⁺ sensor. Thapsigargin- and $Ca^{2+}$-sensitive (SERCA) ATPase activity was measured at 5 μM free $Ca^{2+}$ and 80 mM Na⁺ ions. The total ionic strength was kept constant at 180 mM with 100 mM KCl and 80 mM of (NaCl or choline-chloride) (16). The SERCA-ATPase activity of [n/c]CC was regulated by Na⁺ ions, while the activity of CCC, CNC, CC[c/n], or mouse endogenous enzyme was not affected (only the data on CCC and Ltk are shown in the figure). The ED50 value for Na⁺ stimulation of [n/c]CC was ~50 mM, which is within the physiological local concentrations of Na⁺ ions after an action potential in spite of the slightly larger ED50 value compared to that obtained in the wild-type Na⁺/K⁺-ATPase (16).

## The carboxyterminal 161 amino acids of the Na⁺/K⁺-ATPase α1-subunit are responsible for K⁺ sensitivity.

K⁺ ions activated the wild-type SERCA-ATPase (CCC) in a biphasic fashion; the initial slight activation had an ED50 value of ~15 mM and subsequent full activation, an ED50 ~70 mM, resulting in an overall ED50 value of ~50 mM. The activities of [n/c]CC, and CNC were regulated by K⁺-ions in a manner similar to the wild-type SERCA-ATPase. However, the effect of K⁺ ions on CC[c/n] was the same as that on the wild-type Na⁺/K⁺-ATPase (16), being activated in a monophasic fashion with a single ED50 value of ~5 mM. Some of these results are illustrated in Figure 6, in which a ratio of SERCA-ATPase activities in the presence of 20 mM K⁺ vs. 100 mM K⁺ was taken as a parameter of K⁺ activation of the enzyme. The results demonstrated that only CC[c/n] can be activated at low concentration of K⁺ comparable to the ED50 for the wild-type Na⁺/K⁺-ATPase. This indicates that the carboxy-terminal 161 amino acids of the Na⁺/K⁺-ATPase confers the sodium-pump type K⁺-sensitivity onto the SERCA1, thus acting as a K⁺ sensor.

The Na⁺/K⁺-ATPase tightly binds K⁺ ions in the absence of Na⁺ (22). This phenomenon has been termed "K⁺ occlusion" and thought to be important for Na⁺ and K⁺ transport. A recent report (6) suggested that a 19 kDa fragment derived from the carboxy-terminus of the Na⁺/K⁺-ATPase may be the site for K⁺-occlusion. This finding complements our results that indicate the importance of the carboxy-terminal of the Na⁺/K⁺-ATPase for its activation by K⁺ ions. It remains to be resolved how the Na⁺ and K⁺ effects mediated by the sensors are related to the transport of Na[+] and K[+] ions. It is interesting to note that the K[+]-sensitive region (Figure 6) and the β subunit assembly domain (25) reside within the same short stretch of amino acids at the carboxy terminus of the Na[+]/K[+]-ATPase. A recent report using chemical modification techniques has indicated that a [86]Rb-occlusion property of the 19 kDa fragment of the α-subunit can be affected by the integrity of the β-subunit (28).

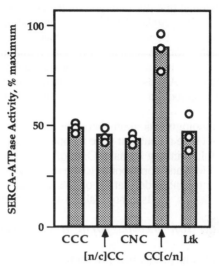

**Figure 6.** Carboxy-terminal 161 amino acids of the Na,K-ATPase can form a K[+] sensor. A SERCA-ATPase activities in transfected cells were measured at 5 μM free Ca[2+] and different concentrations of K[+]. The total ionic strength was kept at 100 mM with KCl plus choline-chloride. The SERCA-ATPase activity at 100 mM K[+] was taken as 100 %, and the % maximum activity at 25 mM K[+] was calculated. For transfected cells, the % maximam was culculated after subtracting the background SERCA-ATPase activity observed in Ltk cells.

## D. Chimeras between the Na[+]/K[+]-ATPase and the PM Ca[2+]-ATPase provide unique information about the molecular mechanism of regulation by Ca[2+]/calmodulin.

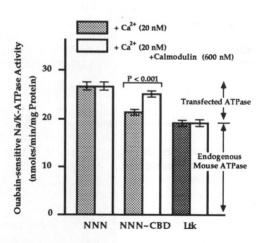

**Figure 7.** Carboxy-terminal 150 amino acids of the PM Ca[2+]-ATPase confer Ca[2+]/calmodulin-sensitivity to the Na[+]/K[+]-ATPase α-subunit. The ouabain-sensitive Na[+]/K[+]-ATPase activity was measured in cell homogetes including mouse endogenous and chicken exogenous (wild-type or chimeric) ATPases. (n = 3)

As a complementary approach to the functional chimeric SERCA-ATPase, functional chimeric Na[+]/K[+]-ATPases was constructed. We used the Na[+]/K[+]-ATPase as a parental molecule, and monitored the Na[+]/K[+]-ATPase functions modified with the characteristics

271

of the PM Ca$^{2+}$-ATPase. We have constructed a chimeric cDNA that encodes a chimeric protein in which the carboxy terminal 165 amino acids of the rat brain PM Ca$^{2+}$-ATPase II was added to the carboxy terminus of the ouabain-sensitive chicken Na$^+$/K$^+$-ATPase α1 subunit. Equal level of expression of this molecule and the wild-type Na$^+$/K$^+$-ATPase in ouabain-resistant mouse L cells was assured by the binding of $^3$H-ouabain and monoclonal IgG specific to the chicken Na$^+$/K$^+$-ATPase α1 subunit. Functional analysis demonstrated that this chimeric molecule is an ouabain-sensitive Na$^+$/K$^+$-ATPase which can be further regulated by Ca$^{2+}$ and calmodulin in a dose-dependent manner, consistent with the current consensus that the carboxy terminal region of the PM Ca$^{2+}$-ATPase is responsible for the regulation by Ca$^{2+}$/calmodulin (9). These results indicate that the presence of structurally homologous domains within the Na$^+$/K$^+$-ATPase α1 subunit and the PM Ca$^{2+}$-ATPase, which, via interaction with the calmodulin-binding domain, can mediate calmodulin activation of ATPase activity.

Five homologous domains between the rat brain PM Ca$^{2+}$-ATPase II and the chicken Na,K-ATPase α1 subunit were identified as candidates for the sites that interact with the calmodulin binding domain. One of such domains, Ile206-Val271, is located between M2 and M3, and corresponds to the region identified in the PM Ca$^{2+}$-ATPase by using a photo-activatable derivative of the synthetic calmodulin binding domains (9). (For further details, see Ishii and Takeyasu in this volume.)

## E. Can ion pumps form a channel- (pore-) like structure?

So far, we have identified several critical domains of the Na$^+$/K$^+$-ATPase α subunit (Figures 1,3-7), including the ouabain-binding domain and ion sensors. However, one of the major questions has remained unanswered; how can specific ions be transported across the membranes? Are the ion sensors the same as the sites for ions to be transported? Is there a pore in the sodium pump? We have begun to address these questions by applying a new physical method, atomic force microscopy (AFM), to elucidation of molecular structure of the Na$^+$/K$^+$-ATPase in membrane. This technique allowed us to monitor some structural aspects of the ATPase without crystallization.

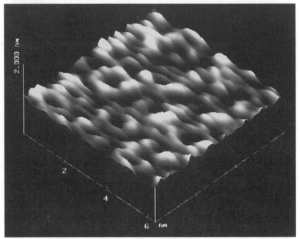

**Figure 8.** AFM imaging of the Na$^+$/K$^+$-ATPase in purified dog kidney membranes.

We used purified kidney membranes that was prepared according to the method of Jorgensen. Close to 95 % of the proteins in this membrane preparation are the Na$^+$/K$^+$-

ATPase. Figure 8 shows an AFM image, illustrating channel-like structures (see Paul et al. in this volume). These observations suggest that the Na+/K+-ATPase may form a pore, and strongly support the future of AFM as a powerful technique in the field of structural biology.

Ion channels and pumps in cell membranes contain multiple transmembrane segments that are thought to be critical for transport of ions. Channel structures supported by these transmembrane segments are characteristic of ion channels (7) whereas such pore-like structures have not been identified in ion pumps. By using AFM working under tapping mode, we have identified a channel-like conformation of the Na+/K+-ATPase.

## F. Summary and Future Directions

We have addressed some fundamental issues concerning the structure and function of P-type ATPases (see model in Figure 1). Concurrently, several new issues have been brought to surface. How do the Na+ and K+ sensors in amino- and carboxy-terminal regions contribute to the property of Na+- and K+-ion transport? For addressing this fundamental question, it will be necessary to develop assay systems that allow one to monitor specific ion-transport properties of expressed chimeric ATPase molecules. Is the K+ sensor the same as the assembly site with the β subunit? To answer this question, further construction and expression of additional chimeras (including those reported by Lemas et al. in this volume) will be required. Are the transmembrane domains M3 - M6 essential for the Na+/K+-ATPase activity as seen in the SERCA-ATPases? To answer this question, expression in non-animal systems (such as yeast or plant systems) will be required. Chimeric ATPases between the PM Ca²⁺-ATPase and the SERCA-ATPase will provide a clue for targeting mechanisms. Detailed structural analysis using AFM will be utilized for 3-dimensional localization of the amino and carboxy termini and the β subunit. ·

*References*
1. Askari, A., Huang, W.-H., McCormick, P.W. (1983) J. Biol. Chem. 258: 3453-3460.
2. Blostein, R., Zhang, R.-P., Gottardi, C.J., Caplan, M.J. (1993) J. Biol. Chem. 268: 10654-10658.
3. Brent, R., Ptashne, M. (1985) Cell 43: 729-736.
4. Canessa, C.M., Horisberger, J.-D., Louvard, D., Rossier, B.C. (1992) EMBO J. 11: 1681-1687.
5. Cantley, L.G., Zhou, X.-M., Cunha, M.J., Epstein, J., Cantley, L.C. (1992) J. Biol. Chem. 267: 17271-17278.
6. Capasso, J.M., Hoving, S., Tal, D.M., Goldshleger, R., Karlish, S.J.D. (1992) J. Biol. Chem. 267: 1150-1158.
7. Catterall, W. (1988) Science 242: 50-61.
8. Clarke, D.M., Loo, T.W., Inesi, G., MacLennan, D.H. (1989) Nature 339: 476-478.
9. Carafoli, E. (1992) J. Biol. Chem. 267: 2115-2118.
10. Eakle, K.A., Kim, K.S., Kabalin, M.A., Farley, R.A. (1992) Proc. Natl. Acad. Sci. U.S.A. 89: 2834-2838.
11. Fukuda, K., Kubo, T., Maeda, A., Akiba, I., Bujo, H., Nakai, J., Mishina, M., Higashida, H., Neher, E., Marty, A., Numa, S. (1989) Trends Pharmacol. Sci. (Suppl.): 4-10.
12. Goldshleger, R., Tal, D.M., Moorman, J., Stein, W.D., Karlish, S.J.D. (1992) Proc. Natl. Acad. Sci. U.S.A. 89: 6911-6915.
13. Hansen, O. (1984) Pharmacol. Rev. 36: 143-163.
14. Horowitz, B., Eakle, K.A., Scheiner-Bobis, G., Randolph, G.R., Chen, C.Y., Hitzeman, R.A., Farley, R.A. (1990) J. Biol. Chem. 265: 4189-4192.

15. Ishii, T. Takeyasu, K. (1993) Proc. Natl. Acad. Sci. U.S.A. In press.
16. Jewell, E.A., Lingrel, J.B. (1991) J. Biol. Chem. 266: 16925-16930.
17. Jorgensen, P.L., Andersen, J.P. (1988) J. Membrane Biol. 103: 95-120.
18. Jorgensen, P.L., Perterson, J. (1985) Biochim. Biophys. Acta 821: 319-333.
19. Jorgensen, P.L. (1974) Biochim. Biophs. Acta 356: 36-52.
20. Karin, N.J., Kaprielian, Z., Fambrough, D.M. (1989) Mol. Cell Biol. 9: 1978-1986.
21. Kaprielian, Z., Fambrough, D.M. (1987) Dev. Biol. 124: 490-503.
22. Karlish, S.J.D., Goldshleger, R., Stein, W.D. (1990) Proc. Natl. Acad. Sci. U.S.A. 87: 4566-4570.
23. Kijima, Y., Ogunbunmi, E., Fleischer, S. (1991) J. Biol. Chem. 266: 22912-22918.
24. Kobilka, B.K., Kobilka, T.-S., Daniel, K., Regan, J.W., Caron, M.G., Lefkowitz, R.J. (1988) Science (Wash. DC). 240: 1310-1316.
25. Lemas, M.V., Takeyasu, K., Fambrough, D.M. (1992) J. Biol. Chem. 267: 20987-991.
26. Luckie, D.B., Boyd, K.L., Takeyasu, K. (1991) FEBS Lett. 281: 231-234.
27. Luckie, D.B., Lemas, M.V., Boyd, K.L., Fambrough, D.M., Takeyasu, K. (1992) Biophys. J. 62: 220-227.
28. Lutsenko, S., Kaplan, J.H. (1993) Biochemistry 32: 6737-6743.
29. Lytton, J., Westlin, M., Hanley, M.R. (1991) J. Biol. Chem. 266: 17067-17071.
30. Maeda, M., Oshiman, K., Tamura, S., Futai, M. (1990) J. Biol. Chem. 265: 9027-9032.
31. McDonough, A.A., Geering, K., Farley, R.A. (1990) FASEB. J. 4: 1598-1605.
32. Mounsbach, A.B., Skriver, E., Herbert, H. (1990) The sodium pump: Recent developments, 44th annual symposium 160-172.
33. Noguchi, S., Mishina, M., Kawamura, M., Numa, S. (1987) FEBS Lett. 225: 27-32.
34. Pedemonte, C.H., Kaplan, J.H. (1990) Am. J. Physiol. 258: (cell Physiol. 27) C1-C23.
35. Price, E.M., Lingrel, J.B. (1988) Biochemistry 27: 8400-8408.
36. Sagara, Y., Fernandez-Belda, F., Meis, L.D., Inesi, G. (1992) J. Biol. Chem. 267: 12606-12613.
37. Schultheis, P.J., Lingrel, J.B. (1993) Biochemistry 32: 544-550.
38. Sumbilla, C., Cantilina, T., Collins, J.H., Malak, H., Lakowicz, J.R., Inesi, G. (1991) J. Biol. Chem. 266: 12682-12689.
39. Sumbilla, C., Lu, L., Sagara, Y., Inesi, G., Ishii, T., Takeyasu, K., Feng, Y., Fambrough, D.M. (1993) J. Biol. Chem. 268: 21185-21192.
40. Takemura, H., Hughes, A.R., Thastrup, O., Putney, J.W.Jr. (1989) J. Biol. Chem. 264: 12266-12271.
41. Takeyasu, K., Tamkun, M.M., Siegel, N., Fambrough, D.M. (1987) J. Biol. Chem. 262: 10733-10740.
42. Takeyasu, K., Tamkum, M.M., Renaud, K.J., Fambrough, D.M. (1988) J. Biol. Chem. 263: 4347-4354.
43. Tanabe, T., Akams, B.A., Numa, S., Beem, K.G. (1991) Nature 352: 800-803.
44. We thank Dr. A. Askari for providing purified dog kidney membranes, and Dr. G.K. Shull for a gift of cDNAs encoding PM $Ca^{2+}$-ATPase isoforms. We also thank Dr. D.M. Fambrough for chicken-specific monoclonal antibodies and a cDNA encoding SERCA1 ATPase. M.G. is a recipient of a Post-doctoral Fellowship from the American Heart Association, Ohio Affiliate. K.T. is an Established Investigator of the American Heart Association, and has been supported by the National Institutes of Health.

## Appendix: Detailed Methods

1.　　　Transfection and cloning of permanent cell lines - The chimeric chicken cDNA constructs encoding CCC (20), [n/c]CC (15), CNC (39), and CC[c/n] (25) (C and c stand for portions of the SERCA1, and N and n for portions of the $Na^+/K^+$-ATPase) were introduced into mouse L cells that had been transfected with a cDNA encoding the chicken $Na^+/K^+$-ATPase $\beta$1 subunit (41). The cells were selected for G418 resistance and high level of expression of the encoded chimeric ATPase under control of the RcCMV promoter by immunofluorescence microscopy and immunoprecipitation using monoclonal antibodies, IgG 5D2 specific to the chicken SR $Ca^{2+}$-ATPase (SERCA1) (21), and IgG 5 specific to the chicken $Na^+/K^+$-ATPase $\alpha$ subunit (42).

**2.** $Ca^{2+}$-ATPase Activity - $Ca^{2+}$-ATPase activity was measured using crude membrane preparation (15) by monitoring the release of inorganic $^{32}P$ from $\gamma$-$[^{32}P]ATP$ (Amersham, PB218) over time at 37 °C. The assay system (0.5 ml) consisted of membranes (~0.1 mg), 100 mM KCl, 50 mM Tris-HCl (pH 7.4), 2 mM ATP (containing 100 μM KOH and $\gamma$-$[^{32}P]ATP$), 3 mM MgCl2, 2μM A23187, 1 mM NaN3, and appropriate amounts of CaCl2 and EGTA (pH adjusted to 7.4 with ~29 μM KOH) to produce the required free $Ca^{2+}$ concentration. The reaction was terminated at intervals by the addition of 0.5 ml of 8% perchloric acid. The cleaved $^{32}P$ was converted to phosphomolybdate, extracted into 2-methyl-propanol, and its radioactivity counted in a liquid scintillation counter. The $Ca^{2+}$- and thapsigargin-sensitive ATPase (SERCA-ATPase) activity was defined as a difference in the $Ca^{2+}$-ATPase activities measured in the presence and absence of 500 nM thapsigargin. The $Ca^{2+}$-ATPase activity in the absence of thapsigargin was determined as a difference in the activities in the presence of a given concentration of free $Ca^{2+}$ or 5 mM EGTA without $Ca^{2+}$. The $Ca^{2+}$-ATPase activity in the presence of thapsigargin was determined as a difference in the activities in the presence of a given concentration of free $Ca^{2+}$ and 500 nM thapsigargin or 5 mM EGTA (no free $Ca^{2+}$) and 500 nM thapsigargin. All assays were started with 5 min preincubation at 37 °C in the presence or absence of thapsigargin.

**3.** $Na^+/K^+$-ATPase Activity - $Na^+/K^+$-dependent ATPase activity was measured by the same way as that used in the $Ca^{2+}$-ATPase Activity, except that the assay medium containing crude membrane preparation (~0.1 mg protein), 100 mM NaCl, 5 mM KCl, 50 mM Tris-HCl (pH 7.4), 2 mM ATP (containing 100 μM NaOH and $\gamma$-$[^{32}P]ATP$), 3 mM MgCl2, 5 mM NaN3 and 1 mM EGTA (pH adjusted to 7.4 with ~260 μM NaOH. Ouabain-sensitive $Na^+/K^+$-ATPase activity was defined as a difference in the $Na^+/K^+$-ATPase activities measured in the presence and absence of 5 mM ouabain.

**4.** Sample preparation for AFM, imaging and analysis - 20 μl of the purified dog kidney membrane preparation (1) (2 mg protein/ml) was directly applied to freshly cleaved mica. The membrane was allowed to attach to the substrate for 5 minutes and fixed with 1% uranyl acetate. The fixed samples were dried overnight under ambient conditions, and were imaged using a type D scanner (Digital instruments inc.) under tapping mode. Cantilevers made of silicon (Digital Instruments inc.) were used. Enough time was allowed for the vibrational characteristics of the tip to stabilize before imaging. Scanning rates were fixed after trial scans starting at 1.5 scans/sec. The vibrational amplitude of the scanning tip was chosen to be around 9 V and the setpoint was set at 6.75V (3/4th of the free vibrating amplitude). The setpoint was later changed according to hardness of the sample as observed from initial scans. An analog two pole butterworth filter with a cut-off frequency of 25 kHz was applied to the feedback loop controlling the piezo movement in case of atomic scans to avoid spurious signals arising from accoustic vibrations. Since subsequent scans in the same area seemed to distort the sample (although to a very small degree), images were collected immediately upon locating the point of interest. The images were collected in both height and deflection modes and stored in 256X256 pixel format. The captured images were fitted into planes at which surface features were most visible. Image acquisition and data analysis were performed using Nanoscope III software (Digital Instruments inc.).

# Na,K-ATPase: Cardiac Glycoside Binding and Functional Importance of Negatively Charged Amino Acids of Transmembrane Regions

Jerry B Lingrel, James Van Huysse, Elizabeth Jewell-Motz, Patrick Schultheis, Earl T. Wallick, William O'Brien, and G. Roger Askew

Department of Molecular Genetics, Biochemistry and Microbiology, University of Cincinnati College of Medicine, 231 Bethesda Avenue, Cincinnati, Ohio 45267-0524

## Introduction

$Na^+/K^+$-ATPase is the receptor for cardiac glycosides, a class of drugs used to treat congestive heart failure and arrhythymias. Binding of these compounds to the enzyme is antagonized by $K^+$ ions suggesting that the binding sites for these ligands may either overlap or that binding of one may effect the other. Thus, defining the binding site for this class of drugs may help in understanding how $Na^+/K^+$-ATPase transports cations. An approach for defining the cardiac glycoside binding site is to use site-directed mutagenesis coupled with expression and selection systems. In addition, these techniques can be used to determine the role of specific amino acid residues in catalytic functions of the enzyme such as cation binding. Utilizing this approach we have identified amino acid residues which act as determinants of ouabain sensitivity as well as investigated the role specific transmembrane amino acids play in the catalytic activity of the enzyme. A functional approach is also being developed to determine if a naturally occurring ligand for $Na^+/K^+$-ATPase exists and whether it is physiologically significant.

### Determinants of Ouabain Sensitivity

The approach initially utilized to identify determinants of ouabain sensitivity involved the expression of chimeras formed between sensitive and insensitive isoforms. This was followed by the introduction of mutations into a cDNA encoding a cardiac glycoside sensitive α isoform, subcloning this altered cDNA into a eukaryotic expression vector, and transfection into sensitive HeLa cells. If resistance to ouabain is conferred on the transfected cells, the amino acid substitution must have altered the ability of the drug to inhibit $Na^+/K^+$-ATPase. Utilizing this approach, multiple amino acid residues have been identified which affect ouabain sensitivity. These are shown in Figure 1.

276

EXTRACELLULAR

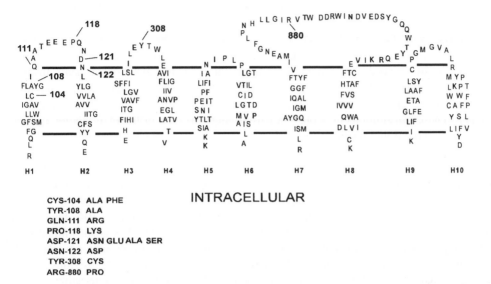

INTRACELLULAR

CYS-104 ALA PHE
TYR-108 ALA
GLN-111 ARG
PRO-118 LYS
ASP-121 ASN GLU ALA SER
ASN-122 ASP
TYR-308 CYS
ARG-880 PRO

Figure 1 Amino acid residues involved in ouabain sensitivity. The ten transmembrane model of the sheep α1 subunit of Na$^+$/K$^+$- ATPase is shown and amino acid residues which have been implicated in determining cardiac glycoside sensitivity are numbered. The Tyr to Cys substitution (317 in dog α1 corresponding to 308 in sheep) has been identified by Canessa et al (4) in ouabain resistant dog MDCK cells.

This figure depicts the 10 transmembrane model of the α subunit of Na$^+$/K$^+$-ATPase which is similar to that proposed for other P- type ATPases, i.e. Ca$^{2+}$-ATPase and H$^+$/K$^+$-ATPase. In the chimera studies, advantage was taken of the existence of both high affinity and low affinity forms of the Na$^+$/K$^+$-ATPase in nature. The rat α1 isoform, as well as those from other rodents and certain species of frogs (11) and butterflies (10), are much less sensitive to cardiac glycosides than the α1, α2 and α3 isoforms from most mammals and other organisms possessing a Na$^+$/K$^+$-ATPase. Initially, chimeras between the sensitive sheep α1 and the insensitive rat α1 subunits were prepared and the ability of these chimeras to confer resistance to otherwise sensitive HeLa cells determined. It was observed that the N- terminal half of the α subunit is responsible for the differential sensitivity (16). These studies were followed up by site-directed mutagenesis which demonstrated that two residues of the sheep α1 isoform, Gln 111 and Asn 122, when substituted with Arg and Asp respectively (the amino acids located at the equivalent positions in the low affinity rat α1 isoform), yielded enzyme which conferred resistance comparable to that observed with the rat α isoform (16). This indicated that these two amino acids were responsible for the differential ouabain sensitivity observed between the ouabain sensitive sheep α1 subunit and the relatively insensitive rat α1 subunit. Subsequently, these two amino acids were substituted with a variety of charged amino acids and in each case, resistant HeLa cells were observed following transfection with the sheep α1 isoform carrying these substitutions (18). Resistance was not obtained when these residues were substituted with an uncharged amino acid such as Ala. Because any combination of charged amino acid substitutions at these two positions appears to give resistance and because these residues are located at the membrane interface, it is possible that the charge prevents this region from interacting with the membrane upon drug binding. These substitutions may prevent partial internalization of the hydrophobic cardiac glycoside into the lipid bilayer.

Asp 121, when substituted with a variety of amino acids such as Asn, Glu, Ala, and Ser also confers resistance when sheep α1 cDNAs carrying these substitutions are transfected into sensitive HeLa cells (17). It appears that this amino acid is critical for ouabain sensitivity. Two other amino acids in this region, Tyr 108 and Cys 104, have been implicated as determinants of ouabain sensitivity (3,19). When Tyr 108 is substituted with Ala or Cys 104 is substituted with Ala or Phe, ouabain resistant colonies are obtained. Interestingly, these amino acids are located within the transmembrane regions. This finding suggests that cardiac glycosides may be completely or partially internalized in the lipid bilayer. This is not unreasonable as this class of compounds is hydrophobic, and in particular, it may be the steroid moiety of the molecule which is inserted into the membrane. Of course, it is possible that these mutations indirectly affect the structure of the binding site through changes in conformation.

Interestingly, Arg 880 is also implicated in ouabain sensitivity (20). This amino acid was identified by selecting clones following random mutagenesis. In this instance, a cassette of the sheep α1 cDNA corresponding to the extracellular regions between membrane regions 7 and 8 was mutagenized with formic acid and this cassette was used to replace the corresponding region of the wild type cDNA. Following the introduction of the resulting cDNA into an expression vector and transfection into HeLa cells, several ouabain resistant colonies were obtained. Sequencing of the H7-H8 region of the cDNA from one of these colonies revealed that Arg 880 had been replaced by Pro. This replacement was shown by direct substitution to be responsible for the ouabain resistance observed. Pro may disrupt the conformation of this region and therefore a general structural change may be responsible for the effect of this substitution on ouabain inhibition. Tyr 317, corresponding to 308 in the sheep α1 subunit, was found by Canessa et al (4) to confer decreased sensitivity to ouabain in mutant MDCK cells. This amino acid is located in the second extracellular region. The Pro and Tyr substitutions indicate that regions outside the first transmembrane and extracellular regions alter ouabain binding.

Because much of the α subunit has not yet been explored, it is likely that additional substitutions will be identified which affect ouabain sensitivity. This makes the development of a model for ouabain binding to the Na,K-ATPase difficult at this point. Nevertheless, the concentration of substitutions which effect ouabain inhibition in the $NH_2$-terminal part of the molecule suggest that this region may contain at least a portion of the binding site. However, in no case is it known whether these residues are located at the binding site or whether they act at a distance to alter the conformation of the binding site.

**Table 1.  Kinetic constants for binding of [$^3$H]ouabain in the presence of magnesium and phosphate**

| Sheep α1 Substitution | $K_1$ $(min^{-1}\mu M^{-1})$ | $K_{-1}$ $(min^{-1}x10^3)$ | $K_D$ (nM) |
|---|---|---|---|
| Wild Type Sheep α1 | 2.83 | 4.40 | 1.55 |
| E116Q | 3.65 | 4.86 | 1.33 |
| Y124F | 3.85 | 5.48 | 1.42 |
| C104A | 3.05 | 13.8 | 4.52 |
| P118K | 1.75 | 10.8 | 6.17 |
| Y108A | 2.69 | 18.1 | 6.73 |
| R880P | 4.79 | 38.6 | 8.05 |

Methods and calculations are described in reference 20.

Two approaches have been utilized to better understand the role these amino acid residues play in inhibition. The first of these is the analysis of on and off rates for [$^3$H] ouabain binding to each of these substituted enzymes. Some mutations, such as those at the border of the first extracellular region, exhibit such low affinity that accurate analysis cannot be carried out. However, those for which this type of analysis has been performed (20) are listed in Table 1. From the measured association and dissociation rate constants, the kD, which is a measure of the overall affinity of the enzyme for ouabain, can be calculated. The on rate is only modestly effected by the amino acid substitutions. For example, Arg 880 Pro shows the greatest change but the on-rate is less than two fold faster than that of the native enzyme. Other substitutions show a much more modest increase from the wild type sheep α1, the only exception being Pro 118 Lys which exhibits an approximately 40% slower on-rate. In contrast, the off rates vary to a much greater extent. Arg 880 Pro, which exhibits a faster on rate than the wild type, exhibits an approximately 10 fold increase in off rate. This increased dissociation rate off sets the faster on rate giving rise to the higher $K_D$ and thus to the relative insensitivity to ouabain of enzyme carrying this mutation. Unfortunately, there is no particular pattern exhibited by these substitutions which would help to better define the role of each of these amino acids in ouabain binding. Based on previous studies (22,24,25), it was envisioned that mutations in a particular part of the molecule would effect on rate much more than off rate or vice versa. However, this is not the case with the amino acid substitutions studied.

Another approach for understanding ouabain sensitivity is to determine whether specific amino acids bind to a particular moiety of cardiac glycosides. This class of drugs are composed of a lactone ring, a steroid portion and a carbohydrate. Ouabain and ouabagenin, for example, vary only in the carbohydrate portion, with ouabagenin lacking the sugar. The $I_{50}$ values of these two compounds vary by 20 fold, i.e. a 20 fold higher concentration of ouabagenin is required to give the same level of inhibition as ouabain. This implies that the enzyme contains a binding site for the sugar moiety. If one of the known amino acid determinants of ouabain sensitivity were interacting with the carbohydrate portion, then substitution of this amino acid residue with one that cannot bind the sugar would result in an enzyme with similar ouabain and ouabagenin $I_{50}$ values. When such analyses were carried out with the following substitutions: C104F, D121E, N122D, Q111K, N122K, Q111R, and A1128, the 20 fold difference in $I_{50}$ was maintained indicating that none of the amino acid substitutions studied interact with the sugar (15). These studies suggest that amino acids which interact with the sugar portion of cardiac glycosides are not among the residues identified to date which alter ouabain sensitivity. Similar studies can be carried out with other paired analogs such as ouabain and dihydroouabain which vary in the lactone ring and digoxin and digitoxin which vary in a steroid hydroxyl group. It is hoped that as additional amino acid determinants of ouabain sensitivity are identified that a model can be developed of the binding site which will be amenable to testing with further amino acid substitutions.

**Cation Binding Sites**

A site-directed mutagenesis-expression-selection system can also be used to identify amino acids required for catalytic activity, such as those which bind cations. This is accomplished by introducing mutations into the cDNA of an $\alpha$ subunit which is relatively resistant to cardiac glycosides, followed by expression of the altered $\alpha$ subunit in cells that are sensitive to these drugs. If the encoded amino acid substitution permits the altered enzyme to function, then cells expressing the modified $Na^+/K^+$-ATPase will grow and produce colonies in 1 $\mu$M ouabain, since the exogenous enzyme is resistant to cardiac glycosides (23). On the other hand, if the amino acid substitution occurs at a residue which is critical to the catalytic cycle of the enzyme and if the $Na^+/K^+$-ATPase activity of the altered enzyme is abolished by the replacement, the transfected cells will not survive in media containing a 1 $\mu$M concentration of the cardiac glycoside (13).

Of course, the absence of colonies can also be a result of the inability of the cell to synthesize the altered $\alpha$ subunit, of its failure to transport the altered protein to the plasma membrane, or it may be due to an impairment of the ability of the modified $\alpha$ chain to interact with the endogenous ß subunit. To determine whether the altered α subunit is synthesized, specific antibodies can be used to estimate the amount of exogenous α protein which is present. In addition, specific antibodies to the exogenous α and endogenous ß subunits can be used in conjunction with confocal microscopy to visualize the co-localization of the two within the plasma membrane (7).

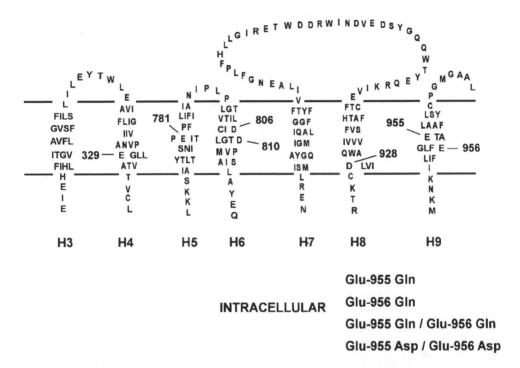

Figure 2. Negatively charged amino acid residues located in transmembrane regions of the ten transmembrane model of the rat α1 subunit. Only that portion of the α subunit containing negatively charged amino acids is shown (spanning transmembrane regions H4-H9). The amino acid substitutions at E955 and E956 are indicated.

The above selection scheme was used to determine if candidate cation-binding amino acids are critical for enzyme function. It has been postulated that negatively charged residues located within the plasma membrane may be involved in cation binding (5,6,12). Based on the 10 transmembrane model for the α subunit described in Figure 2, seven negatively charged amino acids occur in membrane-spanning regions. Studies combining chemical modification with trypsin digestion suggest that two of these, E955 and E956 of the rat α1 isoform, are cation binding residues (8). In the present work, the cDNA for the rat α1 isoform was altered by site-directed mutagenesis to encode substitutions at these two sites. When the rat α1 isoform carrying a Glu955Gln, Glu956Gln, Glu955Gln-Glu956Gln or Glu955Asp-Glu956Asp substitution was expressed in HeLa cells, colonies were produced in numbers equivalent to those seen with the wild-type cDNA, indicating that each substituted α1 isoform was active. Transfection with a rat α3 cDNA construct (which encodes a ouabain-sensitive isoform) did not produce colonies, demonstrating the absence of spontaneous ouabain resistance with this system.

**Table 2. Kinetic constants (mM) derived from a cooperative model of $Na^+$ and $K^+$ Binding**

| Transfected cDNA | $K_{0.5}(Na^+)$ | $K_{0.5}(K^+)$ |
|---|---|---|
| Wild Type rat α1 | $3.45 \pm 0.22$ | $0.78 \pm 0.07$ |
| E955Q | $2.76 \pm 0.22$ | $0.86 \pm 0.06$ |
| E956Q | $4.92 \pm 0.43$ | $0.78 \pm 0.09$ |
| E955Q-E956Q | $4.16 \pm 0.11$ | $1.45 \pm 0.09$ |
| E955D-E956D | $6.21 \pm 0.52$ | $1.16 \pm 0.08$ |

Methods and calculations are described in reference 23.

Due to a nearly 1000-fold difference in cardiac glycoside sensitivity between the endogenous (HeLa) and exogenous enzymes, the $Na^+/K^+$-ATPase activities of the mutant and wild type rat α1 isoforms could be distinguished from that of the HeLa protein. This was achieved by performing studies at 5.0 μM ouabain, which inhibits the HeLa enzyme, but not the ouabain-resistant isoform. When the cation dependence properties of the exogenous enzymes were examined, there was either no difference or only a small difference between the apparent cation affinity of the wild-type rat α1 isoform and that of each mutant rat α1 protein (Table 2). Thus, while the $Na^+$ and $K^+$ dependence of $Na^+/K^+$-ATPase activities were slightly altered in some cases, these studies clearly indicate that Glu (and hence a negatively charged side chain) is not required at positions 955 and 956 of the rat α1 isoform of $Na^+/K^+$-ATPase. Studies of the effects of additional substitutions at these two sites are ongoing, as well as studies examining the effects of substitutions at the remaining five negatively charged transmembrane amino acids.

**Do Endogenous Cardiotonic Steroids Regulate the Na,K-ATPase In vivo?**

Cardiac glycosides have a long history of use as herbal medications and have been used specifically for the treatment of congestive heart failure since William Withering first conceived of this therapy in 1785. These compounds act directly on the $Na^+/K^+$-ATPase by blocking its enzymatic activity, subsequently increasing intracellular sodium levels which have a profound effect on the regulation of intracellular calcium levels. The resulting increased flux in intracellular calcium levels is responsible for the positive inotropic action and thus the therapeutic value of these drugs. Recently, several reports have demonstrated the existence of endogenous cardiac glycosides in several mammalian species including human (9,21) which have the potential for regulating a variety of physiological processes (2). In an attempt to test the physiological significance of endogenous cardiac glycosides, we have developed a genetic approach to modify the responsiveness of a $Na^+/K^+$- ATPase $\alpha$ isoform to cardiac glycosides in the mouse. Using a gene targeting strategy, the $\alpha2$ subunit gene has been modified such that the encoded enzyme has a significantly reduced affinity for cardiac glycosides. The plan is to develop mice from these modified ES cells so that the effect of this mutation on the physiology of the whole animal can be tested.

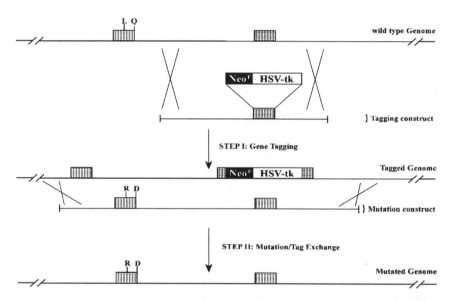

Figure 3   Tag-and-exchange strategy for genetically converting the $\alpha2$ subunit of the cardiac glycoside $Na^+/K^+$-ATPase from a high affinity to a low affinity receptor for cardiac glycosides. Shaded boxes represent exon 4 (on the left) and 8 of the $\alpha2$ $Na^+/K^+$-ATPase gene. The neomycin resistance gene (Neo$^r$) and thymidine kinase gene of Herpes Simplex Virus (HSV-tk) were inserted in tandem into exon 8 of the tagging construct. Point mutations, encoding the substitutions Leu (L)  Arg (R) and Glu (Q)  Asp (D), were introduced by in vitro mutagenesis into exon 4 of the mutation construct. Crossover points of homologous recombination are represented by X.   (Reprinted by permission of Blackwell Scientific Publications, Inc.)

The $\alpha2$ subunit gene of murine embryonic stem (ES) cells was modified to encode a low affinity receptor by introducing point mutations into the codons of the border residues of the first extracellular domain. Nucleotide substitution at these positions result in an enzyme in which the wild type border residues, Leu and Asn, are replaced with the charged amino acids, Arg and Asp, respectively. These border residue substitutions have been shown to convert the $\alpha 2$ Na$^+$/K$^+$-ATPase to a low affinity receptor (16). This genetic manipulation was achieved by a two step gene targeting strategy in which the target gene is tagged with selectable markers and then exchanged for a sequence encoding the desired point mutations (1). This tag-and-exchange strategy is illustrated in Fig. 3 (14). The tagging construct is comprised of a portion of mouse genomic sequence of $\alpha 2$ containing exons 8-10. Two selectable markers, the neomycin resistance gene (Neo$^r$) and the thymidine kinase gene of Herpes Simplex Virus (HSV-tk) have been inserted into exon 8. These selectable markers confer resistance to the broad spectrum antibiotic neomycin and sensitivity to the nucleotide analog gancyclovir, respectively. The genomic sequences act by directing these selectable markers to the target gene via homologous recombination. In the first targeting step, the tagging construct was electroporated into ES cells and neomycin resistant cells were further screened by PCR and Southern blot analysis for correct targeting. Once obtained, correctly tagged cells were expanded and subsequently retargeted with the exchange vector. The exchange vector is comprised of a genomic sequence overlapping that of the tagging construct and contains site-directed point mutations, in exon 4, encoding the substitutions described above. In addition, functionally silent point mutations were introduced into the exchange vector which create unique restriction sites to aid in subsequent screening for products of the exchange event. Since exchange at the tagged locus results in loss of the selectable markers, products of this exchange are resistant to gancyclovir. Thus, following transfection with the exchange vector, cells were selected in gancyclovir and resistant colonies were analyzed for correct targeting with the exchange vector.

Using this strategy, we developed four lines of ES cells which carry the site-directed mutations at the target locus thereby encoding a low affinity $\alpha 2$ Na$^+$/K$^+$-ATPase. These cell lines have been injected into blastocysts and chimeric animals have been developed from these lines and are now being tested for germline transmission of the mutation. Transmission of the mutation to the offspring of chimeric animals will allow development of a mouse line which will be used to determine whether endogenous cardiac glycosides play a significant physiological role. There are several possible consequences of these mutations on the physiology of the whole animal. If high affinity for cardiac glycosides by $\alpha 2$ Na$^+$/K$^+$-ATPase is physiologically important, then a phenotype proportional to the significance of the cardiac glycoside $\alpha 2$ interaction should be observed. If, on the other hand, this interaction is not essential, then no abnormal phenotype would be expected. Finally, it is possible that induction of limited physiological conditions would be necessary to observe the effects of this mutation on the whole animal. While the absolute physiological phenotype of animals with this mutation is difficult to predict, possible consequences of this mutation include effects on vascular tone and cardiac function.

## Acknowledgments

This work was supported by National Institutes of Health Grants P01 HL 41496, RO1 HL 28573 and AHA grant SW-92-01-1.

## References

1.    Askew GR, Doetschman T, Lingrel JB (1993) Site-directed point mutations in embryonic stem cells: a gene-targeting tag- and-exchange strategy.  Mol Cell Biol 13: 4115-4124

2.    Blaustein MP  (1993)  Physiological effects of endogenous ouabain: control of intracellular $Ca^{2+}$ stores and cell responsiveness.  Am J Physiol 264: C1367-C1387

3.  Canessa CM, Horisberger J-D, Louvard D, Rossier BC (1992) Mutation of a cysteine in the first transmembrane segment of Na,K-ATPase $\alpha$ subunit confers ouabain resistance. Embo J 11: 1681-1687

4.  Canessa CM, Horisberger J-D, Rossier BC (1993)  Mutation of a tyrosine in the H3-H4 ectodomain of Na,K-ATPase $\alpha$ subunit confers ouabain resistance.   J Bio Chem 268: 17722-17726

5.  Clarke DM, Loo TW, Inesi G, MacLennan DH (1989) Location of high affinity $Ca^{2+}$-binding sites within the predicted transmembrane domain of the sarcoplasmic reticulum $Ca^{2+}$-ATPase. Nature 339: 476-478

6.  Clarke DM, Loo TW, MacLennan DH (1990) Functional consequences of alterations to polar amino acids located in the transmembrane domain of the $Ca^{2+}$-ATPase of sarcoplasmic reticulum. J Biol Chem 265: 6262-6267

7.  DeTomaso AW, Xie ZJ, Guoquan L, Mercer RW (1993) Expression, targeting, and assembly of functional Na,K-ATPase polypeptides in Baculovirus-infected insect cells. J Biol Chem 268: 1470-1478

8.    Goldshleger R, Tal DM, Moorman J, Stein WD, Karlish SJD (1992) Chemical modification of Glu-953 of the $\alpha$ chain of $Na^+,K^+$- ATPase associated with inactivation of cation occlusion. Proc Natl Acad Sci USA 89: 6911-6915

9.  Hamlyn JM, Blaustein MP, Bova S, DuCharme DW, Harris DW, Mandel F, Mathews WR, Ludens JH (1991)  Identification and characterization of a ouabain-like compound from human plasma. Proc Natl Acad Sci USA 88: 6259-6263

10.  Holzinger F, Frick C, Wink M (1992) Molecular basis for the insensitivity of the Monarch (Danaus plexippus) to cardiac glycosides. FEBS 314: 477-480

11.  Jaisser F, Canessa CM, Horisberger JD, Rossier BC (1992)  Primary sequence and functional expression of a novel ouabain- resistant Na,K-ATPase.  J Biol Chem 267: 16895-16903

12.  Karlish SJD, Goldshleger R, Stein WD (1990) A 19-kDa C- terminal tryptic fragment of the $\alpha$ chain of Na/K-ATPase is essential for occlusion and transport of cations.  Proc Natl Acad Sci USA 87: 4566-4570

13.  Lane LK, Feldmann JM, Flarsheim CE, Rybczynski CL  (1993)  Expression of rat $\alpha$1

Na,K-ATPase containing substitutions of "essential" amino acids in the catalytic center. J Biol Chem 268: 17930-17934

14. Lingrel JB, Van Huysse J, O'Brien W, Jewell-Motz E, Askew GR, Schultheis P (1993) Structure-function studies of the Na,K- ATPase. Kidney Intl, in press

15. O'Brien WJ, Wallick ET, Lingrel JB (1993) Amino acid residues of the Na,K-ATPase involved in ouabain sensitivity do not bind the sugar moiety of cardiac glycosides. J Biol Chem 268: 7707-7712

16. Price EM, Lingrel JB (1988) Structure-function relationships in the Na,K-ATPase $\alpha$ subunit: Site-directed mutagenesis of Glutamine-111 to Arginine and Asparagine-122 to Aspartic Acid generates a ouabain-resistant enzyme. Biochemistry 27: 8400-8408

17. Price EM, Rice DA, Lingrel JB (1989) Site-directed mutagenesis of a conserved extracellular aspartic acid residue affects the ouabain sensitivity of sheep Na,K-ATPase. J Biol Chem 264: 21902-21906

18. Price EM, Rice DA, Lingrel JB (1990) Structure-function studies of Na,K-ATPase: Site-directed mutagenesis of the border residues from the H1-H2 extracellular domain of the $\alpha$ subunit. J Bio Chem 265: 6638-6641

19. Schultheis PJ, Lingrel JB (1993) Substitution of transmembrane residues with hydrogen-bonding potential in the $\alpha$ subunit of Na,K-ATPase reveals alterations in ouabain sensitivity. Biochemistry 32: 544-550

20. Schultheis PJ, Wallick ET, Lingrel JB (1993) Kinetic analysis of ouabain binding to native and mutated forms of Na,K- ATPase and identification of a new region involved in cardiac glycoside interactions. J Bio Chem, in press

21. Shaikh IM, Lau BWC, Siegfried BA and Valdes R Jr (1991) Isolation of Digoxin-like immunoreactive factors from mammalian adrenal cortex. J Biol Chem 266: 13672-13678

22. Thomas R, Gray P, Andrews J (1990) Digitalis: Its mode of action, receptor, and structure activity relationships. In: B. Testa (eds) Advances in Drug Research. Academic Press, New York, 19: 313-562

23. Van Huysse JW, Jewell EA, Lingrel JB (1993) Site-directed mutagenesis of a predicted cation binding site of Na,K-ATPase. Biochemistry 32: 819-826

24. Wallick ET, Pitts BJR, Lane LK, Schwartz A (1980) A kinetic comparison of cardiac glycoside interactions with $Na^+,K^+$- ATPase from skeletal and cardiac muscle and from kidney. Arch Biochem Biophys 202: 442-449

25. Yoda A (1974) Association and dissociation rate constants of the complexes between various cardiac monoglycosides and Na,K- ATPase. Ann NY acad Sci 242: 598-618

# Functional Analysis of the Disulfide Bonds of Na+/K+-ATPase β-subunit by Site-directed Mutagenesis

M. Kawamura[1], S. Noguchi[2], S. Ueno[1], M. Kusaba[2], K. Takeda[1]

[1]Faculty of Medicine, University of Occupational and Environmental Health, Kitakyushu, 807 Japan and [2]Department of Biochemical Engineering and Science, Faculty of Computer Science and Systems Engineering, Kyushu Institute of Technology, Iizuka, 820 Japan

**Introducton**

The ion transporting $Na^+/K^+$-ATPase of animal plasma membranes consists of the catalytic $\alpha$- and the glycosylated $\beta$-subunit.    All the functional roles in the catalytic activity of the enzyme so far known belong to the $\alpha$-subunit, whereas those of the $\beta$-subunit remain still unknown.    However, the $\beta$-subunit has been revealed to play important roles in the biogenesis of the enzyme.    It acts as a stabilizer of the nascent $\alpha$-subunit within the endoplasmic reticulum [9, 10] and takes part in targeting the resulting $\alpha\beta$-complex to the plasma membrane [2,12].

The $\beta$-subunit spans the membrane once, leaving a short N-terminal portion in the cytoplasmic side, and therefore a large portion of the $\beta$-subunit including the C-terminus exists outside of the cell.    Within the extracellular domain of the $\beta$-subunit, there are three highly conservative disulfide bonds ($Cys^{127}$-$Cys^{150}$, $Cys^{160}$-$Cys^{176}$, $Cys^{215}$-$Cys^{278}$ [7] ; Numbering refers to *T. californica* $Na^+,K^+$-ATPase $\beta$-subunit.) , whose reduction results in the loss of the enzyme activity [4,5,8].    Interestingly, the reduced and inactivated enzyme is solubilized with a diluted detergent such as sodium dodecylsulfate at a concentration usually used for the purification of $Na^+,K^+$-ATPase [6].    This suggests that the disulfide bonds of the $\beta$-subunit may play significant roles in the folding of the whole enzyme within the membrane.

In the present study, cysteine residues of the $\beta$-subunit were converted to serine by site-directed mutagenesis leading to the disruption of any one of the three disulfide bonds.    The mutant in which the $Cys^{127}$-$Cys^{150}$ disulfide bond was disrupted, formed a stable complex with the $\alpha$-subunit.    The resulting $\alpha\beta$ complex showed, however, little ATPase activity.    The disruption of either the $Cys^{160}$-$Cys^{176}$ or $Cys^{215}$-$Cys^{278}$ disulfide bond resulted in the $\beta$-subunit lacking the ability to assemble with the $\alpha$-subunit.    The structure of the extracellular domain of the $\beta$-subunit including the latter two disulfide bonds, should be important to correctly assemble with the $\alpha$-subunit.

The primary structures of the α and β-subunit of Na⁺/K⁺-ATPase are homologous to those of the corresponding subunits of gastric H⁺/K⁺-ATPase. In the next place of this study we examined whether the β-subunit of H⁺/K⁺-ATPase could replace the β-subunit of Na⁺/K⁺-ATPase during the assembly of the α subunit of Na⁺/K⁺-ATPase in the plasma membrane. We found that the α subunit of *Torpedo californica* Na⁺/K⁺-ATPase was assembled in the microsomal membrane of oocytes with the H⁺/K⁺-ATPase β- subunit. This result suggests that the β-subunit of H⁺/K⁺-ATPase replaces the β-subunit of Na⁺/K⁺-ATPase when assembled with the α-subunit.

### Mutation of a cysteine in the transmembrane segment of the β-subunit

The β-subunit of the Na⁺/K⁺-ATPase contain seven well-conserved cysteine residues ; one is located within the single transmembrane segment and the other six, which are disulfide-bonded in sequence, are located in the extracellular domain. First, we introduced the mutation to the free cysteine in the transmembrane segment (βC46S).

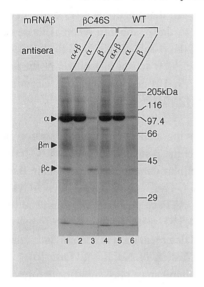 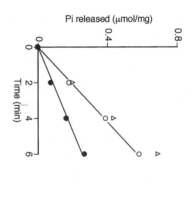

Fig. 1. Functional expression of Cys46 mutant. The mRNA β (10ng/oocyte) of βC46S (lanes 1-3) or wild-type (lanes 4-6) were injected into *Xenopus* oocytes together with mRNA α (10ng/oocyte) of *T. californica* Na⁺/K⁺-ATPase. After incubation at 19°C for 3 days with [¹⁴C]leucine, the labeled translation products were immunoprecipitated with antiserum to the α-subunit (lanes 2 and 5) or β-subunit (lanes 3 and 6) or a mixture of both (lanes 1 and 4). βm and βc represent fully glycosylated and core-glycosylated β-subunit, respectively. WT represents wild-type β-subunit. In the right panel, the ouabain-sensitive ATPase activity of the pellets from the oocytes injected with mRNAβ of βC46S (△) or wild type β-subunit (○) together with mRNA α are shown. The pellet from oocytes into which none of mRNA was injected was used to assess the endogeneous Na⁺/K⁺-ATPase activity (●).

The mRNAβ of the βC46S was coinjected with the wild-type mRNAα into *Xenopus* oocytes, which were followed by a 3 day-incubation at 19°C. Then the translation products produced in the oocytes were assayed by immunoprecipitation and fluorography. As shown in Fig.1, the mutant β-subunits (both the fully glycosylated form, βm, and the core-glycosylated form, βc) and the α-subunit were synthesized in oocytes (lanes 1-3) in nearly the same fashion as when the wild-type mRNAβ was injected (lanes 4-6). In accordance with our previous observations [10], the amounts of the α-subunit synthesized in oocytes coinjected with mRNAβ were much larger than when mRNAα was injected alone (data not shown). Moreover, anti-β-subunit antiserum precipitated the α-subunit(Fig.1, lanes 3 and 6). These results suggested that the mutant β-subunit could assemble with and thereby stabilize the α-subunit in the membranes as the wild-type β-subunit. Next, we assayed the ouabain-sensitive ATPase activity of the microsomal fractions obtained from the mRNA-injected oocytes. The result is shown in the right panel of Fig.1. The membrane fraction from oocytes injected with βC46S mRNAβ plus mRNAα exhibited nearly the same activity as that of oocytes injected with the wild-type mRNAβ together with mRNAα. We concluded that the βC46S mutant was active as the wild-type β-subunit. The isoforms of the β-subunit (β2 and β3) that lack corresponding cysteine residue in the transmembrane segment have been shown to assemble with the α-subunit leading to the active αβ complex [3,11], which is in an accordance with the results described above.

**Disruption of disulfide bonds by site-directed mutagenesis**

Next, we introduced mutations into the cysteine residues in the extracellular domain. These cysteine residues are bonded in sequence resulting in the formation of 3 disulfide bonds ($Cys^{127}$-$Cys^{150}$, $Cys^{160}$-$Cys^{176}$ and $Cys^{215}$-$Cys^{278}$). Either the N-terminal-side one or both (double mutant) of two cysteine residues of the respective disulfide bond was changed to serine (βC127S; βC127,150S; βC160S; βC160,176S; βC215S and βC215,278S). The results of immunoprecipitation of the translation products in the oocytes injected with these mutant mRNAβs together with wild-type mRNAα are shown in Fig.2. In oocytes injected with mRNAβ of βC127S or βC127, 150S together with mRNAα, anti-β-subunit antiserum precipitated the α-subunit (lanes 3 and 6) and the amount of the α-subunit immunoprecipitated with anti-α-subunit antiserum were large and nearly the same (lanes 2 and 5) as in oocytes coinjected with wild-type mRNAβ (lanes 20). These results indicated that the products of the double mutant βC127,150S as well as of the mutant βC127S assembled with and stabilized the α-subunit nearly in the same fashion as the wild-type β-subunit. In contrast to these results, the products of the remaining four mutants (βC160S; βC160,176S; β C215S and βC215,278S) neither assemble with nor stabilize the α-subunit (lanes, 7-18). The faint bands of the α-subunit found in the lanes of these four mutants (lanes 7, 8, 10, 11, 13, 14, 16 and 17) were likely due to the unassembed form of the α-subunit as observed in the oocytes injected with mRNAα alone (data not shown but see lane 10 in Fig.3).

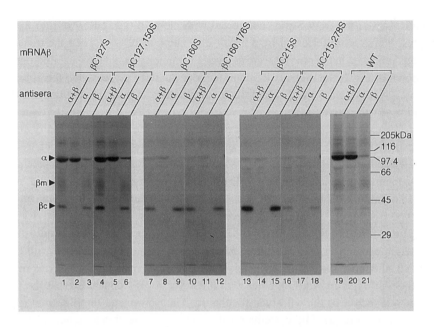

Fig. 2. Translation products in *Xenopus* oocytes injected with mutant mRNAβ. The mRNAβ
(10ng/oocyte) of βC127S (lanes 1-3); βC127S, 150S (lanes 4-6); βC160S (lanes 7-9); βC160S,
176S (lanes 10-12); βC215S (lanes 13-15); βC215, 278S (lanes 16-18) or wild-type β-subunit
(lanes 19-21) were injected into *Xenopus* oocytes together with mRNAα (10ng/oocyte). The
translation products were immunoprecipitated and assayed as described for Fig. 1.

Indeed, as shown in Fig.3, limited trypsinolysis revealed that the unassembled α-subunit
in the oocytes injected with mRNAα plus mRNAβ of βC160S was digested completely
even when treated at a trypsin/protein ratio of 1/100 (w/w)(Fig. 3 lanes 7-12), whereas
the α-subunit in the oocytes injected with mRNAα together with mRNAβ from βC127S
or βC127, 150S was resistant to trypsin (lanes 1-6). From these results, it was
concluded that the β-subunit could not assemble with the α-subunit if either one of the
two C-terminal-side disulfide bonds was broken.

These results indicate that the C-terminal-side portion of the β-subunit, including loops
of $Cys^{160}$-$Cys^{176}$ and $Cys^{215}$-$Cys^{278}$, constituted some specific steric structures which
may play indispensable roles in the assembly with the α-subunit, and that the loop of
$Cys^{127}$-$Cys^{150}$ is not necessarily essential in this assembly.

It is interesting whether the stable complexes of the α-subunit with the β-subunit
mutated in the $Cys^{127}$-$Cys^{150}$ bridge are catalytically active or not. As seen in Table I,
the oocytes injected with either mRNAβ of βC127S or βC127,150S together with
mRNAα showed only a little increment, if any, in the ouabain-sensitive ATPase activity
from that of uninjected oocytes. When these mutants were assayed for ouabain-binding

290

activity, similar results were obtained (data not shown). The result that the stable and trypsin-resistant complexes between $\beta$C127S or $\beta$C127,150S mutant $\beta$-subunit and the $\alpha$-subunit could not exhibit apparent catalytic activity suggests the possibility that the interaction of the loop of Cys$^{127}$-Cys$^{150}$ with the $\alpha$-subunit may be critical in the functional expression of this enzyme.

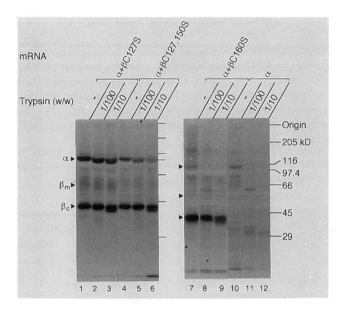

Fig. 3. Trypsin sensitivity of the $\alpha$-subunit expressed in oocytes injected with mRNA$\alpha$ together with mutant mRNA$\beta$. mRNA$\alpha$ (10ng/oocyte) was injected into *Xenopus* oocytes alone (lane 10-12) or together with mRNA$\beta$ (10ng/oocyte) of $\beta$C127S (lane 1-3); $\beta$C127,150S (lanes 4-6) or $\beta$C160S (lanes 7-9). The membrane pellets from injected oocytes were digested with trypsin at trypsin: protein ratios (w/w) of 0(lanes 1, 4, 7 and 10), 0.01 (lanes 2, 5, 8 and 11), and 0.1 (lanes 3, 6, 9 and 12) for 60 min on ice.

## Assembly of the $\alpha$-subunit of Na$^+$/K$^+$-ATPase with the $\beta$-subunit of H$^+$/K$^+$-ATPase

Recombinant plasmids carrying the gene for the $\beta$-subunit of H$^+$/K$^+$-ATPase were designed for their efficient expression in oocytes. The mRNA$\beta$ of H$^+$/K$^+$-ATPase was injected into oocytes in combination with mRNA$\alpha$ of *Torpedo* Na$^+$/K$^+$-ATPase. The protein synthesized dependent on the mRNA was immunoprecipitated and subjected to polyacrylamide gel electrophoresis. The amount of the $\alpha$ subunit detectable was very low when only mRNA$\alpha$ was injected(Fig.4, lanes1-3 ), but increased greatly by coinjection with mRNA for H$^+$/K$^+$-ATPase $\beta$-subunit (lanes 4-6): a similar amount of the $\alpha$-subunit was detected when the mRNAs for both the Na$^+$/K$^+$-ATPase $\alpha$- and Na$^+$/K$^+$-ATPase

291

β-subunit were injected (lanes 7-9 ).　　Furthermore, the Na⁺/K⁺ α-subunit could be precipitated with anti-Na⁺/K⁺ β-subunit antiserum or anti-H⁺/K⁺-ATPase antiserum, indicating that the Na⁺/K⁺-ATPase α- subunit assembled with the H⁺/K⁺-ATPase β-subunit.

Table 1　　Na⁺,K⁺-ATPase activity of the oocytes injected with mutant mRNAβ together with wild-type mRNAα

*Xenopus* oocytes were injected with mRNAβ (10ng/oocyte) shown in the table together with wild-type mRNAα (10ng/oocyte) and incubated for 3 days at 19°C.　Na⁺/K⁺-ATPase activity of the microsome preparations (pellets after centrifugation at 160,000 x g for 1 h) which had been treated with 1M NaSCN, were assayed in the presence and absence of 1mM ouabain.　The figures in parentheses are the activities after subtraction of the activity of uninjected oocytes.

| mRNAβ injected | Na⁺/K⁺-ATPase activity ($\mu$mol Pi/mg/h) | | |
|---|---|---|---|
| | exp-1 | exp-2 | exp-3 |
| βC46S | 4.9 (2.8) | 4.0 (1.7) | 3.2 (1.4) |
| βC127S | 2.4 (0.3) | 2.3 ( 0 ) | 1.6 (-0.2) |
| βC127,150S | 2.6 (0.5) | 2.4 (0.1) | 1.8 ( 0 ) |
| WT | 4.2 (2.1) | 3.3 (1.0) | 2.5 (0.7) |
| uninjection | 2.1 ( - ) | 2.3 ( - ) | 1.8 ( - ) |

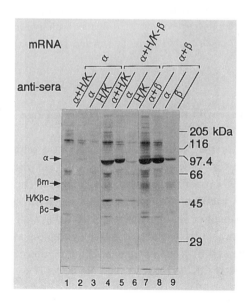

Fig. 4.　Assembly of the H⁺/K⁺-ATPase β-subunit and Na⁺/K⁺-ATPase α-subunit.　The mRNAα (10ng/oocyte) was injected alone (lanes 1-3) or together with mRNAβ (5ng/oocyte) of H⁺/K⁺-ATPase (lanes 4-6) or Na⁺/K⁺-ATPase (lanes 7-9).　The antisera used for immunoprecipitation are shown at the top.

The Na$^+$/K$^+$-ATPase α-subunit assembled with the H$^+$/K$^+$-ATPase β-subunit in microsomes showed similar typsin sensitivity to the Na$^+$/K$^+$-ATPase α-subunit assembled with the Na$^+$/K$^+$-ATPase β-subunit. The α-subunit present in microsomes without β-subunit was digested completely with trypsin (trypsin : membrane protein ratio, 1:100) (Fig.5, lanes1-3). On the other hand, the α-subunit expressed with the H$^+$/K$^+$-ATPase (lanes 4-6 ) or Na$^+$/K$^+$-ATPase (lanes 7-9) β-subunit was resistant to higher concentrations of trypsin (trypsin: membrane protein ratio, 1:10).

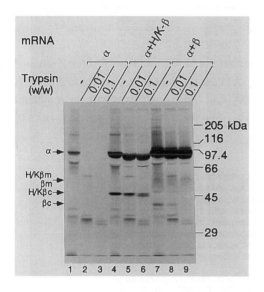

Fig. 5. Trypsin-sensitivity of the α-subunit in oocytes injected with mRNA α together with mRNAβ of the H$^+$/K$^+$-ATPase. The mRNAs were injected as descrived for Fig. 4. The pellets from injected oocytes were digested with trypsin as descrived for Fig. 3.

Table 2    Na$^+$,K$^+$-ATPase activity of the oocytes injected with H$^+$/K$^+$-ATPase mRNAβ together with wild-type mRNAα

Injection of mRNAs into *Xenopus* oocytes and assay of Na$^+$/K$^+$-ATPase activity were carried out as described in Table 1. For exp. 4 and exp. 5, mRNA of pig H$^+$/K$^+$-ATPase β-subunit was used instead of rabbit H$^+$/K$^+$-ATPase mRNAβ which was used for all other experiments reported here.

| mRNAβ injected | Na$^+$/K$^+$-ATPase activity (μmolPi/mg/h) | | | | |
|---|---|---|---|---|---|
| | exp.1 | exp.2 | exp.3 | exp.4 | exp.5 |
| α | 2.4(0.2) | 3.3(-0.1) | 3.1(0.3) | - | - |
| α + β H,K | 2.9(0.7) | 3.7(0.3) | 2.8( 0 ) | 1.8(0.1) | 2.4(-0.1) |
| α + β Na,K | 4.6(2.4) | 5.1(1.7) | 4.9(1.9) | 7.8(6.1) | 7.2(4.7) |
| uninjection | 2.2( - ) | 3.4( - ) | 2.8( - ) | 1.7( - ) | 2.5( - ) |

These results suggest that the α-subunit assembled with the H⁺/K⁺-ATPase β-subunit had a similar conformation in membranes to that with the Na⁺/K⁺-ATPase β-subunit. However, the assemblies of the Na⁺/K⁺-ATPase α- and H⁺/K⁺-ATPase β-subunits were not exactly the same as the assembly of the two subunits in Na⁺/K⁺-ATPase. Very little activity of Na⁺/K⁺-ATPase of the heterogeneous assembly could be detected in the injected oocytes, as shown in Table II. These results suggest that the H⁺/K⁺-ATPase β-subunit has a functional role in the assembly of the α-subunit, but does not form a catalytic complex similar to that of Na⁺/K⁺-ATPase.

**Discussion and Conclusion**

The observations described above indicate that the Na⁺/K⁺-ATPase β-subunit mutated in the loop of Cys$^{127}$-Cys$^{150}$ and the H⁺/K⁺-ATPase β-subunit share common properties in the assembly with the α-subunit; both can form stable complexes with the α-subunit but the resulting αβ complexes are inactive. We compared amino-acid sequences of

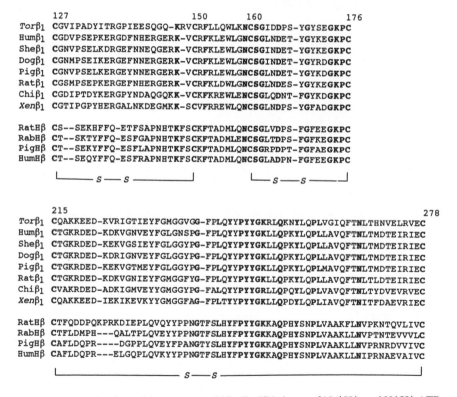

Fig. 6. Alignment of amino acid sequences within disulfide loops of Na⁺/K⁺- and H⁺/K⁺-ATPase β-subunit. Identity in all species is indicated by bold letters.

294

the two $\beta$-subunits with special references to the sequences of disulfide loops. Amino-acid sequences of $Na^+/K^+$-ATPase $\beta 1$- and $H^+/K^+$-ATPase $\beta$- subunit are aligned in Fig. 6. The sequences of the loops of $Cys^{160}$ - $Cys^{176}$ and $Cys^{215}$ - $Cys^{278}$ are well-conserved. Seven out of 15 and 16 out of 62 are identical for the former and the latter loops, respectively. On the other hand, the sequence of the loop of $Cys^{127}$ - $Cys^{150}$ is quite different in $Na^+/K^+$-ATPase $\beta 1$- and $H^+/K^+$-ATPase $\beta$-subunits and $Lys^{147}$ is solely aligned at an identical position within this loop between the two subunits.

The findings that the $\beta$-subunit of $H^+/K^+$-ATPase, whose sequence of the loop of $Cys^{127}$ - $Cys^{150}$ is quite different from that of the $\beta$-subunit of $Na^+/K^+$-ATPase, assembles with the $\alpha$-subunit of $Na^+/K^+$-ATPase but that the resulting hybrid complex is inactive suggest that there should be a specific sequence within the loop of $Cys^{127}$-$Cys^{150}$ which is required for the functional assembly between $Na^+/K^+$-ATPase $\alpha$- and $\beta$-subunits. This together with the results of $Cys^{127}$-$Cys^{150}$ mutant draw a conclusion that the first loop of disulfide plays a critical role in the functional assembly of the $\alpha$- and $\beta$-subunit. The reduction of disulfide bond(s) of the $\beta$-subunit results in the inactivation of $Na^+/K^+$-ATPase[4,8] and $H^+/K^+$-ATPase [1]. Upon inactivation, the least stable disulfide bond is $Cys^{127}$-$Cys^{150}$ [8], which also suggests that $Cys^{127}$-$Cys^{150}$ is important to keep the $\alpha\beta$ complex active. The $\beta$-subunit of the $Na^+/K^+$-ATPase not only facilitates the correct folding of the $\alpha$-subunit during the biogenesis of the enzyme but also may be essential in the functional expression of the enzyme.

### Acknowledgements

This work was supported by Grant-in-Aid for Scientific Research on Priority Areas from the Ministry Education, Science and Culture of Japan.

### REFERENCES

[1] Chow DC, Browning CM, Forte JG (1992) Gastric $H^+K^+$-ATPase activity is inhibited by reduction of disulfide bonds in $\beta$-subunit. Am J Physiol 263: C39-C49

[2] Geering K (1991) The functional role of the $\beta$-subunit in the maturation and intracellular transport of Na,K-ATPase. FEBS Lett 285:189-193

[3] Jaunin P, Horisberger JD, Richter K, Good PJ, Rossier BC, Geering K. (1992) Processing, intracellular transport, and functional expression of endogeneous and exogeneous $\alpha$-$\beta_3$ Na,K-ATPase complexes in *Xenopus* oocytes. J Biol Chem 267:577-585

[4] Kawamura M, Nagano, K (1984) Evidence for essential disulfide bonds in the $\beta$-subunit of $(Na^+,K^+)$-ATPase. Biochim Biophys Acta 774:188-192

[5] Kawamura M, Ohmizo K, Morohashi M, Nagano K (1985) Protective effect of $Na^+$ and $K^+$ against inactivation of $(Na^+,K^+)$-ATPase by high concentrations of 2-mercaptoethanol at high temperatures. Biochim Biophys Acta 821:115-120

[6] Kawamura M, Ohta T, Nagano K (1980) Effect of reducing agents on the solubilization of renal sodium and pottassium dependent ATPase with detergent. J Biochem 87:1327-1333

[7] Kirley TL (1989) Determination of three disulfide bonds and one free sulfhydryl in the β-subunit of (Na,K)-ATPase. J Biol Chem 264:7185-7192

[8] Kirley TL (1990) Inactivation of $(Na^+,K^+)$-ATPase by β-mercaptoethanol. J Biol Chem 265:4227-4232

[9] Noguchi S, Higashi K, Kawamura M (1990) A possible role of the β-subunit of (Na,K)-ATPase in facilitating correct assembly of the α-subunit into the membrane. J Biol Chem 265:15991-15995

[10] Noguchi S, Mishina M, Kawamura M, Numa S (1987) Expression of functional $(Na^+,K^+)$ATPase from cloned cDNAs. FEBS Lett. 225:27-32

[11] Schmalzing G, Kroner S, Schachner M, Gloor S (1992) The adhesion molecule on glia (AMOG/β2) and $α_1$ subunits assemble to functional sodium pump in *Xenopus* oocyte. J Biol Chem 267:20212-20216

[12] Takeyasu K, Tamkun MM, Renaud KJ. Fambrough DM (1988) Ouabain-sensitive $(Na^+,K^+)$-ATPase activity expressed in mouse Ltk cells by fransfection with DNA encoding the α-subunit of avian sodium pump. J Biol Chem 263:4347-4354

# Purified Renal Na$^+$/K$^+$-ATPase; Subunit Structure and Structure-Function Relationships of the N-Terminus of the α1-Subunit

P.L. Jørgensen

Biomembrane Research Centre, August Krogh Institute, Copenhagen University, 2100 Copenhagen OE, Denmark

## Introduction

This article is concerned with two aspects of the structure of Na$^+$/K$^+$-ATPase: the overall quaternary structure of the transport molecule and the function of the N-terminus of the α1-subunit. Selective cleavage of bonds in the α-subunit are important for identifying structure-function relationships of the protein. The N-terminus can be cleaved selectively with trypsin at bond K30-E31 (13,17) and the effects of this truncation on ligand binding, phosphorylation-dephosphorylation, enzymatic (14,19) and transport reactions (21) have been studied extensively. More recently truncated cDNAs have been expressed in *Xenopus* oocytes (4,29) for electrophysiological analysis. In this article, this data are evaluated to identify parameters that are relevant for characterization of mutant Na,K-pumps.

The quaternary structure problem is central to all considerations of the transport mechanism. If a single αβ-unit of Na$^+$/K$^+$-ATPase can carry out all catalytic reactions and bind, occlude, and translocate Na$^+$ and K$^+$, the implication is that the protein must be porous and form a cation transport pathway through the membrane. If oligomerization to (αβ)$_2$ units or higher oligomers is required, the possibility exists, that the cation pathway may be formed in the interface between subunits. The subunit structure of the purified Na$^+$/K$^+$-ATPase has been studied in the membrane bound and the soluble state. It is important to consider the difference between the conditions for studying the pump protein in the two states. In the tightly packed fragments of the purified membrane-bound Na$^+$/K$^+$-ATPase the difference in free energy required for self association of the αβ-units is several fold less than that required for association of protein molecules in solution. Even minor changes of intermolecular forces may therefore precipitate association of the membrane embedded particles that are not necessarily of functional importance for Na,K-pumping.

## The αβ-unit is the minimum asymmetric unit in membrane crystals of Na$^+$/K$^+$-ATPase

In the disc shaped membrane fragments, 1000-3000 Å in diameter, the densely packed protein particles with diameters of 30-50 Å represent αβ-units. They are arranged in irregular clusters or stands and appear to be free to move in the plane of the membrane without formation of well defined oligomeric structures (7,28). The predominant crystal form, induced in the presence of vanadate or other ligands that stabilize the protein in the E$_2$-conformation has the two-sided plane group symmetry, p1, and contains one protomeric $\alpha\beta$-unit per unit cell. Crystals with two sided plane group symmetry, p21, with two $\alpha\beta$-units occupying one unit cell, are transient and less frequent (7,28). The data show that the monomeric $\alpha\beta$-unit is the minimum asymmetric unit of Na$^+$/K$^+$-ATPase.

## The $\alpha\beta$-unit is the minimum functional unit of purified soluble Na$^+$/K$^+$-ATPase

After gradual refinement of techniques for solubilization and chromatography, reliable data defining the mass and subunit structure of purified renal Na$^+$/K$^+$-ATPase have become available (2,3,5,15). Conditions with respect to concentration of detergent, salt and protein can be found where the Na$^+$/K$^+$-ATPase is soluble, has a molecular weight corresponding to the sum of the $\alpha$-subunit and $\beta$-subunit and yet is enzymatically active (2,3) also during sedimentation in the analytical ultracentrifuge (3,31). Earlier observations of oligomeric $\alpha_2\beta_2$-units (5) or even $\alpha_2\beta_4$-units (6) are probably explained by artifactual aggregation due to covalent associations between detergent molecules (15). Molecular weights of the soluble renal Na$^+$/K$^+$-ATPase as determined by sedimentation equilibrium and sedimentation velocity studies (2,3,31) agree with those calculated from amino acid composition within +/- 10% (147,000 for $\alpha\beta$-unit of Na$^+$/K$^+$-ATPase). Sedimentation coefficients are 6-7 S for the protomeric $\alpha\beta$-unit of Na$^+$/K$^+$-ATPase (2,3). The soluble Na$^+$/K$^+$-ATPase is shown to undergo the spectral changes indicative of a conformational change as in the membrane embedded Na$^+$/K$^+$-ATPase (20). Soluble Na$^+$/K$^+$-ATPase undergoes time dependent denaturation and aggregation as a function of temperature and cation composition of the medium (20). Nucleotides were shown to be able to counteract the ion-induced changes again just as in the membrane embedded forms of Na$^+$/K$^+$-ATPase. Cation occlusion experiments showed that the two forms of soluble and membrane embedded Na$^+$/K$^+$-ATPase were equally capable of [86]Rb and [22]Na occlusion (30). It can therefore be concluded that the subunit structure ($\alpha\beta$) of the fully active and soluble Na$^+$/K$^+$-ATPase is the same as that of the minimum assymmetric unit in the membrane crystals.

## Active site concentration in purified membrane bound Na$^+$/K$^+$-ATPase

In the original publication on purification after incubation with NaDodSO$_3$ the specific activity of Na$^+$/K$^+$-ATPase in the fractions from the first 14 zonal centrifugations varied from 25 to 42 U/mg protein, Fig. 1 (10). The maximum ATP binding capacity was 4 nmoles/mg protein for a preparation with specific activity of 36 U/mg protein (11). Following the demonstration of a binding capacity for ouabain of 5.2 nmoles/mg protein (23), the determination of the maximum specific activity and ligand binding was repeated on a series of preparations with exceptionally high specific activities of Na$^+$/K$^+$-ATPase that were selected among several series of zonal centrifugations. In several of these preparations right-side-out vesicles from metrizamide gradients (16) were used as samples for incubation with NaDodSO$_3$ and zonal centrifugation. Protein determination was based on quantitative aminoacid analysis. To control the protein determination, aliquots of 2 mg of the purified preparation were transferred to ammmoniumcarbaminate buffer and the dry weight of the preparation was compared to the weight of its components. The sum of the weight of protein plus total lipid plus total carbohydrate determined as in Ref. 3 was equal to more than 95% of the dry weight of the preparation thus eliminating the possibility of a major error in determination of the protein. It is seen from Fig. 2 that there is a correlation between specific acticity of Na$^+$/K$^+$-ATPase and the amount of inbibitior or substrate sites in these purified

**Figure 1.** Specific activity of Na$^+$/K$^+$-ATPase in samples (■) and in fractions at the peak (●) of Na$^+$/K$^+$-ATPase after zonal centrifugation of a crude membane fraction that had been incubated with NaDodSO$_3$ in the presence of ATP as described before (10,12,16)

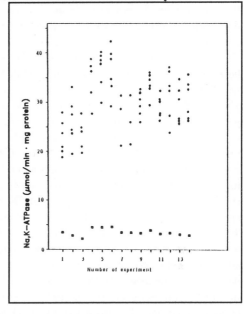

preparations with a molecular activity of 7.857 P$_i$/min and a maximum specific activity approaching 50 U/mg protein. The capacities for substrate or inhibitor binding are in the range of 4-6.4 nmol/mg protein and approaching the value (6.8 nmol/mg protein) for one substrate or inhibitor molecule per $\alpha\beta$-unit. In these purified preparations of Na$^+$/K$^+$-ATPase from outer renal medulla the ion binding capacities were 10 nmol/mg protein for $^{86}$Rb and 15-17 nmol/mg protein for $^{22}$Na. The $\alpha$1-subunit forms 65-70% of the total protein and the molar ratio of $\alpha$ to $\beta$ is 1:1, corresponding to a mass ratio of about 3:1 (15).

**Figure 2.** Relation between the capacities for ligand bining and the specific activity of Na$^+$/K$^+$-ATPase in fractions selected among several zonal centrifugations. Procedures for determination of binding of [$^3$H]ouabain (△), [$^{48}$V]vanadate (○), and [$^{32}$P]CrATP (▲) as described before (16,30).

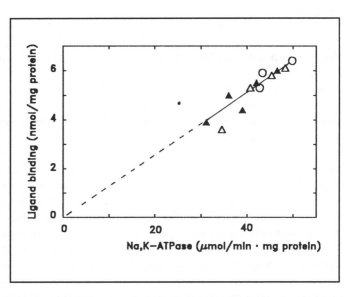

In a series of conference proceedings from 1983-1991, Nørby and Jensen (for ref. se 24) and Reynolds (27) raised objections to the determinations of protein and site concentrations and they found support for oligomeric $\alpha2\beta2$ or $\alpha2\beta4$ structures. In the proceedings from the Conference on $Na^+/K^+$-ATPase in Fuglsø, Reynolds (27) argues that our articles do not provide quantitative data of the content of $Na^+/K^+$-ATPase in homogenates, crude membrane preparations and calculations of the recovery of $Na^+/K^+$-ATPase in the purified preparations. In direct contrast to this, inspection of our articles (8-16) show that data on the amounts and the activities of $Na^+/K^+$-ATPase in homogenates, crude membranes and purified preparations from the outer renal medulla were reported repeatedly. Nørby and Jensen have summarized their objections in 1991 (24). They illustrate the discrepancy by reporting on a series of five preparations produced on my equipment for zonal centrifugation, but using procedures that are clearly distinct from ours. This in particular applies to the dissection of the pig kidney tissue resulting in a lower specific activity of $Na^+/K^+$-ATPase in the crude membrane preparation that form the sample for zonal centrifugation. The protein determinations on the five purified preparations were controlled by 5 different laboratories with expertise in $Na^+/K^+$-ATPase at the Institute of Biophysics, Aarhus University. The average specific activity was $33.5 \pm 1.3$ U/mg protein and the binding capacity for ADP was $3.63 \pm 0.03$ nmol/mg protein. Nørby and Jensen conclude that their data favor a teory with one ADP site per $(\alpha\beta)_2$. To characterize this preparation completely, it should be added that gel electrophoresis in $NaDodSO_3$ was performed in three laboratories but the data remain unpublished. The gels show a dense packing of contaminant protein bands in the region between the $\beta$-subunit and the position of the tracking dye. The contamination can be estimated to correspond to 30% of the total protein. After correction for these impurities, the specific activity of $Na^+/K^+$-ATPase would approach 48 U/mg protein with a site concentration of 5.2 nmoles/mg protein. With this in mind the data do not favour any particular subunit structure in the membrane bound $Na^+/K^+$-ATPase.

Cooperative phenomena in kinetic data on ATP or ADP binding (24) to $Na^+/K^+$-ATPase from outer renal medulla, inhibition studies (1) and cross-linking studies of red cell $Na^+/K^+$-ATPase (26) have been conveniently explained by an oligomeric pump model. This interpretation is forwarded as arguments against the competency of a functional monomer $\alpha\beta$-unit. Alternative explanations for the kinetic data on $Na^+/K^+$-ATPase from outer renal medulla should be considered in view of the high concentration of sites in the purified membrane fragments where the density of $\alpha\beta$-units is 12.000 per $\mu^2$ or 0.5-1 g protein per ml of lipid phase. In contrast to observations on kidney membranes (26) recent experiments by Martin and Sachs provided no evidence for $\alpha$-$\alpha$ subunit cross-linking in red cell membranes (22). The red cell membrane provides a natural dilution of the $Na^+/K^+$-ATPase as it represents less than 0.01 % of the membrane protein. There are about 3000 band-3 anion exchange proteins for each $Na^+/K^+$-ATPase . The most favored and only significant interaction in the presence of o-phenantroline-Cu complex is between $\alpha$-subunits and band 3 (22). Thus, in the red cell membrane, $Na^+/K^+$-ATPase exists and functions as a single $\alpha\beta$-unit. Alternative explanations for the kinetic data may bave to be considered which are perhaps less convenient than oligomers but more consistent with these data on red cell membranes (22), as well as with the direct determination of the molecular weight of soluble Na$^+$/K$^+$-ATPase and the demonstration of the $\alpha\beta$-unit as the minimum asymmetric unit of membrane bound Na$^+$/K$^+$-ATPase from outer renal medulla.

## Selective Cleavage of the N-terminus (T2) of the $\alpha$1-subunit

The N-terminus of the $\alpha$-subunit is strongly hydrophilic with clusters of alternating positive and negative residues between residue 15 and 60. It is a flexible structure with a strong propensity for $\alpha$-helix formation and several predicted turns, notably at residues 14-16, 35-36, 49-50 and 70-72. The removal of residues 1-30 by selective tryptic cleavage at Lys$^{30}$ (T$_2$) is possible because the rate of cleavage of this bond is up to 60 fold higher than the rate of tryptic cleavage at Arg$^{262}$ (T$_3$) in the second slow phase of inactivation (13,14,17,21).

**Figure 3.** Amino acid sequence of the N-terminus of the $\alpha$-subunit with the $\alpha$-helices and $\beta$-turns suggested from analysis using the Chou Fassmann algoritm.

Trypsin cleaves the E$_1$-form rapidly at Lys$^{30}$ (T2) and more slowly at Arg$^{262}$ (T3) to produce the characteristic biphasic pattern of inactivation. Localization of these splits was determined by sequencing N-termini of fragments after isolation on high resolution gel filtration columns (17). In the E$_2$-form, inactivation of K-phosphatase is delayed because cleavage of T1 and T2 in sequence is required for inactivation of K-phosphatase activity. Thus, transition from E1 to E2 involves a change in position of T2 in the N-terminus relative to the central domain (T1) so that cleavage of T2 becomes secondary to cleavage of T1 within the one $\alpha$-subunit (14).

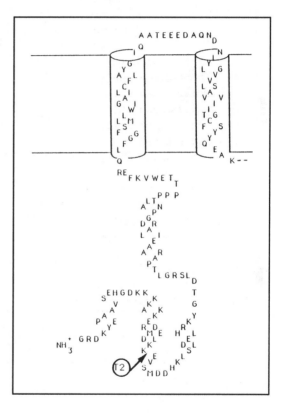

## Truncation of the N-terminus does not affect ATP and cation binding at equilibrium

It is seen from Fig. 4 that T2 cleavage reduces Na$^+$/K$^+$-ATPase activity by 50-60%. There is a parallel loss of Na,K-transport in reconstituted phospholipid vesicles (21). The

301

Na-ADP-ATP exchange activity is increased to 150-160% of control and the K-phosphatase activity is reduced to 15%, while the number of ATP binding sites remain within 80-90% of control. After cleavage to 45% of Na$^+$/K$^+$-ATPase remaining, Fig. 2 shows that the 83 kDa fragment of the $\alpha$-subunit (Res. 263-1016) forms less than 5% of the protein thus demonstrating that the rate of cleavage of bond T3 is very low (14). Fig. 5 shows that the capacities for binding of ATP and $^{86}$Rb remains within 80-90% of control and that the disssociation constants for binding of these ligands are the same for trypsinized and control preparations. Neither the number of ATP binding and cation sites nor the affinities of the sites for direct ligand binding are affected by the removal of the N-terminal 30 residues of the $\alpha$-subunit.

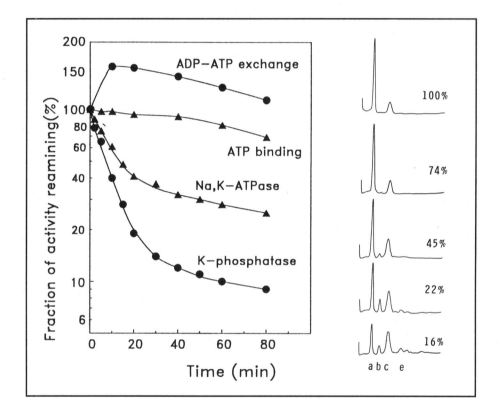

**Figure 4.** Semilogaritmic plots of tryptic modification of ADP-ATP exchange (●), [$^{14}$C]ATP binding (▲), Na$^+$/K$^+$-ATPase (▲) and K-phosphatase (●) activity. To the right electrophoretograms of purified Na$^+$/K$^+$-ATPase and the proteins remaining in the membrane after tryptic digestion in the presence of 150 mM NaCl as described before (13,14).

## Effects of N-terminal truncation of the α1-subunit on the phosphoenzymes of Na⁺/K⁺-ATPase

The reciprocal changes of ADP-ATP-exchange and K-phosphatase activities (Fig. 4) with intact number of sites for ATP binding and phosphorylation suggested that T2 cleavage may affect the $E_1P$-$E_2P$ equilibrium or the dephosphorylation reactions. The fraction of ADP-sensitive phosphoenzyme and the K-sensitivity of dephosphorylation were therefore determined as shown in Fig. 6. It is seen from the left hand part of Fig. 6 that the fraction of ADP sensitive phosphoenzyme is increased from 14% in control enzyme to 55% in trypsinized Na⁺/K⁺-ATPase. Truncation of the N-terminus thus results in a 2-3 fold increase of the relative amount of ADP-sensitive phosphoenzyme. The right hand side of Fig. 6 shows that the K-dependent increment of the rate constant for dephosphorylation was less than half that of the control. In Fig. 7, the reactivity of the phosphoenzyme to KCl was examined over a wider range of concentrations. Is is seen that KCl depresses the phosphoenzyme level of the trypsinated enzyme to about

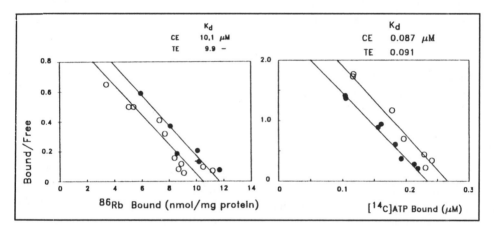

**Figure 5.** ⁸⁶Rb and [¹⁴C]ATP binding at equilibrium after selective tryptic cleavage at T2 (●) and in control (○). Procedures as described before (14,16).

50%, while the level was about 15% for the control enzyme. These values agree well with the estimate in Fig. 4 of a ratio of E1P/E2P of 15/85 in control enzyme as compared to 55/45 in trypsinated enzyme. Two Na-occluded phosphoenzyme intermediates (E1P[3Na] and E2P[2Na]) have been demonstrated in Na⁺ medium. The $E_2P$[2Na] intermediate is sensitive to ADP, because binding of one Na⁺ allows it to return to the $E_1P$[3Na] form for reaction with ADP and formation of ATP. After addition of K⁺, exchange of Na⁺ for K⁺ at the extracellular surface would lead to dephosphorylation. The $E_2P$[2Na]-form appears in both the ADP-sensitive and in the K-sensitive fraction of the phosphoenzyme and its amount can be determined as the sum of the ADP-sensitive and K-sensitive minus the total phosphoenzyme. The data in Fig. 6 and 7 show that the sum of the ADP-sensitive and K-sensitive phosphoenzymes is

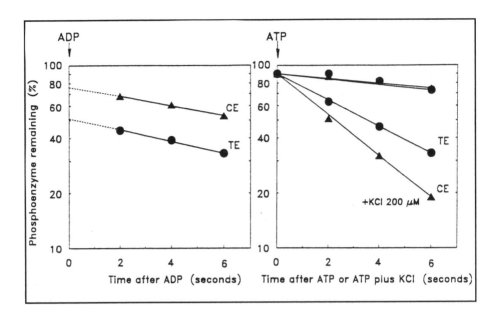

**Figure 6.** Effect of selective tryptic cleavage at T2 on ADP dependent dephosphoryl-ation (left) and dephosphorylation after addition of ATP or ATP plus KCl (right).
TE (●) is treated with trypsin and CE (▲) is a control preparation. Procedures as described before (19).

close to unity indicating that the steady state concentration of the $E_2P[2Na]$ is quite low both in the trypsinized and in the control preparations. In the trypsinized $Na^+/K^+$-ATPase, the fraction of ADP-sensitive $E_1P[3Na]$ is increased and the $E_1P[3Na]$-$E_2P[2Na]$ conformational equilibrium is shifted towards the $E_1P$ form. The $E_1P$-$E_2P$ transition releases a single $Na^+$-ion at the extracellular surface and this transition is the voltage sensitive Na transloation step. The $E_2P[2Na]$ form represents an occluded state in transition to $E_2P$-2Na with $Na^+$ leaving the sites making them accessible for binding of $K^+$ from the extracellular phase. The reduction of the rate of $Na^+/K^+$-ATPase and Na,K-transport at physiological ATP concentrations (2-3 mmoles/l) is adequetely explained by the reduced rate of the $E_1P[3Na]$-$E_2P[2Na]$ transition.

The weak effect of $K^+$ on the rate constant for dephosphorylation and the low K-phosphatase activity also reflect the abnormal transformation from $E_1P$ to $E_2P$ with inadequate exposure of the cation sites to the extracellular medium. Fig. 7. shows that the Km for KCl for depression of the steady state level of the phosphoenzyme was the same for control and trypsinized enzyme. The apparent affinity of the $E_2P$ forms for $K^+$ has therefore not been altered by truncation of the N-terminus. In contrast, the Km value for potassium activation of overall ATP hydrolysis measured by Pi release was 3-4-fold higher for trypsinized enzyme than for the control. Thus, in this truncated preparation where ion binding sites are intact, the apparent $K^+$ sensitivity of reactions involving a series of reactions including $E_1$-$E_2$ transitions is reduced by several fold.

**Figure 7.** Effect of KCl on (top) the steady state levels of phosphoenzyme formed by control preparations (▲) and preparations that had been cleaved at T2 (●)

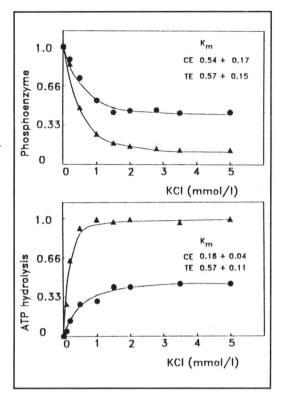

## Conformational transitions after N-terminal truncation

Intrinsic protein fluorescence is a particularly valuabe tool for monitoring conformational transitions in the native and truncated $Na^+/K^+$-ATPase . Recordings of the steady-state levels of fluorescence intensity in Fig. 8 illustrate the antagonism between $K^+$ and ADP or $Na^+$ (13). Titration of the response to $K^+$ in medium with 10 mM NaCl showed that the $K_{0.5}$ for $K^+$ was about 2-fold lower for the control than for the truncated $Na^+/K^+$-ATPase. In contrast, titration of the response to ADP in presence of a low concentration of KCl (0.3 mmoles/l) showed that the truncated $Na^+/K^+$-ATPase was 2-3-fold more sensitive to ADP than the control. The apparent affinity of the truncated $Na^+/K^+$-ATPase to $K^+$ is reduced and the apparent affinity to ADP is increased as compared with the control enzyme. An adequate explanation is that the equilibrium between the $E_1$- and $E_2$- forms is shifted in direction of $E_1$ (18). This change in K-nucleotide antagonism after truncation of the α-subunit is reflected in the stimulation by $K^+$ at low ATP concentrations and in an increase of the rate of K-deocclusion in trypsin treated enzyme (32). We also observed that the two phosphoenzyme forms have different fluorescence intensities, $E_2P$ having the same fluorescence intensity as $E_2K$, while the intensity of $E_1P$ is similar to that of $E_1Na$ (18). The fraction of phosphoenzyme present as $E_2P$ can therefore be determined as the amplitude of the fluorescence change accompanying phosphorylation in the absence of $K^+$ divided by the amplitude of the full response to $K^+$.

In the trypsinized $Na^+/K^+$-ATPase , the amplitude of the fluorescence change accompanying addition of ATP to a medium containing $Mg^{2+}$, $Na^+$ and $K^+$ was smaller than in the control. This defective fluorescence responses of the trypsinized protein are therefore in agreement with the shift in equilibrium between $E_1P$-$E_2P$ forms in direction of the $E_1P$ form.

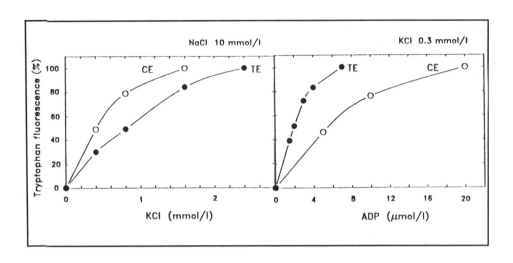

**Figure 8.** Concentration dependence of changes in tryptophan fluorescence. Left: increase in fluorescence intensity following addition of K$^+$ to control ( o ) and trypsinized ( ● ) Na$^+$/K$^+$-ATPase in 10 mM NaCl medium. Right: decrease in fluorescence following addition of ADP to control ( o ) and T2 cleaved ( ● ) Na$^+$/K$^+$-ATPase. Procedures as before (18).

**Functional expression of N-terminal truncated α-subunits in *Xenopus laevis* oocytes**

Deletion mutants of α1-subunit isoforms from A6 cells, (ΔM29 and ΔM34 in Fig. 9) were expressed by cRNA injection in *Xenopus laevis* oocytes by Rossier and coworkers (4). The truncation was reported to result in change of ouabain binding sites

---

```
Pig kidney α1-subunit
CE   GRDKYEPAAVSEHGDKKK-AK-K--ERDMDELKK-EVSMDDHKLS-
TE-ΔE31                                 EVSMDDHKLS-

Xenopus laevis
α1    GRDKYEPAARSEQGGKKKKGKGKGKEKDMDELKK-EVTMEDHKLS-
α1-ΔM29                      MDELKK-EVTMEDHKLS-
α1-ΔM38                             MEDHKLS-

Torpedo
α1    ASEKYQPAATSENAKNSKKSKSKT--TDLDELKK EVSMEDHKLS-
α1-ΔK28                      KT--TDLDELKK EVSMEDHKLS-
α1-ΔT29                       T--TDLDELKK EVSMEDHKLS-
α1-ΔK37                                K EVSMEDHKLS-
```

---

**Figure 9.** N-terminal amino acid sequences of α1-subunit from pig kidney (17), *Xenopus laevis* (4), and *Torpedo* (29).

paralleled by a concomitant increase in Na,K-pump current from $252 \pm 18$ nA/oocyte in the control to $412 \pm 39$ in $\Delta M29$ and $219 \pm 9$ nA/oocyte in $\Delta M34$. A two-fold increasase in the $K_{1/2}$ (from 1.7 to 2.3 mmol/l for $\Delta M34$) for K-activation at the extracellular surface was observed. These results are in contrast with our observation of an unchanged number of Na,K-pump sites and a reduced rate of Na,K-pumping in the $\Delta E31$ truncated renal $Na^+/K^+$-ATPase. However, the two-fold increase in $K_{1/2}$ for K-activation is in agrement with our observation of a weak effect of $K^+$ on dephosphorylation and a reduced apparent affinity for $K^+$ in reactions involving the conformational transitions.

Truncation of the N-terminus of the *Torpedo* Na,K-pump as shown in Fig. 9 removing part of the lysine cluster ($\Delta K28$ and $\Delta T29$) lead to an increase of the apparent affinities at 0 mV for external $Na^+$ and $K^+$ and of the apparent effective charges for the interaction of the ions with the external binding sites. The expressed $\Delta K37$ protein supports ouabain binding and uptake, but it is unable to generate significant current at physiological potentials; the resting potential being more negative than -200 mV in Na-free and 5 mM K-containing solution (29). There is no obvious relationship between these data and the present measurements on the truncated renal $\alpha1$-subunit of $Na^+/K^+$-ATPase.

In conclusion, the N-terminus is not diretly involved in formation of ligand binding sites as the number of sites and the dissociation constants for binding of $^{86}Rb$ or $^{14}C$-ATP are untaltered after the $\Delta E31$ truncation of the $\alpha1$-subunit. The properties of the selectively cleaved derivative suggest that charged residues in the N-terminus engage in ionic interactions of importance for the $E_1$-$E_2$ transition. The shift in $E_1$-$E_2$ equilibrium towards the $E_1$-form can explain the reduced apparent affinities for $K^+$ and ADP when reactions are monitored that involve conformaitonal transitions in the protein. The data illustrate that determination of $K_{1/2}$ values for activation by $K^+$, $Na^+$ or ATP of reaction sequences involving conformational transitions does not lead to adequate parameters for estimating the functional consequences of mutations to the $\alpha$-subunit of $Na^+/K^+$-ATPase.

**References**

1 Buxbaum E, Schoner W (1990) Eur J Biochem 193:355-360
2 Brotherus JR, Jacobsen L, Jørgensen PL (1983) Biochim Biophys Acta 731: 290-303
3 Brotherus JR, Møller JV, Jørgensen PL (1981) Biochem Biophys Res Commun 100: 146-154
4 Burgeneer-Kairuz P, Horisberger JD, Geering K, Rossier BC (1991) FEBS Lett 290: 83-86
5 Esmann M, Christiansen C, Karlsson KA, Hansson GC, Skou JC (1980) Biochim Biophys Acta 603:1-12
6 Hastings DF, Reynolds JA 81979) Biochemistry 18:817-821
7 Hebert H, Jørgensen PL, Skriver E, Maunsbach AB (1982) Biochim Biophys Acta, 689: 571-574
8 Jørgensen PL, Skou JC (1971) Biochim Biophys Acta 233: 366-380
9 Jørgensen PL, Skou JC, Solomonson L (1971) Biochim Biophys Acta 233: 381-394
10 Jørgensen PL (1974) Biochim Biophys Acta 356: 36-52
11 Jørgensen PL (1974) Biochim Biophys Acta 356: 53-67
12 Jørgensen PL (1974) Methods Enzymol 32: 277-290

13  Jørgensen PL (1975) Biochim Biophys Acta 401: 399-415

14  Jørgensen PL (1977) Biochim Biophys Acta 466: 97-108

15  Jørgensen PL (1982) Biochim Biophys Acta 694: 27-68

16  Jørgensen PL (1988) Methods Enzymol 156: 29-43

17  Jørgensen PL, Collins JH (1986) Biochim Biophys Acta 860: 570-576

18  Jørgensen PL, Karlish SJD (1980) Biochim Biophys Acta 597: 305-317

19  Jørgensen PL, Klodos I (1978) Biochim Biophys Acta 507: 8-16

20  Jørgensen PL, Andersen JP (1988) J Membr Biol 103: 95-120

21  Karlish SJD, Pick U (1981) J Physiol 312:505-529

22  Martin DW, Sachs JR (1992) J Biol Chem 267: 23922-23929

23  Moczydlowsky EG, Fortes PAG (1981) J Biol Chem 256: 2346-2356

24  Nørby JG, Jensen J (1991) In: Kaplan JH, De Weer P (eds) The Sodium Pump: Structure Mechanism and Regulation. Rockefeller Univ. Press, New York, pp. 173-188

25  Ohta T, Noguchi S, Nakanishi M, Mutoh Y, Hirata H, Kagawa Y, Kawamura M (1991) Biochim Biophys Acta 1059: 157-164

26  Periyasamy SM, Huang WH, Askari A (1983) J Biol Chem 258: 9878-9885

27  Reynolds JA (1988) In: Skou JC et al. (eds) The Na,K-pump, Part A: Molecular Aspects. A.R. Liss, Inc. New York, pp 137-148

28  Skriver E, Maunsbach AB, Jørgensen PL (1981) FEBS Letters 131: 219-222

29  Vasilets M, Ohta T, Noguchi S, Kawamura M, Schwarz W (1993) Eur Biophys J 21: 433-443

30  Vilsen B, Andersen JP, Petersen J, Jørgensen PL (1987) J Biol Chem 262: 10511-10517

31  Ward DG, Cavieres JD (1993) Proc Natl Acad Sci 90: 5332-5336

32  Wierzbicki W, Blostein R (1993) Proc Natl Acad Sci 90: 70-74

# Organization of the Cation Binding Domain of the Na$^+$/K$^+$-pump

R. Goldshleger, D.M. Tal, A. Shainskaya, E. Or, S.Hoving, and S.J.D. Karlish

Biochemistry Department, Weizmann Institute of Science, Rehovot 76100, Israel.

## Introduction

Knowledge of the structural organization of the cation occlusion domain of the Na$^+$/K$^+$-pump is essential for an understanding of active transport. In the absence of detailed molecular structure, cation sites can be studied by a variety of techniques (14). For example one can manipulate the genes to create point mutations, truncated forms, and chimeric molecules, or manipulate the protein using proteolytic enzymes and chemical modification. We have used, chemical modification with N-N'-dicyclohexylcarbodiimide, DCCD, to investigate the role of carboxyl groups (16, 32), extensive proteolysis of renal Na$^+$/K$^+$-ATPase in order to define the minimal peptide components capable of cation occlusion (4, 20), and proteolysis of the α-chain in native membrane vesicles and in reconstituted proteoliposomes as an approach to define the trans-membrane topology (21). Topological organization has become a crucial issue, because of the evidence that cation sites are located in membrane-spanning segments. Topological models are usually based on hydropathy analysis. Hydropathy plots for alpha chains of P-type ATPases are similar, and indicate clearly the four N-terminal trans-membrane segments, but there is significant ambiguity in the C-terminal region. The result is that several models have been proposed, having as few as 6 and as many as 12 segments, and differing mainly in the C-terminal half. Models with 8 or 10 segments are the most likely, but it is clear that direct determination of topology is necessary. This can be looked at by antibody binding, or by proteolysis and chemical labeling. This paper summarizes our recent work on the cation occlusion domain and α-chain topology.

## Voltage Dependence of the Na$^+$/K$^+$-pump

A unique if indirect approach is the study of functional aspects, such as voltage sensitivity or charge carrying properties, for these studies lead to interesting structural implications, which may be testable using direct methods. In the models in Fig 1, with corresponding kinetic intermediates, cation sites face inwards or are occluded in E$_1$ forms, and face outwards or are occluded in E$_2$ forms. In occluded states barriers are present at both surfaces, and in states with accessible cation sites, barriers are open at one surface i.e this is a Moving Barrier model. Some years ago we proposed that cation sites contain two negatively charged sites to which 2Na$^+$ or 2K$^+$ could bind. The third Na$^+$ site could be neutral. The basis for this proposal was an inference, from experiments with reconstituted vesicles, that the conformational change with 3Na$^+$ ions bound is a voltage sensitive step, while that with 2K$^+$ ions bound is not (15). Analysis of current-voltage relations in whole cells (8) and the use of styryl dyes to detect local changes in electric field in purified membrane-bound enzyme, suggest a similar conclusion (22). In addition, there is now strong evidence, that dissociation of Na$^+$ ions or binding of K$^+$ ions at the extracellular surface are charge carrying steps, suggesting the existence of a high resistance access path or "cation well" at that surface (12, 22).

309

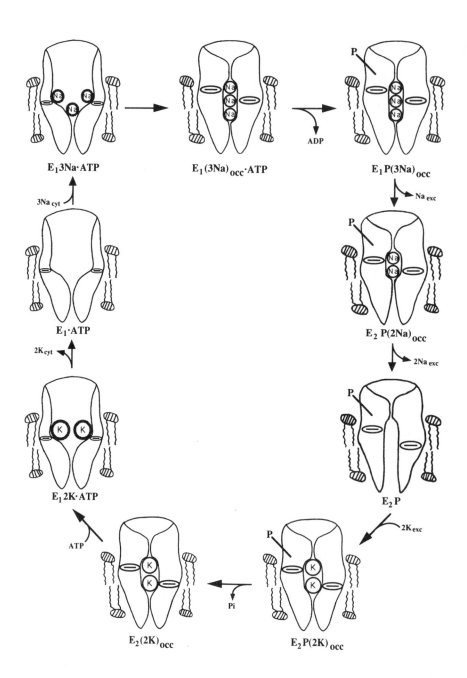

Fig. 1. Models illustrating Structural Implications for Cation sites based on Studies of Function.

Dissociation of occluded $Na^+$ ions is thought to be ordered, release of $1Na^+$ ion accompanying the transition of $E_1P(3Na^+)$ to $E_2P(2Na^+)$ (previously referred to as $E^*$, see 18) , and release of $2Na^+$ ions accompanying the $E_2P(2Na^+)$ to $E_2P$ transition (14).

310

Thus, both the conformational change and dissociation of 1Na$^+$ and the subsequent dissociation of 2Na$^+$ ions are charge carrying steps (34). Deocclusion of 2K$^+$ or Rb$^+$ ions at the extracellular surface is ordered (11, 13), a fact compatible with the channel concept.

Fig. 1 illustrates a new proposal (28) that, at the cytoplasmic surface, cation occlusion occurs in two steps. The evidence is based on effects of organic guanidinium derivatives, acting as competitive Na$^+$ antagonists (7). The first step consists of non-specific recognition of the carboxyls, in a vestibule, by partially hydrated Na$^+$ or K$^+$ ions or the guanidinium derivatives which are not shown. It is followed by the conformational change or closing of the barrier behind either K$^+$ or Na$^+$ ions which are dehydrated. The guanidinium derivatives bind in the vestibule, but are not occluded, and block (Na$^+$/K$^+$)-ATPase activity. There is additional evidence, from vesicle experiments (15) and the styryl dye signal (22), that binding of Na$^+$ to the neutral site at the cytoplasmic surface is voltage sensitive, and lies in a cation well, located deepest into the membrane.

What are the structural implications? First, negatively charged residues in cation sites are, of course, being actively sought by chemical modification and site-directed mutagenesis. The functional evidence for the deep extracellular channel, and cytoplasmic vestibule or channel must eventually find its counterpart in the structure and organization of the membrane spanning helices of the protein. Channels need not be regular as depicted. A strong implication of access channels at both surfaces is that the occlusion sites themselves are located deep within the membrane domain and occupy a restricted region. This is suggestive of the Moving Barrier concept as opposed to a Moving Site mechanism, and furthermore that these structural changes are of limited magnitude.

## Chemical Modification of Carboxyl Groups

The question arose as to the role, number and identity of carboxyl groups in cation occlusion, and in addition whether Na$^+$ and K$^+$ bind to the same sites. Carboxyl reagents which have been used include DCCD (16, 30, 32) and a diazomethane derivative 4-(diazomethyl)-7-(diethylamino)-coumarin DEAC (2).

We have shown previously that DCCD inactivates both Rb$^+$ and Na$^+$ occlusion, with identical kinetics and protection by Rb$^+$ or Na$^+$ (32). This led to the hypothesis that one or a small number of carboxyls in a hydrophobic environment were modified and either Na$^+$ or Rb$^+$ were bound to these same residues. This hypothesis was tested (16) by comparing the Rb$^+$-protected covalent incorporation of $^{14}$C-DCCD and inactivation of Rb$^+$ occlusion. A linear relationship was observed between inactivation and specific labeling of the isolated a chain, or into a C-terminal 19kDa fragment, which is described below, suggesting, with a high probability, that the chemical modification is the cause of inactivation. Incorporation of 2 molecules of DCCD per α chain, or one molecule of DCCD per 19kDa fragment, is associated with full inactivation. The labeled residue in the 19kDa peptide was identified in a CNBr fragment, as Glu$^{953}$. The other labeled residue was not identified.

This work led to the proposal that the cation occlusion domain consists of ligating groups donated by different trans-membrane segments, and includes two carboxyl groups

311

( such as Glu$^{953}$ and perhaps Glu$^{327}$) as well as neutral ligating groups in two sites, to which two Na$^+$ or two K$^+$ ions bind, and only neutral groups in the third Na$^+$ site (16).

Despite the simplicity of the conclusion, and it's consistency with that based on the voltage dependence (Fig. 1), the identity of the relevant carboxyl residues is not certain. As pointed out in (16), in the relevant part of the sequence of pig, brine shrimp, and drosophila α-chains, whereas Glu$^{954}$ (*) is conserved, phenylalanine replaces Glu$^{953}$ in brine shrimp and drosophila.

```
                                       *
Pig         K N K I L I F G L F E E T A L A A F L
Artemia     K N G T L N F A L V F E T C V A A F L
Drosophila  R N W A L N F G L V F E T V L A A F L
```

Thus a charged residue may not be essential in position 953, and it cannot be excluded that Glu$^{953}$ lies near but not in the site, and DCCD interferes with occlusion by a steric effect. On a trans-membrane alpha helix, Glu$^{953}$ might be oriented towards the lipid, where it reacts with the DCCD. In this case the adjacent Glu$^{954}$ could be oriented towards the other helices. Kaplan and Argüello (3) have proposed that the reagent DEAC labels a different residue, Glu$^{779}$ in the fifth trans-membrane segment, and suggested that it is essential for activity.

Predictions based on chemical modification are beginning to be tested by site-directed mutagenesis. Van Huysse et al. (36) mutated Glu$^{953}$ and Glu$^{954}$ singly or together, to glutamine and aspartic acid, and showed that the Glu$^{953}$ and Glu$^{954}$ are not essential for ATPase activity, and observed small or moderate effects of the mutations on apparent activation constants for Na$^+$ or K$^+$ ions. Schwartz and Vasilets have reported (at this conference) that mutations to Ala of the homologous Glu$^{959}$ and Glu$^{960}$ residues in the *Torpedo* enzyme ,expressed in *Xenopus* oocytes, induce significant changes in apparent K$^+$ affinity at the extracellular surface. More recently Lingrel and colleagues ( reported at this conference) have mutated other carboxyls assumed to lie in the membrane domain. Mutations Glu$^{327}$Leu or Asp, Glu$^{953}$Gln or Asp, Glu$^{954}$Asn or Asp and Asp$^{925}$Asn or Leu do not inactivate, while mutations Glu$^{327}$Asp, Glu$^{779}$Leu, Asp$^{803}$Glu or Asn and Asp$^{807}$Leu inactivate ATPase activity.

The interpretation of all these findings is not simple. It is hazardous to exclude a role for residues which can be mutated without complete loss of function or assign a role in cation binding of residues mutated with loss of activity, on the basis of ATPase measurements. Since each site probably consists of 6-8 ligating groups, replacement of one residue might not be lethal, and might lead to only moderate changes of binding affinity, but these might be hard to detect in ATPase assays due to kinetic factors. Occlusion measurements are necessary, but the making of such measurements will not be straightforward, due to the low density of pumps expressed in the cell systems currently available. Even if a way can be found around the technical problem, it may be very difficult to distinguish direct from indirect effects of point mutations. The evidence presented below for a multitude of interactions between occluded cations, and essentially all trans-membrane segments, and loops and tails of α and β chains outside the membrane, implies that mutations which affects the mutual interactions will affect occlusion, even if the mutated residues are far from a cation site.

312

# Extensive Proteolytic Digestion of Renal Na$^+$/K$^+$-ATPase

As to the question of the minimal structure able to occlude cations, we have described a procedure for extensive tryptic digestion of renal Na$^+$/K$^+$-ATPase in the presence of Rb$^+$ and absence of Ca$^{2+}$ ions (4, 20). This digestion leads to so-called "19kDa-membranes"" containing a stable 19kDa and smaller membrane-embedded fragment of the α-chain, and a β-chain which is intact or is partially split to a smaller glycosylated fragment and a 16 or 14 kDa fragments. Most importantly, Rb$^+$ occlusion is intact, even though about 50% of the protein has been removed. The smaller membrane-embedded fragments were resolved on long 16.5% Tricine gels and reveal the presence of 7 peptides of ≈8-12kDa. Digestion of renal Na$^+$/K$^+$-ATPase with non-specific proteases, such as pronase or proteinase-K, in the same conditions removes up to 70% of membrane protein, and the specific activity of the occlusion rises up to 2.5-fold (4). The 19kDa or an 18.5kDa fragment remain as the largest membrane associated fragment while the β-chain is removed. Thus the 19kDa fragment is resistant to virtually all proteases. In the absence of Rb$^+$ or presence of Ca$^{2+}$ ions the 19kDa fragment is digested and occlusion is lost. The enzyme digested with trypsin has been named "19kDa-membranes" because a controlled digestion in a Ca$^{2+}$-containing medium showed a perfect correlation between the amount of the fragment and capacity for Rb$^+$ occlusion (20). Thus, the 19kDa fragment is essential for Rb$^+$ occlusion, being either directly involved or indirectly supporting the occlusion structure. The DCCD labeling (16) suggested a direct role. Of course, this does not exclude participation of other fragments (see below).

Functional properties of the "19kDa-membranes" have been studied in some detail (20). The specific activity of both Rb$^+$ and Na$^+$ occlusion is roughly double that of native enzyme, and the ratio is maintained. Thus all 3Na$^+$ and 2K$^+$ sites are preserved. By contrast all ATP- and Pi-dependent functions are destroyed. Recently it has been demonstrated, with measurements of the styryl dye RH421, that in "19kDa-membranes", the ATP-dependent electrical field changes are absent, but the field change associated with binding of one cytoplasmic Na$^+$ ion is intact (31). In addition, binding of ouabain is intact for, within error, the affinity for ouabain as an inhibitor of styryl dye responses and Rb$^+$ occlusion is the same in native enzyme and "19kDa-membranes" (31).

Strong conclusions from these studies are that **a.** In "19kDa-membranes", the intra-membrane and extracellular domains are intact, but the cytoplasmic domain is removed.**b.** Cation occlusion sites are located within the membrane spanning segments. **c.** Cation (intra-membrane) and ATP (cytoplasmic) are separate domains. Interactions between cation and ATP sites during active transport are indirect.

For identification of all the fragments of "19kDa-membranes", it has been necessary to develop reliable methods for purification, concentration and resolution of the hydrophobic fragments of 8-12 kDa. The method combines size-exclusion HPLC in SDS solution, and long 16.5% Tricine gels. The peptides of 8-12 kDa were resolved on this system (4). After transfer to PVDF paper all peptides of "19kDa-membranes" were sequenced, see Table 1 (Intact).

**Table 1. Tryptic Fragments in Intact and Digested "19kDa-membranes"**

| Intact | | Digested | |
|---|---|---|---|
| Peptide | Calculated $M_r$ kDa | Peptide $M_r$ kDa | Calculated |
| *Alpha Chain Fragments* | | | |
| $Asp^{68}$--$Arg^{166}$ | 10.98 | $Ala^{72}$--$Lys^{146}$ | 8.31 |
| $Ile^{263}$--$Lys^{347}$ | 9.08 | $Ile^{263}$--$Lys^{342}$ | 8.44 |
| $Gln^{737}$--$Arg^{830}$ | 10.32 | $Thr^{772}$--$Lys^{826}$ | 5.93 |
| $Asn^{831}$--$Tyr^{1016}$ | 21.82 | $Leu^{842}$--$Arg^{880}$ | 4.23 |
| | | $Ile^{946}$---$Arg^{972}$ | 2.90 |
| | | $Met^{973}$--$Arg^{998}$ | 3.29 |
| *Beta Chain Fragments* | | | |
| $Ala^{5}$--$Arg^{142}$ | 16.05 | $Thr^{28}$--$Lys^{112}$ | 9.77 |
| $Gly^{143}$--$Ser^{302}$ | 18.52 (protein) | not found | |

The data includes the experimentally determined N-terminal, the likely C-terminus based on tryptic split sites and the appropriate $M_r$ value, respectively. In the case of the 19kDa fragment with N-terminus $Asn^{831}$, the fragment is known to extend to the true C-terminus of the alpha chain, because it binds an antibody to the final 4 residues Glu Thr Tyr Tyr (4). Therefore the true $M_r$ is close to 22kDa. The groups of peptides of 8-12kDa, are long enough to contain two trans-membrane helices, and include peptides corresponding to all other putative trans-membrane segments, namely $Asp^{68}$, M 1+2, $Ile^{263}$, M3+4, and $Gln^{737}$, M5 or M5+6 depending on the topological model. There are in fact three fragments with N-terminal $Asp^{68}$ and slightly different weights, of which the one in the Table is the smallest. Fragments of the β-chain include the 16kDa N-terminal domain, beginning at $Ala^{5}$, and including the single trans-membrane segment, and the C-terminal glycosylated fragment with N-terminal $Gly^{143}$.

Evidently the 19kDa and these other fragments constitute the best defined minimal occlusion structure. Pronase, proteinase-K and other proteases can truncate several fragments further without loss of occlusion, but the products have not been defined. The true minimum structure consists essentially of trans-membrane segments and extracellular loops.

Which trans-membrane segments are involved in the cation occlusion? We have no direct answer as yet, but some interesting insights emerge from analysis of the protection of 19kDa fragment against trypsin by monovalent cations and its destabilization by $Ca^{2+}$ ions. As reported in (4), either $Rb^+$ or $Na^+$ ions protect against trypsin and lead to appearance of "19kDa-membranes" with maintenance of $Rb^+$ occlusion. This lack of selectivity of alkali metal cations can be compared with the classical finding that $Na^+$ and $K^+$ induce different conformation of native enzyme, which are distinguishable by trypsin at initial stages of digestion (17). The finding seems to imply that, near to the membrane, any conformational differences between $Na^+$ and $K^+$-bound forms are too small to be detected by the trypsin. This is compatible with the Moving Barrier notion discussed in connection with Fig 1. A surprising point, and the one that should be emphasized in the present context, is that the cations protect not only the 19kDa fragment of the $\alpha$-chain but also the 16kDa fragment of the $\beta$-chain. The latter finding fits well with recent results on destabilization of "19kDa-membranes" by $Ca^{2+}$ (33). $Ca^{2+}$ and $Rb^+$ ions have been shown to compete directly in occlusion sites, but upon displacement of the $Rb^+$ an irreversible thermal inactivation follows. The rate of loss of occlusion depends only on the presence and concentration of $Ca^{2+}$ ions, while trypsin is without effect. Thus, trypsin does not inactivate the occlusion, but it digests thermally inactivated membranes. We have observed that **all** fragments in the "19kDa-membranes" are digested to smaller pieces, or conversely, one may conclude, **all** fragments are protected by occluded $Rb^+$. The limit tryptic membrane-embedded fragments have now been analyzed, using the HPLC-Tricine gel combination (Table 1, digested). Note that all fragments have been digested. In most cases they are truncated at N- or C-terminals or both. The 19kDa fragment is cut into at least 3 pieces which are long enough to include only a single trans-membrane segment. A hydrophobic fragment of the 19kDa peptide, $Ile^{906}$-$Lys^{931}$, which might have been predicted to be associated with the membrane, was not found .

There appear to be three important implications: **a**. Occluded cations stabilize a complex of **all** membrane-embedded fragments in "19kDa-membranes". All the fragments interact directly, or indirectly via other fragments, with the occluded cations. Presumably, the cation occlusion "cage" consists of several trans-membrane segments. **b**. Occluded cations within trans-membrane segments stabilize the loops and tails of 19kDa peptide outside the membrane at both the extracellular and cytoplasmic surface. These sections might be buried in the membrane, as suggested by increased labeling of hydrophobic probes (19), or become more compact. **c**. Occluded cations protect the 16kDa fragment of the $\beta$-chain, presumably via indirect interactions between the $\alpha$ and $\beta$ fragments. A necessary corollary is that the reciprocal interactions between the 16kDa and the $\alpha$-chain fragments affect $Rb^+$ occlusion. Segments of the 16kDa fragment released by digestion, $Ala^5$--$Arg^{27}$ and $Asp^{113}$--$Arg^{142}$, are candidates for mediating this interaction.

**Trans-membrane Topology of the Alpha Chain**

With this notion of a complex of trans-membrane segments, the building of a model requires knowledge as to the trans-membrane topology, and organization of the trans-membrane segments in the plane of the membrane. The results of chemical modification of the $Na^+$/$K^+$-ATPase (16), and site-directed mutagenesis of $Ca^{2+}$-ATPase (5, 6) as well as molecular model building, suggest the notion that cations are bound to co-ordinating groups, including carboxyl and neutral oxygen containing ligating groups, oriented towards the cavity between several trans-membrane a-helices. Molecular models of the spatial organization of 8 or 10 alpha helices of the $\alpha$ chain with one of the $\beta$ chain, suggest that a four helix bundle could be involved (10). The models are based on calculations of a molecular hydrophobicity potential of single and pairs of helices and docking together of such pairs. A necessary prerequisite for models of spatial relations between helices, is prior knowledge of the trans-membrane topology. Previously, labeling work, using photoactivated hydrophobic probes such as 5-[$^{125}$I]-iodonapthyl-1-azide, INA, or a ouabain site affinity label, led to prediction of the topology of the N-terminal domain even before the primary sequence was available (Jørgensen, et al., 1982). For the purpose of distinguishing different topological models, identification of sites of proteolytic splits is a more definitive technique.

## A. Proteolysis at the Cytoplasmic Surface

Fig. 5 in ref. (4) depicted four possible topological arrangements of tryptic fragments; one with 7 segments, two with 8 segments as in current models of Na$^+$/K$^+$-ATPase (see ref. 23), and a 10 segment model, similar to that for Ca$^{2+}$-ATPase (24). Recently, the C-terminal of the Na$^+$/K$^+$-ATPase α-chain was shown to be cytoplasmic by antibody binding and labeling (1, 25, 36). Since the N-terminal is also cytoplasmic, models with uneven numbers can be excluded (1). In Fig 5 of ref (4), the N-terminal half, with four trans-membrane segments, are the same in all models. A difference between the 8 segment and 10 segment models is that the former two predict that, in the hydrophobic section following the central cytoplasmic loop, there is only one trans-membrane segment M5, while the latter predicts that there are two segments M5+ 6. That is to say, the N-terminal Asn$^{831}$ of the 19kDa fragment should be extracellular according to the 8 segment models but cytoplasmic according to the 10 segment model.

The prediction above has been tested using intact right-side-out renal membrane vesicles and reconstituted proteoliposomes (21). The renal vesicles are right-side-out as judged by ouabain binding, inaccessibility of the ATP site, and necessity of adding detergents to unmask ATPase activity. Thus the question arose whether one can digest these vesicles to produce the 19kDa fragment or whether it is necessary to add a detergent to expose the cytoplasmic face. The result is that it is necessary to add the detergent deoxycholate in order to observe fragmentation to the 19kDa fragment, and the simple interpretation is that the detergent made the cytoplasmic bond, Arg$^{830}$--Asn$^{831}$, accessible to trypsin. Because one might argue that the detergent induced a conformational change, or postulate a more complex series of proteolytic splits, we also looked at digestion of reconstituted proteoliposomes. These consist of pumps in both right-side-out and inside-out orientation in roughly equal proportions. Here we were able to detect the 19kDa fragment in the absence of detergents, and a population of α-chains, determined by functional studies to be right-side-out, were trypsin resistant. The combination of these approaches provides clear evidence that Asn$^{831}$ is cytoplasmic and hence that there are two segments M5+6 in the hydrophobic domain following the central loop and not one segment, M5, as in previous models for the Na$^+$/K$^+$-ATPase.

The result appears compatible with the 10 segment model as proposed for Ca$^{2+}$-ATPase. Nevertheless there are a number of paradoxical findings in the literature. First, the model predicts that only 7.5% of the protein is extracellular, while a number of labeling studies suggest that a significantly higher proportion must be outside (9, 27). Second, there are reports that the hydrophobic peptide Thr$^{913}$-Cys$^{930}$, corresponding to the putative M8 trans-membrane segment, is released into the medium upon proteolysis, and is not labeled with the photoactivated hydrophobic probe 3-trifluoromethyl-3-(m-[$^{125}$I]Iodophenyl)Diazirine, TID, (26, 29). As mentioned above, this fragment was not found as a limit tryptic fragment in the membrane (see Table 1). Third, the putative

M10 helix is not hydrophobic by comparison with other trans-membrane helices . The crucial issue is therefore the topology of the 19kDa fragment itself. The fragment must have an even number of segments, either four as in the 10 segment model or perhaps only two (see Fig 4A or 4B respectively in ref. 21).

## B. Proteolysis at the Extracellular Surface.

Although the $\alpha$-chain is highly resistant to proteolysis at the extracellular surface, digestion has now been achieved, using the right-side out renal vesicles, following a perturbation of the structure. This involves heating of the vesicles to 55°C for 30 min, before cooling to room temperature for the proteolysis. The $\alpha$-chain is cut to a fragment of apparent $M_r$ 91kDa. In $Rb^+$-loaded vesicles the $\alpha$-chain is protected even after this treatment. The N-terminal of the 91kDa fragment is $Gly^1$ of the $\alpha$–chain, and therefore the split must have occurred near the C-terminal. None of the many other proteins in the vesicles are digested. Thus, one may conclude that the vesicle structure is intact even after the heating.

Why did $Rb^+$ protect the $\alpha$-chain? It turns out that the heating inactivates ATPase activity and $Rb^+$ occlusion, in the $Rb^+$-free vesicles, but not in the $Rb^+$-loaded vesicles. Thus the $\alpha$-chain is digested at the outside only when it is denatured. The 91kDa fragment is not digested further because the vesicle is impermeable to trypsin, and presumably the cytoplasmic domains of the 91kDa fragment are inaccessible. $Rb^+$ protects against trypsin, because it protects against thermal denaturation.

Where in the C-terminal region was the $\alpha$-chain cut? The true $M_r$ of the 91kDa fragment should be about 95-97kDa. Thus we were expecting to observe a fragment of about 15kDa, present in small quantities. The C-terminal of such a fragment should bind an antibody against the final 4 residues of the $\alpha$-chain Glu Thr Tyr Tyr, since it is cytoplasmic and should be inaccessible to proteases. This prediction was tested in immunoblots of the digested microsomal vesicles, and produced a surprising result. In the $Rb^+$-protected sample the antibody bound to the $\alpha$-chain and essentially no other proteins. In the $Rb^+$-free digested sample, the antibody bound to a fragment of apparent $M_r$ 14.5 kDa but, in addition, it bound to three closely related bands of $M_r$ 12.1, 9.9 and 7.4kDa. Chymotrypsin digests also produced four bands, of slightly different sizes, $M_r$ 15.7, 13.1, 9.6 and 7.8 kDa, as expected for the different specificity. No fragment larger than 15.7kDa or smaller than 7.4kda was observed. The chymotryptic fragments have also been separated by size exclusion chromatography, showing that they are truly different in size, and not only in electrophoretic mobility. The results seems to imply that there are four points of proteolytic digestion at the extracellular surface, which lie on a piece of polypeptide chain of $M_r$ about 7kDa i.e. about 60 residues long.

After partial purification on the TSK-3000 column, the N-terminal of a $\approx$15kDa tryptic fragment, produced in $Rb^+$-free but not in $Rb^+$-loaded vesicles, was identified as $Trp^{887}$. A $\beta$-chain fragment of about 10kDa has N-terminal $Ala^1$, indicating again that trypsin did not enter the vesicles after the heating. The identification of the other fragments is now being undertaken. This is difficult due to the small quantities present. It will require a further step of purification, probably immunoprecipitation with the anti-Glu Thr Tyr Tyr antibody.

A tentative topological model to account for these findings would consist of 8 segments (similar to the hypothetical model Fig 4B in ref. 21). The model suggests the existence of an extracellular loop of about 80 residues, that about 11 % of the α-chain is located outside, and a cytoplasmic tail of about 50 residues. By comparison with the 10 segment model, M7 is the same, the hydrophobic sequence Thr$^{910}$--Cys$^{930}$, is placed outside the membrane, M9 becomes M8, and M10 is no longer a trans-membrane segment but is associated with the membrane. The latter assumption is needed to account for the findings that the smallest fragment produced by the extracellular split is 7.4kDa and a 3kDa limit membrane-associated tryptic fragment with N-terminal Met$^{973}$ is produced in open membranes (Table 1). Two trans-membrane segments M8 and M10, which by other criteria appear problematical, lie outside the membrane.

There are two important questions. **a.** Does the 8 segment model describe the native state? If this is the case, the postulated extracellular loop could unfold upon heating and become accessible to proteases. The alternative hypothesis is that the denaturation altered the topology. Clearly, it will be necessary to distinguish between these possibilities and obtain independent evidence for the topology in the native state, for example by labeling work. **b.** If the topology is correct, it is clear that the hydrophobic sequence (residues 910-930) cannot be free in the aqueous medium. It might be stabilized by association with the lipid, or with hydrophobic residues on the β-chains (see the report by Geering and colleagues at this conference, that a sequence of hydrophobic residues at the C-terminal of the β-chain are involved in stabilization of the α–β complex), or associate with hydrophobic residues of the α-chain.

*Acknowledgments*

This work was supported by the Weizmann Institute Renal Research Fund. We are greatly indebted to Prof. Jack Kyte for the gift of the anti Glu Thr Tyr Tyr antibody.

**References**

1.  Antolovic, R., Brüller, H-J., Bunk, S., Linder, D., and Schöner, W. (1991). Epitope mapping of amino-acid-sequence-specific antibodies reveals that both ends of the α subunit of Na$^+$/K$^+$-ATPase are located on the cytoplasmic side of the membrane. Eur. J. Biochem. 199: 195-202.
2.  Argüello, J.M. and Kaplan, J.H. (1991). Evidence for Essential Carboxyls in the Cation-binding Domain of the Na$^+$/K$^+$-ATPase. J. Biol. Chem. 266: 14627-14635.
3.  Argüello, J.M. and Kaplan, J.H. (1993). Localization of an Essential Carboxyl in the Cation Binding Site of the Na$^+$, K$^+$-ATPase. Biophys. J. 64: 352a.
4.  Capasso, J.M., Hoving, S., Tal, D.M., Goldshleger, R., and Karlish, S.J.D. (1992). Extensive digestion of Na$^+$, K$^+$-ATPase by Specific and Non-specific Proteases with Preservation of Cation Occlusion sites. J. Biol. Chem. 267: 1150-1158.
5.  Clarke, D.M., Loo, T.W., Inesi, G., and MacLennan, D.H. (1989). Location of high affinity Ca$^{2+}$-binding sites within the predicted transmembrane domain of the sarcoplasmic reticulum Ca$^{2+}$-ATPase. Nature. 339: 476-478.
6.  Clarke, D.M., Loo, T.W., Inesi, G., and MacLennan, D.H. (1990). Functional consequences of Alterations to Polar Amino Acids ~Located in the trans-membrane Domain of the Ca$^{2+}$-ATPase of sarcoplasmic reticulum. J. Biol.

Chem. 265: 6262-6267.

7.	David, P., Mayan, H., Cohen, H., Tal, D.M., and Karlish, S.J.D. (1992) Guanidinium Derivatives act as High Affinity Antagonists of $Na^+$ ions in Occlusion sites of $Na^+$, $K^+$-ATPase. J. Biol. Chem. 267: 1141-1149.

8.	De Weer, P., Gadsby, D.C., and Rakowski, R.F. (1988). Voltage dependence of the $Na^+$, $K^+$ pump. Ann. Revs. Physiol. 50: 225-241.

9.	Dzhandzhugazyan, K.N., and Jorgensen, P.L. (1985). Asymmetric orientation of amino groups in the $\alpha$-subunit of $(Na^++K^+)$-ATPase in tight right-side-out vesicles of basolateral membranes membranes from outer medulla. Biochim. Biophys. Acta. 817: 165-173.

10.	Efremov, R.G., Gulyaev, D.I., and Modyanov, N.N. (1993). Application of a Three-Dimensional Molecular Hydrophobicity Potential to the Analysis of Spatial Organization of Membrane Domains in Proteins: III Modeling of Intramembrane Moiety of $Na^+$, $K^+$-ATPase. J. Protein Chem. 12: 143-151.

11.	Forbush, B. III. (1987). Rapid release of $^{42}K^+$ or $^{86}Rb^+$ from two distinct transport sites on the $Na^+$, $K^+$-pump in the presence of Pi or $VO_4$. J. Biol. Chem. 262: 11116-11127.

12.	Gadsby, D.C., Rakowski, R.F., and De Weer, P. (1993). Extracellular Access to the $Na^+/K^+$-pump: Pathway Similar to Ion Channel. Science 260: 100-103.

13.	Glynn, I.M., Howland, J.L., and Richards, D.E. (1985). Evidence for the ordered release of rubidium ions occluded within the $Na^+$, $K^+$-ATPase of mammalian kidney. J. Physiol. 368: 453-469.

14.	Glynn, I.M., and Karlish S.J.D. (1990). Occluded Cations in Active Transport. Ann. Revs. Biochem. 59: 171-205.

15.	Goldshleger, R., Karlish, S.J.D., Rephaeli, A., and Stein, W.D. (1987). The Effect of Membrane Potential on the Mammalian Sodium/Potassium Pump Reconstituted into Phospholipid Vesicles. J. Physiol. (Lond.) 387: 331-355.

16.	Goldshleger, R., Tal, D.M., Moorman, J., Stein, W.D., and Karlish, S.J.D. (1992). Chemical Modification of Glu 953 of the Alpha Chain of $Na^+$, $K^+$-ATPase Associated with Inactivation of Cation Occlusion. Proc. Natl. Acad. Sci. 89: 6911-6915.

17.	Jorgensen, P.L. (1975). Purification and Characterisation of $Na^+$, $K^+$-ATPase. V. Conformational changes in the enzyme. Transitions between the Na-from and K-form Studied with tryptic digestion as a tool. Biochim. Biophys. Acta. 401: 399-415.

18.	Jorgensen, P.L. (1991). Conformational Transition in the $\alpha$-Subunitand Ion Occlusion. In: The sodium pump: structure, mechanism and regulation, (J.H. Kaplan and P. De Weer, Eds.), Rockefeller University Press pp. 189-200.

19.	Jorgensen, P.L., Karlish, S.J.D., and Gitler, C. (1982). Evidence for the Organization of the Trans-membrane Segments of $Na^+$, $K^+$-ATPase Based on Labeling Lipid-embedded and Surface Domains of the Protein. J. Biol. Chem. 257: 7435-7442.

20.	Karlish, S.J.D., Goldshleger, R., and Stein, W.D. (1990). Identification of a 19kDa C-terminal Tryptic Fragment of the Alpha Chain of $Na^+$, $K^+$-ATPase, Essential for Occlusion and Transport. Proc. Natl. Acad. Sci. 87: 4566-4570.

21.	Karlish, S.J.D., Goldshleger, R., and Jorgensen, P.L. (1993). Location of Asn-831    of the Alpha Chain of $Na^+/K^+$-ATPase at the Cytoplasmic Surface. Implication for Topological Models. J. Biol. Chem. 268: 3471-3478.

22.    Läuger, P. (1991). Kinetic Basis of Voltage Dependence of the $Na^+/K^+$-pump. In: The sodium pump: structure, mechanism and regulation, (J.H. Kaplan and P. De Weer, Eds.), Rockefeller University Press pp. 304-315.

23.    Lingrel, J.B., Orlowski, J., Shull, M.M. and Price, E.M. (1990). Molecular Genetics of $Na^+$, $K^+$-ATPase. Prog. Nucl. Acid Res. and Mol. Biol. 38: 37-89.

24.    MacLennan, D.H., Brandl, C.J., Korczak, B., and Green, N.M., (1985). Amino acid sequence of a $Ca^+$-ATPase from rabbit muscle sarcoplasmic reticulum deduced from its complementary DNA sequence. Nature 316: 696-700.

25.    Modyanov, N.N., Lutsenko, S., Chertova, E., Efremov, R., and Gulyaev, D. (1992). Trans-membrane Organization of the $Na^+$, $K^+$-ATPase molecule. Acta. Physiol. Scand. 146: 49-58.

26.    Modyanov, N.N., Vladimirova, N.M.,Gulyaev, D.I., and Efremov, R.G. (1992). Architecture of the sodium pump molecule. Vectorial labeling and computer modeling. Ann. N.Y. Acad. Sci. 671: 134-146.

27.    O'Connell, M.A. (1982). Exclusive labeling of the extracytoplasmic surface of the sodium ion and potassium ion activated adenosine triphosphatase and a determination of the distribution of surface area across the bilayer. Biochemistry 21: 5984-5991.

28.    Or, E., David, P., Shainskaya, A., Tal, D.M., and Karlish, S.J.D. (1993). Effects of Competitive Sodium-like Antagonists on $Na^+/K^+$-ATPase Suggest that Cation Occlusion from the Cytoplasmic Surface Occurs in Two Steps. J. Biol. Chem. 268: 16929-16937

29.    Ovchinnikov, Y.A., Arzamazova, N.M., Arystarkhova, E.A., Gevondyan, N.M., Aldanova, N.A., and Modyanov, N.N. (1987). Detailed structural analysis of exposed domains of membrane bound $Na^+$, $K^+$-ATPase. A model for trans-membrane arrangement. FEBS Lett. 217: 269-274.

30.    Pedemonte, C.H., and Kaplan, J.H. (1986). Carbodiimide Inactivation of $Na^+$, $K^+$-ATPase. A consequence of internal cross-linking and not carboxyl modification. J. Biol. Chem. 261: 3632-3639.

31.    Schwappach, B., Stürmer, W., Apell, H-J., and Karlish, S.J.D. (1993). Interaction between cytoplasmic sodium binding sites and cardiotonic steroids. This meeting.

32.    Shani-Sekkler, M., Goldshleger, R., Tal, D.M., and Karlish, S.J.D. (1988). Inactivation of $Rb^+$ and $Na^+$ Occlusion on $(Na^+, K^+)$-ATPase by modification of carboxyl groups. J. Biol. Chem. 263: 19331-19342.

33.    Shainskaya, A., and Karlish, S.J.D. (1993). Evidence that the Cation Occlusion Domain of $Na^+/K^+$-ATPase Consists of a Complex of Membrane-spanning Segments. Analysis of Limit Membrane-embedded Tryptic Fragments. J. Biol. In Press.

34.    Stürmer, W., and Apell, H-J. (1992) Fluorescence study of cardiac glycoside binding to the $Na^+/K^+$-pump. Ouabain binding is associated with movement of electrical charge. FEBS Lett. 300: 1-4.

35.    Thibault, D. (1993). The Carboxy Terminus of Sodium and Potassium Ion Transporting ATPase is Located on the Cytoplasmic Surface of the Membrane. Biochemistry 32: 2813-2821.

36.    Van Huysse, J.W., Jewell, J.A., and Lingrel, J.B. (1993). Site-directed Mutagenesis of a Predicted Cation binding site of $Na^+$, $K^+$-ATPase. Biochemistry 32: 819-826.

# The Stabilization of Cation Binding and its Relation to $Na^+/K^+$-ATPase Structure and Function

Jack H. Kaplan, José M. Argüello, Graham C.R. Ellis-Davies, Svetlana Lutsenko

Department of Physiology, University of Pennsylvania, School of Medicine, Philadelphia, Pennsylvania 19104-6085 USA

During the transport cycle of the sodium pump, monovalent cations are somehow smuggled through the low dielectric medium of the plasma membrane as $Na^+$ ions are expelled from the cell and $K^+$ ions taken up. These ions also have specific and well-studied effects on the partial catalytic reactions of the $Na^+,K^+$-ATPase activity that accompanies transport. These two aspects of the roles of $Na^+$ and $K^+$ ions in active transport form the subject of the present article. It is an implicit assumption in most studies that the $Na^+$ ions which activate phosphorylation from ATP are indeed the $Na^+$ ions transported and the extracellular $K^+$ ions which catalyze dephosphorylation are taken up by the cell. The central issues in these two aspects are which parts (amino acid residues) of the protein interact directly with the cation and how are the processes involved in ATP hydrolysis linked to transport. Obviously, this last feature also involves cation recognition by the protein but adds to this the question of "coupling" or "transduction". These are terms used to connote information transfer between different regions of the protein which are directly involved in cation binding, ATP binding, and phosphorylation.

## Stabilization of Cation Binding

Proteins which mediate the transport of ionic species across membranes provide pathways for these ions through an intrinsically unfavorable environment. The amino acid side chains which are in close proximity to the ions and coordinate directly with them, substitute for solvent molecules found in the more hospitable aqueous compartments. Little is presently known about which specific residues in the $Na^+$ pump protein are involved, however, some progress is being made.

Two plausible assumptions that have been made about the amino acids at cation binding and transport sites have obtained support from recent protein chemistry studies. These are (i) that the residues would reside in the intramembrane domains of the $\alpha$-subunit structure and (ii) that carboxyl-bearing amino acids (Asp or Glu) would be directly involved in coordination.

Evidence in support of this first assumption comes from recent studies which have shown that following extensive tryptic digestion of the purified protein in the presence of Rb ions, a so-called "19 kDa membrane" preparation is obtained. This preparation is still able to occlude Rb ions(14). The preparation when examined by SDS-PAGE contained a 19 kDa-22 kDa major component and Karlish et al. focused on this component as being specifically and essentially involved in cation occlusion. The extensive tryptic digestion results in the removal of most of the extra-membraneous segments of the α-subunit of the Na⁺ pump protein. Thus, the idea that the important elements of the cation binding sites were localized in the intramembrane domains was supported. It has also become clear that the organization of the cation occlusion site is far more complex than was initially apparent. Examination of the composition of the "19 kDa membranes", however, reveals (as shown in Fig. 1) that many components remain. When post-tryptic membranes obtained in the presence of Rb (normal occlusion levels) were compared with these obtained under phosphorylation conditions (20% occlusion recovered), the following structural differences were observed (Fig. 1); (i) the length of the M1-M2 domain was different; (ii) the extent of digestion of the β-subunit and the size of the N-terminal fragment of the β-subunit were different; and (iii) the 21 kDa fragment was digested under phosphorylation conditions(18).

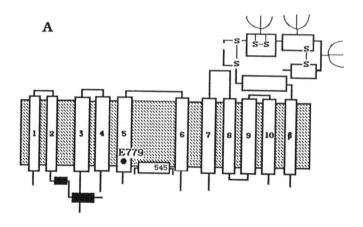

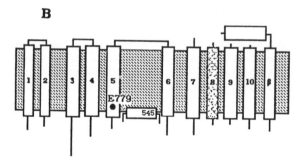

Fig. 1 Composition of post-tryptic membranes obtained in the presence of Rb(A) or after phosphorylation (B)

322

Thus, the conclusion that 19 kDa peptide is the essential element of structure for cation binding is an overstatement. It is not possible, as yet, to eliminate any of the components of this post-tryptic preparation from consideration as being essential or directly involved in cation occlusion. The notion that the presence of the occluded cation itself stabilized the post-tryptic residue also made it tempting to speculate that this supported the idea that cation occlusion specifically stabilizes the intramembrane structure. However, our recent finding that the same post-tryptic preparation is obtained after carrying out proteolysis with trypsin in the presence of ouabain also makes this speculation less plausible(18).

The second assumption that cation binding sites would involve carboxyl-bearing amino acids was attractive for several reasons. If positively charged ions are to cross biomembranes, then the positive charge might best be stabilized by charge neutralization with an anionic moiety on the protein itself. Furthermore, early studies on chemical modification (reviewed in Ref. 22) had shown that a carboxyl-selective reagent DCCD could inactivate the $Na^+,K^+$-ATPase and that such inactivation was prevented by the simultaneous presence of $K^+$ ions. Thus, the protection suggested that $K^+$ ions, by occuping their binding site protected the carboxyl group involved in complexation from DCCD modification.

There have been a series of subsequent studies using carbodiimides to attempt to confirm the importance of carboxyl-bearing amino acids and to identify the essential Asp or Glu residues at the cation binding sites. Much of this work has been reviewed previously(13). Studies in the laboratory of Dr. S.J.D. Karlish led to the identification of $Glu^{953}$ as an essential amino acid residue of the $\alpha$-subunit at the $K^+$ binding site of the sodium pump(10,15). However, studies from our laboratory suggested that cation-protectable inactivation by carbodiimides was a result of intramembrane cross-linking which rendered the identification of a specific carboxyl using these reagents highly questionable (21).

An alternative approach for carboxyl-group modification has emerged with the use of DEAC, 4-(diazomethyl)-7-(diethylamino)-coumarin), a modified fluorescent diazomethane analog which esterifies carboxyl moieties in proteins. We have shown that the reagent modifies only 1 or 2 carboxyls in the $\alpha$-subunit in a cation ($Na^+$ or $K^+$)-preventable fashion. The modified protein retains many of the attributes of native enzyme (ATP binding, $E_1 \rightleftharpoons E_2$ conformational flexibility) but cannot bind or occlude $Na^+$ or $K^+$ ions and is inactive as an ATPase(1). Recently we have described the identification of the site of modification by DEAC which proved to be $Glu^{779}$, a residue in the fifth transmembrane segment(2). $Glu^{779}$ is conserved in all reported primary structures of $Na^+$ pump $\alpha$-subunit and the equivalent residue is also conserved in the sarcoplasmic reticulum Ca pump and gastric proton pump (see Table I).

Table 1. Comparison of the sequences surrounding Glu[779] and the corresponding residue in three different P-type ATPases. Taken from Ref. 2

| ATPase | ISOFORM and SPECIES | SEQUENCE | | |
|--------|---------------------|----------|---|---|
| Na, K-ATPase | sheep α1 | L[765] KKSIAYTLT | SNIPEITPFL | IFIIANIPLP |
| | rat α2 | L[764] KKSIAYTLT | SNIPEITPFL | LFIIANIPLP |
| | rat α3 | L[762] KKSIAYTLT | SNIPEITPFL | LFIMANIPLP |
| | Drosophila | L[797] KKSIAYTLT | SNIPEISPFL | ASILCDIPLP |
| | Artemia | L[747] KKSIAYTLT | ANIPEISPFL | MYILFDLPLA |
| | Hydra | L[776] KKSIVYTLT | SNIPEISPFL | MFILFGIPLP |
| H, K-ATPase | rat | L[782] KKSIAYTLT | KNIPELTPYL | IYITVSVPLP |
| | pig | L[781] KKSIAYTLT | KNIPELTPYL | IYITVSVPLP |
| Ca-ATPase | rabbit SERCA1 | M[756]KQFIRYLIS | SNVGEVVCIF | LTAALGLPEA |
| | rabbit SERCA2 | M[757]KQFIRYLIS | SNVGEVVCIF | LTAALGFPEA |
| | rat SERCA3 | M[757]KQFIRYLIS | SNVGEVVCIF | LTAILGLPEA |
| | Drosophila | M[757]KQFIRYLIS | SNIGEVVSIF | LTAALGLPEA |

The identification of Glu[779] as an essential part of the Na$^+$ and K$^+$ binding site of the Na pump has several interesting consequences. One is that the result focusses attention on transmembrane segments which are not part of the C-terminal 19 kDa fragment as components of the cation complexation site. Another is that since both Na$^+$ ion occlusion (facilitated by oligomycin) and K$^+$ ion occlusion are lost, that Glu[779] is involved in complexation of both cations. This supports the parsimonious hypothesis that the sites which coordinate Na$^+$ ions on the exit pathway via the pump reorganize to stablilize K$^+$ binding on its entry pathway into the cell. This result also has consequences for the mechanism of coupling or transduction (discussed below). Mutagenesis studies of rat α-subunit transfected into HeLa cells have confirmed the prediction that Glu[779] is essential for activity. Mutation of Glu[779] renders the pump inoperative(12). Interestingly similar mutations of Glu[953] (a residue not conserved in all Na pump α-subunits) produce a functional pump showing that Glu[953] is not essential(25). Mutagenesis studies on the sarcoplasmic reticulum Ca$^{2+}$-ATPase by MacLennan and collaborators have also suggested that several intramembrane carboxyl residues in this ion pump are also important for cation binding functions. The residue which is equivalent to Glu[779] is essential for Ca$^{2+}$ binding functions(5,6).

In summary, chemical modification studies and site-directed mutagenesis have begun to identify amino acid residues which are likely candidates for the coordinating ligands directly involved in complexation of Na$^+$ or K$^+$ ions. The current barriers to further rapid advances exist because of severe limitations in the quantity of expressed functional Na$^+$ pump protein. Further advances from the chemical modification approach can be expected but ultimately rely on the availability of new reagents and rigorous applications of functional assays. Both approaches, however, suffer from the same limitations in interpretation. We are attempting to make structure-function correlates in a macromolecule where our structural knowledge is rudimentary

It is often instructive to examine other proteins which have evolved for similar functions or with similar structures. This comparative approach has been very fruitful in the modeling studies of Green and his colleagues on P-type ATPases and their nucleotide binding domains(11,23). However, few relevant clues exist for the monovalent cation binding sites. Recently, an interesting study appeared on a dialkylglycine decarboxylase(24). This protein is bifunctional and carries out decarboxylation and transamination reactions. It is of particular interest to us because it normally binds either 1 $K^+$ or 1 $Na^+$ ion at site 1 and a $Na^+$ ion at a second site. Occupation of site 1 by $Na^+$ leads to inhibition and by $K^+$ leads to activation. The crystal structure of this protein has been solved and the closest ligands at site 1 identified. This site involves complexation by an Asp residue. There are profound changes in Ser and Tyr residues when $Na^+$ replaces $K^+$ at this site and a conformational change in this region is evident. This conformational change is transmitted to the catalytic center where substrate binds. The second $Na^+$ binding site is more reminiscent of alkali metal ion binding to natural ionophores and synthetic macrocyclic ligands where charge compensation is probably effected by the helix macrodipole. This example of a protein utilizing two such different modes of cation complexation is particularly instructive. Perhaps a close examination of site 1 geometry and model building based on similar distances among candidate intramembrane residues of the P-type ATPases might be a promising starting point. The observation that the same residues are closest to $Na^+$ or $K^+$ ions at the site 1 when they are complexed but lead to either active or inactive enzyme is reminiscent of effects of $Na^+$ or $K^+$ on ATP binding in the sodium pump. Our results with DEAC show that Na and K occlusion by the sodium pump have at least one residue in common, namely Glu[779].

### The Involvement of the β-subunit

The heavily glycosylated β-subunit of the Na pump until recently has not been thought to play a major role in the catalytic or transport cycles of the enzyme. Many studies had suggested that the important biological role for this subunit was in the correct insertion and/or assembly with the α-subunit in the membrane. Recently, the role of the β-subunit in membrane insertion has been questioned. In recent studies, we have provided evidence for an essential role for the β-subunit and for its extracellular domain, in particular, in stabilizing the occluded cation intermediates which are essential for transport and activity(19,20). The studies came about as extensions of earlier work which had observed that high concentrations of DTT or mercaptoethanol resulted in inactivation of the enzyme. Such inactivation could be prevented by the presence of high concentrations of $Na^+$ or $K^+$ ions(17,16). We wondered if this protection was specific and whether or not it was indicative of an important role for S-S bridges in maintaining appropriate cation-protein interactions. Initial studies indicated that extensive reduction of S-S bridges in the β-subunit was accompanied by a loss of enzyme activity. The loss of activity was prevented by $K^+$ or $Na^+$ ions (but not by tris or choline)(19). The phenomenon is complicated by the simultaneous presence of some heat-inactivation but by using appropriate controls it is clear that the major cation-protectable effect is due to S-S bridge reduction. The appearance of extra-SH residues after reduction was monitored using a coumarin-maleimide which is highly fluorescent and specific for sulfhydryls (Fig. 2). The increase in sulfhydryl content and loss of activity were both prevented by the presence of monovalent cations which are occluded by the pump. Furthermore, the loss of activity was the result of a loss of the ability of the enzyme to occlude the cations. While this work was in progress, similar studies were reported on the gastric $H^+,K^+$-ATPase. Here, although, occlusion couldn't be directly measured similar cation-preventable reduction of the β-subunit and associated loss of activity was reported(4).

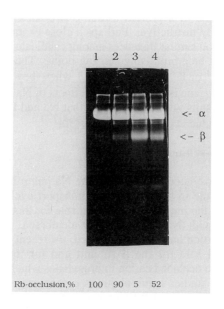

Fig. 2  Correlation between reduction of the β-subunit (monitored using fluorescent SH-group directed coumarin-maleimide) and loss of Rb-occlusion. (Taken from Ref. 19 with permission)

In our studies, attention  was then focussed on the post-tryptic "19 kDa membranes" described above.  These preparations are greatly reduced in structural complexity compared with the native intact enzyme.  Earlier reports had concluded, from work with these preparations, that the β-subunit was not involved in stabilizing occlusion(3).  However, the conclusions had been based on occlusion measurements of proteolyzed preparations and comparisons with the profile of peptides in SDS-PAGE after running gels under reducing conditions (i.e. after treatment with DTT).  This comparison revealed that in preparations that had been treated with pronase, occlusion was maintained but the β-subunit was apparently cleaved.  Our studies demonstrated that DTT reduction of the "19 kDa membrane" preparation resulted in a loss of occlusion and greater fluorescent labeling (i.e. an increase in free sulfhydyls) of the β-subunit.  Repeating the studies with pronase revealed that, if, following pronase digestion, DTT reduction was performed then these membranes also lost the ability to occlude cations(20).  Furthermore, evidence of cleavage of the β-subunit was not observed if the gels were run in non-reducing conditions.

326

Our interpretation of these phenomena is that reduction of the S-S bridges in the extracellular domain of the β-subunit results in a loss of structural integrity and the necessary involvement of the β-subunit in stabilizing the occluded cation state is lost. In the absence of DTT, although pronase does clip the extracellular domain of the β-subunit, the three extracellular S-S bridges hold the cleaved peptides in the natural conformation and function (occlusion) via correct interaction with the α-subunit is maintained. It has not been possible, by selective reduction, to determine which of the three S-S bridges are essential. A variety of reducing agents have been tested but until now high concentrations or extreme conditions are required for reduction and inactivation.

A corrolary of the involvement of the β-subunit in cation occlusion has recently been seen in studies of extensive proteolysis of the enzyme(18). The idea was to see whether or not the binding of particular ligands of the sodium pump which produce different enzyme conformations result in altered digestion patterns after extensive degradation. Since such studies on purified enzyme inevitably lead to initial extensive digestion of the α-subunit, effects on the β-subunit would be difficult to interpret. Right-side-out vesicles from canine kidney medulla were employed; since little of the α-subunit is exposed to the extracellular surface and the enzyme is insensitive to external trypsin(9). These studies showed that when the protein is incubated with either $Rb^+$ or $MgP_i$ ouabain, producing either $E_2(Rb)$ or $E_2P$ ouabain forms, different cleavage sites are produced in the extracellular domain of the β-subunit. These studies show that the well known conformational changes occurring in the α-subunit during the transport and catalytic cycles affect the conformation of the extracellular domain of the β-subunit, presumably close to contact points between the subunits. It is interesting that the cleavage points on the β-subunit, for $E_2(Rb^+)$ and for $E_2P$ ouabain are contained within the first extracellular S-S bridge, a region identified as being important for correct α- and β-subunits association from other studies.

**Proteins Rearrangements and the Mechanism of Enzyme Transduction**

It is extremely likely that distinct regions (domains) of the protein perform distinct functional roles in the ATPase reaction cycle. A clear example of evidence which supports this notion comes from chemical modification studies which show that modification of a single residue (say $Lys^{501}$) can effectively remove high affinity ATP binding without greatly affecting cation occlusion. One such reagent which does this is NIPI (N-(2-nitro-4-isothiocyanophenyl)-imidazole)(7). The opposite case is seen say with DEAC, which as discussed above, modifies $Glu^{779}$, removing the ability to occlude monovalent cations but leaving the nucleotide binding domain intact(1). However, several pieces of well established evidence suggest that these domains are not completely independent. These include: the observation that $Na^+$ supports phosphorylation from ATP while $K^+$ does not, that $K^+$ binding to the protein eliminates high affinity ATP binding, etc. The way in which this kind of information (occupancy of one domain) is transmitted to another domain of the protein is an essential element of the mechanism of enzyme transduction. In other words, how the conformational changes occurring when $Na^+$ or $K^+$ bind and are transported

327

across the membrane are coordinated with ATP binding and β-γ phosphate bond cleavage. Chemical studies are beginning to provide some clues to such a description.

Lys$^{501}$ is a residue which appears to be particularly reactive towards aryl isothiocyanates including FITC, NIPI, and SITS. All of these reagents inactivate the enzyme in an ATP-preventable fashion. Interestingly, when Na$^+$ ions are bound to the enzyme, reaction of Lys$^{501}$ occurs about ten times more readily than when K$^+$ ions are bound(8). Thus, in the E$_1$Na form (associated with high affinity ATP binding) Lys$^{501}$ is more readily accessible or reactive to NIPI than in E$_2$K when K$^+$ ions are bound. This change in reactivity towards NIPI mirrors changes in ATP binding site affinity. Since Glu$^{779}$ is in the 5th transmembrane segment, a direct coupling mechanism seems plausible. The ATP binding domain is contained within the large cytoplasmic loop between transmembrane segments 4 and 5 (see Ref. 13 for discussion). Thus, spatial reorganization in the vicinity of Glu$^{779}$ likely occurs as either Na$^+$ or K$^+$ are complexed by this carboxyl-bearing residue. These changes are immediately transferred to the ATP binding site (and Lys$^{501}$) via the direct connection between this region and transmembrane segment 5 (see Fig. 3).

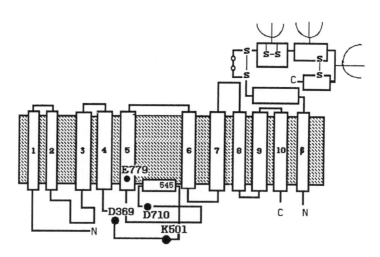

Fig. 3 Two-dimensional model of spatial organization of Na-pump molecule

Are any other segments of the protein involved in the spatial rearrangements which occur when either Na$^+$ or K$^+$ ions bind or when ATP binds to and phosphorylates the enzyme? Analysis of extensive proteolytic digestion with trypsin carried out in the presence of various pump ligands is beginning to extend our knowledge of the changes in intra-protein interactions which occur in various conformational states(18). For example, evidence has been obtained for interactions between the cytoplasmic loop between M2 and M3 and the phosphorylation domain of the $\alpha$-subunit. The binding of $Na^+$ or $K^+$ ions to their sites results in a tighter interaction between these cytoplasmic loop regions, close to the membrane surface. Phosphorylation of the protein causes a loosening of this interaction as different cleavage sites are exposed. From an analysis of the change in susceptibility of different regions of the $\alpha$-subunit to extensive digestion, we have been able to identify likely intra-protein interactions which alter as the protein undergoes conformational changes associated with different steps in the catalytic cycle.

**Conclusions**

Chemical approaches to the questions of intra-protein interactions, inter-subunit interactions, and localization of ligand binding domains continue to contribute to our understanding of the sodium pump protein. Although structural knowledge of this membrane protein is in its infancy, the combination of rigorous functional assays with precisely monitored modifications promises to provide more insight into how the sodium pump achieves its complex active transport function.

**Acknowledgment** The work described in this article was supported by NIH Grants GM39500 and HL30315

**References**

1.    Argüello, J.M., Kaplan, J.H. (1991) Evidence for essential carboxyls in the cation binding domain of the $Na^+,K^+$-ATPase. J Biol Chem, 266:14627-14635.

2.    Argüello, J.M., Kaplan, J.H. (1993) Glutamate 779, an intramembrane carboxyl, is essential for monovalent cation binding by the Na,K-ATPase. J Biol Chem, in press.

3.    Capasso, J.M., Hoving, S., Tal, D.M., Goldshleger, R., Karlish, S.J.D. (1992) Extensive digestion of $Na^+,K^+$-ATPase by specific and non-specific proteases with preservation of cation occlusion sites. J Biol Chem, 267:1150-1158.

4.    Chow, D.L., Browning, C.M., Forte, J.G. (1992) Gastric $H^+,K^+$-ATPase activity is inhibited by reduction of disulfide bands in β-subunit. Am J Physiol, 263:C39-C44.

5.    Clarke, D.M., Loo, T.W., MacLennan, D.H. (1990) Functional consequences of alterations to polar amino acids located in the transmembrane domain of the $Ca^{2+}$-ATPase of sarcoplasmic reticulum. J Biol Chem 265, 6262-6267.

6.    Clarke, D.M., Loo, T.W., Inesi, G., MacLennan, D.H. (1989) Location of high affinity $Ca^{2+}$-binding sides within the predicted transmembrane domain of the sarcoplasmic reticulum $Ca^{2+}$-ATPase. Nature 339,476-478.

7.    Ellis-Davies, G.C.R., Kaplan, J.H. (1990) Binding of $Na^+$ ions to the $Na^+,K^+$-ATPase increases the reactivity of an essential residue in the ATP binding domain. J Biol Chem, 265:20570-20576.

8.    Ellis-Davies, G.C.R., Kaplan, J.H. (1993) Modification of lysine-501 in the $Na^+,K^+$-ATPase reveals coupling between cation occupancy and changes in the ATP binding domain. J Biol Chem, 268:11622-11627

9.    Forbush, B. III (1982) Characterization of right-side-out membrane vesicles rich in (Na,K)-ATPase and isolated from dog kidney outer medulla. J Biol Chem, 257:12678-12684.

10.   Goldshleger, R., Tal, D.M., Moorman, J., Stein, W.D., Karlish, S.J.D. (1992) Chemical modification of Glu-953 of the a-chain of $Na^+,K^+$-ATPase associated with inactivation of cation occlusion. Proc Natl Acad Ssi, 89:6911-6915.

11.   Green, N.M., Stokes, D.L. (1992) Structural modeling of P-type ion pumps. Acta Physiol Scand, 146:59-68

12.   Jewell-Motz, E.A., Lingrel, J.B. (1993) Site-directed mutagenesis of the Na,K-ATPase: consequences of substitutions to negatively-charged amino acids localized in the transmembrane domains. *Biochemistry* in press.

13.   Kaplan, J.H. (1991) Localization of ligand binding from studies of chemical modification. In Kaplan, J.H., De Weer, P. (Eds.) The Sodium Pump: Structure, Mechanism and Regulation Rockefeller Univ. Press, NY pp 117-128.

14.   Karlish S.J.D., Goldshleger, R., Stein. W.D. (1990) A 19 kDa C-terminal fragment of the α-chain of Na,K-ATPase is essential for occlusion and transport of cations. Proc Natl Acad Ssi, 87:4566-4570.

330

15.     Karlish, S.J.D.,Goldshleger, R., Tal, D.M., Capasso, J.M., Hoving, S., Stein, W.D. (1992) Identification of the cation binding domain of the Na/K-ATPase. Acta Physiol Scand, 146:69-76.

16.     Kawamura, M., Nagano, K. (1984) Evidence for essential disulfide bonds in the β-subunit of $(Na^+,K^+)$-ATPase. Biochim Biophys Acta, 774:188-192.

17.     Kirley, T.L. (1990) Inactivation of $(Na^+,K^+)$-ATPase by β-neocaptoethanol. Differential sensitivity to reduction of the three β-sunitunit disulfide bonds. J Biol Chem, 265:4207-4232.

18.     Lutsekno, S., Kaplan, J.H. (1993) Molecular events in close proximity to the membrane associated with the binding of ligands to the Na,K-ATPase. J Biol Chem, in press

19.     Lutsenko, S., Kaplan, J.H. (1992) Evidence of a role for the $Na^+,K^+$-ATPase β-subunit in active cation transport. Ann NY Acad Sci, 671:147-155.

20.     Lutsenko, S., Kaplan, J.H. (1993) An essential role of the extra-cellular domain of the $Na^+,K^+$-ATPase β-subunit in ion-occlusion. Biochemistry, 32:6737-6743.

21.     Pedemonte, C.H., Kaplan, J.H. (1986) Carbodiimide inactivation of $Na^+,K^+$-ATPase. A consequence of internal cross-linking and not carobxyl group modification. J Biol Chem, 261:3632-3639.

22.     Pedemonte, C.H., Kaplan. J.H. (1990) Chemical modification as an approach to the elucidation of sodium pump structure-function relations. Am J Physiol, 258 (Cell Physiol) C1-C23.

23.     Taylor, W.R., Green, N.M. (1989) The predicted secondary structure of the nucleotide-binding sites of six cation-transporting ATPases lead to a probable tertiary fold. Eur J Biochem, 179:241-248.

24.     Toney, M.D., Hohenester, E., Cowan, S.W., Jansonius, J.N. (1993) Dialkylglycine decarboxylase structure: bifunctional active site and alkali metal sites. Science, 261:756-759.

25.     Van Huysse, J.W., Jewell, E.A., Lingrel, J.B. (1993) Site-directed mutagenesis of a predicted cation binding site of Na,K-ATPase. Biochemistry, 32-819-826.

# Is the Sodium Pump a Functional Dimer ?

W. Schoner, D. Thönges, E. Hamer, R. Antolovic, E. Buxbaum, M. Willeke, E.H. Serpersu, G. Scheiner-Bobis

Institute of Biochemistry & Endocrinology, Justus-Liebig-University, Frankfurter Str. 100, D-35392 Giessen, Germany

## Introduction

It is presently unclear whether the membrane-embedded sodium pump works as an $(\alpha\beta)$ monomer (4) according to the single site model (15,32) or as an $(\alpha\beta)_2$ diprotomer of interacting $\alpha$ subunits (21-23,33). The single site model includes as an essential step the conversion of the low affinity ATP binding site ($E_2ATP$ site) to the high affinity ATP binding site ($E_1ATP$ site). This model implies that high and low affinity ATP binding sites cannot exist simultaneously, since they are two different forms of the same ATP binding site that can exist either in the $E_1ATP$ or the $E_2ATP$ conformational state. There are a number of data which are difficult to reconcile with the idea that the membrane-embedded sodium pump works according to the single site model as an $(\alpha\beta)$ monomer: (a) a negative cooperativity in ATP hydrolysis is seen (3,24) which may indicate an interaction of ATP binding sites; this is also seen under conditions when the catalytically active $(\alpha\beta)$ protomer of the sodium pump is existent as shown by active-enzyme analytical centrifugation of detergent solubilized $Na^+/K^+$-ATPase (37). (b) radiation inactivation reveals a bigger target size for the overall reaction than for partial reactions of the pump (7,16,17); (c) the enzyme shows the phenomenon of superphosphorylation at high concentrations of ATP (20); (d) simultaneous binding of TNP-ADP and phosphate to the FITC-modified enzyme have been recorded (27) (FITC binds to the high affinity ATP binding site in the $E_1ATP$ state); (e) Förster energy transfer measurements between the FITC-site and an ATP-protectable erythrosin isothiocyanate-site reveals that both sites are 5.5 nm apart (2); (f) Consistent with the idea that the onset of turnover of the sodium pump might cause $(\alpha\beta)$protomers to form $(\alpha\beta)_2$ diprotomers is the observation that various concentrations of ATP, protein and phosphatidylserine affect the $(\alpha\beta)$ protomer $\rightleftharpoons$ $(\alpha\beta)_2$ diprotomer equilibrium (14,10).

Whether pumping of $Na^+$ and $K^+$ ions through the cell membrane is accomplished by an $(\alpha\beta)$ monomer or by an $(\alpha\beta)_2$ diprotomer of the sodium pump can be clarified with ATP analogs which specifically modify exclusively either the $E_1ATP$ site or the $E_2ATP$ site. Such ATP analogs should affect the overall reaction of a sodium pump working according to the protomeric model to the same extent. We therefore synthesized such ATP analogs which inactivate the $Na^+/K^+$-ATPase through their specific interaction with either the high affinity or the low affinity ATP binding site, and investigated the partial reactions of the inactivated enzymes.

## Are $E_1ATP$ and $E_2ATP$ sites in equilibrium or do they coexist?

Substitution-inert metal complexes of ATP are known to yield tight complexes with glycolytic enzymes leading thereby to an inactive enzyme (8,31). This principle is also valid for P-type transport ATPases (5,6,9,19,25,26,29,30,35,36). $Cr(H_2O)_4ATP$ inactivates $Na^+/K^+$-ATPase in a slow reaction by formation of a stable phosphointermediate (18,19). This reaction is activated by $Na^+$ and $Mg^{2+}$ but hindered by the presence of $K^+$ (19). Since micromolar concentrations of ATP protect $Na^+/K^+$-ATPase from inactivation by $Cr(H_2O)_4ATP$, one should assume that the (ATP-analog·enzyme) complex is formed within the $E_1ATP$ site (18,19). In agreement with this is also the observation that $Na^+$ ions can be occluded by the $Cr(H_2O)_4ATP$-inactivated enzyme (35). $Cr(H_2O)_4ATP$ also inactivates $Ca^{2+}$-ATPase from sarcoplasmic reticulum as well and leads to the stable occlusion of $Ca^{2+}$ into the membrane enzyme (29,36).

**Table:** Apparent affinities of high and low affinity ATP binding sites for MgATP and $MgPO_4$ complex analogs

The data are compiled from references (5,6,9,18,19,25,26,29,30)

| Substrate | High Affinity Site (E₁ATP) $K_d$ [$\mu$M] | Substrate | Low Affinity Site (E₂ATP) $K_d$ [$\mu$M] |
|---|---|---|---|
| $Cr(NH_3)_4ATP$ | 0.62 | $Cr(NH_3)_4ATP$ | 500 |
| $Cr(NH_3)_3(H_2O)ATP$ | 1.71 | $Co(NH_3)4\text{-}8\text{-}N_3\text{-}ATP$ | 210 |
| $Cr(NH_3)_2(H_2O)_2ATP$ | 1.12 | $Co(HN_3)_4ITP$ | 1400 |
| $Cr(H_2O)_4ATP$ | 8 | | |
| $Cr(H_2O)_4AMP\text{-}PCP$ | 26 | $Co(NH_3)_4PO_4$ | 300 |
| | | Fluorescent ATP Analogs: | |
| $Rh(H_2O)_nATP$ | 1.8 | $Co(NH_3)_4\text{-}\epsilon\text{-etheno-ATP}$ | 1250 |
| $Co(NH_3)_4ATP$ | 0.25 not inactivating) | $Co(NH_3)_4MANT\text{-}ATP$ | 19 |

All metal complexes of ATP analogs studied inactivate $Na^+/K^+$-ATPase (Table), but they differ from each other with respect to their mode of action. Chromium complexes of ATP inactivate the enzyme by their interaction with the high affinity ATP site ($E_1ATP$ site), as does $Rh(H_2O)_nATP$ (30). However, $Co(NH_3)_4ATP$ and the $MgPO_4$ complex analog $Co(NH_3)_4PO_4$ inactivate the sodium pump by binding to the $E_2ATP$ site, since the slow inactivation of the enzyme by these substances is only hindered at high concentrations of ATP (5,6,25,26). It is observed, however, that low concentrations of $Co(NH_3)_4ATP$ bind to the $E_1ATP$ site without inactivating the enzyme (25). It

333

is presently unclear which mechanisms are responsible for the interaction of the $E_2ATP$ site exclusively with the cobalt but not with the chromium or rhodium complexes of ATP.

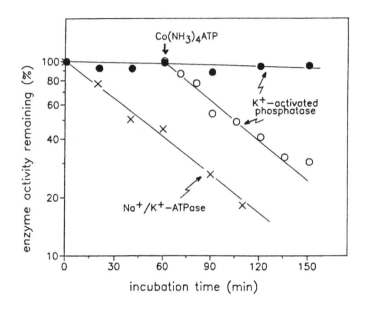

**Figure 1.** Effect of 200 μM $Cr(H_2O)_4$AMP-PCP on the activities of $Na^+/K^+$-ATPase (x) and $K^+$-activated p-nitrophenylphosphatase (●). After 60 min 1 mM $Co(NH_3)_4$ATP was added (o) and both enzyme activities were recorded (from ref. (9)).

Adenosine 5'(ß,γ-methylene) triphosphate, AMP-PCP does not transfer its terminal phosphate group to aspartyl residue 369 to yield a phosphointermediate. In the absence of $Na^+$ and $K^+$ the ATP analog $Cr(H_2O)_4$AMP-PCP causes a slow inactivation of the sodium pump (9). The $K^+$-activated p-nitrophenylphosphatase, a partial reaction associated with the $E_2$ conformation, however, remains unchanged (Fig. 1). The finding is in agreement with the observation that modification of the $E_1ATP$ site by FITC affects the overall $Na^+/K^+$-ATPase reaction but not the phosphatase reaction (13). The $Cr(H_2O)_4$AMP-PCP-modified enzyme has an unaltered capacity to occlude $^{86}Rb^+$ and to form the ouabain enzyme complex II in the presence of $Mg^{2+}$ + inorganic phosphate, but its ability to form the ouabain enzyme complex I in the presence of ATP, $Mg^{2+}$ + $Na^+$ has been lost (9).

This finding might be taken as an indication for the simultaneous presence of high and low affinity ATP binding sites. The experiment, however, cannot exclude the possibility that a rather small molecule like p-nitrophenyl phosphate can still bind to and become hydrolyzed at the same ATP binding site that is already occupied by

334

$Cr(H_2O)_4AMP$-PCP. The $p$-nitrophenylphosphatase activity, however, is lost upon addition of $Co(NH_3)_4ATP$, which is known to inactivate the sodium pump by its binding to the $E_2ATP$ site (Fig. 1) (25,26). Therefore these data favor the idea of coexisting $E_1ATP$ and $E_2ATP$ sites and are inconsistent with the single site model as the mechanism for active transport through the plasma membrane.

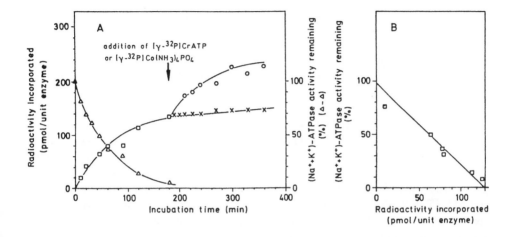

**Figure 2:** $Na^+/K^+$-ATPase was inactivated ($\triangle$) and labeled ($\square$) by ($\gamma$-$^{32}$P)$Co(NH_3)_4ATP$. Addition of ($\gamma$-$^{32}$P)$Cr(H_2O)_4ATP$ after 200 minutes results in excess phosphorylation of the already inactive enzyme (o). Additional phosphorylation is not obtained when $Co(NH_3)_4{}^{32}PO_4$ is used (x) (from ref. (6))

Provided the idea that $Na^+/K^+$-ATPase works as an $(\alpha\beta)_2$ diprotomer of interacting catalytic $\alpha$ subunits is correct, it should be possible to demonstrate the coexistence of $E_1ATP$ and $E_2ATP$ binding sites not only in the $E_1ATP$-modified enzyme but also in the $E_2ATP$-modified enzyme. Inactivation of $Na^+/K^+$-ATPase by $Co(NH_3)_4ATP$ at the $E_2ATP$ site leads to an enzyme whose $Na^+$ dependent phosphorylation from ATP and ADP-ATP exchange reactions are preserved (25,26). These reactions are partial activities of the $E_1ATP$ site. Similar findings are also obtained when the $E_2ATP$ site is modified with the stable $MgPO_4$ complex analog $Co(NH_3)_4PO_4$ (5,6). This latter analog seems to arrest $Na^+/K^+$-ATPase in a stable $E_2 \cdot Co(NH_3)_4PO_4$ form (5,6). When radioactive $Co(NH_3)_4ATP$ is used to inactivate $Na^+/K^+$-ATPase, 130 picomoles of the ATP analog were incorporated per unit of enzyme to achieve complete inactivation (Fig. 2B). Subsequent addition of radioactive $Cr(H_2O)_4ATP$ to the $Co(NH_3)_4ATP$-inactivated enzyme results in an excess incorporation of an equivalent amount of $Cr(H_2O)_4ATP$ into the enzyme (Fig. 2A). These results are independent of the purity of the enzyme preparation used (6). Similar observations are made when the $Co(NH_3)_4PO_4$-inactivated enzyme is exposed to radioactive $Cr(H_2O)_4ATP$ (data not

335

shown). In summary, the incorporation of $Co(NH_3)_4ATP$ into the $Cr(H_2O)_4AMP$-PCP-inactivated enzyme (Fig. 1) and the incorporation of $Cr(H_2O)_4ATP$ into the $Co(NH_3)_4ATP$-inactivated enzyme (Fig. 2) both lead to the conclusion that $E_1ATP$ and $E_2ATP$ sites coexist and that they are not easily interconvertible. The simultaneous binding of TNP-ADP and phosphate to the FITC modified enzyme are consistent with these results (27).

### Do $E_1ATP$ and $E_2ATP$ sites cooperate during catalysis?

As demonstrated by eosin fluorescence, modification of the $E_2ATP$ site by $Co(NH_3)_4ATP$ or $Co(NH_3)_4PO_4$ shifts the $E_1ATP$ site to a $Na^+$ form (5,26). Additionally it is well established that $Na^+/K^+$-ATPase shows negative cooperativity towards the substrate ATP (3,24) and that ATP binds with much higher affinity to the $E_1ATP$ site than to the $E_2ATP$ site.

The fluorescent 3'-O-methylanthranyloyl-ATP (MANT-ATP) binds, however, with higher affinity to the $E_2ATP$ than to the $E_1ATP$ site. The 8-azido derivative of MANT-ATP (MANT-8-$N_3$-ATP) binds almost with equal affinity to both binding sites (34). Both ATP analogs are hydrolyzed by $Na^+/K^+$-ATPase. In contrast with the negative cooperativity obtained with ATP, MANT-ATP displays a positive cooperativity and MANT-8-$N_3$ATP shows Michaelis-Menten kinetics (Fig. 3). We suggest that the conformation of the ribose of the ATP and the ATP analogs is important for the ATP site selectivity and for the cooperativity of the $E_1ATP$ and $E_2ATP$ sites during catalysis (34).

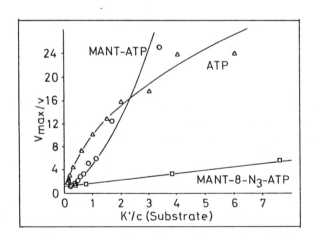

**Figure 3:** Lineweaver Burk plot of the hydrolysis of either ATP (△), MANT-ATP (o), or MANT-8-$N_3$-ATP (□) by $Na^+/K^+$-ATPase. Ordinate and abscissa were normalized. $Na^+/K^+$-ATPase activity was determined by the coupled optical test (28)

## Are $E_1$ATP and $E_2$ATP sites on the same or on different $\alpha$ subunits?

Tryptic digestion of $Na^+/K^+$-ATPase in the presence of $Na^+$ yields a 80 kDa carboxyterminal peptide of the $\alpha$ subunit (11,12). Trypsinolysis in the presence of $K^+$ results in the formation of an aminoterminal 40 kDa and a carboxyterminal 60 kDa fragment (11,12) (Fig. 4). $Na^+$ induces the $E_1$-conformation and $K^+$ the $E_2$-conformation (11,12), and it is thought that the tryptic digestion pattern obtained in the presence of $Na^+$ corresponds to the $E_1$ conformational state, while the one obtained in the presence of $K^+$ represents the $E_2$ conformational state of the enzyme (11,12). Mixed $E_1/E_2$ tryptic digestion patterns in the presence of $Na^+$ and $K^+$ have not been described.

Electrophoresis and immunostaining of tryptic peptides derived from $Na^+/K^+$-ATPase previously inactivated by either $Cr(H_2O)_4$ATP or $Co(NH_3)_4$ATP results in peptide patterns indicating the simultaneous presence of $E_1$ and $E_2$ conformational states. These tryptic digestions, shown in figure 4, were performed in the absence of $Na^+$ and $K^+$ ions. Trypsinolysis of the $Co(NH_3)_4$ATP-inactivated enzyme yields primarily a 80 kDa fragment corresponding to the $Na^+$-form and a minor amount of 40 kDa and 60 kDa peptides corresponding to the $K^+$-form of the enzyme. These latter fragments become rapidly degraded to smaller peptides. An $E_2$ATP-directed ATP analog would be expected to induce a $K^+$ conformational state of the enzyme, which would result in the formation of a 60 kDa and a 40 kDa peptide upon trypsinolysis. The simultaneous presence of the 80 kDa fragment along with the 60 kDa and 40 kDa peptides, and therefore of $E_1$ and $E_2$ forms, is not in accordance with an enzyme working as a protomer. This finding can easily be explained by a diprotomeric enzyme.

Trypsinolysis of a $Cr(H_2O)_4$ATP-inactivated enzyme reveals a rapidly formed 60 kDa carboxyterminal fragment like the one seen in tryptic digestions in the presence of $K^+$. According to the single site model, one would have expected to see the formation of a carboxyterminal 80 kDa peptide of the $E_1$ conformation for a $E_1$ATP site-modified enzyme. Prolonged tryptic digestion, however, leads to the formation of a 80 kDa peptide. Since the 80 kDa carboxyterminal peptide ($Na^+$ form) cannot be formed from a 60 kDa carboxyterminal peptide ($K^+$ form) one has to conclude that the 80 kDa peptide is formed from the undigested $\alpha$ subunit (100 kDa). Apparently, the $Cr(H_2O)_4$ATP-inactivated $Na^+/K^+$-ATPase exists in two different conformations whose cleavage site for the 60 kDa carboxyterminal peptide ($K^+E_2$ form) is more susceptible for trypsin than the one for the 80 kDa carboxyterminal peptide ($Na^+E_1$ form). These findings are thus consistent with the possibility that the $Cr(H_2O)_4$ATP-inactivated $Na^+/K^+$-ATPase contains within an $(\alpha\beta)_2$ dipromer one $\alpha$ subunit in the $E_1Cr(H_2O)_4$ATP form and the other one in the $E_2$ form. If this is correct, the tryptic digestion patterns in the presence of $Na^+$ and $K^+$ must then indicate the existence of a $Na^+E_1/Na^+E_1$ dimer or a $K^+E_2/K^+E_2$ dimer. Binding of ATP to those homogeneous enzyme forms would then induce a conformational $E_1/E_2$ heterogeneity.

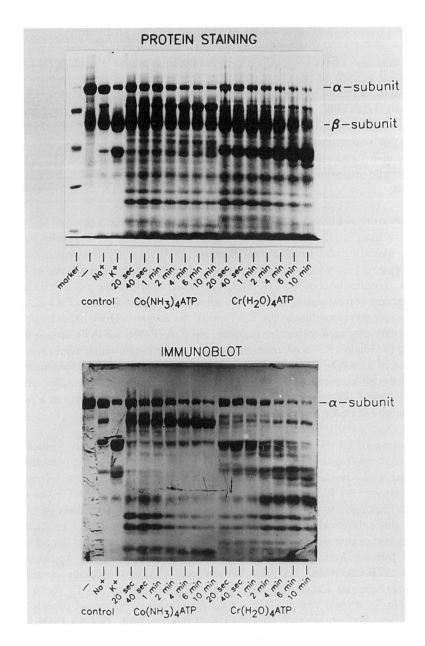

**Figure 4.** Kinetics of the formation of tryptic peptides from $Na^+/K^+$-ATPase of pig kidney. The enzyme was completely inactivated prior to trypsinolysis with either $Co(NH_3)_4ATP$ or $Cr(H_2O)_4ATP$. Tryptic digestion was performed in the absence of $Na^+$ and $K^+$. Protein staining was performed with coomassie blue (upper half). Detection of cleavage products of the catalytic $a$ subunit (lower half) was performed with a polyclonal antibody raised against the carboxy terminal amino acid sequence 991-1006 of the $a$ subunit of pig kidney sodium pump (1).

338

The data presented here greatly favour the assumption that the sodium pump works as a functional dimer. In order to obtain conclusive evidence that the sodium pump works as a diprotomer it will be, however, required to isolate and determine the primary structure of the ATP binding site of the $E_2$ATP and the $E_1$ATP conformational state of the catalytic $\alpha$ subunit. Work addressing this topic is in progress.

*Acknowledgements*

This work has been supported by grants of the Deutsche Forschungsgemeinschaft, Bonn, the Fonds der Chemischen Industrie, Frankfurt/Main, the Giessener Graduiertenkolleg "Molekulare Biologie und Pharmakologie" and the Gießener Hochschulgesellschaft e.V.

# References

1. Antolovic R, Brüller HJ, Bunk S, Linder D, Schoner W (1991) Epitope mapping by amino-acid-specific antibodies reveals that both ends of the $\alpha$-subunit of $Na^+/K^+$-ATPase are located on the cytoplasmic side of the membrane. Eur J Biochem 199: 195-202

2. Amler E, Abott A, Ball Jr W (1992) Structural dynamics and oligomeric interactions of $Na^+/K^+$-ATPase as monitored using fluorescence energy transfer. Biophys J 61: 553-568

3. Askari A (1988) Ligand binding sites of $(Na^+ + K^+)$-ATPase: Nucleotides and cations. Progr Clin Biol Res 268A: 149-165

4. Brotherus JR, Jacobsen L, Jørgensen PL (1983) Soluble and enzymatically stable $(Na^+ + K^+)$-ATPase from mammalian kidney consisting predominantly of protomer $\alpha\beta$-units. Preparation, assay and reconstitution of active $Na^+,K^+$ transport. Biochim Biophys Acta 781: 290-303

5. Buxbaum E, Schoner W (1990) Blocking of $Na^+/K^+$-transport by the $MgPO_4$ complex analogue $Co(NH_3)_4PO_4$ leaves the $Na^+/Na^+$-exchange reaction of the sodium pump unaltered and shifts its high affinity ATP-binding site to a $Na^+$-like form. Eur J Biochem 193: 355-360

6. Buxbaum E, Schoner W. (1991) Phosphate binding and ATP-binding sites coexist in $Na^+/K^+$-transporting ATPase, as demonstrated by the inactivating $MgPO_4$ complex analogue $Co(NH_3)_4PO_4$. Eur J Biochem 195: 407-419

7. Buxbaum E, Schoner W (1992) Investigation of subunit interaction by radiation inactivation: The case of $Na^+/K^+$-ATPase. J theor Biol 155: 21-31

8. Danenberg KD, Cleland WW (1975) Use of chromium adenosine triphosphate and lyxose to elucidate the kinetic mechanism and coordination state of the nucleotide substrate for yeast hexokinase. Biochemistry 14: 28-38

9. Hamer E, Schoner W (1993) Modification of the $E_1$ATP binding site of $Na^+/K^+$-ATPase by the chromium complex of adenosine 5'-($\beta,\gamma$-methylene)triphosphate blocks its overall reaction but not the partial activities of the $E_2$ conformation. Eur J Biochem 213: 734-748

10. Hayashi Y, Kobayashi T, Nakajima T (1993) Protomer association of solubilized Na$^+$/K$^+$-ATPase induced by ATP. Biol Chem Hoppe-Seyler 374: 589

11. Jørgensen PL (1975) Purification and characterization of (Na$^+$,K$^+$)-ATPase. V. Conformational changes in the enzyme. Transitions between the Na-form and the K-form studied with tryptic digestion as a tool. Biochim Biophys Acta 401: 399-415

12. Jørgensen PL, Farley RA (1988) Proteolytic cleavage as a tool for studying structure and conformation of pure membrane-bound Na$^+$,K$^+$-ATPase. Meth Enzymol 156: 291-301

13. Karlish SJD (1980) Characterization of conformational changes in (Na,K)ATPase labeled with fluorescein at the active site. J Bioenerg Biomembr 12: 111-136

14. Mimura K, Matsui H, Takagi T, Hayashi Y (1993) Change in oligomeric structure of solubilized Na$^+$/K$^+$-ATPase induced by octaethylene glycol dodecyl ether, phosphatidylserine and ATP. Biochim Biophys Acta 1145: 63-74

15. Moczydlowski EG, Fortes PAG (1981) Inhibition of sodium and potassium adenosine triphosphatase by 2',3'-0-(2,4,6-trinitrocyclohexadienylidene)adenine nucleotides. J Biol Chem 256: 2357-2366

16. Nørby JG, Jensen J (1991) Functional significance of the oligomeric structure of Na,K-Pump from radiation inactivation and ligand binding. In: The Sodium Pump: Structure, Mechanism and Regulation (Kaplan JH, De Weer P, eds) The Rockefeller University Press, New York, pp 173-188

17. Ottolenghi P, Ellory JC (1983) Radiation inactivation of (Na,K)-ATPase, an enzyme showing multiple radiation sensitive domains. J Biol Chem 258: 14895-14907

18. Pauls H, Bredenbröcker B, Schoner W (1980) Inactivation of (Na$^+$ + K$^+$)-ATPase by chromium(III) nucleotide triphosphates. Eur J Biochem 109: 523-533

19. Pauls H, Serpersu EH, Kirch U, Schoner W (1986) Chromium(III)ATP inactivating (Na$^+$ + K$^+$)-ATPase supports Na$^+$-Na$^+$ and Rb$^+$-Rb$^+$ exchanges in everted red blood cells but not Na$^+$,K$^+$-transport. Eur J Biochem 157: 585-595.

20. Peluffo RD, Garrahan PJ, Rega AF (1992) Low affinity superphosphorylation of the Na,K-ATPase by ATP. J Biol Chem 267: 6596-6601

21. Plesner IW (1987) Application of the theory of enzyme subunit interactions to ATP hydrolyzing enzymes. The case of Na,K-ATPase. Biophys J 51: 69-78

22. Repke KRH, Schön R (1973) Flip-Flop model of (NaK)ATPase function. Acta Biol Germ 31: K19-K30

23. Repke KRH, Dittrich F (1979) Subunit-subunit interaction: Determinant of reactivity and cooperativity of Na,K-ATPase. In: Na$^+$,K$^+$-ATPase, Structure and Kinetics (Skou JC, Nørby JG, eds) pp 487-502, Academic Press, London, New York and San Francisco.

24. Robinson JD, Flashner MS (1979) The (Na$^+$ + K$^+$)-activated ATPase. Enzymatic and transport properties. Biochim Biophys Acta 549: 145-176.

25. Scheiner-Bobis G, Fahlbusch K, Schoner W (1987) Demonstration of cooperating α-subunits in working (Na$^+$ + K$^+$)-ATPase by the use of the MgATP complex analogue cobalt tetrammine ATP. Eur J Biochem 168: 123-131

26. Scheiner-Bobis G, Esmann M, Schoner W (1989) Shift of the Na$^+$-form of (Na$^+$ + K$^+$)-ATPase due to modification of the low affinity ATP binding siite by Co(NH$_3$)$_4$ATP. Eur J Biochem 183: 173-178

27. Scheiner-Bobis G, Antonipillai J, Farley RA (1993) Simultaneous binding of phosphate and TNP-ADP to FITC-modified $Na^+$,$K^+$-ATPase. Biochemistry 32: 9592-9599

28. Schoner W, v Ilberg C, Kramer R, Seubert W (1976) On the mechanism of $Na^+$- and $K^+$-stimulated hydrolysis of adenosine triphosphate. 1. Purification and properties of a $Na^+$- and $K^+$-activated ATPase from ox brain. Eur J Biochem 1: 334-343

29. Serpersu EH, Kirch U, Schoner W (1982) Demonstration of a stable occluded form of $Ca^{2+}$ by the use of the chromium complex of ATP in the $Ca^{2+}$-ATPase of sarcoplasmic reticulum. Eur J Biochem 122: 347-354

30. Serpersu EH, Bunk S, Schoner W (1990) How do MgATP analogues differentially modify high- and low-affinity ATP-binding-sites in Na/K-ATPase. Eur J Biochem 191: 397-404

31. Serpersu EH, Summitt LL, Gregory JD (1992) Inactivation of yeast phosphoglycerate kinase by Cr-ATP complexes and its implications on the conformation of the enzyme active site. J Inorg Chem 48: 203-215

32. Smith RL, Zinn K, Cantley LC (1980) A study of the vanadate-trapped state of (Na,K)-ATPase. J Biol Chem 255: 9852-9859

33. Stein WD, Lieb WR, Karlish SJD, Eilam Y (1973) A model for active transport of sodium and potassium ions as mediated by a tetrameric enzyme. Proc Natl Acad Sci, USA 70: 275-278.

34. Thönges D, Hahnen J, Schoner W (1994) The ribose conformation of ATP affects the kinetic properties of $Na^+$/$K^+$-ATPase (this volume)

35. Vilsen B, Andersen JP, Petersen J, Jørgensen PL (1987) Occlusion of $^{22}Na^+$ and $^{86}Rb^+$ in membrane-bound and soluble protomeric αß-units of Na,K-ATPase. J Biol Chem 262: 10511-10517

36. Vilsen B, Andersen JP (1992) CrATP-induced $Ca^{2+}$ occlusion in mutants of the $Ca^{2+}$-ATPase of sarcoplasmic reticulum. J Biol Chem 267: 25739-25743.

37. Ward DG, Cavieres JD (1993) Solubilized αß Na,K-ATPase remains protomeric during turnover yet shows apperent negative cooperativity toward ATP. Proc Natl Acad Sci USA 90: 5332-5336

# Atomic Force Microscopy of Crystallized Na$^+$/K$^+$-ATPase

A.B. Maunsbach and K. Thomsen

Department of Cell Biology, Institute of Anatomy, University of Aarhus, DK-8000 Aarhus C, Denmark

## Introduction

The atomic force microscope, AFM (3), images hard surfaces at atomic resolution but its application to soft biological samples is still a great challenge (4, 5, 6, 7, 8). An important characteristic of the AFM is that it allows imaging of biological material not only in air or vacuum but also in aqueous solutions without any preceeding fixation or staining. We have applied the AFM to the analysis of the two-dimensional crystals of Na$^+$/K$^+$-ATPase which we have previously characterized by electron microscopy following negative staining (11) or cryo-electron microscopy (10). The present observations demonstrate that the lipid regions of Na$^+$/K$^+$-ATPase membranes can be imaged unstained in aqueous solutions at molecular resolution.

## Materials and Methods

Na$^+$/K$^+$-ATPase was purified in membrane-bound form from the outer medulla of pig kidney by zonal centrifugation (9). Two-dimensional p21 crystals were induced in the membrane fragments with the aid of magnesium and vanadate (14). The crystallized enzyme was then adsorbed on freshly cleaved mica and imaged with a Rasterscope$^{TM}$ 3000 Atomic Force Microscope (Danish Micro Engineering A/S, Denmark) using a 9 μm scanning range. The membrane crystals were analyzed in air or in aqueous solutions either completely untreated or following fixation with glutaraldehyde. For analyses in solutions we designed a fluid cell consisting of an ordinary thin cover-glass fastened over the AFM cantilever. For comparison enzyme samples were also negatively stained with uranyl acetate and analyzed by electron microscopy.

## Results

Na$^+$/K$^+$-ATPase membranes imaged with the AFM in aqueous solutions corresponded in shape and dimensions to parallel samples observed by electron microscopy following negative staining (Figs. 1, 2). The diameter of the membrane crystals ranged from 0.1 - 0.5 μm. Height measurements in the AFM were compatible with regions of lipid bilayer (about 5 nm), single membrane crystals (11-14 nm), or in places folded or stacked membranes (20-26 nm or more).

A large number of membrane fragments were scanned with the different cantilevers and in different aqueous solutions (destilled water, phosphate-buffered saline, pH between 6.0-7.5, crystallization buffer, 500 mM potassium chloride) and with different scan speeds (range 1000-40000 nm/sec) and with variations in the force (0.1-6 nN). None of

342

the various combinations of these parameters provided AFM images showing distinct crystalline arrays within the 10-12 nm or 22-26 nm thick protein-rich regions. In some areas a wavy surface pattern was observed but individual protein complexes or arrays similar to those observed by high resolution electron microscopy were not identified. Further understanding is necessary of the interactions between the cantilever tip and the soft proteins of the biological specimen (15).

Repeated scans over membrane fragments with forces around 1 nN resulted in a gradual displacement and removal of the thick membrane regions representing protein, while the lipid regions remained stationary (Figs. 3, 4). Eventually all the protein-rich regions were removed and the lipid regions of the membranes remained as uniform discs with a thickness of about 5 nm (Fig. 5). Further imaging of these regions revealed distinct crystalline arrays with uniform elevations with center-to-center distances of about 7-8 nm (Fig. 6). Imaging of the adjacent mica surface revealed atomic resolution (Fig. 7).

**Discussion**

The present observations demonstrate that two-dimensional crystals of $Na^+/K^+$-ATPase can be imaged unfixed and unstained in aqueous solution with the atomic force microscope. We did not in the present investigation observe individual protein complexes as reported by others (1, 2) despite the fact that atomic resolution was demonstrable on adjacent mica surfaces. On the other hand the lipid regions of the membranes exhibited molecular resolutions and showed highly ordered arrays with dimensions similar to Langmuir-Blodgett (LB) films (12, 13). Since the thickness of these regions was constant and around 5 nm and since the imaging was performed in aqueous solutions we conclude that the ordered arrays represent the hydrophilic upper surface of the lipid bilayer of the membranes. Thus, the biomembrane lipids have been imaged in aqueous solution at molecular resolution without prior solubilization and reconstitution of the cell membranes.

*Acknowledgements*

Supported by grants from the Danish Medical Research Council and by the Danish Biomembrane Research Center, University of Aarhus.

**References**

1.  Apell HJ, Colchero J, Linder A, Marti O (1993) Investigation of the Na,K-ATPase by SFM. In: Marti O, Amrein M (eds) STM and SFM in Biology. Academic Press, San Diego, pp. 275-308
2.  Apell HJ, Colchero J, Linder A, Marti O, Mlynek J (1992) Na,K-ATPase in crystalline form investigated by scanning force microscopy. Ultramicroscopy 42-44: 1133-1140
3.  Binnig G, Quate CF, Gerber C (1986) Atomic force microscope. Phys. Rev. Lett. 56: 930-933
4.  Butt HJ, Downing KH, Hansma PK (1990) Imaging the membrane protein bacteriorhodopsin with the atomic force microscope. Biophys. J. 58: 1473-1480
5.  Egger M, Ohnesorge F, Weisenhorn AL, Heyn SP, Drake B, Prater CB, Gould SAC,

Hansma PH, Gaub HE (1990) Wet lipid-protein membranes imaged at submolecular resolution by atomic force microscopy. J. Struct. Biol. 103: 89-94

6.  Engel A (1991) Biological applications of scanning probe microscopes. Annu Rev Biophys Biophys Chem 20: 79-108

7.  Hansma PK, Elings VB, Marti O, Bracker CE (1988) Scanning tunneling microscopy and atomic force microscopy: Application to biology and technology. Science 242: 209-216

8.  Hoh JH, Lal R, John SA, Revel JP, Arnsdorf MF (1991) Atomic force microscopy and dissection of gap junctions. Science 253: 1405-1408

9.  Jørgensen PL (1974) Purification and characterization of (Na⁺+K⁺)-ATPase. III. Purification from outer medulla of mammalian kidney after selective removal of membrane components by sodium dodecyl sulphate. Biochim. Biophys. Acta 356: 36-52

10. Maunsbach AB, Hebert H, Kavéus U (1992) Cryo-electron microscope analysis of frozen-hydrated crystals of Na,K-ATPase. Acta Histochem. Cytochem. 25: 279-285

11. Maunsbach AB, Skriver E, Hebert H (1991) Two-dimensional crystals and three-dimensional structure of Na,K-ATPase analyzed by electron microscopy. In: Kaplan JH, De Weer P (eds) The Sodium Pump: Structure, Mechanism, and Regulation. The Rockefeller University Press, pp. 159-172.

12. Peltonen JPK, He P, Rosenholm JB (1992) Order and defects of Langmuir-Blodgett films detected with the atomic force microscope. J Am Chem Soc 114: 7637-7642

13. Schwartz DK, Garnaes J, Viswanathan R, Zasadzinski JAN (1992) Surface order and stability of Langmuir-Blodgett films. Science 257: 508-511

14. Skriver E, Maunsbach AB, Jørgensen PL (1981) Formation of two-dimensional crystals in pure membrane-bound Na⁺,K⁺-ATPase. FEBS Lett 131: 219-222

15. Yang J, Tamm LK, Somlyo AP, Shao Z (1993) Promises and problems of biological atomic force microscopy. J Microscopy 171: 183-198

## Figure Legends

Fig. 1.   Electron micrograph of Na⁺/K⁺-ATPase membrane showing two-dimensional Na⁺/K⁺-ATPase crystal at P and adjacent lipid region at L. Bar: 0.1 μm.

Fig. 2.   Atomic force microscope image of crystallized Na⁺/K⁺-ATPase membrane. Height measurements in the AFM reveal that the thickness of the membrane is about 5 nm at L and 11-14 nm at P. Bar: 0.1 μm.

Fig. 3.   AFM image of Na⁺/K⁺-ATPase fragment following repeated scans at a force of 0.7 nN and a scanning speed of 25,000 nm/sec. Protein regions (light) are gradually being displaced by the horizontal scans of the cantilever. Bar: 0.1 μm.

Fig. 4.   AFM image of the same region following extended scanning. After removal of the thick protein regions, thin uniform areas remain at locations corresponding to the lipid regions of the crystallized Na⁺/K⁺-ATPase membranes. Height measurements in the AFM were carried out along the line between A and B. Bar: 0.1 μm.

Fig. 5.   Thickness measurements along the line between A and B in Fig. 4. The lipid layers at arrowheads have a thickness relative to the mica surface (M) close to 5 nm.

Fig. 6.   AFM image of the region labelled x in Fig. 4. The surface was scanned with a speed of 1300 nm/sec and with a force of 5.7 nN and shows ordered arrays interpreted as the hydrophilic ends of the lipids in the membrane bilayer. Bar: 5 nm.

Fig. 7.   AFM image of mica surface immediately adjacent to the membrane in Fig. 4 shows atomic resolution. Bar: 5 nm.

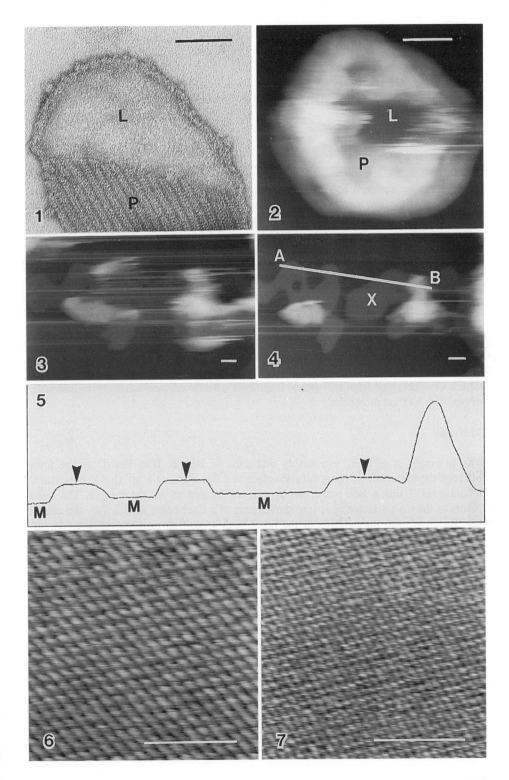

# Scanning force microscopical studies of the $Na^+/K^+$-ATPase

Linder, A., Apell, H.-J.
Department of Biology, University of Konstanz, D-78434 Konstanz, Germany

## Introduction

Scanning Force Microscopy (SFM) is an imaging technique, that can resolve features on a nanometer scale on both conducting and non-conducting surfaces. This advantage was almost immediately used to scan organic surfaces like crystals of amino acids (1) or organic monolayers (2). To extend the method onto biological material was close at hand and first examples have been blood cells and bacteria (3). Especially the purple membrane which consists of crystalline arrays formed by a single proton pumping protein, bacteriorhodopsin, was investigated extensively. Its crystalline arrangement has been characterized by various techniques and it was known how the structure "looks like". Therefore, these preparations are crucial to get information about the capabilities of SFM. Bacteriorhodopsin has been scanned in dried form on lysine-coated glass and on mica (3, 4) and in buffer solutions (5). The observed lattice constant was in agreement with results of electron diffraction experiments. A recent review of the application of the SFM technique is provided in Ref. 6. A second integral membrane protein, investigated by SFM, was the $Na^+/K^+$-ATPase (7). In this paper, results are presented which were obtained from membrane fragments with two-dimensional crystalline arrays of $Na^+/K^+$-ATPase molecules.

## Methods

The principle of SFM is extensively discussed in Ref. 6. It is based on the forces between the molecules of the scanned surface and the atoms of a tip, that is mounted on a cantilever. Using a tube piezo, the tip can be scanned across the sample. The movements of the tip are transduced into movements of a laser beam reflected by the cantilever. By measuring the position of the beam with a position sensitive photodiode, subnanometer movements of the lever can be detected. Up and down movements of the cantilever provides information on the topography of the surface, twisting of the cantilever contains information on lateral forces ("friction") which are a property of the substance that forms the surface (6).

## Results

Two-dimensional crystals of $Na^+/K^+$-ATPase were raised in a buffer containing 1 mM $MgCl_2$ and 5 mM phosphate as described elsewhere (8). The crystals were adsorbed to mica, that had been rinsed briefly with the crystallization buffer after cleavage. The images were taken in the same buffer at room temperature and without fixation. As in electron microscopy, at low magnification, membrane fragments with a size of 0.25 - 1 µm can be seen. Fig. 1 shows a pseudo three dimensional presentation of two membrane fragments. Two different domains are clearly resolved: a lower, flat one in

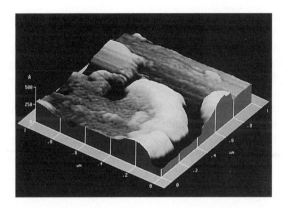

**Figure 1:** Image of membrane fragments containing crystallized Na⁺/K⁺-ATPase. The image size is 1 × 1 μm.

the upper left of Fig. 1, which consists of a pure lipid phase and a higher structure at the bottom, containing a protein phase. The height is given in arbitrary units and corresponds to 5 nm for the lipid phase and about 14 nm for the protein phase. At higher resolution, rows could be identified in the protein domain, as shown in Fig. 2.

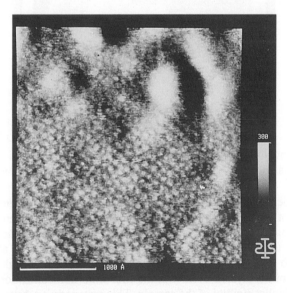

**Figure 2:** Scanning force microscopy image of membrane fragments containing crystallized Na⁺/K⁺-ATPase in buffer at 20 °C. The size of the scanned area is 250 × 250 nm.

347

The distance between the rows, that take their course from the upper left to the lower right was 10 nm. Along the other axis, in an angle of approximately 75° to the first one, the distance between the rows was about 8 nm. The size of the corrugations between the rows depended strongly on the quality of the tips. With the sharpest tips we have used so far, corrugations of about 1 nm have been observed. This is half the value that has been described for the extracellular protrusions of the protein by electron microscopic investigations. This discrepancy can be explained by the blunt shape of the tips. The dimensions of the features in Fig. 2 corresponded well to the predicted values. The image has not been filtered in any way. A further zoom (Fig. 3) shows single bumps which are assumed to represent single pumps. Distances between the particles are in the range of 14 nm to 17 nm.

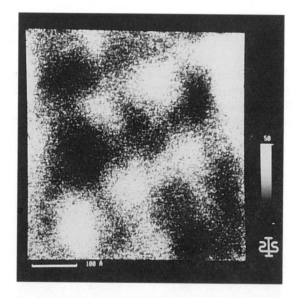

**Figure 3:** Zoom into the crystallized domain of a membrane fragments containing $Na^+/K^+$-ATPase as shown in Fig. 2. The size of the scanned area is $50 \times 50$ nm.

In the present experiments, the lateral resolution is limited to about 5 nm. This is due to the deformation of the sample by the imaging forces. On the fragments, the net force between sample and tip is repulsive. Therefore, we have to push the tip against the sample, in order to get in contact. Unfortunately, at a distance of about 1 nm, the repulsive coulomb forces between tip and sample become high enough to decrease the long range order of the crystals even with the first scan. Immobilization of the proteins should lead to a higher rigidity of the sample. This may permit a better structural conservation and resolutions in the sub-nanometer range, that are needed to answer structural questions. Moreover, new silicon nitride tips with a curvature radius of less than 10 nm have bcome available last year. This enhances the fractionof suitable tips from less than 10 % to almost 100 %, leading to higher reliability of the results and faster progress in the experimental work

At the present state of development of the SFM method arrays of crystalline Na$^+$/K$^+$-ATPase in membrane fragments can be resolved on a molecular level under almost physiological conditions (at room temperature, in aqueous buffers). The resolution is almost comparable to that of electron-microscopical studies. Further progress depends on the accessibility of even sharper tips and on the immobilization of the Na$^+$/K$^+$-ATPase molecules in their (liquid) membrane environment.

*Acknowledgments*

This work has been financially supported by the Deutsche Forschungsgemeinschaft (Sonderforschungsbereich 156).

## References

1   Gould SAC., Marti O, Drake B, Hellemans L, Bracher CE, Hansma PK, Keder NL, Eddy MM, Stucky GD (1988) Molecular resolution images of amino acid crystals with the atomic force microscope. Nature 332:332-334.
2   Marti O, Ribi HO, Drake B, Albrecht TR, Quate CF, Hansma, PK (1988) Atomic force microscopy of an organic monolayer. Science 239:50-52.
3   Gould SAC, Drake B, Prater CB, Weisenhorn AL, Maune S, Hansma HG, Hansma PK, Massie J, Longmire M, Elings V, Dixon Northern B, Mukergee B, Peterson CM, Stoeckenius W, Albrecht TR, Quate CF (1990) From atoms to integrated circuit chips, blood cells, and bacteria with the atomic force microscope. J. Vac. Sci. Technol. A8:369-373.
4   Worcester DL, Kim HS, Miller RG, Bryant PJ (1990) Imaging bacteriorhodopsin lattices in purple membranes with atomic force microscopy. J. Vac. Sci. Technol. A8:403-405.
5   Butt H-J, Downing KH, Hansma PK (1990) Imaging the membrane protein bacteriorhodopsin with atomic force microscopy. Biophys. J. 58:1473-1480.
6   Marti O, Amrein M (1993) STM and SFM in Biology. Academic Press, San Diego.
7   Apell H-J, Colchero J, Linder A, Marti O, Mlynek J (1992) Na,K-ATPase in crystalline form investigated by scanning force microscopy. Ultramicroscopy 42-44:1133-1140.
8   Apell H-J, Colchero J, Linder A, Marti, O (1993) Investigation of the Na,K-ATPase by SFM. In Marti O, Amrein M (eds.) STM and SFM in Biology. Academic Press, San Diego, pp. 275-308

# Molecular Imaging of Purified Na$^+$/K$^+$-ATPase Molecules in Canine Kidney Membranes Using Atomic Force Microscopy

Jose K. Paul, Mehdi Ganjeizadeh, Huiying Yu, Mamoru Yamaguchi* and Kunio Takeyasu

Department of Med. Biochem. and Biotechnology Center, and *Department of Veterinary Anatomy and Cell Biology, The Ohio State University, Columbus, Ohio 43210 USA

## Introduction

In order to ultimately elucidate the structure-function relationships of macromolecules, it is necessary to deduce detailed structural features by high resolution imaging in parallel to other techniques. In the case of membrane proteins this task has proven to be typically challenging due to limited availability of suitable specimen-preparation techniques and narrow applicability of conventional imaging techniques that require at least 2-dimensional crystals for high resolution. So far, a combination of electron microscopy (EM) and X-ray diffraction techniques, has been the most successful in the analysis of some of membrane proteins such as bacteriorhodopsin (7), nicotinic acetylcholine receptor (21) and Ca$^{2+}$-ATPase (5). Atomic force microscopy (AFM) is emerging as a powerful tool for analyzing biological samples at molecular resolutions (3,6). Recent successful application of AFM to the channel structure of gap-junction molecules has proven the potential applicability of this technique to the structural (and, thereby, functional) analysis of membrane proteins in general (8). In this report, using AFM working under tapping mode, we imaged Na$^+$/K$^+$-ATPase molecules in membranes at molecular resolution:

## Biological Application of AFM under Tapping Mode

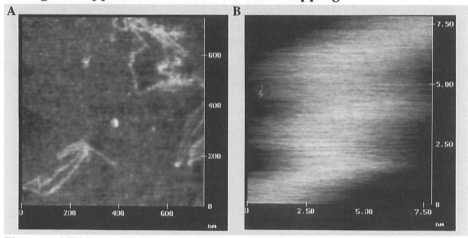

**Figure 1.** AFM image of DNA molecules containing the promoter region of the Na$^+$/K$^+$-ATPase α-subunit gene. A: Low resolution images of circular DNA molecules. B: Molecular imaging of DNA reveals a double helix of with pitch~ 3.4 nm. For detailed methods, see Takeyasu et al. in this volume.

We, all biologists, are most interested in what we can learn about biological processes using this physical technique rather than in how this technique works. Some of our preliminary results have demonstrated the powerful ability in structural analysis of biological molecules (Figure 1). The stuctures of hard materials such as plant polymers can be easily studied by AFM, revealing even atomic interactions at 3-5 Å (Figure 1). On the other hand, the molecular structures of softer materials such as DNAs can be achieved at slightly lower resolutions. We have been investigating the molecular mechanisms of Na+,K+-ATPase gene regulation using AFM. Figure 1b shows an example of DNA images at the highest resolution. Application of AFM to such studies on DNA-protein interaction has just begun (7), and holds great promise in elucidation of the mechanism of Na+/K+-ATPase gene regulation at the molecular level.

## Molecular Imaging of Na+/K+-ATPase Molecules

AFM under tapping mode is also applicable to structural analysis of membrane proteins. The Na+,K+-ATPase molecules in purified dog kidney membranes are well characterized by conventional biochemical techniques (8) and high resolution EM (9-11; also see figure 2). AFM at comparable resolution to EM (figure 3) also showed membrane disquettes, which, on detailed examination, were found to have protein molecules aggregated on their surface. Vesicles were also observed.

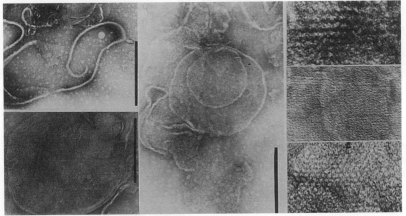

**Figure 2.** EM imaging of purified kidney membranes rich in Na+,K+-ATPase molecules shows membrane patches (scale bar= 335 nm). At higher magnifications, arrays of protein-aggregates and tubular formations are observed as reprorted (9-11).

Figure 3b shows molecular imaging of the Na+,K+-ATPase molecules in purified kidney membranes using AFM. The figure suggests that the protein forms a channel-like structure with a central pore. The average depth of the pore was found to be 12 Å. The entire protein complex had a diameter of 20 Å and the pore diameter was 6 Å. The height of the protein was typically 15 Å. The size of the protein [~20 Å] was found to be comparable to that deduced from the EM analysis (Figure 2), but smaller than the previously reported size (~40 Å). This may be in part attributed to the shrinkage in the fixation step with uranyl acetate. The slight flattening of the sample caused by the tip may also in part contribute to the height of the protein imaged by AFM. Results from fluorescence energy transfer studies (20) show that the vertical distance between ouabain binding site (extracellular) and FITC binding site (intracellular) is 72.5 Å . The total

351

site (intracellular) is 72.5 Å . The total width of the lipid bilayer is about 50Å. Hence the height of the protein protruding out of the bilayer in the intracellular side (22.5 Å) is similar to, but slightly larger than, the values obtained from AFM studies (15 Å). Although many different specimens were imaged, the dimensions and the overall morphology of the protein seemed to be consistent (Figure 3). In spite of repeated experiments, the intracellular side of the membrane seemed to be exposed to the scanning tip consistently. This "preference" could occur due to the difference in charge distribution in the two sides; e.g., extensive glycosylation of the β-subunit in the extracellular region.

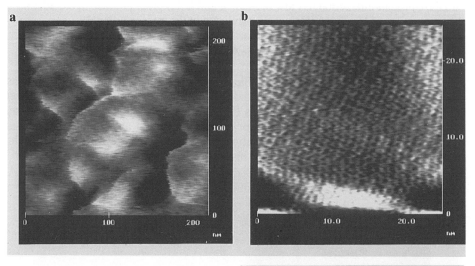

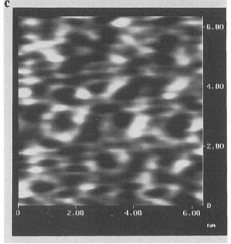

**Figure 3.** AFM images of purified dog kidney membranes at different resolutions.

Ion-motive ATPases including $Na^+/K^+$-ATPase are thought to undergo a series of conformational changes including a phosphorylation-dephosphorylation cycle (11). During these events specific ions are transported across membranes. Recent studies employing recombinant DNA techniques have identified distinct ion-sensitive domains for $Ca^{2+}$, $Na^+$ and $K^+$ (1, 9, 20), but the actual mechanism of transport of these ions across the membrane remains to be solved. It has been a fascinating question whether the

352

ion pumps possess a pore or not, since the structural studies on other types of ion-transporting proteins (gap junction proteins and ion channels) have revealed the existence of "channel-like structures" (4). A marine toxin, palytoxin, is known to increase ouabain-sensitive $Na^+$ conductance to 9.6 pS which is comparable to the property of known ion channels (12). This suggests that the $Na^+/K^+$-ATPase may take a specific channel-like conformation under a certain condition (12, 17, 22). Our present data extends this view further to the idea that native $Na^+/K^+$-ATPase may form a channel-like structure without a specific ligand.

In summary, the shape of the protein was found to be very similar to that of connexons at gap junctions (8), except that the dimensions were smaller. Both proteins seemed to have a channel-like structure, with a well-defined pore in the middle. It may be possible that all membrane proteins which facilitate transport of substances across the membrane exhibit similar morphology. Investigation into this possibility using AFM will provide us with valuable general information about the structure and function of transporting proteins in membranes.

### References

1. Andersen, J.P. (1993) In this volume.
2. Askari, A., Huang, W.-H., McCormick, P.W. (1983) J. Biol. Chem. 258: 3453-3460.
3. Bustamante, C., Keller, D. Yang, G. (1993) Curr. Opinion in Structural Biol. 3-3: 363-372.
4. Catterall, W. A. (1986) Science 242: 50-61.
5. Green, N. M., Stokes, D. L. (1990) J. Mol. Biol. 213: 529-538.
6 Hansma, P. K., Elings, V. B., Marti, O., Bracker, C. E. (1988) Science, 242: 209-216.
7. Henderson, R., Unwin, P. N. T., (1975) Nature, 257: 28-32
8. Hoh, J. H., Sosinsky, G. E., Revel, J-P. & Hansma, P. K. (1993) Biophys. J. 65: 149-163.
9. Ishii, T. & K. Takeyasu (1993) Proc. Natl. Acad. Sci. USA. (In Press).
10 Jorgensen, P.L. (1974) Biochim Biophys Acta 356: 36-52.
11 Jorgensen, P.L. & J.P. Andersen (1988) J. Memb. Biol 103: 95-120.
12 Kim, S. Y., Wu, C. H., Beress. L. (1990) In "The Sodium Pump: Recent Developments" (Ed. J. Kaplan & P. De Weer) p.p. 505-508.
13 Maunsbach, A. B., Skriver, E., Søderholm, M., Hebert, H. In "The Na,K-Pump, Part A: Molecular Aspects, pp 39-56 (1988)
14 Mohraz. M., Smith, P.R. (1984) J. Cell Biol.105: 1-8.
15 Yang, J., Takeyasu, K., Shao, Z. (1992) FEBS Lett. 301: 173-176.
16 Rees, W. A., Keller, R. W., Vesenka, J. P., Yang, G., Bustamante, C. (1993) Science, 260: 1646-1649
17 Scheiner-Bobis, G., Farley, B.A. (1993) In this volume.
18 Shinoguchi, E, Ito, E, Kudo, A, Nakamura, S, Taniguchi, K. (1990) In " The Sodium Pump: Recent Developments" (Society of General Physiologists- 44th Annual Symposium) pp. 363-367.
19 Skriver, E., Hebert, H., Kavéus, U., Maunsbach, A.B. (1990) In "The Sodium Pump: Recent Developments" (Society of General Physiologists- 44th Annual Symposium) pp. 243-247.
20 Takeyasu, K., T. Ishii, J.K. Paul, M.V. Lemas, M. Ganjeizadeh, H.-Y. Yu, S.-S. Wang & T. Kuwahara (1993) In this volume.
21 Unwin, N., Toyoshima, C., Kubalek, E. (1988) J. Cell Biol. 107: 1123-1138.
22 Yoda,A., M. R. Rosner, P. Morrison, S. Yoda (1990) J. Gen. Physiol. 96: 74a.
23. We thank Dr. Askari for providing purified dog kidney membranes. This work was supported by grants from NIH (GM44373) and AHA (901107). M.G. is a recipient of Postdoctoral Fellowship from the American Heart Association, Ohio Affiliate. K.T. is an Established Investigator of the American Heart Association.

# Cytoplasmic and Extracellular Domains in Two-dimensional Crystals of Na+/K+-ATPase

E. Skriver, R. Antolovic, J.V. Møller

Department of Cell Biology, Institute of Anatomy and Institute of Biophysics, University of Aarhus, DK-8000 Aarhus C, Denmark and Institute of Biochemistry and Endocrinology, Justus-Liebig University, W-6000 Giessen, Germany

## Introduction

Information about the topology of Na+/K+-ATPase subunits in the cell membrane has been obtained from their primary sequences and hydropathy plots combined with site-specific labelling, immunological and proteolytic digestion studies. The N- and C-termini have been located on the cytoplasmic side by Antolovic et al. (1) on right side-out vesicles using immunological techniques. The number and location of membrane spanning regions is still a matter of dispute as the hydropathy plots in the C-terminal area do not give any clear answer (4).

In order to identify the cytoplasmic and extracellular regions of Na+/K+-ATPase from pig kidney we prepared two-dimensional (2-D) Na+/K+-ATPase crystals with p21 symmetry induced by vanadate (3,6,9). Surface replicas of freeze-dried membranes from such preparations are suited for high resolution electron microscopy studies including immunogold labelling (8). Therefore, we used this method in the present study to locate the N-terminus and the C-terminus as well as amino acids 815-828 and 889-903 with sequence-specific antibodies.

## Structural identification of extracellular and cytoplasmic sides of Na+/K+-ATPase membrane crystals

Following Ta/W shadowing of freeze-dried membranes with vanadate-induced Na+/K+-ATPase 2-D crystals two distinct surface patterns were observed (Fig. 1). While the sides of the membranes were covered with Ta/W and carbon, revealing the structural details on the electron micrograph, the reverse sides were exposed to labelling with antibodies. The sidedness of the membranes was identified by labelling the non-replicated/exposed side with antibodies against epitopes located in the C-terminal part of the β1-subunit at the extracellular side of the membranes (Fig. 2). Rabbit polyclonal antibodies against the α-subunit (1) bound mostly to the opposite (cytoplasmic) side of the membranes. Thus, the extracellular surface pattern showed marked rows with metal deposition separated by distinct linear arrays without metal. On the cytoplasmic side the rows appeared rather blurred. The average spacing was about 140 Å. The surface patterns were identical to those observed following freeze-etching and probably reflect a more extended space filling polypeptide mass on the cytosolic side of the membrane.

## Localization of C-terminus and N-terminus as well as peptides 815-828 and 889-903

The surface-replicated membranes were floated on a PBS buffer and immunolabelled with sequence-specific polyclonal rabbit antibodies (1,7) and subsequently with gold-conjugated anti-rabbit immunoglobulin (BioCell). Some of the samples of membrane-crystals (0.45 mg/ml protein) were treated with neuraminidase (0.2 U/ml in 25 mM imidazole, pH 6.7). After surface-replication some of the specimens were treated with trypsin, 0.3 mg/ml in 50 mM BisTris, 50 mM NaCl, 10 mM $MgCl_2$ and 0.1 mM $CaCl_2$ (pH 6.5), before immunolabelling. Application of 2 sequence-specific antibodies against the N-terminus (1-12 and 1-13) and 3 against the C-terminus (991-1005, 1005-1016 and 1002-1016) showed that both the N-terminus and the C-terminus were located on the cytoplasmic side (Figs. 3 and 4). Sequence-specific antibodies against amino acid 815-828 reacted with the intracellular side (Fig. 5) while sequence-specific antibodies against the peptide with amino acids 889-903 bound to the extracellular side (Fig. 6). These findings support a model of α-chain topology with either 8 or with 10 transmembrane segments (4), similar to the model proposed for sarcoplasmic reticulum $Ca^{2+}$-ATPase (2,5). Controls with non-immune IgG instead of primary antibody, or with no primary antibody, showed almost no labelling over membrane areas. This was also the case using an antibody raised against β2-subunit (Upstate Biotechnology, Inc, New York). Tryptic digestion of the free side of the $Na^+/K^+$-ATPase membranes before immunolabelling prevented binding of the antibodies against the N-terminus as well as against β1. No changes in labelling of the other sequence-specific antibodies were observed. Treatment with neuraminidase caused no changes in the surface pattern and did not influence the immunolabelling.

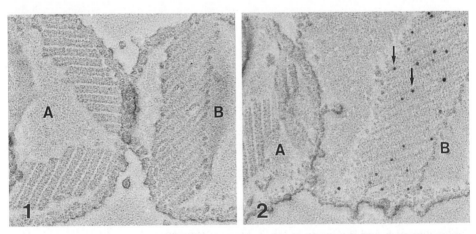

Fig. 1. Following metal-shadowing of freeze-dried membrane two distinct surface patterns of $Na^+/K^+$-ATPase crystals were observed. One side showed marked rows of heavy metal deposition separated by distinct linear arrays without metal. This was identified as the extracellular side of the membrane. Membranes with the extracellular side covered with Ta/W and carbon and the intracellular sides exposed to the buffer/medium are marked with an **A**. Membranes with the intracellular side metal shadowed and extracellular side exposed are marked with a **B**. Most membranes adsorbed to the mica with the cytoplasmic side. Mag.: × 160,000.

Fig. 2. Isoform-specific antibodies raised against amino acids 152-340 of β1-subunit in rat (Upstate Biotechnology, Inc, New York) selectively labelled the extracellular the side of 2-D crystalline arrays. Mag.: × 160,000.

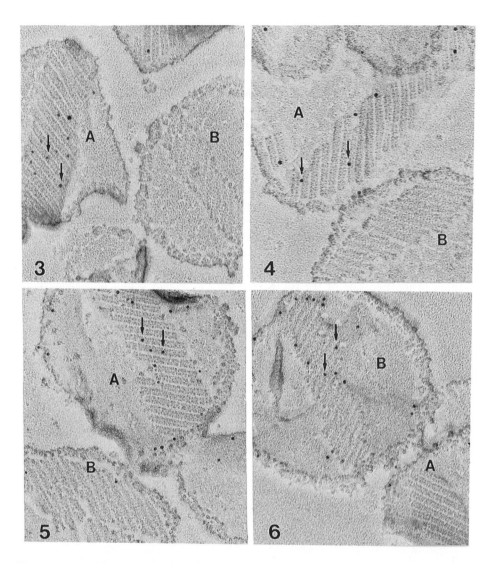

Fig. 3. Immunolabelling with sequence-specific antibodies against peptide (1-13) of the N-terminus showing selective labelling of the cytoplasmic side of the Na$^+$/K$^+$-ATPase crystals. Explanation of **A** and **B** is given in Fig. 1. Mag.: $\times$ 160,000.

Fig. 4. Immunolabelling with sequence-specific antibodies against the C-terminus peptide (1002-1016) showing labelling confined to the intracellular part of the Na$^+$/K$^+$-ATPase crystals although a small number of membranes also had a few gold particles attached to their extracellular side. Mag.: $\times$ 160,000.

Fig. 5. Immunolabelling with sequence-specific antibodies against peptide (815-828) showing significant labelling of the cytoplasmic side of the Na$^+$/K$^+$-ATPase crystals (about 90% of the gold particles). Mag.: $\times$ 160,000.

Fig. 6. Immunolabelling with sequence-specific antibodies against peptide (889-903) showing labelling of the extracellular side of the Na$^+$/K$^+$-ATPase crystals (about 90% of the gold particles). Mag.: $\times$ 160,000.

# Conclusions

Our electron microscopy immunolocalization study on Na$^+$/K$^+$-ATPase membrane crystals with p21 symmetry using sequence-specific antibodies confirms that the N-terminus and the C-terminus of the $\alpha$-subunit is located on the cytoplasmic side of the membrane. This implies an equal number (8 or 10) of transmembrane segments. The hydrophilic region containing amino acids 815-828 is localized on the cytoplasmic side, indicating the presence of an M5-M6 loop in agreement with proteolytic data (4). By contrast, the hydrophilic region comprising amino acids 889-903 is located on the extracellular side, supporting a membrane traverse by the intervening, predominantly hydrophobic stretch of amino acid residues 842-879.

## Acknowledgments

Excellent technical assistance by Annette Blak Rasmussen is gratefully acknowledged. This work was supported by grants from the Danish Biomembrane Research Center, University of Aarhus, the Danish Medical Research Council (SSVF 12-0780-1) and the Deutsche Forschungsgemeinschaft through SFB 249 Giessen "Pharmakologie biologischer Makromoleküle".

## References

1. Antolovic R, Bruller HJ, Bunk S, Linder D, Schoner W (1991) Epitope mapping by amino-acid-sequence-specific antibodies reveals that both ends of the $\alpha$ subunit of Na$^+$/K$^+$-ATPase are located on the cytoplasmic side of the membrane. Eur J Biochem 199:195-202.
2. Brandl CJ, Green NM, Korczak B, MacLennan DH (1986) Two Ca$^{2+}$ ATPase genes: homologies and mechanistic implications of deduced amino acid sequences. Cell 44:597-607.
3. Hebert H, Jørgensen PL, Skriver E, Maunsbach AB (1982) Crystallization patterns of membrane-bound (Na$^+$+K$^+$)-ATPase. Biochim Biophys Acta 689:571-574.
4. Karlish SJ, Goldshleger R, Jørgensen PL (1993) Location of Asn$^{831}$ of the $\alpha$ chain of Na/K-ATPase at the cytoplasmic surface. Implication for topological models. J Biol Chem 268:3471-3478.
5. MacLennan DH, Brandl CJ, Korczak B, Green NM (1985) Amino-acid sequence of a Ca$^{2+}$+Mg$^{2+}$-dependent ATPase from rabbit muscle sarcoplasmic reticulum, deduced from its complementary DNA sequence. Nature 316:696-700.
6. Mohraz M, Yee M, Smith PR (1985) Novel crystalline sheets of Na,K-ATPase induced by phospholipase A$_2$. J Ultrastruct Res 93:17-26.
7. Møller JV, Juul B, Lee Y-J, le Maire M, Champeil P (1993) The topology of sarcoplasmic reticulum Ca$^{2+}$-ATPase and Na$^+$,K$^+$-ATPase in the M5/M6 region. this volume (In Press)
8. Nermut MV (1989) Strategy and tactics in electron microscopy of cell surfaces. Electron Microsc Rev 2:171-196.
9. Skriver E, Maunsbach AB, Jørgensen PL (1981) Formation of two-dimensional crystals in pure membrane-bound Na$^+$,K$^+$-ATPase. FEBS Lett 131:219-222.

# Membrane Topology of the α Subunit of Na,K-ATPase

M. Mohraz[1], E. Arystarkhova[2] and K.J. Sweadner[2]

[1]Dept. of Cell Biology, New York University School of Medicine, 550 First Ave., New York, NY 10016, and [2]Massachusetts General Hospital, Fruit St., Boston, MA 02114

INTRODUCTION

Although the amino acid sequence of the Na,K-ATPase is known, the transmembrane folding of its α subunit is the subject of considerable controversy. Several lines of evidence (8) support hydropathy plots that predict 4 transmembrane spans in the N-terminal third of the sequence. There is also general agreement that the middle third of the polypeptide chain is entirely cytoplasmic. The confusion lies in the C-terminal third of the sequence where hydropathy plots are ambiguous and models with three, four or six spans have been proposed. Rigorous identification of sites exposed on one side of the membrane or the other is essential for the development of a reliable folding model.

Immunoelectron microscopy is a powerful technique for the study of transmembrane folding of a protein whose amino acid sequence is known. Use of monoclonal antibodies allows unambiguous localization of defined epitopes relative to the membrane bilayer. We probed the transmembrane folding of the α subunit by immunoelectron microscopy and double-labeling techniques, using monoclonal antibodies $VG_4$, $VG_2$, and $IIC_9$. The epitopes of these antibodies were previously determined (1,7). $VG_4$ binds to a fragment encompassing the H1-H2 transmembrane hairpin, presumably at the short extracellular stretch. $VG_2$ binds to a cyanogen bromide fragment (M)VPAISLAYEQAESDI(M) after H5, and $IIC_9$ binds to a typtic fragment (R)WINDVEDSYGQQWTYEQR further down the sequence. The epitopes of all three antibodies were localized on the extracellular surface of the membrane, leading to new models for folding of the α subunit (6).

IMMUNOELECTRON MICROSCOPY

Membrane-associated Na,K-ATPase, purified by the method of Jorgensen (4), consists mostly of broken vesicles and membrane fragments. Since both the extracellular and the cytoplasmic sides of the membrane are accessible, immunoelectron microscopy and double-labeling techniques can be used to directly determine the membrane orientation of epitopes. One label is the antibody of interest and the other a control with well-established membrane orientation.

We used wheatgerm agglutinin, conjugated to colloidal gold, as the control for the extracellular surface and the monoclonal antibody M10-P5-C11 (2) as the reference for the cytoplasmic side of the membrane. Protein A conjugated to colloidal gold was used to label the antibodies. Double-labeling was done with two different sizes of colloidal gold particles. After labeling, the samples were embedded in epon and thin sections were prepared for electron microscopy.

Figure 1 shows double-labeling of $VG_4$ and $IIC_9$ with the controls. In the upper panels the small gold particles ($VG_4$ and $IIC_9$) and large gold particles (wheatgerm agglutinin) on the same side of the membrane. In the lower panels the small gold particles ($VG_4$ and $IIC_9$) and large gold particles (M10-P5-C11) are on the opposite sides of the lipid bilayer. These results indicate that epitopes for both $VG_4$ and $IIC_9$ are located on the extracellular surface. Examples of double-labeling with different combinations of the three antibodies $VG_4$, $VG_2$, and $IIC_9$ are show in figure 2. In all, small and large gold particles appear on the same side of the membrane, confirming the extracellular disposition of all three epitopes.

Competition binding studies were also performed to probe the sidedness of the binding sites for $VG_4$, $VG_2$, and $IIC_9$ (6). All three antibodies bound to sealed, right-side-out kidney vesicles in solution. This provided further support for the extracellular location of

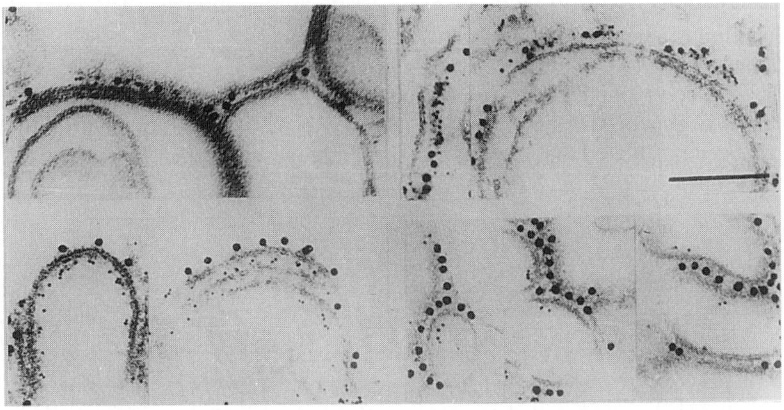

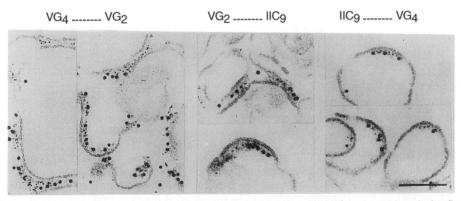

Figure 1. Upper Panels show $VG_4$ and $IIC_9$ (small gold particles) and the control for the extracellular side of the membrane (wheatgerm agglutinin, large gold particles) on the same side of the membrane. In the lower panels $VG_4$ and $IIC_9$ and the control for the cytoplasmic side (M10-P5-C11, large gold particles) are on opposite sides of the membrane. Scale bar: 200 nm.

Figure 2. Combinations of double-labeling with $VG_4$, $VG_2$, and $IIC_9$. In each pair the first marker was the small gold particle (5 nm) and the second was the large gold particle (15 nm). In all cases both labels are on the same side of the membrane. Scale bar: 200 nm.

359

the epitopes (6). The binding of IIC9 depended on experimental conditions, which explained why in previous studies it had not bound to intact cells (7).

An extracellular disposition for $VG_4$ and $VG_2$ binding sites is in agreement with previously published results (1,7). However, extracellular localization of the $IIC_9$ epitope contradicts the previous findings, which were based on its failure to stain intact cells.(7).

FOLDING MODELS

It is not possible to fit all the available data from studies of Na,K-ATPase and the homologous H,K-ATPase into a single model. Figure 3 illustrates four models for the folding of $\alpha$ subunit. All are consistent with the extracellular location of epitopes for $VG_2$ and $IIC_9$. Transmembrane spans in the C-terminal end have been given letter designations (A-G) to facilitate comparison among models. The asterisk (∗) denotes the location of $Asn^{831}$.

Span A corresponds to H5 in previously published models and there is general agreement that it begins on the cytoplasmic surface. Recently, based on tryptic digestion studies, it has been postulated that there is a transmembrane span (span B) immediately after span A

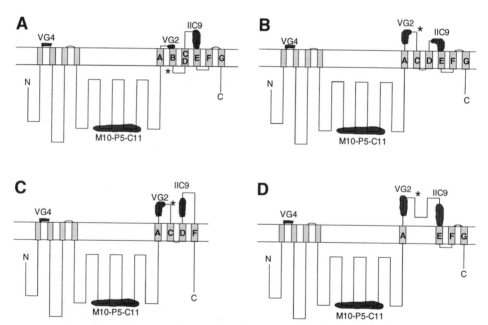

Figure 3. Models compatible with an extracellular location of $VG_4$, $VG_2$, and $IIC_9$ epitopes. The first 4 transmembrane spans (unlabeled) are the H1, H2, H3, and H4 spans found in all models of P-type ATPases. Spans at the C-terminal end are given letter designations which remain constant among models. Note that span B appears only in model A. The lengths of the lines outside the membrane are proportional to the number of amino acids predicted to lie there. The locations of epitopes for the 4 antibodies are shown. Asterisks mark the location of $Asn^{831}$, which lies on the cytoplasmic surface only in model A.

360

in Na,K-ATPase (5) and H,K-ATPase (3). These studies concluded that cleavage at Asn$^{831}$ (pig Na,K-ATPase) and Leu$^{853}$ (rat H,K-ATPase) occurred at the cytoplasmic surface.

These results are not compatible with our studies and an earlier report (7) that located the VG$_2$ epitope on the extracellular surface of the membrane. Model A in figure 3 shows a folding pattern that accommodates both our observations and the digestion results. However, in this model only a small part of the binding sequence is exposed for reaction with VG$_2$ and the transmembrane span B is uncharacteristically short and highly polar.

Our results are best accommodated by the other models in fig. 3. In these, the long hydrophobic stretch between the binding sites of VG$_2$ and IIC$_9$ (span C/D in model A) either crosses the membrane twice as in models B and C, or is entirely extracellular as in model D. In these models the tryptic cleavage site at Asn$^{831}$ is extracellular.

It is not yet established whether the polypeptide chain spans the membrane only once or three times after the IIC$_9$ site to arrive at the cytoplasmic location of the C-terminus. Figure 3 shows both possibilities. Models A, B and D have three and Model C has one transmembrane segment(s) after the IIC$_9$ site.

ACKNOWLEDGMENTS

We thank Dr. W.J. Ball Jr. for his generous gift of Mab M10-P5-C11. This work was supported by grants GM35399 (MM) and HL36271 (KJS). Figures 1 and 2 are from (6) and are printed by permission from the Journal of Biological Chemistry.

REFERENCES

1. Arystarkhova, E, Gasparian, M, Modyanov, NN and Sweadner, KJ (1992) Na,K-ATPase extracellular surface probed with a monoclonal antibody that enhances ouabain binding. J. Biol. Chem. 267:13694-13701
2. Ball, Jr., WJ (1986) Uncoupling of ATP binding to Na$^+$,K$^+$-ATPase from its stimulation of quabain binding: studies of the inhibition of Na$^+$,K$^+$-ATPase by a monoclonal antibody. Biochemistry 25:7155-7162
3. Besancon, M, Shin, JM, Mercier, F, Munson, K, Miller, M, Hersey, S and Sachs, G (1993) Membrane topology and omeprazole labeling of the gastric H$^+$,K$^+$-adenosine triphosphatase. Biochemistry 32:2345-2355
4. Jorgensen, PL (1988) Purification of Na$^+$,K$^+$-ATPase: enzyme sources, preparative problems, and preparation from mammalian kidney. Methods Enzymol. 156:29-43
5. Karlish, SJD, Goldshleger, R and Jorgensen, PL (1993) Location of Asn$^{831}$ of the alpha chain of Na,K-ATPase at the cytoplasmic surface: implication for topological models. J. Biol. Chem. 268:3471-3478
6. Mohraz, M, Arystarkhova, E and Sweadner, KJ (1993) Immunoelectron microscopy of epitope on Na,K-ATPase catalytic subunit. Implications for the transmembrane organization of the C-terminal domain. J. Biol. Chem. In press
7. Ovchinnikov, YuA, Luneva, NM, Arystarkhova, EA, Gevondyan, NM, Arzamazova, NM, Kozhich, AT, Nesmeyanov, VA and Modyanov, NN (1988)Topoloy of Na$^+$,K$^+$-ATPase . Identification of the extra- and intracellular hydrophilic loops of the catalytic subunit by specific antibodies. FEBS Lett. 227:230-234
8. Sweadner, KJ and Arystarkhova, E (1992) Constraints on models for the folding of the Na,K-ATPase. Ann. N.Y. Acad. Sci. 671:217-227

# Identification of epitopes recognized by Na$^+$/K$^+$-ATPase-directed antibodies

H.Homareda, Y.Nagano, H.Matsui

Department of Biochemistry, Kyorin University School of Medicine, Mitaka, Tokyo 181, JAPAN

## Introduction

It would be expected that the surface of the Na$^+$/K$^+$-ATPase molecule is responsible for the antigenic properties of the enzyme than the interior. Therefore, we examined the surface of the intact Na$^+$/K$^+$-ATPase molecule to identify the epitopes recognized by polyclonal antibodies directed against it. The data obtained suggested that K508*, a putative ATP binding site, is located on the surface of the enzyme molecule, whereas other putative ATP binding sites, and the phosphorylation site at D376*, are located in the interior of the molecule.

## Methods

Polyclonal antibodies against the Na$^+$/K$^+$-ATPase holoenzyme, and its $\alpha$- and $\beta$-subunits which were isolated after denaturation of the holoenzyme with SDS, were prepared as described elsewhere (2).

The purified Na$^+$/K$^+$-ATPase (20 $\mu$g) was preincubated for 1 h at 37°C with 4 $\mu$l antiserum or 0-30 $\mu$g IgG in the reaction mixture without ATP (100 mM NaCl, 10 mM KCl, 5 mM MgCl$_2$, 1 mM EDTA and 50 mM imidazole-HCl, pH 7.4). A portion of the preincubation mixture containing 2 $\mu$g enzyme was used for measurement of ATPase activity.

The cDNAs for the $\alpha$- and $\beta$-subunits of human Na$^+$/K$^+$-ATPase (6,7) were cleaved with suitable restriction enzymes, and then, ligated into pSP64, pSP65 or pGEM-3Zf(+) plasmids. The linearized plasmids were expressed in a cell-free transcription and translation system supplemented with [$^{35}$S]methionine. The synthesized peptides were immunoprecipitated with 4 $\mu$l antiserum and then subjected to SDS-PAGE by the methods of Anderson and Blobel (1) and Laemmli (8), respectively. The detailed methods have been reported previously (2).

---

*Amino acids positions quoted in the present data are based on those of the human Na$^+$/K$^+$-ATPase $\alpha$- and $\beta$-subunits

## Results and Discussion

Anti-D and anti-PK, which are polyclonal antibodies directed against canine and porcine $Na^+/K^+$-ATPase, respectively, became bound to the intact $Na^+/K^+$-ATPase (Fig. 1) and inhibited its activity (Fig. 2a). On the other hand, anti-$\alpha$ and anti-$\beta$, which are polyclonal antibodies directed against the SDS denatured $\alpha$- and $\beta$-subunits, respectively, did not bind to the intact enzyme (Fig. 1) or inhibit its activity (Fig. 2b). Since it was expected that the epitopes recognized by anti-D and anti-PK would be present on the surface of the intact $Na^+/K^+$-ATPase molecule, and that some of them would be essential for ATPase activity, we attempted to identify the epitopes by immunoprecipitation of peptide fragments corresponding to various regions of the $\alpha$- and $\beta$-subunits with the antibodies. Anti-D and anti-PK bound to the M32-D75, M158-D197 and M470-L534 regions of the $\alpha$-subunit and the C-terminal one-half of the $\beta$-subunit (Fig. 3). These regions must be present on the surface of the intact $Na^+/K^+$-ATPase molecule, and some of them must be involved in ATPase activity. The antibodies employed did not appear to bind to the M267-I442 and M615-Y1023 (C-terminal) regions of the $\alpha$-subunit or to the N-terminal one-third of the $\beta$-subunit (Fig. 3). These results suggest that K508, which is one of the putative ATP binding sites proposed on the basis of results obtained by the chemical modification technique (4), is located on the surface of the molecule, whereas other putative ATP sites, e.g., D716, D721 and K726, the phosphorylation site at D376, the C-terminal one-third of the $\alpha$-subunit which contains the region resistant to trypsin digestion (5), the region responsible for association with the $\beta$-subunit (9), and the N-terminal one-third of the $\beta$-subunit are located in its interior.

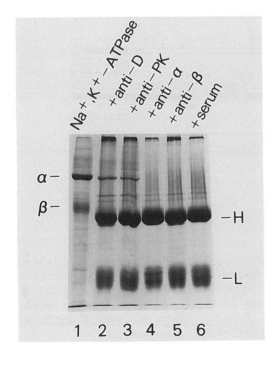

Figure 1. Binding of antibodies with the intact $Na^+/K^+$-ATPase molecule. Twenty micrograms of $Na^+/K^+$-ATPase was incubated with 4 µl anti-D, anti-PK, anti-$\alpha$, anti-$\beta$ and control serum for 1 h at 37°C in a reaction mixture without ATP. The antibody-enzyme complex was collected and subjected to electrophoresis by the methods of Anderson and Blobel (1) and Laemmli (8), respectively. The gel was stained with coomassie brilliant blue. $\alpha$ and $\beta$ indicate $\alpha$- and $\beta$-subunits of $Na^+/K^+$-ATPase, and H and L indicate the heavy and light chains of IgG, respectively.

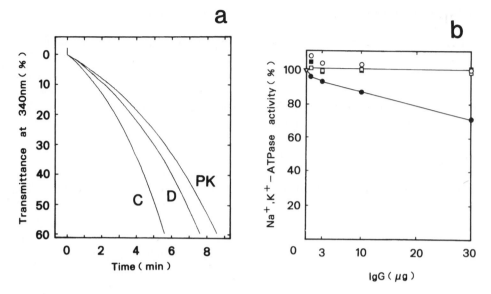

Figure 2. Effect of antibodies on ATPase activity. Twenty micrograms of Na$^+$/K$^+$-ATPase was incubated with a) 4 $\mu$l control serum (C), anti-D (D) or anti-PK (PK), b) 0-30$\mu$g IgG from control serum (O) anti-D (●) anti-$\alpha$ (□) or anti-$\beta$ (■) for 1 h at 37°C in the reaction mixture without ATP. A portion of the incubation mixture containing 2 $\mu$g enzyme was used for measurement of ATPase activity by a) a coupled assay system or b) a standard method.

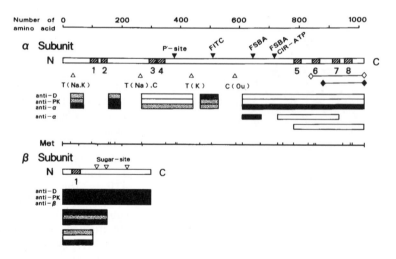

Figure 3. Summary of immunoprecipitation reactions. Black, dotted and white areas denote strong, weak and no antibody binding sites. P-site denotes the phosphorylation site. FITC, FSBA, and CIR-ATP denote putative ATP binding sites (4). (◇) and (◆) show the region responsible for association with $\beta$-subunit (9) and the region resistant to trypsin digestion (5), respectively. T and C denote the sites of tryptic and chymotryptic cleavage in the presence of NaCl or KCl, respectively (3).

Anti-α bound to the M615-Q677 region as well as the regions that bound to anti-D and anti-PK, but showed little or no binding to the M267-I442 and M734-Y1023 regions, respectively (Fig. 3). Anti-β bound to the N-terminal one-third as well as the C-terminal two-thirds of the β-subunit (Fig.3). These results suggest that the M617-Q677 region of the α-subunit and the N-terminal one-third of the β-subunit are exposed by denaturation of the enzyme with SDS, whereas the M267-I442 and M734-Y1023 regions are usually located in the enzyme interior.

*Acknowledgments*

We wish to thank Drs. Kiyoshi Kawakami, Osamu Urayama and Yutaro Hayashi for suppling the cDNA of human Na$^+$/K$^+$-ATPase, antibody against porcine kidney Na$^+$/K$^+$-ATPase and purified Na$^+$/K$^+$-ATPase, respectively, and Ms. Nobuko Shinji for excellent technical assistance.

**References**

1. Anderson DJ, Blobel G (1983) Immunoprecipitation of proteins from cell-free translations. Methods Enzymol 96:111-120
2. Homareda H, Nagano Y, Matsui H (1993) Immunochemical identification of exposed regions of the Na$^+$/K$^+$-ATPase α-subunit. FEBS Lett 327:99-102
3. Jørgensen PL, Farley RA (1988) Proteolytic cleavage as a tool for studying structure and conformation of pure membrane-bound Na$^+$/K$^+$-ATPase. Methods Enzymol 156:291-301
4. Kaplan JH (1990) Localization of ligand binding from studies of chemical modification. In Kaplan JH, Weer PD (eds) The Sodium Pump: structure, mechanism, and regulation. The Rockefeller University Press, New York, pp117-128
5. Karlish SJD, Goldshleger R, Stein WD (1990) A 19-kDa C-terminal tryptic fragment of the α chain of Na/K-ATPase is essential for occlusion and transport of cations. Proc Natl Acad Sci USA 87:4566-4570
6. Kawakami K, Nojima H, Ohta T, Nagano K (1986) Molecular cloning and sequence analysis of human Na/K-ATPase β-subunit. Nucleic Acids Res 14:2833-2844
7. Kawakami K, Ohta T, Nojima H, Nagano K (1986) Primary structure of the α-subunit of human Na/K-ATPase deduced from cDNA sequence. J Biochem 100:389-397
8. Laemmli UK (1970) Cleavage of structural proteins during the assembly of the head of bacteriophage T4. Nature 227:680-685
9. Lemas MV, Takeyasu K, Fambrough DM (1992) The carboxyl-terminal 161 amino acids of the Na/K-ATPase α-subunit are sufficient for assembly with the β-subunit. J Biol Chem 267:20987-20991

# Probing the α-subunit folding by dot-sandwich immunochemical analysis

K.N. Dzhandzhugazyan, N.N. Modyanov

Shemyakin Institute of Bioorganic Chemistry, Russian Academy of Sciences, ul. Miklukho-Maklaya, 16/10, 117871, Moscow V-437, Russia

## Introduction

Folding of the $Na^+/K^+$-ATPase α-subunit still remains uncertain, particularly in C-terminal portion of the polypeptide chain. Among other methods immunochemical approaches are widely employed in studies of the $Na^+/K^+$-ATPase topography. As a common such studies include comparison of purified enzyme with that in sealed membrane vesicles or cells, where $Na^+/K^+$-ATPase is a minor component and crossreactivity with other proteins can be hardly ruled out. Disruption of the vesicles or cells integrity for epitope sidedness determination might influence antigenic structure and antibodies (ABs) access to the epitope in many ways and not only providing penetration of ABs to the opposite side of the plasma membrane. Due to association of purified $Na^+/K^+$-ATPase with open membrane fragments dot-sandwich-ELISA as more direct method was chosen to probe the recognition sidedness of different enzyme-specific ABs in this work.

## Localization of $VG_2$, $VG_4$ and $IIC_9$ epitopes

For dot-sandwich-ELISA assay ABs of known sidedness of recognition were applied on nitrocellulose filter, and sandwich (AB-ATPase-AB*) was formed by asymmetric immunoadsorption of $Na^+/K^+$-ATPase membrane fragments followed by incubation with set of $Na^+/K^+$-ATPase-specific, second ABs* to be probed in comparison with the first applied AB. The second ABs* were peroxidase-conjugated, as indicates asterisk.

All poly- and monoclonal ABs to pig kidney ATPase were isolated by protein A - Sepharose affinity chromatography, except pigeon AB (β), isolated by adsorption on $Na^+/K^+$-ATPase (3). For dot-sandwich-ELISA indicated in the figures the first ABs (0.2-2 μg) were applied on 0.1 μm pore size nitrocellulose filter sheet (Millipore) in 50 μl TBS by BIO-DOT apparatus. After blocking by 5% nonfatty dry milk/TBS sheet was incubated overnight at 4°C with $Na^+/K^+$-ATPase, 0.2 μg/ml in 100 ml TBS. Washed nitrocellulose sheet was placed on filter paper to soak wet and then its pieces carrying complete set of applied ABs were incubated for 30 min in 5% dry milk/TBS/0.25M sucrose with one of peroxidase-conjugated anti-ATPase ABs* diluted 80-300 fold. ABs* binding was visualized in peroxidase-catalyzed reaction with 4-chloro-1-naphtol. Dilution of applied ABs was chosen from serial dilutions in direct ELISA to produce similar binding of $Na^+/K^+$-ATPase per dot, tested by fluorescamine- or [$^{125}$I]-Bolton-Hunter-labelled ATPase.

Remarkable asymmetry of $Na^+/K^+$-ATPase immobilization (i.e. low AB* binding, when the same AB, or similar one in sidedness, was used as the first AB) was observed, when at constant enzyme amount $Na^+/K^+$-ATPase concentration was reduced to 10 μg/50ml TBS. Probably, binding from diluted solutions provides higher asymmetry due to better immobilization along all the surface of the membrane sheet rather then its attachment.

Polyclonal rabbit ABs to the enzyme and pigeon ABs to the $\beta$-subunit were shown to be specific to the intra- and extracellular portions of $Na^+/K^+$-ATPase, respectively (3). Sequence including epitope for monoclonal $VG_4$ was localized at the extracellular surface of the membrane (2). Relationship in the recognition sidedness for these marker ABs was tested in dot-sandwich-ELISA, upon control of similar binding of fluorescamine- or $[^{125}I]$-Bolton-Hunter-labelled ATPase per dot. Results were consistent with previous marker ABs mapping. Therefore, $VG_4$ was employed as the marker of extracellular localization.

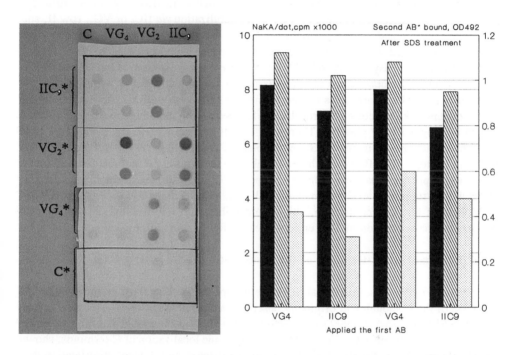

Figure 1 (left). Mapping of $VG_2$ and $IIC_9$ epitopes. Eight dots were applied vertically of each of the four ABs - C (control IgG), $VG_4$, $VG_2$ and $IIC_9$ on the nitrocellulose filter. After incubation with $Na^+/K^+$-ATPase the filter was cut into four pieces each containing two dots of each of the applied ABs. The filter pieces were incubated with peroxidase-conjugated (*) $IIC_9^*$, $VG_2^*$, $VG_4^*$ and $C^*$, respectively marked.

Figure 2 (right). Effect of $Na^+/K^+$-ATPase SDS treatment on the $VG_2^*$ and $\alpha$-p999* binding. 12 dots of each of $VG_4$ and $IIC_9$ were applied on two nitrocellulose filters. The filters were incubated with $[^{125}I]$-Bolton-Hunter-labelled ATPase, one with purified and one with additionally SDS treated enzyme. Each filter was cut in two (containing 6 dots of each of applied ABs) and incubated with peroxidase-conjugated ABs* - one half with $VG_2^*$ (dashed), the other half with $\alpha$-p999* (dotted bars). Bound ATPase (black bars, cpm/dot) and the second bound AB* ($OD_{492}$ in peroxidase-catalized reaction with o-phenylenediamine) were compared.

Fig. 1 shows localization of epitopes for $VG_2$ and $IIC_9$ in comparison with $VG_4$. Amount of dotted ABs, as above, was chosen to produce comparable $Na^+/K^+$-ATPase binding. Therefore, comparison of peroxidase-catalyzed product intensity for each row reveals in highest staining in pairs $IIC_9^*$-$VG_2$, $VG_2^*$-$IIC_9$, $VG_2^*$-$VG_4$, $VG_4^*$-$VG_2$. Taking $VG_4$ as the

367

marker of the outer surface, the obtained result means, that epitope for $VG_2$ is localized at opposite from $VG_4$ side of the membrane, i.e. on the cytoplasmic surface, and epitope for $IIC_9$ - on the extracellular side. The same disposition of $VG_2$ and $IIC_9$ epitopes was observed, when pigeon ABs to the $\beta$-subunit with the same sidedness of recognition as $VG_4$, were applied as support ABs. Conformational changes affecting epitope structure and (or) accessibility could provide an alternative explanation of results obtained - different binding capacity for conjugated second AB* dependent on the first bound AB. Nevertheless, strong and uniform conformational response, mutually induced by three different monoclonal ABs (increased binding of $VG_2$* at immobilization via $IIC_9$ or $VG_4$, row II, and vice versa, rows I and III) seems rather doubtful. The fact, that the same recognition pattern was observed with ABs to the $\beta$-subunit used as support instead of $VG_4$, further stressed the role of asymmetry of $Na^+/K^+$-ATPase immobilization in sidedness of AB binding. In addition, total change in binding capacity for conjugated ABs* requires reaction of majority of ATPase molecules on the membrane fragments with the first ABs - taking into account net-like structure of nitrocellulose filters carrying immobilized ABs quantitative antigen binding is hardly possible. Therefore, results on Fig. 1 place fragment 810-825 of the $\alpha$-chain, including epitope of $VG_2$, in cytoplasmic domain of the $\alpha$-subunit, and 887-904 including epitope for $IIC_9$ - in the extracellular region of the polypeptide chain. It contradicts to previous mapping of these epitopes (2,9). Nevertheless, former result on $VG_2$ agrees with cytoplasmic localization of Asp-831 (6), and sidedness of $IIC_9$ epitope, revised by Mohraz et al., this conference, (8) is compatible with our finding. Discrepancy in immunochemical detection of $VG_2$ epitope localization remains to be elucidated.

### Disposition of the C-terminus of the $\alpha$-subunit

Polyclonal rabbit ABs, raised against synthetic peptide matching the C-terminal sequence 999-1008 of the $\alpha$-subunit and originally described as $\alpha$-p999 (9), were tested also in dot-sandwich-ELISA. When $\alpha$-p999 were applied on the filter as the first ABs, strong staining with $IIC_9$* was observed. It is consistent with cytoplasmic localization of C-terminus, shown immunochemically by Antolovic et al. (1), and confirmed by chemical modification data (7). When, on the contrary, enzyme was immobilized via extracellular side, i.e. on applied $IIC_9$ or $VG_4$ (Fig 2), strong recognition by $VG_2$* and more weak by $\alpha$-p999* was observed at comparable binding of the $Na^+/K^+$-ATPase - the difference could be explained by originally described lower affinity of $\alpha$-p999 to the enzyme. Treatment of $Na^+/K^+$-ATPase at 0.09% SDS in the presence of 3 mM ATP resulted in loss of ~ 90% of initial enzymatic activity and shifted pattern of recognition. $[^{125}I]$-enzyme was immunoadsorbed by dots $VG_4$ and $IIC_9$ and recognized by $VG_2$* with almost the same efficiency, as before SDS treatment (compare black and dashed bars in the first group with the third one, and in the second group with the fourth one, Fig.2), but more strong response was observed with $\alpha$-p999* (dotted bars). Quantification, performed by measurement of soluble product in peroxidase-catalyzed reaction with o-phenylenediamine and determination of the $Na^+/K^+$-ATPase bound per dot (cpm of $[^{125}I]$-labelled enzyme) resulted in increase of specific binding at C-terminus by about 50% as a result of additional detergent treatment. It indicates, that C-terminus might be shadowed by association with other fragments of the $\alpha$ or N-terminus of the $\beta$, or (and) embedded in cytoplasmic monolayer. This is consistent with both conflicting results on hydrophobic modification and tryptic release of the C-terminal fragment from plasma membrane after Trp residues modification procedure.

368

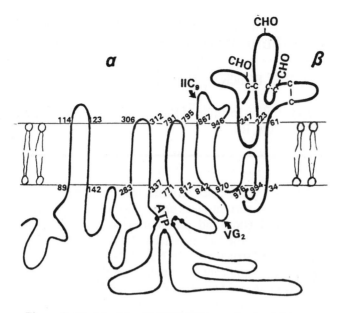

Figure 3. Model of the $Na^+/K^+$-ATPase subunits folding.

Fig. 3 represents one of the probable models of the $Na^+/K^+$-ATPase subunits folding accommodating extracellular localization of epitopes for $IIC_9$ and $VG_4$, intracellular - for $VG_2$ and embedded into cytoplasmic monolayer hairpin of C-terminal fragment of the $\alpha$-subunit. Results on chemical modification in intra- and extracellular domains of $Na^+/K^+$-ATPase, as well as relative distribution of the label among extracellular regions of the $\alpha$- and $\beta$-subunits, are taken into consideration (4,5) and tight, versatile association between $\alpha$- and $\beta$-subunits thought to be important for coordinated stabilization of the enzyme structure is drawn. Possible functional role of accessibility of the C-terminus in $Na^+/K^+$-ATPase $\alpha$-subunit, proposed for some of P-type ATPases, remains to be established.

## References

1.  Antolovic, R., Brüller, H.-J., Bunk, S., Linder, D., Schoner, W. (1991) Eur. J. Biochem., 199: 195-202.
2   Arystarkhova, E., Gasparian, N., Modyanov, N.N., Sweadner, K.J. (1992) J. Biol. Chem., 267: 13694-13701.
3.  Dzhandzhugazyan, K.N., Modyanov, N.N., Ovchinnikov, Yu.A. (1981) Bioorg. Khimia, 7: 847-857.
4.  Dzhandzhugazyan, K.N., Modyanov, N.N. (1985a) In The sodium pump. (eds. Glynn, J., Ellory, C.) The Company of Biologists Ltd., Cambridge, 129-134.
5.  Dzhandzhugazyan, K.N., Jorgensen, P.L. (1985b) Biochim. Biophys. Acta, 817: 165-173.
6.  Karlish, S.J.D., Goldshleger, R., Jørgensen, P.L. (1993) J. Biol. Chem., 268: 3471-3478.
7.  Modyanov, N.N., Vladimirova, N.M., Gulaev, D.J., Efremov, R.G. (1992) Annals N.J.Acad.Sci.USA, 671: 134-145.
8.  Mohraz, M., Arystarkhova, E., Sweadner, K.J. (1993) Biol. Chem. Hoppe-Seyler, 374:580.
9.  Ovchinnikov, Yu.A., Luneva, N.M., Arystarkhova, E.A., Gevondyan, N.M., Arzamazova, N.M., Kozhich, A.T., Nesmeyanov, V.A., Modyanov, N.N. (1988) FEBS Lett., 227: 230-234.

# Determination of sidedness of COOH-terminus of Na$^+$/K$^+$-ATPase alpha subunit by vectorial labeling

N.M. Vladimirova, R.E. Efendiyev, N.A. Potapenko, N.N. Modyanov

Shemyakin and Ovchinnikov Institute of Bioorganic Chemistry Russian Academy of Sciences, 117871 Moscow, Russia

## Introduction

Extensive studies have been performed on structure-function relationship of the Na$^+$/K$^+$-ATPase. Nevertheless, current ideas on basic processes underlying coupled ATP hydrolysis, energy transduction and cation transport are deficient in molecular details. This is mainly due to the lack of reliable data on the spatial protein structure (1). The three-dimensional structure of the Na$^+$/K$^+$-ATPase cannot yet be solved by X-ray analysis. However, significant information on the spatial organization of the protein molecule may be derived by other experimental approaches such as limited proteolysis, affinity modification, labeling by permeable and impermeable reagents, analysis with specific antibodies and spectroscopic methods. A combination of these techniques has been used to identify functional domains and to probe the folding of subunit polypeptide chains in the membrane. It is generally accepted that the beta subunit spans the membrane only once (1,4). In contrast, the transmembrane folding of the catalytic alpha subunit has been the subject of considerable controversy. Several topological models differ in the number and location of the membrane-spanning segments within the COOH-terminal half of the polypeptide chain and therefore the orientation of the COOH-terminus (2,3,5,7).

We describe here a biochemical approach to the spatial location of the COOH-terminus of Na$^+$/K$^+$-ATPase alpha subunit. The method compares the extent of the lactoperoxidase (LPO)-catalyzed radioiodination of the two C-terminal tyrosine residues exposed on the surface of the enzyme molecule in intact cells or exposed in broken membrane preparation or purified, membrane bound enzyme. It is shown that the COOH-terminal tyrosine residues of the alpha subunit are accessible to modification only when LPO has access to the inner side of the plasma membrane as is the case in the membrane preparations but not in intact cells. Thus the COOH-terminus of the alpha subunit is located on the cytoplasmic surface of the pump molecule, and the polypeptide chain must have an even number of transmembrane segments. Based on these data we have corrected the model for the alpha subunit proposed earlier (6).

### Spatial localization of the alpha subunit COOH-terminus

The location of the COOH-terminal fragment of the Na$^+$/K$^+$-ATPase alpha subunit has been the subject of considerable controversy as pointed out above using either theoretical or experimental methods.

For example probing the sidedness of the COOH-terminus of the alpha subunit by sequence-specific antibodies against synthetic peptides produced conflicting results. This region in mammalian enzymes has the following sequence: (991)IFVY-DEVRKLIIRRRPGGWVEKETYY(1016) (5,7). Although apparently providing different results, these antibody studies show that the C-terminal region of the molecule is accessible in the membrane bound form. This suggest that iodination of the exposed C-terminal tyrosines would be successful as a method for resolving the location of the C-terminal amino acids with respect to the plane of the membrane.

The optimal conditions for enzymatic reactions and reliable procedures for isolation of the alpha subunit and for analysis of labeled amino acids were determined in experiments on membrane bound $Na^+/K^+$-ATPase. Table 1 presents data from experiments on labeling of the alpha subunit within the membrane-bound $Na^+/K^+$-ATPase with lactoperoxidase/$^{125}$I at different molar ratio iodide to protein. In the first set of experiments in order to achieve surface-specific labeling and to avoid possible membrane damage especially in case of intact cells, the extent of modification was limited to a small part of the iodinatable sites. In this case only a very small fraction of tyrosines in the alpha subunit was labeled, and this raises the possibility that an unrepresentative population consisting of denaturating molecules of the $Na^+/K^+$-ATPase preparation might be a preferential target for LPO-mediated iodination. To answer this question two additional experiments on the labeling with the excess of iodide were performed (Table 1 b and c). The alpha subunit was isolated by electrophoresis and transferred to the PVDF-membrane. A selected cleavage of the C-terminus of the immobilized protein was achieved by the two-step hydrolysis with carboxypeptidases B and Y. The set of cleaved amino acids completely corresponded to the COOH-terminus sequence of the alpha subunit.

Table 1. Labeling of the alpha subunit within the membrane-bound $Na^+/K^+$-ATPase with lactoperoxidase /$^{125}$I at molar ratio iodide to protein: a) 0.5 : 1, b) 22.5 : 1, c) 45 : 1.

| $\alpha$-subunit band on PVDF membrane cpm* | Iodine incorp. pmol* | cpB + cpY treatment | | Released tyrosine pmol* | COOH-terminal label % |
|---|---|---|---|---|---|
| | | Released label cpm* | PVDF membrane cpm* | | |
| a) 6.72x10⁵ | 0.129 | 1.61x10⁵ | 5.05x10⁵ | 152 | 32 |
| 8.01x10⁵ | 0.154 | 1.51x10⁵ | 6.44x10⁵ | 136 | 28 |
| 7.30x10⁵ | 0.140 | 1.90x10⁵ | 5.33x10⁵ | 168 | 31 |
| | | | | | |
| b) 5.46x10⁵ | 52 | 1.32x10⁵ | 3.36x10⁵ | 146 | 33 |
| 4.92x10⁵ | 46 | 1.08x10⁵ | 3.80x10⁵ | 150 | 29 |
| | | | | | |
| c) 1.93x10⁶ | 183 | 0.40x10⁶ | 1.50x10⁶ | 132 | 32 |
| 2.12x10⁶ | 200 | 0.36x10⁶ | 1.72x10⁶ | 124 | 27 |

* - Data are given per 100 pmol of protein.

Analysis of dansyl-derivatives of cleaved amino acids revealed that radioactivity removed from the protein belonged only to iodinated tyrosine. Taking into account the yield of carboxypeptidase hydrolysis, incorporation of the radioactive label into COOH-terminal tyrosines was 27-33%. Data clearly show that one or both COOH-terminal tyrosine residues of the alpha subunit are proper substrates for labeling by $^{125}$I and lactoperoxidase. If the level of labeling of the targets is almost equal, then only 3-7 tyrosine residues (including the COOH-terminal ones) from 24 in the alpha subunit are accessible to modification. It was shown that up to two atoms of iodide could be incorporated into the alpha subunit molecule, when molar ratio of reagent/protein was increased to 45/1. Even in this case, unspecific labeling was not detected in control experiments without LPO or peroxide. About 30% of the label was again found in COOH-terminal tyrosine residues of the alpha subunit. The extent of modification of the particular residues in these experiments was about 60%, if one tyrosine was iodinated or 30%, if both were labeled. It is improbably that this high specific activity preparation contains this level of denaturated enzyme. Incidentally it was found in this experiment that the LPO-mediated iodination of tyrosine residues located on the surface of the $Na^+/K^+$-ATPase molecule led to substantial inhibition (up to 70%) of ATP-hydrolysing activity. The effect was not detected, when LPO or peroxide were omitted from the reaction mixture. Determination of the location of labeled residues in subunit polypeptide chains, now in progress, could provide new information on the topology and activity of the $Na^+/K^+$-ATPase molecule. To establish orientation of the alpha subunit C-terminal domain relative to the plasma membrane, a series of experiments on vectorial labeling of intact cells of the pig kidney embryonic (PKE) cell line were performed along with modification of the plasma membrane fraction (PMF) isolated from the same cells. If these residues are exposed on the cell outer surface, their level of labeling should then be almost the same in both experiments. On the contrary, if incorporation of the label increases significantly on modification of the open plasma membrane fragments, the COOH-terminus is oriented cytoplasmically. The viability of the cells being labeled must be maintained as close to 100% as possible, because labeling of dead cells is not restricted to external plasma membrane components. In our case, the viability of cells, determined by Trypan Blue exclusion, was about 94-95%, and did not change after limited iodination.

Labeling of the intact cells and PMF was performed under the conditions developed in the first set of experiments for the isolated $Na^+/K^+$-ATPase. To isolate the individual alpha subunit, a mixture of plasma membrane proteins was separated by PAAG electrophoresis and transferred onto PVDF membrane. The precise position of the alpha subunit band was established using the visualization procedure previously described (8) and Western blotting with alpha-p999 antibodies. Homogeneity of isolated samples was verified by $NH_2$-terminal analysis. The content of the radioactive label in C-terminal tyrosines comprised 27-28% and 33-35% of the total for PMF and cells, respectively (Table 2). At the same time, the extent of alpha subunit labeling in the experiments with intact cells and PMF differs drastically. In the first case the specific incorporation of the label comprises only about 6% of the latter. This value correlates well with the percentage content of disrupted cells in the cell samples used (viability 94-95%).

372

Table 2. Labeling of the $Na^+/K^+$-ATPase alpha subunit within the plasma membrane fraction and viable cells with lactoperoxidase /$^{125}$I.

| | $\alpha$-subunit band on PVDF membrane cpm* | cpB + cpY treatment | | Released tyrosine pmol* | COOH-terminal label % |
| | | Released label cpm* | PVDF membrane cpm* | | |
|---|---|---|---|---|---|
| PMF | $2.53 \times 10^5$ | $0.40 \times 10^5$ | $2.08 \times 10^5$ | 120 | 27 |
| | $2.27 \times 10^5$ | $0.41 \times 10^5$ | $1.80 \times 10^5$ | 130 | 28 |
| Viable cells | $1.52 \times 10^4$ | $0.38 \times 10^4$ | $1.10 \times 10^4$ | 140 | 35 |
| | $1.40 \times 10^4$ | $0.29 \times 10^4$ | $1.06 \times 10^4$ | 135 | 33 |

* - Data are given per 100 pmol of protein.

These findings led us to conclude that upon limited enzymatic iodination of viable cells the $Na^+/K^+$-ATPase, alpha subunit can be labeled only within disrupted cells, i.e. modification of the alpha subunit takes place only when lactoperoxidase has an access to the inner side of the plasma membrane. Thus, all targeted tyrosine residues (including the COOH-terminal ones) are exposed on the cytoplasmic surface of the pump molecule. This means that the alpha subunit polypeptide chain has an even number of transmembrane segments.

**References**

1. Horisberger J-D, Lemas V, Kraechenbuhl JP, Rossier BC (1991) Structure-function relationship of $Na^+/K^+$-ATPase. Ann Rev. Physiol 53: 565-584
2. Kawakami K, Noguchi S, Noda M, Takahashi H, Ohta T, Kawamura M, Nojima H, Nogano K, Hi rose T, Inayama S, Hayashida H, Miyata T, Numa S (1985) Primary structure of the $\alpha$-subunit of *Torpedo Californica* $Na^+/K^+$-ATPase deduced from cDNA sequence. Nature (London) 316: 733-736
3. Karlish SJD, Goldshleger R, Tal D-M, Capasso JM, Hoving S, Stein WD (1992) Identificaton of the cation binding domain of $Na^+/K^+$-ATPase. Acta Physiol Scand 146: 69-76
4. Modyanov N, Lutsenko S, Chertova E, Efremov R (1991) Architecture of the sodium pump molecule: probing the folding of the hydrophobic domain. In: Kaplan J, De Weer P (Eds) The Sodium Pump: Structure, Mechanism, and Regulation. pp 99-115
5. Ovchinnikov Yu, Arsenyan S, Broude N, Petrukhin K, Grishin A, Aldanova N, Arzamazova N, Arystarkhova E, Melkov A, Smirnov Yu, Guryev S, Monastyrskaya G, Modyanov N (1985) Primary structure of the $\alpha$-subunit of the pig kidney $Na^+/K^+$-ATPase deduced from cDNA sequence. Proc Acad Sci USSR 285: 1491-1495
6. Ovchinnikov YuA, Arzamazova NM, Arystarkhova EA, Gevondyan NM, Aldanova NA, Modyanov NN (1987) Detailed structural analysis of exposed domains of membrane-bound $Na^+/K^+$-ATPase. A model of transmembrane arrangement. FEBS Letters 217: 269-274
7. Shull GE, Schwartz A, Lingrel JB (1985) Amino-acid sequence of the catalytic subunit of the $Na^+/K^+$ ATPase deduced from a complementary DNA. Nature (London) 316: 691-695
8. Vladimirova (Arzamazova) NM, Potapenko NA, Levina NB, Modyanov NN (1991) $Na^+/K^+$-ATPase isoforms in different areas of calf brain. Biomedical science 2: 68-78

# Approaches to Disulfide Bonds Identification in the $Na^+/K^+$-ATPase Alpha Subunit

N.M. Gevondyan, N.N. Modyanov

Shemyakin and Ovchinnikov Institute of Bioorganic Chemistry Russian Academy of Sciences, 117871 Moscow, Russia.

## Introduction

$Na^+/K^+$-ATPase is known to be a thiol-dependent enzyme (1,2). Available data indicate to the presence of essential SH groups in the enzyme active site (10). Disulfide bonds have been identified and their reduction was shown to lead to inactivation of the enzyme (7,8). The $Na^+/K^+$-ATPase beta subunit was found to contain 3 disulfide bonds (8,9), while their number in the alpha subunit of the enzyme is still the subject of discussion. Evidently, the discrepancy in the reported results be due to the presence in $Na^+/K^+$-ATPase of a large number of masked SH groups, which complicates differentiating them from true cysteine residues involved in the formation of disulfide bonds (3). For understanding of the domain organization of the catalytic subunit, localization of S-S bonds in the polypeptide chain is of prime significance and therefore has been chosen as a focus of our present investigations.

## Analysis of cysteine residues in $Na^+/K^+$-ATPase

Ammetric titration with silver nitrate revealed in pig kidney $Na^+/K^+$-ATPase 20 free SH groups (6,56 of which were readily accessible and 13,4 masked) and 5 disulfide bonds (2 bonds in the alpha subunit and 3 bonds in the beta subunit).

The protocol for localization disulfide bonds in the alpha subunit involved the following steps: limited digestion of the membrane-bound $Na^+/K^+$-ATPase under conditions excluding the regrouping of disulfide bonds; isolation of cystine-containing peptides from the tryptic digest both alkylated and without prior alkylation; analysis of cystine-containing peptides and cysteinyl peptides obtained in the result of S-S bonds cleavage.

## Isolation and analysis of cysteinyl peptides from alkylated digest

In view of the large content of masked sulfhydryl groups in the native enzyme that are inaccessible for alkylation, the enzyme was subjected to trypsinolysis. However, low reactive potency of SH groups required very rigid conditions for alkylation, which resulted in the strong substance aggregation (4). As a consequence, HPLC separation of water-soluble fragments of the alkylated digest failed to yield homogeneous disulfide-containing peptides. It was only after reduction of disulfide bonds with excess 2-mercaptoethanol and alkylation of released SH groups with radiolabeled iodoacetic acid that 3 homogeneous modified peptides were isolated and identified. The N-terminal amino acid residues of these were Cys, Ile and Val, respectively. One of the fractions was detected to have a 2-fold higher radioactivity level which evidences the presence in the peptide of 2 modified cysteine residues. The amino acid analysis of the isolated peptides (4) thus confirmed the existence in their contents of cysteine residues modified after reduction and involved in disulfide bonds.

### Isolation and analysis of cystine-containing and cysteinyl peptides of tryptic digest without prior alkylation

Three disulfide-containing fractions obtained in the result of HPLC separation of water-soluble fragments in the tryptic digest of $Na^+/K^+$-ATPase were found to have S-S bonds with differential sensitivity to reduction (5).

Because of the simultaneous presence in the fractions of both S-S bonds and free SH groups, we undertook a step-by-step analysis of cysteine residues. The procedure included the determination of the amino acid sequences of cystine-containing peptides without prior reduction as well as using various methods of chemical modification of SH groups (alkylation by 4-(aminosulfonyl)-7-fluoro-2,1,3-benzoxadiazole (ABD-F) or 4-vinilpyridine) in the presence or absence of a reducing agent during the sequence analysis. As a result, 3 fragments of the alpha-subunit polypeptide chain consisting of the amino acids 452-461, 507-519, 545-558, respectively, were identified. All of them were demonstrated to contain cysteine residues involved in disulfide bonds which could be identified merely after isolation and analysis of cysteinyl peptides.

Localization of disulfide bonds in the $Na^+/K^+$-ATPase alpha-subunit was made possible in the result of amino acid sequence analysis of modified cysteinyl peptides isolated by fractionation of disulfide-containing peptides, successively subjected to alkylation with radiolabelled iodoacetic acid and then ABD-F prior to and following the reduction, respectively. The experimental conditions used were described before (6). Incorporation of the selective fluorescent label allowed the identification of cysteine residues involved in the formation of disulfide bonds.

The comparative analysis of all experimental results obtained allows a conclusion that the disulfide-containing peptides of the alpha subunit isolated from the $Na^+/K^+$-ATPase tryptic digest (both after preliminary alkylation and without prior alkylation) are identical and covalently bonded at positions: 452 and 456; 511 and 549 (Fig. 1). The presence of two covalent bonds in the cytoplasmatic domain of $Na^+/K^+$-ATPase suggests their functional significance.

Figure 1. The amino acid sequence of cystine-containing peptides of the $Na^+/K^+$-ATPase alpha subunit.

*Acknowledgements*

We are thankful to the Organizing Committee of the VII International Conference on The Sodium Pump: Structure, Mechanism, Hormonal Control and its Role in Disease, as well as the International Science Foundation for the offered opportunity to publish the materials. The studies were supported by Russian Foundation for Fundamental Investigations, grant 93-04-20327 and Protein Engineering grant 64.

### References

1.  Esmann M (1982) Sulfhydryl groups of $Na^+$ / $K^+$-ATPase from rectal glands of *Squalus acanthias*. Biochim Biophys Acta 688: 251-259
2.  Esmann M, Norby J (1985) A kinetic model for N-ethylmaleimide inhibition of the $Na^+/K^+$-ATPase from rectal glands of *Squalus acanthias*. Biochim Biophys Acta 812: 9-20
3.  Gevondyan NM, Gevondyan VS, Gavrilyeva EE, Modyanov NN (1989) Analysis of free sulfhydryl groups and disulfide bonds in $Na^+/K^+$-ATPase. FEBS Lett 255: 265-268.
4.  Gevondyan NM, Gevondyan VS, Modyanov NN (1993) Analysis of disulfide bonds in the $Na^+/K^+$- ATPase alpha subunit. Biochem and Molec Biol Int 29: 327-337
5.  Gevondyan NM, Gevondyan VS, Modyanov NN (1993) Sequence analysis of cystine-containing peptides in the $Na^+/K^+$-ATPase $\alpha$-subunit. Biochem and Molec Biol Int 30: 337-346
6.  Gevondyan NM, Gevondyan VS, Modyanov NN (1993) Location of disulfide bonds in the $Na^+/K^+$-ATPase alpha-subunit. Biochem and Molec Biol Int 30: 347-355
7.  Kawamura M, Ohmizo K, Morohashi M, Nagano K (1985) Protective effect of $Na^+$ and $K^+$ against inactivation of $Na^+/K^+$-ATPase by high concentrations of 2-mercaptoethanol at high temperatures. Biochim Biophys Acta 821: 115-120
8.  Kyrley TL (1989) Determination of three disulfide bonds and one free sulfhydryl in the $\beta$ subunit of $Na^+/K^+$-ATPase. J Biol Chem 264: 7185-7192
9.  Miller RP, Farley RA (1990) $\beta$-Subunit of $Na^+/K^+$-ATPase contains three disulfide bonds. Biochemistry 29: 1524-1532
10. Patzelt-Wenczler R, Schoner W (1975) Disulfide of thioinosine triphosphate. An ATP-analog inactivating $Na^+/K^+$-ATPase. Biochim Biophys Acta 403: 538-543

# Is the Cysteine$_{104}$ Part of the Sugar Binding Subsite of the Cardiac Glycoside Receptor of Na$^+$/K$^+$-ATPase?

R. Antolovic[1], J. Hahnen[2], D. Linder[3] and W. Schoner[1]

[1]Institute of Biochemistry & Endocrinology, Faculty of Veterinary Medicine; [2]Institute of Microbiology & Molecular Biology; [3]Institute of Biochemistry, Human Medical Faculty; Justus-Liebig-University, D-35392 Giessen, Germany.

## Introduction

Na$^+$/K$^+$-ATPase contains at its extracellular side of the catalytic $\alpha$ subunit the molecular receptor for cardiac glycosides. The cardiac glycoside receptor complex is stabilized when a phosphointermediate is formed either by the forward reaction in the presence of (Na$^+$ + Mg$^{2+}$ + ATP) yielding complex I or by "backdoor phosphorylation" in the presence of (Mg$^{2+}$ + P$_i$) yielding complex II. K$^+$ ions lower the affinity of the receptor for cardiac glycosides. From sequence comparisons of the $\alpha$ subunits of Na$^+$/K$^+$-ATPase of cardiac glycoside sensitive and insensitive species and from site-directed mutagenesis experiments the cardiac glycoside binding site has been localized to the extracellular loop between the 1st and 2nd membrane spanning region [1,2]. Cys$_{104}$ in this loop has been proposed to participate in ouabain binding [2]. An interaction with the rhamnose of the ouabain has been excluded, because the aglycone ouabagenin bound equally well [3]. There are indications that also the extracellular loop region between the third and the fourth membrane spanning region are part of the receptor [1]. Affinity labeling of the cardiac glycoside receptor site could be a means to realize whether the conclusions drawn from molecular biological experiments are justified. This was the reason why the action of two protein-reactive digoxigenin derivatives was studied.

## Materials and Methods

Right-side-out vesicles were prepared by the method of Forbush [4]. Na$^+$/K$^+$-ATPase with a specific activity of 28-34 U/mg was isolated from pig kidney by modification of Jørgensen's method [5]. Labeling of the enzyme with N-hydroxysuccinimidyl digoxigenin-3-O methyl-carbonyl-$\epsilon$-aminocaproate ester (HDMA) and digoxigenin-3-O succinyl-[2-(N-maleimido)] ethylamide (DSME) was done at 37°C in 2 hours in 20 mM Imidazol/HCl pH 7.2, 3 mM ATP, 3 mM MgCl$_2$ and 150 mM NaCl. Labeling of the proteins with HDMA and DSME was detected by antibodies against digoxigenin. Modeling of digoxigenin derivatives was performed using the chemnote 2-dimensional construction tool for generating 3-dimensional structure according to the CHARM force field. HDMA labeled material was cut out from PVDF membrane and analysed in a pulsed liquid phase sequencer for its amino acids sequence.

## Results and Discussion

As expected, incubation of purified $Na^+/K^+$-ATPase with increasing concentration of both, HDMA and DSME, at 37°C inhibited the enzyme. HDMA is a protein reactive digoxigenin derivative which modifies $\epsilon$-amino groups and DSME reacts with SH groups. After 60 min of incubation under the conditions forming complex I, HDMA inhibited $Na^+/K^+$-ATPase half-maximally at $5,7 \times 10^{-7}$ M and DSME at $5,7 \times 10^{-6}$ M. Presence of $K^+$ ions greatly decreased the inactivation. With HDMA ($Na^+$ + $Mg^{2+}$ + ATP)-supported labeling of the $\alpha$ subunit was seen which was hindered in the presence of an excess of ouabain (Fig. 1). Similar results were obtained with right-side-out vesicles which were incubated with HDMA. Therefore one must conclude that HDMA labels covalently the cardiac glycoside receptor at the extracellular side of the $\alpha$ subunit.

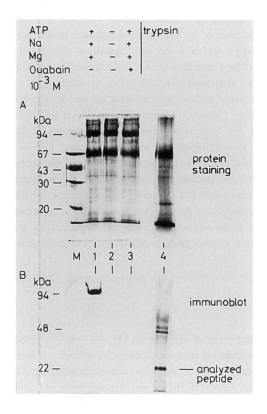

**Figure 1.** Western blot of an SDS-PAGE after incubation of $Na^+/K^+$-ATPase with $10^{-7}$ M HDMA for 60 min at 37°C under different conditions. Upper part of figure shows coomasie blue staining and lower part the immunostaining with digoxigenin-antibodies on PVDF membrane.

It is evident that the digoxigenin part of both derivatives is recognized by the cardiac glycoside binding site but the different substituents at the C3-OH group of ring A affect the interaction. A possible explanation for the differing reactivity of both digoxigenin derivatives is that both substituents have different length and orientation of their reactive groups. It may be therefore that they do not reach a reactive group within the cardiac glycoside binding site. To test this, the 3 dimensional structures of digoxi-

378

genin-monodigitoxoside, HDMA and DSMA and their energy minima under vacuum conditions were calculated. The results are shown in figure 2. It is evident from these calculations that in DSME the sulfhydryl-reactive maleimido group shows a completely different orientation than the aminoreactive N-hydroxy-succinimidyl group. This residue fills the space of the digitoxose substituent of digitoxigenin and is close to its 3'-OH and 4'-OH groups.

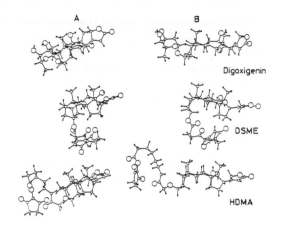

Digoxigenin

DSME

HDMA

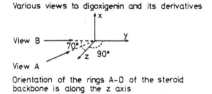

Various views to digoxigenin and its derivatives

View B

View A

Orientation of the rings A–D of the steroid backbone is along the z axis

**Figure 2.** Computer modeling of the structure of HDMA, DMSE and digoxigenin-1-digitoxoside with the Quanta modeling package in combination with the CHARM molecule mechanism program.

Tryptic digestion of the HDMA-labeled $\alpha$ subunit resulted in the formation of 20 kDa (Fig. 1), 12,5 kDa and 11,2 kDa fragments. The amino acid sequence analysis of the aminoterminal end of these peptides gave (doubtful analyses in brackets):

      Peptides 11.2 and 12.5 kDa:
      Found    (D)-(G)-P-N-A-L-T-(P)
      Pig $\alpha$    $^{68}$D - G -P-N-A-L-T- P

      Peptide 20 kDa:
      Found    (D)-M-(D)-E-L-K-K-?-V-S-M
      Pig $\alpha$    $^{24}$D -M- D -E-L-K-K-E-V-S-M

All proposals on the folding structure of the $\alpha$ subunit agree with respect to the arrangement for the first 4 transmembranal portions [6]. According to all these folding models the tryptic cleavage site at the aminoterminal end of the 20, 12.5 and 11.2 kDa peptides of these tryptic peptides is on the cytosolic side prior to the first transmembranal helix. Their molecular size does not exceed the 1st and 2nd transmembranal

379

domains. Since in Forbush vesicles affinity labeling with HDMA occurs, the cardiac glycoside binding site must reside in the loop area between these both transmembranal stretches. HDMA reacts by forming esters with amino group residues. However, there are neither lysine nor argine residues in the loop area but $Cys_{104}$ which could react with HDMA by forming a thioester bond. If a thioester should be formed then this bond should be cleaved by treatment of the peptide with 2 M hydroxylamine. This was the case, indeed. This finding is consistent with the finding of others [2] that this residue is involved in cardiac glycoside binding. The comparison of the 3-dimensional structures of HDMA and of digoxigenin-monodigoxide shows that the reactive hydroxy-succinimidyl group is in the glycoside subsite. A systematic modification of gomphoside is bound to the receptor through the 3'(axial) OH-group and the 5'(equatorial) $CH_3$-group [7]. Although site-directed mutagenis experiments seem to have ruled out an interaction of $Cys_{104}$ with the cymarose moiety in ouabain [3], the comparison of 3 dimensional structures rather suggests this.

*Acknowledgements*

This work has been supported by grants of the Deutsche Forschungsgemeinschaft / Bonn through Scho 139/19-1, the Fonds der Chemischen Industrie, Frankfurt and by the Giessener Hochschulgesellschaft.

**References**

1.  Price EM, Rice DA and Lingrel J (1990) Structure-function studies of Na,K-ATPase. J Biol Chem 265: 6638-6641
2.  Canessa CM, Horisberger J-D, Louvard D and Rossier BC (1992) Mutation of a cysteine in the first transmembrane segment of Na,K-ATPase $\alpha$ subunit confers ouabain resistence. EMBO J 11: 1681-1687
3.  O`Brien JW, Wallick TE and Lingrel J (1993) Amino acids residues of the Na,K-ATPase involved in ouabain sensitivity do not bind the sugar moiety of cardiac glycosides. J Biol Chem. 268: 7707-7712
4.  Forbush B (1982) Characterisation of right-side-out membrane vesicles rich in (Na,K)- ATPase and isolated from dog kidney outer medulla. J Biol Chem. 257: 12678-12684
5.  Jørgensen PL (1974) Purification and characterisation of $(Na^+/K^+)$-ATPase. Biochim Biophys Acta 356: 36-52
6.  Sweadner KJ and Arystarkhova E (1992) Constraints of models for the folding of the Na,K-ATPase. Ion-reactive ATPases: Structure, function and regulation. Ann NY Acad Sci 671: 217-227
7.  Höltje HD and Anzali S (1992) Molecular modeling studies on the digitalis binding site of the $Na^+/K^+$-ATPase. Pharmazia 47: 691-697

# Towards Understanding the Molecular Mechanism of the Inhibitory Cardiac Glycoside Interaction with the Receptor Enzyme $Na^+/K^+$-Transporting ATPase

R. Schön, J. Weiland, R. Megges, K.R.H. Repke

Energy Conversion Unit, Max Delbrück Centre for Molecular Medicine,
Robert-Rössle-Str. 10, D-13122 Berlin-Buch, Germany

## Introduction

Characterization by protein-analytical methods of the cardiac glycoside binding matrix in the $\alpha$-subunit of sensitive $Na^+/K^+$-ATPase as to the identification of amino acid residues directly involved in inhibitory cardiac glycoside interaction has been elusive (1). Using the relations between inhibitor structure, inhibitory potency, and isoenzyme sensitivity accumulated in the authors' laboratory for over 600 congeneric compounds [for a review see (2)] should help to overcome this difficulty. The following is a first attempt to integrate some of the data into a tentative model of the digitalis binding site.

## Results and Discussion

Though biological activity of cardiac glycosides has long been known to reside in the aglycon moiety, it has not been until 1985, that the steroid nucleus itself was identified as the pharmacophoric lead structure (3). As shown in Table 1, the same differential inhibitor susceptibility of four isoenzyme preparations is demonstrated as for the naked steroid **16** as for the genins and glycosides **6-14** with differing side chains at C-3 and C-17, respectively. Even steroids with A/B-*trans* and C/D-*cis* ring junctions **18**, A/B-*flexible* and C/D-*cis* **15** and A/B-*flexible* and C/D-*trans* junctions **19-21** all inhibit the isoenzymes in a ouabain-typical differential gradation, in spite of a more than 100,000fold change in absolute inhibitory potency. In contrast, there is no such differentiation by compounds **2-5** inhibiting $Na^+/K^+$-ATPase by mechanisms different from that of the cardiac glycosides.

Experiments with $Na^+/K^+$-ATPase-$Ca^{2+}$-ATPase chimerae have restricted the cardiac glycoside binding site to the N-terminal half of $Na^+/K^+$-ATPase, containing the H1-H2 and H3-H4 extracellular junctions (4). Of the amino acid substitutions probed in these regions, only variations in the H1-H2 junction region have been shown to be connected with differences in ouabain susceptibility of $Na^+/K^+$-ATPase (5). Thus, this region is

Table 1 The huge differences in the affinity of Na/K-ATPase isoenzymes are sensed by the steroid nucleus. The isoenzyme preparations were from mouse kidney, guinea-pig heart, human heart, and human brain cortex. Their inhibitory susceptibility is expressed by the μmolar inhibitor concentrations required for 50% of maximum inhibition at equilibrium ($IC_{50} \sim K'_D$).

| | Systematic (trivial) name of inhibitor | mouse kidney | guinea-pig heart | human heart | human brain |
|---|---|---|---|---|---|
| 1 | 3β-O-Rhamnosyl-ouabagenin (ouabain) | 67 | 0.80 | 0.027 | 0.033 |
| 2 | Progesterone 3,20-bisguanylhydrazone | 5.0 | 0.69 | 2.4 | 1.9 |
| 3 | Prednisolone 3,20-bisguanylhydrazone | n.d. | 0.27 | 14 | 22 |
| 4 | trans-α'-[4-(2-Dimethylamino)-ethoxyphenyl]-α-ethylstilbene (tamoxifen) | 5.6 | n.d. | 7.0 | 4.5 |
| 5 | Cibacron blue F3GA | n.d. | 15 | 30 | 16 |
| 6 | 3β-O-Tridigitoxosyl-digoxigenin (digoxin) | 46 | 0.55 | 0.014 | 0.057 |
| 7 | 3β-O-Tridigitoxosyl-digitoxigenin (digitoxin) | 45 | 0.19 | 0.0079 | 0.013 |
| 8 | 3β-O-Rhamnosyl-digitoxigenin | 13 | 0.12 | 0.0033 | 0.0068 |
| 9 | 3β-O-(4'ξ-Amino-4'-deoxy-rhamnosyl)-digitoxigenin | 3.6 | 0.019 | 0.0016 | 0.0031 |
| 10 | 3β,14-Dihydroxy-5β,14β-card-20(22)-enolide (digitoxigenin) | 163 | 1.9 | 0.053 | 0.086 |
| 11 | 3β-O-Rhamnosyl-bufalin | 1.2 | 0.0068 | 0.00043 | 0.00039 |
| 12 | 3β,14-Dihydroxy-5β,14β-bufa-20,22-dienolide (bufalin) | 16 | 0.057 | 0.0034 | 0.0059 |
| 13 | 3β-Amino-3-deoxy-bufalin | 32 | 0.055 | 0.0067 | 0.0086 |
| 14 | 3β-Digitoxosyloxy-17β-(pyridazin-4'-yl)-5β,14β-androstane-12β,14-diol | 130* | n.d. | 0.025 | 0.093 |
| 15 | 17β-(Pyrid-3'-yl)-14β-androst-4-ene-3β,14-diol | 152 | 1.4 | 0.13 | 0.15 |
| 16 | 5β,14β-Androstane-3β,14,17β-triol | 714* | 2400* | 180* | 180* |
| 17 | 3β-Rhamnosyloxy-5β,14β-androstane-14,17β-diol | n.d. | 170* | 2.4 | 9.4 |
| 18 | 3β,14-Dihydroxy-5α,14β-card-20(22)-enolide (uzarigenin) | n.d. | 9.2 | 0.38 | 0.66 |
| 19 | 17α-Acetoxy-6-methyl-pregna-4,6-diene-3,20-dione (megestrol acetate) | n.d. | 30 | 7.7 | 16 |
| 20 | 17α-Acetoxy-6-chloro-3β-hydroxy-pregna-4,6-dien-20-one (chlormadinol acetate) | n.d. | 160* | 4.1 | 7.2 |
| 21 | 3-Oxo-17α-pregna-4,6-diene-21,17-carbolactone (canrenon) | n.d. | 590* | 120* | 85* |

* extrapolated value, owing to limited solubility.   n.d. not determined

382

first choice for providing interacting amino acid side chains. The steroid skeleton is known to be surrounded by a positive electrostatic field (6), the charge of which stems from the C-H bonds (7). Hence, a negatively charged surface area in the protein chain would help in attracting, orienting, and binding the approaching steroid. The Glu-Glu-Glu 115-117 sequence seems to be especially suited for that purpose.

When we accept this amino acid sequence as the primary binding site for the steroid nucleus, the lactone substituent at C-17$\beta$ also should interact with neighbouring components of the H1-H2 junction. The large contribution of the lactone moiety to the composite Gibbs interaction energy, calculated from $\Delta G^{0'} = RT\ln K'_D$ (-20.7 of -43.3 kJ·mol$^{-1}$ for 10 and -27.7 of -50.3 kJ·mol$^{-1}$ for 12), calls for strong binding forces. The marked electronegativity of both oxygens points to their capability to serve as acceptors in hydrogen bonding (8). Of the amino acid residues in the H1-H2 region especially fitted to participate in strong hydrogen bonding, Gln 119, Asn 120, Asn 122, replacement of the latter by alanine does not change ouabain sensitivity (5). This leaves Asn 120 and Gln 119 as potential hydrogen bonding partners. Mutagenic replacement of Asp 121 reduces ouabain affinity by nearly 100fold (9). Its anionic carboxylic residue could have its countercharge in a positive charge located in the butenolide plane [cf. (7)].

The increase in inhibitory potency by glycosylation at C-3$\beta$ is already fully realized by introducing only one sugar residue. This suggests that the sugar binding subsite covers only the sugar proximate to the steroid skeleton. Its proper location has been rendered possible by the outcome of energy transfer experiments with anthroylouabain (10), which indicated the presence of a tryptophan residue near to the rhamnose. The only tryptophan present (Trp 310) is in the H3-H4 junction which, accordingly, should provide for or at least participate in the sugar binding subsite. In any case, both the H1-H2 and the H3-H4 extracellular junctions have to come close enough together to allow reaction of the anthroyl residue with Trp 310.

The functional mechanism of Na$^+$/K$^+$-ATPase, as treated in (11), comprises phosphorylation-induced conformation changes which not only alter cation locations and affinities, but also change digitalis affinities. Thus, there has to be 'conformational coupling' between the active centre in the intracellular H4-H5 loop and the H1-H2 extracellular steroid binding domain. We therefore propose that steroid binding to H1-H2 induces and stabilizes a conformational state in which H1-H2 and H3-H4 are fixed together, thus interrupting through the energy transduction pathway H4 (12,13) the conformational cycle neccessary for Na$^+$/K$^+$-ATPase function.

*Ackowledgements*

This work has been rendered possible by grants from the Deutsche Forschungsgemeinschaft (Re 878/1-2) and the Fonds der Chemischen Industrie.

## References

1.  Lingrel JB, Orlowski J, Shull MM, Price EM (1990) Molecular genetics of Na,K-ATPase. Progr Nucleic Acid Res Mol Biol 38:37-89
2.  Repke KRH, Weiland J, Megges R, Schön R (1993) Approach to the chemotopography of the digitalis recognition matrix in $Na^+/K^+$-transporting ATPase as a step in the rational design of new inotropic steroids. Progr Med Chem 30:135-202
3.  Schönfeld W, Weiland J, Lindig C, Masnyk M, Kabat MM, Kurek A, Wicha J, Repke KRH (1985) The lead structure in cardiac glycosides is $5\beta,14\beta$-androstane-$3\beta,14$-diol. Naunyn-Schmiedeberg's Arch Pharmacol 329:414-426
4.  Luckie DB, Lemas V, Boyd KL, Fambrough DM, Takeyasu K (1992) Molecular dissection of functional domains of the E1E2-ATPase using sodium and calcium pump chimeric molecules. Biophys J 62:220-227
5.  Lingrel JB, Orlowski J, Price EM, Pathak BG (1991) Regulation of the $\alpha$ subunit genes of the Na,K-ATPase and determinant of cardiac glycoside sensitivity. In: Kaplan JH, DeWeer P, eds, The Sodium Pump: Structure, Mechanism and Regulation. Society of Physiologists Series, vol. 46, The Rockefeller University Press, New York, pp 1-16
6.  Repke KRH (1985) New developments in cardiac glycoside structure-activity relationships. Trends Pharmacol Sci 6:275-278
7.  Scrocco E, Tomasi J (1978) Electronic molecular structure, reactivity, and intermolecular forces: an euristic interpretation by means of electrostatic molecular potentials. Adv Quantum Chem 11:115-193
8.  Repke KRH, Schönfeld W, Weiland J, Megges R, Hache A (1989) Steroidal Inhibitors of $Na^+/K^+$-ATPase. In: Sandler M, Smith J, eds, Design of Enzyme Inhibitors as Drugs, Oxford University Press, Oxford, pp 435-502
9.  Price EM, Rice DA, Lingrel JB (1989) Site-directed mutagenesis of a conserved, extracellular aspartic acid residue affects the ouabain sensitivity of sheep Na,K-ATPase. J Biol Chem 264:21902-21906
10. Fortes PAG (1977) Anthroylouabain: a specific fluorescent probe for the cardiac glycoside receptor of the Na-K ATPase. Biochemistry 16:531-540
11. Repke KRH, Schön R (1992) Role of protein conformation changes and transphoshorylations in the function of $Na^+/K^+$-transporting adenosine triphosphatase: an attempt at a unifying model. Biol Rev 67:31-78
12. Repke KRH (1982) On the mechanism of energy release, transfer, and utilization in Na,K-ATPase transport work: old ideas and new findings. Ann N Y Acad Sci 402:272-286
13. Shull GE, Schwartz A, Lingrel JB (1985) Amino acid sequence of the catalytic subunit of the $(Na^++K^+)$ATPase deduced from a complementary DNA . Nature (London) 316:691-695

# Palytoxin induces $K^+$ release from yeast cells expressing $Na^+,K^+$-ATPase.

D. Meyer zu Heringdorf, E. Habermann, M. Christ*, and G. Scheiner-Bobis*

*Institut für Biochemie und Endokrinologie, Frankfurter Str. 100, and Rudolf-Buchheim-Institut für Pharmakologie, Frankfurter Str. 107. 35392 Giessen, Germany.

## Introduction

Palytoxin is unique in many respects (2). It is synthesized by corals (Palythoa *caribaeorum*) or by a therewith associated, unknown microorganism. It is the most toxic animal toxin, with an $LD_{50}$ (i.v.) of around 100 ng/kg in rabbits, mice, and guinea-pigs. It is a food poison for man because it occurs, perhaps secondarily, also in çrabs. In contrast to most of the other animal toxins, it is not composed of typical amino acids, but of three sequential components connected by peptide links: namely, a polyhydroxy omega-amino acid of about 2500 Da, dehydro beta-alanine, and n-aminopropanol. Palytoxin forms pores of high selectivity for alkali ions in all mammalian cell membranes investigated so far. The resulting membrane depolarization promotes $Ca^{2+}$ entry, which leads to contraction of smooth and striated muscles and to release of neurotransmitters from the brain.

Efflux of $K^+$ ions from erythrocytes is a sensitive parameter of palytoxin action ($EC_{50}$ about 1 pM) that can easily be monitored against a low and constant background efflux. Release of $K^+$ proceeds to completion when palytoxin is used at a sufficiently high concentration. The concentration versus effect plot is shifted to the left in the presence of $Ca^{2+}$ and borate. The action of palytoxin is inhibited, except in rat erythrocytes, by ouabain in a concentration-dependent manner. Rat $Na^+,K^+$-ATPase is known to be largely resistant to ouabain, although rat erythrocytes do not differ from those of man in their sensitivity to palytoxin. Except at very high concentrations ouabain fails to antagonize palytoxin actions in rat tissues. Palytoxin, at comparatively high concentrations inhibits $Na^+,K^+$-ATPase prepared from pig kidney. It decreases [$^3$H]ouabain binding, and ouabain decreases [$^{125}$I]palytoxin binding (1). Both the binding and the action of palytoxin differ from those of ouabain. Unlike palytoxin, ouabain does not promote pore formation. Although all data accumulated thus far implicate $Na^+,K^+$-ATPase as the only target of palytoxin (2), other membrane constituents have also been advocated (4).

To test the hypothesis that palytoxin acts via its binding to the $Na^+,K^+$-ATPase (2), a comparison between mammalian cells with and without $Na^+,K^+$-ATPase would have been helpful. Such cells, however, are not known to exist. Therefore we used a heterologous expression system of the $Na^+,K^+$-ATPase in the yeast Saccharomyces *cerevisiae* to address the above question. Yeast cells do not contain endogenous $Na^+,K^+$-ATPase whose presence could interfere with the intended measurements. It is furthermore known that yeast can express $Na^+,K^+$-ATPase $\alpha$ and $\beta$ subunits in a functional form that binds ouabain and displays all other activities of mammalian sodium pumps (3). Using this system, four questions were assessed:

1.      Is Saccharomyces a target for palytoxin?
2.      Does expression of $Na^+,K^+$-ATPase make Saccharomyces sensitive to palytoxin?
3.      Are both $\alpha$ and $\beta$ subunits of $Na^+,K^+$-ATPase required for toxin action?
4.      How does Saccharomyces respond to substances known to modify the interaction of palytoxin with erythrocytes?

**Results and discussion**

In order to investigate our hypothesis, yeast cells were transformed with the yeast vector pCGY1406$\alpha\beta$ that codes for sheep $\alpha$1 subunit and dog $\beta$ subunit (3). Other cells were transformed with the vector YEP$\alpha$3, carrying the cDNA for the rat $\alpha$3 subunit, or with the yeast vector pCGY1406$\beta$ that codes for the dog $\beta$ subunit. The latter was simply constructed from the pCGY1406$\alpha\beta$ vector after removal of the sheep $\alpha$1 by Kpn I digestion and back ligation of the cohesive Kpn I overhangs of the vector. The auxotrophic marker for all vectors was tryptophan. Nontransformed cells served as controls.

The answers to the four questions addressed above are consistent with our hypothesis that $Na^+,K^+$-ATPase is the only target for palytoxin.

ad 1: Nontransformed yeast is not a target for palytoxin over the whole concentration range investigated (Fig. 1a). This observation rules out any nonspecific membrane action comparable to that of amphotericin B or nystatin. Such antibiotics are known to increase the potassium efflux from erythrocytes in a manner different from that observed with palytoxin.

ad 2: When both $\alpha$ and $\beta$ subunits of the $Na^+,K^+$-ATPase are expressed, yeast becomes sensitive to palytoxin (Fig. 1a). As with erythrocytes, the yeast response to palytoxin can be monitored by potassium release. Concentration dependence is observed between 5 and 1000 nM and efflux response never reaches 100 % (Fig. 1b,c,d). Some of the yeast potassium might be located in intracellular compartments not accessible to palytoxin. The action of palytoxin on yeast cells as compared with its action on red cells (2) is slow (Fig. 1b).

ad 3: Both $\alpha$ and $\beta$ subunits of $Na^+,K^+$-ATPase are required in order to obtain palytoxin-sensitive yeast cells. Palytoxin-induced $K^+$ release from cells expressing either the catalytic $\alpha$3 subunit or the $\beta$ subunit alone could not be detected (Fig. 1a). Expression of the subunits was confirmed by Western blots with specific antibodies (not shown). This experiment rules out the possibility that transformation itself sensitizes the cells to palytoxin, and is in agreement with the observation that both $\alpha$ and $\beta$ subunits are required in order to obtain ouabain binding in yeast (3).

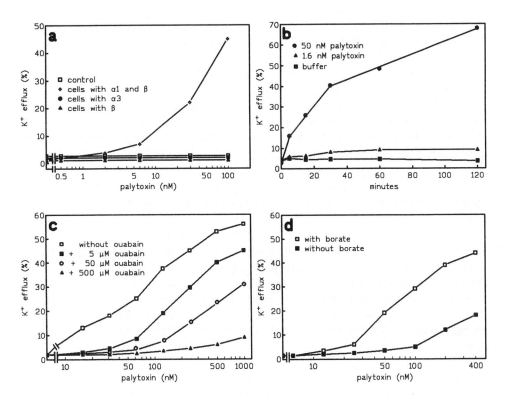

**Figure 1.** Action of palytoxin on yeast cells expressing $Na^+,K^+$-ATPase.

Yeast cells expressing $\alpha$ and/or $\beta$ subunits of $Na^+,K^+$-ATPase were diluted to $10^8$ cells/ml in GHBC (300 mM glucose, 0.5 mM boric acid, 1 mM $CaCl_2$, 10 mM Hepes, pH 7.4) and incubated with various concentrations of palytoxin for 60 min at 30$^\circ$C. After centrifugation (10 min at 8000 x g), supernatant $K^+$ was determined by flame photometry. The ordinate represents the fractional $K^+$ loss from the cells. Total $K^+$ content of the suspension was between 200 and 300 $\mu$M as determined after lysis with 0.02 % SDS for 10 min at 95$^\circ$ C.

a)    Dependence on expression of $\alpha$ and $\beta$ subunits.

b)    Dependence on incubation time.

c)    Inhibition by ouabain. The cells were preincubated with ouabain for 30 min at 30$^\circ$ C, followed by treatment with palytoxin for 60 min.

d)    Activation by borate (0.5 mM).


ad 4: The response of mammalian erythrocytes to palytoxin is inhibited when the cells are preincubated with ouabain. This does not apply for rat erythrocytes, which are known to carry a ouabain-insensitive $\alpha$1 isoform of $Na^+,K^+$-ATPase. Palytoxin actions on yeast cells expressing the $Na^+,K^+$-ATPase are also very sensitive to ouabain (Fig.

1c). Thus, the ouabain binding site is involved in the action of palytoxin, probably by mediating the binding of the toxin(1).

Borate potentiates palytoxin-induced $K^+$ efflux from yeast cells (Fig. 1d), as it does with every other palytoxin action investigated (2). In the absence of palytoxin borate does not induce $K^+$ efflux from erythrocytes (2) or from yeast cells expressing $Na^+,K^+$-ATPase (not shown). It is thought that borate does not interact with the cell surface but with some of the numerous hydroxyl groups of palytoxin. Borate complexes with carbohydrates are well known.

As in erythrocytes, $Ca^{2+}$ also promotes the action of palytoxin in yeast expressing the sodium pump (not shown). The ion is known to promote palytoxin binding (1), and may potentiate the so-called Gardos-effect.

The experiments presented here clearly show that yeast cells expressing both $\alpha$ and $\beta$ subunits of the mammalian sodium pump recognize palytoxin in a way comparable with that observed for erythrocytes. Palytoxin-associated effects, however, cannot be obtained with nontransformed yeast cells or with cells expressing either $\alpha3$ or $\beta$ subunits alone. These observations, and the fact that ouabain prevents the effects of palytoxin on cells expressing the $\alpha1\beta$-heterodimer, strongly suggest that palytoxin acts via the sodium pump.

## References

(1)     Böttinger, H., Beress, L., and Habermann, E. (1986) Biochim. Biophys. Acta 861, 165-176
(2)     Habermann, E. (1989) Toxicon 27, 1171-1187
(3)     Horowitz, B., Eakle, K.A., Scheiner-Bobis, G., Randolph, G.R., Chen, C.Y., Hitzemann, R.A., and Farley, R.A. (1990) J. Biol. Chem. 265, 4189-4192
(4)     Van Renterghem, C., and Frelin C. (1993) Brit. J. Pharmacol. 109, 859-865

*Acknowledgments*

The authors thank Ms. P. Kronich for excellent technical assistance. This work was supported by the Deutsche Forschungsgemeinschaft, Bonn, through SFB 249, and by the Fonds der Chemischen Industrie, Frankfurt.

388

# Bafilomycin and Concanamycin Derivatives as Specific Inhibitors of P- and V-Type ATPases

S. Dröse[1], K.U. Bindseil[2], E.J. Bowman[3], A. Zeeck[2], K. Altendorf[1]

[1] Fachbereich Biologie/Chemie, Universität Osnabrück, 49069 Osnabrück Germany
[2] Institut für Organische Chemie, 37077 Göttingen Germany
[3] Department of Biology, University of California, Santa Cruz, California 95064 USA

## Introduction

The bafilomycins, unusual macrolide antibiotics with a 16-membered lactone ring, were isolated from *Streptomyces griseus* (6). Recently, in a detailed analysis we compared the effects of bafilomycin $A_1$ (1) (Figure 1) on representative enzymes of the three classes of membrane bound ion-translocating ATPases (2). $F_1F_0$ ATPases (F-type) from bacteria and mitochondria are not affected by this antibiotic. In contrast, P-type ATPases are moderately sensitive (micromolar concentrations) to this inhibitor. However, V-type ATPases from *Neurospora* vacuoles, chromaffin granules and plant vacuoles are extremely sensitive (nanomolar concentrations). For the first time these results showed convincingly that bafilomycin $A_1$ (1) is useful for distinguishing among the different types of ATPases and that it is a potent, relatively specific inhibitor of the vacuolar ATPases (2). Here, we show that the concanamycins, which contain an 18-membered lactone ring (5), and which are structurally related to the bafilomycins (Figure 1), are also useful for distinguishing among the different ATPases and are even more specific inhibitors of V-type ATPases.

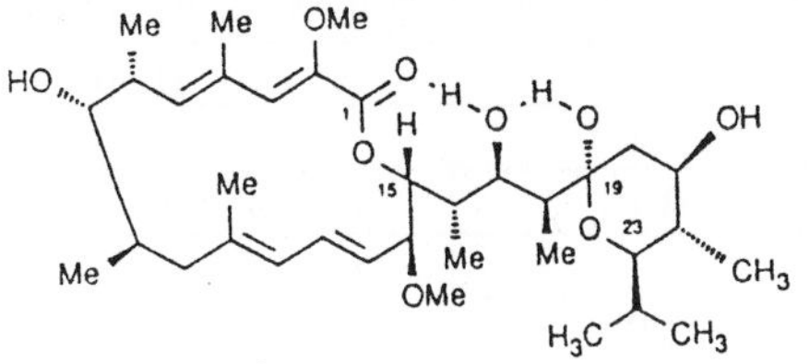

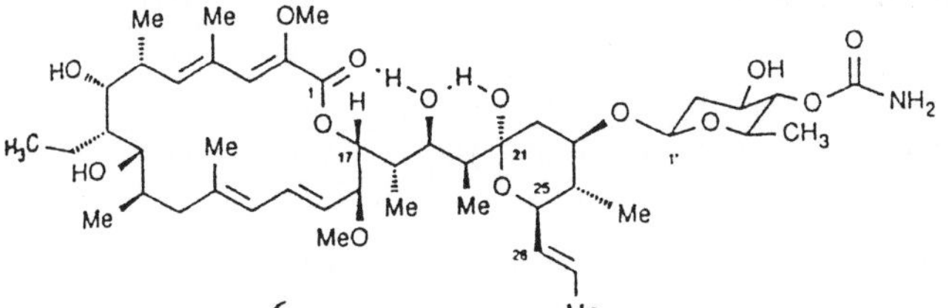

**Figure 1.** Structure of bafilomycin $A_1$ (1) and concanamycin A (6)

## Results and Discussion

We tested the effects of the concanamycins in comparison to bafilomycin A$_1$ (**1**) on representative enzymes of the three classes of ATPases. In addition, we tested the inhibitory effects of modified bafilomycins and concanamycins with the aim of probing structure-activity relationships and of developing a strategy for synthesizing modified antibiotics able to bind covalently to the enzyme complex. The measurements revealed that the ATP synthase (F-type) from *Escherichia coli* was not affected at concentrations of concanamycin A (**6**) up to 50 μM (equivalent to 150 μmol of **6**/mg of protein; data not shown). By contrast, the Kdp-ATPase (P-type) of *E.coli* showed intermediate sensitivity (Table 1). The most striking result, however, was found with the vacuolar ATPase (V-type) from *Neurospora crassa*. The membrane-bound ATPase exhibited high sensitivity to this antibiotic, with an $I_{50}$ value of 0.002 x 10$^{-3}$ μmol mg$^{-1}$. Thus, concanamycin A (**6**), compared to bafilomycin A$_1$ (**1**), shows an even greater specificity for the different kinds of ATPases because of its more than 20 times improved inhibitory effect on the V-type ATPase. In concanamycin C (**7**), the carbamoyl group attached to the sugar hydroxy group at position C-4′ is missing in comparison to **6**. This naturally occurring derivative shows a very weak inhibitory effect on the P-type ATPase (Table 1). However, due to solubility problems of **7** in water, only concentrations up to 10 - 20 μM could be tested. In contrast, concanamycin C (**7**) has retained its effect on the V-type ATPase (Table 1).

**Table 1.** Sensitivity of two ATPases to macrolide antibiotics

| macrolide antibiotic | H$^+$-ATPase (*N.crassa*) $K_i$ (nM) | $I_{50}$ (nmol mg$^{-1}$) | Kdp-ATPase (*E.coli*) $K_i$ (μM) | $I_{50}$ (μmol mg$^{-1}$) |
|---|---|---|---|---|
| bafilomycin A$_1$ (**1**) | 0.5 | 0.05 | 1.3 | 0.36 |
| bafilomycin A$_2$ (**2**) | 1.0 | 0.1 | | |
| 21-*O*-acetylbafilomycin A$_1$ (**3**) | 1.3 | 0.13 | 1.2$^a$ | 0.33$^a$ |
| bafilomycin A$_1$ 21-ketone (**4**) | | | 1.7 | 0.47 |
| bafilomycin D (**5**) | 20.0 | 2.0 | 20.8 | 5.64 |
| concanamycin A (**6**) | 0.02 | 0.002 | 1.8 | 0.5 |
| concanamycin C (**7**) | 0.02 | 0.002 | $b$ | $b$ |
| concanolide (**8**) | 0.06 | 0.006 | 8.4 | 2.33 |
| 23-*O*-methylconcanolide (**9**) | | | 4.7$^a$ | 1.3$^a$ |
| 21,23-di-*O*-methylconcanolide (**10**) | 3.0 | 0.3 | 7.1$^a$ | 1.97$^a$ |

$^a$ Due to solubility problems, 100% inhibition could not be achieved. $^b$ Due to solubility problems, only 30-40% inhibition could be obtained, making the determination of the $K_i$ and $I_{50}$ values impossible. Materials and methods are detailed in (1) and (3).

To obtain information about structural elements of the concanamycins and bafilomycins, which are important for the inhibitory activity toward P- and V-type ATPases, we tested chemically modified forms of the macrolides. In concanolide (8), the sugar residue is absent, moving this derivative closer to bafilomycin $A_1$ (1). In comparison to 6, the inhibitory action towards both ATPases is reduced by a factor of 3-4, but the inhibition of the V-type ATPase remains nearly 10-fold better than for 1. An additional methylation at position C-21 and C-23 of concanolide as in 9 or 10 caused a reduction of the inhibitory effect towards the V-type ATPase (Table 1). Similar but not as drastic losses can be observed for the bafilomycin $A_1$ derivatives bafilomycin $A_2$ (2), 21-O-acetyl-bafilomycin $A_1$ (3), and bafilomycin $A_1$ 21-ketone (4). Because the acetylated compounds are less soluble in water, measurements with concentrations above 5 - 10 µM with the P-type ATPase were difficult (Table 1). In bafilomycin D (5) the six-membered hemiketal ring has been opened. This modification has a pronounced effect on the inhibitory action on both P- and V-type ATPases. Both enzymes, compared to 1, are more then 20-fold less sensitive towards 5 (Table 1).

The almost identical inhibitory pattern between bafilomycin $A_1$ (1) and concanamycin A (6) is based on structural and conformational similarities between the two compounds (Figure 1). The improved inhibitory properties of 6 towards the V-type ATPase could be caused by the additional carbohydrate residue or the increased size of the macrolide ring of the concanamycins, compared to the bafilomycins. Since this effect is maintained in concanolide (8), the aglycon of 6, the additional carbohydrate residue probably does not play an important role in the inhibitory mechanism. This is in accord with the observation that the removal of the carbamoyl group at position C-4' of the sugar in concanamycin C (7) does not change the effect of this compound on the V-type ATPase. The hemiketal ring of the macrolides seems to play an important role in the biological activity, since bafilomycin D (5) is 20 times less active than bafilomycin $A_1$ (1) against both P- and V-type ATPases. This would lead to the conclusion that the macrolide ring together with the adjacent sixmembered ring system play a key role in biological activities.

Besides some circumstantial evidence that bafilomycin $A_1$ (1) binds to the $V_O$-part of the V-type ATPases (4), the actual binding site of the antibiotic within the P- and V-type ATPases is still unknown. In order to obtain information about that, the synthesis of modified compounds (e.g. nitrene-generating derivatives) able to form a covalent bond with the enzyme complex while retaining their specific inhibitory properties should be rewarding. Therefore, we have set out to introduce chemical modifications at different positions of the macrolides. Since acetylation or even oxidation (as in 3 and 4, respectively) of the C-21 hydroxy group of 1 has no significant effect on the inhibitory properties of these modified compounds, this position seems to be a promising site for further chemical modifications. Since concanamycin A (6) is a chemically more stable compound than bafilomycin $A_1$ (1), the former might be more suited for further modifications. Recently, we synthesized an 3′-azido derivative of concanamycin A (6), which still retained its inhibitory effect on the Kdp-ATPase (P-type) of *E.coli*.

*Acknowledgements*

This work was supported by a research grant from the Deutsche Forschungsgemeinschaft (SFB 171/B5) to K.A., by the Fonds der Chemischen Industrie to K.A. and A.Z., and by a research grant, GM-28703, from the National Institutes of Health to B.J.B. and E.J.B.

**References**

1.  Bindseil KU, Zeeck A (1993) The chemistry of unusual macrolides. Part 1 The preparation of the aglycones of concanamycin A and elaiophylin. J Org Chem, in press

2.  Bowman EJ, Siebers A, Altendorf K (1988) Bafilomycins: A class of inhibitors of membrane ATPases from microorganisms, animal cells, and plant cells. Proc Natl Acad Sci USA 85: 7972-7976

3.  Dröse S, Bindseil KU, Bowman EJ, Siebers A, Zeeck A, Altendorf K (1993) Inhibitory effects of modified bafilomycins and concanamycins on P- and V-type adenosine-triphosphatases. Biochemistry 32: 3902-3906

4.  Hanada H, Moriyama Y, Maeda M, Futai M (1990) Kinetic studies of chromaffin granule $H^+$-ATPase and effects of bafilomycin $A_1$. Biochem Biophys Res Commun 170: 873-878

5.  Kinashi H, Someno K, Sakaguchi K (1984) Isolation and characterization of concanamycin A, B and C. J Antibiot 37: 1333-1342

6.  Werner G, Hagenmaier H, Drautz H, Baumgartner A, Zähner H (1984) Metabolic products of microorganisms. 224 Bafilomycins, a new group of macrolide antibiotics. J Antibiot 37: 100-117

# Tertiary Amines as Cytosolic $K^+$-Antagonists of Gastric $H^+/K^+$-ATPase

H.G.P. Swarts, C.H.W. Klaassen, F.M.A.H. Schuurmans Stekhoven, and J.J.H.H.M. De Pont

Department of Biochemistry, University of Nijmegen, P.O. Box 9101, 6500 HB Nijmegen, The Netherlands

## Introduction

The reported properties of the overall $H^+/K^+$-ATPase activity vary with the conditions in which they are analyzed. The affinity of $K^+$ for $H^+/K^+$-ATPase ranges from 0.3 to 5.0 mM. Such variations in $K^+$-affinity can be due to differences in experimental conditions like the ATP and $Mg^{2+}$ concentration, temperature and pH. Another prominent variable is the type of buffer used.
Since preliminary experiments indicated that buffers, similarly as with $Na^+/K^+$-ATPase, have a marked effect on the $K^+$-affinity of $H^+/K^+$-ATPase, we investigated this effect in more detail with the use of leaky and ion-tight vesicles.

## Materials and Methods

Fresh (ion-tight) and leaky $H^+/K^+$-ATPase vesicles were isolated from pig gastric mucosa. The steady-state ATP-phosphorylation level, the dephosphorylation rate, the $K^+$ activated ATPase and p-nitrophenylphosphatase activities were determined as reported previously [4].

## Results and Discussion

The $K^+$-dependent ATP hydrolysis of $H^+/K^+$-ATPase.
Figure 1A shows that the $K^+$ activation of the overall $H^+/K^+$-ATPase activity is not only depending on the pH [1] but also on the type of buffer used. In the presence of Tris or Hepes the $K_{0.5}$ for $K^+$ varied from 0.3 mM at pH 8.0 to 2.0 mM at pH 6.0. In the presence of imidazole the $K_{0.5}$ value increased to 13.2 mM at pH 6.0. The concentration of Hepes or Tris had virtually no effect on the $K^+$-affinity at pH 7.0 (Fig 1B), but imidazole increased the $K_{0.5}$ value linearly with increasing concentrations of this buffer. This indicates that imidazole acts as a $K^+$-antagonist in the ATPase reaction. From these values a $K_i$ value of 13 mM was calculated.
Comparable effects of imidazole were observed when we studied the

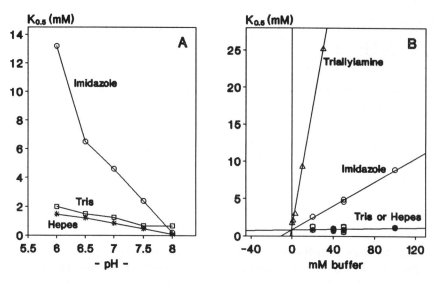

**Figure 1.** The effect of pH, type of buffer (50 mM) or the concentration of buffer at pH 7.0 (B) on the $K^+$-activation of the $H^+/K^+$-ATPase activity.

reaction mechanism of $Na^+/K^+$-ATPase [2,3]. In those studies we demonstrated that the effect of imidazole was also found with other tertiary amines, with triallylamine as the most potent compound. We therefore also tested the effect of triallylamine on the $H^+/K^+$-ATPase activity. The latter compound also decreased

competitively the $K^+$-affinity in the overall $H^+/K^+$-ATPase reaction and a $K_i$-value of 1.6 mM for triallylamine was determined at pH 7.0 (Fig. 1B).

The effect of amines on the ATP-phosphorylation level and dephosphorylation rate.

The steady-state ATP-phosphorylation level itself was not influenced by either imidazole or Hepes or Tris but was about 30% inhibited by triallylamine, but the $K^+$ sensitivity of the steady-state ATP phosphorylation level depended on the type and concentration of the buffer used. Similar effects of the ligands Tris, imidazole and triallylamine were observed when the $K^+$ activation of the dephosphoryla-

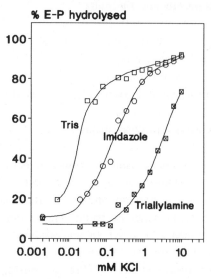

**Figure 2.** The effect of 100 mM buffer amine (pH 7.0) on the $K^+$ sensitivity of the phosphointermediate. The amount of E-P hydrolysed in 3 s is plotted.

tion reaction was studied (Fig. 2). At low buffer concentrations, with 100 choline chloride or with 100 mM Tris the $K^+$ activation of the dephosphorylation reaction already occurred in the micromolair range ($K_{0.5} = 25$ $\mu M$) while in the presence of 100 mM imidazole or 100 mM triallylamine the $K_{0.5}$ was found to be 0.15 and 3 mM, respectively. Thus tertiary amines reduce the $K^+$ affinity of the phospho-enzyme and thus the $E_2$-P to $E_2$.K transition.

<u>Experiments with ion-tight vesicles.</u>

By using freshly prepared ion tight $H^+/K^+$-ATPase vesicles, where the cytosolic side faces the extra vesicular medium [4], it is possible to study the effects of tertiary amines exclusively at the cytosolic side of the membrane. The stimulatory $K^+$-site for p-nitrophenylphosphatase activity

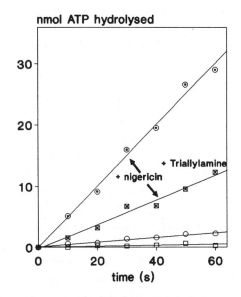

**Figure 3.** Time dependence of the inhibition by 20 mM triallylamine (□,◙) of the $K^+$-activated ATP hydrolysis in ion-tight vesicles, studied with (●,◙) and without nigericin (○,□).

is located at the cytosolic side of the membrane [1]. As the $K^+$-activation of phosphatase activity was mixed-type inhibited by 20 mM triallylamine, the $K_{0.5}$ for $K^+$ increased from 1.1 mM to 4.5 and the $V_{max}$ decreased to 78 %, the site of inhibition is probably located at the outside of these vesicles.

A possible effect of transmembrane diffusion of triallylamine is unlikely since the $pK_a$ value of triallylamine is 8.3 and at pH 7.0 the uncharged form, which can diffuse directly over the membrane, is only 5% of total triallylamine. This is further supported by the observation of Fig. 3 where the degree of inhibition of the $H^+/K^+$-ATPase activity did not change in time. This indicates that triallylamine reacts immediately with inside-out $H^+/K^+$-ATPase vesicles.

<u>Dephosphorylation studies in ion-tight vesicles.</u>

The phosphoenzyme formed in ion-tight vesicles preparations reacted only with $K^+$ when nigericin was also present, indicating that $K^+$ stimulated at the luminal side. If the enzyme was phosphorylated in the presence of 50 mM triallylamine, the basal dephosphorylation rate decreased from 0.04 to 0.02 $s^{-1}$. In order to stimulate the $K^+$-activated dephosphorylation reaction, the presence of nigericin was also necessary, but the activating effect of 1 mM KCl on the dephosphorylation rate was quite lower than in the absence of triallylamine (0.11 vs 0.28 $s^{-1}$). This indicates that triallylamine antagonises the $K^+$-stimulated dephosphorylation process in both leaky (Fig. 2) and ion-tight vesicles.

## The ATP-phosphorylation level in ion-tight vesicles

The reduction of the steady-state ATP-phosphorylation level by $K^+$, observed in leaky vesicles, must at least in part be due to a stimulation of the dephosphorylation reaction ($E_2$-P to $E_2$.K). However, a shift in the $E_1$.H to $E_1$.K equilibrium could also contribute to this effect.

In ion-tight vesicles where no activation by luminal $K^+$ can occur it was possible to study the latter possibility. When the $K^+$ concentration during the 5-s phosphorylation period was increased the affinity for ATP concentration-dependently decreased, indicating that $K^+$ shifts the $E_1$.H to $E_1$.K equilibrium to the right. When enzyme was preincubated with

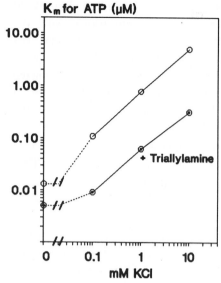

**Figure 4.** The effect of $K^+$ or in combination with 10 mM triallylamine on the ATP affinity of the steady-state phosphorylation level measured in ion-tight vesicles.

triallylamine the decrease of the ATP affinity by $K^+$ was counteracted. So there is also a $K^+$-triallylamine antagonism in the $E_1$.H to $E_1$.K transition.

The above results indicate that tertiary amines decrease the affinity for $K^+$ at both luminal and cytosolical binding sites by interaction at the cytosolic side of the membrane. This results in shifts of both the $E_1$.H$\leftrightarrow$$E_1$.K transition and in the dephosphorylation reaction, $E_2$-P$\rightarrow$$E_2$.K.

## References

1.  Ljungström M, Vega FV, Mardh S (1984) Effects of pH on the interaction of ligands with the $H^+/K^+$-ATPase purified from pig gastric mucosa. Biochim Biophys Acta 769:220-230
2.  Schuurmans Stekhoven FMAH, Swarts HGP, 't Lam GK, Zou YS, De Pont JJHHM (1988) Phosphorylation of $Na^+/K^+$-ATPase; stimulation and inhibition by substituted and unsubstituted amines. Biochim Biophys Acta 937:161-176
3.  Schuurmans Stekhoven FMAH, Zou YS, Swarts HGP, Leunissen L, De Pont JJHHM (1989) Ethylenediamine as active site probe for $Na^+/K^+$-ATPase. Biochim Biophys Acta 982:103-114
4.  Swarts HGP, Van Uem TJF, Hoving S, Fransen JAM, De Pont JJHHM (1991) Effect of free fatty acids and detergents on $H^+/K^+$-ATPase - The steady-state ATP phosphorylation level and the orientation of the enzyme in membrane preparations. Biochim Biophys Acta 1070:283-292

# Probing the cation binding site of $Na^+/K^+$-ATPase with competitive $Na^+$ antagonists

E. Or, P. David, A. Shainskaya, R. Goldshleger, D. M. Tal, and S. J. D. Karlish

Biochemistry Department, Weizmann Institute of Science, Rehovot 76100, Israel

## Introduction

Organic guanidium derivatives have been characterized as competitive $Na^+$-like antagonists in $Na^+/K^+$-ATPase (2). They compete for $Na^+$ and $Rb^+$ occlusion by $Na^+/K^+$-ATPase, and inhibit ATPase activity and $Na^+$-dependent phosphorylation from ATP. Like $Na^+$, they inhibit phosphorylation from inorganic phosphate and stabilize the $E_1$ conformation of FITC-labeled enzyme. Two compounds m- and p- xylylene bisguanidinium , pXBG and mXBG, were found to compete for cation occlusion with the highest affinity (~8µM). This work utilizes pXBG and other $Na^+$-like antagonists to investigate structural and functional properties of the cation binding and transport site of renal $Na^+/K^+$-ATPase. The first section summarises experiments which compare the ability of pXBG and other $Na^+$-like antagonists with that of transported alkali metal cations to protect the enzyme against covalent modification and structural perturbations of the cation occlusion site (9). The second section describes experiments with reconstituted proteoliposomes which define the sidedness of inhibition and effects of electrical diffusion potentials on inhibition by $Na^+$ antagonists.

## Effects of $Na^+$-like antagonists on Structural Modifications

The carboxyl modifying reagent DCCD inactivates the capacity of $Na^+/K^+$-ATPase to occlude $Rb^+(K^+)$ and $Na^+$. This inactivation can be protected against by $Rb^+$ or $Na^+$ (10). We find also that $Na^+$ antagonists like pXBG, guanidinium ions, and ethylene-diamine, which are not occluded by the enzyme, can protect $Na^+/K^+$-ATPase against inactivation by DCCD (9).

$Na^+/K^+$-ATPase is digested by trypsin in selective conditions ($Rb^+$ present, $Ca^{2+}$ absent), to so-called "19kDa-membranes" (6). The treatment removes 50% of the original protein, leaving a 19kDa fragment and smaller membrane-embedded fragments (8-11.6 kDa) of the $\alpha$-chain, together with an intact $\beta$-chain (1). "19kDa-membranes" are devoid of any ATP- and Pi- dependent functions but cation occlusion remains intact. We find that unlike $Rb^+(K^+)$ and $Na^+$ (1), $Na^+$-like antagonists do not protect the 19kDa fragment against further tryptic digestion and loss of $Rb^+$ occlusion (9).

A new phenomenon of thermal inactivation of $Rb^+$ occlusion by "19kDa-membranes" is also described (9). "19kDa-membranes" rapidly loose their ability to occlude $Rb^+$ when they are incubated at 37°C. Nevertheless they do retain full capacity to occlude $Rb^+$ at 37°C if $Na^+$, $Rb^+(K^+)$ or congeners are present in the medium. We find that, unlike alkali metal cations, the $Na^+$-like antagonist, guanidinium, pXBG and ethylene-diamine, do not protect "19kDa-membranes" against thermal inactivation. It is suggested that in the absence of occluded cations the transmembrane segments comprising the cation binding site do not interact tightly, the structure is loose and thermo-labile, and tryptic-sensitive bonds are exposed to the medium.

As described below, the $Na^+$ antagonists are selective for the cytoplasmic surface. On the basis of this fact and the experiments reported above, we have proposed to a two-step model for cation occlusion from the cytoplasmic surface of $Na^+/K^+$-ATPase (9). The first step is governed by direct binding of the cations to two charged carboxyl residues, which reside within transmembrane segments, but are accessible from the medium. This initial recognition is not very selective, since in addition to two $Rb^+(K^+)$ and two $Na^+$ ions, pXBG and other $Na^+$-like antagonists recognize these carboxyls (the 3rd $Na^+$ ion binds at a neutral site lying inside an ion-well). This step is detected by the ability of the cations to protect against inactivation of these essential carboxyls by DCCD. The second step is more selective, and is suggested to involve a conformational change to an occluded form with a compact structure, which is resistant to thermal motions and is inaccessible to trypsin. During this step, the remaining water molecules forming the hydration shell around the transported cations are shed and are replaced in the occluded state by additional contacts with polarized ligating residues. The guanidinium ion and its organic derivatives such as pXBG are not occluded presumably because the guanidinium group is too large, and hence they block cation occlusion and $Na^+/K^+$-ATPase activity.

### Sidedness of Effects of $Na^+$-antagonists and Effects of Diffusion Potentials

A $Na^+$-like antagonist, such as pXBG, is expected to bind preferentially at the cytoplasmic side of $Na^+/K^+$-ATPase. The sidedness of it's action can be studied conveniently in liposomes reconstituted with $Na^+/K^+$-ATPase. pXBG competes with $Na^+$ at cytoplasmic sites and inhibits active $Na^+/K^+$-exchange on inside-out oriented pumps (see below). However, a more definitive measure of sidedness is obtained in experiments examining its effects on passive $Rb^+$-activated $Rb^+$-$K^+$ exchange (5), sustained by pumps in both orientations. Vanadate-sensitive $^{86}Rb$ uptake into $K^+$-loaded vesicles represents the flux via in-side-out oriented pumps, while the saturable component of the vanadate-insensitive $^{86}Rb$ uptake represents the flux via right-side-out oriented pumps. Indeed, pXBG shows a pronounced selectivity for the cytoplasmic side competing for $Rb^+$ with an apparent $K_i$ of 570μM, while at the extracellular side pXBG (0.7mM) hardly affected $Rb^+$-$K^+$ exchange (9).

In previous work, inside negative diffusion potentials ($\Delta\psi$ up to -150mV, produced by ionophores and outwardly directed cation gradients) were shown to increase the $Na^+$ affinity for activation of ATP-dependent $Na^+/K^+$-exchange via in-side-out oriented pumps (4). This observation was interpreted as being compatible with the existence of a high resistance access pathway or "Na-well" connecting the cytoplasmic $Na^+$ binding site to the bulk medium (4). As a result the probability of a $Na^+$ ion to bind at its site is increased, and thus the apparent binding affinity of $Na^+$ is increased too (7). Subsequently, more direct evidence for a cytoplasmic "Na-well" has been obtained in studies with the styryl dyes (8).

398

The question arose whether the apparent affinity at the cytoplasmic side of $Na^+$ antagonists is affected by an electrical diffusion potential, as found for $Na^+$ ions. Apparent binding affinities of $Na^+$ antagonists at the cytoplasmic face of $Na^+/K^+$-ATPase were calculated from inhibition of ATP-dependent $Na^+/K^+$-exchange, at a fixed and low Na concentration, by increasing concentrations of the antagonists. We have found that, unlike the case of $Na^+$, inside negative diffusion potentials do not affect the binding affinities of pXBG ($758\pm55\mu M$ vs. $674\pm1\mu M$), guanidinium ($23.3\pm3.1mM$ vs. $19.6\pm2.4mM$), ethylenediamine ($151\pm29\mu M$ vs. $167\pm28\mu M$), and $Rb^+$ ($3.42\pm0.02mM$ vs. $3.9\pm0.8mM$). The raw data can also be calculated as an enhancement ratio ($V_{\Delta\psi<0}/V_{\Delta\psi=0}$) versus concentration of the inhibitor. The enhancement ratio is constant and independent of the concentration of the inhibitor. This is the predicted behavior if binding of $Na^+$ antagonists at the cytoplasmic side is unaffected by a transmembrane potential.

The cation binding site of $Na^+/K^+$-ATPase has been suggested to contains two negatively charged residues, which bind either two $Na^+$ or two $K^+$ ions, while the third $Na^+$ ion binds at a neutral site (3). The present results suggest that the neutral site lies inside an ion-well and that this site alone confers the observed voltage-dependence of $Na^+$ binding at the cytoplasmic side. The two charged sites to which $Na^+$-like antagonists also bind, lie outside the ion-well and therefore binding at these sites is unaffected by potential.

A more detailed resolution of the effects of the diffusion potential on the three $Na^+$ binding constants at the cytoplasmic side could be obtained by studying inhibition by $Na^+$ of passive $Li^+_{Exc}/Rb^+_{Cyt}$ exchange, activated by very low cytoplasmic $Rb^+$ ($12\mu M$). This exchange is defined as the saturable component of the ouabain-insensitive $^{86}Rb$ uptake into $Li^+$-loaded vesicles (5). Inhibition of the exchange by $Na^+$ was described best by a model for $Na^+$ binding at two sites rather than three ($K_1=1899\pm516\mu M$, $K_2=263\pm103\mu M$). Probably, the third $Na^+$ ion binds with a much lower affinity, while the exchange is already inhibited by the first two $Na^+$ ions bound. Imposition of a diffusion potential with a $Li^+$-ionophore (-76mV) increased the binding affinity of the first $Na^+$ ion 2.4-fold to $781\pm190\mu M$, but did not affect the binding affinity of the second $Na^+$ ion ($361\pm158\mu M$). As expected the diffusion potential did not significantly affect inhibition by pXBG at the cytoplasmic side assayed under identical conditions ($K_i=82\pm4\mu M$ vrs. $K_i=69\pm3\mu M$). The fractional depth of the ion-well for $Na^+$, calculated from our data, is 0.29. We propose that, of the three cytoplasmic $Na^+$ sites, the neutral $Na^+$ site, which lies inside the ion-well, binds the first $Na^+$ ion.

Assuming that in the form $E'P(Na_2)$ the two $Na^+$ ions are bound at the two negatively charged sites, then the first $Na^+$ ion released to the extracellular medium during the conformational change $E_1P(Na_3)\rightarrow E'P(Na_2)$, is released from the neutral site. One may speculate that $Na^+$ transport by $Na^+/K^+$-ATPase is of FIFO type; The neutral $Na^+$ site at the cytoplasmic side of the enzyme binds the first $Na^+$ ion, and following $Na^+$ occlusion it is the first site from which $Na^+$ is released to the extracellular milieu. The neutral $Na^+$ site lies at the end of a cytoplasmic ion-well deeper into the protein's interior than the two negatively charged sites. This structural feature is compatible with a moving barrier model (11) as the basis for FIFO type semi-ordered transport of $Na^+$ by $Na^+/K^+$-ATPase.

## References

1. Capasso, J.M., Hoving, S., Tal, D.M., Goldshleger, R., and Karlish, S.J.D. (1992). Extensive digestion of $Na^+,K^+$-ATPase by Specific and Non-specific Proteases with Preservation of Cation Occlusion sites. J. Biol. Chem. 267: 1150-1158.

2. David, P., Mayan, H., Cohen, H., Tal, D.M., and Karlish, S.J.D. (1992). Guanidinium Derivatives act as High Affinity Antagonists of $Na^+$ ions in Occlusion sites of $Na^+,K^+$-ATPase. J. Biol. Chem. 267: 1141-1149.

3. Glynn, I.M. and Karlish, S.J.D. (1990). Occluded Cations in Active Transport. Ann. Revs. Biochem. 59: 171-205.

4. Goldshleger, R., Karlish, S.J.D., Rephaeli, A. and Stein, W.D. (1987) The Effect of Membrane Potential on the Mammalian Sodium/Potassium Pump Reconstituted into Phospholipid Vesicles. J. Physiol. (Lond.) 387: 331-355.

5. Karlish, S. J. D., and Stein, W. D. (1982) Passive Rubidium Fluxes Mediated by $Na^+/K^+$-ATPase Reconstituted into Phopholipid Vesicles when ATP- and Phophate-Free. J. Physiol. 328: 295-316.

6. Karlish, S.J.D., Goldshleger, R., and Stein, W.D. (1990). Identification of a 19kDa C-terminal Tryptic Fragment of the Alpha Chain of $Na^+,K^+$-ATPase, Essential for Occlusion and Transport. Proc. Natl. Acad. Sci. 87: 4566-4570.

7. Laüger, P. (1984) Thermodynamic and kinetic properties of electrogenic ion pumps. Biochim. Biophys. Acta. 779: 307-341.

8. Läuger, P. (1991) Kinetic Basis of Voltage Dependence of the $Na^+/K^+$-pump. In: The sodium pump: structure, mechanism and regulation, (J.H. Kaplan and P. de Weer, Eds. ), Rockefeller University Press pp.304-315.

9. Or, E., David, P., Shainskaya. A., Tal, D.M., and Karlish, S.J.D. (1993). Effects of Competitive Sodium-like Antagonists on $Na^+/K^+$-ATPase Suggest that Cation Occlusion from the Cytoplasmic Surface Occurs in Two Steps. J. Biol. Chem. 268: 16929-16937.

10. Shani-Sekkler, M., Goldshleger, R., Tal, D. M., and Karlish, S.J.D. (1988). Inactivation of $Rb^+$ and $Na^+$ Occlusion on $(Na^+,K^+)$-ATPase by modification of carboxyl groups. J. Biol. Chem. 263: 19331-19342.

11. Tanford, C. (1982) Simple model for the chemical potential change of a transported ion in active transport. Proc. Natl. Acad. Sci. 79: 2882-2884.

400

# Structure of the Cation Occlusion Domain of $Na^+/K^+$-ATPase Studied by Proteolysis, Interaction with $Ca^{2+}$ ions and Thermal Denaturation

A.M.Shainskaya and S.J.D.Karlish

Biochemistry Department, Weizmann Institute of Science, Rehovot 76100, Israel

## Introduction

Knowledge of the structure of cation binding sites and of organization of trans-membrane segments of cation pumps, such as $Na^+/K^+$-ATPase, is essential for an understanding of the active transport mechanism. Renal $Na^+/K^+$-ATPase extensively digested with trypsin, in the presence of $Rb^+$ and absence of $Ca^{2+}$ ions, to a C-terminal 19kDa and smaller membrane-embedded fragments (8-11kDa) of the alpha chain, and intact beta chain or beta split into two fragments, is a valuable tool for establishing structure-function relationships. In these "19kDa membranes", cation occlusion sites are intact, and although it is evident that the sites are located within trans-membrane segments, the identity and number of segments involved is unknown. The question as to which or how many of the trans-membrane segments actually occlude the ions is unanswered, and is the major focus of the present work. The first section of the present paper analyses the nature of the antagonism between $Ca^{2+}$ and $Rb^+$ ions in "19kDa-membranes", and explains the previous observation that the presence of $Ca^{2+}$ during digestion destabilises the 19kDa fragment to trypsin and occlusion is destroyed. The second section describes the purification and identification of the limit tryptic membrane-embedded fragments, produced by digestion of "19kDa-membranes" in the presence of $Ca^{2+}$.

### Binding of $Ca^{2+}$ ions in $Rb^+$ occlusion sites on "19kDa-membranes"

We provide here evidence for direct competition between $Ca^{2+}$ and $Rb^+$ ions in occlusion sites on "19kDa membranes". In experiments examining $Rb^+$ occlusion at different $Rb^+$ concentrations, at various fixed concentrations of $Ca^{2+}$ ions, sigmoid $Rb^+$ activation curves are observed. $Ca^{2+}$ ions behave competitively in that, $Rb^+$ affinity is reduced in the presence of $Ca^{2+}$ ($K_m$=20.5 µM increased to 603.6 µM at 1mM $Ca^{2+}$), while maximal binding (5.25±0.25 nmol $Rb^+$/mg protein) is unchanged. Experiments with fixed $Rb^+$ and variable $Ca^{2+}$ concentrations suggest that binding of a single $Ca^{2+}$ antagonises binding of the two $Rb^+$ ions. Assuming a simple one-site model for $Rb^+$ occlusion and the measured $K_{0.5}$ value, the intrinsic dissociation constant for $Ca^{2+}$ is calculated to equal 2.8µM. The findings are compatible with previous work describing $K^+$-like effects of $Ca^{2+}$, and $Ca^{2+}$ occlusion on intact $Na^+/K^+$-ATPase (4,12,14), and add to the previous information in that the possibility of indirect allosteric antagonism between $Ca^{2+}$ and $Rb^+$ on "19kDa-membranes" can be excluded. Measurement of the rate of dissociation of occluded [86]Rb, when displaced by $Ca^{2+}$ ions, provides a more rigourous test for true competition. Deocclusion of [86]Rb was compared in media containing concentrations of Rb$^+$ or Ca$^{2+}$ ions sufficient to completely displace the occluded $^{86}$Rb. The deocclusion into the Rb$^+$ medium is markedly biphasic, as reported previously (7) with rate constants of 0.16 sec$^{-1}$ and 0.002 sec$^{-1}$ and almost equal amplitudes of the two phases. By contrast, displacement of $^{86}$Rb by Ca$^{2+}$ ions is monoexponential with a rate constant (0.06 sec$^{-1}$), about half that of the faster of the two rate-constants for dissociation of $^{86}$Rb into an excess of Rb$^+$. The monophasic dissociation observed with Ca$^{2+}$ is the result one might expect for an ion that competes with Rb$^+$ in the site, and does not affect the dissociation of the first $^{86}$Rb, but is unable to block release of the second deeply bound $^{86}$Rb.

Further evidence as to the nature of the interaction of Ca$^{2+}$ ions with the sites is provided by the pH dependence of the competition. Elevation of inhibitory potency of Ca$^{2+}$ from 5% to about 80% as pH is raised from 6.0 to 8.5 suggests competition between Ca$^{2+}$ and protons in the binding site, with a pKa of around 7.5. The apparent Rb$^+$ affinity for occlusion is about the same at pH 7.0 and pH 8.5, and therefore the greater potency of Ca$^{2+}$, as the pH is raised, reflects an increased intrinsic binding affinity. Similar effects of pH have been observed on the modification of carboxyl residues by DCCD (13), on electrogenicity of reconstituted pumps sustaining so-called "uncoupled Na$^+$-flux" (5), and on the potency of guanidinium derivatives acting as competitive Na$^+$ antagonists (3).

**Protection by Ca$^{2+}$ ions against inactivation by DCCD and Thermal Inactivation of Rb$^+$ Occlusion in the presence of Ca$^{2+}$.**

Ca$^{2+}$ ions resemble Rb$^+$, Na$^+$ and Na$^+$ antagonists (6,9) in respect of protection against inactivation of Rb$^+$ occlusion by DCCD. Inactivation of "19kDa membranes", with DCCD (1 μmol/mg protein) at 30 $^{\circ}$C for 3 minutes resulted in 65% loss of Rb$^+$ occlusion. Ca$^{2+}$ decreases the extent of DCCD inactivation, but less so than Rb$^+$ which is fully effective (6). Protection by Ca$^{2+}$ may be incomplete, due to an accelerated thermal inactivation of Rb$^+$ occlusion in the presence of Ca$^{2+}$ ions. We have reported recently that incubation of "19kDa-membranes" at 37$^{\circ}$C, in the absence of occluded cations, leads to irreversible inactivation of Rb$^+$ occlusion. Rb$^+$, Na$^+$ and other alkali metal cations protect against this thermal inactivation process (9). Therefore, the question arose whether the reversible displacement of Rb$^+$ by Ca$^{2+}$ ions is followed by the irreversible process. This was addressed by looking at the ability of EGTA to reverse the inhibition of Rb$^+$ occlusion by Ca$^{2+}$ ions. Initially, addition of EGTA after Ca$^{2+}$ was able to largely reverse inhibition of occlusion, but subsequently, this ability was lost because of irreversible thermal inactivation of Rb$^+$ occlusion. In the presence of Rb$^+$ ions and absence of Ca$^{2+}$ ions, the occlusion was quite stable. The effects of Ca$^{2+}$ ions suggest that they, like the Na$^+$ antagonists, could interact with carboxyl residues in the initial recognition complex, but fail to be occluded and induce the conformational changes necessary to confer resistance to proteases and thermal stability.

**Tryptic digestion of "19kDa-membranes" in the presence of Ca$^{2+}$. Inactivation of Rb$^+$ occlusion**

The evidence discussed above for occupation of the sites by Ca$^{2+}$ ions and inhibition of Rb$^+$ occlusion, and subsequent irreversible inactivation, suggests that the destabilisation towards trypsin is related to these phenomena. Incubation of "19kDa-

402

membranes" at 37°C in the presence of $Rb^+$ ions with and without $Ca^{2+}$ ions and with or without trypsin, leads to rapid and irreversible loss of $Rb^+$ occlusion. Occlusion is lost at a rate which depends only on the presence and concentration of $Ca^{2+}$ ions, and trypsin itself is entirely without effect. Therefore, trypsin must have digested the "19kDa-membranes" which had been inhibited or irreversibly inactivated by $Ca^{2+}$ ions, and digestion is not the cause of loss of occlusion. It appears finally that trypsin gains access to and digest only "19kDa-membranes" from which $Rb^+$ is displaced by $Ca^{2+}$, but which has not yet been thermally inactivated, in addition to the thermally inactivated state.

## Identification of Tryptic Fragments

The limit membrane-embedded peptides, resulting from tryptic digestion of "19kDa-membranes" in the presence of $Ca^{2+}$, have been partially purified and concentrated by size-exclusion HPLC, resolved on 16.5% Tricine gels and identified by sequencing, as described in (1). In digested "19kDa-membranes" we have detected truncated peptides corresponding to M1+2, M3+4 and M5+6, and three fragments derived from the 19kDa peptide, $L^{842}$--$R^{880}$, $I^{946}$--$R^{972}$, and $M^{973}$--$R^{998}$. A predicted hydrophobic tryptic peptide $I^{906}$--$K^{931}$ was not found in the membranes ( see also 10). The 16kDa fragment of the beta chain is truncated at both N- and C-terminals upon displacement of the $Rb^+$. The fragment $T^{28}$--$K^{112}$ with Mr 16.05kDa has been identified. The glycosylated fragment of the beta chain of $M_r \approx 50$kDa is resistant to cleavage. It is a striking fact that, in the presence of $Ca^{2+}$ ions, trypsin digests the 19kDa, 16kDa and also the family of 8-12kDa fragments to smaller fragments, while all of these peptides are protected from trypsin in the presence of $Rb^+$ (and absence of $Ca^{2+}$) during preparation of the "19kDa-membranes". The 19kDa fragment is known to extend to the C-terminal of the alpha chain as judged by binding of an antibody to the final four amino acid residues E T Y Y. After tryptic digestion in the presence of $Ca^{2+}$ ions no binding of the anti-E T Y Y could be observed. The are several points of interest.

1. Occlusion of cations ($Rb^+$ or $Na^+$) protects all the fragments in "19kDa-membranes" against proteolysis. Thus all fragments are in a complex, and interact directly or indirectly via other fragments, with occluded cations. The cation-binding domain consists of several membrane-spanning segments, extra-membrane loops and, as discussed below, the C-terminal of the alpha chain. When occluded cations are displaced (e.g. $Rb^+$ by $Ca^{2+}$) the complex relaxes and is digested by proteases.

2. Occluded cations stabilise the 19kDa peptide. Loops and tails at both extracellular and cytoplasmic surface are protected against proteolysis by cations occluded within trans-membrane segments.

3. Occluded cations protect the 16kDa fragment of the beta-chain against proteolysis, presumably via interaction between the alpha chain and beta fragments. Interaction between the alpha chain and 16 kDa fragment must affect $Rb^+$ occlusion.

4. Although analysis of tryptic fragments cannot provide conclusive evidence on their trans-membrane topology, it may provide useful supportive information for models suggested on the basis of definitive techniques. For example the lack of evidence for a membrane associated hydrophobic peptide $I^{906}$--$K^{931}$ is consistent with an new 8 segment model (see Goldshleger et al, this conference). The fragment $M^{973}$-- $R^{998}$ may be associated with the membrane without spanning it.

## A Regulatory Role for the C-terminal of the Alpha Chain?

Digestion of 19kDa fragment (to about 18.5kDa) by Carboxypeptidase Y occurs only in conditions when $Rb^+$ occlusion is lost, in the presence of $Ca^{2+}$ ions, or at pH 5.5. The family of smaller peptides (8-12kDa) appeared to be clipped in all conditions. Loss of the anti-E T Y Y binding confirm that digestion of the C-terminal occurs only in the inactivated samples. Evidently, occlusion of $Rb^+$ induces a conformation in which the C-terminal tyrosines are hidden from the Carboxypeptidase. A necessary corollary is that the C-terminal must interact either directly or indirectly with $Rb^+$ occlusion domain, and affect it. Precedents for regulatory interactions between C-terminal and other domains of P-type pumps, include (a) the calmodulin binding domain of cell membrane $Ca^{2+}$-ATPase (2) and (b) the C-terminal of plant $H^+$-ATPase (11). It would be interesting if this turned out to be a general feature of P-type cation pumps.

## References

1. Capasso JM, Hoving S, Tal DM, Goldshleger R, and Karlish SJD (1992) Extensive digestion of $Na^+,K^+$-ATPase by Specific and Non-specific Proteases with Preservation of Cation Occlusion sites. J Biol Chem 267: 1150-1158

2. Carafoli E (1992) The $Ca^{2+}$ pump of the plasma membrane. J Biol Chem 267: 2115-2118

3. David P, Mayan H, Cohen H, Tal D, and Karlish SJD (1992) Guanidinium Derivatives act as High Affinity Antagonists of $Na^+$ ions in Occlusion sites of $Na^+,K^+$-ATPase. J Biol Chem 267: 1141-1149

4. Forbush B, III (1988) Rapid Release of $^{45}Ca$ from an Occluded State of the $Na^+,K^+$-ATPase. J Biol Chem 236: 7970-7978

5. Goldshleger R, Shachak Y, and Karlish SJD (1990) Electrogenic and electroneutral Transport of Renal Na/K-ATPase Reconstituted into Proteoliposomes. J Membr Biol 113: 139-154

6. Goldshleger R, Tal D, Moorman J, Stein WD, and Karlish SJD (1992) Chemical Modification of Glu 953 of the Alpha Chain of $Na^+,K^+$-ATPase Associated with Inactivation of Cation Occlusion. Proc Natl Acad Sci U.S.A. 89: 6911-6915

7. Karlish SJD, Goldshleger R, and Stein WD (1990) Identification of a 19kDa C-terminal Tryptic Fragment of the Alpha Chain of $Na^+,K^+$-ATPase, Essential for Occlusion and Transport. Proc Natl Acad Sci U.S.A. 87: 4566-4570

8. Karlish SJD, Goldshleger R, and Jorgensen PL (1993) Location of Asn-831 of the Alpha Chain of Na/K-ATPase at the Cytoplasmic Surface. Implication for Topological Models. J Biol Chem 268: 3471-3478

9. Or E, David P, Shainskaya A, Tal DM, Karlish SJD (1993) Effects of Competitive Sodium-like Antagonists on Na/K-ATPase Suggest that Cation Occlusion from the Cytoplasmic Surface Occurs in Two Steps. J Biol Chem 268: 16929-16937

10. Ovchinnikov YuA, Arzamazova NM, Arystarkhova EA, Gevondyan NM, Aldanova NA, and Modyanov NN (1987) Detailed structural analysis of exposed domains of membrane bound $Na^+,K^+$-ATPase. A model for trans-membrane arrangement. FEBS Letters 217: 269-274

11. Palmgren GM, Sommarin M, Serrano R, and Larson Ch (1991) Identification of an Autoinhibitory Domain in the C-terminal Region of the Plant Plasma Membrane $H^+$-ATPase. J Biol Chem 266:20470-20475

12. Post RL, and Stewart HB (1985) Occupancy of a monovalent cation-binding centre in $Na^+,K^+$-ATPase by calcium ion. In: Glynn I, Ellory C (eds) The Sodium Pump. The Company of Biologists , Cambridge, pp. 429-441

13. Shani-Shekler M, Goldshleger R, Tal D, and Karlish SJD (1988) Inactivation of $Rb^+$ and Na Occlusion on (Na,K)-ATPase by modification of carboxyl groups. J Biol Chem 263: 19331-19341

14. Vassalo PM, Post RL (1986) Calcium Ion as a Probe of the Monovalent Centre of Sodium Potassium ATPase. J Biol Chem 261: 16957-16962

# Functional consequences of alterations to Glu329 and other residues in the transmembrane sector of Na$^+$/K$^+$-ATPase

B. Vilsen

The Danish Biomembrane Research Centre, Institute of Physiology, University of Aarhus, DK-8000 Aarhus C, Denmark

## Cloning of a Glu329→Gln variant of the rat Na$^+$/K$^+$-ATPase

I have previously described the cloning of a cDNA encoding the ouabain-resistant rat kidney $\alpha_1$-isoform of Na$^+$/K$^+$-ATPase from a cDNA library (3). Further examination of the library revealed, in addition to this wild-type cDNA, a variant Na$^+$/K$^+$-ATPase cDNA containing a single G to C base substitution, which on amino acid level gave rise to a glutamate to glutamine substitution at position 329 within the motif 328-PEGL in the predicted 4th transmembrane helix. As Glu329 has recently been attributed a role in cation binding (1), it was pertinent to examine whether the Glu329→Gln variant was able to carry out active transport of Na$^+$ and K$^+$ and Na$^+$/K$^+$-activated ATP hydrolysis. The cDNAs encoding either the Glu329→Gln variant or the wild-type rodent Na$^+$/K$^+$-ATPase were, therefore, transfected into COS-1 cells, which were grown in the presence of 5 $\mu$M ouabain i.e. conditions under which only COS-1 cells expressing functional exogenous ouabain-resistant Na$^+$/K$^+$-ATPase are able to survive.

Surprisingly, the Glu329→Gln variant of the ouabain resistant rodent Na$^+$/K$^+$-ATPase was able to confer ouabain resistance to COS-1 cells, as revealed by the appearance of ouabain resistant colonies after 2-3 weeks of growth in the presence of 5 $\mu$M ouabain. PCR was used to verify that the Glu329 to Gln substitution was still present in the cDNA stably integrated into the chromosome of the transfectants grown in presence of ouabain.

It can thus be concluded that the expressed Glu329→Gln modified enzyme is functional in Na$^+$/K$^+$-transport.

A number of other mutants made by in vitro mutagenesis were also analyzed in the same way as the Glu329→Gln variant (Table 1).

## Determination of the phosphorylation capacity and molecular turnover numbers

Phosphorylation from AT$^{32}$P was carried out under conditions where the dephosphorylation rate is low and phosphorylation therefore nearly stoichiometric (0°C, 150 mM Na$^+$ present, absence of K$^+$). Since the expressed Na$^+$/K$^+$-ATPase constitutes only a minute fraction of the total protein present in the COS-cell plasma membranes, an accurate determination of the amount of phosphorylated Na$^+$/K$^+$-ATPase required separation of the phosphorylated Na$^+$/K$^+$-ATPase from other phosphorylated proteins. This was accomplished by SDS-polyacrylamide gel electrophoresis under acid conditions (4). All the phosphoprotein migrating corresponding to 100 kDa on the gel was demonstrated to be Na$^+$/K$^+$-ATPase, since it disappeared upon addition of K$^+$. Preincubation with ouabain and MgCl$_2$ in absence of Na$^+$ and K$^+$ for 30 min at 20 °C, to inhibit specifically the ouabain-sensitive endogenous Na$^+$/K$^+$-ATPase prior to initiation of the phosphorylation reaction, was without significant effect on the phosphoenzyme levels formed in membranes from cells expressing the rat wild-type enzyme or the Glu329→Gln variant. This indicates that the endogenous enzyme was inactive or not present at all in these membranes. A likely explanation is that the ouabain-sensitive endogenous enzyme had already formed a stable ouabain-bound complex during propagation of the COS-1 cells in the presence of ouabain. By contrast, phosphorylation was completely prevented by preincubation with ouabain in membranes isolated from mock-transfected cells that expressed only the endogenous ouabain-sensitive enzyme and which had been grown in the absence of ouabain.

To determine the turnover numbers the concentrations of active Na$^+$/K$^+$-ATPase sites present in the membrane preparations isolated from the transfectants were measured by the phosphorylation method. The specific Na$^+$/K$^+$-ATPase activities were measured at 10 $\mu$M ouabain, since at this ouabain concentration the difference between the ATPase activities of the ouabain-sensitive endogenous enzyme and the ouabain-insensitive rat Na$^+$/K$^+$-ATPase is near maximal (3).

There was no significant difference between the maximum turnover numbers of 28,600 min$^{-1}$ $\pm$ 3200 and 26,900 min$^{-1}$ $\pm$ 3150 calculated for the wild-type enzyme and the Glu329→Gln variant, respectively, at saturating Na$^+$, K$^+$, and ATP concentrations (average values of 4 experiments $\pm$ S.D.).

By contrast, the turnover number of the Leu332→Ala mutant was found to be reduced to around 35% that of the wild-type enzyme (Table 1). This may be explained by a reduction of the rate of a partial reaction following the phosphorylation reaction, in accordance with the finding that the E$_1$P-E$_2$P conversion is blocked in the homologous Ca$^{2+}$-ATPase mutants Pro312→Ala, Gly and Leu (4).

## Na$^+$, K$^+$, and ATP dependencies of Na$^+$/K$^+$-ATPase activity

Despite the fact that the Glu329→Gln variant was functional, it did not respond to variation of the concentrations of Na$^+$, K$^+$, and ATP in the same way as the wild-type enzyme. As can be seen in Table 1, the Glu329→Gln variant displayed a 15-fold increase in apparent affinity for ATP relative to the wild type and a reduced apparent affinity for either of the cations, corresponding to a 2-fold increase in K$_{0.5}$ for cytoplasmic Na$^+$ and a 6-fold increase in K$_{0.5}$ for extracellular K$^+$. The increased apparent affinity for ATP displayed by the Glu329→Gln variant suggests that the Glu to Gln substitution displaced the E1-E2 equilibrium in favor of the E1 conformation possessing high affinity for ATP. The reduced K$^+$ affinity may be related to this displacement of the E1-E2 equilibrium. The finding that the Glu329→Gln variant exhibited a 2-fold reduced affinity for Na$^+$ can, however, not be understood by reference to the conformational equilibrium, since the displacement in favor of E1 would tend to increase the apparent affinity for cytoplasmic

Na$^+$. It is necessary, then, to postulate that in addition to the displacement of the equilibrium in favor of the E1 conformation, the intrinsic affinity of the E1 conformation for Na$^+$ was reduced in the Glu329→Gln variant. Hence, it seems that the Glu329→Gln substitution disturbed one or more of the high affinity cytoplasmic Na$^+$ sites. The reduction of intrinsic Na$^+$ affinity may have amounted to much more than 2-fold, since the observed 2-fold decrease in apparent affinity may be the net effect resulting from a combination of reduced intrinsic affinity with the displacement of the conformational equilibrium towards the Na$^+$ form.

**Table 1. $K_{0.5}$ for Na$^+$, K$^+$ and ATP and turnover numbers of the mutants**

| Mutant | $K_{0.5}$ (mM) | | | Turnover (min$^{-1}$) |
|---|---|---|---|---|
| | Na$^+$ | K$^+$ | ATP | |
| Wild type | 7.13 | 0.78 | 0.279 | 26,900 |
| *M4 segment:* | | | | |
| Pro328→Ala | 13.04 | 2.46 | 0.086 | 24,300 |
| Glu329→Gln | 13.79 | 4.79 | 0.019 | 28,600 |
| Leu332→Ala | 3.92 | 1.97 | 0.042 | 10,700 |
| *M5 segment:* | | | | |
| Pro780→Ala | 7.10 | 4.60 | 0.110 | 27,600 |
| *M6 segment:* | | | | |
| Pro813→Ala | 10.72 | 2.60 | 0.130 | 25,200 |

## pH-dependence of Na$^+$/K$^+$-ATPase activity of the Glu329→Gln variant and the wild-type enzyme expressed in COS cells

The figure shows the pH-dependencies of the maximum turnover rates of the Glu329→Gln variant and the wild type. The turnover rate of the Glu329→Gln variant was almost unaffected by varying pH above 8.0. Assuming that Glu329 is located within a hydrophobic region in Na$^+$/K$^+$-ATPase leading to an increase in the pK of Glu329 relative to its isolated state, the simplest possibility would be that Glu329 is the residue whose deprotonation is responsible for the decrease in activity observed with the wild-type enzyme at high pH values. Alternatively, the titration of an amino group is responsible for the decrease in enzyme activity in the wild-type enzyme at alkaline pH.

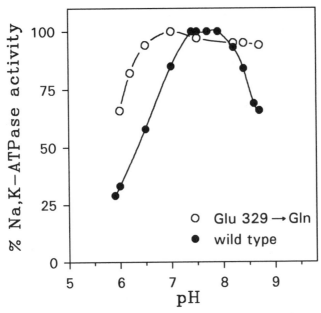

Maximal Na$^+$/K$^+$-ATPase activity may require that Glu329 is neutralized by formation of a salt-bridge with the protonated amino group. An increase in pH above 8.0 would lead to deprotonation of the amino group, breaking the salt-bridge and leaving the negative charge of Glu329 unshielded. In the Glu329→Gln variant, there is no negative charge to be neutralized, explaining why the activity of the Glu329→Gln variant remained high at alkaline pH.

In the lower pH range, the activity of the Glu329→Gln variant was also enhanced relative to that of the wild type, with a downward shift of the pH optimum. At low pH, the E2(K)→E1(Na) transition is rate limiting for the overall turnover rate in the wild type (2). Therefore, the observed difference between the Glu329→Gln variant and the wild type in the lower pH range is in line with the finding that the Glu329→Gln variant displayed a higher apparent affinity for ATP than the wild-type enzyme.

**References**

1. Goldshleger R, Tal DM, Moorman J, Stein WD, Karlish SJD (1992) Chemical modification of Glu-953 of the α chain of Na$^+$/K$^+$-ATPase associated with inactivation of cation occlusion Proc Natl Acad Sci USA 89:6911-6915
2. Skou JC (1983) On the mechanism behind the ability of Na$^+$/K$^+$-ATPase to discriminate between Na$^+$ and K$^+$. Current topics in membranes and transport 19:323-341
3. Vilsen B (1992) Functional consequences of alterations to Pro328 and Leu332 located in the 4th transmembrane segment of the α-subunit of the rat kidney Na$^+$/K$^+$-ATPase FEBS Lett 314:301-307
4. Vilsen B, Andersen JP, Clarke DM, MacLennan DH (1989) Functional consequences of proline mutations in the cytoplasmic and transmembrane sectors of the Ca$^{2+}$-ATPase of sarcoplasmic reticulum. J Biol Chem 264:21024-21030

# Identification of Glu[779] as an Essential Part of the Cation Binding Site of the Sodium Pump

José M. Argüello and Jack H. Kaplan

Department of Physiology, School of Medicine, University of Pennsylvania, Philadelphia, Pennsylvania 19104-6085, USA

## Introduction

The chemical modification of purified renal $Na^+/K^+$-ATPase with the fluorescent carboxyl-selective reagent, 4-(diazomethyl)-7-(diethylamino)-coumarin (DEAC), results in enzyme inactivation via disruption of the monovalent cation binding sites and loss of $K^+$ and $Na^+$ binding capacity. Modification of 1 or 2 carboxyl residues in the α-subunit in a $K^+$ or $Na^+$-preventable manner leaves the ATP binding unaltered and the enzyme is still able to undergo $E_1 \leftrightarrow E_2$ transitions. These characteristics of the DEAC-modified enzyme are evidence that the modified residues are located at the cation binding site (1, 2). Furthermore, the specificity of DEAC to react with carboxyl groups supports the early idea that carboxyl residues, in or near the membrane, would be involved in the cation oclusion and transport by the $Na^+/K^+$-ATPase.

We now describe the localization of the modified carboxyl within the primary structure of the α-subunit and propose that its position in the putative 5th transmembrane fragment provide a link between the nucleotide binding site in the cytoplasmic loop and the cation binding site in the membrane segments of the $Na^+/K^+$-ATPase.

## Methods

$Na^+/K^+$-ATPase ($\approx$ 20-25 μmol $P_i$ $mg^{-1}$ $min^{-1}$) was purified from dog kidney outer medulla according to Jørgensen (5) with the modifications of Liang and Winter (7). The enzyme, 10-20 mg (0.5 mg/ml), was treated with 250 μM DEAC, in 50 mM imidazole, pH 6.5, 1 mM EDTA, 3 mM $MgCl_2$, 3 mM $P_i$, 10% dimethylsulfoxide during 2 h at 37°C. The modified protein (DEAC-enzyme) retained 10% of its initial activity. 100 mM KCl or 100 mM NaCl was included in the media when it was desired to obtain K-protected enzyme (90% active) or Na-protected enzyme (75% active) respectively (1, 2). The α-subunits from DEAC-enzyme, K-protected, and Na-protected enzymes were isolated by SDS-PAGE in 7.5% acrylamide gels (6). The α-subunit bands were cut out of the gel, eluted with 0.1% SDS, 0.1 M $NH_4HCO_3$ pH 8.5, concentrated using Centricon 30, and precipitated with methanol (-20°C, overnight). Purified α-subunits were cleaved with protease $V_8$ (1:10; w:w) in 0.05% SDS, 20 mM phosphate buffer pH 7, 6 hs, 37°C followed by overnight at room temperature. Tricine-urea SDS-PAGE of $V_8$-treated α-subunits was performed in 16.5% polyacrylamide gels (9). Two fluorescent DEAC-labeled cation-protectable bands with apparent molecular weights (MW) of 15-17 kDa (Peptides I and II) and 5-6 kDa (Peptides IIIa and IIIb) were observed. The band corresponding to peptides IIIa and IIIb was cut out of the gel, eluted with 0.1% SDS, 0.1 M $NH_4HCO_3$ pH 8.5, concentrated in Centricon 3 filters and precipitated with acetone (-20°C, overnight). The peptides were resuspended in 0.1 M $NH_4HCO_3$ pH 8.5 and spotted onto polyvinylidene fluoride (PVDF) membranes for sequencing. The DEAC-labeled cation-protectable peptides with an apparent MW $\approx$ 15-17 kDa (Peptides I and II) were resolved in 15% polyacrylamide tris-glycine-2 M urea SDS-PAGE (6). These peptides were recovered from the gel as indicated for peptides III.

Peptides IIIa and IIIb were digested with thermolysin (1:10, w:w) in 1 mM $CaCl_2$, 20 mM $NH_4HCO_3$ pH 8.5, 2 h, 45°C and the resulting digest separated in Tricine-urea gels (9). The only fluorescent band observed in this gel (Peptide IV) was cut out, eluted with 20 mM $NH_4HCO_3$ pH 8.5, concentrated by lyophilization, and spotted onto PVDF membranes. For sequencing of Peptide IV in particular, the PVDF membrane was positioned on top of a Polybrene-conditioned filter disk (10).

In a set of experiments designed to prove the length of Peptide IV, DEAC labeled Peptides IIIa and IIIb and Peptide IV were treated with 20 μM 2,5-dimethoxystilbene-4'-maleimide (DMSM) in 50 mM Tris pH 8.2 at room temperature, in the dark, during 1 h. The fragments were then run in a non-reducing Tricine-urea gel (9). DEAC and DMSM residues were detected by their fluorescence emission after equilibrating the gel for 20 min either in 0.1 M Tris pH 8.25 or in 10% acetic acid pH 2.5 respectively.

## Results

For the localization of the labeled residues, we purified DEAC labeled α-subunits free of non-covalently bound coumarines. The α-subunits were cleavaged with $V_8$ protease which, besides producing fragments large enough to be separated by tricine gels, yielded identical peptide maps for the protected and the unprotected α-subunits (Coomassie Brilliant Blue staining Fig. 1A). Fluorescence visualization of these gels, gave clear differentiation between cation-protectable (responsible for inactivation) and non-specific unprotectable labeling. $Na^+$ or $K^+$ protected $V_8$ digested α-subunit showed the same

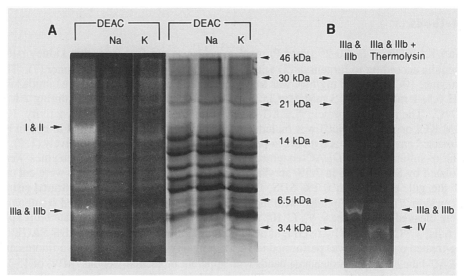

Figure 1: SDS-PAGE (tricine-urea-16.5% polyacrylamide) of: A) V8 protease treated α-subunits from DEAC-enzyme, Na-protected enzyme and K-protected enzyme. The fluorescent bands (3 left lanes) in the unstained gel were visualized by illumination with a UV lamp. The 3 lanes on the right show the coomassie brilliant blue staining of the gel. B) Peptides IIIa and IIIb and Peptide IV obtained by thermolysin treatment of Peptides IIIa and IIIb. Peptides contained in each fluorescent band and the position of molecular weight marker are indicated.

fluorescent labeling pattern, indicating that both cations protect the labeling of the same carboxyl residue. Two cation-protectable bands with apparent MW 5-6 kDa and 15-17 kDa were observed (Fig. 1A). Isolation of these fragments and sequencing showed the presence of four peptides in these bands (Peptides I and II were separated by Laemmli gels, not shown, and Peptides IIIa and IIIb were co-sequenced). Peptides I (apparent MW ≈ 17 kDa), II (≈ 15 kDa) and IIIa (≈ 5-6 kDa) started at Gly[758] while Peptide IIIb (≈ 5-6 kDa) started at Gly[561] (sequence assignments are based on the $\alpha1$ subunit from dog (11)). Since Peptides I, II, and IIIa have the same starting sequence they were the result of differences in the position of $V_8$ cleavage at their carboxyl termini and they contain in common the same DEAC labeled carboxyl residue. Subsequent proteolysis of peptides IIIa and IIIb (which co-migrate in tricine-urea gels) with thermolysin followed by electrophoresis, revealed a single smaller fluorescent band (Fig. 1B). N-terminal sequence analysis of this material showed a single peptide extending from Leu[773] to Leu[784] with a single carboxyl residue, Glu[779], which thus, is the amino acid modified by DEAC.

Since the DEAC ester bond is hydrolyzed during sequencing (and therefore Glu[779] is detected) it is necessary to demonstrate that Peptide IV contained only one carboxyl residue (Glu[779]) and did not extend to the next carboxyl beyond Leu[784]. We took advantage of the fact that before the next carboxyl residue down-stream in the sequence (Asp[804]) there is a single thiol group (Cys[802]), and we exploited the pH sensitivity of DEAC fluorescence (DEAC is not fluorescent at low pH) and the pH insensitivity of DMSM, a fluorescent -SH reagent. After treatment with DMSM, Peptide IV showed no fluorescence at pH 2.5, hence it was not labeled with DMSM, did not include Cys[802], and therefore did not reach Asp[804] (Fig. 2). As a positive control Peptides IIIa & IIIb (which contain cysteine residues) were subjected to similar treatment and they proved to be fluorescent at pH 2.5 when treated with DMSM.

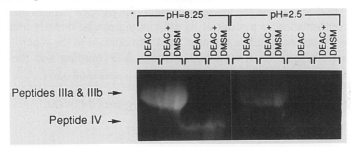

Figure 2: SDS-PAGE (tricine-urea-16.5% polyacrylamide-nonreducing conditions) of DMSM treated DEAC-labeled Peptide IV and Peptides IIIa and IIIb. DEAC-labeled Peptides without DMSM treatment were run as controls. DMSM and DEAC fluorescence was visualized by illumination with a UV lamp.

Further pieces of evidence also suggested that the DEAC-labeled fragment was a very small peptide with a single acidic amino acid. (i) The dramatic fall in yield on sequencing when sequencing was attempted without a second support (the Polybrene disk) (not shown). This specific effect was previously reported for the attempted sequencing of model decapeptides (10). (ii) The modified peptide starting at Leu[773] and reaching Asp[804] would have a molecular weight of 3691 Da. Peptide IV runs with our 3 kDa standard, but mobility in tricine-urea gels is unreliable for estimating the molecular weight of small peptides (9). However, Peptide IV was not retained and was

411

centrifuged through a 3 kDa cut-off filters, but it was retained by 1 kDa cut-off filters. These results and the double labeling experiment with DMSM strongly suggest $Glu^{779}$ is the only carboxyl in Peptide IV and is the residue labeled by DEAC.

## Discussion

We have shown that $Glu^{779}$ is the amino acid modified by DEAC in a $Na^+$- or $K^+$- protectable faction. Based on the characteristics of the DEAC modified $Na^+/K^+$-ATPase, we can conclude that this glutamate is part of the cation binding site of the enzyme, and essential for $Na^+$ and $K^+$ active transport by the sodium pump.
Several other observations point to the importance of this residue. This glutamic residue is conserved in all the sequences we examined for $Na^+/K^+$-ATPase, gastric $H^+/K^+$-ATPase, and the sarco(endo)plasmic reticulum $Ca^+$-ATPase; although it is not present in other P-type ATPases such as plasma membrane Ca pump, yeast H pump, or E. Coli K pump. Furthermore, Lingrel et al. [8] using heterologously expressed rat enzyme in HeLa cells, showed that the mutation of this residue yields HeLa cells unable to live in presence of ouabain, suggesting that the mutation of $Glu^{779}$ produced non-functional pumps. In addition, the mutations of the equivalent residue in the Ca pump (SERCA2), $Glu^{771}$ ($Glu^{771} \rightarrow$ Gln and $Glu^{771} \rightarrow$ Asp), have produced pumps unable to support $Ca^{++}$-dependent functions in that enzyme [3, 4].
According to the different topological models proposed for the $Na^+/K^+$-ATPase (10 or 8 transmembrane segments) $Glu^{779}$ is located in the $5^{th}$ transmembrane segment. There are important functional implications that result from the position of this residue. The $5^{th}$ transmembrane segment is directly connected to the cytoplasmic loop which contains the phosphorylation site and ATP binding domain. The localization of $Glu^{779}$ as part of the cation binding site reveals how information can be transfered between the catalytic domain (ATP hydrolysis) and the transport domain (ion passage through the membrane). Changes in conformation as a result of ligand binding can be directly transmitted, via $Glu^{779}$ at the $5^{th}$ transmembrane segment from one domain to the other. In this way, the antagonistic effects of $K^+$ on high affinity ATP binding and the ATP or $P_i$ stimulated deocclusion of $K^+$ ions are rationalized on a structural basis.

*Acknowledgement* This work was supported by NIH grant GM39500

## References

1. Argüello, J.M. and Kaplan, J.H. (1991) *J. Biol. Chem.* **266**, 14627-14635.
2. Argüello, J.M. and Kaplan, J.H. (1991) In *The Sodium Pump: Recent Developments* (Kaplan, J.H. and De Weer, P. Eds.) Rockefeller University Press,. New York, pp 199-203
3. Clarke, D.M., Loo, T.W., Inesi, G. and MacLennan, D.H.(1989) *Nature* **339**,476-478.
4. Clarke, D.M., Loo, T.W. and MacLennan, D.H. (1990) *J. Biol. Chem.* **265**, 6262-6267.
5. Jørgensen, P.L. (1974) *Biochim. Biophys. Acta* **356**,36-52.
6. Laemmli, U.K. (1970) *Nature* **224**,680-685.
7. Liang, S.M.L. and Winter, C.G. (1976) *Biochim. Biophys. Acta* **452**,552-565.
8. Lingrel, J.B., Schultheis, P.J., Jewell, E.A., Van Huysse, J.W., Askew, G.R., O'Brien, W.J., Kuntzweiler, T.A., Wallick, E.T. (1993) *Biol. Chem. Hoppe-Seyler* **374**:545-546.
9. Shagger, H. and Von Jagow, G. (1987) *Anal. Biochem.* **166**, 368-379.
10. Simpson, R.J., Movitz, R.L., Begg, G.S., Rubira, M.R. and Nice, E.C. (1989) *Anal. Biochem.* **177**, 221-236.
11. Xie, Z.J., Li, H., Liu, Q., Wang, Y., Askari, A. and Mercer, R.W. (1993) This Volume

# Rearrangements of the membrane-associated segments of $Na^+/K^+$-ATPase upon phosphorylation, ion-occlusion and ouabain binding

Svetlana Lutsenko, Jack H. Kaplan

Department of Physiology, University of Pennsylvania, Philadelphia, PA 19104-6085
USA

## Introduction

The reaction cycle of the $Na^+/K^+$-ATPase includes several intermediate stages, which are produced by the binding of different physiological ligands accompanied by conformational changes of the enzyme (2). Recently we have shown that extensive proteolytic digestion in the presence of various physiological ligands could be a useful tool in studies of rearrangements of membrane-associated segments of $Na^+/K^+$-ATPase (10). Distinctive conformational changes were revealed by N-terminal amino acid sequence analysis of the digests following SDS PAGE (10). The changes in cleavage patterns result from alterations in domain-domain interactions of the protein. Here, we exploit this approach to shed light on events which occur following ouabain binding. We addressed this question by a comparative study of the posttryptic products obtained after extensive degradation of $Na^+/K^+$-ATPase after phosphorylation, in the presence of potassium or Mg,Pi,ouabain.

## Materials and Methods

$Na^+/K^+$-ATPase was purified from dog kidney according to Jørgensen (3) with the average specific activity of 26-30 mmol Pi mg protein$^{-1}$ min$^{-1}$ assayed according to Brotherus et al (1). Protein concentration was determined by the method of Lowry et al. (8) . Rb occlusion was measured essentially as described previously (12).

Proteolytic digestion. $Na^+/K^+$-ATPase (1.5 mg/ml) was suspended in media containing 25 mM Imidazole-HCl, 1 mM EDTA-Tris, pH 7.5 and one of the following ligands: 1) 5 mM $MgCl_2$, 5 mM Trizma-Pi; 2) 5 mM $MgCl_2$, 5 mM Trizma-Pi, 1 mM ouabain; 3) 10 mM RbCl or 10mM KCl; 4) 10 mM Tris-HCl; 5) 10 mM NaCl; 6) 7 mM Na-ATP, 3 mM $MgCl_2$. After incubation of the samples at room temperature for 30 min, TPCK-Trypsin (1:10 w/w with respect to $Na^+/K^+$-ATPase) was added and tubes were incubated at 37$^O$C for 1 h. Soybean trypsin inhibitor was added (5-7:1 w/w with respect to trypsin) to stop digestion; samples were incubated an additional 10 min at 37$^O$C and they were then diluted with 1 ml of 25 mM Imidazole, pH 7.5, 1 mM EDTA-Tris, 2 mM RbCl, and the pelleted membranes were collected by centrifugation at 353000g for 30 min, at 4$^O$C. The membranes were homogenized in the latter buffer and centrifugation was repeated twice. Pellets were suspended in buffer containing 25 mM Imidazole-HCl, pH 7.5, 1 mM EDTA-Tris, 2 mM RbCl; and aliquots were taken for Rb-occlusion measurements or mixed with an equal volume of 10% SDS and modified with CPM as described in (9). Samples were separated in Tricine gels according to Schagger & Von Jagow (13), fragments were then transferred onto PVDF-membranes by electroblotting in 10mM CAPS, 10% methanol, pH 11 (11) and sequenced.

**Results**

The following considerations were made when conditions for proteolytic digestion were chosen: 1) the presence of ouabain stimulates phosphorylation from Mg,Pi; 2) the binding of ouabain prevents Rb-occlusion; 3) posttryptic membranes (obtained in the presence of Rb) have about the same level of Rb-occlusion as native enzyme. The Na$^+$/K$^+$-ATPase was digested in the presence of Mg,Pi,ouabain, K(Rb) or after phosphorylation (Mg,Pi or Na,Mg,ATP), membrane-associated fragments were separated by SDS-PAGE and they were then characterized by N-terminal amino-acid sequencing and compared.

The most extensive degradation was found under phosphorylation conditions, while protection of the 21 kD fragment can be seen in both Rb or Mg,Pi,ouabain media (Fig.1).

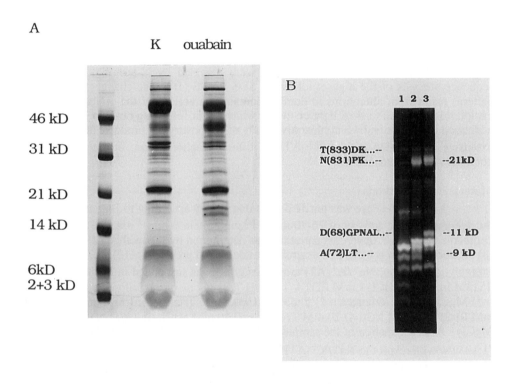

Fig.1 Composition of posttryptic membranes after extensive proteolytic digestion of Na$^+$/K$^+$-ATPase in the presence of K/Rb (3), MgPi,ouabain (2) or Mg,Pi /NaMgATP (1) A) 15% Laemmli gel - Coomassie staining; B) Tricine gel, nonreducing conditions - fluorescence visualization.

N-terminal amino-acid sequencing and staining with anti-ETYY (C-terminus) antibody revealed that in the presence of Rb or Mg,P$_i$,ouabain the same C-terminal fragment is protected against tryptic digestion (table).

When Rb-occlusion properties were compared, essential differences between all three samples were found (table). As expected the level of Rb-occlusion of the most extensively digested preparation was low. On the other hand, inspite of complete preservation of the 21kD fragment (which earlier had been claimed to be essential for Rb-occlusion (5)) ouabain-stabilized posttryptic membrane have markedly reduced Rb-occlusion capacity. Control experiments on digestion in the presence of radioactive ouabain explained the apparent reduction of Rb-occlusion. Ouabain remained bound to the membranes after digestion and even after several washings and thus the Rb-ouabain antagonism is preserved (see also table).

Table Comparison of posttryptic membrane preparations, obtained in the presence of Rb; MgPiouabain or Mg,Pi (Na,Mg,ATP).

|  | Rb | Mg,Pi,ouabain | Mg,Pi | i |
|---|---|---|---|---|
| Rb-occlusion nmol/mg prot | 7.17 | 3.22 | 1.7 | |
| presence of 21kD fragment | + | + | - | |
| N-terminus of 21kD | N(831)PKTDK | T(833)DKLVN | - | |
| Staining with anti ETYY-AB | + (21kD) | + (21kD) | - | |
| M.w. of M1-M2 fragment | 11 kD | 9 kD | 9 kD | |
| N-terminus of M1-M2 | D(68)GPNALT | A(72)LTPP | A(72)LTPP | |
| Protection of the b-subunit against digestion | + | - | - | |

Along with the striking similarities between K and ouabain with respect to stabilization of the C-terminal portion of a-subunit, distinct differences were seen upon comparison of the low molecular weight fragments and the $\beta$-subunit fragmentation pattern (Fig. 1). Digestion of the $\beta$-subunit is less in the presence of potassium than in the presence of ouabain. The different fragmentation of the $\beta$-subunit was also seen when $Na^+/K^+$-ATPase was digested in right-side-out microsomal vesicles (not shown). In addition, an 11kD fragment ($Asp^{68}$GPNA- ) which includes the M1-M2 transmembrane domain was obtained in Rb-stabilized membranes, whereas a 9kD fragment is produced from the same domain in ouabain media. When Mg,Pi or Na media were used a 9kD fragment was found under the phosphorylating conditions ($Ala^{72}$LTP- Fig.1B) and a 10.5kD fragment in the presence of Na ions ($Asp^{68}$GPNA- not shown).

**Discussion and Conclusions**

Our data indicate that the binding of ouabain at the extracellular surface is followed by rearrangement of the membrane-associated portion of the $Na^+/K^+$-ATPase to form a very stable ion-occluded-like state. Ouabain remains bound to the posttryptic membrane and apparent Rb-occlusion is drastically reduced inspite of complete preservation of C-terminal 21kD fragment. Moreover, the intracellular C-terminus of the M1-M2 domain is protected against digestion by the presence of monovalent cations, but is exposed after phosphorylation. Ouabain binding in the presence of Mg,Pi does not change its accessibility with respect to trypsin. These findings lead to the conclusion that remote effects of ouabain binding on the cytoplasmic portion of Na-pump, which have been reported (6, 4), are likely to be secondary effects and they are unlikely to be major structural changes, producing inhibition of the Na-pump. Trapping of the enzyme in one particular conformation seems to be the structural basis of the inhibitory action of ouabain and the transmembrane fragments and extracellular loops play the major role in this structural rearrangement. Our data also illustrate the involvement of the M2-M3 cytoplasmic loop in conformational changes, which take place upon phosphorylation and ion-binding.

*Acknowledgements.* This work was supported by NIH grant GM39500

## References

1.  Brotherus, J.B., Moller, J.V., Jørgensen, P.L. (1981) Soluble and active renal Na,K-ATPase with maximum protein molecular mass 170,000 ± 9,000 daltons; formation or larger units by secondary aggregation. Biochem Biophys Res Commun 100, 146-154.

2.  Jørgensen, P.L. (1975a) Purification and characterization of $(Na^+,K^+)$-ATPase V. Conformatinal changes in the enzyme. Transitions between the Na-form and the K-form studied with tryptic digestion as a tool. Biochim Biophys Acta 401, 399-315.

3. J  ørgensen, P.L. (1975b) Purification and characterization of $(Na^+ + K^+)$-ATPase III. Purification from the outer medulla of mammalian kidney after selective removal of membrane components by sodium dodecyl sulphate. Biochim Biophys Acta 356, 36-52.

4.  Jørgensen, P.L., Farley, R.A. (1988) Proteolytic cleavage as a tool for studying structure and conformatin of pure membrane-bound $Na^+,K^+$-ATPase. Methods in Enzymology 156, 291-301.

5.  Karlish, S.J.D., Goldshleger, R., Stein, W.D. (1990) A 19 kDa C-terminal fragment of the $\alpha$-chain of Na,K-ATPase is essential for occlusion and transport of cations. Proc Natl Acad Sci USA 87, 4566-4570.

6.  Kirley T.L., Peng, M. (1991) Identificatin of cysteine residues in lamb kidney (Na,K)-ATPase essential for ouabain binding. J Biol Chem 266, 19953-19957.

7.  Laemmli, U.K. (1970) Preparation of slab gels for one- or two-dimensional polyacrylamide sodium dodecyl sulfate gel electrophoresis. Nature 227, 680-685.

8.  Lowry, O.H., Rosebrough, N.J., Farr, A.L., Randall, R.J. (1951) Protein measurement with the folin phenol reagent. J Biol Chem 193, 265-275.

9.  Lutsenko, S., Kaplan, J.H. (1993a) An essential role of the extra-cellular domain of the $Na^+,K^+$-ATPase $\beta$-subunit in ion-occlusion. Biochemistry 32, 6737-6743.

10. Lutsenko, S., Kaplan, J.H. (1993b) Molecular events in close proximity to the membrane associated with the binding of ligands to the Na,K-ATPase. J Biol Chem (in press)

11. Matsudaria, P. (1987) Sequence from picomole quantities of proteins electroblotted onto polyvinylidene difluoride membranes. J Biol Chem 262, 10035-10038.

12. Shani, M., Goldshleger, R, Karlish, S.J.D. (1987) $Rb^+$ occlusion in renal $(Na^+ + K^+)$-ATPase characterized with a simple manual assay. Biochim Biophys Acta 904, 13-21.

13. Schagger, H., Von Jagow, G. (1987) Tricine-sodium dodecyl sulfate-polyacrylamide gel electrophoresis for the separation of proteins in the range from 1 to 100 kDa. Analyt Biochem 166, 368-379.

# Na$^+$/K$^+$-ATPase: Regulation of K$^+$ Access Channels by K$^+$

W.-H. Huang, A. Askari

Department of Pharmacology, Medical College of Ohio, P.O. Box 10008, Toledo, OH 43699-0008 USA

## Introduction

Based on studies with the purified canine kidney enzyme (1), we concluded recently that Rb$^+$ (or K$^+$) occlusion sites of the unphosphorylated E$_2$ state of the enzyme are confined within the protein matrix and connected to the medium by narrow and heterogeneous access channels, and that channel heterogeneity is distinct from any differences that may exist among the affinities of the occluded sites that are deep within these channels. More importantly, the same studies indicated that regulation of these channels is also distinct from the regulation of the events at the occluded sites, and identified allosteric regulatory effects of ATP, Na$^+$, and Rb$^+$ on the "width" of the access channels. Here, we present further findings on the interactive allosteric effects of ATP and Rb$^+$ (or K$^+$) on these channels. All experiments were done at 4°C by methods and procedures described before (1).

## Results and Discussion

ATP accelerates both Rb$^+$ binding and release (1). The present experiments focus on ATP effects on the release of occluded Rb$^+$ from the enzyme. As described before, experiments of Fig. 1 show that (a) release of occluded $^{86}$Rb$^+$ into a medium of low

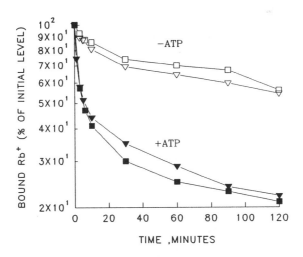

Fig. 1. Effects of 2 mM ATP on the time-course of $^{86}$Rb$^+$ release into media containing 20 mM Rb$^+$ ($\square$,$\blacksquare$) or no Rb$^+$ ($\triangledown$,$\blacktriangledown$). $^{86}$Rb$^+$ was bound to the enzyme at 3 $\mu$M Rb$^+$ (1).

ionic strength at 4°C is slow and not affected by the presence of 20 mM unlabeled $Rb^+$ in the medium; and (b) 2 mM ATP accelerates $^{86}Rb^+$ release with or without unlabeled $Rb^+$ being in the medium. In experiments of Fig. 2 we compared the

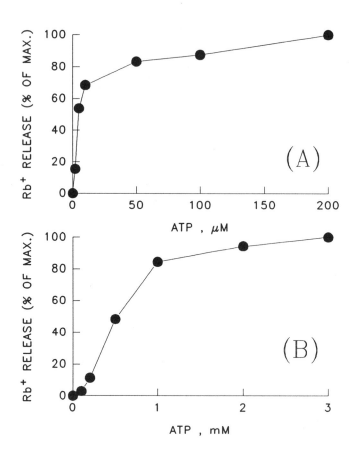

Fig. 2. Effects of varying ATP concentrations on $^{86}Rb^+$ release into media containing 20 mM $Rb^+$ (A) or no $Rb^+$ (B).

effects of varying concentrations of ATP on $^{86}Rb^+$ release under two conditions: in the presence of 20 mM unlabeled $Rb^+$, and without medium $Rb^+$. Under each condition, $^{86}Rb^+$ release was measured after 20 min of incubation. In Fig. 2 the effect of each concentration of ATP is expressed as per cent of the maximal increase in $^{86}Rb^+$ release that was induced by 2-3 mM ATP. As the data show, $K_{0.5}$ of ATP in the presence of medium $Rb^+$ (Fig. 2A) is about two orders of magnitude lower than $K_{0.5}$ in the absence of medium $Rb^+$ (Fig. 2B). We should point out immediately that there is uncertainty about the absolute magnitudes of these $K_{0.5}$ values (5 $\mu$M and 0.5 mM). As evident from data of Fig. 1, and as discussed in detail before (1), $Rb^+$ release is not monoexponential, and ATP accelerates both the fast and the slow phases of release, and increases the ratio of fast/slow components of release. Therefore, in experiments such as those of Fig. 2,

418

where ATP effects are noted at one time point, the resulting $K_{0.5}$ values are influenced by the choice of this point. For better characterization of the effects of ATP, it is necessary to repeat experiments similar to those of Fig. 2 and obtain complete dissociation curves at different ATP concentrations. Nevertheless, the present data are sufficient to show that when medium $Rb^+$ is present, much lower concentrations of ATP are required to widen the access channels.

In experiments where ATP was kept constant (0.2 mM), and the increase in $^{86}Rb^+$ release was measured at one time point (20 min) as a function of medium $Rb^+$ concentration, the $K_{0.5}$ of $Rb^+$ was about 6 mM (Fig. 3). While this value is subject to the same kind of uncertainty mentioned above, it is evident that this is a "low-affinity" effect of medium $Rb^+$. $K^+$ also had a similar "low-affinity" effect on ATP-induced $Rb^+$ release (data not shown).

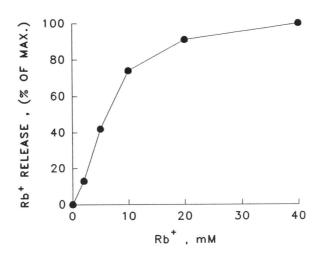

Fig. 3. Effects of varying concentrations of medium $Rb^+$ on $^{86}Rb^+$ release into a medium containing 0.2 mM ATP.

Because of the well-entrenched concept of $K^+$-ATP antagonism, the present findings may seem unexpected. However, an increase in the rate of transition of $E_2$ to $E_1$ by a regulatory effect of $K^+$ at a low-affinity site has already been shown in studies of eosin interaction with the enzyme (3). Our findings simply confirm the existence of this allosteric $K^+$ site, and show that its occupancy increases the affinity of $E_2$ for ATP significantly, but to a value that is still lower than the affinity of $E_1$ for ATP. The value of the intermediate $K_d$ for ATP binding to $E(K^+)K^+$ must be about 5-20 $\mu$M (Fig. 2A of this paper and Fig. 9 of reference 2); in contrast to the well-known $K_d$ values of about 0.2 $\mu$M ATP and greater than 0.2 mM ATP for $E_1Na^+$ and $E_2(K^+)$ respectively. The possible physiological significance of this allosteric effect of $K^+$ (presumably at an internal site of the pump) was discussed by Skou and Esmann (3).

Although the full characterization of the allosteric effects of ATP, $K^+$, and $Na^+$ on the access channels requires a good deal of additional work, we think that what we have presented here and elsewhere (1) is enough to suggest the necessity of the inclusion of interactions between the transport sites, the regulatory sites, and the access channels in the reaction cycle of the pump. We have already pointed out the structural bases of these allosteric interactions (1).

Acknowledgements

This work was supported by NIH grant HL-36573 awarded by National Heart, Lung, and Blood Institute, United States Public Health Service, DHHS.

References

1.    Hasenauer J, Huang W.-H, Askari A (1993) Allosteric regulation of the access channels to the $Rb^+$ occlusion sites of $(Na^+ + K^+)$-ATPase. J Biol Chem 268:3289-3297
2.    Kapakos JG, Steinberg M (1986) Ligand binding to (NaK)-ATPase labeled with 5-iodoacetamidofluorescein. J Biol Chem 261:2084-2089
3.    Skou JC, Esmann M (1983) The effects of $Na^+$ and $K^+$ on the conformational transitions of $(Na^+ + K^+)$-ATPase. Biochim Biophys Acta 746:101-113

# The Ribose Conformation of ATP Affects the Kinetic Properties of Na$^+$/K$^+$-ATPase

D. Thönges, J. Hahnen and W. Schoner

Institute of Biochemistry & Endocrinology and SFB 272/Z1, Justus-Liebig-University, Frankfurter Str. 100, D-35392 Giessen, Germany

## Introduction

The transport of sodium and potassium across the cell membrane occurs by oscillation between two different enzyme conformations. These sodium exporting $E_1$- and potassium importing $E_2$-conformations have differing affinities for ATP and its analogs. There are indications that Na$^+$/K$^+$-ATPase acts as an $(\alpha\beta)_2$-diprotomer [1-3]. In such a dimer the kind of interaction between $E_1$ATP- and $E_2$ATP binding-sites is usually described by the model of Adair [4]. It formulates the interactions between catalytic centres as a function of the relative affinities of the binding-sites for the substrate as positive, negative or no cooperativity. In the case of positive cooperativity the enzyme changes upon the increase of substrate concentration to a conformation with better substrate affinity. The inverse is true in the case of negative cooperativity. No cooperativity is seen, however, if both substrate binding-sites have equal affinities. Since the structure of ATP may affect the cooperativity between binding-sites, we studied the behaviour of ATP-analogs on the interaction of substrate sites and correlated it with the structural parameters.

## Experiments

Enzyme activity of Na$^+$/K$^+$-ATPase as a function of substrate concentration has been determined by an optical assay. The intrinsic affinities for the $E_1$ATP binding-site were analyzed by the competitive interaction against Cr(H$_2$0)$_4$-ATP whilst those of the $E_2$ATP binding-site against Co(NH$_3$)$_4$-ATP. The K$_d$ values of ATP or its analogs were determined from their protective effects against the inactivation by the metal derivatives [5]. As it is evident from the Adair model of two cooperating binding-sites the Hill coefficient is described by eqn 1 as $n_H = 2/(1 + (K_2/K_1)^{1/2})$.

## Results

In Figure 1 substrate hydrolysis by Na$^+$/K$^+$-ATPase was studied as a function of its concentration using various ATP analogs. The interaction of the ATP analogs with the $E_1$ATP- and $E_2$ATP binding-sites reveal the K$_d$ values listed in Table 1. It furthermore shows as to whether the substrate is hydrolyzed, the cooperativity observed and the Hill coefficient calculated according to eqn. 1.

421

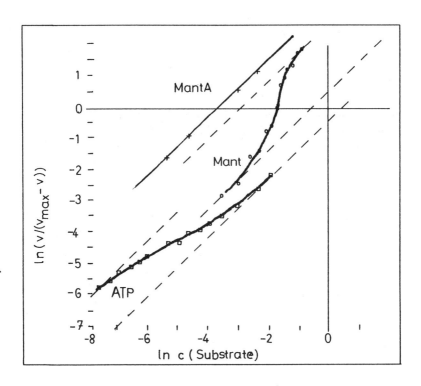

**Figure 1.** Hill-Plot of ATP-analogs hydrolyzed by $Na^+/K^+$-ATPase showing all types of cooperativity possible in a dimer.

**Table 1.** Variation of kinetic properties of $Na^+/K^+$-ATPase with ATP-analogs

| ATP-analogs | $K_d$-values [$\mu M$] $E_1/E_2$ATP | Substrate function cooperativity, $n_H$ | Calculated $n_H$ |
|---|---|---|---|
| TNP-ATP | 0.6/N.D. | no hydrolysis | --- |
| ATP | 2.5/360 | negative $n_H = 0.7$ | 0.2 |
| MANT-8-$N_3$-ATP | 37/40 | no, $n_H = 1.0$ | 1.0 |
| MANT-ATP | 200/35 | positive, $n_H = 2.3$ | 1.4 |
| DANS-8-$N_3$-ATP | ---/15 | no hydrolysis | --- |

Abbreviations. DANS: 2'(3')-Dimethylaminonaphthalenesulfonyl; MANT: 2'(3')-Methylanthraniloyl; TNP: 2',3'-Trinitrophenyl; N.D.: not determined

## Discussion

The data available so far suggest the structure/function-relationships of the various ATP-analogs to $Na^+/K^+$-ATPase as it is shown in Figure 2. The observation of changing cooperativity in substrate's hydrolysis is explicable only by changes in the interaction of two binding-sites. In any case, substrate hydrolysis follows under turnover conditions the predictions of the Adair-model. The calculated Hill-coefficients correspond fairly well to the experimentally obtained data. It is furthermore evident from Table 1 that the relative affinities of the substrate analogs for the $E_1ATP$- and $E_2ATP$ binding-sites affect the kind of cooperativity. Those ATP analogs which bind to one ATP binding-site only are not hydrolyzed by $Na^+/K^+$-ATPase.

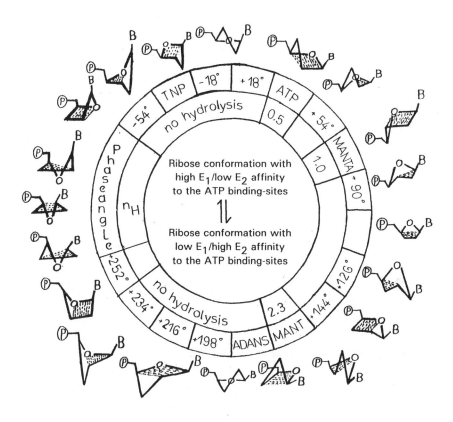

**Figure 2.** Relationship between the structure of ATP-analogs and the kinetic properties of $Na^+/K^+$-ATPase.

It is furthermore evident from the data in Table 1, that the ATP binding-sites are able to discriminate between the ATP analogs. This may have structural and conformational reasons. The ribose in nucleotides is substituted by the purine base, the phosphate chain and in the case of an analog also by a fluorophore. The energy-barrier of free pseudorotation (puckering) between ribose conformations is increased by these substituents [6]. The prefered structures in nucleotides are the $C_2$·exo-$C_3$·endo (North) or the $C_2$·endo-$C_3$·exo (South) conformations. ATP and its analogs were subjected to energy minimization by the Quanta Molecular Modeling Program. The equilibrium between North- and South conformations was studied by molecular dynamic simulations. These data are suported by $^1$H-NMR-spectroscopy as the protons $H_{1'}$ and $H_{2'}$ have different coupling constants.

We found that ATP has energy minima in the North and $J(H_{1'},H_{2'}) = 5$ Hz. This coupling constant is too high for only equatorial-equatorial orientation like in the North conformation. In dynamic simulation ATP changes from the North- to the South conformation. Therefore, the coupling constant represents a time average.

DANS-8-$N_3$-ATP exhibits the South conformation and $J(H_{1'},H_{2'}) = 8$ Hz. This value stands only for axial-axial orientation like in the South conformation. Obviously, a charge-transfer complex between DANSyl and adenine restricts puckering of the ribose.

TNP-ATP prefers the North conformation. The dynamic simulation shows a very high energy barrier of pseudorotation, because puckering is restricted by a Meisenheimer complex.

*Acknowledgements*

This work has been supported by grants of the Fonds der Chemischen Industrie, Frankfurt/Main and by the Giessener Graduiertenkolleg "Molekulare Biologie und Pharmakologie".

**References**

1.  Askari A (1988) Ligand Binding Sites of (Na$^+$+K$^+$)-ATPase: Nucleotides and Cations. Progr Clin Biol Res 268A: 149-165

2.  Nørby JG, Jensen J (1991) Functional Significance of the Oligomeric Structure of Na,K-Pump from Radiation Inactivation and Ligand Binding. in: The Sodium Pump: Structure, Mechanism and Regulation (Kaplan JH, De Weer P, eds) The Rockefeller University Press, New York, pp 173-188

3.  Buxbaum E, Schoner W. (1991) Phosphate binding and ATP-binding sites coexist in Na$^+$/K$^+$-transporting ATPase, as demonstrated by the inactivating MgPO$_4$ complex analogue Co(NH$_3$)$_4$PO$_4$. Eur J Biochem 195: 407-419

4.  Ricard J, Cornish-Bowden A (1987) Cooperative and allosteric enzymes: 20 years on. Eur J Biochem 166: 255-277

5.  Serpersu EH, Bunk S, Schoner W (1990) How do MgATP analogues differentially modify high- and low-affinity ATP binding sites of Na$^+$/K$^+$-ATPase? Eur J Biochem 191: 397-404

6.  Altona C, Sundaralingam M (1972) Conformational analysis of the sugar ring in nucleotides. A new description using the concept of pseudorotation. JACS 94, 23: 8205-8212

# ATP accelerates phosphorylation of the Na$^+$/K$^+$-ATPase acting with low apparent affinity

R.D. Peluffo, R.C. Rossi, A.F. Rega and P.J. Garrahan

Departamento de Química Biológica. Facultad de Farmacia y Bioquímica. Universidad de Buenos Aires. Junín 956 (1113) Buenos Aires. Argentina

## Introduction

In a previous paper (2) we showed that as the concentration of ATP tends to infinity, the maximum level of phosphorylation tends to a value that is at least twice the maximum binding of ouabain, a phenomenon that we called *superphosphorylation*. Superphosphorylation was transient and so fast that made it difficult to measure its rate. The present work was aimed to study the effect of ATP on the rate of phosphorylation under conditions that make the Na$^+$/K$^+$-ATPase able to undergo superphosphorylation.

## Materials and Methods

Na$^+$/K$^+$-ATPase (specific activity 25.7 µmol P$_i$/(mg protein·min)) from pig kidney purified by zonal centrifugation (1) was kindly provided by Dr. J. Jensen (Århus University, Denmark). Phosphorylation was performed at 25°C in a rapid mixing apparatus (Intermekron, Uppsala). The incubation buffer contained 150 mM NaCl; 0.2 mM EDTA; 40 mM imidazole-HCl (pH 7.5); and enough MgCl$_2$ to give 0.5 mM free Mg$^{2+}$. The reaction was started by mixing the enzyme with [γ-$^{32}$P]ATP and quenched with an ice-cold solution of 11% (w/v) TCA, 50 mM H$_3$PO$_4$ and 10 mM ATP. The mixture was filtered through a 0.45-µm Millipore filter and the insoluble material washed three times with 10 ml of an ice-cold solution of 7% (w/v) TCA and 32 mM H$_3$PO$_4$. The filters were dried and their radioactivity was measured. Phosphoenzyme (*EP*) was estimated from the difference of the radioactivity incorporated into the enzyme in incubation buffers with and without MgCl$_2$.

*The Meaning of r$_m$*: A reaction involving phosphoenzymes includes at least the following steps:

$$E + ATP \underset{k_{-1}}{\overset{k_1}{\longleftrightarrow}} EATP \overset{k_2}{\longrightarrow} EP \overset{k_3}{\longrightarrow} E + P_i$$

Analytical solution of this scheme yields:

$$[EP]_{(t)} = \frac{k_1 k_2 [ATP][E]_T}{K_1 K_2 (K_2 - K_1)} \left( K_2 - K_1 - K_2\, e^{-K_1 t} + K_1\, e^{-K_2 t} \right) \qquad (1)$$

where K$_1$ and K$_2$ are functions of the rate constants and the concentration of ATP. Eqn (1) predicts that $v_0 = (d[EP]/dt)_{t=0} = 0$ except when formation of *EATP* is instantaneous, that is if the nucleotide binds in rapid equilibrium or its concentration tends to infinity. For this reason we estimated the rate of phosphorylation measuring $r_m = (d[EP]/dt)_{max}$, *i.e.*, the maximum slope of $[EP] = f$ (t). For the experimental results, $r_m$ was obtained from the equation that best fits the data and for the simulations, $r_m$ was obtained from the maximum value of $\Delta[EP]/\Delta t$.

425

## Results and Discussion

$r_m$ *as a Function of ATP Concentration*: Time courses of phosphorylation were measured in media containing from 1 to 1000 μM ATP. At low ATP concentrations a lag was apparent that tended to disappear at higher concentrations of the nucleotide. When the concentration of ATP was 500 μM or more superphosphorylation became apparent. The values of $r_m$ were plotted against the concentration of ATP. Results in Fig. 1 show that at low ATP concentrations, $r_m$ increased along a curve that tended to saturation (inset in Fig. 1) and that at about 300 μM ATP, $r_m$ started to increase again along a parabolic curve that showed no signs of saturation.

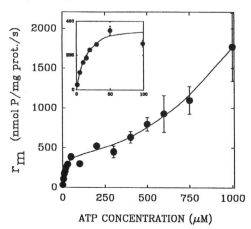

Fig. 1. $r_m$ as a function of the ATP concentration. The curve represents the solution of $r_m = (r_m)_{max}$ [ATP]/ $(K_m + [ATP]) + A$ [ATP]$^n$ for the values of the parameters that provided the best fitting to the experimental data. These (± standard error) were: $(r_m)_{max}$ = 459.5 ± 37.6 nmol P/(mg protein·s); $K_m$ = 15.7 ± 2.3 μM; $A$ = 7.9·10$^{-4}$ ± 1.7·10$^{-4}$ nmol P/(mg protein·s·μM$^n$); n = 2.07 ± 0.33

*Effect of Turnover on the Time Course of EP:* The enzyme was preincubated with 10 μM unlabeled ATP in the incubation buffer at 25°C during 2 min and then phosphorylated in media with either 15 or 505 μM [γ-$^{32}$P]ATP. $r_m$ values at 15 and 505 μM ATP (74.5 ± 4.9 and 93.5 ± 31.0 nmol P/(mg protein·s) respectively), were not significantly different. When preincubation was performed in the absence of Mg$^{2+}$ the low-affinity effect of ATP on $r_m$ persisted. Hence the disappearance of the effect of high ATP on $r_m$ required the Na$^+$/K$^+$-ATPase to undergo turnover.

*A Reaction Scheme to Account for the Low-Affinity Effect of ATP on $r_m$:* To explain the results, the reaction scheme we had used to account for superphosphorylation (2) was modified as shown in Scheme 1. The basic assumptions are: (a) in the Na$^+$/K$^+$-ATPase at rest ($E_r$) both high- and low-affinity sites for ATP coexist and become phosphorylated, and (b) the transition of the enzyme from the resting to cycling state requires phosphorylation which is faster if $E_r$ simultaneously binds one molecule of ATP with high affinity and two with low affinity. Scheme 1 was developed only to show that it is feasible to describe the results.

426

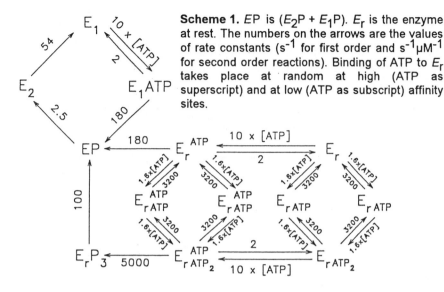

**Scheme 1.** $EP$ is $(E_2P + E_1P)$. $E_r$ is the enzyme at rest. The numbers on the arrows are the values of rate constants ($s^{-1}$ for first order and $s^{-1}\mu M^{-1}$ for second order reactions). Binding of ATP to $E_r$ takes place at random at high (ATP as superscript) and at low (ATP as subscript) affinity sites.

Simulations were performed assuming that at the start of the reaction the enzyme was a mixture of equal amounts of $E_r$ and $E_1$. Scheme 1 correctly reproduced the experimental time courses of phosphorylation. In Fig. 2 the curve of $r_m$ as a function of the concentration of ATP (*dashed line*) is plotted together with the curve in Fig. 1 (*continuous line*). It can be seen that the experimental and simulated curves were very similar to each other. The additional curve allows to compare the experimental and simulated curves with those generated by the Albers-Post model (*dotted line*). This model predicted a Michaelis-Menten like dependence with $K_m$ near 20 $\mu M$. The inset in Fig. 2 shows that up to 30 $\mu M$ ATP the three curves were not distinguishable from each other.

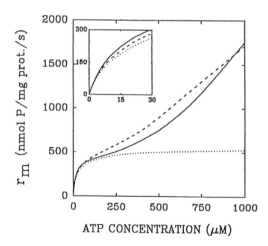

Fig. 2 Dashed line: simulation of Scheme 1; dotted line: simulation of the Albers-Post model; full line: the function that best fits the experimental data of $r_m$ in Fig. 1. Initial conditions were: Scheme 1: $E_r = 50\%$, $E_1 = 50\%$; Albers-Post model: $E_1 = 90\%$, $E_2 = 10\%$.

*Effects of The Initial Amount of $E_r$ on the Predictions of Scheme 1:* Results in Fig. 3 indicate that increasing the proportion of $E_r$ increased both the low-affinity effect of ATP on $r_m$, and superphosphorylation. However, if the fraction of $E_r$ is sufficiently lowered superphosphorylation disappears, but the low-affinity effect on velocity, albeit reduced, persists.

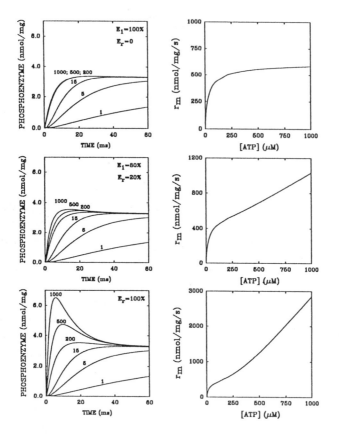

**Fig. 3.** Left: Time courses of $EP$ for the indicated initial proportions of $E_r$ and ATP concentrations (µM, numbers near the curves).
Right $r_m = f([ATP])$ for the initial proportions of $E_r$ indicated in the corresponding figure on the left.

## Conclusions

The low-affinity effect of ATP on $r_m$ (and superphosphorylation) is not explained by the Albers-Post model and may be a property of a conformer of the $Na^+/K^+$-ATPase that disappears when the enzyme is undergoing turnover. Scheme 1 illustrates such a hypothesis, preserving the possibility of using the Albers-Post model to describe steady-state $Na^+/K^+$-ATPase activity.

## References

1.  Jørgensen, PL (1974) Purification and characterization of (Na⁺ + K⁺)-ATPase III Purification from the outer medulla of mamalian kidney after selective removal of membrane components by SDS *Biochim. Biophys. Acta* 356: 36-52
2.  Peluffo, RD, Garrahan, PJ and Rega, AF (1992) Low affinity superphosphorylation of the Na,K-ATPase by ATP. *J Biol Chem* 267: 6596-6601

# Effects of ATP on the Steady-State Level of the Phosphoenzyme of Pig Kidney Na$^+$/K$^+$-ATPase

P.J. Schwarzbaum, R.C. Rossi, S. B. Kaufman and P.J. Garrahan

Departamento de Química Biológica. Facultad de Farmacia y Bioquímica. Universidad de Buenos Aires. Junín 956 (1113) Buenos Aires. Argentina

## Introduction

The Albers-Post model for the hydrolysis of ATP by the Na$^+$/K$^+$-ATPase postulates that the release of Pi from ATP is preceded by the formation of at least two conformers of a phosphoenzyme as shown in Scheme I

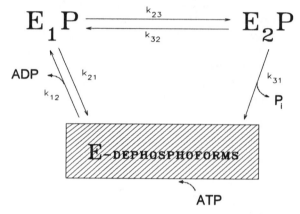

Scheme 1. The Albers Post model for ATP hydrolysis by the Na$^+$/K$^+$-ATPase postulates that ATP binds to the dephosphoforms of the ATPase and that release of inorganic phosphate is preceded by the sequential formation of at least two phosphoenzymes.

During steady-state, the rate of ATP hydrolysis ($v_i$) will be equal to the net rate of any of the elementary steps of the reaction. Hence in the absence of Pi:

$$v_i = k_{31} \times E_2P = k_{23} \times E_1P - k_{32} \times E_2P \tag{1}$$

Since $EP_T = E_1P + E_2P$, it follows that: $EP_T = \dfrac{E_2P \times (k_{31} + k_{32} + k_{23})}{k_{23}}$

and therefore $\dfrac{v_i}{EP_T} = \dfrac{k_{31} \times k_{23}}{k_{31} + k_{32} + k_{23}} \tag{2}$

Equations essentially similar to Eqn (2), will apply for reaction schemes including any amount of additional phosphoenzymes. Eqn (2) evinces one of the fundamental features of the class of models shown in Scheme 1, *i.e.*, that the ratio between steady-state activity and steady-state level of total phosphoenzyme will be independent of the rate of phosphorylation and of the transitions among the different dephosphoforms of the ATPase.

Moreover if, as it is proposed by the Albers-Post model, the values of the rate constants involved in Eqn. (4) are not affected by ATP, then the ratio $v_i/EP_T$ will be also independent of the concentration of the nucleotide. Hence the functions relating ATPase activity and total phosphoenzyme level to the concentration of ATP should be identical in shape. This prediction was submitted to experimental test.

## Materials and Methods

$Na^+/K^+$-ATPase from pig kidney (specific activity of 9-12 units/mg) was prepared by the procedure of Jensen et al. (1). Steady-state level of phosphoenzyme was measured at 25°C in media containing 150 mM NaCl, 0.2 mM EDTA, 10 mM of either KCl, RbCl or $NH_4Cl$, enough $MgCl_2$ to give 0.5 mM free $Mg^{2+}$, and 25 mM imidazole-HCl (pH=7.45 at 25°C). The reaction was started by mixing 1 ml of $[\gamma^{-32}P]ATP$ with 1 ml of the enzyme suspension and quenched, after 1-3 seconds, by adding 3 ml of an ice-cold solution of 11% (w/v) trichloroacetic acid (TCA) and 50 mM $H_3PO_4$. The denatured phosphoenzyme was washed three times with 10 ml of 7% (w/v) TCA and 32 mM $H_3PO_4$ on either Whatman GF/C or Millipore filters (Type GS, 0.22 μm pore size). Phosphoenzyme was estimated from the difference between the $^{32}P$ incorporated in the assay media described above, and in the same media without $MgCl_2$. Ouabain-sensitive ATPase activity was measured in media of identical composition and at the same temperature as in phosphorylation experiments.

## Results and Discussion

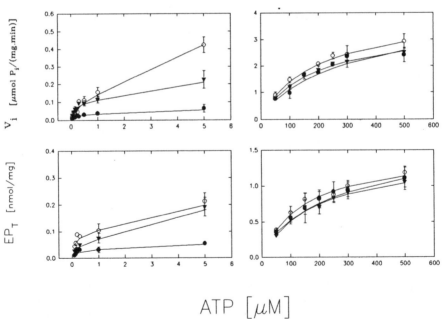

ATP $[\mu M]$

**Figure 1.** Ouabain-sensitive ATPase activity and phosphoenzyme level in media with 10 mM of either $K^+$ (▼), $Rb^+$ (•) or $NH_4^+$ (○). For the sake of clarity the curves were divided into two segments.

In media without $K^+$ or its congeners for the whole range of ATP concentrations tested (0-50 μM), $Na^+$-ATPase activity and phosphoenzyme level were single hyperbolic functions of [ATP] with identical Km values ($0.16 \pm 0.07$ μM for $EP_T$ and $0.16 \pm 0.01$ μM for $v_i$). Hence under these conditions $v_i/EP_T$ remained independent of the concentration of ATP (average value $0.8 \pm 0.05$ s$^{-1}$) and the prediction of Eqn. (2) was fulfilled.

As shown in Fig. 1, in media with either $K^+$, $Rb^+$ or $NH_4^+$, both $EP_T$ and $v_i$ were biphasic functions of ATP.

In Fig. 2 the ratio $v_i/EP_T$ for all the cations tested is plotted as a function of the concentration of ATP. It can be seen that the points could be fitted by a rectangular hyperbola (continuous line in the figure) that goes from 19.8 to 40.08 s$^{-1}$ as ATP concentration goes from zero to infinity. The effect of ATP was half-maximal at $9.52 \pm 5.07$ μM.

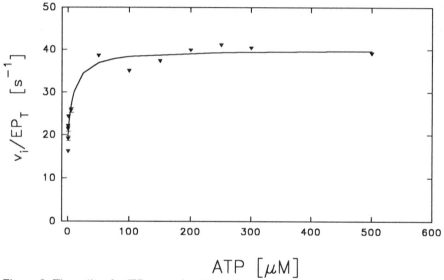

ATP [μM]

Figure 2. The ratio of $v_i/EP_T$ as a function of the concentration of ATP. Each point is the average value of the ratio for $K^+$, $Rb^+$ and $NH_4^+$ at each concentration of ATP taken from the experiments in Fig. 1.

Hence, the constancy of $v_i/EP_T$ was only fulfilled at concentrations of the nucleotide above 100 μM.

To test if the decrease in $v_i/EP_T$ was caused by partial inactivation of the enzyme at low concentrations of ATP, $Na^+/K^+$-ATPase was preincubated in media with 5 μM ATP with and without 10 mM $K^+$. After 10 min, ouabain-sensitive activity was measured under optimal conditions. No effect of preincubation was detected on maximal activity of $Na^+/K^+$-ATPase.

increases with the concentration of ATP. To obtain the observed effect $k_{31}$ should increase from 50 to 300 s$^{-1}$ or $k_{23}$ from 75 s$^{-1}$ to 240 s$^{-1}$ as ATP concentration goes from zero to infinity. These values of the rate constants are compatible with those found in the literature.

Up to now we have only tested the effect of ATP on $k_{31}$. To do this the enzyme was phosphorylated during 3 seconds at 20°C and then K$^+$ was added to give a final concentration of 1 mM together with enough ATP to give final concentrations ranging from 5 to 200 μM. EP$_T$ was measured at 15 and 25 ms after the addition of K$^+$. Results showed that K$^+$ induced a decrease in EP$_T$ that was independent of [ATP], confirming previous experiments at 0°C by Klodos and Nørby (2).

This seems to indicate that Scheme 1 would hold if ATP acted on the distribution between conformers of the phosphoenzyme. Since no experimental evidence is yet available for this, it still remains an open question whether such an effect explains the results in Fig. 2 or whether a more profound modification of the reaction scheme is necessary.

## Acknowledgments

This work was supported by grants from CONICET, Fundación Antorchas and University of Buenos Aires.

## References

1.  Jensen J, Nørby JG, Ottolenghi P (1984) Sodium and potassium binding to the sodium pump of pig kidney: stoichiometry and affinities evaluated from nucleotide-binding behaviour. J Physiol 346:219-241
2.  Klodos I, Nørby JG (1988) Does ATP affect the interconversion and the dephosphorylation of the phosphoenzymes of the Na,K-pump?. In: Skou JC, Nørby JG, Maunsbach AB, Esmann M (eds) The Na$^+$, K$^+$-pump. Part A: Molecular aspects, Allan R. Liss, New York, pp.321-326.

# A Phase-Change Model of Transient Kinetics of Phosphoenzyme of Na$^+$/K$^+$-ATPase

I. Klodos, R. L. Post, B. Forbush III

Institute of Biophysics, University of Aarhus, DK-8000 Aarhus C, Denmark; Department of Physiology, School of Medicine, University of Pennsylvania, Philadelphia, Pennsylvania 19104-6085, USA; Department of Cellular and Molecular Physiology, Yale University School of Medicine, New Haven, Connecticut 06510, USA.

An extraordinary kinetic response of the phosphoenzyme was reported by Nørby et al. (6) and was interpreted with difficulty. The phosphoenzyme responded rapidly and partially to a jump in [NaCl]. Klodos and Forbush (3) demonstrated the nature of the effect and Martin and Sachs (5) proposed a model of 5 components functioning in parallel. Klodos et al. (4) examined the effect in detail and proposed a phase-change model. This article outlines their phase-change model.

The nature of the response is illustrated by the following experiment. It is well known that the phosphoenzyme consists of two principal components. One is sensitive to ADP and resistant to K$^+$, "E1P", and the other is sensitive to K$^+$ and resistant to ADP, "E2P". A high [NaCl] increases the proportion of E1P and a low [NaCl] increases the proportion of E2P (2). In this experiment the phosphoenzyme was formed from [$\gamma$-$^{32}$P]ATP in 600 mM NaCl. At zero time further formation was terminated and also E2P was dephosphorylated by addition of KCl. The remaining K-resistant E1P disappeared slowly (Fig. 1). But when the [NaCl] was diluted subsequently, a part, and only a part, of E1P disappeared rapidly, whereas the other part continued to disappear slowly (Fig. 1). The amount of the slow component was a little larger than the amount that would have been found if the initial phosphorylation had been done in the final [NaCl]. Klodos et al. (4) showed that the rapid phase is a conversion of E1P to E2P. They also showed a similar rapid partial conversion of E2P to E1P by an upward jump in [NaCl].

These kinetics raise a problem. During the slow phase of dephosphorylation E1P converts to E2P slowly, since E2P is dephosphorylated immediately by the K$^+$. But during the rapid phase it converts rapidly and only partially. How can a sustained change in a parameter produce a transient change in a rate? Martin and Sachs (5) used heterogeneity of molecules of Na$^+$/K$^+$-ATPase; Klodos et al. (4) used heterogeneity of the environment of homogeneous molecules (phases).

The phase-change model begins with interconversion as an exchange between two pools. Exchange takes place at an interface between the pools. To effect a slow exchange and a rapid net conversion the model introduces a constraint on mixing within the pools. Exchange of molecules is rapid but slowness of mixing restricts interconversion to those molecules that are at the interface. Slowness of mixing does not affect net conversion. A computer program illustrates the principle.

433

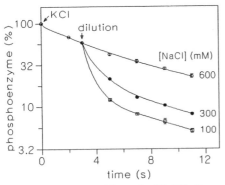

**Figure 1.** Downward jump in [NaCl] after KCl chase. Redrawn from (3) by copyright permission of the Rockefeller University Press.

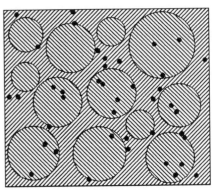

**Figure 3.** Section of model membrane with two lipid phases, (///) (\\\), containing Na⁺/K⁺-ATPase (●).

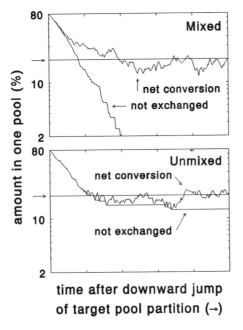

**Figure 2.** Model of effect of partition jump on mixed and unmixed pools.

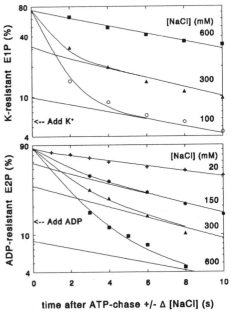

**Figure 4.** Fit of model (*solid lines*) to data from references (3) above and (6) below.

*Comment on Fig. 4.* The lines that extrapolate from the fitted curves back to the vertical axis show fits of the model to chases at constant [NaCl]'s. From the corresponding partition values at zero time a phase diagram was constructed as a function of [NaCl].

434

The data in this program consist of elements divided into two pools, A and B. The partition between the pools changes by one element with each iteration; either 1 element of A is converted to an element of B or *vice versa* at random, subject to a bias. The bias is the difference between the partition and a target value. The bias keeps the partition close to the target value.

To show the difference of exchange between unmixed phases and mixed phases, pool A was divided into two parts, A1 and A2. At the start of a series of iterations (time = 0) all of A was A1 and A2 was zero. A2 represented elements of A that were formed by conversion from B. Every iteration that converted B to A decreased B and increased A2. Conversion of A to B was different for mixed and unmixed pools. For mixed pools A1 and A2 were converted to B in proportion to their relative amounts, since mixing permitted either A1 or A2 to be at the interface for exchange. For unmixed pools only A2 was converted to B unless its amount was zero and only then was A1 converted. Since no mixing took place, A2 remained at the interface and was converted preferentially.

Fig. 2 shows net and exchange conversions generated by the program. The initial value of A was 80%. The target value of A was 20%. Net conversion proceeded rapidly to the target value. For mixed phases exchange proceeded monoexponentially. For unmixed phases exchange ceased shortly after the target was reached. Thus the model showed rapid net conversion and slow exchange.

What sort of molecular structure can provide a suitable phase separation and changes? We thought of a phase separation in the lipid of the membrane in which $Na^+/K^+$-ATPase is embedded. Fig. 3 shows a section of such a membrane. The phase boundaries are shown by circles; one phase is inside the circles and the other phase is outside. $Na^+/K^+$-ATPase molecules are distributed at random. In a steady state the phase boundaries and the locations of the $Na^+/K^+$-ATPase molecules are relatively stable. The lipid phase controls the form of the phosphoenzyme. In one phase E1P predominates and in the other phase E2P predominates. Thus the relative amounts of E1P and E2P vary with the partition of the phases. Although the phase boundaries are stable, the individual lipid molecules at these boundaries interconvert between the phases rapidly. The [NaCl] controls the partition between the phases. Thus when the [NaCl] changes rapidly, the phase boundaries move rapidly to the new partition. As the boundaries sweep past some of the $Na^+/K^+$-ATPase molecules, these molecules experience a phase transition and change their conformation rapidly.

The model was expressed as algebraic equations and was fitted to the data in two experiments. The equations assigned a rapid or a slow value to rate constants for equilibration of E1P and E2P in each phase ($k_{12}$ and $k_{21}$). In phase A $k_{12}$ was slow (0.089 s$^{-1}$) and $k_{21}$ was rapid (1.025 s$^{-1}$) so that E1P predominated. In phase B $k_{12}$ was rapid (0.54 s$^{-1}$) and $k_{21}$ was slow (0.022 s$^{-1}$) so that E2P predominated. In addition to two values of rate constants, the equations specified a partition of the phases that was dependent on [NaCl] (Fig. 4).

435

A phase diagram was drawn for each experiment and an empirical equation was fitted to the data. From the K-chase data in Fig. 4 the percent of phase A was $8 + 92/(((430 \text{ mM})/[\text{NaCl}])^3 + 1)$. From the ADP-chase data in Fig. 4 the percent of phase B was $84/(([\text{NaCl}]/(230 \text{ mM}))^2 + 1)$. The sum of phase A plus phase B was as low as 65-70% at $[\text{NaCl}] = 200\text{-}300$ mM. Thus there was a third phase between them, phase C. When a $Na^+/K^+$-ATPase molecule moved from either phase A or phase B into phase C, its rate constants for interconversion in either direction became rapid ($k_{12} = 1.025 \text{ s}^{-1}$ and $k_{21} = 0.54 \text{ s}^{-1}$ from Fig. 4). Thus in this phase the phosphoenzyme was sensitive to ADP or to $K^+$ and was equivalent to Yoda and Yoda's "E*P" (7). Unlike their model this E*P is not an obligatory intermediate between E1P and E2P but represents phosphoenzyme in a phase that is intermediate between phases in which E1P and E2P predominate. In the structure of a membrane, phase C may correspond to $Na^+/K^+$-ATPase molecules at or close to phase boundaries (Fig. 3) that jump about slightly.

This phase-change model has implications for $Na^+/K^+$-ATPase activity. Nørby et al. found too little slowly-converting E1P to support activity in the presence of $K^+$ (6). In the phase-change model, rapidly-converting E1P can be formed directly from ATP in phases B and C and so can support a much higher ATPase activity in the presence of $K^+$ than slowly-converting E1P in phase A can. Comparison of $Na^+/K^+$-ATPase activity predicted from the phase diagram above with data in Fig. 10 of Nørby et al. (6) showed that the model can support up to 2.7-fold more activity than is found in the data. Froehlich and Fendler also used parallel pathways to interpret their data (1).

## References

1.  Froehlich JP, Fendler K (1991) The partial reactions of the $Na^+$- and $Na^+ + K^+$-activated adenosine triphosphatases. In: Kaplan JH, De Weer P (eds) The Sodium Pump: Recent Developments. Rockefeller University Press, New York, pp 227-247

2.  Glynn IM, Karlish SJD (1990) Occluded cations in active transport. Annu Rev Biochem 59: 171-205

3.  Klodos I, Forbush B III (1991) Transient kinetics of dephosphorylation of Na,K-ATPase after dilution of NaCl. In: Kaplan JH, De Weer P (eds) The Sodium Pump: Recent Developments. Rockefeller University Press, New York, pp 327-331

4.  Klodos I, Post RL, Forbush B III (in press) Kinetic heterogeneity of phosphoenzyme of Na,K-ATPase modeled by unmixed lipid phases. Competence of the phosphointermediate. J Biol Chem

5.  Martin DW, Sachs JR (1991) Simulation of dephosphorylation curves by an ensemble of enzyme molecules with different rate constants. J Gen Physiol 98: 419-426

6.  Nørby JG, Klodos I, Christiansen NO (1983) Kinetics of Na-ATPase activity by the Na,K pump. Interactions of the phosphorylated intermediates with $Na^+$, $Tris^+$, and $K^+$. J Gen Physiol 82: 725-759

7.  Yoda S, Yoda A (1986) ADP- and $K^+$-sensitive phosphorylated intermediate of Na,K-ATPase. J Biol Chem 261: 1147-1152

# The $E_2$ Conformation of $Na^+/K^+$-ATPase During the $Na^+$-ATPase Activity

Marta Campos and Luis Beaugé

Instituto M. y M. Ferreyra, Casilla de Correo 389, 5000 Córdoba, Argentina

## Introduction

Two main reactions can lead to ATP hydrolysis by the $Na^+/K^+$-ATPase (EC 3.6.I.37) (see ref. 1). In the presence of $K^+(o)$, the $(Na^+/K^+)$-dependent activity associated to $Na^+(i)$-$K^+(o)$ exchange. $K^+(o)$ accelerates the breakdown of the phosphoenzyme and in its transit across the membrane is trapped by the dephosphoenzyme in an "occluded" $(E_2(K))$ form. The spontaneous release of $K^+$ into the cell is slow but that rate is markedly increased by ATP acting with low affinity in a non phosphorylating role. Without $K^+(o)$ there is what we know as $Na^+$-ATPase activity where the return path, beginning with enzyme dephosphorylation is drastically altered. In the absence of $Na^+(o)$, dephosphorylation is spontaneous and goes with an "uncoupled" $Na^+$ efflux. When there is $Na^+(o)$, this cation acts like $K^+$ stimulating dephosphorylation and being transported inwardly in exchange for $Na^+(i)$. Little is known about the after dephosphorylation steps of the $Na^+$-ATPase. Based on phosphorylation from Pi at $0°$ C Post et al. (2) suggested the existence of an $E_2$ form following dephosphorylation.

## Results and Discussion

In the present work we aimed to further characterize the dephosphointermediate/s during the cycle of $Na^+$-ATPase activity of the $Na^+/K^+$-ATPase but at 20 ° C. By working a that temperature we expected to use rate constants available from the literature and make comparisons between the prediction of kinetic model/s with the actual experimental data. For theoretical calculations we used the Albers-Post scheme (see ref. 1) assuming the existence of a $E_2$ form that can return to $E_1$ without or with previous binding of ATP to a low affinity regulatory site. (Figure 1). Note that the path leading to $E_2P^*$ is an addition that does not modify the model and was used in order to compute at the same time $E_2P$ formed from ATP ($E_2P$) and from Pi ($E_2P^*$). The rate constants f1, b1, b2, b3, f4, f5 and f6 were taken from ref. 3 and 4; b4 is the maximal value that predicts no EP formation without ATP; f3 and b6 by assuming the Ks for ATP and Pi equal to 100 $\mu$M; f2 from the restrictions of the cycle. The experiments were carried out with partially purified pig kidney $Na^+,K^+$-ATPase. For technical details see ref. 5 and legends to Figure 2 and Table 1.

The working model predicts: (i) without Pi the $Na^+$-ATPase activity has high affinity for ATP (single michaelian) but becomes double michaelian (with high and low ATP affinity) in the presence of Pi; (ii) the apparent affinity for Pi is substantially lower than its Kd and in the mM range; (iii) a steady state phosphorylation from Pi is attained in about 2 s. (iv) Pi incorporation can be fully inhibited by ATP acting with low affinity.

437

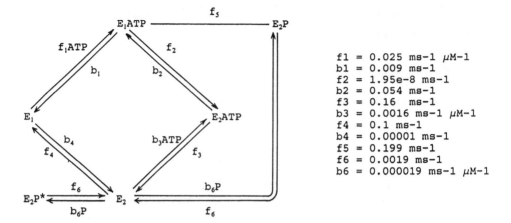

Figure 1. Model for the Na$^+$-ATPase activity of Na$^+$/K$^+$-ATPase.

```
f1 = 0.025 ms-1 µM-1
b1 = 0.009 ms-1
f2 = 1.95e-8 ms-1
b2 = 0.054 ms-1
f3 = 0.16  ms-1
b3 = 0.0016 ms-1 µM-1
f4 = 0.1 ms-1
b4 = 0.00001 ms-1
f5 = 0.199 ms-1
f6 = 0.0019 ms-1
b6 = 0.000019 ms-1 µM-1
```

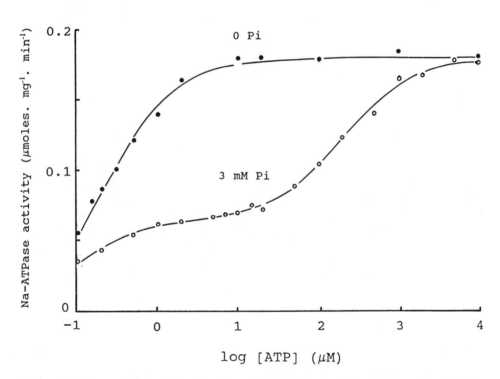

Figure 2. Effects of Pi on ATP activation of Na$^+$-ATPase activity. See text for details.

The effects of Na$^+$ cannot be predicted from the model.

Figure 2 illustrates the ATP stimulation of Na$^+$-ATPase activity in the absence and presence of 3 mM Pi. The lines through the data points are the best fit to a single michaelian in the absence of Pi and to the sum of two michaelians in the presence of 3 mM Pi. Without Pi we observe single site kinetics which are transformed in double michaelian when inorganic phosphate is added. On the basis of the unidirectional rate constants the true affinities for ATP (Ks) are 0.36 $\mu$M and 100 $\mu$M for the catalytic and regulatory sites respectively (see Figure 1). The fit of the experimental points gives a Km of 0.22 $\pm$ 0.02 $\mu$M for the catalytic site in the absence of Pi. With 3 mM Pi that Km is reduced to 0.09 $\pm$ 0.008 $\mu$M; the fractional reduction coincides with the fit of theoretical ATP activation curves from the model fitted to the same equations (not

Table 1. Effect of Na$^+$ and ATP concentrations on steady state Pi incorporation during Na$^+$-ATPase turnover of Na$^+$/K$^+$-ATPase at 20° C

| | Fractional Pi incorporation | | | |
|---|---|---|---|---|
| ATP ($\mu$M) | 10mM Alone | 10mM NaCl + Tris.HCl | 10mM NaCl + NMG.HCl | 150mM NaCl |
| 2 | 0.52$\pm$0.05 | 0.55$\pm$0.03 | 0.58$\pm$0.04 | 0.24$\pm$0.03 |
| 100 | 0.24$\pm$0.03 | 0.24$\pm$0.03 | 0.25$\pm$0.04 | 0.12$\pm$0.03 |

All solutions contained 30 mM Imidazole, pH (20° C) 7.4 and 1.8 mM Mg$^{++}$. The entries immediately below 10 mM Na mean no further additions (alone) or enough Tris.HCl or n-methyl glucamine.HCl (NMG.HCl) to give the same ionic strength as in the 150 mM NaCl media. Following 1 sec. of turnover in Pi-free conditions 1 mM [$^{32}$P]Pi and the required MgCl$_2$ were added; the reaction was stopped after 3 sec.

shown). The Km$^{\text{ATP}}$ of the regulatory site in Figure 2 is 187 $\pm$ 15 $\mu$M (low affinity) while the V$_{max}$'s are identical.

Table 1 summarizes the effects on Pi incorporation during Na$^+$-ATPase turnover of two concentrations of Na$^+$ (10 mM and 150 mM) at constant and variable ionic strength and of ATP (2 $\mu$M and 100 $\mu$M). The [Pi] was 1 mM in these cases. Increasing [Na$^+$] from 10 mM to 150 mM Na$^+$ reduces Pi incorporation to about one half. This is a genuine Na$^+$ and not due to changes in ionic strength. Table 1 also shows that ATP inhibits Pi

439

incorporation and that, Na$^+$ and ATP sum their effects.

In agreement with predictions from the model in Figure 1, we found: (i) The $t_{1/2}$ of Pi phosphorylation was about 400 ms. (ii) At constant [Mg$^{++}$] of 1.8 mM, the Km$^{Pi}$ was 2020 $\pm$ 80 $\mu$M and the maximal expected incorporation 100 percent. (iii) ATP can fully inhibit Pi incorporation along an hyperbolic curve with an apparent affinity of 136 $\pm$ 24 at 1 mM Pi. In addition, both in 150 mM Na$^+$-K$^+$-free and 150 mM Na$^+$-0.5 mM K$^+$, the dephosphorylation rate of the EP formed from Pi during Na$^+$-ATPase turnover was indistinguishable from that obtained from ATP and Na$^+$ .

The most economical way to account for these results is that, even in the absence of K$^+$(o), the cycle of ATP hydrolysis by the Na$^+$/K$^+$-ATPase enzyme go through an E$_2$ conformation that has the same low affinity ATP site observed in the E$_2$(K) state. In turn this would mean that, if not all, at least many of the steps following enzyme dephosphorylation are the same with and without external K$^+$.

Acknowledgements

Supported by Grants from CONICET, CONICOR, Volkswagen-Stiftung I/68 788 and Fundación Pérez Companc. We wish to thank Myriam Siravegna for her skillful technical assistance.

References

1. Glynn, I.M. (1985) The Na$^+$/K$^+$-transporting adenosine triphosphatase. In The enzymes of biological membranes (ed. A.N. Martonosi) pp. 35-114. Plenum Press. New York.

2. Post, R.L., Toda, G. and Rogers, F.N. (1975) Phosphorylation by inorganic  phosphate of sodium plus potassium ion transport adenosine triphosphatase, J. Biol. Chem. 250, 691-701.

3. Rossi, R.C. and Garrahan, P.J. (1889) Steady-state kinetic analysis of the Na$^+$/K$^+$-ATPase. The effects of adenosine 5'[beta,gamma-methylene]triphosphate on substrate kinetics, Biochim. Biophys. Acta 981, 85-94.

4. Hobbs, A., Albers, R.W. and Froehlich, J.P. (1988) Complex time dependence of phosphoenzyme formation and decomposition in electroplax Na$^+$/K$^+$-ATPase, Progr. Clin. Biol. Res. 268A, 307-314.

5. Campos, M. and Beaugé, L. (1992) Effects of magnesium and ATP on pre-steady state phosphorylation kinetics of the Na$^+$/K$^+$-ATPase, Biochim. Biophys. Acta 1105, 51-60.

# PARALLEL PATHWAY MODELS FOR ELECTRIC ORGAN Na$^+$- AND Na$^+$/K$^+$-ATPases

J. P. Froehlich, A. S. Hobbs and R. W. Albers

National Institute on Aging, National Institues of Health, Baltimore, Maryland, 21224; National Institute of Neurological Disease and Stroke, National Institutes of Health, Bethesda, Maryland, 20892

The Na$^+$- and Na$^+$/K$^+$-ATPases exhibit a variety of complex kinetic and thermodynamic properties that are incompatible with the Albers-Post mechanism and simple modifications thereof (1-5). Specific examples of this behavior are shown in Fig. 1 which depicts the multiphasic time courses of phosphorylation and dephosphorylation measured in native Electrophorus membranes using the acid quench-flow technique. In the case of the phosphorylation reaction, it is necessary to add Na$^+$ to the enzyme last in order to demonstrate the biphasicity which consists of fast and slow components in a ratio of 2:1 (1). Addition of Na$^+$ prior to ATP and Mg$^{2+}$ yields monophasic behavior without the slow phase (1). The biphasicty resulting from late Na$^+$ addition cannot be attributed to an equilibrium mixture of the E$_1$ and E$_2$ conformational states because pre-incubation with 0.01-0.1 mM ATP, which favors the high-affinity E$_1$ conformer, does not affect the relative amounts of the fast and slow components.

The same 2:1 ratio of fast and slow phases is also observed in the dephosphorylation reaction following the addition of a chase containing KCl + EDTA. This reaction also contains a stable component (not shown) which arises predominantly from sealed vesicles with sequestered K$^+$ sites (1) and contributes to the fast and slow decay phases in the same 2:1 ratio as the phosphoenzyme in unsealed vesicles. If the K$^+$ + EDTA chase is added during the pre-steady state of phosphorylation, the ratio of fast and slow decay phases remains at 2:1 (4), implying that the slowly decaying component accumulates rapidly. A detailed kinetic analysis of this behavior revealed that the fast and slow components of dephosphorylation are produced by a parallel pathway phosphorylation mechanism (1). Such behavior could arise from alternate (uncoupled) pathways for Na$^+$ binding as depicted by the mechanism shown in Fig. 2. This model assumes that the transport sites for Na$^+$ are in a low-field access channel and that the alternate pathways for loading these sites manifest different kinetics which give rise to the multiphasic patterns of phosphoenzyme formation and decomposition. Single and double occupation of the transport sites leads to equally probable states, whereas the third Na$^+$ binds slowly to one-third of the states because of a conformational constraint imposed by occupation of the other sites. The principal drawback of this scheme is that it requires that the fully occupied states be conformationally and functionally distinct depending on their loading history and that this distinction be maintained (by hysteresis)

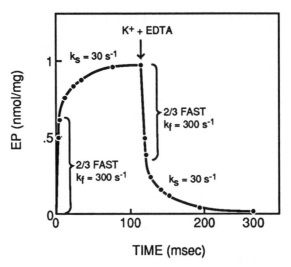

**Figure 1**. Facsimile of multiphasic phosphorylation and dephosphorylation in Na$^+$-ATPase.

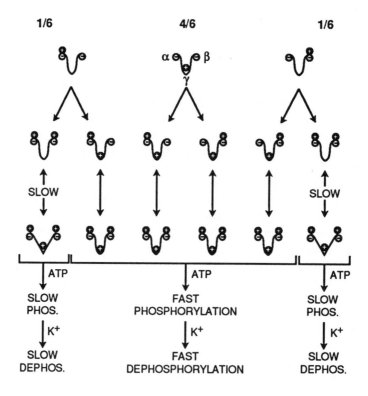

**Figure 2**. Statistical partitioning model for heterogeneous Na$^+$ binding at the cytoplasmic transport sites.

442

throughout phosphorylation, phosphoenzyme conversion and dephosphorylation. It also does not explain how oligomycin and nonionic detergents act to change the apparent stoichiometry of Na$^+$ binding reflected in the multiphasic patterns of phosphorylation (see below).

Biphasic phosphorylation similar to that in Fig. 1 has also been observed in the presence of oligomycin (3) which blocks the conversion of $E_1P$ to $E_2P$. As in the previous example, Na$^+$ must be excluded during the pre-incubation with oligomycin to observe biphasic behavior. A salient difference is that the proportions of fast and slow components are 1:1, suggestive of an $E_1/E_2$ dimer ($\alpha_2\beta_2$ diprotomer). Because only half of the enzyme ($E_1$) binds oligomycin in the absence of Na$^+$ and K$^+$, the slow accumulation of phosphoenzyme subsequent to the fast phase probably corresponds to inhibition of the remaining adjacent enzyme ($E_2$) subunits. The different proportions of rapid and slow phases seen in the native and oligomycin-treated membranes can be reconciled by assuming that in the former population about one-third of the enzyme is in a rapidly phosphorylating, monomeric state. In the absence of Na$^+$, oligomycin interacts strongly with the $E_1/E_2$ dimer, shifting the ratio of fast and slow components from 2:1 to 1:1. An important distinction between this scheme and that shown in Fig. 2 is that the parallel catalytic pathways in the dimer are coupled as opposed to being completely independent. Coupling imposes "staggered" operation on the subunits such that phosphorylation (or dephosphorylation) occurring in one subunit precedes phosphorylation (or dephosphorylation) in the neighboring subunit.

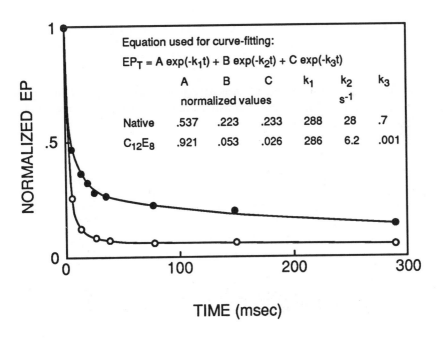

**Figure 3**. Dephosphorylation of native and C$_{12}$E$_8$-solubilized Electrophorus membranes by KCl + EDTA.

Evidence in support of the dimer is presented in Fig. 3 which shows the time dependence of dephosphorylation of native and $C_{12}E_8$-solubilized Electrophorus membranes produced by chasing the phosphoenzyme with 10 mM KCl + 10 mM EDTA. Detergent treatment virtually eliminates the slow (B) decay component which, in the context of the dimer mechanism, results from phosphorylation of the lagging ($E_2$) subunit. Low (less-than-solubilizing) detergent concentrations also transform the biphasic phosphorylation pattern to monophasic kinetics (not shown). A challange for the future will be to distinguish subunit uncoupling effects from detergent-induced changes in secondary and tertiary protein structure and to determine the specific pattern of cataytic coupling that prevents the $\alpha\beta$-protomeric units from behaving independently.

## References

1. Froehlich, JP, Fendler, K (1991) The partial reactions of the $Na^+$- and $Na^+/K^+$-activated adenosine triphosphatases. In: De Weer, PD, Kaplan, JH (eds) The Sodium Pump; Structure, Mechanism and Regulation. The Rockefeller University Press, New York, pp 227-247

2. Hamer, E. Schoner, W (1993) Modification of the $E_1$ATP binding site of $Na^+/K^+$-ATPase by the chromium complex of adenosine 5'-[$\beta,\gamma$-methylene]adenosine blocks the overall reaction but not the partial activities of the $E_2$ conformation. Eur J Biochem 213:743-748

3. Hobbs, AS, Albers, RW, Froehlich, JP (1983) Effects of oligomycin on the partial reactions of the sodium plus potassium-stimulated adenosine triphosphatase. J Bio Chem 258:8163-8168

4. Hobbs, AS, Albers, RW, Froehlich, JP (1985) Quenched-flow determination of the $E_1$P to $E_2$P transition rate constant in electric organ $Na^+/K^+$-ATPase. In: Glynn, IM, Ellory, C (eds) The Sodium Pump. The Company of Biologists, Cambridge, pp 355-361

5. Ottolenghi, P, Jensen, J (1983) The $K^+$-induced apparent heterogeneity of high-affinity nucleotide binding sites in $Na^+/K^+$-ATPase can only be due to the oligomeric structure of the enzyme. Biochim Biophys Acta 727:89-100

# Enzymic activities of solubilised and dissociated (α + ß) chains and of soluble αß protomers of pig kidney Na$^+$/K$^+$-ATPase

C. S. Madden, D. G. Ward, T. J. H. Walton, R. F. Washbrook, A. J. Rowe[*] and J. D. Cavieres.

*Department of Cell Physiology and Pharmacology and (\*) National Centre for Macromolecular Hydrodynamics, Leicester University, Leicester LE1 9HN, England.*

Two important issues about the sodium pump concern the subunit organisation of the active form(s) of the enzyme. One would like to know whether the membrane-bound enzyme is protomeric or dimeric, and also to decide whether the intact ß chain has a functional role in the reaction. As a first step, it seems necessary to define the catalytic potential of different subunit arrangements and detergent solubilisation of Na$^+$/K$^+$-ATPase in active form has been a fruitful approach to the question. This offers not only the possibility of manipulating the subunit composition but also of being able to characterise the resulting particles with the analytical ultracentrifuge on the basis of sound physical principles[14].

Soluble and active αß protomers were first obtained by Jørgensen and colleagues by incubating zonal preparations of kidney Na$^+$/K$^+$-ATPase with C$_{12}$E$_8$, a non-ionic detergent, at 3 mg detergent/mg protein[1,2]. The protomeric preparation has Na$^+$/K$^+$-ATPase and K$^+$-phosphatase activities, experiences conformational transitions and occludes Na and Rb ions[15]. Not surprisingly, we have found, besides, that the protomeric enzyme catalyses ATP-ADP exchange at 20°C (0.13 U/mg), albeit at a lower rate than the membrane-bound enzyme (0.71 U/mg) (at least, after N-ethyl-maleimide treatment; R.F. Washbrook & J.D. Cavieres, unpublished observations). Notwithstanding this convincing array of functions, and perhaps because of them, it was important to obtain direct experimental evidence about two questions: i) whether the soluble protomer had to form (αß)$_2$ dimers to be active and ii) whether the apparent negative co-operativity[13] towards ATP was abolished in the protomeric state. We approached the first question using active-enzyme analytical ultracentrifugation (AEC) and the second by measuring the Na$^+$/K$^+$-ATPase activity of the soluble protomers between 50 nM and 200 μM [$\tau^{32}$P]-ATP and by using a non-phosphorylating ATP analogue (ADPNP)[17].

AEC is a technique that has been available for some time[6,7], and which Martin[10] has put to elegant use to show that the monomer of the sarcoplasmic-reticulum calcium pump in C$_{12}$E$_8$ solution is active. The method takes advantage of a change in optical density caused by the enzymic reaction, in order to establish a footprint of the distribution of active particles in the ultracentrifuge cell. We used Phenol Red, whose absorbance at 550 nm decreases because of the H$^+$ release associated with ATP hydrolysis. To do an AEC run, Phenol Red together with ATP, Mg$^{2+}$, Na$^+$, K$^+$, dilute buffer and detergent go in the main chamber and 10 μl of solubilised enzyme in a side chamber of the centrepiece. As the rotor speeds up, the solubilised protein is layered on top of the solution in the main chamber. Thus, the band of active enzyme particles bleach the dye as they hydrolyse ATP during their migration towards the base of the chamber. The sedimentation coefficient ($s_{20,w}$) of the active enzyme species could then be calculated from scans at 550 nm taken at various times during the centrifugation. This was compared to the $s_{20,w}$ of the protein, determined separately in sedimentation-velocity runs (enzyme in main chamber), scanning at 280 nm. Our results showed that the $s_{20,w}$ values were indistinguishable, each averaging $6.5 \pm 0.2$ S. This gave a molecular weight of $151,000 \pm 9,000$. Therefore, the soluble $\alpha\beta$ protomer of $Na^+/K^+$-ATPase does not have to dimerise to be active[17].

Contrary to expectations, the ATP-concentration dependence of the $Na^+/K^+$-ATPase activity of the soluble $\alpha\beta$ protomer gave a non-Michaelian pattern[17], Lineweaver-Burk plots showed a downward curvature as had been found with the membrane-bound enzyme[13]. The apparent negative co-operativity with respect to ATP was confirmed in experiments using ADPNP: whereas at high or low ATP concentrations ADPNP inhibited the $Na^+/K^+$-ATPase reaction, the analogue could stimulate the rate at intermediate ATP concentrations[17]. These results are compatible with either 2 ATP sites per $\alpha$ chain or a change of ATP affinity round the reaction cycle. We concluded that the apparent negative co-operativity was intrinsic to the $\alpha\beta$ protomer and that it could not, therefore, be thought of as a diagnostic test for $\alpha$ subunit interactions.

To achieve the dissociation of $\alpha$ from $\beta$ chains in mild conditions, we decided to try CHAPS, a zwitterionic detergent that has been used to solubilise active membrane receptors[5] and to solubilise and reconstitute $Na^+/K^+$-ATPase in phospholipid vesicles[18]. At 10 mM, CHAPS solubilised the zonal $Na^+/K^+$-ATPase preparation and the high-speed supertatants of the solubilisate showed two bands in CHAPS PAGE, either in 7% Laemmli gels or in 5-20% gradient gels. Treatment[12] of the solubilised enzyme with N-glycosidase F displaced the faster band but did not affect the slower band. Fig. 1 shows a sedimentation-velocity run of the CHAPS supernatant in the

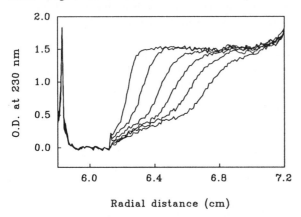

FIGURE 1. Analytical ultracentrifugation of CHAPS-solubilised Na+/K+-ATPase. The high-speed supernatant was spun at 40,000 rpm in a Beckman Optima XLA and scanned at 230 nm every 30 min. The $s_{20,w}$ estimates were 4.6 S (greater component) and < 2 S (smaller component).

analytical ultracentrifuge and two components are readily apparent: a faster one, with a $s_{20,w}$ of 4.6 S and a slow particle with $s_{20,w} < 2$ S. As the lipid, carbohydrate and detergent contents of these protein micelles are not known, it is not possible to obtain molecular weights. However, the ratio of optical densities at 230 nm (fast/slow) is about 3.5, which compares favourably with an $\alpha/\beta$ protein mass[11] ratio of 3.2.

Taken together with the electrophoresis results, these data strongly suggest that $\alpha$ and $\beta$ chains have been dissociated during solubilisation in CHAPS.

The CHAPS high-speed supernatant was tested for enzymic activity at 20°C. Fig. 2 shows that, in the absence of K ions, the CHAPS enzyme presents ATP-ADP exchange and $Na^+$-ATPase activities. The specific activities can reach values of 0.26 and 0.37 U/mg, respectively, and this "$\alpha+\beta$" enzyme is reasonably stable at 20°C over the observation period, just as the solubilised $\alpha\beta$ protomers[9]. The possibility that the observed exchange could be due to a minor contamination with undissociated $\alpha\beta$ protomers seems unlikely, as the ATP-ADP exchange activity of the $\alpha\beta$ protomers is lower than that of the CHAPS-solubilised enzyme.

The result of assaying the ATPase activity of the "$\alpha+\beta$" enzyme in the presence of K ions is shown in Fig. 3. The enzyme initially splits ATP at a rate faster than

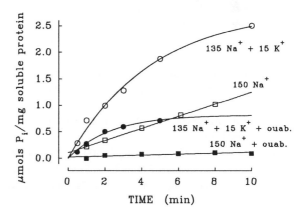

FIGURE 2. ATP-ADP exchange and Na+-ATPase activities of the high-speed supernatant of the CHAPS-solubilised Na+/K+-ATPase. Simultaneous reactions[4] at 20°C, in K+-free medium.

in the absence of $K^+$, but it is then rapidly inactivated. This contrasts with the stable behaviour (linear time course) obtained with the $Na^+$-ATPase activity over the same time span, in conditions such that the *total* $P_i$ release was greater than in the presence of K ions[16].

FIGURE 3. ATPase activity of Na+/K+-ATPase solubilised with 10 mM CHAPS and measured[3] in the presence and in the absence of K+, with and without 0.1 mM ouabain.

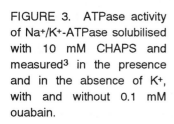

Therefore, the "$\alpha+\beta$" enzyme seems capable of catalysing two reactions that do not involve $K^+$ occlusion or low-affinity ATP effects, and can also bind ouabain (we do not yet know about the binding affinity). The results shown in Figs. 2 and 3 suggest

447

that the ATP-ADP exchange and the $Na^+$-ATPase activities might be catalysed by the $\alpha$ chain on its own. However, preliminary estimates indicate that $\alpha$ and $\beta$ chains might still recombine fast enough to match each ATP-ADP exchange or $Na^+$-ATPase cycle, although they might not do so vis-à-vis the faster $Na^+/K^+$-ATPase cycle. It is then possible that the $\beta$ chain be required to stabilise the $\alpha$ chain during steps involving activation by or occlusion of $K^+$ ions, as proposed recently by Lutsenko & Kaplan[8], or to stabilise low-affinity ATP binding. We are trying to achieve a mild separation of the dissociated $\alpha$ and $\beta$ subunits to be able to provide a definitive answer about the role of the $\beta$ chain.

We thank Mrs. P. Mistry for technical assistance and The Wellcome Trust, the Medical Research Council, the Science & Engineering Research Council and The Royal Society for financial support.

1. Brotherus, J.R., Møller, J.V. & Jørgensen, P.L. (1981). *Biochem. Biophys. Res. Commun.* **100**, 146-154.
2. Brotherus, J.R., Jacobsen, L. & Jørgensen, P.L. (1983). *Biochim. Biophys. Acta* **731**, 290-303.
3. Cavieres, J.D. (1987). *FEBS Lett.* **225**, 145-150.
4. Cavieres, J.D. (1987). *Biochim. Biophys Acta.* **899**, 83-92.
5. Cavinato, A.G., Macleod, R.M., Ahmed, M.S. (1988). *Preparative Biochemistry* 18, 205-216.
6. Cohen, R., Giraud, B. & Messiah, A. (1967). *Biopolymers* **5**, 203-225.
7. Kemper, D.L. & Everse, J. (1973). *Methods Enzymol.* **28**, 67-81.
8. Lutsenko, S. & Kaplan, J.H. (1993). *Biochemistry* **32**, 6737-6743.
9. Madden, C.S. & Cavieres, J.D. (1993). *J. Physiol.* **459**, 380P.
10. Martin, D.W. (1983). *Biochemistry* **22**, 2276-2282.
11. Ovchinnikov, Yu.A., Modyanov, N.N., Broude, N.E., Petrukhin, K.E., Grishin, A.V., Arzamazova, N.M., Aldanova, N.A., Monastyrskaya, G.S. & Sverdlov, E.D. (1986). *FEBS Lett.* **201**, 237-245.
12. Pedemonte, C.H., Sachs, G. & Kaplan, J.H. (1990). *Proc. Natl. Acad. Sci. USA* **87**, 9789-9773.
13. Robinson, J.D. (1976). *Biochim. Biophys. Acta* 429, 1006-1019.
14. Tanford, C., Nozaki, Y., Reynolds, J.A. & Makino, S. (1974). *Biochemistry* **13**, 2369-2376.
15. Vilsen, B., Andersen, J.P., Petersen, J. & Jørgensen, P.L. (1987). *J. Biol. Chem.* **262**, 10511-10517.
16. Walton, T.J.H. & Cavieres, J.D. (1993). *J. Physiol.* **467**, 346P.
17. Ward, D.G. & Cavieres, J.D. (1993). *Proc. Natl. Acad. Sci. USA* **90**, 5332-5336.
18. Yoda, A., Clark, A.W. & Yoda, S. (1984). *Biochim. Biophys. Acta* **778**, 332-340.

# One Phosphorylation Site per αβ-Protomer in Reconstituted Shark Na⁺/K⁺-ATPase.

Flemming Cornelius
Institute of Biophysics. University of Aarhus. DENMARK.

## INTRODUCTION

The oligomeric structure of the Na⁺/K⁺-ATPase in the native state is still a controversial issue. Several lines of evidence suggest the minimal functional unit to be a dimer, an $(\alpha\beta)_2$-unit. These include radiation inactivation studies, ligand binding -and kinetic studies, and structural studies of two-dimensional crystals. However, equally compelling evidence are present for an αβ-protomer as the minimal functional unit.

From studies of ligand binding capacity and phosphorylation, it is generally found that the Na⁺/K⁺-ATPase contains an equal number of sites when measured as phosphorylation capacity, or ouabain- and vanadate binding, or high-affinity ATP binding (4), whereas apparent disagreements exist as to the number of sites per mass of protein: In detergent disrupted, partially purified membrane preparations between 2-3 nmoles of sites are found per mg of protein, whereas in highly purified enzyme preparations between 4-6 nmoles per mg using detergent solubilization (5,2,7), or employing zonal centrifugation are found (8). Therefore, in both highly purified membrane-bound and in solubilized preparations the site number is higher than 1 per $(\alpha\beta)_2$-dimer and approaches in some preparations 1 per αβ-protomer.

## RESULTS

*Phosphorylation.* By solubilization with the detergent $C_{12}E_8$ the number of phosphorylation sites (EP) after phosphorylation from ATP increases from 2.5 to 4.2 nmoles per mg of protein, **Fig. 1** (protein determined by the Lowry method (9) or its modification by Peterson (11)), in accordance with previous results (12), and is probably the result of a purification. In the solubilized Na⁺/K⁺-ATPase this corresponds to about one phosphorylation site per $(\alpha\beta)_2$-unit, assuming a molecular weight for the α and β polypeptide chains of 112 Kda and 35 Kda, respectively. After reconstitution (1) the EP-site number increases to 6.9±0.6 (n=13) nmol per mg protein, and phosphorylation by the 'direct route' from $Mg^{2+}$ (3mM) plus Pi (1 mM) gives 6.1±0.2 (n=9) phosphorylation sites per mg protein. Both values are close to the theoretical value of 6.8 calculated for 1 site per αβ chain.

The increase in phosphorylation site number following reconstitution is, however, followed by a parallel decrease in molar activities measured both during Na⁺/K⁺-exchange and during Na⁺/Na⁺-exchange. This is probably ascribed to a preferential detergent inactivation of hydrolytic activity, since it can be partially lifted by phospholipid addition.

Since the fraction of the total enzyme with extracellular side exposed (i.e. the right-side out) does not participate in the phosphorylation reaction after reconstitution, very

449

accurate determinations of enzyme orientation is necessary in order to evaluate the specific EP-site number. In the present investigation the fraction of i:o-oriented enzyme in the different proteoliposome preparations used was very consistent, 0.128±0.023 (mean±SD, n=10). Moreover, identical EP-site number is obtained in the reconstituted preparation in the absent of ouabain inhibition of enzyme with both sides exposed (non-oriented, n-o), and the calculation based on the fractions of i:o+n-o enzyme, lending further confidence to the results.

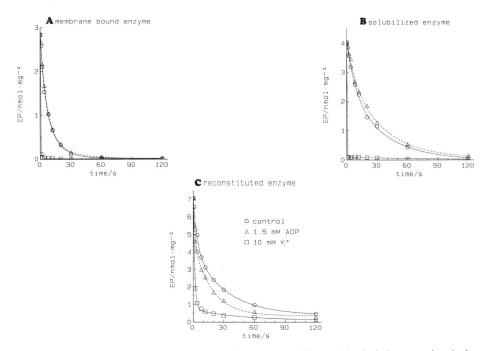

**Fig. 1.** Comparison of phosphorylation site number (EP) and hydrolytic capacity during Na$^+$/K$^+$ - or Na$^+$/Na$^+$ exchange for three preparations of shark Na$^+$/K$^+$-ATPase: A. membrane bound, B. solubilized, and C. reconstituted. Mean±SEM (n>10).

*Dephosphorylation.* The results given in **Fig. 2** show dephosphorylation (spontaneous) at 0°C of purified membrane-bound, solubilized, and reconstituted Na$^+$/K$^+$-ATPase, together with ADP- or K$^+$-activated dephosphorylation. The K$^+$-activated dephosphorylation for membrane-bound and solubilized enzyme is too fast even at 0°C to follow with the present technique. The dephosphorylation curves are all biexponential with a rapid and a slow component. The rate of the slower component is in all cases insufficient to account for the measured hydrolytic activities measured at comparable conditions (sub-optimal, see below), if the phosphoenzymes are assumed consecutive intermediates in a single reaction pathway. Only if the two fractions, rapid and slow, represents phosphoenzyme species that are dephosphorylated in parallel (3, 10) does the rate of dephosphorylation match the rate of hydrolysis. Alternatively, the chase conditions either induces a redistribution in enzyme species from that of steady-state, or forms new reactants (ATP, Mg$^{2+}$, or MgATP could react with EP-forms). An indication of the latter possibility is that the $Mg^{2+}$ concentration in these ATP chase experiments must be kept high enough to effectively depress the concentration of free ATP, which otherwise inhibits the dephosphorylation. Therefore the conditions during dephosphorylation is not optimal for hydrolysis.

As shown in **Fig. 2**, ADP accelerates dephosphorylation only in the reconstituted preparation, indicating that the phosphoenzyme distribution here is less in favour of the $E_2P$-form than for membrane-bound and solubilized enzyme, where the distribution is apparently strongly poised towards the $E_2P$-form.

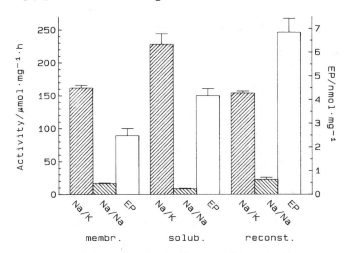

**Figure 2.** Dephosphorylation of membrane bound, solubilized, and reconstituted $Na^+/K^+$-ATPase at 0 °C. Dephosphorylation (spontaneous) was followed by adding stopping solution at different time intervals after chasing of radioactive ATP with cold ATP (1 mM) in the presence of high $Mg^{2+}$ concentration (10 mM). ADP- or $K^+$-supported dephosphorylations were measured by adding 1.5 mM ADP or 10 mM $K^+$ to the chasing solution. In the case of proteoliposomes nigericin (0.5 µg/ml) was included in the $K^+$-dephosphorylation to equilibrate the liposomes. The curves are computer-fit to the data using biexponential decay, with the following rate-constants ($\tau$) and fractions ($f$) for reconstituted enzyme, control: $\tau_1$=0.23 s$^{-1}$, $\tau_2$=0.03 s$^{-1}$, $f_1$=3.04 nmol/mg, and $f_2$=3.63 nmol/mg; +ADP: $\tau_1$=0.75 s$^{-1}$, $\tau_2$=0.05 s$^{-1}$, $f_1$=2.44 nmol/mg, $f_2$=4.25 nmol/mg; +$K^+$: $\tau_1$=0.78 s$^{-1}$, $\tau_2$=0.03 s$^{-1}$, $f_1$=6.4 nmol/mg, $f_2$=0.69 nmol/mg. The distribution of EP-forms after reconstitution was calculated according to a "three EP-pool model", from the fractions with rapid decay of the biexponential computerfits to the data. At 0°C the distributions were: $E_1P:E^*P:E_2P$ = 0.09:0.74:0.18, and at 10°C: 0.09:0.69:0.22.

### DISCUSSION

From the present study it is concluded that reconstitution of solubilized $Na^+/K^+$-ATPase from shark rectal glands into lipid vesicles increases the number of phosphorylation sites per mg protein from about one per $\alpha\beta$-dimer to one per $\alpha\beta$-protomer. Concomitantly, the molar activity decreases proportionally, an effect which can probably be ascribed to a preferential and partial inactivation by the detergent $C_{12}E_8$ of hydrolytic activity over that of phosphorylation.

451

The increase in EP-site number by reconstitution is probably not the result of a further purification after solubilization, since the protein recovery in the reconstitution is near 100%, and the pattern in SDS gels are identical for solubilized and reconstituted enzyme (results not shown), nor can it be explained by an underestimation of protein (6).

Interference in the measurements of EP-sites in the solubilized state from the detergents is a possibility. Alternatively, the lower site number in solubilized enzyme as compared to reconstituted enzyme could be due to inactive enzyme or contaminating protein, which is removed by reconstitution.

Another explanation could be, that in the native membranes and after solubilization negative cooperativity between protomers takes place such that phosphorylation of only one of two protomers is supporting turnover of a dimer. By reconstitution monomerization could result, as a consequence of dilution in the lipid phase, and each protomer must be phosphorylated in order to turn over.

## REFERENCES

1. Cornelius, F (1991) Biochim. Biophys. Acta 1071;19-66
2. Esmann, M (1984) Biochim. Biophys. Acta 787:71-80
3. Froehlich, JP and Fendler, K (1991) In: The Sodium Pump: Structure, Mechanism, and Regulation (Kaplan, JH and De Weer, P, eds.) pp. 227-247. The Rockefeller University Press, New York.
4. Glynn, IM (1985) In; The Enzymes of Biological Membranes (Martonosi, AM, ed.) Vol. 3 pp. 35-114. Plenum Press, New York.
5. Hastings, DF and Reynolds, JA (1979) Biochemistry 18:817-821
6. Jensen, J (1992) Biochim. Biophys. Acta 1110:81-87
7. Jensen, J and Ottolenghi, P (1983) Biochim. Biophys. Acta 731:282-289
8. Jørgensen, PL (1988) Meth. Enzymol. 156:29-43
9. Lowry, OH, Rosebrough, NJ, Farr, AL, and Randall, RJ (1951) J. Biol. Chem. 193:265-275
10. Martin, DW and Sachs, JR (1991) J. Gen. Physiol. 98:419-426
11. Peterson, GL (1977) Anal. Biochem. 83:346-356
12. Skou, JC and Esmann, M (1979) Biochim. Biophys. Acta 567:436-444

# Protomer Association of Solubilized Na$^+$/K$^+$-ATPase Induced by ATP

Y.Hayashi, T.Kobayashi, T.Nakajima and H.Matsui

Department of Biochemistry, Kyorin University School of Medicine, Mitaka, Tokyo 181, JAPAN

## Introduction

We have shown by means of low-angle laser light-scattering photometry coupled with high-performance gel chromatography (HPGC/LALLS method) that solubilized Na$^+$/K$^+$-ATPase is in a dissociation-association equilibrium ($2\alpha\beta \rightleftharpoons (\alpha\beta)_2$) between the $\alpha\beta$-protomer (M$_r$ = 1.5 x 10$^5$) and ($\alpha\beta$)$_2$-diprotomer (3.0 x 10$^5$)(3). The association constant (K$_a$) of the equilibrium in the presence of 0.1 M KCl was about 50 times larger than that in the presence of 0.1 M NaCl (3,5). Thus, it was strongly suggested that two protomers in the E$_2$ conformation were associated more strongly than in the E$_1$ conformation. It this study, the solubilized enzyme was incubated with ATP under conditions expected to produce a phosphorylated intermediate in the E$_2$-conformational state (E$_2$-P), and was subjected to chromatography to estimate the weight ratio of the protomer to the diprotomer. The content of diprotomer was markedly increased whereas that of the protomer was decreased. These results strongly suggest that ATP induces the association of the protomer to form the diprotomer through E$_2$-P formation.

## Methods and Results

Membrane-bound Na$^+$/K$^+$-ATPase was purified from dog kidney, and solubilized with octaethylene glycol n-dodecyl ether (C$_{12}$E$_8$), as described elsewhere (3). The solubilized enzyme was incubated with 5 μM to 4.8 mM ATP in the presence of 83 mM NaCl and 4.8 mM MgCl$_2$, as well as reagents used for solubilization including 5 mg/ml C$_{12}$E$_8$ at pH 7.0 and 0°C for 10 min, and then subjected to chromatography in a TSKgel G3000SW$_{XL}$ column equilibrated with an elution buffer containing 0.3 mg ml$^{-1}$ C$_{12}$E$_8$/0.1 M NaCl/1 mM EDTA/10 mM imidazole/13 mM Hepes (pH 7.0) at -0.2°C. As shown in Fig.1, the protomer and diprotomer accounted for 56% and 37% (w/w) of all the protein eluted, respectively, when the enzyme was incubated without ATP. With increasing ATP concentration, the content of the protomer decreased while that of the diprotomer increased, and their relative proportions reached 26% and 64% (w/w), respectively, when more than 1 mM ATP was present. Thus, ATP induced association of the protomer into the diprotomer. The ATP concentrations necessary for half-maximal (K$_{0.5}$) increase of the diprotomer and decrease of the protomer were 0.44 and 0.45 mM, respectively.

In another series of experiments, incubation of the enzyme with ATP was followed by

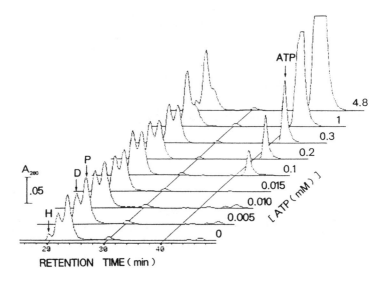

**Figure 1.** ATP-dependent increases of $(\alpha\beta)_2$-diprotomer (D) and decreases of $\alpha\beta$-protomer (P) revealed by high-performance gel chromatography developed at -0.2°C using a TSKgel G3000SW$_{XL}$ column. H means oligomers with a M$_r$ about 4-fold higher than that of the protomer.

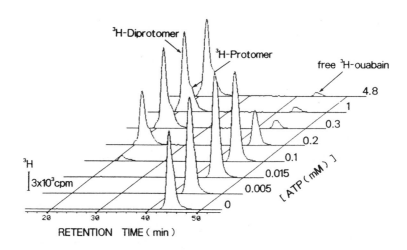

**Figure 2.** Dependence of [³H]ouabain binding of solubilized Na⁺/K⁺-ATPase upon ATP concentration. The enzyme was incubated with ATP under the same conditions as those used for the experiments described in Fig.1. The incubation of the enzyme was followed by incubation with a stoichiometric amount of [³H]ouabain for 5 min at 0°C, and the resulting enzyme was subjected to chromatography as described in the legend for Fig. 1. Elution patterns of A$_{280}$ are not shown, but the conversion between the diprotomer and protomer can be observed from the patterns like those in Fig.1.

454

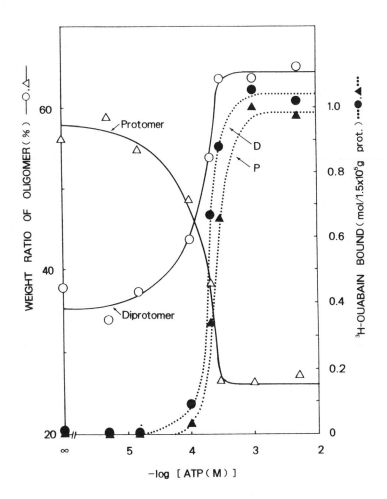

**Figure 3.** Coincidence between ATP-dependence of conversion of the protomer to the diprotomer and that of ouabain binding to the diprotomer. The data were taken from the results shown in Fig. 2.

incubation with a stoichiometric amount of [³H]ouabain for 5 min at 0°C. It was shown by the elution patterns monitored at $A_{280}$ that the conversion of protomer to diprotomer occurred as in the case without [³H]ouabain (data not shown). The elution patterns monitored by [³H]ouabain radioactivity are shown in Fig.2. The amount of [³H]ouabain bound to the diprotomer increased with increasing ATP concentration in parallel with the conversion of the protomer to the diprotomer, and reached 1 mol per mol protomeric unit in the presence of more than 0.3 mM ATP. $K_{0.5}$ values of ATP for increase of the diprotomer and decrease of the protomer were 0.16 and 0.17 mM, respectively, while those for [³H]ouabain binding to the diprotomer and protomer were 0.19 and 0.25 mM, respectively (Fig. 3).

## Conclusions and Discussion

ATP induced oligomeric conversion of the solubilized enzyme from the protomer into the diprotomer. The conversion was brought about without addition of [$^3$H]ouabain (cf. Fig.1) or with ouabain. According to Hara *et al.* (2), membrane-bound Na$^+$/K$^+$-ATPase produces the intermediate of E$_2$-P in the presence of a low concentration of NaCl like that used here. Ouabain is known to bind to the enzyme in its E$_2$-P form (1, 4). Since K$_{0.5}$ of ATP for change of the oligomers was consistent with that for ouabain binding to the diprotomer, the intermediate of E$_2$-P would have been produced, followed by association of the protomer into the diprotomer, and simultaneously by binding of ouabain to the diprotomer. Although ouabain bound to the protomer, this would be attributable to dissociation of the diprotomer bound to [$^3$H]ouabain upon chromatography. Thus, it was strongly suggested that ATP caused association of the two protomers into a diprotomer through formation of E$_2$-P. We have already shown that phosphatidylserine-induced associated solubilized Na$^+$/K$^+$-ATPase was dissociated by ATP with a K$_{0.5}$ value of 0.16 mM (5). It is concluded that ATP has double aspects of dissociation and association of the solubilized enzyme. These results are consistent with our conclusion obtained so far that the conformational states of E$_1$ and E$_2$ correspond to a loosely associated diprotomer and a tightly associated diprotomer *in situ,* respectively.

*Acknowledgements*

We are grateful to Miss M.Ishihira and Miss N.Shinji for technical assistance in the purification of the enzyme. This work was supported in part by a Grant-in-Aid for Scientific Research (No.05259222 to Y.H.) from the Ministry of Education, Science and Culture of Japan.

## References

1. Charnock JS, Post, RL (1963) Evidence of the mechanism of ouabain inhibition of cation activated adenosine triphosphatase. Nature 199: 910-911
2. Hara Y, Nakao M (1981) Sodium ion discharge from pig kidney Na$^+$/K$^+$-ATPase. Na$^+$-dependency of the E$_1$-P $\rightleftharpoons$ E$_2$-P equilibrium in the absence of KCl. 90:923-931
3. Hayashi Y, Mimura K, Matsui H, Takagi T (1989) Minimum enzyme unit for Na$^+$/K$^+$-ATPase is the αβ-protomer. Determination by low-angle laser light scattering photometry coupled with high-performance gel chromatography for substantially simultaneous measurement of ATPase activity and molecular weight. BiochimBiophys Acta 983: 217-229
4. Matsui H, Schwartz A (1968) Mechanism of cardiac glycoside inhibition of the (Na$^+$-K$^+$)-dependent ATPase from cardiac tissue. Biochim Biophys Acta 151:655-663
5. Mimura K, Matsui H, Takagi T, Hayashi Y (1993) Change in oligomeric structure of solubilized Na$^+$/K$^+$-ATPase induced by octaethylene glycol dodecyl ether, phosphatidylserine and ATP. 1145:63-74

# Arginine Appears not to be Located in the Active Site of $Na^+/K^+$-ATPase

O.D. Lopina, M. Vachova

Department of Biochemistry, School of Biology, Moscow State University, 119899, Moscow, Russia

## Introduction

It was found that modification of $Na^+/K^+$-ATPase by arginine specific reagents produced inhibition of the enzyme activity (1,2,4). ATP (1,2,4) and ADP (1,2) protected $Na^+/K^+$-ATPase against the inactivation. These data supported the idea that an arginine residue is located in nucleotide binding site of $Na^+/K^+$ATPase. We have compared inhibition of $Na^+K^+$ATPase from duck salt gland by two arginine specific reagents 2,3-butanedione and phenylglyoxal in the presence of three essential ligands - $Na^+$, $K^+$ and ATP.

## Experimental procedure

$Na^+/K^+$ATPase was prepared from duck salt glands by the method of Smith (5). The rate of ATP hydrolysis by $Na^+/K^+$-ATPase was determined as described earlier (3). Modification of $Na^+/K^+$-ATPase activity with 2,3-butanedione and phenylglyoxal were carried out at $20^{\circ}C$ in a medium containing 50 mM $Na^+$ or $K^+$-borate buffer (pH 8.0 and 7.6 for butanedione and phenylglyoxal respectively) and appropriate concentrations of the modificators. Reaction was started by the addition of the enzyme at final concentration 1 mg/ml. Solutions of 2,3-butanedione and phenylglyoxal were freshly prepared for each experiment, pH of solutions were readjusted with 1 N NaOH or KOH and concentrations of modificators were determined from absorbance measurements.

## Results

Treatment of purified $Na^+/K^+$-ATPase from duck salt glands by 2,3-butanedione or by phenylglyoxal inhibits enzyme activity. The inhibition of $Na^+/K^+$-ATPase by 2,3-butanedione follows pseudo-first order kinetics.

457

However in the presence of phenylglyoxal (2-10 mM) the $Na^+/K^+$-ATPase exhibits a two-exponential decay curve indicating more than one class of arginine residues modificated for complete inhibition. Inactivation of $Na^+/K^+$-ATPase by phenylglyoxal as well as by 2,3-butanedione in the presence of borate buffer was irreversible. A storage of the enzyme at -20°C results in a significant decrease of the rate of $Na^+,K^+$-ATPase inactivation by 2,3-butanedione. However the enzyme activity during the storage practically does not change (Table 1). This finding suggests that arginine residues are not essential for the catalytic activity of $Na^+/K^+$-ATPase.

Table 1. Effect of a storage of $Na^+/K^+$-ATPase preparation at -20°C on the enzyme inactivation by 2,3-butanedione

| Time of storage (days) | Enzyme activity (mmol/mg h) | Pseudo-first-order rate constants (min$^{-1}$) |
| --- | --- | --- |
| 1 | 1135 | 0.088 |
| 3 | 1090 | 0.050 |
| 25 | 1167 | 0.027 |

The pseudo-first-order rate constant of $Na^+/K^+$-ATPase inactivation by 2,3-butanedione practically does not change when $Na^+$ in the solution is substituted by the same amount of $K^+$ (Table 2). In Na-containing medium ATP (0.2-1 mM) decreases the rate of enzyme inactivation by 2,3-butanedione demonstrating considerable protection. Effect of ATP on the inactivation in K-containing medium is more complex: the loss of $Na^+/K^+$-ATPase activity in the presence of ATP and $K^+$ is two-exponential process. The rate constant of the enzyme inactivation in the fast phase is 10 times higher and in a slow phase - 10 times less than in a medium without ATP (Table 2).

Table 2. Effect of $Na^+$, $K^+$ and ATP on the $Na^+/K^+$-ATPase inactivation by 2,3-butanedione

| KCl (mM) | NaCl (mM) | ATP (mM) | Pseudo-first-order rate constants fast phase | slow phase |
| --- | --- | --- | --- | --- |
| 50 | | | | 0.018 |
| | 50 | | | 0.016 |
| 50 | | 1 | 0.13 | 0.0032 |
| | 50 | 1 | | 0.0047 |

$Na^+$ and $K^+$ ions have opposite effects on the inhibition rate of the enzyme activity by phenylglyoxal (Table 3). ATP demonstrates protective effect against the inactivation by phenylglyoxal only in the presence of $K^+$. On the contrary in Na-containing medium ATP increases the rate of the inactivation (Table 3).

458

Because protective effect of ATP against the inactivation by phenylglyoxal does not observed in the presence of Na$^+$ (under conditions when ATP can binds with the enzyme) one can suggest that protective effect of nucleotide is due to

Table 3. Effect of Na$^+$, K$^+$ and ATP on the Na$^+$/K$^+$-ATPase inactivation by phenylglyoxal

| KCl (mM) | NaCl (mM) | ATP (mM) | Pseudo-first-order rate constants | |
|---|---|---|---|---|
| | | | fast phase | slow phase |
| 50 | | | 0.300 | 0.0081 |
| | 50 | | 0.184 | 0.0046 |
| 50 | | 1 | 0.087 | 0.0018 |
| | 50 | 1 | 0.207 | 0.0079 |

the conformational transition of the enzyme induced by ATP-binding but not direct interaction of ATP with arginine residue.

## References

1.  De Pont JJHHM, Shoot BM, Van Proooijen AVE, Bonting SL (1977) An essential arginine residue in ATP-binding centre of (Na +K )-ATPase. Biochim Biophys Acta 482: 213-227
2.  Grisham CM (1979) Characterization of essential arginyl residues in sheep kidney (Na$^+$ + K$^+$)-ATPase. Biochem Biophys Res Commun 88: 229-236.
3.  Lopina O, Skackova D, Baranova L, Khropov Yu (1991) Inhibition of Na,K-ATPase by new ATP analog, adenosine-5-N-(2,4-dinitro-5-fluorophenyl)phosphohydrazide. FEBS Lett 282: 228-230.
4.  Pedemonte CH, Kaplan JH (1988) Modification of arginine residues in the Na$^+$,K$^+$-ATPase. Biophys J 53: 222a.
5.  Smith TW (1988) Purification of Na$^+$,K$^+$-ATPase from the supraorbital salt glands of the duck. Methods in Enzymol 156: 46-47.

# Does The α-Subunit of the Red Cell Membrane Na⁺/K⁺-ATPase Exist as a Homodimer?

**D.W. Martin, J.R. Sachs**

Hematology Division, State University of New York, Stony Brook, New York 11794-8151 USA

## Introduction

Chemical cross-linking techniques have been used in prior studies to demonstrate associations between the subunits of the $Na^+/K^+$-ATPase in purified kidney microsomes. Both alpha-alpha and alpha-beta subunit interactions have been observed depending upon the cross-linking reagent and enzyme substrate conditions. The $Na^+/K^+$-ATPase represents the predominant protein species in purified kidney preparations and subunit associations could be directly assayed by densitometric scans of polyacrylamide gels of cross-linked preparations (1).

Periyasamy et al. (5) in related studies using red blood cell membrane vesicles concluded that Cu and $o$-phenanthroline (CuP) also induced alpha-alpha homodimers of the red cell $Na^+/K^+$-ATPase. The $Na^+/K^+$-ATPase is a minor component of the red cell membrane (~200-400 copies/cell) and subunit interactions could not be directly followed using densitometric scans. Radioactive "scans" of the phosphorylated $Na^+/K^+$-ATPase obtained by counting successive slices of polyacrylamide gels were used to resolve the cross-linked species.

If true, the above conclusions would be significant. Given the natural high dilution and presumable random distribution of active $Na^+/K^+$-ATPase in the red cell membrane, cross-linked alpha-alpha dimers present a strong argument for the pre-existence of closely associated alpha chains in the native membrane.

Band 3 is the predominant protein of the red cell membrane (~1,200,000 copies/cell) and has a molecular weight similar to that of the alpha subunit of the $Na^+/K^+$-ATPase (~100,000 Da). We explored the possibility that cross-linking reagents used to generate alpha-alpha and alpha-beta dimers in kidney preparations may in the red cell induce hetero-cross-links between the $Na^+/K^+$-ATPase and unrelated membrane protein(s), with band 3 being a likely candidate (4).

The extracellular treatment of intact red cells with Pronase in a manner known to proteolyze Band 3 did not effect the electrophoretic migration of phosphorylated α-chains of the Na$^+$/K$^+$-ATPase isolated from inside-out vesicles (IOV) (3). We found (4) that while Pronase treatment did not effect the migration of Na$^+$/K$^+$-ATPase α-chains it reduced the apparent M$_r$ of the cross-linked species induced by CuP treatment as illustrated in Figure 1. This observation was inconsistent with CuP induction of α-α homodimers and throughout the study we saw no evidence of α-α homodimer formation. We concluded that CuP induced heterodimer formation between the α-chain and another membrane protein which displayed the identical proteolytic and H$_2$DIDS annealing sensitivities as that of Band 3 (4).

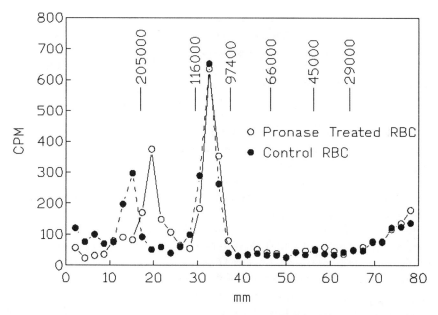

**Figure 1.** The M$_r$ of the CuP cross-linked species was sensitive to pronase treatment of the extracellular surface of intact red blood cells. Procedures as described in (2).

We also determined (4) that α-chain-Band 3 heterodimer formation was dependent upon the conformational state of the α-chain with cross-linking being favored by the E$_2$ state of the ATPase similar to that observed for α-α homodimer formation in purified kidney enzyme preparations (1,5). Phosphorylation of red cell (Na$^+$,K$^+$)-ATPase at 0°C by ATP results in predominantly E$_1$P formation (2). Extended co-incubation (60 sec, 0°C) of $^{32}$P-ATP, CuP and red cell IOV obtained from pronase treated and control red cells resulted in the formation of a small amount of cross-linked product, consistent with the small amount of E$_2$P formed under these conditions (Figure 2).

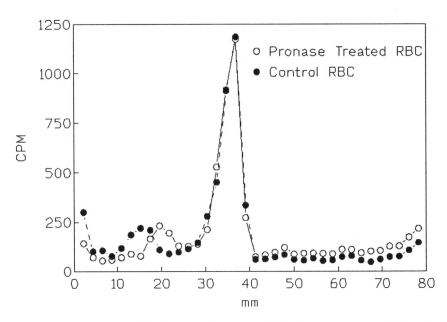

**Figure 2.** Phosphorylation by ATP produced similar pronase sensitive CuP cross-linked species as with phosphorylation by $P_i$ (Fig.1). Procedures as in (2) and described in text.

Although the amount of cross-linked product was substantially less than that observed when $E_2P$ was formed directly at room temperature with $^{32}P_i$ (Fig. 1), the cross-linked species showed an identical pronase induced shift in the apparent $M_r$. The conformational dependence of the cross-linking reaction was further demonstrated by conducting the phosphorylation and cross-linking reaction in sequential stages. When IOV were first phosphorylated with $^{32}P$-ATP (5 s), followed by addition of strophanthidin (5 s) which favors the $E_2P$ state, followed by CuP (5 s), a substantial increase in the amount of labeled cross-linked species was produced resulting in quanities of label in the heterodimer bands comparable to that seen in Fig. 1 (see figure 7 in (4)).

There are two ways to account for these data:

1) $Na^+,K^+$ pumps of red cell membranes are organized as diprotomers. In that case, one must believe that Cu and $o$-phenanthroline, which cross-link $\alpha$-chains in a conformationally dependent manner in preparations in which it is known that $Na^+,K^+$ pumps exist as $(\alpha\beta)_2$ dimers (1,5), do not form $\alpha$-$\alpha$ cross-links from similar dimers in red cell membranes when the pumps are in a comparable conformational state. Nevertheless, $\alpha$-chains are cross-linked to band 3 molecules in a conformationally dependent manner. One is

462

then forced to the absurd conclusion that the casual association of $\alpha$-chains with band 3 molecules is closer than the association of $\alpha$-chains in a pre-existing $(\alpha\beta)_2$ diprotomer.

2) Cu and $o$-phenanthroline cross-link $\alpha$-chains in purified membrane preparations in which $(\alpha\beta)_2$ diprotomers are known to exist. Cu and $o$-phenanthroline do not cross-link $\alpha$-chains in red cell membrane even though they do cross-link $\alpha$-chains to band 3 molecules. The conclusion is that $Na^+,K^+$ pumps in red cell membranes must exist as monomers, not diprotomers.

We believe the second explanation is by far the most credible.

*Acknowledgements*

This work was supported by United States Public Health Service Grant DK-19185.

**References**

1.  Askari A, Huang W-H, Antieau, JM (1980) $Na^+,K^+$-ATPase: ligand-induced conformational transitions and alterations in subunit interactions evidenced by cross-linked studies. Biochemistry 19:1132-1140
2.  Kaplan JH, Kenney LJ (1985) Temperature effects on sodium pump phosphoenzyme distribution in human red cells. J Gen Physiol 85:123-136
3.  Knauf PA, Proverbio F, Hoffman JF (1974) Chemical characterization and pronase susceptibility of Na:K pump-associated phospho-protein of human red blood cells. J Gen Physiol 63:305-323
4.  Martin DW, Sachs JR (1992) Cross-linking of the erythrocyte $(Na^+,K^+)$-ATPase: Chemical cross-linkers induce $\alpha$-subunit-Band 3 Heterodimers and do not induce $\alpha$-subunit homodimers. J Biol Chem 267:23922-23929
5.  Periyasamy SM, Huang W-H, and Askari A (1983) Subunit associations of $(Na^++K^+)$-dependent adenosine triphosphatase: Chemical cross-linking studies. J Biol Chem 258:9878-9885

463

# Functional Diversity of Tissue-Specific $Na^+/K^+$-Pumps Delivered from Exogenous Sources into Erythrocytes

J. S. Munzer and R. Blostein

Montreal General Hospital Research Institute, 1650 Cedar Avenue, Montreal, Quebec, Canada H3G 1A4

## Introduction

In recent years considerable insight into the structure and tissue-specific expression of $Na^+,K^+$-ATPase isoforms has been obtained. However, the functional basis for the distinct forms is largely unknown. This paper describes experiments aimed to shed light on the tissue-specific kinetic behavior of the pump, in particular the activation by intracellular $Na^+$ and extracellular $K^+$. We have examined two issues: one deals with the question of whether pumps of the same isoform ($\alpha_1\beta_1$), but derived from different tissues or species, behave similarly; the other concerns the behavior of pumps from tissues of distinct isoform composition.

To address these questions, we developed a system for delivering $Na^+/K^+$-pumps of exogenous (microsomal) membrane sources into the structurally and metabolically simple mammalian erythrocyte using polyethylene glycol-mediated fusion (8). As discussed in that report, delivery of functional pumps was apparent in experiments showing that pumps of dog kidney microsomes fused into dog erythrocytes are fueled by intracellular ATP, and ATP added to inside-out vesicles prepared from the fused cells stimulates strophanthidin-sensitive $Na^+$ influx.

### Rat kidney pumps delivered into human erythrocytes: comparative behavior of rat kidney and human erythrocyte pumps.

Previous studies have indicated that the isoform composition of the erythrocyte $Na^+/K^+$-pump, like that of the kidney, is $\alpha_1\beta_1$ (2,4). To compare the kinetic behavior of these two kinds of pumps, we have delivered the rat kidney enzyme into human erythrocytes. The activity of the exogenous kidney pumps was distinguished from that of endogenous erythrocyte pumps using low (5 $\mu$M) and high (5 mM) ouabain concentrations, respectively, as described earlier (8). The two pumps were compared with respect to the effect of extracellular $K^+$ ($K_{ext}$) on ouabain-sensitive $K^+$ influx using the congener $^{86}Rb^+$ as tracer. As shown in Fig. 1, kidney and erythrocyte pumps have indistinguishable behavior; for both, the apparent affinity for $K_{ext}$ ($K_{K(ext)}$) was 0.16 mM. [Extracellular $Na^+$ was maintained at a constant low (5 mM) concentration].

The effect of intracellular $Na^+$ concentration ($Na_{in}$) on ouabain-sensitive $^{22}Na^+$ efflux was tested using a modification (7) of the nystatin permeabilization method (1) to alter $Na_{in}$ concentration ($[Na]_{in}$). These experiments were complicated by the fact that exogenous pumps are delivered into only a small fraction of the cells (8). As a result, the relatively high pump-mediated $Na^+$ efflux from the pump-rich cells decreases their $[Na]_{in}$ to concentrations lower than those of the whole cell population. Therefore, values for the

464

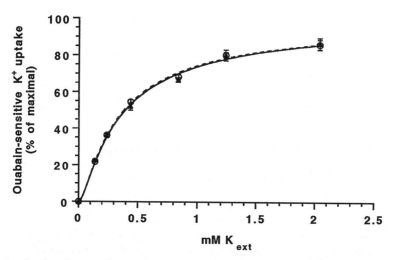

Figure 1. **Activation of endogenous human erythrocyte pumps and exogenous rat kidney pumps by $K_{ext}$.** Rat kidney microsomes were fused with human erythrocytes and assays were carried out using $^{86}Rb^+$ as a congener of $K^+$ as described previously (8). Endogenous human pump activity (open circles) was distinguished from exogenous rat pump activity (solid triangles) by the addition of 5 μM or 5 mM ouabain, respectively, during a 5-minute preincubation.

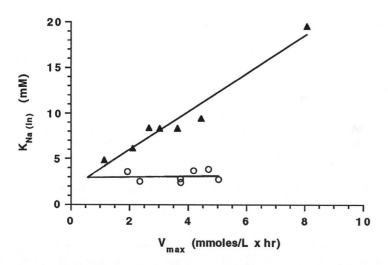

Figuire 2. **Activation of endogenous human erythrocyte pumps and exogenous rat kidney pumps by $Na_{in}$: relationships between $K_{Na(in)}$ and $V_{max}$.** The $Na_{in}$ concentration of the fused cells was varied using nystatin with constant (10 mM) KCl and varying choline chloride to maintain osmolarity, and then equilibrated with $^{22}NaCl$ essentially as described by Mairbaurl and Hoffman (7). Endogenous pumps (open circles) and exogenous pumps (solid triangles) were distinguised as described in Fig. 1. $^{22}Na^+$ efflux assays were carried out as described previously (8). Kinetic parameters were determined as described in Table 1.

465

apparent affinity for $Na_{in}$ ($K_{Na(in)}$) obtained from plots of pump rate versus (measured) $[Na]_{in}$ must be larger than the 'true' value. Discrepancies between observed and 'true' values should, however, diminish as $V_{max}$ decreases. Thus, in 7 independent experiments summarized in Fig. 2, values of $K_{Na(in)}$ of the exogenous kidney pumps decreased directly with $V_{max}$; $K_{Na(in)}$ for the erythrocyte pump was 2.8 mM, independent of $V_{max}$. At low $V_{max}$, the plot of $K_{Na(in)}$ values for the kidney pumps intersects that of the erythrocyte pumps. We conclude, therefore, that the $Na^+/K^+$-pumps of the rat kidney and the human erythrocyte have similar affinities for intracellular $Na^+$.

### Comparison of rat axolemma and rat kidney pumps.

In experiments aimed to examine $K_{ext}$ and $Na_{in}$ activation of pumps of tissues of distinct isoform composition, rat kidney and axolemma microsomes were fused into dog erythrocytes which are almost devoid of pumps. Since the levels of exogenous pump activity were relatively low ($V_{max} \leq 1$ mmoles/L/hr; c.f. activity of the rat kidney pump fused into human erythrocytes shown in Fig. 2), we assumed that deviations of $K_{Na(in)}$ from the 'true' $K_{Na(in)}$ were minimal. The results of these experiments indicate a 3-fold lower $K_{K(ext)}$ and a $\approx$3-fold higher $K_{Na(in)}$ for the axolemma compared to kidney pumps. Based on measurements of the proportion of total axolemma $Na^+/K^+$-ATPase which was relatively insensitive to ouabain, we concluded that the $\alpha_1$ isoform constitutes $\approx$10% of the total axolemma $Na^+/K^+$-ATPase. From comparisons of the immunological reactivity (quantitative densitometry of Western blots) of rat axolemma, kidney and muscle, the latter comprising $\alpha_1$ and $\alpha_2$ isoforms quantified from ouabain-sensitivity profiles of EP formation, we estimated that $\alpha_3$ comprises $\geq$70% of the remaining ouabain-sensitive activity in axolemma.

### Conclusion

In the present study, we show that pumps of the same isoform composition but derived from at least two distinct species (human versus rat) and tissues (erythrocyte versus kidney) have similar cation activation kinetics.

The kinetic differences between axolemma and kidney pumps probably reflect the distinct behavior of the $\alpha_1$ and $\alpha_3$ isoforms since $\alpha_3$ is the predominant isoform in axolemma. These flux experiments are consistent with earlier studies of the cation activation kinetics of the membrane $Na^+/K^+$-ATPase of isoform-specific transfected HeLa cells. In those experiments, a 3-fold lower $K_{Na}$ and a $\approx$1.5-fold increase in $K_K$ were observed (5). More recent studies concerned with the transport characteristics of these transfected cells have revealed similar differences, albeit of greater magnitude, i.e. a >5-fold lower $K_{Na(in)}$ and $\approx$3-fold higher $K_{K(ext)}$ for $\alpha_3$ compared to $\alpha_1$ or $\alpha_2$[1].

In contrast to the aforementioned differences between axolemma and kidney pumps, studies of $Na^+/K^+$-ATPase activities of rat kidney, brain and pineal gland (predominantly $\alpha_3\beta_2$ pumps) led Shyjan et al. (9) to conclude that $\alpha_3$ has a lower affinity for $Na^+$ than $\alpha_1$ pumps, and similar affinities for $K^+$. Whether these dichotomies reflect an effect of the (distinct) $\beta$ subunit, consistent with a kinetic effect of the $\beta$ subunit described recently (6), remains to be determined. Even though $\beta_2$ is present in rat axolemma and pineal

---

[1] Munzer JS, Daly SE, Jewell EA, Lingrel JB, Blostein R. Manuscript in preparation.

glands, we cannot rule out a difference in $\alpha$–$\beta$ pairing, for example; mainly $\alpha_3\beta_1$ in axolemma and $\alpha_3\beta_2$ in the pineal gland.

*Acknowledgements*

We thank Dr. John C. Parker, University of North Carolina, for helpful discussions. This study was supported by a Medical Research Council of Canada grant (MT-3876).

# References

1. Cass A, Dalmark M (1973) Equilibrium dialysis of ions in nystatin-treated red cells. Nat New Biol 244: 47-49
2. Dhir R, Nishioka Y, Blostein R (1990) Na,K-ATPase isoform expression in sheep red blood cell precursors. Biochim Biophys Acta 1026: 141-146
3. Garay RP, Garrahan PJ (1973) The interaction of sodium and potassium with the sodium pump in red cells. J Physiol 231: 297-325
4. Inaba M, Maeda Y (1986) Na,K-ATPase in dog red cells. Immunological identification and maturation-associated degradation by the proteolytic system. J Biol Chem 261: 16099-16105
5. Jewell EA and Lingrel JB (1991) Comparison of the substrate dependence properties of the rat Na,K-ATPase $\alpha_1$, $\alpha_2$, and $\alpha_3$ isoforms expressed in HeLa cells. J Biol Chem 266: 16925-16930
6. Lutsenko S, Kaplan JH (1993) An essential role for the extracellular domain of the Na,K-ATPase $\beta$-subunit in cation occlusion. Biochemistry 32: 6737-6743
7. Mairbaurl H, Hoffman JF (1992) Internal magnesium, 2,3-diphosphoglycerate, and the regulation of the steady-state volume of human red blood cells by the Na/K/2Cl cotransport system. J Gen Physiol 99: 721-746
8. Munzer JS, Silvius JR, Blostein R (1992) Delivery of ion pumps from exogenous membrane-rich sources into mammalian red blood cells. J Biol Chem 267: 5202-5210
9. Shyjan AW, Cena V, Klein DC, Levenson R (1990) Differential expression and enzymatic properties of the $Na^+,K^+$-ATPase $\alpha_3$ isozyme in rat pineal cells. Proc Natl Acad Sci USA 87: 1178-1182

# Two-sided bi-directional $Na^+/K^+$-ATPase-liposomes. Structure and function

M. Moosmayer, D. Lacotte, B. Volet, B.M. Anner

The Laboratory of Experimental Cell Therapeutics, Geneva University Medical School, CH-1211 Geneva 4, Switzerland

## Introduction

The minimal functional unit of the $Na^+/K^+$-ATPase or sodium pump system, which contains also a hormonal and pharmacological receptor, is composed of a 110 kD $\alpha$-subunit, a $\beta$-glycoprotein, membrane phospholipids and cholesterol; its primary function consists in mediating the transmembrane movements of Na and K ions with energy derived from ATP hydrolysis (13,14). It is well known that the link between the chemical reaction and the vectorial process is made in a first step by transfer of the terminal phosphate of the ATP molecule to an aspartyl residue located on the intracellular protein loops of the $\alpha$ subunit and that, secondly; the phosphorylation of the pump induces a conformational change priming the transmembrane Na/K-exchange. However, the molecular mechanism of the vectorial ion movement, in particular its allosteric modulation by the receptor for cardioactive steroids is still poorly understood.

### Two-sided bi-directional NKA-liposomes

To correlate the receptor functions of the $\alpha$-$\beta$ complex with vectorial ion movements and to investigate its control by the extracellular receptor, we concentrated our efforts on the controlled incorporation of purified functional Na,K-ATPase molecules into the artificial membranes of liposomes (5,6). On the basis of extensive numerical analyses (2,3,9,10) a model was predicted and experimentally verified: two-sided bidirectional NKA-liposomes, i.e., liposomes containing on the average 4 randomly oriented pumps per vesicle and an internal reservoir of 10,000 ATP molecules, 1000 Mg and 44,000 Na ions to drive the pump population in cell-like orientation upon activation by external K or Rb addition; the pump population in reversed orientation remains silent in this condition and can then be activated also by externally added ATP to extrude the previously accumulated Rb or K pool (7,8). Fig. 1 shows the ultrastructure of thin-sectioned ATP-filled NKA-liposomes prepared as described (7) except for a higher initial protein/lipid ratio (800 $\mu$g protein/100 $\mu$l) and the use of lamb instead of rabbit kidney as source for NKA. Statistical analysis showed that the preparation contained about 96% single-walled vesicles; they are of relatively homogenous 100 nm size if the random sectioning of the vesicles is taken into account, in agreement with previous determinations by freeze fracture (9,10) and laser light scattering (11).

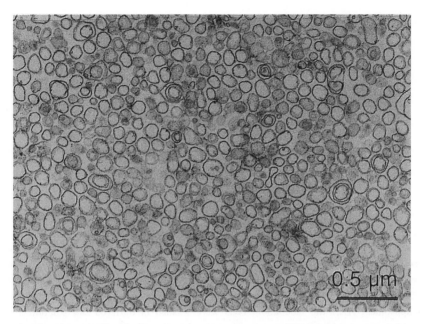

**Figure 1.** Electron micrographs of NKA-liposomes. For electron microscopy, liposome pellets were prefixed in 2% glutaraldehyde in cacodylate buffer, postfixed in 2% $OsO_4$-collidine, dehydrated in ethanol and embedded in epon. Ultrathin sections stained in uranyl acetate and lead citrate were examined in a Philips 400 electron microscope. Magnification: 32,500 x.

The stable and reproducible structure of the NKA-liposomes is the results of a self-assembling and self-organising process initiated by cholate-removal which leads finally to a stable 100 nm sphere with minimal energy. For these thermodynamic reasons it is no coincidence that the artificially formed NKA-vesicles are a perfect copy of naturally formed intracellular vesicles which must also be in a minimal energy state. In further analogy to the self-assembled NKA-liposomes, most natural vesicle contain ATPases in their membrane and are filled often with high concentrations of ATP; hence, it was tempting to verify experimentally whether the artificially formed NKA-vesicles could be introduced into cells as a model or substitute for natural vesicles as proposed (7). To monitor vesicle uptake by cells, NKA-liposomes were prepared as described previously (8) except for the addition of 200 mM of 5,6-carboxyfluorescein (CF, Fluka, Switzerland) to label the water phase during dialysis. After dialysis the liposomes were washed four times by 100,000 x g centrifugation in 1 mM EDTA, 30 mM L-histidine, 5 mM $MgCl_2$, 50 mM NaCl and 50 mM KCl, pH 7.4). Fluorescence-activated cell sorting (FACS) was chosen as a method to follow the internalisation of fluorescent liposomes by freshly isolated human lymphocytes (9); this technique has the advantage of taking into account the individual cell fluorescence of 10,000 cells from each sample for statistical calculation.

469

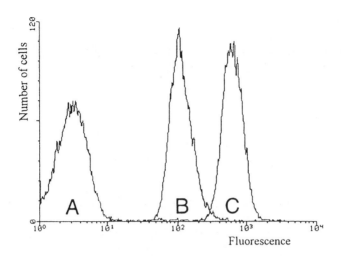

**Figure 2.** Effect of temperature on uptake of carboxyfluorescein-containing NKA-liposomes by isolated human lymphocytes; $1.5 \times 10^6$ human lymphocytes were incubated for 30 min in 60 $\mu$l of RPMI-1640 at 37°C (A, C) or at 4 °C (B) without (A) or with (B,C) 3 $\mu$l (about $6 \times 10^{10}$) NKA-liposomes. After the incubation, cells were washed in buffered salt solution and their fluorescence measured by fluorescence-activated cell sorting (Becton-Dickinson).

Fig. 2 shows that the cells become all fluorescent when incubated either at 4°C or at 37°C with fluorescent liposomes in contrast to control experiments showing background fluorescence of the cells (A) or low fluorescence when free CF was added at concentrations which could be theoretically released from the aqueous liposome phase (not shown). The cells were more fluorescent after an incubation at 37°C than at 4°C. If we take the mean value of cell fluorescence of the populations analysed, we can calculate that the fluorescence at 37°C is 5.3 times more elevated than at 4°C. These results show that liposome uptake can be followed by FACS analysis, that isolated human lymphocytes incubated with fluorescent liposomes become all fluorescent by a temperature-dependent process. Despite a possibly important uptake of NKA-liposomes, cell survival was not affected as measured by trypan blue exclusion and visualised by electron microscopy (data not shown).

In conclusion, we have developed a new model system, two-sided bi-directional NKA-liposomes, for resolution of structure-function relationships of NKA, in particular for investigating the distinct roles of the extracellular receptors for ouabain (1), isomeric ouabain (15) and palytoxin (12) on ion fluxes. The ultrastructure of the self-assembled vesicles was indistinguishable from naturally formed intracellular vesicles; hence, we speculated that NKA-liposomes might mimic natural vesicles (4); NKA-liposomes were indeed able to enter isolated human lymphocytes spontaneously without affecting cell survival.

470

*Acknowledgements*

We are grateful to Dr. Dominique Wohlwend for help with the fluorescent activated cell sorter. Supported by SNSF grant No. 31-25666.88 (BMA).

## References

1.  Anner BM (1985) The receptor function of the $Na^+,K^+$-activated adenosine triphosphatase system. Biochem. J. 227: 1-11
2.  Anner, BM (1985) Interaction of (Na+/K+)-ATPase with artificial membranes. I. Formation and structure of (Na + K)-ATPase-liposomes. Biochim. Biophys. Acta 822: 319-334
3.  Anner BM. (1985) Interaction of (Na+/K+)-ATPase with artificial membranes. II. Expression of partial transport reactions. Biochim. Biophys. Acta 822: 335-353
4.  Anner, BM (1987) Electrical sorting: the missing link between membrane potential and intracellular vesicle traffic? Perspect. Biol. Med. 30: 527-545
5.  Anner BM, Lane LK, Schwartz, A, Pitts BJR (1977) A reconstituted Na,K-pump in liposomes containing purified $(Na^+/K^+)$-ATPase from kidney medulla. Biochim. Biophys. Acta 467: 340-345
6.  Anner, BM. and Moosmayer, M. (1981) Preparation of Na,K-ATPase-containing liposomes with predictable transport properties by a procedure relating the $Na^+/K^+$-transport capacity to the ATPase activity. J. Biochem. Biophys. Methods 5: 299-306
7.  Anner BM. Moosmayer M (1985) Right-side-out pumping $Na^+/K^+$-ATPase-liposomes: a new tool to study the enzyme's receptor function. Biochem. Biophys. Res. Commun. 129: 102-108
8.  Anner, BM, Rey, HG, Moosmayer, M, Meszoely, I, Haupert, GT (1990) A hypothalamic sodium pump inhibitor characterized in two-sided liposomes containing Pure renal $(Na^+/K^+)$ATPase. Am. J. Physiol., 258: F144-F153
9.  Anner BM, Robertson JD, Ting-Beall, HP (1984) Characterization of $(Na^+/K^+)$-ATPase liposomes. II. Effect of alpha-subunit digestion on intramembrane particle formation and Na,K-transport. Biochim. Biophys. Acta 773: 262-270
10. Anner BM, Ting-Beall HP. Robertson JD (1984) Characterization of $(Na^+/K^+)$-ATPase liposomes. I. Effect of enzyme concentration and modification on liposome size, intramembrane particle formation and Na,K-transport. Biochim. Biophys. Acta 773: 253-261
11. Apell HJ, Marcus MM, Anner BM, Oetliker H, Läuger P. (1985) Optical study of active ion transport in lipid vesicles containing reconstituted Na,K-ATPase. J. Membrane Biol. 85: 48-65
12. Habermann E (1989) Palytoxin acts through $Na^+/K^+$-ATPase Toxicon 27: 1171-1187
13. Jørgensen PL (1974) Isolation of $Na^+/K^+$-ATPase. Methods Enzymol. 32: 277-290
14. Jørgensen PL (1982) Mechanism of the Na,K-pump. Protein structure and conformations of the pure $Na^+/K^+$-ATPase. Biochim. Biophys. Acta 694: 27-69
15. Tymiak AA, Norman JA, Bolgar M, DiDonato GC, Lee H, Parker WL, Lo L-C, Berova N, Nakanishi K, Haber E, Haupert GT (1993) Physicochemical characterization of a ouabain isomer isolated from bovine hypothalamus. Proc. Natl. Acad. Sci. 90: 8189-8193
16. Volet B, Anner BM (1993) Analysis of liposome-lymphocyte interaction by fluorescent technologies Experientia 49: A53

471

# Voltage Sensitivity of the $Na^+/K^+$ Pump: Structural Implications

Paul De Weer, R.F. Rakowski, David C. Gadsby

Department of Physiology, University of Pennsylvania School of Medicine, Philadelphia PA 19104; Department of Physiology and Biophysics, University of Health Sciences/ The Chicago Medical School, North Chicago IL 60064; and Laboratory of Cardiac and Membrane Physiology, Rockefeller University, New York NY 10021, USA.

## Introduction

It is now firmly established that the $Na^+/K^+$ pump in its normal forward mode exports three sodium ions and imports two potassium ions per cycle (13,22,25). The enzyme thus produces a transmembrane electric current as it hydrolyzes ATP and must, like any chemical battery, for thermodynamic reasons be sensitive to the voltage difference across the cell membrane. For a given $Na^+:K^+:ATP$ stoichiometry and known conditions (ion gradients; free energy of ATP hydrolysis) only the equilibrium (zero-current or "reversal") potential can be computed (6). In typical animal cells the reversal potential is expected to be quite negative ($\leq$ -200 mV), i.e. probably beyond experimental reach (4). The *shape* of the steady-state current-voltage relationship (I-V curve) away from the reversal potential is determined by the kinetic properties of the pump, in particular by the voltage sensitivity of individual steps in the pump's reaction cycle. Were every step and its voltage sensitivity known, the I-V curve could be deduced from first principles. While such detailed knowledge is lacking, several researchers have examined the predictions of more or less elaborate models. The most comprehensive of these (e.g. 17,18) incorporate possible voltage sensitivity of ion binding, ion translocation, and protein conformational steps. We review here the converse problem: *given* an experimental voltage dependence curve, what information can be extracted regarding the kinetics of various steps in the pump cycle? Several groups including ours have exploited the pump's voltage sensitivity to discover during *which* step(s) a charge is moved across an electric field, or a field is moved across a charge (17).

## Voltage sensitivity of electrogenic forward and backward pumping

Investigations on a variety of preparations including cardiac myocytes (2,10,11,21), *Xenopus* oocytes (16,26,27), renal $Na^+/K^+$ ATPase reconstituted into phospholipid vesicles (14,28), and squid giant axons (25) bathed in normal (i.e. high-Na) extracellular fluids, showed that forward pump turnover is accelerated as the cell interior is made more positive. Over most of the experimentally accessible voltage range, pump current is well described by a section of a sigmoid curve rising from a low value at negative potentials to an approximate plateau near zero transmembrane potential. Two additional features of some of the findings deserve emphasis. First, the familiar inhibitory effect of external [Na] on forward pumping is more pronounced at negative membrane potentials. That is, in the absence of external Na the pump's voltage dependence is much less steep than in its presence. Second, it appears that *at low external [K]* (27) the forward pump is also inhibited by sufficiently *positive* potentials, which suggests that there may be more than one voltage-sensitive step in the cycle (15).

By reversing the normal ionic gradients and exposing the cytoplasmic side of the membrane to ADP + inorganic phosphate, the $Na^+/K^+$ pump can be made to run backwards and produce an inward current. This current, too, is voltage sensitive. It is enhanced along a sigmoid curve by negative potentials in cardiac myocytes (1), squid giant axons (24), and *Xenopus* oocytes (8).

## Voltage sensitivity of electroneutral (ADP-requiring) Na/Na exchange

Early work in the laboratories of Karlish (28), Bamberg (9), and Läuger (3) strongly pointed to a major charge-translocating step in the Na leg of the transport cycle. In addition, Gadsby and coworkers were able to elicit, in response to a step change in transmembrane voltage, transient charge movements in cardiac myocytes whose pump cycle was restricted to Na translocation by removal of K (20) but not in those restricted to K translocation by removal of Na (1). Because of these earlier observations (7), we undertook a study (12) of the voltage dependence of electroneutral, ADP-requiring Na/Na exchange to gain insight into the nature of the step(s) which confer(s) voltage sensitivity to that part of the $Na^+/K^+$ pump cycle.

The internally dialyzed squid giant axon is an ideal preparation for accurate measurements of isotope fluxes and transmembrane current under close control of electrical, ionic, and metabolic conditions. We restricted the pump's operation to Na/Na exchange by excluding potassium from both sides of the membrane and including ATP + ADP in the dialyzate, which also contained 50 mM $Na^+$ (labeled with $^{22}Na$), sufficient to saturate internal Na-binding sites. Pump-mediated $^{22}Na$ efflux was defined as the component inhibitable by dihydrodigitoxigenin ($H_2DTG$), a rapidly reversible cardiotonic steroid. By measuring $H_2DTG$-sensitive transmembrane current, we verified that over 90% of the $^{22}Na$ efflux was, in fact, electroneutral and required external Na.

For the purpose of the kinetic analysis that follows, the Na/Na exchange sequence can, without loss of generality, be reduced to a two-step reaction:

$$nNa_i \underset{k_{-1}}{\overset{k_1}{\rightleftharpoons}} E\text{-}P\cdot Na_n \underset{k_{-2}}{\overset{k_2}{\rightleftharpoons}} E\text{-}P \quad nNa_o$$

$$E$$

where $n$ is the apparent molecularity of the ion binding and $k_1$ and $k_2$ are first-order, and $k_1$ and $k_{-2}$ pseudo-first-order, rate constants. Because the exchange involves *both* steps in *both* directions, it is clear that the reaction velocity (measured as $^{22}Na$ efflux) must vanish if *any one* of the four rate constants approaches zero. Less obvious is that the reaction velocity will also vanish if $k_{-1}$ and/or $k_2$ become very large, as the reaction intermediates would then become "trapped" at the left and/or right end of the sequence.

473

One can now examine the kinetic consequences of assigning voltage sensitivity to any or all rate constants. Since in the forward direction a positive charge is translocated from left to right (3,9,20) in the above reaction scheme, a negative membrane potential will decelerate $k_1$ and/or $k_2$ exponentially: $k_i = k_i^o \exp[(1 - \delta_i)z_i V_m F/RT]$, and accelerate $k_{-1}$ and/or $k_{-2}$ exponentially: $k_{-i} = k_{-i}^o \exp(-\delta_i z_i V_m F/RT)$ with a steepness determined by $z_i$ (an equivalent charge) and by $\delta_i$ ($0 \leq \delta_i \leq 1$), an asymmetry factor that describes the apportioning of a given step's voltage dependence between forward and reverse transition. These voltage effects are schematically depicted in Figure 1.

## All four rate constants V-sensitive

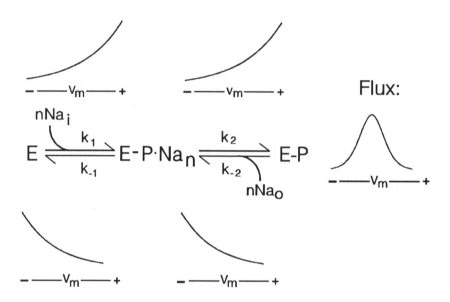

**Figure 1.** Effect of membrane potential on electroneutral exchange flux through the sodium pump, when all four rate constants of the reduced model are voltage-sensitive. Exponentials adjacent to each rate constant show the shape of their voltage sensitivity. The flux vs. voltage curve vanishes at extreme negative potentials because $k_1$ and $k_2$ vanish; the flux also vanishes at extreme positive potentials because $k_1$ and $k_2$ vanish. Further description in the text.

In the most general case (all four rate constants voltage-sensitive) the exchange reaction clearly must vanish at extreme positive or negative membrane potentials (hence pass through a maximum) because either $k_{-i}$ or $k_i$ become very small. Inspection of Fig. 1 shows, furthermore, that *any combination of at least two* voltage-sensitive rate constants will cause Na/Na exchange as a function of membrane voltage to display a maximum. What happens if a *single* rate constant (out of four) retains voltage sensitivity? When either $k_1$ or $k_2$ is the sole voltage-sensitive step, the exchange reaction rate vs. membrane potential curve is *still biphasic* because the rate constant either vanishes or traps the reaction sequence in one or the other corner at extreme voltages. Only the remaining two simple cases ($k_1$ or $k_2$ as the sole voltage-sensitive rate constant) display *monotonic* (saturating) flux vs. voltage curves. Figure 2 shows that voltage sensitivity *of $k_1$ only* predicts a curve that saturates at positive potentials, while voltage sensitivity *of $k_2$ only* predicts a curve that saturates at negative potentials (Fig. 3).

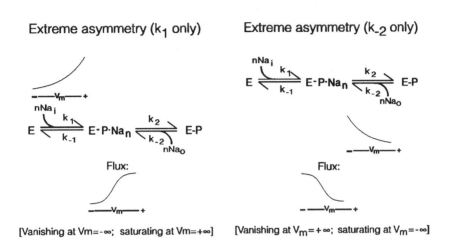

Extreme asymmetry ($k_1$ only)

Extreme asymmetry ($k_2$ only)

Flux:

Flux:

[Vanishing at $V_m = -\infty$; saturating at $V_m = +\infty$]    [Vanishing at $V_m = +\infty$; saturating at $V_m = -\infty$]

**Figure 2.** Effect of membrane potential on electroneutral Na/Na exchange flux in the limiting case where *only* $k_1$ is voltage sensitive. Flux vanishes at extreme negative potentials because $k_1$ vanishes, but *saturates* at extreme positive potentials because the remaining three, voltage-insensitive, rate constants become rate-limiting.

**Figure 3.** Effect of membrane potential on electroneutral Na/Na exchange flux in the limiting case where *only* $k_2$ is voltage sensitive. Flux vanishes at extreme positive potentials because $k_2$ vanishes, and *saturates* when the three voltage-insensitive rate constants become rate-limiting.

We examined the voltage dependence of Na/Na exchange against these theoretical predictions. Figure 4 shows that *electroneutral* Na/Na exchange is indeed voltage sensitive and that the experimental data conform best to a simple model in which $k_2$ is the *sole* voltage-sensitive rate constant while the others, *in particular $k_2$*, remain voltage-independent. In other words, it appears as if step 2 in the reaction sequence is voltage sensitive in a highly asymmetric way, i.e. $\delta_2 \simeq 1$. [A least-squares fit yielded $\delta_2 = 0.95 \pm 0.05$. We will return below to the possibility that the curve might have a shallow maximum.] If the voltage sensitivity of step 2 resides predominantly (perhaps exclusively) in its backward pseudo-first-order rate constant $k_2 = k_{-2}^o \exp(-z_i V_m F/RT)$ while forward $k_2$ remains unaffected, the inescapable conclusion follows that the voltage sensitivity must reflect a property of the Na concentration term contained in $k_2$ (Na$_o$ rebinding) but not in $k_2$ (Na release).

475

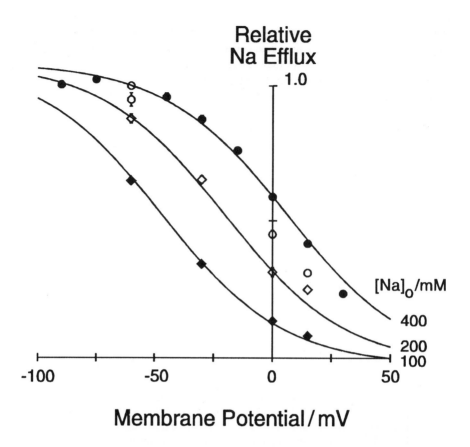

**Figure 4.** Electroneutral Na/Na exchange flux in squid giant axon sodium pump, as a function of membrane potential, normalized to that in 400 mM Na$_o$ seawater at -60 mV. Data are from ref. 12 plus additional points. Four external [Na] were used: 400, 300, 200, and 100 mM, but for clarity only three curves are shown. The least-squares flux ($\phi$) curves obey the expression $\phi = \phi_{max}/\{1 + [K_{0.5}^o \exp(\lambda V_m F/RT)]^n/[Na]_o^n\}$ with Hill coefficient $n = 1.8$, fractional channel depth $\lambda = 0.69$, and half-maximal activating concentration at 0 mV : $K_{0.5}^o = 0.34$ *M*. Parallel shift between curves is 26 mV.

**An external access channel model for Na/Na exchange**

Voltage dependence of an apparent Na$_o$ concentration is readily explained if the pump's reaction with external Na$^+$ occurs *within* the transmembrane electric field, that is, *at the bottom of a narrow, high-field, access channel*. As a result, the probability of an individual external Na ion being present at the bottom of the access channel (where it can interact with the external "intake" site of the pump) is multiplied by $\exp(-\lambda V_m F/RT)$, a Boltzmann term where $\lambda$ is the fractional depth of the channel, in such a way that negative $V_m$ values enhance the effectiveness of a given [Na]$_o$ at stimulating Na/Na exchange, and positive $V_m$ values diminish it.

476

A sketch of the "access channel" model of the pump is shown in Fig. 5, which makes explicit the absence of charge translocation in the steps that take place across the remaining fraction, 1-λ, of the transmembrane field. A straightforward prediction of this model is that *external ion concentration* and *membrane potential* should be *kinetically equivalent*, i.e. the effect, on Na/Na exchange velocity, of a change in concentration can be overcome by a change in potential, and *vice versa*. The data of Fig. 4 verify this prediction: successive twofold reductions in $Na_o$ were equivalent, within experimental error, to successive 26-mV leftward shifts of the sigmoid flux vs. voltage curve, which reflects a fractional access channel depth of 0.69. Stated another way: each 26 -mV hyperpolarization causes a doubling of the apparent affinity of the pump for external Na.

# Model for $Na_o$/$Na_i$ exchange

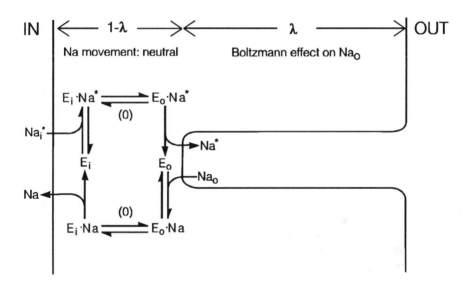

**Figure 5.** Sketch of high-field access channel model for electroneutral Na/Na exchange by the sodium pump. Asterisks mark the "forward" path followed by internal [22]Na. Channel depth (as a fraction of the transmembrane field) is λ. In this first-approximation model, no voltage-sensitive steps are assumed in the remainder, 1-λ, of the field. Through a Boltzmann effect, negative membrane potentials will enhance the probability of (unlabeled) $Na_o$ ions being present at the bottom of the channel, and thus accelerate the exchange rate at all but saturating [Na]$_o$ levels. They will also in theory retard the escape of transported [22]Na to the external solution; but since the escape rate is still expected to be orders of magnitude faster than the pump cycle kinetics, overall turnover should not be affected by this mechanism.

**Discussion**

The highly asymmetric voltage sensitivity of the forward and reverse rate constants in isotopic Na/Na exchange, described here, echoes the asymmetric kinetics of voltage jump-induced transient charge movements seen in cardiac myocyte (20) and *Xenopus* oocyte (23) pumps restricted to Na translocation. An access channel model (19,30) readily predicts such asymmetry. Within the framework of the "pure" access channel model of Fig. 5 (i.e. with no voltage-sensitive steps in the $1-\lambda$ part of the field), voltage-jump experiments such as first carried out by Nakao and Gadsby (20) are kinetically equivalent to *concentration*-jump experiments.

As stated earlier, the forward-running pump is inhibited along a sigmoid curve as membrane potential is hyperpolarized (i.e. the cell interior made more negative), but this inhibition is very much attenuated in the absence of external Na. Here again, the model of Fig. 5 provides a more straightforward interpretation than previously (5,21) entertained: if external Na inhibits forward pumping, then negative membrane potentials, through a Boltzmann enhancement of the effective [Na] at the bottom of the external access channel, should potentiate this inhibition.

Another verifiable prediction of the access channel model is that external Na, the "substrate" for backward pumping, should accelerate that mode in a voltage-dependent manner, i.e. more strongly at negative potentials. This is indeed observed, as is the expected lateral shift of the current-voltage curve when $[Na]_o$ is reduced or increased.

A corollary of the simple model of Fig. 5, where no charge is translocated in the Na leg over the $1-\lambda$ fraction of the field, is that a $-1$ elementary charge must be carried over $1-\lambda$ of the field in the K leg (if we assume $K_o$ uptake to occur at the bottom of the same or similar-depth access channel). This corollary appears difficult to reconcile with the lack of evidence for charge translocation in the potassium leg of the pump's transport cycle (1,14,28,29). The question arises whether the charge translocated over $1-\lambda$ of the field in the sodium leg is truly zero (as in Fig. 5) or rather a (small) positive charge, which would account for the slight droop seen in the leftmost data points of Fig. 4. In fact, a somewhat better fit to our data was obtained (unpublished results) with a more elaborate model which allows $\sim +0.5$ elementary charge to be carried over $1-\lambda$ in the sodium leg and, consequently, a $\sim -0.5$ elementary charge over $1-\lambda$ in the K leg. The predicted voltage dependence of these charge translocations is so shallow that it could easily have escaped earlier attention, especially if obscured by Boltzmann effects in access channels.

Finally, if the sodium pump delivers transported Na ions to the bottom of an external channel, does it also take up external K ions from the same or a similar channel? Such an economical model is very attractive because it provides an explanation (2,27) for the observation that, at low $[K^+]_o$, where the pump's $K^+$ binding sites are not saturated, membrane potential changes in the negative direction increase pump turnover rate -- presumably through a Boltzmann enhancement of $K^+$ binding probability -- whereas changes towards positive potentials slow down pump turnover by lowering the presence of $K^+$ at the bottom of the access channel.

# Conclusion

Kinetic analysis of the voltage dependence of electroneutral (ADP-requiring) Na/Na exchange through the sodium pump of squid giant axon, at saturating internal $[Na^+]$ levels, shows that a very limited range of models will accommodate the data. The voltage sensitivity of the reaction under these conditions resides exclusively (or nearly so) in the *external* $Na^+$ *rebinding* step, the physical substrate for which must be a narrow access channel at the external face of the membrane, that traverses at least half of the transmembrane electric field. Such an access channel model is compatible with several established properties of the pump.

*Acknowledgments*

Our research was supported by US NIH grants NS11223, NS22979, and HL36783, and the Irma T. Hirschl Trust (to DCG).

# References

1.      Bahinski A, Nakao M, Gadsby DC (1988) Potassium translocation by the $Na^+/K^+$ pump is voltage insensitive. Proc Natl Acad Sci USA 85:3412-3416

2.      Bielen FV, Glitsch HG, Verdonck F (1993) $Na^+$ pump current-voltage relationships of rabbit cardiac Purkinje cells in $Na^+$-free solution. J Physiol 465:699-714

3.      Borlinghaus R, Apell H-J, Läuger P (1987) Fast charge translocations associated with partial reactions of the Na,K-pump. I. Current and voltage transients after photochemical release of ATP. J Membr Biol 97:161-178

4.      De Weer P (1984) Electrogenic pumps: theoretical and practical considerations. In: Blaustein MP, Lieberman M (eds) Electrogenic Transport: Fundamental Principles and Physiological Implications. Raven, New York, pp 1-15

5.      De Weer P (1990) The Na/K pump: a current-generating enzyme. In: Reuss L, Szabo G, Russell JM (eds) Regulation of Potassium Transport Across Biological Membranes. University of Texas Press, Austin, pp 5-28

6.      De Weer P, Gadsby DC, Rakowski RF (1988) Voltage dependence of the Na-K pump. Ann Rev Physiol 50:221-241

7.      De Weer P, Gadsby DC, Rakowski RF (1988) Stoichiometry and voltage dependence of the Na/K pump. In: Skou JC, Nørby JC, Maunsbach AB, Esmann M (eds) The $Na^+,K^+$-Pump, Part A: Molecular Aspects. Alan Liss, New York, pp 421-434

8.      Efthymiadis A, Schwarz W (1991) Conditions for a backward-running $Na^+/K^+$ pump in *Xenopus* oocytes. Biochim Biophys Acta 1068:73-76

9.  Fendler K, Grell E, Haubs M, Bamberg E (1985) Pump currents generated by purified Na,K-ATPase from kidney on black lipid membranes. EMBO J 4:3079-3085

10. Gadsby DC, Kimura J, Noma A (1985) Voltage dependence of Na/K pump current in isolated heart cells. Nature 315:63-65

11. Gadsby DC, Nakao M (1989) Steady-state current-voltage relationship of the Na/K pump in guinea pig ventricular myocytes. J Gen Physiol 94:511-537

12. Gadsby DC, Rakowski RF, De Weer P (1993) Extracellular access to the Na,K pump: pathway similar to ion channel. Science 260:100-103

13. Glynn IM (1984) The electrogenic sodium pump. In: Blaustein MP, Lieberman M (eds) Electrogenic Transport: Fundamental Principles and Physiological Implications. Raven, New York, pp 33-48

14. Goldshlegger R, Karlish SJD, Rephaeli A, Stein WD (1987) The effect of membrane potential on the mammalian sodium-potassium pump reconstituted into phospholipid vesicles. J. Physiol 387:331-355

15. Hansen U-P, Gradmann D, Sanders D, Slayman CL (1981) Interpretation of current-voltage relationships for "active" ion transport systems: I. Steady-state reaction kinetic analysis of class-I mechanisms. J Membr Biol 63:165-190

16. Lafaire AV, Schwarz W (1986) The voltage dependence of the rheogenic $Na^+/K^+$ ATPase in the membrane of oocytes of *Xenopus laevis*. J Membr Biol 91:43-51

17. Läuger P (1991) Electrogenic Ion Pumps. Sinauer, Sunderland, 313 pp

18. Läuger P, Apell H-J (1986) A microscopic model for the current-voltage behaviour of the Na,K-pump. Eur Biophys J 13:309-321

19. Läuger P, Apell H-J (1988) Transient behaviour of the $Na^+/K^+$-pump: microscopic analysis of nonstationary ion-translocation. Biochim Biophys Acta 944:451-464

20. Nakao M, Gadsby DC (1986) Voltage dependence of the Na translocation by the Na/K pump. Nature 323:628-930

21. Nakao M, Gadsby DC (1989) [Na] and [K] dependence of the Na/K pump current-voltage relationship in guinea pig ventricular myocytes. J Gen Physiol 94:539-565

22. Post RL, Jolly PC (1957) The linkage of sodium, potassium, and ammonium active transport across the human erythrocyte membrane. Biochim Biophys Acta 25:118-128

23. Rakowski RF (1993) Charge movement by the Na/K pump in *Xenopus* oocytes. J Gen Physiol 101:117-144

24. Rakowski RF, De Weer P, Gadsby DC (1988) Current-voltage relationship of the backward-running Na/K pump in voltage-clamped internally-dialyzed squid giant axons. Biophys J 53:233a

25.     Rakowski RF, Gadsby DC, De Weer P (1989) Stoichiometry and voltage dependence of the sodium pump of squid giant axon. J Gen Physiol 93:903-941

26.     Rakowski RF, Paxson CL (1988) Voltage dependence of Na/K pump current in *Xenopus* oocytes. J Membr Biol 106:173-182

27.     Rakowski RF, Vasilets LA, LaTona J, Schwarz W (1991) A negative slope in the current-voltage relationship of the $Na^+/K^+$ pump in *Xenopus* oocytes produced by reduction of external $[K^+]$. J Membr Biol 121:177-187

28.     Rephaeli A, Richards DE, Karlish SJD (1986) Electrical potential accelerates the $E_1P(Na) \rightarrow E_2P$ conformational transition of (Na,K)-ATPase in reconstituted vesicles. J Biol Chem 261:12437-12440

29.     Stürmer W, Apell H-J, Wuddel I, Läuger P (1989) Conformational transitions and charge translocation by the Na,K pump: Comparison of optical and electrical transients elicited by ATP-concentration jumps. J Membr Biol 110:67-86

30.     Stürmer W, Bühler R, Apell H-J, Läuger P (1991) Charge translocation by the Na,K-pump: II. Ion binding and release at the extracellular face. J Membr Biol 121:163-176

# Electrogenic Properties of the Endogenous and of Modified *Torpedo* Na$^+$/K$^+$-Pumps in *Xenopus* Oocytes: The Access Channel for External Cations

W.Schwarz, L.A.Vasilets[*]), H.Omay, A.Efthymiadis, J.Rettinger, S.Elsner

Max-Planck-Institut für Biophysik, D-60596 Frankfurt/M, Germany
[*]) Permanent Address: Institute of Chemical Physics in Chernogolovka, Russian Academy of Sciences, Chernogolovka, Moscow region 142432, Russia

## Introduction

During the last years we have examined in voltage-clamp experiments the influence of extracellular Na$^+$ and K$^+$ on the current generated by the Na$^+$/K$^+$-pump. As a model system we used the *Xenopus* oocytes, and we analysed in these cells the endogenous *Xenopus* pump as well as expressed wild-type and mutated pumps of *Torpedo* electroplax by means of two-microelectrode and giant-patch voltage clamp. The results allowed us to draw conclusion about possible structure-function relationships of the Na$^+$/K$^+$-ATPase.

Variations in the voltage dependence of pump current were described on the basis of a "high-field" access channel (19) that has to be passed by the external cations to reach their binding sites within the transport protein. For the description of the transport of Na$^+$ out of the cell and of K$^+$ into the cell we used the Albers-Post reaction cycle (1,21). Since we investigated only variations in effects of external cations

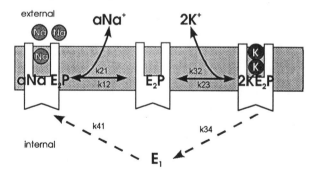

**Figure 1.** Reduced Albers-Post reaction scheme illustrating external cation interaction with the pump molecule in its E$_2$P form through a narrow access channel. The states following K$^+$ binding and preceding Na$^+$ release are lumped together in E$_1$. The external cation interactions can be described by voltage-dependent binding rates

$$k = k^*_0 [\text{cation}]^n e^{-zVF/RT}$$

and voltage-independent unbinding (9,25). n represents a Hill coefficient, and z an effective valence of the charge moved during the corresponding binding step.

and their consequences for pump stimulation and inhibition, we will restrict ourselves
to the reduced diagram illustrated in Fig. 1.

## Voltage dependence of the forward-running pump

It is now generally accepted that the reaction cycle of the pump can be modulated by at
least two voltage-dependent steps (see contribution by DeWeer). These steps are represented by extracellular binding of $K^+$ and $Na^+$ as illustrated in Fig. 1. For *Xenopus*
oocytes (28,29), this interpretation is based on the finding that the voltage dependence
of pump activity changes dramatically if the extracellular cation composition is altered
(Fig. 2). In $Na^+$-free and high-$K^+$ containing solutions, pump activity shows

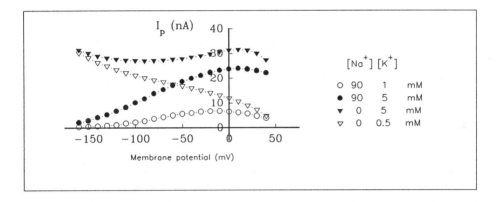

**Figure 2.** Current-voltage dependencies of the endogenous *Xenopus* pump current
under conditions of different external cation compositions as indicated (Data taken from
(26)).

no significant voltage dependence over a wide range of potentials. Reduction of external
$[K^+]$ to lower concentrations (<2 mM) results in a clear negative slope in the voltage
dependence, which can be explained by a voltage-dependent stimulation of the pump
due to voltage-dependent $K^+$ binding. On the other hand, adding external $Na^+$ results
in a positive slope, which can be explained by a voltage-dependent inhibition of the
pump due to voltage-dependent $Na^+$ rebinding. At physiological $[K^+]$ and $[Na^+]$
stimulation by $K^+$ becomes rate determining at positive potentials, and inhibition by
$Na^+$ at negative potentials leading to a maximum in the current-voltage dependence at
about 0 mV (16). Since charge translocations in steps following $K^+$ occlusion or
preceding $Na^+$ occlusion have been excluded (see contribution by DeWeer), a
straightforward explanation for these dependencies is the above mentioned voltage-
dependent access of the cations to their occlusion sites through a narrow high-field
access channel (17). We have performed detailed analysis of the dependencies of pump
current on membrane potential, and on external $[Na^+]$ and $[K^+]$ by experiments of the
type shown in Fig. 3A and B (25-27). If the binding of $K^+$ and $Na^+$ are the dominating

voltage-dependent steps in the reaction cycle the potential and concentration dependencies can be fitted on the basis of the reduced diagram of Fig. 1 (see Fig. 3) by (25):

$$I = \frac{k_{12}k_{23}^{*}e^{-z_{K}VF/RT}k_{34}k_{41}}{(k_{12}+k_{21}^{*}e^{-z_{Na}VF/RT})(k_{32}+k_{34})k_{41}+k_{23}^{*}e^{-z_{K}VF/RT}(k_{41}k_{34}+k_{41}k_{12}+k_{34}k_{12})} \tag{0}$$

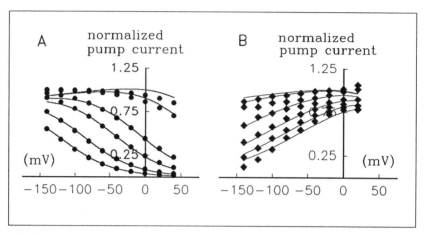

**Figure 3.** Voltage dependence of normalised and averaged endogenous pump current in *Xenopus* oocytes for **(A)** different $K^+$ concentrations (5, 2.5, 0.5, 0.25, 0.1, 0.05 mM) in absence of external $Na^+$, and **(B)** different $Na^+$ concentrations (100, 75, 50, 25, 5 ,1 mM) in presence of 5 mM $K^+$ externally. The solid lines represent a simultaneous fit of equ. (0) to the data. The fitted parameters for the $K^+$ and $Na^+$ interactions are for $K^+$: $k_{23}^{*}$ = 106 $s^{-1}mM^{-n}$ (n=1.3), $k_{32}$ = 34 $s^{-1}$, $z_K$ = 0.66; for $Na^+$: $k_{12}$ = 169 $s^{-1}$, $k_{21}^{*}$ = 2 $s^{-1}mM^{-n}$ (n=1.3). $z_{Na}$ = 1.1.

The fitted rate coefficients (see legend to Fig. 3) are compatible with those obtained in measurements of enzyme activity (see e.g. (12,18)), and the values for $z_K$ and $z_{Na}$ are nearly identical to those obtained form separate fits (see equ. (2) and (3) below) of the potential dependencies of $K_m$ for pump stimulation by external $K^+$ and $K_I$ for pump inhibition by external $Na^+$, respectively (compare Table). In this model the current can be written in the form (25):

$$I(V,[K^+],[Na^+]) = I_{max}\frac{[K^+]^m}{K_m^m+[K^+]^m}\frac{K_{1/2}^n}{K_{1/2}^n+[Na^+]^n} \tag{1}$$

with voltage-dependent apparent $K_m$ values for pump stimulation by $K^+$ and $K_{1/2}$ values for pump inhibition by $Na^+$. m and n represent corresponding Hill coefficients. The

$K_m$ value for pump stimulation can be determined in the absence of external $Na^+$ and has in this description an exponential potential dependence (23,27):

$$K_m = K_m(0mV)\, e^{\frac{z_K VF}{mRT}} \qquad .$$

(2)

$z_K$ represents an effective valence for charge that is moved in the electrical field during steps associated with the $K^+$ binding. In terms of the access channel it is a measure for the apparent dielectric length of the channel. The voltage dependence of the apparent $K_m$ were determined for the endogenous *Xenopus* pump and for the *Torpedo* pump expressed in the oocytes (Fig. 4A). In contrast to the voltage dependence of the $K_m$ value for the *Xenopus* pump, the voltage dependence for the *Torpedo* pump has to be described by the sum of two exponentials. This has been interpreted by a sequential binding of the two $K^+$ ions (27), the component with the higher effective valence representing the binding of the first $K^+$ ion, the component with the lower effective valence the binding of the second ion that senses less of the electrical field. In terms of the access channel, this could mean that the *Xenopus* pump has a less pronounced high-field access channel than the *Torpedo* pump with a dielectric geometry where the $K^+$ ion have to move in single file to reach their binding sites.

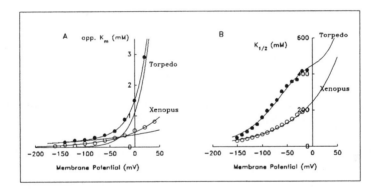

**Figure 4.** Voltage dependence of apparent $K_m$ for pump stimulation by external $K^+$ **(A)** and of $K_{1/2}$ values for pump inhibition by external $Na^+$ in presence of 5 mM $[K^+]$ **(B)**. Open circles represent data for the endogenous *Xenopus* pump, filled circles for the pump of *Torpedo* electroplax expressed in the oocytes. Lines in (A) represent fits of equ. (2) (note that for the *Torpedo* pump the sum of two exponentials is necessary, see text), in (B) fits according to equ. (3). The fitted parameters are listed in the Table. Data are taken from (25,27)

The $K_{1/2}$ value for pump inhibition by $Na^+$ shows a more complex dependency (see Fig. 4B) demonstrating voltage-dependent competition between $Na^+$ and $K^+$ in their interaction with the $E_2P$ form (comp. Fig. 1). Assuming this competition, a voltage-dependent apparent $K_I$ can be calculated from the $K_{1/2}$ value. It also described by an exponential corresponding to equ. (2) and is related to $K_{1/2}$ by (see (25)):

$$K_I = K_I(0mV)\, e^{\frac{z_{Na}VF}{nRT}} = \frac{K_{1/2}}{\left(1 + \frac{\left[K^+\right]^m}{K_m^m}\right)^{1/n}} \tag{3}$$

with $K_{1/2}$ determined from

$$I = I_{(Na=0mM)} \frac{K_{1/2}^n}{K_{1/2}^n + \left[Na^+\right]^n} .$$

The nominator in equ. (3) illustrates the dependence of $K_I$ on $[K^+]$ and shows that for the $K_{1/2}$ value additional voltage dependence is introduced by the voltage dependence of $K_m$. This contribution is particularly apparent for the *Torpedo* pump with the more prominent voltage dependence of $K_m$ (Fig. 4).

### Voltage dependence of inward-directed pump current

The $Na^+/K^+$ pump can be forced to run in a reversed mode transporting 3 $Na^+$ ions into the cell and 2 $K^+$ ions out of the cell under synthesis of 1 ATP and thus generating an inward-directed current. This current has been detected in several preparations (2,22) including the oocytes (8) under conditions of elevated outward-directed gradient for $[K^+]$ and inward-directed gradient for $[Na^+]$ and of drastically reduced cytoplasmic [ATP]/[ADP] ratio. The voltage dependence of the inward pump current (Fig. 5A) demonstrates pump stimulation with more negative potentials, which is in line with the idea that, as for the forward-running pump, external $Na^+$ binding into an access channel is voltage-dependent and rate determining. The observation of a negative slope cannot be expected since there is no external $K^+$ present that could rebind at extreme negative potentials and inhibit backward pumping.

In addition to the current generated by the backward-running pump a small inward-directed and ouabain-sensitive current can also be detected if not only external $K^+$ but also external $Na^+$ is absent (23). The ouabain sensitivity demonstrates that the current is generated by the $Na^+,K^+$-pump (see Fig. 5B). Moreover, this current can be activated only if internal ATP is present. The mode of transport has not yet been determined with certainty. The pump-mediated current is nearly absent when $Mg^{2+}$ instead of $Ba^{2+}$ is the only external divalent cation (comp. filled triangles and squares), it is stimulated by external $H^+$ (comp. filled and open squares) and low $[Na^+]$ (<10 mM) (not shown), and does not depend on whether $Na^+$ or $K^+$ is present internally (comp. filled circles and open triangles). It has been suggested (7) that in the low-$Na^+$ and $K^+$-free solution the

Na$^+$,K$^+$-pump can either form a conductive pathway that is permeable to Ba$^{2+}$ or protons or can operate in its conventional transport mode accepting Ba$^{2+}$ as K$^+$ congener.

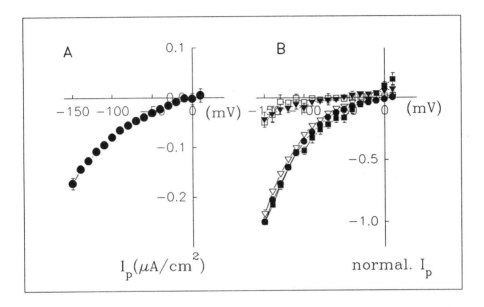

**Figure 5: (A)** Voltage dependence of endogenous *Xenopus* pump current generated by the reversed operating pump in K$^+$-free solution. Determined as strophantidin-sensitive current by two-electrode voltage clamp in oocytes with reduced cytoplasmic [Na$^+$] and [ATP]/[ADP] ratio (data taken from (8)). **(B)** Voltage dependence of *Torpedo* pump current determined in giant patches in absence of extracellular [Na$^+$] and [K$^+$] (for methods see contribution by Rettinger et al.). The currents were normalised to the value at -150 mV; the averaged absolute value roughly equals the maximum outward current of the forward running pump. At pH=6.5 and 5 mM Ba$^{2+}$: filled circles (ATP-activated current in presence of Na$^+$), open triangles (ATP-activated current in presence of K$^+$); filled squares (ouabain-sensitive current). Ouabain-sensitive current at pH=7.8 (open squares). Ouabain-sensitive current at pH=6.5 but 5 mM Mg$^{2+}$ instead of Ba$^{2+}$ (filled triangles).

### Effects of mutation on cation interactions

The data on the voltage dependencies of K$_m$ for pump stimulation by external K$^+$ and of K$_{1/2}$ or K$_I$ for pump inhibition by external Na$^+$ demonstrate clear species differences between the *Xenopus* and *Torpedo* pump (Fig. 4 and Table). Structurally, the α-subunits of different isoforms and from different species show a high degree of homology in their amino acid sequence; major differences are predominantly found in the N-terminal part (see (28,29). In particular, the cytoplasmic N-terminus carries a large number of charged residues and the number of charges may vary considerably among different α-subunits. In addition, differences among isoforms of the same

487

species have been classified with respect to their ouabain sensitivity. Therefore, we addressed the question whether truncations within a cytoplasmic lysine-rich cluster (see Fig. 6) and mutation that leads to ouabain insensitivity can affect external cation interaction. The *Xenopus* oocytes are particularly suited for this purpose since they allow to compare electrogenic properties of wild-type and mutated pumps in the same microenvironment.

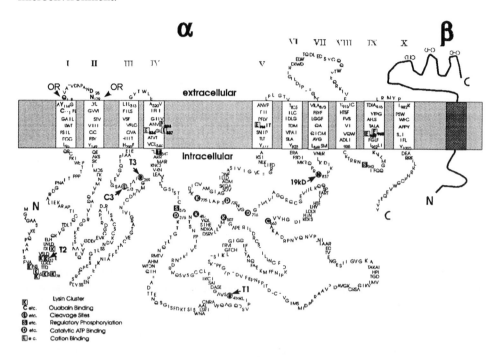

**Figure 6.** Amino acid sequence of the α-subunit of the Na⁺,K⁺-ATPase of *Torpedo* electroplax (15) and possible orientation of the pump in the membrane. Amino acids that have been identified to be functionally relevant are indicated, as well as trypsin (T1-T3, 19kD) and chymotrypsin (C3) cleavage sites (13,14). Mutation of the amino acids labelled with OR to charged one leads to ouabain resistance. Based on Fig. 3 from (28).

To approach this question, we performed experiments with mutants of the *Torpedo* pump that were truncated at different positions within the K-cluster (25,26). Fig. 7A shows that truncation after Lys-28 (ΔT29) leads to a pronounced increase of the effective valencies (see Table) and to an increase of the apparent affinity for K⁺ over a wide range of potentials. In terms of the access channel this would mean an increase of the dielectric length as if the binding sites become less accessible. On the other hand, the effective valence ($z_{K2}$=4.3) larger than 2 is not compatible with a pure access channel and, therefore, some additional voltage-dependent step other than binding of external K⁺ has to be postulated. Leaving Lys-28 attached has a much less pronounced effect suggesting that particularly Lys-28 effects the apparent binding of external K⁺. Since the cytoplasmic N-terminus influences extracellular K⁺ sensitivity, an allosteric effect seems most likely. A qualitatively similar result is obtained for the inhibition by $Na^+$ (Fig. 7B, compare also Table). The similar variations in $K_m$ and $K_I$ with respect to species and N-terminus differences support not only the view that differences in the N-terminus can contribute to the species differences in cation sensitivity but also that both $Na^+$ and $K^+$ interact with binding sites within the same access channel.

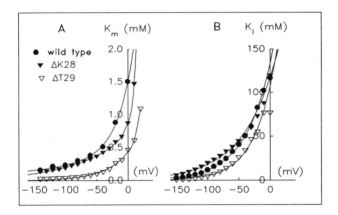

**Figure 7:** Voltage dependencies **(A)** of apparent $K_m$ value for pump stimulation by external $K^+$ (data taken from (26)), and **(B)** of the $K_I$ value for pump inhibition by external $Na^+$ (data taken from (25)). Data are shown for the wild-type *Torpedo* pump and truncated mutants expressed in the oocytes. The solid lines represent fits of one or the sum of two exponentials to the data based on exponential dependence of binding rates for external $K^+$ and $Na^+$ (see Table).

**Table:** Fitted parameter of voltage dependence of apparent $K_m$ and $K_I$ values

$$K_m = \sum_i K_{mi}(0mV)e^{Z_{Ki}VF/RT} \qquad K_I = K_I(0mV)e^{Z_{Na}VF/RT}$$

with i=1 for the endogenous pump and i=2 for the Torpedo pumps.

| | X | wT | Δ28K | Δ29T | wT | OR |
|---|---|---|---|---|---|---|
| $Z_{Na}$ | 1.10 | 0.73 | 0.84 | 1.50 | | |
| $K_I(0mV)$ (mM) | 44 | 133 | 117 | 107 | | |
| n | 1.6 | 1.2 | 1.8 | 2.0 | | |
| $Z_{K1}$ | 0.61 | 0.16 | 0.33 | 0.66 | 0.15 | 0.10 |
| $K_{m1}(0mV)$ (mM) | 0.63 | 0.40 | 0.73 | 0.42 | 0.13 | 0.17 |
| $Z_{K2}$ | | 1.03 | 3.4 | 4.3 | 0.53 | 0.78 |
| $K_{m1}(0mV)$ (mM) | | 1.11 | 0.16 | 0.02 | 0.25 | 0.11 |
| m | 1.3 | 1 | 1 | 1 | 1 | 1 |

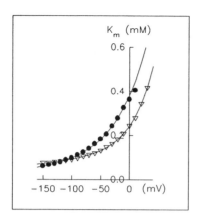

**Figure 8:** Voltage dependence of apparent $K_m$ for wild-type *Torpedo* pump and ouabain-resistant mutant (OR). The solid lines represent fits of the sum of two exponentials to the data based on exponential dependence of binding rates for external $K^+$ (see Table).

In addition to the experiments with truncated mutants, we analysed an ouabain-resistant mutant of the *Torpedo* pump. Ouabain-resistance was achieved by double mutation of the two bordering amino acids of the first extracellular loop to charged amino acids (Gln-118 to Arg and Asn-129 to Asp, see Fig. 6). Since there are often pronounced seasonal variations in the electrogenic properties, it is always necessary to perform control measurements for the wild-type *Torpedo* pumps in parallel; only then conclusions about the effects of mutation or modification are reliable. The variations may reflect possible influences of different intracellular conditions like different states of regulatory phosphorylation or interactions with the cytoskeleton. Comparison of the $K_m$ values for the wild-type *Torpedo* pump shown in Fig. 7A and Fig. 8 gives an example for the variability of voltage dependence (for the endogenous pump see Fig. 9). The data in Fig. 8 illustrate that the loss of ouabain sensitivity results in slightly modified voltage dependence with higher sensitivity for $K^+$ at 0 mV (see also Table).

It was proposed that during cation translocation two positive charges are compensated by negative charges (6,10). Based on their measurements with the carboxyl reagent DCCD Karlish and coworkers (11) suggested for possible candidates the glutamic acid residues E-334 and E-959 (in *Torpedo* numbering, see Fig. 6). On the other hand, expression of pump molecules with point mutations of either Glu-334, Glu-959, or Glu-960 to alanine leads to fully functional enzymes (see contribution by Vasilets et al.). This finding is in line with the results obtained by Lingrel and coworkers (24) who demonstrated expression of functionally active Glu-mutants in transfected HeLa cells (see also contribution in this volume). Since, at least in the oocytes, the apparent $K_m$ values of the Glu-mutants clearly differ from those for the wild type, these results imply that the residues, nevertheless, may be involved in cation interaction (see contribution by Vasilets et al.).

**Regulation by protein kinases**

In the regulation of membrane transport phosphorylation of transport molecules by protein kinases has turned out to play an important role. For the α-subunit of the

Na$^+$/K$^+$-ATPase phosphorylation by protein kinase A (PKA) and C (PKC) could be demonstrated *in vitro* and also in homogenates of *Xenopus* oocytes (3-5). In addition, modulation of pump activity was observed after stimulation of both protein kinases in *Xenopus* oocytes. Stimulation of PKA leads to an increase of maximum transport rate, stimulation of PKC to a decrease (27). Not only maximum transport activity is regulated by protein kinases, also the voltage-dependent stimulation of pump activity by external K$^+$ is modulated. Figure 9A illustrates for the endogenous *Xenopus* pump that stimulation of PKA by cAMP leads to a slight increase of the apparent affinity for external K$^+$ at 0 mV and to a reduced potential dependence which may be interpreted as a reduction of the apparent dielectric length of the access channel. As for the maximum transport activity, stimulation of PKC has the opposite effect (Fig. 9B).

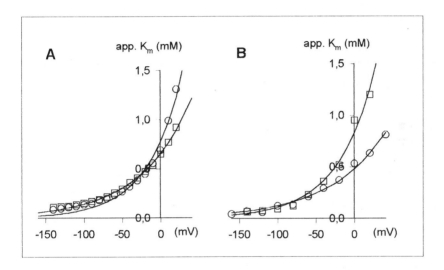

**Figure 9:** Voltage dependencies of apparent K$_m$ values of the endogenous *Xenopus* pump. Effect of **(A)** stimulation of PKA by microinjection of cAMP, and **(B)** stimulation of PKC by microinjection of diC$_8$ into the oocyte (Data taken from (27)). Open circles represent control; open squares-microinjection of activators. Solid lines represent exponential fits to the data (equ. (2)). Fitted parameters are:

|  | control A | cAMP | control B | diC$_8$ |
|---|---|---|---|---|
| $z_K$ | 0.73 | 0.49 | 0.38 | 0.66 |
| $K_m$(0mV) (mM) | 0.78 | 0.62 | 0.48 | 0.83 |

Based on the localisation of fragments of the $\alpha$-subunit phosphorylated by protein kinases serines and threonines could be identified as putative phosphorylation sites (4, 5) according to consensus sequences (see Fig. 6). Since they are located on cytoplasmic loops but have their effect on extracellular cation access. As for the lysine cluster this phenomenon suggests an allosteric influence on the access channel.

## Selectivity of K$^+$ congeners

Not only K$^+$ can be used as an external ligand for activation of the Na$^+$/K$^+$-pump, but also a variety of other small cations can act as K$^+$ congeners like Tl$^+$, Rb$^+$, NH$_4^+$ , or Cs$^+$(20). The selectivity is of this order with Tl$^+$ (K$_m$(0mV)= 0.1 mM) being two orders of magnitude more effective than NH$_4^+$ (K$_m$(0mV)= 13 mM) for the endogenous pump. Interestingly, the ratios for the apparent affinities are identical to the selectivity sequence of delayed rectifier K$^+$ channels suggesting functional similarities where the cations have to pass a series of energy barriers to reach their occlusion sites.

Above we illustrated that external cation sensitivity shows clear species differences and that differences in the cytoplasmic N-terminus may contribute. Figure 10 shows for Tl$^+$ and Rb$^+$ that also the selectivity exhibits species differences and that N-terminal truncation modifies the selectivity. The largest difference in K$_m$ values or highest selectivity is found for the *Xenopus* pump (open circles), the *Torpedo* pump (filled circles) exhibits a slightly lower selectivity than the *Xenopus* pump, which is further reduced in the truncated mutant Δ29T (open triangles)

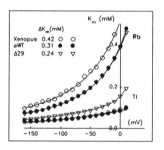

**Figure 10:** Voltage dependence of apparent K$_m$ values for pump stimulation by external Tl$^+$ and Rb$^+$.

## Summary

The dominating voltage dependent steps in the reaction cycle of the Na$^+$,K$^+$-ATPase can be attributed to voltage-dependent binding of external Na$^+$ and K$^+$ via a high-field access channel, and the access can be described by an effective valence, which reflects a dielectric length of the channel. Differences in the cytoplasmic amino-acid composition of the N-terminus and in ouabain sensitivity may contribute to species and isoform differences of the α-subunit by effecting the apparent affinities and dielectric length of the access channel. Putative intramembraneous glutamic acid residues are not necessary but may be involved in the formation of the channel. The activity of the Na$^+$/K$^+$-ATPase can be regulated by protein kinases; stimulation of protein kinases modulates maximum transport activity and the dielectric geometry of the access channel.

*Acknowledgements*

We thank Drs. M.Kawamura, T.Ohta and K.Takeda for providing cDNAs for the Torpedo pumps. The work was supported by Deutsche Forschungsgemeinschaft (SFB 169).

**References**

1. Albers RW. (1967) Biochemical aspects of active transport. Ann. Rev. Biochem., 36, 727-756.
2. Bahinski A, Nakao M, Gadsby DC. (1988) Potassium translocation by the $Na^+/K^+$ pump is voltage insensitive. Proc. Natl. Acad. Sci. USA, 85, 3412-3416.
3. Chibalin AV, Lopina OD, Petukhov SP, Vasilets LA. (1991) Phosphorylation of Na,K-ATPase by protein kinase C and cAMP-dependent protein kinase. Biol. Mem., 8, 1440-1441.
4. Chibalin AV, Lopina OD, Petukhov SP, Vasilets LA. (1993) Phosphorylation of the Na,K-ATPase by Ca,phospholipid-dependent and cAMP-dependent protein kinases: Mapping of the region phosphorylated by Ca,phospholipid-dependent protein kinase. J. Bioenerg. Biomembr., 25, 61-66.
5. Chibalin AV, Vasilets LA, Hennekes H, Pralong D, Geering K. (1992) Phosphorylation of Na,K-ATPase α subunits in microsomes and in homogenates of Xenopus oocytes resulting from the stimulation of protein kinase A and protein kinase C. J. Biol. Chem., 267, 22378-22384.
6. DeWeer P, Gadsby DC, Rakowski RF. (1988) Voltage dependence of the Na-K pump. Ann. Rev. Physiol., 50, 225-241.
7. Efthymiadis A, Rettinger J, Schwarz W. (1994) Inward-directed current generated by the $Na^+,K^+$ pump in $Na^+$- and $K^+$-free medium. Cell Biology International, (submitted),
8. Efthymiadis A, Schwarz W. (1991) Conditions for a backward-running $Na^+/K^+$ pump in Xenopus oocytes. Biochim. Biophys. Acta, 1068, 73-76.
9. Gadsby DC, Rakowski RF, DeWeer P. (1993) Extracellular access to the Na,K pump - Pathway similar to ion channel. Science, 260, 100-103.
10. Goldshleger R, Karlish SJD, Rephaeli A, Stein WD. (1987) The effect of membrane potential on the mammalian sodium-potassium pump reconstituted into phospholipid vesicles. J. Physiol., 387, 331-355.
11. Goldshleger R, Tal DM, Moorman J, Stein WD, Karlish SJD. (1992) Chemical modification of Glu-953 of the alpha-chain of $Na^+,K^+$- ATPase associated with inactivation of cation occlusion. Proc. Natl. Acad. Sci. USA, 89, 6911-6915.
12. Gschwendt M, Kittstein W, Marks F. (1991) Protein kinase C activation by phorbol esters: do cystein-rich regions and pseudosubstrate motifs play a role?. TIBS, 16, 167-169.
13. Jorgensen PL, Andersen JP. (1988) Stuctural basis for E1-E2 conformational transitions in Na,K-pump and Ca-pump proteins. J. Membrane Biol., 103, 95-120.

493

14. Karlish SJD, Goldshleger R, Stein WD. (1990) A 19-kDa C-terminal tryptic fragment of the alpha-chain of Na/K-ATPase is essential for occlusion and transport of cations. Proc. Natl. Acad. Sci. USA, 87, 4566-4570.

15. Kawakami K, Noguchi S, Noda M, et al. (1985) Primary structure of the $\alpha$-subunit of Torpedo californica $(Na^{+}+K^{+})$ATPase deduced from cDNA sequence. Nature, 316, 733-736.

16. Lafaire AV, Schwarz W. (1986) Voltage dependence of the rheogenic $Na^{+}/K^{+}$ ATPase in the membrane of oocytes of Xenopus laevis. J. Membrane Biol., 91, 43-51.

17. Läuger P. (1991) Kinetic basis of voltage dependence of the Na,K-pump. In: The Sodium Pump: Structure, Mechanism, and Regulation (ed. JH Kaplan, P DeWeer), pp 303-315. Rockefeller Univ. Press, New York.

18. Läuger P. (1991) Electrogenic Ion Pumps. Sinauer Associates Inc., Sunderland.

19. Läuger P, Apell H-J. (1986) A microscopic model for the current-voltage behaviour of the Na, K-pump. Europ. Biophys. J., 13, 309-321.

20. Omay HS, Schwarz W. (1992) Voltage-dependent stimulation of $Na^{+}/K^{+}$ pump current by external cations: Selectivity of different $K^{+}$ congeners. Biochim. Biophys. Acta, 1104, 167-173.

21. Post RL, Kume S, Tobin T, Orcutt B, Sen AK. (1969) Flexibility of an active centre in sodium-plus-potassium adenosine triphosphatase. J. Gen. Physiol., 54, 306s-326s.

22. Rakowski RF, DeWeer P, Gadsby DC. (1988) Current-voltage relationship of the backward-running Na/K pump in voltage-clamped internally-dialyzed squid giant axons. Biophys. J., 53, 223a

23. Rakowski RF, Vasilets LA, LaTona J, Schwarz W. (1991) A negative slope in the current-voltage relationship of the $Na^{+}/K^{+}$ pump in Xenopus oocytes produced by reduction of external $[K^{+}]$. J. Membrane Biol., 121, 177-187.

24. Vanhuysse JW, Jewell EA, Lingrel JB. (1993) Site-directed mutagenesis of a predicted cation binding-site of Na, K-ATPase. Biochemistry, 32, 819-826.

25. Vasilets LA, Ohta T, Noguchi S, Kawamura M, Schwarz W. (1993) Voltage-dependent inhibition of the sodium pump by external sodium: Species differences and possible role of the N-terminus of the $\alpha$-subunit. Europ. Biophys. J., 21, 433-443.

26. Vasilets LA, Omay H, Ohta T, Noguchi S, Kawamura M, Schwarz W. (1991) Stimulation of the $Na^{+}/K^{+}$ pump by external $[K^{+}]$ is regulated by voltage-dependent gating. J. Biol. Chem., 266, 16285-16288.

27. Vasilets LA, Schwarz W. (1992) Regulation of endogenous and expressed $Na^{+}/K^{+}$ pumps in Xenopus oocytes by membrane potential and stimulation of protein kinases. J. Membrane Biol., 125, 119-132.

28. Vasilets LA, Schwarz W. (1993) Structure-function relationships of cation binding in the $Na^{+}/K^{+}$-ATPase. Biochim. Biophys. Acta, (in press),

29. Vasilets LA, Schwarz W. (1994) The $Na^{+}/K^{+}$ pump: Structure and function of the alpha-subunit. Cell. Physiol. Biochem., 4, 81-95.

□

# Electrogenic and electroneutral partial reactions in Na$^+$/K$^+$-ATPase from eel electric organ

K.Fendler,S.Jaruschewski,J.P.Froehlich,W.Albers and E.Bamberg

Max-Planck-Institut für Biophysik,D-60596 Frankfurt, Germany.
National Institute on Aging, NIH, Baltimore,MD 21224, USA.
National Institute of Neurological Desease and Stroke, NIH, Bethesda, MD 20892, USA.

## Introduction

The generally accepted reaction scheme of the Na$^+$/K$^+$-ATPase is the Albers-Post cycle. It defines the reaction mechanism of the enzyme by a sequence of intermediates which are characterized by their chemical or structural properties, namely, conformation and state of phosphorylation. While much information has been accumulated about the reaction mechanism, much less is known about the transport mechanism of the Na$^+$/K$^+$-ATPase. An obvious strategy to address the latter problem is to correlate charge movement to the partial reactions in the Albers-Post scheme. The underlying assumption in this approach is that the charge movement reflects the movement of the transported Na$^+$or K$^+$ ions inside the protein.

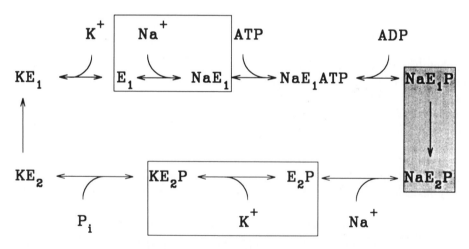

Figure 1: Reaction cycle of the Na$^+$/K$^+$-ATPase. The reactions discussed below are enclosed in boxes. Electrogenic reaction: shaded box, electroneutral reaction: open box. A Na$^+$/K$^+$ transport stoichiometry of 3/2 is assumed but not given in the figure.

Conformational transitions are obvious candidates for electrogenic steps since in these transitions the ions are thought to be translocated across the dielectric barrier. In addition, high resistance access channels leading to the ion binding sites have been discussed as a possible source of the electrogenicity of the pump (13,10). The number and assignment of electrogenic steps in the $Na^+K^+$-ATPase reaction cycle is controversial although there seems to be agreement about an electrogenic step in the $Na^+$-transporting part of the reaction cycle (4,16,2).

Models for the transport activity of the $Na^+/K^+$-ATPase with a single electrogenic step have been proposed previously (11). There is, however, experimental evidence that at least two electrogenic transitions are required in the presence of low concentrations of $K^+$ (14,18). In addition, electrogenic $Na^+$-binding at the cytoplasmic side of the enzyme was proposed from experiments with the electrochromic dye RH421 (20).

## Bilayer measurements

Microsomal membranes containing $Na^+/K^+$-ATPase were prepared from the electric organ of the eel, *Electrophorus electricus* .The activity of the enzyme was 2-3 µmol Pi/mg protein/minute at 21 $^O$C. Optically black lipid membranes (BLM) were formed in a thermostated teflon cell as described elsewhere (4). Each of the two compartments of the cell was filled with 1.5 ml of electrolyte containing 3 mM $MgCl_2$, 1 mM DTT, 25mM imidazole (pH 6.2) and various amounts of NaCl and KCl. The temperature was kept at 24 $^O$C. The membrane-forming solution contained 1.5%(w/v) diphytanoyl-phosphatidylcholine and 0.025%(w/v) octadecylamine dissolved in n-decane. 15 µl of a suspension containing ca. 5 mg/ml protein were added to one compartment of the cuvette and stirred for 30 minutes. Then caged ATP at concentrations between 300 and 600 µM was added under stirring to the cuvette.

To photolyse the caged ATP, light pulses of an excimer laser ($\lambda$ = 308 nm) were focused onto the lipid bilayer membrane. Typically, a release of 17 % of ATP from caged ATP was obtained at the membrane surface. The membrane was connected to an external measuring circuit via polyacrylamid gel salt bridges and Ag/AgCl or platinized Pt electrodes. The signal was amplified, filtered and recorded with a digital oscilloscope.

The recorded signals were fitted with a sum of n exponentials. This corresponds to a system of n+1 intermediates, in which first-order reactions may take place between any two components (6,13). The number of electrogenic steps is arbitrary. Each data set was fitted with 3 to 5 exponential components. The solution yielding a satisfactory fit to the data with the minimum number of exponentials was used for further evaluation. In the absence of $K^+$,3 and in its presence 4 components were sufficient to describe the data in the time range t > 3ms. The differential equations describing the kinetic models shown in Fig. 5 were solved numerically and fitted to the dataset using the program MLAB (Civilized Software, Bethesda, USA).

As shown previously (2,6) the capacitive coupling introduces an additional time constant into the electrical signal. This so-called system time constant $\tau_0$ depends on the capacitances $C_p$ and $C_m$ and the conductivities $G_p$ and $G_m$ of the membrane fragments and the underlying bilayer: $\tau_o^{-1}=(G_p+G_m)/(C_p+C_m)$. Under saturating ATP and $Na^+$ concentrations $\tau_o$ corresponds to the slowest time constant $\tau_3$ of the signal (6). Since $\tau_o$ is not related to enzymatic activity it will not be discussed in the following.

## Quenched-flow measurements

Rapid mixing experiments were performed using a chemical quenched-flow device described elsewhere (7). The temperature of the syringes and the flow path was kept at $24^\circ C$. The time course of phosphoenzyme formation was measured in a medium with ionic composition identical to that used in the electrical measurements. Electric organ microsomal membranes (protein concentration: 0.66 mg/ml) suspended in 3 mM $MgCl_2$, 0.1 mM EDTA, 25 mM imidazole at pH 6.2, 130 mM NaCl, and 10 mM KCl were mixed with an equal volume of an identical medium containing 20 $\mu M$ $[\gamma^{32}P]ATP$. After a brief time delay (2.5 to 300 msec) the reaction was quenched by the addition of 3% perchloric acid and 2 mM $H_3PO_4$ (final concentrations). The phosphoprotein and the released $P_i$ in a 2 ml sample of the quenched reaction mixture were determined as described elsewhere (7).

## An electrogenic $Na^+$-dependent reaction

Because of their probable association with charge translocation, electrical signals generated by the $Na^+/K^+$-ATPase have been used to obtain information about the $Na^+$ translocation and release steps. From measurements of membrane potential-dependent steady state currents in $Na^+/K^+$-ATPase, it was concluded that at least one of the $Na^+$ dependent reactions of the enzymatic cycle is electrogenic (17). Rate constants for charge movement within the $Na^+$-dependent limb of the Albers-Post cycle have also been obtained from current transients measured in cardiac myocytes after a voltage step (16).

Current transients generated by the purified $Na^+/K^+$-ATPase on lipid bilayers as described in this work have previously been measured for enzyme prepared from pig kidney (4,5), rabbit kidney (2), and eel electric organ (6). A typical current trace is shown in figure 2.

The exponential components related to the rising phase, the decaying phase and the slowest phase were labeled $\tau_1$, $\tau_2$, and $\tau_3$. The rising phase of these signals with a relaxation time of $\tau=10ms$ was attributed to an electrogenic step in the $Na^+$-dependent limb of the $Na^+/K^+$-ATPase reaction cycle (5). Subsequent analysis on the basis of the reduced photolysis rate constant of caged ATP in the presence of $Mg^{2+}$ (21) now shows that the rise of the electrical signal in these measurements is partially limited by reactions other than the electrogenic step (6). Using a chymotrypsin-modified enzyme, which is able to occlude $Na^+$ ions and catalyze phosphorylation and ADP/ATP exchange, but is unable to undergo the $E_1P \rightarrow E_2P$ transition Borlinghaus et al. (2) demonstrated loss of the electrical signal following activation of the enzyme by photolytic release of ATP from caged ATP. These results clearly implicate a reaction downstream from phosphorylation, most likely $E_1P \rightarrow E_2P$ transition. In the following, the electrical signal will be analysed under the assumption that the $E_1P \rightarrow E_2P$ transition is the electrogenic step of the $Na^+$-part of the $Na^+/K^+$-ATPase reaction cycle (16,2). However, a possible alternative interpretation, as e.g. an electroneutral $E_1P \rightarrow E_2P$ transition followed by a rapid electrogenic reaction in which $Na^+$ dissociation or rebinding occurs, is kinetically indistinguishable from this assumption. The term "$E_1P \rightarrow E_2P$ transition" will therefore be used in the sense as to include all kinetically indistinguishable alternatives.

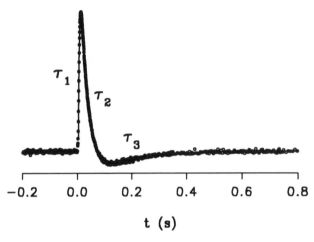

Figure 2: Typical electrical signal obtained with $Na^+/K^+$-ATPase from eel elctric organ in the presence of 130 mM $Na^+$. The solid line is a 3-exponential fit curve. For fitting, only data points with $t > 3$ms were used. The time constants $\tau_1$, $\tau_2$ and $\tau_3$ corresponding to the 3 different phases of the signal are shown in the figure.

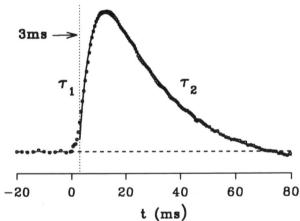

### Is intracellular $Na^+$-binding electrogenic ?

At low $Na^+$ concentration, only a fraction of the enzyme is in the $E_1Na$-state and therefore ready for rapid phosphorylation by released ATP. By reducing the $Na^+$ concentration we can therefore probe the $E_1 \rightarrow E_1Na$ reaction. Figure 3 shows the $Na^+$-

498

dependence of the electrical signal. The fast reciprocal time constant $\tau_1^{-1}$ decreases with decreasing $Na^+$ concentration while $\tau_1^{-1}$ and $\tau_2^{-1}$ remain constant. Only at very low $Na^+$ concentrations do $\tau_2^{-1}$ and $\tau_3^{-1}$ also decrease. However, this is only due to the fact that here the $Na^+$-dependent time constant becomes so slow that it determines the decay of the signal rather than the rising phase. A similar effect has been found for $\tau_2^{-1}$ in the ATP-dependence of the electrical signal (6). In a simple kinetic model, which takes into account the effects of caged ATP and ATP, the $Na^+$-dependence is described by:

$$NaE_1 cagedATP \leftrightarrow NaE_1 \leftrightarrow NaE_1 ATP \xrightarrow{k_p} NaE_1 P \rightarrow NaE_2 P$$
$$\updownarrow \qquad\qquad \updownarrow \qquad\qquad \updownarrow$$
$$E_1 cagedATP \leftrightarrow E_1 \leftrightarrow E_1 ATP$$

A similar model was introduced previously to explain the ATP-dependence of the electrical signals (6) and is here expanded to take into account $Na^+$ binding. The intermediates with and without bound $Na^+$ are assumed to be in rapid equilibrium with each other described by the dissociation constant $K_{Na}$. For simplicity, only a single $Na^+$ ion is included in the model although 3 $Na^+$ ions are translocated during the transport cycle. As shown below, this simple model is sufficient to describe the experimental results.

It can be shown that this reaction sequence is formally equivalent to:

$$NaE_1 cagedATP \leftrightarrow NaE_1 \leftrightarrow NaE_1 ATP \xrightarrow{k_p'} NaE_1 P \rightarrow NaE_2 P$$

This is the same model as at saturating $Na^+$ concentrations except that the rate constant for phosphorylation is replaced by an effective rate constant $k_p' = k_p c_{Na} / (c_{Na} + K_{Na})$. Based on this model, a hyperbolic saturation dependence of the apparent rate constant for phosphorylation is expected with a half-saturation value equal to the $Na^+$-dissociation constant.

A hyperbolic dependence of $\tau_1^{-1}$ on $Na^+$ concentration is indeed observed (Fig. 3) with a half-saturation value of $K_{Na}$=10mM. This is surprising since the $Na^+$ transport stoichiometry is 3. Sigmoidal behaviour indicative for cooperative binding of 2 or 3 $Na^+$ ions has been found in previous studies (19,9). In contrast, the electrical activity of the pump was shown to be stimulated by $Na^+$ with a Hill coefficient close to 1 (17). According to the kinetic model, the half saturation value for $k_p'$ represents the $Na^+$-dissociation constant $K_{Na}$. It is similar to a half saturation concentration of 4 mM found for $Na^+$ activation in steady-state phosphorylation experiments (19,9) and 11 mM reported for patch-clamp measurements (17).

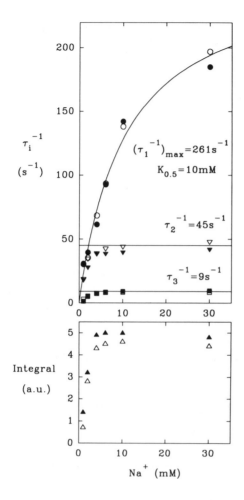

Figure 3: Upper panel: $Na^+$-dependence of the reciprocal time constants of the electrical signal. A hyperbolic saturation curve was fitted to $\tau_1^{-1}$. Lower panel: Integral of the pump current $I_p(t)$ over the range of $0 < t < 0.1$ s.

A problem in interpreting $\tau_1$, the time constant of the rising phase of the signal, is that several processes contribute to this phase (6). Major contributions come from phosphorylation, the $E_1P \rightarrow E_2P$ transition and the release of ATP from caged ATP. All these reactions have rate constants in the range of 400-1000 $s^{-1}$ (6). The saturating behavior of $\tau_1$ may therefore not reflect the $Na^+$-dependence of the rate constant for phosphorylation but partly show the effect of an additional $Na^+$ independent reaction that is rate limiting at high $Na^+$ concentrations.

In conclusion, the kinetic model which explains the $Na^+$ dependence of $\tau_1^{-1}$ by an apparent $Na^+$-dependent rate constant for phosphorylation reproduces the data quite well. 10 mM seems to be a good approximation to the $Na^+$-dissociation constant although the saturation value of $\tau_1^{-1}$ certainly contains contributions from reactions other than phosphorylation.

When no $K^+$ is present, dephosphorylation is slow and the experiments shown in Fig. 3 are essentially single turnover experiments. In this case, integration over the pump current yields the charge transported during the first turnover. The pump current $I_p(t)$ can be calculated from the measured current $I(t)$ using an equation derived in reference 2 which takes into account the capacitive coupling of the ion pumps to the measuring circuit. Subsequently, the integral of the pump current was determined by numerical integration. This integral is shown in the second panel of Fig. 3. It is seen that the charge remains constant over a region where $\tau_1^{-1}$ changes drastically. Only at very low concentrations, where the electrical signal becomes so slow that it exceeds the upper integration limit (100ms), does the integral decrease. This demonstrates that even in a situation where most of the protein starts the reaction without $Na^+$ in the binding site, the amount of charge transported during the first turnover does not change. This suggests that $Na^+$ binding at the cytoplasmic surface is not electrogenic.

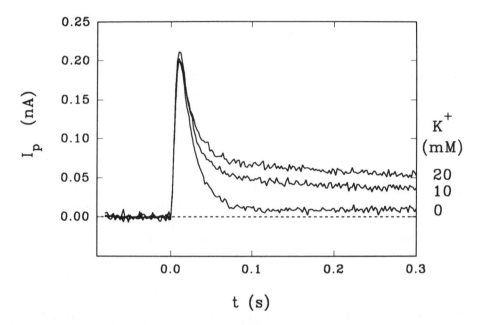

Figure 4: Transient current at different $K^+$ concentrations. The pump current $I_p$ (t) was calculated from the measured current $I(t)$ as described in the text.

## Is extracellular $K^+$ binding electrogenic ?

Figure 4 shows the current generated by the $Na^+/K^+$-ATPase at different $K^+$ concentrations. The pump current $I_p(t)$ was calculated from the measured current $I(t)$ according to reference 2. The stationary current changes with $K^+$ concentration while the current transient during the first 30 ms remains unchanged. The increased stationary current reflects the increased turnover of the $Na^+/K^+$-ATPase in the presence of $K^+$. The fact that the current transient does not change after addition of $K^+$ demonstrates that with and without $K^+$ the amount of charge transported during the first turnover is the same. This implies that $K^+$ binding to the extracellular binding site is electroneutral.

## Correlation of charge transport, phosphoenzyme formation and $P_i$ release

Electrical and phosphorylation experiments were performed under the same conditions (130 mM $Na^+$, 5 mM $K^+$, 3 mM $Na^+$, pH 6.2 and 24$^o$C) using the same enzyme preparation. The only difference between the two techniques was that, in the electrical experiments, the reaction was initiated by photolytic release of ATP from caged ATP while, for the phosphorylation measurements, ATP was rapidly mixed with the enzyme-containing solution. This required a different reaction model for the two experiments. The white region in Fig. 5 shows the reaction sequence used to describe the phosphorylation measurements. For the electrical experiments, additional intermediates given in the shaded region of Fig. 5 were required.

The symbol caged ATP* in Fig. 5 represents a caged ATP molecule that has absorbed a photon and is ready to decay to ATP. As indicated in the figure, the photolytic reaction is possible for molecules in solution as well as for molecules bound to the enzyme ($E_1$cagedATP*). The reaction in solution was measured under the conditions of our experiments using photometric detection according to reference 15. The rate constant determined was $\lambda = 510$ s$^{-1}$. Whether the photolytic process occurs in the binding site and proceeds with the same efficiency and rate constant is unknown. In a previous publication (6) release of ATP in the binding site was rejected since the electrical signal could not be simulated with this assumption. However, based on the rather low value ($\lambda = 510$ s$^{-1}$) determined for photolytic ATP release in solution we have to conclude that both possibilities have to be taken into account. For our analysis we assumed the same rate constant and efficiency in solution and in the binding site. We investigated two limiting cases (dashed arrows in Fig. 5): a) no release of ATP in the binding site (1 → 2), b) release in the binding site (1 → 4).

The kinetic models of Fig. 5 were solved numerically. The table in Fig. 5 shows the results of a fit with the corresponding kinetic model to the data of a phosphorylation (phos) and an electrical (el) experiment assuming release of ATP in the binding site (ris) or no release in the binding site (nris). For some of the rate constants good experimental values determined separately under the same conditions were available. For these fixed values were chosen. Variable parameters were $k_a^+$, $k_{56}$ ($E_1P \rightarrow E_2P$), $k_{83}$ ($E_2 \rightarrow E_1$), and for the electrical experiments $k_0$.

Good agreement between data and the calculated fit curve was found as demonstrated in Fig. 6. Both models, ris as well as nris, gave good quality fits to the electrical data. The parameters determined in the electrical and the phosphorylation experiments were

502

consistent except for $k_{56}$ which depended strongly on the model (ris or nris) chosen. Since the value determined in the phosphorylation measurement lies in between the values determined in the ris and nris kinetic model it is possible, that neither of the limiting cases describes the actual situation and ATP is released in the binding site but with a somewhat reduced efficiency or rate. Alternatively, it is conceivable that electrical measurements and phosphorylation experiments, indeed, probe different partial reactions of the reaction cycle. This will be discussed in the following paragraph.

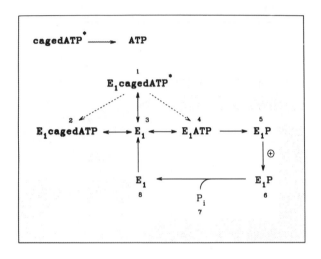

Figure 5: Kinetic model for the fit to electrical and phosphorylation measurements. The Table shows the parameters obtained from the fit to the phosphorylation (phos) and electrical (el) measurements. nris = no release of ATP in the binding site, ris = release in the binding site.

Eel electric organ: [Adjustable] and fixed parameters

| | | | phos | el nris | el ris |
|---|---|---|---|---|---|
| ATP: | $c_a^0$ | $/\mu M$ | 10 | | |
| caged ATP: | $c_c^0$ | $/\mu M$ | | 270 | 270 |
| | $\eta$ | | | 0.19 | 0.19 |
| | $\lambda$ | $/s^{-1}$ | | 510 | 510 |
| | $k_c^+$ | $/s^{-1}M^{-1}$ | | $2.9 \cdot 10^7$ | $2.9 \cdot 10^7$ |
| | $k_c^-$ | $/s^{-1}$ | | 1000 | 1000 |
| ATP: | $k_a^+$ | $/s^{-1}M^{-1}$ | [$1.1 \cdot 10^7$] | [$0.87 \cdot 10^7$] | [$1.0 \cdot 10^7$] |
| | $k_a^-$ | $/s^{-1}$ | 94 | 94 | 94 |
| | $k_{4,5}$ | $/s^{-1}$ | 340 | 340 | 340 |
| | $k_{5,6}$ | $/s^{-1}$ | [510] | [2000] | [270] |
| | $k_{6,8}$ | $/s^{-1}$ | 600 | 600 | 600 |
| | $k_{8,3}$ | $/s^{-1}$ | [10] | [13] | [16] |
| Network: | $k_0$ | $/s^{-1}$ | | [3.5] | [3.9] |
| | $k_m$ | $/s^{-1}$ | | 0 | 0 |

It is important to note that the rate constant for the $E_1P \rightarrow E_2P$ transition determined from the phosphorylation experiments represents a lower limit for this transition. The Albers-Post model in Fig. 5 is a minimal model that has been stripped to the bare essentials needed to explain ATP binding and transphosphorylation. What is missing from this scheme are the steps involving $Na^+$ dissociation and deocclusion which lie between the phosphoenzyme transition and $K^+$-activated hydrolysis of $E_2P$.

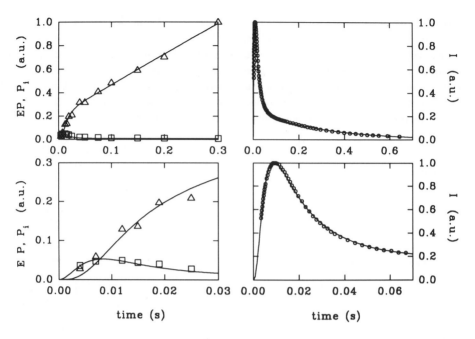

Figure 6: Time-dependence of phosphoenzyme (EP) formation, phosphate ($P_i$) production and current (I) measured under the same conditions (see text). For calculation of the fit curve the parameters of the table (ris) in Fig. 5 were used.

For a transport mechanism involving 3 $Na^+$ ions, there are 3 $Na^+$ dissociation steps and 2 conformational changes situated between the release of the first and second and third ions (assuming that transport does not occur at 3 independent sites). In addition to this, there is the $Na^+$ deocclusion reaction at the extracellular surface which may involve more than one step (the reactions involved in $K^+$ binding and occlusion are non-contributory in this argument because their kinetic behavior is reflected in the rate constant for $K^+$-activated dephosphorylation measured by rapid acid quenching). Although the rate constants for these reactions are unknown, it is evident that the introduction of any additional steps in the reaction sequence will necessitate an increase in the simulated rate constant for the $E_1P \rightarrow E_2P$ transition. Recent measurements of mammalian $Na^+/K^+$-ATPase pump current transients using the giant patch technique indicate that $Na^+$ deocclusion might be quite slow with a time constant of ca. 6 ms (12). This raises the possibility that $Na^+$ deocclusion in the eel enzyme may proceed at rates comparable to the other partial reactions in the ATPase mechanism and therefore contribute significantly to rate limitation of $P_i$ release during the first turnover. At the same time, the electrical signal could possibly be not affected by this rate limitations and therefore yield a higher value as shown in the table for the nris model.

504

# Conclusions

Electrical as well as phosphorylation measurements on $Na^+/K^+$-ATPase from eel electric organ could be described with a consistent set of rate constants. Because of an ambiguity in the choice of the appropriate model for the electrical measurements we have to rely on the quenched-flow experiments for the determination of the rate constant of the $E_1P \rightarrow E_2P$ transition. Using this technique, values of $500$ $s^{-1}$ (Fig. 5) to $1000$ $s^{-1}$ (8) were obtained at $24^oC$ and pH 6.2. However, measurements performed on enzyme from different sources seem to yield a slower rate constant (6). This may indicate a species difference or the necessity for a more complex reaction model.

The electrogenic step is the $E_1P \rightarrow E_2P$ transition or a rapid step following the conformational transition (6). Electrical experiments performed at different $Na^+$ and $K^+$ concentrations indicate that $Na^+$ binding at the intracellular side of the enzyme and $K^+$ binding at the extracellular side are electroneutral. This is not in agreement with measurements on $Na^+/K^+$-ATPase from canine kidney using electrochromic dyes (20) and voltage clamp experiments on oocytes (18). These data seem to require electrogenic $Na^+$ and $K^+$ binding, respectively, at their corresponding binding sites. The reason for this discrepancy is unknown and may be related to the sensitivity of the different techniques to the electrogenicity of certain partial reactions

# References

1.    Apel H-J, Borlinghaus R, Läuger P (1987) Fast charge translocations associated with partial reactions of the Na,K-Pump: II. Microscopic analysis of transient currents. J. Memb. Biol. 97:179-191.

2.    Borlinghaus R, Apell H-J, Läuger P (1987). Fast charge translocations associated with partial reactions of the Na,K-Pump: I. Current and voltage transients after photochemical release of ATP. J. Memb. Biol. 97:161-178.

3.    Fahr A, Läuger P, Bamberg E (1981). Photocurrent kinetics of purple-membrane sheets bound to planar bilayer membranes. J. Memb. Biol. 60:51-62.

4.    Fendler K, Grell E, Haubs M, Bamberg E (1985). Pump currents generated by the purified $Na^+K^+$ATPase from kidney on black lipid membranes. EMBO J. 4:3079-3085.

5.    Fendler K, Grell E, Bamberg E (1987). Kinetics of pump currents generated by the $Na^+,K^+$-ATPase. FEBS Lett. 224:83-88.

6.    Fendler K, Jaruschewski S, Hobbs AS Albers RW, Froehlich JP (1993) Presteady state charge translocation in NaK-ATPase from eel electric organ. J. Gen. Physiol. 102: 631-666.

7.    Froehlich JP, Hobbs AS, Albers RW (1983) Evidence for parallel pathways of phosphoenzyme formation in the mechanism of ATP hydrolysis by Electrophorus Na,K-ATPase. Curr. Top. Memb. Transp. 19:513-535.

8. Froehlich JP, Fendler K (1991). The Partial Reactions of the Na- and Na+K- activated Adenosine Triphosphatase. In J.H. Kaplan und P. De Weer (eds): The Sodium Pump: Structure, Mechanism and Regulation. New York: Rockefeller University Press: 227-248.

9. Froehlich JP, Hobbs, AS Albers RW (1993) Parallel pathway models for electric organ Na,K-ATPase. This volume.

10. Gadsby DC, Rakowski RF, DeWeer P (1993). Extracellular Access to the Na, K Pump: Pathway Similar to Ion Channel. Science 260:100-103.

11. Goldshlegger R, Karlish SJD, Rephaeli A, Stein WD (1987) The effect of membrane potential on the mammalian sodium-potassium pump reconstituted into phospholipid vesicles. J. Physiol. 387:331-355

12. Hilgeman DW (1993) Steady state and transient Na/K pump currents in giant excised, inside-out membrane patches. This volume.

13. Läuger P, Apell H-J (1988) Voltage dependence of partial reactions of the Na,K-pump: predictions from microscopic models, Biochi. Biophys. A. 945:1-10

14. Lafaire AV, Schwarz W (1986). Voltage dependence of the rheogenic Na+/K+ ATPase in the membrane of Oocytes of Xenopus Laevis. J. Memb. Biol. 91:43-51.

15. McCray JA, Herbette L, Kihara T, Trentham DR (1980). A new approach to time-resolved studies of ATP-requiring biological systems: Laserflash photolysis of caged ATP. PNAS 77:7237-7241.

16. Nakao M, Gadsby DC (1986). Voltage dependence of Na translocation by the Na/K pump. Nature 323:628-630.

17. Nakao M, Gadsby DC (1989) [Na] and [K] Dependence of the Na/K Pump Current-Voltage Relationship in Guinea Pig Ventricular Myocytes. J. Gen. Physiol. 94:539-565.

18. Rakowski RF, Vasilets LA, LaTona J, Schwarz W (1991). A Negative Slope in the Current-Voltage Relationship of the Na/K-Pump in Xenopus Oocytes Produced by Reduction of External [K]. J. Memb. Biol. 121:177-187.

19. Siegel GJ, Albers RW (1967) Sodium-Potassium-activated Adenosine Triphosphatase of Electrophorus Electric Organ. J. Biol. Chem. 242:4972-4979.

20. Stürmer W, Bühler R, Apell HJ, Läuger P (1991) Charge translocation by the Na,k-pump.2. ion binding and release at the extracellular face. J. Memb Biol. 121:163-176.

21. Walker JW, Reid GP, McCray JA, Trentham DR (1988). Photolabile 1-(2-Nitrophenyl)ethyl Phosphate Esters of Adenine Nucleotide Analogues. Synthesis and Mechanism of Photolysis. JACS 110:7170-7177.

# FLEXIBILITY AND CONSTRAINT IN THE INTERPRETATION OF Na$^+$/K$^+$ PUMP ELECTROGENICITY: WHAT IS AN ACCESS CHANNEL?

Donald W. Hilgemann

Department of Physiology, University of Texas Southwestern
Medical Center at Dallas, 5323 Harry Hines Boulevard,
Dallas, TX 75235-9040 USA

**Introduction**

A central question which electrophysiologists ask about the Na$^+$/K$^+$ pump often appears esoteric: How and when do ions cross the membrane electrical field during the pump cycle? From the electrophysiologist's view point, this question challenges profoundly our understanding of the molecular basis of ion transport. The reasons are apparent when pump function is thought of as a selective, stoichiometric movement of ions across the cell membrane barrier, and therefore across membrane electrical field. Several electrophysiological approaches have raised questions and suggested hypotheses about molecular structure-function relations of the Na$^+$/K$^+$ pump. Recently, it has been proposed that the major charge-carrying reaction of the Na$^+$/K$^+$ pump is the diffusion of sodium through a 'high resistance access channel', open to the extracellular side in the E$_2$ pump state (1). The present article will describe some of the flexibilities and constraints encountered by the author in accounting for recent experimental results on Na$^+$/K$^+$ pump electrogenicity with the help of simulations. The results considered will include refined charge movement measurements in giant cardiac membrane patches.

**The channel analogy: Can 'narrow access channel' behavior be distinguished from fast ion occlusion?**

To one degree or another, analogies can be made between transporter and channel function, whereby the transporter is viewed as a channel closed at one end. Figure 1 illustrates two general mechanisms of electrogenicity which can be evoked (2). The first mechanism (A) is the 'narrow access channel', whereby ions approach their binding sites through a channel-like region of the transporter along which membrane field falls off. If the electrical (dielectric) properties of the protein are more like membrane than like water, then the 'access channel' would presumably have to be so narrow as to allow structuring of water within it. Simplistically, the charge movement of ion binding would be directly related to the ion charge and the

507

'electrical distance' that the ion travels. Ion passage through the access channel is to some extent analogous to ion permeation in channels. Selectivity would be another essential property of the pore to explain transporter function.

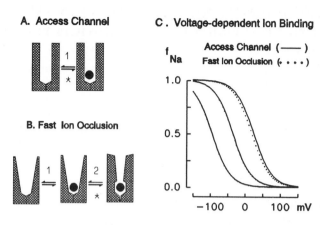

**A. Access Channel**

**B. Fast Ion Occlusion**

**C. Voltage-dependent Ion Binding**

Access Channel (——)

Fast Ion Occlusion (· · · ·)

$f_{Na}$

1.0

0.5

0.0

−100    0    100 mV

**Figure 1. Hypothetical Mechanisms of Electrogenicity in Ion Transporters (\*) .**

The second mechanistic possibility is that charge movements arise during the actual transport of ions, which is assumed to involve significant conformational changes related to the 'occlusion' of ions. As illustrated in Figure 1B, charge movements would arise when binding sites and bound ions rearrange with respect to membrane field. The magnitude of the charge movement will depend both on the magnitude of the net charge of binding sites plus bound ions and the extent of rearrangement of membrane field. A fast occlusion/deocclusion process would bear analogy to the opening and closing of a channel, albeit only at one channel end. The time course of channel openings and closings, as observed in single channel recording, has not been resolved down to a methodological limit to date of a few microseconds. The electorgenic occlusion/deocclusion process would be expected to be similarly fast and involve conformational changes of similar magnitude. Such fast ion occlusion reactions might be called 'microscopic' occlusion reactions. In the Na$^+$/K$^+$ pump they would constitute partial reactions of the overall ('macroscopic') ion occlusion process which generates stable ion-occluded, enzyme intermediates.

From much experience with simple (Markovian) transport models, it appears remarkable how similar the predicted behaviors of a narrow access channel and a fast ion occlusion reaction might be for the Na$^+$/K$^+$ pump. As shown in Figure 1C, simulations of ion binding (as well as transport behaviors) for the two mechanisms can be entirely indistinguishable. '$f_{Na}$' gives the fraction of sites occupied by sodium for the access channel (solid results) and the fraction of sites with an occluded ion for the occlusion reaction (dotted results). The fast ion occlusion reaction, consisting of a voltage-independent ion binding (recognition) step and a voltage-dependent occluding reaction, has a much greater explanatory range than does the access channel. To mimic the access channel, two constraints are needed. The 'recognition' state with a bound-but-not-occluded ion is a transitional state, as might be expected for any ion chelating reaction, and the total binding site charge be small. *Membrane voltage then acts precisely as if it is changing the effective ion concentration, as in the narrow access channel !* Only if the bound-but-not-occluded state accumulates significantly, then voltage will still have an effect (via ion occlusion) when the maximum effect of increasing ion concentration has been achieved. In fact, this is the case for current-relations of the cardiac $Na^+/Ca^{2+}$ exchanger (3) with respect to sodium concentration, a behavior which cannot be explained by an access channel model. Furthermore, for the $Na^+/Ca^{2+}$ exchanger, calcium binding to the cytoplasmic side appears to move negative charge into membrane field. This is explained easily by an ion occlusion reaction involving negatively charged binding sites; the patter cannot be explained by the simple access channel fails.

**A narrow or a wide access channel?**

Another concern in interpreting electrophysiological data in structural terms is whether a 'narrow' access channel can be distinguished from a 'wide' access channel. Figure 2 illustrates the possible problem. As shown in cartoon form in panel A, membrane electrical field is suggested to pass through the aqueous region of the narrow access channel (2), such that an ion experiences membrane field during binding. As shown in panel B, however, ion binding might also be strongly electrogenic in a wide access channel. Here, it is imagined that the

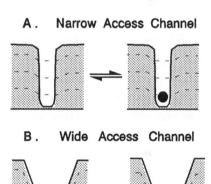

**Figure 2**

access channel has a conical shape. The field lines do not cross into the aqueous region in the absence of a bound ion. When an ion binds, however, the field lines in the tip of the pore may change (e.g. as a result of dehydration). A shifting of membrane field across both the ion and binding sites is easily imagined. Small conformational changes, resulting in a 'tightening' of the binding site, would strongly enhance this effect.

## Electrogenic ion binding versus other electrogenic mechanisms

Considerations such as these raise caution about structural interpretations of data on the electrogenicity of transporters. The narrow access channel concept is very attractive in its simplicity for the $Na^+/K^+$ pump, and contradictory experimental results have not been identified, as in the case of the cardiac $Na^+/Ca^{2+}$ exchanger. For the time being, however, the conservative interpretation of data on the electrogenicity of the $Na^+/K^+$ pump is that electrogenicity has been closely associated with extracellular ion binding, which may or may not involve significant binding site conformational changes. The concept of ion binding justifiably includes all of the possibilities of access channels and ion occlusion. The identification of ion binding as the major course of electrogenicity in the $Na^+/K^+$ pump is contrasted to alternatives where reactions other than ion binding are electrogenic (e.g. rearrangements of empty binding sites). For the $Na^+/K^+$ pump, this does not appear to occur, and details of $Na^+/K^+$ pump electrogenicity will be considered in this conservative format in the following.

## Electrogenic ion binding as the basis of $Na^+/K^+$ pump electrogenicity

Multiple experimental approaches (1,4,5,9) lend support to the idea that the major electrogenic events in the $Na^+/K^+$ pump cycle are related to extracellular ion binding, primarily of sodium. Figure 3 shows some predictions of the simplest possible equivalent transport cycle with (instantaneous) extracellular ion binding as the only voltage-dependent step. First, as simulated in Figure 3B for the simple

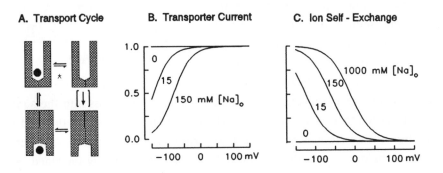

**Figure 3. I/V Relations and Ion Self Exchange Predicted by Access Simple Channel Transport Model.**

510

transporter, the forward running (sodium-extruding) $Na^+/K^+$ pump current shows almost no voltage dependence in the absence of extracellular sodium. Addition of extracellular sodium induces voltage dependence, corresponding to progressively strong inhibition of forward pump activity at more negative potentials. This behavior is precisely that expected if extracellular sodium is causing a voltage-dependent block of the pump by rapidly binding to sites which must be empty to allow subsequent forward pump reaction steps. The block is logically removed by depolarization. Second, Figure 1C shows behaviors of ion 'self exchange' under the condition that no transporter cycling is possible; the returning rate in the cycle (given by the arrow in parenthesis in 3A) is zero. Sodium self exchange for the $Na^+/K^+$ pump is defined as ouabain-sensitive sodium efflux dependent on the presence of extracellular sodium (see ref. 1). The flux increases monotonically with hyperpolarization, and the relationship with voltage shifts in a parallel fashion to more positive potentials with increasing extracellular sodium concentration. This behavior, like the changes of current-voltage relations, is exactly predicted for voltage-dependent ion binding from the extracellular side.

Another type of data which can be interpreted in support of extracellular sodium binding as the basis of $Na^+/K^+$ pump electrogenicity is the recording of charge movments (current transients) related to sodium translocation in the absence of potassium. Figure 4 shows the predictions for the simple model of Figure 3. When the transported ion is bound on either side, it can be transported. Ion binding/unbinding is assumed to be instantaneous, and all electrogenicity occurs at the extracellular ion binding reaction (*). As binding sites are gained or lost from the extracellular side, via the voltage-independent translocation reaction, a 'slow' charge movement is generated from the binding/unbinding of extracellular ion. The 'slow' charge movements track the ion translocation reaction, and they are in part consistent with experimental data. In particular, the rate of the 'on' charge movement increases with hyperpolarization, as more sites are occupied by extracellular ion, until all

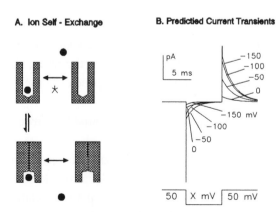

**Figure 4. Charge Movements Predicted by Simple Access Channel Model.**

511

sites are occupied. The charge movement rate becomes limited by the voltage independent translocation reaction as larger and larger voltatge steps are made in the depolarizing direction.

In addition to these features, however, the 'access channel hypothesis' implicitly makes several important predictions about $Na^+/K^+$ pump charge movements. Figure 4B presents the expected current transients for applying voltage pulses from a holding potential of +50 mV to various negative potentials ('on' charge movement) and back to +50 mV ('off' charge movement). The voltages given for the return transients are those *from which* membrane potential is stepped: 1) If it is true that ion binding-unbinding is electrogenic, then it must possible to monitor a fast charge movement related to the binding reaction, at least under appropriately chosen conditions. In general, a fast charge movement of ion binding should precede the 'slow' transient. 2) The charge movement of ion binding should be largest when the entire $Na^+/K^+$ pump is forced into the $E_2$ configuration with binding sites facing the extracellular side, and membrane potential is stepped quickly to a negative potential. 3) With voltage steps to very negative potentials, it should be possible to saturate binding sites, in a fast phase, and thereby preclude slow charge transients during the actual translocation reaction. And 4), as membrane potential is stepped to extreme negative potentials, the slow charge movement must become smaller, and an increasing discrepancy should appear between the magnitudes of charge moved in the 'on' and 'off' directions.

Over the last three years, I have tested extensively for these patterns in giant cardiac membrane patches (6,7), where $Na^+/K^+$ pump transients can be monitored at 4 $\mu$s resolution and from +150 to -300 mV. On the one hand, fast charge movements have been identified which appear to be related to extracellular sodium binding. On the other hand, the fast charge movements revealed are smaller in magnitude and have a more shallow voltage dependence than expected from the simple access channel model. Furthermore, the specific patterns just described have not been found (e.g. discrepancies between 'on' and 'off' of slow charge).

**The access channel, if it exists, must be a transitional state.**

Figure 5A shows a model of sodium release from the $Na^+/K^+$ pump which can account for most, if not all, available data available on $Na^+/K^+$ pump electrogenicity. In this model, the phosphorylated $E_1$ pump with 3 occluded sodium ions undergoes a voltage-independent conformational change (a/b reaction) which enables the binding sites to open to the extracellular side. Electrogenicity (*) can

512

arise from either a fast deocclusion reaction, as drawn here, or from the subsequent unbinding and release of the first sodium to the outside. The states occuring during release of the first sodium are highly unstable, transitional states. Therefore, they never accumulate to any significant amount, and it is never possible to evoke a large charge movement via the immediate binding/unbinding of sodium. For this reason, the loss of 'slow' charge movement with increasing hyperpolarization, just described for the simple access channel model, cannot occur. Finally, the pump opens further to allow the release of the last two sodium ions (reactions c/d). These binding/unbinding reactions are weakly electrogenic, as will also be the case for the subsequent binding of potassium.

Figure 5B shows charge movements of the $Na^+/K^+$ pump in a cardiac giant membrane patch (6,7). To quantify adequately fast and slow components of charge movements, the charge movements are recorded as 'charge' per se, directly from an integrating patch clamp. Thus, the records given are the time integral of membrane current, and membrane current is the first derivative of these records. The records presented here are low-pass filtered at 5 kHz, they are from a 7.5 pF patch, and they are with 0.4 mM cytoplasmic ATP, 10 mM cytoplasmic sodium and 120 mM extracellular sodium. Stable voltage clamp was achieved in 4 $\mu$s using pipettes with <100 kilo-ohm resistance. (Details of these measurements are submitted for publication with *Science*).

From top to bottom in Figure 5B, $Na^+/K^+$ pump charge transfer records are given for 5 ms voltage pulses from 0 mV to +150 mV, +100 mV, +50 mV, 0 mV, -50 mV, -100 mV, -150 mV, and for 2.5 ms voltage pulses from 0 mV to -200 mV and 250 mV. The calibration assumes a specific membrane capacitance of 0.8 $\mu$F/cm$^2$. Note the presence of fast and slow components in the records, whereby the fast components are larger for pulses to and from positive potentials. The charge movements are entirely blocked by 10 $\mu$M extracellular ouabain. The initial charge jumps are identifed as extracellular sodium binding reactions on the basis of their dependence on extracellular sodium and on pump configuration (i.e. $E_1$ versus $E_2$, as induced by application of sodium and MgATP from the cytoplasmic side).

Figure 5C shows a reconstruction of the data via the model of Figure 5A. The calibration of the simulation gives the number of unitary charges which have apparently moved the entire distance through membrane field. In the depolarizing direction, the 'slow' charge movement becomes rate-limited by the voltage-independent reaction 'a'. In the hyperpolarizing direction, the charge movement becomes rate-limited by the voltage independent reaction 'd'. The charge jumps

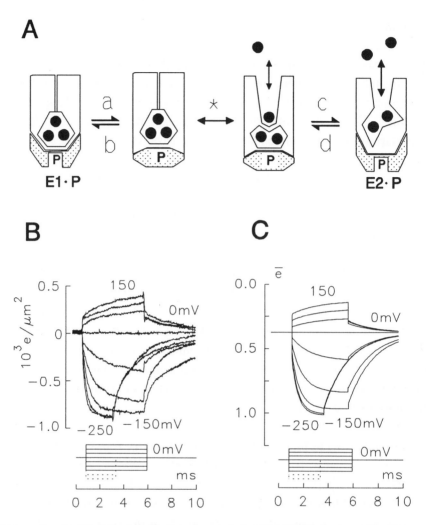

**Figure 5.** A. Model of sodium release from the Na/K pump. B. Charge movement of Na release in a giant membrane patch from a guinea pig myocyte. C. Simulation of Na/K pump charge movements.

upon changing potential correspond largely to the binding/unbinding of the 2 sodium ions in $E_2 \cdot P$. In part, however, they arise from the major electrogenic step (*), constituting the movement of a single charge over 80% of the membrane field. Although all charge movement is taking place in quasi-instantaneous reactions, the slow charge transfer reactions completely dominate the recordings. These represent the appearance and disappearance of binding sites available for electrogenic ion binding.

## Discussion

Electrophysiological measurements provide the most detailed available functional information on the translocation of ions by ion transporters, and the improved electrophysiological methods provide much more powerful tools to extract such information. However, the need for caution in interpreting this data is impressive. The interpretation of electrophysiological data in terms of 'narrow' and 'wide' access channels in a real structural sense may be compromised. The meaning of interpretive jargon such as the 'depth of an ion well' is compromised at best. A major lesson from work with the $Na^+/K^+$ pump is that reactions being monitored kinetically as a charge movement may in fact not be electrogenic reactions at all, but may be electroneutral reactions which track faster electrogenic reactions.

The refined results on charge movement of the $Na^+/K^+$ pump, using the giant patch clamp technique, appear to solidify and, to a remarkable extent, to unify the results and interpretations of many groups. The results to date are entirely consistent with the 'narrow access channel' concept (1), when it is allowed that the access channel is a transitional state which never comprises more than a small fraction of the total '$E_2$' pump state. *The access channel, if it exists, cannot constitute the entire $E_2$ state.* For those who have been inclined to interpret ion deocclusion as the major electrogenic step (8,9), the final fast reactions of sodium release upon deocclusion to the extracellular side can be viewed as an extension of the deocclusion process. It is perhaps not surprising that the release of three sodium ions from the $Na^+/K^+$ pump might ultimately be subdivided into multiple partial reactions. It is strongly suspected that the voltage-independent reactions involve the cytoplasmic domain of the pump, and that these reactions serve to unlock the microscopic electrogenic reactions. That the $E_2 \cdot P$ state with 3 sodium binding sites is a transitional state, is consistent with a lack of evidence for its existence as a significant $E_2$ state in other studies (e.g. ref. 9). The scheme given in Figure 5A can explain why the overall deocclusion process appears like a single reaction with asymmetrical voltage-dependence, the backward occlusion step being voltage-dependent and the releasing step being voltage-independent. The model of Figure 5A can be seen as a refined version of the recent electrostatic model of Apell (9), proposed on the basis of other types of data. The model can account easily for a relatively small voltage dependence of extracellular potassium binding. For the time being, the idea of a 'voltage-dependent gating' of the pump, in some way related to extracellular potassium binding, (10) is not included in the model. It is expected that the convergence of electrophysiological data to a functional detailed model of sodium translocation will facilitate the development of molecular structure-function models of the $Na^+/K^+$ pump in the coming years.

## Acknowledgement

I am indebted to Dr. David C. Gadsby for his many discussions and ecouragement of this work.

## References

1. D.C. Gadsby, R.F. Rakowski, and P. DeWeer (1993) Extracellular access to the Na,K pump: pathway similar to ion channel. Science 260, 100-103.
2. P. Läuger (1991) Electrogenic Ion Pumps 313 pp. Sinauer Associates (1991).
3. S. Matsuoka and D.W. Hilgemann (1993) Steady state and dynamic properties of cardiac Na/Ca exchange: Ion and voltage dependencies of the transport cycle. J Gen Physiol 100, 963-1001.
4. M. Nakao and D.C. Gadsby (1986) Voltage dependence of Na translocation by the Na/K pump. Nature 323, 628-630.
5. R.F. Rakowski (1993) Charge movement by the Na/K pump in Xenopus oocytes. J. Gen. Physiol. 101, 117-144.
6. D.W. Hilgemann (1989) Giant cardiac membrane patches: Na- and Na/Ca exchange current. Pfluegers Arch 415, 247-249.
7. A. Collins, A.V. Somlyo, and D.W. Hilgemann, The giant cardiac membrane pathc method: Stimulation of outward $Na^+$-$Ca^{2+}$ exchange current by MgATP. J. Physiol. 454, 27-57 (1992).
8. R. Goldshleger, S.J.D Karlish, A. Rephaeli and W.D. Stein (1987) The effect of membrane potential on the mammalian sodium-potassium pump reconsituted into phospholipid vesicles. J. Physiol. 387, 331-355.
9. S. Heyse, I. Wuddel, H.-J. Apell, and W. Sturmer, J. Gen. Physiol. [in press].
10. L. Vasilets, H.S. Omay, S. Ohata, M. Noguchi, M. Kawamura, and W. Schwarz (1991) J. Biol. Chem. 266, 16285-16288.

# Partial Reactions in $Na^+/K^+$- and $H^+/K^+$-ATPase Studied with Voltage-sensitive Fluorescent Dyes

I. Klodos

Institute of Biophysics, Aarhus University, 8000 Aarhus C, Denmark

The present paper discusses the results of a study of the partial reactions of $H^+/K^+$-ATPase (EC 3.6.1.36) in which we have used a special type of chemical compounds, the voltage sensitive, fluorescent, amphiphilic RH-dyes (aminostyrylpyridinium dyes). For comparison, some results with $Na^+/K^+$-ATPase (EC 3.6.1.37) will also be reported.

The RH-dyes were originally developed for optical recording of changes in the transmembrane potential of cells (9,11,13,21) and they were designed on the basis of the principle of electrochromism (see below) (19-22). Therefore, in contrast to earlier voltage sensitive probes which respond to the electric field by motions of their molecules, the response of RH-dyes is very rapid.

In 1988 we reported (16) that these dyes not only respond to transmembrane potentials, but also to events accompanying the hydrolysis of ATP by $Na^+/K^+$-ATPase in broken membranes, i. e. membranes where the electric potential is the same on both sides. The dyes also respond to events accompanying the hydrolysis of ATP by several other P-type cation transport ATPases -- gastric $H^+/K^+$-ATPase, sarcoplasmic reticulum $Ca^{2+}$-ATPase (EC 3.6.1.38) (18) and $H^+$-ATPase from *Neurospora Crassa* (EC 3.6.1.34) (25,26), and they do so without interfering with the enzymic activity of these ATPases. To better the understanding of the nature of the molecular events reported by RH-dyes we first consider the chemical structure, and a probable mechanism of response of RH-dyes.

**Aminostyrylpyridinium dyes -- their chemical structure, localization in membranes and mechanism of the response.**

The *p*-aminostyrylpyridinium chromophore of RH-dyes carries a positive charge centered in the pyridinium end of the molecule. Upon absorption of light an excited state is formed and the charge moves to the aniline end of the molecule.

RH-dyes partition into the lipid bilayer, and their orientation in the bilayer is determined by hydrophobic aliphatic chains on the aniline nitrogen of the chromophore and a hydrophilic charged group bound to the pyridinium nitrogen.

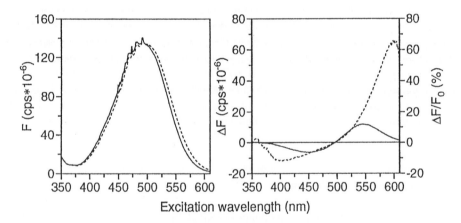

**Zwitterionic dye RH-421**

The dyes are presumably oriented perpendicular to the membrane surface with hydrophilic groups at the aqueous interface and lipophilic groups in the low dielectric region. The redistribution of charge within the chromophore of such an oriented dye molecule is influenced by the electric field, which affects the energy, i. e. the required wavelength of absorbed light for the transition from the ground state to the excited state.

The partitioning of the dye molecules into the lipid bilayer is connected with a significant increase in their ground state fluorescence and causes some shift in their absorption spectrum (2,13,18). Under the experimental conditions described below, more than 95% of RH-160 (18) and 90% of RH-421 (2) was bound to the membranes. When exposed to an electric field the bilayer bound dyes demonstrate a shift in both the absorption spectrum and the fluorescence excitation and emission spectrum. Such spectral changes were observed both in hemispherical lipid bilayers, lipid vesicles, vesicular preparations and in whole cells in response to the transmembrane potential changes.

Figure 1. Effect of TPB⁻ on the fluorescence excitation spectrum of membrane bound RH-421. Stippled line in the left panel shows the spectrum recorded in the presence of 2.5 μM TPB⁻. Right panel shows the absolute difference in fluorescence excitation spectra $\Delta F = F_{TPB} - F_{-TPB}$ and the fractional fluorescence change $\Delta F/F_{-TPB} = (F_{TPB} - F_{-TPB})/F_{-TPB}$ (stippled line).

The dyes also respond to modifications of the intramembrane charge caused by the insertion of lipophilic ions like tetraphenylphosphonium ($TPP^+$) or tetraphenylboron ($TPB^-$) (Fig. 1, and Fig. 2 in ref. 17). $TPB^-$ caused a red-shift in the fluorescence excitation spectrum of the membrane bound RH-421, seen both as absolute difference between the spectra ($\Delta F = F_{TPB} - F_{-TPB}$) and more clear-cut as the fractional fluorescence change $\Delta F/F_{-TPB}$. The direction of fluorescence excitation spectrum shift in response to lipophilic ions shows that the chromophore is localized between the surface and the adsorption plane for the lipophilic ions in the membrane which is 2-4 Å below the hydrophilic heads of membrane lipids (8). This is in agreement with the calculated total length of RH-dyes which is more than 23 Å, the length of the chromophore being about 15 Å.

According to the published data the mechanism of the response of the dyes to changes in electrical field is not purely electrochromic, but their response is modified  by a variety  of different factors, such as the orientation of the dyes in the membrane, the existence of *cis* and *trans* isomers and di- and oligomerization of the dyes in the lipid bilayers (3,4,11). Thus, since the exact mechanism of the dyes response to the electric field is still unknown the spectrum shift elicited by the lipophilic ions will be used in the following as a model for the response of the dyes to changes of local electric field.

Although a number of questions about the response mechanism are still unresolved, the RH-dyes have found increasing application in investigations of electrogenic properties of $Na^+/K^+$-ATPase (2,3,6,10,15,17,25,26,30,32,37,38) and other cation transport ATPases. This interest appear not only to be due to the fact that the dyes apparently report on charge movements by the ATPases, but also because a) most RH-dyes have no effect on the hydrolytic activity of the ATPases, b) large fractional fluorescence responses can be obtained, c) the dyes probably do not bind to ATPases, but partition into the bilayer. The latter point was documented in our study by the following observations: a) the fractional fluorescence response, defined as $(F_{event} - F_0)/F_0$ is independent of the dye/protein ratio within the range of 0.1 to more that 10 molecules of dye per 1 molecule of enzyme (18), and b) removal of the membranous structure, by solubilization of the enzyme, eliminates most of the fluorescence responses without affecting the activity of the shark $Na^+/K^+$-ATPase (7).

### Investigations of $Na^+/K^+$-ATPase

Fig. 2A shows a typical recording of RH-421 fluorescence responses to additions of $Na^+$, ATP and $K^+$. The main conformational states of $Na^+/K^+$-ATPase and the corresponding fractional fluorescence levels of RH-dyes are summarized in Fig. 2B. The magnitude of the fractional responses, especially to ATP addition in the presence of $Na^+$ and $Mg^{2+}$, i. e. to the phosphorylation of the enzyme, depends on the RH-dye used (17) and on the purity and the source of the enzyme preparation. With RH-421 the responses obtained with brain $Na^+/K^+$-ATPase (specific activity 5 U/mg protein) were much smaller than those recorded with dog kidney enzyme (10 U/mg) which gave a lower response than that obtained with pig kidney and shark rectal glands enzymes (20-25 U/mg protein). Enzyme purity is not the only factor determining the magnitude of RH-response. The RH-421 response to phosphorylation was at least 3 times higher with shark gland enzyme than with pig kidney enzyme, although both the phosphorylation levels and the specific activity of these enzymes were the same. We have at present no explanation for the dependency of the magnitude of the response on the degree of purification of the enzyme or on the enzyme source.

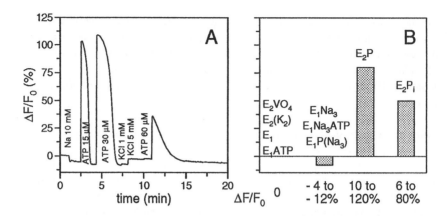

Figure 2. **A** Fractional fluorescence responses of membrane-bound RH-421 recorded with shark rectal glad $Na^+/K^+$-ATPase. The medium contained 40 μg of the enzyme (specific activity 25 U/mg protein), 1 μM RH-421, HEPES/MES (10 mM of each) buffer, pH 7.5, and 1 mM $MgCl_2$. The fluorescence was recorded in a SPEX fluorometer using an excitation wavelength 580 nm and a 630 nm emission cut-off filter.
**B** Conformational states of $Na^+/K^+$-ATPase and the fluorescence response of RH-dyes. The fluorescence was recorded as in **A** with dog kidney, pig kidney, bovine brain and shark rectal gland enzymes using zwitterionic dyes RH-160, RH-421, RH-237, and positively charged dyes RH-414, RH-461, and RH-795 ( ). The numbers below the figure show the range of responses.

The two main molecular events reported by the RH-dyes are the binding of $Na^+$ to the dephosphoenzyme, and the formation of phosphointermediate. The binding of $Na^+$ to $Na^+/K^+$-ATPase is registered by RH-dyes with a slight, but significant fall in the fluorescence (2,6,10,16,17,38). Ouabain inhibition or heat inactivation of the enzyme resulted in disappearance of the response (18). In the absence of $Na^+$ the binding of $K^+$ to the $E_1$ form of the enzyme did not elicit any fluorescence response (6,16) showing that RH-dyes are not indicators of $E_1$ to $E_2$ transitions. In the presence of $Na^+$, $K^+$ returned the fluorescence to "pre-$Na^+$ level" (6,16), showing once more that low fluorescence represents the Na-bound form.

ATP by itself, in the absence of $Na^+$ and/or $Mg^{2+}$ had no effect on the fluorescence (16,18). The phosphorylation of the enzyme elicited a large increase in fluorescence. The response was not a simple consequence of the phosphorylation. This was proved in experiments with a) chymotrypsin treated enzyme, where the transition of $E_1 P$ to $E_2P$ is inhibited (18,38), and b) by the Na-dependency of the response. a) With chymotrypsin treated enzyme the fluorescence response to the phosphorylation decreased in parallel with the activity of the enzyme whereas ATP binding, phosphorylation level, and the fluorescence response to $Na^+$ binding remained almost unaffected by the proteolysis (18). b) At high NaCl concentrations, where the total amount of the phosphoenzyme is constant but the ratio of $E_1P/E_2P$ is increased, the fluorescence response to phosphorylation of the enzyme was also diminished (16,18,38). Both experiments show that RH-dye response requires the transition of $E_1P$ to $E_2P$ and is probably due to $Na^+$ release from the phosphoenzyme. This conclusion finds support in the work of Pratap and Robinson (30). The phosphorylation with $P_i$ leads to a somewhat lower increase in the fluorescence than can be obtained by phosphorylation from ATP (17). Binding of $AsO_4$, but not of vanadate, also elicited an increase in the fluorescence similar to that of $P_i$ (18). The formation of enzyme-ouabain complex either in the presence of $Na^+$, $Mg^{2+}$ and ATP, or of $Mg^{2+}$ and $P_i$ also gives rise to a fluorescence change (18,37).

The fluorescence excitation spectrum shifts in response to $Na^+$ binding to the enzyme and to the formation of $E_2P$, i. e. a release of $Na^+$, are very similar to the shifts caused by local charge of lipophilic ions (Fig. 1 in ref. 6, and Fig. 2 in ref. 17). This suggests that the dyes are signalling changes in the electrostatic field surrounding the pump molecule as $Na^+$ is bound and released. The direction of the shift accompanying the binding of $Na^+$ corresponds to an appearance of positive charge in the vicinity of the chromophore. The response the release of $Na^+$ is consistent with an appearance of a negative charge.

### Investigations of $H^+/K^+$-ATPase

As discussed above, the molecular events in the $Na^+/K^+$-ATPase cycle reported by the RH-dyes are $Na^+$ binding and the formation of $E_1Na$ and the release of $Na^+$ accompanying the transition of $E_1P$ to $E_2P$. These are the steps in the Na-limb of the transport cycle, in which $Na^+/K^+$-ATPase moves one positive charge out of the cell (1,12). $H^+/K^+$-ATPase moves charge across the membrane in both H- and K-limbs of the transport cycle (1,23,33,35,40,41), and the formation of cation bound dephosphoforms ($E_1H_2$, $E_2K_2$ and $E_2(NH_4)_2$), as well as the formation of phosphoenzyme are accompanied by a significant change in the RH-dyes fluorescence.

The experiments were performed with vesicular fraction of hog gastric $H^+/K^+$-ATPase permeabilized by freeze-drying. Thus, as in the experiments with $Na^+/K^+$-ATPase described above, the RH-dyes and the ligands had access to both sides of the membrane and the transmembrane potential was zero.

The $H^+$ binding to $H^+/K^+$-ATPase was evaluated from the difference in the fluorescence levels of membrane bound RH-421 at pH 8.5 and pH 6.0. As seen from Fig. 4 the fluorescence level at pH 6.0 was 17% lower than at pH 8.5. This fluorescence change could reflect some unspecific effects of pH on the fluorescence of membrane bound RH-421 and/or an unspecific binding of protons to membrane proteins. The experiments performed with heat inactivated $H^+/K^+$-ATPase showed a much smaller pH dependent fall the fluorescence of the membrane bound dye (Fig. 3, left panel). Similar small response was seen with pig kidney $Na^+/K^+$-ATPase (Fig. 3, right panel). Taking into account that heat inactivation causes a disruption of hydrogen bonds and thus might increase proton binding, and that $Na^+/K^+$-ATPase binds $H^+$ specifically at low pH in the absence of $Na^+$ and $K^+$ (14,28), these results show that at least a part of the pH response is due to the specific binding of $H^+$ to $H^+/K^+$-ATPase. The binding of $H^+$ to the enzyme caused a shift in the fluorescence excitation spectrum, shown in Fig. 3, similar to those induced by binding of $Na^+$ to $Na^+/K^+$-ATPase or of $TPP^+$ to the membranes (compare Fig. 1 and Fig. 3, and Fig. 2 in ref. 17), indicating that the cation bound $E_1$ form of $H^+/K^+$-ATPase, $E_1H_2$, is also charged.

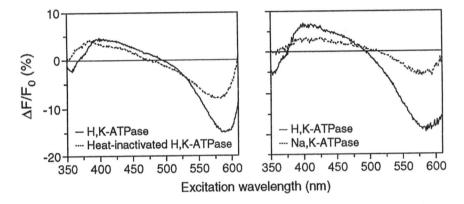

Figure 3. Shift in the fluorescence excitation spectrum of membrane bound RH-421 induced by $H^+$ binding to $H^+/K^+$- and $Na^+/K^+$-ATPase. Responses to $H^+$ binding to the native and to the heat inactivated $H^+/K^+$-ATPase are shown in the left panel. Right panel shows responses recorded with $H^+/K^+$-and $Na^+/K^+$-ATPase (equal amounts of protein with a similar specific activity). The displacement of fluorescence excitation spectrum is shown as a fractional change $\Delta F = (F_{pH\,6.0} - F_{pH\,8.5})/F_{pH\,8.5}$. The experiments were performed in HEPES/MES (10 mM of each) buffer system. The buffer was adjusted with N-methyl-D-glucamine (NMG) to pH 8.5 and, to eliminate unspecific effects of ionic strength and $NMG^+$ on the $H^+$ binding, NMGCl was added to the same concentration at pH 6.0.

Binding of $K^+$ or its congener $NH_4^+$ (23) to $H^+/K^+$-ATPase was accompanied by a decrease in the fluorescence. At pH 8.5 saturating $[K^+]$ or $[NH_4^+]$ induced ~13% decrease, similar to the fall elicited by pH change. $K_{0.5}$ for $K^+$ and $NH_4^+$, $4 \pm 0.5$ mM and for $21 \pm 1.3$ mM, respectively, are comparable with the values reported in the literature (5,31). SCH 28080, which inhibits $H^+/K^+$-ATPase by preventing $K^+$ binding to the enzyme (24,34,39,42), partially blocked the response to $K^+$ (not shown). Binding of $Na^+$, which substitutes for $H^+$ in the $H^+/K^+$-ATPase reaction at high pH (5,29,31), was also followed by a decrease in the fluorescence with $K_{0.5}$ of $10 \pm 0.6$ mM, but the response was only half of that to $K^+$ (not shown). The fluorescence excitation spectrum shift induced by the binding of $K^+$ or $NH_4^+$, but not $Na^+$, displayed the same characteristics as the response to $H^+$ binding, indicating that both $E_2K_2$ and $E_2(NH_4)_2$ are charged.

At pH 6.0 both $K^+$ and $NH_4^+$, but not $Na^+$, induced fluorescence decrease although the maximal response was only one-third of that at high pH (not shown). Both responses were blocked by SCH 28080, and the affinity of the enzyme to the inhibitor was much higher, indicating that the enzyme is inhibited by the protonated form of SCH 28080

(42). The analysis of the fluorescence excitation spectrum did not shown any distinct shift of the spectrum upon cation binding. This could indicate that the binding of $K^+$ or $NH_4^+$ and the transition from $E_1H_2$, predominant at the low pH, to $E_2K_2$ (or $E_2(NH_4^+)_2$) does not change the charge of the intermediates, i. e. both $E_1H_2$ and $E_2K_2$ are charged intermediates.

Figure 4 shows a typical recording of RH-421 fluorescence responses to sequential additions of ATP, $K^+$, and then to ATP in the presence of $K^+$ at pH 6.0 and pH 8.5. The presence of $Mg^{2+}$ was essential for the response to ATP, demonstrating that it was due to phosphorylation of the enzyme. SCH 28080 added *prior* to ATP completely inhibited the response at pH 6.0, but not at pH 8.5. At high pH the response was still present even with 200 μM SCH 28080 in the medium, although the activity, measured as the time necessary for the return to the starting fluorescence level, was significantly decreased (not shown). Some acceleration of ATP hydrolysis by $K^+$ was seen in both experiments, and in both experiments an increase in [ATP] increased the response, but $K^+$ had more pronounced effect on the magnitude of response at low pH.

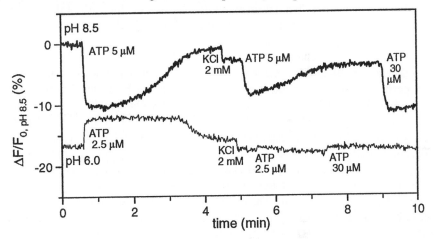

Figure 4. Fractional fluorescence responses of membrane bound RH-421 recorded with $H^+/K^+$-ATPase from hog stomach. The medium contained 1 μM RH-421, 1 mM $MgCl_2$, HEPES/MES (10 mM of each) buffer, pH 8.5 (upper trace) or 6.0 (lower trace) and ~ 200 μg of freeze-dried vesicular fraction (specific activity 2.3 U/mg protein measured in the presence of 60 mM $NH_4Cl$, 5 mM $MgCl_2$, 5 mM ATP in 30 mM imidazole buffer pH 7.1 at 37 °C). The level of fluorescence at pH 6.0 is expressed in % of that at pH 8.5.

Phosphorylation by $P_i$ in the absence of $K^+$ induced changes in fluorescence in the same direction as those caused by the phosphorylation from ATP. Also the effects of SCH 28080 and $K^+$ were the same. Moreover, the magnitude of $P_i$ induced fluorescence changes was at least the same as with ATP in agreement with the previous data showing that the levels of phosphoenzyme formed from ATP and from $P_i$ are the same (36,39).

The fact that the characteristics of responses induced by phosphorylation of the enzyme from ATP or $P_i$ are identical implies, in agreement with data in the literature (36,39,41), that the same phosphoenzyme is formed by the two routes of phosphorylation. And since the phosphoenzyme formed from $P_i$ is $E_2P$ (36,39,41) it proves that also with $H^+/K^+$-ATPase the ATP response of RH-421 is due to the formation of the $E_2P$ form.

The question is then why does the formation of $E_2P$ induce a decrease in the fluorescence at high pH, but an increase at low pH. Can these responses be interpreted in terms of charge of the phosphointermediate? The excitation spectrum shifts upon the formation of $E_2P$ from either ATP or $P_i$ displayed an electrochromic characteristics, but with opposite directions of the shift at low and at high pH. In the experiments performed with $Na^+/K^+$-ATPase under identical conditions the formation of $E_2P$ induced only a red-shift independently of pH.

Let us consider the following: At high pH, where the concentration of proton bound $E_1H_2$ form is low, the fluorescence level of the membrane bound RH-421 is high. At low pH and thus a high concentration of $E_1H_2$, the fluorescence level is low. The formation of $E_2P$ is accompanied by an increase in the fluorescence in comparison with the $E_1H_2$ level (a red-shift of the spectrum), but a decrease in fluorescence (a blue-shift of the spectrum) in comparison with the $E_1$ level. Thus the charge of the $E_2P$ form appears to be in between that of the $E_1$ and the $E_1H_2$ forms.

In the following we assume that the net charge of $H^+/K^+$-ATPase molecule is increased by two positive charges upon the binding of 2 $H^+$ to the $E_1$ form, denoted here as $^{+2}H_2E_1$, where the number of charges refers to the relative charge with respect to that of $E_1$ form, which is assigned a zero charge, $^0E_1$. Furthermore it is assumed that all intermediates with 2 $H^+$ bound carry the same charge. Since it has been suggested that the transition of $H_2E_1P$ to the K-sensitive phosphoenzyme is not accompanied by a release of $H^+$ (36,41), both $H_2E_1P$ and $H_2E_2P$ will have the same positive charge, viz. $^{+2}H_2E_1P$ and $^{+2}H_2E_2P$ in our notation. Only in a step after transition to $H_2E_2P$ form are protons released and $E_2P$ formed. The charge of $E_2P$ will be the same as that of ligand-free enzyme, i. e. $^0E_2P$.

It has been previously shown that the ratio of $E_1P$ to $E_2P$ intermediates is almost independent of pH and that the K-sensitive $E_2P$ is a predominant phosphoenzyme species (36,39,41). If this K-sensitive species is $^0E_2P$ then the phosphorylation of the enzyme should induce an increase in the fluorescence. The increase should be much greater at pH 6.0 than at pH 8.5, because of the concentration of charged $^{+2}H_2E_1$ is higher at this pH and thus the change of charge upon formation of $^0E_2P$ will be greater. And this is clearly not the case as can be seen from fig. 4. If a predominant phosphointermediate is $^{+2}H_2E_2P$ then no fluorescence response, or even a decrease in the fluorescence, should be seen at pH 6.0, which is again contrary to the experimental observation. Thus under the steady state conditions neither $^0E_2P$ nor $^{+2}H_2E_2P$ can be predominant. One of the possibilities is illustrated in the left panel of the scheme. In this case the $E_2P$ pool consists of both $^{+2}H_2E_2P$ and $^0E_2P$ and the fluorescence response of RH-dye reflects an average charge of $E_2P$-pool. Since the phosphorylation at high pH will result in the formation of $^0E_2P$ but also of $^{+2}H_2E_2P$ the average charge of the enzyme species will increase and the fluorescence will decrease in comparison with pre-phosphorylation level.

524

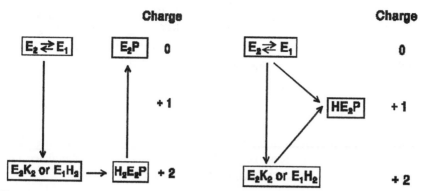

The opposite will happen at low pH, which is exactly the behavior seen in the Fig. 4. There is, however, one observation which remains unaccounted for - the identity of the fluorescence responses to the phosphorylation from ATP and $P_i$. It is difficult to imagine that $^{+2}H_2E_2P$ will be formed through a low affinity $H^+$ binding to $^0E_2P$ especially at high pH. This means that the description of the $E_2P$ pool as consisting of only two phosphoenzyme forms, $^{+2}H_2E_2P$ and $^0E_2P$, appears to be insufficient. We propose therefore an additional form of the $E_2$-phosphoenzyme with a charge in between that of $^{+2}H_2E_2P$ and $^0E_2P$, i. e. $^{+1}HE_2P$, as illustrated in the right panel of the scheme.

Thus the assumption that both $H_2E_1$, $H_2E_1P$ and $H_2E_2P$ carry 2 positive charges more than $E_1$ and $E_2P$ allows to simulate the observed fluorescence response to the phosphorylation of the enzyme and leads to the following reaction scheme for $H^+/K^+$-ATPase:

$$^0E_1 \rightleftharpoons {}^{+2}H_2E_1 \rightleftharpoons {}^{+2}H_2E_1P \rightleftharpoons {}^{+2}H_2E_2P \rightleftharpoons {}^{+1}HE_2P \rightleftharpoons {}^0E_2P \rightleftharpoons {}^{+2}K_2E_2P \rightleftharpoons {}^{+2}K_2E_2 \rightleftharpoons {}^0E_1$$

The above analysis of the correlation between the RH-dyes fluorescence responses and changes of the charge of the enzyme induced by cation binding to and release from the enzyme finds support in the similar analysis of the results obtained with $Na^+/K^+$-ATPase.

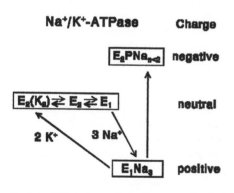

It has been suggested that binding of 2 $K^+$ is electroneutral, whereas the binding of 3 $Na^+$ results in $E_1Na_3$ form with 1 positive charge (27), and consistently the decrease in RH-dye fluorescence is induced by binding of $Na^+$, but not of $K^+$. The phosphorylation of the enzyme results in the formation of the ADP sensitive $E_1P(Na_3)$ form and subsequently the K-sensitive $E_2P$. The transition is accompanied by a stepwise release of $Na^+$, and thus a change of charge. The dyes do not respond to the phosphorylation of the enzyme when the transition of $E_1P(Na_3)$ to $E_2P$, and thus $Na^+$ release, is hindered. The release of $Na^+$ upon the conversion of $E_1P(Na_3)$ to $E_2P$ induces an increase in the fluorescence, associated with the red-shift of the fluorescence excitation spectrum indicating the appearance of negative charge in the vicinity of the dye. Release of the first $Na^+$ should result in a decrease of charge from +1 to 0 and of the next $Na^+$ to -1 and -2. Thus the direction of response of the RH-dyes to $Na^+$ release should show an increased negative charge in comparison with $E_1Na_3$ or $E_1P(Na_3)$ level.

**Summary and conclusions**

1. RH-dyes respond to changes in the charge of the cation binding and release sites, i. e. to cation binding to the $E_1$ form and cation release from the phosphorylated intermediates, as shown by an electrochromic shift of the excitation spectrum.

2. The fluorescence responses of RH-dyes indicates that the two ATPases differ in their cation binding. With $Na^+/K^+$-ATPase only the binding of three $Na^+$, but not of two $K^+$, is connected with the appearance of a positively charged intermediate. With $H^+/K^+$-ATPase both binding of two $H^+$ or of two $K^+$ leads to a positively charged state.

3. Based on the analysis of RH-421 response to the phosphorylation of $H^+/K^+$-ATPase we suggest that the net charge of $H^+/K^+$-ATPase molecule is increased by two positive charges upon binding of 2 $H^+$, and that the phosphorylated intermediates, both $H_2E_1P$ and $H_2E_2P$ carry the same charge. The response of the dye to the phosphorylation appears to be elicited by a release of $H^+$ upon the formation of $E_2P$ and/or $HE_2P$.

*Acknowledgements.* The author wishes to thank Flemming Cornelius, Carin Briving and Elisabeth Skriver for their helpful collaboration, Jan Joop H.H.M. De Pont and Herman G.P. Swarts for teaching how to prepare $H^+/K^+$-ATPase, Karl A. Jørgensen for calculations on the structure of RH-dyes, and Jens G. Nørby for his encouragement, valuable comments and critical reading of the manuscript. The author wishes to express her gratitude to Biff Forbush, with whom the studies with $Na^+/K^+$-ATPase and RH-dyes began, and to Natasha U. Fedosova for her collaboration, criticism and help. The work was supported by the NIH (GM-31782), The Danish Medical Research Foundation, The Danish Natural Science Foundation, The Danish Research Academy, The Novo Nordic Foundation, and The Biomembrane Research Center, Aarhus University.

**References**

1.      Bamberg E, Butt H-J, Eisenrauch A, Fendler K (1993) Charge transport of ion pumps on lipid bilayer membranes. Quarterly Reviews Biophys 26: 1-25
2.      Bühler R, Stürmer W, Apell H-J, Läuger P (1991) Charge translocation by the Na,K-pump: I. Kinetics of local field changes studied by time-resolved fluorescence measurements. J Membrane Biol 121: 141-161
3.      Clarke RJ, Schrimpf P, Schöneich M (1992) Spectroscopic investigations of the potential-sensitive membrane probe RH421. Biochim Biophys Acta 1112: 142-152
4.      Ephardt H, Fromherz P (1989) Fluorescence and photoisomerization of an amphiphilic aminostilbazolium dye as controlled by the sensitivity of radiationless

deactivation to polarity and viscosity. J Phys Chem 93: 7717-7725

5.     Faller LD, Diaz RA, Scheiner-Bobis G, Farley RA (1991) Temperature dependence of the rates of conformational changes reported by fluorescein 5'-isothiocyanate modification of $H^+,K^+$- and $Na^+,K^+$-ATPases. Biochemistry 30: 3503-3510

6.     Fedosova NU, Klodos I (1993) Conformational states of $Na^+/K^+$-ATPase probed with RH-421 and eosin. This volume

7.     Fedosova NU, unpublished

8.     Flewelling RF, Hubbell WL (1986) The membrane dipole potential in a total membrane potential model. Application to hydrophobic ions interactions with membranes. Biophys J 49: 541-552

9.     Fluhler E, Burnham VG, Loew LM (1985) Spectra, membrane binding, and potentiometric responses of new charge shift probes. Biochemistry 24: 5749-5755

10.    Forbush B III, Klodos I (1991) Rate-limiting steps in Na translocation by the Na/K pump. Soc Gen Physiol Ser 46: 211-225

11.    Fromherz P, Müller CO (1993) Voltage-sensitive fluorescence of amphiphilic hemicyanine dyes in neuron membranes. Biochim Biophys Acta 1150: 111-122

12.    Glynn IM (1993) All hands to the sodium pump. J Physiol (London) 462: 1-30

13.    Grinvald A, Hildesheim R, Farber IC, Anglister L (1982) Improved fluorescent probes for the measurement of rapid changes in membrane potential. Biophys J 39: 301-308

14.    Hara Y, Nakao M (1986) ATP-dependent proton uptake by proteoliposomes reconstituted with purified $Na^+,K^+$-ATPase. J Biol Chem 261: 12655-12658

15.    Heyse S, Apell H-J, Stürmer W (1993) Electrogenic partial reactions of the sodium pathway of the Na,K-ATPase. this volume

16.    Klodos I, Forbush B III (1988) Rapid conformational changes of the Na/K pump revealed by a fluorescent dye, RH-160. J Gen Physiol 92: 46a

17.    Klodos I, Cornelius F, Fedosova, NU (1993) Screening of different potential sensitive fluorescent RH-dyes as probes for $Na^+/K^+$-ATPase reaction. this volume.

18.    Klodos I, Schoppa NL, Forbush B III, unpublished

19.    Loew LM (1982) Design and characterization of electrochromic membrane probes. J Biochem Biophys Methods 6: 243-260

20.    Loew LM, Bonneville GW, Surow J (1978) Charge shift optical probes of membrane potential. Theory. Biochemistry 17: 4065-4071

21.    Loew LM, Cohen LB, Salzberg BM, Obaid AL, Bezanilla F (1985) Charge-shift probes of membrane potential. Characterization of aminostyrylpyridinium dyes on the squid giant axon, Biophys J 47: 71-77

22.    Loew LM, Simpson LL (1981) Charge-shift probes of membrane potential. A probable electrochromic mechanism for $p$-aminostyrylpyridinium probes on a hemispherical lipid bilayer. Biophys J 34: 353-365

23.    Lorentzon P, Sachs G, Wallmark B (1988) Inhibitory effects of cations on the gastric $H^+,K^+$-ATPase. A potential-sensitive step in the $K^+$ limb of the pump cycle. J Biol Chem 263: 10705-10710

24.    Mendlein J, Sachs G (1990) Interaction of a $K^+$-competitive inhibitor, a substituted imidazol[1,2a]pyridine, with the phospho- and dephosphoenzyme forms of $H^+,K^+$-ATPase. J Biol Chem 265: 5030-5036

25.    Nagel G, Slayman CL, Klodos I (1989) Fluorescence probing of a major conformational change in the plasma membrane $H^+$-ATPase of *Neurospora*. Biophys J 55: 338a

26.    Nagel G, Klodos I, Xu J-C, Slayman CL (1990) Evidence for binding of ATP to the $E_2$-conformation of the plasma membrane ATPase from *Neurospora*. Biophys J 57:

349a

27. Or E, David P, Shainskaya A, Tal DM, Karlish SJD (1993) Effects of competitive sodium-like antagonists on Na,K-ATPase suggest that cation occlusion from the cytoplasmic surface occurs in two steps. J Biol Chem 268: 16929-16937

28. Polvani C, Blostein R (1988) Protons as substitutes for sodium and potassium in the sodium pump reaction. J Biol Chem 263: 16757-16763

29. Polvani C, Sachs G, Blostein R (1989) Sodium ions as substitutes for protons in the gastric H,K-ATPase. J Biol Chem 264: 17854-17859

30. Pratap PR, Robinson JD (1993) Rapid kinetic analyses of the $Na^+/K^+$-ATPase distinguish among different criteria for conformational change. Biochim Biophys Acta 1151: 89-98

31. Rabon EC, Bassilian S, Sachs G, Karlish SJD (1990) Conformational transitions of the H,K-ATPase studied with sodium ions as surrogates for protons. J Biol Chem 265: 19594-19599

32. Ruf H, Lewitzki E, Grell E (1993) Na,K-ATPase: probing of a high affinity binding site for $Na^+$ by spectrofluorometry. this volume

33. Sachs G, Chang HH, Rabon E, Schackman R, Lewin M, Saccomani G (1976) A nonelectrogenic $H^+$ pump in plasma membranes of hog stomach. J Biol Chem 251: 7690-7698

34. Scott CK, Sundell E, Castrovilly L (1987) Studies on the mechanism of action of the gastric microsomal $(H^+ + K^+)$ATPase inhibitors SCH 32651 and SCH 28080. Biochem Pharmacol 36: 97-104

35. Stegelin M, Fendler K, Bamberg E (1993) Kinetics of transient pump currents generated by the (H,K)-ATPase after an ATP concentration jump. J Membrane Biol 132: 211-227

36. Stewart B, Wallmark B, Sachs G (1981) The interaction of $H^+$ and $K^+$ with the partial reactions of gastric $(H^+ + K^+)$-ATPase. J Biol Chem 256: 2682-2690

37. Stürmer W, Apell H-J (1992) Fluorescence study on cardiac glycoside binding to the Na,K-pump. Ouabain binding is associated with movement of electrical charge. FEBS Lett 300: 1-4

38. Stürmer W, Bühler R, Apell H-J, Läuger P (1991) Charge translocation by the Na,K-pump: II. Ion binding and release at the extracellular face. J Membrane Biol 121: 163-176

39. Van der Hijden HTWM, Koster HPG, Swarts HGP, De Pont JJHHM (1991) Phosphorylation of $H^+/K^+$-ATPase by inorganic phosphate. The role of $K^+$ and SCH 28080. Biochim Biophys Acta 1061: 141-148

40. Van der Hijden HTWM, Grell E, De Pont JJHHM, Bamberg E (1990) Demonstration of the electrogenicity of proton translocation during the phosphorylation step in gastric $H^+$ $K^+$-ATPase. J Membrane Biol 114: 245-256

41. Van Uem TJF, De Pont JJHHM (1992) Structure and function of gastric H,K-ATPase. New Comprehensive Biochemistry 21: 27-55

42. Wallmark B, Briving C, Fryklund J, Munson K, Jackson R, Mendlein J, Rabon E, Sachs G (1987) Inhibition of gastric $H^+,K^+$-ATPase and acid secretion by SCH 28080, a substituted pyridyl(1,2a)imidazole. J Biol Chem 262: 2077-2084

# Changes of membrane capacitance coupled with electrogenic transport by the $Na^+/K^+$-ATPase

V.S. Sokolov[†], K.V. Pavlov[†], K.N. Dzhandzhugazyan[‡]

Frumkin Institute of Electrochemistry, Russian Academy of Science, Moscow, 117071, Russia[†] and Shemyakin and Ovchinnikov Institute of Bioorganic Chemistry, Russian Academy of Sciences, Moscow, 117871, Russia[‡]

## Introduction

In a recent publication we have shown that changes of capacitance of lipid bilayer membrane with adsorbed membrane fragments containing $Na^+/K^+$-ATPase can be induced by photodissociation of "caged"-ATP or addition of ATP to the solution [8]. In our subsequent work we tried to identify the nature of the effect and to investigate its dependence upon frequency under different conditions.

## Measurement of small changes of membrane capacitance

Bilayer membranes were created on the 1 mm. orifice in the wall between two compartments of a teflon cell. To measure the capacitance changes a sine wave voltage of ±50 mV was applied to the electrodes submerged into the cell. Initial capacitive current was compensated by a special circuit consisting of the voltage inverter and RC-equivalent of the cell. To obtain abrupt increase of ATP concentration we used its photo-activated form, so called "caged"-ATP [3]. Immediately after the laser flash of ultraviolet light about 10% of its initial concentration released in the form of free ATP, thus activating the ATPase and causing a change of the capacitive current. Changes of capacitance were measured by Lock-in-Amplifier. In order to get rid of experimental errors caused by frequency dependence of the instrumentation the rate of the signal measured under each frequency was set by the value of full capacitance. In some series of experiments bathing solution contained ATP consuming enzyme, apyrase. It was added to the solution in order to speed up expiration of free ATP and hence the ATPase recovery time [1]. All the following results were obtained in potassium-free solution where capacitance has been found to increase on activation of the ATPase.

## Kinetics of the decay of capacitance with time

Fig. 1a represents records of the change of capacitance after light flash ($\Delta C$) in two experiments at the same frequency of the alternating voltage (20 Hz), in presence and absence of apyrase. It is clearly visible that in the presence of apyrase this change decays much faster. This result shows that some fraction of the capacitance increment decays simultaneously with removal of ATP from the solution. The value of the decay depended on the frequency of voltage applied. Fig. 1b shows three records of the change of capacitance made with the same membrane in the same concentration of apyrase, but with three different frequencies. At higher frequencies the capacitance increment is much smaller and does not decay so fast.

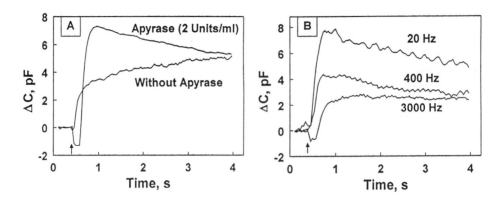

**Figure 1:** Time dependence of capacitance increment at different conditions. The lipid bilayer was formed by a mixture of diphytanoyllecithin (20 mg/ml) and octadecylamine (2 mg/ml) dissolved in n-decane. The absolute value of capacitance after adsorption of membrane fragments with $Na^+/K^+$-ATPase was 420 pF. The water solution contained 130 mM NaCl, 25 mM imidazol, 10 mM $MgCl_2$, 5 mM dithiothreitol, pH 6.5 and apyrase. The time moments of UV light flash are marked by arrows. **A:** Influence of apyrase (2 units/ml) on the decay of capacitance at 20 Hz, **B:** Influence of frequency on the decay of capacitance in the presence of apyrase (2 units/ml).

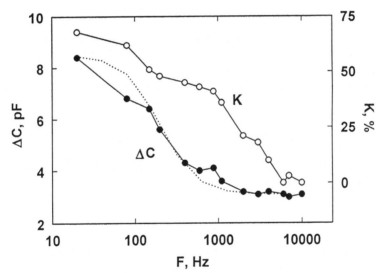

**Figure 2:** Dependence of the increment of capacitance $\Delta C$ and of the relaxation coefficient K on the frequency of alternating voltage. The values of $\Delta C$ measured at each frequency were normalised by dividing this value to the absolute value of capacitance measured at the same frequency and multiplying the result to the capacitance measured at 20 Hz. The dashed line is the function $\Delta C = \Delta C_g + \Delta C_e \, F_o^2/(F^2+F_o^2)$ with the parameters equal to: $\Delta C_g = 3.1$ pF, $\Delta C_e = 5.4$ pF, $F_o = 200$ Hz.

## Dependence of the capacitance change on frequency

The frequency dependence of the effect is shown in the figure 2, where open circles represent values of $\Delta C$ and filled circles - a characteristic of the signal decay for which we have chosen the ratio $K = (\Delta C_{max} - \Delta C_\infty)/\Delta C_\infty$ , where $\Delta C_{max}$ is the maximal signal and $\Delta C_\infty$ the signal 5 seconds after the UV flash (5 seconds was the time range of the experimental score). The value of K which we named relaxation coefficient exhibits distinct correlation with the values of capacitance increment $\Delta C$. The capacity increment itself is diminished with increasing frequency and reaches a plateau at frequencies higher then 1 kHz. The value of K decreases with frequency and becomes virtually equal to zero at frequencies higher then 1 kHz. All these data allow to suggest that the measured signal consists of different terms which are caused by different processes.

The first one is ATP dependent, it disappears instantaneously when ATP is expired, and can be measured only at low frequencies. The second process appears to be independent upon the actual concentration of ATP and upon frequency (in the range under investigation) and can be related with a real change of the system capacitance that does not cease immediately when ATP concentration drops to zero. Such changes can be caused for example by osmotic swelling of the gap between absorbed membranes or the compression of the membrane in the electric field generated by the $Na^+/K^+$-ATPase. The ATP dependent process seems to be more interesting and to contain more information about electrogenic transport by ATPase. It can be ascribed to the so called pseudo capacitance caused by displacement of charged groups or ions inside the membrane. If a the simple model of the charge moving between two states is applied [4,5] one can show that the frequency dependence of pseudo-capacitance can be described by the function $F_0^2/(F^2+F_0^2)$, where F is the frequency and $F_0$ a kinetic parameter similar to the rate constant of the exponential decay of the current in voltage-clamp experiments. The fit of experimental points by this function is shown in figure 2 by the dotted curve. It can be seen that this function describes satisfactory the experimental data at high frequencies, but there is some disagreement in the low frequency range. This disagreement may indicate on the existence of other relaxation processes which are also dependent on the concentration of ATP. The disagreement of the simplest two state model with experiment was found also in other works [2,5,7].

Although there are no comprehensive data on the temperature dependence of the effect, some preliminary results show the trend that the transition frequency $F_0$ increases with temperature. A similar behaviour of current relaxation rate constant with temperature has been reported in the work with voltage-clamp technique [4]. This result is in agreement with the hypothesis on the nature of the effect for the mobility of charged particles and hence that the frequency, up to which pseudo capacitance effects can be measured, should increase with temperature.

*Acknowledgements*

We thank H.-J. Apell for kindly provided preparations of apyrase and fruitful discussion. The work was supported by Russian Foundation of Fundamental Research (Grant #93-04-7652) and partially by "Cygnus Therapeutic" company.

**References**

1. Borlinghaus, R., Apell, H.-J., Läuger, P.(1987) Fast charge translocation associated with partial reactions of the Na,K pump: I. Current and voltage transients after photochemical release of ATP. J Membrane Biol 97: 161-178
2. De Weer P., Gadsby D.C, Rakowski R.F. (1988) Voltage dependence of the Na-K pump. Annu Rev Physiol 50: 25-241
3. Fendler, K., Grell, E., Haubs, M., Bamberg, E. (1985) Pump currents generated by the purified $Na^+,K^+$-ATPase from kidney on black lipid membranes. EMBO J 4:3079-3085
4. Gadsby, D.C., Nakao M., Bahinski A. (1991) Voltage-induced Na/K pump charge movements in dialysed heart cells. In: J. H. Kaplan and P. De Weer (eds)The Sodium Pump: Structure, mechanism, and Regulation. Rockfeller University Press, New York, pp 356-371.
5. Läuger P., Apell H.-J. (1988) Transient behaviour of the Na+/K+ pump: Microscopic analysis of nonstationary ion translocation. Biochim Biophys Acta 944: 461-464
6. Nakao, M., Gadsby, D.C. (1986) Voltage dependence of Na translocation by the Na/K pump. Nature 323:628-630.
7. Rakowski R.F. (1993) Charge movement by the Na/K pump in Xenopus oocytes. J Gen Physiol 101:117-144.
8. Sokolov V.S., Pavlov K.V., Dzhandzhugazyan K.N., Bamberg E. (1992) Capacitance and conductivity changes of the model membrane due to Na,K ATPase action. Biologicheskie Membrany 9: 961-969 (in Russian).

# Reversal potential of $Na^+$ pump current in cardioballs at various $\Delta G$ of ATP hydrolysis

H.G. Glitsch, A. Tappe

Department of Cell Physiology, Ruhr–University, 44780 Bochum, Germany

## Introduction

The $Na^+/K^+$ pump of animal cells extrudes $3\ Na^+$ from the cell interior and takes up $2\ K^+$ per ATP molecule hydrolysed. Thereby the transport generates the pump current $I_p$ which is an outward current under physiological conditions. The chemical and electrical work, which can be performed by the pump, depends on the free energy of the pump's ATP hydrolysis ($\Delta G_{ATP}$). If intracellular $Na^+$ and external $Cs^+$ are pumped the following equation should hold:

$$\Delta G_{ATP} = F \times (E_{rev} - 3\ E_{Na} + 2\ E_{Cs}) \qquad \text{(eqn. 1)}.$$

F is the Faraday constant; $E_{Na}$ and $E_{Cs}$ stand for the $Na^+$ and $Cs^+$ equilibrium potential, respectively. Thus at fixed $Na^+$ and $Cs^+$ gradients across the cell membrane $E_{rev}$, the membrane potential where $I_p$ changes its sign, is expected to vary with $\Delta G_{ATP}$. To test this expectation we have carried out experiments on isolated, cultured cardiac cells. Some results of these experiments are described below.

## Methods

The membrane current of cardioballs prepared from sheep Purkinje fibres (5) was measured by means of whole cell recording (7) at various clamp potentials (V) and 30 to 34°C. The holding potential was $-20$ mV throughout. The measurements were started 10 to 15 min after getting access to the cell interior. $I_p$ was identified as current blocked by $2 \times 10^{-4}$ M dihydroouabain (DHO), a specific inhibitor of the $Na^+/K^+$ pump. The $[ADP]x[P_i]/[ATP]$ quotient of the pipette solution was calculated to give two different $\Delta G_{ATP}$ values ($-60$ kJ/mole and $-39$ kJ/mole, respectively) according to:

$$\Delta G_{ATP} = \Delta G^{\bullet} + R \times T \times \ln \frac{[ADP] \times [P_i]}{[ATP]} \qquad \text{(eqn. 2)}.$$

The appropriate $\Delta G^{\bullet}$ value, which varies with free $[Mg^{2+}]$ of the medium, was taken from Guynn & Veech (6). Free $[Mg^{2+}]$ and binding constants at 32 °C were derived by means of a program written by Fabiato (2). The internal (pipette) solution for $\Delta G_{ATP} = -60$ kJ/mole ($-39$ kJ/mole) contained (mM): 110(60) CsCl; 30(50) CsOH; 10(10) NaOH;

10(10) MgATP; 0.3(20) TrisADP; 0.15(10) $Cs_2HPO_4$, 0.15(10) $CsH_2PO_4$, 3(3) $MgCl_2$; 6(6) EGTA; 56(36) HEPES. The pH of both media was adjusted to 7.4 with CsOH. The external (bathing) solution contained (mM): 140 NaCl; 10 NaOH; 6CsCl; 0.9 $CaCl_2$; 1.5 $MgCl_2$; 20 HEPES (pH 7.4 with HCl); 2 $BaCl_2$ and 5 $NiCl_2$ (to block the K and Ca conductances, and the Na/Ca exchange of the sarcolemma, respectively). Thus the $Na^+$ and $Cs^+$ gradients across the cell membrane and the [MgATP] of the pipette solution remained always constant. The initial resistance of the patch pipettes (when filled with one of the pipette solutions) varied between 1 and 3 MΩ. The surfaces of the cells studied were estimated from microscopic measurements (5).

## Results

Fig.1 A displays an original record of the membrane current at various clamp potentials starting from the holding potential of –20 mV. The internal (pipette) solution designed for $\Delta G_{ATP}$ = –60 kJ/mole is used. First a control current–voltage relationship is recorded.

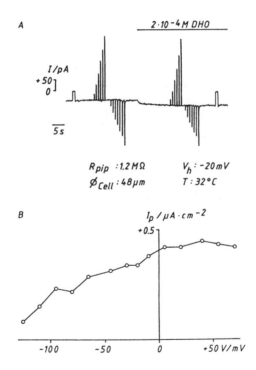

**Figure 1.** $I_p$–V relation of a cardioball at $\Delta G_{ATP}$ = – 60 kJ/mole. **A.** Original record of I–V curves before and during inhibition of the $Na^+/K^+$ pump by DHO (bar). The two rectangular signals at the start and the end of the current trace indicate zero current level. **B.** $I_p$–V relationship calculated from the difference between the two I–V curves displayed in A. No $E_{rev}$ in the potential range studied.

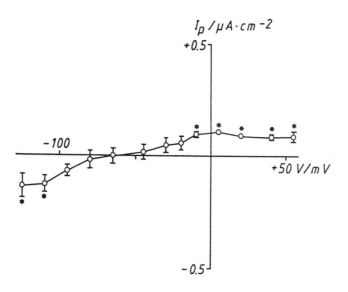

$I_p / \mu A \cdot cm^{-2}$

+0.5

−100

+50 V/mV

−0.5

**Figure 2.** Mean Ip–V relationship of sheep Purkinje cardioballs at $\Delta G_{ATP}$ = −39 kJ/mole (n = 5). Note $E_{rev}$ near −65 mV. Asterisks mark $I_p$ densities significantly different from 0 $\mu A$ x cm$^{-2}$.

Thereafter a DHO containing bathing fluid is applied to the cell. The inhibition of the $Na^+/K^+$ pump by the drug causes an inward shift of the holding current. A second current–voltage curve is measured during the blockade of the active $Na^+$ transport. The current amplitude recorded at the end of a 200 ms pulse to each clamp potential during the second run was subtracted from the the corresponding control current. The resulting $I_p$–V relationship is shown in Fig. 1B. The pump current density is plotted versus the membrane potential. The $I_p$–V curve exhibits a positive slope at negative potentials and little change at positive voltages. $E_{rev}$ of $I_p$ is not located within the accessible range of membrane potentials. By the way of contrast $E_{rev}$ can be found in this voltage range if the pipette solution for $\Delta G_{ATP}$ = −39 kJ/mole is used for the intracellular perfusion. Fig. 2 shows the mean $I_p$–V relationship (mean ± S.E.M.) of five cardioballs under the latter condition. The plot suggests $E_{rev}$ of $I_p$ to be near −65 mV. However, it should be realized that the mean $I_p$ densities do not differ significantly from 0 $\mu A$ x cm$^{-2}$ between −20 and −95 mV. The current densities differ from zero at more negative and positive membrane potentials (asterisks; $p < 0.05$). Compared to Fig. 1B the general slope of the $I_p$–V curve seems to be essentially unchanged though the $I_p$ densities are much smaller at the potentials positive to $E_{rev}$.

535

## Discussion

The shape of the $I_p$–V relationship depicted in Fig. 1B is typical for cardiac cells under physiological conditions. $E_{rev}$ of $I_p$ is negative to the range of membrane potentials accessible (3,5). In contrast, if $\Delta G_{ATP}$ is reduced to –39 kJ/mole $E_{rev}$ is clearly situated within this voltage range (Fig. 2). The finding of an inwardly directed $I_p$ at potentials negative to $E_{rev}$ is not unexpected in view of the early report by Garrahan & Glynn (4) on backwards running of the $Na^+/K^+$ pump under appropriate conditions. Furthermore, Bahinski et al. (1) already demonstrated that a substantial inward $I_p$ can be recorded from cardiac ventricular cells in adequately designed experiments. Our results on cultured Purkinje cells confirm their observations and show that $I_p$ can reverse its direction at physiological membrane potentials upon lowering $\Delta G_{ATP}$ even if the gradients of the transported ions across the sarcolemma remain unchanged.

## References

1. Bahinski A, Nakao M, Gadsby DC (1988) Potassium translocation by the $Na^+/K^+$ pump is voltage insensitive. Proc Natl Acad Sci USA 85: 3412–3416
2. Fabiato A (1988) Computer programs for calculating total from specified free or free from specified total ionic concentrations in aqueous solutions containing multiple metals and ligands. Methods Enzymol Biomembr 157: 378–417
3. Gadsby DC, Kimura J, Noma A (1985) Voltage dependence of Na/K pump current in isolated heart cells. Nature 315: 63–65
4. Garrahan PJ, Glynn IM (1966) Driving the sodium pump backwards to form adenosine triphosphate. Nature 211: 1414–1415
5. Glitsch HG, Krahn T, Pusch H (1989) The dependence of sodium pump current on internal Na concentration and membrane potential in cardioballs from sheep Purkinje fibres. Pflügers Arch 414: 52–58
6. Guynn RW, Veech RL (1973) The equilibrium constants of the adenosine triphosphate hydrolysis and the adenosine triphosphate–citrate lyase reactions. J Biol Chem 248: 6966–6972
7. Hamill OP, Marty A, Neher E, Sakmann B, Sigworth FJ (1981) Improved patch–clamp techniques for high–resolution current recording from cells and cell–free membrane patches. Pflügers Arch 391: 85–100

# Characteristics of sodium pump current in guinea–pig and rat dorsal root ganglion neurons

A.N. Hermans, H.G. Glitsch[*] and F. Verdonck

Interdisciplinary Research Centre, Catholic University of Leuven, Campus Kortrijk, B-8500 Kortrijk, Belgium and [*]Department of Cell Physiology, Ruhr University, D-44780 Bochum, Germany.

## Introduction

The properties of the $Na^+$ pump current ($I_p$) in mammalian neurons are largely unknown. We have studied some fundamental characteristics of electrogenic $Na^+/K^+$ pumping in guinea–pig and rat dorsal root gangion cells (DRG neurons) and report below on a few results of this study.

## Methods

DRG neurons were dissected in principle as described by Delree et al. (2) and kept in primary culture for 1–2 days at 37°C in an incubator. They were grown on coverslips in dishes which were transferred to the stage of an inverted microscope for whole cell recording. The superfusion medium contained (mM): 144 NaCl, 0–5.4 KCl, 0.5 $MgCl_2$, 1.8 $CaCl_2$, 10 glucose, 10 HEPES (adjusted by NaOH to pH 7.4 at 35°). $BaCl_2$ (2mM)and $NiCl_2$ (5mM) were included in order to suppress the $K^+$ and $Ca^{2+}$ conductances and the $Na^+/Ca^{2+}$ exchange of the sarcolemma. Dihydroouabain (DHO), a specific inhibitor of the $Na^+/K^+$ pump, was added from an aqueous stock solution. All K containing media were applied to the neuron studied via multibarrelled pipettes nearby. Prewarmed solution was continuously perfusing the dish. The solution within the patch pipettes (pipette solution) contained (mM): 110 caesium aspartate or tetraethylammonium chloride, 40 NaOH, 5 $MgCl_2$, 5 glucose, 5 Mg-ATP, 5 sodium creatine phosphate, 40 HEPES (with HCl to pH 7.35). Thus the Na concentration of the pipette solution was 50 mM throughout. The $Na^+$ pump current of the neurons was identified as DHO inhibited current which is activated by extracellular K ($K_o$). The cells were clamped to preset membrane potentials and the resulting membrane current was measured by means of whole cell recording (3) as described previously in detail (1). Holding potential was –20 mV. The initial resistance of the patch electrodes filled with one of the pipette solutions varied between 2 and 3 MΩ. The experiments were carried out at 35°C. n indicates the number of cells studied. Whenever possible data are presented as means ± S.E.M.

## Results

Table 1 shows $I_p$ densities in guinea–pig and rat DRG neurons at 2 mM and 5.4 mM $K_o$, respectively. The measurements were performed 1–2 min after getting access to the cell interior. As can be seen from the table $I_p$ densities measured at 5.4 mM $K_o$ are slightly higher than at 2 mM. Pump current densities from the corresponding isolated cardiac ventricular cells are smaller by a factor of two at both $[K]_o$ (data not shown ).

**Table 1.** $I_p$ densities and $K_{0.5}$ values for $I_p$ inhibition by DHO in guinea–pig and rat DRG neurons (holding potential: –20 mV).

| species | $K_o$ (mM) | $I_p$ density (Axcm$^{-2}$) | n | $K_{0.5}$ value (M DHO) |
|---|---|---|---|---|
| guinea–pig | 5.4 | $(3.3 \pm 0.3) \times 10^{-6}$ | 48 | $1.9 \times 10^{-5}$ |
| | 2.0 | $(3.1 \pm 0.1) \times 10^{-6}$ | 65 | $9.1 \times 10^{-6}$ |
| rat | 5.4 | $(4.8 \pm 0.4) \times 10^{-6}$ | 44 | $1.7 \times 10^{-3}$ |
| | 2.0 | $(3.7 \pm 0.4) \times 10^{-6}$ | 41 | $9.4 \times 10^{-4}$ |

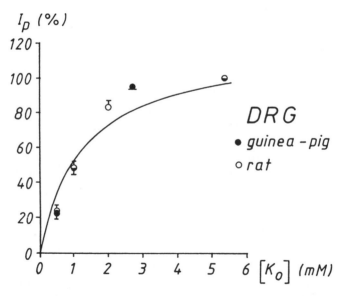

**Fig. 1.** $I_p$ activation by $K_o$ in guinea-pig (n=9) and rat (n=12) DRG neurons. $I_p$ normalized to the respective $I_p$ value measured at 5.4 mM $K_o$. Holding potential –20 mV. The curve fitted to the data obeys simple saturation kinetics with half maximum $I_p$ activation at 1.3 mM $K_o$ and $I_{pmax}$=120%; $r^2$=0,99 for both data sets. Assuming more complicated kinetics did not improve the fit.

538

Half maximum inhibition of $I_p$ by DHO occurs in guinea–pig neurons (n=8) at a concentration ($K_{0.5}$ value) two order of magnitude lower than in rat DRG cells (n=17). Lowering $K_0$ to 2 mM reduces the $K_{0.5}$ value for $I_p$ inhibition in both species (n=12) by about 50 per cent. Nearly identical $K_{0.5}$ values are obtained with corresponding ventricular cells (data not shown).

The activation of $I_p$ in the neurons by $K_0$ is depicted in Fig.1. Obviously, there is little species difference. Half maximum $I_p$ activation is observed at 1.3 mM $K_0$, both in guinea–pig and rat DRG neurons. Again, similar values are obtained with guinea–pig and rat single cardiac ventricular cells. Fig.2 shows the voltage dependence of $I_p$ in guinea–pig and rat DRG neurons. The $I_p$ values were normalized to their respective values at the holding potential. The neurons of both species display similar $I_p$–V relationships. $I_p$ increases with depolarization at negative membrane potentials and remains essentially unchanged at positive voltages. Corresponding cardiac ventricular cells exhibit nearly identical $I_p$–V relations.

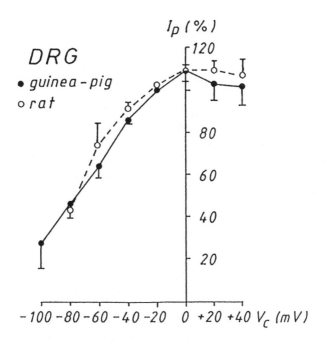

**Fig. 2.** Ip–V relationships of guinea–pig (n=7) and rat (n=6) DRG neurons at 2 mM $K_0$. $I_p$ normalized to the respective $I_p$ value measured at the holding potential of –20 mV. Error bars shown only if larger than symbols. Lines drawn to connect the data points.

## Discussion

Apart from a slightly higher current density, $I_p$ of DRG neurons in guinea–pigs and rats has properties similar to those known for $I_p$ of the corresponding cardiac ventricular cells. For example, half maximum $I_p$ activation by $K_o$ occurs at 1.3 mM in the neurons and at 1.1 mM in rat ventricular myocytes (Hermans et al., unpublished results) or at 1.5 mM $K_o$ in guinea–pig ventricular cells (4). Furthermore, there is also a close correspondence of the data on the voltage dependence of $I_p$ and on $I_p$ inhibition by DHO from neurons and cardiac cells (Hermans et al., unpublished results). The well–known insensitivity of rat cardiac cells against cardiac glycosides is present in rat DRG neurons too. These observations might be surprising at first sight since different isozymes of the $Na^+/K^+$–ATPase may be expressed in DRG neurons and cardiac ventricular cells (see: 5). It is, however, conceivable that the existence of different isozymes in DRG neurons and cardiac cells is not reflected in variations of the $I_p$ properties studied here.

*Acknowledgements*

The study was supported by the Belgian National Research Fond (NFWO).

## References

1.  Bielen FV, Glitsch HG, Verdonck F (1991) Dependence of $Na^+$ pump current on external monovalent cations and membrane potential in rabbit cardiac Purkinje cells. J Physiol 442: 169–189
2.  Delree P, Leprince P, Schoenen J, Moonen G, (1989) Purification and culture of adult dorsal root ganglia neurons. J Neuroscience Res 23: 198–206
3.  Hamill OP, Marty A, Neher E, Sakmann B, Sigworth FJ (1981) Improved patch–clamp techniques for high–resolution current recording from cells and cell–free membrane patches. Pflügers Arch 391: 85–100
4.  Nakao M, Gadsby DC (1989) [Na] and [K] dependence of the Na/K pump current–voltage relationship in guinea pig ventricular myocytes. J Gen Physiol 94: 539–565
5.  Sweadner, KJ (1989) Isozymes of the $Na^+/K^+$– ATPase. Biochim Biophys Acta 988: 185–200

# Kinetic Analysis and Voltage Dependency of Alpha-3 $Na^+/K^+$-Pump

Y. Hara, T. Furukawa[*], O. Urayama, Y. Ito[+], T. Kojima,
K. Hirakawa[+], M. Hiraoka[*] and Y. Ikawa

Departments of Biochemistry and [+]Neurosurgery, Faculty of Medicine, and
Department of [*]Cardiovascular disease, Medical Research Institute,
Tokyo Medical and Dental University, 1-5-45, Yushima, Bunkyo-Ku, Tokyo
113 Japan

## Introduction

The $Na^+/K^+$-pump generates $Na^+$ and $K^+$ gradients across the plasma
membrane, driving both various $Na^+$-dependent co-transporting systems
and the action potentials in excitable cells. There are three types of $Na^+/K^+$-
pump in mammalian tissues. A general one, designated as $\alpha1$ $Na^+/K^+$-
ATPase, is ubiquitously distributed in all tissues. The second one, $\alpha2$
$Na^+/K^+$-ATPase, exists in nerve, muscle, heart and adipocytes. The third
type is $\alpha3$ $Na^+/K^+$-ATPase, existing in central nervous tissues and the
Purkinje fibers of mature animals. Despite the distinct tissue-specificity, no
differences in enzyme characteristics among those three types have been
found except in ouabain sensitivity. In order to get more information, we
examined the kinetic characteristics of the $\alpha1$ and the $\alpha3$ enzymes. As the $\alpha3$
enzyme is localized in the CNS with the $\alpha1$ and $\alpha2$ enzymes and the
separation of $\alpha3$ from $\alpha1$ and $\alpha2$ is very difficult, the precise nature of $\alpha3$
has remained unclear. So, we established an $\alpha3$ cDNA-transfected cell line,
expressing 5-7 fold excess of $\alpha3$ enzyme activity in relation to the
endogenous $\alpha1$ enzyme activity. We previously reported that $\alpha3$ $Na^+/K^+$-
ATPase expressed in the transformant showed a very high ouabain-sensitivity
($Ki=1x10^{-7}M$), like that of rat brain $Na^+/K^+$-ATPase (4). Here we describe
a difference in rate-limiting step between the $\alpha1$ and $\alpha3$ enzymes and suggest
that $\alpha3$ $Na^+/K^+$-ATPase may be regulated by the membrane potential.

## Methods
*Cell and Cell culture*
All cells were maintained in DMEM medium containing 10% FCS. T333, $\alpha3$
cDNA transfected cells, were treated with 10 mM sodium butylate for 48
hours before use (3).
*Pump current measurement*
$Na^+$ pump currents of cultured cells were measured by using the patch clamp

method. The extracellular solution contained 145 mM NaCl, 5 mM KCl, 2.3 mM MgCl$_2$, 2 mM BaCl$_2$, 5.5 mM dextrose and 5mM Hepes, pH 7.2. Ionic compositions of the intracellular solution (pipette solution) are shown in the figure legends. Membrane potential was held at -40 mV and step pulses with 800msec duration were applied from -120 mV to 40 mV in 20 mV intervals. The $\alpha$1 pump activity was measured as $10^{-3}$ M ouabain-sensitive pump current in Balb/c 3T3 cells, while $\alpha$3 pump activity was obtained as the $10^{-5}$ M ouabain-sensitive pump current in T333 cells. All pump currents were measured at 37°C.

## Results
### Kinetic parameters of Na$^+$/K$^+$-ATPase
$\alpha$3 Na$^+$/K$^+$-ATPase activity was obtained from the membrane fraction of T333 cells as described previously (4). The $\alpha$1 Na$^+$/K$^+$-ATPase activity was obtained from BALB/c 3T3 cell membrane. As shown in Table 1, the Km for ATP of the $\alpha$3 enzyme was half that of the $\alpha$1 enzyme, suggesting that the E1-E2 equilibrium of the $\alpha$3 enzyme is shifted more to the E1 form than that of the $\alpha$1 enzyme. The Kis of $\alpha$3 for both ADP and NEM were smaller than those of $\alpha$1. As ADP and NEM bind to E1P rather than E2P, those low Ki values also support the hypothesis that the E1P-E2P equilibrium of $\alpha$3 enzyme is shifted much more to the E1P form much more than that of the $\alpha$1 enzyme.

The shift of E1P-E2P equilibrium was also apparent from the measurement of the E1P/E2P ratio. It was reported that brain enzyme containing the $\alpha$3 enzyme showed a higher E1P/E2P ratio (30/70) than that (15/85) of kidney enzyme containing only the $\alpha$1 enzyme (5,1). The equilibrium shifts of the E1-E2 and the E1P-E2P transition steps in the $\alpha$3 enzyme raise the possibility that the rate-limiting step of the $\alpha$3 enzyme may be the E1P--E2P step rather than the E2--E1 step, though the latter transition step is known to be rate-limiting in the $\alpha$1 enzyme at neutral pH (2).

**Table 1    Kinetic parameters of Na$^+$/K$^+$-ATPase $\alpha$ isoforms**
Standard Assay condition: 140 mM NaCl, 14 mM KCl, 3 mM ATP, 5 mM MgCl$_2$, 0.5 mM EDTA and 25 mM Imidazole, pH 7.4

|  | $\alpha$1 | $\alpha$3 |
|---|---|---|
| Km for Na$^+$ | 17 mM | 14 mM |
| Km for K$^+$ | 1.6 mM | 2.8 mM |
| Km for ATP | 0.36 mM | 0.18 mM |
| Ki for NEM | 1.0 mM | 0.1 mM |
| Ki for ADP | 2.0 mM | 1.0 mM |
| Ki for ouabain | 45 $\mu$M | 80 nM |

*Effects of membrane potential on Na$^+$ pump activity*

It is generally accepted that the E1P--E2P transition step is electrogenic, while the E2K--E1 transition step is electroneutral at saturating concentration of extracellular K$^+$ (6). If the E1P--E2P transition step is rate-limiting, Na$^+$/K$^+$-pump activity should be influenced by membrane potential. So, we examined the effects of membrane potential on $\alpha$1 and $\alpha$3 pump activities. As shown in Fig.1, $\alpha$3 pump activity increased with increasing membrane potential (-120 to +40mV), while $\alpha$1 pump current increased, reached at maximum and then decreased with increasing membrane potential at 80 mM intracellular K$^+$. At higher K$^+$ concentration, the pump current of $\alpha$1 decreased with increasing membrane potential, while $\alpha$3 pump activity increased with increasing membrane potential at various intracellular K$^+$ concentration (data not shown).

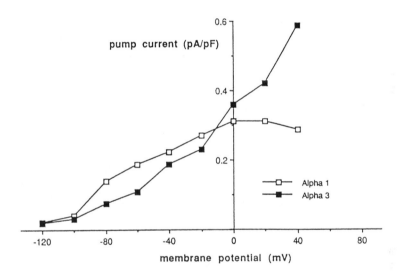

**Fig. 1** Effects of membrane potential on $\alpha$1 and $\alpha$3 pump current. [Na$^+$]$_o$=145 mM, [K$^+$]$_o$=5 mM, [Na$^+$]$_i$=10 mM, [K$^+$]$_i$=80 mM

## Discussion

The results can be interpreted as follows:
1) From the kinetic parameters, $\alpha$1 enzyme favors E2K or E2P forms while $\alpha$3 enzyme favors E1ATP or E1P forms, as those four states are stable and major enzyme forms as shown in Fig.2. Then it is suggested that the E2K--E1ATP step or E1P--E2P step is a rate-limiting in $\alpha$1 or $\alpha$3 enzyme, respectively.
2) $\alpha$3 enzyme activity increased with increasing membrane potential in wide range of potential, suggesting that the E1P--E2P transition step is rate-

543

limiting. α1 pump current also increased with potential but reached at maximum at around zero mV in our experimental condition. This indicates that E1P--E2P step limits overall $Na^+/K^+$-pump activity partially but E2K--E1ATP step may become a dominant rate-limiting step at depolarized membrane potential in α1 pump.

3) α3 $Na^+/K^+$-pump activity can be influenced strongly by the membrane potential. The increase in enzyme activity with depolarization might be important for a quick restoration of cellular ionic condition after the membrane excitation and at least in part for a generation of membrane excitation.

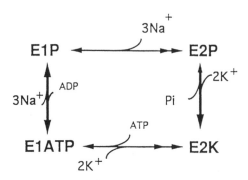

**Fig. 2** Minimal reaction cycle of $Na^+/K^+$-ATPase. The two occluded states and two major forms of the enzyme are shown.

References

1.      Fukushima Y, Nakao M (1980) Change in affinity of $Na^+$- and $K^+$-transporting ATPase for divalent cations during its reaction sequence. J Biol Chem 255: 7813-7819

2.      Glynn I, Hara Y, Richard D E, and Steinberg M (1987) Comparison of Rates of Cation Release and of Conformational Change in Dog Kidney $Na^+,K^+$-ATPase. J Physiol 383: 477-485

3.      Gorman C M, Howard B H (1983) Expression of recombinant plasmids in mammalian cells is enhanced by sodium butyrate. Nucleic Acids Res 11: 7631-7648

4.      Hara Y, Nikamoto A, Kojima T, Matsumoto A, Nakao M (1988) Expression of sodium pump activities on BALB/c 3T3 cells transfected with cDNA encoding α3-subunit of rat brain $Na^+,K^+$-ATPase. FEBS Lett 238: 27-30

5.      Hara Y, and Nakao M (1981) Sodium Ion Discharge from Pig Kidney $Na^+,K^+$-ATPase. J Biochem 90: 923-931

6.      Lauger P (1991) Kinetic basis of voltage dependence of the NaK-pump. in: Kaplan J H, De Weer P(eds)The sodium Pump. The Rockefeller University

# Access Channel Model for Na$^+$ and K$^+$ Translocation by the Na$^+$/K$^+$-pump

R.F. Rakowski, A. Sagar, M. Holmgren

Department of Physiology and Biophysics, University of Health Sciences/ The Chicago Medical School, 3333 Green Bay Road, North Chicago, Illinois 60064 USA

## Introduction

Recent studies (6, 1) have strongly suggested that external Na$^+$ and K$^+$ reach their binding sites deep within the Na$^+$/K$^+$-pump via a high-field access channel. Pre-steady-state transient currents measured under Na$^+$/Na$^+$-exchange conditions are dependent on external [Na$^+$] ([Na]$_o$) and are asymmetrically voltage dependent (4, 3) as expected for an access channel. The voltage dependence of the apparent affinity for external [K$^+$]- ([K]$_o$) activation of steady-state pump current (5) suggests that external K$^+$ binding sites are also located within the membrane field. Since there must be at least two voltage-dependent steps in the pump cycle to produce steady-state pump current-voltage (I-V) relationships that have regions of both positive and negative slope (5), we decided to ask how well a simple access channel model with only two voltage-dependent rate coefficients can explain the observed voltage dependence of pump current.

### Pseudo-two-state kinetic model

Any electrogenic cyclic reaction scheme with only two oppositely-directed voltage-dependent rate coefficients can be reduced to an equivalent pseudo-two-state model in which the steady-state current ($I$) is given by

$$I = FN\rho \frac{ac - bd}{a + b + c + d} \tag{1}$$

where $a$ and $d$ are collections of forward and reverse voltage-independent rate constants, and $b$ and $c$ are voltage-dependent forward and reverse rate coefficients respectively. $N$ is the pump site density, $\rho$ is a voltage-independent reserve factor (2) and $F$ is Faraday's constant. For an external access channel the effective concentration of a positive ion is increased by hyperpolarization according to a Boltzmann relationship. If we adopt the Hill approximation for multi-site binding, the forward and reverse pseudo-first order rate coefficients for external Na$^+$ and K$^+$ binding may be written as follows :

$$b = \bar{b} \; [Na]_o^{\gamma_N} \exp\left(-\gamma_N \lambda_N U\right) \tag{2}$$

$$c = \bar{c} \; [K]_o^{\gamma_K} \exp\left(-\gamma_K \lambda_K U\right) \tag{3}$$

where $\bar{b}$ and $\bar{c}$ are the respective rate constants at 0 mV, $\gamma_N$ and $\gamma_K$ are Hill coefficients for Na$^+$ and K$^+$ respectively, $\lambda_N$ and $\lambda_K$ are the fractional dielectric distances (well-depths) at which Na$^+$ and K$^+$ bind, and $U=FV/RT$, where $V$ is membrane potential and

*F/RT* has its usual meaning. For an irreversible forward-going pump cycle far from equilibrium we may assume $d = 0$ and substitute Eqs. 2 and 3 in 1 to obtain

$$I = \frac{FN\rho a}{1 + \dfrac{a + \bar{b}\,[\mathrm{Na}]_o^{\gamma_N} \exp(-\gamma_N \lambda_N U)}{\bar{c}\,[\mathrm{K}]_o^{\gamma_K} \exp(-\gamma_K \lambda_K U)}} \qquad (4)$$

**Fit of the model to steady-state pump I-V data**

Fig. 1 shows steady-state pump I-V measurements made at various $[\mathrm{K}]_o$ in 15 mM $[\mathrm{Na}]_o$. The solid lines are calculated from Eq. 4 using the parameters determined by the overall fit to 18 different experimental conditions, three of which are shown in Fig. 1.

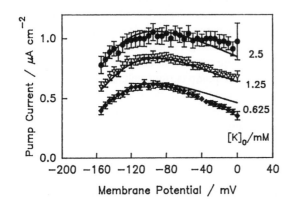

**Figure 1.** Steady-state Na⁺/K⁺-pump current measured in 15 mM $[\mathrm{Na}]_o$ at various $[\mathrm{K}]_o$. Mean values ($\pm$SEM) of $[\mathrm{K}]_o$-activated current from twelve oocytes are shown. Oocytes were voltage clamped using a two-micro-electrode technique. Pump current was determined by subtraction of control I-Vs in K⁺-free solution from I-Vs recorded in the presence of the indicated $[\mathrm{K}]_o$. From Sagar and Rakowski (manuscript).

The model provides a good fit to the data. The values of fractional dielectric distance ($\lambda_N$ and $\lambda_K$) obtained from the overall fit were $0.486 \pm 0.010$ and $0.256 \pm 0.009$ respectively. The meaning of the other kinetic parameters is obscured by the adoption of the Hill approximation and the reduction to a pseudo-two-state model.

**Measurements of pre-steady-state charge movement**

Fig. 2 shows $\mathrm{Na}_o$-sensitive pre-steady-state charge movement measured using the cut-open oocyte technique (7) under Na⁺-Na⁺-exchange conditions (K⁺-free solutions, 5 mM ATP and ADP). The solid line in Fig. 2A is a least squares fit of Eq. 5 to the steady-state charge

$$Q = Q_{\min} + \{Q_{\mathrm{tot}}/(1 + \exp(z_q(V_q - V)F/RT))\} \qquad (5)$$

Fig. 2B shows the voltage dependence of the exponential relaxation rate of the $\mathrm{Na}_o$-sensitive transient current. The solid line is drawn according to Eq. 6.

$$k = a_k \{1 + \exp(z_k(V_k - V)F/RT)\} \qquad (6)$$

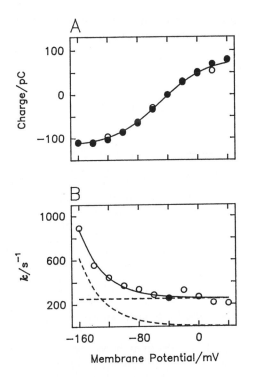

**Figure 2.** Voltage dependence of pre-steady-state charge movement. A. Voltage dependence of the $[Na]_o$-sensitive chage movement ($Q$) in response to voltage pulses from a holding potential of –40 mV to test voltages from –160 to +40 mV. Filled circles: $Q$ measured from the *on* transient current. Open circles (obscured by filled circles): $Q$ measured from the *off*. The solid line is a least-squares fit of Eq. 5 with $Q_{min} = -117 \pm 6$ pC, $Q_{tot} = 200 \pm 13$ pC, $z_q = 0.81 \pm 0.10$ and $V_q = -49.9 \pm 3.5$ mV. B. Voltage dependence of the transient current relaxation rate. The solid line is a least squares fit of Eq. 6 with $a_k = 252 \pm 2$ s$^{-1}$, $z_q = 0.76 \pm 0.11$ and $V_k = -130 \pm 7$ mV. The dashed lines are the forward and backward rate coefficients from Eq. 6. From Holmgren and Rakowski (manuscript).

The least-squares parameters are given in the legend. Note that the exponential steepness for charge ($z_q$) and relaxation rate ($z_k$) are similar, but that the mid-point voltages ($V_q$ and $V_k$) are different. Also note the asymmetry of the voltage dependence of the forward and reverse rate coefficients (dashed lines). This behavior is consistent with that expected for an ion well (at low occupancy) and provides independent support for the use of Eqs. 2 and 3 in the pseudo-two-state model.

### An electrostatic model of charge translocation

It may not be immediately obvious how it is possible for one net charge to be translocated in each pump cycle if charge translocating steps only occur in external-facing access channels. Fig. 3 shows how this is possible. 1) Three Na$^+$ bind to three negative charges at the inside face of the enzyme. 2) Electroneutral translocation occurs over a fractional dielectric distance $1 - \lambda_N$. 3) The three Na$^+$ are released within an access channel resulting in the net outward movement of three positive charges over the fractional dielectric distance $\lambda_N$. 4) An electroneutral conformational change takes place in which two of the three negative charges are translocated outward by $\lambda_N - \lambda_K$, and one moves inward $1 - \lambda_N$. For this step to be electroneutral requires that $2(\lambda_N - \lambda_K) = 1(1 - \lambda_N)$, that is, $3\lambda_N - 2\lambda_K = 1$. 5) Two K$^+$ move over $\lambda_K$ within an external-facing access channel and bind to the 2 negative charges. 6) Electroneutral translocation of K$^+$ occurs over a dielectric distance $1 - \lambda_K$. 7) The two K$^+$ are released at the internal face of the

547

enzyme making two negative charges available together with the one negative charge from step (4) to provide the three charges required for the electroneutral $Na^+$ translocation step (1).

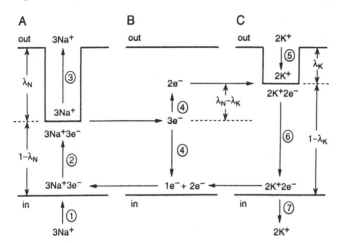

**Figure 3.** Electrostatic model of charge translocation by the $Na^+/K^+$-pump. A. $Na^+$ translocation. B. Electroneutral conformational change. C. $K^+$ translocation. See text for details. From Sagar and Rakowski (manuscript).

**Calculation of the dielectric coefficients for $Na^+$ and $K^+$ translocation**

It is instructive to see whether $\lambda_N$ and $\lambda_K$ obtained from the fit to the steady-state current data are consistent with the requirement that the sum of the dielectric coefficients must equal 1 if one net charge is translocated in each pump cycle. We calculate the dielectric coefficient for $Na^+$ ($\alpha_N = 3\lambda_N$) to be $+1.46 \pm 0.03$ and for $K^+$ ($\alpha_K = -2\lambda_K$) to be $-0.51 \pm 0.02$. The sum $\alpha_N + \alpha_K$ is $0.95 \pm 0.05$, close to the theoretical value of 1.0.

*Acknowledgement*

Supported by NIH grant NS-22979.

**References**

1.  Gadsby DC, Rakowski RF, De Weer P (1993) Extracellular access to the Na,K pump: pathway similar to ion channel. Science 260: 100-103
2.  Hansen U-P, Gradmann D, Sanders D, Slayman CL (1981) Interpretation of current-voltage relationships for "active" ion transport systems: I. Steady-state reaction-kinetic analysis of class-I mechanisms. J Membr Biol 63: 165-190
3.  Nakao M, Gadsby DC (1986) Voltage dependence of Na translocation by the Na/K pump. Nature 323: 628-630
4.  Rakowski RF (1993) Charge movement by the Na/K pump in *Xenopus* oocytes. J Gen Physiol 101: 117-144
5.  Rakowski RF, Vasilets LA, LaTona J, Schwarz W (1991) A negative slope in the current-voltage relationship of the $Na^+/K^+$ pump in *Xenopus* oocytes produced by reduction of external $[K^+]$. J Membr Biol 121: 177-187
6.  Stürmer W, Bühler R, Apell H-J, Läuger P (1991) Charge translocation by the Na,K - pump: II. Ion binding and release at extracellular face. J Membr Biol 121: 163-176
7.  Taglialatela M, Toro L, Stefani E (1992) Novel voltage clamp to record small, fast currents from ion channels expressed in *Xenopus* oocytes. Biophys J 61: 78-82

# Na$^+$/K$^+$-pump mutant with a slow charge translocating step

J.-D. Horisberger, F. Jaisser, C.Canessa, B.C. Rossier

Institut de Pharmacologie et de Toxicologie, Bugnon 27, CH-1005 Lausanne, Switzerland

## Introduction

During its transport cycle the Na$^+$/K$^+$-pump transport 3 Na$^+$ ions out of and to K$^+$ ions into the cells. The translocation of the Na$^+$ ions is associated with a net movement of charge across the membrane. Charge translocation by the Na$^+$/K$^+$-pump can be studied as ouabain-sensitive transient current after fast voltage perturbation under Na/Na exchange conditions (5,6). The translocation of charges has been associated with the release of Na$^+$ ions across an external high field access channel (2,7). The structure responsible for Na$^+$ ion translocation is not known.

We have observed that an aminoterminus-truncated form of the $\alpha$ subunit of the Bufo Na$^+$/K$^+$-pump presented slower ouabain-sensitive transient current than wild type.

## Methods

Synthetic cRNAs coding for wild type (WT) and truncated mutant (Tr) of the *Bufo* Na$^+$/K$^+$-ATPase $\alpha$1 subunit and wild type ß1 subunit (4) were prepared as described (3). The 40 N-terminal amino-acids of the $\alpha$ subunit were deleted so that the truncated mutant started at the methionine M41 of the wild type sequence. Stage V-VI *Xenopus* oocytes were injected with 10 ng of $\alpha$ subunit cRNA and 1 ng ß subunit. We have shown previously that this results in a large expression of ouabain-resistant (K$_i$ 50 µM) *Bufo* Na$^+$/K$^+$-pumps (4). Electrophysiological measurements were perfromed 3 to 7 days after cRNA injection. Na$^+$/K$^+$-pump activity measurements were restricted to the *Bufo* pumps by previous inhibition of the *Xenopus* Na$^+$/K$^+$-pumps endogenous to the oocyte by exposure to 1 µM ouabain. Whole cell currents were measured using the two-electrode voltage clamp technique, using large current-passing electrodes (resistance around 0.2 MOhms). Steady-state Na$^+$/K$^+$-pump currents at -60 mV were measured as described earlier (3) by recording the outward current induced by addition of 10 mM K$^+$ to a K-free solution, in Na$^+$-loaded oocytes, in the presence of 5 mM Ba$^{++}$ and 10 mM TEA. Ouabain-sensitive transient currents were obtained by recording in a K-free solution the current induced by 50 ms voltage steps of various amplitudes (from +100 to -100 mV) starting from a -60 mV command potential, before and after the addition of 2 mM ouabain.

## Results

Figure 1 shows that the relaxation rate of the ouabain-sensitive current was slower in Tr than in WT at all membrane potential less more positive than -60 mV.

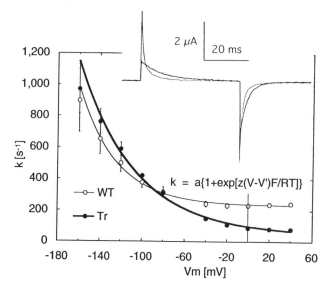

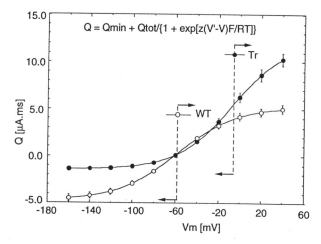

Figure 1.

**Top Panel.** Voltage dependence of the relaxation rate (k) of the ouabain-sensitive transient current obtained with oocytes injected with wild type (WT) and truncated (Tr) form of the a subunit of the *Bufo* Na,K-pump. The solid lines are computed from the equation (see ref. 6).

**Inset** : Original tracings of the ouabain-sensitive current in a K-free solution during a 50 ms voltage step from -60 to +40 mV. The figure shows one WT (thin line) and one Tr (thick line) tracing superimposed. Both tracings show an approximately mono-exponential relaxation. The relaxation rate of the Tr form is much slower .

**Bottom Panel.** Voltage dependence of the ouabain-sensitive charge translocation (Q). The solid lines are calculated from the Boltzman equation (shown in the figure). The values of the mid-point potential (V') are indicated by the dotted lines. The values of Qmin and Qmax are indicated by the solid half-lines. Note the right shift in the mid-point voltage (V', from -58 mV, WT, to -7 mV,Tr) for the truncated a subunit. The z values were 0.92 and 0.99 for WT and Tr, respectively.

Values are mean ± SE of 7 measurements with wild type (WT) and 8 measurements with truncated (Tr) form of the Na,K-pump a subunit.

The values in WT and Tr were similar at high membrane potential. The relaxation rate is the sum of the forward ($k_f$) and backward ($k_b$) rate constant of the charge translocating step (5). Using the model of Rakowski (6) with a voltage independent $k_f$ and an exponentially voltage-dependent $k_b$ (see equation in figure 1, top panel) our results are consistent with a large (4 to 5 fold) decrease of $k_f$ in Tr and no or little change of $k_b$.

The steady state Na,K-pump current at -50 mV and under $V_{max}$ conditions (10 mM external $K^+$) was $468 \pm 49$ nA (n=7) and $190 \pm 17$ nA (n=8) in WT and Tr, respectively. This difference was increasing at high negative membrane potentials. The difference of pump current could be attributed to 1) a smaller number of active pumps expressed at the plasma membrane, 2) to a change of stoichiometry, or 3) to a slower turnover rate of each pump unit. The data of figure 2 support the last hypothesis.

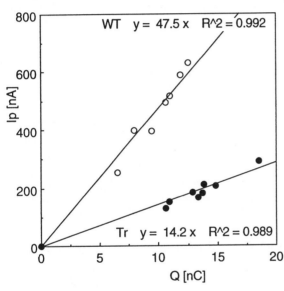

Figure 2

Relation of the steady-state pump current in 10 mM $K^+$ (Ip) to the maximal translocated charge obtained as the best fitting value of the maximal translocated charge ($Q_{tot}$) of the Boltzman equation (see fig 1). Each paramter was measured in 7 oocytes of the wild type (WT, black circles) group and 8 oocytes of the truncated (Tr, open circles) $\alpha$ subunit group.

Assuming a ratio of one translocated charge per pump in both groups, the turnover rate would be 47 $s^{-1}$ in the wild type group and 14 $s^{-1}$ in the truncated a subunit group. This difference in turnover rate can be explained by the slower forward rate constant of the $Na^+$ translocating step which is rate limiting at negative membrane potentials.

The N-terminal end of the $\alpha$-subunit is the most variable region of the $Na^+/K^+$-pump sequence. It contains a large number of positive charges (figure 2). Amino-terminal truncated mutants have been associated with altered potassium activation kinetics, either increase (8) or decrease (1) of apparent affinity depending on the exact location of the truncation and the species used.

## Conclusions

The Na,K-pump of Bufo marinus in which the N-terminus of the $\alpha$ subunit has been truncated 40 amino acids has a slower turnover rate due to a large reduction of the forward rate of the voltage-sensitive $Na^+$ translocating step. The N-terminal end of the $\alpha$ subunit of the Na,K-pump appears to be directly involved in the conformational change that allows the translocation of $Na^+$ ions. This "slow" mutant may make easier the analysis of charge translocation kinetics.

## References

1.    Burgener-Kairuz P, Horisberger J-D, Geering K, Rossier BC (1991) Functional expression of N-terminus truncated a-subunits of Na,K-ATPase in Xenopus laevis oocytes. FEBS Lett 290:83-86

2.    Gadsby DC, Rakowski RF, De Weer P (1993) Extracellular access to the Na,K pump: Pathway similar to ion channel. Science 260:100-103

3.    Horisberger J-D, Jaunin P, Good PJ, Rossier BC, Geering K (1991) Coexpression of a1 with putative ß3 subunits results in functional Na-K-pumps in Xenopus oocyte. Proc Natl Acad Sci USA 88:8397-8400

4.    Jaisser F, Canessa C, Horisberger J-D, Rossier BC (1992) Primary sequence and functional expression of a novel ouabain-resistant Na,K-ATPase. J Biol Chem 267:16895-16903

5.    Nakao M, Gadsby DC (1986) Voltage dependence of the Na translocation by the Na/K pump. Nature 323:628-630

6.    Rakowski RF (1993) Charge movement by the Na/K pump in *Xenopus* oocytes. J Gen Physiol 101:117-144

7.    Stürmer W, Bühler R, Apell H-J, Läuger P (1991) Charge translocation by the Na,K-pump: II. Ion binding and release at the extracellular face. J Membrane Biol 121:163-176

8.    Vasilets LA, Omay HS, Ohta T, Noguchi S, Kawamura M, Schwarz W (1991) Stimulation of the $Na^+/K^+$ pump by external $[K^+]$ is regulated by voltage-dependent gating. J Biol Chem 266:16285-16288

# Analysing the $Na^+/K^+$-pump in Outside-out Giant Membrane Patches of *Xenopus* Oocytes

J. Rettinger, L.A. Vasilets*, S.Elsner, W. Schwarz

Max-Planck-Institut für Biophysik, Kennedyallee 70, 60596 Frankfurt/Main, Germany
*Permanent address: Institute of Chemical Physics, Chernogolovka, Moscow region, 142432, Russia

## Introduction

To obtain further insight into electrogenic steps of the $Na^+/K^+$-pump cycle, current transients can be analysed that result from the perturbance of a steady-state distribution by application of a voltage pulse. Recently a giant patch method was developed (2,4) that allowed detection of pump-generated membrane currents in inside-out patches of cardiac cells and *Xenopus* oocytes. Application of the two-electrode voltage-clamp method is questionable in this respect since space clamp cannot be achieved within the first milliseconds. A drawback of the giant-patch method was that outside-out orientation could be obtained only occasionally. Here we apply a modified procedure that allows routinely also the formation of outside-out membrane patches of oocytes with diameters of up to 45 μm.

## Methods

Conventional patch pipettes were pulled from borosilicate glass (Hilgenberg (Germany) 1.5 mm o.d./0.86 mm i.d.). The desired tip diameter of 20-40 μm was generated as illustrated in Fig. 1. With these electrodes it is possible to get stable and long-lasting giga seals without any coating of the electrode tip. Seals were formed by applying negative pressure of no more than 3 mbar. After the seal resistance has reached its

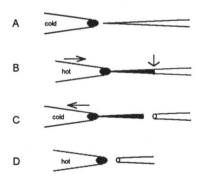

**Figure 1.** Fabrication of large diameter electrodes. A bead of solder glass is melted on a V-shaped platinum wire **(A)**. The filament is heated and liquid soft glass is brought into contact with the tip of a standard-shaped patch pipette. The melted soft glass will fill the tip of the pipette **(B)**. When the heating is switched off the shrinking wire will lead to a cut of the pipette at the contact rim between soft and hard glass **(C)** After heatpolishing of the tip **(D)** the electrode is ready for formation of in-side- as well as outside-out patches.

maximum value, the membrane patch was destroyed by application of a +800 mV voltage pulse of 5 ms duration. A slight positive pressure was applied to blow intracellular material out of the pipette tip. After adjusting the pressure difference to zero the electrode was slowly withdrawn from the oocyte, and the membrane in the vicinity of the tip closed in most cases to the outside-out conformation. For rapid solution changes without affecting the oocyte the patch was transferred from the oocyte chamber to another chamber with a built-in liquid filament solution exchange device (3) connected to 6 different solutions. All experiments were performed at 22 °C. The currents were digitised and stored on hard disc via an EPC-9/Atari Mega ST 4 setup (HEKA Lambrecht, Germany). Pump currents were determined as the ouabain-sensitive component of total membrane current or as $K^+$-sensitive current (6). To improve the signal-noise ratio all experiments were performed with oocytes with expressed ouabain-resistant $Na^+/K^+$-pumps of *Torpedo* electroplax (cDNAs were kindly provided by Drs. M.Kawamura and K.Takeda). The endogenous *Xenopus* pump was inhibited by 1 µM ouabain.

## Measurements of steady state pump currents

In order to demonstrate that identical results can be obtained with the two-electrode voltage-clamp and the giant outside-out patch clamp, we determined the dependence of steady-state pump current on membrane potential and external $[K^+]$. Figure 2A shows

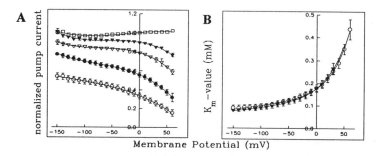

**Figure 2.** Voltage dependence of pump current **(A)** and $K_m$-values **(B)**. The current - voltage dependencies were determined from the difference of total current in absence of external $K^+$ and in presence of different external $K^+$-concentrations (in mM).(A) Data from giant patches: Open squares 5, filled triangles 1, open triangles 0.5, filled circles 0.25, open circles 0.1. (B) Open circles: mean of 5 giant patch experiments, filled triangles: mean of 8 voltage-clamp experiments. Solutions for giant patch experiments (in mM): Pipette: 10 $MgCl_2$, 5 $Na_2ATP$, 5 EGTA, 25 NaOH, 5 MOPS pH=7.8. Bath: 20 TEA-Cl, 5 $BaCl_2$, 5 MOPS/TRIS pH=7.3 and different $K^+$-concentrations. The bathing solution for the two-electrode voltage-clamp experiments contained in addition 5 mM $NiCl_2$ and to keep osmolalic balance with the oocyte 100 mM TMA-Cl.

the voltage dependence of pump currents at different external $[K^+]$ in outside-out giant patches; Fig. 2B demonstrates that the voltage dependencies of the half-activating $[K^+]$ (apparent $K_m$-value) obtained with the two techniques do not differ significantly.

## Measurements of transient pump currents

Analysis of electrogenic steps in the $Na^+$-$Na^+$ or $K^+$-$K^+$ halfcycle is important for kinetic characterisation of the $Na^+/K^+$-pump, and can be done by measurements of current transients. This kind of experiments was previously performed on ventricular myocytes with the whole-cell patch-clamp technique (1) and on oocytes using the two-electrode voltage-clamp technique (5). With the giant-patch clamp technique it is possible to perform relaxation experiments with a greatly improved time resolution (compare Fig. 3A and B). To demonstrate the advantage of the giant patch technique over the two-electrode method both methods were compared by analysing transient currents mediated by the $Na^+/K^+$-pump operating in the $Na^+$-$Na^+$-halfcycle. The rates

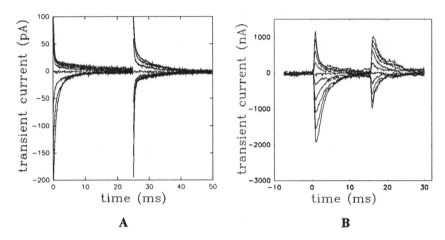

**A**                              **B**

**Figure 3. (A)** Transient currents generated by the sodium pump in a giant outside-out patch experiment. Currents were obtained by subtracting total membrane currents in presence of 5 mM extracellular ouabain from those in absence of ouabain. The currents are elicited by voltage jumps from a holding potential of -60 mV to potentials ranging from -150 to +60 mV. The extracellular solution contained (in mM): 100 NaCl, 20 TEACl, 5 $BaCl_2$, 5 MOPS/TRIS pH=7.8. The pipette solution contained: 80 NaCl, 25 NaOH, 5 $Na_2ATP$, 10 $MgCl_2$, 5 EGTA, 5 MOPS pH=7.3. The signal was filtered at 10 kHz and sampled at 20 kHz, averaged three times. **(B)** Transient pump currents measured by two-electrode voltage-clamp at identical extracellular conditions as in (A) except for additional 5 mM $NiCl_2$. The filter frequency was 1 kHz, the pulse protocol and the number of averages is also identical to (A).

of the exponential current decline during on-responses are qualitatively the same in the two types of experiments. Their voltage dependence shows an exponential decline with less negative potentials and reaches a plateau at positive potentials that can be described by $k=k_\infty(1+\exp(z_a(V_d-V)R/FT)$ (see Table A). For the two-electrode clamp, the parameters depend on the time where the original signal was cut off coming closer to the patch value the larger the cut-off interval. Figure 4 shows for the two methods voltage dependencies of the charges moved during the on-responses. As for the rate constants the results for two-electrode clamp depend on the type of analysis giving again more reliable data with increasing cut-off time. Nearly identical results are obtained for the total charge calculated from numerical integration of the current response.

In conclusion, the two-microelectrode techniques can successfully be applied in the analysis of transients of pump current if total charges are calculated by numerical integration. For more detailed analysis the gaint-patch clamp technique with its much higher time resolution provides more accurate results.

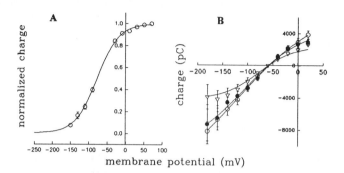

membrane potential (mV)

**Figure 4.** Voltage dependence of charge moved during the on responses using the giant patch **(A)** or two-electrode technique **(B)**. The charge in (A) was calculated as the time integral of the fitted e-function as $Q=I_0/k$, which was identical to the numerical integration. The charge in (B) was calculated as the time integral by cutting off after 2ms (open circles) or 4ms (closed circles) or by numerical integration (triangles). The charge data were fitted by a Boltzmann-equation: $Q=Q_{min}+(Q_{tot}/(1+\exp(z_q(V_m-V)F/RT)))$ (see (5)). The fit-parameters are shown in the Table B).

**Table:** Fitted parameters for **(A)** the exponential voltage dependence of the rate constants for on responses (see text) and for **(B)** the Boltzmann distribution of moved charges (see legend to Fig. 4).

|   |   | cut-off 2 ms | cut-off 4 ms | num.-integrated | giant-patch |
|---|---|---|---|---|---|
| **A** | $k_\infty$ (s$^{-1}$) | 180 | 150 | | 140 |
| | $V_d$ (mV) | -131 | -119 | | -60 |
| | $z_a$ | 0.41 | 0.38 | | 0.45 |
| **B** | $V_m$ (mV) | -123 | -107 | -78 | -78 |
| | $z_q$ | 0.29 | 0.42 | 0.7 | 0.9 |

# References

1.  Bahinski A, Nakao M, Gadsby DC (1988) Potassium translocation by the Na/K pump is voltage insensitive. PNAS 85: 3412-3416
2.  Collins A, Somlyo AV, Hilgemann DW (1992) The giant cardiac membrane patch method: Stimulation of outward $Na^+$-$Ca^{2+}$ exchange current by MgATP. J. Physiol. 454: 27-57
3.  Franke C, Hatt H, Dudel J (1987) Liquid filament switch for ultra-fast exchanges of solutions at excised patches of synaptic membrane of crayfish muscle. Neuroscience Letters 77: 199-204
4.  Hilgemann DW (1989) Giant excised cardiac sarcolemmal membrane patches: sodium and sodium-calcium exchange currents. Pfluegers Arch. 415: 247-249
5.  Rakowski RF (1993) Charge movement by the Na/K pump in Xenopus oocytes. J.Gen.Physiol. 101: 117-144
6.  Rakowski RF, Vasilets LA, LaTona J, Schwarz W (1991) A negative slope in the current-voltage relationship of the $Na^+/K^+$ pump in Xenopus oocytes produced by reduction of external [$K^+$]. J.Membrane Biol. 121: 177-187

# The Role of Putative Intramembraneous Glutamic Acid Residues of the α-Subunit of the Sodium Pump in External Cation Binding

L.A.Vasilets[*)+)], M.Kawamura[#)], K.Takeda[#)], T.Ohta[°)] and W.Schwarz[+)]

[+)] Max-Planck-Institut für Biophysik, D-60596 Frankfurt/M, Germany
[#)] Dept. of Biology, University of Occupational and Environmental Health, Kitakyushu 807, Japan
[°)] Institute of basic Medical Sciences, Univ. of Tsukuba, Tsukuba, Ibaraki 305,Japan
[*)] Permanent Address: Institut of Chemical Physics in Chernogolovka, Russian Academy of Sciences, Chernogolovka, Moscow region 142432, Russia

## Introduction

During the cation translocation of the $Na^+/K^+$ transport cycle two positive charges have been suggested to be compensated by negative charges (3) It was proposed that like in other cation binding proteins the cations are ligated with six to eight oxygens of carboxyl groups and that glutamic acid residues in intramembranous domains are involved (1,2,4). Based on their measurements with the carboxyl reagent DCCD Karlish and coworkers (4) suggested for possible candidates the glutamic acid residues E-334 and E-959 (in *Torpedo* numbering). We addressed this question in voltage-clamp experiments on *Xenopus* oocytes with expressed mutated pumps of *Torpedo* electroplax by analysing the influence of external $K^+$ on transport activity. In absence of extracellular $Na^+$ an apparent $K_m$ value for pump stimulation by external $K^+$ can be determined that shows exponential dependence on membrane potential. This has been interpreted by a voltage-dependent access of the $K^+$ ions to their occulsion site (5). For the *Torpedo* pump the voltage dependence of $K_m$ had to be described by the sum of two exponentials

$$K_m = K_{m1}(0mV) \, e^{z_1 VF/RT} + K_{m2} \, e^{z_2 VF/RT} \tag{1}$$

and has been explained by a sequential binding of the two $K^+$ ions (6) where the effective valencies $z_i$ reflect the dielectric length of the access for the two cations.

## Results and Discussion

To elucidate the possible role of the glutamic acid residues Glu-334, -959, and -960 in interaction with external $K^+$, wild-type and mutated pumps of *Torpedo* electroplax were expressed in *Xenopus* oocytes. Under two-microelectrode voltage clamp pump-generated currents were determined as ouabain-sensitive currents or $K^+$-activated currents (5). Figure 1A shows maximum pump currents of the endogenous *Xenopus* pump (control) and of wild-type (wT) and the mutated *Torpedo* pumps E334A and E960A where the corresponding glutamic acid residues were replaced by alanine. The inset shows [86]Rb uptake indicating that control, wT and E334A pumps are operating in their conventional mode. Figure 1B gives the amount of bound ouabain that is a measure for the number of functionally expressed pumps. Comparison of the two parameters demonstrates that maximum transport activity of the single pump molecule is not significantly affected by mutating Glu-334 or Glu-960 to alanine indicating that these amino acids are not essential for cation occlusion. On the other hand, the apparent $K_m$ values are clearly influenced.

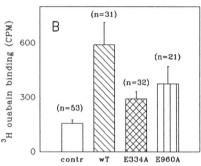

**Figure 1: (A)** Currents generated by ouabain-sensitive pumps in *Xenopus* oocytes under optimal conditions (0 Na$^+$, 5 mM K$^+$, 0 mV). Inset shows amount of $^{86}$Rb-uptake by oocytes without voltage clamp but in the same external solution. **(B)** Amount of bound $^3$H-labelled ouabain from the same batches of oocytes. Non-injected oocytes with endogenous *Xenopus* pumps (contr), oocytes with expressed *Torpedo* pumps: wild type (wT) and *Torpedo* mutants with Glu mutated to Ala (E334A, E959A, E960A).

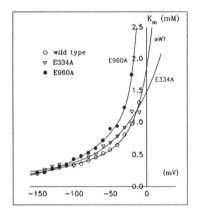

**Figure 2:** Voltage dependence of apparent $K_m$ values for pump stimulation by external K$^+$. Data for ouabain-sensitive wild-type and mutated *Torpedo* pumps are shown as indicated. Solid lines represent fits of equ. (1) to the data and fitted parameters are given in the Table.

Figure 2 shows the voltage dependence of apparent $K_m$ values for the wild-type *Torpedo* pump and for the mutants E334A and E960A. The effective valencies for the mutant E334A are significantly reduced (see Table) though the actual $K_m$ values differ only slightly from the wild-type over a wide range of potentials. The mutation of Glu-960, on the other hand, leads to a pronounced increase in the effective valencies and a clear reduction in apparent affinity (see also Table), which becomes particularly apparent in the change of $K_m$ at 0 mV.

To investigate expressed *Torpedo* pumps the contribution of the endogenous *Xenopus* pump to total pump current was subtracted numerically (see e.g. (6)). Another possibility is to use ouabain-resistant mutants (OR) of the *Torpedo* pump and to block the endogenous pump by 1 μM ouabain, which is without effect on the OR pump. The two kinds of subtraction of endogenous pump current give the same result as shown for the ouabain-resistant mutant of E960A (OR-E960A) in Fig. 3. As for the ouabain-sensitive mutants OR-E959A and OR-E960A are fully active (Fig. 4).

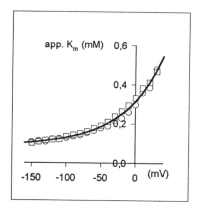

**Figure 3:** Voltage dependence of $K_m$ values of the ouabain-resistant *Torpedo* mutant OR-E960A determined from pump currents where the contribution of the endogenous pump was subtracted from total pump current by numerical subtraction of the current determined in control oocytes of the same batch (circles) or by blockage with i μM ouabain (squares). Ouabain resistance was achieved by mutation of Gln-118 to Arg and Asn-129 to Asp.

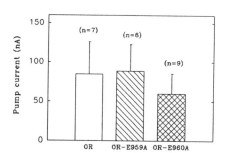

**Figure 4:** Currents generated by ouabain-resistant pumps in *Xenopus* oocytes under optimal conditions (0 $Na^+$, 5 mM $K^+$, 0 mV).

Interestingly, transferring to ouabain resistance completely changes the effect of Glu-mutation. Figure 5 shows that the $K_m$ values for the OR-E960A mutant are nearly indistinguishable from the normal OR pump and were fitted by the same voltage dependence of $K_m$. For the OR-E959A a clear difference is detectable with the component of higher effective valency dominating (see Table).

In conclusion, the glutamic acid residues Glu-334, -959 and 960 are not essential for cation binding but their mutation appears to modify external cation interaction by changing the apparent dielectric length of the access channel. Effects of mutation can be detected by determination of the potential dependence of apparent $K_m$ values that will depend on whether ouabain-sensitive or -resistant mutants are investigated.

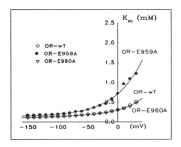

**Figure 5:** Voltage dependence of apparent $K_m$ values for pump stimulation by external $K^+$. Data for ouabain-resistant *Torpedo* pumps are shown as indicated without and with mutated glutamic acid residues.

**Table:** Fitted parameters for voltage dependence of $K_m$ values according to equ. (1). Parameters for ouabain-sensitive (S) and ouabain-resistant (OR) Torpedo pumps and the respective mutants are given

|  | wT | E334A | E960A | OR/E960A | OR-E959A |
|---|---|---|---|---|---|
| $z_1$ | 0.22 | 0.04 | 0.35 | 0.11 | 0.08 |
| $K_1(0mV)$ (mM) | 0.82 | 0.23 | 1.65 | 0.21 | 0.16 |
| $z_2$ | 1.56 | 0.49 | 3.4 | 0.84 | 0.64 |
| $K_2(0mV)$ (mM) | 1.03 | 1.25 | 7.56 | 0.09 | 0.54 |

*Acknowledgements*

The work was supported by Deutsche Forschungsgemeinschaft (SFB 169).

**References**

1. Arguello J, Kaplan JH. (1993) Identification of Glu[779] as an essential part of the cation binding site of the sodium pump. Biol. Chem. Hoppe-Seyler, 374, 585
2. Arguello JM, Kaplan JH. (1991) Evidence for essential carboxyls in the cation-binding domain of the Na,K-ATPase. J. Biol. Chem., 266, 14627-14635.
3. DeWeer P, Gadsby DC, Rakowski RF. (1988) Voltage dependence of the Na-K pump. Ann. Rev. Physiol., 50, 225-241.
4. Goldshleger R, Tal DM, Moorman J, Stein WD, Karlish SJD. (1992) Chemical modification of Glu-953 of the alpha-chain of $Na^+,K^+$- ATPase associated with inactivation of cation occlusion. Proc. Natl. Acad. Sci. USA, 89, 6911-6915.
5. Rakowski RF, Vasilets LA, LaTona J, Schwarz W. (1991) A negative slope in the current-voltage relationship of the $Na^+/K^+$ pump in Xenopus oocytes produced by reduction of external $[K^+]$. J. Membrane Biol., 121, 177-187.
6. Vasilets LA, Schwarz W. (1992) Regulation of endogenous and expressed $Na^+/K^+$ pumps in Xenopus oocytes by membrane potential and stimulation of protein kinases. J. Membrane Biol., 125, 119-132.

# Conformational States of Na$^+$/K$^+$-ATPase Probed with RH-421 and Eosin

N. U. Fedosova and I. Klodos

Institute of Biophysics, University of Aarhus, 8000 Aarhus C, Denmark

## Introduction

RH-dyes partition into the lipid bilayer and respond to binding of Na$^+$ to Na$^+$/K$^+$-ATPase (EC.3.6.1.37) by a decrease in their fluorescence (1,5,7). Subsequent addition of K$^+$ returns the fluorescence to the "pre-Na$^+$ level". In an attempt to define steps of the reaction cycle or conformational transitions of Na$^+$/K$^+$-ATPase associated with changes in RH-dyes fluorescence we applied one of the RH-dyes, RH-421, and eosin, a well known conformational probe, to characterize the cation binding to the intracellular sites. Increase in eosin fluorescence reflects a high affinity binding of eosin to the substrate binding site of the enzyme, identifying the E$_1$-dephosphoforms of Na$^+$/K$^+$-ATPase (6). RH-421 appeared to be able to distinguish between the Na-bound E$_1$ form and any other E$_1$-form in the pool of dephosphoenzymes.

## Methods

The experiments were performed with membrane bound Na$^+$/K$^+$-ATPase from pig kidney prepared as previously described (4). The specific activity was 25 U(mg protein)$^{-1}$. The buffer system was 10 mM HEPES, 10 mM MES, 10 mM EDTA, adjusted with N-methyl-D-glucamine to pH 7.5. Measurements of RH fluorescence were performed in a SPEX fluorometer using an excitation wavelength of 580 nm and a 630 nm emission cut-off filter as described by Klodos and Forbush (5). Eosin fluorescence was measured according to Skou and Esmann (6). The protein concentration was 30-50 µg/ml in all experiments.

## Results and Discussion

Eosin fluorescence did not increase upon addition of Na$^+$ to Na$^+$/K$^+$-ATPase enriched membranes, showing that the starting enzyme form (in the presence of buffer alone) belonged to the high-eosin-affinity E$_1$-pool. In parallel experiments with RH-421 an addition of Na$^+$ was followed by a decrease in the fluorescence, but in spite of the fact that the only enzyme form present was E$_1$, the decrease of RH-421 fluorescence was at double-exponential. The amplitude of the response in equilibrium experiments was dependent on [Na$^+$] with K$_{0.5}$ equal to 0.45 mM NaCl (Fig. 1A). Since no fluorescence change was observed upon addition of other monovalent cations, the decrease was due to a specific Na$^+$-binding to Na$^+$/K$^+$-ATPase. The RH-421 response to Na$^+$ binding was due to an electrochromic spectrum shift indicating the appearance of a charge in the vicinity of the dye (Fig. 1B). Addition of $K^+$ in the absence of $Na^+$ caused no change in fluorescence (Fig. 1A) showing that the dye does not respond to formation of the $E_2K$-form in the absence of $Na^+$.

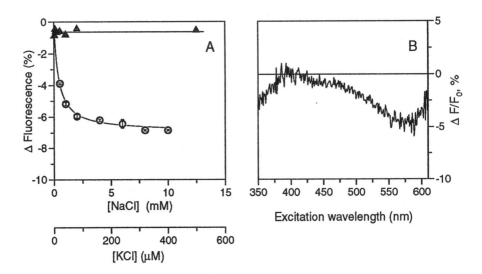

Figure 1. A - Changes in RH-421 fluorescence as a function of the cation concentration. *Triangles* - $K^+$-titration, *circles* - $Na^+$-titration. Fluorescence changes are expressed as percentage of the starting level. The values are the mean of three experiments ± SEM. B - Fractional change in the excitation spectrum of RH-421 upon addition of 4 mM NaCl.

An investigation of transitions between Na- and K-bound forms of the enzyme was performed in two sets of experiments: a) the transition of $E_1Na$ to $E_2K$ was studied in the presence of 4 mM NaCl adding varying $[K^+]$; and b) the transition from $E_2K$ to $E_1Na$ was investigated in the presence of KCl adding varying $[Na^+]$. Changes in fluorescence of both eosin and RH-421 were under these conditions monoexponential. They were fitted with monoexponential functions and the parameters of these functions, rates and amplitudes ($k_{obs}$ and $\Delta F/\Delta F_{max}$), were compared. The data were adequately fitted to the minimal model $E_2K \rightleftharpoons E_1K \rightleftharpoons E_1 \rightleftharpoons E_1Na \rightleftharpoons E_1Na_2$ which implies exchange of 2 $Na^+$ for 1 $K^+$ in one site.

The parameters derived from the first set of experiments were largely independent of the dye used as indicator. $k_{obs}$ values showed linear dependence on $[K^+]$ (not shown). $\Delta F/\Delta F_{max}$ *vs.* $[K^+]$ were hyperbolic functions (Fig. 2A) with slightly different $K_{0.5}$ - 0.14 mM, measured with RH-421, and 0.4 mM, obtained with eosin. These results are in agreement with the previously reported data on the $K^+$-binding to the $E_1$-form (2,3).

In the second set of experiments where $Na^+/K^+$-ATPase was initially in the $E_2K$ form, the equilibrium titration with $Na^+$ gave evidence that more than one $Na^+$ was bound to the enzyme in the $E_1$-form (Fig. 2B). Again the data on RH-421 fluorescence were in agreement with those obtained with eosin.

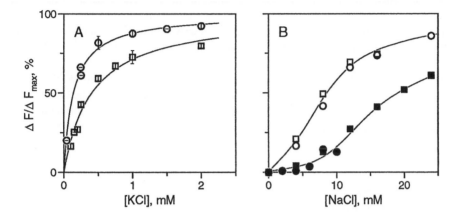

Figure 2. A - Equilibrium fluorescence change as a function of [$K^+$] in the presence of 4 mM NaCl. Changes in eosin fluorescence (*squares*) and in RH-421 fluorescence (*circles*) are normalized to the maximal fluorescence change observed with each probe. The values are the mean of three experiments ± SEM.
B - Equilibrium fluorescence change as function of $Na^+$ in the presence of 0.5 mM KCl (*open symbols*) or 2 mM KCl (*filled symbols*). Symbols are the same as in Fig. 2A.

Comparison of the data on RH-421 fluorescence with those from eosin experiments shows that both probes can be used for the investigation of the $E_1Na \rightleftharpoons E_2K$ interconversions. RH-dyes, however, have several advantages. One of them is that they distinguish Na-bound $E_1$ form from any other form of the enzyme. The dyes do not affect activity of the enzyme and do not interfere with the cation-enzyme equilibrium. In contrast, eosin binding to the $E_1$-conformation shifts the equilibrium in favor of the $E_1$-form and thus hinders $K^+$-binding. That is probably why the $K_{0.5}$ value for $K^+$, obtained with RH-421 is slightly lower, than that obtained with eosin (Fig. 2A). This kind of effects of a probe adds to the complexity of the kinetic study of the system. Drawbacks of the RH-dyes in this context are their relatively small response to $Na^+$-binding and their sensitivity to ionic strength of the medium.

The double-exponential time course observed with RH-dye could be a reflection of sequential $Na^+$-binding, where a rapid formation of the $E_1Na$-form is followed by a slow transition to an another Na-bound form (e. g. $E_1(Na)$ form). Such a sequential scheme implies that the rate of the K-induced reversal should change with the time of incubation of the enzyme with $Na^+$ *prior* to $K^+$-addition, i. e. when a slow transition to the next Na-bound form is completed, a relatively slow (or double-exponential) response to $K^+$ should be observed, whereas when $K^+$ is added together with $Na^+$ the rate of K-induced reversal should be relatively fast. It appeared, however, that the rate coefficient of the monoexponential fluorescence change was independent of the time of $K^+$-addition. The $k_{obs}$ values were the same when $K^+$ was added after the equilibrium with $Na^+$ was reached or when both cations were added simultaneously. These results rule out the sequential model. Eosin fluorescence in parallel experiment showed the same $k_{obs}$ for the $E_1Na \rightarrow E_2K$ transition giving additional evidence for similar behavior of these probes with respect to transitions between the dephosphoenzymes.

The apparent heterogeneity of $Na^+$-binding to the $E_1$-form, reported by the RH-421 as a double-exponential decrease in the fluorescence implies heterogeneity of forms comprising the $E_1$-pool and needs further investigation. It could reflect multisite cation binding but until now we do not have experimental evidence for this hypothesis.

*Acknowledgements*

We thank The Danish Natural Science Council, The Danish Research Academy and The Biomembrane Research Center, University of Aarhus, for financial support.

**References**

1.      Bühler R, Stürmer W, Apell H-J, Läuger P (1991) Charge translocation by the Na,K-pump: I. Kinetics of local field changes studied by time-resolved fluorescence measurements. J Membrane Biol 121: 141-161

2.      Faller LD, Diaz R, Scheiner-Bobis G, Farley RA (1991) Temperature dependence of the rates of conformational changes reported by fluorescein-5'-isothiocyanate modification of H,K- and Na,K-ATPases. Biochemistry 30: 3503-3510

3.      Glynn IM, Karlish SJD (1982) Conformational changes associated with $K^+$ transport by the Na,K-ATPase. Membrane and Transport (Martonosi AN, ed.), Plenum Press, New York, 1: 529-536

4.      Jensen J, Nørby JG, Ottolenghi P (1984) Binding of sodium and potassium to the sodium pump of pig kidney evaluated from nucleotide-binding behaviour. J Physiol (London) 346: 219-241

5.      Klodos I, Forbush B III (1988) Rapid conformational changes of the Na/K pump revealed by a fluorescent dye, RH-160. J Gen Physiol 92: 46a

6.      Skou JC, Esmann M (1981) Eosin, a fluorescent probe of ATP binding to $(Na^++K^+)$-ATPase. Biochim Biophys Acta 647: 232-240

7.      Stürmer W, Bühler R, Apell H-J, Läuger P (1991) Charge translocation by the Na,K-pump: II. Ion binding and release at the extracellular face. J Membrane Biol 121: 163-176

# Do Lipophilic Anions Affect the Spatial Organization of the ATP Binding Site of the Na$^+$/K$^+$-ATPase?

N.U. Fedosova and J. Jensen

Institutes of Biophysics and Physiology, University of Aarhus, 8000 Aarhus C, Denmark

## Introduction

It has been shown that the fluorescence of eosin is increased upon its binding to the substrate site of Na$^+$/K$^+$-ATPase (EC 3.6.1.37) (3,8). Studying the effect of lipophilic ions on Na$^+$/K$^+$-ATPase we observed that these ions affect eosin fluorescence (5). Both tetraphenylphosphonium (TPP$^+$) and tetraphenylboron (TPB$^-$) increased the eosin fluorescence. While TPP$^+$ influenced non-specific eosin fluorescence, TPB$^-$ affected exclusively fluorescence of eosin bound to high affinity sites on the Na$^+$/K$^+$-ATPase. A more detailed study of the TPB$^-$-effect on both eosin fluorescence and ADP-binding is presented below.

## Methods

The experiments were performed with membrane bound Na$^+$/K$^+$-ATPase from pig kidney and rectal glands of *Squalus acanthias* prepared as previously described (4,7). The specific activity was 17 and 27 U(mg protein)$^{-1}$ respectively. Eosin fluorescence was measured according to Skou and Esmann (8). Specific eosin binding was induced by the presence of 10 mM NaCl in 10 mM HEPES, 10 mM MES, 10 mM EDTA, pH 7.5. Protein concentration in fluorescence experiments was 30-50 µg/ml. NaTPB dissolved in dimethyl sulfoxide was added to a final concentration of 50 µM. $^{14}$C-ADP binding was measured by a centrifugation assay as described by Nørby and Jensen (6). The protein to TPB$^-$ ratio was kept constant in all experiments as the TPB$^-$ concentration in the lipid phase appeared to be of importance.

## Results and Discussion

The fluorescence of eosin specifically bound to the Na$^+$/K$^+$-ATPase increased instantaneously upon addition of TPB$^-$, the effect being concentration-dependent and saturating at [TPB$^-$] ~ 100 µM. Change in specific fluorescence could occur for the following reasons: a) a change in affinity for the fluorescent compound, b) a change in the number of binding sites, and finally c) a change in the quantum yield of the fluorescence.

Titration of Na$^+$/K$^+$-ATPase with eosin in the absence and in the presence of 50 µM TPB$^-$ revealed no difference in eosin dissociation constant (0.32 and 0.29 µM respectively). So, either TPB⁻ induced additional eosin binding with the same affinity or changed the quantum yield of fluorescence of eosin already bound to the enzyme.

Direct measurements of [14]C-ADP binding showed that the number of ADP-binding sites (2.10 nmoles per mg protein) was not affected by the lipophilic anion (Fig.1, inset). Therefore, the TPB⁻-induced increase in the number of eosin binding sites appears to be unlikely. The assumption that more eosin molecules bind to the same site with unchanged affinity is also unlikely.

We are thus left with the third explanation of the TPB⁻ effect: increase in quantum yield of bound eosin. The ability of eosin to change the fluorescence upon the transfer into the environment with different polarity is one of the main characteristics that allows to use it as a conformational probe. Eosin dissolved in ethyl alcohol and eosin specifically bound with Na⁺/K⁺-ATPase have similar spectrum characteristics (7). On this basis it is reasonable to suggest that a further increase in quantum yield of bound eosin in the presence of TPB⁻ is due to a further decrease in the polarity of its environment, i.e. of the ATP binding site. The fact that TPB⁻ increased $K_{0.5}$ for ADP obtained in fluorescence experiments by substitution of eosin in different concentrations (Fig.1) and in direct ADP-binding experiments (without eosin) fits the hypothesis about alterations in ATP-binding site.

The decreased affinity for ADP (but not for eosin) in the presence of the lipophilic anion could be explained by an electrostatic repulsion of the two negatively charged molecules. Note that at pH 7.5 the negative charge of ADP is about 3 times higher than that of eosin. We studied therefore the effect of TPB⁻ on the fluorescence of eosin analogues - 5-carboxy-eosin and 6-carboxy-eosin, which have an additional carboxyl group fully deprotonated at the pH used, i. e. the net charge of these molecules is twice that of eosin. If the reason for decreased ADP affinity is the interaction of charges one could expect a decrease in the affinity of these eosin analogues to Na⁺/K⁺-ATPase in the presence of lipophilic anion.

TPB⁻ did not affect the affinity of any of the analogues to Na⁺/K⁺-ATPase. Moreover, the increase in quantum yield of 5-carboxy-eosin in the presence of lipophilic anion was much less than that of eosin, and 6-carboxy-eosin fluorescence was completely insensitive to TPB⁻. These results suggest that TPB⁻ inhibition of ADP-binding is not due to simple electrostatic interactions.

It appears however, that the additional carboxyl group in eosin analogues affects their sensitivity to the TPB⁻-induced alterations in the ATP-binding site and that the position of this carboxyl group is crucial.

It has been shown that TPB⁻ can be washed out of the membrane and all its effects on Na⁺/K⁺-ATPase are reversible (5). Solubilization of Na,K-ATPase with $C_{12}E_8$ eliminated TPB⁻ influence on both ADP-affinity and quantum yield of 5-carboxy-eosin fluorescence. The reason for the use of 5-carboxy-eosin in solubilization experiments was that its fluorescence is not affected by $C_{12}E_8$ (2). Therefore, we conclude that the effect of lipophilic ion was mediated by the lipid bilayer.

The fact that a lipophilic anion localized in the lipid phase of the membrane affects the ADP-binding site in the cytoplasmic loop of $Na^+/K^+$-ATPase is surprising. Since $TPB^-$ is located so deep in the membrane that its charge is not screened by increasing ionic strength (1) any long-distance electrostatic interactions between $TPB^-$ and charged group(s) of the ATP-binding site in the cytoplasmic loop of $Na^+,K^+$-ATPase appear to be unlikely.

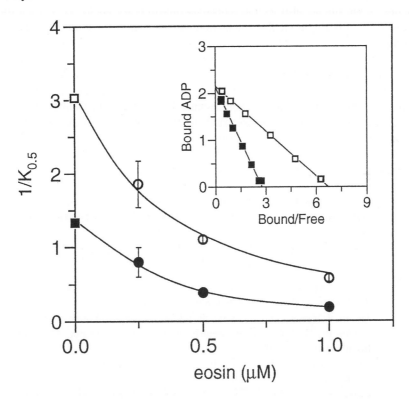

Figure 1. The ADP affinity to Na,K-ATPase at different eosin concentrations. The data were obtained by fluorescence measurements (circles) and by estimation of [14]C-ADP-binding (squares) in the absence (open symbols) and in the presence (filled symbols) of $TPB^-$. All values are the mean of at least three experiments ± SEM. The sets of data were fitted according to equation

$$1/K_{0.5} = 1/K_D / (1 + K_{Eo} / [Eo]),$$

where $K_D$ is dissociation constant for ADP, $K_{Eo}$ - dissociation constant for eosin, [Eo] - eosin concentration. The following values for the parameters were obtained: $K_D$ in the absence and in the presence of $TPB^-$ - 0.33 µM and 0.75 µM respectively, $K_{Eo}$ was 0.32 µM in the absence and 0.29 µM in the presence of lipophilic anion.
Scatchard plot for ADP binding in the absence and in the presence of lipophilic anion is shown in the *inset*.

An alteration in the structure of the ATP-binding site could be due to a TPB⁻ effect on intramembrane amino acid residues, which in consequence elicit small changes in the folding of the cytoplasmic loop. The changes in the ATP-binding site probably occur in the region of hydrophobic interactions between the nitrogenous base of the substrate and corresponding amino acid residues. Eosin responds to the changes by increased quantum yield of fluorescence. As an additional carboxyl group in eosin analogues changes the orientation of the molecule in the site, moving these analogues out of the region with decreased polarity, the sensitivity of the analogues decreases. The changes in the structure of the site are also seen as a decreased affinity for nucleotide but not for eosin, suggesting that the hydrophobic interactions are of major importance for the substrate binding.

*Acknowledgements*

We thank Edith B. Møller for excellent technical assistance, The Danish Natural Science Council and The Biomembrane Research Center, University of Aarhus, for financial support.

**References**

1.      Andersen OS, Feldberg S, Nakadomari H, Levy S, McLaughlin S (1978) Electrostatic interactions among hydrophobic ions in lipid bilayer membranes. Biophys J 21: 35-70

2.      Esmann M (1991) Conformational transitions of detergent-solubilized Na,K-ATPase are conveniently monitored by the fluorescent probe 6-carboxy-eosin. Biochem Biophys Res Commun 174: 63-69

3.      Esmann M (1992) Determination of rate constants for nucleotide dissociation from Na,K-ATPase. Biochim Biophys Acta 1110: 20-28

4.      Jensen J, Nørby JG, Ottolenghi P (1984) Binding of sodium and potassium to the sodium pump of pig kidney evaluated from nucleotide-binding behaviour. J Physiol (London), 346: 219-241

5.      Klodos I, Fedosova NU, Plesner L (1993) Influence of intramembrane electric charge on Na,K-ATPase (in preparation)

6.      Nørby JG, Jensen J (1988) Measurement of binding of ATP and ADP to Na⁺,K⁺-ATPase. Methods Enzymol 156: 191-210

7.      Skou JC, Esmann M (1979) Preparation of membrane-bound and of solubilized (Na⁺+K⁺)-ATPase from rectal glands of *Squalus acanthias*. The effect of preparative procedures on purity, specific and molar activity. Biochim Biophys Acta 567: 436-444

8.      Skou JC, Esmann M (1981) Eosin, a fluorescent probe of ATP binding to (Na⁺+K⁺)-ATPase. Biochim Biophys Acta 647: 232-240

# Na$^+$/K$^+$-ATPase: Probing for a High Affinity Binding Site for Na$^+$ by Spectrofluorometry

H. Ruf, E. Lewitzki, E. Grell

Max-Planck-Institute of Biophysics, Kennedy-Allee 70, D-60596 Frankfurt, FRG

## Introduction

The transfer of Na$^+$ and K$^+$ across membranes by the Na$^+$/K$^+$-pump is coupled to the transition between two conformational states of the enzyme denoted E$_1$ and E$_2$ (1,12). In the E$_1$ state the enzyme binds preferentially Na$^+$, while in the E$_2$ state it binds preferentially K$^+$. Fluorescent probes like fluorescein-isothiocyanate (FITC) covalently bound to the protein (8) or potential-sensitive styryl dyes such as RH 421 incorporated into the membrane (2,9,13) change their fluorescent properties in the presence of Na$^+$ or K$^+$, and thus offer the possibility of studying the binding of these ions and conformational transitions of the enzyme spectroscopically. FITC-labeled Na$^+$/K$^+$-ATPase is strongly fluorescent in the presence of high Na$^+$ concentrations, but exhibits a much lower fluorescence if K$^+$ or one of its congeners is bound. Accordingly the two fluorescence emission intensity states have been assigned to the conformational states E$_1$ and E$_2$ (6,8,11). Na$^+$ binding alone can also be studied by means of the fluorescence changes of RH 421, which has been used to study phosphorylation of the native enzyme by ATP in the presence of Na$^+$ and Mg$^{++}$ under conditions where no K$^+$ is present (2,9,13). Titrations of the enzyme with Na$^+$ in the presence of these dyes showed that the two different fluorescent labels report binding of Na$^+$ to two different sites (4). The implications of these findings to the assignment of fluorescent states to conformations of the enzyme will be discussed here in more detail.

## Spectrofluorometric titrations with NaCl

A typical Na$^+$ titration curve obtained with the FITC-enzyme isolated and labeled from pig kidney according to (4,7,8) is shown in Fig. 1a. The data are described well by a binding model based on 1:1 stoichiometry of complex formation. The pK value of 2 compares well with those determined before (11). The spectrofluorometrically determined affinities of the Na$^+$ complex of the FITC-enzyme, however, are much smaller than those determined by Matsui et al. (10) using a filtration method with $^{22}$Na$^+$. Moreover, spectrofluorometric titrations with a series of mono and divalent cations other than K$^+$ and its congeners showed similar titration curves as with Na$^+$ (3,4). It was found that the affinities of these compounds depend practically only on the charge of the ions and not on their chemical nature. For all monovalent ions the pK values are similar to that of Na$^+$, which suggests that Na$^+$ binding indicated by the high fluorescent state of the FITC-enzyme is a non-selective one.

On the other hand, RH 421 fluorescence changes its intensity at rather low Na$^+$ concentrations indicating high affinity binding of Na$^+$ to the enzyme, whereas this is practically not changed in the concentration range where Na$^+$ binding is reported by

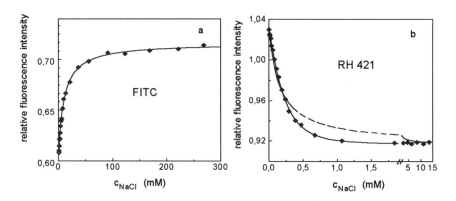

**Figure 1.** Fluorometric titrations of Na$^+$/K$^+$-ATPase with NaCl in the presence of 10 mM imidazole-HCl, 0.1 mM EDTA at pH = 7.5 (25 °C). (a) FITC-labeled enzyme (exc. 495 nm, em. 518 nm): the solid line represents the theoretical titration curve based on 1:1 stoichiometry of complex formation. (b) Unmodified enzyme in the presence of 0.7 μM RH 421 (exc. 580 nm, em. 640 nm): the data are well described by 2:1 stoichiometry for Na$^+$ binding (solid line). The titration curve based on 1:1 stoichiometry (dashed line) is also given for comparison.

FITC fluorescence (Fig. 1b). In addition, a quantitative description of the experimental titration curve requires a more complex stoichiometry. A good fit is obtained here by a model that assumes a 2:1 stoichiometry of binding (pK$_1$ = 3.4, pK$_2$ = 3.5). The pK values indicate here a more than one order of magnitude higher binding affinity for Na$^+$. The decrease in RH 421 fluorescence upon binding of Na$^+$ is quite specific. RH 421 doesn't exhibit pronounced fluorescence changes when the other cations including K$^+$ are bound. For example, practically no intensity changes are observed with K$^+$ up to 10 mM and with choline up to 50 mM. As an exception Mg$^{++}$ causes an increase in fluorescence. From titrations with Mg$^{++}$ a pK$_{0.5}$ value of about 4.0 was determined. Although the characteristics of these titrations are influenced by ouabain, we doubt whether this really reflects binding to specific sites of the protein, since the same behaviour was observed after heat-inactivation of the enzyme that abolished the Na$^+$ effect. From this specific probing of Na$^+$ binding by RH 421, from the marked differences in the affinities and from the fact that RH 421 reports only high affinity binding while FITC reports only low affinity binding, it is concluded that two different types of binding sites exist for Na$^+$ on the enzyme.

To check whether the FITC label, which prevents phosphorylation of the enzyme by ATP, will have an affect on high affinity Na$^+$ binding, titrations were carried out with both the labeled and the unmodified enzyme in the presence of RH 421 at different pH values (Fig. 2a). There are practically no differences which indicate that the high affinity binding sites are also present on the modified enzyme and unaffected by the FITC label. The latter is consistent with results obtained with the FITC-enzyme from titrations with K$^+$ in the presence of different amounts of Na$^+$ (5). It was found that Na$^+$ lowers the affinity for K$^+$ at rather low concentrations, where Na$^+$ does not affect the FITC fluorescence yet. This competitive effect is indicative of the existence of additional high affinity binding sites for Na$^+$. The pK$_{0.5}$ value of about 3.8 obtained

570

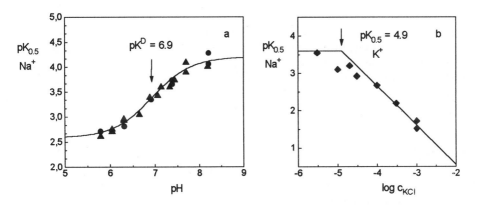

Figure 2. Na$^+$ affinities characterized by pK$_{0.5}$ obtained from titrations of Na$^+$/K$^+$-ATPase in the presence of 0.7 μM RH 421. (a) Dependence on pH of unmodified (♦ ) and FITC-enzyme (● ). The same behaviour is found with both enzymes, which indicates protonation of an amino acid residue in the binding region (pK$^D$ = 6.9). (b) Dependence on KCl concentration at pH 7.5. A pK$_{0.5}$ value of about 4.9 is determined from the intersection of the limiting slope with the pK$_{0.5}$ value of Na$^+$. Other conditions: 10 mM imidazole-HCl, 0.1 mM EDTA, 25 °C.

from these competition titrations for Na$^+$ compares very well with that directly determined from Na$^+$ titrations with the unmodified enzyme in the presence of RH 421 and with the data reported by Matsui et al. (10). In addition, the high affinity Na$^+$ binding shows a pronounced pH dependence. The affinity increases by roughly two orders of magnitude as the pH increases from 6 to 8, which suggests that a charged, protonable amino acid residue, possibly aspartic or glutamic acid, is part of the high affinity Na$^+$ binding region (pK$^D$ ≈ 6.8). This finding points to a possible role of pH in controlling Na$^+$ binding and thus turnover numbers of the native enzyme.

High affinity binding of K$^+$ is not directly reported by changes in RH 421 fluorescence. However, just as Na$^+$ at low concentrations starts to affect high affinity binding of K$^+$ to the FITC-enzyme (5), K$^+$ also reduces high affinity binding of Na$^+$. This is shown in Fig. 2b, where the pK$_{0.5}$ values of Na$^+$ are plotted against the logarithm of the KCl concentration. The affinity of this competing cation can be determined from the intersection of the limiting slope and the K$_{0.5}$ value of Na$^+$ in the absence of K$^+$. The pK$_{0.5}$ value of 4.9 for K$^+$ binding extrapolated in this way compares well with that determined directly from titrations with the FITC-labeled enzyme (4). These results demonstrate again the usefullness of competition titrations for revealing binding sites that are not directly obvious from changes in fluorescence.

The data presented here show that each of the fluorescent probes reveals directly one high affinity binding site of one of the two species transported by the Na$^+$/K$^+$-pump, and indirectly, via competition titrations, the existence of high affinity binding sites for the other ion, respectively. In view of the presently accepted models for the action of the pump, the low fluorescent state of the FITC-enzyme in the presence of K$^+$ and its congeners can be assigned to the $E_2$ conformation of the enzyme, where $K^+$ is bound in an occluded form. The competitive effects and the magnitude of the affinity itself suggest that high affinity binding of $Na^+$ reported by RH 421 can be associated with the $E_1$ conformation. On the other hand, the high fluorescent state of the FITC enzyme caused by low affinity, non-specific binding of $Na^+$ as well as by a variety of other cations has to be assigned to a different state denoted as $F_1$. A detailed discussion on this issue is given elsewhere in this volume (5).

## Acknowledgements

We thank Mrs. A. Schacht and Mr. G. Schimmack for expert technical assistance and the German Research Foundation (SFB 169) for support.

## References

1.  Albers RW (1967) Biochemical aspects of active transport. Ann Rev Biochem 36: 727-756
2.  Bühler R., Stürmer W, Apell H-J, Läuger P (1991) Charge translocation by the Na,K-pump: I. Kinetics of local field changes studied by time-resolved fluorescence measurements. J Membrane Biol 121: 141-161
3.  Grell E, Warmuth R, Lewitzki E, Ruf H (1991) Precision titrations to determine affinity and stoichiometry of alkali, alkaline earth and buffer cation binding to Na,K-ATPase. In: Kaplan JH, de Weer P (eds) The Sodium Pump: Recent Developments. Rockefeller University Press, New York, pp 441-445
4.  Grell E, Warmuth R, Lewitzki E, Ruf H (1992) Ionics and conformational transitions of Na,K-ATPase. Acta Physiol Scand 146: 213-221
5.  Grell E, Lewitzki E, Ruf H (1994) Reassignment of cation-induced population of main conformational states of FITC-$Na^+/K^+$-ATPase as detected by fluorescence spectroscopy and characterized by equilibrium binding studies. This volume.
6.  Hegyvary C, Jørgensen PL (1981) Conformational changes of renal sodium plus potassium ion-transport adenosine triphosphatase labeled with fluorescein. J Biol Chem 256: 6296-6303
7.  Jørgensen PL (1974) Purification and characterization of $(Na^+ + K^+)$-ATPase III. Biochim Biophys Acta 336: 36-52
8.  Karlish SJD (1979) Cation induced conformational states of Na,K-ATPase studied with fluorescent probes. In: Skou JC, Norby JG (eds) Na,K-ATPase: structure and kinetics. Academic Press, New York, pp 115-128
9.  Klodos I, Forbush III B (1988) Rapid conformational changes of the Na/K pump revealed by a fluoresent dye. J Gen Physiol 92: 46a
10. Matsui H, Homareda H, Hayashi Y (1985) Characteristics of $Na^+$ and $K^+$ binding to Na,K-ATPase. In: Glynn I, Ellory C (eds) The Sodium Pump. The Company of Biologists, Cambridge, pp 243-249
11. Mezele M, Lewitzki E, Ruf H, Grell E (1988) Cation selectivity of membrane proteins. Ber der Bunsenges 92: 998-1004
12. Post RL, Seiler SM, Vasallo PM (1987) Action of hydrogen ions on sodium, potassium adenosine triphosphatase labelled with fluorescein. In: Mukohata Y, Morales MF, Fleischer S (eds) Perspectives of biological energy transduction. Academic Press, Tokyo, pp 306s-326s
13. Stürmer W, Bühler R, Apell H-J, Läuger P (1991) Charge translocation by the Na,K-pump: II. Ion binding and release at the extracellular side. J Membrane Biol 121: 163-176

# Interaction between cytoplasmic sodium binding sites and cardiotonic steroids

B. Schwappach, W. Stürmer, H.-J. Apell, S.J.D. Karlish

Department of Biology, University of Konstanz, D-78434 Konstanz, Germany, and Department of Biochemistry, Weizmann Institute, Rehovot 76100, Israel

## Introduction

Information on structure-function relations of a protein can be obtained by looking at effects of specific structural modifications on defined functions. In this paper we present data obtained with native $Na^+,K^+$-ATPase from rabbit kidney, and so-called "19 kD-membranes", produced by a specific trypsin digestion. 19 kD-membranes consist of a 19 kD and smaller ($\approx$ 8-12 kD) fragments of the $\alpha$-chain, and an intact $\beta$-chain. (1). Cation occlusion is maintained but ATP-dependent functions are destroyed. We have now compared the ability of native enzyme and 19 kD-membranes to bind $Na^+$ ions at the cytoplasmic surface in an electrogenic fashion and to bind ouabain and inhibit the pump from the external side.

## Results

Purified membrane fragments or trypsin modified membrane preparations, the 19 kD membranes, have been investigated by a fluorescence method, using the styryl dye RH 421, which allows detection of charge movements within membrane bound proteins, which are associated with changes of a local electric field (2). A standard experiment is shown in Fig. 1.

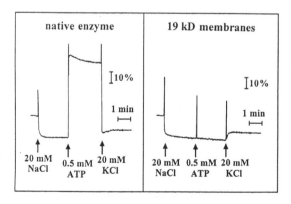

**Figure 1:** 10 µg/ml protein was equilibrated in buffer containing 30 mM imidazole, 1 mM EDTA, 5 mM $MgCl_2$ and 200 nM RH 421 at pH 7.2. Successively NaCl (final concentration 20 mM), ATP (0.5 mM) and KCl (20 mM) were added.

In both preparations an electrogenic $Na^+$ binding has been observed. Phosphorylation by ATP and the successive reaction to state $P\text{-}E_2$ was observed only with native enzyme. Addition of $K^+$ led to $K^+$ binding and dephosphorylation in the case of native enzyme. In the case of 19 kD-membranes there appears to be a partial binding of $K^+$ ions, presumably reflecting the different affinities of $Na^+$ and $K^+$ at the cytoplasmic surface.

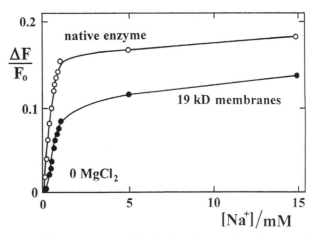

**Figure 2:** Titration of the cytoplasmic Na binding sites in the absence of $Mg^{2+}$ ions

The electrogenic binding of $Na^+$ ions has been investigated as a function of the $Na^+$ concentration in the presence and absence of 10 mM $MgCl_2$. The $K_M$ of the $Na^+$ binding was ~ 0.7 mM in the absence of Mg in both protein preparations (Fig. 2). In the presence of Mg the $K_M$ was shifted by a factor of 10 to 7 mM for native enzyme and the 19 kD membranes. It can be concluded that cytoplasmic ion binding sites are hardly affected by trypsin induced removal of the cytoplasmic protein parts. This is consistent with the conclusion drawn on the basis of direct measurements of cation occlusion (1).

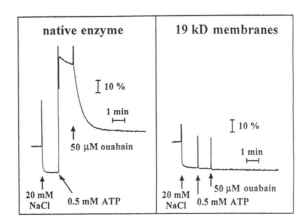

**Figure 3:** 10 µg/ml protein was equilibrated in buffer containing 30 mM imidazole, 1 mM EDTA, 5 mM $MgCl_2$ and 200 nM RH 421 at pH 7.2. Successively NaCl (final concentration 20 mM), ATP (0.5 mM) and ouabain (50 µM) were added.

As has been shown elsewhere (3, 4) that cardiac glycosides inhibit the Na,K-ATPase by binding to one conformational form, $E_2$-P with 2 cations bound ($Na^+$, $K^+$). The inhibition of the native enzyme can be detected by a significant fluorescence decrease after addition of ouabain as shown in Fig. 3. This corresponds to the transition $E_2$-P $\rightarrow$ $E_2$-P($Na_2$)-ouabain. In case of the 19 kD membranes neither ATP nor ouabain produce any effect in the presence of $Na^+$.

Effects of phosphate and ouabain in the absence of $Na^+$ are shown in Fig. 4. The interpretation of these signals is not completely clear. Yet they provide evidence for binding of ouabain under different conditions.

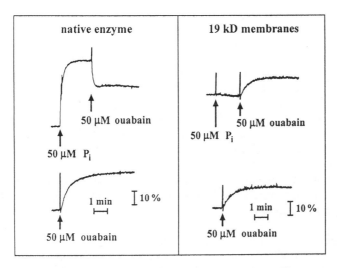

**Figure 4:** Change of RH 421 fluorescence induced by binding of ouabain in the absence of $Na^+$ ions. 10 μg/ml protein was equilibrated in buffer containing 30 mM imidazole, 1 mM EDTA, 5 mM $MgCl_2$ and 200 nM RH 421 at pH 7.2. Successively $P_i$ (final concentration 0.25 mM) and for ouabain (50 μM) were added.

Phosphorylation by inorganic phosphate, $P_i$, and inhibition by ouabain has been observed with native enzyme. A repetition with 19 kD membranes did not show an effect of $P_i$, but an effect of binding of ouabain was observed. Control experiments with the fluorescent analog Anthroyl ouabain demonstrated binding to the protein. The lower traces of Fig. 4 demonstrate that binding of ouabain occurred similarly for both preparations in the absence of $Na^+$ ions and $P_i$. Under these condition ouabain binding could be reversed by addition of 20 mM $Na^+$ or $K^+$. Binding and reversal of binding occurred 6 - 7 times faster for the 19 kD membranes compared to the native enzyme. (data not shown). Binding affinities for strophanthidin have been determined. Varying the strophanthidin concentration led to a Michaelis-Menton type of binding with $K_I$ values of 0.75 μM for the native enzyme and 1.2 μM for the 19 kD membranes .

Ouabain has also been found to inhibit $Rb^+$ occlusion on native enzyme and 19 kD-membranes with similar affinities ( 1.7 μM and 2.1 μM respectively, data not shown.).

Since in native enzyme and 19 kD membranes, $Na^+$ binding and steroid inhibition are antagonistic, the question has been raised as to how both properties are directly connected via protein structure. Therefore, heat inactivation of 19 kD fragments has been measured. At 37 °C and 25 °C, after the given time of incubation, the $Na^+$ induced and the strophanthidin induced RH fluorescence change has been measured in parallel. In both experiments the corresponding effects disappeared with the same time course. The half-life times of $Na^+$ and strophanthidin binding have been 10 min. at 37 °C and greater than 1 hour at 25 °C.

## Conclusions

1. Cytoplasmic $Na^+$ binding and cardiotonic steroid binding and 'inhibition' have the same characteristics in the 19 kD membranes with as in native protein in membrane fragments. Thus, it appears that both the transmembrane domain and extracellular loops are intact in 19 kD-membranes, whereas, of course, the cytoplasmic domain has been removed.
2. $Na^+$ binding to the cytoplasmic sites and ouabain binding to the extracellular face are antagonistic even in 19 kD membranes, in which in which the usual distinction between an $E_2$ or $E_2$-P state cannot be made.
3. The parallel disappearance of $Na^+$ and ouabain binding upon thermal inactivation suggests that the fragments of 19 kD-membranes form a complex which is disrupted in a concerted fashion.

*Acknowledgments*

This work has been financially supported by the Deutsche Forschungsgemeinschaft (Sonderforschungsbereich 156).

## References

1  Karlish SJD, Goldshleger R, Stein WD (1990) A 19-kDa C-terminal tryptic fragment of the a-chain of the Na,K-ATPase is essential for occlusion and transport of cations. Proc. Natl. Acad. Sci. USA 87: 4566-4570
2  Bühler R, Stürmer W, Apell H-J, Läuger P (1991) Charge translocation by the Na,K-pump: I. Kinetics of local field changes studied by time-resolved fluorescence measurement. J Membr Biol 121: 141-161
3  Jørgensen PL, Andersen PJ (1988) Structural basis for E1-E2 conformational transitions in Na,K-pump and Ca-pump proteins. J Membr Biol 103: 95-120
4  Stürmer W, Apell H-J (1993) Fluorescence study on cardiac glycoside binding to the Na,K-pump. FEBS Lett. 300: 1-4

# Dielectric Coefficients of the Extracellular Release of Sodium Ions

I. Wuddel, W. Stürmer, H.-J. Apell

Department of Biology, University of Konstanz, D-78434 Konstanz, Germany

## Introduction

The pumping mechanism of the $Na^+/K^+$-ATPase can be described on the basis of a kinetic reaction model which is generally accepted as Post-Albers cycle. To describe pump functions mathematically this scheme is used with its characteristic parameters, rate constants and dielectric coefficients (1). The dielectric coefficients describe the charge displacements of the corresponding partial reactions and reflect the electrogenicity of the process (2).

Experiments with fluorescence dyes, which are sensitive to changes of the local electric field, have been used to identify electrogenic reaction steps of the $Na^+/K$ pump (3, 4, 5): (1) Binding of the third $Na^+$ ion to the cytoplasmic binding site $Na_2E_1 \rightarrow Na_3E_1$, (2) the conformational change, deocclusion and release of the first $Na^+$ ion to the extracellular side, $(Na_3)E_1\text{-}P \rightarrow P\text{-}E_2(Na_2)$, (3) the subsequent release of the remaining two $Na^+$ ions, $P\text{-}E_2(Na_2) \rightarrow P\text{-}E_2Na \rightarrow P\text{-}E_2$, (4) binding of two $K^+$ ions from the extracellular aqueous phase, $P\text{-}E_2 \rightarrow P\text{-}E_2K \rightarrow P\text{-}E_2K_2$. Independent evidence of the electrogenicity of steps (2) and (4) can be found elsewhere (6, 7, 8).

To obtain quantitative information on the corresponding dielectric coefficients Na,K-ATPase containing membrane fragments were adsorbed to planar lipid bilayers. Activation of the pump was triggered by a fast ATP-concentration jump by a UV flash given to the buffer which contained caged ATP (9, 10). Due to the principle of capacitive coupling the charge movement in the pumps can be detected as current transient in the external measuring circuit (10).

To restrict the reaction steps to the Na branch of the Post-Albers cycle, the buffers did not contain $K^+$ ions. The resulting pump current were in the range of 10 pA to 5 nA. As a specific inhibitor of the $Na^+/K^+$-ATPase strophanthidin has been used, which binds and blocks the enzyme in the state $E_2\text{-}P(Na_2)$ (5, 11). This property allows to lock the enzyme in a defined state and to discriminate between the dielectric coefficients ß and $\delta_1 + \delta_2$.

## Results

### Experiments in the absence and presence of strophanthidin

The membrane with the adsorbed membrane fragments was formed in the presence of the indicated $Na^+$ concentration in the absence of an inhibitor. After an equilibration time the current transient upon enzyme phosphorylation by ATP was measured. Then 1 mM of strophanthidin were added and after an incubation time of 10 min the experiment was repeated. Two typical experiments in the presence of 150 mM and 2 M NaCl are shown in Figure 1.

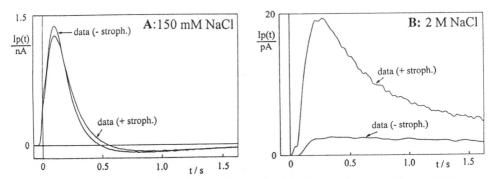

**Figure 1**: Comparison of ATP induced current transients buffers containing (A) 150 mM NaCl and (B) 2 M NaCl, in the absence and presence of 1 mM strophanthidin. Buffer composition: 30 mM imidazole, 10 mM $MgCl_2$, 1 mM EDTA, pH 7.2 and indicated concentration of NaCl, T = 20 °C.

At 150 mM NaCl the currents in the absence and presence of the inhibitor do not change significantly, the amplitude is slightly reduced and the kinetic of the falling phase of the signal a bit slower in the presence of strophanthidin. In the presence of high NaCl concentrations (1.5 M and 2 M) significant changes of the current transients can be observed. While in the absence of the inhibitor the current was small, the presence of strophanthidin increases the amplitude. In the absence of inhibitor a stationary pump-current component can be observed which is caused by steady state electrogenic 3Na:2Na exchange. In the presence of 1 mM strophanthidin the steady state current disappears. By control experiments with double flash technique it has been proven that 1 s after the first flash the enzyme was completely blocked.

Integrating the electrical currents over up to 8 s (until the current transient has vanished), the charge Q is obtained which is moved by the transient. It can be determined with and without inhibitor present and compared by calculation of the ratio S of the charges

$$S = \frac{Q(+\text{strophanthidin})}{Q(-\text{strophanthidin})}$$

The concentration dependence of the ratio S is shown in Fig. 2.

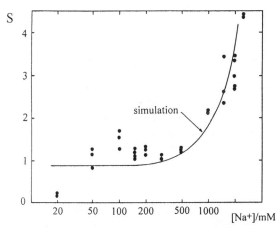

**Figure 2:** Concentration dependence of the ratio S of transferred charges in the presence and absence of the pump inhibitor strophanthidin.

At low concentrations (< 25 mM) the incubation with 1 mM strophanthidin causes a strong reduction of S. This effect is caused by an inhibition of the enzyme before phosphorylation by ATP. This has been proven by independent experiments (data not shown). In a concentration range of NaCl between 40 mM and 400 mM the incubation of the enzyme with the inhibitor changed the amount of transferred charge hardly. Only the quasi-stationary current disappeared (which was subtracted before the charge Q(-strophanthidin) was determined. In high concentrations of NaCl (> 500 mM) the ratio S of transferred charges increased significantly indicating that in the presence of the pump inhibitor more charge is moved than in the absence.

**Discussion**

As has been shown earlier, enzyme phosphorylation is electroneutral (10). Therefore, the charge carrying steps have to be $(Na_3)E_1$-P $\rightarrow$ P-$E_2(Na_2)$ and P-$E_2(Na_2)$ $\rightarrow$ P-$E_2(Na)$ $\rightarrow$ P-$E_2$, to which the dielectric coefficients ß, $\delta_1 + \delta_2$ have been assigned respectively. In Na concentrations up to 400 mM and in the absence of strophanthidin the pump relaxes after the ATP jump preferentially into state P-$E_2$ or P-$E_2(Na)$. In the presence of strophanthidin it will end up quantitatively in state P-$E_2(Na_2)_{inh}$. Since S is close to 1 under these conditions only a small amount of charge can be displaced between states P-$E_2(Na_2)$ $\rightarrow$ P-$E_2(Na)$ $\rightarrow$ P-$E_2$.

In Na concentrations above 500 mM the ratio S became larger than 1. This indicates that in the presence of the inhibitor more charge is moved across the membrane than in the absence. (*5) have shown that in approx. 500 mM NaCl no release or binding of Na ions can be observed by addition of inhibitor, while at higher concentrations of Na ouabain induces dissociation of Na from the pump. At 2 M NaCl most of the pumps remain in state $(Na_3)E_1$-P after phosphorylation by ATP.

These results lead to the conclusion that significantly more charge is moved within the protein during the first step involving conformational change and release of the first Na ion than in the subsequent steps in which the remaining two Na ion dissociate from the

579

protein. Since the electrogenic contributions are parametrized in the dielectric coefficients, ß and $\delta_1 + \delta_2$, have been determined by numerical simulations.

A quantitative determination of the dielectric coefficients, ß and $\delta_1 + \delta_2$, has been obtained by numerical simulations on the basis of an extended Post-Albers model, which takes into account all reaction pathways and rate constants of the Na,K-pump. A rather exact description of the experimental findings which allows no significant variation of the values is shown as solid line in the figure representing S as function of the sodium concentration. The corresponding values are:

$$\text{ß} = 0.75, \ \delta_1 + \delta_2 = 0.4 \ (\text{or } \delta_1 = \delta_2 = 0.2).$$

*Acknowledgments*

This work has been financially supported by the Deutsche Forschungsgemeinschaft (Sonderforschungsbereich 156).

# References

1. Stürmer W, Apell H-J, Wuddel I, Läuger P (1989) Conformational transitions and charge translocation by the Na,K pump: Comparison of optical and electrical transients elicited by ATP-concentration jumps. J Membr Biol 110: 67-86

2  Läuger P (1991) Electrogenic Ion Pumps. Sinauer Associates Inc. Sunderland, MA

3. Bühler R, Stürmer W, Apell H-J, Läuger P (1991) Charge translocation by the Na,K-pump: I. Kinetics of local field changes studied by time-resolved fluorescence measurement. J Membr Biol 121: 141-161

4. Stürmer W, Bühler R, Apell H-J, Läuger P (1991) Charge translocation by the Na,K-pump: II. Ion Binding and Release at the Extracellular Face. J Membr Biol 121: 163-176

5  Stürmer W, Apell H-J (1993) Fluorescence study on cardiac glycoside binding to the Na,K-pump. FEBS Lett. 300: 1-4

6  Nakao M, Gadsby DC (1986) Voltage dependence of Na translocation by the Na/K pump. Nature 323: 628-630

7  Rakowski RF, Vasilets LA, LaTona J, Schwarz W (1991) A negative slope in the current-voltage relationship of the $Na^+/K^+$ pump in Xenopus oocytes produced by reduction of external $[K^+]$. J Membr Biol 121: 177-187

8  Gadsby DC, Rakowski RF, DeWeer P (1993) Extracellular Access to the Na,K Pump: Pathway Similar to Ion Channel. Science 260: 100-103

9  Fendler K, Grell E, Haubs M, Bamberg E (1985) Pump currents generated by the purified $Na^+K^+$-ATPase from kidney on black lipid membranes. EMBO J 4: 3079-3085

10 Borlinghaus R, Apell H-J, Läuger P (1987) Fast charge translocations associated with partial reactions of the Na,K-pump: I. Current and voltage transients after photochemical release of ATP. J Membr Biol 97: 161-178

11 Jørgensen PL (1991) Conformational transitions in the $\alpha$-subunit and ion occlusion. In Kaplan JH, DeWeer P (eds.) The Sodium Pump: Structure, Mechanism and Regulation. The Rockefeller University Press, New York, pp. 189-200

580

# Changes in the Conformational State of Probe-Labeled Na+/K+-ATPase in Real Time

K. Taniguchi , D. Kai*, S. Inoue*, E. Shinoguchi*, K.Suzuki*, Y. Nakamura, Y.Adachi, S. Kaya

Department of Chemistry, Faculty of Science, School of Dentistry*, Hokkaido University, Sapporo, Japan

## Introduction

Conformational changes during $Na^+,K^+$-ATPase reaction have been suggested (1,16,23) and shown from kinetical experiments of phosphorylation, reactivity of sulfhydryl groups under various ligand conditions, sensitivity to proteolytic digestion, and intrinsic and extrinsic fluorescence probes (4,8,9,10,15-18,30).To understand the mechanism of energy transduction in $Na^+,K^+$-ATPase, a detailed knowledge of its conformational changes is essential. The study of conformational changes in real time seems to be especially useful for this purpose. The free energy of the ATP molecule appears to be converted to a change in the enzyme conformation. It induces transport of $Na^+$ and $K^+$ against the electrochemical gradients accompanying sequential appearance of reaction intermediates (1,4,16,17) according to the Post-Albers Mechanism (Scheme I)). In this paper, we would like to summarize what we have done using $Na^+,K^+$-ATPase preparations labeled with fluorescence probes (11-13,21,24-34).

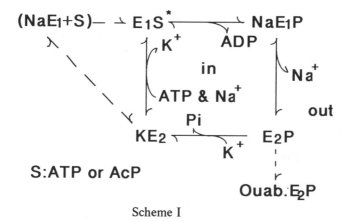

Scheme I

Pig kidney enzyme preparations purified with sodiumdeoxycholate and NaI were specifically labeled with fluorescence probes such as N-(p-(2-benzimidazolyl) phenyl) maleimide (BIPM) at Cys-965, fluorescein 5'-isothiocyanate (FITC) probe at Lys- 501 and pyridoxal probes at Lys-480 (9,11,12,15). To phosphorylate the probe labelled enzyme, radioactive ATP, acetylphosphate (AcP) and p-nitrophenylphosphate have been used (11,24,32,34). To accumulate ADP- or acetate or p-nitrophenol sensitive phosphoenzyme $(E_1P)$ , 2 M NaCl was added to the phosphorylation reaction mixture containing 4 mM $MgCl_2$ to shift the equilibrium between $E_1P$ and $K^+$-sensitive phosphoenzyme ($E_2P$) to the former (25,27). To accumulate $E_2P$, 16 mM NaCl was added. In some experiments, 1984 mM choline chloride + 16 mM NaCl were used to accumulate $E_2P$ to keep the ionic strength constant.

## Multiple conformational states of phosphoenzymes and dephosphoenzymes

To investigate the decrease of BIPM fluorescence induced by ATP in the presence of 2 M NaCl with $Mg^{2+}$, the time course of both fluorescence intensity and the amount of phosphoenzyme were measured (Fig.1A). The addition of ATP induced a monophasic increase in the amount of phosphoenzyme and a monophasic decrease in the fluorescence intensity. To investigate whether the decrease in BIPM fluorescence is simultaneous with the accumulation of $E_1P$, the time course of the fluorescence change and the amount of phosphoenzyme were measured after addition of ATP in the presence of 2 M NaCl with $Ca^{2+}$ to form $E_1P$ at a steady state (Fig.1B). The amount of $E_1P$ increased to give 80% of active sites ($t_{1/2}$=1.2 s), while the fluorescence intensity showed a more rapid monophasic decrease ($t_{1/2}$= 0.1 s). These data clearly shows that the decrease in the BIPM fluorescence occur preceding $E_1P$ formation (28).

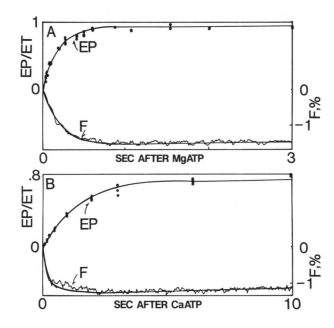

Figure 1. Change in fluorescence and phosphorylation induced by ATP in the presence of 2M NaCl with $Mg^{2+}$ (A) or $Ca^{2+}$ (B). Procedure are described (28).

Reversible changes in light scattering accompanying $E_1P$ formation were detected in BIPM modified enzyme preparations (29). Stopped flow measurements showed that the fluorescence change accompanying the $E_1P$ formation ($t_{1/2}$=0.1 s) occurred preceding the light scattering change ($t_{1/2}$= 1 s) as shown (Fig.2). The phosphoenzyme formed was split by a chase of ADP rapidly at a rate not affected by the relative intensity of light scattering of $E_1P$ ( ). The data show that the change in the light scattering develops even after the completion of $E_1P$ formation and suggest the sequential formation of species of $E_1P$ with low and high relative light scattering, because the enzyme had been fully phosphorylated.

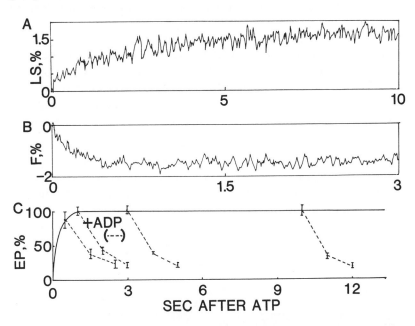

Figure 2. Time course of ATP-induced changes light scattering (LS) and BIPM fluorescence (F) in the presence of 2 M $Na^+$ with $Mg^{2+}$ (29).

The presence of a tight $Na^+$ binding state has been reported as the occlusion of $Na^+$ in $E_1P$ by using chymotrypsin- or N-ethyl maleimide-treated enzymes to stabilize $E_1P$ and later in an enzyme-$Na^+$-oligomycin complex in the absence of phosphorylation (3,5). However, the relationship between these occluded $Na^+$ forms of the enzyme and their conformational states remain to be elucidated. Such investigations could be very important for the understanding of the mechanism of energy transduction in $Na^+,K^+$-ATPase. Oligomycin reduced the fluorescence intensity of BIPM probe at Cys-964 of the alpha-chain of pig kidney $Na^+,K^+$-ATPase with increases in the concentration of $Na^+$ with a Hill coefficient of $n_h$=0.77 with $K_h$ =231 mM. The data suggested that the presence of different conformational states in $(Na)E_1$ rather than the presence of enzyme-oligomycin complexes that did not bind enough $Na^+$ to saturate the binding (26). The relative fluorescence intensity was measured in the presence of 2 M $Na^+$ and 0.43

mM $Mg^{2+}$ while changing the order of additions of ATP and oligomycin, each of which presumably induces $Na^+$ occlusions with or without phosphorylation (3,5). The addition of ATP to $NaE_1$ immediately decreased the fluorescence intensity to a -4.6% level to accumulate $E_1P$. When oligomycin was added to $E_1P$, the fluorescence intensity decreased from a -4.6 to a -5.2% level. The addition of oligomycin to $NaE_1$ reduced the fluorescence intensity to a -4% level to form, $Na^+$ occluded enzyme, $(Na)E_1$. The addition of ATP further reduced the fluorescence to a -5% level to accumulate $E_1P$. The data show that the effect of ATP (-4.6%) and that of oligomycin (-4%) were not additive but that the presence of both gave the maximum decrease (-5 or -5.2%), demonstrating that nearly 80% of the decrease was dependent on $Na^+$ occlusion but independent of phosphorylation (26).

To compare the ability of $NaE_1$ and $(Na)E_1$ to act as precursors of $E_1P$, ATP or radioactive ATP were added to follow BIPM fluorescence change and the amount of phosphoenzyme. The apparent rate constants for the $E_1P$ formation from $NaE_1$ and $(Na)E_1$ were 20 and 13/s,respectively and those for fluorescence change were 9 and 10/s. These data clearly show the species of $(Na)E_1$ were more competent than $NaE_1$ as precursors of $E_1P$, irrespective of their conformational states. The data also show that the phosphorylation of species of $NaE_1$ and $(Na)E_1$ by ATP + $Mg^{2+}$ was completed before the fluorescence change was. When $Mg^{2+}$ was replaced with $Ca^{2+}$, the gross fluorescence change induced by ATP + $Ca^{2+}$ was completed before the phosphorylation was (26). From these results we propose that the ATP-dependent BIPM fluorescence changes reflect conformational events accompanying the change in $Na^+$ binding states presumably related to $Na^+$ migration in the enzyme (Fig.3). The data also indicate that changes in the $Na^+$ binding state and the formation of $E_1P$ are not tightly coupled.

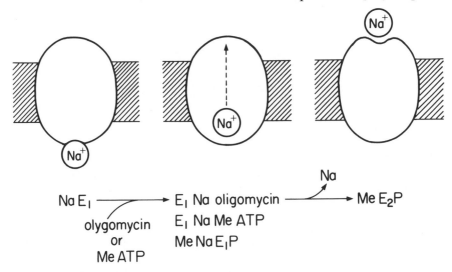

Figure 3. A model for change in the $Na^+$ binding state or $Na^+$ migration detected by BIPM probe at Cys-964 preceding formation of $E_2P$. Ellipses designate $Na^+$ pump embedded in membranes. Me designayes $Mg^{2+}$ or $Ca^{2+}$.

# Reversible changes in the fluorescence energy transfer between probes of BIPM at Cys-964 to FITC at Lys-501 accompanying sequential formation of reaction intermediates in probe-labeled $(Na^+, K^+)$-ATPase

BIPM-FITC doubly labeled enzymes showed little $Na^+, K^+$-ATPase activity with retention of nearly 90% of phosphorylation capacity from AcP (32). The addition of AcP to the preparation in the presence of 16 mM $Na^+$ with 4 mM $Mg^{2+}$ induced reversible fluorescence intensity changes of both probes as shown (Fig. 4). The addition of ouabain completely stabilized the fluorescence changes to accumulate ouabin$E_2P$: the BIPM fluorescence gave the highest level and the FITC fluorescence gave the lowest level. Both levels are obtained also by adding higher concentrations of AcP to accumulate $E_2P$ fully without ouabain.

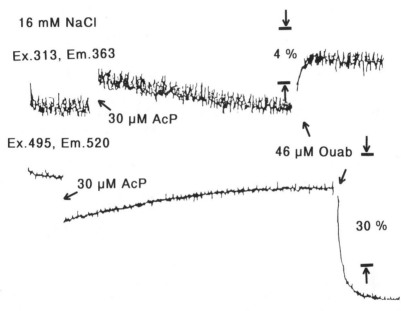

Fig. 4. Acetyl phosphate dependent reversible fluorescence intensity changes of BIPM and FITC probes in BIPM-FITC enzyme. The time course runs from left to right.

The addition of AcP to the enzyme in the presence of 2 M $Na^+$ with $Mg^{2+}$ induced acetate sensitive but not ADP sensitive phosphoenzyme formation with a decrease in the fluorescence intensity (32):the BIPM fluorescence decreased with a simultaneous increase in the amount of phosphoenzyme; there was a significant delay in a decrease in the FITC fluorescence. The extent of the decrease in the BIPM fluorescence and the increase in the amount of phosphoenzyme both showed monophasic kinetics with a similar dependence on the concentration of AcP, while that of FITC fluorescence showed a biphasic decrease. The data suggest the presence of at least two conformationally different $E_1Ps$; one gave little and the other gave large FITC fluorescence decrease.These data and others (24,32) are consistent with a hypothesis that both fluorescence probes sense the change in $Na^+$ binding states presumably related to $Na^+$ migration in the enzyme.

Excitation of BIPM-FITC doubly labeled enzyme gave three fluorescence peaks (Fig.5a). Excitation (305 nm) of the BIPM probe gave both BIPM at 360 nm (B) and FITC fluorescence at 520 nm (BF) and excitation (470 nm) of the FITC probe, gave different FITC fluorescence at 520 nm (F). The addition of 0.12% SDS to denature the enzyme increased the fluorescence intensity of B to 115 % and decreased that of BF and F to 22 and 40%, respectively; SDS decreased BF more than F. As SDS only slightly influenced the BIPM fluorescence intensity of the BIPM enzyme, the increase in B after the addition of SDS to the BIPM-FITC enzyme suggested that fluorescence energy transfer from the BIPM to the FITC probe was decreased by denaturation (24). The data also suggested some possibility of change in the energy transfer accompanying formation of reaction intermediates because the microenvironments of various probes changed out of phase (24,27,29,32).

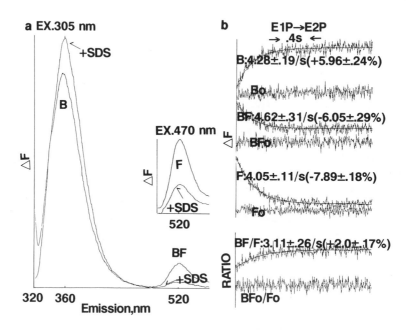

Fig.5. Emission spectrum and time course of fluorescence intensity changes from $E_1P$ to $E_2P$ (24).

Change in the fluorescence energy transfers from BIPM to FITC probes were estimated from the time course of change in relative fluorescence intensities of BF and F using the ratio of BF to F (BF/F) calculated. Details are already described (24). The increase and the decrease of the ratio from the starting point reflect the increase and the decrease of the fluorescence energy transfer, respectively.

Addition of acetyl phosphate to a $Na^+$ bound enzyme ($NaE_1$) to accumulate acetate-sensitive phosphoenzyme ($E_1P$) induced three different single exponential fluorescence decreases. The decrease of the fluorescence intensity of B occurred prior to that of F. A faster and greater of FITC fluorescence (F) decrease occurred when excited at 470 nm than at 305 nm. The BF/F ratio increased to a steady level, which indicated the increase in fluorescence energy transfer. The transition from $E_1P$ to $E_2P$ accompanied a single exponential increase of B and decreases of both BF and F (Fig.5b). The decrease of BF was 77% that of F, which demonstrated an increase in the fluorescence energy transfer accompanying the transition. The BF/F ratio increased to a steady level. While the ratio of the transition from $E_2P$ to $E_1P$ decreased to a steady level as expected.

Similar experiments were done in the presence of oligomycin. However, only small changes in the fluorescence and ratio were detected (Fig. 5b, Bo,BFo,Fo and BFo/Fo). This indicated that oligomycin strongly inhibited the conformational transition in the presence of 160 mM $Na^+$.

The addition of $K^+$ to $[^{32}P]E_2P$ to form a $K^+$ bound enzyme ($KE_2$) induced a rapid dephosphorylation (140 /s) and induced only a slight decrease in B but clear single exponential increases in BF and F. The increase in BF was 61% that of F, which demonstrated a decrease in fluorescence energy transfer (24).

The addition of $Na^+$ to $KE_2$ induced a single exponential decrease in B with both a slow increase of the FITC fluorescence and a larger fluorescence increase when excited at 470 nm. The BF/F ratio decreased to a staedy state while the ratio of the transition from $NaE_1$ to $KE_2$ increased to a seady level.

The data are summarized in Fig. 6. Transition of $NaE_1$ to, $Na^+$ occluded enzymes, $(Na)E_1$ was accompanied by a slight decrease in the ratio with a decrease in both B and F. Transition of $NaE_1$ to $E_1P$ occurred together with an increase in the ratio with decreases in both types of fluorescence. Transition of $E_1P$ to $E_2P$ was accompanied by a large increase in the ratio with both the maximum increase in B and the maximum decrease in F, which was strongly inhibited by oligomycin. Transition of $E_2P$ to $KE_2$

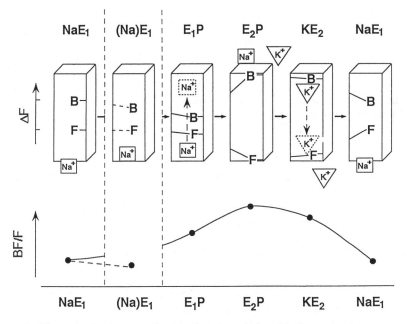

Figure 6. Diagram of a hypothesis showing the relationship between the enzyme state for ion migration and the change in the fluorescence energy transfer from the BIPM probe to the FITC probe.

occurred together with a decrease in the ratio with a decrease in B but an increase in F. Transition of $KE_2$ to $NaE_1$ accompanied the maximum decrease in the ratio with both the maximum decrease in B and the maximum increase in F. The reversible change in the ratio during the sequential appearance of reaction intermediates reflected a reversible change in fluorescence energy transfer and demonstrated another aspect of conformational change which has not been reported before. Transitions of $E_1P$ to $E_2P$ and of $KE_2$ to $NaE_1$ are supposed to occur accompanying the liberation or deocclusion of $Na^+$ and $K^+$, respectively (4,5,17,18). The data suggest that change in fluorscence energy transfer from BIPM to FITC probes accompanies the processes of migration of $Na^+$ and $K^+$ in the pump molecules. The reversible change in the fluorescence energy transfer may be a reflection of the change in the distance between the probes due to movement of both $Na^+$ and $K^+$ (33).

## Phospholipids dependence of BIPM probe at Cys-964 but not FITC probe at Lys-501

It has been assumed from a hydropathy analysis that Cys-964 is in the transmembrane segment and Lys-501 is in soluble domain (15,22). However there seems to be little direct biochemical evidence which shows Cys-964 is in the transmembrane segment, except that BIPM containing peptides became soluble only after extensive trpsin digestion (12). It has already been shown that $Na^+$, $K^+$-ATPase requires phospholipids for its activity (14,20,31). Lipid removal from $Na^+$, $K^+$-ATPase preparations by detergents or phospholipase treatment leads to partial or complete inactivation of the enzyme and the addition of PS and/or PI or asolectin restores the activity. The treatment induced different inhibition of the partial reactions in $Na^+$, $K^+$-ATPase (14,25,31). These data prompted us to investigate the phospholipids dependence on the dynamic fluorescence intensity changes of the BIPM probe at Cys-964 and of the FITC probe at Lys-501: the role of phospholipids in the conformational change seems not to have been studied yet.

Phospholipase $A_2$ (EC.3.1.1.4) treatment of probe labeled enzymes reduced the extent of fluorescence intensity changes of the BIPM probe to below one-third of the initial level with only a slight decrease of that of the FITC probe, and reduced the rates of change of both probes accompanying accumulation of various enzyme forms (13). The anisotropy of each probes was nearly the same in various enzyme forms accumulated and little influenced by the treatment. The addition of phosphatidyl serine (PS) or phosphatidyl inositol (PI) to phospholipase-treated BIPM-FITC-labeled enzyme increased the rate of the FITC fluorescence change. Phospholipase treatment of BIPM-enzyme strongly reduced $Na^+$, $K^+$-ATPase activity. The addition of PS or PI to the treated enzyme induced reactivation. The treatment showed a clear difference in the susceptibility of the fluorescence intensity changes between BIPM and FITC probes. Differences in the rates of conformational changes in different domains of the enzyme were clearly shown from the difference in the initial rates and the extents of fluorescence changes of both probes of the BIPM and of the FITC. The data also show that the breakdown of phospholipids reduced the rate of conformational change after formation of $E_2P$, which reflects changes in the binding states of $Na^+$ and $Mg^+$.

These data suggest that phospholipase A treatment reversibly disrupted protein-phospholipid interactions in such a way as to quench the BIPM fluorescence at Cys-964 with little influence on FITC fluorescence at Lys-501 (13). Cys-964 is believed to be present in the transmembrane segment near the C-terminal of the alpha-chain (15,22). The present data indicate that PS molecules somehow participate in the construction of domain(s) in such a way as to manifest a BIPM fluorescence change at Cys-964, to increase the rate of conformational changes detected clearly by the increase in the rate of FITC fluorescence change at Lys-501(Fig.3B,C) and to accelerate the rates of partial reactions (13,31).

The treatment affected neither the microenvironment of the FITC probe at Lys-501 nor the domain structure of the active site containing Asp-369, which is assumed to be present in the cytoplasmic side of the membrane as the largest soluble domain between transmembrane segments (22). The reason why the phospholipase treatment only slightly influenced the amount of phosphoenzyme and the extent of FITC fluorescence compared with the BIPM fluorescence may be that phospholipid molecules are scarcely present in the vicinity of the FITC probe at Lys-501 and are not required for the appearance of FITC

fluorescence: the slow conformational change detected via FITC fluorescence change is due to the breakdown of phospholipids which are present in areas other than the vicinity of Lys-501(13). ATP and FITC binding to the enzyme occur mutually exclusively (9) and delipidated enzymes have normal nucleotide binding sites (14). Quite recently, a possibility of the cardiac glycoside binding site in intramembrane rather than the extracellular is reported (2,19). The distance between the FITC probe at Lys-501 and the ouabain binding site labelled with anthroylouabain has been assumed to be 64 to 83 Å by several groups (see the references of 33) which suggests that the FITC binding site is not in the membrane. The distance between the BIPM probe to FITC probe has been calculated to be 35Å (33).

Figure 7 show a herical wheel representation (left) and the space-filling model (right) of an alpha-Helix of the hydrophobic transmembrane peptide from Ile-946 to leu-971 containing BIPM probe at Cys-964, and an $Rb^+$ protectable dicyclohexyl carbodiimide binding site Glu-953 (6, and see the reference of 33). The model suggest that the dynamic fluorescence change of the BIPM probe accompanying formation of reaction intermediates occurred rather restricted region (see the BIPM probe in the Fig. 7 right ) and that the hydrophobic fluorophore of BIPM probe is near to the hydrocarbon chain(s) of phospholipid molecules , in either leaflet of the membrane (Fig. 7 center). The probe would be sensitive enough to the change in the microenvironment in a rather rotationally restricted location. The BIPM fluorescence was more sensitive than the Trp fluorescence to both the hydrophilic quencher acrylamide and non-ionic detergent $C_{12}E_8$ in the state of $E_2P$ rather than $NaE_1$. These considerations favor the hypothesis that the interaction between the BIPM probe and the hydrocarbon chains of phospholipid molecules changes dependent on the change in the binding state of $Na^+$ and $K^+$ and also $Mg^{2+}$ in the enzyme (13).

The model also indicates that Glu-953 and -954 and the BIPM probe at cys-964 lie rather one side of the helix (Fig.7 left) but conservative H-bonding amino acids, Thr-955 and Ser-962 lie the other side (Fig.7 left). BIPM probe at Cys-964 influences little on $Na^+$, $K^+$-ATPase activity and senses various conformational states of the enzyme (7,12,13,21,24-34). The single or double replacements of Glu-953 and -954 were essentially devoid of effect on the enzymatic activity (19) and Glu-953 is replaced with Phe in *Artemia* and *Drosophila* (see the reference of 6). These data suggest that the side containing Cys-964, Glu-953 and Glu -954 (Fig. 7 bottom) is located in the vicinity of the surface area and faces to phospholipids molecules such as to be influenced as described above and to be modified by BIPM probe (12) and dicyclohexyl carbodiimide

(6). The other side containing conserved H-bonding amino acids (19,22), Thr-955 and Ser-962 (Fig. 7 top) may interact other hydrophobic transmembrane segments to construct the cation migration route or the cation binding cage rather than to lie on the vicinity of the surface area (13,33).

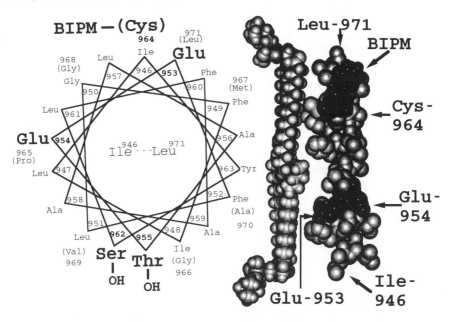

Figure 7. Model illustrating localization of the BIPM probe at Cys-964 in the trans membrane segment (24).

A definitive answer for the arrangement of transmembrane segments of $Na^+$, $K^+$-ATPase in the plane of the membrane seems not to have been obtained yet. The orientation of the segment containing the BIPM probe is dependent on the model proposed (6). PS molecules have been known to be preferentially found in the cytoplasmic leaflet . If we accept that the role of PS in the enzyme in *vivo* is to directly construct some essential domain(s) around which Cys-964 is present , it follows that Cys-964, which is 7 residues from Leu-971 would be near the cytoplasmic face rather than the exoplasmic face. The length of the fluorophore of the BIPM probe (- 5Å) would be 1/4 of that of the hydrocarbon chains of phospholipid molecules (Fig. 7 left). Such considerations suggest that Ile-946 and Leu-971 are at the extracellular and cytoplasmic face, respectively. The distances between BIPM at Cys-964 and FITC at Lys-501 estimated by fluorescence energy transfer ranged around 36Å (33), which seem to support the model.

*Acknowledgements*

We thank Drs S. Mårdh for the collaboration of rapid phosphorylation experiments and helpful discussions (24) and I. Tanaka and Y. Kawano for helpful discussions for the Fig.7. The studies were supported by The Naito Foundation, The Akiyama Foundation, Nagase Science and Technology foundation, The Uehara Memorial

Foundation, a Grant-in-Aid for Science Research on Priority areas of Bioenegy and Grants-in-Aid Scientific Research (03558026, 03454543).

## References

1   Albers R W (1968) Studies on the interaction of ouabain and other cardioactive steroids with sodium-potassium-activated adenosine triphosphatase. Mol Pharmacol 4: 324-336

2   Arystarkhova E, Gasparian M, Modyanov N N, Sweadner K (1992) Na,K-ATPase extracellular surface probed with a monoclonal antibody that enhances ouabain binding. J Biol Chem 267:13694-13701

3   Esmann M, Skou J C (1985) Occlusion of Na by the Na,k-ATPase in the presence of oligomycin. Biochem Bipys Res Comm 127:857-863

4   Glynn I M (1985) The Na,K-transporting adenosine triphosphatase. In: Martonosi A N (ed) The enzymes of Biological Membranes. 2nd edition Plenum press New York,pp35-114

5   Glynn I M, Karish S J D (1990) Occluded cations in active transport. Ann Rev Biochem 59:171-205

6   Goldschleger R, Tal D M, Scainskaya A, Hoving S, Or E, Karlish S J D (1993) Organization of the cation occlusion domain of the Na/K pump. Biol Chem Hoppe-Seyler 374:551

7   Inoue S, Taniguchi K, Shimokobe H, Iida I (1987) Effect of peptide bond splitting on ouabain sensitive conformational changes in Na, K -ATPase treated with N-[p-benzimidazolyl)phenyl]maleimede. Jap Jour Pharmacol43 :107-111

8   Jørgensen P (1975) Purification and characterization of $(Na^{+}+K^{+})$-ATPase. V. Conformational changes in the enzyme, transition between the Na-form and the K-form studied with tryptic digestion as a tool. Biochim Biopys Acta 401: 399-415

9   Karlish S J D (1980) Characterization of conformational changes in (Na,K)ATPase labeled with fluorescein at the active site. J Bioenerg Biomembr 12:111-135

10   Karlish S J D, Yates D W (1978) Tryptophan fluorescence of $(Na^{+}+K^{+})$-ATPase as a tool for study of the enzyme mechanism. Biochim Biopys Acta 527:115-130

11   Kaya K,Tsuda T, Hagiwara K, Fukui T ,Taniguchi K (1994) Pyridoxal-5'-phosphate Probes at Lys-480 Can Sense the Binding of ATP and the Formation of Phosphoenzymes in $Na^{+}$, $K^{+}$- ATPase. J Biol Chem (in press)

12   Nagai M, Taniguchi K, Kangawa K, Matsuo H, Nakamura S, Iida S (1986) Identification of N-[p-(2-benzimidazolyl]maleimide modified residue participating in dynamic fluorescenc change accompanying Na and K -dependent ATPase. J Biol Chem 261:1319-13202

13   Nakamura Y, Kai D, Kaya S, Adachi Y, Taniguchi K (1994) Different susceptibility to phospholipaseA treatment of the fluorescence intensity changes in the vicinity of Cys-964 and Lys-501 in the alpha chain of probe labeled $Na^{+}/K^{+}$- ATPase. J Biochem (in press)

14   Ottolenghi P (1979) The repidation of delipidatedNa,K-ATPase. An analysis of complex formation with dioleoylphosphatidylcholine and with dioleoylphosphatidylethanolamine. Eur J Biochem 99:113-131

15   Pedemonte C H, Kaplan J H (1990) Chemical modification as an approach to elucidation of sodium pump structure-function relations. Am J Physiol 258:C1-C23

16   Post R L, Kume S, Tobin T, Orcutt B, Sen A K (1969) Flexibility of an active center in sodium-potassium-activated adenosine triphosphatase. J Gen Physiol  54:306s-326s

17   Repke K R H, Schon R (1992) Role of protein conformation changes and transphosphorylations in the function of $Na^{+}/K^{+}$-transporting adenosine tiphosphatase ; an attempt at an integration into the $Na^{+}/K^{+}$pump mechanism. Biol Rev  67:31-78

18   Robinson J D, Pratap P R (1993) Indicators of conformational changes in the $Na^{+}/K^{+}$- ATPase and their interpretation. Biochim Biopys Acta 1154: 83-104

591

19   Schulthesis P J, Lingrel J P (1993) Substitution of transmembrane residues with hydrogen- bonding potential in the A subunit of Na$^+$/K$^+$- ATPase reveals alterations in ouabain  sensitivity. Biochemistry 32:544- 550

20   Schuurmans Stekhoven F,  Bonting S L (1981) Transport adenosine triphosphatase: properties and functions. Physiol Rev  61:1-76

21   Shinoguchi E, Ito E, Kudo A, Nakamura S, Taniguchi K (1988) Fluorescence energy transfer between   N-(p-(2-benzimidazolyl)phenyl) maleimide,N-(1-anilinonaphtyl-4)maleimide, fluorescein 5'- isothiocyanate, and Anthroylouabain in  Na$^+$/K$^+$- ATPase. The Sodium Pump: Recent developments. The Rockefeller University  press, 363-367

22   Shull G E A, Schwarts A, Lingrell J B (1985) Amino acid sequence of the catalytic subunit of the  Na$^+$/K$^+$- ATPase deduced from a complementary  DNA. Nature 316:691-695

23   Somogyi J (1968) The effect of proteases on the (Na$^+$+K$^+$)-activated adenosine triphosphatase system of rat braint.  Biochim Biopys Acta 151:421-428

24   Taniguchi K, Mårdh S (1993) Reversible changes in the fluorescence energy transfer accompanying formation of reaction intermediates in  probe-labeled  Na$^+$/K$^+$- ATPase. J Biol  Chem 268:15588-15594

25   Taniguchi  K, Post R L (1975) Synthesis of adenosine thiphosphate and exchange between inorganic phospate and adenosine triphosphate in sodium and potassium ion transport adenosine triphosphatase. J Biol Chem 250:3010-3018

26   Taniguchi K, Sasaki T, Shinoguchi E, Kamo Y,Ito E (1991) Conformational change accompanying formation of oligomycib-induced Na-bound forms and their conversion to ADP-sensitive phosphoenzymes in  Na$^+$-K$^+$- ATPase. J Biochem 109:299-306

27   Taniguchi K, Suzuki k, Iida S (1982) Conformational change accompanying transition of ADP-sensitive phosphoenzyme to potassium sensitive phosphoenzyme of Na$^+$, K$^+$ -ATPase modified with N-[p(2benzimidazolyl)phenyl]maleimide. J BiolChem 257: 10659-10667

28   Taniguchi K, Suzuki K, Kai D, Matsuoka I,  Tomita K, Iida S (1984) Conformational change of sodium and potassium dependent adenosine triphosphatase. Conformational evidence for the Post-Albers mechanism in Na$^+$ , K$^+$ -dependent hydrolysis of ATP. J Biol Chem  259:15228-15233

29   Taniguchi K,  Suzuki K, Sasaki T, Shimokobe H, Iida S (1986) Reversible change in the light scattering following the formation of ADP-sensitive phosphoenzyme in Na, K - ATPase modified with N-[p-(2-benzimidazolyl)phenyl]maleimede.  J Biol Chem261:3272- 3281

30   Taniguchi K , Suzuki K, Shimizu J, Iida S (1980) ATP dependent reversible conformational  change of  Na$^+$+K$^+$ -ATPase modified with N-(p-(2-benzimidazolyl) phenyl)Maleimide. J  Biochem 88:609-612

31   Taniguchi K, Tonomura Y (1971) Inactivation of Na,K-dependent ATPase by phospholipase- treatment and its reactivation by phospholipids. J Biochem 69:543-557

32   Taniguchi K, Tosa H, Suzuki K, Kamo Y (1988) Microenvironment of two different extrinsic fluorescence probes in Na, K -ATPase changes out of phase during sequential appearance of reaction intermediates. J  Biol Chem263:12943-12947

33   Tosa H, Sawa A, Nakamura N, Takizawa S, Kaya S, Taniguchi K, Araiso T, Koyama T, Fortes P A G (1993) Estimation of distance changes between Cys-964 and Lys-501 in Na/K-ATPase intermediates and the orientation of transmembrane segment containing Cys-964. in this volume

34    Yamazaki Y, Kaya S,Araki Y, Scimada A, Taniguchi K (1993) Phosphorylation of Na$^+$+K$^+$-ATPase by p-nitrophenylphosphate and other phosphatase substrates. in this volume

# Mechanism of the Conformational Change in Fluorescein 5'-Isothiocyanate-Modified Na⁺/K⁺-ATPase

Larry D. Faller, Irina N. Smirnova, Shwu-Hwa Lin & Martin Stengelin

Center for Ulcer Research and Education, Department of Medicine, University of California at Los Angeles School of Medicine, and Department of Veterans Affairs Medical Center Wadsworth Division, West Los Angeles, California 90073 USA

## Introduction

It is generally agreed that conformational changes explain both coupling of transport to ATP hydrolysis (16) and the physical translocation of $Na^+$ and $K^+$ ions by sodium pump (12). Karlish reported 15 years ago at the IInd Annual Conference on the Sodium Pump (10) that a conformational change can be studied by chemically modifying the enzyme with fluorescein 5'-isothiocyanate (FITC). He showed that the time course of the reaction can be followed with a stopped-flow instrument, and he proposed that the reaction reported by fluorescein is a rate-limiting conformational change in dephosphoenzyme (11). Subsequent studies of the conformational change in dephosphoenzyme using fluorescein and other reporter groups have been reviewed by Glynn (6).

When I began comparing the FITC-modified proton and sodium pumps in collaboration with Robert Farley and Georgios Scheiner-Bobis (4), we assumed that the only reaction reported by fluorescein is the conformational change in dephosphoenzyme. We also assumed that the mechanism of that reaction is known for sodium pump. Neither assumption has turned out to be entirely correct. Therefore, in this paper two questions are addressed. First, what else does FITC-modification of sodium pump report? Second, what is the mechanism by which ions cause the conformational change reported by fluorescein?

## Experimental

Two types of experiments will be described. In one, $K^+$ is added to FITC-modified enzyme causing a decrease in emission intensity, or fluorescence quench. In the other, $Na^+$ is added to enzyme preincubated with $K^+$ causing an increase in emission intensity, or fluorescence enhancement. For convenience, the reaction with $K^+$ is considered the forward reaction, so the two reactions will be referred to as the $K^+$ quench and the $Na^+$ reversal. Two numbers can be estimated from each type of experiment, an exponential time coefficient and the magnitude of the fluorescence change.

What do we do differently that has caused us to question, first of all, what fluorescein

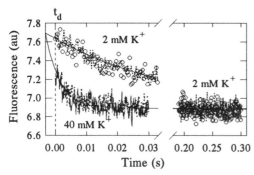

**Figure 1.** The reproducibility of K$^+$ quench data at two K$^+$ concentrations is shown by different symbols. The solid lines are fits of a single exponential to the data. The parameter estimates are $1/\tau = 25$ s$^{-1}$, $\Delta F = 0.76$ at 2 mM K$^+$ and $1/\tau = 225$ s$^{-1}$, $\Delta F = 0.41$ at 40 mM K$^+$. The instrument dead-time ($t_d$) is 3 ms. The corrected amplitudes are 0.82 and 0.81 arbitrary units, respectively. The design of K$^+$ quench experiments has been described elsewhere (20).

is reporting? The principal innovation is that we estimate the change in fluorescence level, as well as the time constant, from stopped-flow experiments. Earlier investigators obtained the kinetic constant and the fluorescence intensity at equilibrium from separate experiments, because of concern about the reproducibility of stopped-flow amplitudes. Figure 1 shows that stopped-flow amplitudes and rates are both reproducible. Two quench experiments at each of two different K$^+$ concentrations are superimposed. The concentrations were purposely chosen to illustrate another important point. The rate increases with increasing [K$^+$], but the amplitude appears to be inversely related. Does this mean that a plot of fluorescence change versus [K$^+$] is bell-shaped (18)? The answer is no, because stopped-flow amplitudes must be corrected for the dead-time of the instrument. In our machine we see 93% of the total fluorescence change at the lower [K$^+$], but only 51% at the higher [K$^+$]. The reason is that the half-time of the quench caused by 40 mM K$^+$ is approximately equal to the dead-time of our instrument. The amplitudes of the fitted curves at each [K$^+$] are almost the same.

When we compared stopped-flow titrations of FITC-modified sodium pump with K$^+$ as a function of Mg$^+$ concentration (20) with previously published titrations (8,11), a significant discrepancy was noted. In the published experiments a decrease in magnitude of the fluorescence change with increasing [Mg$^{2+}$] was observed and interpreted as evidence for a different conformation of the magnesium, or metalloenzyme (8,11). Since the amplitude did not depend on [Mg$^{2+}$] in the stopped-flow titration, we drew the opposite conclusion that only two limiting conformations are needed to explain the transition reported by fluorescein. The published experiments (8,11) are equilibrium titrations in which aliquots of K$^+$ were sequentially added to enzyme in a stationary-state fluorometer. The stopped-flow experiment is a "kinetic" titration in which the percentage change in fluorescence is estimated from the amplitude of stopped-flow traces at each [K$^+$]. To understand why equilibrium and kinetic titrations give different results, we posed the question "What does fluorescein report?".

A novel feature of the On Line Instrument Systems software interfaced with our

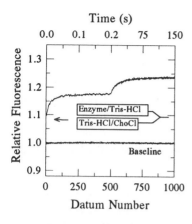

Figure 2. The fluorescence enhancement when 200 mM ChoCl is mixed with FITC-enzyme in low ionic strength Tris-HCl buffer is resolved into three phases. The first 500 data points were collected in 0.2 s and last 500 data points in 150 s. The arrow indicates the start of the faster exponential increase. The baseline was obtained by mixing enzyme with buffer at the same instrument settings. See (14) for details.

mechanical mixer is that reactions occurring on different time scales can be resolved by acquiring data from a single mix with two different collection times. Figure 2 shows that the fluorescence change observed when choline chloride (ChoCl) is mixed with FITC-modified sodium pump in low ionic strength Tris-HCl buffer can be resolved into three reactions on order of magnitude different time scales. Exactly the same triphasic curve is observed when the titrant is NaCl (14). Why do we conclude there are at least three reactions when only two exponential increases in fluorescence are visible? The reason is that we can calculate from the dead-time of our instrument that the reaction on the millisecond time scale began at the arrow. Therefore, an approximately 8% baseline shift must already have occurred within the dead-time of the instrument.

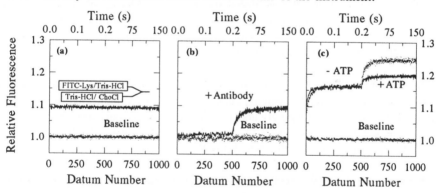

Figure 3. The fluorescence changes observed when ChoCl is mixed with (a) FITC-lysine, (b) FITC-modified enzyme in the presence of anti-FITC antibody and (c) enzyme that was labeled with FITC either in the presence (solid) or absence (stippled) of ATP are compared. The stippled baselines were obtained by mixing the corresponding fluorescein-containing solution with buffer.

What are the three phases? Figure 3 explains by showing that each of the phases in a "reversal" by ChoCl can be selectively eliminated or reduced. The rationale for these experiments was provided by an important study with anti-fluorescein antibody in which Ball and coworkers (1) showed that FITC labels two classes of sites. Fluorescein

595

incorporation into "antibody-accessible" sites is slowly reversible. Only fluorescein at "antibody-inaccessible", ATP-protectable sites reports the conformational change between $E_1$ and $E_2$.

In panel (a) FITC reacted with lysine was substituted for enzyme. Only the baseline shift is observed. In panel (b) ChoCl was mixed with FITC-enzyme plus anti-fluorescein antibody. Only the slowest enhancement with several second half-time is seen. Conversely, when enzyme that was labeled in the presence of ATP was mixed with ChoCl in panel (c), the two fastest phases are practically unaffected, but the magnitude of the slowest enhancement is greatly reduced.

What do these experiments mean? The titrant concentration is halved in a mixing experiment, so when 200 mM ChoCl is mixed with FITC-lysine there is a 100 mM ionic strength jump. There is no protein in panel (a), so we attribute the baseline shift to an effect of ionic strength on the fluorescence quantum yield of fluorophore that either did not react, or came off the labile antibody-accessible sites.

The second phase is not observed with FITC-lysine, so it must be reporting an effect of ionic strength on fluorescein bound to protein. Since the enhancement on the millisecond time scale is eliminated in panel (b) by anti-fluorescein antibody, the fluorescein must be located at antibody-accessible sites.

The only phase in panel (c) that is clearly affected by labeling in the presence of ATP is the slowest enhancement, the reaction that is not affected by anti-fluorescein antibody. Therefore, it must result from a change in the fluorescence quantum yield of fluorescein at ATP-protectable, antibody-inaccessible sites. These are the sites that report a change in fluorophore environment caused by a conformational change.

The titrant in panel (c) is ChoCl, so we conclude that non-specific ions like ChoCl can cause the slow protein conformational change that has been implicated in transport, as well as the two faster and relatively nonspecific effects of an ionic strength jump on free fluorophore and fluorescein bound to antibody-accessible sites. Grell and coworkers have independently suggested this possibility (7).

In what sense are the transported ions specific? This question is answered by Figure 4. The set of bars on the left shows that if labeled enzyme preincubated with $K^+$ is mixed with $Na^+$, all three stopped-flow phases are still observed. In sharp contrast, the middle set of bars shows that there is no enhancement corresponding to the slow conformational change when ChoCl is the titrant. The simplest explanation of this result is that ChoCl cannot compete with $K^+$. In other words, the specificty of the transported ions results from tighter binding.

Figure 4 also shows that the changes observed in a stopped-flow experiment add up to the change seen in an equilibrium titration In every case, the resultant of the time-resolved enhancements indicated by the narrow, diagonally cross-hatched bars is equal within experimental error to the relative fluorescence change measured in a conventional

596

stationary-state fluorometer that is denoted by the rectangularly cross-hatched bar.

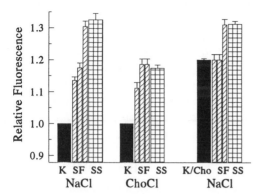

**Figure 4.** Stopped-flow (SF) and stationary-state (SS) titrations of FITC-enzyme in KCl (K) with NaCl and ChoCl are compared. The filled bars are normalized to 1.0 without ChoCl. From left to right the narrow diagonally hatched bars represent the baseline shift, faster and slower exponential enhancements in Figure 2. The ionic strength was kept constant in the titration with NaCl on the right by including ChoCl in the enzyme syringe.

Finally, Figure 4 demonstrates that the change on the slowest time scale is not an ionic strength effect. The set of bars on the right shows that when enzyme plus $K^+$ in ChoCl is mixed with $Na^+$, so that there is no change in ionic strength, the slowest enhancement is still observed. This is how we design our experiments, so that all we see is the fluorescence change reported by fluorescein at antibody-inaccessible, ATP-protectable sites that results from the conformational change we are interested in studying.

## Observations

Turning now to the question "How do ions cause the conformational change reported by fluorescein?", what observations have we made that have caused us to question the mechanism in the literature? We use a single exponential equation with offset to estimate the time constant, which we designate $1/\tau$, and amplitude expressed as percentage change in fluorescence from 1000 point kinetic curves. In Figures 5 & 6 the average estimates are shown as solid circles $\pm$ standard deviation plotted against the independent variable in the two types of experiments.

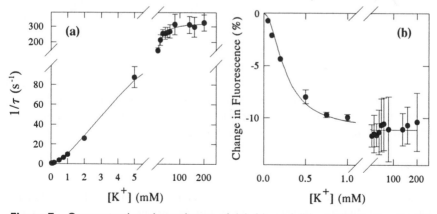

**Figure 5.** Concentration dependence of (a) $1/\tau$ and (b) amplitude for the $K^+$ quench reaction at 22 °C. The theoretical curves were calculated with Equations 2 & 3, as explained in the text.

597

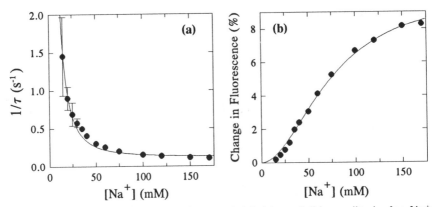

**Figure 6.** Concentration dependence of (a) $1/\tau$ and (b) amplitude for $Na^+$ reversal of the quench caused by 30 mM $K^+$ at 15 °C. The theoretical curves were calculated with Equations 2 & 4, as explained in the text.

Karlish (11) reported hyperbolic dependence of the time constant for the $K^+$ quench on $[K^+]$. Figure 5a shows the time constants we estimate for the $K^+$ quench as a function of $[K^+]$. The axes are broken, so the reader can see clearly that the functional dependence is sigmoidal, not hyperbolic. The $Na^+$ reversal reaction was not studied as a function of $[Na^+]$ by previous investigators. Figure 6a shows that the time constant for the $Na^+$ reversal is inversely related to $[Na^+]$. The percentage change in fluorescence estimated from kinetic curves is shown in Figure 6b to depend sigmoidally on $[Na^+]$. Therefore, the experimental observations we must explain are a sigmoidal dependence of the time constant for the $K^+$ quench on $[K^+]$, an inverse relationship between the time constant for the $Na^+$ reversal and $[Na^+]$, and a sigmoidal dependence of the amplitude of the $Na^+$ reversal on $[Na^+]$.

## Model

Equation 1 is the simplest formal mechanism that can explain our results.

$$E_1Na_2 \rightleftharpoons E_1Na \rightleftharpoons E_1 \overset{k_f}{\underset{k_r}{\rightleftharpoons}} E_1K \rightleftharpoons E_1K_2 \rightleftharpoons E_2K_2$$
$$\searrow \qquad \nearrow$$
$$E_1NaK$$
$$\tag{1}$$

We propose that 2 $K^+$ ions must bind to cause the conformational change in dephosphoenzyme reported by fluorescein and that at least 2 $Na^+$ ions bind to shift the equilibrium back from the $E_2$ to the $E_1$ conformation.

## Test of Model

What is different about how we have tested our model? The principal innovation is that we treat stopped-flow measurements as relaxation experiments. The concentration jump in a mixing experiment is large and most textbooks say relaxation methods are restricted

598

to small perturbations, so how can we analyze our stopped-flow traces as relaxation effects? The answer is that the reaction we are following is unimolecular. Since the relaxation expression for a unimolecular reaction is derived from a perfectly general rate equation without discarding any terms involving the square of deviations from equilibrium concentrations, the result is correct for any size perturbation (20). There are two advantages to thinking of stopped-flow experiments as relaxation effects. First, referring to the time constant estimated from kinetic traces as a reciprocal relaxation time $(1/\tau)$ emphasizes that it is not a rate constant. Even in the simplest case of a unimolecular reaction, $1/\tau$ is the sum of two rate constants. Second, derivation of the relaxation expression is reduced to a problem in algebra.

There are 7 states, or differently liganded enzyme forms, in Equation 1, so as many as 6 relaxation times could be measured. The fact that we only need one exponential to fit our stopped-flow data justifies *a posteriori* assuming the ion binding steps are fast and using equilibrium dissociation constants to calculate the fraction of the $E_1$ conformation with 2 $K^+$ ions bound. $E_1K_2$ changes conformation with a probability given by the forward rate constant $(k_f)$.

$$\frac{1}{\tau} = k_f \left[ \frac{[K^+]_o^2}{[K^+]_o\{[K^+]_o + K_{1K_2}(appNa)\} + K_{1KK}(appNaNa)} \right] + k_r \quad (2)$$

The derivation of Equation 2 assumes that $K^+$ and $Na^+$ bind competitively and that the binding of each ion can be described by a single intrinsic, or microscopic dissociation constant. The "apparent" terms are series expansions involving $Na^+$ to the first power (appNa) and $Na^+$ to the first plus second power (appNaNa).

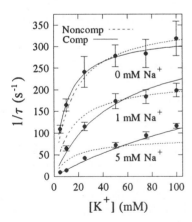

**Figure 7.** Fits of competitive (Equation 2) and noncompetitive binding models to $1/\tau$ estimates as a function of $K^+$ and $Na^+$ concentration are compared.

Once analytical expressions for the independent variable have been derived, different models can be tested by comparing how well they fit the experimental data. For example, if noncompetitive binding of $Na^+$ and $K^+$ is assumed, the "apparent" expression involving $Na^+$ multiplies both terms in the denominator of Equation 2. Figure 7 compares the fit of competitive and noncompetitive binding models to data for the dependence of $1/\tau$ on $K^+$ concentrations above the inflection point in Figure 6a as a function of $[Na^+]$ concentration. The better fit of the competition model to the $1/\tau$ values estimated from rate measurements is more convincing evidence for competitive binding of $Na^+$ and $K^+$ than competitive binding studies (11), because of the uncertainty about what is measured in equilibrium titrations.

Figure 7 illustrates another important point. There are 4 parameters, 2 microscopic equilibrium constants and 2 rate constants, in the "simplest" mechanism for Na$^+$ and K$^+$ interaction with the enzyme in Equation 1. The number of parameters increases rapidly, if K$^+$ and Na$^+$ are not assumed to bind to identical and independent sites. Is the situation hopeless? The answer is no for two reasons. First, the number of data points can be increased by "globally" fitting the data. In Figure 7 the best simultaneous fits of the competitive and noncompetitive models to the data for $1/\tau$ versus [K$^+$] at all three Na$^+$ concentrations are shown. That is why the fitted curves to the zero Na$^+$ data are not exactly the same, even though both models reduce to the same equation when Na$^+$ is absent. Second, two reactions can be studied. Therefore, different models can be judged by whether consistent numerical estimates of the equilibrium and rate constants explain the concentration dependence of both $1/\tau$ and amplitude for both K$^+$ quench and Na$^+$ reversal experiments.

Perfectly general expressions for the amplitude of the fluorescence change can also be derived. The expression for the corrected amplitude ($\Delta F_o$) of a K$^+$ quench is

$$\Delta F_o = \frac{[K^+]_o^2}{[K^+]_o\left[[K^+]_o + \dfrac{K_{1K_2}}{(1 + K_c)}\right] + K_{0.5KK}} \cdot \frac{K_c}{1 + K_c}\, \Delta F_{max} \qquad (3)$$

and the corrected amplitude of a Na$^+$ reversal is

$$\Delta F_o = \left[\frac{[Na^+]_o\{[Na^+]_o + K_{1Na_2}(appK)\}}{[Na^+]_o\{[Na^+]_o + K_{1Na_2}(appK)\} + K_{0.5NaNa}(appKK)}\right] Ampl KKQ \qquad (4)$$

where *AmplKKQ* is given by Equation 3. Again the apparent terms are series expansions. This time in K$^+$ concentration. In each case sigmoidal dependence of amplitude on titrant concentration is predicted with a half maximum given by the positive root of a quadratic equation (21).

The theoretical lines in Figures 5 & 6 were calculated with Equations 2-4. They show that the 2K$^+$/2Na$^+$ model in Equation 1 can satisfactorily explain the experimentally observed dependence of $1/\tau$ and amplitude on titrant concentration in both K$^+$ quench and Na$^+$ reversal experiments.

The theoretical line in Figure 5a is the nonlinear least squares fit of Equation 2 to the K$^+$ quench data. There is no Na$^+$ in this experiment, so the parameter estimates are independent of the number and magnitude of the Na$^+$ dissociation constants. Neither does derivation of Equation 2 with [Na$^+$] = 0 require that the two K$^+$ ions bind to identical and independent sites. The shape of the curve is determined by the constants for dissociation of the first ($K_{1K_1}$) and second ($K_{1K_2}$) K$^+$ ions from the E$_1$ conformer of the enzyme. The curve extrapolates to the sum of the forward and reverse rate constants as the K$^+$ concentration is increased giving a reliable estimate of $k_f$, since $k_r$ is near zero. $K_{1K_1}$, $K_{1K_2}$ and $k_f$ were fixed in fitting Equation 3 to the amplitude data in Figure 5b.

Conversely, plots like Figure 6a of $1/\tau$ versus [Na$^+$] extrapolate to the reverse rate constant at high [Na$^+$], so that a good estimate of $k_r$ can be obtained from Na$^+$ reversal of quenches by low K$^+$ concentrations that are not shown. The values of $K_{1K_1}$, $K_{1K_2}$, $k_f$

and $k_r$ were fixed at the values reported previously (20) in fitting Equations 2 and 4 to the Na$^+$ reversal data in Figure 6. In addition, the ratio of the macroscopic Na$^+$ dissociation constants was constrained to be 4, since independent binding was assumed in deriving Equation 2. The constants used to fit the data at 15 °C were $k_f = 142 \pm 2$ s$^{-1}$, $k_r = 0.12 \pm 0.01$ s$^{-1}$, $K_{1K_1} = 2.7 \pm 1.0$ mM, $K_{1K_2} = 10.0 \pm 1.2$ mM, $0.14 \leq K_{1Na_1} \leq 0.19$ mM and $0.55 \leq K_{1Na_2} \leq 0.77$, fulfilling the criterion that a consistent set of constants explain the variation of both parameters, relaxation time and amplitude, in both reactions, K$^+$ quench and Na$^+$ reversal.

## Conclusions

What have we learned? By analyzing stopped-flow amplitudes and by treating mixing experiments as relaxation experiments, we have been able to make the study of the conformational change in unphosphorylated sodium pump quantitative.

What does fluorescein report? We have shown that fluorescein reports at least three reactions. First, there is a baseline shift that we attribute to free fluorophore. Second, there is an enhancement on the millisecond time scale reported by fluorescein bound to antibody-accessible sites. The first two fluorescence increases are effects of ionic strength. Finally, there is a slow enhancement with a half-time of several seconds reported by fluorescein bound to antibody-inaccessible, ATP-protectable protein sites. Only this last enhancement reports the conformational change in dephosphoenzyme. ChoCl reverses the conformational change in low ionic strength buffer, but only Na$^+$ reverses the quench caused by K$^+$. Therefore, the specificity of the transported ions results from tighter binding.

The reason for discrepancies between equilibrium and kinetic titrations of FITC-modified Na$^+$,K$^+$-ATPase is that the fluorescence change observed in a stationary-state measurement is the resultant of the reactions reported by fluorescein. Our results also show that it is not easy to avoid artifacts in equilibrium experiments, even if their cause is known. For example, the conclusion that ChoCl also causes the conformational change means that equilibrium titrations cannot be corrected for ionic strength effects by a control titration in which an equivalent aliquot of ChoCl is added (19).

The influence of nonspecific ions on the conformational equilibrium and the heterogeneity of FITC labeling make kinetic titrations the method of choice when working with FITC-modified sodium pump. There is no change in ionic strength in a properly designed stopped-flow experiment. Other complications that occur in the time required for a sequential, equilibrium titration are also avoided (14). Most important of all, the contribution of a conformational change to the observed fluoresence change can be isolated by resolving the reactions reported by fluorescein along the time axis.

Given that FITC-modification reports a conformational change, is it a conformational change involved in transport? The conformational change reported by fluorescein is not an artifact of FITC-modification, because Na$^+$ also causes a fluorescence change with several second half-time in unmodified enzyme (9). This fluorescence change in native enzyme reported by intrinsic tryptophan fluorescence is accelerated by millimolar ATP, so that it has the same properties as the slow conformational change originally inferred from studies of the catalytic cycle by Post and coworkers (17).

Two results of our research strengthen the conclusion that fluorescein reports the conformational change implicated in transport. First, we have demonstrated that 2 $K^+$ ions are required to cause the conformational change reported by fluorescein (21). Therefore, the stoichiometry of the conformational change in dephosphoenzyme agrees with the consensus transport stoichiometry (6). Second, we showed earlier that reversal of the $K^+$ quench of FITC-modified $H^+,K^+$-ATPase is two orders of magnitude faster than the corresponding reaction of $Na^+,K^+$-ATPase (3,4). This discovery was subsequently confirmed by measurements in another laboratory (13). Therefore, the rate of the conformational change in $H^+,K^+$-ATPase reported by fluorescein correctly predicts failure to observe a $K^+$-occluded form of unphosphorylated gastric proton pump (2).

What is the mechanism by which ions cause the conformational change reported by fluorescein in sodium pump? Our results may explain the stoichiometry of ion transport. At least 2 $K^+$ ions must bind to cause the conformational change. This is the first kinetic evidence that 2 $K^+$ ions must be bound before the conformational change occurs. At least 2 $Na^+$ ions are required to explain reversal of the $K^+$ quench. It should be emphasized that Equation 1 is the simplest mechanism that can explain our experimental results. Not surprisingly expansion of the model to include binding of a third $Na^+$ ion fits the data in Figure 6 better. However, we are not sure yet whether the fit of a family of curves like those in Figure 6 at different $K^+$ concentrations is enough better to justify introducing an additional parameter into Equations 2 & 4.

It may be possible to learn the mechanism of the conformational change by studying FITC-modified enzyme. Table 1 summarizes the $K^+$ dissociation constants estimated from plots of $1/\tau$ versus $[K^+]$ for $K^+$ quench experiments at two temperatures with two different enzyme preparations (A & B). What is intriguing is the ratio of the constants for dissociation of the first and second $K^+$ ions. $Na^+$ was not present in these experiments, so derivation of the equation that was used to estimate the ratio 4 did not require any assumptions about whether ion binding occurs anticooperatively to a single site, or to two indentical and independent sites.

**Table 1.** Estimated potassium dissociation constants

| TEMP(°C) | ENZYME | $K_{1K_1}$ (mM) | $K_{1K_2}$ (mM) | $K_{1K_2}/K_{1K_1}$ |
|---|---|---|---|---|
| 15 | A | $2.7 \pm 1.0$ | $10.0 \pm 1.2$ | $3.7 \pm 1.8$ |
|  | B | $2.8 \pm 0.9$ | $14.0 \pm 1.4$ | $5.0 \pm 2.1$ |
| 22 | B | $2.2 \pm 1.5$ | $10.1 \pm 1.6$ | $4.6 \pm 3.9$ |

A ratio of 4 could result from ordered binding to two sites coincidentally differing in their affinity for $K^+$ by a factor of four. There is evidence both from deocclusion experiments (5) and from electrical measurements (13) for anticooperative binding of two $K^+$ ions to a single "pocket" or "ion well".

Alternatively, the ratio 4 could mean that $K^+$ binds to two identical and independent sites. The macroscopic dissociation constants for binding to two sites with the same intrinsic affinity differ by four, because the first ion has two ways to go on the protein but only one way to come off, while the second ion can bind to only one unoccupied site but dissociate from either of two occupied sites. One test of an independent binding model has already been carried out. If $K^+$ binds to two independent and identical sites and $Na^+$ competes with $K^+$, then 2 $Na^+$ ion must also bind independently and the ratio of the $Na^+$ dissociation constants must be 4. Figure 6 demonstrates satisfactory agreement between theory and experiment when the ratio $K_{1Na_1}$ to $K_{1Na_2}$ is constrained to be 4.

One advantage of an independent binding model is that it is simpler. Since there is a statistical relationship between the macroscopic constants for dissociation of an ion from independent and identical sites, only four parameters were assumed in the derivation of Equation 1, which can satisfactorily explain our data. At least two more parameters would be required for an anticooperative binding model. Another advantage is that there is precedent for independent binding of ligands causing conformational changes in proteins. One thinks immediately of the concerted model proposed by Monod, Wyman and Changeux (15). In this model binding occurs with the same microscopic dissociation constant to protomers constrained to the same conformation within an oligomer until the equilibrium favors the alternative conformation of all the subunits. An obvious disadvantage of the Monod-Wyman-Changeux mechanism for the sodium pump is that it assumes an oligomeric structure. A paper in this Symposium by Professor Schoner and coworkers deals with the issue of whether sodium pump is a functional dimer.

*Acknowledgements*

This work was supported by National Science Foundation Grant MCB9106338, National Institutes of Health Grant DK36873, and a Veterans Administration Merit Review Award.

# References

1. Abbott, AJ, Amler, E, Ball, WJ (1991) Immunochemical and spectroscopic characterization of two fluorescein 5'-isothiocyanate labeling sites on Na,K-ATPase. Biochemistry 30: 1692-1701
2. De Pont, JJHHM, Helmich-de Jong, ML, Bonting, SL (1985) K,H-ATPase and Na,K-ATPase: similarities and differences. In: Glynn, I, Ellory, C (eds) The Sodium Pump. The Company of Biologists, Cambridge, pp 733-738
3. Faller, LD, Diaz, RA, Scheiner-Bobis, G, Farley, RA (1990) $E_1E_2$-ATPase conformational energetics. FASEB J 4: A1962
4. Faller, LD, Diaz, RA, Scheiner-Bobis, G, Farley, RA (1991) Temperature dependence of the rates of conformational changes reported by fluorescein 5'-isothiocyanate modification of $H^+$,$K^+$- and Na,K-ATPases. Biochemistry 30: 3503-3510

5. Forbush, B (1987) Rapid release of $^{42}$K or $^{86}$Rb from two distinct transport sites on the Na,K-pump in the presence of $P_i$ or vanadate. J Biol Chem 259: 11116-11127

6. Glynn, IM (1985) The Na$^+$,K$^+$-transporting adenosine triphosphatase. Enzymes Biol Membr 3: 35-114

7. Grell, E, Warmuth, R, Lewitzki, E, Ruf, H (1991) Precision titrations to determine affinity and stoichiometry of alkali, alkaline earth, and buffer cation binding to Na,K-ATPase. In: Kaplan, JH, De Weer, P (eds) The Sodium Pump: Recent Developments. The Rockefeller University Press, New York, pp. 441-445

8. Hegyvary, C, Jorgensen, PL (1981) Conformational changes of renal sodium plus potassium ion-transport adenosine triphosphatase labeled with fluorescein. J Biol Chem 256: 6296-6303

9. Karlish, SJD, Yates, DW (1978) Tryptophan fluorescence of (Na$^+$ + K$^+$)-ATPase as a tool for study of the enzyme mechanism. Biochim Biophys Acta 527: 115-130

10. Karlish, SJD (1979) Cation induced conformational states of Na,K-ATPase studied with fluorescent probes. In: Skou, JC, Norby, J (eds) (Na,K) ATPase Structure and Function. Academic Press, New York, pp 115-128

11. Karlish, SJD (1980) Characterization of conformational changes in Na,K-ATPase labeled with fluorescein at the active site. J Bioenerg Biomembr 12: 111-135

12. Kyte, J (1975) Structural studies of sodium and potassium ion-activated adenosine triphosphatase. J Biol Chem 250: 7443-7449

13. Läuger, P, Apell, H-J (1986) A microscopic model for the current-voltage behaviour of the Na,K-pump. Eur Biophys J 13: 309-321

14. Lin, S-H, Faller, LD (submitted) Time resolution of fluorescence changes observed in titrations of fluorescein 5'-isothiocyanate-modified Na$^+$,K$^+$-ATPase with monovalent cations. Biochemistry

15. Monod, J, Wyman, J, Changeux, J-P (1965) On the nature of allosteric transitions: a plausible model. J Mol Biol 12: 88-118

16. Post, RL, Kume, S, Tobin, T, Orcutt, B (1969) Flexibility of an active center in sodium-plus-potassium adenosine triphosphatase. J Gen Physiol 54: 306s-326s

17. Post, RL, Hegyvary, D, Kume, S (1972) Activation by adenosine triphosphate in the phosphorylation kinetics of sodium and potassium ion transport adenosine triphosphatase. J Biol Chem 247: 6530-6540

18. Pratap, PR, Robinson, JD, Steinberg, MI (1991) The reaction sequence of the Na,K-ATPase: rapid kinetic measurements distinguish between alternative schemes. Biochim Biophys Acta 1069: 288-298

19. Rabon, EC, Bassilian, S, Sachs, G, Karlish SJD (1990) Conformational transitions of the H,K-ATPase studied with sodium ions as surrogates for protons. J Biol Chem 265: 19594-19599

20. Smirnova, IN, Faller, LD (1993) Role of Mg$^{2+}$ ions in the conformational change reported by fluorescein 5'-isothiocyanate modification of Na$^+$,K$^+$-ATPase. Biochemistry 32: 5967-5977

21. Smirnova, IN, Faller, LD (1993) Mechanism of K$^+$ interaction with fluorescein 5'-isothiocyanate modified Na$^+$,K$^+$-ATPase. J Biol Chem 268: 16120-16123

# Relaxation Spectroscopy Applied to Determination of Rate Constants for Ligand Binding and Dissociation

M. Esmann

Institute of Biophysics, Ole Worms Allé 185, University of Aarhus, DK-8000 Aarhus, Denmark

## Introduction

Application of relaxation methods for determining individual rate constants for ligand binding to a receptor is presented. The method requires that a ligand is available which has a spectroscopic property that changes upon binding to the receptor. The rate constants for binding ($k_2$) and dissociation ($k_{-2}$) for such a "labelled" ligand (Y) can be determined using conventional relaxation methods (3). The purpose of the present paper is to show that kinetic parameters $k_1$ and $k_{-1}$ for an unlabelled ligand (X) also can be determined, using the influence of the unlabelled ligand on the binding of the labelled ligand to the receptor (R).

$$R \cdot Y \underset{k_2 \cdot [Y]}{\overset{k_{-2}}{\rightleftharpoons}} R \underset{k_{-1}}{\overset{[X] \cdot k_1}{\rightleftharpoons}} R \cdot X \tag{1}$$

The method has been used to determine rates of nucleotide binding reaction of the $Na^+/K^+$-ATPase, using eosin as the labelled ligand. Please refer to reference (2) for experimental and numerical procedures.

## Simulation of perturbation experiments and analysis using relaxation theory

Small changes (i.e. perturbations) of the concentration of a ligand will lead to changes in the concentration of the three complexes R, RX and RY, cf. scheme 1. It will be assumed that the concentration of RY can be followed with time (in the example with eosin and Na,K-ATPase (2), the fluorescence yield of the eosin·Na,K-ATPase complex is much higher than that of free eosin, which allows the concentration of the eosin·Na,K-ATPase complex to be monitored). The time-dependence of [RY] after perturbation will according to perturbation theory (1,6) follow a bi-exponential time-course, which is characterized by two relaxation rate constants $\lambda_1$ and $\lambda_2$ :

$$[RY]_t = A_1 \cdot \exp(-\lambda_1 \cdot t) + A_2 \cdot \exp(-\lambda_2 \cdot t) + [RY]_{t=\infty} \tag{2}$$

The magnitude of $\lambda_1$ and $\lambda_2$ depend on the rate constants and the concentrations of X and Y after the perturbation (1,6):

$$\lambda_1 = 0.5 \cdot \left\{ k_1 \cdot [X] + k_{-1} + k_2 \cdot [Y] + k_{-2} - \sqrt{(k_1 \cdot [X] + k_{-1} + k_2 \cdot [Y] + k_{-2})^2 - 4 \cdot (k_{-1} \cdot k_{-2} + k_1 \cdot [X] \cdot k_{-2} + k_{-1} \cdot k_2 \cdot [Y])} \right\} \tag{3}$$

$$\lambda_2 = 0.5 \cdot \left\{ k_1 \cdot [X] + k_{-1} + k_2 \cdot [Y] + k_{-2} + \sqrt{(k_1 \cdot [X] + k_{-1} + k_2 \cdot [Y] + k_{-2})^2 - 4 \cdot (k_{-1} \cdot k_{-2} + k_1 \cdot [X] \cdot k_{-2} + k_{-1} \cdot k_2 \cdot [Y])} \right\} \tag{4}$$

605

Figure 1 shows the calculated time-courses of such perturbations, either with a net reaction towards an increase in [RY] (upper curve, the perturbation is a 30% increase in [Y]) or with a net displacement of the equilibrium towards RX due to an increase in [X]. The bi-exponential curve-fits are characterized by two relaxation rate constants (5.31 and 50.6 $s^{-1}$ for the upper curve) which are in close agreement with (or identical to) the relaxation rate constants that can be calculated from equations 3 and 4 (5.24 and 50.1 $s^{-1}$, respectively, using the rate constants given in the legend to Figure 1). Note that the amplitudes for both exponentials characterizing the upper curve have the same (negative) sign, i.e. both terms give rise to an increase in [RY] with time. The lower curve is also fitted by two exponentials, and here there is a clear lag-period (corresponding to the negative sign of the amplitude for the second term) before the decrease in [RY]. Both relaxation rate constants determined from the bi-exponential fit of the lower curve are also in agreement with those calculated from equations 3 and 4 (see legend to Figure 1), showing that a change in [RY] of about 15% is acceptable as a "perturbation".

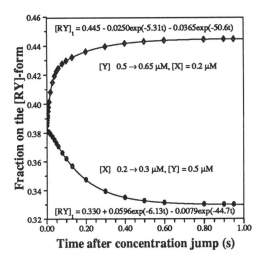

**Figure 1. Simulation of the time-dependence of the RY-complex after a perturbation of the ligand concentrations.** Rate constants are : $k_1 = 20.5 \cdot 10^6$ $M^{-1} \cdot s^{-1}$, $k_{-1} = 4$ $s^{-1}$, $k_2 = 45 \cdot 10^6$ $M^{-1} \cdot s^{-1}$, $k_{-2} = 18$ $s^{-1}$, [X] = 0.2 μM and $[R]_{TOT}$= 0.1 μM. The upper curve gives the perturbation when - at time zero - [Y] is increased from 0.5 to 0.65 μM, and the lower curve shows the displacement from the RY-form when [X] is increased to 0.3 μM. The points are calculated concentrations of RY at different times, and the full lines represent non-linear least-squares curve fits of the form $[RY]_t = [RY]_{t=\infty}$ + $A_1 \cdot exp(-\lambda_1 \cdot t) + A_2 \cdot exp(-\lambda_2 \cdot t)$, values are given in the curve.

Classical interpretations of the concentration dependence of relaxation rates [1, 3-6] rely on for example a plot of $(\lambda_1 + \lambda_2)$ versus e.g. [X], which will yield a slope of $k_1$ and an intercept of $(k_2 \cdot [Y] + k_{-1} + k_{-2})$. A plot of the product $\lambda_1 \cdot \lambda_2$ versus [X] will similarly give values for $k_1 \cdot k_{-2}$ and $k_{-1} \cdot (k_2 \cdot [Y] + k_{-2})$, from which the individual rate constants may be calculated (e.g. Table 4 in (1)). It is of interest to perform the experiments in such a way that the rate constants $k_{-1}$ and $k_1$ are determined relatively accurately. Analysis of the concentration dependence of $\lambda_1$ and $\lambda_2$ on [X] (at a fixed [Y], see equations 3 and 4) shows that:

$$\lambda_1 \rightarrow k_{-1} \quad \text{and} \quad \lambda_2 \rightarrow k_2 \cdot [Y] + k_{-2} \quad \quad \text{for } [X] \rightarrow 0 \quad \quad (5)$$

$$\lambda_1 \rightarrow k_{-2} \quad \text{and} \quad \lambda_2 \rightarrow k_1 \cdot [X] \ (\sim \infty) \quad \quad \text{for } [X] \rightarrow \infty \quad \quad (6)$$

Therefore, an analysis of the relaxation process (notably of $\lambda_1$) over a wide range of [X] (from very high concentrations to those approaching zero) will give estimates of the individual rate constants for the dissociation reaction of the ligands.

A optimal experimental design for a reliable determination of $\lambda_1$ depends on the relative magnitudes of $k_{-1}$ and $k_{-2}$ ($k_1$ and $k_2$ are to be multiplied by [X] and [Y], respectively, and the magnitude of terms in equations 3 and 4 containing these constants can be adjusted by choosing an appropriate concentration of X and Y). Figure 2A shows the concentration dependence of $\lambda_1$ as [X] is varied over a wide range, calculated using equation 3. Values for $k_2 \cdot [Y]$ and $k_{-2}$ are fixed (at 30 and 20 $s^{-1}$, respectively) and the equilibrium dissociation constant for binding of X ($K_X = k_{-1}/k_1$) is kept constant. The large range of values for $k_{-1}$ (and thus $k_1$) give an idea of the applicability of the method to systems with very different kinetic properties, here is shown a range from relatively slow binding and dissociation of X ($k_{-1}=1\ s^{-1}$, $k_1=2 \cdot 10^6\ M^{-1} \cdot s^{-1}$) to more rapid processes ($k_{-1} = 50\ s^{-1}$ and $k_1 = 10^8\ M^{-1} \cdot s^{-1}$). The "titration" of $\lambda_1$ with [X] follows a smooth curve (see equation 3), and the "midpoint" of the titration (the [X] corresponding to a relaxation rate constant of $\lambda_1 = 0.5 \cdot (k_{-1} + k_{-2})$ indicated by an arrow as an example in Figure 2A) is related to the binding rate constant $k_1$ :

$$k_1 = \{2 \cdot k_2 \cdot [Y] + k_{-2} - k_{-1}\}/\{2 \cdot [X]_{0.5}\} \tag{7}$$

where $[X]_{0.5}$ denotes the concentration of X corresponding to the midpoint of the titration. Since is it assumed that $k_2$ and $k_{-2}$ can be determined independently, and since $k_{-1}$ is the limiting value at low [X], the value for $k_1$ can be readily calculated. This also allows the equilibrium dissociation constant ($K_X = k_{-1}/k_1$) to be calculated.

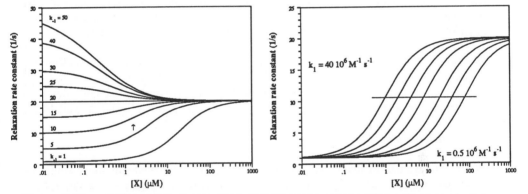

**Figure 2. The relationship between the relaxation rate constant for a decrease in the concentration in RY and the concentration of unlabelled ligand.** In *Panel A* (left), the lines are calculated for values of $k_{-1}$ between 1 and 50 $s^{-1}$, keeping the ratio $k_{-1}/k_1$ constant at 0.5 µM. For Y, the constants chosen were $k_2 = 30 \cdot 10^6\ M^{-1} \cdot s^{-1}$ and $k_{-2} = 20\ s^{-1}$, with [Y]= 1 µM. In *Panel B* (right), the lines show the dependence of the relaxation rate constant $\lambda_1$ calculated in the same manner as for panel A, but fixing the rate constant $k_{-1} = 1\ s^{-1}$ and allowing $k_1$ to vary between 0.5 and $40 \cdot 10^6\ M^{-1} \cdot s^{-1}$.

Figure 2B shows a set of simulations of the concentration dependence of $\lambda_1$ on [X]. Here the equilibrium dissociation constant for binding of X is allowed to vary. As predicted, the midpoint of titration ($[X]_{0.5}$-values corresponding to the horizontal line in Figure 2B) follows the value of $k_1$ inversely, here is shown a range from $0.5 \cdot 10^6$ to $40 \cdot 10^6\ M^{-1} \cdot s^{-1}$.

The amplitude of the perturbation in terms of the change in [RY] is important for determination of the rate constants (i.e., there must be a measurable change in [RY]). The fraction of receptor on the RY-form is given by:

$$[RY]/[R]_{TOT} = [Y]/(K_Y \cdot \{1+[X]/K_X+[Y]/K_Y\}) \tag{8}$$

If it is assumed that the experiment is performed at $[Y]=K_Y$, then the change in [RY] can - as an example - be calculated to $0.05 \cdot [R]_{TOT}$ when [X] is perturbed from $0.8 \cdot K_X$ to a final concentration of $1.26 \cdot K_X$ (i.e. a fractional change in [RY] of 10%). As the final concentration of X is lowered considerably relative to $K_X$, the response is diminished. The maximal change which can be observed (i.e. with a jump in [X] from zero to a given $[X]_{final}$) obeys the following relation (again taking $[Y]=K_Y$):

$$([RY]_{[X]=0} - [RY]_{[X]final})/[R]_{TOT} = 0.5/(1+2 \cdot K_X/[X]_{final}) \tag{9}$$

A change of $0.01 \cdot [R]_{TOT}$ thus occurs at $[X]\sim0.04 \cdot K_X$, increasing to about 0.42 at $[X]=10 \cdot K_X$.

It is clear from the above that the simple model describing binding of the ligands must be followed, i.e. Y must exhibit a competitive binding for X. The perhaps simplest case is that the site for X and for Y is the same site. In this case the method can be used to test a hypothesis of site-identity for X and Y, since there should be no (or only accidental) consistency between rate constants obtained by this method and parameters obtained by other kinetic observations if the sites for X and Y were different.

In conclusion, relaxation spectroscopy in conjunction with for example the rapid-mixing stopped-flow method (by which relaxation rate constants up to 300 s$^{-1}$ can be measured) can be applied to determination of receptor binding and dissociation rate constants for a given ligand in a simple manner, provided a suitable competitive ligand is available.

*Acknowledgements*

The author wishes to thank Jens G. Nørby for a critical reading of the manuscript. Financial support was received from the Danish Medical Research Council, Aarhus University Research Foundation and the Danish Biomembrane Research Center.

**REFERENCES**

1.      Eigen, M. (1968) New looks on physical enzymology. Q. Rev. Biophys. **1**, 3-33.
2.      Esmann, M. (1992) Determination of rates of nucleotide binding and dissociation from Na,K-ATPase. Biochim. Biophys. Acta **1110**, 20-28.
3.      Gutfreund, H. (1972) Enzymes: Physical Principles, Wiley Interscience, London.
4.      Hague, D.N. (1971) "Fast reactions", Wiley Interscience, pp. 135-142.
5.      Hammes, G.G. and Schimmel, P.R. (1970) in "The Enzymes" (ed. P.D. Boyer), Academic Press, vol. 2, pp. 67-114.
6.      Matsen, F.A. and Franklin, J.L. (1950) A general theory of coupled sets of first order reactions. J. Am. Chem. Soc. **72**, 3337-3341.

608

# Does Eosin Treat all P-type ATPases Equally?

Craig Gatto and Mark A. Milanick

MA415 Medical Sciences Bldg., Department of Physiology, University of Missouri-Columbia, Columbia, Missouri  65212  USA

## Introduction

A detailed chemical and molecular mechanism by which the hydrolysis of ATP is coupled to the uphill movement of ions remains to be elucidated, even though a variety of studies have provided a wealth of information about structural and kinetic aspects of P-type pump mechanisms.  Identification of the amino acid residues of the ATP site would provide important information for determining this chemical mechanism.  Many results are consistent with the idea that fluorescein-5-isothiocyanate (FITC) labels a lysine that is part of the ATP site of the $Na^+/K^+$-, $H^+/K^+$-, sarcoplasmic reticulum (SR) $Ca^{2+}$-, and plasma membrane (PM) $Ca^{2+}$-pumps.

The primary evidence that FITC labels a lysine in the ATP site is that ATP protects against inactivation by FITC.  The stoichiometry of FITC labeling is generally 1 FITC per pump monomer.  In the $Na^+$-pump, Farley *et al.* (2) and Kirley *et al.* (5) report that Lys-501 is the only residue labeled by FITC.  In contrast, Xu (13) found that FITC modified 3 different lysine residues, even when the stoichiometry of FITC to $Na^+$-pump alpha subunit was 1:1.  She suggests that all 3 lysines are part of the ATP binding pocket.

If the fluorescein moiety directs FITC to a reactive lysine that is part of the ATP site, then one would expect that fluorescein itself should compete with ATP for the P-type pumps.  Indeed, tetrabromofluorescein (eosin) competes with ATP for the $Na^+/K^+$-pump (12) and for the $H^+/K^+$-pump (4).  Alternatively, it is possible that a reactive lysine exists in the ATP site which will react with any isothiocyano containing compound (i.e. the fluorescein moiety is unnecessary).  For example, 4-sulfophenyl isothiocyanate is an ATP protectable inactivator of the PM $Ca^{2+}$-pump (Gatto and Milanick, unpublished).  Also, N-(2-nitro-4-isothiocyanophenyl)-imidazole (NIPI) and phenyl isothiocyanate are ATP protectable inactivators of the $Na^+/K^+$-pump (1).

One approach to determine if these reactive lysine residues are part of the ATP site in P-type pumps is site-directed mutagenesis.  Mutation of the FITC lysine in the SR $Ca^{2+}$-pump reduced but did not eliminate $Ca^{2+}$ transport activity (70% inhibition); there was no decrease in ATP-dependent phosphoenzyme production (7).  In the $Na^+/K^+$-pump, mutating all three FITC lysines did not significantly alter the $K_m$ for ATP or the $V_{max}$ for ATPase activity (6).  These results argue against the essential involvement of these lysine residues in ATP binding.  Assuming these lysine residues are the only sites for FITC labeling, an intriguing prediction is that FITC would <u>not</u> inhibit the residual activity of the mutated enzymes.

If these lysines are not essential parts of the ATP site, why does ATP protect against FITC inhibition?  An obvious explanation is that ATP binding results in a conformational change that decreases the accessibility of the reactive lysines.  Another possibility is that these lysines are close to the ATP site.  In this case, the ATP molecule itself sterically prevents FITC from contacting the lysine residue.

The present study characterizes eosin inhibition of the calmodulin-sensitive PM $Ca^{2+}$-pump.  Unlike other P-type pumps, we found that eosin does not compete with ATP for the PM $Ca^{2+}$-pump.  We also dicuss the successes and failures of the assumption that ATP, eosin, and FITC go to the same site in the P-type pumps.

## Methods

Inside-out vesicles (IOVs) were prepared from pig red cells according to standard procedures (10). ATP-dependent $^{45}Ca^{2+}$ uptake into IOVs was measured as described previously (3). Briefly, assays were started by adding IOVs (final [protein] = 20-80 µg/ml) to reaction tubes containing (in mM) 0.02 $CaCl_2$, 0.1 $MgCl_2$, 130 NaCl, 30 HEPES, and MgATP as indicated in Fig. 1 (assays contained 0.1-0.5 µCi $^{45}Ca^{2+}$/ml). Assays were performed in a total volume of 1.1 ml at 37°C. After 30 s, fluxes were stopped by removing 400-µl aliquots from reaction tubes (in duplicate) and mixing them with 400-µl of ice-cold stop solution containing (in mM) 2 EGTA, 20 HEPES, and 150 KCl. The total volume was vacuum filtered (pore size = 0.45 µm) and rinsed 4-times with 165 mM NaCl. The filter discs were combined with scintillation cocktail for liquid scintillation counting.

## Results and Discussion

If eosin binds exclusively at the ATP site, then increasing [ATP] should displace eosin from the binding site and the same $V_{max}$ will be attained in the presence or absence of eosin. As shown in Fig. 1, eosin clearly decreased the $V_{max}$ for $Ca^{2+}$ uptake as ATP was increased. The fact that eosin and ATP did not compete for the $Ca^{2+}$-pump was surprising given the results on other pumps (4, 12) and the model in which FITC goes to the ATP site (2, 5, 11). A simple explanation for the data in Fig. 1 is that eosin binds to the ATP site as well as an additional site on the PM $Ca^{2+}$-pump.

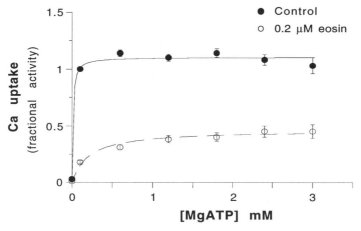

**Fig. 1.** Eosin (0.2 µM) inhibition of ATP-dependent $Ca^{2+}$-uptake in pig red cells. Data were fit to the Michaelis-Menten equation; control, $V_{max}$= 1.10; eosin treatment, $V_{max}$= 0.46. Data (mean $\pm$SEM) from 5 separate experiments were normalized to the control uptake rate at 0.1 mM ATP. Increasing [MgATP] did not decrease the inhibition by eosin suggesting that eosin does not compete with ATP. We also have observed that eosin does not compete with ATP in the human red cell Ca-pump when either ATP-dependent $Ca^{2+}$ uptake or $Ca^{2+}$-dependent ATP hydrolysis were measured (3).

At this time it may be helpful to define the common but vague term, "site". In the chemical sense, "site" refers to the specific amino acid residues which are in contact with a particular ligand. The problem then becomes determining when two compounds share the same site. For example, assume ATP contacts 8 amino acid residues when bound.

610

How many of these 8 residues would have to contact eosin or FITC before these compounds can be called ATP site probes? For simplicity, many assume that two compounds bind to the same site if they share at least one contact point. Unfortunately, this assumption may not be sufficient (see below). Under this definition, ATP and ADP bind to the same site (although defined kinetically as different sites). In this case, the importance of the phosphorylated aspartate residue is de-emphasized because it is not part of the true ADP site. Nonetheless, we will use this definition of "site" to discuss a model where eosin, FITC, and ATP bind to the same site, i.e. a one site model.

The following findings are consistent with this model (for review see 9): a) ATP protects against FITC inactivation in all of the P-type pumps. b) The major FITC-reactive lysine is in a conserved region among the P-type pumps (a.a. sequence KGAPE for the $Na^+/K^+$-, SR $Ca^{2+}$-, and $H^+/K^+$-pumps and KGASE for the PM $Ca^{2+}$-pump). c) Eosin competes with ATP for the $H^+/K^+$- and $Na^+/K^+$-pumps. d) We find that eosin completely eliminates inactivation by FITC in the PM $Ca^{2+}$-pump. A 3.5 min incubation with 30 μM FITC irreversibly inhibits 90% of the $Ca^{2+}$ uptake activity; in the presence of 20 μM eosin, 30 μM FITC was unable to inhibit pump activity.

In contrast, the data shown in Fig. 1 are not easily reconciled with this model. However, we propose two possible explanations for eosin not competing with ATP for the PM $Ca^{2+}$-pump with the constraints of a one site model. In both explanations, the ATP-binding site must be slightly different for the PM $Ca^{2+}$-pump than in the other P-type pumps (e.g. $S$ replacing $P$, above). 1) Eosin and ATP can both bind to the site at the same time. For example, assume that eosin and ATP each bind to 8 residues; ATP binds to residues 1-**8** and eosin binds to residues **8**-15. While eosin is bound, ATP binds at residues 1-7 without displacing eosin from residues 8-15. Although this model is allowed by the above definition of site, we hesitate to call them the same site (since 14 of 15 residues are not shared by both ligands). This model puts severe limits on the usefulness of eosin as an ATP site probe in the PM $Ca^{2+}$-pump. 2) Eosin binds equally well to both the non-phosphorylated and the phosphorylated enzyme (see Fig. 2).

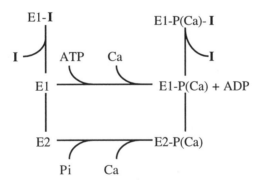

**Fig. 2.** Kinetic model depicting eosin as a mixed-type inhibitor of the PM $Ca^{2+}$-pump.

In this model, eosin and ATP (and ADP) share several of the same contact points but they do not compete. For example, assume that eosin binds to residues 1-7 and ATP binds to residues 1-8 (8 is the phosphorylated aspartate). After enzyme phosphorylation, the affinity for ATP significantly decreases (because residue 8 is significantly altered). On the other hand, the affinity for eosin does not change because residues 1-7 remain unchanged; these residues also comprise the ADP binding site. According to the above definition, eosin is binding to the same site before and after phosphorylation. However,

611

this site behaves kinetically as two different sites which results in a mixed-type inhibition for eosin. Interestingly, this model predicts that eosin will compete with ATP for Ca-dependent phosphorylation (E1 to E1-P) in the forward direction as well as compete with ADP for pump dephosphorylation in the reverse direction (E1-P to E1). These predictions are currently under investigation.

Above we discussed eosin inhibition with the premise that eosin binds to the ATP site. However, there is no definitive evidence proving that eosin binds only to the ATP site in all of the P-type pumps. An alternative explanation for eosin not competing with ATP for the PM $Ca^{2+}$-pump is that this enzyme contains a high affinity inhibitory site which is separate from the ATP site. This model allows for possible endogenous regulation of the $Ca^{2+}$-pump. Structurally, eosin appears to be at least as similar to steroids and opiates as it is to ATP; eosin is the most potent inhibitor known for this PM $Ca^{2+}$-pump ($IC_{50}$ <0.2 µM; 3, 8).

**Acknowledgements**
We gratefully thank Betty Jo Wilson for excellent technical assistance. This work was supported by National Institute of Health Grant DK-37512. M.A. Milanick is a recipient of a Research Career Development Award from the NIDDK.

**References**
1.   Ellis-Davies GCR, Kaplan JH (1990) Binding of $Na^+$ ions to the $Na^+/K^+$-ATPase increases the reactivity of an essential residue in the ATP binding domain. J Biol Chem 265: 20570-20576
2.   Farley RA, Tran CM, Carilli CT, Hawke D, Shivly JE (1984) The amino acid sequence of a fluorescein-labeled peptide from the active site of $Na^+/K^+$-ATPase J Biol Chem 259: 9532-9535.
3.   Gatto C, Milanick MA (1993) Inhibition of the red blood cell calcium pump by eosin and other fluorescein analogues. Am J Phys 264: C1577-C1586.
4.   Helmich-de Jong ML, van Duynhoven JPM, Schuurmans Stekhoven FMAH, DePont JJHHM (1986) Eosin, a fluorescent marker for the high-affinity ATP site of $H^+/K^+$-ATPase. Biochim Biophys Acta 858: 254-262.
5.   Kirley TL, Wallick ET, Lane LK (1984) The amino acid sequence of the fluorescein isothiocyanate reactive site of Lamb and Rat kidney Na+/K+-ATPase. Biochem Biophys Resch Comm 125(2): 767-773.
6.   Kuntzweiler TA, Lingrel JB (1993) Altering the nucleotide binding site of Na+/K+-ATPase by site-directed mutagenesis. This volume
7.   Maruyama K, Maclennan DH (1988) Mutation of aspartic acid-351, lysine-352, and lysine-515 alters the $Ca^{2+}$ transport activity of the $Ca^{2+}$-ATPase expressed in COS-1 cells. Proc Natl Acad Sci USA 85: 3314-3318.
8.   Mugica H, Rega AF, Garrahan (1984) The inhibition of the calcium dependent ATPase from human red cells by erythrosin B. Acta Physiol Pharm Latinoam 34: 163-173.
9.   Pedemonte CH, Kaplan JH (1990) Chemical modification as an approach to elucidation of sodium pump structure-function relations. Am J Physiol 258: C1-C23.
10.  Sarkadi B, Szasz I, Gardos G (1980) Characteristics and regulation of active calcium transport in inside-out red cell membrane vesicles. Biochim Biophys Acta 598: 326-338.
11.  Shull GE, Greeb J (1988) Molecular cloning of two isoforms of the plasma membrane $Ca^{2+}$-ATPase from rat brain. J Biol Chem 263: 8646-8657.
12.  Skou JC, Esmann M (1981) Eosin, a fluorescent probe of ATP binding to the $Na^+/K^+$-ATPase. Biochim Biophys Acta 647: 232-240.
13.  Xu K-Y (1989) Any of several lysines can react with 5'-isothiocyanofluorescein to inactivate sodium and potassium ion activated adenosine triphosphatase. Biochemistry 28: 5764-5772.

# Kinetic Analyses of Conformational Changes and Charge Transfer by $Na^+/K^+$-ATPase

P. R. Pratap and J. D. Robinson

Department of Pharmacology, SUNY Health Science Center, Syracuse, NY 13210 USA

## Introduction

How does the $Na^+/K^+$-ATPase couple ATP hydrolysis with transport? According to common interpretations of the Albers-Post reaction scheme, the enzyme cycles through phosphorylated and unphosphorylated forms of two conformations, $E_1$ and $E_2$; ion transport is mediated by transitions between these conformations. Evidence for these two conformations include differences in: (a) affinity for $Na^+$ and $K^+$, (b) tryptic digestion patterns, (c) fluorescence quantum efficiencies of various probes, and (d) sensitivity to dephosphorylation induced by ADP ($E_1P$) and by $K^+$ ($E_2P$).

Discrepancies with this model included measurements of $E_1P$ and $E_2P$ that summed to more than 100% of total EP. Attempts to deal with this discrepancy (by Nørby et al. [3] and by Yoda and Yoda [7]) have resulted in the partial or complete dissociation of ion transport from the $E_1$ - $E_2$ transition. Based on these studies, Skou [5] presented a detailed scheme which defined $E_1$ and $E_2$ in terms of the number of cation binding sites: $E_1$ having three, and $E_2$, two, implying that the conformational change occurs after ion transport. Earlier work from our laboratory [4] was consistent with this model: those data indicated that the fluorescence of IAF-labeled enzyme (and thus the conformation) changed at or after the release of the second and/or third $Na^+$ from the enzyme to the extracellular side; however, this conclusion was based on the assumption that the fluorescence change in the IAF-enzyme indicates the $E_1$-$E_2$ conformational change.

In this work, we tested this and the underlying assumption in the Albers-Post model that a single conformational change mediates $Na^+$ transport by identifying the steps

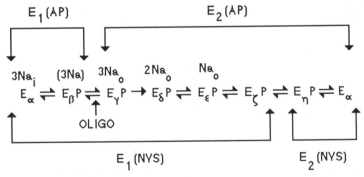

**Figure 1.** General model for Na+ transport. Each step in the reaction sequence is shown as a separate conformation (denoted by Greek subscripts). Conformations corresponding to $E_1$ and $E_2$ according to the Albers-Post (AP) scheme and the scheme proposed by Skou (NYS) are indicated. The release of the first Na+ is assumed to be irreversible under the conditions used here. Oligomycin inhibits Na+ deocclusion.

indicated by various fluorescent probes (BIPM, IAF, and RH421) and affected by various inhibitors (oligomycin and $Ca^{2+}$). To interpret our data, we have put forward a scheme for $Na^+$ transport that departs from the traditional concept of two conformations (Fig. 1): we propose that each reaction step involves a conformational change, and that the various indicators examined here indicate different transitions in the reaction scheme.

## Materials and Methods

Isolation of enzyme from the outer medulla of frozen dog kidneys and labeling with IAF have been described earlier [4]. BIPM-labelled enzyme was prepared as described by Taniguchi and coworkers [6]. RH421 enzyme was prepared by mixing IAF enzyme in the appropriate buffer with RH421 in ethanol, as described by Apell and coworkers [1]; the final RH421 concentration was 2 μM.

Where used, oligomycin in ethanol was added to the enzyme at 10 μg/ml when the enzyme suspension was prepared, and an equal volume of ethanol was added to the control enzyme.

## Results and Discussions

Figure 2 shows the rate constant (k) and magnitude (%ΔF) of ATP-induced change in the fluorescence of RH421-enzyme. The saturating value of k ($k_{max}$) is 100 s$^{-1}$, slightly higher than that seen for IAF (80 s$^{-1}$ [4]). Further, $K_{1/2}$ for $Na^+$ is comparable to that for the IAF fluorescence change. These results indicate that the rate-limiting step seen with RH421 is the same as that seen with IAF, viz., the release of the second and/or

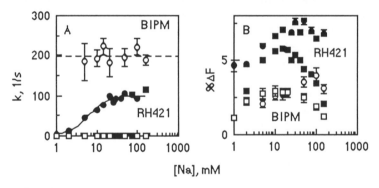

[Na], mM

**Figure 2.** k and %ΔF of RH421 (●,■)) and BIPM (○,□) fluorescence change as a function of [$Na^+$] in the absence (○,●) and presence (□,■) of oligomycin. Error bars: standard error of the mean. Panel A: solid line -- fit of k(–oligo,RH421) to the Hill equation ($k_{max}$ = 100 ± 2 s$^{-1}$, $K_{1/2}$ = 6.0 ± 0.3 mM, and $n_H$ = 1.70 ± 0.07); dashed line -- average k(–oligo,BIPM) (199 ± 8 s$^{-1}$). Panel B: %ΔF(–oligo) was fitted to the Hill equation: for RH421 (%ΔF$_{max}$ = 7.2 ± 0.1, $K_{1/2}$ = 1.0 ± 0.1 mM, and $n_H$ = 1.1 ± 0.2), and for BIPM (%ΔF$_{max}$ = 3.5 ± 0.4, $K_{1/2}$ = 3.5 ± 1.3, and $n_H$ = 1.0 ± 0.6). %ΔF(+oligo) was fitted to the Hill equation + a straight line: for RH421 (%ΔF$_{max}$ = 6.03 ± 0.03, $K_{1/2}$ = 2.02 ± 0.03 mM, $n_H$ = 2.1 ± 0.1, slope = -0.0261 ± 0.0004), and for BIPM (%ΔF$_{max}$ = 3.0 ± 0.2, $K_{1/2}$ = 1.3 ± 0.1, $n_H$ = 2.2 ± 0.7, slope = -0.0118 ± 0.0018).

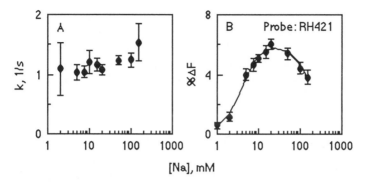

**Figure 3.** k (A) and %ΔF (B) of RH421-enzyme fluorescence change vs [Na+] with Mg$^{2+}$ replaced by Ca$^{2+}$. Panel B: Line is a fit to Hill equation + a straight line (%ΔF$_{max}$ = 6.4; K$_{1/2}$ = 4 mM; n$_H$ = 1.8).

third Na$^+$ [4].

In oligomycin-treated enzyme, k$_{max}$ is reduced by a factor of 100, while %ΔF$_{max}$ is reduced only by 20%. These observations indicate that the step reported by RH421 is separated from the step inhibited by oligomycin (Na$^+$-deocclusion) by an irreversible step. At higher [Na$^+$], %ΔF decreases, in accordance with the higher Na$^+$-affinity of oligomycin-treated enzyme [2].

Figure 2 also shows the Na$^+$-dependence of k and %ΔF for BIPM-labelled enzyme. k is independent of [Na$^+$] over the range examined, and is higher than k$_{max}$ for RH421 or IAF. Oligomycin reduces k to the same level as with RH421 and IAF. %ΔF increases with [Na$^+$], and %ΔF$_{max}$ is approximately the same with and without oligomycin. These results imply that: BIPM reports a step before the release of the second and/or third Na$^+$ and after Na$^+$ deocclusion, and is separated from the latter by an irreversible step. These observations imply that BIPM reports the release of the first Na$^+$.

Figure 3 shows k and %ΔF of RH421-enzyme fluorescence change when Mg$^{2+}$ is

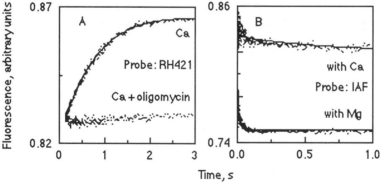

**Figure 4.** Effect of oligomycin on the stopped-flow responses of RH421-enzyme in the presence of Ca$^{2+}$ (A) and the effect of Ca$^{2+}$ on IAF-enzyme response (B). In panel A, solid line is a fit to a single exponential: k = 1.5 s$^{-1}$; %ΔF = 4.6. In panel B, solid lines represent fits of data (2 ms to 16 s) to a single exponential: for enzyme with Mg$^{2+}$ (lower trace) k = 82 s$^{-1}$, %ΔF = -5.9; for enzyme with Ca$^{2+}$, k = 1.5 s$^{-1}$, %ΔF = -2.8.

replaced by $Ca^{2+}$. k has been reduced by two orders of magnitude and does not vary with $[Na^+]$ over the range examined, while $\%\Delta F_{max}$ remains approximately the same as with $Mg^{2+}$. At higher $[Na^+]$, $\%\Delta F$ decreases, similar to that seen in enzyme treated with oligomycin. The decrease in that case was attributed to the increase in $Na^+$ affinity of the oligomycin-treated enzyme. In the presence of $Ca^{2+}$, oligomycin completely abolishes the RH421 response (Fig. 4A). These results implies that oligomycin and $Ca^{2+}$ inhibit the same step in the enzyme cycle, viz., $Na^+$ deocclusion.

For the IAF enzyme (Fig. 4B), $Ca^{2+}$ reduces both k and $\%\Delta F$ under conditions where $Ca^{2+}$ reduces only k for the RH421-enzyme. For this enzyme, k is consistent with that seen for RH421 in the presence of $Ca^{2+}$ (Fig. 4A, upper trace). This result implies that (a) in addition to inhibiting $Na^+$ deocclusion, $Ca^{2+}$ inhibits the step reported by IAF, and (b) this step occurs after the step reported by RH421.

Based on these results, the general model of $Na^+$ transport shown in Fig. 1 has been

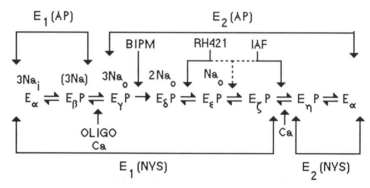

**Figure 5.** General model of $Na^+$ transport by $Na^+/K^+$-ATPase (as in Fig. 1) indicating steps reported by IAF, RH421 and BIPM, and steps inhibited by $Ca^{2+}$ and oligomycin.

expanded (Fig. 5) to show (a) the steps reported by IAF, RH421 and BIPM, and (b) the steps inhibited by $Ca^{2+}$. We conclude that a simple model with two conformations ($E_1$ and $E_2$) is insufficient to explain the results of transient kinetic experiments. Further, it is likely that such experiments can yield information regarding transitions between enzyme conformations involved in ion transport.

**References**

6. Bühler, R., Stüurmer, W., Apell, H.-J., and Läuger, P. (1991) J. Membr. Biol. 121, 141-161.
7. Esmann, M. and Skou, J.C. (1985) Biochem. Biophys. Res. Commun. 127, 857-863.
1. Nørby, J.G., Klodos, I. and Christiansen, N.O. (1983) J. Gen. Physiol. 82, 725-759.
4. Pratap, P.R., Robinson, J.D. and Steinberg, M.I. (1991) Biochim. Biophys. Acta, 1069, 288-298.
3. Skou, J.C. (1990) FEBS Lett. 268, 314-324.
5. Taniguchi, K., Suzuki, K., and Iida, S. (1982) J. Biol. Chem. 257, 10659-10667.
2. Yoda, S. and Yoda, A. (1986) J. Biol. Chem. 261, 1147-1152; Yoda, S. and Yoda, A. (1987) J. Biol. Chem. 262, 103-109; Yoda, A. and Yoda, S. (1986) J. Biol. Chem. 262, 110-115; Yoda, A. and Yoda, S. (1986) J. Biol. Chem. 263, 10320-10325.

# Reassignment of cation-induced population of main conformational states of FITC-Na$^+$/K$^+$-ATPase as detected by fluorescence spectroscopy and characterized by equilibrium binding studies

E. Grell, E. Lewitzki, H. Ruf, M. Doludda

Max-Planck-Institute for Biophysics, Kennedy Allee 70, D-60596 Frankfurt, Germany

## Introduction

To link proposed features of the Na$^+$/K$^+$-ATPase reaction cycle to the molecular structure, spectroscopic studies mainly concerning the elucidation of partial reactions are carried out. This can be done by introducing fluorescence labels such as the FITC-group (1) which enables the study of alkali ion binding as a key feature of this ion pump. According to earlier studies, a correlation between the fluorescence emission intensity and the position of the E$_1$/E$_2$ equilibrium is suggested: low fluorescence intensity in the presence of K$^+$ has been attributed to E$_2$, high intensity to E$_1$(1-5). Evidence will be presented that this correlation concerning E$_1$ is not fulfilled. Membrane-bound Na$^+$/K$^+$-ATPase has been prepared from pig kidney according to (6), the FITC-enzyme according to (1); details are given in (7).

## Assignment of FITC fluorescence intensity levels to main conformational states

**High fluorescence intensity** state: Essentially the same high fluorescence intensity state is reached in the presence of very high concentrations of Na$^+$ and also upon binding of a large number of cations, such as choline (Ch$^+$), guanidinium (G$^+$), tetramethylammonium (TMA$^+$), N-methylglucammonium (NMG$^+$), Li$^+$, buffer cations (e.g. imidazolium ImH$^+$, TrisH$^+$), organic divalent cations (e.g. protonated diamines like ethylene- (EnH$_2$$^{++}$), n-propane- (PH$_2$$^{++}$), n-butane- (BuH$_2$$^{++}$), n-pentane diamine (PeH$_2$$^{++}$) or the bis-guanidinium compounds m-xylene- (XBG$^{++}$) and n-pentane-bis-guanidinium (PBG$^{++}$) (originally introduced by (8)) and alkaline earth cations. As indicated earlier (5,7), in a given buffer system of constant pH all these monovalent cations exhibit essentially the same binding affinity (Fig. 1a) with a pK value around 2.3 (based on a 1:1 binding model). The originally reported course of the Na$^+$ titration curve (5) could not be confirmed by later measurements and is attributed to the enzyme preparation at that time. Accordingly, the binding of all divalent cations is also characterized approximately by a single pK value around 4.4 (Fig. 1b). Evidently, the binding affinity does not depend on size or the chemical nature of the cation; it mainly depends on cation charge. For this reason, this type of binding is denoted as non-selective cation binding. To explain these results, a model has been suggested (5,7) by assuming that these cations are preferentially bound electrostatically to a double negatively charged site probably on the enzyme's cytoplasmic side. As a consequence of charge neutralisation, non-selective binding must be expected to be sensitive to the electric field. The binding of choline and $Na^+$ is essentially pH independent provided the ionic strength of the buffer is kept constant over the whole pH range.

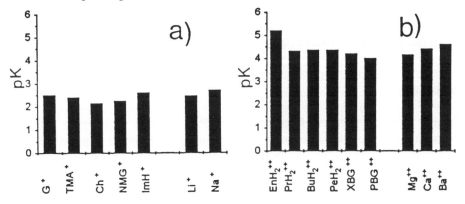

Figure 1. Non-selective cation binding to FITC-$Na^+$/$K^+$-ATPase, as determined in 25 mM histidine- or 10 mM imidazole-HCl pH 7.5 (25°C). Symbols are explained in the text.

Since these direct cation titrations on the FITC-enzyme leading to the high intensity fluorescence state provide only evidence for low affinity $Na^+$ binding and no evidence for $Na^+$ selectivity, the properties of this $Na^+$ complex indicated by the FITC-enzyme cannot fulfill the principal selectivity requirements for the $E_1$ state according to the Albers-Post model. For this reason, this high intensity fluorescence state, which is also populated by all other non-selectively bound cations, is denoted $F_1$ (7). This concept is supported by observations, that a selective, high affinity $Na^+$ binding site can be detected with RH 421 not only for the unmodified, but also for the FITC-enzyme (7,9) as well as from the conclusions of corresponding kinetic studies (10). According to these kinetic studies, non-selective cation binding is characterized by a two-step process where the main fluorescence intensity change can be attributed to the rate-limiting reaction step, which is due to a slow conformational rearrangement ($k_{23}$ in eq. 1 of (10)) subsequent to the initial binding process. Since the turn-over number of the unmodified enzyme under optimal conditions is more than $10^3$ times higher than the rate constant of this rearrangement process, this binding process cannot reflect $Na^+$ binding leading to $E_1$.

A further analysis of this concept is possible by carrying out spectrofluorometric competition titrations, for example of $K^+$, in the presence of different concentrations of choline chloride or NaCl. Fig. 2 shows the experimentally observed dependence of the $K^+$ affinity (expressed as the pK value related to the 1:1 binding model) as a function of increasing choline chloride and NaCl concentrations. The $K^+$ affinity decreases with increasing concentrations of both salts, which provides evidence for a competing choline site ($K^+$ and choline binding is assumed to be mutually exclusive) with a pK value around 2.2 for choline (Fig. 2a), which corresponds to the value obtained independently by direct titration (Fig. 1a). If the effect of $Na^+$ were the same as that found for choline, a similar pK value would be expected for this alkali ion (cf. broken line in Fig. 2b). However, in the presence of $Na^+$ a $K^+$ affinity decrease is found at much lower concentrations than for choline leading to a pK value for $Na^+$ of about 3.8 ($K^+$ and $Na^+$ binding is assumed to be mutually exclusive), which is consistent with the value obtained for $Na^+$ by direct titration in the presence of RH421 (7,9). These results clearly indicate that the FITC-enzyme is capable of adopting the predicted, selective , high affinity $E_1$-$Na^+$ state, but it can not display the population of this state by a fluorescence intensity change. Formation of $E_1$ can affect the thermodynamic parameters, but does not appreciably influence the fluorescence properties of the FITC-enzyme. Thus, this new assignment of the high intensity fluorescence state to $F_1$ is consistent with all currently available data.

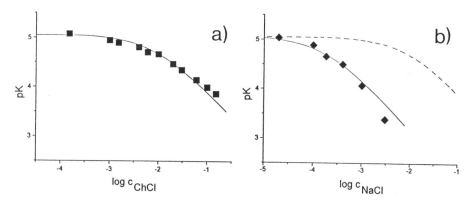

Figure 2. $K^+$ affinities determined at different concentrations of choline chloride (a) and NaCl (b) in 10 mM imidazole-HCl pH 7.5 (25°C): Plot of pK versus logarithm of corresponding salt concentrations. Based on 1:1 stoichiometry, the theoretical curves (solid lines) have been calculated for pK values of 5.1 for $K^+$, 2.2 for $Ch^+$ and 3.8 for $Na^+$ binding. Details are given in the text.

**Low intensity** fluorescence state: Spectrofluorometric titrations of the FITC-enzyme with $K^+$, $Rb^+$, $Cs^+$, $NH_4^+$ and $Tl^+$ lead to the low fluorescence intensity state which has been denoted $E_2$ (1,2,5,7). The affinities are characteristic of a size-selective site which excludes, for example, protonated primary amines, such as n-propylammonium. This selective type of alkali ion binding is typical for inner sphere coordination to a neutral cavity as has been found for the $K^+$-selective membrane carrier valinomycin (11). Thus the experimentally found properties are consistent with the expected ones according to the Albers/Post model implying that the low fluorescence intensity state of the $K^+$ complex indicates the formation of the occluded $E_2(K^+)_2$ state, which is also consistent with the interpretation of kinetic results (10).

## Acknowledgements

We thank Mrs. A. Schacht and Mr. G. Schimmack for expert technical assistance and the German Research Foundation (SFB 169) for support.

## References

1. Karlish SJD (1979) Cation induced conformational states of Na,K-ATPase studied with fluorescent probes. In: Skou JC, Norby JG (eds) Na,K-ATPase: structure and kinetics, Academic Press, New York, pp 115-128
2. Hegyvary C, Jørgensen PL (1981) Conformational changes of renal sodium plus potassium ion-transport adenosine triphosphatase labeled with fluorescein. J Biol Chem 256: 6296-6303
3. Post RL, Seiler SM, Vasallo PM (1987) Action of hydrogen ions on sodium, potassium adenosine triphosphatase labelled with fluorescein. In: Mukohata Y, Morales MF, Fleischer S (eds) Perspectives of biological energy transduction, Academic Press, Tokyo, pp 306s-326s
4. Mezele M, Lewitzki E, Ruf H, Grell E (1988) Cation selectivity of membrane proteins. Ber der Bunsenges 92: 998-1004
5. Grell E, Warmuth R, Lewitzki E, Ruf H (1991) Precision titrations to determine affinity and stoichiometry of alkali, alkaline earth, and buffer cation binding to Na,K-ATPase. In: Kaplan JH, de Weer P (eds) The Sodium Pump: Recent Developments, Rockefeller University Press, New York, pp. 441-445
6. Jørgensen PL (1974) Purification and characterization of $(Na^+ + K^+)$-ATPase III. Biochim Biophys Acta 336: 36-52
7. Grell E, Warmuth R, Lewitzki E, Ruf H (1992) Ionics and conformational transitions of Na,K-ATPase. Acta Physiol Scand 146: 213-221
8. David P, Mayan H, Cohen H, Thal DM, Karlish SJD (1992) Guanidinium Derivatives act as High Affinity Antagonists of Na ions in Occlusion sites of Na,K-ATPase. J Biol Chem 267: 1141-1149
9. Ruf H, Lewitzki E, Grell E (1994) Na,K-ATPase: Probing of a high affinity binding site for $Na^+$ by spectrofluorometry. This volume
10. Doludda M, Lewitzki E, Ruf H, Grell E (1994) Kinetics and mechanism of cation binding to Na,K-ATPase. This Volume
11. Grell E, Oberbäumer I (1977) Dynamic aspects of carrier mediated cation transport through membranes. In: Rigler R, Pecht I (eds) Chemical Relaxation in Molecular Biology, Springer, Heidelberg, pp 371-413

# Caged Na$^+$ and K$^+$ ligands: photochemical properties, application for membrane transport studies and selective fluorimetric detection of alkali ions

R. Warmuth[*], B. Gersch[*], F. Kastenholz[*], J.-M. Lehn[+], E. Bamberg[*], E. Grell[*]

[*]Max-Planck-Institut für Biophysik, Kennedyallee 70, 60596 Frankfurt, Germany
[+] Institut Le Bel, Université Louis Pasteur, 4, rue Blaise Pascal, 67000 Strasbourg, France

## Introduction

In many biological studies the generation of fast cation concentration jumps without mechanical disturbance of the biological system is of considerable interest. We have recently reported the synthesis of photo-cleavable cryptands **1** and **2** substituted with a 2-nitrophenyl group at one of the cryptand bridges as alkali ion selective ligands (2,3,5). These cryptands are suitable for binding alkali as well as alkaline earth cations (caged-cations) and releasing the complexed cation after a short pulse of UV light, due to the cleavage of the substituted cryptand bridge. This results in a drastical decrease of the stability of the cation/ligand complex (2,5). Cryptands **3** and **4** were synthesized as fast, selective and sensitive indicators for alkali cations in aqueous solution. Fluorescence and cation binding properties of these ligands are discussed.

1  n = 1  NC221  caged Na$^+$ ligand
2  n = 2  NC222  caged K$^+$ ligand

3  n = 1  F221  indicator for Na$^+$
4  n = 2  F222  indicator for K$^+$

## Cation complexation studies and photochemical properties

In aqueous solution the protonation of the cryptand bridge nitrogens affects cation complexation and thus a pH dependent apparent dissociation constant $K_{app}$ is observed. The protonation constants of the cryptands and the dissociation constants of the cryptates were determined by potentiometric titrations. As expected, the results obtained for compounds **1** and **2** differ only slightly from the values of the non substituted cryptands reported by Lehn and Sauvage (4). NC222 **2** shows peak selectivity for K$^+$ with a selectivity of approximately 40 over Na$^+$. The smaller NC221 **1** exhibits a peak selectivity for Na$^+$ in the alkali ion series (cf. Table 1).

Table 1. Protonation constants (pK$^D$) and cryptate dissociation constants (pK) of cryptands **1** and **2** in aqueous solution (T = 25 °C)

| Cryptand | pK$^D_1$ | pK$^D_2$ | pK | |
|---|---|---|---|---|
| | | | Na$^+$ | K$^+$ |
| NC221 **1** | 7.40 | 10.50 | 5.35 | 3.80 |
| NC222 **2** | 7.20 | 9.70 | 3.80 | 5.45 |

The quantum yields of photolysis were determined spectrometrically and by applying HPLC. The quantum yields of NC221 **1** and NC222 **2** and their Na$^+$ and K$^+$ cryptates are approximately 0.3 at 266 nm.

The rates of the UV-flash induced benzylether cleavage are pH-dependent; the rates are in the time range of milliseconds at pH 7 and in the 10 millisecond range at pH 12 for cryptands **1** and **2**. The rates of photolysis of the alkali ion cryptates of **1** and **2** are of the same order of magnitude as the rates of the corresponding free cryptands.

**Applications**

To demonstrate the suitability of photo-cleavable cryptands for biochemical studies the kinetics of cation binding to FITC-Na$^+$/K$^+$-ATPase was investigated by flash photolysis of a solution containing the K$^+$ cryptate of NC222 at pH 8.35 and 25 °C (3). The values obtained for the association rate (k$_f \approx 10^4$ M$^{-1}$ sec$^{-1}$) and dissociation rate (k$_d \approx 1$ sec$^{-1}$) using a one-step binding model (1:1 stoichiometry) are of similar magnitude as those obtained from stopped flow studies when evaluated according to the same model.

The applicability of photo-cleavable cryptands for cation transport studies was demonstrated by electrical measurements on black lipid membranes containing adsorbed Na$^+$/K$^+$-ATPase or the cation carrier valinomycin (Fig. 1 and 2). Electrogenic Na$^+$ transport by Na$^+$/K$^+$-ATPase could be demonstrated after photolysis of caged-ATP in one of the compartments (Fig. 1B) by electric current measurements across black lipid membranes containing absorbed enzyme in the same compartment (1) at pH 8.3. An identical transient current could be induced by photolysis of the Na$^+$ complex of NC221 **1** in the presence of ATP at the same pH (Fig. 1D). This result is considered direct proof that the transient signal is due to electrogenic Na$^+$ transport. A K$^+$ concentration jump induced by applying NC222 **2** leads to an increased membrane conductivity in the presence of valinomycin, as shown in Fig. 2.

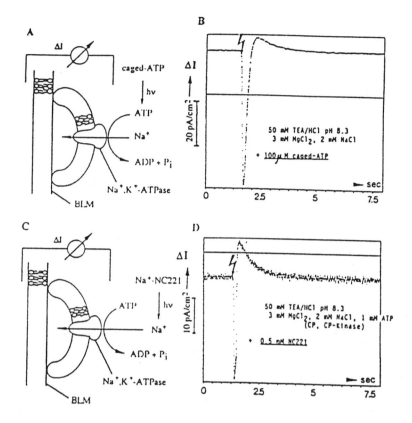

Figure 1. Schematic illustration of concentration jump experiments on Na$^+$/K$^+$-ATPase (purified from pig kidney) adsorbed to a black lipid membrane (BLM) consisting of diphytanoyllecithin with current detection performed at 25 °C by photolysis of caged-ATP (A) and the Na$^+$ cryptate of NC221 1 (C) in the presence of ATP, creatine phosphate (CP) and its kinase to maintain constant ATP concentration. Transient short circuit current (B,D) due to electrogenic Na$^+$ transport by the enzyme induced by UV-flash (start marked by arrow: 0.125 sec in B, 0.5 sec duration in D). TEA: triethanolamine.

## Fluorescent cryptands as alkali cation indicators in aqueous solution

Cryptands 3 and 4 with an trifluoromethylcoumarino residue annelated to one of the bridges are suitable as cation selective alkali ion indicators in aqueous solution. Upon excitation at 360 nm both cryptands exhibit a broad fluorescence emission band at 470 nm at pH 8.0. The Na$^+$-sensitive F221 3 displays a marked fluorescence intensity decrease, whereas the K$^+$-sensitive F222 4 shows an intensity increase of the fluorescence emission band upon complexation. The negative logarithms of the dissociation constants of the Na$^+$ and K$^+$ crpytates of F221 3 and F222 4 at pH 8.0 are approximately 3.5 as determined by spectrofluorometric titrations.

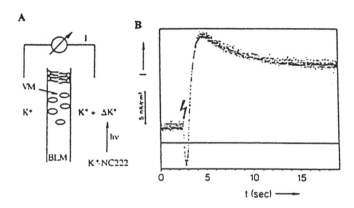

Figure 2. (A) Schematic illustration of a black lipid membrane (BLM) containing valinomycin (VM) and of the light induced $K^+$ concentration jump. (B) Time dependent current change of a black diphytanoyllecithin membrane in the presence of 0.7 μM valinomycin after a light induced $K^+$ concentration jump on one side of the membrane at a potential of 10 mV; one compartment contained 0.5 mM KCl and 0.5 mM NC222 2 in 100 mM triethanolamine/HCl pH 8.3, the other 0.5 mM KCl in the same buffer (25 °C). The start of irradiation (0.25 sec duration) is marked by the arrow.

## Acknowledgement

We thank Prof. G. Quinkert (Universität Frankfurt) for stimulating discussions, Mrs. A. Schacht and Mrs. A. Hüby for helpful assistance. This work was supported by the Deutsche Forschungs Gesellschaft (DFG, SFB 169). One of us (B. G.) thanks the Fonds der Chemischen Industrie for a Kekulé grant.

## References

1.  Fendler K, Grell E, Haubs M, Bamberg E (1985) Pump currents generated by the purified $Na^+K^+$-ATPase from kidney on black lipid membranes. EMBO J 4: 3079-3085
2.  Grell E, Warmuth R (1993) Caged cations. Pure Appl Chem 65: 373-379
3.  Grell E, Warmuth R, Lewitzki E, Ruf H (1992) Ionics and conformational transitions of Na,K-ATPase. Acta Physiol Scand 146: 213-221
4.  Lehn J-M, Sauvage J-P (1975) [2]-Cryptates: Stability and Selectivity of Alkali and Alkaline-Earth Macrobicyclic Complexes. J Am Chem Soc 97: 6700-6707
5.  Warmuth R, Grell E, Lehn J-M, Bats JW, Quinkert G (1991) 66. Photo-Cleavable Cryptands: Synthesis and Structure. Helv Chim Acta 74: 671-681

# Kinetic and spectroscopic characterisation of the cardiac glycoside binding site of Na$^+$/K$^+$-ATPase employing a dansylated ouabain derivative

E. Lewitzki, U. Frank, E. Götz, E. Grell

Max-Planck-Institute for Biophysics, Kennedy-Allee 70,
D-60596 Frankfurt, Germany

## Introduction

Fluorescence spectroscopy is a useful tool for studying equilibrium and dynamic properties related to structural aspects of biological systems. Dansylated compounds have been widely used as fluorescent probes to characterize protein structure. Their wide use is a result of a favourable lifetime (~10 nsec), promising spectral characteristics, including the high sensitivity of the emission spectrum of the dansyl moiety to solvent polarity, and suitability for energy transfer studies with tryptophan residues in proteins (2).
(N-(Dansyl)-N`(ouabain)-ethylenediamine (DEDO), a chemically stable, novel fluorescent derivative of ouabain (Fig.1) was used to characterize the cardiac glycoside binding site of Na$^+$/K$^+$-ATPase, purified from kidney tissue of different animal species (6). The hydroxy groups of the sugar moiety are no longer present after the chemical modification.

## DEDO binding to Na$^+$/K$^+$-ATPase

Spectrofluorometric equilibrium titrations of Na$^+$/K$^+$-ATPase with DEDO in the presence of Mg$^{++}$ and P$_i$ provide evidence for 1:1 complex formation between the enzyme and the inhibitor with an affinity of around 30 nM in the case of rabbit kidney enzyme.

The kinetics of DEDO complex formation (k$_{on}$) with the purified enzymes was initiated by rapid mixing in a medium containing either MgCl$_2$ and P$_i$ (complex II) or Mg$^{2+}$-ATP and NaCl (complex I) and monitored spectrofluorimetrically. Dissociation rate constants (k$_{off}$) were determined by adding a 1000 fold excess of unlabeled ouabain to the pre-formed DEDO complex. Under these experimental conditions the dissociation of the fluorescent DEDO is rate limiting. DEDO dissociation could always be characterized by a single exponential function. If k$_{on}$ and k$_{off}$ are known, K can be calculated according to K= k$_{off}$/k$_{on}$.
As reported earlier (4, 9) the measured rate of formation of complex I and II depends only

slightly on the tissue origin of the enzyme (dog, pig, sheep, rabbit). The determined rate constants of about $3 \times 10^4$ $M^{-1}sec^{-1}$ for DEDO binding are in good agreement with those reported for the unmodified inhibitor ouabain (1,3,6). On the other hand, the rate constant of DEDO dissociation is about 10 times higher ($1 \times 10^{-3}$ $sec^{-1}$) compared with that of ouabain. An exception is found for the rabbit kidney enzyme, where the values of the dissociation rate constants of the DEDO as well as of the ouabain complex are similar; complex I ($6.5 \times 10^{-4}$ $sec^{-1}$) and complex II ($3.5 \times 10^{-4}$ $sec^{-1}$).

In the case of DEDO complex formation in the presence of $Mg^{++}$ only (denoted here as complex III), the formation rate constant is 100 times lower compared with that for complex I + II, whereas the DEDO dissociation constant is comparable to that of complex II.

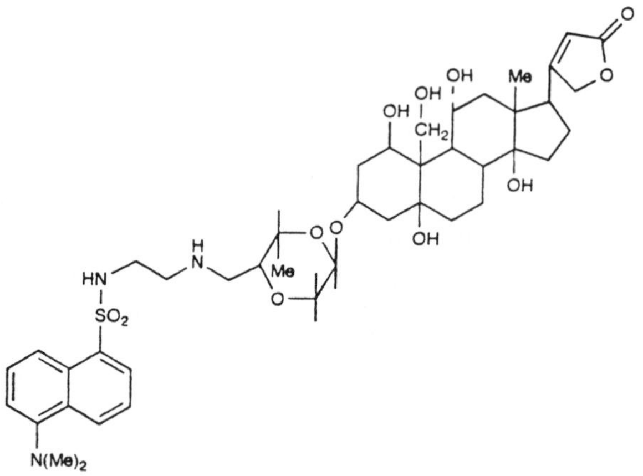

**Figure 1.** Structure of DEDO

**Spectroscopic properties**

Upon binding of DEDO to $Na^+/K^+$ATPase, isolated from rabbit kidney, a large blue shift (~60 nm) and a 8 times higher fluorescence intensity is observed (Fig.2). This is indicative of a very hydrophobic binding microenvironment (3-10 Debye) of the dansyl group. The enzymes isolated from the other species exhibit only blue shifts between 10 -20 nm and an 1.5 to 2 fold increase of the maximum emission intensity.

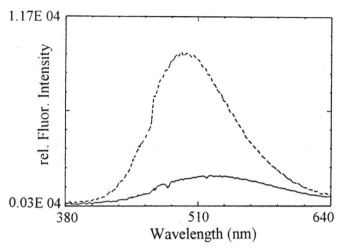

**Figure 2.** Fluorescence emission spectra of free DEDO (——) and in the presence of 0.3 µM Na$^+$/K$^+$-ATPase (- - - ), isolated from rabbit kidney, in 25 mM imidazole/HCl pH 7.5 containing 3 mM MgCl$_2$, 3 mM Tris-P$_i$, 25°C.

In addition, fluorescence emission intensity of the tryptophan residues (Trp) of Na$^+$/K$^+$-ATPase is quenched by 20 - 30 % upon binding of DEDO (Fig.3), which is indicative of energy transfer. Similar observations were reported also for anthroylouabain (7). Differences in energy transfer efficiency are found between complex I and II. Except for the sheep enzyme, the energy transfer efficiency of complex I always appears to be larger than that of complex II.

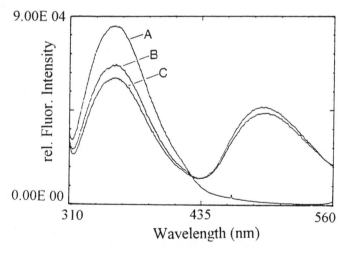

**Figure 3.** Emission spectra indicating energy transfer from Trp in rabbit kidney Na$^+$/K$^+$-ATPase (0.3 µM) to DEDO in 25 mM imidazole/HCl pH 7.5 (25°C, $\lambda$ exc =295 nm) :
A.) Emission spectra of rabbit kidney Na$^+$/K$^+$-ATPase without DEDO,
B.) of complex II ; 3 mM Tris-P$_i$, 3 mM MgCl$_2$, 0.6 µM DEDO;
C.) of complex I  ; 100 mM NaCl, 0.5 mM MgCl$_2$, 0.5 mM Tris-ATP, 0.6 µM DEDO

Estimations according to Förster`s theory (5) for the interaction of DEDO with a single Trp residue would indicate a distance of about 26 Å in the case of the sheep enzyme. Since both sequences of this enzyme (8) contain 16 Trp residues ( 4 are localized in the β subunit, 2 within putative transmembrane segments of the α chain), it is not possible to determine such distances for multible Trp sites without any detailed structural features. A model consideration implies that if all Trp residues were equally separated by 50 Å from the dansyl group the calculated transfer efficiency would lead to a decrease of the Trp - emission intensity of only about 9 %. Transfer efficiencies of 20-30 % suggest that one or two Trp residues are closely  located to the dansyl group of the bound DEDO.

The observed results suggests that DEDO acts as a very sensitive probe molecule for the investigation of correlations of structural aspects associated with specific inhibitor binding to $Na^+/K^+$-ATPase. Several of these aspects could be investigated in a detailed way by employing site-directed mutagenesis on a preparative scale.

*Acknowledgements*

We thank Prof. W. J. Richter for analytical support, Mrs. A. Schacht and Mr. G. Schimmack for their help and expert technical assistance and the Deutsche Forschungsgemeinschaft (SFB 169) for support.

**References**

1.  Askari A, Kakar SS, Huang Wh (1988) Ligand binding sites of the ouabain-complexed Na,K-ATPase. J Bio Chem 263: 235-242
2.  Chen RF (1967) Fluorescence of dansyl amino acids in organic solvents and protein solutions. Arch of Biochem and Biophys 120: 609-620
3.  De Pover A (1990) Dissociation kinetics of ouabain Na,K-ATPase isoform complexes. J of Gen Physiol 96: 80a
4.  Erdmann E, Schoner W (1973) Ouabain-receptor interactions in Na,K-ATPase preparations from different tissues and species. Determination of kinetic constants and dissociation constants. Biochim Biophys Acta 307: 386-398
5.  Förster Th (1948) Intermolecular energy migration and fluorescence. Ann Phys (Leipzig) 2: 55-75
6.  Grell E, Lewitzki E, Ifftner A (1985) Dynamics and mechanism of cardiac glycoside binding to Na,K-ATPase. In: Glynn I, Ellory C (eds) The sodium pump.The Company of Biologists, Cambridge, pp 289-294
7.  Moczydlowski EG, Fortes PAG (1980) Kinetics of cardiac glycoside binding to Na,K-ATPase studied with a fluorescent derivative of ouabain. Biochem 19: 969-977
8.  Shull GE, Schwartz A, Lingrel JB (1985) Amino acid sequence of the catalytic subunit of the Na,K-ATP-ase deduced from a complementary DNA. Nature 316: 691-695
9.  Wallick ET, Pitts BJR, Lane LK, Schwartz A (1980) A kinetic comparsion of cardiac glycoside interactions with Na,K-ATPase from skeletal and cardiac muscle and from kidney. Arch of Biochem and Biophys 202: 442-449

# Kinetics and mechanism of cation-binding to $Na^+,K^+$-ATPase

by M. Doludda, E. Lewitzki, H. Ruf, E. Grell

Max-Planck-Institute of Biophysics, Kennedyallee 70, D-60596 Frankfurt, Germany

## Introduction

Most mechanistic aspects of cation binding and transport by $Na^+,K^+$-ATPase are still unknown. The kinetics of cation binding with the FITC-enzyme can be studied using the stopped flow technique (1,3,4) taking advantage of the large fluorescence intensity changes caused by these reactions (1,2). In contrast to equilibrium titration experiments, kinetic studies can provide evidence for the existence of intermediate states.

Two types of cation binding (4,5,6) are distinguished: non-selective binding, occurring probably on the cytoplasmic side, by buffer cations, choline, $Na^+$ (low affinity site), divalent organic or alkaline earth cations, etc., and selective binding of $Na^+$ (high affinity site) or $K^+$ (and congeners), occurring with highest affinity on the extracellular side. Non-selective binding depends essentially on the cation charge (2) and is indicated by a fluorescence intensity increase; whereas selective binding depends on the size of the preferred alkali ion and is indicated by a decrease in the fluorescence emission of the FITC-enzyme. Binding of $K^+$ is cooperative and has been shown to follow 1:2 stoichiometry (2 $K^+$ per enzyme unit) (2,4,7).

All experiments were performed with the purified enzyme isolated from pig kidney (8). The kinetic studies were carried out with a SF.17MV stopped-flow set-up from Applied Photophysics equipped with fluorescence detection.

## Non-selective binding of cations to $Na^+,K^+$-ATPase

A representative stopped-flow trace for the non-selective binding of a monovalent cation (choline) is shown in Figure 1.A. Besides a fast, non-resolvable amplitude change, a single, resolvable kinetic phase is observed that is described by a single exponential function. The dependence of the reciprocal time constant as a function of cation concentration exhibits saturation behaviour (Figure 1.B, 2.A) for all investigated cations. This suggests that at least a two-step process (based on 1:1 stoichiometry) exists:

$$E + C^{n+} \underset{k_{21}}{\overset{k_{12}}{\rightleftharpoons}} E\bullet C^{n+} \underset{k_{32}}{\overset{k_{23}}{\rightleftharpoons}} F_1\text{-}C^{n+} \qquad (1)$$

where $K_{12}=k_{21}/k_{12}$; $K_{23}=k_{32}/k_{23}$; $K=K_{12}K_{23}$. Here E is assumed to represent the conformational state $E_1$ whereas the high fluorescence state of the FITC enzyme is denoted $F_1$ (4,5). If the binding step in eq. (1) is fast compared to the isomerisation step and if $[C^{n+}]_0 \gg [E]_0$,

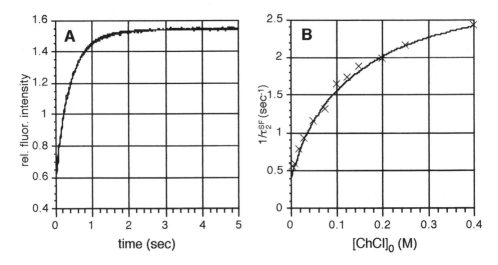

**Figure 1.** Kinetics of choline binding to FITC-Na$^+$,K$^+$-ATPase in 25 mM histidine-HCl, pH 7.5, containing 0.4 M choline chloride, 0.1 mM EDTA and 0.1 mM dithiothreitol (25°C); $\lambda_{ex}$ = 466nm; $\lambda_{em}$ > 495nm. **A.** Stopped flow trace of a typical experiment. The solid line represents an monoexponential fit with a decay time of 0.4 sec$^{-1}$. **B.** Plot of $1/\tau_2^{SF}$ versus total choline chloride concentration. Theoretical curve is fitted according eq. (3); results are shown in Table 1.

where the index stands for total concentrations, we expect a biphasic reaction, characterized by the time $\tau_1^{SF}$ and $\tau_2^{SF}$ according to (2) and (3):

$$\frac{1}{\tau_1^{SF}} = k_{12}[C^{n+}]_0 + k_{21} \qquad (2) \qquad\qquad \frac{1}{\tau_2^{SF}} = \frac{k_{23}[C^{n+}]_0}{K_{12} + [C^{n+}]_0} + k_{32} \qquad (3)$$

The fast binding process (eq. 1) cannot be temporally resolved by the stopped-flow technique. Its existence is indicated solely by the presence of a fast, initial amplitude change. The reciprocal decay time of the slower process $1/\tau_2^{SF}$ shows saturation behaviour (Figure 1.B), as would be expected for the reaction mechanism predicted above (eq. 3). The parameters obtained according to eq. (3) are given in Table 1 for the cations investigated.

The experimentally observed kinetic process is assigned to an equilibration step between the intermediate state E•C$^{n+}$ and the final complex F$_1$-C$^{n+}$ coupled to a faster electrostatic binding process. Since all determined $k_{23}$ values are of the same magnitude, it is assumed that the equilibration step in eq. (1) represents a slow conformational change of the enzyme, induced by the non-selective cation binding process. Good correspondence is found between the equilibrium constants K determined by spectrofluorimetric equilibrium titrations and those calculated from the parameters obtained from kinetic experiments (Table 1).

**Table 1.** Kinetic parameters of non-selective cation binding to FITC-Na$^+$,K$^+$-ATPase by fitting eq. (3) to a plot of $1/\tau^{SF}$ versus $[C^{n+}]_0$ as determined in 25 mM histidine-HCl, pH 7.5, containing 0.1 mM dithiothreitol, 0.1 mM EDTA and the indicated cations in form the of their chloride salts (25°C).

| Cation[a] | $K_{12}$ (M) | $k_{23}$(sec$^{-1}$) | $k_{32}$(sec$^{-1}$) | $K_{23}=\dfrac{k_{32}}{k_{23}}$ | pK$_{kin}$[b] | pK$_{tit}$[c] |
|---|---|---|---|---|---|---|
| Ch$^+$ | 0.13 | 3.0 | 0.4 | $1.3\times10^{-1}$ | 1.8 | 2.2 |
| Gua$^+$ | 0.08 | 0.6 | 0.1 | $1.6\times10^{-1}$ | 1.9 | 2.3 |
| Na$^+$ | 0.11 | 0.8 | 0.2 | $1.8\times10^{-1}$ | 1.7 | 1.9 |
| BuH$_2^{2+}$ | $1.3\times10^{-3}$ | 9.3 | 0.6 | $6.5\times10^{-2}$ | 4.1 | 4.2 |
| BGP$^{2+}$ | $4.5\times10^{-5}$ | 0.3 | 0.2 | $6.6\times10^{-1}$ | 4.6 | 4.9 |

[a] Ch$^+$ choline; Gua$^+$ guanidinium; BuH$_2^{2+}$ protonated 1,4 diamino butane; BGP$^{2+}$ 1,5-bis-guanidinium-n-pentane
[b] calculated from kinetic data: $K = K_{12} K_{23}$
[c] Dissociation constants obtained from spectrofluorimetric equilibrum titrations (2)

### Selective binding of cations to the Na,K-ATPase

The selective binding of cations (eg. M$^+$=K$^+$) to the FITC-enzyme can be resolved by employing the stopped-flow technique (3,7). The experimental data observed in 10 mM imidazole-HCl pH 7.5, containing 0.1 M choline chloride are characterized by a single exponential ($\tau^{SF}$). The dependence of $1/\tau^{SF}$ as a function of the total KCl concentration exhibits sigmoidal behaviour (Figure 2.B), as reported by (7), and can be described by a kinetic model that predicts two fast, non-resolvable K$^+$ binding steps preceding the slow conformational rearrangement of the enzyme, which leads to the occluded state E$_2$(M$^+$)$_2$:

$$2M^+ + E \underset{}{\overset{K_1}{\rightleftharpoons}} M^+ + E\text{-}M^+ \underset{}{\overset{K_2}{\rightleftharpoons}} E\text{-}M^+M^+ \underset{k_2}{\overset{k_1}{\rightleftharpoons}} E_2(M^+)_2 \qquad (4)$$

E is assumed to represent the state E$_1$. In addition, if $[M^+]_0 \gg [E]_0$, the parameters can be obtained by fitting eq. (5) to the data (Figure 2.B):

$$\frac{1}{\tau_2^{SF}} = k_1 \frac{[M^+]_0^2}{K_1K_2 + K_2[M^+]_0 + [M^+]_0^2} + k_2 \qquad (5)$$

leading to $K_1 = 2 \times 10^{-4}$ M, $K_2 = 7 \times 10^{-4}$ M, $k_1 = 90$ sec$^{-1}$, $k_2 = 2$ sec$^{-1}$. These parameters are consistent with those reported by (7) and confirm earlier findings concerning K$^+$ binding stoichiometry (2,4).

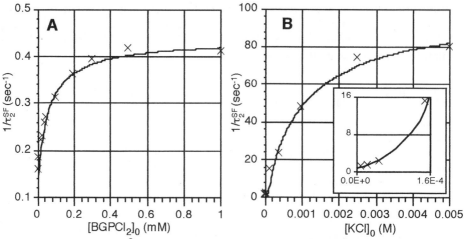

**Figure 2. A.** Kinetics of $BGP^{2+}$ binding to FITC-Na,K-ATPase ($4 \times 10^{-7}$ M) in 25 mM histidine-HCl pH 7.5 containing 0.1 mM EDTA and 0.1 mM dithiothreitol (25°C). Plot of $1/\tau^{SF}$ versus total $BGP^{2+}$ concentration. The solid line represents a fit according to eq. (2); results are shown in Table 1. **B.** Kinetics of $K^+$ binding to FITC-Na$^+$,K$^+$-ATPase ($4 \times 10^{-7}$ M) in 10 mM imidazole-HCl pH 7.5 containing 100 mM choline chloride, 0.1 mM EDTA and 0.1 mM dithiothreitol (25°C). Plot of $1/\tau^{SF}$ versus total KCl concentration. The solid line represents a fit according eq. (3), results are given in the text. The insert represents the results obtained at low KCl concentrations.

*Acknowledgements*

We thank Prof. G. Ilgenfritz for helpful discussions, Mrs. A. Schacht for her help and expert technical assistance and the Deutsche Forschungsgemeinschaft (SFB 169) for support.

**References**

1. Karlish, SJD (1979) Cation induced conformational states of Na,K-ATPase studied with fluorescent probes. In: Skou JC, Nørby JG (eds) Na,K-ATPase: Structure and Kinetics. Academic Press, London pp 115-128

2. Grell E, Warmuth R, Lewitzki E, Ruf H (1991) Precision titrations to determine affinity and stoichiometry of alkali, alkaline earth and buffer cation binding to Na,K-ATPase. In: Kaplan JH, Weer P (eds) The Sodium Pump: Recent Developments. The Rockefeller University Press, New York pp 441-445.

3. Faller LD, Diaz RA, Scheiner-Bobis G, Farley R (1991) Temperature dependence of the rates of conformational changes reported by fluorescein 5'-isothiocyanate modification on $H^+$,$K^+$- and Na$^+$,K$^+$-ATPases. Biochem 30:3503-3510.

4. Grell E, Warmuth R, Lewitzki E, Ruf H (1992) Ionics and conformational transitions of Na,K-ATPase. Acta Physiol Scand 146: 213-221.

5. Grell E, Lewitzki E, Ruf H, Doludda M, Brand K, Schneider FW, Zachariasse K, Reassignment of cation induced population of main conformational states of FITC-Na,K-ATPase as detected in fluorescence spectroscopy and characterized by eqilibrium binding studies. Published in this edition.

6. Ruf H, Lewitzki E, Grell E, Na,K-ATPase: Probing of a high affinity binding site for Na$^+$ by spectrofluorometry. Published in this edition.

7. Faller LD, Smirnova IN (1993) Mechanism of $K^+$ interaction with fluorescein 5'-isothiocyanate-modified Na,K-ATPase. J Bio Chem 268/22: 16120-16123.

8. Jørgensen PL (1974) Purification and characterization of Na,K-ATPase III. Purification from the outer medulla of mammalian kidney after selective removal of membrane components by sodium dodecylsulphate. Biochim Biophys Acta 336: 36-52.

# Comparison of $E_1$ and $E_2$ Conformers of Membrane Bound Na$^+$/K$^+$-ATPase Using Time-Resolved Phosphorescence Anisotropy Measurements

A. Boldyrev, O. Lopina, A. Rubtsov, D. McStay, L. Yang and P. Quinn

Moscow State University, 119899, Moscow, Russia and King's College London, London W8 7AH, UK

**Introduction**

Na$^+$/K$^+$-ATPase is known to form oligomeric complexes in the membrane bilayer. On the other hand, there are strong evidence that monomeric form possesses all of the main hydrolytic and transport properties of the Na$^+$-pump. This raises the question as to whether formation of oligomeric complexes is a the necessary step in the biochemical processes associated with the translocation of cations across the membrane or simply an architectural feature of the intercalation of the proteins into the phospholipid bilayer. We have addressed this question by measuring the rotational relaxation of the ATPase in different conformational states of the pump cycle using flash-photolysis methods.

**Experimental procedure**

In experiments duck salt gland microsomes have been used with specific activity of Na$^+$/K$^+$-ATPase of 2-2.5 units (37°C). Microsomes have been labelled with 5'-eosin isothiocyanate which resulted in attachment of 0.7-0.9 moles eosin per α-subunit and 25-30% inhibition of Na$^+$/K$^+$-ATPase (3). The decay of laser flash induced phosphorescence anisotropy was analyzed according to a double exponential process. The rapidly rotating component has been ascribed to rotation of the monomeric form of the enzyme about its axis perpendicular to the plane of the membrane and the slowly rotating species to that of oligomeric forms of Na$^+$/K$^+$-ATPase.

**Rotational motion of the enzyme in the presence of Na$^+$ and K$^+$**

The effect of monovalent ions on rotational parameters of monomeric and oligomeric enzyme forms is clearly distinct. Whereas fast rotational motion of the enzyme at different temperatures is not altered by the presence of monovalent cations, the slow rotational component is increased in the presence of K$^+$ and decreased in the presence of Na$^+$, which is especially clear at low temperature. It is noteworthy that at physiological temperatures the sizes of the different oligomeric forms, represented by slower rotational correlation times, in the presence of sodium or potassium were indistinguishable.

Study of temperature effects on rotational correlation time of the labelled enzyme (Fig.1) demonstrated a linear dependence on temperature, showing increased rotational motion as the temperature increases. The associated weighting functions which represent the contribution of the respective motion to the overall anisotropy signal show a decrease in the amount of monomers with subsequent increase in oligomers of the enzyme with a raise in temperature. One possible interpretation of this data is that monomers tend to associate into oligomeric forms of the enzyme with increasing temperature. Another explanation could be the dissociation of very large aggregates into smaller oligomers.

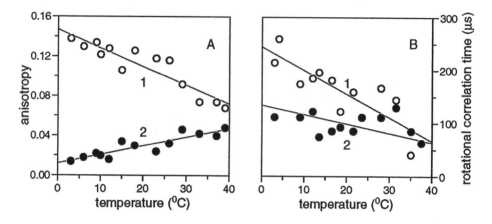

Figure 1. Anisotropy values (A) for fast (1) and slow (2) rotational components of labelled $Na^+/K^+$-ATPase and rotation correlation time (B) for the slow component in the presence of $K^+$ (1) or $Na^+$ (2).

### pH effect

Rotational motion of the fast rotating species is similar in both $E_2$ and $E_1$ forms of the enzyme and does not change significantly with pH. The rotational correlation times of the slower rotating species, however, showed the opposite tend as a function of pH. Thus the size of the rotating species in $K^+$ containing media decreases with increase in pH whereas the size of the rotating species in $Na^+$ containing media increases under these conditions. At pH 7.5, optimal for $Na^+/K^+$-ATPase the sizes of the rotating species in both conformers are similar.

The rate of rotational motion is inversely related to the radius of the rotating cylinder and the viscosity of the lipid. Estimation of the viscosity of the lipid matrix of membrane preparations using fluorescence probe techniques and assuming contact of the rotating protein throughout the thickness of a lipid bilayer have enabled us to estimate the size of the rotating proteins. The results of these calculations provide a value of 2.2-2.4 nm for the radius of the fast rotating species which is consistent with rotation of monomers of the enzyme. The slow rotating species of protein presumably consist of oligomeric complexes the size of which can again be calculated from the above parameters (Table

1). Accordingly, the size of these complexes formed in $K^+$ containing media decreased and that in $Na^+$ containing media increased with increase in pH. As expected from the similarity of $t_0$ values for the potassium and sodium forms of the enzyme at physiological pH values, the apparent radius of rotation of complexes is close to 9 nm. It is not possible to provide an accurate estimation of the number of monomers in these complexes because changes in the orientation of the probe on the protein induced by interaction with other membrane proteins.

Table 1. Apparent radius of rotation (nm) of $Na^+/K^+$-ATPase calculated from rotational motion of fast and slow components of the $E_1$ and $E_2$ conformers at 20 °C (calculated according to membrane microviscosity measurements, $\rho$, being equal to 7.8 Poise).

| pH | $E_2$-form | | $E_1$-form | |
|---|---|---|---|---|
| | fast | slow | fast | slow |
| 6.0 | 2.4 | 13.2 | 2.2 | 6.9 |
| 6.5 | 2.5 | 12.2 | 2.4 | 7.9 |
| 7.0 | 2.4 | 11.0 | 2.4 | 7.7 |
| 7.5 | 2.4 | 9.8 | 2.3 | 8.4 |
| 8.0 | 2.4 | 10.3 | 2.4 | 9.4 |
| 8.5 | 2.3 | 5.9 | 2.4 | 9.7 |

**ATP effect**

In order to isolate the effects of changes in conformation that take place during the pumping cycle the relative proportion of monomers present was determined in the presence of different combination of ligands including different conformations of the enzyme. These include $E_1Na$, $E_1NaATP$, $E_2K$ and $E_2KATP$. The result of the calculations derived from these studies is that interaction between protomers undergoes a change during the pumping cycle. This leads to association - dissociation reactions that are illustrated in the descriptive terms of a dance in Fig. 2. The number of partners at each stage in the cycle corresponds to the best estimates from experimental evidence available from published data (1, 2).

**Conclusions**

The transient formation of oligomeric complexes of $Na^+/K^+$-ATPase is supported by the existence of functional aggregates of different size distribution when the enzyme is induced to form one or other conformer. Such studies will assist in our understanding of the mechanism of ion translocation which, on the basis of the evidence currently available, appears to involve transient associations between individual protomers during the pumping cycle.

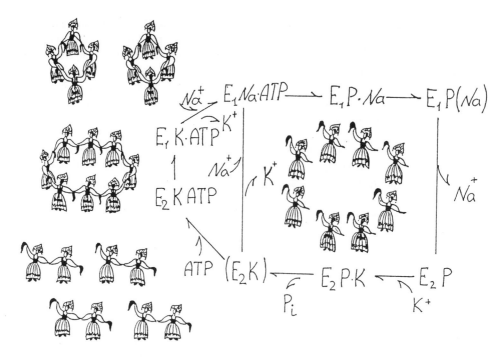

Figure 2. Schematic representation of the $Na^+/K^+$-ATPase operation with indications of size of oligomeric ensembles at different steps of the cycle.

**References**

1.  Boldyrev A, Fedosova N, Lopina O (1991) The mechanism of the modifying effect of ATP on $Na^+/K^+$-ATPase. Biomed Sci 2: 450-454
2.  Mimura K, Matsui H, Takagi T, Hayashi Y (1993) Change in oligomeric structure of solubilized $Na^+/K^+$-ATPase induced by octaethylene glycol dodecyl ether, phosphatidylserine and ATP. Biochim Biophys Acta 1145: 63-74
3.  Yang L, Lopina O, Rubtsov A, McStay D, Boldyrev A, Quinn P (1993) Rotational dynamics of membrane-bound $Na^+/K^+$-ATPase revealed by the time-resolved phosphorescence anisotropy. Biochem J (in press)

# Application of Time-Resolved Phosphorescence Anisotropy Measurement for the Study of Rotational Mobility of Eosin-Labelled $Na^+/K^+$-ATPase and $Ca^{2+}$-ATPase

A.M. Rubtsov, O.D. Lopina, [*]L. Yang, [*]D. McStay, A.A. Boldyrev, [**]P.J. Quinn

Department of Biochemistry, School of Biology, Moscow State University, 119899 Moscow, Russia, [*]School of Applied Sciences, The Robert Gordon University, Aberdeen AB1 1HG, Scotland and [**]Biochemistry Section, Division of Life Sciences, King's College London, London W8 7AH, UK

## General approaches

To study the rotational mobility of $E_1$-$E_2$-type ATPases under different conditions the membrane preparations of $Na^+/K^+$-ATPase from duck salt glands and $Ca^{2+}$-ATPase from rabbit skeletal muscle sarcoplasmic reticulum were specifically labelled at single lysine residue in or near putative ATP-binding site by eosin-5'-isothiocyanate (EITC) (1,4). Samples were excited into the triplet state by a short laser pulse from a frequency-doubled Nd:YAG laser. The resulting time-resolved phosphorescence depolarization was then measured by a "T"-format phosphorimeter and anisotropy was calculated and analyzed using a non-linear exponential function curve fitting method (Marquardt procedure) (5). Incorporation of EITC into $Ca^{2+}$-ATPase protein provides for a complete inhibition of enzyme hydrolytic activity when level of dye incorporation reaches 1 mole per mole $Ca^{2+}$-ATPase; incorporation of the same amount of EITC into $Na^+/K^+$-ATPase preparations provides only 30% inhibition of enzyme activity (data not shown).

### Rotational mobility of ion-transport ATPases

The initial anisotropy value of EITC-labelled $Ca^{2+}$-ATPase preparations is equal to 0.11, whereas the initial anisotropy value of EITC-labelled $Na^+/K^+$-ATPase is much higher and reaches 0.27 that probably reflects some differences in the orientation of dye molecule(s) bound to these proteins. Phosphorescence

637

anisotropy decay data obtained from both enzymes are satisfactory described by a sum of two exponents with residual constant term. The fast and slow rotating components have a rotational correlation times in a range of 10-20 $\mu s$ and 100-400 $\mu s$ respectively.

Study of the influence of different agents modifying the lipid bilayer viscosity (diethyl ether and nonionic detergent $C_{12}E_9$) on the relevant weighting of the fast and slow rotating components and residual anisotropy shows that the decrease of overall membrane viscosity provides for a significant decrease of residual anisotropy and correspondent increase of weighting functions for fast and slow rotating components (Tables 1 and 2). An addition of high glycerol concentrations and short-term heating of sarcoplasmic reticulum preparations induce protein aggregation in the membranes (1,2); after such treatment residual anisotropy is increased

Table 1. Influence of glycerol and diethyl ether on normalized anisotropy (A) and rotational correlation time ($\phi$) values of EITC-labelled $Ca^{2+}$-ATPase (pH 7.0, EITC:$Ca^{2+}$-ATPase ratio 1:1)

| conditions | $A_1$ | $\phi_1$ | $A_2$ | $\phi_2$ | A |
|---|---|---|---|---|---|
| control | 0.460 | 13.0 | 0.265 | 182.9 | 0.285 |
| 20% glycerol | 0.462 | 14.3 | 0.206 | 205.6 | 0.331 |
| 40% glycerol | 0.398 | 14.9 | 0.201 | 230.3 | 0.401 |
| 60% glycerol | 0.323 | 18.6 | 0.183 | 314.1 | 0.496 |
| 2% diethyl ether | 0.531 | 14.0 | 0.266 | 162.0 | 0.202 |
| 4% diethyl ether | 0.539 | 14.2 | 0.269 | 210.0 | 0.192 |
| 6% diethyl ether | 0.530 | 14.1 | 0.333 | 220.0 | 0.139 |

drastically and weighting functions of fast and slow rotating components are decreased (Tables 1, 2, and 3). In addition protein aggregation in the membranes increases significantly the rotational correlation time of slow rotating component. These results allow to suggest that fast rotating component

Table 2. Influence of glycerol and nonionic detergent $C_{12}E_9$ on normalized anisotropy (A) and rotational correlation time ($\phi$) values of EITC-labelled $Na^+/K^+$-ATPase (pH 7.4, EITC:$Na^+/K^+$-ATPase ratio 0.9:1)

| conditions | $A_1$ | $\phi_1$ | $A_2$ | $\phi_2$ | A |
|---|---|---|---|---|---|
| control | 0.666 | 13.8 | 0.171 | 123.2 | 0.163 |
| 25% glycerol | 0.680 | 11.4 | 0.170 | 134.8 | 0.150 |
| 50% glycerol | 0.546 | 14.6 | 0.156 | 188.3 | 0.298 |
| 2% $C_{12}E_9$ | 0.686 | 9.4 | 0.314 | 67.1 | 0.000 |

638

of phosphorescence anisotropy decay signal belongs to a monomeric form of enzymes, the slow rotating component - to the oligomeric complexes of $Ca^{2+}$-ATPase and $Na^+/K^+$-ATPase, and residual anisotropy reflects the presence in membranes of large protein aggregates.

Table 3. Influence of short-term heating on normalized anisotropy (A) and rotational correlation time ($\phi$) values of EITC-labelled Ca2+-ATPase in the presence of different $Ca^{2+}$-ATPase ligands (pH 6.0, EITC:$Ca^{2+}$-ATPase ratio 1:1)

| conditions | $A_1$ | $\phi_1$ | $A_2$ | $\phi_2$ | A |
|---|---|---|---|---|---|
| EGTA ($E_2$) | 0.398 | 13.4 | 0.218 | 167.8 | 0.385 |
| + heating | 0.397 | 13.2 | 0.202 | 213.2 | 0.401 |
| $Ca^{2+}$ ($E_1$) | 0.482 | 13.5 | 0.301 | 141.8 | 0.217 |
| + heating | 0.478 | 12.7 | 0.260 | 303.7 | 0.262 |
| $Mg^{2+}$ ($E_1$) | 0.493 | 14.0 | 0.274 | 136.7 | 0.233 |
| + heating | 0.475 | 15.3 | 0.216 | 186.4 | 0.309 |

Sarcoplasmic reticulum $Ca^{2+}$-ATPase protein was transformed into either of its putative conformeric state, designated $E_1$ and $E_2$, by the addition of appropriate ligands to the medium (3). In all cases the rotational correlation time of slow rotating component of phosphorescence from molecules in the $E_2$ conformation registered a lager value than that of molecules in $E_1$ conformation (Tables 3 and 4). The residual anisotropy of EITC-labelled $Ca^{2+}$-ATPase in $E_2$ conformation was also relatively higher than that of enzyme in $E_1$ conformation. Therefore, transition of $Ca^{2+}$-ATPase from $E_1$ into $E_2$ conformation is probably connected with protein oligomerization and/or aggregation. Moreover, the molecular size of oligomeric complexes of $Ca^{2+}$-ATPase in $E_2$ conformation seems to be relatively higher than that in E1 conformation.

Table 4. Influence of $Ca^{2+}$-ATPase conformational state on normalized anisotropy (A) and rotational correlation time ($\phi$) values of EITC-labelled $Ca^{2+}$-ATPase in the presence of different $Ca^{2+}$-ATPase ligands (pH 6.0, EITC:$Ca^{2+}$-ATPase ratio 0.5:1)

| conditions | $A_1$ | $\phi_1$ | $A_2$ | $\phi_2$ | A |
|---|---|---|---|---|---|
| EGTA ($E_2$) | 0.376 | 16.7 | 0.277 | 174.1 | 0.314 |
| EGTA + $P_i$ ($E_2$) | 0.478 | 15.4 | 0.234 | 284.6 | 0.288 |
| $Ca^{2+}$ ($E_1$) | 0.483 | 13.0 | 0.310 | 140.1 | 0.207 |
| ATP + $Ca^{2+}$ ($E_1$) | 0.484 | 13.6 | 0.329 | 147.8 | 0.188 |

The overall results indicate that phosphorescence depolarization could be a useful method for the study of the conformational transitions and oligomeric state of $Na^+/K^+$-ATPase and $Ca^{2+}$-ATPase.

## References

1.  Birmachu W, Thomas DD (1990) Rotational dinamics of the Ca-ATPase in sarcoplasmic reticulum studied by time-resolved phosphorescence anisotropy. Biochemistry 29: 3904-3914
2.  Geimonen ER, Rubtsov AM, Boldyrev AA (1993) Mechanism of thermal uncoupling of sarcoplasmic reticulum Ca-pump: change in functional activity and structural state of the enzyme molecule. Biokhimiya 58: 1318-1328
3.  Jorgensen PL, Andersen JP (1988) Structural basis for E1-E2 conformational transition in Na,K-pump and Ca-pump proteins. J Membr Biol 103: 95-120
4.  Yang L, Lopina OD, McStay D, Rogers AJ, Quinn PJ (1993) Phosphorescence depolarization and the rotational mobility of Na+,K+-ATPase in crude microsomal membranes. Optical Engineering 32(2): 347-353
5.  Yang L, McStay D, Sharma A, Rogers AJ, Quinn PJ (1992) A novel dual-channel time domain phosphorimeter for time-resolved spectroscopic study of biomolecules. Proc SPIE (Biomed Optics) 1640: 513-519

# Pyridoxal-5'-phosphate Probe at Lys 480 can Monitor Conformational Events Induced by Acetyl Phosphate in Na+/K+-ATPase

S.Kaya, T.Tsuda, K.Hagiwara, A.Shimada, K.Taniguchi, T.Fukui*

Department of Chemistry, Faculty of Science, Hokkaido University, Sapporo 060, Japan. *The Institute of Scientific and Industrial Research, Osaka University, Ibaraki, Osaka 567, Japan.

## Introduction

The hydrolysis of ATP by Na+/K+-ATPase is accompanied by conformational changes of the enzyme. Several conformational states of the enzyme have been characterized with intrinsic and extrinsic fluorescence probes (2-4,8-12). Pyridoxal-5'-phosphate (PLP) or its analogue, adenosine diphosphopyridoxal (AP2PL) have been used to probe the ultrastructure of catalytic sites of nucleotide binding enzymes. Modification of the ion transport ATPase with these reagents have been shown to inactivate the enzyme activity (1,4,6,14). However there appears to be no reported evidence of these probes sensing the conformational events of transport ATPase. We have prepared specifically modified Na+/K+-ATPase and monitored conformational events accompanying the formation of phosphoenzyme by AcP.

## Modification of Na$^+$/K$^+$-ATPase by PLP and AP$_2$PL

When Na+/K+-ATPase was modified with PLP or AP2PL in the presence of 25 mM Na+ (Fig. 1A) or 30 mM K+ (Fig. 1B), the ATPase activity decreased to around 30 % with only a slight decrease in K+-pNPPase activity. The data show that NaE1 has a

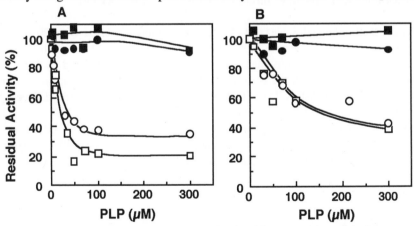

**Figure 1**. Effect of PLP treatment on Na+/K+-ATPase activity and the amount of phosphoenzyme. The Na+/K+-ATPase was treated with various concentrations of PLP in the presence of 25 mM NaCl (A) or 30 mM KCl (B). ATPase (O), K+-pNPPase (●), phosphoenzyme from ATP (□), and from AcP (■) were assayed as described (9-11).

higher affinity to those probes than KE2 as is already known for the higher affinity of NaE1 to ATP (7). The decrease in ATPase activity is attributed to a reduction in the formation of phosphoenzyme from ATP. However the enzyme can still be phosphorylated from AcP as was the case of FITC modification (11, 12). The inhibition of ATPase activity with PLP or AP2PL was suppressed in the presence of ATP {Fig. 2). ADP had a similar effect and AMP and pNPP had a slightly protective effect but neither AcP nor p-nitrophenylphosphate had any effect. These results suggest that these PLP or AP2PL probes were bound to the residue in or near the nucleotide-binding region.

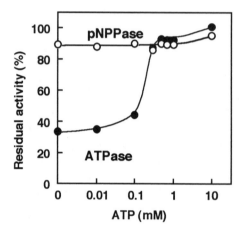

**Figure** 2. Protection of Na+/K+-ATPase by ATP from modification with PLP. Na+/K+-ATPase was treated with 100 $\mu$M PLP in the presence of various concentrations of ATP. The ATPase activity (●) and K+-pNPPase activity (O) were assayed.

## Identification of a residue modified with PLP and AP2PL

To identify the binding sites, Na+/K+-ATPase preparations modified with 20 $\mu$M PLP or AP2PL were solubilized with SDS, and the $\alpha$-chain was isolated. After extensive digestion of $\alpha$-chains by trypsin, the fluorescence peptide were separated by HPLC. The amino acid sequence of both the main peptides were shown to have the same common sequence Asn-Ser-Thr-Asn-X-Tyr. From a comparison of the sequence deduced from cDNA (6), X was shown to correspond to Lys-480 of the $\alpha$-chain in Na+/K+-ATPase. The site is the same as the binding site protected by the ATP on PLP or AP2PL in the presence of both Na+ and K+ (1). Our data indicate that the Lys-480 is not directly involved in the catalytic reaction as has been reported (13).

## Fluorescence change induced by acetyl phosphate

It has been shown that the addition of AcP to FITC-treated enzyme preparations reduce the FITC fluorescence intensity. Fig. 3B shows that AcP induced a single exponential fluorescence increase in PLP-enzyme in the presence of 16 mM $Na^+$ and 4 mM $Mg^{2+}$ where the enzyme preparations may accumulate E2P (9-11). Addition of AcP to the PLP-enzyme in the presence of 2 M $Na^+$ to accumulate E1P, seemed to show a biphasic fluorescence change, namely a slight increase preceding a decrease. The opposite fluorescence change induced by AcP in different concentrations of $Na^+$ reflects the accumulation of conformationally different phosphoenzymes E1P and E2P. On the other hand, ATP also induced a single exponential fluorescence increase in both $Na^+$ concentrations (data not shown). This suggests the formation of a Mg-Na-ATP-Enzyme complex, because the PLP treatment reduced the amount of phosphoenzyme from ATP as shown in Fig. 1 and the fluorescence increase by ATP required the presence of both $Na^+$ and $Mg^{2+}$. These data suggest that the introduction of the PLP probes to Lys-480 inhibits the transphosphorylation reaction to form E1P from ATP without affecting that from AcP, and that these probes can sense the binding and formation of phosphoenzyme in $Na^+/K^+$-ATPase. The investigation presented here indicates the usefulness of PLP probe analogues and [$^{32}$P]AcP to investigate the mechanisms of energy transduction in $Na^+/K^+$-ATPase and other P-type ATPases as well as the mechanism of other enzyme affinity labeled with these analogues.

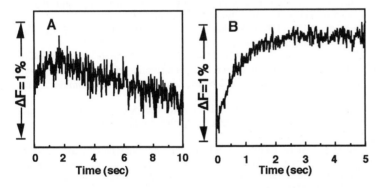

**Figure 3**. Acetyl phosphate induced fluorescence change in the PLP-labeled enzyme. Fluorescence change induced by acetyl phosphate was analyzed by the stopped flow technique. The experimental procedures are described elsewhere (12). The labeled enzyme was mixed with 10 mM acetyl phosphate-tris in the presence of 2 M NaCl (A) or 16 mM NaCl (B). The data shown is the ratio of accumulated data in the presence of the acetyl phosphate to that in the presence of the same concentration of tris-HCl.

*Acknowledgements*

This work was supported in part by The Naito Foundation, The Akiyama Foundation, The Nagase Science and Technology foundation, The Uehara Memorial Foundation, a Grant-in-Aid for Science Research on Priority areas of Bioenegy and Grants-in-Aid Scientific Research (03558026, 03454543).

## References

1.  Hinz HR , Kirley TL (1990) Lysine 480 is an essential residue in the putative ATP site of lamb kidney (Na,K)-ATPase J Biol Chem 265:10260-10265

2.  Hegyvary C, Jørgensen PL (1981) Conformational change of renal sodium plus potassium ion-transport adenosine triphosphatase labeled with fluorescein J Biol Chem 256:6296-6303

3.  Karlish SJD, Yates DW (1978) Tryptphan fluorescence of (Na$^+$, K$^+$)-ATPase as a tool for study of the enzyme mechanism Biochim Biophys Acta 527:115-130

4.  Karlish SDJ (1980) Characterization of conformational changes in (Na,K)ATPase labeled with fluorescein at the active site J. Bioenerg Biomembr 12: 111-136

5.  Maeda M, Tagaya M, Futai M (1988) Modification of gastric (H$^+$+K$^+$)-ATPase with pyridoxal 5'-phosphate J Biol Chem 263:3652-3656

6.  Ovchinikov YA, Modyanov NN, Broude NE, Petrukhin KE, Grishin AV, Arzamazova NM, Aldanova NA, Monastyrskaya GS, Sverdlov ED (1986) Pig kidney Na$^+$,K$^+$-ATPase FEBS Lett 201:237-245

7.  Skou JC (1982) The effect of pH, of ATP and of modification with pyridoxal 5-phosphate on the conformational transition between the Na$^+$-form and the K$^+$-form of the (Na$^+$+K$^+$)-ATPase Biochim Biophys Acta 688:369-380

8.  Steinberg M, Karlish SJD (1989) Studies on conformational changes in Na,K-ATPase with 5-iodoacetoamidofluorescein J Biol Chem 264:2726-2734

9.  Taniguchi K, Suzuki K, Iida S (1982) Conformational change accompanying transition of ADP-sensitive phosphoenzyme to potassium-sensitive phosphoenzyme of (Na$^+$,K$^+$)-ATPase modified with N-[p-(2-benzimidazolyl)phenyl]maleimide J Biol Chem 257:10659-10667

10. Taniguchi K, Suzuki K, Kai D, Matsuoka I, Tomita K, Iida S (1984) Conformational change of sodium- and potassium-dependent adenosine triphosphatase J Biol Chem 259:15228-15233

11. Taniguchi K,Tosa H,Suzuki K, Kamo Y (1988) Microenvironment of two different extrinsic fluorescence probes in Na$^+$, K$^+$-ATPase changes out of phase during sequential appearance of reaction intermediates J Biol Chem. 263:12943-12947.

12. Taniguchi K, Mårdh S (1993) Reversible change in the fluorescence energy transfer accompanying formation of reaction intermediates in probe-labeled (Na$^+$,K$^+$)-ATPase J Biol Chem 268:15588/15594

13. Wang K, Farley R A (1992) Lysine 480 is not essential residue for ATP binding or hydrolysis by Na,K-ATPase J Biol Chem 267:3577-3580

14. Yamamoto H, Imamura Y, Tagaya M, Fukui T, Kawakita M (1989) Ca$^{2+}$-dependent conformational change of the ATP-binding site of Ca$^{2+}$-transporting ATPase of sarcoplasmic reticulum as revealed by an alteration of the target site specificity of adenosine triphosphopyridoxal J Biochem (Tokyo) 106:1121-1125

# Phosphorylation of Na$^+$/K$^+$-ATPase by p-Nitrophenylphosphate and Other Phosphatase Substrates

A.Yamazaki, S.Kaya, Y.Araki, A.Shimada, and K. Taniguchi

Department of Chemistry, Faculty of Science, Hokkaido University, Sapporo 060, Japan.

## Introduction

Na$^+$/K$^+$-ATPase can hydrolyze some phosphatase substrates such as acetyl phosphate (AcP), p-nitrophenylphosphate (pNPP), and carbamylphosphate (CarbP) in the presence of Mg$^{2+}$ and K$^+$. When K$^+$ is replaced with Na$^+$, AcP is known to induce conformational changes and accumulate E$_1$P and E$_2$P in the presence of 2 M and 16 mM Na$^+$, respectively (1,5,7,8). It was shown that some nucleotides and K$^+$-phosphatase substrates can support some transport reactions (3). However there seems to be little evidence that pNPP or other phosphatase substrates induce conformational change and/or accumulate phosphoenzymes. To investigate this, the enzyme treated with N-(p-(2-benzimidazolyl)phenyl)maleimide (BIPM) (6) or fluoroscein 5'-isothiocyanate (FITC) (4-8) were used to follow conformational changes. To measure the amount of phosphoenzyme, [$^{32}$P]pNPP was synthesized.

### Conformational change induced by phosphatase substrates

As reported elsewhere, the addition of AcP to probe-labeled Na$^+$/K$^+$-ATPase induces different fluorescence changes accompanying the formation of E$_1$P and E$_2$P (5-8). Addition of AcP, CarbP, but not phenyl phosphate to BIPM-labeled enzyme induced a similar fluorescence decrease and increase, respectively, in the presence of 2 M and 16 mM Na$^+$ as observed by the addition of ATP (Fig. 1). Similar results were obtained when FITC-labeled enzymes were used. These results indicate that CarbP also phosphorylates the enzyme as AcP. Addition of pNPP to FITC-labeled enzyme in the presence of 16 mM Na$^+$ caused a decrease of fluorescence intensity like other phosphatase substrates, suggesting the formation of E$_2$P. The K$_{0.5}$ of pNPP for the fluorescence change was about 640 μM and is much higher than that of AcP (60 μM) or CarbP (100 μM). In the presence of 2 M Na$^+$ to accumulate E$_1$P, the fluorescence change induced by pNPP was smaller than that observed in the presence of 16 mM Na$^+$.

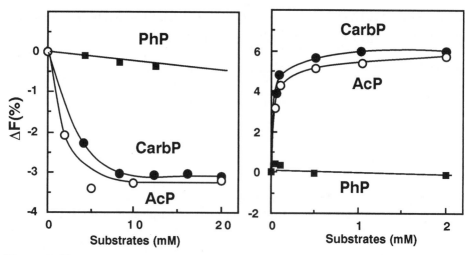

**Figure 1.** Fluorescence changes of the BIPM-labeled Na$^+$/K$^+$-ATPase induced by phosphatase substrates   Fluorescence changes of the BIPM-labeled enzyme was determined  in the presence of 2 M (left) and 16 mM (right) NaCl.

## Phosphorylation of Na$^+$/K$^+$-ATPase by [$^{32}$P]pNPP

To investigate the Na$^+$-dependent phosphorylation by pNPP, [$^{32}$P] pNPP was synthesized (2). The purity of [$^{32}$P]pNPP was determined by paperchromatography (Fig. 2). The specific activity was about 90 mCi/mmol and the content of inorganic phosphate

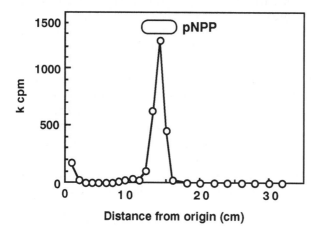

**Figure 2**. Analysis of [$^{32}$P]pNPP by paperchromatography.  [$^{32}$P]pNPP  was applied on paperchromatography in the solvent,1 M ammonium acetate (pH 3.8):EtOH,=2:5.

646

in the substrate was 10 %. The extent of phosphorylation formation in the presence of 16 mM $Na^+$ is similar to the decrease of fluorescence intensity of FITC-labeled enzyme (Fig.3). The $K_{0.5}$ of pNPP was about 1.5 mM in the presence of 16 mM NaCl. The enzyme is also phosphorylated by pNPP in the presence of 2 M $Na^+$ but the apparent affinity to pNPP was more than 10-fold lower than that observed in the presence of 16 mM NaCl.

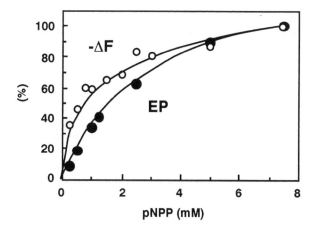

**Figure 3**. Formation of phosphoenzyme from [32P]pNPP and fluorescence changes of FITC-labeled enzyme induced by pNPP. Assay of phosphoenzyme was performed as described (7). Phosphorylation reaction was performed in a reaction mixture containing 25 mM imidazole-HCl, (pH 7.4) 4 mM $MgCl_2$, 25 mM sucrose, 16 mM NaCl, 0.1 mM EDTA-tris and various concentration of pNPP.

## Sensitivity of ADP-sensitive phosphoenzyme (E1P) to p-nitrophenol

The ADP-sensitive phosphoenzyme (E1P) can accept ADP to form ATP. We investigated the sensitivity of E1P to p-nitrophenol (pNP) and the formation of pNPP from E1P and pNP. The $Na^+/K^+$-ATPase was first phosphorylated by [32P]ATP in the presence of 2 M $Na^+$ to accumulate E1P, then various concentrations of pNP was added. Parts of the reaction mixtures were withdrawn and applied to paperchromatography and the phosphoenzyme remaining was also measured. [32P]pNPP was detected on paperchromatography and the amount of pNPP formed is dependent on the concentration of pNP added and is similar to the decrease in the amount of phosphoenzyme. About 50 % of phosphoenzyme was dephosphorylated by 1.5 M pNP within 10 seconds. These results indicate that E1P is dephosphorylated by pNP to form pNPP. Thus, it is concluded $Na^+/K^+$-ATPase accepts pNPP to accumulate E1P and E2P accompanying conformational changes.

647

*Acknowledgements*

This work was supported in part by The Naito Foundation, The Akiyama Foundation, The Nagase Science and Technology foundation, The Uehara Memorial Foundation, a Grant-in-Aid for Science Research on Priority areas of Bioenegy and Grants-in-Aid Scientific Research (03558026, 03454543).

## References

1       Campos M, Berberián G, Beaugé L (1988) Some total and partial reactions of $Na^+/K^+$-ATPase using ATP and acetyl phosphate as a substrate Biochim Biophys Acta 938:7-16

2.      Guan K, Dixon JE (1991) Evidence for protein-tyrosine-phosphatase catalysis proceeding via a cystein-phosphate internediate J Biol Chem 266:17026-17030

3.      Guerra M, Robinson JD, Steinberg M (1990) Differntial effects of substrates on three transport modes of the $Na^+/K^+$-/ATPase Biochim Biophys Acta 1023:73-80

4.      Hegyvary C, Jørgensen PL (1981) Conformational change of renal sodium plus potassium ion-transport adenosine triphosphatase labeled with fluorescein J Biol Chem 256:6296-6303

5.      Rephaeli A, Richards DE, Karlish SDJ (1986) Electrical potential accelerates the E1P(Na)-E2P conformational transition of (Na,K)-ATPase in reconstituted vesicles J Biol Chem 261:12437-12440

6       Taniguchi K, Suzuki K, Iida S (1982) Conformational change accompanying transition of ADP-sensitive phosphoenzyme to potassium-sensitive phosphoenzyme of $(Na^+,K^+)$-ATPase modified with N-[p-(2-benzimidazolyl)phenyl]maleimide J Biol Chem 257:10659-10667

7.      Taniguchi K,Tosa H,Suzuki K, Kamo Y (1988) Microenvironment of two different extrinsic fluorescence probes in $Na^+$, $K^+$-ATPase changes out of phase during sequential appearance of reaction intermediates J Biol Chem. 263:12943-12947.

8.      Taniguchi K, Mårdh S (1993) Reversible change in the fluorescence energy transfer accompanying formation of reaction intermediates in probe-labeled $(Na^+,K^+)$-ATPase J Biol Chem268:15588-15594

# Estimation of Distance Changes Between Cys-964 and Lys-501 in Na$^+$/K$^+$-ATPase Intermediates and the Orientation of Transmembrane Segment Containing Cys-964

H. Tosa*, A. Sawa, K. Nakamura, N. Takizawa, S. Kaya, K. Taniguchi , T. Araiso**, T. Koyama** and P.A.G.Fortes***

Department of Chemistry, Faculty of Science, School of Dentistry*, Research Institute for Electronic Science** Hokkaido University, Sapporo, Japan and Department of Biology***, University of California, San Diego,La Jolla,CA, USA

## Introduction

Na$^+$/K$^+$-ATPase (EC.3.6.1.3) from pig kidney was modified with N-(p-(2-benzimidazolyl)phenyl)- maleimide (BIPM) and fluorescein 5'-isothiocyanate (FITC), which react primarily with Cys-964 and Lys-501 (11,14). The resulting (BIPM-FITC) enzymes showed little Na$^+$/K$^+$-ATPase activity with retention of nearly 90% of phosphorylation capacity from acetyl phosphate: phosphoenzymes from acetyl phosphate in the presence of 2 M and 16 mM NaCl have been already shown to be conformationally E$_1$P (15) and E$_2$P (16), which showed sensitivity to acetate and K$^+$, respectively. Recently, it was clearly shown that BIPM-FITC enzymes showed reversible changes in fluorescence energy transfer from the BIPM to the FITC probe accompanying the formation of reaction intermediates (16). The transition of NaE$_1$ to E$_1$P occurred with an increase in energy transfer. The transition of E$_1$P to E$_2$P was accompanied also by an increase. The transition of E$_2$P to KE$_2$ occurred with a decrease in energy transfer. The transition of KE$_2$ to NaE$_1$ showed the maximum decrease. The reversible change in fluorescence energy transfer may be a reflection of the change in the distance between the probes due to movement of both Na$^+$ and K$^+$ (16). In this paper, we have estimated the extent of the change in the distance between the probes during the sequential appearance of reaction intermediates of the enzyme.

## Results and Discussion

Comparison of the emission intensity of the BIPM enzymes and the BIPM-FITC enzymes in NaE$_1$ indicated a decrease in the intensity of the BIPM fluorescence (donor quenching), and an increase in the intensity of the FITC fluorescence (sensitized emission). Thus, the energy transfer efficiencies (E) were determined by measuring the decrease in the BIPM (donor) fluorescence and the increase in the FITC (acceptor) fluorescence, under incubation conditions that stabilized different forms of the enzyme such as E$_1$P, E$_2$P, KE$_2$ and ouabainE$_2$P (15,16). Estimates of the efficiencies of energy transfer from lifetime measurements were unsuccessful because they were the same within experimental error, around 1 ns, irrespective of the intermediates accumulated. The quantum yield of BIPM enzyme (Q) and the overlap integrals (J) between the emission spectrum of the BIPM probe and the absorption spectrum of the   FITC probe were determined also. The apparent distance ($R$) between BIPM and FITC probes ranged from 30 to 42Å in different preparations, assuming a random orientation factor of 2/3. Table 1 shows data of typical experiments in which the values of E were obtained from the increase in the FITC fluorescence. The anisotropies of the BIPM probe were around 0.29  irrespective of the enzyme forms accumulated (10). The minimum and maximum values of the orientation factor were calculated to be 0.15 and 3.0 from the anisotropies of the probes (4). Thus, the actual distance is within 0.78xR to 1.28xR of that shown in Table 1, calculated with an orientation factor of 2/3. Relative apparent distance changes of the different enzyme forms in each enzyme preparation were within the experimental error (100±2%) of the measurements. The data suggest that the continuous increase in fluorescence energy transfer detected accompanying formation of $E_2P$ from $NaE_1$ via $E_1P$ and the decrease accompanying formation of $NaE_1$ from $E_2P$ via $KE_2$ reflected rather small distance changes by less than 1Å.

## Table 1  Estimation of distance between BIPM-FITC

|  | E | Q | J | R |
|---|---|---|---|---|
|  |  |  | $\times 10^{-15}\ cm^3\ M^{-1}$ | Å |
| NaE1 | 0.088±0.028 | 0.129±0.032 | 13.6 | 36.2±4.1 |
| E1P | 0.089±0.012 | 0.118±0.027 | 13.6 | 35.6±2.3 |
| E2P | 0.074±0.032 | 0.123±0.005 | 13.6 | 37.1±4.2 |
| KE2 | 0.097±0.023 | 0.123±0.030 | 13.6 | 35.3±3.2 |
| Ouabain-E2P | 0.101±0.014 | 0.131±0.020 | 13.6 | 35.4±1.9 |

Data of distances between various probes in $Na^+/K^+$-ATPase are reported (1,3,5,7,8,9,13). Some of these data with the data of ouabain binding (2,12) are summarized in Fig.1. The distance between the BIPM probe at Cys-964 and the FITC probe at Lys-501 is around 36Å (the theoretical minimum and maximum limits of the distance are 29 and 47Å, respectively ), which is the mean of several experiments using three different enzyme preparations. The membrane thickness is around 40Å and the distance between anthroylouabain (A-O) and FITC is reported to be around 72Å (1,3,5,13). The ouabain binding domain is supposed to be in the hydrophobic transmembrane or perimembrane region (2,12) and the FITC binding domain to be in the cytoplasmic region (11,14), which suggest that the ATP binding sites must be at least 30Å away from the cytoplasmic face (1,3,5,13). A model of the transmembrane peptide containing BIPM probe at Cys-964 (10,14) suggests that the fluorophore  is very close to the membrane face (- 5Å). These considerations support an idea that Cys-964 would be near the cytoplasmic face (around 6Å=36-30Å) rather than the exoplasmic face. A definitive answer for the arrangement of transmembrane segments of the enzyme in the plane of the membrane have not been obtained yet. Our data coincides with a model (10) that the transmembrane segment containing Cys-964 begins from Ile-946 at the extracellular and ends to Leu-971 at the cytoplasmic face. Quite recently, experimental evidences were reported to suggest that the alpha chain has 8 transmembrane segments (6) in which the segment containing Cys-964 is the same arrangement to our model.

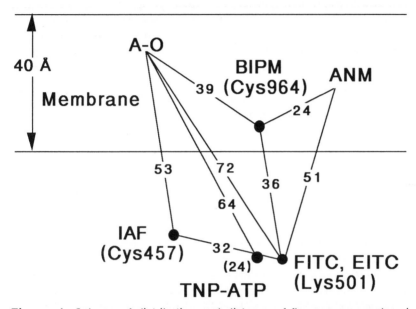

**Figure 1.** Scheme of distribution and distance of fluorescence probes in the alpha chain in Na/K-ATPase. The binding sites of BIPM, FITC and 5-iodoacetamidefluorescein (IAF) were determined already (11).The site of erythrosin isothiocyanate (EITC) seems to be the same to that of FITC (1). The site of N-(1-anilinonaphtyl-4)maleimide (ANM) is under investigation in our laboratory. Distances shown except between BIPM probe to FITC probe are from published data; A-O to FITC (1,3,5,13), BIPM to ANM (13), A-O to BIPM (13), ANM to FITC (13), A-O to IAF (1), IAF to EITC (1), A-O to IAF (5), IAF to 2',3'-o-(2,4,6-trinitrocyclohexadienylidine (TNP-ATP) (5). The binding site of A-O was taken to be the same to that of ouabain. Recently, independent experimental evidences were reported to suggest the possibility that the cardiac glycoside binding site may be intramembrane or perimembrane rather than extracellular (2,12).

*Acknowledgements*

This work was supported in part by The Naito Foundation, The Akiyama Foundation, Nagase Science and Technology foundation, The Uehara Memorial Foundation, a Grant-in-Aid for Science Research on Priority areas of Bioenegy, Grants-in-Aid Scientific Research (03558026, 03454543) and NIH Grant GM 47165.

# References

1.  Amler E, Abbott A, Ball W J Jr (1993)Structural dynamics and oligomeric interactions of Na$^+$/K$^+$- ATPase as modified using fluorescence energy transfer. Biopys J 61:553-568

2.  Arystarkhova E, Gasparian M, Modyanov N N, Sweadner K (1992) Na,K-ATPase extracellular surface probed with a monoclonal antibody that enhances ouabain binding. J Biol Chem 267:13694-13701

3.  Carilli C T, Farley R A, Perlman D M,Cantley L C (1982) The active site structure of Na- and K-stimulated ATPase. J Biol Chem 257:5601-5606

4.  Dale R E, Eisinger J, Blumberg W E (1979) The orientational freedom of molecular probes:The orientation factor in intramolecular energy transfer. Biophys J 26:161-194

5.  Fortes G P A, Aguilar R (1988) Distance between 5-iodoacetamide fluorescein and the ATP and ouabain sites of Na$^+$/K$^+$- ATPase determined by fluorescence energy transfer. Prog Clin Biol Res 268A:197-204

6.  Goldshleger R, Tal D M, Shainskaya A, Hoving S, Or E, Karlish S J D (1993) Organisation of the cation occlusion domain of the Na$^+$/K$^+$- ATPase. in this volume

7.  Jesaitis A J, Fortes P A G (1980) Fluorescence studies of sodium and potassium adenosine triphosphatase labeled with fluorescein mercuric acetate and anthroylouabain. J Biol Chem 255:459-467

8.  Lee J A, Fortes P A G (1986) Spatial relationship and conformational changes between the cardiac glycoside site and beta-subunit oligosaccarides in (Na,K)-ATPase. Biochemistry 25:8133-8141

9.  Moczydlowski E G, Fortes P A G (1981) Characterization of 2',3'-O-(2,4,6-trinitrocyclohexdienylidene)adenosine triphosphate as a fluorescent probe of th ATP site of sodium and potassium adenosine triphosphatase. J Biol Chem 256:2346-2356

10. Nakamura Y, Kai D, Kaya S, Adachi Y, Taniguchi K (submitted to J Biochem) Different susceptibility to phospholipaseA treatment of the fluorescence intensity changes in the vicinity of Cys-964 and Lys-501 in the alpha chain of probe labeled Na$^+$/K$^+$- ATPase.

11. Pedemonte C H, Kaplan J H (1990) Chemical modification as an approach to elucidation of sodium pump structure-function relations. Am J Physiol 258:C1-C23

12. Schulthesis P J, Lingrel J P (1993) Substitution of transmembrane residues with hydrogen- bonding potential in the A subunit of Na$^+$/K$^+$- ATPase reveals alterations in ouabain sensitivity. Biochemistry 32:544- 550.

13. Shinoguchi E, Ito E, Kudo A, Nakamura S, Taniguchi K (1988) Fluorescence energy transfer between N-(p-(2-benzimidazolyl)phenyl) maleimide,N-(1-anilinonaphtyl-4)maleimide, fluorescein 5'- isothiocyanate, and Anthroylouabain in Na$^+$/K$^+$- ATPase. The Sodium Pump: Recent developments. The Rockefeller University press, 363-367

14. Shull G E A, Schwarts A, Lingrell J B (1985) Amino acid sequence of the catalytic subunit of the Na$^+$/K$^+$- ATPase deduced from a complementary DNA. Nature 316:691-695

15. Taniguchi K, Tosa H, Suzuki K, Kamo Y (1988) Microenvironment of two different extrinsic fluorescence probes in Na$^+$/K$^+$- ATPase changes out of phase during sequential appearance of reaction intermediates. J Biol Chem 263:12943-12947

16. Taniguchi K, Mårdh S (1993) Reversible changes in the fluorescence energy transfer accompanying formation of reaction intermediates in probe-labeled Na$^+$/K$^+$- ATPase. J Biol Chem 268:15588-15594

# Heterogeneity of distribution and of regulation by corticosteroids of transport ATPases along the rat collecting tubule

C. Khadouri, C. Barlet-Bas, S. Marsy, L. Cheval, A. Doucet

Laboratoire de Physiologie Cellulaire, Collège de France, 11 Place M. Berthelot, 75231 Paris Cedex 05, France

## Introduction

In rodents, the pH of the tubular fluid decreases as it flows along the inner medullary collecting duct (IMCD) towards the tip of the papilla (17, 18, 30), indicating that this nephron segment is an important site of urinary acidification. This acidification primarily results from active proton secretion rather than bicarbonate reabsorption since it is accompanied with a luminal acidic disequilibrium pH (8, 16). However, there is little information concerning the molecular mechanism of apical proton secretion in this nephron segment.

Although studies revealed the presence of a $Na^+/H^+$ exchanger in the IMCD (23) as well as in cultured IMCD cells (2, 24), this transporter cannot be involved in proton secretion because it is located at the basolateral pole of the cells (23, 24). Alternately, primary secretion of protons along the IMCD could be accounted for by proton-ATPases. However, the involvement and even the presence in IMCD of two possible candidates, i.e., the electrogenic N-ethylmaleimide (NEM)-sensitive ATPase and the electroneutral $H^+/K^+$-ATPase, are not clearly established as yet. This may result in part from the axial heterogeneity of the IMCD since it consists of two successive subsegments - called initial ($IMCD_i$) and terminal IMCD ($IMCD_t$) - which are morphologically (5, 21, 22) and functionally distinct (27, 28). Using polyclonal antibodies against different subunits of bovine kidney electrogenic $H^+$-ATPase, Brown et al. have revealed the presence of an apical labelling in intercalated cells of the rat $IMCD_i$ whereas no labelling was found along the $IMCD_t$ (3). In contrast, Garg and Narang have measured NEM-sensitive ATPase activity in both rat and rabbit IMCD (11, 13). However, these last studies were performed before recognition of the axial heterogeneity of the IMCD, and it is unclear whether the results apply to the $IMCD_i$, the $IMCD_t$ or a combination of both. Sabatini et al. also reported the presence of NEM-sensitive ATPase in the rat $IMCD_t$ (26). Concerning $H^+/K^+$-ATPase, two studies have revealed its presence along the CCD and OMCD of both rat and rabbit (7, 12), but none of them has questioned whether it might be also present along the IMCD.

The first aim of this study was therefore to determine whether the electrogenic $H^+$-ATPase and/or the electroneutral $H^+/K^+$-ATPase are present in the subsegments of the IMCD. Because corticosteroids control proton secretion in the collecting duct, the second aim of this study was to determine whether these ATPases might be under the control of these hormones by comparing the activities of NEM-sensitive ATPase and of $H^+/K^+$-ATPase along the IMCD of normal and adrenalectomized rats.

## Methods

*Animals.* Experiments were carried out on normal and adrenalectomized male Wistar rats weighing 150-200 g and fed the normal laboratory diet ad libitum. Bilateral adrenalectomy was performed under light ether anesthesia seven days before study. Adrenalectomized (ADX) rats had free access to a 0.9 % NaCl solution for drinking.

*Renal function.* Acid-base balance was studied in 5 control and 6 ADX rats which were thereafter used for ATPase measurements. Animals were anesthetized with Inactin (Promonta, Hamburg, FRG, 10 mg/100g body wt ip) and were laparotomized. A urine sample was taken for pH measurement by bladder puncture. After a 20-30 min recovery period, the abdominal aorta was punctured for blood sampling. Blood pH, $PCO_2$ and $HCO_3$ concentration were determined with an automatic blood gas analyzer (ABL 30, Radiometer, Copenhagen, DK).

*Tubule microdissection and ATPase measurement.* Tubule microdissection was carried out after collagenase treatment of the kidney as previously described (6). The successive subsegments of the inner medullary collecting duct were characterized according to the criteria defined by Sands et al. (27, 28) and their origin was confirmed by measuring $Na^+/K^+$-ATPase activity which is higher in the $IMCD_i$ than in the $IMCD_t$. The length of each tubule (0.3-1.1 mm) was measured by automatic image processing and served as reference for enzymatic activities. $Na^+/K^+$-ATPase, NEM-sensitive ATPase and $H^+/K^+$-ATPase activities were measured as previously described (1, 6, 7).

654

# Results

*Acid-base balance.* As summarized in table 1, ADX rats excreted an alkaline urine with a pH more than 1.5 unit higher than control rats, whereas plasma pH was similar in the two groups of animals, which indicates that renal proton excretion was inhibited in ADX rats. In addition, ADX rats displayed a metabolic acidosis, as assessed by the decreased plasma bicarbonate concentration, which was compensated by respiratory alkalosis revealed by increased $PO_2$.

| | Control | P | ADX |
|---|---|---|---|
| Urinary pH | $6.07 \pm 0.08$ | <0.001 | $7.68 \pm 0.07$ |
| Plasma pH | $7.37 \pm 0.01$ | NS | $7.33 \pm 0.02$ |
| Plasma $PCO_2$ | $6.9 \pm 0.5$ | NS | $6.2 \pm 0.5$ |
| Plasma $[HCO_3]$ | $30.0 \pm 2.0$ | <0.025 | $23.8 \pm 1.0$ |
| Plasma total $CO_2$ | $31.6 \pm 2.1$ | <0.025 | $25.2 \pm 1.1$ |
| Plasma $PO_2$ | $12.2 \pm 0.3$ | <0.005 | $13.7 \pm 0.3$ |

**Table 1.** Acid-base balance in control and adrenalectomized rats. Values are means $\pm$ SE obtained in 5 normal rats (control) and 6 rats adrenalectomized 7 days before study (ADX). Plasma $PCO_2$ and $PO_2$ are expressed in kPa ; bicarbonate and total $CO_2$ concentrations are in mM. Statistical significance was assessed by Student's t test. NS, not significant.

*Distribution of transport ATPases along the collecting duct.* $Na^+/K^+$-ATPase was detected in all the subsegments of the collecting duct of normal rats, but its activity was higher in the CCD and $IMCD_i$ than in the OMCD and $IMCD_t$ (table 2). Differences between mean $Na^+/K^+$-ATPase activities in the successive subsegments were statistically significant (table 2), indicating a true segmentation of the collecting duct with regard to $Na^+/K^+$-ATPase. NEM-sensitive ATPase activity was also measurable in all the subsegments of the collecting duct. However, there was no clearcut axial segmentation, except

655

for a statistically significant decrease in activity from the CCD to the OMCD. Although absolute activities were lower, the profile of distribution of $H^+/K^+$-ATPase along the collecting duct was quite similar to that of $Na^+/K^+$-ATPase, except that the activity was almost undetectable (and not statistically different from zero) in the $IMCD_t$. Here again, differences between mean ATPase activities in the successive collecting duct subsegments were statistically significant.

| | CCD | P | OMCD | P | $IMCD_i$ | P | $IMCD_t$ |
|---|---|---|---|---|---|---|---|
| $Na^+/K^+$-ATPase (n = 6) | $1238 \pm 129$ | <0.001 | $576 \pm 22$ | <0.001 | $1008 \pm 74$ | <0.01 | $668 \pm 70$ |
| $H^+$-ATPase (n = 8) | $326 \pm 20$ | <0.01 | $254 \pm 13$ | NS | $268 \pm 30$ | NS | $300 \pm 20$ |
| $H^+/K^+$-ATPase (n = 6) | $289 \pm 23$ | <0.05 | $165 \pm 20$ | <0.001 | $307 \pm 24$ | <0.001 | $30 \pm 11$ |

**Table 2.** Activities of $Na^+/K^+$-ATPase, NEM-sensitive ATPase ($H^+$-ATPase) and $H^+/K^+$-ATPase in the successive subsegments of the collecting duct from normal rats. ATPase activities are expressed as $pmol.mm^{-1}.h^{-1} \pm SE$. Statistical significance between the successive subsegments of the collecting duct was assessed by variance analysis (NS, not significant). CCD, cortical collecting duct ; OMCD, outer medullary collecting duct ; $IMCD_i$ and $IMCD_t$, initial and terminal parts of the inner medullary collecting duct.

*Effect of adrenalectomy.* In all the subsegments of the collecting duct, $Na^+/K^+$-ATPase was markedly decreased (> 50 %) in ADX rats (figure 1). Similarly, adrenalectomy decreased by 70-85 % $H^+/K^+$-ATPase activity in the CCD, OMCD and $IMCD_i$ of ADX rats, and this activity remained undetectable in the $IMCD_t$. Conversely, adrenalectomy decreased NEM-sensitive ATPase activity (> 60 %) only in the CCD and the OMCD whereas it had no effect along the IMCD.

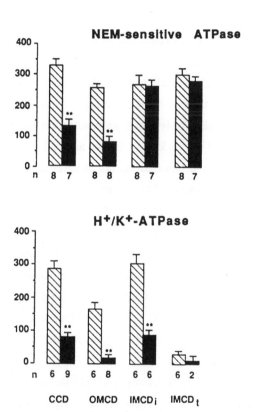

**Figure 1.** Activity of Na⁺/H⁺-ATPase (top), NEM-sensitive ATPase (middle) and $H^+/K^+$-ATPase along the collecting duct of normal (hatched bars) and adrenalectomized rats (dark bars). Values (in $pmol.mm^{-1}.h^{-1}$) are means ± SE from n animals. CCD, cortical collecting duct ; OMCD, outer medullary collecting duct ; $IMCD_i$ and $IMCD_t$, initial and terminal part of the inner medullary collecting duct. Differences statistically significant between normal and ADX rats were determined by Student's t test : *, p<0.005 ; **, p<0.001.

657

## Discussion

Present results demonstrate the existence of both NEM-sensitive ATPase activity and $H^+/K^+$-ATPase activity along the IMCD of the rat : NEM-sensitive ATPase is present along the whole IMCD whereas $H^+/K^+$-ATPase is detected only in the initial part of the IMCD. Since it is now clearly established that these two ATPase activities are the enzymatic counterparts of the electrogenic V-type $H^+$-pump and the electroneutral P-type $H^+/K^+$-pump, respectively (4, 19, 25), our findings suggest that these two pumps may account in part for active proton secretion along the IMCD.

The presence of NEM-sensitive ATPase activity both in the $IMCD_i$ (which consists of principal and type A intercalated cells) and in the $IMCD_t$ (which only contains IMCD cells) indicates that both intercalated cells and IMCD cells possess a membrane-bound electrogenic $H^+$-ATPase. Conversely to intercalated cells, which are known to have an active $H^+$-pump, the presence of NEM-sensitive ATPase in IMCD cells is somehow controversial. Indeed, using polyclonal antibodies raised against subunits of the electrogenic $H^+$-ATPase of bovine kidney, Brown and colleagues could not find labelling in rat $IMCD_t$ (3). Opposite to that finding, but in agreement with results of this study, Sabatini et al. previously reported NEM-sensitive ATPase activity in the terminal IMCD of the rat (26). The discrepancy concerning the presence of $H^+$-ATPase in the $IMCD_t$ may reveal a molecular heterogeneity of $H^+$-ATPase along the collecting duct : The enzyme would be sensitive to NEM along the whole collecting duct but the form present in IMCD cells would not be recognized by the antibodies used by Brown et al. conversely to the form present in intercalated cells. In fact, molecular heterogeneity of renal $H^+$-ATPase has been reported (17, 31). In contrast to NEM-sensitive ATPase, it is likely that $H^+/K^+$-ATPase is exclusively present in intercalated cells since no activity was detected in the terminal IMCD.

Within one week, bilateral adrenalectomy modified the activities of transport ATPases along the collecting duct. $Na^+/K^+$-ATPase activity, which likely originates from principal cells in CCD, OMCD and $IMCD_i$ and from IMCD cells in $IMCD_t$, was markedly reduced along the whole collecting duct of adrenalectomized rats. This observation extends to the inner medullary collecting duct what was already known for the CCD and OMCD (9). In the IMCD, it confirms the previous report by Terada and Knepper who observed that both low-NaCl diet and chronic deoxycorticosterone treatment increased Na-

K-ATPase activity (29). Such findings suggest the presence of mineralocorticoid receptors in IMCD cells. $H^+/K^+$-ATPase activity was also decreased markedly by adrenalectomy in whole the subsegments of the collecting duct in which it is detectable. It suggests that $H^+/K^+$-ATPase activity might be controled by corticosteroids in type A intercalated cells, in agreement with a preliminary report by Garg showing that chronic aldosterone administration increased $H^+/K^+$-ATPase activity in the rabbit collecting duct (10). As $H^+/K^+$-ATPase is known to secrete proton and to reabsorb potassium (4, 14, 25, 32) it should be mentioned that the stimulatory action of corticosteroids on $H^+/K^+$-ATPase goes against the well documented kaliuretic action of mineralocorticoids.

Inhibition of NEM-sensitive ATPase in both CCD and OMCD of adrenalectomized rats confirms previous findings (20). However, the lack of effect of adrenalectomy in the IMCD was unexpected, especially in the $IMCD_i$. Indeed, in the $IMCD_t$, the lack of inhibitory action of corticosteroid depletion might result from the expression in IMCD cells of a distinct molecular form of $H^+$-ATPase (see above) which is insensitive to corticosteroids. In the $IMCD_i$, however, one would have expected the same molecular form of $H^+$-ATPase as in the OMCD since this ATPase likely originates from the same type of cells (type A intercalated cells) in both $IMCD_i$ and OMCD. Nevertheless, it remains possible that NEM-sensitive ATPase is controled by corticosteroids and that its inhibition during steroid-depletion is counterbalanced by an as yet unknown stimulatory process.

# References

1. Ait-Mohamed AK, Marsy S, Barlet C, Khadouri C, Doucet A (1986) Characterization of N-ethylmaleimide-sensitive proton pump in the rat kidney. Localization along the nephron. J Biol Chem 261 : 12526-12533

2. Alexander EA, Schwartz JH (1991) Regulation of acidification in the rat inner medullary collecting duct. Am J Kidney Dis 18 : 612-618

3. Brown D, Hirsch S, Gluck S (1988) Localization of a proton-pumping ATPase in rat kidney J Clin Invest 82 : 2114-2126

4. Cheval L, Barlet-Bas C, Khadouri C, Féraille E, Marsy S, Doucet A (1991) K-ATPase-mediated Rb transport in the rat collecting tubule : modulation during K-depletion. Am J Physiol 260 : F800-F805

5. Clapp WL, Madsen KM, Verlander JW, Tisher CC (1987) Intercalated cells of the rat inner medullary collecting duct. Kidney Int 31 : 1080-1087

6. Doucet A , Katz AI , Morel F (1979) Determination of Na-K-ATPase activity in single segments of the mammalian nephron. Am J Physiol 237 : F105-F113

7. Doucet A , Marsy S (1987) Characterization of K-ATPase activity in distal nephron : stimulation by potassium depletion. Am J Physiol 253 : F418-F423

8. Dubose TDJr (1982) Hydrogen ion secretion by the collecting duct as a determinant of the urine to blood $PCO_2$ gradient in alkaline urine. J Clin Invest 69 : 145-156

9. El Mernissi G, Doucet A (1983) Short-term effect of aldosterone on renal sodium transport and tubular Na-K-ATPase in the rat. Pflügers Arch 399 : 139-146

10. Garg LC (1991) Effects of aldosterone on H-K-ATPase activity in rabbit nephron segments. In : Aldosterone : Fundamental Aspects, edited by J.P. Bonvalet, N. Farman, M. Lombès and M.E. Rafestin-Oblin. Paris, John Libbey Eurotext,p.328 (Abst.)

11. Garg LC, Narang N (1985) Stimulation of an N-ethylmaleimide-sensitive ATPase in the collecting duct segments of the rat nephron by metabolic acidosis. Can J Physiol Pharmacol 63 : 1291-1296

12. Garg LC, Narang N (1988) Ouabain-insensitive K-adenosine triphosphatase in distal nephron segments of rabbit. J Clin Invest 81 : 1204-1208

13. Garg L C, Narang N (1988) Effects of aldosterone on NEM-sensitive ATPase in rabbit nephron segments. Kidney Int 34 : 13-17

14. Gifford JD, Rome L, Galla JH (1992) $H^+$-$K^+$-ATPase activity in rat collecting duct segments. Am J Physiol 262 : F692-F695

15. Graber ML, Bengele HH, Mroz E, Lechene C, Alexander EA (1981) Acute metabolic acidosis augments collecting duct acidification rate in the rat. Am J Physiol 241 : F669-F676

16. Graber ML, Bengele HH, Schwartz JH, Alexander EA (1981) pH and $pCO_2$ profiles of the rat inner medullary collecting duct. Am J Physiol 241 : F659-F668

17. Hemken P, Guo XL, Wang ZQ, Zhang K, Gluck S (1992) Immunologic evidence that vacuolar $H^+$-ATPases with heterogeneous forms of Mr = 31.000 subunit have different membrane distributions in mammalian kidney. J Biol Chem 267 : 9948-9957

18. Hierholzer K (1961) Secretion of potassium and acidification in collecting ducts of mammalian kidney. Am J Physiol 201 : 318-324

19. Khadouri C, Cheval L, Marsy S, Barlet-Bas C, Doucet A (1991) Characterization and control of proton-ATPases along the nephron. Kidney Int 40 : S71-S78

20. Khadouri C, Marsy S, Barlet-Bas C, Doucet A (1987) Effect of adrenalectomy on NEM-sensitive ATPase along the rat nephron and on urinary acidification. Am J Physiol 253 : F495-F499

21. Madsen KM, Tisher CC (1986) Structural-functional relationship along the distal nephron. Am J Physiol 250 : F1-F15

22. Madsen KM, Clapp W.L, Verlander JW (1988) Structure and function of the inner medullary collecting duct. Kidney Int. 34 : 441-454

23. Matsushima Y, Yoshitomi K, Koseki C, Kawamura M, Akabane S, Imanishi M, Imai M (1990) Mechanisms of intracellular pH regulation in the hamster inner medullary collecting duct perfused in vitro. Pflügers Arch 416 : 715-721

24. Nord EP, Hart D (1991) Polarized distribution of $Na^+/H^+$ antiport and $Na^+/HCO_3$ cotransport in primary cultures of rat inner medullary collecting duct cells. J Biol Chem 266 : 2374-2382

25. Planelles G, Anagnostopoulos T, Cheval L, Doucet A (1991) Biochemical and functional characterization of H-K-ATPase in the distal amphibian nephron. Am J Physiol 260 : F806-F812

26. Sabatini S, Laski ME, Kurtzman NA (1990) NEM-sensitive ATPase activity in rat nephron : effect of metabolic acidosis and alkalosis. Am J Physiol 258 : F297-F304

27. Sands J M, Knepper M A (1987) Urea permeability of mammalian inner medullary collecting duct system and papillary surface epithelium. J Clin Invest 79 : 138-147

28. Sands JM, Nonoguchi H, Knepper MA (1987) Vasopressin effects on urea and $H_2O$ transport in inner medullary collecting duct subsegments. Am J Physiol 253 : F823-F832

29. Terada Y, Knepper MA (1989) $Na^+-K^+$-ATPase activities in renal tubule segments of rat inner medulla. Am J Physiol 256 : F218-F223

30. Ullrich KJ, Eigler FW (1958) Sekretion von Wasserstoffionen in den Sammelrohren der Sungetierniniere. Pflügers Arch 267 : 491-496

31. Wang ZQ, Gluck S (1990) Isolation and properties of bovine kidney brush border vacuolar $H^+$-pump : A proton pump with enzymatic and structural differences from kidney microsomal $H^+$-ATPase. J Biol Chem 265 : 21957-21965

32. Wingo CS (1989) Active proton secretion and potassium absorption in the rabbit outer medullary collecting duct. Functional evidence for proton-potassium-activated adenosine triphosphatase. J Clin Invest 84 : 361-365

# Hormonal regulation of Na$^+$,K$^+$-ATPase activity

A. Aperia, J. Fryckstedt, G. Fisone, U. Holtbäck, S. Cheng, M-L. Syrén, D. D.Li, H.C. Hemmings Jr, A.C. Nairn, P. Greengard

Department of Pediatrics, St Görans´ Childrens´ Hospital, Karolinska Institutet, 11281 Stockholm and Department of Molecular and Cellular Neuroscience, The Rockefeller University, 1230 York Ave., New York, N Y10021-6399

## Introduction

The kidney maintains electrolyte and fluid balance with great precision. Under basal conditions, almost all filtered Na$^+$ is reabsorbed in the renal tubules. The Na$^+$,K$^+$-ATPase pump, localized to the basolateral membranes, is the active step of this process, providing the energy for the vectorial transport of electrolytes and solutes across the epithelial cells at the cost of ATP hydrolysis. Na$^+$ reabsorption is a dynamic process under the influence of local and humoral factors which can cause several-fold variations in Na$^+$ excretion. Imperfect regulation of Na$^+$ homeostasis results in disturbances in electrolyte balance of significant clinical importance. Revealing the molecular mechanisms underlying the regulation of natriuresis is of prime importance in understanding the pathogenesis o,f and directing specific therapy for, diseases such as hypertension. Here we present a model in which a cascade of intracellular events results in a bidirectional modulation of Na$^+$,K$^+$-ATPase activity through regulation of its phosphorylation and dephosphorylation.

## Studies on phosphorylation of the α 1 subunit of Na$^+$,K$^+$-ATPase

The α subunit of Na$^+$,K$^+$-ATPase, in preparations from rat renal cortex and shark rectal gland, was phosphorylated by cAMP dependent protein kinase (PKA) or by protein kinase C (PKC). Phosphorylation by PKA or PKC was accompanied by a decrease in enzyme activity (Figure 1). Phospho amino acid analysis and peptide mapping revealed that the phosphorylation sites for PKA and PKC were distinct. PKA phosphorylated the α subunit with a stoichiometry of 1:1 at serine 943. PKC phosphorylated the α subunit with a stoichiometry of 2:1 at both serine and treonine. Protein phosphatases 1, 2A and 2B (PP-1,PP-2A, PP-2B) each dephosphorylated Na$^+$,K$^+$-ATPase in vitro (5). These results demonstrated that the phosphorylation of

662

Na+,K+-ATPase in vitro is associated with a decrease in enzyme activity.

**A.**

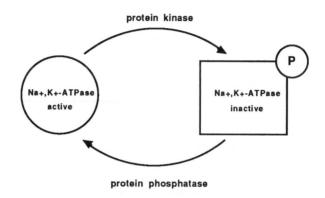

**B.**

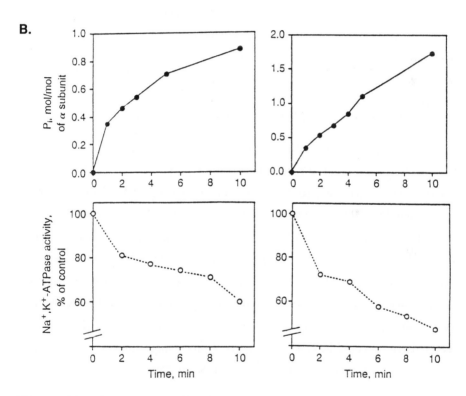

Figure 1. Phosphorylation of Na+,K+-ATPase is associated with inhibition of enzyme activity. .A. Schematic presentation. B. A purified preparation of shark rectal gland was phosphorylated in the presence of either PKA (left) or PKC (right) for the indicated times, and the stoichiometry of phosphorylation (upper) and the Na+,K+-ATPase activity (lower) were determined (from ref. (5) with permission)

In order to assess the possible physiological role of the phosphorylation of Na+,K+-ATPase in intact cells by PKA, COS cells were transfected with cDNA corresponding to either of 2 forms of the $\alpha$ subunit of the Na+,K+-ATPase, namely the wild-type form or a mutated form in which an alanine was substituted for serine 943. Na+,K+-ATPase activity was assayed at various concentrations of Na+ using membranes isolated from the transfected cells. Under control conditions, the concentration of Na+ required for half-maximal activation of the enzyme ($K_{0.5}$) was similar in the two cell types. Incubation of wild-type cells with forskolin plus 3-isobutyl-1-methylxanthine (IBMX) caused a significant decrease in Na+,K+-ATPase activity, measured at non-saturating Na+ concentrations. The decrease in Na+,K+-ATPase activity was associated with a decrease in the affinity of the enzyme for Na+. In contrast, in membranes from cells containing the mutated $\alpha$ subunit, no significant difference in Na+,K+-ATPase activity at non-saturating Na+ concentration or in Na+ affinity was observed following forskolin plus IBMX treatment (Fisone et al, unpublished).

### The role of DARPP-32 and PP-1 in the regulation of Na+,K+-ATPase activity

The intracellular signalling systems regulating Na+,K+-ATPase activity have been extensively studied in intact renal tubule cells. The studies were carried out in the proximal convoluted tubule (PCT), responsible for the bulk of Na+ reabsorption, and in the medullary thick ascending limb of Henle (mTAL), responsible for the concentrating capacity of the kidney. Na+,K+-ATPase activity was measured as ouabain-sensitive hydrolysis of $\gamma^{32}P$-ATP in isolated permeabilized tubular segments (6). This procedure allowed ATP to enter the cells and made the effect of changes in Na+ entrance negligible since intracellular Na+ concentration could be equilibrated with the medium. In most studies, the Na+ concentration was either 20 mM, where observed effects could be considered to be of physiological relevance, or 70 mM, which represented Vmax conditions for the Na+,K+-ATPase.

The role of phosphorylation in the regulation of Na+,K+-ATPase activity was initially indicated by the effect of the endogenous phosphatase inhibitor, DARPP-32. DARPP-32 is a dopamine- and cAMP-regulated phosphoprotein present in cells containing the dopamine-1 ($D_1$) receptor in the central nervous system and in a few peripheral tissues. DARPP-32 is phosphorylated by PKA and by cGMP-dependent protein kinase (PKG) which, in renal tissue, are activated by the first messengers dopamine and atrial natriuretic peptide (ANP). In its phosphorylated state, DARPP-32 is a potent inhibitor of PP-1 DARPP-32 is dephosphorylated and inactivated by the $Ca^{2+}$ regulated PP-2B, calcineurin (8). DARPP-32 was found to be localized to renal tubule cells by immunohistochemistry and in situ hybridization (10). The possible involvement of DARPP-32 in the regulation of the activity of $Na^+,K^+$-ATPase activity was examined by incubating permeabilized tubular segments in the presence of a synthetic DARPP-32 peptide. This 31 amino acid peptide contained the phosphorylation site for PKA and was active as a phosphatase inhibitor. The phosphorylated peptide inhibited $Na^+,K^+$-ATPase activity in a dose-dependent manner, while the unphosphorylated peptide was without effect (2) (Figure 2).

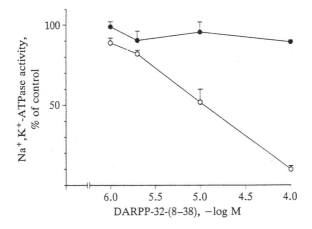

Figure 2. Effects of various concentrations of phospho- (unfilled circle) or dephospho- (filled circle) DARPP-32 on $Na^+,K^+$-ATPase activity in single permeabilized mTAL segments (Reproduced from ref. (2) with permission).

Okadaic acid and calyculin A, exogenous inhibitors of PP-1 and 2A, also inhibited $Na^+,K^+$-ATPase activity (9, Li et al unpublished). Calyculin A is ten times more potent as an inhibitor of PP-1 than PP-2A and inhibited $Na^+,K^+$-ATPase activity with a ten times greater potency than okadaic acid, confirming that $Na^+,K^+$-ATPase activity in renal tubule cells is regulated by PP-1.

**Hormonal regulation of natriuresis by modulation of Na$^+$,K$^+$-ATPase activity .
I. Natriuretic agents**

Dopamine, an intrarenal natriuretic hormone, is commonly used in clinical practice to maintain diuresis when renal function is compromised. Dopamine decreases tubular Na$^+$ reabsorption by inhibiting Na$^+$,K$^+$-ATPase activity in several nephron segments (1, 10). In mTAL segments, the DA$_1$ receptor agonist fenoldopam and the cAMP analogue, dBcAMP, inhibited Na$^+$,K$^+$-ATPase activity to a similar extent and this effect was abolished by an inhibitor of PKA (10, 7, 2). These findings support the pathway of intracellular signalling depicted in Figure 3.

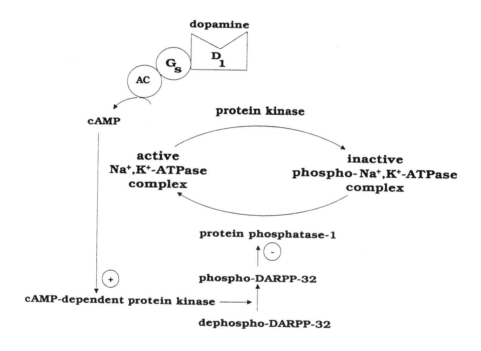

Figure 3. Proposed mechanism for the regulation of Na+,K+-ATPase activity by phosphorylation and dephosphorylation. Dopamine, acting at the DA1 receptor, causes an increase in the level of cAMP, activation of cAMP dependent protein kinase, and phosphorylation of DARPP-32. Phospho-DARPP-32 inhibits PP-1 leading to an increased phosphorylation state, and thereby decreased activity, of Na+,K+-ATPase. Gs, stimulatory G-protein. (Reproduced from ref. (2) with permission).

ANP stimulates cGMP accumulation and inhibits Na+,K+-ATPase activity in mTAL (11, Syrén et al, manuscript in preparation). These observations suggest that ANP may decrease Na+,K+-ATPase activity by stimulating PKG, thus leading to the phosphorylation of DARPP-32. Phosphorylated DARPP-32 would then inhibit Na+,K+-ATPase activity as discussed above.

## II. Antinatriuretic agents

Regulation of Na+ excretion by renal sympathetic nerve activity and by the renin-angiotensin system plays a major role in the maintenance of Na+ homeostasis. These agents might promote Na+ reabsorption by modulating the DARPP-32/PP-1 pathway. The $\alpha$-adrenergic agonist, oxymetazoline, stimulated Na+,K+-ATPase activity measured at non-saturating Na+ concentration in PCT segments. The stimulation was dose-dependent, with a maximal stimulation of 70 % at $10^{-6}$M oxymetazoline. It decreased the $K_{(0.5)}$ for Na+ from 13 to 5 mM. The effect of oxymetazoline may be mediated by an increase in intracellular $Ca^{2+}$ and activation of PP-2B, since the $Ca^{2+}$ ionophore A23187 mimicked this effect and oxymetazoline increased fura-2 fluorescence in PCT. This hypothesis was supported by the use of the immunosuppressant drug FK506, a specific inhibitor of PP-2B. Incubation of PCT segments with FK506 abolished the effect of oxymetazoline, whereas rapamycin, a structurally related analogue of FK506 which does not inhibit PP-2B, had no effect. (4).

Angiotensin (AII) stimulated Na+,K+-ATPase activity in PCT segments in a bimodal manner with a maximal effect (27%) at 5 x $10^{-11}$M.This result was in concord with AII stimulation of Na+ reabsorption only within a certain concentration range reported earlier. Since AII increased intracellular $Ca^{2+}$ in renal tubular cells, the mechanism of action was postulated to be similar to that of oxymetazoline, i.e. activation of PP-2B and a resulting decrease in the phosphorylation state of Na+,K+-ATPase (3).

### Interactions between hormonal systems

Correlations exist between the interactions of natriuretic and anti-natriuretic agents observed at the whole organ level and interactions between pharmacological agents at the cellular level. Dopamine and its second messenger, cAMP, as well as ANP and cGMP, abolished the stimulatory effect of oxymetazoline on $Na^+,K^+$-ATPase activity. Similarly, AII stimulation of $Na^+,K^+$-ATPase activity was abolished by ANP/cGMP, dopamine/cAMP, as well as FK506 (3)(Figure 4).

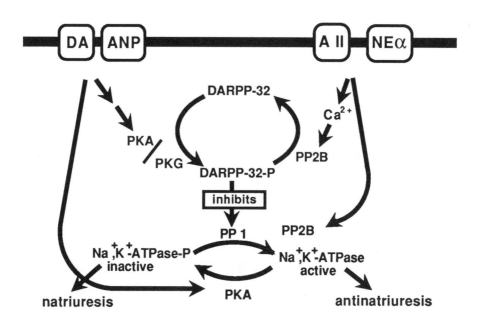

Figure 4. Postulated pathways for regulation of $Na^+$ metabolism in renal tubule cells by natriuretic and antinatriuretic factors. For further details see text. (Modified from ref. (3)).

## Conclusion

These results point to a common pathway for the regulation of $Na^+$ balance, in which the ultimate step is the phosphorylation/dephosphorylation of $Na^+,K^+$-ATPase and an associated change in its activity. Hormones and second messengers modulate $Na^+,K^+$-ATPase activity by regulating protein kinase and phosphatase activities. The principal concept in this model is that each tubule cell possesses a balanced system for regulation of $Na^+,K^+$-ATPase activity. Extracellular signals shift the balance in

668

the direction of natriuresis or antinatriuresis. This model provides stability as well as versatility in the regulation of Na+,K+-ATPase activity, since the system is capable of integrating different extra- and intracellular signals. Further exploration of the regulation of Na+,K+-ATPase activity should increase our understanding of diseases that involve altered Na+ homeostatic mechanisms in renal cells.

## References

1. Aperia A, Bertorello A and Seri I (1987) Dopamine causes inhibition of Na+-K+-ATPase activity in rat proximal convoluted tubule segments. Am. J. Physiol. 252: F39-F45.
2. Aperia A, Fryckstedt J, Svensson L-B, Hemmings Jr HC, Nairn AC and Greengard P (1991) Phosphorylated Mr 32,000 dopamine- and cAMP-regulated phosphoprotein inhibits Na+-K+-ATPase activity in renal tubule cells. Proc. Natl. Acad. Sci. USA. 88: 2798-2801.
3. Aperia A, Holtbäck U, Syrén M-L, Svensson L-B, Fryckstedt J and Greengard P (1993) Activation/Deactivation of renal Na+-K+-ATPase: A final common pathway for regulation of natriuresis. FASEB: submitted
4. Aperia A, Ibarra F, Svensson L-B, Klee C and Greengard P (1992) Calcineurin mediates α-adrenergic stimulation of Na+-K+-ATPase activity in renal tubule cells. Proc. Natl. Acad. Sci. USA 89: 7394-7397.
5. Bertorello AM, Aperia A, Walaas SI, Nairn AC and Greengard P (1991) Phosphorylation of the catalytic subunit of Na+-K+-ATPase inhibits the activity of the enzyme. Proc. Natl. Acad. Sci. U.S.A. 88: 11359-11362.
6. Doucet A, Katz AI and Morel F (1979) Determination of Na+-K+-ATPase activity in single segments of the mammalian nephron. Am. J. Physiol. 237: F105-F113.
7. Fryckstedt J and Aperia A (1992) Sodium-dependent regulation of sodium, potassium, adenosine-tri-phosphatase (Na+-K+-ATPase) activity in medullary thick ascending limb of Henle segments. Effect of cyclic adenosine-monophosphate, guanosine-nucleotide-binding-protein activity and arginine vasopressin. Acta Phys. Scand. 144: 185-190.
8. Hemmings Jr HC, Nairn AC and Greengard P (1984) DARPP-32, a dopamine- and adenosine 3':5'-monophosphate-regulated neuronal phosphoprotein. II. Comparison of the kinetics of phosphorylation of DARPP-32 and phosphatase inhibitor 1. J. Biol.Chem. 259: 14491-14497.
9. Ibarra F, Aperia A, Svensson L-B, Eklöf A-C and Greengard P (1993) Bidirectional regulation of Na+-K+-ATPase activity by dopamine and an α-adrenergic agonist. Proc. Natl. Acad. Sci. USA 90: 21-24
10. Meister B, Fryckstedt J, Schalling M, Cortés R, Hökfelt T, Aperia A, Hemmings Jr HC, Nairn AC, Ehrlich ME and Greengard P (1989) Dopamine- and cAMP-regulated phosphoprotein (DARPP-32) and dopamine DA$_1$ agonist-sensitive Na+-K+-ATPase in renal tubule cells. Proc. Natl. Acad. Sci. USA 86: 8068-8072.
11. Syrén M-L, Fryckstedt J (1993) Atrial natriuretic peptide-induced cGMP accumulation. in the rat kidney. A comparative study of glomeruli and outer medullary tubules. Acta Phys. Scand. submitted

# Regulation of the Na$^+$,K$^+$-pump by Insulin

Jonathan Lytton, Janet Lin, Luisa DiAntonio, Jeff Brodsky, Julie McGeoch, Diana McGill, and Guido Guidotti

Department of Biochemistry and Molecular Biology, Harvard University, 7 Divinity Avenue, Cambridge, Massachusetts 02138

**Introduction.**

   Insulin administration reduces the serum glucose and K$^+$ concentrations (6) by promoting their entry into skeletal muscle (20). The acute function of insulin, thus, is to clear from the blood two of the major nutrients ingested during a meal. Clausen and Kohn (13) and Resh et al. (33) showed that insulin stimulates the activity of the Na$^+$,K$^+$-pump in rat skeletal muscle and adipocytes, respectively, by a direct effect on the pump rather than by an increase in the intracellular Na$^+$ concentration. On the other hand, there are instances in which insulin does stimulate a Na$^+$/H$^+$ exchanger resulting in an increase in intracellular Na$^+$ and increased activity of the (Na$^+$,K$^+$)-pump through an increase in the substrate concentration (15,26,34). The central question here is the mechanism of the direct activation of the (Na$^+$,K$^+$)-pump by insulin.

**Isoforms of the (Na$^+$,K$^+$)-ATPase and insulin action.**

   Direct activation of the Na$^+$,K$^+$-pump by insulin is mediated by the α2 isoform of the Na$^+$,K$^+$-ATPase. In the first place, rat skeletal muscle contains predominantly this isoform of the pump on the basis of ouabain binding (11) and ouabain-dependent Rb$^+$ uptake (18) with intact muscle cells, and ouabain-inhibitable phosphatase activity (28). In isolated sarcolemma, the α1 and α2 isoforms of the Na$^+$,K$^+$-ATPase are present in a ratio of 3 to 7 on the basis of immunoblots and enzymatic activity (18); since the recovery of plasma membrane is at best 8-10 %, it is not clear whether the α1 isoform in these preparations derives from the sarcolemma or the plasma membrane of other cells (fibroblasts, neurons, blood vessels).

   The adipocyte, on the other hand, has both isoforms of the Na$^+$,K$^+$-ATPase in the plasma membrane. However, the results obtained by Lytton (22,23) and McGill (25), the latter shown in Fig. 1, indicate that the major effect of insulin is on the activity of the α2 isoform. The increase in activity appears to result from an apparent decrease in the K$_{0.5}$ for Na$^+$ and an increase in Vmax of the α2 isoform of the Na$^+$,K$^+$-pump, measured as the ability to pump Rb$^+$ into adipocytes equilibrated with different concentrations of Na$^+$.

**Plasma membrane density of Na$^+$,K$^+$-pumps.**

   In view of the evidence that the effect of insulin is in part to increase the Vmax of the α2 isoform of the Na$^+$,K$^+$-pump, one can inquire whether this effect is due to an increase in the number of pumps in the plasma membrane. This possibility is an important consideration because it is accepted that the increase in glucose transport brought about by insulin in both skeletal muscle and adipocytes is a result of the translocation of GLUT 4 glucose transporters in intracellular vesicles to the plasma membrane (5). It is our opinion that insulin does not change the number of Na$^+$,K$^+$-pumps in the plasma membrane. The number of ouabain binding sites in intact adipocytes (12,33) and rat skeletal muscle (12) is not affected by insulin; the number of pumps in adipocytes determined by back-door phosphorylation is not affected by insulin (32); and the number of pumps determined by immunogold electron microscopy is unaltered by insulin in adipocytes (35).

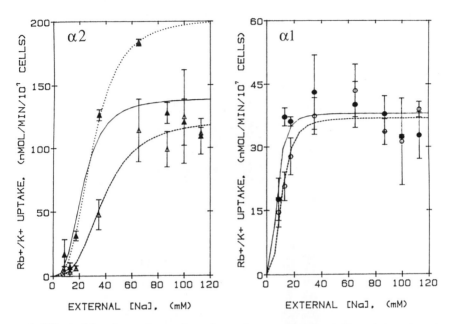

**Figure 1.** Effect of insulin on the sodium dependences of the activities of the α1 and α2 isozymes. Parameters have the units of nmol/min/$10^7$ cells for $V_{max}$ and mM for $K_{0.5}Na^+$. O---O: α1, control; $V_{max}$= 37; $K_{0.5}$=11. ●——●: α1, insulin; $V_{max}$=38; $K_{0.5}$= 8. Δ——Δ: α2, control; $V_{max}$=122; $K_{0.5}$= 39. ▲——▲: α2, insulin (all points used); $V_{max}$=139; $K_{0.5}$= 23. ▲······▲: α2, insulin (three points omitted); $V_{max}$=203; $K_{0.5}$=30. (Reprinted with permission from reference 25)

On the other hand, the number of $Na^+,K^+$-pumps in the sarcolemma of frog (30) and rat (19) skeletal muscle has been reported to be increased by exposure of the muscles to insulin. Our view is that the yield of sarcolemma is so low (approximately 5-8 %) that the change in the number of pumps is related to the effect of insulin on the relative recovery of sarcolemma as compared to that of the plasma membrane of contaminating tissues (blood vessels, neurons, fibroblasts).

Furthermore, the work of both Lytton and McGill suggests that a substantial effect of insulin is on the $K_{0.5}$ for $Na^+$ of the pump, an effect which is independent of a change in the number of pumps. We have suggested that one interpretation of this effect of insulin is the presence of a $Na^+$-dependent inhibitor of the α2 isoform of the $Na^+,K^+$-pump; insulin action would result in an increase in the affinity of the inhibitor for $Na^+$ and a release of the inhibition of the pump (25). This interpretation is compatible with the observation that the $Na^+$-dependence of $Rb^+$ uptake remains highly cooperative in spite of the apparent shift in the $K_{0.5}$ for $Na^+$.

### Phosphorylation of the Na⁺,K⁺-ATPase.

Some evidence suggests that in kidney (1) and brain (3) dopamine causes a decrease in Na⁺,K⁺-ATPase activity, possibly through phosphorylation of the α chain (2,4). Since insulin is known to regulate the activity of glycogen synthase through phosphorylation of protein phosphatase 1 (14), one must inquire whether insulin affects the phosphorylation of the α chain of the Na⁺,K⁺-ATPase.

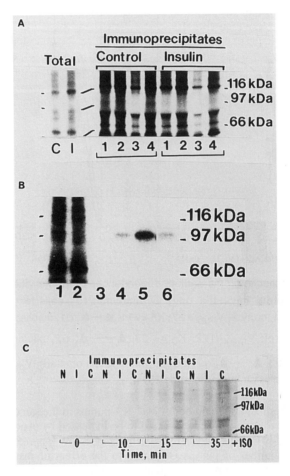

**Figure 2.** Immunoprecipitation of the adipocyte α chains with antiserum 620. **A**. Adipocytes were metabolically labeled with $^{32}$Pi and then incubated with 130 nM insulin for 20 min. The cells were washed and lysed into 2 % SDS, 5 mM EDTA, 1 mM NaVO₄, and 40 mM NaF before addition of the antiserum. Total: C, control; I, insulin. Immunoprecipitates: Lane 1: Nonimmune serum. Lane 2: Immune serum. Lane 3: Immune serum + unlabelled denatured kidney (Na⁺,K⁺)ATPase for competition. Lane 4: Immune serum + unlabelled purified α2 subunit for competition. **B**. Adipocyte membranes were incubated with γ[$^{32}$P]ATP and either 100 mM NaCl, or 100 mM NaCl and 20 mM KCl for 15 s before immunoprecipitation. Lanes 1 and 2 : Total membranes labelled in the absence and presence of KCl, respectively. Lane 3: Nonimmune serum with the reaction in lane 1. Lane

4: Immune serum with the reaction in lane 2. Lane 5: Immune serum and the reaction in lane 1. Lane 6: Immune serum and the reaction in lane 1 + unlabelled denatured kidney (Na,K)ATPase for competition. **C.** Adipocytes were metabolically labeled with $^{32}$Pi, treated with isoproterenol, and subjected to immunoprecipitation. For each time point, there are three conditions: N, nonimmune serum; I, immune serum; C, immune serum + unlabelled denatured α1 subunit of rat kidney (Na,K)ATPase for competition.

With polyclonal antibodies to the α chain, Lytton examined the effect of insulin on the phosphorylation of this polypeptide in rat adipocytes incubated with [$^{32}$P]phosphate to achieve labeling of ATP. The results, shown in Figure 2A, are that there is no observable phosphorylation of the α chain; on the other hand, when the active site aspartate is labeled with γ[$^{32}$P]ATP, a labeled band with apparent Mr=96,000 is clearly visible in the immunoprecipitate, even though most of the label is lost during the SDS PAGE (Fig. 2B). The conclusion is that insulin does not affect the state of phosphorylation of the α chain in adipocytes; since the isoform affected principally by insulin in these cells is α2, the activity of the α2 isoform is not regulated by phosphorylation.

### Activation of the Na$^+$,K$^+$-ATPase by epinephrine.

In most cases insulin and epinephrine have opposing effects on metabolic processes, for instance glycogen metabolism (8). The work of Clausen and Flatman (9), however, shows that in skeletal muscle epinephrine also activates the Na$^+$,K$^+$-pump and causes hyperpolarization of the membrane.

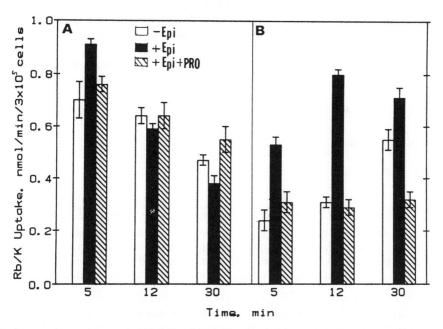

**Figure 3.** Effect of epinephrine on $^{86}$Rb$^+$ uptake by the α1 and α2 isoforms of the (Na,K)ATPase. Adipocytes were incubated for 5, 12, and 30 min with no addition (- Epi), 3 μM epinephrine (+ Epi), or epinephrine and 29 μM propranolol. Rb$^+$ uptake was measured as described in reference 12. **A** : α1 isoform. **B**: α2 isoform.

We asked whether epinephrine also activates the $Na^+,K^+$-pump in rat adipocytes and affects mainly one isoform of the pump. The results of experiments by Janet Lin are shown in Figure 3. The rates of $Rb^+$ uptake by the $\alpha1$ and $\alpha2$ isoforms of the pump are shown after several times of exposure of the cells to epinephrine, in the absence or the presence of propranolol. It is evident that the major effect of the catecholamine is to stimulate the activity of the $\alpha2$ isoform of the $Na^+,K^+$-pump, as is the case with insulin. We inquired, therefore, whether in this situation activation of the pump might be accompanied by phosphorylation of the $\alpha$ chain. Figure 2C shows that there is no detectable phosphorylation of this polypeptide either before or after exposure of the cells to epinephrine. This result supports the previous conclusion: the activity of the $\alpha2$ isoform of the pump is not regulated by phosphorylation of the $\alpha$ chain.

**The $\alpha2$ isoform is necessary but not sufficient for direct activation of the pump by insulin.**

The evidence in the preceding sections indicates that insulin activates the $\alpha2$ isoform of the $Na^+,K^+$-pump. The question is whether the presence of the insulin receptor and this isoform of the pump in a cell is sufficient for activation of the pump.

Luisa DiAntonio studied the effect of insulin on the activation of glucose transport, glycogen synthetase, and $Rb^+$ uptake in guinea pig adipocytes. The reason for this peculiar choice was the report of Horuk et al. (17) that in these adipocytes insulin did not stimulate glucose uptake because there was no redistribution of glucose transporters to the plasma membrane.

Table 1

**Effect of insulin on $Rb^+$ uptake, deoxyglucose uptake, and glycogen synthetase activity in guinea pig and rat adipocytes.**

|  | Guinea pig adipocytes | | | Rat adipocytes | | |
|---|---|---|---|---|---|---|
|  | Fold ± se | n | P value | Fold ±se | n | P value |
| $Rb^+$ uptake | 1.13 ± 0.056 | 8 | 0.050 | 1.42 ± 0.085 | 4 | 0.010 |
| 2-Deoxyglucose uptake | 1.34 ± 0.094 | 5 | 0.010 | 8.60 ± 1.5 | 4 | 0.010 |
| Glycogen synthetase activity | 1.9 | 1 |  | 2.3 | 1 |  |

As is shown in Table 1, insulin does not increase either glucose transport or $Rb^+$ uptake in guinea pig adipocytes; on the other hand, activation of glycogen synthetase is similar to that in rat adipocytes. Surprisingly, there is only the $\alpha2$ isoform in the guinea pig adipocyte plasma membrane as determined by immunoblotting (Figure 4A, lane 4), by ouabain titration of $Rb^+$ uptake (Figure 4B), and by ouabain inhibition of the $Na^+,K^+$-ATPase activity of the adipocyte membranes (Figure 4C). In addition, the apparent $K_{0.5}$ for $Na^+$ of this isoform is the same in intact cells as well as in isolated plasma membranes and comparable to that of the $\alpha1$ isoform of the rat adipocyte. Thus, the activity of the $\alpha2$ isoform of the $Na^+,K^+$-pump in the guinea pig adipocyte is not regulated: there is no inhibition of the activity in the absence of insulin and no relief of the inhibition brought about by insulin. The conclusion is that the guinea pig adipocyte lacks the mechanism to repress the activity of the $\alpha2$ isoform. Thus, the presence of the α2 isoform is necessary but not sufficient for the activation of the $Na^+,K^+$-pump by insulin.

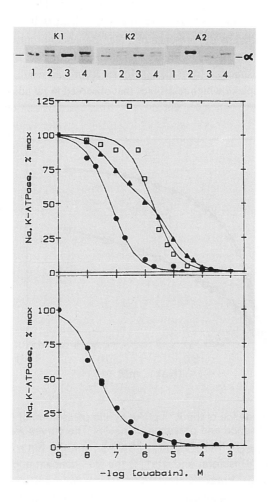

**Figure 4.** Isoforms present in tissues of the guinea pig. **Top panel**: Immunoblots with antibodies K1, K2, and A2 (from K. Sweadner) were done as in reference 23. Lane 1: Guinea pig kidney $Na^+,K^+$-ATPase. Lane 2: Guinea pig brain membranes. Lane 3: Dog kidney $Na^+,K^+$-ATPase. Lane 4: Guinea pig adipocyte membranes. **Middle panel**: Ouabain inhibition of $^{86}Rb^+$ uptake by guinea pig adipocytes (--●--, $K_i = 6.5 \times 10^{-8}$ M), of guinea pig brain membrane $Na^+,K^+$-ATPase activity (--▲--, $K_{i1} = 7.6 \times 10^{-8}$ and $K_{i2} = 7.4 \times 10^{-6}$ M), and of purified guinea pig kidney $Na^+,K^+$-ATPase activity (--□--, $K_i = 1.8 \times 10^{-6}$ M). **Bottom panel** : Ouabain inhibition of guinea pig adipocyte membrane $Na^+,K^+$-ATPase activity (--●--, $K_i = 2.1 \times 10^{-8}$ M).

Accordingly, we propose that signal transduction from the insulin receptor to the α2 isoform of the $Na^+,K^+$-pump involves at least four participants : the insulin receptor, the α2 isoform of the pump, a repressor that inactivates the pump, and an insulin-dependent messenger that derepresses the pump.

675

**The repressor of the α2 isoform of the Na⁺,K⁺-pump.**

The effect of insulin on the activity of the α2 isoform of the Na⁺,K⁺-pump in rat adipocytes (Figure 1) appears to be mediated by an increase in Vmax and a decrease in the apparent $K_{0.5}$ for Na⁺. We consider that these changes in the activity of the pump can be simply explained by the presence of a Na⁺ dependent repressor of the α2 isoform. As can be seen from the curves in Figure 5, a change in the Na⁺ affinity of the repressor, caused by signal transduction from the insulin receptor, can result in a change in the Na⁺ dependence of Rb⁺ uptake which resembles that observed in rat adipocytes (Figure 1).

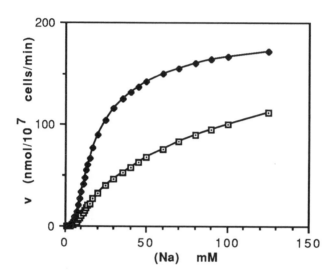

**Figure 5.** Na⁺ dependence of the K⁺ uptake by the α2 isoform of the Na⁺,K⁺-ATPase in adipocytes in the absence and presence of insulin. The curves were calculated with the equation : $v = (Vmax_{app} \times [Na^+]^3)/(Km^3 + [Na^+]^3)$ , in which Km = 10 mM is the apparent $K_{0.5}$ for Na⁺ of the α2 isoform and [Na⁺] is the Na⁺ concentration in mM. $Vmax_{app}$ is defined by : $Vmax_{app} = Vmax (K+[Na^+])/(Ka+[Na^+])$, where K is a constant that defines the Na⁺ dependence of the inhibition of Vmax (200 nmol/10⁷ cells/min) by the inhibitor I present at a concentration $[I]/K_I = 50$, and $Ka = K(1+[I]/K_I)$. The values of K are 2 mM for the curve -insulin and 0.4 mM for the curve +insulin.

The existence of a repressor of the α2 isoform of the pump is supported by the following observations. Brodsky (7) showed that in synaptosomes the relative pumping activities of the α1 and α2 isoforms are in a ration of 3:1; the ATPase activities of the α1 and α2 isoforms in synaptosomal membranes, however, are in a ratio of 1:4, which corresponds to the relative amounts of the α1 and α2 polypeptides (Table 2). Obviously the activity of the α2 isoform in synaptosomes is repressed by a mechanism that is not present in the synaptosomal membranes: both the $K_{0.5}$ for Na⁺ and the Vmax of the α2 isoform are affected in synaptosomes. In the second place, the relationship between the rate of Rb⁺ uptake and the intracellular Na⁺ concentration of the α2 isoform in rat adipocytes shows the same degree of positive cooperativity in the unstimulated and insulin-stimulated state. This can occurr only if all the molecules of the α2 isoform are

676

in rapid equilibrium with a dissociable repressor; if there were two populations of α2 molecules with different Na⁺ affinities and Vmax values, the curves would show evidence of negative cooperative behavior.

**Table 2**

**Ouabain affinities, $K_{0.5}$ for Na⁺ and fractional pumping activities of the α1 and α2 isoforms in rat brain synaptosomes.**

| | $K_{0.5}$ (ouabain) (μM) | Vmax[a] | Pump rate Basal[b] | Fractional | $K_{0.5}$ (Na⁺) (mM) |
|---|---|---|---|---|---|
| α1 | 110 | 0.26 | 73 | 1.0 | 12 |
| α2 | 1.5 | 1.04 | 27 | 0.09 | 36 |

a, μmol ATP hydrolyzed/mg/min;  b, nmol Rb⁺ uptake/mg/5min.

Finally, it appears appropriate to imagine that the affinity of the repressor should be dependent on the [Na⁺]; as the [Na⁺]ᵢ increases, the repressor is released and the activity of the α2 isoform increases. This arrangement would allow the cell to store a substantial part of the Na⁺,K⁺-pump in the plasma membrane in an inactive state until additional pumping activity is required, as happens with insulin stimulation and might happen with a sudden Na⁺ influx. The advantage of this device is shown in Figure 6, which depicts the activity of the Na⁺,K⁺-pump as a function of the [Na⁺]ᵢ in three situations.

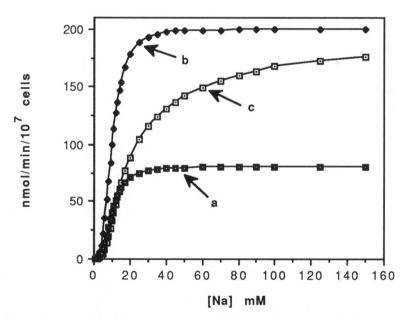

**Figure 6.** Sodium dependence of K⁺ transport activity of thr Na⁺,K⁺-pump under three conditions. In **a**, the pump has a Vmax= 80 nmol K/10⁷ cells/min and a $K_{0.5}$ for Na⁺ of 10

mM. In **b**, the pump Vmax is 200 nmol/$10^7$ cells/min and $K_{0.5}$ for $Na^+$ =10 mM. In **c**, the conditions are as in case **b**, except that Vmax $_{app}$ is a function of $[Na^+]i$ : as in Figure 5, Vmax app= Vmax(K+[Na^+])/(Ka+[Na^+]), where Vmax=200 nmol/$10^7$cells/min, K=2 mM.

In **a**, there is one isoform of the pump with Vmax for $K^+$ entry of 80 nmol/$10^7$ cells/min and a $K_{0.5}$ for $Na^+$ of 10 mM. In **b**, is shown the activity vs $Na^+$ concentration curve for cells in which the Vmax is increased to 200 nmol/$10^7$ cells/min with the same $K_{0.5}$. Increasing the amount of enzyme in the cell results in an increase in the amount of pumping activity at 10 mM $[Na^+]_i$; the pumping activity can be mantained at the level observed for 10 mM $[Na^+]_i$ in case **a** by a decrease in the $[Na^+]_i$ to 5 mM. In **c**, there is also one isoform of the pump with $K_{0.5}$ for $Na^+$ of 10 mM and a Vmax of 200 nmol/$10^7$ cells/min which varies with the $[Na^+]i$ according to Vmax $_{app}$ = Vmax (K+[Na^+])/(Ka+[Na^+]). The curves illustrate the point that the ion metabolism of a cell need not be perturbed by increasing the amount of the pump in the membrane if the activity of the additional pumps can be regulated. Thus, the reason for the existence of the α2 isoform is to provide the ability to activate inactive $Na^+,K^+$-pumps on demand.

## $Na^+$ metabolism during insulin action.

Insulin activation of the $Na^+,K^+$-pump in skeletal muscle and adipocytes is not a transient event; it persists for as long as insulin is present. In this new steady state, there is increased active transport of $Na^+$ and $K^+$ by the pump and increased movement of these ions down their concentration gradients. In the case of $Na^+$, which is a principal substrate of the pump, the entry of $Na^+$ into the cell must be increased sufficiently to maintain the increased activity of the pump. The question is whether $Na^+$ entry is also regulated by insulin.

In 1977, Clausen and Kohn (13) showed that insulin stimulates the entry of $Na^+$ in rat skeletal muscle; this activity was not affected by amiloride or bumetanide (10,18), suggesting that neither a $Na^+/H^+$ exchanger or a $Na,K,Cl_2$ cotransporter were involved. The view that a $Na^+$ conductance is affected by insulin is supported by two other observations. The skeletal muscles of hypokalemic rats are depolarized by insulin in media containing low $[K^+]$ (21,29,31). In the muscles of these animals, which have only 10 % of the α2 isoform present in normal muscle (18,27), there is insufficient $Na^+,K^+$-pump activity at the low $[K^+]_o$ to pump out of the cell the $Na^+$ entering through the insulin-activated channel; thus the plasma membrane depolarizes rather than hyperpolarizes as happens with normal rats. In the second place, insulin decreases the hyperpolarization of the membrane of skeletal muscle caused by epinephrine (16) even though both hormones activate the α2 isoform of the $Na^+,K^+$-pump; the reason is that only insulin increases the $Na^+$ conductance of the membrane.

## Insulin-stimulated Cation Channel.

Julie McGeoch has identified an insulin-sensitive cation channel by patching rat skeletal muscle plasma membrane reconstituted in a phospholipid bilayer (24). The properties of the channel are shown in Figures 7, which illustrates the dependence of the insulin- and GTP-stimulated current on voltage, and on the concentration of insulin, cGMP, μ-conotoxin, and $Ca^{2+}$.

The channel is relatively nonspecific allowing the movement of $Na^+$, $K^+$, $Mg^{2+}$, and $Ca^{2+}$; the $Na^+$ conductance is 10-30 pS at 10 mM $Na^+$. The channel is activated by insulin on one side of the patch and GTP on the other side, or by cGMP alone; the Km for insulin is 5 nM and for cGMP 0.4 μM. $Ca^{2+}$ on the cytoplasmic surface of the channel inhibits the current with a Ki of 2 μM, as does μ-conotoxin on the extracellular surface of the channel with a Ki of 50 nM. This cation channel is the principal portal of entry of $Na^+$ into skeletal muscle cells, and possibly other cells as well.

It has been possible recently to monitor the $[Na^+]i$ of rat skeletal muscle cells with the fluorescent dye SBFI (sodium-binding benzofuran isophthalate). The results (data not shown) support the view that insulin increases $Na^+$ entry into muscle cells. In the presence of ouabain to block the $Na^+,K^+$-pump, insulin causes an increase in $[Na^+]i$, which is blocked by μ-conotoxin a specific inhibitor of the channel. However, in the absence of ouabain, the increase in $Na^+$ entry is balanced by activation of the $Na^+,K^+$-pump with the result that the $[Na^+]i$ decreases, as has been previously shown by other methods (13,25).

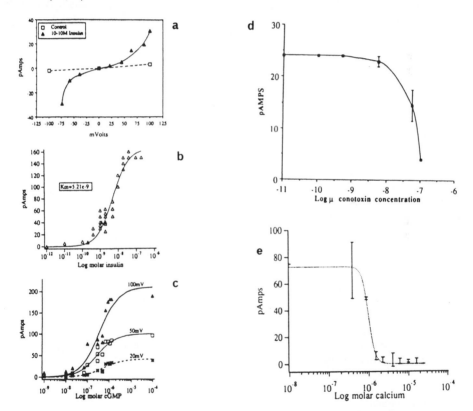

**Figure 7**. Characteristics of the cation channel. **a**, I/V plot with 10 mM $Na^+$ on both sides of the patch; insulin was added to the bath to a final concentration of $10^{-10}$ M. **b**, channel current as a function of the insulin concentration (Km=5.2 nM). **c**, $Na^+$ current as a function of [cGMP] at three voltages (Km=0.37 μM). **d**, inhibition of channel current by μ-conotoxin (Ki=0.5 nM). **e**, calcium inhibition of the channel current (Ki=1 μM). (Reprinted with permission from reference 24).

679

## Conclusions.

Insulin increases the uptake of $K^+$ from the blood into skeletal muscle, presumably to clear dietary $K^+$ from the blood. The mechanism of this effect involves the $\alpha2$ isoform of the $Na^+,K^+$-pump, which is the principal isoform in skeletal muscle. The critical property of this isoform is that it can be regulated by interaction with a repressor that modulates its Vmax and its $K_{0.5}$ for $Na^+$. The repressor-pump interaction is affected by the $[Na^+]i$ and by signal transduction from the insulin repressor, leading to stimulation of the $Na^+,K^+$-pump by increased entry of $Na^+$ and by insulin action. In skeletal muscle, insulin also activates a cation channel which allows $Na^+$ to enter the cell at an increased rate and to fuel the activated $Na^+,K^+$-pump; in this way a new steady state is obtained with a slightly lowered $[Na^+]i$ and a raised $[K^+]i$ which persists until the insulin concentration decreases.

The situation is described by the cartoon shown in Figure 8. The hypothesis is that there are at least four components required for the activation of the $Na^+,K^+$-pump by insulin: the $\alpha2$ isoform of the $Na^+,K^+$-ATPase, a repressor of the activity of the $\alpha2$ isoform, an effector that releases the repression, and the insulin receptor that activates the effector. The pathway that leads to activation of the cation channel is still obscure.

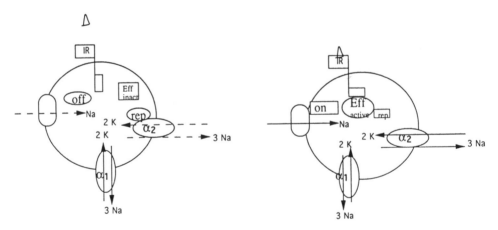

**Figure 8**. Cartoon depicting the putative mechanism of activation of the a2 isoform of the $Na^+,K^+$-pump and the $Na^+$ channel by insulin.

Acknowledgements.

The work was supported by grants from the NIH.

680

# References.

1. Aperia, A., Bertolello, A., and Seri, I. (1987) Am. J. Physiol. 252, F39-F45.
2. Aperia, A., Ibarra, F., Svensson, L.B., Klee, C., and Greegard, P. (1992) Proc. Natl. Acad. Sci. 89, 7394-7397.
3. Bertolello, A.M., Hopfield, J.F., Aperia, A., and Greengard, P. (1990) Nature 347, 386-388.
4. Bertolello, A.M., Aperia, A., Walaas, S.I., Nairn, A,C., and Greengard, P. (1991) Proc. Natl. Acad. Sci. 88, 11359-11362.
5. Birnbaum, M.J. (1989) Cell 57, 305-315.
6. Briggs, A.P., Koechig, I., Doisy, E.A., and Weber, C.J. (1924) J. Biol. Chem. 58, 721-730.
7. Brodsky, J., and Guidotti G. (1990) Am. J. Physiol. 258, C803-C811.
8. Cohen, P. (1986) in The Enzymes, 3rd. edn. (Boyer, P.D. and Krebs, E,G,, eds) Vol 17 pp361-397, Academic Press, Orlando, FL.
9. Clausen, T., and Flatman, J.A. (1977) J. Physiol. 270. 383-414.
10. Clausen, T., and Flatman, J.A. (1987) Am. J. Physiol. 252, E492-E499.
11. Clausen, T., and Hansen, O. (1974) Biochim. Biophys. Acta 345, 387-404.
12. Clausen, T., and Hansen, O. (1977) J. Physiol. 270, 415-430.
13. Clausen, T., and Kohn, P.B. (1977) J. Physiol. 265, 19-42.
14. Dent, P., Lavoinne, A., Nakielny, S., Caudwell, F.B., Watt, P., and Cohen, P. (1990) Nature 348, 301-308.
15. Fehlmann, M., and Freychet, P. (1981) J. Biol. Chem. 256, 7449-7453.
16. Flatman, J.A., and Clausen, T. (1979) Nature 281, 580-581.
17. Horuk, R., Rodbell, M., Cushman, S.W., and Wardzala, L.J. (1983) J. Biol. Chem. 258, 7425-7429.
18. Hsu, Y.M., and Guidotti, G. (1991) J. Biol. Chem. 266, 427-433.
19. Hundal, H.S., Marette, A., Mitsumoto, Y., Ramlal, J., Blostein, R., ans Klip, A. (1992) J. Biol. Chem. 267, 5040-5043.
20. Kamminga, C.E., Willebrands, A.F., Groen, J., and Blickman, J.R. (1950) Science 111, 30-31.
21. Kao, I., and Gordon, A.M. (1975) Science 188, 740-741.
22. Lytton, J. (1985) J. Biol. Chem. 260, 10075-10080.
23. Lytton, J., Lin, J.C., and Guidotti, G. (1985) J. Biol. Chem. 260, 1177-1184.
24. McGeoch, J.E.M., and Guidotti, G. (1992) J. Biol. Chem. 267, 832-841.
25. McGill, D. and Guidotti, G. (1991) J. Biol. Chem. 266, 15824-15831.
26. Moore, R.D. (1981) Biophys. J. 33, 203-210.
27. Norgaard, A., Kjeldsen, K., and Clausen, T. (1981) Nature 293, 739-741.
28. Norgaard, A., Kjeldsen, K., and Clausen, T. (1984) Biochim. Biophys. Acta 770, 203-209.
29. Offerijns, F.G.J., Westerink, D., and Willebrands, A.F. (1958) J. Physiol. 141, 377-384.
30. Omatsu-Kambe, M., and Kitasato, H. (1990) Biochem. J. 272, 723-733.
31. Otsuka, M., and Ohtsuki, I. (1970) Am. J. Physiol. 219, 1178-1182.
32. Resh, M.D. (1982) J. Biol. Chem. 257, 11946-11952.
33. Resh, M. D., Nemenoff, R., and Guidotti, G. (1980) J. Biol. Chem. 255, 10438-10445.
34. Rosic, N.K., Standaert, M.L., and Pollet, R.J. (1985) J. Biol. Chem. 260, 6206-6212.
35. Voldstedlund, M., Tranum-Jensen, J., and Vinten, J. (1993) J. Membrane Biol. 136, 63-73.

# Phosphorylation of Na,K-ATPase by Protein Kinases: Structure-Functions Relationship

P. Beguin, A.T. Beggah, A.V. Chibalin, L.A. Vasilets, B.C. Rossier, F. Jaisser, K. Geering

Institut de Pharmacologie et de Toxicologie de l'Université, CH-1005 Lausanne, Switzerland

## Introduction

In view of the variety of functions that depend on $Na^+,K^+$-ATPase in different tissues, it is essential that the activity of this enzyme is regulated in response to changing physiological demands. Recent experimental evidence suggest that $Na^+,K^+$-ATPase might be a candidate for regulatory phosphorylation by protein kinases. Physiological agonists and drugs which modulate protein kinase A and C, or intracellular phosphatases, have been shown to affect $Na^+,K^+$-ATPase activity (1,4,5,9.10). In addition, in vitro (2,3) and in vivo (8) phosphorylation of the $\alpha$-subunit of $Na^+,K^+$-ATPase upon protein kinase C stimulation has been reported. However, so far no evidence has been provided that a direct link exists between phosphorylation of the $\alpha$-subunit and modulation of $Na^+,K^+$-pump activity. To approach this question, we have attempted to identify potential phosphorylation sites on the $\alpha$-subunit for protein kinase A and C. Threonine (T939) and serine (S943) residues in a consensus phosphorylation sequence (939TRRNS943) in the C-terminus of the $\alpha$-subunit of Bufo marinus $Na^+,K^+$-ATPase (6) were mutated. The mutants were expressed in Xenopus oocytes and phosphorylation of the $\alpha$-subunit was studied in oocyte homogenates upon stimulation of oocyte protein kinases A or C (3).

## Results and Discussion

To verify that $\alpha$ wild type ($\alpha$ wt) and $\alpha$ mutants are similarly expressed in Xenopus oocytes, the $\alpha$-subunits were immunoprecipitated from Triton extracts of oocytes previously injected with Bufo $\beta$ cRNA (6), together with $\alpha$ wt or a mutant cRNA and metabolically labelled with $^{35}$S-methionine. Compared to non-injected oocytes (Fig 1A, lane 1), the $\alpha$ wt (lane 2), the single mutant S943A (lane 3) and the double mutant T939A/S943A (lane 4), led to a similar and significant increase in the cellular accumulation of $\alpha$-subunits. For the phosphorylation studies, non-injected or cRNA injected oocytes were incubated for 3 days to permit expression of functional $Na^+,K^+$-pumps at the cell surface. Homogenates were prepared and supplemented with $[\gamma\text{-}^{32}P]ATP$ and cAMP or phorbol 12-myriastate 13-acetate

(PMA) before immunoprecipitation with an anti α-serum was performed. Fig 1B shows that the endogenous oocyte α-subunit was not phosphorylated in the absence of stimulators (Fig 1B, lane 1). The phosphorylation of the oocyte α-subunit could be increased by cAMP (lane 2), however, only in the presence of 0.2 % Triton (3). The oocyte α-subunit was not phosphorylated upon stimulation of protein kinase C with PMA (lane 3). In cRNA injected oocytes, the signal of phosphorylated α-subunit in the presence of cAMP increased several folds over the signal observed in non-injected controls (compare lane 5 to lane 2), which reflects the expression of exogenous Na$^+$,K$^+$-pumps. In contrast to endogenous oocyte α-subunits, the exogenous Bufo α-subunits became phosphorylated upon stimulation of protein kinase C with PMA (compare lane 6 to lane 3). The phosphorylation signal in cAMP stimulated oocytes expressing the single mutant S943A (lane 8) or the double mutant T939A/S943A (lane 11) was much lower than in oocytes expressing the α wt (compare lanes 8 and 11 to lane 5) and fell to the level of non-injected oocytes (compare lanes 8 and 11 to lane 2). In contrast, incubation of oocyte homogenates with PMA did not influence the phosphorylation state of the mutated α-subunits compared to α wt (compare lanes 9 and 12 to lane 6).

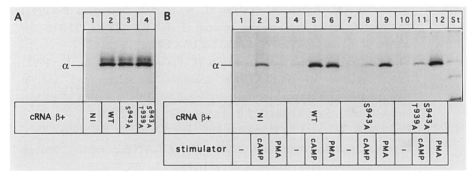

**Figure 1.** Expression and phosphorylation by protein kinase A and C stimulation of wild type and mutated α-subunits of Na$^+$,K$^+$-ATPase. **A. Cellular expression of α-subunits.** Oocytes were either not injected (NI, lane 1) or injected with 0.3 ng β cRNA, together with 7 ng α wt (WT, lane 2), S943A mutant (lane 3) or T939A/S943A mutant cRNA (lane 4). Oocytes were metabolically labelled with $^{35}$S-methionine for 24h and after a chase period of 2 days, the α-subunits were immunoprecipitated from Triton extracts, as previously described (7). **B. Phosphorylation of α-subunits.** Oocytes were injected as described in A and incubated for 3 days. Homogenates were prepared and supplemented with (g-$^{32}$P) ATP in the absence (lanes 1,4,7,10) or presence of 50 μM cAMP and 0.2 % Triton X-100 (lanes 2,5,8,11) or 100 nM PMA and 1.5 mM CaCl$_2$ (lanes 3,6,9,12) and after 50 min of incubation, the α-subunit was immunoprecipitated as described (3).

In conclusion, the data indicate that the α-subunit of Na$^+$,K$^+$-ATPase can indeed be phosphorylated under certain experimental conditions upon stimulation of protein kinase A or C. The results obtained with mutants affected in a highly conserved phosphorylation consensus sequence in the C-terminus of the α-subunit rule out that threonine 939 or serine 943 are target residues for phosphorylation by protein kinase C. On the other hand, a unique phosphorylation site for protein kinase A appears to be located at serine 943 in the α-subunit of Bufo Na$^+$,K$^+$-ATPase. Since phosphorylation of α-subunits upon stimulation of protein kinase A can only be revealed in the presence of detergent, its physiological relevance remains, however, questionable.

## Acknowledgements

The studies were supported by the Swiss National Fund for Scientific Research (Grant No 31-26241-89 and 31-33676-92).

## References

1.  Bertorello A, Aperia A (1989) Na$^+$-K$^+$-ATPase is an effector protein for protein kinase C in renal proximal tubule cells. Am J Physiol 256: F370-F373

2.  Bertorello AM, Aperia A, Walaas SI, Nairn AC, Greengard P (1991) Phosphorylation of the catalytic subunit of Na$^+$,K$^+$-ATPase inhibits the activity of the enzyme. Proc Natl Acad Sci USA 88: 11359-11362

3.  Chibalin AV, Vasilets LA, Hennekes H, Pralong D, Geering K (1992) Phosphorylation of Na,K-ATPase a-subunits in microsomes and in homogenates of Xenopus oocytes resulting from the stimulation of protein kinase-A and protein kinase-C. J Biol Chem 267: 22378-22384

4.  Horiuchi A, Takeyasu K, Mouradian MM, Jose PA, Felder RA (1993) D$_{1A}$ dopamine receptor stimulation inhibits Na$^+$/K$^+$-ATPase activity through protein kinase-A. Mol Pharmacol 43: 281-285

5.  Ibarra F, Aperia A, Svensson LB, Eklof AC, Greengard P (1993) Bidirectional regulation of Na$^+$,K$^+$-ATPase activity by dopamine and an a-adrenergic agonist. Proc Natl Acad Sci USA 90: 21-24

6.  Jaisser F, Canessa CM, Horisberger JD, Rossier BC (1992) Primary sequence and functional expression of a novel ouabain-resistant Na,K-ATPase - The b-subunit modulates potassium activation of the Na,K-pump. J Biol Chem 267: 16895-16903

7.  Jaunin P, Horisberger JD, Richter K, Good PJ, Rossier BC, Geering K (1992) Processing, intracellular transport and functional expression of endogenous and exogenous a-b$_3$ Na,K-ATPase complexes in Xenopus oocytes. J Biol Chem 267: 577-585

8.   Middleton JP, Khan WA, Collinsworth G, Hannun YA, Medford RM (1993) Heterogeneity of protein kinase-C-mediated rapid regulation of Na/K-ATPase in kidney epithelial cells. J Biol Chem 268: 15958-15964

9.   Smart JL, Deth RC (1988) Influence of a1-adrenergic receptor stimulation and phorbol esters on hepatic $Na^+/K^+$-ATPase activity. Pharmacology 37: 94-104

10.  Vasilets LA, Schwarz W (1992) Regulation of endogenous and expressed $Na^+/K^+$ pumps in Xenopus oocytes by membrane potential and stimulation of protein kinases. J Membr Biol 125: 119-132

# Stimulation by Phorbol 12-Myristate 13-Acetate (PMA) of Na,K-Pump Mediated $^{86}$Rb$^+$-Influx Into Human Lymphocytes

J.G. Nørby and N. Obel

Institute of Biophysics, University of Aarhus, DK-8000 Aarhus C, Denmark

## Introduction

The Na,K-pump in cells is a target for both long-term (latency of 1-24 hours) and acute (seconds-minutes) regulation (5,6). In both cases stimulation or decrease in pumping activity is observed. The subject of this paper is the acute up-regulation of the pump, for which a molecular explanation is lacking. We report an acute, stimulating effect of the protein kinase C activator PMA on the pump mediated K$^+$-influx into lymphocytes. Ionomycin has a similar effect, and amiloride does not abolish the stimulation.

## Methods

Preparation of lymphocytes: Buffy-coat from 0.5 l freshly drawn human blood was diluted 1:1 with 0.9% NaCl, layered over Lymphoprep (Nycomed, density 1077 kg/m$^3$) and fractionated by centrifugation (3). The lymphocytes were harvested from a layer above the Lymphoprep and washed 3 times in 0.9% NaCl by centrifugation (14). If the cells were not used immediately, they were suspended in the growth medium RPMI 1670 (Gibco BRL, with 10% serum, L-glutamin, streptomycin and penicillin) and kept in an atmosphere of 5% CO$_2$ at 37 °C for not more than 3 days before use.

Measurement of K$^+$-influx: All fluxes were measured as cellular uptake of $^{86}$Rb (about 2000 cpm/µl assay) at 37 °C from a medium (11) containing (mM): NaCl 135, KCl 5, MgSO$_4$ 1, CaCl$_2$ 1, Tris base 12.5, HCl 8.5, D-glucose 5, and bovine serum albumin 1 mg/ml. pH was 7.4 and osmolarity 290 mOsm/l. The assay volume was 500-600 µl containing about 5·10$^6$ lymphocytes, which were washed in the medium and counted prior to the assay. For each time point a 200 µl sample was transferred to a small plastic tube (Fig. 1) containing 5 µl Lymphoprep and 100 µl "oil", i.e. a 3:1 mixture of dibutyl- and dinonylphthalate. The tube was centrifuged for 2-3 min. at 6000·g and the tip containing the cells was cut off and transferred to vials for counting of $^{86}$Rb.

## Results

Reproducibility of fluxes: To ascertain that the lymphocytes were in steady-state at the beginning of the assay they were preincubated for 45-60 min. at 37 °C before addition of $^{86}$Rb. It was found (not shown) that the K$^+$-influx was higher and less reproducible if the cells in stead were kept on ice with only a short preincubation period at 37 °C. This treatment will lead to an increase in cytosolic [Na$^+$] and thereby an increased Na,K-pump activity (10). This was abolished by the longer acclimatization period at 37 °C.

686

Characteristics of the K$^+$-influx: The flux-curve on Fig.1 has an Y-axis intercept which corresponds to a contamination of the cell pellet with $1.27 \pm 0.04$ µl flux medium/10$^7$ cells (n = 31). Assuming a cellular volume of 1.8 µl/10$^7$ cells (13), the contamination amounts to 40% of the pellet volume, in agreement with earlier observations (1,10). It is equivalent to a 0.6 µm thick layer of medium around a cell with a radius of 3.5 µm.

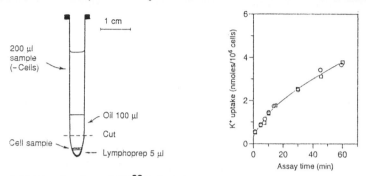

**Figure 1. Determination of K$^+$($^{86}$Rb$^+$)-influx into human lymphocytes.** The lymphocytes were separated from the flux medium by centrifugation through "oil" (1) with a density about 1030 kg/m$^3$ (see Methods). The figure (left) shows the tube after centrifugation. The 5 µl Lymphoprep, density 1077 kg/m$^3$, is essential to prevent sticking of the cells to the bottom after centrifugation. The curve (right) represents "total influx" and is obtained after 45 minutes preincubation in the flux medium at 37 °C before addition of $^{86}$Rb$^+$. It illustrates the typical Y-axis intercept at t = 0 min. and the often, but not always, encountered biphasic time course. See the text for further explanation and calculations.

The flux curve in Fig.1 is biphasic: The intial, more rapid phase is seen in several (but not all, see Fig.2) experiments. Since it disappears in the presence of ouabain, it is not a leak flux. If we assume that this rapid flux reflects a Na,K-pump mediated flux into cells with a volume of 1.8 µl/10$^7$ cells and an intracellular [K$^+$] $\sim$ 120 mM (10,12), and that these cells attain isotope equilibrium within 30 min., then we can estimate that the fast flux only represents the pump in $5.1 \pm 0.8$ % of the cells (n = 11). Note that the magnitude of the rapid phase is not affected by PMA (Fig.2) or ionomycin (not shown).

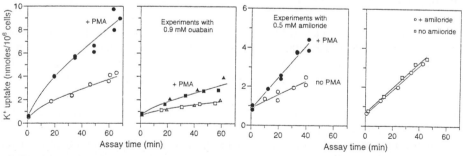

**Figure 2. Stimulation of K$^+$-influx by PMA and lack of effect of amiloride.** The cells were preincubated in the flux medium at 37 °C for 50 min. before addition of PMA, 120 ng/ml (filled symbols). 3 min. later 0.9 mM ouabain (panel 2) or buffer (panels 1,3,4) was added and 5 min. later the flux assay was started by addition of $^{86}$Rb. Panel 3 shows the lack of effect of 0.5 mM amiloride which was added 2 min. before the PMA.

Stimulation of $K^+$-influx by PMA (Fig.2) and ionomycin (not shown): The fluxes reported below are all estimated from the (approximately) linear portion of the flux curve, for $t > 20$ min., which according to the above discussion presumably represents more than 90 % of the cells. The fluxes recorded are listed in Table 1.

**Table 1. Stimulation of $K^+$-influx, nmol/($10^6$cells·hr), into lymphocytes by PMA**

| | no PMA | | | with PMA | | |
|---|---|---|---|---|---|---|
| | Total flux | Ouab. insens. | Pump flux | Total flux | Ouab. insens. | Pump flux |
| | 3.4±0.3 | 1.3±0.4 | 2.1 | 5.9±0.5 | 1.7±0.4 | 4.2 |
| n = | 15 | 4 | | 8 | 3 | |

The normal fluxes (no PMA) are comparable to earlier measurements of $K^+$-influxes in lymphocytes (e.g. ref. 2). Both ouabain insensitive and pump fluxes were approximately doubled by addition of PMA, as seen also in paired experiments (Fig. 2, panels 1,2). The major portion (80%) of the increase in $K^+$-influx was due to a stimulation of the pump-flux by PMA. Amiloride, a $Na^+/H^+$-exchange inhibitor, had no effect on the fluxes or on the stimulation by PMA (Fig.2, panels 3,4). This indicates that the effect of PMA on the pump flux is not due to a stoichiometric stimulation by an increased intracellular $[Na^+]$ obtained via the $Na^+/H^+$ antiporter.

The $Ca^{2+}$-ionophore ionomycin also stimulated the pump $K^+$-influx rapidly and to the same extent as PMA (not shown). The mitogen phytohaemagglutinin (PHA), however, had no short-term effect on the $K^+$-fluxes. It should be noted that the PHA stimulation of Na,K-pump fluxes found by others have always been after a preincubation period with PHA of at least 1 hr (and usually much longer).

**Discussion**

Phorbol 12-myristate 13-acetate (PMA) is among the many compounds known to activate protein kinase C (PKC), and although this may not be its only effect on the cell, it is tempting to speculate that the acute stimulation of Na,K-pump mediated $K^+$-influx by PMA, which is observed in the present study, occurs via a protein kinase C dependent pathway. It has been shown that the catalytic $\alpha$-subunit of Na,K-ATPase in purified and microsomal preparations (4,9) can serve as a substrate for PKC, but no ATPase activities were reported in these studies. In two reports (7,8) there is evidence that activation of PKC may lead to stimulation of Na,K-ATPase activity, like it has been observed here. To the contrary it was found in a recent study with oocytes, that activation of PKC by phorbol esters induces downregulation of the Na,K-ATPase (15).

The activating effect of ionomycin (not shown) could in this study also be via PKC, since this enzyme is presumed to be stimulated by a rise in cellular $[Ca^{2+}]$.

688

The conclusion at this point must be that the mechanism for the acute regulation of cellular Na,K-ATPase and Na,K-pump activity is still unknown. There is evidence that regulatory mechanisms involving protein kinases could be important, but at the moment there is no agreement as to whether PKC stimulation results in activation or inhibition of the pump, or whether the regulatory mechanisms are tissue specific or modulated by factors not controlled in the experiments discussed here.

*Acknowledgements*

The excellent technical assistance by Susanne Pedersen is gratefully acknowledged. This work was supported in part by the Aarhus University Research Foundation and by the Biomembrane Research Centre, Aarhus University.

## References

1. Andreasen PA, Schaumburg BP, Østerlind K, Vinten J, Gammeltoft S, Glieman J (1974) Anal Biochem 59:610-616
2. Averdunk R, Lauf PK (1975) Exptl Cell Res 93:331-342
3. Bøyum A (1984) Methods Enzymol 108:88-102
4. Chibalin AV, Vasilets LA, Hennekes H, Pralong D, Geering K (1992) J Biol Chem 267:22378-22384
5. Clausen T (1986) Physiol Rev 66:542-580
6. Gick GG, Ismail-Beigi F, Edelman IS (1988) In: Skou JC, Nørby JG, Maunsbach, AB, Esmann, M (eds) The Na$^+$,K$^+$-Pump, Part B: Cellular Aspects. Alan R Liss, New York, pp 277-295
7. Gupta S, Ruderman NB, Cragoe EJ, Sussman I (1991) Am J Physiol 261:H38-H45
8. Lattimer SA, Sima AAF, Green DA (1989) Am J Physiol 256:E264-E269
9. Lowndes ML, Hokin-Neaverson M, Bertics PJ (1990) Biochim Biophys Acta 1052:143-151
10. Pedersen KE, Klitgaard NA, Jest P (1987) Scand J Clin Lab Invest 47:801-811
11. Petersen RH, Knudsen T, Johansen T (1991) FEBS Lett 279:153-156
12. Segel GB, Lichtman MA, Hollander MM, Gordon BR, Klemperer MR (1976) J Cell Physiol 88:43-48
13. Stevenson HC (1984) Methods Enzymol 108:242-249
14. Thorsby E, Bratlie A (1970) In: Terasaki P (ed) Histocompatibility Testing, Munksgaard, København, pp 655-656
15. Vasilets LA, Schmalzing G, Mädefessel K, Haase W, Schwarz W (1990) J Membrane Biol 118:131-142

# Regulation of the Sodium Pump by Fatty Acids and Acylglycerols

Z. Xie, M. Jack-Hays, Y. Wang, W.-H. Huang, A. Askari

Department of Pharmacology, Medical College of Ohio, P.O. Box 10008, Toledo, Ohio 43699-0008 USA

## Introduction

Several years ago we noted that some common laboratory detergents, at low concentrations that do not solubilize the purified membrane-bound $Na^+/K^+$-ATPase, lower the apparent $K_m$ of ATP without affecting maximal velocity (4). Because this seemed to be a nonspecific effect of these amphiphiles, at the time we suggested the possibility that some endogenous amphiphiles may regulate the pump by allowing it to operate at full capacity when ATP is in short supply (4). Our subsequent studies on the interactions of the enzyme with fatty acids and their CoA esters established that these endogenous amphiphiles do indeed activate the pump when ATP is suboptimal, clarified the mechanism of this activation, and showed that there was some measure of specificity in these interactions (5-8). More recently, it occurred to us that receptor-linked regulation of the pump in some tissues may be related to direct interaction of $Na^+/K^+$-ATPase with the products of receptor-linked hydrolysis of membrane phospholipids, and we showed that monoacylglycerol (MAG), but not inositoltrisphosphate and diacylglycerol (DAG), also activates the pump when ATP is not optimal (1). Here, we present further observations on the specificities of the effects of these hydrophobic ligands on $Na^+/K^+$-ATPase.

## Results and Discussion

Data of Fig. 1 are typical of the effects of a long-chain fatty acid (FA) and the corresponding DAG and MAG. The FA has a biphasic effect (activation at lower levels, followed by inhibition at higher concentrations), DAG has no effect, and MAG has a pronounced activating effect. The experiments were done with the purified canine kidney enzyme at 50 $\mu$M ATP, with the assay period of 30 s. The short duration of assay is necessary, because as we (7) and others (9) have noted, FA and derivatives have time-dependent irreversible inhibitory effects on the enzyme. The activating effects described here are readily reversible (7).

To explore the possibility of interactive effects of FA, MAG, and DAG, we examined the combined effects of two activators (FA and MAG) on the enzyme. When the two activators were used at suboptimal concentrations, their effects were additive but not synergistic (data not shown). The results of further experiments, exemplified by those of Fig. 2, indicated that DAG which is ineffective when used alone, caused blockade of FA activation but had little or no effect on MAG activation.

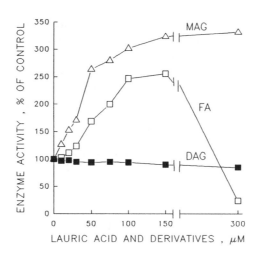

Figure 1. Effects of varying concentrations of lauric acid, monolauroylglycerol and dilauroylglycerol on $Na^+/K^+$-ATPase activity at 50 $\mu$M ATP. Methods and procedures as described before (1).

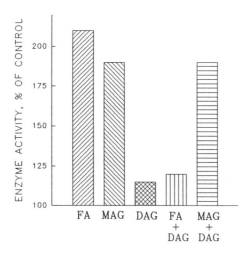

Figure 2. Effects of oleic acid (30 $\mu$M), monoolein (50 $\mu$M) and diolein (100 $\mu$M) on $Na^+/K^+$-ATPase activity at 50 $\mu$M ATP.

The second messenger DAG is released by the agonist-induced breakdown of either phosphatidylinositol or phosphatidylcholine (3). In turn, DAG is converted to MAG and FA. The regulatory roles of arachidonic acid and its metabolites that are generated by this route are well established. Our findings now suggest that three products of this pathway may act as direct regulators of the sodium pump: MAG and FA as activators, and DAG as an inhibitor of activation by FA. It is evident from the work of others,

691

however, that there may be additional components to the receptor-linked regulation of the pump within this pathway. The DAG-activated protein kinase C phosphorylates the isolated $Na^+/K^+$-ATPase (2), and the properties of both endogenous and expressed sodium pumps in Xenopus oocytes are altered by phorbol esters and DAG (10). The intricate relationships between the direct regulatory effects of the second messengers on the pump, and regulation through protein kinase C (Fig. 3) remain to be determined.

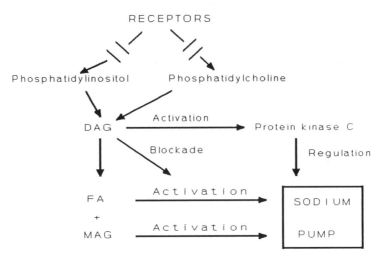

Figure 3. Suggested scheme for the direct and indirect regulation of the sodium pump by fatty acids and acylglycerols.

Acknowledgement

This work was supported by NIH grants HL-45554 and HL-36573 awarded by National Heart, Lung, and Blood Institute, United States Public Health Service/DHHS.

References

1.  Askari A, Xie Z, Wang Y, Periyasamy S, Huang W.-H (1991) A second messenger role for monoacylglycerols is suggested by their activating effects on the sodium pump. Biochim Biophys Acta 1069:127-130
2.  Chibalin AV, Lopina OD, Petukhore SP, Vasilets LA (1993) Phosphorylation of the Na,K-ATPase by Ca, phospholipid-dependent and cAMP-dependent protein kinases. J Bioenerg Biomembr 25:61-66
3.  Exton JH (1990) Signaling through phosphatidylcholine breakdown. J Biol Chem 265:1-4
4.  Huang W.-H, Kakar SS, Askari A (1985) Mechanism of detergent effects on membrane-bound $(Na^++K^+)$-ATPase. J Biol Chem 260:7356-7361
5.  Huang W.-H, Kakar SS, Askari A (1986) Activation of $(Na^++K^+)$-ATPase by long-chain fatty acids and fatty acyl coenzyme A. Biochem Int 12:521-528

692

6.    Huang W.-H, Kakar SS, Askari A (1988) Control of the sodium pump by liponucleotides and unsaturated fatty acids: side-dependent effects in red cells.  Prog Clin Biol Res 268B:401-407

7.    Huang W.-H, Wang Y, Askari A (1989) Mechanism of control of (Na$^+$+K$^+$)-ATPase by long-chain acyl coenzyme A.  J Biol Chem 264:2605-2608

8.    Kakar SS, Huang W.-H, Askari A (1987) Control of cardiac sodium pump by long-chain acyl coenzyme A.  J Biol Chem 262:42-45

9.    Swarts HGP, Schuurmans Stekhoven FMAH, DePont JJHHM (1990) Binding of unsaturated fatty acids to Na$^+$,K$^+$-ATPase leading to inhibition and inactivation.  Biochim Biophys Acta 1024:32-40

10.   Vasilets LA, Schwarz W (1992) Regulation of endogenous and expressed Na$^+$/K$^+$-ATPase in Xenopus oocytes by membrane potential and stimulation of protein kinases. J Membrane Biol 125:119-132

# Relations of $Na^+/K^+$-Pump Kinetics to Polyunsaturated Molecular Species of Plasmalogen-Phospatidylethanolamine in Intact Human Erythrocytes

Jochen Duhm, Bernd Engelmann

Physiologisches Institut, Universität München, Pettenkoferstr. 12, D-80336 München, Germany

Recent studies indicate that the percentages of certain phospholipid classes and of distinct molecular species of phosphatidylcholine (PC) and phosphatidylethanolamine (PE) of the red blood cell membrane exhibit considerable variations among normolipidaemic and hyperlipidaemic donors (1,2). The dependence of the function of the $Na^+/K^+$-pump on its lipid environment is well established. The transport system prefers anionic phospholipids (3) and is sensitive to small variations in the molecular species composition of PC (4-6). Also the function of the $Na^+/K^+$ cotransport system is dependent on the lipid environment (7). In addition, the $Na^+/K^+$-cotransport system and, to a lesser extent, the $Na^+/K^+$-pump are known to exhibit interindividual differences in activity. These findings provided the background of the present study on possible relations of the kinetic properties of the two transport systems on the membrane phospholipid composition in normolipidaemic (n = 7) and hyperlipidaemic individuals (n = 17, 5 with hypercholesteraemia (type IIa) and 12 with hypertriglyceridaemia (9 type IV, 2 type V, 1 type III)).

The methods applied to analyse $Na^+/K^+$-pump kinetics and membrane phospholipids are described in the legends to Figs. 1 and 2. The kinetics of the $Na^+/K^+$-cotransport system were evaluated using the Eadie plot from data of ouabain-resistant $Rb^+$ uptake in $Na^+$ media sensitive to inhibition by 10 μM bumetanide (1 to 10 mM $Rb^+_o$).

## Results

No significant relations were observed between the apparent $V_{max}$ and $K_m$ values of the $Na^+/K^+$-cotransport system and the percentages of phospholipid classes, the cholesterol content and any of the 26 molecular species analysed (9 in diacyl-PC, 9 in diacyl-PE and 8 in plasmalogen-PE (2)).

The kinetic parameters of the $Na^+/K^+$-pump, in contrast, exhibited a number of significant relations to membrane lipids. Notably, the maximal rate of both $Rb^+$ uptake and $Na^+$ extrusion rose significantly with increasing percentage of the molecular species 16:0/20:4 (1-palmitoyl,2-arachidonoyl) in plasmalogen-PE. The data for $Na^+$ uptake are shown in Fig. 2. The relation was significant even when the individuals with hyperlipedaemias were considered alone (r = 0.62, 2p < 0.05). The species 16:0/20:4 in diacyl-PE was not related to the maximal velocity of the $Na^+/K^+$-pump (2p > 0.7), indicating that the nature of the bond in $sn_1$ of PE plays a decisive role. The relation of 16:0/20:4 in diacyl-PC to the pumping rate was not significant (r = 0.32, p = 0.069).

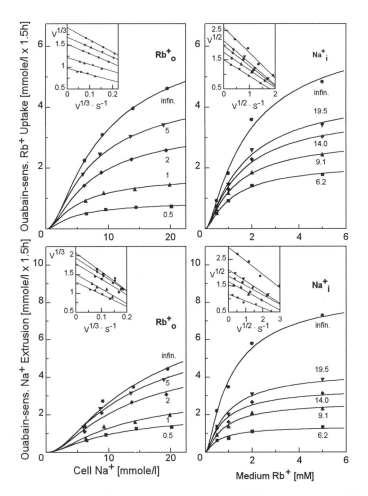

Figure 1: Ouabain-sensitive $Rb^+$ uptake and $Na^+$ extrusion of human erythrocytes are plotted as a function of cell $Na^+$ and medium $Rb^+$ (1.5 h of incubation in 137 mM $Na^+$ media containing 10 mM MOPS, 5 mM glucose, 1 mM inorganic phosphate, pH 7.4, $37^\circ C$, haematocrit 2.5%, $Rb^+$ replaced by choline). Cell $Na^+$ was varied at the expense of $K^+$ by preincubation in isotonic salicylate media containing varying proportions of $Na^+$ and $K^+$ ($0^\circ C$, 3h) (8,9). Cellular $Na^+$ and $Rb^+$ contents were measured by atomic absorption. The four lower curves of each panel are drawn according to the kinetic constants obtained by analysis of the experimental data using a modified Eadie plot:

$$V^{1/n} = V_{max}^{1/n} - (K_S \cdot V^{1/n}) / [S],$$

where V is the actual velocity at the cation concentration [S]. $V_{max}$ and $K_S$ are the apparent maximum velocity and dissociation constant, and n is the number of cations bound (three $Na^+_i$ and two $Rb^+_o$). The most upper curves are plots of the $V_{max}$ values obtained from the data of the neighbouring panel, yielding kinetic parameters pertaining to infinite (infin.) concentrations of $Na^+_i$ and $Rb^+_o$. In the Eadie plots (inserts), the mean standard deviations of the intercepts at the ordinate ($V_{max}^{1/n}$) and the slopes ($-K_S$) were greater for ouabain-sensitive $Na^+$ extrusion ($\pm 4.0 \pm 1.4\%$ and $\pm 14.4 \pm 8.7\%$) than for ouabain-sensitive $Rb^+$ uptake ($\pm 1.5 \pm 1.2\%$ and $\pm 6.6 \pm 4.1\%$). The apparent dissociation constants $K_S$ increased from 3.3 to 5.0 mmole $Na^+$/l cells and from 0.55 to 0.87 mM $Rb^+_o$ with rising concentration of the other cation, as evidenced by the increasing negative slopes of the Eadie plots. Two sets for $Rb^+$ uptake and $Na^+$

695

extrusion pertaining to infinite concentrations of the two cations were obtained, respectively. The pairs of kinetic constants were rather similar (e.g. $V_{max}$ of $Rb^+$ uptake 4.61 versus 4.55 and of $Na^+$ extrusion 5.96 versus 5.87 mmole/l per h). In Results, the mean values of the kinetic constants are evaluated.

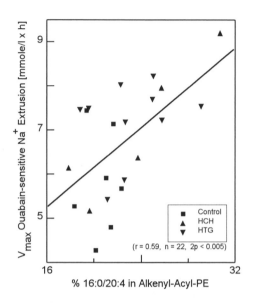

Figure 2: Relation of the maximum velocity of ouabain-sensitive $Na^+$ extrusion to the percentage of the molecular species 16:0/20:4 in red blood cell membrane plasmalogen-PE. The determination of $V_{max}$ is described in Fig. 1. Analysis of membrane phospholipid classes (diacyl-PC, lyso-PC, diacyl-PE, alkenylacyl(plasmalogen)-PE, sphingomyelin, phosphatidic acid and phosphatidylserine plus phospatidylinositol) and of the molecular species composition was performed as described elsewhere (1,2). In brief, phospholipids were extracted according to Rose and Oklander and separated as proposed Broekhyuse. Molecular species analysis of diacyl-PC, diacyl-PE and plasmalogen-PE was performed as described by Takamura et al. (10). Subsequent to one-dimensionall thin-layer chromatography the ester bond in $sn_3$ of the phospholipids was cleaved using phospholipase C. The diradylglycerol species were derivatized using 3,5-dinitro-benzoylchloride and the diacyl-PC, diacyl-PE and alkenylacyl-PE derivatives were separated on HPTLC plates and the molecular species were quantitated by HPLC with detection at 254 nm. - HCH = hypercholesterolaemia, HTG = hypertriglyceridaemia

The apparent dissociation constant of the $Na^+/K^+$-pump for cell $Na^+$ fell with rising percentage of phophatidylserine plus phosphatidylinositol (r = - 0.55, 2p < 0.01), indicating that the pump may be activated by anionic phospholipids not only in isolated (3) but also in intact membranes at non-saturating $Na^+$ concentrations. The dissociation constant of the pump for $Na^+$ showed, in addition, a positive relation to 16:0/20:4 in plasmalogen PE (r = 0.54, 2p < 0.01), and a negative one to the species 18:0/22:4 (1-stearoyl,2-docosatetraenoyl) in the same phospholipid subgroup (r = - 0.47, 2p < 0.05). In multiple regression analysis (using the SPSS/PC+ program) only 16:0/20:4 in plasmalogen PE was significantly related to $K_S Na^+_i$ (multiple r = 0.54, significance F = 0.012). The affinity of the pump for $Rb^+$ was inversely related to 18:0/22:4 in plasmalogen-PE (r = - 0.52, 2p < 0.05).

## Conclusion

On the basis of the present data no conclusion can be drawn as to the causal nature of the relations observed. The results are consistent with the view that certain polyunsaturated molecular species of plasmalogen-PE (especially 16:0/20:4) preferentially interact with the $Na^+/K^+$-pump protein. In this respect it is worthwhile to mention that recently polyunsaturated plasmalogen-PE has been identified as a major component of lipids (together with anionic phospholipids) tightly bound to the sarcoplasmatic $Ca^{2+}$ pump (11). In addition, it has been shown that phospholipid species with 16:0 in $sn_1$ and 20:4 in $sn_2$ activate the $Na^+/K^+$-pump in erythrocytes (6) and lymphocytes (4).

*Acknowledgements*

The authors thank Dr. W.O. Richter (II. Medizinische Klinik, Universität München) for selection of the patients and J. Kronauer for technical assistance. This work is part of the doctoral thesis of S. Streich and U.M. Schönthier. The studies were supported in part by a grant of the Wilhelm-Sander-Stiftung to B.E.

## References

1. Engelmann B, Streich B, Schönthier UM, Richter WO, Duhm J (1992) Changes of membrane phospholipid composition of human erythrocytes in hyperlipidemias. I. Increased phosphatidylcholine and reduced sphingomyelin in patients with elevated levels of triacylglycerol-rich lipoproteins. Biochim Biophys Acta 1165:32-37
2. Engelmann B, Schönthier UM, Richter WO, Duhm J (1992) Changes of membrane phospholipid composition of human erythrocytes in hyperlipidemias. II. Increases in distinct molecular species of phosphatidylethanolamine and phosphatidylcholine containing arachidonic acid. Biochim Biophys Acta 1165:38-44
3. Roelofsen B (1981) The (non)specificity in the lipid-requirement of calcium- and (sodium plus potassium)-transporting adenosine triphosphatases. Life Sci 29:2235-2247
4. Szamel M, Resch K (1981) Modulation of enzyme activities in isolated lymphocyte plasma membranes by enzymatic modification of phospholipid fatty acids. J Biol Chem 256:11618-11623
5. Marcus MM, Apell H-J, Roudna M, Schwedener RA, Weder H-G, Läuger P (1986) ($Na^+$ + $K^+$)-ATPase in artificial lipid vesicles: Influences of lipid structure on pumping rate. Biochim Biophys Acta 854:270-278
6. Engelmann B, Op den Kamp JAF, Roelofsen B (1990) Replacement of molecular species of phosphatidylcholine: influence on erythrocyte Na transport. Am J Physiol 258:C682-C691
7. Wiley JS, Cooper RA (1975) Inhibition of cation cotransport by cholesterol enrichment of human red cell membranes. Biochim Biophys Acta 413:425-431
8. Wieth JO (1970) Paradoxical temperature dependence of sodium and potassium fluxes in human red cells. J Physiol (London) 207:563-580
9. Duhm J, Engelmann B, Schönthier UM, Streich S (1993) Accelerated maximal velocity of the red blood cell $Na^+/K^+$ pump in hyperlipidemia is related to increase in 1-palmitoyl,2-arachidonoyl-plasmalogen phosphatidylethanolamine. Biochim Biophys Acta 149:185-188
10. Takamura H, Narita H, Park HJ, Tanaka K, Matsuura T, Kito M (1987) Differential hydrolysis of phospholipid molecular species during activation of human platelets with thrombin and collagen. J Biol Chem 262:2262-2269
11. Bick RJ, Younker KA, Pownall HJ, Van Winkle WB, Entman ML (1991) Unsaturated aminophospholipids are preferentially retained by the fast skeletal muscle CaATPase during detergent solubilisation. Evidence for a specific association between aminophospholipids and the calcium pump protein. Arch Biochem Biophys 286: 346-352

# Inhibitory Actions of Monoclonal Antibody to Phosphatidylserine on Na$^+$/K$^+$-ATPase

F.M.A.H. Schuurmans Stekhoven, M. Umeda, K. Inoue and J.J.H.H.M. de Pont

Department of Biochemistry, The University of Nijmegen, Nijmegen, The Netherlands and Department of Health Chemistry, Faculty of Pharmaceutical Science, The University of Tokyo, Tokyo, Japan

## Introduction

Previously (5) one of us (F.S.S.) had found that phosphatidylserine (PtdSer) is involved in cation activation of Na$^+$/K$^+$-ATPase. During that study monoclonal antibodies against PtdSer have been developed in a separate laboratory (6). These antibodies, due to their specificity, appeared to be ideal tools for the confirmation of PtdSer as aid in cation activation.

## Materials and Methods

Na$^+$/K$^+$-ATPase was isolated and purified from rabbit kidney outer medulla (zonal centrifugation mode, Ref. 2). Monoclonal antibodies to PtdSer (PS4A7, PS1G3) were raised as described before (6). PS1G3 was stored in 4 M guanidine-Cl because of aggregation in media of low ionic strength. Prior to use 0.4 ml was dialysed twice overnight against 1 l 300 mM choline-Cl + 20 mM triethanol-NHCl, pH 7.4. Antibody/antigen (Na$^+$/K$^+$-ATPase) interaction was tested kinetically by tracing inhibition of the monoclonal antibodies on Na$^+$/K$^+$-ATPase and p-NPPase activities, which are reflections of the forward and backward reaction steps of the title enzyme (4). Due to antibody/antigen interaction being ruled by ionic strength and antibody/antigen ratio, and the hydrolytic activity (including inactivation) by time and temperature of (pre)incubation, we have chosen for a low ionic strength, relatively high molecular antibody/antigen ratio (~ 10), moderate (23°C) preincubation and assay temperature and an ample (30 min) preincubation time. Assay media (controls) contained (Na$^+$/K$^+$-ATPase): 20 mM NaCl, 1 mM KCl, 1 mM ATP, 1.15 mM MgCl$_2$, choline-Cl (equivalent to antibody addition) $\leq$ 30 mM and 9.4 $\mu$g Na$^+$/K$^+$-ATPase/ml 25 mM triethanol-NHCl, pH 7.0; (p-NPPase): 10 mM KCl, 5 mM p-NPP, 6 mM MgCl$_2$, choline-Cl (equivalent to antibody addition) $\leq$ 18 mM and 9.4 $\mu$g Na$^+$/K$^+$-ATPase/ml 25 mM triethanol-NHCl, pH 7.0. Hydrolysis was run for 10-15 min and stopped by addition of 2 ml 5% trichloroacetic acid to 0.1 ml assay medium. Pi was determined in 1 ml aliquots by means of the Malachite Green procedure (7).

## Results and Discussion

Na$^+$/K$^+$-ATPase activity is inhibited to any appreciable extent (40%) by the IgG PS1G3, but not by the bulkier IgM PS4A7 (Fig. 1). Inhibition of the p-NPPase activity by PS1G3 is stronger ($\sim$ 70%) than of the ATPase activity, due to induction of a K$^+$/substrate antagonism (Fig. 2A), not seen in the overall ATPase reaction in the presence of PS1G3 (Fig. 2B). The antibody does not affect the half-maximally activating cation and substrate concentrations in the ATPase reaction, but reduces the Hill coefficient for Na$^+$ and K$^+$ (Table 1). In the p-NPPase reaction the half-maximally activating K$^+$ concentration is strongly increased, but the Hill coefficient hardly affected.

This PtdSer antagonism is strongly resemblant to the effect of detergent, inducing the E1 conformation (1). The reduced overall Na$^+$/K$^+$-ATPase reaction may be due to a shift to the left in the E1P $\rightleftharpoons$ E2P equilibrium or a consequence of a reduced cooperation between subunits or cation activation sites or both. The data confirm earlier findings for a role of PtdSer in conformational changes of the enzyme (3).

**Table 1.** Effects of PS1G3 on the kinetic parameters of Na$^+$/K$^+$-ATPase
Kinetic parameters were determined via Eady-Scatchard and Hill plots. PS1G3 was present in maximally inhibitory concentrations (31-40 $\mu$g/ml).
Data ± S.E.M. are averages of duplicate determinations. Na$^+$/K$^+$-ATPase: [Na$^+$] 2-20 mM, [K$^+$] 0.1-4 mM, [Mg ATP] 0.1-1 mM; p-NPPase: [K$^+$] 1-10 mM.

| Na$^+$/K$^+$-ATPase activity | -PS1G3 | +PS1G3 | V$_{max}$ ratio (+/-PS1G3) |
|---|---|---|---|
| [Na$^+$]$_{0.5}$ (mM) | 1.8 ± 0.15 | 2.4 ± 0.4 | |
| nH Na$^+$ | 3.0 ± 0.4 | 1.6 ± 0.3 | |
| | | | 0.61 ± 0.01 |
| [K$^+$]$_{0.5}$ (mM) | 0.35 ± 0.05 | 0.35 ± 0.05 | |
| nH K$^+$ | 1.6 ± 0.1 | 0.8 ± 0.1 | |
| | | | 0.50 ± 0.005 |
| [MgATP]$_{0.5}$ (mM) | 0.085 ± 0.015 | 0.105 ± 0.005 | |
| nH MgATP | 1.5 ± 0.3 | 1.5 ± 0.2 | |
| | | | 0.65 ± 0.12 |
| p-NPPase activity | | | |
| [K$^+$]$_{0.5}$ (mM) | 0.63 | 6.4 | |
| nH K$^+$ | 0.9 | 1.4 | |
| | | | 0.43 |

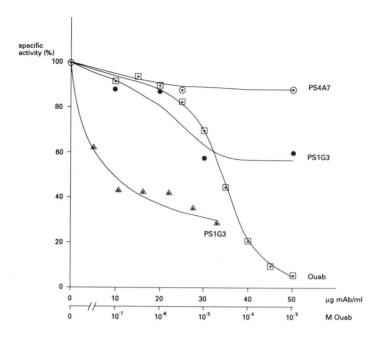

**Figure 1.** Monoclonal PtdSer-antibody inhibition of overall $Na^+/K^+$-ATPase and p-NPPase activity. Meaning of symbols: (o), (●), residual ATPase activity; (▲), residual p-NPPase activity; for comparison inhibition of the ATPase activity by ouabain (□) is shown.

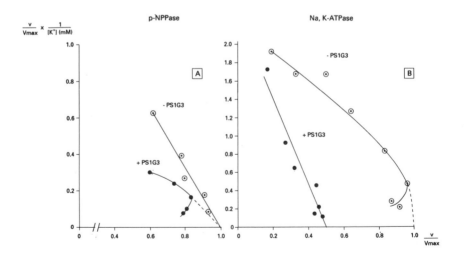

**Figure 2.** Eady-Scatchard plots of p-NPPase (A) and $Na^+/K^+$-ATPase activity (B) in dependence on the $K^+$ concentration with (●) and without (o) PS1G3. PS1G3 is 11 $\mu$g/ml in A and 31 $\mu$g/ml in B.

# References

1. Huang W-H, Kakar SS, Askari A (1985) Mechanisms of detergent effects on membrane-bound ($Na^+ + K^+$)-ATPase. J Biol Chem 260: 7356-7361
2. Jørgensen PL (1974) Purification and characterization of ($Na^+ + K^+$)-ATPase. III. Purification from the outer medulla of mammalian kidney after selective removal of membrane components by sodium dodecylsulphate. Biochim Biophys Acta 356: 36-52
3. Mimura K, Matsui H, Takagi T, Hayashi Y (1993) Change in oligomeric structure of solubilized $Na^+/K^+$-ATPase induced by octaethylene glycol dodecyl ether, phosphatidylserine and ATP. Biochim Biophys Acta 1145: 63-74
4. Schuurmans Stekhoven FMAH, Bonting SL (1981) Sodium-potassium-activated adenosinetriphosphatase. In: Bonting SL, De Pont JJHHM (eds) Membrane Transport (New Compr Biochem 2). Elsevier/North-Holland Biomedical Press, Amsterdam, pp 159-182
5. Schuurmans Stekhoven FMAH, Tesser GZ, Ramsteyn G, Swarts HGP, De Pont JJHHM (1992) Binding of ethylenediamine to phosphatidylserine is inhibitory to $Na^+/K^+$-ATPase. Biochim Biophys Acta 1109: 17-32
6. Umeda M, Igarashi K, Nam KS, Inoue K(1989) Effective production of monoclonal antibodies against phosphatidylserine: stereo-specific recognition of phosphatidylserine by monoclonal antibody. J Immunol 143: 2273-2279
7. Van Veldhoven PP, Mannaerts GP (1987) Inorganic and organic phosphate measurements in the nanomolar range. Anal Biochem 161: 45-48

# Regulation of sodium pump with physiologically active ligands (prostaglandins, galanin)

T.Kullisaar, T.Vihalemm, A.Valkna, M.Zilmer

Department of Biochemistry, Tartu University, Tartu EE2400, Estonia

## Introduction

Due to the multifunctional nature (catalytic, receptoric, transport) the regulation of the $Na^+/K^+$- ATPase must be versatile. Hence, several direct (ATP,$Na^+$,$K^+$,EDLF) and indirect (hormones, neuropeptides etc.) modulators exist. Action of the latter agents is mediated via receptors (signal transduction system) and receptor-independent way. A definite role of lipid-protein interactions in indirect (receptor-independent) regulation of $Na^+/K^+$-ATPase is quite possible.

## Lipid-protein interaction in $Na^+/K^+$-pump regulation

The $Na^+/K^+$-ATPase preparations (EP) were incubated ($37^0$) at different concentrations of prostaglandins ($PGE_2$,$PGF_{2\alpha}$) and neuropeptide galanin (GAL). Their effect on enzyme activity was incomplete (Table 1). Such incompleteness, a definite correlation between inhibitory effects and the lipid status of EP (lipid peroxydation degree, the total content of phopholipids and cholesterol) refer to possibility, that regulation of sodium pump by PGs and GAL has, evidently, a receptor-independent character, i.e. may be mediated via lipid-protein interaction. Such suggestion is supported by our and

Table 1. Maximum inhibitory effect (%) of $PGE_2$ (3mM), $PGF_{2\alpha}$ (9mM) and GAL ($10^{-4}$M) on $Na^+/K^+$-ATP- ase from human brain, human brain tumor (glioma), rat kidney and $KBrO_3$-treated rat kidney and the lipid status of EP from these sources

|  | $PGE_2$ | $PGF_{2\alpha}$ | GAL | TBARS[c] | PL[d] | CH[e] |
|---|---|---|---|---|---|---|
| Brain | 50±8 | 44±8 | 17±3 | 2±0.2 | 1.1±0.1 | 12±2 |
| Brain tumor | 35±5[a] | 30±5[a] | n.d.[b] | 1±0.1[a] | 0.7±0.1 | 22±3[a] |
| Kidney | 30±5 | 37±4 | 2±1 | 116±14 | 1.5±0.1 | 15±2 |
| Kidney +$KBrO_3$ | 19±3[a] | 22±4[a] | 18±3[a] | 203±19[a] | 1.3±0.2 | 16±2 |

[a]$p<0.05$ compared to normal sources; [b]nondedectable;[c]TBA-reactive substances: nmol/g wet tissue (kidney; nmol/mg protein (brain,tumor); [d]phospholipids (μmol Pi/mg protein); [e]cholesterol (arbitrary units)

**Table 2. The parameters of nontransformed and transformed by oncoprotein E5 (PVO) C127 cells**

| | AC activity[e] | TBARS[b] | PL[c] | Relative capacity[d] |
|---|---|---|---|---|
| C127 | 3 | 120±10 | 1.2±0.1 | 55±3 |
| PVO | 6 | 35±4[a] | 0.8±0.1[a] | 46±4 |

[a]$p<0.05$ vs C127; [b](nmol/$10^9$ cells); [c]$\mu$mol $P_i$/mg protein; [d]arbitrary capacity of $Na^+/K^+$-ATPase; measured via outtransported $Na^+$ (4); [e]pmol cAMP/mg/min

literature facts: 1) recently, a receptor-independent action way for some peptides is suggested (1); 2) PGs ($E_2$,$F_{2\alpha}$) are lipophilic allosteric ($n_H$=1.4) modulators for $Na^+/K^+$-ATPase, 3) the regulative effect of neuropeptide GAL on basal adenylatecyclase (AC) in brain (frontal cortex, hypothalamus, hippocampus) was also limited,12-23% (2); 4) the lack of protection (ATP) and competition ($Na^+$,$K^+$) effects in the case of GAL's influence on transport ATPase (3); 5) similar and minimal basal activity of AC, but different lipid status, plasma membrane content of PGs ($E_2$, $F_{2\alpha}$) and relative capacity of the $Na^+/K^+$-pump in the case of nontransformed and transformed cells (Table 2).

### References

1. Mousli M, Bueb JL, Bronner C, Rouot B, Landrey Y (1990) G protein activation: a mode of action for cationic amphiphilic neuropeptides and venom peptides. TIBS 11:358-362
2. Karelson E, Zilmer M,Laasik J (1993) Possible mechanism of galanin neurotropic activity. Transactions of Tartu University 961:58-62
3. Laasik J, Sillard R, Langel Ü, Zilmer M(1992) Lipid-protein interactions in neurotropic Na-pump regulation. Acta Physiol Scand 146:249-251
4. Kalinina LM, Cheschevik AB (1991) Permeability of eryhtrocyte membranes for monovalent ions in patients with stable stenocardia Quest Medical Chem 37:39-40

# Characterization of the molecular isoforms of Na+/K+-ATPase subunits along the rat nephron by polymerase chain reaction

L. Cheval, C. Barlet-Bas, A. Doucet

Laboratoire de Physiologie Cellulaire, Collège de France, 11 Place M. Berthelot, 75231 Paris cedex 5, France.

## Introduction

It is commonly admitted that the kidney only expresses the $\alpha_1$ and $\beta_1$ isoforms of Na+/K+-ATPase subunits. However, previous results from our laboratory indicate an heterogeneity of Na+/K+-ATPase along the successive segments constituting the rabbit nephron : The collecting tubule displays a higher affinity for ouabain (4) and for sodium (2) than more proximal segments such as the proximal tubule and the thick ascending limb of Henle's loop. Furthermore, Na+/K+-ATPase from rabbit collecting duct is recognized by an anti-$\alpha_3$ antibody but not by an anti-$\alpha_1$ antibody, whereas the opposite is observed in nephron segments located upstream (1). These findings suggest that the rabbit collecting duct preferentially expresses an $\alpha_3$-like isoform of Na+/K+-ATPase catalytic subunit, whereas other nephron segments would express the $\alpha_1$ isoform.

In the rat nephron, we have also characterized two distinct forms of Na+/K+-ATPase which differ by their sensitivity to ouabain and to the anti-$\alpha_1$ and the anti-$\alpha_3$ antibodies (7, Féraille et al. unpublished results). However, conversely to the rabbit, these two distinct forms are present along the whole nephron, although their respective proportions vary from one segment to an other. The aim of this study was therefore to characterize the different isoforms of Na+/K+-ATPase subunits ($\alpha_1$, $\alpha_2$, $\alpha_3$, $\beta_1$, $\beta_2$) expressed in the different segments of the rat nephron. For this purpose, the presence of specific mRNAs encoding for these isoforms was searched at the level of microdissected segments of rat nephron using reverse transcription followed by polymerase chain reaction (RT-PCR).

## Material and methods

*Microdissection of nephron segments and RNA extraction.* Single segments of rat nephron were microdissected in Hank's sterile solution (Sigma) under stereomicroscopic observation according to the technique previously described (5). Pools (20-50 mm) of nephron segments (see nomenclature in the legend of figure 2) free of contaminating cells were submitted to RNA extraction using a microadaptation (6) of the method described by Chomczynski and Sacchi (3). Tubules were denaturized in a solution containing 4 M guanidium thiocyanate, 25 mM sodium citrate (pH 7), 0.5% sarcosyl and 20 μg polyinosinic acid (Pharmacia) used as carrier (8). Then, phenol-chloroform-isoamylalcool (25:24:1) extraction was performed. After precipitation with ice-cold isopropanol, the RNA pellet was rinced two times with ethanol 75%, redissolved in 30-40 μl of RNA dilution buffer (Tris-HCl pH 8 10 mM, RNAse inhibitor -Pharmacia- 29 U in DEPC-treated water) and stored at -20°C. The same procedure was used to extract total RNAs from entire brain and kidney.

*Design and synthesis of primers.* The primers were designed with the help of Oligo Software (Med Probe, Norway). They were chosen according to criteria of minimal self-priming and upper/lower dimer formation, and specificity for each isoform. The sequences and the localization of primers for each isoform are indicated figure 1. Even if the sequence of some primers displays a high degree of homology with the corresponding sequences of other isoforms (e.g., the $\alpha_1$ upstream primer, see fig. 1) they all revealed as specific under the conditions of RT-PCR used. The primers were synthetized by Bioprobe Systems. Working solutions were stored at -20°C in Tris-EDTA pH 8.

*RT-PCR of RNAs.* The RT and PCR steps were performed in the same reaction tube. RT was performed with the specific downstream primer in a final volume of 50 μl containing : 5 μl of 10xRT mix (200 mM Tris-HCl pH 8.3, 15 mM $MgCl_2$ and 500 mM KCl), 100 ng of RNA extracted from entire brain or kidney or RNA extracted from 2-5 mm of tubule, 2-20 pmol of specific downstream primer (concentration varied for each primer), 200 μM of each dNTP, 2 mM $MgCl_2$, 8 mM DTT and 200 U of Moloney Murine Leukaemia virus reverse transcriptase (Gibco). The final volume was adjusted with DEPC-treated water. The RT reaction lasted for 45 min at 42°C. Then, the temperature was raised to 96°C for 30 sec to inactivate the enzyme and denaturize the DNA-RNA hybrids. The amplification reaction was initiated by adding to each tube 50 μl of a solution containing : 5 μl 10xPCR mix (200 mM Tris-HCl pH 8.3, 5 mM $MgCl_2$, 500 mM KCl and 0.1% gelatin), 2-20 pmol of specific upstream primer and 5 U of Taq Polymerase (Gibco). The samples were then overlaid with 50 μl of mineral oil and submitted to the following cycles : 96°C, 30 sec ; Tm ($\alpha_3$ : 62°C, $\alpha_2$ and $\beta_1$ : 68°C, $\beta_2$ : 70°C), 30 sec ; 74°C, 1 min. For the $\alpha_1$ isoform, the cycle was : 96°C, 30 sec ; Tm (73°C), 1.5 min. Generally, 30-35 cycles were realized. Then, a final elongation step was performed at 74°C for 10 min.

*Analysis of RT-PCR products.* 10-30 μl of each reaction mixture was deposited on a 2% agarose gel. Bands were detected with ethidium bromide under UV illumination. The expected lengths of amplification products for each isoform are indicated in figures 2 and 3.

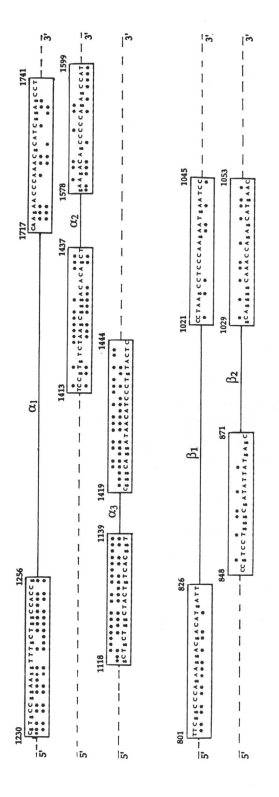

**Figure1.** Sequence and localization of oligonucleotide primers used in this study. Positions of the upstream and downstream primers are numbered from the ATG initiation codon of the different cDNAs. ✱ represent homologies between the corresponding sequences of the different isoforms.

706

## Results

Figure 2 displays the results of a typical RT-PCR experiment carried for the two well established renal isoforms of $Na^+/K^+$-ATPase subunits, namely $\alpha_1$ and $\beta_1$. Single bands of expected sizes (511 bp and 244 bp for $\alpha_1$ and $\beta_1$ respectively) were consistently generated from RNAs extracted from either whole kidney or any constitutive nephron segment tested.

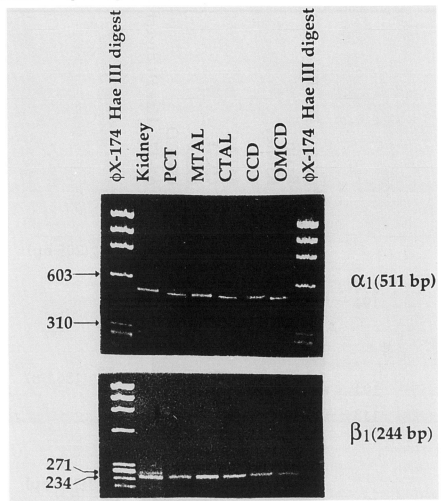

**Figure 2.** Detection by RT-PCR of $\alpha_1$ and $\beta_1$ mRNAs in whole kidney and in microdissected nephron segments (proximal convoluted tubule, PCT ; medullary and cortical part of the thick ascending limb of Henle's loop, MTAL and CTAL ; cortical and outer medullary collecting duct, CCD and OMCD). Left and right lanes correspond to the molecular weight marker ($\phi$X-174 HaeIII digest, Pharmacia).

Figure 3 pictures results obtained for β2, α2 and α3 isoforms. Brain RNA preparation was used as a positive standard for these isoforms, and indeed, bands of expected sizes (205 bp, 186 bp and 326 bp for β2, α2 and α3, respectively) were consistently observed in that tissue. A product corresponding to the β2 isoform was most often (but not always) obtained with whole kidney RNAs as well as RNAs from microdissected segments of nephron. Conversely, α2 and α3 isoforms were never detected in preparations from kidney or microdissected segments.

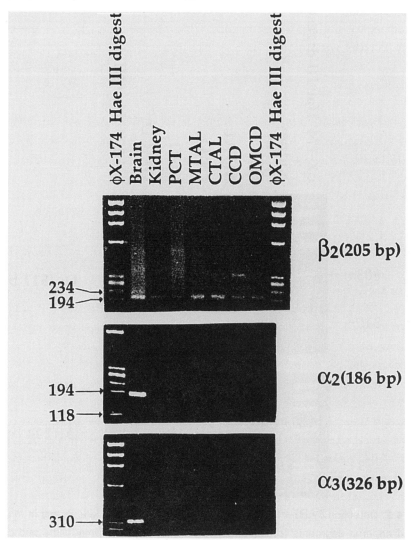

**Figure 3.** Detection by RT-PCR of β2, α2 and α3 mRNAs in whole brain and kidney and in microdissected nephron segments. Left and right lanes correspond to the molecular weight marker.

# Conclusion

These results confirm the presence of mRNAs encoding for the $\alpha_1$ and $\beta_1$ isoforms of Na$^+$/K$^+$-ATPase subunits along the whole rat nephron. They suggest that the $\beta_2$ isoform might be also expressed in all nephron segments of the rat. However, nether $\alpha_2$ nor $\alpha_3$ mRNAs could be detected in any segment of the nephron. Together with the pharmacological and immunological data mentioned above, this last finding suggests that another isoform of Na$^+$/K$^+$-ATPase catalytic subunit, distinct from $\alpha_1$, $\alpha_2$ and $\alpha_3$ (although it displays some immunological resemblance with $\alpha_3$) and of high affinity for ouabain is expressed along the rat nephron.

*Acknowledgements*

The authors are endebted to Drs J.M. Buhler and J.M. Elalouf for their expert advices during the development of RT-PCR procedure.This work was supported by a grant from the Institut National de la Santé et de la Recherche Médicale (CRE 910201).

# References

1. Barlet-Bas C, Arystarkhova E, Cheval L, Marsy S, Sweadner K, Modyanov N, Doucet A (1993) Are there several isoforms of Na,K-ATPase $\alpha$ subunit in the rabbit kidney ? J Biol Chem 268 : 11512-11515

2. Barlet-Bas C, Cheval L, Khadouri C, Marsy S, Doucet A (1990) Difference in the affinity of Na$^+$-K$^+$-ATPase along the rabbit nephron : modulation by K. Am J Physiol 259 : F246-F250

3. Chomczynski P, Sacchi N (1987) Single step method of RNA isolation by acid guanidium thiocyanate-phenol-chloroform extraction. Anal Biochem 162 : 156-159

4. Doucet A, Barlet C (1986) Evidence for differences in the sensitivity to ouabain of NaK-ATPase along the nephrons of the rabbit kidney. J Biol Chem 261 : 993-995

5. Doucet A, Katz AI, Morel F (1979) Determination of Na-K-ATPase activity in single segments of the mammalian nephron. Am J Physiol 237 : F105-F113

6. Elalouf JM, Buhler JM, Tessiot C, Bellanger AC, Dublineau I, De Rouffignac C (1993) Predominant expression of $\beta_1$-adrenergic receptor in the thick ascending limb of rat kidney. J Clin Invest 91: 264-272

7. Féraille E, Vogt B, Rousselot M, Barlet-Bas C, Cheval L, Doucet A, Favre H (1993) Mechanism of enhanced Na-K-ATPase activity in cortical collecting duct from rats with nephrotic syndrome. J Clin Invest 91 : 1295-1300

8. Winslow SG, Henkart PA (1991) Polyinosinic acid as a carrier in the microscale purification of total RNA. Nucleic Acids Res 19 : 3251-3253

# Increase in Na Pump Activity of Brain-type Isoforms *via* Increased Turnover Rate after Glutamate Excitation of Cerebral Neurons

H.Matsui, H.Homareda, Y.Hayashi, N.Inoue*

Department of Biochemistry, Kyorin University School of Medicine, Mitaka, Tokyo 181, *Department of Biochemistry, Yokohama City University, School of Medicine, Kanazawa-ku, Yokohama 236, JAPAN

## Introduction

Na$^+$/K$^+$-ATPase (=Na pump) is abundant in nervous tissue and plays an important role in neuronal excitation by producing sodium and potassium ion-gradients across the cell membrane. Isoforms of Na$^+$/K$^+$-ATPase are known to exist in brain. Highly ouabain-sensitive isoform (brain-type; $\alpha$2 and $\alpha$3) appeared and increased during the maturation of primary cultured rat cerebral neurons besides the weakly ouabain-sensitive isoform pre-existing (common-type; $\alpha$1) [1]. We demonstrated that the Na pump activity of brain-type isoform in the matured neurons increased several folds after glutamate excitation while the activity of common-type isoform decreased slightly [2,3]. To discriminate whether the increase in Na pump activity is due to increase in number or turnover rate of the pump, the number of Na pump was determined by [$^3$H]ouabain binding experiment.

## Materials and Methods

Cerebral neurons generated from 17 day-old rat embryos were cultured on 24-well plates for 13~17 days as described already [1,2]. The neurons were incubated at 37°C for 30 min with 1 $\mu$mol [$^3$H]ouabain in a standard assay medium, which is the same as used for K$^+$ uptake experiments, containing 130mM NaCl, 2mM KCl, 0.8mM MgCl$_2$, 1.8mM CaCl$_2$, 5mM glucose, and 25mM Hepes-Imidazol buffer (pH 7.3). The ouabain-binding reaction was terminated by removal of the medium with suction. The cells were washed twice with cold assay medium and dissolved in NaOH solution, and the radioactivity of [$^3$H]ouabain recovered were counted in a liquid scintillation spectrometer. Amount of specific binding of ouabain was calculated as the difference between the amounts of [$^3$H]ouabain obtained in the absence and presence of 0.5mM non-radioactive ouabain.

Simultaneous measurements of ouabain binding and K uptake activity of the neurons were performed in a double labeling experiment with [$^3$H]ouabain and $^{42}$KCl as follows. Neurons were preincubated in the assay medium containing 1$\mu$M [$^3$H]ouabain without or with 0.5mM non-radioactive ouabain at 37°C for 30 min, and then the reaction mediums were changed to those containing 100$\mu$M glutamate. After 12 min stimulation with glutamate, the assay mediums were again changed to those containing not glutamate but $^{42}$K labeled KCl, and the incubation for $^{42}$K$^+$ uptake was lasted for 12 min. Termination of the reaction and treatment of samples were the same as above. Other details of measurements of [³H]ouabain binding and ⁴²K uptake were described elsewhere (2,4,6).

## Results and Discussion

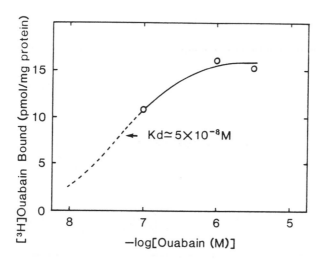

**Figure 1.** Concentration dependency of ouabain binding to primary cultured cerebral neurons. The neurons were incubated with various concentrations of [³H]ouabain in the standard assay medium at 37°C for 30 min.

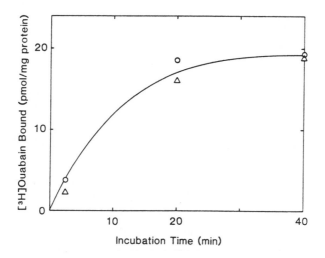

**Figure 2.** Effects of glutamate stimulation on ouabain binding. After 20 min preincubation in the standard assay medium, the neurons were incubated for 12 min in the absence ( ○ ) or presence ( △ ) of 100mM glutamate at 37°C. Then, the incubation for ouabain binding continued in the assay medium containing 1mM [³H]ouabain or 1mM [³H]ouabain + 0.5mM non-labeled ouabain for the periods indicated at abscissa.

711

High-affinity ouabain binding saturated at about 1~3 μM ouabain (Fig. 1). The concentration-dependency of ouabain binding corresponds with the inhibition of brain-type Na pump activity by ouabain in the absence and presence of glutamate (2,3). $Kd$ for ouabain was around $5 \times 10^{-8}$ M, which was comparable with 0.1μM of $Ki$ for ouabain of brain-type Na pump. The level of saturation binding of 20 pmol/mg protein of cell, was also comparable with the brain-type $Na^+/K^+$-ATPase activity of 0.16 μmol/min/mg protein in the homogenate of neurons (1, and cf 6). It is natural from these results to conclude that the high-affinity ouabain binding is ouabain binding to the site of brain-type isoform of Na pump.

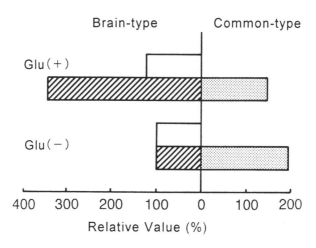

**Figure 3.** Simultaneously measured Na pump activity and ouabain binding capacity of the cultured neurons after glutamate stimulation. Assay procedure was described in Methods. ▭ ,Ouabain binding; ▨ , $K^+$ uptake by brain-type Na pump; ▦ , $K^+$ uptake by common-type Na pump.

**Table I.** Turnover rate of Na pump calculated from $K^+$ uptake and ouabain binding with or without glutamate stimulation. The turnover rate of Na pump was calculated from the results of experiments shown in Fig. 3 and the turnover rate of $Na^+/K^+$-ATPase as a reference was calculated from the following dada; ATPase activity, 45 μmol/min/mg protein, and ouabain bound, 6.1 nmol/mg protein (5).

|  |  | -Glutamate | +Glutamate |
|---|---|---|---|
|  | K+ Uptake (nmol/min/mg) | 5.8 | 16. |
|  | Ouabain bound (pmol/mg) | 13.8 | 16.4 |
| Turnover rate | Na Pump | 3.5 | 8.6 |
| (sec⁻¹) | $Na^+/K^+$-ATPase | 123 |  |

As shown in Fig. 2, glutamate stimulation did not change the level of ouabain bound although $K^+$ uptake by the neurons had increased 3 folds by the stimulation in previous results (2,3). To study this relationship, ouabain binding and $K^+$ uptake activity of the neurons stimulated with glutamate were simultaneously measured in a double labeling experiment with [$^3$H]ouabain and $^{42}$KCl (Fig 3). No significant change of ouabain binding and 3 fold increase of the brain-type Na pump activity by glutamate stimulation were confirmed and at the same time the common-type Na pump activity was observed to decrease a little similarly to the previous results (2, 3). Turnover rate of the pump calculated from the data of pump activity and bound ouabain (= pump number) in the experiment of Fig 3 were shown in Table 1. The rate increased from 3.5 to 8.6 /sec correspondingly with the increase of $K^+$ uptake by the stimulation of glutamate. Turnover rate of $Na^+/K^+$-ATPase activity of the highly purified enzyme from canine kidney was calculated to be 123 from the data of ATPase activity of 45 μmol/min/mg protein and bound ouabain of 6.1 nmol/mg/protein in a previous paper (5). Then, relative values of the pump turnover rate are only 2.6% (with glutamate) and 6.2% (after glutamate stimulation) of the $Na^+/K^+$-ATPase turnover rate under optimal conditions. The results indicate that the Na pump in neuron has ample scope for its activity by increasing its turnover rate without increasing number of pump.

*Acknowledgments*

We are grateful to Mrs. Y. Nagano, Miss N. Shinji and Miss M. Ishihira for their technical assistance. This study was supported in part by a Grant-in-Aid for Scientific Research (No. 04266220 to H.M.) from the Ministry of Education, Science and Culture of Japan.

# References

1. Inoue N, Matsui H, Tsukui H, Hatanaka H (1988) The appearance of a highly digitalis-sensitive isoform of $Na^+,K^+$-ATPase during maturation *in vitro* of primary cultured rat cerebral neurons. J Biochem 104:349-364
2. Inoue N, Matsui H (1990) Activation of a brain type Na pump after glutamate excitation of cerebral neurons. Brain Res 634:309-312
3. Inoue N, Matsui H (1991) Changes in responses of Na pump isoforms to glutamate stimulation of cerebral neurons during maturation in culture. In: Kaplan JH, De Weer P (eds) The Sodium Pump : Recent Developments. The Rockefeller University Press, New York, pp 597-600
4. Matsui H, Homareda H (1982) Interaction of sodium and potassium ions with $Na^+,K^+$-ATPase. I. Ouabain-sensitive alternative binding of three $Na^+$ or two $K^+$ to the enzyme. J Biochem 92:193-217
5. Matsui H, Hayashi Y, Homareda H, Taguchi M (1983) Stoichiometrical binding of ligands to less than 160 kilodaltons of Na,K-ATPase. In: Current Topics in Membrane Transport. Vol 19, Hoffman JF, Forbush III B (eds) Structure, Mechanism, and Function of Na/K Pump. Academic Press, New York, pp 146-148
6. Nagamatsu S, Inoue N, Matsui H (1986) Evaluation of sodium and potassium pump activity and number in diabetic erythrocytes. Acta Endoclinol 111:69-74

# The $\alpha 3$ isoform protein of the $Na^+/K^+$-ATPase is associated with the sites of neuromuscular and cardiac impulse transmission

R. Zahler, W. Sun, T. Ardito, M. Brines, M. Kashgarian

Dept. of Medicine, Yale University School of Medicine, New Haven, CT 06510 USA

## Introduction

In heart and skeletal muscle the sodium pump is crucial for maintenance of membrane potential and cellular integrity. Of the three $\alpha$-subunit isoforms, $\alpha 2$ and $\alpha 3$ mRNA's are concentrated in cardiac conduction system (CCS) of normal adult rat (6). We thus studied expression of pump isoform proteins in normal rat heart and skeletal muscle, and in vascular endothelial tissue, using immunofluorescence, immuno-EM, and Western blotting with isoform-specific antibodies.

## Methods

Tissues were obtained from normal newborn and adult male Sprague-Dawley rats. Specimens were also obtained from Wistar rats that had undergone right-sided sciatic nerve transection 7 days prior to sacrifice. The isoform-specific antibodies McK1 and 6H (anti-$\alpha 1$), McB2 (anti-$\alpha 2$), F9G10 and CM (both, anti-$\alpha 3$) were generously supplied by Drs. K. Sweadner, K. P. Campbell, and M. Caplan. For immunofluorescence, primary antibodies McK1 and McB2 were used at a 1:5 dilution, F9G10 at 1:15, and CM at 1:10. For the neuromuscular junction (NMJ) studies, double-labeling with FITC-$\alpha$-bungarotoxin at 1:50 dilution and Texas-Red-conjugated $Fab^2$ was used. Immuno-electron microscopy was performed as previously described (2). Western blots were done on crude microsomal fractions of cultured cells.

## Results and Discussion

Endothelium of large vessels such as pulmonary vein expressed $\alpha 1$. Capillary endothelial cell membranes, however, stained with McB2, while immunoreactivity to $\alpha 2$ was also present in endothelium of adult and neonatal coronary arteries. Endothelium of these smaller vessels had no detectable $\alpha 1$ or $\alpha 3$ immunoreactivity. In fact, no immunoreactive $\alpha 3$ was detected in any vascular tissue (adult or neonatal). To confirm these endothelial expression results, we performed immunoblots with isoform-specific antibodies on microsome preparations from cultured endothelial cells of various types. Human umbilical vein endothelial cells (HUVEC) expressed $\alpha 1$ protein, but did not react with the McB2 antibody to $\alpha 2$. Although McB2 is known to recognize the bovine $\alpha 2$ isoform, BAEC also was not immunoreactive with $\alpha 2$ on immunoblots. Neither of two isoform-specific antibodies against $\alpha 1$ reacted with BAEC, even when large amounts of microsomal protein were loaded. Rat epidydimal fat pad EC, however, reacted strongly with the McB2 antibody to $\alpha 2$, but no signal for $\alpha 1$ was detected. Thus, in all cases we studied, EC from large vessels expressed mainly $\alpha 1$, while those from small vessels and capillaries expressed $\alpha 2$.

In adult cardiac tissue, immunofluorescence demonstrated α1 at the surface membranes of myocytes throughout the atrial and ventricular myocardium. Control slides incubated with secondary antibody did not show membrane-associated fluorescence. In addition to this sarcolemmal expression, transverse tubules in ventricular myocardium were also immunoreactive with anti-α1 antibody McK1, although not with McB2 or F9G10. Ventricular myocardium was also reactive with the anti-α2 antibody McB2; the α2 signal was especially strong in papillary muscles. Immunostaining for α1 was diffusely present in both atria, although it appeared more intense in the left atrium than in the right atrium. The α2/α3 isoforms, however, were undetectable in adult atria. The cardiac impulse spreads from one working myocardial cell to the next via the junctional complex. Although there was no detectable signal for α3 in the sarcolemmal membrane of working ventricular myocytes, the junctional complex of these cells reacted strongly with both anti-α3 antibodies, at both the light microscopic and EM levels.

The cardiac conduction system transmits impulses from the sinoatrial (SA) node through the atrioventricular (AV) node, His bundle, and Purkinje fibers to the working myocardium. We have previously reported that there is markedly increased expression of α2 and α3 isoform mRNA (6) in conduction system of adult heart. Although adult ventricular myocardium generally had undetectable α3 signal, there was specific membrane-associated fluorescence in the conduction system when sections were stained with each of the two anti-α3 antibodies. While F9G10 labeled the sarcolemma uniformly, however, CM stained discrete linear regions of membrane. This pattern is similar to that observed at the NMJ, and thus may represent signal in neural structures innervating the AV node.

In neonatal heart, both the F9G10 and CM antibodies yielded very intense staining in areas associated with the conduction system, which was much stronger than that in adjacent myocardium. Unlike adult heart, however, there was no detectable immunoreactivity of either working myocardium or conduction system with McB2 in the neonatal hearts. In neonatal heart sections stained for α1, conduction tissue had fluorescence no greater than that of adjacent working myocardium. Also, regions of heart that expressed significant amounts of α2 in adult animals, such as papillary muscle, were negative for α2 in neonatal heart. Signal for the α3 isoform was present diffusely in RV and LV of neonatal heart. Discrete areas of the LA were positive for α3 with both CM and F9G10 antibodies. In addition, F9G10 reacted strongly with junctional complex, as in adult heart. Isoform expression patterns in neonatal vascular tissue, however, were similar to those in adult heart.

The NMJ is the interface between motor neuron and skeletal muscle. The Na,K-ATPase is important for the generation of the membrane potential in both of these excitable tissues, although the biochemical and electrical characteristics of pre- and post-synaptic membrane are markedly different. Skeletal muscle expresses predominantly the α2 isoform, while α3 is found in both cell bodies and axonal processes of neurons(1,3-5). This suggests that at the NMJ, a structure expressing predominantly α2 may be adjacent to one expressing predominantly α3.

Although F9G10 was negative on Western blots of rat heart ventricle and skeletal muscle, histologically it stained discrete linear regions of skeletal muscle cell surface membrane. Double-labeling showed that the linear α3-positive areas in muscle were adjacent to α-bungarotoxin-positive regions which had the typical morphology of NMJ's. Double-labeling indicated that α3 protein is expressed at the NMJ in close proximity to the postsynaptic motor endplate. Colocalization with bungarotoxin (a specific marker for the acetycholine receptor) was seen with both the F9G10 and the CM antibodies to α3. The α1 isoform, however, was not immunoreactive with NMJ. The α2 signal was present throughout the muscle cell membrane; in some cases this fluorescence was more intense at the NMJ, but most NMJ's did not have increased α2 signal. The fluorescence for α3 was offset slightly from that for bungarotoxin, suggesting that α3 is on the presynaptic side of the NMJ. To confirm that the α3 isoform is presynaptic, we studied lower extremity muscle obtained from rats 7 days after unilateral denervation, when there is predictable degeneration of the nerve terminal but not of the postsynaptic structure. Although endplates marked by FITC-α-bungarotoxin fluorescence persisted in the denervated limb, those in the distal muscle groups had markedly decreased to absent Texas red fluorescence for α3; the difference from the contralateral limb was highly statistically significant. Fluorescence for the McB2 antibody to α2 was unaffected in the denervated limb, however. Furthermore, right- and left-sided proximal muscles had identical staining patterns for α3, consistent with the expected reinnervation at this level at day 7.

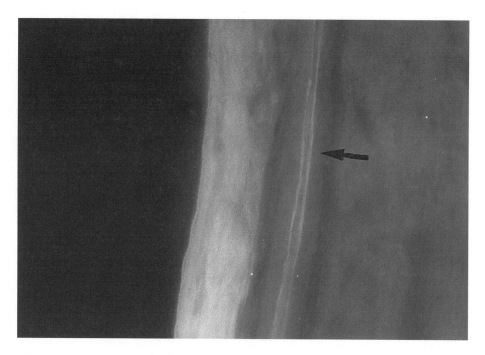

Figure 1. Purkinje fiber in subendocardium, left side of interventricular septum of adult rat heart, showing linear staining (arrow) with anti-α2 antibody McB2.

716

Thus, the complex spatial and developmental distribution of sodium pump isoforms in cardiovascular tissues may provide clues to distinct physiologic roles of these transporters. In particular, the α3 isoform is found in the cardiac conduction system, cardiac junctional complex, and presynaptic NMJ, suggesting that this isoform may be adopted to maintenance of membrane potential and/or intracellular ion concentrations required for impulse transmission. Alternatively, differences in stability between the isoforms, or distinct roles in cellular adhesion, could underlie these findings.

*Acknowledgements*

We thank Dr. E.J. Benz for his advice and support, Drs. F. Chen, M. Caplan, and K. Sweadner for helpful comments, Drs. Sweadner, Caplan, and K.P. Campbell for providing antibodies, Drs. Jeffrey R. Bender and Steven Pfau for providing HUVEC, Dr. Bauer Sumpio for providing BAEC, and Dr. J. Madri for providing rat epididymal fat pad EC.

# References

1. Brines M.L. and R.J. Robbins. Cell-type specific expression of Na,K-ATPase catalytic subunits in cultured neurons and glia: evidence of polarized distribution in neurons. *Brain Research* (in press).

2. Kashgarian, M., Biemesderfer, D., Forbush, B., and Caplan, M. (1985) Monoclonal antibody to Na,K-ATPase: immunocytochemical localization along nephron segments. *Kidney Int.* **28**, 899-913

3. McGrail, K.M., J.M. Phillips, and K.J. Sweadner. Immunofluorescent localization of three Na,K-ATPase isozymes in the rat central nervous system. *J. Neuroscience* **11**: 381-391, 1991.

4. Pietrini, G., M. Matteoli, G. Banker, and M.J. Caplan. Isoforms of the Na,K-ATPase are present in both axons and dendrites of hippocampal neurons in culture. *Proc. Natl. Acad. Sci.* **89** : 8414-8418, 1992.

5. Sweadner, K.J. Isozymes of the Na,K-ATPase. *Biochim. Biophys. Acta* **988** : 185-220, 1989 .

6. Zahler R, Brines M, Kashgarian M, Benz, E.J. Jr., Gilmore-Hebert M: The cardiac conduction system in the rat expresses the α2 and α3 isoforms of the Na,K-ATPase; Proc. Natl. Acad. Sci. **89** (1992) 99-103.

# Direct Correlation of Na$^+$/K$^+$-ATPase -Isoform Abundance and Myocardial Contractility in Mouse Heart

G. R. Askew*, J. B Lingrel*, I. L. Grupp[§], and G. Grupp[§]

*Departments of Molecular Genetics, Biochemistry and Microbioloty ML 0524 and [§]Pharmacology and Cell Biophysics ML 0575, University of Cincinnati College of Medicine Cincinnati, Ohio 45267 USA

## Introduction

In earlier experiments on rat heart ventricles (1, 2 and 7) we found that high- and low-affinity positive inotropic effects of ouabain (PIE) were correlated with the abundance of $\alpha2$ and $\alpha1$ Na$^+$/K$^+$-ATPase isoforms respectively, whereas rat atria showed only low affinity PIE and only $\alpha1$ isoforms. We also reported (3) that rat ventricular strips at high affinity concentrations (0.1 to 5 µM) accumulated a$^i$Na significantly by 2mM. We now report the characteristics of the hearts of 2 mouse strains "FVB/N" and "129" with regard to contractile response and Na$^+$/K$^+$-ATPase isoform abundance. We measured myocardial contractility in atrial and ventricular strips under isometric conditions. Resting tension (final 0.5 to 1.0 g) was increased stepwise close to the point of maximally developed force (L$_{max}$). At this level cumulative ouabain dose-response curves (10 nM to 200 µM) were obtained. The strips were suspended in oxygenated Krebs-Henseleit solution at 35°C and stimulated at 1Hertz, pH 7.4. Relative abundance of $\alpha1$ and $\alpha2$ isoform mRNAs and enzyme activities were determined by northern blot and enzyme coupled spectrophotometric assays, respectively.

## Contractile Response of Mouse Heart Strips to Increasing Concentration of Ouabain.

Figure 1 contains the responses of **ventricular strips** of two mouse species: FVB/N ($\blacklozenge$, n=22) and 129 ($\square$, n=7). The high affinity (30 nM to 3 µM) contractility responses to ouabain of the FVB/N ventricles are greater than 50% of the total response whereas the low affinity (3 to 200 µM) response is less than 50%. The estimated ratio is approximately 55 to 45% high to low affinity distribution. It is of interest that the 129 ventricles (open squares) have a lower contractility response and the high affinity response begins only at 1 µM ouabain.

The **atrial response** to ouabain ($\blacksquare$, n=15) is fundamentally different from the ventricular: There is no or very little high affinity response. The major response is low affinity, beginning at 20 µM and peaking at 400 µM. In addition the maximal response is significantly higher (by about 33%) than the ventricular. Although the mouse atrial response is very similar to the rat atrial response, the ventricular responses of the two species are very different (see table 1). Whereas the mouse ventricular high versus low affinity response is about 55:45%, respectively, the rat response is about 29:71% (1).

718

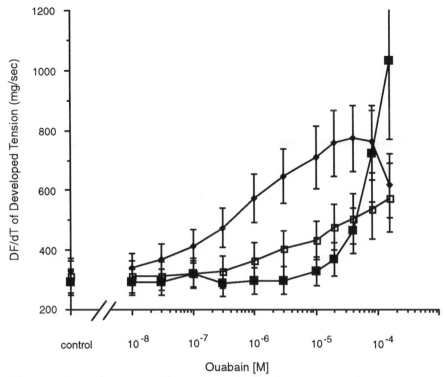

**Figure 1.** Effect of ouabain on ventricular and atrial tissue of different strains (FVB/N and 129) of mice. Ventricular response of FVB/N (♦) and 129 strain (□ mice and combined (FVB/N and 129) atrial response (■) was measured as described in the text.

## Relative Abundance of α1 and α2 Isoform Subunits in Rat and Mouse Myocardium

To quantitate expression of subunit isoform mRNAs, northern blots were prepared with total RNA from adult atria and ventricles and several control tissues (brain, lung, skeletal muscle and kidney). Blots were hybridized with isoform specific probes (7), labeled, by the random oligo priming method, to equivalent specific activities, and signals were quantitated using a phosphorimager (Molecular Dynamics). Relative mRNA levels indicate that the mouse ventricle has roughly a 1:1 proportion of α1:α2 transcripts (figure 2). In contrast, the ratio of α1 to α2 mRNA levels for rat ventricle is roughly 2:1. Comparison of the rat and mouse ratios reveals a shift towards greater expression of the "high affinity" α2 than the "low affinity" α1 subunit mRNA in the mouse ventricle.

Measurements of atrial α1 and α2 mRNA levels indicate that α1 represents the majority (>95% ±6%) of $Na^+/K^+$-ATPase transcripts in both rat and mouse and that the ratios of mRNA expression are roughly equivalent in this tissue (figure 2).

To determine relative contribution of α1 and α2 isoforms to total $Na^+/K^+$-ATPase activity of mouse and rat ventricle, microsomal preparations (6) from each tissue were assayed by differential ouabain inhibition using the coupled enzyme spectrophotometric method as previously described (4). In this assay 100 % $Na^+/K^+$-ATPase activity is defined as the proportion of ATPase activity which is inhibited by

719

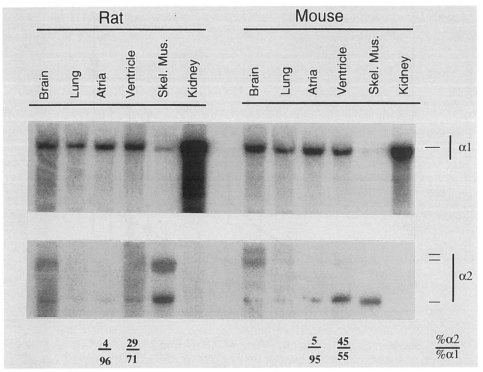

**Figure 2.** Northern Blot of rat and mouse tissue total RNAs. The representative blot shown was sequentially hybridized with cDNA fragment $^{32}$P-dCTP-labeled probes which are specific for α1 and α2, top and bottom exposures respectively. Signals of bands representing mRNA species of each isoform were quantitated with a phosphorimager and relative levels of expression are provided, as % of total NKA mRNA, at the bottom of the figure. The image shown is a digital reproduction of the northern blot exposure, prepared directly from the phosphorimager data.

10 mM ouabain, α2 activity is defined as the proportion of total Na$^+$/K$^+$-ATPase activity which is inhibited by 100 nM ouabain and α1 activity is the portion of total Na$^+$/K$^+$-ATPase activity which is inhibited by 10 mM ouabain but not by 100 nM ouabain. The results of these measurements demonstrate colinearity of isoform mRNA expression with isoform enzyme activity levels. In the mouse ventricle α1 accounts for 40% (±4%) and α2 for 60 % (±4%) of total Na$^+$/K$^+$-ATPase activity. By contrast, in the rat ventricle α1 accounts for the majority (78% ±4%) and α2 only 22% (± 4%) of the total Na$^+$/K$^+$-ATPase activity, which is in general agreement with previous observations (5).

In conclusion, we have found that the relative proportion of α2:α1 Na$^+$/K$^+$-ATPase expression is significantly greater in the mouse ventricle than in the rat ventricle in terms of mRNA abundance and enzyme activity. Since the α2 isoform of Na$^+$/K$^+$-ATPase displays a greater affinity for ouabain than the α1 isoform, the greater abundance of the more sensitive α2 isoform is directly correlated with and may contribute to, or account for, the greater low dose inotropic response to ouabain

**Table 1. Summary of contractility and Na⁺/K⁺ATPase expression data for rat and mouse myocardium**

| Ventricles | Rat α1 | α2 | ratio | Mouse α1 | α2 | ratio |
|---|---|---|---|---|---|---|
| Relative mRNA levels (± 6%) | 65% | 35% | ($\approx$2:1) | 47% | 53% | ($\approx$1:1) |
| Microsomal Na⁺/K⁺ATPase (±4%) | 78% | 22% | ($\approx$3:1) | 40% | 60% | ($\approx$2:3) |
| Contractility | 70% | 30% | ($\approx$2.5)@ | 45% | 55% | ($\approx$1:1) |
| **Atria** | | | | | | |
| Relative mRNA levels (± 6%) | 95% | 5% | ($\approx$20:1) | 96% | 4% | ($\approx$20:1) |
| Contractility | 99% | 1% | ($\approx$100:1)@ | 90% | 10% | ($\approx$9:1) |

@ Contractility data for rat myocardium was previously reported in reference (1).

(compared to that of the rat) which we observed in mouse ventricular myocardium. The atria of rat and mouse exhibit similar inotropic response profiles as well as comparable proportional expression of α1 and α2 isoform mRNAs with the consistent feature being predominant expression of the low affinity α1 Na⁺/K⁺-ATPase which is directly correlated with primarily a high dose inotropic response.

*Acknowledgments*
The authors wish to thank Gilbert Newman, Jay Slack, and Mehtap Tosun for excellent technical assistance. This work was supported by NIH grant PO1 HL 41496 and AHA grant SW-92-01-1.

**References**

1. Adams RJ, Schwartz A, Grupp G, Grupp IL (1982) High affinity binding site and low-dose inotropic effect in rat myocardium. Nature 296:167-169.
2. Grupp IL, Grupp G, Schwartz, A (1981) Digitalis receptor desensitization in rat ventricle: Ouabain produces two inotropic effects. Life Sci 29:2789-2794.
3. Grupp I, Im WB, Lee CO, Lee SW (1985) Relation of sodium pump inhibition to positive inotropy at low concentrations of ouabain in rat heart muscle. J Physiol 360:149-160.
4. Jewell EA and Lingrel JB (1991) Comparison of the substrate dependence properties of the rat Na,K-ATPase α1, α2 and α3 isoforms expressed in HeLa cells. J Biol Chem 266 (25) 16925-16930.
5. Lucchesi PA, Sweadner KJ (1991) Postnatal changes in Na,K-ATPase isoform expression in rat cardiac ventricle. J. Biol. Chem. 266(14) 9327-9331.
6. Van Alstyne E, Burch RM, Knickelbein RG, Hungerford RT, Gower FJ, Webb JG, Poe SL, Lindenmayer GE (1980) Isolation of sealed vesicles highly enriched with sarcolemma markers from canine ventricle. Biochem Biophys Acta 602:131-143.
7. Young RM and Lingrel JB (1987) Tissue distribution of mRNAs encoding α isoforms and β subunit of rat Na⁺/K⁺-ATPase. Biochem Biophys Res Commun 145:52-58.

# Discovery of Endogenous Ouabain - A New Mammalian Hormone

J.M. Hamlyn

Department of Physiology, University of Maryland, 655 West Baltimore St, Baltimore MD 21201 USA

## Introduction

Cardiac glycosides and their aglycones have been used widely in the therapy of shock and congestive heart failure (13,23). It is generally believed that the therapeutic actions of this class of compounds are secondary to their specific interaction with a conserved receptor site on the $Na^+/K^+$-pump. Speculation that there may be an endogenous mammalian counterpart to the cardiac glycosides can be traced back more than a century and has been repeated often (26). However, most of the evidence for the existence of such a factor has been fragmentary and based primarily upon bioassay or other indirect data. Despite the fact that enormous progress has been made in the chromatographic and analytical sciences in the last thirty years, nothing was known about the general class of compounds to which this factor may belong or the manner in which this inhibitor interacts with the $Na^+/K^+$-pump. The following is an account of the purification and identification of an $Na^+/K^+$-pump inhibitor from human plasma that was carried out in collaboration with The Upjohn Company.

## General Considerations

In the search for new compounds, the integrity of the screening assay is a critical element that influences outcome. At the start of this endeavour, it was decided that any biologically relevant material must inhibit the transport activity of the $Na^+/K^+$-pump in intact cells. This aspect was emphasized because other techniques such as ATPase activity or ouabain binding methods, although more suited for mass screening, were considered to be indirect and were already known to be more susceptible to false positives (17). To circumvent the disadvantage of using ion fluxes as a screening assay, 96 well plates and semiautomated techniques were so employed so that several hundred assays could performed daily when required. Another key factor concerned the choice of starting material. It is generally believed that the therapeutic actions of the cardiac glycosides are secondary to their presence in the circulation. Therefore, it was assumed that an endogenous biologically active counterpart would circulate; a presumption supported by the numerous observations of an $Na^+/K^+$-pump inhibitor in the circulation of volume expanded as well as hypertensive animals and man (16,34,35). The decision to use plasma as a source material was also reinforced by the lack of consensus in the literature regarding a tissue source and also by the uncertainty that active material would be present in urine.

Bioassay techniques had suggested that the circulating concentration of the $Na^+/K^+$-pump inhibitor in the pig was in the subnanomolar range (18). Simple calculations assuming an overall yield of 10-16% suggested that a total of seven metric tons of human plasma would be required to produce 50 µg of pure material suitable for conclusive structural analysis. This raised four issues. First, the identification of a new compound present in such low amounts using plasma as a starting source had not been previously accomplished. Second, there was no reliable information on the stability of the endogenous compound in stored blood. Thus, large volumes of fresh plasma would be required; a need that was met by routine plasmapheresis. Third, the routine use of large volumes of human blood from multiple donors raised significant safety issues. Therefore all donors were screened for HIV and special handling procedures were required to ensure the safety of those handling these materials. Fourth, none of the individual plasma donors could be taking cardiac glycosides or any other medications that might interfere with the screening assay. When these issues were satisfied, the purification of active materials commenced.

**Purification and Characterization**

The first attempts used 10 ml of plasma while the final methods that were developed after 4-5 years of experience, processed 85 liter batches at a time. Overall, approximately two metric tons of plasma were used; three quarters of which went for methods development. The following is an outline of the general features of the purification scheme used and, where convenient, some rationale is given for each of the steps used and the manner in which certain critical decisions were made.

The first step involved dialysis because preliminary experiments showed that inhibitory material passed easily through a membrane with a 6-8 kDa cut off. Dialysis had the disadvantage that the dialysate volume was considerably greater that the plasma itself but this was offset to some extent by the fact that the bulk of the plasma proteins remained behind. Initially, the dialysate was lyophilized, methanol extracted, dried and subjected to preparative HPLC. Lyophilization proved time consuming and the extracts often led to column blockage despite attempts to minimize this problem. However, when acceptable runs were performed and the fractions were screened, a polar and highly active area of interest was observed. This material was taken through five sequential steps of HPLC involving different solvents and column packings. In the last chromatographic step, involving a Waters semipreparative phenyl column developed with a biphasic gradient program of isopropanol in 0.1% heptafluorobutyric acid, a single highly active fraction eluting at 26 minutes was obtained. This material was purified by a factor in excess of $10^9$-fold and exhibited strong UV absorbance at 220 nm (20). Unfortunately, only 0.8 µg of this material were recovered from the 125 liters of plasma used to start the purification and given the unknown nature of the material, no structural work was feasible. However, the amount of material was sufficient to characterize the general nature of the compound and to determine its interaction with the $Na^+/K^+$-pump. The following paragraph summarizes the results from those studies.

The inhibitory activity was found to be an uncharged organic molecule that was soluble in low order alcohols and water and that was inactivated by heating to $150^{\circ}C$ in air. It was resistant to proteases, peptidases and esterases, but lost activity with prolonged boiling in water or in strong acid or alkaline solutions. The material eluted mid way

723

between the 318 and 900 MW markers in high performance gel-sieving chromatography. It was a specific inhibitor of the dog kidney $Na^+/K^+$-ATPase, no effects were seen on a variety of other ATPases tested. Enzyme titration experiments using a purified dog kidney $Na^+/K^+$-ATPase suggested that the apparent binding affinity of the inhibitor was very high (approximately 0.3 nM). However, the [3]H-ouabain binding capacity of the $Na^+/K^+$-ATPase used in those experiments was 10-fold lower than expected from activity measurements. Subsequent experiments indicated that the binding affinity of the inhibitor was actually around 3 nM; a value that is similar to that for commercial ouabain under the same conditions. Dose-response relationships and Scatchard analysis indicated that the inhibitor competed with ouabain for reversible binding to a common class of receptor sites in dog kidney $Na^+/K^+$-ATPase. Studies of the ligand dependence showed that inhibitor binding was potassium-sensitive and was of highest affinity in MgPi or Na,MgATP solutions, i.e. conditions associated with the $E_2$-P form of the $Na^+/K^+$-pump. Like ouabain, the binding of the inhibitor increased the apparent affinity for Pi and suported the acid stable incorporation of $^{32}Pi$ into the $Na^+/K^+$-ATPase. In addition, the dissociation rates of ouabain and the human compound from the preformed inhibitor-dog kidney $Na^+/K^+$-ATPase complex were strikingly similar under different conditions as shown in Table 1. At that time we were unaware that the rate at which cardiac glycosides dissociate from the $Na^+/K^+$-pump is critically dependent upon various structural elements such as the saturation of the lactone ring, the stereochemical configuration of the steroid moiety and the nature and orientation of the sugars at C-3 (37,43). Thus, a prepared mind could have recognized from the data in Table 1 that the endogenous human material was probably ouabain itself.

**Table 1. Dissociation Rate Constants for Ouabain and the Human $Na^+/K^+$-pump Inhibitor from $Na^+/K^+$-ATPase.**

|  | $K_{-1}$ ( x $10^{-3}$ $min^{-1}$) | |
|---|---|---|
|  | EDTA | $MgCl_2$ |
| Ouabain | 7.6±0.7 | 0.2±0.02 |
| Human Inhibitor | 6.8±0.8 | 0.3±0.06 |

Preformed inhibitor-$Na^+/K^+$-ATPase (dog kidney) complexes were allowed to dissociate at $37^oC$ in 7.5 mM imidazole-HCl pH 7.4 containing either 5mM EDTA or 5 mM $MgCl_2$. See reference (20) for details.

Therefore, the importance of the aforementioned studies was the demonstration that human plasma contained a compound that could be purified in active form and that showed remarkable similarities to the cardiac glycosides in its interactions with the $Na^+/K^+$-ATPase. These observations, along with the indications that the material was probably not peptidic, validated the screening philosophy and fueled enthusiasm to generate more material for structural analyses.

Although a complete HPLC based purification sequence was already in place, the 5-10% yield was unacceptably low. Therefore efforts were directed to remedy this problem. First the dialysate was passed over a large Amberlite XAD-2 column. Following extensive washing, bound materials were eluted with methanol and then subjected to preparative scale HPLC. The introduction of the XAD-2 step bypassed the time consuming requirement for lyophilization, greatly reduced the tendency of the preparative HPLC to clog, and improved the working resolution and yield of the HPLC preparative column so that gradient elution could be performed. When the fractions from this column were screened, four distinct peaks of $Na^+/K^+$-pump inhibitory activity were usually observed (22). Of these, there was a prominent polar material that consistently accounted for >70% of the total inhibitory activity in the run (see Figure 1 in ref. 29 and Figure 4 in ref. 22).

However, the presence of four active peaks now required a process to determine which was most worthy of pursuit. As the levels of circulating sodium pump inhibitory activity change in response to physiological alterations of sodium balance, small amounts (5-10ml) of plasma were obtained from pigs and man before and following oral or intravenous salt loads. Analytical scale HPLC runs of those samples generally showed a fraction that decreased the red cell $Na^+/K^+$-pump activity by 10%. The polarity of this inhibitory activity was similar to the most polar highly active fraction observed in the preparative scale HPLC and the magnitude of inhibitory activity was greater under conditions of salt loading (Hamlyn, Harris and Ludens, unpublished). No other peaks of inhibitory activity were noted from the analytical column, probably because their concentration was below the threshold for detection. Therefore, a decision was made to pursue the large polar activity from the preparative scale HPLC and the eventual success or failure of the project would hinge upon those small changes in $Na^+/K^+$-pump flux. That such an important decision was taken in the absence of more conclusive evidence may appear quite extraordinary. However, it emphasizes one aspect of the difficulties that are encountered in this type of exploratory work especially when experimental materials are in short supply.

In order to purify the active material to homogeneity in good yield, a reduction in the number of HPLC steps appeared desirable and the natural affinity of the material for its native receptor was exploited in a batch purification. Therefore, the polar active fraction from the preparative HPLC was incubated with a two fold molar excess of purified lamb kidney $Na^+/K^+$-ATPase under conditions (Mg+Pi) previously shown to result in high affinity binding of the human material. The enzyme-inhibitor complexes were pelleted by centrifugation, washed, and incubated with buffer containing EDTA to induce dissociation (c.f. Table 1). Following centrifugation, the resultant supernatant was lyophilized and prepared for additional HPLC. The results from the affinity purification step were such that two steps of HPLC sufficed to complete the purification and 31μg of pure material were obtained at 30% yield from 300 liters of plasma (29).

**Identification**

Initially, HPLC fractions that either contained or were immediately adjacent to the inhibitory material were subjected to low resolution fast atom bombardment mass spectroscopy (FABMS). A single unique protonated molecular ion was observed at m/z 585 in the active but not adjacent fractions. Subsequently an accurate protonated mass of 585.292 Da was determined consistent with the elemental formula ($C_{29}H_{45}O_{12}$) of ouabain. To obtain additional information, linked scan tandem mass spectrometry was performed to obtain a fragmentation fingerprint. The daughter ion spectra for ouabain and the human compound were identical and exhibited a major ion at 439.2 Da corresponding to the diprotonated form of ouabagenin. These data reveal that the sugar moiety in the human material was a deoxyhexose. In order to assess the number and reactivity of the OH groups expected in the human compound, acetylated derivatives were prepared. Under conditions designed to prevent complete acetylation, six primary and secondary OH groups were detected in ouabain and the human compound by FABMS (33). When more rigorous acetylation conditions were used, FABMS spectra revealed molecular ions consistent with the acetylation of a seventh and eighth OH in both sources of ouabain (19). Thus, these astounding results indicated that the human plasma $Na^+/K^+$-pump inhibitor was indistinguishable from ouabain by mass spectroscopy.

Follow up studies were performed to compare the behaviour of structural grade human ouabain with commercial ouabain in various systems (Table 2).

**Table 2. Comparison of the Behaviour of Human and Commercial Ouabain in Various Systems.**

| SYSTEM | CONDITIONS | RESULT | REF |
|---|---|---|---|
| Chromatography | Bakerbond C-18* Cyclobond II γ** | Coelution | 33 |
| Receptor Interactions | Human RBC $Na^+/K^+$-pump | No difference in affinity | 19 |
| | Dog Kidney $Na^+/K^+$-ATPase | No difference in affinity | 19 |
| | Binding studies. Dog kidney $Na^+/K^+$-ATPase | Competitive, single class of sites, no difference in affinity. | 19 |
| | Rat RBC $Na^+/K^+$-pump | No difference in affinity | † |
| | Dissociation rates | See table 1 | 20 |
| Cardiotonic Action | Guinea pig atrium. Onset and wash out of inotropic effect | No difference | 6 |
| Antibody Interactions*** | Competition with ouabain or $^3$H-ouabain in ELISA or RIA | No difference | 22,‡ |

* Shown to separate sugar isomers of ouabain that differ only in the stereochemistry of a single sugar -OH (9). ** Routinely used for the separation of racemic drug mixtures in the pharmaceutical industry. *** The recognition epitopes involve the groups attached to the C/D rings, the cis fusion of these rings, and the lactone ring area. †Ludens, Harris & Hamlyn (unpublished). ‡ Hamlyn and Manunta (unpublished).

The amounts of material available for the studies indicated in Table 2 were limited so that the primary purpose was to gain information about the biological responses to the

726

human compound. However, a significant difference in any response would indicate that the human ouabain has a unique stereochemical configuration. Thus, there is as yet no indication that the chromatographic, receptor, cardiotonic or immunological techniques used distinguish significantly between the human and commercial forms of ouabain.

## Is Ouabain Endogenous to Mammals?

Several criteria suggest that the circulating ouabain may be endogenous to man and other mammals. First, unlike digoxin, ouabain is not effective in the treatment of congestive heart failure unless given intravenously. The explanation lies in the fact that the bioavailability of oral ouabain is minimal and this was the first indication that the circulating ouabain did not originate from a dietary source (40). Second, patients maintained by total parenteral nutrition for periods of seven days had similar plasma levels to those observed on their normal diets (19). As the half time for the urinary clearance of injected ouabain in man is approximately 19-24 hours (39), the plasma concentration would have been expected to decline by at least 7 half lives assuming that dietary intake were the immediate source. Third, the adrenal is highly enriched in ouabain in many mammals and the tissue content remains remarkably constant under different conditions. Fourth, plasma levels of ouabain fall in adrenalectomized rats but do not in rats whose adrenal medulla has been removed. Thus, the adrenal cortex is a source of circulating ouabain (30). Fifth, the ouabain content of the adrenal venous effluent is 3-5 fold greater than that of the arterial blood entering the gland in conscious chronically cannulated afebrile dogs (5). Sixth, elevated plasma levels of ouabain have been found in hypertensive patients with adrenocortical tumors (31). Removal of the tumors was associated with the normalization of plasma ouabain and the remission of hypertension. These results are compatible with those in other patients with adrenocortical tumors (32) and may represent the first example of ouabain secreting adrenocortical tumors causing hypertension. Seventh, cultured human and bovine adrenocortical cells secrete ouabain into the culture fluid (19,28). In addition, there is some preliminary evidence that the secretion is increased by vasopressin as well as alpha-1 adrenoceptor agonists (28). As these cells raise the ouabain content of the culture medium without altering the cellular content of ouabain, it appears there is *de novo* synthesis of ouabain in the adrenal. Collectively, the aforementioned data suggest that the adrenal cortex is an important  source of plasma ouabain in man and the rat. Thus, the endogenous production and existence of specific receptors throughout the body indicates that  ouabain is a new mammalian hormone.

## Physiopathology

There is accumulating evidence that ouabain has significant physio-pathological roles in mammals. The body of data are too large to be reviewed in detail here and the interested reader is referred to articles that describe actions of ouabain in the cardiovascular and renal area, the central nervous system, and the visual system (1-4, 6-8, 10-12, 14-16, 24, 25, 27, 36, 38, 41-45). However, given the manner in which speculation regarding endogenous counterparts to the cardiac glycosides arose, it is appropriate to consider one particularly intriguing set of observations. In 1953, Szent Gyorgyi suggested that some patients developed congestive heart failure because they lacked an endogenous cardiac glycoside. Approximately four decades later, it has been shown that most patients with

727

congestive heart failure have plasma levels of endogenous ouabain that are elevated in proportion to the degree of failure while other patients have low circulating levels (12). Thus, as anticipated by Szent Gyorgyi, it will be of great interest to determine whether these latter individuals benefit from therapy with exogenous cardiac glycosides.

**Future Directions**

The evidence presented shows that the endogenous human counterpart to the cardiac glycosides is not only a member of this class of compounds but is indistinguishable from ouabain. Clearly, there are many implications that follow from these statements. The wide distribution of the sodium pump and the recognition of multiple isoforms that may differ in their sensitivity to ouabain suggests the potential for involvement in many fundamental biological processes. Indeed, the systemic and cellular actions of high concentrations of ouabain appear, for the most part, to be well understood. Not known are the physiological effects of chronic exposure to the concentrations of ouabain that circulate under normal and abnormal conditions. This, along with the details of the biosynthetic pathway for ouabain, remains a key frontier for future work. Moreover, elucidation of the biological significance of ouabain will likely be reflected in the recognition of new and unexpected roles for sodium pumps.

*Acknowledgements*

I especially thank the members of the Cardiovascular Diseases and Analytical Services sections of the Upjohn Company, Kalamazoo, Michigan who made critical contributions in the discovery of human ouabain. I thank Drs. Mordecai Blaustein, Bruce Hamilton, and Paolo Manunta and the many other authors cited for their interest, contributions and support of this work over the years. Portions of this work were supported specifically by the National and Local Maryland Chapters of the American Heart Association and the UMAB Bressler Research Foundation. JH is an Established Investigator of the American Heart Association.

**References**

1. Abarquez RF (1967) Digitalis in the treatment of hypertension, a preliminary report. Acta Medica Philippina, Series 2, 3:161-170.

2. Baldy-Moulinier M, Arias LP, Passouant P (1973) Hippocampal epilepsy produced by ouabain. Eur Neurol 9:333-348.

3. Blanco G, Berberian G, Beauge L (1990) Detection of a highly ouabain-sensitive isoform of rat brainstem Na,K-ATPase. Biochim Biophys Acta 1027:1-7.

4. Blaustein MP (1993) The physiological effects of endogenous ouabain: control of cell responsiveness. Am J Physiol 264:C1367-C1387.

5. Boulanger BR, Lilly MP, Hamlyn JM, Laredo J, Shurtleff D, Gann DS (1993) Ouabain is secreted by the adrenal gland in the awake dog. Am J Physiol 264:E413-E419.

6. Bova S, Blaustein MP, Ludens JH, Harris DW, DuCharme DW, Hamlyn JM (1991) Effects of an endogenous ouabainlike compound on heart and aorta. Hypertension 17:944-950.

7. Brender D, Vanhoutte PM, Shepherd JT (1969) Potentiation of adrenergic venomotor responses in dogs by cardiac glycosides. Circ Res 25:597-606.

8. Caldwell RW, Songu-Mize E, Bealer SL (1985) The vasopressor response to centrally administered ouabain. Circ Res 55:773-779.

9. Cassels BK (1985) Analysis of a Maasai arrow poison. J Ethnopharmacol 14:273-281.

10. Donaldson J, st-Pierre T, Minnich J, Barbeau A (1971) Seizures in rats associated with divalent cation inhibition of Na,K-ATPase. Can J Biochem 49:1217-1224.

11. Gillis RA, Quest JA (1980) The role of the nervous system in the cardiovascular effects of digitalis. Pharmacol Rev 31:19-97.

12. Gottlieb SS, Rogowski AC, Weinberg M, Krichten CM, Hamilton BP, Hamlyn JM (1992) Elevated concentrations of endogenous ouabain in patients with congestive heart failure. Circulation 86:420-425.

13. Greeff K, Schadewaldt H (1981) Introduction and remarks on the history of cardiac glycosides. In:  K. Greeff, ed. Cardiac Glycosides, Part I: Experimental Pharmacology. New York, Springer Verlag, p 1-12.

14. Gupta JD, Harley GD (1975) Decreased adenosine triphosphatase activity in human senile cataractous lenses. Exp Eye Res 20:207-209.

15. Gutman Y, Chaimovitz M, Bergmann F, Zerachia A (1971) Hypothalamic implantation of ouabain and electrolyte secretion: evidence for a central effect on sodium balance. Physiol Behav 6:399-401.

16. Haddy FJ, Pamnani MB (1985) Evidence for a circulating endogenous $Na^+$-$K^+$ pump inhibitor in low renin hypertension. Fed Proc 44:2789-27845.

17. Hamlyn JM (1988) Endogenous digitalis: where are we? ISI Atlas of Pharmacology 2:339-344.

18. Hamlyn JM (1989) Increased levels of a humoral digitalis-like factor in deoxycorticosterone acetate-induced hypertension in the pig. J Endocr 122:409-420.

19. Hamlyn JM, Blaustein MP, Bova S, DuCharme DW, Harris DW, Mandel F, Mathews WR, Ludens JH (1991) Identification and characterization of a ouabain-like compound from human plasma. Proc Natl Acad Sci (USA)88:6259-6263.

729

20. Hamlyn JM, Harris DW, Ludens JH (1989) Digitalis-like activity in human plasma: purification, affinity and mechanism. J Biol Chem 64:7395-7404.

21. Hamlyn JM, Manunta P (1992) Ouabain, Digitalis-like factors and hypertension. J. Hypertension 10 (suppl 7):S99-S111.

22. Harris DW, Clark MA, Fisher JF, Hamlyn JM, Kolbasa KP, Ludens JH, DuCharme DW (1991) Development of an immunoassay for endogenous digitalis-like factor. Hypertension 17:936-943.

23. Horton JAG, Davison MHA (1955) Ouabain in the treatment of shock. Br J Anaesth 27:139-144.

24. Huang BS, Harmsen E, Yu H, Leenen F (1992) Brain ouabain-like activity and the sympathoexcitatory and pressor effects of central sodium in rats. Circ Res 71:1059-1066.

25. Jacomini LCL, Elghozi JL, Dagher G, Devynck MA, Meyer P (1984) Central hypertensive effect of ouabain in rats. Arch Int Pharmacodyn Ther 267:310-318.

26. Kim RS, LaBella FS (1981) Endogenous ligands and modulators of the digitalis receptor: some candidates. Pharmacol Ther 14:391-399.

27. Kumar R, Yankopoulos NA, Abelman WH (1973) Ouabain-induced hypertension in a patient with decompensated hypertensive heart disease. Chest 63:105-107.

28. Laredo J, Hamlyn JM (1992) Modulation of the release of endogenous ouabain from human adrenal CRL 7050 cells by peptidergic and non-peptidergic secretagogues. Endocrine Soc Abstracts, 74th Annual Meeting, San Antonio, Texas, p.71, abstr 79.

29. Ludens JH, Clark MA, DuCharme DW, Harris DW, Lutzke BS, Mandel F, Mathews WR, Sutter DM, Hamlyn JM (1991) Purification of an endogenous digitalislike factor for structural analysis. Hypertension 17:923-929.

30. Ludens JH, Clark MA, Robinson FG, DuCharme DW (1992) Rat adrenal cortex is a source of circulating ouabainlike compound. Hypertension 19:721-724.

31. Manunta P, Evans G, Hamilton BP, Gann D, Resau J, Hamlyn JM (1992) A new syndrome with elevated plasma ouabain and hypertension secondary to an adrenocortical tumor. J. Hypertension 10(4):S27, Abstr. P36.

32. Masugi F, Ogihara T, Hasegawa T, Sakaguchi K, Kumahara Y (1988) Normalization of high plasma level of ouabain-like immunoreactivity in primary aldosteronism after removal of adrenals. J Hum Hypertens 2:409-420.

33. Mathews WR, DuCharme DW, Hamlyn JM, Harris DW, Mandel F, Clark MA, Ludens JH (1991) Mass spectral characterization of an endogenous digitalislike factor from human plasma. Hypertension 17:930-935.

34. Moreth K, Renner D, Schoner W (1987) A quantitative receptor assay for digitalis-like compounds in serum. Demonstration of raised concentrations in essential hypertension and correlation with arterial pressure. Klin Wochenschr 65:179-184.

35. Poston L, Sewell RB, Wilkinson SP, Richardson PJ, Williams R, Clarkson EM, MacGregor GA, de Wardener HE (1981) Evidence for a circulating sodium transport inhibitor in essential hypertension. Brit Med J 282:847-849.

36. Saxena PR, Bhargava K (1975) The importance of a central adrenergic mechanism in the cardiovascular response to ouabain. Eur J Pharmacol 31:332-346.

37. Schwartz A, Lindenmayer GE, Allen JC (1978) The sodium potassium adenosine triphosphatase: pharmacological, physiological and biochemical aspects. Pharmacol Rev 29:187-220.

38. Sekihara H, Yazaki Y, Kojima T (1992) Ouabain as an amplifier of mineralocorticoid-induced hypertension. Endocrinology 131:3077-3082.

39. Seldin R, Margolies MN, Smith TW (1974) Renal and gastrointestinal excretion of ouabain in dog and man. J. Pharmacol Exp Ther 188:615-623.

40. Strobach H, Wirth KE, Rojsathaporn K (1968) Absorption, metabolism and elimination of strophanthus glycosides in man. Naunyn-Schmiedebergs Arch Pharmacol 334:496-500.

41. Szalay KS (1993) Ouabain - a local paracrine aldosterone synthesis regulating hormone. Life Sci 52:1777-1780.

42. Szent-Gyorgyi A. *Chemical Physiology of Contraction in Body and Heart Muscle.* New York, Academic Press. 1953, pp 86-91.

43. Taddei S, Salvetti A, Pedrinelli R (1988) Ouabain vasoconstricts human forearm arterioles through alpha-adrenergic stimulation. J Hypertens 6(suppl 4):S357-S359.

44. Yoda A (1974) Association and dissociation rate constant of the complexes between various cardiac monoglycosides. Ann NY Acad Sci 242:598-616.

45. Yuan C, Manunta P, Hamlyn JM, Chen S, Bohen E, Yeun J, Haddy FJ, Pamnani MB (1993) Long term ouabain administration produces hypertension in rats. Hypertension 22: 178-187.

# Structure and Biological Activity of the Na$^+$/K$^+$-ATPase Inhibitor Isolated from Bovine Hypothalamus: Difference from Ouabain

Garner T. Haupert, Jr.

Renal Unit, Massachusetts General Hospital, Harvard Medical School, 149 13th Street, Charlestown, Massachusetts 02129 U.S.A.

## Introduction

The search for an endogenous inhibitor of Na$^+$/K$^+$-ATPase has been supported by the occurrence of digitalis-like cardiotonic steroids in plants and the presence of a highly conserved binding site for ouabain in mammalian tissues (5). The only precedent for cardiotonic steroids outside of the plant kingdom are the bufodienolides isolated from toads, although no glycosylated analogs were described (23). A substantial amount of indirect evidence supporting the existance of Na$^+$/K$^+$-ATPase inhibitory activity in human tissues has continued to fuel the search for a molecule of mammalian origin which would share at least some of the properties of the plant-derived glycosides, particularly as this compound(s) has been implicated in sodium and water homeostasis and the pathogensis of a prevalent human disease, hypertension (8).

Much effort has been expended in many laboratories to identify and characterize a plasma, urinary, or tissue inhibitor of Na$^+$/K$^+$-ATPase. These efforts have been hampered by the very low concentrations of putative candidates in mammals, and the difficulty of obtaining stringently purified samples for analysis. For recent developments in this field, the reader is referred to the excellent review by Goto and colleagues (7).

Many compounds of known structure have been have been studied as possible candidates, but as yet no unique molecule possessing the stringent biochemical criteria for a physiologic regulator of the Na$^+$/K$^+$-ATPase has been isolated and structurally identified. Hamlyn and co-workers described a ouabain-like inhibitor of Na$^+$/K$^+$-ATPase from human plasma that was indistinguishable from ouabain by several criteria including chromatography, mass spectrometry and immunoreactivity, but the analytical methods employed could not permit assignment of precise structure such as the nature of the putative sugar moiety or the stereochemistry of the steroidal portion (11). However, physiological testing of this purified plasma inhibitor again gave results indistinguishable from ouabain (2). These results raised interesting speculation about the ability of humans to synthesize ouabain, previously known only as a plant product, or perhaps a closely related mammalian analogue.

In this vein, we have pursued for more than a decade a high affinity, reversible inhibitor of the Na$^+$,K$^+$-ATPase, isolated from bovine hypothalamus, which has biological properties similar to but not identical to those of ouabain (4, 13, 14). Although tested at various degrees of purity in biochemical and physiological assays over the years, the trend of the results remained consistent in revealing significant differences from ouabain tested in parallel experiments, and the distinctive biological activity of the Hypothalamic Inhibitory Factor, HIF, predicted different, even if related, structures for the respective plant- and animal- derived inhibitors.

Using affinity chromatography coupled with preparative HPLC steps, HIF was recently purified to a single compound. Following development of microanalytical methods applicable to submicrogram amounts of natural product, HIF was shown to be an isomer of ouabain, differing in the regio- or stereochemistry of the aglycone portion of the molecule. This structural difference is presumed to account for the significant differences in biological activity observed for HIF and ouabain. The purpose of this communication is to briefly review the biological properties and structural characteristics which distinguish the pharmacologic inhibitor, ouabain, from the putative physiological mammmalian analogue, HIF.

## Reversibility of HIF Inhibition

The first important biological evidence that HIF differs from ouabain was found in a study of the binding and dissociation characteristics of the two compounds in intact renal tubular epithelial cells (LLC-PK$_1$). These experiements were undertaken since an endogenous inhibitor of the Na$^+$/K$^+$-ATPase had been proposed as a natriuretic hormone (5). Using the active influx of $^{86}$Rb$^+$ as a measure of Na$^+$ pump activity in the intact renal cell, HIF was found to completely inhibit this pump activity. The more important finding was that the HIF-induced inhibition was rapidly reversible following a brief washout period, while ouabain-induced inhibition in the cells was not (3). This more rapid reversibility of Na$^+$ pump inhibition (and thus binding of HIF to the intact enzyme) was later found also in the study of HIF-induced positive inotropy using neonatal rat cardiac myocytes (9). In these cardiac cells, a 5 min washout period was necessary for the reversal of ouabain-induced inotropic effects (28), while HIF effects were completely reversed following a rapid rinse of under 1 min (9)

In the renal tubular cells, the dose response curves using HIF showed a sigmoidial shape consistent with an allosteric binding reaction (3), again different from ouabain binding interactions with LLC-PK$_1$. This positive cooperativity in HIF binding was confirmed in subsequent studies of HIF inhibition of highly purified Na$^+$/K$^+$-ATPase incorporated into liposomes (1). Such "allosteric" binding reactions are common in biology and provide the potential advantage of sensitive control of reaction rates of enzymatic processes (22). Taken together, the binding and dissociation reactions of HIF in both intact cell and purified enzyme preparations were different from those of ouabain and consistent with physiological regulation *in vivo*.

## Effects of HIF on Isoenzymes of the Na$^+$/K$^+$-ATPase and on Na$^+$/K$^+$-ATPase of Ouabain-resistant Species

The rationale for target organ selectivity for an endogenous Na$^+$/K$^+$-ATPase inhibitor was boosted with the discovery that the α (catalytic) subunit exists as a family of isoforms whose principal functional distinction is differning sensitivity to ouabain inhibition (25). In rat and man there are three expressed α isoforms corresponding to three separate genes. Their expression is tissue specific and developmentally regulated (23, 16).

Within the rat species, inactivation of the ouabain-resistant $\alpha_1$ isoform (obtained from rat kidney) requires concentrations of ouabain $10^2$-$10^3$ greater than for the ouabain-sensitive $\alpha_2/\alpha_3$ isoforms (obtained from rat brain axolemma) (25). Such is however not the case for HIF. When Ferrandi and co-workers prepared HIF from bovine and rat hypothalamus (6), they found that a mere 5-fold increase in HIF dose was required to achieve the $IC_{50}$ for rat renal $Na^+/K^+$-ATPase compared to rat brain synaptosomal $Na^+/K^+$-ATPase (Fig. 1).

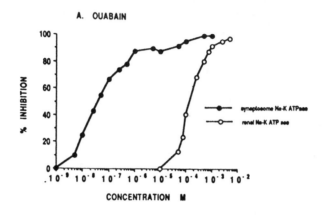

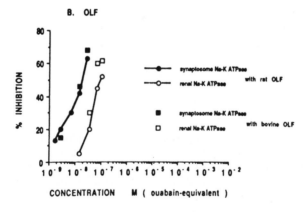

**Figure 1.** Inhibition of rat renal $Na^+/K^+$-ATPase ($\alpha_1$ isoform, open symbols) and rat synaptosomal $Na^+/K^+$-ATPase ($\alpha_2/\alpha_3$ isoforms, closed symbols) by ouabain (panel A) and HIF (OLF, ouabain-like factor) extracted from rat and bovine hypothalamus (panel B). Reproduced from (6) with permission.

It is felt that the amino acid sequence comprising the extracellular domain between the H1 and H2 transmembrane domains determines the ouabain sensitivity of the particular isoform (21). The ability of HIF to effectively inhibit the ouabain-resistant rat isoform is consistent with earlier preliminary work in which HIF was found to inhibit ouabain-resistant $Na^+/K^+$-ATPase expressed in primate cells transfected with a murine ouabain resistance gene (15).

734

Ferrandi and colleagues also found that renal Na$^+$/K$^+$-ATPase prepared from young Milan strain hypertensive and normotensive rats was less inhibited by HIF than enzyme prepared from mature Milan rats, that enzyme from young MHS rats was more sensitive than that of young MNS rats, while sensitivity to ouabain was the same for all enzymes irrespective of strain or age (6).

While the existence of isoforms showing different sensitivities to ouabain and HIF (albeit of a widely different magnitude) provides a seed of rationale for target organ selectivity, it is clear that such selectivity in any eventual endogenous regulation must be more complex since tissues expressing the same $\alpha_1$ isoform across species show widely different sensitivities to ouabain. Thus canine renal Na$^+$/K$^+$-ATPase, $\alpha_1$, is ouabain sensitive, but rat renal Na$^+$/K$^+$-ATPase, also $\alpha_1$, is highly resistant as noted. Consistent with functional differences concerning isoform sensitivities, HIF was also shown to not distinguish between ouabain-sensitive and ouabain-resistant renal Na$^+$/K$^+$-ATPase forms (Fig. 2). This finding may be important since it is further evidence that HIF interacts with target organ enzymes differently from ouabain, and may consequently have a different regulatory role *in vivo*.

## % Inhibition of Purified Renal Na+-K+ ATPase

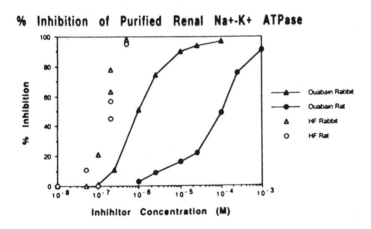

**Figure 2**. Inhibition of purified Na$^+$/K$^+$-ATPase from rabbit renal outer medulla ($\alpha_1$ isoform, ouabain-sensitive species) and from rat renal outer medulla ($\alpha_1$ isoform, ouabain-resistant species) reconstituted into liposomes, by ouabain (closed symbols) and HIF (open symbols) (B. Anner and G. Haupert, unpublished data)

While much remains to be understood about HIF regulation of Na$^+$/K$^+$-ATPase and possible target organ selectivity, one concludes from the foregoing data that HIF interactions with the enzyme differ from those of ouabain, again predicting structural difference for the two compounds.

## Effects of HIF in Cardiac Myocytes

It is generally agreed that to the extent that an endogenous inhibitor of Na$^+$/K$^+$-ATPase is truly "digitalis-like", it should manifest the cardinal pharmacologic feature of the digitalis glycosides which is positive cardiac inotropy. HIF and ouabain were therefore tested in parallel in spontaneously contracting cultured neonatal rat cardiac myocytes. The presence of $\alpha 2/\alpha 3$ isoforms in cardiac cells in the neonatal period provides a ouabain-sensitive preparation. Like ouabain, HIF was found to inhibit the sodium pump in these cells, to produce a dose-related increase in intracellular free $Ca^{2+}$ concentration ($[Ca^{2+}]_i$), and to produce an inotropic response equal to the maximal response caused by an optimal, non-toxic dose of ouabain (9). But the approximate dose of HIF ($1 \times 10^{-9}$ M) to achieve this degree of inotropic effect was about 500-fold less than the respective dose of ouabain ($5 \times 10^{-7}$ M). Concentrations of HIF were approximated based on determination of an apparent affinity constant using an enzyme titration technic (13) and application of this result to a standardized human erythrocyte $^{86}Rb^+$ uptake assay where 1 unit HIF = 0.75 pmol ouabain-equivalent bioactivity (1). Now that the precise molecular weight of HIF is known (*vide infra*), these earlier calcuations turn out to be remarkably accurate.

As noted previously, HIF effects in the cardiac cells were more readily reversible than those of ouabain. The unexpected finding was the dissociation of ouabain-induced cardiotoxicity from simple progressive elevation in $[Ca^{2+}]_i$ since HIF, at a dose producing similar pump inhibition and greater elevations in $[Ca^{2+}]_i$ than the toxic dose of 1 µM ouabain, did not produce this toxicity (9). As partially purified HIF inhibited the $Ca^{2+}$-ATPase of sarcoplasmic reticulum (SR) (but not the membrane $Ca^{2+}$-ATPase) (4), the different effects on $[Ca^{2+}]_i$ and toxicity in the myocytes might be due in part to blockade of $Ca^{2+}$ reuptake in the myocyte SR, particularly since HIF was shown to be membrane permeable while ouabain is not (1). It should be emphasized, however, that effects of pure HIF on $Ca^{2+}$-ATPase activity of SR have not yet been determined.

### Effects of HIF in Isolated Blood Vessels

Infusions of cardiac glycosides into blood vessels increases vascular resistance (27), while incubation of vessels with ouabain in vitro has been found generally to potentiate previously induced tone rather than stimulate significant or sustained de novo vasoconstriction (20). However, many of these studies have been done using rat vessels which are inherently ouabain resistant. Prolonged exposure of human resistance vessels to ouabain did produce concentration-dependent increase in resting tone (29), although chronic treatment of humans with cardiac glycosides does not produce hypertension. Despite these inconsistencies, there continues to be keen interest in the possibility that an endogenous digitalis-like factor may be a pathogenetic factor in hypertension.

Since HIF manifests a number of differences from ouabain in physiological tests, we studied effects of highly purified HIF on vasoconstriction in parallel with ouabain using pulmonary artery (PA) and aortic rings from genetically hypertensive (SHR) and normotensive control rats (18). Pulmonary arteries were studied since SHR have been found to have a degree of primary pulmonary hypertension, independent of the systemic hypertension (19), and we wondered if this pulmonary vasoconstriction could be associated with HIF-induced $Na^+/K^+$-ATPase inhibition.

HIF was found to be a potent constrictor of pulmonary arteries at concentrations in the nanomolar range, while ouabain at much greater concentration had no effect (Fig. 3).

736

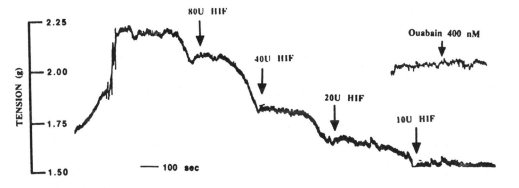

**Figure 3.** Constriction of pulmonary arterial rings mounted in vitro from spontaneously hypertensive rats (SHR) by increasing doses of HIF. Inhibition was completely reversible. Ouabain at 100 times the HIF dose was without effect (inset).

Furthermore, the effects on PA were significantly greater in SHR than in normotensive rats, and within SHR vasocontrictive effects on PA were significantly greater than on aorta. There was no significant difference on aortic ring constriction between SHR and normotensive rats, indicating that the PA of SHR were particularly susceptible to the vasoconstrictive actions of HIF, leaving open the possibility that HIF could play a role in the pathogenesis of primary pulmonary hypertension in these rats (18).

There has been much discussion about the mechanism by which $Na^+/K^+$-ATPase inhibition in vascular tissue could lead to vasoconstriction. Altered $Na^+/Ca^{2+}$ exchange, membrane depolarization with opening of voltage-dependent $Ca^{2+}$ channels, and augmented autonomic neurotransmitter activity have all been proposed. We therefore conducted preliminary experiments to address this issue. HIF-induced vasoconstriction in PA was completely abolished by the $\alpha$ blocker, phentolamine (1 $\mu$M), and greatly attenuated by removing extracellular $Ca^{2+}$ (Fig. 4). The calcium channel blocker, verapamil, had no effect on HIF-induced vasoconstriction (18).

These results suggest that a major mechanism for HIF induced vasoconstriction in these vessels is local modulation of the $Na^+/K^+$-ATPase-adrenergic neuroeffector interaction resulting in potentiation of norepinephrine action at the neuromuscular junction. Exocytotic release of norepinephrine into the junctional cleft is normally triggered by $Ca^{2+}$ entry into the nerve terminal after the propagation of the action potential along the neuronal cell membrane. Norpeinephrine release is therefore inhibited in a zero-calcium milieu, a condition that also eliminated HIF-induced vasoconstriction. This mechanism of action may explain the particular sensitivity of the PA in SHR to HIF, since SHR are known to have enhanced adrenergic innervation in PA (10).

These findings are consistent with the proposed role of an endogenous $Na^+/K^+$-ATPase inhibitor in vasoconstriction and the regulation of blood pressure. The absence of vasoconstriction by ouabain in this model extends the concept that structural difference between HIF and ouabain would be needed to account for differences in biological activity observed for the two inhibitors.

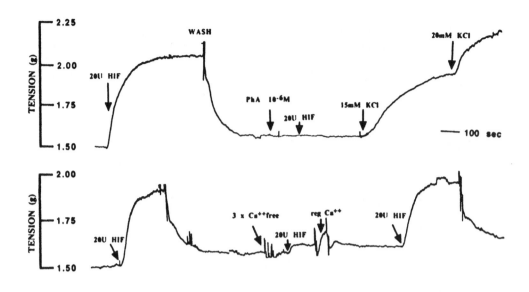

**Figure 4.** Effects of the α-blocker phentolamine (PhA, $10^{-6}$M) **(top)** and nominal zero calcium **(bottom)** on HIF -induced constrictions of pulmonary artery of SHR. Phentolamine abolished the HIF-induced constriction and zero calcium greatly reduced the constriction. Return to normal calcium milieu (reg Ca$^{++}$) restored the vasoconstriction on retreatment with HIF. Reproduced from (18) with permission.

### Physicochemical Characterization of HIF as an Isomer of Ouabain

The principal difficulty in this work was obtaining enough HIF, completely purified, for organic analysis at the microgram or even submicrogram level, since exrapolation to HIF amounts from calculations of HIF bioactivity using a ouabain-standardized assay indicated that no more than 1 µg of HIF could be expected from 5 kg of starting hypothalamic tissue, and yields were found to vary over time. A crucial step in this process was development of a high-yield affinity purification method. Manipulation of samples using purified membranes alone as the binding matrix proved cumbersome. We therefore developed a method whereby SDS-extracted canine kidney Na$^+$/K$^+$-ATPase was coupled to paramagnetic iron particles via glutaraldehyde cross linking of primary amino groups. Separation of magnetic particles holding the enzyme-HIF complex from supernatant containing impurities was achieved with a magnetic bar. The particle-enzyme complex consistently retained active ouabain (and HIF) binding sites, was easily scaled up, and each batch column could be recycled two to three times. HIF recoveries from this step averaged 50%. Using this methodology followed by two HPLC steps HIF was purified to a single peak of bioactivity which ultimately yielded a single compound (26).

The next step was to determine the precise molecular mass of pure HIF. A protocol for analyzing HIF was piloted using submicrogram amounts of authentic ouabain, ionized in the presence of ammonium acetate to form the $(M + NH_4)^+$ ion for ouabain having a m/z of 602. This ion is mass selected in the first quadripole of a triple stage tandem mass spectrometer, then undergoes collision with argon in the second quadripole, splitting off the sugar moiety leaving the steroid portion of ouabain having an m/z = 439, the only ion detected in the third quadripole. Analysis of affinity-purified HIF by this technic yielded the m/z 439 ion, suggesting that HIF is isobaric with ouabain and that it is a glycoside whose sugar is isobaric with rhamnose.

To identify the sugar portion of HIF, it was necessary to cleave the sugar from the aglycone moiety. This was accomplished using naringinase, an enzyme preparation containing α-L rhamnosidase activity. Naringinase hydrolysis reduced the bioactivity of both HIF and ouabain by 100-fold. However, naringinase hydrolysates were unsuitable for sugar analysis since the commercially available preparation was contaminated with sugars, including rhamnose, at the pM level, comparable to the sugar concentration expected in the HIF hydrolysate. But acid hydrolysis was found suitable for generating free sugars for analysis. GC/MS detection of persilylated derivatives then identified the sugar in HIF as rhamnose, and chiral GC/MS analysis of pertrifluoroacetyl derivatives demonstrated that HIF hydrolysates contained only L-rhamnose, the isomer found in ouabain (26).

Since HIF and ouabain are isobaric and both contain rhamnose, it was possible to assign the molecular weight of the aglycone of HIF by difference. However, a direct measure of the molecular weight of HIF aglycone was desirable, and was accomplished by LC/MS analysis of naringinase hydrolysates. Contaminants in naringinase preparation do not interfere with study of the aglycone, and the aglycone of ouabain (and presumably HIF) was destroyed by acid hydrolysis. Both glycosides and aglycones of HIF and ouabain were indistinguishable by HPLC retention times, and full scan ion-spray mass spectra summed over the peak for HIF and ouabain aglycones showed pseudo-molecular ions at m/z 439 $(M + H)^+$ and m/z 456 $(M + NH4)^+$ for both (26).

These characterization experiments identified the sugar in HIF and established a precise molecular mass for HIF aglycone, but obviously did not distinguish HIF from ouabain. To clarify this crucial point, acylaton of reactive hydroxyls was used as a means of amplifying any structural differences. Naphthoyl derivatives were chosen because they are obtained in high yield, give rise to fluorescent products, exhibit intense UV absorptions, and, most importantly, exhibit characteristic exciton-coupled circular dichroism (CD) spectra (17). Reaction conditions were specifically optimized to produce the 1,19,2',3',4' ouabain pentanaphthoate, and then applied to HIF and ouabain in parallel experiments. Treatment of ouabain yielded only the expected pentanaphthoate, while reaction with HIF produced a pentanaphthoate and a more lipophilic product representing a hexanaphthoate derivative of HIF (26).

The pentanaphthoate derivatives of HIF and ouabain showed different retention times on HPLC in coinjection experiments (Fig. 5), and subsequent CD spectroscopy on isolated naphthoylation products of HIF and ouabain confirmed that the two molecules are different. Thus, ouabain pentanaphthoate gave a clearly split CD curve with a positive Cotton effect reflecting exciton coupling between various naphthoate groups, in marked contrast to HIF pentanaphthoate which showed no distinct CD Cotton effect (Fig. 5). The hexanaphthoate derivative of HIF, on further HPLC, was separated into two compounds with opposite Cotton effects on CD analysis, suggesting two isomeric forms of the HIF hexanaphoate derivative formed under conditions where no ouabain hexanaphthoate was produced (26).

The naphthoylation of HIF provided several clues as to how it differs from ouabain. The CD results demonstrarte that the spatial arrangement of chromophores is different for the two derivatives, as would be found for regio- or stereochemical isomers. The formation of hexanaphoates of HIF under conditions where none forms for ouabain suggests that at least one (-OH) acylation site is more accessible in the HIF structure. Furthermore, since two (isomeric) hexanaphthoates of HIF were isolated, it is likely that two hydroxyl groups in HIF are altered relative to ouabain.

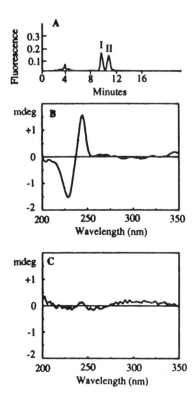

**Figure 5.** (A) Fluorescence-detected HPLC trace for coinjected samples of ouabain pentanaphthoate (I) and HIF pentanaphthoate (II). (B) CD spectrum (acetonitrile) of ouabain pentanaphthoate (I). (C) CD spectrum (acetonitrile) of HIF pentanaphthoate (II). Absence of the signal for II was not due to a lower concentration since amounts of I and II were nearly identical as measured by UV absorbance. mdeg, Millidegrees. Reproduced from (26) with permission.

While the abundance and biological properties of HIF suggest its role as an endogenous inhibitor of the $Na^+/K^+$-ATPase, it is not known whether HIF is truly the product of *de novo* mammalian biosynthesis. Until direct measurement of its production by mammalian cells can be made, the best evidence available bearing on this important question is provided by the studies of Ferrandi and co-workers in Milan hypertensive strain rats. They found the hypothalamic content of HIF in the Milan hypertensive strain to be approximately ten-fold greater than that of the Milan normtensive strain (6). The fact that both strains were housed in identical environments, eating identical rat chow suggests, but does not prove, endogenous production of HIF.

In summary, a consistent body of data from biochemical and physiological experiments in recent years has demonstrated significant differences in the biological activity profiles of HIF and ouabain, suggesting that the two could not be structurally

740

identical. The physical evidence recently obtained confirms this point, showing that HIF is an isomer of ouabain. The structural difference must account for the difference in biological properties which distinguish the pharmacologic inhibitor, ouabain, from the putative physiologic mammalian analogue, HIF. Biological testing of a synthetic candidate will be required to confirm the findings, and to establish definitively the role of HIF in cellular physiology, the regulation of fluid volume status, and the pathogenesis of cardiovascular diseases.

## References

1. Anner BM, Rey HG, Moosmayer M, Meszoely I, Haupert GT Jr (1990) Hypothalamic $Na^+/K^+$-ATPase inhibitor characterized in two-sided liposomes containing pure renal $Na^+/K^+$-ATPase. Amer J Physiol 258:F144-F153
2. Bova S, Blaustein MP, Ludens JH, Harris DW, DuCharme DW, Hamlyn JM (1991) Effects of an endogenous ouabain-like compound on heart and aorta. Hypertension 17:944-950.
3. Cantiello HF, Chen E, Ray S, Haupert GT Jr (1988) Na+ pump ion renal tubular cells is regulated by the endogenous $Na^+/K^+$-ATPase inhibitor from hypothalamus. Proc Natl Acad Sci USA 86:10080-10084.
4. Carilli CT, Berne M, Cantley LC, Haupert GT Jr (1985) Hypothalamic factor inhibits the $Na^+/K^+$-ATPase from the extracellular surface. Mechanism of inhibition. J Biol Chem 260:1027-1031.
5. DeWardener HE, Clarkson EM (1985) Concept of natriuretic hormone. Physiol Rev 65:658-759
6. Ferrandi M, Minotti E, Salardi S, Florio M, Bianchi G, Ferrari P (1992) Ouabainlike factor in Milan hypertensive rats. Am J Physiol 263:F739-F748
7. Goto A, Yamada K, Yagi N, Yoshioka M, Sugimoto T (1992) Physiology and pharmacology of endogenous digitalis-like factors. Pharmacol Rev 44:377-399
8. Haber E, Haupert GT Jr (1987) The search for a hypothalamic $Na^+/K^+$-ATPase inhibitor. Hypertension 9:315-324
9. Hallaq H, Haupert GT Jr (1989) Positive inotropic effets of the endogenus $Na^+/K^+$-transporting ATPase inhibitor from the hypothalamus. Proc Natl Acad Sci USA 86:10080-10084
10. Hallback M, Weiss L (1977) Mechansims of spontaneous hypertension in rats. Med Clin North Am 61:593-609
11. Hamlyn JM, Blaustein MP, Bova S, DuCharme DW, Haris DW, Mandel F, Mathews WR, Ludens JH (1991) Identification and characterization of a ouabain-like compound from human plasma. Proc Natl Acad Sci USA 88:6259-6263
12. Harada N, Nakansihi K (1983) Circular Dichroism Spectroscopy: Exciton Coupling in Organic Stereochemistry. University Science Books, Mill Valley CA
13. Haupert GT Jr, Carilli CT, Cantley LC (1984) Hypothalamic sodium-transport inhibitor is a high-affinity reversible inhibitor of $Na^+/K^+$-ATPase. Am J Physiol 247:F919-F924
14. Haupert GT Jr, Sancho JM (1979) Sodium transport inhibitor from bovine hypothalamus. Proc Natl Acad Sci USA 76:4658-4660
15. Haupert GT Jr, Shulz JT (1989) Hypothalamic factor inhibits ouabain-resistant $Na^+/K^+$-ATPase in primate cells transfected with a ouabain resistance gene. Kidney Int 35:157A
16. Herrera VLM, Emanuel JR, Ruiz-Opazo N, Levenson R, Nadal-Ginard B (1987) Three differentailly expressed $Na^+/K^+$-ATPase alpha subunit isoforms: structural and functional implications. J Cell Biol 105:1855-1865
17. Ikemoto N, Lo L-C, Nakanishi K (1992) Detection of subpicomole levels of compounds containing hydroxyl and amino groups with the fluorogenic reagent, 2-naphthoylimidazole. Angew Chem Int Ed Engl 31:890-891

18.  Janssens SP, Kachoris C, Prker WL, Hales CA, Haupert GT Jr (1993) Hypothalamic Na⁺/K⁺-ATPase inhibitor constricts pulmonary arteries of spontaneously hypertensive rats. J Cardiovasc Pharmacol 22:42-46

19.  Janssens SP, Thompson TB, Spence CR, Hales CA (1991) Acute and chronic hypoxia in the pulmonary and systemic circulation of spontaneously hypertensive and Wistar-Kyoto rats. Am Rev Resp Dis 143:A187

20.  Mulvany MJ (1985) Changes in sodium pump activity and vascular contraction. J Hypeertens 3:429-436

21.  Price EM, Lingrel JB (1988) Structure-function relationships in the Na⁺/K⁺-ATPase a-subunit. Biochemistry 27:8400-8408

22.  Segel IH (1975) Enzyme Kinetics.Wiley, New York, chapt.7, pp 346-384

23.  Shimada K, Fujii Y, Yamashita E, Niizaki Y, Sato Y, Nambara T (1977) Studies on cardiotonic steroids from the skin of Japanese toad. Chem Pharm Bull (Tokyo) 25:714-730

24.  Shull GE, Greeb J, Lingrel JB (1986) Molecular cloning of three distinct forms of Na⁺/K⁺-ATPase alpha-subunit from rat brain. Biochemistry 25:8125-8132

25.  Sweadner KJ (1989) Isoenzymes of the Na⁺/K⁺-ATPase Biochim Biophys Acta 988:185-220

26.  Tymiak AA, Norman JA, Bolgar M, DiDonato GC, Lee H, Parker WL, Lo L-C, Berova N, Nakanishi K, Haber E, Haupert GT Jr (1993) Physicochemical characterization of a ouabain isomer isolated from bovine hypothalamus. Proc Natl Acad Sci USA 90: 8189-8193

27.  Vatner SF, Higgins CV, Franklin D, Braunwald E (1971) Effects of a digitalis glycoside on coronary and systemic dynamics in conscious dogs. Circ Res 28:470-479

28.  Werdan K, Wagenknecht B, Zwissler B, Brown L, Krawietz W, Erdmann E (1984) Biochem Pharmacol 33:1873-1886

29.  Woolfson RG, Hilton PJ, Poston L (1990) Effects of ouabain and low sodium on contractility of human resistance arteries. Hypertension 15:583-590

# Identification and physiological significance of $Na^+/K^+$-pump inhibitors in rats

Masaaki Tamura[*], Thomas M. Harris[$], Sau Kuen Lam[*] and Tadashi Inagami[*]
With the technical assistance of Trinita Fitzgerald and Edward Price, Jr.

Departments of [*]Biochemistry and [$]Chemistry Vanderbilt University School of Medicine, Nashville, TN (U.S.A.)

## Introduction

Recognition of the presence of a receptor on $Na^+/K^+$-ATPase for cardiac glycosides has prompted speculation on the existence of an endogenous ligand for the receptor. In support of this speculation, many investigators have reported the existence of $Na^+/K^+$-ATPase inhibitors in a variety of mammalian tissues (5). The concentrations of these inhibitors have been shown to change in response to several pathophysiological stimuli, particularly evident in hypertension-related conditions (1,3,6). Such inhibitors have been postulated to be involved in the pathogenesis of salt-dependent hypertension (3,6) and some types of essential hypertension (1). Recently, two $Na^+/K^+$-pump specific inhibitors have been isolated from human materials and identified as digoxin (4) and ouabain (13) by mass spectrometric studies. However, it remains controversial as to whether synthetic pathways for these typical plant-derived cardiotonic steroids, digoxin and ouabain, exist in human tissues, whether they are the inhibitors described in many reports (5) and whether they act as regulators of sodium homeostasis. The answers to these questions require further studies.

We have independently attempted to clarify these problems by carrying out the complete purification and structural analysis of the inhibitors. We have demonstrated that the majority of $Na^+/K^+$-ATPase inhibitory activities in the plasma of volume-expanded hogs is attributable to unsaturated fatty acids (20) and lysophosphatidylcholines containing saturated (22) and polyunsaturated fatty acids (21). While we were the first to purify a ouabain-like $Na^+/K^+$-pump specific inhibitor from bovine adrenal glands (23) and a highly potent specific inhibitor structurally different from ouabain from normal pig urine (24), the final purified amounts were not sufficient for the determination of the complete structures. These results indicate that the content of $Na^+/K^+$-pump specific inhibitors in mammals is extremely low, and it is therefore necessary to purify the inhibitors from highly induced tissue. Accordingly, we have investigated the tissue distribution of inhibitors in both normal and pathologically altered rats whose inhibitor contents have been reported to be high (7,15). We have also tried to determine which inhibitors are truly endogenous by switching the diet from regular to pure synthetic chow. Since ouabain or a substance identical to authentic ouabain has been identified from mammalian tissues (8,13), we have also investigated the physiological relevance of ouabain. By this strategy, we have found that the urine from reduced renal mass and renovascular hypertensive rats contains the most $Na^+/K^+$-pump specific inhibitors, although they seem to be derived from the diet. The level of a ouabain-like inhibitor

743

in a variety of tissues, however, is only slightly affected by switching the diets, and it seems that ouabain itself may be the physiological regulator of aldosterone production in adrenal glands. These findings provide evidence for the physiological significance of an endogenous, ouabain-like $Na^+/K^+$-pump inhibitor.

## Methods

<u>Surgical preparation</u>: Male Sprague-Dawley rats weighing 275~300 g were used for the preparation of experimentally hypertensive rats. Under pentobarbital anesthesia (50 mg/kg), the right kidney was removed in both experimental groups. For the preparation of reduced renal mass (RRM) hypertension, two of the three left renal artery branches directly connected to the kidney were completely tied off by silk suture. The other group was used for the preparation of one-kidney one-clip (1K-1C) Goldblatt hypertensive rats, in which the blood flow of the left renal artery was reduced by clamping with a constricting silver clamp (0.2 mm ID). After the surgery the rats were individually housed in metabolic cages and urine volume was monitored every day. Systolic blood pressure was monitored weekly by the tail cuff method without anesthesia.

<u>Extraction of $Na^+/K^+$-pump inhibitors:</u> Blood was drawn from the abdominal aorta under pentobarbital anesthesia (50 mg/kg) and plasma was separated by centrifugation. Urine was collected from conscious rats continuously, 24 hours per day. Solid tissues such as liver, kidney, lung, atria, ventricle, brain, spleen, aorta and adrenal glands were collected after euthanasia by an overdose of pentobarbital. All samples were frozen immediately and stored at -20°C until use.

Heavier weight tissues (>0.2 g) were individually homogenized with 4 volumes (w/v) of phosphate buffered saline (pH 7.4) using a polytron, and the water-soluble fraction was separated by centrifugation at 1000 x g, 4°C for 20 min. The precipitate was re-homogenized and used for re-extraction. Lighter weight tissues (<0.2g) were pooled (total weight: 0.5-1.0g), and inhibitors were extracted by the method described above. Specific $Na^+/K^+$-pump inhibitors in the supernatant of various tissues, plasma and urine were extracted by solid phase extraction using Amberlite XAD-2 mini columns (0.5 x 7.5 cm). The 100% acetonitrile eluate, which contained most of the specific inhibitors, was lyophilized. The inhibitors in this extract were fractionated by HPLC on a YMC ODS column (0.6 x 15 cm, Yamamura, Kyoto, Japan) and quantified by an [86]Rb uptake assay using human erythrocytes and/or a radioimmunoassay for ouabain.

<u>Purification of $Na^+/K^+$-pump inhibitors from rat urine and rat diet:</u> Two specific $Na^+/K^+$-pump inhibitors were prepared from approximately 30 l of RRM hypertensive rat urine and 3 kg of rat diet (# 5001, PMI Feeds, Inc., St. Louis, MO) by the slightly modified method previously described (24). Briefly, rat urine (one batch consisted of approximately 6 l) was directly chromatographed by an Amberlite XAD-2 column (5 x 40 cm). The inhibitors were eluted with 3 column volumes of methanol. The inhibitors in the diet were extracted with 50% methanol at 23°C for 12 hours and then chromatographed by the same column with the same conditions. This column chromatography was repeated appropriately. After evaporating the methanol, the extract was applied to a preparative reverse phase column (2.5 x 60 cm) packed with ODS bonded to silica gel (J.T. Baker, Phillipsburg, N.J.). The elution was performed with a linear gradient of acetonitrile (5-50%) at a flow rate of 5 ml/min. Two major inhibitory activities were then successively purified by 5 steps of high performance liquid

chromatography (HPLC). Reverse phase HPLCs on an Econosphere NH$_2$ column (2.25 x 25 cm, acetonitrile: 0-32%, Alltech, Deerfield, IL), on a YMC ODS column (2 x 30 cm, acetonitrile: 15-40%, Yamamura), on an Econosphere TMS column (1 x 25 cm, acetonitrile: 10-30%, Alltech), on a Vydac phenyl column (1 x 25 cm, acetonitrile: 10-30%, The Nest Group, Southboro, MA) and on a YMC ODS-AQ column (1 x 30 cm, acetonitrile: 15-35%, Yamamura) were run in succession with linear gradients of acetonitrile as indicated in parentheses, at flow rates of 8 ml/min for the preparative columns, at 3 ml/min for the TMS and phenyl columns and at 1 ml/min for the ODS-AQ analytical column.

Structure analysis: $^1$H NMR spectra were recorded on a Bruker AMX-500 spectrometer operated at 500.13 MHz at ambient temperature. The entire purified samples from rat diet, which correspond to approximately 7 μg of each inhibitor, were individually dissolved in 0.5 ml deuterium oxide and then subjected to the NMR studies. The data were acquired over 3 h and 24 h of accumulation for inhibitors A and B, respectively. Liquid secondary ion (LSI) mass spectra of negative ions were obtained on a Kratos Concept HHII tandem mass spectrometer. Approximately 0.25 μg of all four purified samples (two inhibitors each from rat diet and rat urine) were individually desorbed with glycerol (used as matrix) and introduced into the mass spectrometer. Gas-liquid chromatography (GC)-mass spectra (MS) were obtained on a Perkin Elmer 8500 GC-mass spectrometer equipped with a Perkin Elmer ITD ion trap detector using Rtx-1 and Rtx-225 columns with helium as the carrier gas. The purified samples (0.1 μg each) were hydrolyzed by a mixture of TFA and acetic acid at 100°C for 2 h and then subjected to the GC-MS studies. The ultraviolet spectra were recorded on a spectrophotometer (Varian, Palo Alto, CA) using distilled water as the solvent.

In vivo experiments: Male SD rats weighing 275-300 g were divided into two groups. Both groups were fed a normal rat chow diet (#5001). The experimental rats each received an intravenous bolus injection of varying amounts of ouabain (0.6-6000 ng/kg) in 300 μl of saline, and the control group received injections of 300 μl of saline. Injections were made through the tail vein at a rate of 0.1 ml/min. Blood was drawn from the abdominal aorta under pentobarbital anesthesia (50 mg/kg) at appropriate time points. Plasma was quickly separated by centrifugation and stored at -20°C.

Primary culture of bovine adrenal glomerulosa cells: Bovine adrenal glands were obtained within 60 min of sacrifice and placed in ice-cold Ham's F12 medium. Glomerulosa cells were isolated by collagenase digestion and mechanical dispersion of capsular tissue from 15 adrenal glands as described elsewhere (11). Cells were then seeded into 35 mm culture dishes and were grown in a modified Ham's F12 medium. Cells were maintained at 37°C in a humidified air atmosphere. Medium was changed every 48 h. The effect of ouabain ($10^{-10}$-$10^{-6}$ M) on aldosterone secretion was studied during the period 96-144 h after seeding. Aldosterone secreted into the media was directly measured using a commercially available radioimmunoassay kit (Diagnostic Products Corp. Los Angeles, CA).

**Results and Discussion**

Tissue distribution: Before investigating the tissue distribution of the Na$^+$/K$^+$-pump inhibitors, we studied the validity of the solid phase extraction procedure by using both [$^3$H]ouabain and [$^3$H]digoxin, since most of the specific inhibitors detected in various tissues exhibited chromatographic characteristics similar to them. In these preliminary experiments both [³H] compounds, which were mixed with tissue homogenates, urine or plasma, showed more than 85% recovery in every trial. Accordingly, the Amberlite XAD-2 mini column solid phase extraction method was employed for this study. In addition to solid phase extraction, fractionation of the inhibitors by reverse phase HPLC was carried out before quantifying the inhibitor levels, since at least 5 specific inhibitory activity peaks were detected in rat urine. However, four of these inhibitors, which eluted from a YMC ODS column at retention times between those of [³H]ouabain and [³H]digoxin, were not detectable in most of the solid tissues examined. Liver and kidney tissue did contain a small amount (less than 30 pg/g tissue) of inhibitors with retention times similar to those in normal rat urine. The fifth inhibitory activity, however, which eluted at the same position as [³H]ouabain, was consistently detected in most of the tissues. Figure 1 shows levels of this ouabain-like inhibitory activity, expressed as ouabain equivalents, present in plasma and other tissues taken from normal rats maintained on tap water and regular chow *ad libitum*. The highest levels of inhibitor

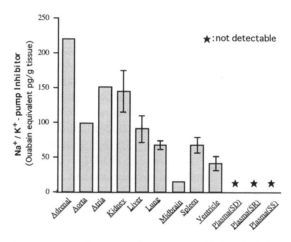

**Fig. 1.** Tissue distribution of the ouabain-like Na⁺/K⁺-pump inhibitor in normal SD rats raised on regular rat chow (Rat Chow 5001). Bars with standard error indications are the means ± SE of five samples. Bars without standard error indications are the means of two pooled samples. For the pooled samples, approximately 0.5-1 g pooled tissues were collected from rats of similar age, and then the inhibitors were extracted as described in the Methods section. Plasma (SS) and plasma (SR) are the plasma obtained from Dahl salt-sensitive and salt-resistant rats, respectively.

were found in adrenals, atria and kidneys. Lower levels were apparent in aorta, liver, lung, spleen and ventricle. The inhibitor levels in plasma obtained from not only normal SD rats but also Dahl salt-sensitive and salt-resistant rats were not detectable. This overall distribution of the ouabain-like inhibitor is confirmatory of the recent publication reported by Ludens *et al.* (12), although the assay method employed in their study was an ELISA for ouabain. A noteworthy point in the present study is that 2 ml plasma was not enough to detect any inhibitory activity by this bioassay, suggesting that the plasma level of the inhibitor is less than 20 pg/ml. In fact, when 10 ml pooled plasma was extracted and assayed by the same method and was also assayed by a ouabain RIA developed in our laboratory, approximately 14 pg/ml ouabain-like inhibitor was detected. This low level of the inhibitor as compared with the levels reported by other groups (8,12,17) can easily be explained by differences in the extraction methodologies. Since the present study employed two column systems for the extraction and fractionation of the inhibitor, a smaller amount of interfering substances contaminated the sample assay tubes as compared with the one column solid-phase extraction method which was used for all of the other studies (8,12,17). Another contradictory finding in the present study is that the specific activity of the ouabain-like inhibitor in adrenal glands is not as high as other groups have reported (12,17). This suggests that the adrenal glands may not be the major production site for the ouabain-like $Na^+/K^+$-pump inhibitor. Further evidence shows that [$^3H$]ouabain accumulates in the adrenal glands after administration (10), and that the plasma immunoreactive ouabain levels in patients after bilateral adrenalectomy are similar to those in normal subjects (14). The content of the inhibitor in midbrain tissue was next to the lowest of all tissues tested, although the midbrain, particularly the hypothalamus, has been considered to be one of the sources of circulating inhibitor. The clarification of this contradiction requires future study.

Among the many tissues, plasma, and urine which were examined, urine was found to contain the most abundant amounts of several different $Na^+/K^+$-pump specific inhibitors (Fig. 2). Accordingly, the relationship between the various inhibitor levels and RRM and 1K-1C hypertension was studied. All of the inhibitors, including the ouabain-like inhibitor and inhibitors A through D, significantly increased in the urine from both groups of hypertensive rats. These results seem to reflect the increased plasma levels of the inhibitors (7,15). The significant increase in inhibitor levels was observed immediately after surgery for the setup of RRM and 1K-1C hypertensions, and this

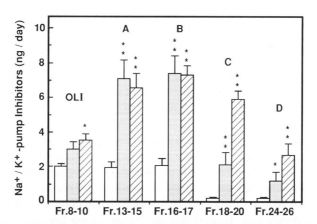

**Fig. 2.** Distribution of $Na^+/K^+$-pump inhibitory activities in urine from normotensive control rats (open bars) and RRM (dotted bars) and 1K-1C (hashed bars) hypertensive rats. Urine from RRM, 1K-1C, and age-matched control rats was collected 3 weeks after the surgeries. The inhibitors were extracted and fractionated as described in the Methods section, and then the fraction combinations designated in the figure were lyophilized and subjected to the assay. All data are the means ± SE of five samples. * $p<0.05$, ** $p<0.01$ compared to the control levels in each fraction

increase was blunted by feeding the high salt diet, although NaCl has been reported to be essential for further increasing the blood pressure in these hypertensive models. These results inspired us to investigate whether the regular rat chow diet may be the source for the inhibitor detected in rat tissues.

Effect of diet on tissue levels of the inhibitors: Some of the commercially available diets, such as rat chow 5001, basal diet 5755, and high and low-salt diets, were tested for this purpose. Each diet (1 g) was extracted with 50% methanol overnight, and the extracts were individually chromatographed by HPLC on a Zorbax ODS analytical column. Rat chow 5001 (regular chow) was found to contain large amounts of a variety of inhibitors, whereas the basal diet and the high and low-salt diets, all of which contain only synthetic nutrient ingredients, had no detectable inhibitory activities. A striking discovery was that the elution pattern of the inhibitors from HPLC of the regular chow was similar to the pattern found in rat urine, strongly suggesting that the majority of the inhibitors detected in normotensive and hypertensive rat urine may be derived from the diet. Accordingly, the effect on tissue distribution of changing from regular to synthetic diet was studied. After four weeks of feeding the basal diet to control, RRM and 1K-1C rat groups, the kidneys, liver, blood and urine were collected. The inhibitory activity in each tissue sample was extracted and fractionated as described in the tissue distribution section, and then the inhibitor levels were quantified with an $^{86}$Rb uptake assay. The overall tissue distribution of the inhibitors in the two solid tissues examined and in plasma was identical to the tissue pattern from rats raised on the regular diet. First, only ouabain-like inhibitory activity was consistently detected in the two tissues, and second, no inhibitory activity was detected in the 2 ml plasma sample. Much of the inhibitory activity detected in the urine of rats raised on the regular diet was absent from the urine of rats raised on the synthetic diet, except for the ouabain-like inhibitor, whose level decreased to approximately one-half of the regular chow rat level. These results clearly demonstrate that the ouabain-like inhibitor may be of endogenous origin but that the majority of the inhibitors detected in the urine of rats raised on the regular chow originate from the diet. In future studies regarding Na$^+$/K$^+$-pump inhibitors, we must all note this discovery, and in fact, we may need to reevaluate papers which have been published in the past. Obviously the tissue levels of inhibitors in regular chow-fed rats cannot be compared with the rats raised on high or low salt diets either, since these diets also consist of synthetic ingredients, like the basal diet. On the other hand, the identity of the ouabain-like inhibitor detected in all tissues is still unclear, since this inhibitor was consistently detected by $^{86}$Rb uptake assay using human erythrocytes, but was not always detectable by an RIA for ouabain. Since this RIA has a higher sensitivity for ouabain (10-20 pg/tube) than the $^{86}$Rb uptake assay, these results are contradictory. Although the possibility that this RIA has been interfered with by some other substances has not been ruled out, these results may suggest that the Na$^+$/K$^+$-pump inhibitory activity and the immunoreactivity in this fraction (with identical retention time to ouabain) are actually due to two different substances.

Purification and identification of Na$^+$/K$^+$-pump inhibitors: Although the study of the dietary effect on tissue levels of the inhibitors revealed that a majority of the urinary inhibitors are derived from the diet, it is still worthwhile to determine whether their chemical natures are identical to each other. Therefore, purification of the major inhibitory activity in RRM rat urine and in the diet was attempted. Crude extracts were individually prepared from a total of approximately 30 $l$ of pooled RRM rat urine and 3.0 kg of regular diet using Amberlite XAD-2 adsorption column chromatography as described under Methods. The extracts were individually chromatographed by reverse phase $C_{18}$ LPLC. The activity peaks corresponding to urinary inhibitors A and B (Fig. 2) were further purified individually by a series of HPLCs. These two peaks were chosen because, as the largest, they were calculated to contain the minimal activity quantities necessary for structural analysis and also because they were characterized as specific for the $Na^+/K^+$-pump and cross-reactive with anti-ouabain antibodies. An Econosphere preparative $NH_2$ column, a YMC preparative ODS column, an Econosphere TMS semipreparative column and a Vydac phenyl semipreparative column were used in succession for the HPLC purification steps. The active substances under peak A and peak B from $C_{18}$ LPLC were finally purified by HPLC on a YMC ODS-AQ column.

The final purified materials obtained by repetitive HPLC were pooled and structurally analyzed by liquid secondary ion (LSI) mass spectrometry and $^1H$ NMR spectroscopy. Approximately 0.25 $\mu g$ each of the purified inhibitor A and inhibitor B, from peak A and peak B on $C_{18}$ LPLC, respectively, were subjected to LSI-mass spectrometry. The mass spectra of negative ions were obtained over a mass range of m/z 50-1650. The nominal mass obtained from LSI mass spectra provided a single de-protonated dominant ion peak at m/z 711 for inhibitor A from both urine and diet and at m/z 535 for inhibitor B from both urine and diet. These LSI mass studies also provided evidence that the four purified inhibitors were homogeneous, since no contaminant ion peaks were observed. In order to analyze the sugar moieties for both compounds, approximately 1/100th of each was also subjected to GC-MS following acid hydrolysis. Identification of the acetylated derivatives was based on their gas-liquid chromatography retention times relative to corresponding standard compounds and on their mass-spectral characteristics. Inhibitor A was found to contain two types of sugars, rhamnose and glucose, whereas the sugar in inhibitor B was identified as rhamnose alone. $^1H$ NMR spectra obtained from the entire purified inhibitors from diet revealed the following: 1) both inhibitors contain one 5-membered lactone ring, 2) inhibitor A has one tertiary methyl group whereas inhibitor B contains two tertiary methyl groups, and 3) inhibitor A contains one aldehyde group.

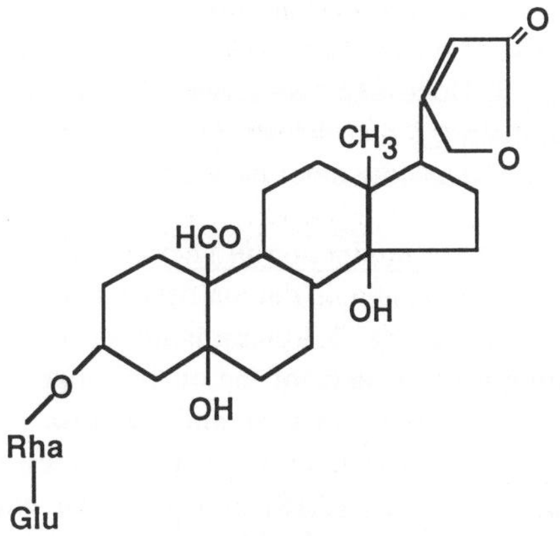

Convalloside (Bogoroside)

**Fig. 3.** Chemical structures of convalloside.

749

The ultraviolet absorption spectra of both inhibitors were identical to spectra of the cardenolides, such as ouabain and digoxin (UV maximum at approximately 218 nm). By studying all of the information and data together, it was concluded that inhibitor A is a stereoisomer of convalloside (Fig. 3). Inhibitor B also shares the common structure of the plant-derived cardenolides, and the structure was assigned tentatively as a mono-rhamnoside of periplogenin.

Figure 4 shows inhibitory activities of the purified inhibitor A from urine against [$^3$H] ouabain binding (A) and $^{86}$Rb uptake into human erythrocytes (B) and immunocross-reactivities with ouabain (C) and digoxin antibodies (D). The purified inhibitor B from rat urine also exhibited almost the identical pattern in the four assays. Both inhibitors A and B are not easily distinguished from ouabain by these assay systems, suggesting that inhibitor A (convalloside stereoisomer) and inhibitor B are equipotent to ouabain and that a more specific assay system for the ouabain-like compound is acutely required for future studies.

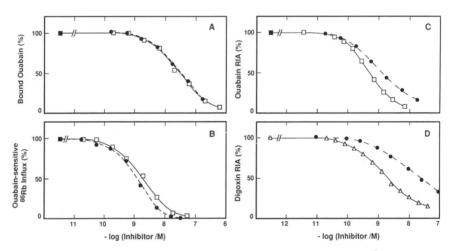

**Fig. 4.** Inhibitory effect of the purified inhibitor (●) on binding of [$^3$H]ouabain to Na$^+$/K$^+$-ATPase (A) and on ouabain-sensitive $^{86}$Rb uptake by human erythrocytes (B), and the cross-reactivities of the inhibitor with anti-ouabain antibodies (C) and with anti-digoxin antibodies (D). Dose response curves of these activities were determined in comparison with ouabain (□) and digoxin (△) under the same assay conditions. Each point is the mean of duplicate determinations.

Effect of ouabain on aldosterone production *in vivo* and *in vitro*: The studies of tissue distribution and the effect of the synthetic diet on tissue and urine levels of the inhibitors suggest that only the ouabain-like Na$^+$/K$^+$-pump inhibitor may be of endogenous origin. Accordingly, the physiological relevance of the ouabain-like inhibitor was investigated by using authentic ouabain, both *in vivo* and *in vitro*. Since relatively higher concentrations of ouabain ($10^{-6}$~$10^{-5}$ M) have been reported to stimulate aldosterone production in rat glomerulosa cells (2,18) and since an endogenous Na$^+$/K$^+$-pump inhibitor has been shown to be associated with salt-dependent hypertension and volume expansion-related high blood pressure (6), we focused our study on investigating the relationship between ouabain and aldosterone production.

Bolus intravenous injections of ouabain at 0.6 μg/kg significantly increased plasma levels of aldosterone 4 hours after injection. The average stimulation of the aldosterone in the experimental group was 2.7 times higher than in the saline-injected control group (Fig 5). However, greater or smaller amounts of ouabain showed no significant effects 4 hours after injection. The time course of the ouabain effect was also investigated using one effective dosage of ouabain (0.6 μg/kg). Measurements were made at time points 0.5 h, 1 h, 2 h, 4 h and 24 h after the bolus injection of ouabain, but only the measurement at 4 h showed a significant increase (p < 0.01) in plasma aldosterone levels. Although the increased plasma aldosterone levels in the experimental rats tended to stay high, the statistical difference between the experimental groups and control groups was no longer observed 24 hours after the ouabain injection. A noteworthy point shown in

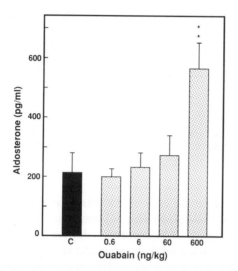

**Fig. 5.** Effect of intravenous injections of ouabain (0.6-600 ng/kg) on plasma levels of aldosterone in normotensive SD rats. Blood was drawn under pentobarbital anesthesia 4 h after the injection and plasma was separated by centrifugation. Results are means ± SE for 5 rats. ** P < 0.01 compared to saline injection.

these results is that the effective dose of ouabain for the increase in plasma aldosterone is approximately 1/10 of the clinical dose of the common cardiotonic steroids, such as digoxin, used for treatment of congestive heart failure and arrhythmia. Szalay has reported the stimulation of aldosterone production in isolated adrenal glands two weeks after a large dose (0.75 mg/kg) of ouabain (19). Her dose is approximately 1000 times higher than the effective dose observed in the present study.

In order to clarify the effect of ouabain on aldosterone production by another method, the dose response and time course of ouabain effect were studied by using primary cultured cells from bovine adrenal glands. These primary cultured cells consisted of more than 95% homogeneous glomerulosa cells with a small amount of contamination

751

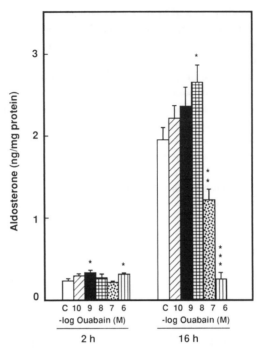

**Fig. 6.** Effect of ouabain on aldosterone production in bovine adrenal glomerulosa cells. Results are means ± SE of triplicate determinations in two separate experiments. *p <0.05, **p<0.01 compared to the control levels at each incubation time.

by fasciculata cells based on the ratio of the marker enzymes. The effects of $10^{-10}$-$10^{-6}$ M ouabain on aldosterone production in glomerulosa cells at different incubation times are presented in Fig. 6. Aldosterone production rose with increasing ouabain concentrations, with maximum production occurring at $10^{-9}$ M for a 2 h incubation with ouabain and at $10^{-8}$ M for a 16 h incubation. Higher concentrations reduced aldosterone production to levels even lower than the control when cells were incubated with ouabain for 16 hours. Contrary to the long time incubations, at 2 h incubation time although the higher concentrations of ouabain reduced the aldosterone production with minimal production occurring at $10^{-7}$ M, an even higher concentration of $10^{-6}$ M ouabain in fact increased aldosterone production. These results obtained from *in vitro* experiments are in excellent agreement with the results from *in vivo* experiments. They perhaps suggest that the lower concentrations of ouabain, approximately equivalent to the concentrations which exhibit less than 50% inhibition of $^{86}$Rb uptake into human erythrocytes, are the stimulatory concentrations for aldosterone production, and likewise, the higher inhibitory concentrations for $^{86}$Rb uptake are also the inhibitory concentrations for aldosterone production. The present studies indicate that only with short incubation times (less than 2 h), higher concentrations of ouabain ($10^{-6}$-$10^{-5}$ M) increase aldosterone production, which has also been reported elsewhere for rat adrenal glomerulosa cells (2,18). However, since the rat Na$^+$/K$^+$-pump is known to consist of the ouabain resistant $\alpha$1 isoform (16), the results obtained from the present studies with bovine cells may not be appropriate to compare with the rat experiments. Nevertheless, the *in vitro* experiments clearly demonstrate that low concentrations of ouabain, which are almost comparable to the tissue levels of ouabain-like inhibitor ($\sim 10^{-10}$ M), stimulate aldosterone production in bovine adrenal glomerulosa cells.

The present study clearly demonstrates that there is one kind of ouabain-like $Na^+/K^+$-pump inhibitor in almost all tissues from the rat, that it may be of endogenous origin, and that this inhibitor may regulate sodium homeostasis through controlling aldosterone production in adrenal glomerulosa cells.

## Acknowledgements

The authors gratefully acknowledge Drs. Ian A. Blair and Dennis Phillips (Vanderbilt University) for their assistance in LSI mass analysis and Drs. Yue Fen Wang and Carl Hellerqvist (Vanderbilt University) for their excellent assistance in GC MS analysis. We are indebted to Mr. Markus Voehler for his excellent technical assistance in $^1$H NMR study. This work was supported by the National Institutes of Health Grants HL14192 and HL35323.

## References

1.  Blaustein MP and Hamlyn JM (1991) Pathogenesis of essential hypertension: a link between dietary salt and high blood pressure. *Hypertension* 18 (Suppl III):III-184-III-195.
2.  Braley LM and Williams GH (1978) The effect of ouabain on steroid production by rat adrenal cells stimulated by angiotensin II, $\alpha$1-24 adrenocorticotropin, and potassium. *Endocrinology* 103:1997-2005.
3.  Buckalew VM Jr and Haddy FJ (1990) Circulating natriuretic factors in hypertension. In: Laragh JH and Brenner BM (eds) *Hypertension: Pathophysiology, diagnosis, and management.* Raven Press Ltd, pp 939-954.
4.  Goto A, Ishiguro T, Yamada K, Ishii M, Yoshioka M, Eguchi C, Shimora M and Sugimoto T (1990) Isolation of a urinary digitalis-like factor indistinguishable from digoxin. *Biochem Biophys Res Commun* 173:1093-1101.
5.  Goto A, Yamada K, Yagi N, Yoshioka M and Sugimoto T (1992) Physiology and pharmacology of endogenous digitalis-like factors. *Pharmacol Rev* 44:377-395.
6.  Haddy FJ (1990) Digitalislike circulating factor in hypertension: potential messenger between salt balance and intracellular sodium. *Cardiovasc Drugs Ther* 4:343-349.
7.  Haddy FJ and Pamnani MB (1983) The role of humoral sodium-potassium pump inhibitor in low-renin hypertension. *Federation Proc* 42:2673-2680.
8.  Hamlyn JM, Blaustein MP, Bova S, Ducharme DW, Harris DW, Mandel F, Mathews WR and Ludens JH (1991) Identification and characterization of a ouabain-like compound from human plasma. *Proc Natl Acad Sci USA* 88:6259-6263.
9.  Haupert GT (1988) Physiological inhibitors of Na,K-ATPase: concept and status. In *The $Na^+,K^+$-Pump, Part B: Cellular Aspects,* Alan R. Liss, Inc., New York, pp. 297-320.
10. Kitano S, Morimoto S, Fukuo K, Yasuda S, Kaimoto T and Ogiwara T (1993) Adrenals accumulate exogenous ouabain and may be origin of plasma ouabain in rats. *Ther Res* 14:201-206.
11. Kramer RE (1988) Angiotensin II causes sustained elevations in cytosolic calcium in glomerulosa cells. *Am J Physiol* 255:E338-E346.
12. Ludens JH, Clark MA, Robinson FG, DuCharme DW (1992) Rat adrenal cortex is a source of a circulating ouabain-like compound. *Hypertension* 19:721-724.

13. Mathews WR, Ducharme DW, Hamlyn JM, Harris DW, Mandel F, Clark MA and Ludens JH (1991) Mass spectral characterization of an endogenous digitalislike factor from human plasma. *Hypertension* 17:930-935.

14. Naruse K, Naruse M, Tanabe A, Yoshimoto T, Watanabe Y, Kurimoto F, Horiba N, Tamura M, Inagami T and Demura H (1993) Does plasma immunoreactive ouabain originate from the adrenal gland? *Hypertension* (in press).

15. Pamnani M, Huot S, Buggy J, Clough D and Haddy F (1981) Demonstration of a humoral inhibitor of the Na$^+$-K$^+$ pump in some models of experimental hypertension. *Hypertension* 3 (Suppl II) II96-II101.

16. Price EM and Lingrel JB (1988) Structure-function relationships in the Na,K-ATPase α subunit: site-directed mutagenesis of glutamine 111 to arginine and asparagine 122 to aspartic acid generates an ouabain-resistant enzyme. *Biochemistry* 27:8400-8408.

17. Rauch A and Buckalew VM Jr (1988) Tissue distribution of an endogenous ligand to the Na,K ATPase molecule. *Biochem Biophys Res Commun* 152:818-824.

18. Schiffrin EL, Grutkowska J and Genest J (1981) Role of Ca$^{2+}$ in response of adrenal glomerulosa cells to angiotensin II, ACTH, K$^+$, and ouabain. Endocrinol Metab 4:E42-46.

19. Szalay KS (1971) The effect of ouabain on aldosterone production in the rat. *Acta Endocrinol* 68:477-484.

20. Tamura M, Kuwano H, Kinoshita T, Inagami T (1985) Identification of linoleic and oleic acids as endogenous Na$^+$, K$^+$-ATPase inhibitors from acute volume-expanded hog plasma. *J Biol Chem* 260:9672-9677.

21. Tamura M, Harris TM, Higashimori K, Sweetman BJ, Blair IA and Inagami T (1987) Lysophosphatidylcholines containing polyunsaturated fatty acids were found as Na$^+$,K$^+$-ATPase inhibitors in acutely volume-expanded hog. *Biochemistry* 26:2797-2806.

22. Tamura M, Inagami T, Kinoshita T and Kuwano H (1987) Identification of lysophosphatidylcholine, gamma-stearoyl as an endogenous Na, K-ATPase inhibitor from acutely volume-expanded hog plasma. *J Hypertension* 5:219-225.

23. Tamura M, Lam T-L and Inagami T (1988) Isolation and characterization of a specific endogenous Na$^+$,K$^+$-ATPase inhibitor from bovine adrenal. *Biochemistry* 27:4244-4253.

24. Tamura M, Harris TM, Konishi F and Inagami T (1993) Isolation and characterization of an endogenous Na$^+$,K$^+$-ATPase-specific inhibitor from pig urine. *Eur J Biochem* 211:317-327.

# Endogenous Digitalis-Like Factor from Umbilical Cord and Ouabain: Comparison of Biochemical Properties

S. Balzan[1], S. Ghione[1], L. Pieraccini[1], P. Biver[2], V. Di Bartolo[3] and U. Montali[4]

CNR Institute of Clinical Physiology[1], Operative Unit of Neonatology[2], CNR Institute of Mutagenesis and Differentiation[3], Institute of Biological Chemistry[4], University of Pisa; via Roma n. 55, 56100 Pisa, Italy.

## Introduction

Digitalis substances (cardenolides and bufodienolides) are potent and specific inhibitors of the cell membrane Na,K-ATPase. Cardenolides are synthetized in a number of plants and bufodienolides are endogenous substances in certain species of amphibians (1,2). The demonstration of cardiotonic steroids in lower vertebrates and of specific receptors for digitalis on the surface of most mammalian cells have suggested the existence of endogenous digitalis-like factors (EDLF) also in mammalians (2,3). Several substances with endogenous digitalis-like activity have been purified from mammalian tissues and body fluids (2,3). Recently, Hamlyn et al. have isolated from human plasma and identified by mass spectroscopy a substance that is indistinguishable from ouabain, proposing that ouabain is an endogenous digitalis-like factor (4). However, this conclusion is still object of controversies (5).

In previous papers we reported on a digitalis-like factor obtained from newborn plasma (6,7). In the present study we compare some properties of this factor with ouabain.

## Materials and methods

Blood plasma from umbilical cord was treated as previously described by lyophilization, three extractions with methanol, prepurification on SepPak C18 cartridges (Millipore, Bedford, MA) and purification by affinity chromatography and HPLC (6,7). Inhibition of erythrocyte $^{86}$Rb uptake was used as an index of digitalis like activity (8).

Ouabain (Sigma) at a final concentration of $10^{-8}$ M and dry amounts of purified EDLF were subjected to the treatments described below. Both gave approximately the same degree of inhibition (60-70%).

1. Proteolytic enzymes: a sample was dissolved in 400 µl of 5mM Tris HCl, 100µM EDTA buffer pH 7.4 and digested with 1.8 units of carboxypeptidase A (Sigma) or 1 mg of protease (Sigma) for 2 h at 37°C.

2. Stability: two samples were heated at 100°C, for 5 and 24 h. A third was treated with HCl 6N at 110°C for 24 h in a tube sealed under vacuum and a fourth was suspended in NH$_3$ 0.5N solution and left at 37 C° for 6 h.

3. Reaction with acetyl chloride: EDLF (or ouabain) was reacted with 10 µl acetyl chloride for 1 h and then acetyl chloride was destroyed by addition of 250 µl H$_2$O.

4. Reaction with sodium metaperiodate: 50µl 2% sodium metaperiodate solution was added to EDLF (or ouabain) and incubated for 30 minutes at 50°C.

5. Reaction with sodium borohydride: 200µl of sodium borohydride (0.5%) was added to EDLF (or ouabain) and incubated for 2 h at 50 C°. The reaction was stopped adding 5 µl of concentrated acetic acid.

6. Incubation with albumin: [86]Rb uptake measurements were done with and without adding human serum albumin (Biagini, Pisa, Italy) to the incubation medium at a final concentration of 4%.

7. Effect of anti-digoxin and anti-ouabain antibodies: the inhibitory activity of EDLF and ouabain was measured also after preincubation with antidigoxin Fab fragments $5 \times 10^{-6}$M (Digibind, Burroughs Wellcome Co., Italy) and antiouabain antiserum prepared according to Harris et al. (9) (raised in rabbits, 1:10).

8. Effect of external K: measurements were done at 0, 1, 5, 10 mM K.

9. Reversibility of inhibition, effect of EDTA: after the preincubation of 3 h with and without inhibitors, erythrocytes were centrifuged and resuspended in presence or absence of EDTA 5 mM for 1 to 4 h. At the end of this incubation, [86]Rb was added and the uptake measured as in all other samples.

10. HPLC chromatographic profile: purified EDLF and [3]H ouabain 17 pmol (specific activity 23.2 Cu/mmol, Amersham) were chromatographed on a C18 300x3.9 mm Boundapak column (Waters), with a mobile phase of acetonitrile/methanol/water (14/14/72 by vol) at a flow 1.5 ml/min and the elution profile was determined for digitalis like activity and radioactivity (for [3]H ouabain).

## Results

The effects of the various chemical treatments on the inhibitory potency of EDLF and ouabain are reported in the Table.

Table. Effect of various treatments on the inhibitory activity of EDLF and ouabain (values are percentage of inhibition of erythrocyte [86]Rb uptake).

| Treatments | Control | Carboxypept A | Protease | 100°/5 hr | 100°/24 hr | HCl 6N | NH3 0.5 N | acetyl. chl. | Na metaperiod. | Na borohydr. | Albumin 4% | Digibind | Anti-ouab. anti-serum |
|---|---|---|---|---|---|---|---|---|---|---|---|---|---|
| EDLF | 71 | 70 | 71 | 69 | 52 | -2 | 72 | 10 | 56 | 70 | 71 | 0 | 0 |
| ouabain | 74 | | | 71 | 74 | 4 | 70 | 0 | 63 | 72 | 73 | 0 | 0 |

When the capacity of EDLF and ouabain to inhibit [86]Rb uptake in presence of increasing concentrations of external $K^+$ was evaluated, the inhibitory activities of EDLF and of ouabain were reduced to the same extent (from about 70% to15%) (Fig. 1a). When EDLF or ouabain were added to the erythrocytes and then removed from the medium, no reversal of the inhibition of Rb uptake could be demonstrated for up to 4 h in control conditions. When, however, EDTA was added to the medium, 25% reversal after 4 h was observed for EDLF and 33% for ouabain (Fig. 1b). Finally, as shown in Fig. 1c, the HPLC elution profiles displayed equal retention times (12.4 min) for the inhibitory activity of EDLF and [3]H ouabain and for the radioactivity of [3]H ouabain.

756

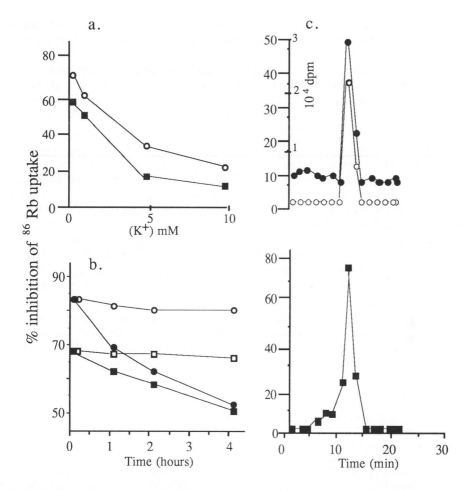

**Figure 1**

**a:** Inhibition of erythrocyte [86]Rb uptake by EDLF (full symbols) and by ouabain $10^{-8}$ M (open symbols) in presence of increasing concentrations of external $K^+$.

**b:** Reversal of the inhibition of erythrocyte [86]Rb uptake by EDLF (squares) and ouabain $10^{-8}$ M (circles) after the inhibitor was removed from the medium. The open symbols: without EDTA 5mM; the full symbols: EDTA 5mM added to the medium.

**c top:** Chromatographic profile of [3]H ouabain (17 pmol) eluted from an HPLC Bondapak C18 column with a mobile phase of methanol, acetonitrile, water (14,14, 72%) flow: 1,5 ml/min. Full circles: inhibition of [86]Rb uptake; open circles: radioactive profile.

**c bottom:** Chromatographic profile of EDLF eluted under the same condition as ouabain.

## Discussion

Our group has purified a factor with digitalis-like activity from the umbilical cord of neonate (6). Recently, Hamlyn et al. identified as ouabain an endogenous digitalis-like factor isolated from human plasma (4). This finding is partially supported by independent evidence that EDLF purified by other Authors have functional and/or structural similarities to ouabain (3,10). On the other hand, recently, Ferrandi et al. reported that an endogenous digitalis-like factor purified from the hypothalamus of hypertensive rats shared some properties with ouabain but differed for other aspects (11). For most of the characteristics studied, the EDLF purified by our group and ouabain were virtually identical. At variance with the results of Hamlyn et al. (6) one should however observe that his compound was inactivated in $NH_3$ 0,2M, whereas our factor, as well as ouabain, were not affected by $NH_3$ at even slightly higher concentrations. $K^+$ reduced the inibitory activity of EDLF and of ouabain to the same extent, suggesting that the binding to Na,K-ATPase is influenced by conformational changes resulting from the interaction of external $K^+$. In addition, the finding that ouabain and EDLF seem not to dissociate from erythrocytes in our medium (as shown in Fig. 1b) and to dissociate slowly after external $Mg^{++}$ has been complexed by EDTA, suggests that the binding of EDLF - as known for ouabain - occurs with highest affinity to the phosphorilated form of the Na,K-ATPase (which requires the presence of $Mg^{++}$).

Our data, although not representing a direct proof, are consistent with the idea that our EDLF may be similar or even identical to ouabain. In addition our data suggest that this factor is an organic non-peptidic compound. Its functional activity seems to depend on some $NH_2$ and/or OH groups (as suggested by the effect of acetyl chloride) but not on vicinal OH groups (effect of metaperiodate) nor on aldehydic groups (sodium borohydride). Finally, the observation that albumin, which strongly binds among other substances also fatty acids, does not reduce the inhibitory activity of EDLF suggests that this factor is not a fatty acid.

## References

1. Heftmann E (1970) Steroid Biochemistry. New York: Academic Press.
2. Lichtstein D, Samuelov S, Gati I, Wechter WJ (1992) Digitalis-like compounds in animal tissue. J Basic Clinical Physiol & Pharmac; 3: 269-370.
3. Goto A, Yamada K, Yagi N, Yoshioka M, Sugimoto T (1992) Physiology and pharmacology of endogenous digitalis-like factors. Pharmacol Rev; 44: 377-399.
4. Mathews WR, DuCharme DW, Hamlyn JM, Harris DW, Mandel F, Clark MA, LudensJH (1991) Mass spectral characterization of an endogenous digitalis-like factor from human plasma. Hypertension; 17: 930-935.
5. Kelly RA, Smith TW (1992) Is ouabain the endogenous digitalis ? Circulation; 86: 694-697.
6. Balzan S, Ghione S, Biver P, Gazzetti P, Montali U (1991) Partial purification of endogenous digitalis-like compound(s) in cord blood. Clin Chem; 37: 277-281.
7. Montali U, Balzan S, Ghione S (1991) Purification of endogenous digitalis-like factor(s) from cord blood of neonate by immunoaffinity chromatography. Biochem Int; 25: 853-859.
8. Balzan S, Montali U, Genovesi-Ebert A, Biver P, Fantoni M, Ghione S (1989) Comparison between endogenous digoxin-like immunoreactivity and 86Rb uptake by erythrocytes in extracts of human plasma. Clin Sci; 77: 375-381.
9. Harris DW, Clark MA, Fisher JF, Hamlyn JM, Kolbasa KP, Ludens JH, DuCharme DW (1991) Development of an immunoassay for endogenous digitalis-like factor. Hypertension; 17: 936-943.
10. Tamura M, Lam TT, Inagami T (1988) Isolation and characterization of a specific endogenous NaK ATPase inhibitor from bovine adrenal. Biochemistry; 27: 4244-4253.
11. Ferrandi M, Minotti E, Salardi S, Florio M, Bianchi G, Ferrari P (1992) Ouabainlike factor in Milan hypertensive rats. Am J Physiol; 263: F739-F748.

# New steroidal digitalis-like compounds in human cataractous lenses

D. Lichtstein[1], I. Gati[1], S. Samuelov[1], D. Berson[2], Y. Rozenman[2], L. Landau[2], J. Deutsch[3]

[1]Department of Physiology, Hebrew University - Hadassah Medical School, Jerusalem, Israel, [2]Department of Ophthalmology Shaare Zedek Hospital, Jerusalem, Israel and [3]Department of Pharmaceutical Chemistry, School of Pharmacy, Hebrew University, Jerusalem, Israel.

## Introduction

Since $Na^+/K^+$-ATPase has a high-affinity receptor for digitalis steroids, it has been postulated that there are endogenous ligands for these receptors which regulate the $Na^+/K^+$-pump activity. Indeed, based on their ability to inhibit $^3H$-ouabain binding and $Na^+/K^+$-ATPase activity, digitalis-like compounds (DLC) have been shown to be present in the brain, heart, adrenal, plasma, cerebrospinal fluid and urine of mammals and in the skin and plasma of toads (for review see 3,6,11). Several substances have been proposed as the DLC, including unsaturated fatty acids, lysophosphatidyl choline, dopamine, dehydroepiandrosterone sulfate, lignans and ascorbic acid (11). However, none of these compounds appears to be the natural ligand of the digitalis receptor of the $Na^+/K^+$-ATPase because of their limited specificity and affinity. Bufodienolides, which resemble the structure of the plant cardiac glycosides, have been identified in the plasma, brain and other tissues of toads (4,12), however their presence in mammals has not yet been demonstrated. Recently, ouabain (7) and digoxin (5) have been identified in mammalian tissues.

The anatomy and biochemistry of the ocular lens as well as its development were extensively studied (for review see 2,8). Studies of lens plasma membrane ion pumps have indicated that the lens $Na^+/K^+$-ATPase could maintain intracellular $Na^+$ and $K^+$ concentrations similar to those in other tissues (16). An abnormal distribution of $Na^+$ and $K^+$ in this tissue can lead to the loss of osmotic regulation and subsequent cataract formation. Loss of $Na^+/K^+$-ATPase activity has been reported to occur during cataract formation in the galactose-fed rat (18), rats with Alloxan-induced diabetes (1) the triparanol-fed rat (13), in certain animal hereditary cataracts (17), and in human senile cataracts (9,13).

## Results and discussion

Recently we have demonstrated that human cataractous lens nuclei extract inhibited in a dose dependent fashion $^3H$-ouabain binding to rat brain synaptosomes and microsomal $Na^+/K^+$-ATPase activity and interacted with anti digoxin antibodies (10). DLC from the human cataractous lenses were purified by a procedure

759

consisting of organic extractions and batch chromatography followed by filtration through 3,000 Dalton cutoff filter and subsequent separations using reverse-phase high performance liquid chromatography. Based on chemical ionization mass spectrometry together with UV spectrometry and biological characterization, it was suggested that new bufodienolides, 19-norbufalin and 19-norbufalin peptide derivatives (Figure 1) are responsible for the endogenous DLC activity (10).

What is the role of these $Na^+/K^+$-ATPase activity inhibitors in the lens and are they involved in cataract formation? In that respect it is important to note that ouabain and other cardiac glycosides are potent cataractogenic factors. As can be seen in Figure 2 the addition of 10 μM ouabain to lens incubation media resulted in a complete lens opacification after one week

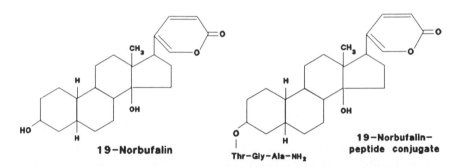

**Figure 1.** Postulated structures of human cataractous lens digitalis-like compounds. The two proposed compounds 19-norbufalin (MW=372) and 19-norbufalin-peptide conjugate (MW=602) are shown.

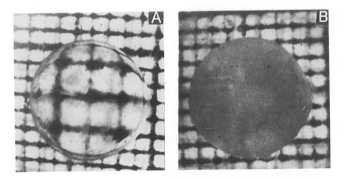

**Figure 2.** Appearance of rat lenses incubated for 7 days in the absence (A) and presence (B) of 10 μM ouabain.

of incubation. The relative high concentration of DLC in human cataractous lenses and the ability of these compounds to induce cataract suggest that the increased levels of DLC in the lens may be, in some cases, a primary cause for the cataract formation. The concentration of these compounds in the cataractous lens vary extensively with an average of 15.05 ± 7.78 nmol/gr wet mass (Mean ± S.E.M ouabain equivalents). This may indicate that either the synthesis of DLC in the cataractous lens is increased or that the compounds are synthesized in other tissue and are accumulated in the lens during cataractogenesis. The variability of the levels of DLC in different pools may signify that the increased levels are obtained only in a small sub-population of cataract patients. At present such a population has not been identified. We suggest that the increased levels of DLC in the lens may cause cataract formation by the following sequence of events: Increased DLC causes the inhibition of $Na^+/K^+$-ATPase in the epithelial and fiber cells of the lens, which results in an increase of intracellular $Na^+$. This ensues a reduction of $Na^+/Ca^{++}$ exchanger activity which in turn produces an elevation in intracellular $Ca^{++}$. This increase of intracellular $Ca^{++}$ results in protein degradation and denaturation which are manifested in lens opacification.

DLC levels have been shown to be increased in the plasma in some forms of hypertension (1-3). Although we have not demonstrated the presence of the lens DLC in the circulation, it is of interest to note that the frequency of cataract in hypertensive patients is significantly higher than in control population (14). Furthermore, Dahl, salt-sensitive hypertensive rats show enhanced frequency of cataract formation whereas the control salt resistant rats do not develop cataracts (15). These observations are consistent with a hypothesis that in sub-populations of hypertensive and cataract patients, cataractogenesis may be related to the hypertensive process. It is possible that increased levels of the DLC lead to a decrease in $Na^+/K^+$-ATPase activity in the lens and in arterial smooth muscle which precede lens opacification and hypertension, respectively.

Acknowledgements

This study was supported in part by grants from the Israeli Ministry of Health and the Yeshaya Horowitz Foundation.

References

1.      Ahmad SS, Tsou KC, Ahmad SI, Rahman MA, Kirmani TH (1985) Studies on cataractogenesis in human and rats with Alloxan-induced diabetes. Ophthal. Res. 17:1-11
2.      Cotlier E (1981) The lens in: Adler's Physiology of the eye (Moses R.A. ed) 277-303, The C. V. Mosby Company, St. Louis
3.      De Wardener HE, Clarkson EM (1985) Concept of natriuretic hormone. Physiological Reviews 65:658-759
4.      Flier J, Edwards MW, Daly JW, Myers CW (1980) Widespread occurrence in frogs and toads of skin compounds interacting with the ouabain binding site of Na,K-ATPase. Science 208:503-505

5.  Goto, A., Ishiguro, T., Yamada, K., Ishii, M., Yoshioka, M., Eguchi, C., Shimura, M. & Sugimoto, T. (1990) Isolation of a urinary digitalis-like factor indistinguishable from digoxin. Biochem. Biophys. Res. Commun. 173:1093-1101

6.  Haber E, Haupert GT (1987) The search for hypothalamic $Na^+$, $K^+$-ATPase inhibitor. Hypertension 9:315-324

7.  Hamlyn JM, Blaustein MP, Bova S, DuCharme DW, Harris DW, Mandel F, Mathews WR Ludens JH (1991) Identification and characterization of ouabain-like compound from human plasma. Proc. Natl. Acad. Sci. U.S.A. 88:6259-6263

8.  Harding JJ, Crabbe J (1984) The lens: development, proteins, metabolism and cataract. In The eye (Davson, H. ed) pp 207-492, Academic Press, London

9.  Kobatashi S, Roy D, Spector A (1983) Sodium/potassium ATPase in normal and cataractous human lenses. Curr. Eye Res. 2:327-334

10. Lichtstein D, Gati I, Samuelov S, Berson D, Rozenman Y, Landau L, Deutsch J (1993) Identification of digitalis-like compounds in human cataractous lenses. Eur. J. Biochem. in press

11. Lichtstein D, Samuelov S, Wechter W J (1992) Digitalis-like compounds in animal tissues.J. Basic Clin. Physiol. Pharmacol. 3:269-292

12. Lichtstein D, Kachalsky S, Deutsch J (1986) Identification of a ouabain-like compound in toad skin and plasma as a bufodienolide derivative. Life Sciences 38:1261-1270

13. Mizuno GH, Chapman CJ, Chipault JR, Pfeiffer DR (1981) Lipid composition and $(Na^+ + K^+)$-ATPase activity in rat lens during Triparanol-induced cataract formation. Biochim. Biophys. Acta 644:1-12

14. Nugent J, Whelan J (1984) Human cataract formation. Ciba Foundation Pitman Press, London, Symposium 106:1-266

15. Rodriguez-Sargent C, Cangiano JL, Caban GB, Marrero E, Martinez-Maldonado M (1987) Cataracts and hypertension in salt-sensitive rats, A possible ion transport defect. Hypertension 9:304-308

16. Sen PC, Pffifer DR (1982) Characterization of partially purified $Na^+$, $K^+$-ATPase from porcine lens. Biochim. Biophys. Acta 693:33-44

17. Takehana M (1990) Hereditary cataract of the Nakano mouse. Exp. Eye Res. 50:671-676

18. Unakar NJ, Tsui J (1980) Sodium-Potassium-dependent ATPase. II. Cytochemical localization during the reversal of galactose cataracts in rat. Invest. Ophtal. 19:378-385

# Large Scale Purification of an Endogenous Na$^+$/K$^+$-pump Inhibitor from Bovine Adrenal Glands

Masaaki Tamura[1], Fumiko Konishi[1], Masayuki Sakakibara[2] and Tadashi Inagami[1]

[1]Department of Biochemistry, Vanderbilt University School of Medicine, Nashville, TN (USA) and [2]Pharmaceutical Laboratory, Kirin Brewery Co. Ltd., Takasaki (Japan)

## Introduction

A Na$^+$/K$^+$-pump specific inhibitor which shows chromatographic and functional characteristics identical to ouabain has been isolated from bovine adrenal glands (5). Similar substances were also purified from human urine (1) and plasma (2), and the plasma inhibitor was identified as ouabain by mass spectrometric studies (4). It remains to be clarified, however, as to whether the chemical structure of the inhibitor is identical to ouabain, particularly its sugar moiety, and whether a synthetic pathway for ouabain exists in mammalian tissues. In order to clarify these questions, we first need a large quantity of the pure inhibitor which will enable us to analyze the detailed chemical structure. Thus, we present herein the improved procedure for large scale purification of a Na$^+$/K$^+$-pump inhibitor from bovine adrenal glands.

## Methods

**Assays:** Four separate assays were used to monitor Na$^+$/K$^+$-pump inhibitory activity and to characterize the purified inhibitor. The assays for Na$^+$/K$^+$-ATPase, ouabain-binding and $^{86}$Rb uptake, and a radioimmunoassay for ouabain were performed as described previously(5).

**Preparation of the Na$^+$/K$^+$-pump inhibitor:** A specific Na$^+$/K$^+$-pump inhibitor, designated as adrexin C, was prepared from approximately 120 kg of bovine adrenal glands obtained from a commercial tissue supplier (Pel-Freeze, Rogers, AR) by the modified method previously described (5). Briefly, trimmed bovine adrenal glands (approx. 8000 adrenal glands, total 120 kg) were ground in a tissue grinder and lyophilized. The dry tissue (2.5 kg wet weight per batch) was delipidated with 7 l of petroleum ether. The Na$^+$/K$^+$-pump inhibitory activities were extracted from the delipidated tissue powder with 4 vol. (w/v) of 50% methanol at 78°C for 20 min. The extract was filtered through Whatman No. 4 filter paper (Whatman, Hillsboro, OR) and was dried by evaporation of the methanol followed by lyophilization. The residue equivalent to 750 g fresh adrenal glands was suspended in 500 ml of distilled water containing 0.1% TFA and was chromatographed by flash chromatography on a preparative-size column packed with octadecylsilane (ODS) bonded to silica gel (5.0 x 18 cm, J.T. Baker, Phillipsburg, NJ). The fraction eluted with 1 liter of 20% acetonitrile containing 0.1% TFA, which contained most of the inhibitory activities, was

763

lyophilized. The dry material obtained from flash chromatography was dissolved in 75 ml of 5 mM ammonium acetate (pH 7.0) and passed through a DEAE-cellulose (Whatman) weak anion exchange column (2.5 x 40 cm). The adrexin C, in the passed through fraction, was preparatively fractionated by low pressure liquid chromatography (LPLC) on a hand made $C_{18}$ column (2.5 x 60 cm) with a linear gradient of acetonitrile (0-24%) at a flow rate of 5 ml/min. The adrexin C-rich fraction from $C_{18}$ LPLC was passed through a sulphoxyethyl (SE)-cellulose (Whatman) strong cation exchange column (2.5 x 40 cm) pre-equilibrated with 0.1% TFA-water. The adrexin C was then purified by four steps of high performance liquid chromatography (HPLC). Reverse phase HPLC, on a preparative YMC ODS column (2 x 30 cm, acetonitrile: 0-35%, Yamamura, Kyoto, Japan), on a semi-preparative Vydac phenyl column (1 x 25 cm, acetonitrile: 9-20%, The Nest Group, Southboro, MA), on a semi-preparative Econosphere trimethylsilane (TMS) column (1 x 25 cm, acetonitrile: 0-20%, Alltech, Deerfield, IL), and on a YMC ODS-AQ column (0.46 x 30 cm, acetonitrile: 10-35%, Yamamura) were run in succession with linear gradients of acetonitrile as indicated in parentheses at flow rates of 8 ml/min for the preparative column, at 3 ml/min for the semi-preparative columns and at 1 ml/min for the final analytical column. All of the solvent systems were acidified with 0.1% TFA. Activity was monitored by measuring the inhibition of [86]Rb uptake activity into human erythrocytes.

### Results and Discussion

**Purification of the $Na^+/K^+$-pump inhibitor:** Crude extracts were prepared by heated methanol extraction from 11.06 kg of delipidated dry tissue powder, which was prepared from a total of 120 kg of bovine adrenal glands. Approximately 3.8 kg of dry material was obtained. Reverse phase $C_{18}$ flash chromatography was used as the second step of extraction, in order to remove lipidic interfering substances. After removing the organic solvent from the 20% acetonitrile fraction by rotary evaporator, the inhibitory activities in the resultant aqueous solution were further cleaned by passage through a DEAE-cellulose column, followed by fractionation by preparative $C_{18}$ LPLC. The activities were separated into seven distinct activity peaks. Since peak VI was chromatographically identified as the same specific $Na^+/K^+$-pump inhibitor purified previously on a small scale from bovine adrenal glands (5), this peak was further purified by a combination of "clean-up" column chromatography and HPLC. A vast majority of the bulky materials in this peak VI fraction from $C_{18}$ LPLC was further removed by passage through an SE-cellulose strong cation exchange column. Since all HPLC columns have only limited capacity and since the resolution ability of successive reverse phase columns becomes greatly improved by using a totally different kind of column, this ion exchange column was very effective in cleaning up the sample. Indeed, this clean-up process drastically minimized the following HPLC purification steps. A preparative ODS column, a semipreparative phenyl column, a semipreparative TMS column and an analytical ODS-AQ column were used in succession for the HPLC purification. Although HPLC on the phenyl column separated the inhibitory activity under peak VI into three separate activity peaks, designated peaks A, B and C, peak C was found to have the predominant activity and to be identical to the inhibitor purified previously (5). The active substance under peak C was then successfully purified by the following two steps of HPLC as mentioned above. The active substance was eluted as an apparently homogeneous, single,

764

symmetrical UV absorption peak devoid of a leading or trailing edge. The total quantity of the purified inhibitor from the present experiment was determined to be ~10 μg by a microbalance. Based on the dry weight obtained from the methanol extraction (3.808 kg), the apparent purification factor for this purification was 380.8 million-fold. During the course of this massive purification process, inconsistent yield of peak VI at the $C_{18}$ LPLC step was noticed. Although this method produced consistently high yield recoveries of radioactivity when [3H]ouabain was added to adrenal glands, only 12 out of 36 batches contained relatively high concentrations of the inhibitor in the peak VI fraction. Accordingly, the relationships between the activity yield and the age and sex of the cows, the season, the geographical location and the morphological appearance of the adrenal glands were investigated. However, no positive correlation was recognized. This observation could be explained by one of the following: 1) the bovine adrenal

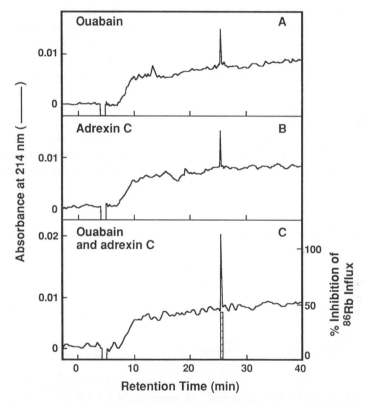

**Figure 1.** Elution profiles of authentic ouabain (A), the purified inhibitor adrexin C (B) and a mixture of ouabain and adrexin C (C). Fifty nanograms each of ouabain and adrexin C and a mixture of 50 ng each of ouabain and adrexin C were individually applied to HPLC on a YMC ODS-AQ column and run with linear gradients of acetonitrile (10-30%) in 0.1% TFA over 40 min at flow rates of 1 ml/min. Each sample was injected at 0 min. One minute fractions were collected from the HPLC of the mixture of ouabain and adrexin C, and monitored for inhibitory activity against 86Rb uptake by erythrocytes (bar in panel C) using 0.25% of the eluates.

gland is not the major source or the storage organ for this inhibitor, 2) this inhibitor is produced only under certain pathophysiological conditions, or 3) this inhibitor is accidentally derived from the diet (contamination). This result is contradictory to a recent report (3), in which the adrenal cortex was concluded to be the major source for the ouabain-like inhibitor in the rat.

**Characterization of the Na$^+$/K$^+$-pump inhibitor:** In order to answer another important question, whether this purified inhibitor is identical to ouabain, immunological and chromatographical characterizations were carried out. The quantity of the inhibitor was determined by RIA using anti-ouabain antibodies, and then 50 ng of the purified inhibitor and the same amount of authentic ouabain determined by a balance were individually chromatographed by HPLC on a YMC ODS-AQ column (0.46 x 30 cm). Both the purified inhibitor adrexin C and ouabain showed exactly the same retention times and peak heights (Fig. 1). These results again demonstrate that the purified inhibitor from bovine adrenal glands is identical to authentic ouabain by immunological and chromatographic means.

The functional characteristics of the purified inhibitor, such as the inhibitory potency against the ATPases, $^{86}$Rb uptake, and ouabain-binding, and the crossreactivity to anti-ouabain antibodies, were identical to ouabain, both in the slopes of the dose response curves and in the ED$_{50}$ values. The physico-chemical characteristics of the purified inhibitor and its sensitivity to various enzymatic digestions were also identical to ouabain. These results support our previous observations (5,6) and indicate that the purified inhibitor from bovine adrenal glands is apparently indistinguishable from authentic ouabain. The true source or induction mechanism of this inhibitor remains to be clarified.

### Acknowledgement

This work was supported by the National Institutes of Health Grants HL14192 and HL35323.

### References

1. Goto A, Yamada K, Ishii M, Yoshioka M, Ishiguro T, Eguchi C, Sugimoto T (1988) Purification and characterization of human urine-derived digitalis-like factor. Biochem Biophys Res Commun 154:847-853
2. Hamlyn JM, Harris DW, Ludens JH (1989) Isolation and characterization of a sodium pump inhibitor from human plasma. Hypertension 13:681-689
3. Ludens JH, Clark MA, Robinson FG, DuCharme DW (1992) Rat adrenal cortex is a source of a circulating ouabain-like compound. Hypertension 19:721-724
4. Mathews WR, Ducharme DW, Hamlyn JM, Harris DW, Mandel F, Clark MA, Ludens JH (1991) Mass spectral characterization of an endogenous digitalis-like factor from human plasma. Hypertension 17:930-935
5. Tamura M, Lam T-T, Inagami T (1988) Isolation and characterization of a specific endogenous Na$^+$,K$^+$-ATPase inhibitor from bovine adrenal. Biochemistry 27:4244-4253
6. Tamura M, Naruse M, Sakakibara M, Inagami T (1993) Isolation of an endogenous Na-pump specific inhibitor from normal pig urine: characterization and comparison with the inhibitor purified from bovine adrenal glands. Biochim Biophys Acta 1157:15-22

# Demonstration of Inhibitors of the Sodium Pump in Human Plasma and Bovine Adrenals Cross-Reacting with Proscillaridin A Antibodies

B. Sich, U. Kirch, R. Antolovic and W. Schoner

Institute of Biochemistry & Endocrinology, Justus-Liebig-University,
Frankfurter Str. 100, D-35392 Giessen, Germany

## Introduction

Since the cardiac glycoside binding site of the sodium pump is conserved over the millenia it was speculated that endogenous cardiac glycosides might exist. The intensive search for such substances led to the isolation of a compound indiscriminable from ouabain from human plasma [1] and of an isomer of ouabain from bovine hypothalamus [2]. Moreover, a substance indiscriminable from digoxin from human urine has been described [3] and recently the existence of bufadienolides in human lenses with cataract was reported [4]. Additionally, a yet unidentified substance of 620 Da with inhibitory activity on the sodium pump has been isolated from pig urine [5].

If vertebrates like toads can synthesize bufadienolides in their skin and parotis glands we wondered whether other vertebrates might be able to do so as well. With this idea in mind we raised antibodies against the bufadienolide proscillaridin A and looked for cross-reactivity in human blood serum. Moreover, we isolated a low molecular weight substance by affinity chromatography on a column containing antibodies against proscillaridin A.

## Materials and Methods

Proscillaridin A was coupled with bovine serum albumin (BSA) and gelatine according to Terano et al. [6]. Antibodies against the BSA-coupled cardiac glycoside were raised in rabbits and detected by an ELISA. The IgG fraction was coupled to Eurocell ONB-carbonat A (Eurochrom/Berlin) to yield a support for affinity chromatography. Inhibition of the sodium pump was measured as inhibition of $^{86}Rb^+$ uptake into human red blood cells. The amount of the inhibitor was calculated in ouabain equivalents assuming that the sodium pump inhibitor has the same affinity as ouabain ($4 \times 10^{-8}$ M). The molecular mass of the inhibitor was determined by FAB mass spectrometry in a FINIGAN MAT 900 mass spectrometer using "magic bullet" as a matrix by Dr. R. Geyer, Biochemistry Department, Medical School, Giessen.

Detection in serum: The material cross-reacting with proscillaridin A antibodies was detected after ethanol precipitation of proteins, evaporation of the supernatant and elution of the water-dissolved residue either from a Bakerbond C18 column with 80%

ethanol or after fractionation on a HPLC C18 column applying a propanol/isopropanol gradient [7].

Purification of the sodium pump inhibitor from bovine adrenals: Adrenals (5 kg) were extracted with methanol and the dried residue was extracted with acetone. The dried acetone extract was dissolved in water and ultrafiltered (exclusion size < 1000 Da). The dried material was dissolved in acetone and fractionated in acetone on Sephadex LH-20. The fraction containing the inhibitor was dried, dissolved in water and further purified by affinity chromatography with a column containing IgG directed against proscillaridin A. After washing the column with 50 mM Tris-HCl (pH 7.8) containing 150 mM NaCl, the retained material was eluted with 1 M propionic acid and the eluate was applied to a C18 HPLC column using a propanol/isopropanol gradient [7]. The fraction eluting at 18-20 min was rechromatographed on a C18 HPLC column using a shallow gradient of acetonitrile in 0.1% TFA.

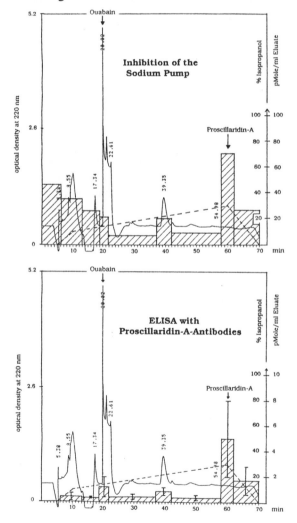

**Figure 1.** Detection of sodium pump inhibitors and of material cross-reacting with proscillaridin A antibodies in human serum of normotensives. To a Lichrospher RP18 (5 $\mu$) (25 x 0.8 cm) an equivalent of 1 ml serum after ethanol extraction was applied. The data are mean values of 10 different sera. The values determined by the ELISA show mean values with standard deviation.

768

## Results and Discussion

The antibodies raised against the bufadienolide proscillaridin A cross-reacted 84% with bufalin, but not with the cardenolides ouabain, ouabagenin, dihydro-ouabain, K-strophanthidin, digoxin or with other steroid hormones. A very weak cross-reactivity (< 0.5%) was found with K-strophanthoside, K-strophanthidine, convallatoxin, gitoxin and digitoxin. When human serum was tested for proscillaridin A immunoreactivity, a median of 7.3 ± 3.5 nmol/l in 30 sera was obtained. Since ouabain-like activity has been detected in plasma [1], we were interested to find out how the cross-reacting material behaved on a HPLC C18 column. Figure 1 shows that 3 distinct peaks could be detected with a retention time of 20, 40 and 60 minutes by the ELISA. These peaks showed also an inhibition of $^{86}Rb^+$ uptake into red blood cells. As is also evident from Figure 1, ouabain eluted under these conditions at 20 min and proscillaridin A at 60 min. The peak, eluting like proscillaridin A, had the highest inhibitory potency.

It is evident from the chromatogram in Figure 1 that the 3 inhibitors of the sodium pump which cross-react with antibodies against proscillaridin A show different polarities.

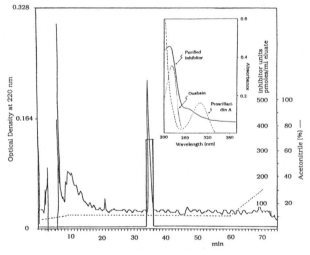

**Figure 2.** Chromatography of the affinity-purified inhibitor of the sodium pump on Lichrospher C18 (5 $\mu$) (25 x 0.4 cm) with 10% acetonitrile in 0.1% TFA. The activity of the inhibitor was quantitated from the inhibition of $^{86}Rb^+$ uptake into human red blood cells. The inset shows a comparison of the UV spectra of the pure inhibitor with the cardenolide ouabain and the bufadienolide proscillaridin A.

To learn more about the nature of these substances, 5 kg of bovine adrenals were extracted with methanol and the extract was processed as described in the Methods section. When the eluate from the proscillaridin A affinity column was given over a

769

Lichrospher C18 (5 $\mu$) column, the material inhibiting the sodium pump from red blood cells and cross-reacting with anti-proscillaridin A IgG eluted with the same polarity like ouabain. Additionally small amounts of a more hydrophobic inhibitor were seen (not shown). When the polar material was rechromatographed on the same column using a shallow gradient of acetonitrile in 0.1% TFA, a sharp peak was obtained which inhibited the sodium pump (Fig. 2). It eluted again with the same retention time as ouabain. Bufadienolides show a maximum of absorbance at 300 nm (Fig. 2, inset). Surprisingly, the UV-spectrum of the peak fraction at 35 min did not show a spectrum like bufadienolides nor cardenolides but rather it resembled the spectrum recorded for the sodium pump inhibitor from pig urine [5]. Since the antibodies against proscillaridin A did not cross-react with ouabain, the polar purified substance could not be ouabain. In fact, when the molecular mass of the compound was determined by FAB mass spectroscopy a mass of 561.2 Da was found, 23 Da smaller than ouabain. Despite the similarity in spectra, the polar inhibitor isolated from bovine adrenals by affinity chromatography cannot be identical with Tamura's inhibitor from pig urine, because its molecular mass is 64 Da bigger. Clearly more effort is necessary to elucidate the structure of this factor.

*Acknowledgements*

This work has been supported by grants of the Deutsche Forschungsgemeinschaft Scho 139/20-1, the Fonds der Chemischen Industrie, Frankfurt/Main and the Giessener Hochschulgesellschaft e.V., Giessen.

**References**

1.  Hamlyn JM, Blaustein MP, Bova S, DuCharme DW, Harris DW, Mandel F, Matthews WR, Ludens JH (1991) Isolation and characterization of a ouabain-like compound from human plasma. Proc Natl Acad Sci USA 88: 6259-6263
2.  Tymiak AA, Norman JA, Bolgar M, Didonato GC, Lee H, Parker WL, Lo LC, Berova N, Nakanishi K, Haber E, Haupert GT (1993) Physicochemical characterization of a ouabain isomer isolated from bovine hypothalamus. Proc Natl Acad Sci USA 90: 8189-8193
3.  Goto A, Ishiguro T, Yamada K, Ishii M, Yoshioka M, Eguchi C, Shimura M, Sugimoto T (1990) Isolation of a urinary digitalis-like factor indistinguishable from digoxin. Biochem Biophys Res Commun 173: 1093-11001
4.  Lichtstein D, Gati I, Samuelov S, Berson D, Rozenman Y, Landau L, Deutsch J (1993) Identification of digitalis-like compounds in human cataractous lenses. Eur J Biochem 216: 261-268
5.  Tamura M, Harris TM, Konishi F, Inagami, T (1993) Isolation and Characterization of an endogenous $Na^+,K^+$-ATPase-specific inhibitor from pig urine. Eur J Biochem 211: 317-327
6.  Terano Y, Tomii A, Masugi F (1991) Production and characterization of antibodies against ouabain. Jpn J Med Sci Biol 44: 123-139
7.  Hamlyn JM, Harris DW, Ludens JH (1988) Digitalis like activity in human plasma. J Biol Chem 264: 7395-7404

# A Labile Endogenous Na-pump Inhibitor in Man

Graves SW[†‡], Soszynski PA[†], Tao QF[†], Williams GH[†], Hollenberg NK[*].
Departments of Radiology[*] and Medicine (Endocrine-Hypertension Division[†]), Harvard Medical School, Brigham and Women's Hospital, 221 Longwood Ave, Boston, MA, USA 02115. ([‡] For correspondence)

A circulating, endogenous sodium pump inhibitor (ESPI) has been implicated in the pathogenesis of essential hypertension (3). Efforts to identify this factor have generated a bewildering array of chemically different candidates (1). Previous efforts to isolate such a factor for further characterization have used techniques, which while appropriate for the purification of stable chemical species, have been too long and/or ineffective in protecting chemically labile species. We have developed a procedure to purify ESPI from human fluids that would be rapid and provide protection for air and thermally sensitive chemical species to ascertain whether such compounds exist and to allow for their study. We had previously carried out a clinical study in which we isolated an ESPI from human peritoneal dialysate (PD) from patients with sustained extracellular fluid volume expansion (6). PD levels of this ESPI were correlated with volume status, blood pressure, and serum $Na^+/K^+$-ATPase inhibitory activity. Additionally, it appeared to be chemically unstable. We have continued to use this same paradigm to insure isolation of a volume-sensitive ESPI and to exploit PD as a source because it is rapidly available and readily processed.

## METHODS

PD, obtained fresh from renal failure patients undergoing sustained volume expansion as part of an approved hospital clinical research protocol, was used as a source for purification. The agent considered to be a candidate ESPI was undetectable in PD when patients were euvolemic and increased in proportion to the rise in body weight that followed liberalization of salt and fluid intake. Blood pressure and serum activity of this factor rose in proportion to the rise in body weight and in PD ESPI levels (6). Ascorbic acid (1 g/L) was added to the PD and the PD placed on ice under an Ar atmosphere. Immediately the PD was filtered through a 1000 dalton exclusion membrane (Centrasette Ultrafiltration System, Filtron, Chicago, IL). ESPI was extracted from filtrate on an $C_{18}$-HPLC column followed by $C_{18}$ HPLC chromatography as described previously (6). Eluate at 19.5 min was previously shown to contain the only volume-sensitive ESPI. An internal standard (trace amounts of $^3$H-ouabain) with a retention time of 13.5 min was included to calibrate elution times.

HPLC eluate from purified PD was collected post chromatography over the region of interest (17.0 to 22.0 min) and each half minute fraction divided into two equal portions. Both halves were taken to complete dryness to remove the organic solvents present in the HPLC mobile phase (ethanol, 0.01% trifluoroacetic acid). Immediately thereafter half was analyzed for its $Na^+/K^+$-ATPase inhibitory activity.

The other half was then i) stored dry overnight under Ar at -70°C or ii) at liquid $N_2$ temperatures; or iii) was redissolved in distilled water and stored overnight frozen under Ar at -70°C. Storage was in silanized glass vials. The following day fractions stored dry specimens were brought up in assay buffer and their $Na^+/K^+$-ATPase inhibitory activity measured. Specimens stored frozen were thawed, dried quickly (rotary evaporator), brought up in assay buffer and inhibitory activity measured. Other less polar cardenolides (strophanthidin G and bufalin) were also processed similarly to assess resolubilization in the aqueous buffer.

ESPI activity was described as the percent inhibition of $Na^+/K^+$-ATPase hydrolysis of 6-labelled [$^{32}$P]ATP by measuring the amount of inorganic phosphate liberated after 30 min of incubation with or without inhibitor (4).

RESULTS

As shown in Figure 1, renal failure patients during sustained volume expansion elaborate an ESPI in their blood which was dialyzed away by their peritoneal dialysis. This factor has an HPLC mobility of $19.5 \pm 0.5$ min on $C_{18}$ HPLC chromatography as described previously (6). ESPI activity, present in this same fraction in the example shown, was undetectable after storage overnight dry under Ar at -20°C.

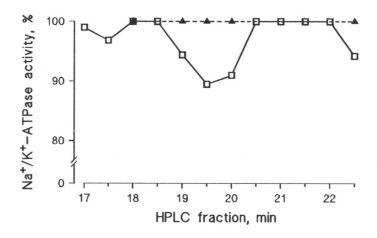

**Figure 1. HPLC fractionated PD's ability to inhibit $Na^+/K^+$-ATPase activity before (□) and after (▲) 20 h storage dry at -20°C under Ar.**

Use of three storage conditions were assessed. These are summarized in Fig. 2. Specimens of the labile ESPI were stored under Ar overnight (20 h). Those stored dry at -70°C and at -196°C (liquid $N_2$) showed comparable results with 75-85% of the activity being lost. About half of these stored materials showed no detectable activity. Those stored frozen in water at -70°C retained about half of their activity and all specimens had some activity. Strophanthidin and bufalin, both stable cardenolides with reduced water solubility, showed no loss of activity when processed similarly (Fig. 2).

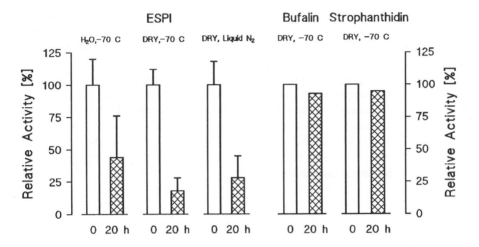

Figure 2. Effect of storage conditions on the labile PD-ESPI, bufalin and strophanthidin. Original activity was normalized to 100% (left bar) and compared to activity 20 h later (right bar).

DISCUSSION

These studies confirm the presence of a chemically labile ESPI in PD from renal failure patients undergoing sustained volume expansion. Its location in the HPLC chromatogram was identical to that previously described for a volume-sensitive ESPI whose levels correlated with blood pressure in other renal failure patients (6). Likewise, it shares the same HPLC retention time with a similar labile ESPI isolated from amniotic fluid (5), whose amniotic fluid levels were elevated in women with preeclampsia and correlated with maternal blood pressure (2). This lability distinguishes this factor from all other candidates for the ESPI.

The instability appears to be chemical in nature and not related to enzymatic degradation. The results show that the storage conditions employed did not overcome the chemical lability, suggesting that the instability may due to more than air sensitivity. The finding that maintenance of an aqueous environment appeared to attenuate the loss of ESPI activity is unusual, since many labile compounds are less stabile in the presence of protic solvents. This pattern of activity loss might also be seen with compounds that cannot be redissolved. However, the fact that the first half of the specimen was also taken to dryness and that less water soluble compounds showed no losses when taken to dryness and stored overnight argue against this explanation. Some compounds, especially proteins, are known to adhere to or be denatured by glass surfaces. However, the use of silanized glass vessels makes this unlikely.

There is still substantial controversy about the origins of some ESPI candidates, e.g. ouabain (7). However, the clinical changes of the labile ESPI while on constant diet (with modifications in NaCl and water intake) plus its instability make it highly unlikely that it originates from an exogenous source and likely to be physiologically relevant. Other preliminary studies have indicated that this same labile ESPI causes direct contraction of vascular smooth muscle (6), suggesting further that it has properties requisite for a hypertensinogenic factor. The labile ESPI is not ouabain or an isomer. Previously we found that the HPLC region containing ouabain, easily distinguished by retention time from the labile ESPI, did not change with sustained volume expansion (6). We suspect that the selection of a protocol that would, inevitably, fail to identify a labile candidate may have contributed to the controversy to isolate an ESPI that fulfills all the criteria required to establish a hormonal mechanism.

REFERENCES

1. Goto A, Yamada K, Yagi N, Yoshioka M, Sugimoto T. Physiology and pharmacology of endogenous digitalis-like factors. Pharmacol Rev 1992;44:377-99.

2. Graves SW, Williams GH. An endogenous ouabain-like factor associated with hypertensive pregnancies. J Clin Endo Metab 1984;59:1070-4.

3. Graves SW, Williams GH. Endogenous digitalis-like natriuretic factors. Ann Rev Med 1987;38:433-44.

4. Graves SW, Eder JP, Schryber SM, Sharma K, Brena A, Antman KH, Peters WP. Endogenous digoxin-like immunoreactive factor and digitalis-like factor associated with the hypertension of patients receiving multiple alkylating agents as part of autologous bone marrow transplantation. Clin Sci 1989;77:501-507.

5. Graves SW, Seely EW, Williams GH, Hollenberg NK. A purified, high affinity inhibitor of $Na^+/K^+$-ATPase causes vasoconstriction. Clin Exp Hypertens 1991;B10:173 (Abstract).

6. Graves SW, Glatter KA, Lazarus JM, Williams GH, Hollenberg NK. Volume expansion in renal failure patients: a paradigm for a clinically relevant $Na^+/K^+$-ATPase inhibitor. J Cardiovas Pharmacol 1993;22(S2):S54-7.

7. Kelly RA, Smith TW. Is ouabain the endogenous digitalis? Circulation 1992;86:794-7.

# Ouabain-like immunoreactivity in the brain and the adrenal medulla is increased in DOCA-salt hypertensive rats

H. Takahashi[1], Y. Terano[2]

1) Department of Clinical Sciences and Laboratory Medicine, Kansai Medical University, Moriguti City, Osaka 570, and 2) Department of Physiology, Osaka City University Medical School, Osaka 545, Japan

## Introduction

The interest in research into hypertension is become focused on the circulating endogenous digitalis-like factor (EDLF). Since EDLF increases when a high-sodium diet was given (7,8) and it has been known that human hypertension is closely associated with an intake of excess salt, the increased level of EDLF caused by an excess sodium, might cause an increase in blood pressure. Because plasma levels of EDLF become elevated when hypertonic saline was intracerebroventricularly infused (9), central nervous system must play a key role in releases of EDLF. On the other hand, Tamura et al. (6) claim that they purified the EDLF from the adrenal tissue. we aimed, in the present study, to search ouabain-like immunoreactive substances in the brain and the adrenal gland by an immunocytochemical technique and radioimmunoassay with mono- and poly-clonal ouabain-antibodies, respectively.

### Immunohistochemistry

Animals were perfused via the left cardiac ventricle with phosphate-buffered saline (0.1M phosphate buffer containing 0.9% NaCl, pH 7.2), followed by 300 ml of a fixative containing 0.5% glutaraldehyde, 4% formaldehyde, and 0.2% picric acid in 0.1M phosphate buffer (pH 7.0) at 4°C; and then 300 ml of 4% formaldehyde, 0.2% picric acid and 0.2% acetic acid fixative; and then 300 ml of 4% formaldehyde fixative. The brains and the adrenals were removed from their bodies and further immersed in the 4% formaldehyde fixative for 24 hours at 4°C. After immersion in 0.1M phosphate buffer containing 20% sucrose at 4°C, the adrenal glands were rapidly embedded in 10% gelatin. The hypothalamic tissues and hardened gelatin blocks of adrenal glands were frozen and cut into 20 mm-thick sections with a cryostat. The sections were stored in 0.1M phosphate buffered saline containing 0.3% Triton X-100 at 4°C. The sections were successively incubated in: 1] mouse monoclonal antibody to ouabain (x3000-x5000 dilution) for 48-72 hours at 4°C; 2] biotin-labeled anti-mouse IgG for 180 minutes; 3] avidin-biotin-peroxidase complex solution for 60 minutes. The immunoreactive substance, colored with 3,3'-diaminobenzidine (DAB), was enhanced with osmication.

The immunoreactivities were observed as dark brown colored deposits, and were diffusely distributed in the cytoplasm of endocrine cells of the adrenal medulla. However, immunoreactive substances were not detected in the adrenal cortex.

Immunoreactive neuronal somata and their proximal processes were found in the paraventricular nucleus (PVN), particularly in its magnocellular part. The intensity of the immunostaining in the proximal processes was greater than in the somata. In the supraoptic nucleus (SON), the very weak immunoreactivity was observed in the neuronal somata and processes. No immunoreactive substance was detected in the other areas of the hypothalamus examined in this study.

**Measurement of ouabain-like immunoreactivity**

Animals: DOCA-salt hypertensive rats were produced by implanting a silicone mold containing DOCA, 35 mg, when unilateral nephrectomy was done under anesthesia with ether. Then 1% saline was substituted as drinking water. Four weeks later when rats were 9 weeks old, they were sacrificed by decapitation.

Preparation and extraction of the material: Tissue was homogenized and sonicated by adding 2 ml of distilled water. It was then boiled for 10 minutes and centrifuged at 15,000 r.p.m. for 30 minutes. The supernatant and urine samples were applied to the Sep-Pak C18 column (Waters Inc., Ltd.) and eluted with 2 ml of 60% acetonitrile.

Radioimmunoassay was performed using the anti-ouabain polyclonal antibody which cross-reacts with digitoxin and digoxin. This antibody was produced by immunizing rabbits with ouabain and bovine serum albumin which was used for the production of monoclonal antibody. Final dilution of the antibody was 25.000 times. Ouabain (Sigma chemicals) was used as the standard. $^{125}$I-Digoxigenin with the buffer solution was supplied from a commercial radioimmunoassay kit (Spac, Daiichi Radioisotope Co. Ltd.). The mixture of sample/standard and anti-ouabain antibody was incubated overnight at 4°C. Then, $^{125}$I-Digoxigenin was added, and incubated overnight at 4°C. Finally, the second antibody (anti-rabbit IgG, Daco Chemicals) was added, and incubated for 4 hours at room temperature. After starring thoroughly, it was centrifuged at 300 r.p.m. for 30 minutes. Radioactivity of the precipitate was counted for 2 minutes.

Urinary excretions of immunoreactivity markedly increased in DOCA-salt hypertensive rats as compared to the control rats (60.5±7.2 vs. 14.3±5.2 µg/day) at 2 weeks after the DOCA-salt treatment, and 43.6±3,1 vs. 16.3±3.8 µg/day at 4 weeks after. Although the hypothalamic content of the immunoreactivity was not different between DOCA-salt and the sham rats, the pituitary content was significantly elevated in DOCA-salt hypertensive rats (431±102 vs. 98±15.2 ng/g-tissue weight). The adrenal content was also significantly elevated in DOCA-salt hypertensive rats (184±22.3 vs. 21.1±5.2 ng/g-tissue weight). However, in the other organs tissues the content was very low as compared to the level in the pituitary and adrenal, and differences in the content between DOCA-salt and the sham rats were not significant.

**Discussion**

We first showed that the ouabain-like immunoreactive materials cross-reacted with the monoclonal antibody to ouabain (MAb-T8B11) exist in the adrenal medulla and the hypothalamic PVN. And, the content of immunoreactivity was significantly increased in the pituitary and the adrenal in a salt-loaded animal model, DOCA-saly hypertensive rats. The production of ouabain-like immunoreactivity was seemed to be increased in DOCA-salt rats as shown by increases in the urinary excretions.

In the previous study (11) with polyclonal anti-digoxin antibody, we find that the immunoreactive substances locate in the PVN and SON, and that their fibers densely distributed in the circumventricular organs of the third ventricle such as the organum vasculosum of the laminae terminalis (OVLT) and the subfornical organ (SFO), the median eminence and the pituitary posterior lobe of macaque and rats. Those areas are known to be closely implicated in regulation of electrolytes and water balance, and consequently in development of hypertension. However, the immunoreactivity was not found in the adrenal cortex and medulla with the similar immunocytochemical technique as used in the present study.

Since we detected the immunoreactivity in the magnocellular part of the PVN even with the monoclonal antibody, the detected material may be the same one of digoxin- and ouabain-like substances as we have reported earlier. However, the missing of the immunoreactivity in the other part of the hypothalamus and the posterior pituitary, may indicate the existence of structurally-different, but very similar substances in the central nervous system. And the monoclonal antibody might have detected one of those substances. Characterization of the immunoreactive substance must be clarified this issue.

The adrenal medulla was stained only with this particular monoclonal antibody. (MAb-T8B11) to ouabain. MAb-T8B11 was characterized according to the method described by Terano et al. [10]. MAb-T8B11 was the murine IgG2b kappa subclass, recognized the entire structure which consists of the cis-linked ring D and the functional groups at positions C13, C14, and C17 on the steroid portion of cardiac glycosidegenins. The partial structure with ouabain-like $Na^+$, $K^+$-ATPase inhibitory activity in a competitive mode against $K^+$ corresponded with that recognized by MAb-T8B11. MAb-T8B11 recognized the pharmacologically active site of cardiac glycosides and neutralized their biological activities. It may, therefore, also indicate the presence of the subtype of the ouabain-like substance, in the adrenal medulla. Since Hamlyn et al. [1,5] first demonstrated presence of ouabain in the adrenal, present findings of the ouabain-like substance in the adrenal medulla is the special interest in this respect. They further found the ouabain-producing adrenal tumors in hypertensive patients [2]. The histological examination revealed that the tumor tissue is of the adrenal cortex origin. After removal of the cortical tumor, plasma concentrations of ouabain significantly decreased and blood pressure was become lowered. Therefore, they thought that the ouabain is produced in the adrenal cortex. Masugi et al. [4] also find that plasma levels of ouabain-like immunoreactivity are elevated not only in patients with essential hypertension but also in those with hyperaldosteronism [3]. In contrast to these findings, our present results demonstrated the immunoreactivity only in the adrenal medulla. There is no proper explanation for this discrepancy.

Our present findings indicate that the subtype of the ouabain-like $Na^+$, $K^+$-ATPase inhibitors, i.e., EDLF, are exclusively produced in the neural tissue such as the hypothalamus and the adrenal medulla.

*Acknowledgment*

This investigation was partly supported by a Grant-in-Aid for Scientific Research from the Ministry of Education, Science and Culture in Japan (H. Takahashi, No. 05671927).

## References

1. Hamlyn JM, Blaustein MP, Bova S et al. (1991) Identification and characterization of a ouabain-like compound from human plasma. Proc Natl Acad Sci, USA 88:6259-6263
2. Manunta P, Evans G, Hamilton BP et al. (1992) A new syndrome with elevated plasma ouabain and hypertension secondary to an adrenocortical tumor. (abstract) 14th Scientific Meeting of the International Society of Hypertension, S27.
3. Masugi F, Ogihara T, Hasegawa T et al. (1986) Circulating factor with ouabain-like immunoreactivity in patients with primary aldosteronism. Biochem Biophys Res Comm 135:41-45
4. Masugi F, Ogihara T, Hasegawa T et al. (1987) Ouabain and non-ouabain-like factors in plasma of patients with essential hypertension. Clin Exp Hypertens A9:1233-1242
5. Mathews WR, DuCharme DW, Hamlyn JM et al. (1991) Mass spectral characterization of an endogenous digitalislike factor from human plasma. Hypertension 17:930-935
6. Tamura M, Lam T-T., Inagami T (1988) Isolation and characterization of a specific endogenous $Na^+$, $K^+$-ATPase inhibitor from bovine adrenals. Biochemistry 27:4244-4253
7. Takahashi H, Matsuzawa M, Okabayashi H, et al. (1986) Evidence for a digitalis-like substance in the hypothalamo-pituitary axis in rats. J Hypertens 4:S317-S320
8. Takahashi H, Matsusawa M, Okabayashi H, et al. (1988) Endogenous digitalis-like substance in an adult population in Japan Am J Hypertens 1:168S-172S
9. Takahashi H, Matsuzawa M, Suga K et al. (1988) Hypothalamic digitalis-like substance is released with sodium-loading in rats. Am J Hypertens 1:146-151
10. Terano Y, Tomii A, Masugi F (1991) Production and characterization of antibodies to ouabain. Jpn J Med Sci Biol, 44, 123-139
11. Yamada H., Ihara N., Takahashi H et al. (1992) Distribution of the endogenous digitalis-like substance (EDLS)-containing neurons labeled by digoxin antibody in hypothalamus and three circumventricular organs of dog and macaque. Brain Res, 584:237-243

# Changes in the Endogenous Ouabain-like Substance in the Brain and Cerebrospinal Fluid after Cold-induced Vasogenic Brain Edema

Z.M. Rap[1], W. Schoner[1], Z. Czernicki[2], G. Hildebrandt[3], W. Mueller[4], O. Hoffmann[4]

Institute of Biochemistry and Endocrinology of the University Giessen Germany[1], Medical Research Center Pol. Acad. Sci., Warsaw, Polen[2], Neurosurgery Clinic Cologne[3], and Giessen[4]

## Introduction

Application of exogenous ouabain into the brain tissue induces cytotoxic brain edema with swelling of glia cells (3,5). This phenomenon is characterized, not only by toxic, but also by ischemic and vasogenic brain edema (2) and is accompanied by reversible decrease of membranal brain ATPase activity (11). The detection of an endogenous ouabain like substance (EOLS) in the brain tissue of animals (1,4,7,9,12) and in the human cerebrospinal fluid (CSF) (6,10) suggests that this factor may participate in the pathogenesis of brain edema. The data to be reported are consistent with this hypothesis.

## Material and Methods

Experiments were carried out on 10 anaesthetized (Nembutal 30 mg/kg) and spontaneosly respirating cats. Animals were continuously hydrated with Ringer solution ($0.09 \ ml \cdot kg^{-1} \cdot min^{-1}$). Blood pressure, heart rate and the blood's gasometric status were continuously recorded. Standardized cold injury of the cortex was produced through the intact dura mater (8). Samples of CSF for the estimation of EOLF were taken before animals were sacrificed at 12 and 24 hrs after cold injury.

Procedures of partial purification and identification of the EOLS in brain tissue.
(a) The edematous and contralaterale brain hemispheres were homogenized and extracted with methanol (1:3). After centrifugation (47000 x g), the supernanants were evaporated and lyophilized. (b) The residues were dissolved in 50% methanol and lipids were extracted into an equal volume of chloroform. The methanol-waterphase was evapored. (c) The residues were dissolved in water, given onto Waters C-18 cartridges and eluted thereof with 80% ethanol. The ethanol fractions were evaporated, dissolved in water and ultrafiltered. (d) Aliquots thereof were subjected to C-18 reverse phase HPLC and eluted isocratic with 30% methanol in water. The fraction with the retention time of ouabain was evaporated. (e) EOLS in crude CSF and brain fraction was estimated by the measuring their inhibitory effect on [86]Rb uptake into

779

human red blood cells, presuming that the sodium pump has the same affinity for EOLS like ouabain ($K_I = 4 \times 10^{-8}$ M) (14).

## Results

As is shown in Figure 1, EOLS in crude CSF increased 12 hrs after cold injury from 27.5 ± 5.4 to 134.2 ± 13.3 nmol/l ouab. equiv. ($p < 0.01$) and fell to 56 ± 28.2 nmol/l ouab. equiv. ($p < 0.05$) at 24hrs. Intravenous Ringer infusion in non-treated control animals did not affect the EOLS concentration in CSF. EOLS concentration in the edematous brain hemisphere rose 12 hrs after injury to 2600 ± 1762 pmol ouab.equiv./g wet tissue ($p < 0.03$) as compared to 612 ± 86 pmol ouab.eqiuv./g of the healthy control hemisphere. It fell in the edematous hemisphere 24 hrs after cold injury to 857 ± 166 pmol ouab.equiv./g which is stil above the concentration of 597 ± 55 pmol ouab eqiuv./g wet tissue in the control hemisphere.

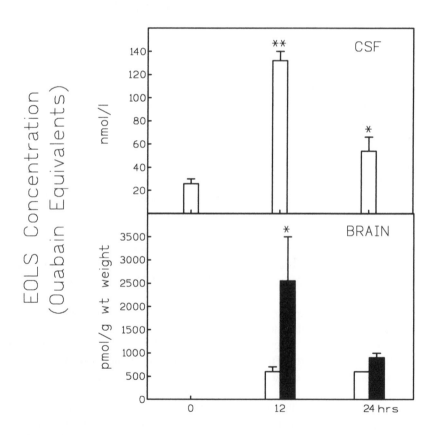

**Figure 1.** EOLS concentration in the CSF and in the control (open bars) and edematous (filled bars) brain hemispheres 12 (n=5) and 24hrs (n=4) after cold injury.

## Discussion

In both, CSF and in edematous brain tissue, alterations of the EOLS concentrations showed the same time course: The EOLS concentration peaked 12 hrs after cold lesion and lowered after 24 hrs. The percental changes in EOLS were approximately the same: After 12 hrs a 425% increase in edematous hemispheres was seen and 260% in CSF and it fell after 24 hrs to 143% in the edematous tissue and 120% in CSF. This pattern correlates well with the dynamics of cold induced vasogenic brain edema (10). The absolute concentrations of EOLS, however, were in the CSF higher than in edematous brain tissue. It remains to be investigated as to whether the EOLS formed in the brain tissue is secreted into CSF.

The concentration of EOLS in edematous brain tissue is in the range of 2.5 - 5 nmol/g tissue. This concentration might suffice to inhibits the cardiac glycoside-sensitive $\alpha_3$ isoform of $Na^+/K^+$-ATPase in astrocytes (13). In conclusion, we propose that increased concentrations of EOLS in CSF and edematous brain tissue may play a role as one of the specific mediators of brain edema.

*Acknowledgements*

We are grateful for the support of this work by the Fonds der Chemischen Industrie, Frankfurt/Main, Germany.

## References

1.   Akagawa K, Hara N, Tsuka Y (1984) Partial purification and the characters of the inhibitors of Na/K-ATPase and ouabain binding in the central nervous system. J Neurochem 42:775-78

2.   Baethmann A, Maier-Hauff K, Kempski O, Unterberg A, Wahl M, Schürer L (1988) Mediators of brain edema and secundary brain damage. Crit Care 16:972 978

3.   Bignami A, Palladini G (1966) Experimentally produced cerebral status spongiosus and continuous pseudorhythmic electroencephalographic discharges with a membrane ATPase inhibitor in the rat. Nature 209:413-414

4.   Fishman MC (1979) Endogenous digitalis-like activity in mammalian brain. Proc Natl Acad Sci (USA) 76:4661-4663

5.   Gazendam J, Go KG, Van der Meer JJ, Zuidetveen F (1979) Changes of electrical impedence in edematous cat brain during hypoxemia and after intracerebral ouabain injection. Exptl Neurol 66:78-87

6.   Halperin JA, Riordan JF, Tosteson DC (1988) Characteristics of an inhibitor of the $Na^+/K^+$ATPase pump in human cerebrospinal fluid. J Biol Chem 263:646-651

7.   Haupert GT, Carilli CT, Cantly LC (1984) Hypothalamic sodium transport inhibitor is a high-affinity reversible inhibitor of $Na^+/K^+$-ATPase. Am J Physiol 247:F919-F924

8.   Klatzo I, Piraux A, Laskowskii EJ (1958) The relationship between edema blood brain barrier and tissue elements in a local brain injury. J Neuropathol exptl Neurol 17:548-564

9.    Lichtstein D, Samuelov (1982) Membrane potential changes induced by the ouabain like compound extracted from mammalian brain. Proc Natl Acad Sci (USA) 79:1435-1456

10.   Lichtstein D, Mine D, Bourrit A, Deutsch J, Karlish SJD, Belmaker H, Rimon R, Palo J (1985) Evidence for the presence of "ouabain like" compound in human cerebrospinal fluid. Brain Res 325:13-19

11.   Rigoulet M, Gerin B, Cohadon E, Vandendreische M (1979) Unilateral brain injury in the rabbit; reversible and irreversible damage of membranal ATPase. J Neurochem 32:535

12.   Takahashi H, Matsusawa M. Ikegaki I, Suga K, Knishimura M, Yoshimura M, Yamada H, Sano Y (1988) Digitalis-like substance is produced in the hypothalamus but not in the adrenal gland in rats. J Hypertension 6:S345-S349

13.   Urayama O, Sweadner KJ (1988) Ouabain sensi-tivity of alpha 3 isozyme of rat Na,K-ATPase. Bioch Bioph Res Commun 156:796-800

14.   Wenzel M, Heidrich E, Kirch U, Moreth-Wolfrat K, Wizemann V, Schoner W (1990) Increased concentrations of a circulating inhibitor of the sodium pump inhibitor in essential hypertension and uremia and its partial purification from hemofiltrate. Physiol Bohemoslovaca 39:79-85

# Modulation of the $Ca^{2+}/Mg^{2+}$-ATPase Activity of Synaptosomal Plasma Membrane Vesicles by the Hypothalamic Hypophysary $Na^+/K^+$-ATPase Inhibitory Factor (HHIF)

M. Ricote, E. Garcia, C. Gutierrez and J. Sancho,

Serv. Endocrinología. Hosp.Ramón y Cajal, 28034 Madrid, Spain and Dept. Bioquímica y Biología Molecular y Genética, Facultad de Ciencias, UNEX, 0680 Badajoz, Spain.

## Introduction

There is a general agreement that in essential hypertension there is a generalized alteration in membrane transport which leads to an increased peripheral resistance, decreased renal sodium excretion and plasma volume expansion (7). A circulating inhibitor of the sodium pump has been implicated in these three phenomena (3). The increased intracellular $Ca^{2+}$ in vascular smooth muscle (VSM) cells is considered the final effect of the inhibitor leading to an increased contractility. It has also been speculated that this substances could be also be involved in the proliferation of VSM cells through the same mechanism. One of these factors is an endogenous compound first isolated from the hypothalamus and hypophysis, HHIF, (8) and later found to be present in other tissues although at lower concentrations (12) that has been shown to increase cell proliferation on rat mesangial cells (10). This nonpeptidic, nonlipidic $Na^+$-pump inhibitor is different from any known cardioglycoside (12). Based in these observation and in the known fact that different ion pumping ATPases, namely $Na^+/K^+$-ATPase and $Ca^{2+}/Mg^{2+}$-ATPases, show little specificity towards inhibitors (11), we have studied the inhibition of HHIF on the $Ca^{2+}$ pump of the plasma membrane of synaptosomes.

## Materials and methods

Synaptosomes have been prepared from Wistar rat brains according to the method of Michaelis, as modified by García-Martín and Gutiérrez-Merino (4). Plasma membrane vesicles were prepared from synaptosomes by hypotonic lysis as previously described (4). HHIF has been purified from bovine hypothalamus and hypophysis as previously reported (8). The inhibition by HHIF of the $Na^+/K^+$-ATPase activity from porcine kidney outer medulla was measured using a coupled assay (2,9) and one unit (1U) was defined as the amount of HHIF required to inhibit by 50% the activity of $8\,\mu g$ of purified $Na^+/K^+$-ATPase (8). The total $Ca^{2+}/Mg^{2+}$-ATPase activity was measured as previously reported (5) using the coupled enzyme system pyruvate kinase-lactate dehydrogenase. The composition of the standard assay medium was the following: 50 mM TES (pH 7.4), 0.1 M KCl, 2 Mm $MgCl_2$, 2 Mm $\beta$-mercaptoethanol, 2 mM ATP, 50 $\mu$M $CaCl_2$, 5 mM sodium azide, 0.42 mM phosphoenol pyruvate, 0.22 mM NADH, 10 IU/ml of pyruvate Kinase, 28 IU/ml of lactate dehydrogenase and 0.02 mg protein/ml. $^{45}Ca^{2+}$ uptake by plasma membrane vesicles

783

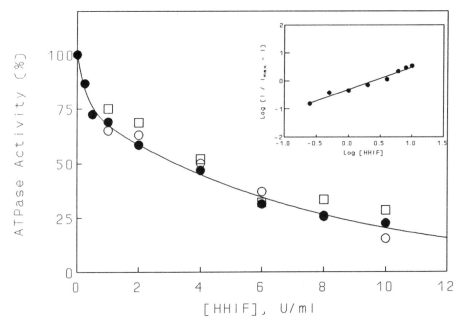

**Figure 1.** Inhibition of the total $Ca^{2+}/Mg^{2+}$-ATPase activity by HHIF. The data was obtained at standard assay conditions, pH 7.4 and $25^0$ C (●), pH 7 and $25^0$ C (□) and pH 7.4 and $37^0$ C (O).

was measured by Millipore filtration through HAWP025000 filters as described by García-Martín and cols. (1990). The reaction medium contained the following: 50 mM TES (pH 7.4), 0.1 M KCl, 2 mM $MgCl_2$, 50 $\mu$M $CaCl_2$ (0.4 $\mu$Ci/ml, free $Ca^{2+}$ 42 $\mu$M) 2 mM $\beta$-mercaptoethanol and 0.12 mg synaptosomal protein per milliliter.

### Results

HHIF inhibits the total $Ca^{2+}/Mg^{2+}$-ATPase of the synaptosomal plasma membrane vesicles. Figure 1 shows that HHIF inhibits the total $Ca^{2+}/Mg^{2+}$-ATPase activity of synaptic plasma membrane vesicles. The dependence of the $Ca^{2+}/Mg^{2+}$-ATPase activity upon the concentration of HHIF is not significantly altered by a change of pH of the assay medium from 7.4 to 7, nor by changing the temperature from $25^0$ C to $37^0$ C. From the Hill plot of these data a $K_{0.5}$ value of 2.45 U/ml and a Hill coefficient of $0.8\pm0.1$ are obtained.

### Inhibition of ATP dependent $Ca^{2+}$ uptake by HHIF

In synaptic membranes $Ca^{2+}$-stimulated ATP hydrolysis is coupled to ATP-dependent calcium uptake (6), therefore we have studied the effect of HHIF on the ATP-dependent $Ca^{2+}$ uptake. The $Ca^{2+}$ uptake by plasma membrane vesicles was measured by Millipore filtration. In the absence of HHIF was $3.9\pm0.6$ and $5\pm0.8$

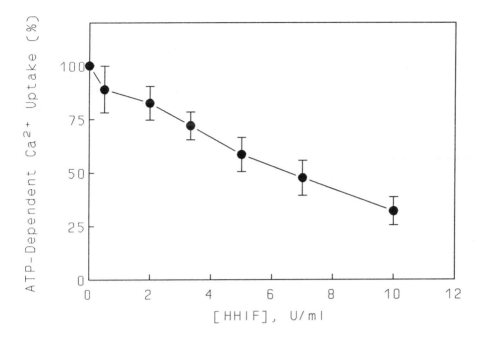

**Figure 2.** Inhibition of the ATP-dependent Ca²⁺ uptake by HHIF. The data was obtained after substraction of the Ca²⁺ uptake observed in the presence of HHIF and absence of ATP in the assay medium.

nmol Ca²⁺ /mg membrane protein at 2 and 5 min, respectively. The results obtained in the presence of different HHIF concentrations are displayed in the Figure 2. From the Hill plot of these data we have obtained a $K_{0.5}$ value of 6.25 U/ml and a Hill coefficient 0f 1.42±0.07. The $K_{0.5}$ values listed above are given as the total HHIF concentration in the assay medium. Calcium uptake measurements were carried out at membrane concentrations of 0.12 mg protein/ml, which makes necessary to correct the value of $K_{0.5}$ for HHIF adsorption to the membrane for both $K_{0.5}$ . The apparent partition coefficient ($K_p$) for HHIF binding to synaptosomal membranes is 0.08 ($\mu$g protein/ml)⁻¹. The corrected value is approximately 0.6 U/ml. This concentration is close to the corrected $K_{0.5}$, 0.9 U/ml, for the inhibition of the Ca²⁺/Mg²⁺-ATPase (see above).

**Discussion**

From the results shown above, it can be deduced that HHIF inhibits the total Ca²⁺/Mg²⁺-ATPase of the synaptosomal plasma membrane and the ATP-dependent Ca²⁺ uptake. This mechanism can account for the increased cytosolic-free Ca²⁺ and proliferation of mesangial cells induced by HHIF (10). In addition to its chemical characteristics, this inhibitory property also distigish HHIF from Digoxin or Ouabain.

The inhibition of the plasma membrane $Ca^{2+}$ pump could produce an increase in $Ca^{2+}$ concentration in synaptic terminals (1), leading to synaptic hyperactivity. This possible property of HHIF is actually under study. Altogether these results suggest that HHIF modulates the $Ca^{2+}$ pump of synaptosomal plasma membrane by direct interaction with hydrophobic sites if this transport site, as suggested earlier for the modulation of local anesthetics of the $Ca^{2+}$ pump of the synaptosomal plasma membrane and the sarcoplasmic reticulum .

*Acknowledgements*

We are indebted to J. Colilla for excellent technical assistance. This work was partially supported by a financial aid from the CAYCIT and FIS and Dr. Ricote fellowship from FIS and Fund. Rich.

**References**

1.      Akerman KEO, Nicholls DG (1983) Ca2+ transport and the regulation of transmitter release in isolated nerve endings. Trends Biochem Sci 8:63-64.
2.      Carilli CT, Berne M, Cantley LCJr, Haupert GT,Jr. (1985) Hypothalamic factor inhibits the (Na/K)ATPase from the extracellular surface. J Biol Chem 260,No.2:102-7-1031.
3.      de Wardener HE, MacGregor GA (1980) Dahl's hypothesis that a saluretic substance may be responsible for a sustained rise in arterial pressure:its possible role in essential hypertension. Kidney Int 18:1-9.
4.      García-Martín E, Gutiérrez-Merino C (1986) Local anesthetics inhibit the Ca2+,Mg2+-ATPase activity of rat brain synaptosomes. J Neurochem 47:668-672.
5.      García-Martín E, Gutiérrez-Merino C (1990) Modulation of the Ca2+,Mg2+-ATPase activity of synaptosomal plasma membrane by the local anesthetics dibucaine and lidocaine. J Neurochem 54:1238-1246.
6.      Gill DL, Grollman EF, Khon LD (1981) Calcium transport mechanisms in memebrane vesicles from guinea pig brain synaptosomes. J Biol Chem 256:184-192.
7.      Haddy FJ, Pamnani MB, Clough DL, Huot S (1982) Role of a humoral sodium-potasium pump inhibitor in experimental low renin hypertension. Life Sci 30:571-575.
8.      Illescas M, Ricote M, Mendez E, G-Robles R, Sancho J (1990) Complete purification of two identical Na(+)-pump inhibitors isolated from bovine hypothalamus and hypophysis. FEBS Lett 261:436-440.
9.      Josephson L, Cantley LCJr (1977) Isolation of a potent (Na+-K+)ATPase inhibitor from striated muscle. Biochemistry 16:4572-4578.
10.     Montero A, Rodriguez-Barbero A, Ricote M, Sancho J, López-Novoa JM (1993) Effect of ouabain and hypothalamic, hypophysary inhibitory factor on rat mesangial cell proliferation. J Cardiovasc Pharmacol 22(Suppl.2):S35-S37.
11.     Pedersen PL, Carafoli E (1987) Ion motive ATPases.I.Ubiquity,properties, and significance to cell function. Trends Biochem Sci 12:146-150.
12.     Sancho J, García-Martín E, García-Robles R, Santirso R, Villa E, Gutiérrez-Merino C, Ricote M (1993) Properties of the purified hypothalamic pituitary Na/K-ATPase inhibitor. J Cardiovas Pharmacol 22 (Suppl.2):S32-S34.

# ENDOGENOUS DIGITALIS IN ACUTE MYOCARDIAL ISCHEMIA

A.Y. Bagrov[1,2*], R.I. Dmitrieva[2], O.V. Fedorova[2], E.A. Kuznetsova[1], N.I. Roukoyatkina[2].

Department of Cardiology, Dzhanelidze Research Institute of Emergency Medicine[1] and Sechenov Institute of Evolutionary Physiology and Biochemistry[2], St. Petersburg, Russia, National Institute on Aging[*], 4940 Eastern Ave, Baltimore, MD 21224, USA

## Introduction

The evidence that endogenous digitalis-like factor (EDLF) may act as an endogenous inotrope includes presence of digitalis-like immunoreactive material in the hearts of mammals (14), high-affinity binding of this material to digitalis receptor sites in ventricular myocardium (15), and digitalis-like positive inotropic effects of EDLF demonstrated in different isolated heart preparations (17, 12). Acute myocardial ischaemia/infarction (AMI) is known to sensitize myocardium to the arrhythmogenic effect of digitalis (13). AMI is associated with both inhibition of $Na^+/K^+$-ATPase in the myocardium (6, 10) and with loss of myocardial digitalis receptors (11). Analyzing these observations one can speculate that these effects observed in AMI at least in part may be explained by the action of one or more circulating $Na^+/K^+$-pump inhibitors, and that EDLF may participate in the genesis of arrhythmias in AMI. Based on this hypothesis we have shown that pretreatment of coronary ligated cats with antidigoxin antiserum leads to a significant increase of the electric threshold of ventricular fibrillation (1). In coronary ligated rats AMI was assocciated with a 2.5 fold increase in plasma level of EDLF, antidigoxin antibody along with the antifibrillatory effect prevented the inhibition of $Na^+/K^+$-ATPase in erythrocytes and myocardium, which were otherwise observed in acute myocardial ischaemia (2, 3). Clinical observations have shown significant inhibition of $Na^+/K^+$-ATPase in erythrocytes in patients during the initial period after first transmural AMI (4). This inhibition was associated with the increase in plasma $Na^+/K^+$-ATPase inhibitory activity and digoxin-like immunoreactivity (4, 5). Here we present new data suggesting that inhibition of the myocardial $Na^+/K^+$-pump due to increased plasma EDLF may contribute to the genesis of arrhythmias in AMI. In order to determine the nature of EDLF acting in AMI, we also have compared antiarrhythmic effects of different antibodies blocking the effects of EDLF. Interactions of cardiac glycosides with sympathetic neural endings in the heart are known to play an important role in cardiac effects of digitalis (16). Therefore, we also looked at a possibility of inhibition of $Na^+/K^+$-ATPase in the neural endings in myocardium in AMI.

## Methods

Coronary ligation was performed as described previously (7) in the anaesthetized (pentabarbital sodium, 75 mg/kg, i/m) and artificially ventilated adult male Wistar rats (180 - 300 g). After 15 min of ECG monitoring, the animals were sacrificed by exsanguination, and the hearts were rapidly excised for the study of myocardial

$Na^+/K^+$-ATPase activity. The pretreatment antibodies or saline in controls were administered 30 minutes before the ligature was tied. Arrhythmias were assessed during a 15 min postligation period. Incidence of ventricular arrhythmias was determined as the total duration of ventricular tachycardia and ventricular fibrillation (VT and VF). Mixed blood samples were collected in cooled polyethylene tubes containing 0.1 M of EDTA and 10 $\mu$M phenylmethylsulphonylfluoride (50 $\mu$l per 10.0 ml of blood). Plasma concentrations of digoxin-like immunoreactivity were measured using DELFIA fluoroimmunoassay (LKB-Wallac, Finland) and expressed as ng/ml of digoxin equivalents. Rabbit polyclonal monospecific antibody to digoxin and to the mixture of bufodienolides was obtained, separated and tested as reported previously (3). Fab fragments of digoxin antibody (Digibind) were obtained from Borroughs Wellcome (Research Triangle, NC, USA). Left ventricular myocardium membranes fraction was prepared as reported previously (8), except 0.15% deoxycholate was substituted by 0.001% sodium dodecyl sulphate, and then exposed to differential centrifugation in the gradient of sucrose (9). Activity of $Na^+/K^+$-ATPase was measured as reported previously (8) in the crude membrane fraction and in two fractions obtained after differential centrifugation and corresponding to sarcolemmal membranes and membranes of neural endings, respectively.

**Results and discussion**

Coronary ligation in the saline-pretreated rats caused the typical ischemic changes in the ECG, and 8 min after ligation, paroxysms of VT and VF. 15 min of acute coronary ligation was associated with no significant changes in the activity of $Na^+/K^+$-ATPase in the crude membrane fraction (8.3 $\pm$ 0.6 vs. 7.7 $\pm$ 1.0, n=8). As shown in Table 1, in control animals, activity of $Na^+/K^+$-ATPase in the fraction corresponding to neural endings was significantly higher than in sarcolemmal membranes. 15 min acute myocardial ischemia was associated with 34% inhibition of $Na^+/K^+$-ATPase activity in sarcolemmal membranes and with 53% inhibition in synaptic membranes (Table 1). Nonequal inhibition of $Na^+/K^+$-ATPase by digitalis has been shown previously in canine heart; the activity of $Na^+/K^+$-ATPase in Purkinje fibers was significantly higher than in sarcolemmal membranes, and, at the same time, $Na^+/K^+$-ATPase in Purkinje fibers was more "inhibitable" by cardiac glycosides (18).

**Table 1. Activity of $Na^+/K^+$-ATPase** ($\mu$moles $P_i$/mg protein/1 hour)   **Plasma EDLF**

| | sarcolemmal membranes | synaptosomal membranes | (ng/ml)[@] |
|---|---|---|---|
| Control (n=8) | 3.32 $\pm$ 0.4 | 7.1 $\pm$ 0.95 | 0.44 $\pm$ 0.08 |
| AMI (n=8) | 2.19 $\pm$ 0.35[*] | 3.34 $\pm$ 0.8[*] | 1.16 $\pm$ 0.29[**] |

Means $\pm$ S.E.M. (*) - p < 0.02, (**) - p < 0.05 as compared with controls, Student's t-test. @ - Results reproduced with permission from Cardiovascular Research, 1993, 27, 1045-1050.

Table 2 compares the effects of different antibodies blocking digoxin/EDLF, on the total duration of VT and VF during 15 min of AMI. Digibind, 5 mg/kg was almost inactive, and demonstrated moderate antiarrythmic effect only at extremely high dose, 40 mg/kg. Polyclonal digoxin antibody and bufodienolide antibody were administered at concentrations blocking the $EC_{75}$ of $Na^+/K^+$-ATPase inhibitory effects of digoxin and mixture of bufodienolides in the intact rat erythrocytes (unpublished data). As shown in Table 2, antibody raised against the mixture of bufodienolides from significantly more active than rabbit polyclonal antidigoxin antibody.

**Table 2. Antiarrhythmic effects of antibodies blocking EDLF in AMI**

|  | AMI[@] | AMI+anti-[@] digoxin Ab 260 $\mu$g/kg | AMI+anti- EDLF Ab 250 $\mu$g/kg | AMI + Digibind 40 mg/kg |
|---|---|---|---|---|
| VT+VF (sec) | 201$\pm$34 | 46$\pm$18[*] | 18.7$\pm$6.5[**] | 74$\pm$34[***] |
| n | 28 | 17 | 10 | 8 |

Means $\pm$ S.E.M. (*) - p < 0.001, (**) - p < 0.002, (***) - p < 0.05 as compared with AMI group, Student's t-test. (@) - Results reproduced with permission from Cardiovascular Research, 1993, 27, 1045-1050

The present observations together with our previous experimental and clinical data demonstrate that acute myocardial ischemia is associated with a significant increase in plasma concentrations of an endogenous substance having digitalis glycosides-like properties, in parallel with the onset of arrhythmias and inhibition of myocardial $Na^+/K^+$-ATPase. Results of the measurement of $Na^+/K^+$-ATPase activity in different membrane fractions prepared from ischemic left venticles suggest, that in AMI symapthetic nervous endings in the heart are likely to be a target for EDLF. In low concentrations, digitalis drugs are known to interact with sympathetic neural endings in the heart causing inhibition of $Na^+/K^+$-ATPase and increasing norepinephrine release (16). These observations are in agreement with our recent data demonstarting that in ischemic rat myocardium $Na^+/K^+$-pump inhibitory effect of EDLF may be overshadowed by $Na^+/K^+$-ATPase activating action of catecholamines (3).

Different antibodies blocking the effects of $Na^+/K^+$-ATPase inhibitors possessed antiarrhythmic effects in AMI. Digibind, antidigoxin antibody having affinity to digoxin in the range of $10^9$ to $10^{10}$ mol $\cdot$ $L^{-1}$ (which is higher than the affinity of digoxin to $Na^+/K^+$-ATPase) showed very weak antiarrhythmic activity. Therefore, although EDLF acting in AMI has "digoxin-like" immunoreactivity, it is likely to be distinct from digoxin itself. At the same time, the fact that antibodies raised against bufodienolide $Na^+/K^+$-ATPase inhibitors have shown the highest antiarrhythmic activity, suggests that rat (and mammalian) EDLF has a bufodienolide structure.

789

# References

1.  Bagrov AY, Ganelina IE, Nikiforova KA, Osipov V, Stolba P (1987) Antifibrillatory effect of antidigoxin antibodies in experimental myocardial infarction. (in Russian). In: Diagnosis and treatment of myocardial infarction. Tbilisi: Metsnierba Eds, 177-179.
2.  Bagrov AY, Fedorova OV, Maslova MN, Roukoyatkina NI, Stolba P, Zhabko EP (1989) Antiarrhythmic effect of antibodies to digoxin in acute myocardial ischemia in rats. Eur J Pharmacol 162: 195-196.
3.  Bagrov AY, Fedorova OV, Roukoyatkina NI, Zhabko EP (1993) Effects of endogenous digoxin like factor and digoxin antibody on myocardial Na,K-pump activity and ventricular arrhythmias in acute myocardial ischaemia in rats. Cardiovasc Res 27: 1045-50.
4.  Bagrov AY, Fedorova OV, Maslova MN, Roukoyatkina NI, Ukhanova MV, Zhabko EP (1991) Plasma Na,K-ATPase inhibitory activity and digoxin like immunoreactivity after acute myocardial infarction. Cardiovasc Res 25: 371-377.
5.  Bagrov AY, Kuznetsova EA, Fedorova OV (1994) Endogenous digoxin like factor in acute myocardial infarction. J Intern Med; 235: 63-67.
6.  Beller GA, Conroy J, Smith TW (1976) Ischemia-induced alteration in myocardial (Na,K)-ATPase and cardiac glycosides binding. J Clin Invest 57: 341-350.
7.  Clark C, Foreman MI, Kane KA, McDonald F, Parrat JR (1980) Coronary artery ligation in anaesthetized rats as a method for the production of experimental disrhythmias and for the determination in infarct size. J Pharmacol Methods 3: 357-368.
8.  Clough DL, Pamnani MB, and Haddy FJ (1983) Decreased myocardial Na,K-ATPase activity in one-kidney, one-clip hypertensive rats. Am J Physiol, 245: H244-H251
9.  Jones DH, Matus A (1974) Isolation of synaptic plasma membrane from brain by combined flotation-sedimentation density gradient centrifugation. Biochim Biophys Acta 356: 276-287.
10. Kim D, Akera T, Weaver LC (1984) Role of sympathetic nervous system in ischemia-induced reduction of digoxin tolerance in anaesthetized cats. J Pharmacol Ther 228: 537-544.
11. Maixent J-M, Lelievre LC (1987) Differential inactivation of inotropic and toxic digitalis receptors in ischemic dog heart. Molecular basis of the deleterious effects of Digitalis. J Biol Chem 262: 12458 - 12462.
12. Navaratnam S, Chau T, Agbanyo M, Bose D, Khatter JC (1990) Positive inotropic effect of porcine left ventricular extract on canine ventricular muscle. Br J Pharmacol 101: 370-374.
13. Okita GT (1977) Dissociation of Na,K-ATPase inhibition from digitalis inotropy. Fed Proc 36: 2225-30.
14. de Pover A, Castsneda-Hernandes G, Godfraind T (1982) Water versus acetone-HCl extraction of digitalis-like factor from guinea-pig heart. Biochem Pharmacol 31: 267-271.
15. de Pover A (1984) Endogenous digitalis-like factor and inotropic receptor sites in the heart. Europ J Pharmacol 99: 365-366.
16. Schwartz A (1993) Positive inotropic action of digitalis and endogenous factors: Na,K-ATPase and positive inotropy; endogenous digitalis. Current Top Membr Transp 19: 825-841.
17. Shimony Y, Gotsman M, Deutsch J, Kachalsky S, Epstein M, Lichtstein D (1986) Further characterization of the inotropic effect of a bufodienolide glycoside: an endogenous ouabain-like compound. Cardiovasc Res 20: 229-239.
18. Somberg JC, Barry WH, Smith TW (1981) Differing sensitivities of Purkinje fibers and myocardium to inhibition of monovalent cation transport by digitalis. J Clin Invest 67: 116-123.

# Sensitivity of the failing and nonfailing human heart to cardiac glycosides

Robert H.G. Schwinger, Michael Böhm, Erland Erdmann

Klinik III für Innere Medizin der Universität zu Köln, Joseph-Stelzmannstr. 9, 50924 Köln, Germany

## Introduction

Cyclic AMP-dependent positive inotropic mechanisms exhibit a reduced effectiveness in failing human myocardium (4), possibly due to a downregulation of myocardial ß-adrenoceptors (4,17) as well as to an increased expression of inhibitory guanine nucleotide binding proteins (6,12) which in turn lead to reduced intracellular cyclic AMP levels (7,14). To investigate the effects of cAMP-independent mechanisms on myocardial contractility, the $Na^+, K^+$ ATPase (EC. 3.6.1.3.) -mediated system in nonfailing myocardium and terminally failing human myocardium due to dilated cardiomyopathy (NYHA IV) was studied.

Cardiac glycosides bind to cell surface sites and thereby inhibit the $Na^+, K^+$-ATPase with subsequent reduction of $Na^+, K^+$ transport activity of the cell membrane (1,20). In consequence, intracellular $Ca^{2+}$ increases via an elevation of intracellular $Na^+$-concentration and subsequent activation of the $Na^+/Ca^{2+}$-exchanger or exchange mechanisms. Consistently, the $Ca^{2+}$ content of the sarcoplasmatic reticulum will be augmented leading to an enhanced $Ca^{2+}$ release during depolarization and therefore increasing force of contraction. Other mechanisms which increase intracellular $Na^+$ may also augment force development, even in the diseased human myocardium (8,15,18,19). In addition, a close relation between intracellular $Na^+$ activity and force of contraction has been documented. Moreover, it has been suggested that increasing $Na^+$ influx is accompanied by an enhanced binding of cardiac glycosides to the membrane bound $Na^+, K^+$-ATPase (10).

Experiments performed with guinea pig myocytes document that the $Na^+$-channel activator BDF 9148 modifies $iNa^+$ (11). BDF 9148 has been shown to inhibit the deactivation of the fast sodium-channel and, thereby, increasing intracellular $Na^+$ (2,11). By this mode of action, the transport rate of cardiac $Na^+, K^+$-ATPase should be enhanced. The aim of the present study was to investigate the influence of ouabain on force of contraction and on the intracellular electrolyte homeostasis in failing and nonfailing human myocardium. We examined the action of the cardiac glycoside ouabain and of ouabain after prestimulation with BDF 9148 on isometric force of contraction of electrically driven left ventricular papillary muscle strips from terminally failing (NYHA IV, heart transplants) and nonfailing (brain death) human myocardium. The interaction of BDF 9148 and ouabain on specific 3H-ouabain binding in myocardial membrane preparations was studied as well.

## Materials and methods

Myocardium from terminally failing human hearts was obtained from patients after cardiectomy during cardiac transplantation (n=25). The preoperative diagnosis was dilated cardio-myopathy in all patients. All patients had been classified as NYHA IV. The pretreatment of the patients consisted of ACE-inhibitors, diuretics and nitrates. None of the patients had received $Ca^{2+}$-channel antagonists nor agonists within 7 days of surgery. None of the patients

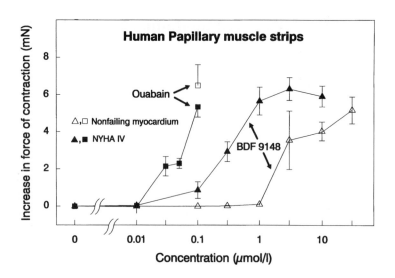

**Fig. 1.** Concentration response curve for the effects on the force of contraction of ouabain (0.01-0.1 µmol/l) and of BDF 9148 (0.01-10 µmol/l) in isolated, electrically driven (1Hz, 37°C) human papillary muscle strips from failing and nonfailing human myocardium.

Ordinate:     Increase in force of contraction in mN

Abscissa:     Concentration in µmol/l

BDF 9148 was more potent to increase force of contraction in failing compared to nonfailing human myocardium.

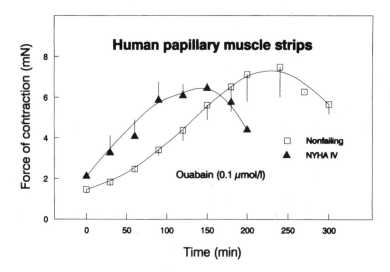

**Fig. 2.** Time-dependency of the positive inotropic effect of ouabain (0.1 µmol/l) in electrically driven human papillary muscle strips from patients with terminally heart failure and in nonfailing human tissue.

Ordinate:     Force of contraction (mN)

Abscissa:     Time in minutes

Ouabain was more potent in failing compared to nonfailing tissue to increase force of contraction.

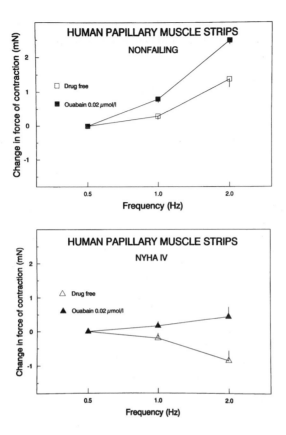

**Fig. 3.** Effect of ouabain (0.02 μmol/l) on the frequency-dependent change in force of contraction in failing and nonfailing human tissue.

Ordinate:     Change in force of contraction (mN)

Abscissa:     Frequency in Hz

Ouabain was effective to partially restore a positive force-frequency-relation in terminally failing human myocardium.

## Discussion

Cardiac glycoside receptors are not subjected to downregulation in human heart failure (16). Even in terminally failing human myocardium with reduced number of ß-adrenoceptors cardiac glycosides remain effective to increase force of contraction as long as there is no significant increase in fibrous tissue. In human myocardium, the Na+-channel activator BDF 9148 shifted the concentration response curve of ouabain to the left. As 3H-ouabain binding to isolated cardiac membranes remains unaffected and only binding to intact cells was facilitated (9), the effects of BDF 9148 on ouabain binding and myocardial inotropy are likely to be indirect via elevation of Na+i. This is in accordance with studies in isolated guinea pig left atria paced at 0.3 Hz (9). There is further evidence, that an increase in Na+i leads to an enhancement of the inotropic response to ouabain. Combined application of the Na+-channel gating modifiers veratridine with dihydro-ouabain, ATX II with ouabain, asebotoxin III with received ß-adrenoceptor agonists 48h prior to operation. Drugs used for general anesthesia were flunitrazepam, fentanyl and pancuroniumbromide with isoflurane. Cardiac surgery was performed on cardiopulmonary bypass with cardioplegic arrest during hypothermia. Nonfailing human myocardium was obtained from five donors with brain death caused by traumatic injury. The nonfailing hearts could not be used for transplantation for technical reasons. Patients gave written informed consent before operation. The cardioplegic solution used was a modified Bretschneider solution containing (in mM): NaCl 15; KCl 10; MgCl2 4; histidine HCl 180; tryptophan 2; mannitol 30; potassium dihydrogen oxoglutarate 1. Immediately after excision, the papillary muscles were placed in ice-cold preaerated Tyrode's solution and delivered to the laboratory within 10 minutes. Contraction experiments as well as binding studies were performed as described previously (5,16,17). Statistical significance was analysed using the Student's t-test for unpaired or paired observations (SPSS PC plus); $p < 0.05$ was considered significant.

## Materials

BDF 9148 (4-(3-(1-Diphenylmethyl-azetidin-3-oxy)-2-hydroxy-propoxy)-1 H-indol-2-carbonitril) was kindly provided from Beiersdorf AG (Hamburg, FRG). Isoprenaline was obtained from Sigma Chemical Co (Deisenhofen, FRG), and ouabain from Boehringer (Mannheim, FRG). The ligand 3H-ouabain was from Amersham-Buchler (Braunschweig, FRG). All other chemicals were of analytical grade or the best grade commercially available. For studies with isolated cardiac preparations stock solutions were prepared and applied to the organ bath. BDF 9148 has been dissolved in 50 % DMSO. The final concentration of DMSO in the bathing solution never exceeded 0.05 %. All other compounds were dissolved in twice distilled water. Applied agents did not change the pH of the medium.

## Results

The receptor density and affinity measured by [$^3$H]-ouabain binding was not significantly different in either group, whereas ß-adrenoceptors ([$^3$H]-CGP 12.177- or [$^{125}$I]-CYP-binding) were significantly downregulated in the studied failing myocardium. Ouabain enhanced force of contraction concentration-dependently in nonfailing and in NYHA IV with a similar effectiveness as an elevation of extracellular $Ca^{2+}$ (Fig. 1).

Time until maximal inotropic effects after ouabain developed were significantly shorter in NYHA IV (mean 150 min) than in nonfailing myocardium (mean 240 min) (Fig.2).

In addition, the $Na^+$-channel activator BDF 9148 was more potent to increase force of contraction in failing compared to nonfailing tissue (Fig. 1). Stimulation of force of contraction by ouabain (0.02 µmol/l) partially antagonized the negative force-frequency-relationship in NYHA IV (Fig.3).

In order to study the effects of ouabain after stimultion of the transport rate of the sarcolemmal $Na^+, K^+$ ATPase, the $Na^+$-channel activator BDF 9148 (0.1 or 1 µM) was applied to increase $Na_i^+$. BDF 9148 enhanced the potency of ouabain to stimulate force of contraction. After pre-stimulation with BDF 9148+ouabain, force of contraction increased after an elevation of frequency of stimulation in NYHA IV, whereas force of contraction gradually declined under control conditions. In myocardial membranes, [$^3$H]-ouabain binding ($B_{max}$, $K_D$) was not affected by BDF 9148, therefore, the influence of BDF 9148 on the effects of ouabain is probably indirect and may be related to cellular $Na^+$ load.

794

ouabain, or the Na$^+$ ionophore monensin with digoxin resulted in an augmented inotropy and accelerated binding of the cardiac glycoside to its membrane bound receptor. This is in accordance to other mechanisms which increase intracellular Na$^+$ (e.g. increased stimulation frequency as well as the combination of extracellular low Ca$^{2+}$-high Na$^+$).

BDF 9148 was as effective as ouabain to increase force of contraction in papillary muscle strips from terminally failing patients due to dilated cardiomyopathy with reduced responsiveness to ß-adrenoceptor agonists. The effectiveness of ouabain and BDF 9148 was preserved in terminally failing human myocardium when compared to nonfailing tissue. Due to different mechanisms, BDF 9148 and ouabain increase intracellular Na$^+$. Both compounds use an identical final pathway to increase force of contraction. In addition, the potency of both compounds, BDF 9148 and ouabain, to increase force of contraction was greater in failing compared to nonfailing myocardium. Therefore, disease related changes may affect the sensitivity to respond to Na$^+$-modulating positive inotropic compounds. This may be due to differences in both, failing and nonfailing myocardium to maintain intracellular electrolyte levels or due to altered basal electrolyte content.

Possible defects are related to intracellular Na$^+$- or Ca$^{2+}$- concentration, altered activity of the Na$^+$-Ca$^{2+}$-exchange mechanism or differences in Na$^+$, K$^+$-ATPase function. Consistently, in myocardium from patients with cardiomyopathy significantly higher resting and enddiastolic Ca$^{2+}$-levels have been measured (3). The intracellular sodium concentration in failing myocardium could be altered as well. As in nonfailing and failing human myocardium the force-frequency-relationship became positive in the presence of low concentrations of ouabain and BDF 9148 the Na$^+$/Ca$^{2+}$-exchanger may be effective in NYHA IV to influence Ca$^{2+}$-handling. The gene expression of the Na$^+$/Ca$^{2+}$ -exchanger in patients with heart failure was observed to be even enhanced compared to nonfailing controls (13). In consequence, changes in intracellular Na$^+$ may be more effective to affect myocardial force of contraction in failing than in nonfailing tissue. Therefore, the Na$^+$,K$^+$ATPase may play an important role for the regulation of the intracellular electrolyte homeostasis. Inhibition of the sodium pump may differently affect electrolyte changes and force development in failing and nonfailing human tissue.

*Acknowledgement*

We thank Mrs Heidrun Villena Hermoza for her excellent technical help. We are indepted to all collegues of the Department of the Cardiothoracic Surgery (Director: Prof. Dr. B. Reichart) for providing us with human myocardial samples. Experimental work was supported by the Deutsche Forschungsgemeinschaft (DFG).

## References

1. Akera T, Brody TM. The role of Na$^+$,K$^+$-ATPase in the inotropic action of digitalis. Pharmacol Rev 1978;29:187-220.
2. Armah BI, Brückner R, Stenzel W. BDF 9148: a novel inotropic agent that transiently prolongs cardiac action potential duration (abstr.). Naunyn Schmiedeberg's Arch Pharmacol 1990;341(Suppl):R50.
3. Beuckelmann DJ, Näbauer M, Erdmann E: Intracellular calcium handling in isolated ventricular myocytes from patients with terminal heart failure. Circulation 1992;85:1046-1055.

4. Bristow MR, Ginsburg R, Minobe W et al. Decreased catecholamine sensitivity and beta-adrenergic-receptor density in failing human hearts. N Engl J Med 1982;307:205-11.
5. Erdmann E, Schoner W. Ouabain-receptor interactions in (Na$^+$,K$^+$)-ATPase preparations from different tissues and species. Determination of kinetic constants and dissociation constants. Biochim Biophys Acta 1973;307:386-98.
6. Feldman AM, Cates AE, Veazey WB et al. Increase of the 40 000-mol wt pertussis toxin substrate (G protein) in the failing human heart. J Clin Invest 1988;82:189-97.
7. Feldman MD, Copelas L, Gwathmey JK et al. Deficient production of cyclic AMP: pharmacologic evidence of an important cause of contractile dysfunction in patients with end-stage heart failure. Circulation 1987;75:331-9.
8. Gwathmey JK, Slawsky MT, Briggs GM, Morgan JP. Role of intracellular sodium in the regulation of intracellular calcium and contractility:effects of DPI 201-106 on excitation-contraction coupling in human ventricular myocardium. J Clin Invest 1988;82:1592-1605.
9. Herzig S, Lilienthal E, Mohr K. The positive inotropic drugs DPI 201-106, BDF 9148, and, veratridine increase ouabain toxicity and 3H-ouabain binding in guinea pig heart. J Cardiovasc Pharmacol 1991;18:182-189.
10. Herzig S, Lüllmann H, Mohr K, Schmitz R. Interpretation of $^3$H-ouabain binding in guinea pig ventricular myocardium in relation to sodium pump activity. J Physiol 1988;396:105-120.
11. Honerjäger P, Dugas M, Wang G. Effects of a new indol-2-carbonitrile, BDF 9148, and its two enantiomers on sodium current in rat heart cells (abstr.). Naunyn Schmiedeberg's Arch Pharmacol 1990;341:(Suppl):R51.
12. Neumann J, Scholz H, Döring V, Schmitz W, v. Meyerinck L, Kalmar P. Increase in myocardial G$_i$-proteins in heart failure. Lancet 1988;II:936-7.
13. Reinecke H, Studer R, Philipson KD, Bilger J, Eschenhagen T, Böhm M, Just H-J, Holtz J, Drexler H. Myocardial gene expression of Na$^+$/Ca$^{2+}$-exchanger and sarcoplasmic reticulum Ca$^{2+}$-ATPase in human heart failure. Circulation 1992;86:I-860.
14. Schmitz W, von der Leyen H, Meyer W, Neumann J, Scholz H. Phosphodiesterase inhibition and positive inotropic effects. J Cardiovasc Pharmacol 1989;14:(Suppl3):S11-4.
15. Scholtysik G. Cardiac Na$^+$ channel activation as a positive inotropic principle. J Cardiovasc Pharmacol 1989;14 (Suppl 3):S24-29.
16. Schwinger RHG, Böhm M, Erdmann E. Effectiveness of cardiac glycosides in human myocardium with and without ''downregulated'' ß-adrenoceptors. J Cardiovasc Pharmacol 1990;15:692-697.
17. Schwinger RHG, Böhm M, Erdmann E. Evidence against spare or uncoupled ß-adrenoceptors in the human heart. Am Heart J 1990;119:889-904.
18. Schwinger RHG, Böhm M, Mittmann C, La Rosée, Erdmann E. Evidence for a sustained effectiveness of sodium-channel activators in failing human myocardium. J Mol Cell Cardiol 1991;23:461-471.
19. Sheu SS, Fozzard HA. Transmembrane Na$^+$ and Ca$^{++}$ electrochemical gradients in cardiac muscle and theire relationship to force development. J Gen Physiol 1982;80:325-351.
20. Smith TW. The fundamental mechanism of inotropic action of digitalis. Therapie 1989;44:431-435.
21. Yajima M, Hotta Y, Takeya K, Sakakibara J. Alternatively potentiative interrelations on theire positive inotropic effects between cardiac glycoside and asebotoxin III. Jpn J Pharmacol 1984;36(Suppl 277P).

# Autoimmune gastritis and H+/K+-ATPase

S.Mårdh, Y.-H.Song and J.-Y.Ma

Department of Cell Biology, Faculty of Health Sciences, S-581 85 Linköping, Sweden

## Introduction

Autoimmune gastritis, which eventually may develop into pernicious anemia (PA), is an organ-specific autoimmune disease characterized by chronic atrophic corpus gastritis and circulating parietal cell autoantibodies (6,13,21). By means of immunofluorescence and complement fixation techniques the antibodies were shown to react with a microsomal component (11,22) and the antigen was further localized to the microvilli of the parietal cell (10). By using a Western blotting technique and a preparation enriched in the H+/K+-ATPase (EC3.6.1.36), the acid pump of the stomach, this enzyme was suggested to be the autoantigen (12). Later the H+/K+-ATPase was shown to comprise two subunits, the $\alpha$- and ß-subunits, which migrate as 94 kDa and 60-90 kDa components on sodium dodecyl sulphate polyacrylamide gel electrophoresis (SDS-PAGE) (9,14,19,23). Both the $\alpha$- and ß-subunits of the H+/K+-ATPase have been identified as autoantigens by immunoblotting and immunoprecipitation using parietal cell autoantibodies (7,8,12). The autoantibodies were reported to bind to the cytoplasmic side of the H+/K+-ATPase (18) and the activity of the H+/K+-ATPase was inhibited by the autoantibodies (2). Removal of the N-linked oligosaccharides from the ß-subunit of pig H+/K+-ATPase eliminated autoantibody binding (8). The intact disulfide bonds in the ß-subunit are essential for gastric H+/K+-ATPase activity (4) and also for autoantibody binding (8). The cDNA of both the $\alpha$- and ß-subunit of the human H+/K+-ATPase were cloned recently (15,16) which made possible the characterization of the antibody epitopes using recombinant DNA technology. The human H+/K+-ATPase ß-subunit comprises a 33 kDa core protein with seven possible N-linked glycosylation sites (15). The sources of the H+/K+-ATPase used as antigen in previous investigations were membrane preparations derived mainly from pig, dog, rat and rabbit. However, by using recombinant DNA technology, we have demonstrated that the *human* H+/K+-ATPase $\alpha$- and ß-subunits are indeed major autoantigens and also that the major epitope of the $\alpha$-subunit is located at, or near the active site of the enzyme (amino acid residues 360-526).

## Patients

Forty-two sera from fasting patients (27 females and 15 males, aged between 34 and 83 years, mean age 66) with atrophic corpus gastritis with PA, examined by gastroendoscopy and biopsy, were selected for investigation. The mean duration of PA was 9.3 (range 0-36) years. The criteria for the diagnosis of PA have been described previously (1). All sera were positive by ELISA for pig H+/K+-ATPase holoenzyme containing both $\alpha$- and ß-subunits. Sera from healthy blood donors served as controls.

**Purification of recombinant human H+/K+-ATPase α-subunit fragments**

Recombinant E. coli producing a fusion protein comprising glutathion S-transferase and H+/K+-ATPase α-subunit fragments were lysed in 1 % Triton X-100 (v/v) in a sonication bath (0 ℃) for 2 x 15 min and then centrifuged at 12.000 x g. The pellet was dissolved in 10 volumes of 8 M urea by sonication (0 ℃) 2 x 0.5 min. The urea was removed by chromatography on Sephadex G-25 followed by dialysis against 150 mM NaCl in 0.01 mM Tris-Cl buffer, pH 8.0. The protein sample was then mixed with glutathione agarose beads and the fusion proteins were purified by affinity chromatography; for the enzyme-linked immunosorbent assay the H+/K+-ATPase α-subunit fragment (Ba) was recovered from the affinity column after thrombin treatment.

**Immunoblotting**

Bacterial extract was solubilized in SDS gel loading buffer comprising 50 mM Tris-HCl, pH 6.8, 100 mM dithiothreitol, 2% SDS, 0.1% bromophenol blue and 10% glycerol. The samples were reduced, alkylated and heated for 3 min at 95 ℃ before loading. SDS-PAGE was carried out by means of a Mini-PROTEAN II slab cell (BioRad, Richmond, CA, USA). After electrophoresis, the proteins were blotted onto nitrocellulose paper (0.2 μm; Schleicher & Schull, Dassel, Germany) in a mini trans-blot electrophoretic transfer cell. After incubation with 1% (w/v) ovalbumin and normal goat serum, the nitrocellulose paper was incubated with human sera diluted 1/20 in PBS (phosphate buffed saline) for 1 h. Detection of IgG binding was carried out using biotinylated goat anti-human IgG, streptavidin and a biotinylated peroxidase system (18).

**Cells and Viruses**

The insect cells, *spodopters frugiperda* 9 (Sf9) cells and the baculovirus, *autographa californica* nuclear polyhedrosis viruses (AcNMPV) were supplied in the MAXBAC™ baculovirus expression system kit (Invitrogen Corp., San Diego, USA). The Sf9 cells were cultured as monolayers in a 25-cm$^2$ Falcon® flask in culture medium (Grace's medium + supplements +10% fetal calf serum) and subcultured 3 times a week. The wild type AcNMPV and recombinant AcNMPV containing human H+/K+-ATPase ß-subunit cDNA were propagated in the Sf9 cells using standard techniques.

**Construction of the recombinant transfer vector pVL1392**

A cDNA fragment of human H+/K+-ATPase ß-subunit (hHKß) was excised from the recombinant phagemid pBluescript II SK+ (15) by digestion with DNA restriction enzymes NotI and EcoRI. The NotI / EcoRI fragment of hHKß was 1089 base pairs (bp) long, which included the 17 bp of the 5' nontranslated leader sequence, the entire open reading frame (873 bp) and the 199 bp of the 3' nontranslated sequence. This NotI / EcoRI fragment of hHKß was purified by electrophoresis on a 1% agarose gel and then inserted into NotI / EcoRI sites of transfer vector pVL1392. After partial sequence analysis for identification, the recombinant transfer plasmid pVL1392 containing hHKß in correct transcriptional orientation was isolated and designated pVL1392-hHKß.

## Co-transfection of Sf9 cells with wild type AcNMPV and pVL1392-hHKß

In order to obtain recombinant AcNMPV, wild type AcNMPV and pVL1392-hHKß were co-transfected into Sf9 cells. Two million Sf9 cells were seeded into a 25-cm$^2$ Falcon® flask and allowed to attach at 27°C for 1 h. The medium was replaced with 0.75 ml of fresh medium and a mixture of 1 µg of wild type AcNMPV DNA and 1.5 µg of pVL1392-hHKß DNA in 0.75 ml of transfer buffer (25 mM Hepes buffer, pH 7.1, 140 mM NaCl, 125 mM CaCl$_2$) was added. The cells were then incubated at 27°C for 4 h. The viral suspension was removed, 5 ml of fresh medium was added and the cells were continuously incubated at 27°C for 6 days.

## Purification of recombinant baculovirus by dot blot hybridization

Purification of recombinant baculovirus was performed by the methods of Pen (20) and the method described in the manual of the MAXBAC™ baculovirus expression system kit. Extracellular viruses were harvested and the titer was determined after 6 days incubation. Ten thousand Sf9 cells were seeded into each well of a 96-well microtiter plate and infected with 10 pfu (plaque form unit) of extracellular viruses in a total volume of 100 µl. After incubation at 27°C for 6 days, the cells were lysed in 200 µl of 0.5 M NaOH, neutralized by the addition of 40 µl of 5 M ammonium acetate, blotted onto a nitrocellulose membrane using a Bio-Dot® Microfiltration Apparatus (BIO-RAD, Richmond, CA, USA) and hybridized with the hHKß probe. The hybridization was performed according to the manufacturer's manual (ECL random prime labelling and detection system, Amersham, U.K.). The culture media from cells which were positive after hybridization were used for a second round of screening, where the Sf9 cells were infected with extracellular viruses at 1 pfu / well. The culture media from cells which were positive after the second hybridization and did not contain occlusion bodies were used for a third round of screening to make certain that pure recombinant baculovirus was obtained. At the third screening, the Sf9 cells were infected with extracellular viruses at 0.5 pfu / well. Pure recombinant baculovirus was obtained after the third round of screening and designated AcNMPV-hHKß.

## Southern blot analysis of AcNMPV-hHKß

Five million Sf9 cells were cultured with viruses, either wild type AcNMPV, or AcNMPV-hHKß, and without viruses in a 25-cm$^2$ Falcon® flask at 27°C for 1 h and then incubated in 5 ml of fresh medium at 27°C for 3 days. The extracellular DNA was pelleted by centrifugation of the medium at 100,000 g for 30 min at 4°C and then resuspended in 270 µl of extraction buffer (0.1 M Tris-HCl buffer, pH 7.5, 0.1 M Na$_2$EDTA, 0.2 M KCl) containing 20 µg of proteinase K. After incubation for 2 h at 50°C, 30 µl of 10% (w/v in H$_2$O) sarkosyl was added (total volume was 300 µl). The incubation was continued for another 2 h at 50°C. The DNA was extracted by phenol /

chloroform (1:1), precipitated in ethanol, digested with NotI / EcoRI and then applied for electrophoresis on a 1% agarose gel. After electrophoresis, the DNA was blotted onto Hybond-N membrane and hybridized with hHKß probe.

## Expression of the human H+/K+-ATPase ß-subunit

A monolayer of Sf9 cells ($3 \times 10^6$ cells / 25-cm$^2$ flask) was infected with purified AcNMPV-hHKß (5-10 pfu / cell) at 27°C for 1-2 h. The viral suspension was removed and the Sf9 cells were further incubated in 5 ml of fresh culture medium at 27°C for 2-3 days. The cells were suspended in the culture medium using a pasteur pipette and harvested by centrifugation at 2000 rpm for 10 min in a low speed centrifuge. The cells were washed in PBS (phosphate-buffer saline) and lysed at 0°C for 5 min in 200 µl of lysis buffer (30 mM Tris-HCl buffer, pH 7.5, 20 mM Mg-acetate, and 2% Nonidet® P 40). After centrifugation, the supernatant (cell extract containing ß-subunit protein) was used for analysis in an enzyme-linked immunosorbent assay (ELISA).

## Tunicamycin-treatment of Sf9 cells

In order to inhibit glycosylation of recombinant human H+/K+-ATPase ß-subunit, the Sf9 cells were treated with tunicamycin (TM). The monolayer of Sf9 cells was infected with AcNMPV-hHKß in a 25-cm$^2$ Falcon® flask and incubated at 27°C for 1-2 h as described above. The incubation was continued at 27°C for 72 h in 5 ml of fresh medium containing 20 µg / ml tunicamycin. The cell extract was prepared as described above and also analyzed by ELISA.

## Enzyme-linked immunosorbent assay

Nunc-Immuno plates were coated with antigen (50 µl / well) at a concentration of 10 µg protein / ml in 50 mM $Na_2CO_3$-$NaHCO_3$ buffer, pH 9.8, and incubated overnight at 4°C. Free binding sites were blocked with PBS containing 0.05% (v/v) Tween 20 (PBS-T) for 1 h at 37°C. After washing the wells three times with PBS-T, 50 µl of sera diluted 1:100 in PBS-T were added. The plates were then incubated at 37°C for 1 h. Antibody binding was detected using a biotinylated goat anti-human IgG, streptavidin and biotinylated alkaline phosphatase system. The substrate, p-nitrophenyl phosphate at 1 mg / ml in 50 mM $Na_2CO_3$-$NaHCO_3$ buffer, pH 9.8, was added to the wells (100 µl / well). The absorbance was measured at 405 nm using a computerized ELISA reader.

## Statistical calculation

Experimental data were evaluated by analysis of variance (ANOVA) using SYSTAT® soft ware (SYSTAT, Inc., Evanston, IL, USA). Quantitative results are given as mean ± SD.

800

# Results

*α-subunit*

The coding region of the H+/K+-ATPase α-subunit cDNA comprises 3105 bp (1035 codons) (Fig. 1). Four fragments were selected for PCR amplification and expression in E. coli. Fragments A (453 bp, 151 codons), Ba (1314 bp, 438 codons), Bb (816 bp, 272 codons) and C (630 bp, 210 codons) encompass 77% of the whole molecule. The PCR-amplified cDNA fragments A, Ba, Bb and C were 0.45, 1.3, 0.8 and 0.63 kb, respectively. Analysis of the recombinant H,K-ATPase fragments by SDS-PAGE and subsequent Coomassie staining revealed four fusion proteins with the approximate molecular weights of 44 kDa, 66 kDa, 50 kDa and 32 kDa.

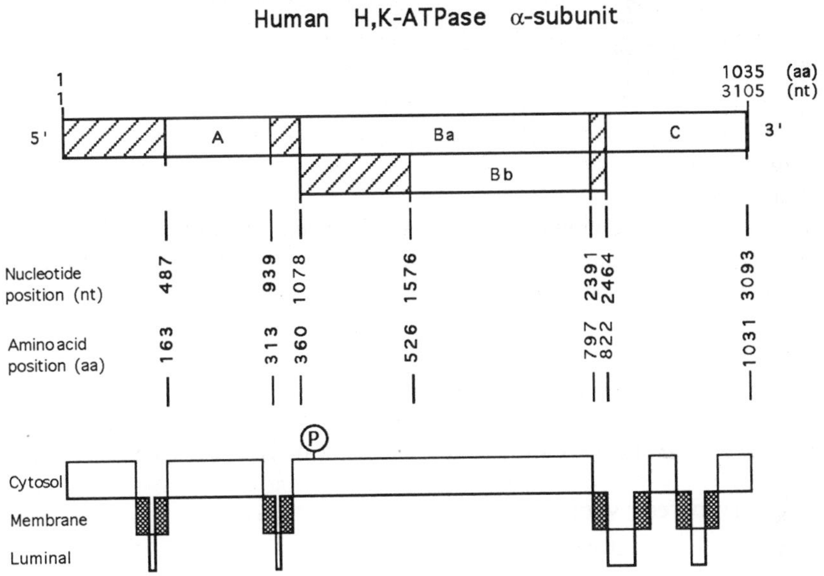

Human  H,K-ATPase  α-subunit

1035  (aa)
3105  (nt)

5'    A    Ba    C    3'

Bb

Nucleotide position (nt): 487  939 1078  1576  2391 2464  3093

Amino acid position (aa): 163  313 360  526  797 822  1031

Ⓟ

Cytosol

Membrane

Luminal

**Figure 1.** Schematic presentation of selected fragments on the human H+/K+-ATPase α-subunit molecule amplified by PCR and expressed in E. coli. *Top*: open boxes represent selected fragments A, Ba, Bb and C. Shaded boxes represent unselected regions. *Bottom*: pattern of transmembrane organization. The active site phosphorylation at residue 387 is indicated by P.

By means of immunoblotting a pool of sera from 10 individual patients with pernicious anaemia was tested for reactivity with the recombinant fusion proteins containing H+/K+-ATPase fragments. The sera bound only to Ba fragments. A pool of sera from 10 individual blood donors was used as control; there was no binding using these controls to any of the fusion proteins. Similar results were obtained by immunoblotting with six individual sera. The antibodies recognised the Ba fragment only. The Ba fragment was purified and further analyzed by enzyme-linked immunosorbent assay (Fig. 2). The mean value of the sera from 42 patients with pernicious anemia and 20 controls were $0.482 \pm 0.252$ (mean $\pm$ SD) and $0.159 \pm 0.083$ (mean $\pm$ SD, $p < 0.001$), respectively. The upper normal limit was 0.325 which was defined as the mean value plus two standard deviations obtained from the control sera. Twentyeight of 42 (67%) patients scored positive.

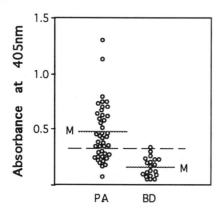

**Figure 2.** Enzyme-linked immunosorbent assay. Binding of antibodies from sera of patients with pernicious anemia (PA) and blood donors (BD) to the purified Ba fragment. The mean values are indicated by M. The broken line indicates the upper normal limit.

*β-subunit*

A recombinant transfer vector was constructed by insertion of the 1089 bp human H+/K+-ATPase ß-subunit cDNA into the transfer vector pVL1392 (Fig. 3).

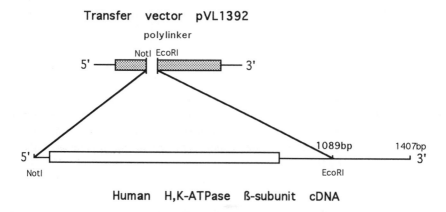

**Figure 3.** Construction of the recombinant transfer vector pVL1392 containing human H+/K+-ATPase β-subunit.

The correct structure was confirmed by digestion of the recombinant pVL1392 (pVL1392-hHKß) with the DNA restriction enzymes NotI and EcoRI and by DNA sequence analysis. The recombinant baculovirus was obtained by co-transfection of Sf9 cells with wild type AcNMPV DNA and pVL1392-hHKß DNA and isolated after three rounds of screening by dot blot hybridization. The DNA fractions prepared from recombinant transfer vector pVL1392-hHKß, AcNMPV-hHKß infected Sf9 cells, wild type AcNMPV infected Sf9 cells and uninfected Sf9 cells were digested with restriction enzymes NotI / EcoRI. The presence of hHKß-DNA in the recombinant baculovirus was confirmed by Southern blot hybridization using hHKß as a probe.

Recombinant human $H^+/K^+$-ATPase ß-subunit, obtained by the baculovirus expression system, was identified as the antigen by ELISA using sera containing autoantibodies against pig $H^+/K^+$-ATPase from patients with PA (Fig. 4). The mean optical density value of the sera from the patients was 0.293, whereas the mean value of the controls was 0.058 ($p < 0.001$). The upper normal limit was 0.104. Thirty-nine out of 42 patients' sera (93%) scored positive and 3 (7%) were negative. The sera from the latter 3 patients were positive for recombinant human $H^+/K^+$-ATPase $\alpha$-subunit when tested as previously described.

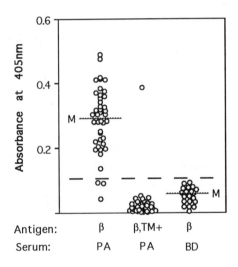

**Figure 4.** Enzyme-linked immunosorbent assay. Cell extracts from Sf9 cells infected with recombinant baculovirus, untreated (β) and treated with tunicamycin (β, TM+) were used as antigen preparations. Sera from patients with pernicious anemia (PA) and 20 blood donors (BD) were assayed. Mean values are indicated by M. The broken line represents the upper normal limit.

The unglycosylated recombinant human $H^+/K^+$-ATPase ß-subunit was expressed in the Sf9 cells in the presence of tunicamycin. Tunicamycin-treatment resulted in the appearance of the core protein of 33 kDa of the ß-subunit after SDS-PAGE which is in accordance with previous results (3,9,15,19). The unglycosylated ß-subunit was used as antigen in ELISA to test the requirement of carbohydrates of human $H^+/K^+$-ATPase ß-subunit on autoantibody binding. The autoantibody binding in the sera of 41 patients (98%) was eliminated (Fig. 4). In one serum sample the absorbance was decreased by 30% only.

## Discussion

Previous studies identified a 65-75 kDa (5,12) and a 92 kDa (12) polypeptide as the major parietal cell autoantigens. However, either a gastric mucosal extract (canine and rodent), or a purified membrane-bound porcine $H^+/K^+$-ATPase was used in those studies. Although the major antigen was suggested to be the $H^+/K^+$-ATPase (12), the autoantibodies may have bound to some other proteins with a similar molecular weight. Using recombinant human $H^+/K^+$-ATPase we now provide a final proof for the $H^+/K^+$-ATPase α-and ß-subunits as major autoantigens in the parietal cell, since 67% of the sera from patients diagnosed pernicious anaemia scored positive against the Ba polypeptide and 93% scored positive against the ß-subunit. All autoantibody epitopes in the pig $H^+/K^+$-ATPase holoenzyme seem to be present also in the human sequences.

In the α-subunit the fragment Bb was part of the Ba fragment. The patient sera, however, bound only to the Ba fragment indicating that the antibody epitope resides in the $NH_2$-terminal part of the Ba fragment (residues 364-526) which contains the active site of the $H^+/K^+$-ATPase (17). The results are in agreement with previous findings that antibodies against parietal cells have a direct inhibitory effect on the $H^+/K^+$-ATPase (2). Furthermore, the antigenic structure was reported localized to the cytosolic side of the $H^+/K^+$-ATPase (18).

Using tunicamycin to inhibit glycosylation of the human $H^+/K^+$-ATPase ß-subunit resulted in elimination of autoantibody binding in 41 of the 42 sera from patients with pernicious anemia. This result is in accordance with previous results (8) indicating that the carbohydrates of the $H^+/K^+$-ATPase ß-subunit are necessary for autoantibody binding. Some epitope(s) of the ß-subunit core protein may also be present in some patients, since the absorbance value of the serum of one patient was decreased by 30% when unglycosylated ß-subunit was used as antigen.

## Future perspectives

Autoimmune gastritis frequently causes insufficient absorption of vitamin $B_{12}$ due to the lack of intrinsic factor production in the gastric mucosa and subsequently pernicious anemia and disturbances from the nervous system develop. A key factor in these processes appears to be insufficient activity of vitamin $B_{12}$ dependent methyl transferase. Chronic atrophic gastritis is a common disease in the elderly as is infection by Helicobacter pylori, a bacterium known to cause peptic ulcer and gastritis. In late stages of atrophy of the gastric mucosa an epithelization by intestinal cell types occurs and Helicobacter pylori disappears. A common feature of the autoimmune gastritis and several other autoimmune diseases, e.g., thyroiditis, insulin dependent diabetes mellitus and sometimes rheumatoid arthritis, is the occurrence of "parietal cell" autoantibodies. Although the major autoantigens appear to be different in these diseases, common B- or

T-cell epitopes and immunological cross-reactivities have been suggested as possible connections.

Future investigations will indicate whether an initial infection by Helicobacter pylori is triggering similar pathogenic mechanisms in the development of the above mentioned diseases. Is an early diagnosis of infection and of the gastritis and an early treatment essential to prevent the development of several autoimmune diseses and even to prevent the degeneration of CNS and mental retardation due to vitamin $B_{12}$ deficiency in the elderly population?

## Acknowledgements

This work was supported by the Swedish Medical Research Council, project 4X-4965, The Swedish Society of Medicine, and AB Astra Hässle.

## References

1.  Borch K, Liedberg G (1984) Prevalence and incidence of pernicious anemia. An evaluation for gastric screening. Scand J Gastroeneterol 19:154-160
2.  Burman P, Mårdh S, Norberg, Karlsson FA (1989) Parietal cell antibodies in pernicious anaemia inhibit H,K-adenosine triphosphatase, the proton pump of t h e stomach. Gastroenterology 96:1434-1438
3.  Callaghan JM, Toh BH, Pettitt JM, Humphris DC, Gleeson PA (1990) Poly-N-acetyllactosamines specific tomato lectin interacts with gastric parietal cells. J Cell Sci 95:563-576
4.  Chow DC, Browning CM, Forte JG (1992) Gastric H+-K+-ATPase activity is inhibited by reduction of disulfide bonds in ß-subunit. Am J Physiol 263:C39-C46
5.  Dow CA, de Aizpurua HJ, Pedersen JS, Ungar B, Toh BH (1985) 65-70 kD protein identified by immunoblotting as the presumptive gastric microsomal autoantigen in pernicious anemia. Clin Exp Immunol 62:732-737
6.  Glass GBJ (1977) Immunology of atrophic gastritis. NY State Med 77:1697-1706
7.  Gleeson PA, Toh BH (1991) Molecular targets in pernicious anaemia. Immunol Today 12:233-238
8.  Goldkorn I, Gleeson PA, Toh BH (1989) Gastric parietal cell antigens of 60-90, 9 2 , and 100-120 kDa associated with autoimmune gastritis and pernicious anaemia. J Biol Chem 264:18768-18774
9.  Hall K, Perez G, Anderson D, Gutierrez C, Munson K, Hersey SJ, Kaplan JH, Sachs G (1990) Location of the carbohydrates present in the H,K-ATPase vesicles isolated from hog gastric mucosa. Biochemistry 29:701-706
10. Hoedemaeker PJ, Ito S (1970) Ultrastuctral location of gastric parietal cell antigen with peroxidase-coupled antibody. Lab Invest 22:184-188
11. Irvine WJ, Davies SH, Delamore IW, Williams AW (1962) Immunological relationship between pernicious anaemia and thyroid disease. Br Med J ii: 454-456
12. Karlsson FA, Burman P, Lööf L, Mårdh S (1988) The major parietal cell antigen in autoimmune gastritis with pernicious anaemia is the acid-producing H,K-ATPase of the stomach. J Clin Invest 81:475-479

805

13. Kaye MD, Whorwell PJ, Wright R (1983) Gastric mucosal subpopulations in pernicious anaemia and normal stomach. Clin Immunol Immunopathol 28:431-440

14. Ljungström M, Norberg L, Olaisson H, Wernstedt C, Vega FV, Arvidson G, Mårdh S (1984) Characterization of proton-transporting membranes from resting pig gastric mucosa. Biochim Biophys Acta 769:209-219

15. Ma J-Y, Song Y-H, Sjöstrand SE, Rask L, Mårdh S (1990) cDNA cloning of the ß-subunit of the human gastric H,K-ATPase. Biochem Biophys ResCommun 180:39-45

16. Maeda M, Osshiman KI, Tamura S, Futai M (1990) Human gastric (H + K)-ATPase gene. J Biol Chem 265:9027-9032

17. Mårdh S, Cabero JL, Song YH (1989) H,K-ATPase, the proton pump of the stomach. Biochem (Life Sci. Adv.) 8:55-59.

18. Mårdh S, Song Y-H (1989) Characterization of antigenic structures in autoimmune atrophic gastritis with pernicious anaemia, the parietal cell H,K-ATPase and the chief cell pepsinogen are the two major antigens. Acta Physiol Scand 136:581-587

19. Okamoto CT, Karpilow JM, Smolka A, Forte JG (1990) Isolation and characterization of gastric microsomal glycoproteins. Evidence for a glycosylated ß-subunit of the H,K-ATPase. Biochim Biophys Acta 1037:360-372

20. Pen P, Welling-Wester S (1989) An efficient procedure for the isolation of recombinant baculovirus. Nucleic Acids Res 17:451

21. Strickland RG, Mackay IR (1973) A reappraisal of the nature and significance of chronic atrophic gastritis. Am J Dig Dis 18:426-440

22. Taylor KB, Roitt IM, Doniach D, Couchman KG, Shapland C (1962) Autoimmune phenomena in pernicious anaemia: Gastric antibodies. Br Med J ii:1347-1354

23. Toh BH, Gleeson PA, Simpson RJ, Moritz RL, Callaghan JM, Goldkorn I, Jones C M, Martinelli TM, Mu FT, Humphris DC, Pettitt JM, Mori Y, Masuda T, Sobieszczuk P, Weinstock J, Mantamadiotis T, Baldwin GS (1990) The 60- to 90-kDa parietal cell autoantigen associated with autoimmune gastritis is a ß subunit of the gastric H,K-ATPase (proton pump). Proc Natl Acad Sci USA 87:6418-6422

# Na$^+$ Modulation of K$^+$ Transport by Na$^+$/K$^+$-Pump in Red Cells of Salt-sensitive and Salt-resistant Dahl Rats

Mitzy Canessa and José R. Romero

Endocrine-Hypertension Division and Department of Medicine Brigham and Women's Hospital and Harvard Medical School, 221 Longwood Avenue, Boston, MA 02115 USA

## Introduction

Most of the kinetics studies of the Na$^+$/K$^+$ pump fluxes have been performed in human red blood cells (RBC) and one tends to consider it as the "normal" pump. Few kinetics studies of pump fluxes have been performed in rat RBCs which have a shorter half-life than human RBCs, a higher pump activity and a different amino acid composition in the segments of the $\alpha$1 isoform involved in ATP hydrolysis and ouabain binding.

We have studied the kinetics of unidirectional Na$^+$ and K$^+$ fluxes of several transport modes in RBCs of two Dahl (D) rat strains selected for susceptibility or resistance to the hypertensive effects of high salt diet (1,2). These studies were motivated by the report of Herrera et al (4) that alleles of the gene encoding the $\alpha$1 subunit of the Na$^+$/K$^+$ pump isoform were identified in Dahl salt-resistant (DR) and Dahl salt-sensitive (DS) rats. Sequencing of the cDNA for these two alleles identified a leucine substitution of glutamine at position 276 for the allele of the DS rat. In microinjection expression experiments performed in Na-loaded *Xenopus laevis* oocytes, this Leu/Gln$_{276}$ substitution resulted in decreased $^{86}$Rb influx. Because rat red cells only express the $\alpha$1 isoform of the Na$^+$/K$^+$ pump, the design of our experiments explored the functional properties of Na$^+$ and K$^+$ transport in both strains.

## Methods

Fifty male DS and fifty male DR rats (Harlan Sprague Dawley Inc. Co., Madison, WI from the in-bred strain of J. Rapp) were studied at 10-11 weeks of age. All rats were fed a low-salt diet (0.05% NaCl, #82049, Teklad, Madison, WI) with tap water to drink *ad libitum*, for 1 week before being sacrificed.

RBC Na$^+$ content was modified by means of a Na-salicylate loading procedure optimized for the DS and DR RBCs as described (2). Initial rates (0-6 min) of unidirectional K$^+$ influx and efflux were measured with use of $^{86}$Rb, and of Na$^+$ efflux and influx with use of $^{22}$Na (2). All flux medium contained (mM): 10 Tris-MOPS (pH 7.4 at 37°C), 1 MgCl$_2$, 10 glucose, and NaCl and KCl in variable concentrations. In addition, net fluxes were determined incubating the cells for one hour w/wo ouabain. Na$^+$ and K$^+$ fluxes by the Na$^+$/K$^+$ pump were defined by the ouabain-sensitive (OS) (2.5 mM ouabain) component of the flux. Flux units (FU) were measured in mmol/l cell x h.

## Results

DS and DR erythrocytes have similar intracellular $Na^+$ content (4.6 ± 0.2 and 4.8 ± 0.1 respectively, Mean ± SE, n=36 rats for each strain). However, the intracellular $K^+$ content was slightly but significantly lower in DS than in DR RBCs (104 ± 1.5 vs. 108 ± 1.2 mmol/l cell, n=29 rats for each strain, p < 0.05). In addition, RBCs of both rat strains have similar ATP, ADP, and Pi content as well as similar $K_d$ for ouabain inhibition of $K^+$ influx (54 $\mu$M).

OS $Na^+$ efflux in fresh RBCs as well as $V_{max}$ (30 FU) and $K_m$ for $Na_i$ (18 mmol/l cell) were similar in both rat strains (2). The high pump activity of rat RBCs cannot be accounted by for the high reticulocyte count of rat blood because they were discarded by suction with the buffy coat prior to the experiments. In contrast, OS $K^+$ influx into fresh cells was significantly lower in DS (3.32 ± 0.18 FU, n=8 rats) than in DR (5.09 ± 0.3 FU, n=7 rats) cells. The similarities in OS $Na^+$ efflux at any intracellular $Na^+$ content, and the different OS $K^+$ influx conferred higher $Na^+:K^+$ coupling ratios for DS than DR pumps.

A kinetic analysis (3) of the $K_o$ dependence of OS $K^+$ influx at $Na_o$ 140 mM was performed using the Enzfitter software program to fit the data to the Garay equation. We found a significantly lower $V_{max}$ and higher $K_m$ for $K_o$ stimulation of $K^+$ influx in DS than in DR pumps (Table 1). When external $Na^+$ was replaced by 75 mM $MgCl_2$ and 85 mM sucrose, OS $K^+$ influx had similar kinetic parameters in both strains (Table 1). Thus, $Na_o$ stimulated $K^+$ influx 3.9-fold in DR but only 1.8-fold in DS cells. Furthermore, the $K_m$ for $K_o$ increased 9-fold in DR cells and only 2.7-fold in DS cells. These results indicate that the $Na^+$-$K^+$ pump in DS and DR cells exhibits marked differences in the modulation of $K^+$ influx by external $Na^+$. Because intracellular $Na^+$ also stimulates OS $K^+$ influx, experiments were conducted to define the $Na_i$-activation of $K^+$ influx in both strains. DS cells showed significantly lower $V_{max}$ of ouabain-sensitive $K^+$ influx than in DR strain but similar $K_m$ for $Na_i$ activation (Table 1).

Fresh DR cells exhibited some OS $K^+$ efflux, as expected for cells with a low $Na^+$ content (4.5 mmol/l cell). In contrast, OS $K^+$ efflux in DS cells was 4-fold higher than in DR cells (Table 1). Kinetic analysis (3) of the inhibition of OS $K^+$ efflux by varying $Na_i$ indicated that the $K_{50}$ for $Na_i$ inhibition of $K^+$ efflux was significantly higher in DS than in DR cells (Table 1). The data demonstrate that the $Na^+:K^+$ coupling ratio in the Dahl strains appear quite different whether unidirectional or net ouabain-sensitive fluxes are considered.

Our findings indicate that Dahl RBCs differ from human RBCs in several ways: $K_{50}$ for ouabain inhibition, the effects of $K_o$ on $Na^+$ and $K^+$ influx and in the $K_m$ for $Na_i$ to activate $K^+$ influx. Quantitative analysis of the intra- and inter- strain variations of unidirectional and net pump fluxes provided support for a higher $Na^+:K^+$ coupling ratio in DS than DR strains.

808

TABLE 1

## KINETIC PARAMETERS OF OUABAIN SENSITIVE $K^+$

## FLUXES IN RED CELLS OF DAHL RATS

| | DR | DS | DS vs DR |
|---|---|---|---|
| **A. $K^+$ Influx, activation by external $K^+$** | | | |
| $Na_o = 140$ mM | | | |
| $V_{max}$ for $K_o$ | 5.70 FU | 2.87 FU | reduced |
| $K_{50}$ for $K_o$ | 2.31 mM | 0.74 mM | reduced |
| $Na_o = 0$ mM | | | |
| $V_{max}$ for $K_o$ | 1.41 FU | 1.65 FU | same |
| $K_{50}$ for $K_o$ | 0.21 mM | 0.23 mM | same |
| **B. $K^+$ Influx, activation by intracellular $Na^+$** | | | |
| $V_{max}$ for $Na_i$ | 14.5 FU | 9.3 FU | reduced |
| $K_{50}$ for $Na_i$ (*) | 8.7 | 7.9 | same |
| **C. $K^+$ Efflux, inhibition by intracellular $Na^+$** | | | |
| $K_{50}$ for $Na_i$ (*) | 3.09 | 9.66 | increased |

FU = mmol/l cell x h. (*) = mmol/l cell

Net $K^+$ uptake, inhibitable by ouabain, was significantly lower in DS than in DR cells but it compensated a lower ouabain-resistant cell $K^+$ lost. Furthermore $Na^+$ modulation of $K^+$ influx and efflux was found altered at both sides of the membrane in DS cells. Our findings suggest that DS pumps may have abnormalities involving $Na^+$ regulation of the $K^+$ de-occlusion step.

**Discussion**

Studies in Dahl rats as in other genetic strains of hypertension face the difficulties of the strain variability. For instance, several sub-strains of the SHR have been selected such as the salt-sensitive Taconic strain, the salt-resistant Charles River SHR, the stroke-prone (Japan) and the obese SHR strains (Cleveland). Highly inbred Dahl rat strains were selected by Rapp (9), but different colonies have been used in studies in USA and Europe without adequate control of the genetic make up. Our RBC transport studies were performed in parallel with the genetic studies of Herrera et al (5) and with another study of our laboratory which tested the salt-sensitivity of the blood pressure in the Dahl strains provided by our commercial source (8).

In contrast, studies carried out by Zicha and Duhm (13) in Munich did not find differences in RBC K$^+$ transport between DS and DR rats. These authors studied 5-week old rats from the colony established in Prague, Czechoslovakia maintained for in 0.06% NaCl diet after weaning (5 weeks). Measurements of net $^{85}$Rb influx at Na$_i$ of 2, 4 and 6 mmol/l cell in the absence of external Mg$^{2+}$ showed no differences in the calculated V$_{max}$ and K$_m$ for ouabain-sensitive $^{85}$Rb influx between the DS and DR strains. The different results of this study might be related to (i) strains differences; (ii) to the pre-treatment of RBCs for 2 hours with a 2.5 mM phosphate media prior to the influx measurements; (iii) or to the lack of external Mg$^{2+}$ in the flux media. In addition, McCormick et al (7) reported no differences in Rb$^+$ influx into RBCs of Dahl rats (28-35 days) obtained from the Brookhaven National Laboratories (Upton, NY) which are not inbred as those selected by J. Rapp (9). Notably, this study reported abnormally high RBC Na$^+$ content (10-15mmol/l cell) in both strains.

The higher coupling ratio between K$^+$ and Na$^+$ transport in DS Na$^+$/K$^+$-pumps appears to be caused by an abnormal Na$^+$ modulation of K$^+$ transport. The dissociation of the intracellular Na$^+$ effects on Na$^+$ and K$^+$ efflux as well as the lack of effect of external Na$^+$ on K$^+$ influx indicate that K$^+$ and Na$^+$ transport are modulated by a different set of internal Na$^+$ sites. This abnormality in DS pumps might be related to the Na$^+$ regulation of the low-affinity ATP binding site that controls K$^+$ deocclusion steps (4). Amino acid sequence analysis of the $\alpha$1 Na$^+$/K$^+$-pump has shown a great deal of variability in the loops involved in ATP binding sites (6).

In addition, the primary amino acid sequence of DS and DR $\alpha$1 Na$^+$-K$^+$ pumps reported by Herrera et al (4) has been questioned by a study of the genomic DNA using the polymerase chain reaction (PCR) technique (11,12). In addition, variations in the salt-sensitivity of DS rats provided by the commercial source Harlan Sprague Dawley Inc. Co., Madison, WI used in that study has been recently identified as a likely source of disagreement between these two molecular biology studies. Herrera et al (5,10) have claimed and confirmed the T/A$^{1079}$ transversion in DS rat genomic DNA and kidney RNA using Taq polymerase error-independent amplification-based techniques. These test consisted of polymerase allele-specific amplification and ligase chain reaction (LCR) analysis that detected mutant T$^{1079}$ in DS genomic DNA and not in DR rat genomic DNA. A third test, thermostable reverse transcriptase-PCR of allele-specific mRNA (allele-specific RT$^{th}$-PCR, detected wild type A$^{1079}$ in DR rat kidney RNA and not in DS rat kidney RNA. Additionally, these investigators (5) have observed a consistent Taq polymerase error that selectively substituted dA at T$^{1079}$ in the PCR amplification and cycle-sequencing of reconfirmed mutant $\alpha$1 Na$^+$/K$^+$ ATPase M13 sub- clones. This Taq polymerase error results in the conversion of the mutant sequence back to the wild type sequence. This identifies a site- and nucleotide-specific Taq polymerase mis-incorporation suggesting that a structural basis

810

might underlie a predisposition to non-random Taq polymerase errors.

In summary, the differences in the kinetic behavior of DS and DR $\alpha 1$ $Na^+/K^+$-pumps might be related to the rat strain selection, alterations in its primary structure or/and by post-translational modifications during maturation of the RBCs in these Dahl strains. From a structural point of view, the $Leu/Gln_{276}$ substitution present in DS pumps reported by Herrera et al (4) may mark an allosteric site involved in the $Na^+$ modulation of $K^+$ transport route because $K^+$ fluxes but not $Na^+$ efflux was found altered in DS cells. The mutant amino acid in DS pumps is in the region identified as a $Na^+$ sensor but it might not be involved in $Na^+$ transport.

## References

1.  Canessa M, Romero JR, Herrera VLM and Ruiz-Opazo N (1990). A mutant $\alpha 1$ $Na^+$ pump in the Dahl salt-sensitive rat has altered allosteric interaction of $Na^+$ ion with $K^+$ transport. J Gen Physiol 96:70a.
2.  Canessa M, Romero JR, Ruiz-Opazo N, Herrera VLM (1993) The $\alpha 1$ $Na^+$-$K^+$ pump of the Dahl salt-sensitive rat exhibits altered $Na^+$ modulation of $K^+$ transport in red blood cells. J Memb Biol 134:107-122.
3.  Glynn IM and Richards DE (1982) Occlusion of rubidium ions by sodium-potassium: its implication for the mechanism of potassium transport. J Physiol (Lond) 330:17-43.
4.  Herrera VLM and Ruiz-Opazo N (1990) Alteration of $\alpha 1$ $Na^+$,$K^+$-ATPase $^{86}Rb^+$ influx by a single amino acid substitution. Science 249: 1023-26
5.  Herrera VLM, Ruiz-Opazo N (1993) Personal communication.
6.  Horisberger JD, Lemas V, Kraehenbuhl JP and Rossier BC (1991) Structure-Function relationship of Na,K-ATPase. Ann Rev Physiol 53:565-84.
7.  McCormick CP, Hennesy JF, Rauch AL, Buckalew VM Jr (1989) Erythrocyte sodium concentration and $Rb^+$ uptake in wealing Dahl rats. Am. J. Hypert 2:604-609.
8.  Pontremoli R, Spalvins S, Menachery A, Torielli L, and Canessa M (1992) Red cell sodium-proton exchange is increased in Dahl salt-sensitive hypertensive rats. Kidney Intern 42:1355-1362.
9.  Rapp JP. Characteristics of Dahl salt-susceptible and salt-resistant rats in Handbook of Hypertension: Experimental and genetic Models of Hypertension (vol 4), edited by De Jong W, Berlin Elsevier Science Publisher, 1984,pp 286-295.
10. Ruiz-Opazo N, Herrera VLM (1992) Analysis of the $Na^+$ pump genes. Hypertension 19: 495.
11. Simonet L, St. Lezin E, Kurtz TW (1991) Sequence analysis of the $\alpha 1$ $Na^+$,$K^+$-ATPase gene in the Dahl salt-sensitive rat. Hypertension 18:689-693.
12. Simonet L, St. Lezin E, Kurtz TW (1992) Analysis of the $Na^+$ pump genes. Hypertension 19: 496.
13. Zicha, J, Duhm J (1990) Kinetics of $Na^+$ and $K^+$ transport in red blood cells of Dahl rats: effect of age and salt. Hypertension 15: 612-627.

# Biochemical characterization of the $Na^+/K^+$-ATPase isoforms in human heart

S. Decollogne[*], J.J. Mercadier[$], Ph. Lechat[&], P. Allen[§], L.G. Lelièvre[*]

[*]Laboratoire de Pharmacologie des Transports Ioniques Membranaires, Université Paris 7, 2, Place Jussieu, 75251 Paris Cedex 05; [$]Dept. de Recherche Médicale, CNRS URA 1159, Hopital M. Lannelongue, Plessis-Robinson, 92350 France; [&]Service de Pharmacologie, Hopital La Pitié Salpetrière, 75013 Paris; [§]Dept. of Anesthesia, Harvard Medical School, Boston, (Mass) 02115, USA.

## Introduction

Adult human heart expresses mRNAs for the three isoforms ($\alpha_1$, $\alpha_2$, $\alpha_3$), products of different genes (8). However, different questions arise: are the three forms expressed as enzymatic activities? are they distinguishable by their contribution as measured by [3H]ouabain-binding levels and by differences in apparent molecular weights? Furthermore, what types of changes can be found in human failing hearts?

So far, two types of $Na^+/K^+$-ATPases expressed as specific ouabain-binding sites seem to be expressed in human cardiac membranes. With [3H]ouabain, De Pover and Godfraind (4) and Shamraj et al. (7) reveal two classes of specific sites. This is evidenced by dissociation kinetics showing two dissociation rate constants ($k_{off}$ values: 0.058 min$^{-1}$ and 0.0092 min$^{-1}$) and by the non-linearity of the Scatchard plot ($K_D$ values: 4.8 nM and 17 nM). By [3H]dihydrodigitoxin-binding assays, Brown and Erdmann (1) reveal two populations of digitalis receptors ($K_{D1}$ = 18 nM, $B_{max1}$ 15 pmoles/mg and $K_{D2}$ = 2 $\mu M$, $B_{max2}$: 150 pmoles/mg). However, [3H]ouabain-binding measurements reveal a unique high affinity site ($K_D$ = 4.5 nm; $B_{max}$ 15 moles/mg).

From an electrophysiological point of view, Rasmussen et al. (6) measure the sensitivity of the $Na^+$ pump activity to acetylstrophanthidin in human atrial tissues. These authors found two inhibitory processes: one of high affinity (half-saturation: 3.9 nM) and low capacity (30-33%) and one of low affinity (half-saturation: 480-610 nM) and high capacity (66-70% of the total inhibition capacity).

Our aim is to compare the sensitivities of the $Na^+/K^+$-ATPase activity to ouabain and to identify the different apparent molecular weights of the $\alpha$-subunits in microsomal vesicles. These vesicles are purified from frozen biopsies ($\pm$ isopentan) of normal auricles (NA), normal (N) and failing (F) (stade IV) human left ventricles. Myocardial tissues have been obtained from adults entering the transplantation programme as organ donors or heart recipients.

## Sensitivities of the $Na^+/K^+$-ATPase activity to ouabain

The $Na^+/K^+$-ATPase activity was measured, as previously described (5), in membranes isolated according to two procedures.

Frozen NA biopsies (200-240 mg wet weights) are homogenized without any previous thawing, for 15 sec with a Polytron PT10, in 10 volumes (w/v) of an isotonic and isoosmotic buffer containing tetrasodium pyrophosphate, sucrose, KCl, $Ca^{2+}$ chelators and imidazole-HCl, pH 7.4. The enzymatic assays are performed with membrane vesicles opened by SDS (Sodium Dodecyl Sulphate) treatments (0.2 mg SDS/mg proteins for 30 min at 20°C, 1 mg protein/ml).

In spite of a low and highly variable specific activity of the $Na^+/K^+$-ATPase [from 7 to 25 $\mu$mol inorganous phosphorus liberated / h x mg prot (unit)], three different sensitivities to ouabain are found in NA biopsies (Table 1).

However, this method largely inactivates the enzyme in left ventricles and another isolation procedure (2) has been used for N and F biopsies. The vesicles are permeabilized with Alamethicin (8 $\mu$g/$\mu$g proteins). This procedure yields a membrane bound $Na^+/K^+$-ATPase with constant specific activities of $10\pm1$ units for N [n > 60 from two biopsies] and $9.6\pm3.4$ units for F [n > 100 with three membrane preparations]. There is no change in specific activities between N and F hearts. Note that in F, there is a much larger scattering of the experimental values than in membranes from N.

In these membranes, log-logit plots of the dose-response curves of $Na^+/K^+$-ATPase to ouabain (Figure 1) reveal two enzyme forms of high and low affinities for ouabain (Table 1). In both N and F, a part of the enzymatic activities is of high affinity for ouabain ($IC_{50}$: 10 and 16 nM, with respective proportional contributions of 33 and 50% of the total activity). The second part of the enzymatic activities is of low affinity for ouabain and differs between N and F (Table 1 - Figure 1).

It is worthy of note that the experimental $IC_{50}$ values are obtained when a stable steady-state level between ouabain and $Na^+/K^+$-ATPase has been reached. Indeed the rate of ATP hydrolysis is linear *versus* time (from 10 to 30 min) indicating that a stabilized equilibrium is achieved at 37°C within the first thirty minutes. Note that the inhibition levels remain unchanged during at least 50 minutes.

**Table 1.** Sensitivities of the $Na^+/K^+$-ATPasse activity to ouabain in human normal auricles (NA) and in human normal (N) and failing (N) left ventricles.

|  | N A | N | F |
|---|---|---|---|
| $IC_{50}$* and (%contribution) | 0.05 nM (35%) 1.7 nM (35%) |  |  |
|  |  | 10 nM (33%) | 16 nM (50%) |
|  |  | 125 nM (66%) |  |
|  | 1.5 $\mu$M (30%) |  | 2 $\mu$M (50%) |

\*$IC_{50}$: concentration of drug causing 50% inhibition.

%Inhibition

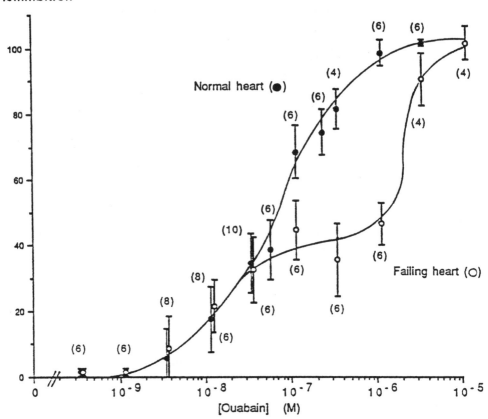

**Figure 1.** Dose-response of Na$^+$/K$^+$-ATPase activity versus ouabain concentrations in microsomal vesicles isolated from normal (●) and failing (o) human hearts. (): number of experiments.

### SDS/Page and immunological detection of the $\alpha$-subunits in N and F.

The samples are resolved on a 6-12% linear acrylamide gradient as described by Sweadner (9). We used specific monoclonal antibodies against rat Na$^+$/K$^+$-ATPase isoforms ($\alpha_1$, $\alpha_2$, $\alpha_3$) from U.B.I. (Upstate Biotechnology, Inc.) and specific antisera against the lamb kidney enzymes raised in rabbits and against fusion protein raised in rats.

With the two lots of polyclonal antibodies, two different molecular forms of the catalytic subunits are resolved:       in N: $105 \pm 3$ kD and $101 \pm 2$ kD (n=4).

in F: $107 \pm 1$ kD and $102 \pm 1$ kD (n=4).

In both N and F, monoclonal anti-$\alpha_2$ and anti-$\alpha_3$ react with a single band [$106 \pm 1$ kD and $107 \pm 1$ kD (n=2), respectively]. The reaction [96 kD] with the monoclonal anti-$\alpha_1$ is only detectable in N. As a general rule, with the monoclonal antibodies, the bands in F are fainter than those found in N.

Two (or three?) functional isoforms of different molecular weights can be revealed in normal and failing human left ventricles. The two apparent affinities for ouabain of the functional $Na^+/K^+$-ATPase as found in N and F are consistent with those reported by Brown and Erdmann (1), and De Pover and Godfraind (4). However, all the apparent affinities reported are low as compared to the non toxic levels of digitalis (in the range of 1 to 3-4 nM), except the very high affinity form found in normal auricles which are not involved in heart functions. A forthcoming approach is to correlate the decreased sensitivity to cardiac glycosides revealed here in human heart failure to the pathological alterations of $Na^+$, $K^+$ (3) and $Ca^{2+}$ homeostasis in these hearts.

*Acknowledgements*
We thank Dr. J. Ball (Dept. of Pharmacology and Cell Biophysics, University of Cincinnati) and Dr. K. Geering (Institut de Pharmacologie et de Toxicologie, Université de Lausanne) for their generous gifts of polyclonal antibodies. This work was supported by grants from la Direction des Recherches et de l'Encadrement des Doctorants (DRED): Action Spécifique: 3201 R10, 1990 - UR: 934-3409 R10-13, 1992,93 (LGL), l'Université Paris 7: 3201 C - 3201 R13, 1993 (LGL), la Fédération Française de Cardiologie (JJM), l'Association Française contre les Myopathies (JJM), the Muscular Dystrophy Association (P.A.) and le Réseau INSERM Inotropie (Ph. L-LGL). S. Decollogne is a recipient of a fellowship from le Ministère de la Recherche et de l'Espace.

**References**
1. Brown L, Erdmann E (1983) Binding of dihydrodigitoxin to beef and human cardiac $Na^+ + K^+$-ATPase: evidence for two binding sites in cell membranes. Biochem Pharmacol 32: 3183-3190.
2. Berrebi-Bertrand I, Maixent JM, Guédé FG, Charlemagne D, Lelièvre LG (1991) Identification of two functional (Na/K)-ATPase isoforms in left ventricular guinea pig heart. Eur J Biochem 196: 129-133.
3. Beuckelmann DJ, Näbauer M, Erdmann E (1993) Alterations of $K^+$ currents in isolated human ventricular myocytes from patients with terminal heart failure. Circ Res 73: 379-385.
4. De Pover A, Godfraind T (1979) Interaction of ouabain with $(Na^+ + K^+)$-ATPase from human heart and from guinea pig. Biochem Pharmacol 28: 3051-3056.
5. Lelièvre LG, Maixent JM,, Lorente P, Mouas C, Charlemagne D, Swynghedauw B (1986) Prolonged responsiveness to ouabain in hypertrophied rat heart: physiological and biochemical evidence. Am J Physiol 250: H923-H931.
6. Rasmussen HH, Ten Eick RE, Okita GT, Hartz RS, Singer DH (1985) Inhibition of electrogenic Na-pumping attributable to binding of cardiac steroids to high-affinity pump sites in human atrium. J Pharm Exptl Ther 236: 629-635.
7. Shamraj OI, Grupp IL, Melvin D, Gradoux N, Kremers W, Lingrel JJ, De Pover A (1993) Characterization of Na,K-ATPase, its isoforms, and the inotropic response to ouabain in isolated failing human hearts. Cardiovasc Res (in press).
8. Shamraj OI, Melvin D, Lingrel JB (1991) Expression of Na,K-ATPase isoforms in human heart. Biochem Biophys Res Comm 179: 1434-1440.
9. Sweadner KJ (1970) Two molecular forms of $(Na^+ + K^+)$-ATPase in brain. Separation and difference in affinity for strophantidin. J Biol. Chem. 254: 6060-6067.

# The Canine Cardiac $\alpha_3$ Isoform of Na$^+$/K$^+$-ATPase is Highly Sensitive to Digitoxigenin

I. Berrebi-Bertrand*, O. Barbey, L.G. Lelièvre* and J.M. Maixent

Laboratoire de Recherches Cardiologiques, Faculté de Médecine Bd Dramard 13916 Marseille,*Laboratoire de Pharmacologie des Transports Ioniques Membranaires, Hall des Biotechnologies, Université P7, 2, Place Jussieu, 75005 Paris, France

## Introduction

Three alpha ($\alpha_1$, $\alpha_2$, $\alpha_3$) and three beta ($\beta_1$, $\beta_2$, $\beta_3$) subunits of Na$^+$/K$^+$-ATPase have been identified from cDNA cloning (9). The expression of only $\alpha_1$ isoform has been shown in the kidney in all the mammalian species considered. In the brain, two other isoforms ($\alpha_2$ and $\alpha_3$) are concomitantly expressed (as reviewed in 15). However, each of them can be expressed separately in other tissues such as : $\alpha_2$ in the cardiac and skeletal muscles and $\alpha_3$ in the pineal gland (14), and macrophages (16). In dog heart, we have previously reported the existence of $\alpha_1$ and $\alpha+$ isoforms. Since it was immunologically demonstrated that the $\alpha^+$ form could contain a mixture of $\alpha_2$ and $\alpha_3$ isoforms (15), it was our objective to further investigate in dog heart and brain the nature and function of these two isoforms. We have used the panel of antibodies specific for the three rat $\alpha$ subunit isoforms (13) in Western blot analysis of canine microsomal fractions from kidney, brain and heart. In these 3 tissues, functional differences were investigated in terms of sensitivity to digitoxigenin (DGN) and Na$^+$ since differences in Na$^+$ affinity have also been associated with isoform properties (4-7, 14).

## Materials and Methods

Microsomal fractions from the renal, cerebral and cardiac tissues were obtained as previously described (10, 11) from sodium pentobarbital anesthetized adult mongrel dogs.

### SDS-PAGE and Western blots

Microsomal preparations from kidney, brain, and cardiac muscles were electrophoresed and immunoblotted as previously described (3) with polyclonal antisera purchased from Upstate Biotechnology Inc., (New-York).

### Enzyme assays

The enzymatic assays were performed with vesicles treated with SDS (0.2-0.3 mg/mg of protein for 30 min at 20°C). The Na$^+$/K$^+$-ATPase activity was measured at 37°C in an ATP regenerating medium by continuously recording NADH oxidation with a coupled assay method using a Beckman DU70 spectrophotometer (8). To ascertain whether there are other changes in the pump characteristics, kinetic studies were performed to estimate the apparent affinity of the Na$^+$/K$^+$-ATPase for Na$^+$. The Na$^+$ requirements of isoenzymes were determined by measuring ATPase activity at a constant $K^+$ level(20 mM) and varying $Na^+$ (0-100 mM) concentrations. To distinguish between the two $\alpha$ forms of the $Na^+/K^+$-ATPase, two different concentrations of DGN have been chosen, 1/ a low concentration (100 nM) which inhibits almost all the high affinity sites (93 %), and a few percent of the low affinity sites and 2/ a high DGN concentrations (50 $\mu$M) which inhibits the whole activity. By difference between the two DGN concentrations, this activity principally yielded the $\alpha_1$ subunit.

### Data analysis

The best fit of the dose-response curves to either $Na^+$ or DGN was analyzed as the sum of one to three saturable and independent sites by the non linear least squares procedures using the cooperative model (for $Na^+$) as previously described (2, 6). The $IC_{50}$ values and proportions result from the analysis with a two site model for heart and brain and a one-site model for kidney. $K_{0.5}$ values are the best fit of the curves according to a model with 3 non-interacting $Na^+$ ions required for activity.

### Results and Discussion

The $\alpha_3$ fusion protein antiserum stained a single polypeptide band of 99 kDa (apparent molecular mass) in dog heart and brain whereas the antiserum $\alpha_2$ reacts with a polypeptide of 99 kDa only in dog brain (Fig 1) We verified (data not shown) the specificity of the $\alpha_3$ antisera by demonstrating that this antibody does not recognize the canine $\alpha_2$ isoform present in skeletal muscle (15). These results indicate that $\alpha_3$-like form is expressed in dog heart muscles and myocytes whereas $\alpha_2$ and $\alpha_3$ -like isoforms coexist in brain.

Microsomes from dog kidney exhibited a single component of DGN-sensitive $Na^+/K^+$-ATPase activity with a contribution of 90 % of the total activity. The $IC_{50}$ value was $0.10 \pm 0.01$ $\mu$M. Data obtained from dog cardiac microsomes (n=3) were best fitted by a two-site model ($IC_{50} = 7\pm2$ nM and $IC_{50} = 0.63\pm0.17$ $\mu$M) respectively, suggesting at least two components of DGN sensitive $Na^+/K^+$-ATPase activity. For cerebral microsomes, the data were best fitted by an equation that assumed the existence of two rather than three inhibitory processes : $IC_{50} = 17\pm5$ nM and $IC_{50} = 1.3\pm0.5$ $\mu$M. Inasmuch as Western blot analysis suggests the coexpression in brain of three $\alpha$ subunits, using the discrimination by DGN of the high and low affinity sites, we found in heart, two $Na^+$ affinities of $6.7 \pm 1.4$ and $10.0 \pm 1.9$ mM attributed to the $\alpha_1$ and $\alpha_3$ subunits respectively. For brain enzymes, the high affinity inhibitory sites to DGN (presumably a mixture of $\alpha_2$ and $\alpha_3$ subunits) exhibit a $K_{0.5}$ for $Na^+$ of $19.6 \pm 4.9$ mM, and a $6.3 \pm 1.2$ mM for the low affinity sites ($\alpha_1$) to DGN.The comparison of the low affinity inhibitory sites for digitalis between organs shows an heterogeneity between the tissues tested, kidney and heart. Parameters other than the $\alpha$ subunit such as $\alpha\beta$ complexes, environment of the pump (4) may have a large influence on the apparent affinity for sodium. One of this is the presence of two possible complexes $\alpha_1\beta_1$ or $\alpha_1\beta_2$ in heart since the $\beta_2$ protein was detected in our canine preparations (JMM, unpublished observation).

Some physiological relevance of $\alpha_2$ and $\alpha_3$ protein expression in cardiac function can be inferred by the analogy with the fact that the switch between 7 and 15 days after birth is synchronous with the shortening of heart action potential duration in rat (9). Fig.

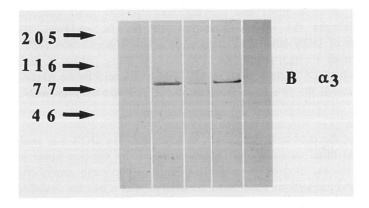

**Figure 1.** Expression of Na$^+$,K$^+$-ATPase $\alpha$3 isoenzyme in canine cardiac microsomes. Solubilized microsomal proteins were fractionated by electrophoresis then transferred to a nitrocellulose filter and probed with the rat $\alpha$ -subunit specific antisera. Rat brain and kidney microsomes were isolated by a previously described procedure (2). Lane 1 : rat kidney, lane 2 : rat brain, lane 3 : dog heart, lane 4 :dog brain, lane 5 : dog kidney.

Thus, the $\alpha$3 expression in dog might be synchronous with a slow action potential. In heart, the $\alpha$1 form has a $K_{0.5}$ for sodium of 6 mM which means that it participated in sodium extrusion under physiological conditions (Na$_{in}$ : 7 mM). The $\alpha$3 on the other hand will be maximally active when (Na$^+$)$_{in}$ reaches its maximun following the depolarisation phase. This could explain the biphasic increase in (Na$^+$)$_{in}$after addition of 1 nM to 100 $\mu$M ouabain to isolated chicken myocytes (1).

Therapeutic concentrations of digitalis resulting in $\alpha$3 inhibition produce their inotropic effects in heart through an enhanced sodium transient and a limited myocardial Na$^+$ accumulation (10).

*Acknowledgments*

We are grateful Dr. R.M. Kawamoto and Dr A. Baggioni (Procter & Gamble Pharmaceuticals Inc., Norwich, NY 13815, USA and France) for helpful discussions.The studies were supported by grants from the Université Paris 7 (n°3409 R10-R11), la DRED (Direction de la Recherche et de l'Encadrement de Doctorants (Action scientifique 1990 n°3201 R10), INSERM (CAR 488002) and by a research contract from Procter & Gamble Pharmaceuticals France.

# References

1. Ahlemeyer B, Weintraut H, Schoner W (1992) Cultured chick-embryo heart cells respond differently to ouabain as measured by the increase in their intracellular $Na^+$ concentration. Biochim Biophys Acta 1137:135-142

2. Berrebi-Bertrand I, Maixent JM, Christe G, Lelièvre, LG (1990) Two active $Na^+,K^+$-ATPase of high affinity for ouabain in adult rat brain membranes. Biochim Biophys Acta 1021:148-156.

3. Berrebi-Bertrand I, Maixent JM, Guede F, Gerbi A, Charlemagne D, Lelièvre, LG (1991) Two functional $Na^+K^+$-ATPase isoforms in left ventricular guinea-pig heart. Eur J Biochem 196:129-133.

4. Brodsky JL, Guidotti G (1990) Sodium affinity of brain $Na^+K^+$-ATPase is dependent on isoenzyme and environment of the pump. Am J Phys 258:C803-C811.

5. Feige G, Leutert T, De Pover A (1988) $Na^+,K^+$-ATPase isoenzymes in rat tissues: differential sensitivities to sodium, vanadate and dihydroouabain. In Prog Clin Biol Res Vol 268A : The $Na^+,K^+$ pump ; Part B : Cellular aspects, eds. Skou JC, Norby JG, Maunsbach AB, Esmann M P. 377-384, New-York, Liss..

6. Gerbi A, Debray M, Maixent JM, Chanez C, Bourre JM (1993) Heterogeneous $Na^+$ sensitivity of $Na^+,K^+$-ATPase isoenzymes in whole brain membranes. J Neurochem 60: 246-252.

7. Jewel EA Lingrel JB (1991) Comparison of the substrate dependence properties of the rat $Na^+,K^+$-ATPase $\alpha1$, $\alpha2$ and $\alpha3$ isoforms expressed in Hela cells. J Biol Chem 266:16925-16930

8. Lelièvre LG, Maixent JM, Lorente P, Mouas C, Charlemagne D, Swynghedauw B (1986) Prolonged responsiveness to ouabain in hypertrophied rat heart : physiological and biochemical evidence. Am J Physiol 250:923-931.

9. Lingrel JB, Orlowski J, Shull MM, Price EM (1990) Molecular genetics of $Na^+K^+$-ATPase. Prog Nucleic Acid Res 38:37-89

10. Maixent JM, Charlemagne D, de la Chapelle B, Lelievre LG (1987) Two $Na^+K^+$-ATPase isoenzymes in canine cardiac myocytes: Molecular basis of inotropic and toxic effects of digitalis. J Biol Chem 262:6842-6848

11. Maixent JM, Fenard S, Kawamoto RM (1991) Tissue localization of active $Na^+,K^+$-ATPase isoenzymes by determination of their profile of inhibition with ouabain, digoxin, digitoxigenin and cordil, a new aminosteroid cardiotonic. J Receptor Res 11:687-698

12. Ng YC, Book LB, (1992) Expression of $Na^+K^+$-ATPase $\alpha1$ and $\alpha3$-isoforms in adult and neonatal ferret hearts. Am J Physiol 263: H1430-H1436

13. Shyjan AW, Levenson R (1989) Antisera specific for the $\alpha1$, $\alpha2$ and $\alpha3$ and $\beta$ subunits of the $Na^+,K^+$-ATPase: Differential expression of $\alpha$ and $\beta$ subunits in rat tissue membranes. Biochemistry 28:4531-4535

14. Shyjan AW, Cena V, Klein DC, Levenson R (1990) Differential expression and enzymatic properties of $Na^+,K^+$-ATPase $\alpha3$ isoenzyme in rat pineal glands. Proc Natl Acad of Sci, USA 87:1178-1182

15. Sweadner KJ (1989) Isozymes of the $Na^+,K^+$-ATPase. Biochim Biophys Acta 988:185-220

16. Vignery A, Wang F, Qian HY, Benz EJ, Gilmore Hebert M, (1991) Detection of the $Na^+,K^+$-ATPase $\alpha3$-isoform in multinucleated macrophages. Am J Physiol 37:1265-1271.

# Heterogeneous Na$^+$ Sensitivities of Na$^+$/K$^+$-ATPase Isoenzymes in Membranes ; Modulation by Anaesthesia and Diets

A. Gerbi, J-M. Maixent, M. Zérouga * and J-M. Bourre*

Laboratoire de Recherches Cardiologiques, Faculté de Médecine Nord, Bd Dramard, 13015, Marseille. *Inserm U26, Hôpital F. Widal, 200 rue du Fg St Denis, 75010 Paris.

## Introduction

Na$^+$/K$^+$-ATPase consists of an active complex of two noncovalently linked subunits (a catalytic $\alpha$ and a glycosilated $\beta$ subunit) surrounded by a lipid bulk of 50 % of the mass molecular (5). In vertebrate (rat, human, chicken), it has been clearly shown that they are three independent genes for the $\alpha$ subunit, known as $\alpha1$, $\alpha2$ and $\alpha3$. If their best characterized difference, ouabain affinity, is not informative of their physiological relevance, their first real physiological significance is their functional difference in response to Na$^+$(4,7,11-13). These specific characteristics can be related to their divergence in aminoacids. However, the specificity of the surrounding environment and its functional implication is not yet understood. This article summarize ouabain and Na$^+$ sensitivities of the brain and kidney isoenzymes under different experimental conditions modulating their membrane environment (anaesthesia and fatty acid diets).

## Material and Methods

The different diets used differed mainly either by their amount of (n-3) polyunsaturated fatty acids (PUFA) (standard and Sunflower oil) or their nature (fish oil diet) (8,9,14). Weaned and adult rats fed diets since two generations (8) were killed by decapitation without anaesthesia, exept for adult rats fed standard diet which were also anaesthetized by intraperitoneal injection of pentobarbital (Sanofi, Santé animale). Ten minutes after injection, rats were decapitated (10).

Whole brain (from the different groups) and kidney membrane fractions were prepared and quantified for their protein amounts (7) and their fatty acid composition of total phospholipids (8,9,12).

Na$^+$/K$^+$-ATPase activity of the different detergent-free membrane fractions were determined by the coupled assay at 10 mM K$^+$. Specific activity varied from 20-35 $\mu$molPi/h/mg Protein whatever the tissue and diets. The proportion of each isoenzyme as a function of ouabain affinity was obtained from dose-response curves. The best fit of the curves was analysed as the sum of two or three saturable and independent sites by non-linear regression with the MKMODEL$^R$ Software (Biosoft, Cambridge, England) (7). Na$^+$ sensitivity were determined by varying NaCl + Choline Chloride = 100 mM

(7). The best fit of the curves was analysed as the sum of saturable and independent sites by non-linear regression using the model that we have written (7). The Hill coefficient as well as the threshold of stimulation (both variable) were not shown in the table. All values are means of 4 animals, experiments were done in triplicate, S.E.M not exceed 5 %.

## Results and Discussion

In whole brain membrane, a model where the three isoenzymes ($\alpha 1$, $\alpha 2$ and $\alpha 3$) are functional (1,2), $Na^+$ response of $Na^+$, $K^+$-ATPase activity was consistent with the existence of more than one reactivity (6). Using two appropriate ouabain concentrations, we were able to discriminate at least five $Na^+$ responses (7). In addition, pentobarbital exhibit a modulatory action on ouabain and $Na^+$ affinities (Table 1-2) of isoenzymes (10). As well for kidney membranes, each isoenzyme presents two $Na^+$ sensitivities in adult rats. This heterogeneous response and pentobarbital effect suggest the involvement of the membrane environment in this functional property. The anaesthetic action of pentobarbital is generally attributed to its membrane disorganising effect (3). In our case, this might result in a modification of the fatty acid composition of the membrane as well as to the binding of pentobarbital on $Na^+$, $K^+$-ATPase (10).

The biological membranes are dependent for their fatty acid composition to nutrition. Two PUFAs (ac. $\alpha$-linolenic C18:3(n-3) and ac. linoleic C18:2(n-6)) are essential. They are precursors of different active molecule, including long chain fatty acid which constitute for 30 % of the cerebral membrane. This long chain fatty acids can be bring by fish oil where they are concentrated.

The diets used modulate significantly the fatty acid composition of the weaned and adult brain membranes, in part by their (n-6)/(n-3) molar ratio (8,9,12). Then, dietary experiments allowed us to investigate the environmental dependence of isoenzymes biochemical properties.

Ouabain affinity and proportion of $Na^+/K^+$-ATPase isoenzymes expressed in the different membrane environments are modulated as shown in Table 1. Our study was restricted at the enzymatic level, so possibly, the change of proportion can reveal changes in mRNA expression not related to the fatty acid composition of the phospholipids. In function of this results, two appropriate concentrations were used to discriminate isoenzyme $Na^+$ sensitivity (summarized in Table 2). No research of heterogeneity has been made for high affinity isoenzyme in membranes of adults rats fed sunflower oil diet. In adults rats, heterogeneous $Na^+$ sensitivity was now well described. In contrast, weaned rats exhibit always for each isoenzyme monophasic $Na^+$ response. $\alpha 1$ isoenzyme $Na^+$ sensitivity was unaffected by the fatty acid modification, except for fish oil at 60-day-old. $\alpha 2$ and $\alpha 3$ isoenzymes $Na^+$ sensitivity was, whatever the age, modulated in different manner such the diet.

These results can explain divergences of these criteria reported in the litterature. They ascribed fatty acid dependence, however, these modulation are not still correlated to specific change occurred in their surrounding environment. No relationship has been found between the (n-6)/(n-3) molar ratio and the changes of affinities. Heterogeneous

Na$^+$ responses can be related to the 6 possible heterodimers and/or environmental heterogeneity. Why a monophasic response exists at 21-day old ? Have isoenzymes specific surrounding environments that could be lipids as well as protein ? What is the correlation membrane-structure-function of isoenzymes ? what is the relevance of threshold of stimulation and Hill coefficient superior to 3 ? and What reveal these altered Na$^+$ sensitivities for neuronal cells ?

Table 1 : ouabain affinity (IC$_{50}$ (M)) and proportion (%) of brain isoenzymes of 21- and 60-day-old rats fed the different diets and anaesthetic effect on brain and kidney isoenzymes of 60-day-old rats. ND ; Not Detected.

| | $\alpha$1 | | $\alpha$2 | | $\alpha$3 | |
|---|---|---|---|---|---|---|
| 21-day-old brain | IC50 | % | IC50 | % | IC50 | % |
| Sunflower oil | 8.4 10$^{-6}$ | 42.7 | 1.0 10$^{-6}$ | 40.6 | 3.9 10$^{-9}$ | 16.7 |
| Fish oil | 2.6 10$^{-5}$ | 26.3 | 2.5 10$^{-6}$ | 19.5 | 1.3 10$^{-7}$ | 54.2 |
| Standard | 8.4 10$^{-6}$ | 41.3 | 3.4 10$^{-7}$ | 25.0 | 2.0 10$^{-9}$ | 33.7 |
| 60-day-old brain | | | | | | |
| Sunflower oil | 5.8 10$^{-5}$ | 28.0 | 5.4 10$^{-6}$ | 36.0 | 3.2 10$^{-8}$ | 36.0 |
| Fish oil | 4.4 10$^{-5}$ | 20.4 | 1.3 10$^{-6}$ | 29.8 | 2.5 10$^{-8}$ | 49.8 |
| Standard (Decap.) | 4.0 10$^{-4}$ | 15.7 | 8.5 10$^{-7}$ | 49.6 | 6.8 10$^{-8}$ | 34.6 |
| Standard (Anaes.) | 3.2 10$^{-4}$ | 21.0 | 4.6 10$^{-7}$ | 38.1 | 2.3 10$^{-8}$ | 41.9 |
| 60-day-old kidney | | | | | | |
| Standard (Decap.) | 2.4 10$^{-5}$ | 100 | ND | ND | ND | ND |
| Standard (Anaes.) | 2.9 10$^{-5}$ | 100 | ND | ND | ND | ND |

Table 2 : Na$^+$ sensitivity (R1 and R2 (mM)) of brain isoenzymes of 21- and 60-day-old rats fed the different diets and anaesthetic effect on brain and kidney isoenzymes of 60-day-old rats. (value), one site fitted, ND ; Not Detected.

| | $\alpha$1 | | $\alpha$2 | | $\alpha$3 | |
|---|---|---|---|---|---|---|
| 21-day-old brain | R1 | R2 | R1 | R2 | R1 | R2 |
| Sunflower oil | 0.98 | ND | 7.21 | ND | 22.50 | ND |
| Fish oil | 1.15 | ND | 9.77 | ND | 12.90 | ND |
| Standard | 1.06 | ND | 2.80 | ND | 43.80 | ND |
| 60-day-old brain | | | | | | |
| Sunflower oil | 0.40 | 6.00 | 8.00 | ND | 17.00 | ND |
| Fish oil | 1.40 | 7.50 | 1.09(16) | 14.70 | 13.1(23) | 26.50 |
| Standard (Decap.) | 0.40 | 6.10 | 2.1(11.6) | 11.50 | 7.8(17.5) | 19.55 |
| Standard (Anaes.) | 3.88 | ND | 4.98 | 28.00 | 3.50 | 20.00 |
| 60-day-old kidney | | | | | | |
| Standard (Decap.) | 1.39 | 11.60 | ND | ND | ND | ND |
| Standard (Anaes.) | 6.00 | ND | ND | ND | ND | ND |

822

# References

1. Berrebi-Bertand I, Maixent JM, Christe G, Lelièvre L G (1990) Two active $Na^+,K^+$-ATPase of high affinity for ouabain in adult rat brain membranes. Biochim Biophys Acta 1021:148-158.

2. Blanco G, Berberian G, Beaugé L (1990) Detection of highly sensitivive isoform of rat brainstem $Na^+,K^+$-ATPase. Biochim Biophys Acta 1027:1-7.

3. Boggs JM, Yoong T, Hsia JC (1976) Site and mechanism of anesthetic action. I. Effect of anesthetics and pressure on fluidity of spin labeled vesicles. Mol Pharmacol 12:127-135.

4. Brodsky J L, Guidotti G (1990) Sodium affinity of brain $Na^+,K^+$-ATPase is dependent on isoenzyme and environment of the pump. Am J Physiol 258:C803-C811.

5. Geering K (1990) Subunit assembly and functional maturation of $Na^+,K^+$-ATPase. J Membr Biol 115:109-1211.

6. Gerbi A, Berrebi-Bertrand I, Leliévre LG (1990) Sensitivities to sodium : Isoenzymes of $Na^+,K^+$-ATPase in brain (Poster), in Satellite Meeting of the XIth International Congress of Pharmacology, Giessen.

7. Gerbi A, Debray M, Maixent JM, Chanez C, Bourre JM (1993) Heterogeneous $Na^+$ sensitivity of $Na^+,K^+$-ATPase isoenzymes in whole brain membranes. J Neurochem 60:246-252.

8. Gerbi A, Zerouga M, Debray M, Durand G, Chanez C, Bourre JM. (1993) Effect of dietary $\alpha$-linolenic acid on functional characteristic of $Na^+,K^+$-ATPase isoenzymes in whole brain membranes of weaned rats. Biochim Biophys Acta 1165: 291-298.

9. Gerbi A, Zerouga, M, Debray M, Durand G, Chanez C, Bourre JM (1993) Effect of fish oil diet on fatty acid composition of phospholipids of brain membranes and on kinetics properties of $Na^+,K^+$-ATPase isoenzymes of weaned and adult rats. (1993) In press in J Neurochem.

10. Gerbi, A, Zerouga M, Debray M, Maixent JM, Chanez C, Bourre JM (1993) Modulation by pentobarbital-induced anaesthetic action of functional properties of $Na^+,K^+$-ATPase isoenzymes. Submitted in Mol. Pharmacol

11. Lytton J (1985) Insulin affects the sodium affinity of the rat adipocyte $Na^+,K^+$-ATPase. J Biol Chem 260:10075-10080.

12. Maixent JM, Berrebi-Bertrand I (1992) Immunological and functional expression of alpha3 isoform of Na,K-ATPase in dog heart. J Mol Cell Cardiol 24: S26.

13. Shyjan AW, Cena V, Klein DC, Levenson R (1990) Differential expression and enzymatic properties of the Na,K-ATPase $\alpha3$ isoenzyme in rat pineal gland. Proc Natl Acad Sci USA 87:1178-1182.

14. Zerouga M, Gerbi A, Debray M, Durand G, Bourre JM (1993) Effect of diet deficient in alpha-linolenic acid on fatty acid composition and enzymatic properties of $Na^+,K^+$-ATPase isoenzymes of brain membranes in the rat. Submitted in J Biochem Nutr.

# Na$^+$/K$^+$-ATPase Isoforms in Human Heart ; Variation with Mammalian Species and Pathophysiological States

Franck Paganelli, Alain Gerbi, Odile Barbey, Alain Saadjian, Isabelle Berrebi-Bertrand, Samuel Lévy and Jean Michel Maixent

Laboratoire de Recherches Cardiologiques, Faculté de Médecine Nord, bd Dramard, 13015 Marseille, France.

## Introduction

The oligomeric structure of Na$^+$/K$^+$-ATPase assumed two polypetides : $\alpha$ catalytic (ouabain receptor) and ß glycoprotein subunits. Each of these subunits are encoded from a multigene family relevant of three $\alpha$ ($\alpha$1, $\alpha$2, $\alpha$3) and two $\beta$ ($\beta$1, $\beta$2) subunit isoforms (7). In adult rat brain, these three different $\alpha$ entities described as express by Northern and Western analysis (7) were considered functional since three inhibitory processes were decomposed from ouabain dose-response curves (3). In the heart, until recently, two classes of binding and inhibitory sites have been demonstrated in human (5), dog (8,9) and guinea pig (4) and well correlated with two $\alpha$ entities (4, 9) except for human. Indeed, the PCR analysis of gene transcription and Northern blot analysis demonstrated that adult human ventricles express mRNA for all three $\alpha$1, $\alpha$2 and $\alpha$3 isoenzymes (1,2,11,12). However these informations need to be correlated to protein properties like Western blot analysis and functional detections. Thus, we have investigated the expression of Na$^+$,K$^+$-ATPase $\alpha$2, $\alpha$3 and $\beta$2 subunits by immunobloting as well as the degree of heterogeneity of Na$^+$,K$^+$-ATPase activity in term of ouabain sensitivity. These characteristics were compared to those we previously described in the dog and guinea pig. Moreover, the understanding of digitalis drugs-induced inotropic action in heart failure requires the study of properties of human heart Na$^+$/K$^+$-ATPase from different states of congestive heart failure (CHF). Thus, displacement of [$^3$H] ouabain (13) has been used to compare the density and affinity for ouabain affinity binding sites in membrane preparations from different pathophysiological states of CHF.

## Materials and Methods

### Tissues

Human hearts were obtained at the time of orthotopic cardiac transplantation from patients with severe CHF or during cardiac surgery using protocols approved by the local hospital institutional committee for the protection of human subjects.

Small pieces of tissue (less than 1g) were excised and frozen in liquid nitrogen within 10 min of heart sample excision and then stored at -80°C until used.

## Preparation of cardiac microsomal membranes

Purified membrane fractions were isolated from the different human source, dog and guinea pig ventricles and also atria for human by differential centrifugation according to a method used for dog tissues described earlier (8). Protein was assayed by the BCA method (Pierce) with a bovine serum albumin standard.

## SDS-PAGE and Immunoblotting

Proteins were electrophoresed with Miniprotean II Cell Apparatus (Biorad) by SDS-PAGE on 4-15 % gradient ready gel transferred to a nitrocellulose membrane (Hybond, Amersham), blocked with 3 % low-fat milk in phosphate buffer saline and probed with isoform specific polyclonal rabbit anti-rat $\alpha$ and $\beta$ antibodies purchased from UBI. ECL detection (Amersham) was carried out with anti-rabbit IgG peroxidase.

## Enzymatic assays

$Na^+/K^+$-ATPase activity was measured in an ATP regenerating medium by continuously recording NADH oxidation with a coupled assay method using a Beckman DU70 spectrophotometer. $Na^+/K^+$-ATPase activity was defined as that activity inhibitable by 0,1 mM ouabain in human and dog and 1mM in guinea pig.

## [$^3$H] ouabain binding assays

Displacement of specifically bound [$^3$H] ouabain by unlabelled ouabain were carried out at 37°C in 1 mM Mg-Pi medium (Binding medium). Microsomes (5 to 12 $\mu$g) were incubated in the binding medium for 2 hours in the presence of 10 nM [$^3$H] ouabain and in the presence of 0.1 nM to 10 $\mu$M ouabain. Binding was terminated by rapid vacuum filtration over whatman GF/C filters and 3 washes with 5 ml of ice cold binding medium.

Bmax and Kd were determined following the equations :

$$Bmax = (b*/l*) \times IC_{50} \ (1) \qquad Kd = IC_{50} - L* \ (2)$$

Where b* is the specific binding expressed in dpm, l* is the constant amount of free [$^3$H] ouabain expressed in dpm, L* is the free concentration of [$^3$H] ouabain and $IC_{50}$ represents the free concentration of ouabain required to inhibit half of the radioligand binding. We used this procedure because the numerical transformation (Scatchard analysis) produce a cumbersome propagation of experimental errors (13).

The best fit of the curves (ouabain dose-response and binding displacement) was analysed as the sum of saturable and independent sites by non-linear-regression with the MKMODEL$^R$ software (Biosoft, Cambridge, England) using the model that we have written (6). The choice of the number of independent sites model to fit the data was made according to the Schwarz criterion (6).

## Results and Discussion

Utilizing commercial isoform specific antibodies, we detect $\alpha$2-, $\alpha$3- and $\beta$2-like isoforms in membrane preparations from human ventricles (data not shown) that confirm the human heart mRNA pattern expression for the $\alpha$ subunit (11,12). Moreover, the comparison of ouabain dose-response curves of the human $Na^+/K^+$-ATPase activity with standard enzyme preparations characterized for their isoform composition: ($\alpha$1 and $\alpha$3 in the dog [9] and, $\alpha$1 and $\alpha$2/$\alpha$3 [4] in the guinea-pig heart) revealed an unique inhibitory shape consistent with three active $Na^+/K^+$-ATPase $\alpha$1,

$\alpha2$ and $\alpha3$ isoenzymes (Table 1). However, no relationship was observed between isoenzymes affinities and their nomenclature in the different species studied so we presents in Table 1 their ouabain characteristics by class of similar affinities. These divergences could be due to aminoacid differences not still reported or a specific membrane environment between species.

**TABLE 1** Ouabain affinity $(IC_{50}$ (M)) and proportion (%) of $Na^+,K^+$-ATPase isoenzymes in dog, guinea pig (GP) and human left ventricular muscles.
**Isoenzyme Class**

| Species | $IC_{50}$ | % | $IC_{50}$ | % | $IC_{50}$ | % | $IC_{50}$ | % | $IC_{50}$ | % |
|---------|-----------|---|-----------|---|-----------|---|-----------|---|-----------|---|
| GP | $3.7\ 10^{-5}$ | 34 | - | - | $2\ 10^{-7}$ | 65 | - | - | - | - |
| Dog | - | - | - | - | $3\ 10^{-7}$ | 34 | $2\ 10^{-9}$ | 66 | - | - |
| Human | - | - | $1.3\ 10^{-6}$ | 32 | - | - | $4.1 10^{-9}$ | 19 | $1\ 10^{-10}$ | 47 |

In heart, the physiological and pharmacological contributions of the receptor with high-affinity for ouabain to inotropy is of interest. So, we studied by displacement of an inotropic concentration (10 nM) their affinities and their density in two pathophysiological states.

In first instance, we improved this procedure at low $[^3H]$ ouabain concentration on different atria sources (n=9). We observed a constant determination of the $IC_{50}$ relevant of the 50 % of $[^3H]$ ouabain displaced by unlabelled ouabain. In relation to equation (2), we determined a Kd of $5.8 \pm 0.41$ nM. However, in spite of this Kd well conserved, we found a large range of different Bmax (18.5-72.7, $\overline{x}$ = $41.3 \pm 17$ pmol/mg protein). Such spreading of ouabain receptor concentrations was also observed in left ventricle samples excised from the same heart obtained from orthotopic cardiac transplantation and valvular heart diseases. All these patients had advanced disease. The transplanted one was specified New York Heart Association (NYHA) class IV and the reported valvular patients were class III. On the first case corresponding to severe CHF, we determine Bmax in the range of 10.3-99.2 ($\overline{x}$ =$42.46 \pm 30.2$ pmol/mg protein, n=7) and in the second case, moderate CHF, Bmax range is 37.5-77.35 ($\overline{x}$ =$55.78 \pm 15.5$ pmol/mg protein, n=4). However, Kd determination is well improved in both two cases : Severe CHF, Kd = $15.2 \pm 0.6$ nM and Moderate CHF, Kd = $37.8 \pm 5.2$ nM. As conclusion of this preliminary result, we show a trend for an increased sensitivity 37.8 vs 15.2 nM in the severe CHF compared to moderate CHF. Although this results seems comparable with finding from (12), we cannot correlated as previously described (2,5) to a decrease in the concentration of total binding sites due to the Bmax spreading. Nevertheless, correlation could be hazardous since we assay only a high ouabain binding sites. In addition, atria and ventricles Kd's suggest that the properties of $[^3H]$ ouabain binding were not similar. Atria seems more sensitive than ventricles. Due to this missing of "normal" ventricles and the divergence between atria and ventricles, we cannot compared moderate CHF to control. Nevertheless, a comparison of our results to human heart without heart disease as (2, 5) seems also hazardous since heterogeneity in level of isoform expression in human heart samples has been reported (11) linked to numerous factor discussed before (11, 12).

In conclusion, we demonstrated that the three isoenzymes are functional in human heart in contrast to other species studied and present ouabain affinities divergence. In addition, we described an apparent sensibilization of the ouabain binding sites with high affinity in relation to the increase of CHF severity.

## References

1. Akopyanz NS, Broude NE, Vinogradova NG, Balabanov YA, Monastyrskaya GS, Sverdlov ED (1991) Differential expression of three $Na^+,K^+$-ATPase catalytic subunit isoforms in human tissues, organs and cell lines. In the Sodium Pump : Recent Developments. Rockfeller University Press (Kaplan, Deweer, ed) p 189-193.
2. Allen PD, Schmidt TA, Marsh JD, Kjeldsen K (1992) $Na^+,K^+$-ATPase expression in normal and failing human heart. Acta Physiologica Scandinavia 146 suppl 607:87-94.
3. Berrebi-Bertrand I, Maixent JM, Christé G, Lelièvre LG (1990) Two active Na,K-ATPase of high affinity for ouabain in adult rat brain membranes. Biochem Biophys Acta 1021:148-156
4. Berrebi-Bertrand I, Maixent JM, Guedé F, Gerbi A, Charlemagne D, Lelièvre LG (1991) Two functional Na,K-ATPase isoforms in left ventricular guinea pig heart. Eur J Biochem 196:129-133
5 De Pover A, Godfraind T (1979) Interaction of ouabain with $Na^+,K^+$-ATPase from human heart and from guinea pig heart. Biochem Pharmacol 28:3051-3056.
6. Gerbi A, Debray M, Maixent JM, Chanez C, Bourre JM (1993) Heterogeneous $Na^+$ sensitivity of $Na^+,K^+$-ATPase isoenzymes in whole brain membranes. J Neurochem 60:246-252.
7. Lingrel JB, Orlowski J, Schull J, Price EM, (1990) Molecular genetics of $Na^+,K^+$-ATPase. Prog Nuclei Acid Res Molec Biol 38:37-89.
8. Maixent JM, Fenard S, Kawamoto RM (1991) Tissue localization of active $Na^+,K^+$-ATPase isoenzymes by determination of their profile of inhibition with ouabain, digoxin, digitoxigenin and Cordil a new aminosteroid cardiotonic. J Receptor Res 11:687-698.
9. Maixent JM, Berrebi-Bertrand I (1992) Immunodetection and functional expression of $\alpha 3$ isoform in dog heart. J Mol Cell Card 24 (VI) 526.
10. Schwinger RGH, Böhm M, Erdmann E, (1990) Effectiveness of cardiac glycosides in human myocardium with and without downregulated $\beta$ adrenoreceptors. J Cardiovasc Pharmacol 15:692-697
11. Shamraj OI, Melvin D, Lingrel JB, (1991) Expression of $Na^+,K^+$-ATPase in human heart. Biochem Biophys Res Commun 179 (3):1434-1440
12. Shamraj OI, Grupp IL, Grupp G, Melvin D, Gradoux N, Kremers W, Lingrel JB, De Pover A (1993) Characterization of Na,K-ATPase its isoforms and the inotropic response to ouabain in isolated failing human hearts. Cardiovasc Res in press.
13. Swillens S (1992) How to estimate the total receptor concentration when the specific racdioactivity of the ligand is unknown. TiPs 13:430-434

827

# Non Effectiveness of Ouabain and Decrease in Na$^+$/K$^+$-ATPase Affinity for Ouabain in Failing Rabbit Heart

Abdellatif Ezzaher*, Renaud Mougenot**, Alain Gerbi, Nour el Houda Bouanani*, Antoine Baggioni**, Bertrand Crozatier* and Jean-Michel Maixent

Laboratoire de Recherches Cardiologiques, Faculté de Médecine Nord, bd Dramard, 13015 Marseille, * Inserm U2, Hôpital H. Mondor, Faculté de Médecine, Av. Gl. Sarrail, 94000 Créteil, ** Procter & Gamble Pharm. France, 1 chemin de Saulxier 91160 Longjumeau, France.

## Introduction

Impaired mechanical perfomance is the usual cause of Congestive Heart Failure (CHF) (2,10). For this reason, digitalis drugs are still the oldest compounds to treat CHF and still remain the mainstay of treatment despite the advent of new inotropes and vasodilators (9). Multiple changes are described in the failing myocardium such as a downregulation of $\beta$ adrenoreceptors, a deficient production in cAMP, a decrease in the function of the sarcoplasmic reticulum, an inhibition in the Na$^+$/Ca$^{++}$-exchange leading to excessive calcium accumulation (2,5,6,10). Several studies have examined the Na$^+$/K$^+$-pump in failing hearts but their conclusions are divergent (1,2,7,11). Some of these discrepancies could be due to the experimental model of heart failure, the utilized mammalian species and enzyme assay using pure sarcolemmal membrane fractions with low recovery.

This study was undertaken to gain further information regarding the molecular mechanism of action of digitalis in CHF. To assess the effectiveness of ouabain in failing rabbit heart produced by a double pressure plus volume overload, we investigated the contractile response to ouabain in Langendorff preparations and the sensitivity to ouabain of Na$^+$/K$^+$-ATPase-enriched preparations from ventricular cardiac muscles.

## Materials and Methods

### Induction of heart failure

Heart failure (HF) was produced in rabbits by a double volume plus pressure overload by creation of an aortic insufficiency (retrograde catheterization of the aortic valve) followed 14 days later by an abdominal aortic stenosis (5,7).

### Physiological studies

The effect of increasing doses of ouabain (0.1 nM-1000nM) was evaluated 14 days later in isolated hearts under a Langendorff apparatus and compared to that of

828

normal hearts. Measured parameters were developed pressure (intraventricular balloon) and peak left ventricular dP/dt. Data were normalized to 100 % of control (see Table 2). Statistical analysis based on values measured during drug infusion for each drug concentration, permitted to compare the response of failing hearts with that of control hearts.

**Biochemical studies : cardiac microsomal membranes.** Purified membrane fractions were isolated from left ventricular tissue according to the method of Maixent et al. (8). The frozen ventricle (1.5 g of left ventricle) was homogenized in 10 volumes of ice-cold buffer with a polytron PT10 (2x20 sec., setting 5) after centrifugation at 7000 x g and 46000 x g. The pellet was resuspended in 100 mM NaCl, 250 mM sucrose and 30 mM imidazole-HCl, pH 7.4 and was quickly frozen in liquid $N_2$. Protein concentration of all membranes was determined by the method of Lowry. The freeze-thaw process was used to render leaky the membranes.

**Enzymes assays.** Activities were measured at 37°C as a function of time (up to 60 min) and amount of proteins from 0.5 to 3 $\mu$g per assay. The $Na^+/K^+$-ATPase activity was measured in an ATP regenerating medium by continuously recording NADH oxidation with a coupled assay method using a Beckman DU70 spectrophotometer. The $Na^+/K^+$-ATPase activity was defined as an activity inhibitable by 1 mM ouabain. Inhibition percentages with varying doses of ouabain were calculated by comparing the activities in the presence or absence of 1 mM of ouabain. Experimental data were fitted using Enzfitter (Biosoft, Elsevier) and the best fit was calculated, using non-linear regression with one or a sum of two functions assuming the presence of one or two inhibitory processes (8).

## Results and Discussion

With this CHF rabbit model, it was shown in our laboratory's previous papers, that $\beta$ adrenoreceptor density was decreased and a larger depressant effect of calcium-blocking agents on contractility with a decreased ventricular response to catecholamine were shown (5, 7).

**Table 1.Anatomic data.(\*\*)** $p < 0.005$ ; (+) $p < 0.001$

| | Left ventricular weight (g) | | Left ventricular weight/body weight (g/kg) | |
|---|---|---|---|---|
| | Physiological studies | Biochemical studies | Physiological studies | Biochemical studies |
| Normal hearts | 5.48±0.30 | 4.47±0.29 | 1.62±1.10 | 1.32±0.14 |
| Failing hearts | 6.58±0.50 | 8.00±0.27[+] | 2.37±1.63[**] | 2.43±0.08[+] |

Left ventricular/body weight ratio was increased in HF (2.43±0.08 g/Kg compared with control (1.32±0.14 g/Kg ; p< 0.001). Left ventricular developed pressure increased significantly in both control and HF at a dose of 1 $\mu$M, but the increase above baseline was significantly larger (p<0.03) in control than in CHF (reaching respectively 143.8±11.6 and 109±6.7 % of baseline, Table 2).

Only $\alpha 3$ isoform was immunostained in control and CHF (data not shown). The presence of $\alpha 2$ isoform was not detected with the polyclonal anti-rat antibody used. This could be due to an absence of crossreactivity between the rat and the rabbit.

**Table 2. Normalized developed pressure (dev. P) (% of control [C]) and peak dP/dt (% of control) with ouabain in control and failing hearts.**

| | Ouabain (M) | | | | | | | |
|---|---|---|---|---|---|---|---|---|
| | $10^{-8}$ | | $10^{-7}$ | | $0.5\ 10^{-6}$ | | $10^{-6}$ | |
| | dev. P | dP/dt | dev. P | dP/dt | dev. P | dP/dt | dev. P | dP/dt |
| Control Heart | 107.7 ±3.7 | 105.8 ±2.9 | 112.8 ±5.6 | 108.7 ±3.4 | 114.5 ±6.4 | 113.0 ±4.6 | 143.8[b] ±11.6 | 169[a] ±10.7 |
| Failing Heart | 95.3 ±3.9 | 94.0 ±6.9 | 102.3 ±2.7 | 111.5 ±7.9 | 99.5 ±4.2 | 98.5 ±9.2 | 109.3 ±6.7 | 133.9 ±13.0 |

Values were normalized to 100 % during baseline. Comparisons between control and failing heart : [a] p<0.05 ; [b] p<0.03.

Receptor density measured by $Na^+/K^+$-ATPase activity in 6 CHF was similar to that of control. The affinity of $Na^+/K^+$-ATPase to ouabain was significantly higher in control (p<0.05) than in CHF ($IC_{50}$ of 2 $\mu M$ vs 90 $\mu M$, p<0.001). Only one inhibitory site to ouabain was detected corresponding to the ubiquitous $\alpha 1$ and the $\alpha 3$ isoenzymes. This could means a similar affinity for both or we cannot rule out the limit of detection of the mathematical analysis.

**Table 3. $Na^+/K^+$-ATPase activity ($\mu$mol Pi/h/mg prot) and ouabain affinity ($IC_{50}$ ($\mu$M) in normal and rabbit heart failure.**

| | Normal | Heart Failure | P |
|---|---|---|---|
| Activity | 41±16 | 36±3.5 | NS |
| $IC_{50}$ | 2±0.1 | 90±2 | <0.05 |

Our results are differrent to those which found a depression of $Na^+/K^+$-ATPase between 25 to 50 %(2). In these stuides, HF was obtained in different animal species (rats{4} and dogs {8,11}) and the aetiologies of the disease were different (4, 8, 11). Our results are similar to those found no change in $Na^+/K^+$-ATPase density in the rat ventricular hypertrophy (1) and in a model of CHF due to myocardial infarction (3). In conclusion, in the present model, the depressed inotropic response to ouabain appears as the consequence of an altered affinity of $Na^+/K^+$-ATPase to ouabain rather than to a decrease in $Na^+/K^+$-ATPase density. Two factors previously identified could cause a change in affinity : one is an isoenzyme change (1, 2), the second is an extrinsic changes such as membrane environment through fatty acid change of the membrane (10) We also can speculate an intrinsic change of the catalytic properties of the enzyme (turnover, $Na^+$ affinity etc..) . Further investigations concerning mRNA expression, affinity for different ligands and fatty acid composition of the sarcolemmal membranes will have to be done.

## References

1. Charlemagne D, Maixent JM, Preteseille M, Lelièvre LG (1986) Ouabain binding sites and $Na^+,K^+$-ATPase activity in rat cardiac hypertrophy. J Bioch Chem 261:185-189.
2. De Pover A, Grupp G, Schwartz A, Grupp IL (1991) Coupling of contraction through effects on $Na^+,K^+$-ATPase : Changes of $Na^+,K^+$-ATPase isoforms in heart desease ? Heart failure 1:201-258.
3. Dhalla NS, Dixon IMC, Rupp H, Barwinsky J (1991) Experimental congestive heart failure due to myocardial infarction. Sarcolemmal receptors and cation transporters. Current Topics in Heart Failure, 13-23.
4. Disow LMG, Hata T, Dhalla MS (1992) Sarcolemmal $Na^+,K^+$-ATPase activity in congestive heart failure due to myocardial infarction. Am J Physiol 262:C664-671.
5. Ezzaher A, Bouanani N, Su JB, Hittinger L, Crozatier B (1991) Increased negative inotropic effect of calcium channel blockers in hypertrophied and failing rabbit heart. J Pharmacol Exp Ther 257:466-471
6. Gerbi A, Maixent JM, Zerouga M, Bourre JM (This issue) Heterogenous $Na^+$ sensitivity of $Na^+,K^+$-ATPase isoenzymes in membranes ; Modulation by anaesthesia and diets.
7. Gilson N, Bouanani N, Corsin A, Crozatier B (1990). Left ventricular function and $\beta$-adrenoceptors in the rabbit failing heart. Am J Physiol 258:H634-H641
8. Khatter JC, Prasad K, (1976) Myocardial sarcolemmal ATPase in dogs with induced mitral insufficiency.Cardiovasc Res 10:637-641
9. Maixent J.M, Fénard S, Kawamoto RM (1991) Tissue localization of active Na,K-ATPase isoenzymes by determination of three profiles of inhibition with ouabain, digoxin, digitoxigenin and LND-796, a new aminosteroid cardiotonic. J Receptor Res 11:687-698
10. Paganelli F, Gerbi A, Barbey O, Maixent JM, Lévy S (1993) Bases pharmacologiques et cliniques de l'emploi des b-bloquants ($\beta$1 et/ou $\beta$2 spécifiques) dans l'insuffisance cardiaque JAMA Suppl Août 4-16
11. Prasad K, Khatter JC, Bharadway B, (1979) Intra and extracellular electrolytes and sarcolemmal ATPase in the failing heart due to pressure overload in dogs Cardiovasc Res 13:95-104

# Digoxin Treatment and Congestive Heart Failure in Light of Human Cardiac and Skeletal Muscle Digitalis Glycoside Receptor Studies

T.A. Schmidt, *H. Bundgaard, +H.L Olesen, +N.H. Secher, *K. Kjeldsen

Departments of Cardiothoracic Surgery RT, +Anaesthesiology AN and *Medicine B, Rigshospitalet, University of Copenhagen, Blegdamsvej 9, 2100 Copenhagen, Denmark.

## Introduction

Previous studies carried out on various in vitro systems (1,3,19) and experimental animals (2,18) have reported an increase in $Na^+/K^+$-ATPase as a result of cardiac glycoside exposure, as have studies performed on human peripheral blood cells (4,11). This has engendered speculation about development of tolerance to the inotropic effect of cardiac glycoside treatment: The idea being that inhibition of $Na^+/K^+$-ATPase obtained initially by digitalization would be counterbalanced by $Na^+/K^+$-ATPase upregulation. However, points of concern may be raised with regard to the methodology of the underlying studies for this hypothesis: 1) The concentrations of cardiac glycoside applied to cell cultures were in micromolar concentration, i.e. toxic to humans 2) In vitro systems, peripheral blood cells and results obtained by digitalization of normal guinea pigs or rats need not mirror the effect of digitalis treatment on human cardiac and skeletal muscle in heart failure. 3) Activity measurements performed on purified membrane fractions may not be applicable for studying quantitative aspects of $Na^+/K^+$-ATPase in muscle tissue (5). On this background it was our aim to evaluate the hypothesis that cardiac glycoside treatment increases digitalis receptors by performing appropriate measurements on the tissue of relevance, i.e. cardiac and skeletal muscle from patients with heart failure. In addition we wished to assess the distribution of specifically bound digoxin to human muscular tissue during digitalization.

## Materials and Methods

Myocardial as well as skeletal muscle $Na^+/K^+$-ATPase may be quantified with high accuracy and precision by vanadate facilitated $^3H$-ouabain binding to intact muscle samples (17); a method which ensures high recovery of enzyme. However, $Na^+/K^+$-ATPase may be underestimated using this method on muscle tissue from digitalized individuals owing to an already existing receptor occupancy with digoxin. This problem, however may be overcome by washing the muscle samples in buffer containing digoxin antibody prior to $^3H$-ouabain binding (16). From patients treated with digoxin for heart failure left ventricular and vastus lateralis specimens were obtained 13 - 16 h after death; muscle tissue unexposed to digoxin was likewise obtained at autopsy (14,15). Vital left ventricular and vastus lateralis samples from digitalized heart failure patients were harvested during heart transplantation (13) and by way of muscle biopsy, respectively. Vital left ventricular control tissue was obtained from brain dead organ donors (13) and by muscle biopsy from heart failure patients prior to digitalization. The studies were conducted in accordance with protocols approved by the local Ethical Committees in accordance with the Helsinki Declaration II.

## Results and Discussion

### Receptor regulation

It was feasible to carry out $^3$H-ouabain binding studies on cardiac and skeletal muscle 13 - 16 h after death based on the following observations: First, it is well established that $^3$H-ouabain binding sites degrade only very slowly postmortem (9,10). Second, the release of specifically bound digitalis glycoside was also found to occur only slowly after death (15). After washing previously bound digoxin off the cardiac glycoside receptors in left ventricular samples from the digoxin treated group (n = 11) and exposing samples from the untreated group (n = 8) to the same procedure, $^3$H-ouabain binding revealed no receptor upregulation in the digoxin treated group but rather a tendency to a 14 % (p > 0.10) reduction of receptor sites compared to the control group.

Skeletal muscle samples from the same patients also showed no receptor upregulation in the digoxin treated group but a 37 % (p < 0.005) reduction of receptor sites compared to the nondigitalized subjects.

Following wash of vital left ventricular samples in digoxin antibody no receptor upregulation was observed in the digoxin treated group (n = 6) compared with the nondigitalized group (n = 5), but rather a trend towards lower (19 %; p < 0.08) $Na^+/K^+$-ATPase concentration in myocardial tissue from patients with heart failure. The observed tendencies to a reduction in myocardial $Na^+/K^+$-ATPase concentration in heart disease seems to be in accord with previous observations of such a reduction in heart failure (7).

Vastus lateralis biopsies were obtained from 10 subjects with clinical heart failure (New York Heart Association class I -II) prior to and 2 days following complete digitalization. No difference in $Na^+/K^+$-ATPase concentration before or after this short term digitalization was observed following wash of samples in digoxin antibody.

Taken together there was no evidence for development of tolerance to digoxin due to upregulation of receptors in relevant human tissues, conversely muscular digitalis glycoside receptor concentration was found to be reduced in heart failure. In order not to inappropriately alter intracellular $Ca^{2+}$ homeostasis in the myocardium or extrarenal K-handling by skeletal muscle $Na^+/K^+$-ATPase, the present data may suggest that it might be a benefit to keep blood digoxin concentration at moderate levels in heart failure patients.

### Digoxin distribution

Within the groups of digitalized patients it is possible to calculate specific cardiac glycoside receptor occupancy with digoxin based on $^3$H-ouabain binding measurements to samples unexposed and exposed to wash in digoxin antibody (Table). Specific cardiac glycoside receptor occupancy with digoxin, thus amounts to 24 - 34 % and 9 - 13 % in heart and skeletal muscle, respectively. This seems in good accord with previous measurements of the reduction in maximum hyperpolarisation in atrial tissue (12) and radioimmunoassay estimation of skeletal muscle digoxin content following digitalization (6).

Table. Receptor occupancy with digoxin

|  | Occupancy (%) | References |
|---|---|---|
| Tissue obtained after death |  |  |
| Left ventricle | 34 | (14) |
| Vastus lateralis | 13 | (15) |
|  |  |  |
| Vital tissue |  |  |
| Left ventricle | 24 | (13) |
| Vastus lateralis | 9 |  |

It may be observed that there is 2.6 - 2.7 fold higher occupancy of receptors with digoxin in the heart than in skeletal muscle. This finding seems to be in good accord with measurements of the apparent $K_D$ for $^3$H-ouabain binding in samples of human muscle and myocardium (8,9).

The distribution of specifically bound digoxin to cardiac and skeletal muscle $Na^+/K^+$-ATPase in life could be calculated and compared to the amount of digoxin in the extracellular volume. This was feasible in heart failure patients based on $^3$H-ouabain binding data from vital left ventricle (13) and skeletal muscle (288 pmol/g wet wt.), an average body mass of around 70 kg and the assumptions that the extracellular volume (ECV), the heart, and skeletal muscle mass constitute 20, 1 and 30% of the body weight, respectively. Hence, following digitalization digoxin specifically bound to the heart and skeletal muscle would amount to 103 and 544 nmol while the ECV would contain 18 nmol digoxin. Thus, it may be appreciated that it would require only a 5% release of specifically bound digoxin from skeletal muscle to more than double the amount of digoxin in the extracelluar space and thereby increase binding to the heart or vice versa. Thus, skeletal muscle $Na^+/K^+$-ATPase may be perceived as an important distribution volume for digoxin during digitalization.

**Acknowledgements**

We thank Grete Simonsen for skilled technical assistance, Stig Haunsø and Erik Sandøe for valuable discussions, and the Wellcome foundation for a gift of digoxin antibody (Digibind$^R$). This work was in part supported by the Danish Heart Foundation, The Faculty of Medicine of Copenhagen University, Nycomed DAK, and Novo's Foundation.

**References**

1.      Bluschke V, Bonn R, Greeff K (1976) Increase in the $(Na^+ + K^+)$-ATPase activity in heart muscle after chronic treatment with digitoxin or potassium deficient diet. Eur J Pharmacol 37:189-191

2.      Bonn R, Greeff K (1978) The effect of chronic administration of digitoxin on the activity of the myocardial (Na + K)-ATPase in guinea-pigs. Arch Int Pharmacodyn Ther 233:53-64

834

3. Brodie C, Sampson SR (1985) Effects of chronic ouabain treatment on [$^3$H]ouabain binding sites and electrogenic component of membrane potential in cultured rat myotubes. Brain Res 347:121-123

4. Erdmann E, Werdan K, Krawietz W (1984) Influence of digitalis and diuretics on ouabain binding sites on human erythrocytes. Klin Wochenschr 62:87-92

5. Hansen O, Clausen T (1988) Quantitative determination of $Na^+$-$K^+$-ATPase and other sarcolemmal components in muscle cells. Am J Physiol 254:C1-C7

6. Joreteg T, Jogestrand T (1984) Physical exercise and binding of digoxin to skeletal muscle - effect of muscle activation frequency. Eur J Clin Pharmacol 27:567-570

7. Nørgaard A, Bagger JP, Bjerregaard P, Baandrup U, Kjeldsen K, Thomsen PEB (1988) Relation of left ventricular function and Na,K-pump concentration in suspected idiopathic dilated cardiomyopathy. Am J Cardiol 61:1312-1315

8. Nørgaard A, Kjeldsen K, Clausen T (1984) A method for the determination of the total number of $^3$H-ouabain binding sites in biopsies of human skeletal muscle. Scand J Clin Lab Invest 44:509-518

9. Nørgaard A, Kjeldsen K, Hansen O, Clausen T, Larsen CG, Larsen FG (1986) Quantification of the $^3$H-ouabain binding site concentration in human myocardium: a postmortem study. Cardiovasc Res 20:428-435

10. Nørgaard A, Kjeldsen K, Stenfatt-Larsen J, Grønhøj-Larsen C, Grønhøj-Larsen F (1985) Estimation of stability of [$^3$H]-ouabain binding site concentration in rat and human skeletal muscle post mortem. Scand J Clin Lab Invest 45:139-144

11. Rapeport WG, Aronson JK, Grahame-Smith DG, Carver JG (1985) Increased specific $^3$H-ouabain binding to lymphocytes after incubation with acetylstrophanthidin for 3 days. Br J Clin Pharmcol 20:277P-278P

12. Rasmussen HH, Okita GT, Hartz RS, Eick RET (1990) Inhibition of electrogenic $Na^{(+)}$-pumping in isolated atrial tissue from patients treated with digoxin. J Pharmacol Exp Ther 252:60-64

13. Schmidt TA, Allen PD, Colluci WS, Marsh JD, Kjeldsen K (1993) No adaption to digitalization as evaluated by digitalis receptor (Na,K-ATPase) quantification in explanted hearts from donors without heart disease and from digitalized patients in endstage heart failure. Am J Cardiol 70:110-114

14. Schmidt TA, Holm-Nielsen P, Kjeldsen K (1991) No upregulation of digitalis glycoside receptor (Na,K-ATPase) concentration in human heart left ventricle samples obtained at necropsy after long term digitalisation. Cardiovasc Res 25:684-691

15. Schmidt TA, Holm-Nielsen P, Kjeldsen K (1993) Human skeletal muscle digitalis glycoside receptors (Na,K-ATPase) - importance during digitalization. Cardiovasc Drugs Ther 7:175-181

16. Schmidt TA, Kjeldsen K (1991) Enhanced clearance of specifically bound digoxin from human myocardial and skeletal muscle samples by specific digoxin antibody fragments. Subsequent complete digitalis glycoside receptor (Na,K-ATPase) quantification. J Cardiovasc Pharmacol 17:670-677

17. Schmidt TA, Svendsen JH, Haunsø S, Kjeldsen K (1990) Quantification of the total Na,K-ATPase concentration in atria and ventricles from mammalian species by measuring $^3$H-ouabain binding to intact myocardial samples. Stability to short term ischemia reperfusion. Basic Res Cardiol 85:411-427

18. Wai Ching Li P, Ho CS, Swaminathan R (1993) The chronic effects of long-term digoxin administration on Na/K-ATPase activity in rat tissues. Int J Cardiol 40:95-100

19. Werdan K, Reithmann C, Erdmann E (1985) Cardiac glycoside tolerance in cultured chicken heart muscle cells - a dose-dependent phenomenon. Klin Wochenschr 63:1253-1264

# Age-Related Changes in $Na^+/K^+$-ATPase and $Ca^{2+}$-ATPase Concentration in Rat Heart Ventricle

J.S. Larsen, *T.A. Schmidt, K. Kjeldsen

Departments of Medicine B and *Cardiothoracic Surgery RT, Rigshospitalet, University of Copenhagen, Blegdamsvej 9, 2100 Copenhagen, Denmark.

## Introduction

Earlier studies of $Na^+/K^+$-ATPase concentration in various excitable tissues and species have shown major variations with development. In rat skeletal muscle $Na^+/K^+$-ATPase concentration has been found to increase from birth to a maximum at the fourth week of life followed by a decrease to a plateau obtained in mature animals (7), in peripheral nerves an increase was reported from the 2nd to 4th week of life followed by a decrease to the 12th week of life (6), in cerebral cortex an increase from birth to a plateau at maturity was seen (11); in mouse skeletal muscle also a raise to a peak value followed by a decrease to a plateau was found (7); in guinea pig skeletal muscle the changes in $^3$H-ouabain binding concentration were different showing a decrease from birth to maturity (7), and also $^3$H-ouabain binding to left ventricle samples decreased (8); in human subjects no change in skeletal muscle $^3$H-ouabain binding site concentration was seen with age (4), in the heart a decrease in $^3$H-ouabain binding site concentration within the first years of age was observed (5). These observations were generally carried out using measurement of $Na^+/K^+$-ATPase by vanadate facilitated $^3$H-ouabain binding to intact samples (2). The benefit of using this method was that it ensured high accuracy and precision of measurements. However, since rat heart has a very low affinity for digitalis glycosides this method has not been applicable to that tissue. Thus, it was of interest to apply another simple method suitable for quantitative determination of $Na^+/K^+$-ATPase to that tissue. Hence, in the present study rat myocardial $Na^+/K^+$-ATPase concentration was assessed by measurement of $K^+$-activated hydrolysis of $p$-nitrophenylphosphate ($p$NPP) (12) in crude homogenates. Furthermore to evaluate whether the age-dependent change in $Na^+/K^+$-ATPase concentrations is selective or reflects a general developmental change in ion-transport proteins it was also of interest to evaluate a putative change in myocardial $Ca^{2+}$-ATPase concentration. This was assessed using in principle the same $p$NPPase assay the exception being activation by $Ca^{2+}$ (10).

## Materials and Methods

Female Wistar rats 0 to 16 weeks old were killed by decapitation and hearts were excised in toto. Heart ventricles were isolated, weighed and homogenized using a Ultra-Turrax T25 homogenisator at $0°C$ in a buffer containing 30 mM histidine, 2 mM EDTA, and 250 mM sucrose (pH 7.2). Subsequently the tissue homogenate was homogenized at $0°C$ in a glass homogenizator with a teflon pestle to a final tissue concentration of 10 mg/ml. $K^+$-dependent $p$-nitrophenyl phosphatase ($p$NPPase) activity was determined

836

incubating 100 $\mu$l of myocardial homogenate (10 mg wet wt./ml) at 37°C in 800 $\mu$l buffer as the difference in values obtained using buffer containing: 31.25 mM histidine, 18.75 mM $MgCl_2$ and 125 mM NaCl (pH 7.4), and buffer containing: 31.25 mM histidine, 6.25 mM $MgCl_2$ and 62.5 mM KCl (pH 7.4). The phosphatase reaction was started by the addition of 100 $\mu$l 100 mM $p$-NPP and stopped after 30 min by the addition of 2 ml 500 mM Tris and 55 mM EDTA. $Ca^{2+}$-dependent $p$NPPase activity was determined after activation with Triton X-100. 100 $\mu$l myocardial homogenate (10 mg wet wt./ml) was preincubated 10 min at 37°C with Triton X-100. $Ca^{2+}$-dependent $p$NPPase was then determined as the difference in values obtained in 800 $\mu$l buffer containing: 25 mM hepes (pH 7.4), 1.25 mM $MgCl_2$, 1.25 mM EGTA, 0.001 % Triton X-100 and 62.5 mM KCl, and buffer containing: 25 mM hepes (pH 7.4), 1.25 mM $MgCl_2$, 1.25 mM EGTA, 0.001 % Triton X-100, 62.5 mM KCl and 0.75 mM $CaCl_2$. The phosphatase reaction was started by the addition of 100 $\mu$l 100 mM $p$-NPP, and stopped after 30 min by the addition of 2 ml 500 mM Tris and 55 mM EDTA. The liberated $p$--nitrophenyl was determined by absorption spectrophotometry using a Hitachi Photometer 4020. The wavelength was 410 nm. On the basis of a molar absorption coefficient of $1.75 \times 10^4$ liter/mol/cm (12) the $p$--nitrophenyl concentration in the assay was calculated. Multiplying this value with the volume of assay (liter) and dividing with time of incubation (min) as well as the amount of tissue employed (g wet wt.) the phosphatase activity was calculated and expressed as $\mu$mol/min/g wet wt. The statistical significance of differences was ascertained by Student's two tailed $t$ test for unparied observations.

## Results and Discussion

Values for $K^+$- and $Ca^{2+}$-dependent $p$NPPase activity in crude homogenates of heart ventricle of 0 - 16 week old rats are given in the Table 1 below.

| Age weeks | $K^+$-dependent $p$NPPase $\mu$mol/min/g wet wt. | $Ca^{2+}$-dependent $p$NPPase $\mu$mol/min/g wet wt. |
|---|---|---|
| 0 | 1.97 $\pm$ 0.01 | 0.099 $\pm$ 0.006 |
| 1 | 2.05 $\pm$ 0.07 | 0.171 $\pm$ 0.001 |
| 2 | 2.11 $\pm$ 0.01 | 0.224 $\pm$ 0.006 |
| 3 | 2.13 $\pm$ 0.09 | 0.287 $\pm$ 0.009 |
| 4 | 1.99 $\pm$ 0.03 | 0.260 $\pm$ 0.017 |
| 5 | 2.03 $\pm$ 0.03 | 0.274 $\pm$ 0.014 |
| 6 | 2.00 $\pm$ 0.08 | 0.235 $\pm$ 0.011 |
| 8 | 2.04 $\pm$ 0.05 | 0.202 $\pm$ 0.007 |
| 12 | 1.87 $\pm$ 0.06 | 0.191 $\pm$ 0.007 |
| 16 | 1.84 $\pm$ 0.05 | 0.180 $\pm$ 0.012 |

**Table 1.** $K^+$-and $Ca^{2+}$-dependent $p$NPPase activity in crude homogenates of heart ventricular tissue of 0 - 16 week-old rats. Values are given as means $\pm$ S.E.M. n = 5 animals in each group.

It can be seen that the $K^+$-dependent $p$NPPase activity varies from 1.84 to 2.13 $\mu$mol/min/g wet wt. Using a molecular activity of 1300 cycles/min (9) it can be calculated that these values correspond to a $Na^+/K^+$-ATPase concentration of around 1500 pmol/g wet wt. This concentration of $Na^+/K^+$-ATPase in rat myocardium seems to be of the same order of magnitude as that earlier determined by $K^+$-dependent 3-$O$-methylfluorescein phosphatse activity in crude homogenates (9).

$Ca^{2+}$-dependent $p$NPPase activity was in the range 0.099 to 0.287 $\mu$mol/min/g wet wt. Using a molecular activity of around 20 cycles/min (3) it can be calculated that this corresponds to a $Ca^{2+}$-ATPase concentration of around 10 nmol/g wet wt. Thus it may be calculated that rat myocardium contains in the magnitude 5-10 times higher concentration of $Ca^{2+}$-ATPase than $Na^+/K^+$-ATPase. This may accord with the observation that $Ca^{2+}$-ATPase uses around 30 % whereas the $Na^+/K^+$-ATPase only 4 % of total basic myocardial energy expenditure (1).

It can be seen from the table that maximum values for $K^+$-dependent as well as $Ca^{2+}$-dependent $p$NPPase activities are obtained at 3 weeks of age. For $K^+$-dependent $p$NPPase activity an increase of 8 % ($p < 0.05$) was observed from birth to the 3rd week of life. This was followed by a decrease by 14 % ($p < 0.001$) from the 3rd to the 16th week of life. Although the changes are small, the pattern in this tissue is in agreement with earlier observations in rat skeletal muscle and peripheral nerves (6,7), whereas it is at variance with the observed monophasic increase in $Na^+/K^+$-ATPase concentration in rat cortex cerebri (11). Thus various excitable tissues within the same animal may show different pattern of development (see Figure 1). The present observed change in myocardial $Na^+/K^+$-ATPase concentration with age may be due to a minor imbalance between synthesis of myocardial $Na^+/K^+$-pumps and tissue mass. That the myocardial $Na^+/K^+$-ATPase concentration is relatively stable may be associated with the need for optimum $Na^+/K^+$-homeostasis within the myocardium already at birth in accordance with the demand for optimum heart rhythm regulation.

From birth to the 3rd week of age the $Ca^{2+}$-dependent $p$NPPase activity increased by 190% ($p < 0.001$), followed by a decrease to the 16th week of age by 37% ($p < 0.001$).

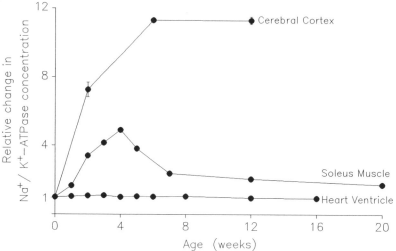

**Figure 1.** Relative change in rat excitable tissue $Na^+/K^+$-ATPase concentration expressed per g wet weight in relation to age. Values from cerebral cortex and soleus muscle obtained from (7,11).

838

This biphasic change in $Ca^{2+}$-dependent $p$NPPase activity indicate a higher synthesis of myocardial $Ca^{2+}$-ATPase than muscle mass during the first weeks of life followed by a period with a larger growth of muscle mass than synthesis of $Ca^{2+}$-ATPase. It was calculated that the $K^+$-dependent $p$NPPase activity constitutes 7-20 times the $Ca^{2+}$-dependent $p$NPPase activity. Although the two ion-pumps show commen pattern of developmental change it is evident that the regulatory changes are dissimilar due to difference in relative amount of the two ion-transporters during development. In the heart $Ca^{2+}$-handling by the $Ca^{2+}$-ATPase is of importance for inotropy. Thus the observed age-dependent changes in myocardial $Ca^{2+}$-ATPase concentration may be of importance for inotropic aspects of heart development.

## Acknowledgements

Stine Laurentzius is thanked for skilled technical assistance and Stig Haunsø for valuable discussions. This work was in part supported by grants from Johan and Hanne Weimann's Foundation, Georg Bestle and Wife's Memorial Foundation, NOVO's Foundation, The Danish Heart Foundation, The Foundation to the Advancement of Medical Science, Leo Pharmaceutical Products and Bristol-Myers Squibb, Denmark.

## References

1. Clausen T, Van Hardeveld C, Everts ME (1991) Significance of cation transport in control of energy metabolisme and thermogenesis. Physiol Rev 71:733-774
2. Hansen O, Clausen T (1988) Quantitative determination of $Na^+/K^+$-ATPase and other sarcolemmal components in muscle cells. Am J Physiol 254:C1-C7
3. Inesi G (1971) p-nitrophenyl phosphate hydrolysis and calcium ion transport in fragmented sarcoplasmic reticulum. Science 171:901-903
4. Kjeldsen K, Grøn P (1989) Skeletal muscle $Na^+, K^+$-pump concentration in children and its relationship to cardiac glycoside distribution. J Pharmacol Exp Ther 250:721-725
5. Kjeldsen K, Grøn P (1990) Age-dependent change in myocardial cardiac glycoside receptor ($Na^+, K^+$-pump) concentration in children. J Cardiovasc Pharmacol 15:332-337
6. Kjeldsen K, Nørgaard A (1987) Quantification of rat sciatic nerve $Na^+/K^+$-ATPase by measurements of $^3$H-ouabain binding in intact nerve samples. J Neurol Sci 79:205-219
7. Kjeldsen K, Nørgaard A, Clausen T (1984) The age-dependent changes in the number of $^3$H-ouabain binding sites in mammalian skeletal muscle. Pflügers Arch 402:100-108
8. Kjeldsen K, Nørgaard A, Hansen O, Clausen T (1985) Significance of skeletal muscle digitalis receptors for $^3$H-ouabain distribution in the guinea pig. J Pharmacol Exp Ther 234:720-727
9. Nørgaard A, Kjeldsen K, Hansen O (1985) $K^+$-dependent 3-O-methylfluorescein phosphatase activity in crude homogenate of rodent heart ventricle: effect of $K^+$- depletion and changes in thyroid status. Eur J Pharmacol 113:373-382
10. Schatzmann HJ (1989) The calcium pump of the surface membrane and of the sarcoplasmic reticulum. Annu Rev Physiol 51:473-485
11. Schmidt TA, Larsen JS, Kjeldsen K (1992) Quantification of rat cerebral cortex $Na^+/K^+$-ATPase: Effect of age and potassium depletion. J Neurochem 59:2094-2104
12. Skou JC (1974) Effect of ATP on the intermediary steps of the reaction of the $(Na^+/K^+)$-dependent enzyme system. III. Effect on the $p$-nitrophenylphosphatase activity of the system. Biochim Biophys Acta 339:258-273

# Digoxin Induced Dysfunction of Skeletal Muscle Potassium Homeostasis During Exercise

H. Bundgaard, *T.A. Schmidt, +H.L. Olesen, +N.H. Secher and K. Kjeldsen

Departments of Medicine B, *Cardiothoracic Surgery RT and +Anaesthesiologi AN, Rigshospitalet, University of Copenhagen, Blegdamsvej 9, 2100 Copenhagen, Denmark.

## Introduction

Digitalization has been shown to cause an occupancy of 24 - 34% of the $Na^+/K^+ATPase$ in human heart (22-24). Furthermore, it has been shown that the $Na^+/K^+$-ATPase concentration in human heart is reduced in dilated cardiomyopathy and heart failure (19,20). Such reductions in capacity for active $Na^+$ and $K^+$ transport may be a potential cause of local electrolyte derangement.

Moreover, skeletal muscle potassium homeostasis is of importance for the heart during exercise. The leak of potassium from the working skeletal muscles can be calculated to 15 $\mu$mol/g wet wt./min. Human skeletal muscles contain around 300 pmol/g wet wt. $Na^+/K^+$-ATPase, and accordingly have a maximum capacity for potassium uptake of around 5 $\mu$mol/g wet wt./min (molecular activity of $Na^+/K^+$-ATPase: 16,000 potassium ions transported per minute) (2,5,7,9,10). At rest only a small percentage of this capacity is used, but during maximum physical stimulation the entire capacity may probably be called for (3,5). It can be calculated that the total leakage of potassium from the entire skeletal muscle pool in healthy human subjects may amount to 420 mmol/min and that the maximum capacity for active reuptake is around 130 mmol/min (10). In comparison the extracellular space contains only around 50 mmol potassium. During exercise the capacity for active potassium uptake in skeletal muscle may be exceeded, and interstitial potassium will equilibrate with plasma-potassium, which thus may rise to at least 8 mmol/l (15).

This extrarenal potassium homeostasis is influenced by changes in activity of existing $Na^+/K^+$-ATPase as well as in concentration of $Na^+/K^+$-ATPase. Thus, catecholamines stimulate $Na^+/K^+$-ATPase mediated potassium uptake in skeletal muscles (2). Consequently, $\beta$ - adrenoceptor agonists reduce plasma-potassium concentration (17,26) and $\beta$ - adrenoceptor antagonists elevate plasma-potassium concentration during exercise (1,13). Also insulin stimulates potassium uptake in skeletal muscle (2,18), and reduces plasma-potassium concentration. In hyperthyroidism the $Na^+/K^+$-ATPase concentration in human skeletal muscle is increased by up to 68% (11) and hyperthyroidism is associated with hypokalaemic attacks (6,16). In physical conditioning in humans plasma-potassium concentration raise during exercise is deminished (12) and sprint training increases skeletal muscle $Na^+/K^+$-ATPase concentration by up to 16% (14). It is of special interest in relation to digitalization that a decrease of 25% was found in skeletal muscle $Na^+/K^+$-ATPase concentration in biopsies from patients with congestive heart failure (20).

With or without a concomitant alteration in myocardial $Na^+/K^+$-ATPase concentration major fluctuations in plasma-potassium level or disturbed extrarenal potassiun handling may pose a risk for devolopment of arrhythmias. On this basis it was of interest to study muscle potassium homeostasis in digitalized human subjects where digitalization has been shown to inhibit 5-13% (8,25) of skeletal muscle $Na^+/K^+$-ATPase as well as it may be reduced by heart failure per se.

## Methods and results

The study was approved by the local ethics committee according to Helsinki Declaration II. Eight in-patients aged $49 \pm 3$ years with congestive heart failure on ischemic basis were studied. All of the patients had acid-base-status, plasma-potassium, -sodium, -creatinine, $-T_3$, $-T_4$ and -TSH within normal range and clinically no other major diseases were found. The study was performed immediately before and two days after digitalization. The digitalization was performed by giving a loading dose of Digoxin® followed by maintenance dose according to standard procedures. After digitalization fasting plasma-digoxin was determined in the morning of the exercise test to $1.3 \pm 0.3$ nmol/l. Heart failure was classified to New York Heart Association class I-II. Angina pectoris was classified to Canadian Cardiovascular Society class I-II. Angiographic examination demonstrated a left ventricular ejection fraction of 25-60 % and coronary artery disease in all patients. Exercise test was performed in the supine position for eight minutes on a cycle ergometer with a workload of 40% of earlier determined maximum work capacity. Catheterisation of the femoral vein and artery was performed. Before and from the 4.-7. minute during the exercise tests blood samples were drawn simultaneously from the femoral artery and vein. Sampling was performed at the same time in the two tests. ABL 4, Radiometer, Copenhagen, was used for plasma-potassium measurements. Results are given as mean value $\pm$ SEM. Significance was ascertained by Students two tailed $t$-test for paired observations.

At rest plasma-potassium concentration decreased in venous blood before digitalization from $4.21 \pm 0.09$ to $4.11 \pm 0.06$ mmol/l after digitalization and in arterial blood from $4.23 \pm 0.10$ to $4.11 \pm 0.07$ mmol/l, i.e. 2-3% (p = 0.08 and 0.12, respectively). After digitalization maximum plasma-potassium concentration during the present exercise load was reduced from $5.15 \pm 0.11$ to $5.08 \pm 0.09$ mmol/l in venous blood compared to before digitalization and from $5.01 \pm 0.12$ to $4.86 \pm 0.09$ mmol/l in arterial blood, i.e. 1-3% (p = 0.45 and 0.20, respectively). Before digitalization plasma-potassium concentration increased in venous blood from $4.21 \pm 0.09$ at rest to $5.15 \pm 0.11$ mmol/l during exercise and in arterial blood from $4.23 \pm 0.10$ to $5.01 \pm 0.12$ mmol/l, i.e. 22-18% (p < 0.001 for both). Thus veno-arterial (V-A)-difference calculated on the basis of the individual data increased from $-0.01 \pm 0.05$ to $0.13 \pm 0.02$ mmol/l, i.e. by 0.14 mmol/l (p = 0.01). After digitalization plasma-potassium concentration increased in venous blood from $4.11 \pm 0.06$ at rest to $5.08 \pm 0.09$ mmol/l during exercise and in arterial blood from $4.11 \pm 0.07$ to $4.86 \pm 0.09$ mmol/l, i.e. 24-18% (p<0.001 for both). Thus V-A-difference increased from $0.01 \pm 0.04$ to $0.22 \pm 0.03$ mmol/l, i.e. by 0.21 mmol/l (p = 0.004). The magnitude of the increase in V-A-difference by digitalization was 50%.

841

| Plasma-K⁺ (mmol/l) | Rest | | Exercise | |
|---|---|---|---|---|
| | ÷ digoxin | + digoxin | ÷ digoxin | + digoxin |
| Venous | 4.21 ± 0.09 | 4.11 ± 0.06 | 5.15 ± 0.11 | 5.08 ± 0.09 |
| Arterial | 4.23 ± 0.10 | 4.11 ± 0.07 | 5.01 ± 0.12 | 4.86 ± 0.09 |
| V-A-difference | −0.01 ± 0.05 | 0.01 ± 0.04 | 0.13 ± 0.02 | 0.22 ± 0.03 |

Figure. Arterial and venous plasma-potassium concentration before and after digitalization measured simultaneously in the femoral vein and artery at rest and during exercise with a constant workload at 40% of maximal work capacity. Values were determined after the same duration of exercise before and after digitalization. Note that V-A-differences were calculated on the basis of the individual pairs of venous and arterial plasma-potassium values.

## Discussion

The present study demonstrates an altered extrarenal potassium homeostasis after digitalization. A tendency to a decreased resting plasma-potassium concentration after digitalization was observed. This accord with earlier studies (4,21) and may be due to a diuretic effect of digoxin. In line with the lower resting values of plasma-potassium concentrations seen after digitalization the present study showed decreased values during the exercise test after digitalization. The V-A-difference reflecting the net potassium leak from the leg increased significantly after digitalization. This probably directly reflects impaired skeletal muscle capacity for active potassium reuptake during exercise due to inhibition of $Na^+/K^+$-ATPase by digoxin. Especially in patients suffering from congestive heart failure or dilated cardiomyopathy with reduced $Na^+/K^+$-ATPase concentration in the heart and an even further reduction in functioning $Na^+/K^+$-ATPase by digitalization, the increased V-A-difference during exercise may pose a risk for development of arrhythmias during exercise.

## Acknowledgements

We thank Stine Laurentzius and Grete Simonsen for skilled technical assistance and Stig Haunsø for valuable discussions. This study was in part supported by grants from The Danish Heart Foundation, The Faculty of Medicine of Copenhagen University, The Danish Sports Research Foundation, NOVO´s Foundation, Nycomed DAK and LEO Pharmaceutical's Foundation.

## References

1.    Carlsson E, Fellenius E, Lundborg P, Svensson L (1978) beta-Adrenoceptor blockers, plasma-potassium, and exercise. Lancet 2:424-425
2.    Clausen T (1986) Regulation of active $Na^+$-$K^+$ transport in skeletal muscle. Physiol Rev 66:542-580
3.    Clausen T, Everts ME, Kjeldsen K (1987) Quantification of the maximum capacity for active

sodium-potassium transport in rat skeletal muscle. J Physiol Lond 388:163-181

4. Edner M, Ponikowski P, Jogestrand T (1993) The effect of digoxin on the serum potassium concentration. Scand J Clin Lab Invest 53:187-189

5. Everts ME, Clausen T (1992) Activation of the Na-K pump by intracellular Na in rat slow- and fast-twitch muscle. Acta Physiol Scand 145:353-362

6. Feldman DL, Goldberg WM (1969) Hyperthyroidism with periodic paralysis. Can Med Assoc J 101:61-65

7. Hansen O, Clausen T (1988) Quantitative determination of Na$^+$/K$^+$-ATPase and other sarcolemmal components in muscle cells. Am J Physiol 254:C1-C7

8. Joreteq T, Jogestrand T (1984) Physical exercise and binding of digoxin to skeletal muscle - effect of muscle activation frequency. Eur J Clin Pharmacol 27:567-570

9. Kjeldsen K (1987) Regulation of the concentration of $^3$H-ouabain binding sites in mammalian skeletal muscle - effects of age, K-depletion, thyroid status and hypertension. Dan Med Bull 34:15-46

10. Kjeldsen K (1991) Muscle Na,K-pump dysfunction may expose the heart to dangerous K levels during exercise. Can J Sport Sci 16:33-39

11. Kjeldsen K, Nørgaard A, Gøtzsche CO, Thomassen A, Clausen T (1984) Effect of thyroid function on number of Na-K pumps in human skeletal muscle. Lancet 2:8-10

12. Kjeldsen K, Nørgaard A, Hau C (1990) Exercise-induced hyperkalaemia can be reduced in human subjects by moderate training without change in skeletal muscle Na,K-ATPase concentration. Eur J Clin Invest 20:642-647

13. Lundborg P (1983) The effect of adrenergic blockade on potassium concentrations in different conditions. Acta Med Scand Suppl 672:121-126

14. McKenna MJ, Schmidt TA, Hargreaves M, Cameron L, Skinner SL, Kjeldsen K (1993) Sprint training increases human skeletal muscle Na,K-ATPase concentration and improves K regulation. J Appl Physiol 75(1):173-180

15. Medbø JI, Sejersted OM (1990) Plasma potassium changes with high intensity exercise. J Physiol Lond 421:105-122

16. Miller D, delCastillo J, Tsang TK (1989) Severe hypokalemia in thyrotoxic periodic paralysis. Am J Emerg Med 7:584-587

17. Montoliu J, Lens XM, Revert L (1987) Potassium-lowering effect of albuterol for hyperkalemia in renal failure. Arch Intern Med 147:713-717

18. Moore RD (1983) Effects of insulin upon ion transport. Biochim Biophys Acta 737:1-49

19. Nørgaard A, Bagger JP, Bjerregaard P, Baandrup U, Kjeldsen K, Thomsen PEB (1988) Relation of left ventricular function and Na,K-pump concentration in suspected idiopathic dilated cardiomyopathy. Am J Cardiol 61:1312-1315

20. Nørgaard A, Bjerregaard P, Baandrup U, Kjeldsen K, Reske-Nielsen E, Thomsen PEB (1990) The concentration of the Na,K-pump in skeletal and heart muscle in congestive heart failure. Int J Cardiol 26:185-190

21. Nørgaard A, Bötker HE, Klitgaard NA, Toft P (1991) Digitalis enhances exercise induced hyperkalaemia. Eur J Clin Pharmacol 41:609-611

22. Rasmussen HH, Okita GT, Hartz RS, Eick RET (1990) Inhibition of electrogenic Na(+)-pumping in isolated atrial tissue from patients treated with digoxin. J Pharmacol Exp Ther 252:60-64

23. Schmidt TA, Allen PD, Colluci WS, Marsh JD, Kjeldsen K (1993) No adaption to digitalization as evaluated by digitalis receptor (Na,K-ATPase) quantification in explanted hearts from donors without heart disease and from digitalized patients in endstage heart failure. Am J Cardiol 70:110-114

24. Schmidt TA, Holm-Nielsen P, Kjeldsen K (1991) No upregulation of digitalis glycoside receptor (Na,K-ATPase) concentration in human heart left ventricle samples obtained at necropsy after long term digitalization. Cardiovasc Res 25:684-691

25. Schmidt TA, Holm-Nielsen P, Kjeldsen K (1993) Human skeletal muscle digitalis glycoside receptors (Na$^+$/K$^+$-ATPase) - importance during digitalization. Cardiovasc Drugs Ther 7:175-181

26. Wang P, Clausen T (1976) Treatment of attacks in hyperkalaemic familial periodic paralysis by inhalation of salbutamol. Lancet 1:221-223

# Na$^+$/K$^+$-ATPase Isoform Expression in Cardiac Muscle from Dahl Salt-Sensitive and Salt-Resistant Rats

P.H. Ogden, G. Cramb

School of Biological and Medical Sciences, University of St. Andrews, St. Andrews, Fife, UK. KY16 9TS

## Introduction

Dahl rats have been selectively bred for their sesitivity (S) or resistance (R) to dietary salt-induced hypertension [1,2]. There is evidence that a number of genetic lesions may be associated with an elevated dietary sodium intake and the pathogenesis of hypertension in the S strain [2-5]. The possibility that these genetic lesions are associated with changes in the expression of Na$^+$/K$^+$-ATPase activities in various tissues are currently being investigated in our laboratory. Here we present our preliminary results in our study of the expression of Na$^+$/K$^+$-ATPase in atrial and ventricular tissues of S and R rats maintained on normal or high salt diets.

## Animals

Dahl resistant (R) and Dahl sensitive (S) rats were maintained by brother x sister mating. At 5 weeks of age, rats from either strain were selected and allowed free access to food (R & M 1 Cube Diet containing 0.8% w/w NaCl) and water. After 10 days the rats from each strain were split into two dietary groups, one of which continued on the normal (N) salt diet (0.8% w/w NaCl) and the other was fed a high (H) salt diet (modified R & M 1 Cube Diet containing 8% w/w NaCl). All groups of animals (designated as NR, HR, NS and HS) were allowed free access to water. The blood pressures of sub-groups of 3 animals of both sexes from each strain and each dietary regime were routinely monitored at two-three day intervals using a tail cuff monitor. After 5 weeks of either dietary regime, all rats were heparinised (5000 units/kg, i.p.) and killed by exsanguination following ether anaesthesia. Atrial and ventricular muscle was dissected free of connective tissue, washed in ice cold phosphate buffered saline, snap frozen in liquid N$_2$ and stored at -90 $^{\circ}$C until required.

## Results

### Blood pressures

Systolic, mean and diastolic blood pressures of subsets of animals from male and female groups at the end of the dietary salt regime were recorded using an Apollo Model 179 Blood Pressure Analyzer (IITC/Life Sciences, California, USA) and are given in Fig. 1. In both Dahl-R and Dahl-S strains, the high salt diet induced a statistically significant increase in systolic blood pressure although only Dahl-S rats showed a significant increase in mean and diastolic pressures when given the high salt diet. Dahl-S rats also exhibited significant increases in all blood pressure measurements compared to their Dahl-R counterparts on the equivalent diet. The highest blood pressures were recorded from the female Dahl-S rats maintained on the high salt diet.

844

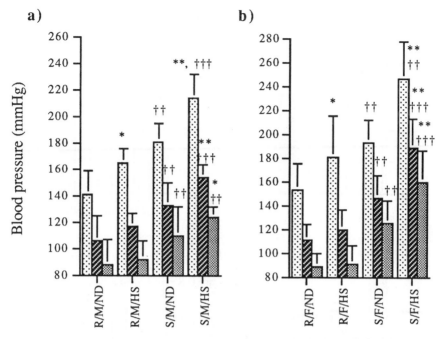

Figure 1 Blood pressure recordings taken from male (a) and female (b) Dahl sensitive (S) and Dahl resistant (R) rats after 5 weeks on normal (ND) and high salt (HS) diets. ( Light stipple, systolic; hash, mean; dark stipple, diastolic pressures).

Values are averages ± SD of 9 or 12 individual recordings from 3 animals in each group measured on three or four separate days. * and † indicate significant differences between R vs R or S vs S and R vs S respectively. (*, † = p < 0.05; **,†† = p < 0.01; ††† = p < 0.001)

## $Na^+/K^+$-ATPase expression

Northern blots for atrial and ventricluar RNAs were probed consecutively with cDNAs specific for $\alpha1$, $\alpha2$, $\alpha3$ and $\beta1$ $Na^+/K^+$-ATPase subunit and $\alpha$-actin mRNAs. In all cases the amount of RNA present in each blot was determined by re-probing with a radiolabelled synthetic antisense oligonucleotide to ribosomal 18S RNA and all mRNA signals expressed as ratios compared to the 18S RNA signal. In the atria, results from both male and female groups indicate small increases in $\alpha1$ and $\beta1$ mRNA expression in the HR, NS and HS groups (results not shown) and increased expression appears to correlate with increased blood pressure. Alpha 2 and $\alpha3$ signals were too low to quantify accurately. Interestingly in the atria, the relative amount of actin mRNA (often used as a control for general mRNA levels) was increased when either group of animals were maintained on the high salt diet (results not shown). Relative increases in actin mRNA abundance were similar in both R and S groups and did not appear to correlate with increases in blood pressure.

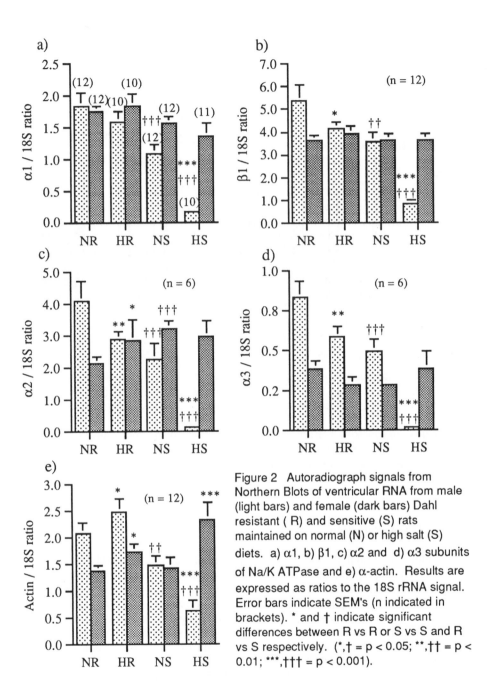

Figure 2 Autoradiograph signals from Northern Blots of ventricular RNA from male (light bars) and female (dark bars) Dahl resistant ( R) and sensitive (S) rats maintained on normal (N) or high salt (S) diets. a) α1, b) β1, c) α2 and d) α3 subunits of Na/K ATPase and e) α-actin. Results are expressed as ratios to the 18S rRNA signal. Error bars indicate SEM's (n indicated in brackets). * and † indicate significant differences between R vs R or S vs S and R vs S respectively. (*,† = p < 0.05; **,†† = p < 0.01; ***,††† = p < 0.001).

In the ventricle there were marked differences in mRNA abundance between male and female groups (Fig. 2). Total ventricluar RNA extracts from male S rats exhibited marked decreases in $\alpha 1$, $\alpha 2$, $\alpha 3$ and $\beta 1$ mRNA's compared to the male R group and these decreases were even more pronounced when animals were fed the high salt diet. This decrease in mRNA abundance was not limited to $Na^+/K^+$-ATPase as actin mRNA showed a similar trend (Fig. 2 e). There were indications of similar but smaller decreases in $\alpha 1$ and $\beta 1$ mRNAs in female animals from both R and S rats when fed high salt, but $\alpha 2$, $\alpha 3$ and actin mRNAs appeared unchanged (Fig. 2 a-e).

## Discussion

The relationship between blood pressure and the potential increases and decreases in $Na^+/K^+$-ATPase mRNA abundance in the atria and ventricle respectively, supports our previous observations with male Milan hypertensive rats (6) and with the earlier reports of Herrera et al. (7) using DOCA-salt and angiotensin II models of hypertension (sex unknown). The associated reduction in actin mRNA (Fig. 2) and the reduced RNA yield from both male Milan (6) and male Dahl rats (results not shown) also suggests that the reduction in mRNA expression is not limited to $Na^+/K^+$-ATPase but may reflect a general decrease in transcription/ mRNA stability in this tissue. A general reduction in mRNA levels is especially difficult to reconcile in Dahl-S rats where there is a marked hypertrophy of the ventricular tissue in the S rats fed high salt (results not shown). The reason(s) for the measured differences in RNA yields and expression between male and female animals remain unresolved, especially since the female Dahl-S rats fed the high salt diet exhibited slightly higher (although not statistically significant) systolic and diastolic blood pressures than the equivalent male group and also show significant ventricular hypertrophy when systolic blood pressures were elevated above 200 mmHg.

## Acknowledgements

This work was funded by the British Heart Foundation (Grant 89/56).

## References

1.      Rapp JP (1982) Dahl salt susceptable and salt-resistant rats: A review. Hypertension 4: 753-763.
2.      Rapp JP, Dene H (1985) Development and characteristics of inbred strains of Dahl salt-sensitive and salt-resistant rats. Hypertension 7:340-349.
3.      Fink GD, Takashita A, Mark AL, Brody MJ (1980) Determinants of renal vascular resistance in the Dahl strain of genetically hypertensive rat. Hypertension 2: 274-280.
4.      Giradin E, Caverzsio J, Iwai J, Bonjour J, Muller AF, Grandchamp A (1980) Pressure natriuresis in isolated kidney from hypertensive-prone and hypertensive-resistant rats (Dahl rats). Kidney Int. 18: 10-19.
5.      Gomez-Sanchez EP, Fort C, Thwaites D (1992) Central mineralocorticoid receptor antagonism blocks hypertension in Dahl S/JR rats. Am J Physiol. 262: E96-E99
6.      Cutler CP, Cramb G (1991) Isoforms of Na,K-ATPase in myocardial tissues of the Milan hypertensive rat. in "The Sodium Pump: Recent Developments" Society of General Physiologists Series Vol 46 (2) pp 635-639 edit JH Kaplan and P DeWeer, Rockefeller University Press.
7.      Herrera VLM, Chobanian AV, Ruiz-Opazo N (1988) Isoform-specific modulation of Na,K-ATPase $\alpha$-subunit gene expression in hypertension. Science 241: 221-223.

# Na$^+$/K$^+$-ATPase from spontaneously hypertensive rats

B.M. Anner, G.T. Haupert, Jr.

The Laboratory of Experimental Cell Therapeutics, Geneva University Medical School, CH-1211 Geneva 4, Switzerland and The Renal Unit, Massachusetts Hospital, Harvard Medical School, Charlestown, MA 02129 USA

## Introduction

The kidney is very rich in Na$^+$/K$^+$-ATPase (NKA) or sodium pump which, being responsible for the Na reabsorption, constitutes about 20% of the membrane protein. In view of its essential role in Na metabolism, the NKA is also suspected to be involved in the pathogenesis of low-renin hypertension, a disease associated with insufficient natriuresis (2). Since blood pressure rises in previously normotensive rats after implantation of kidneys from genetically hypertensive rats, and since similar correlations have been established in men, some pathogenic factors leading to essential hypertension could be renal as formulated by the concept of the "defective kidney" (1). In view of the pivotal role of NKA in renal salt handling we wished to look whether renal NKA of genetically hypertensive rats was altered in such a way as to account for the possibly enhanced renal Na reabsorption. Either qualitative or quantitative defects could cause enhanced Na-transport: normally expressed NKA with hyperaffinity for Na ions or hyperexpressed NKA with normal Na-affinity. To assess the NKA content of genetically hypertensive rats, NKA was isolated simultaneously and quantitatively from renal outer medulla, renal cortex and brain of genetically normo- and hypertensive rats and the protein recovery, specific activity and yield compared.

## Na,K-ATPase activity and yield from kidney and brain

To compare quantity and quality of NKA from hyper- and normotensive animals, the kidneys and brains of Milan hypertensive (MHS) and normotensive rats (MNS) were isolated in parallel, the microsomes were isolated and the NKA first unmasked and then purified by detergent (SDS) treatment. As shown in Table 1, no difference in kidney weight was seen between MHS and MNS whereas slightly less outer medulla was recovered from MHS as compared to MNS and less protein in the microsomal fraction. However, when the microsomal NKA was activated by SDS, 23% more enzyme activity was unmasked in the MHS microsomes indicating possible higher microsomal NKA concentration. The higher NKA concentration in MHS became more apparent by NKA purification as 74% more NKA with 56% higher specific activity could be extracted from MHS microsomal protein as compared to MNS (Table 1).

**Table 1. Comparative activity and yield of outer renal medulla Na,K-ATPase**

| Purification steps | MHS | MNS | MHS/MNS |
|---|---|---|---|
| Kidneys No. | 10 | 10 | 1 |
| weight (g) | 10.06 | 11.6 | 0.86 |
| Outer medulla, weight (g) | 0.57 | 0.73 | 0.78 |
| Microsomes, protein (mg) | 2.3 | 2.9 | 0.79 |
| specific activity + SDS (%) | 123.3 $\pm$ 14.4* | 100 | 1.23 |
| Purified NKA, protein (mg) | 0.208 | 0.151 | 1.38 |
| NKA recovery (% microsome protein) | 9.04 | 5.21 | 1.74 |
| Specific activity ($\mu$mol/mg/h) | 1651 | 1061 | 1.56 |
| Yield ($\mu$mol/h) | 343.4 | 160.21 | 2.14 |

*sem; n=6

Genetically hypertensive Milan rat strains (MHS) and their normotensive controls (MNS) were a gift of Prof. G. Bianchi; their age was 10 to 14 weeks, their weight 250 to 380 g, their systolic arterial pressure (mm Hg) 120 to 140 (MNS) and 160 to 180 mm (MHS). The dissected renal outer medulla was homogenised in a glass potter containing 8 ml solution of 250 mM sucrose, 30 mM histidine, 1 mM EDTA, pH 7.2 and subjected to 15 min centrifugation at 6000 x g, the supernatant, the procedure repeated for 25 min at 6000 x g, the supernatants combined and recovered as a pellet after a 30 min centrifugation at 48,000 x g. To measure detergent activation, the microcosms were incubated in a solution containing 11.2 mg of microsomal protein (1.6 mg/ml) in 7 ml of 3 mM $Na_2ATP$, 25 mM imidazole, 1 mM Tris-EDTA, 2.5 mM SDS for 20 min at 25 °C. To purify NKA, the detergent-treated microsomes were separated by centrifugation in 4 ml 15% sucrose, 16 ml 25% sucrose, 25 mM imidazole, 1 mM Tris-EDTA, pH 7.5, 0 °C for 110 min at 250,000 x g. and the pellet containing purified NKA recovered (3). The linked-enzyme assay was used for the continuous determination the Na,K-ATPase activity: 0.5 to 2 $\mu$g Na,K-ATPase protein was added to 1 ml of 0.3 mM NADH, 2.5 mM phospho-enolpyruvate, 30 mM imidazole, 1 mM Tris-EDTA, 2.5 mM ATP, 5 mM $MgCl_2$, 100 mM NaCl, 10 mM KCl, 8 $\mu$l pyruvate-kinase/lactate-dehydrogenase suspension, pH 7.2. The rate of NADH oxidation was recorded at 340 nm.

In agreement with the augmented NKA observed in MHS outer renal medulla (Table 1), MHS renal cortex seemed also richer in NKA: when the NKA activity was unmasked by SDS, the peak activity was 24 % higher than in MNS in perfect accordance with renal outer medulla (Table 2). Further, 3.5 times more NKA was recovered from MHS microsomes as compared to MNS. Thus, despite lower specific activity of the NKA isolated from MHS cortex, the yield from an equal amount of cortex microsomes was 2.9 higher in MHS as compared to MNS in line with the 2.14 fold increased yield of medulla NKA (Table 1). Along the same line, MHS brain appeared richer in NKA than MNS brain as illustrated by the comparative measurements listed in Table 3: twice as much NKA protein with over twice as much specific activity could be isolated from MHS brain as compared microsomes

**Table 2. Comparative activity and yield of renal cortex Na,K-ATPase**

| Purification steps | MHS | MNS | MHS/MNS |
|---|---|---|---|
| Microsomes, protein (mg) | 6.66 | 6.61 | 1.01 |
|    specific activity + SDS (%) | 124.4 ± 6.8* | 100 | 1.24 |
| Purified NKA, protein (mg) | 1.426 | 0.414 | 3.44 |
| NKA recovery (% microsome protein) | 21.6 | 6.2 | 3.48 |
| Specific activity (μmol/mg/h) | 246 | 293 | 0.84 |
| Yield (μmol/h) | 121.3 | 350.8 | 2.89 |

*sem; n = 12

Na,K-ATPase was isolated from kidney cortex as described in Legend to Figure 1

**Table 3. Comparative activity and yield of brain Na,K-ATPase**

| Purification steps | MHS | MNS | MHS/MNS |
|---|---|---|---|
| Brain No. | 1 | 1 | 1 |
| Microsomes, protein (mg) | 19.46 | 10.24 | 1.90 |
| Purified NKA, protein (mg) | 0.492 | 0.24 | 2.05 |
| NKA recovery (% microsome protein) | 2.51 | 2.34 | 1.07 |
| Specific activity (μmol/mg/h) | 277 | 128 | 2.16 |
| Yield (μmol/h) | 136.28 | 30.72 | 5.32 |

Na,K-ATPase was isolated from 5 MHS brains and 6 MNS brains as described in Legend to Figure 1; for better comparison, values calculated per brain are shown.

That the differences in specific activity and recovery were not due to unequal purity was controlled by separation and analysis of the purified MHS and MNS preparations by gel electrophoresis; the alpha and beta subunit content of MHS and MNS preparations was identical purity (data not shown).

850

The reported procedures and results were the only experimental way we could see to quantify functional, membrane inserted alpha--beta NKA entities in MHS and MNS rat, a ouabain-resistant species precluding NKA-titration by $^3$H-ouabain. From our comparative study it appears clearly that the MHS renal outer medulla contains about twice as much NKA than MNS controls of equal age and weight; renal cortex contained about 3 times more and the brain even 5 times more NKA. Such a tendency was observed also by Parenti et al.(5) who found 1.4 more NKA activity and Na-transport in MHS renal cortex membranes and by Melzi et al. who report 1.78 more NKA activity in PCT and 1.4 more in TAL segments of isolated MHS nephrons as compared to MNS (4). Taken together, the results obtained in kidney consistently indicate 1.4 to 2 fold increased NKA in MHS. In conjunction with the functional normality of NKA purified from MHS (Anner and Haupert, manuscript submitted) the augmented NKA expression might reflect a faulty gene repression mechanism. In support of this hypothesis, augmented NKA gene expression leading to 4 to 5 fold increased NKA mRNA in heart of spontaneously hypertensive SHR rats has been demonstrated and proposed to be an early or even primary pathogenic event (6). The putative altered gene expression must be so stable and predominant that a normal organism receiving a piece of "hypertensive" tissue is unable to correct the defect.

*Acknowledgements*

We are grateful to Mrs M. Moosmayer and E: Meneghini for experimental work and to Prof. G. Bianchi, Drs P. Ferrari, M Ferrandi, G. Barber for Milan rats, advice and many helpful discussion. Supported by the Swiss National Science Foundation (grant No. 31-25666.88 to BMA).

# References

1.  Blaustein MP, Hamlyn JM (1984) Sodium transport inhibition, cell calcium and hypertension. The natriuretic hormone/ Na-Ca exchange/ hypertension hypothesis. Am. J. Med., Oct. 5, 45-59
2.  De Wardener HE, Clarkson EM (1985) Concept of natriuretic hormone. Physiol. Rev. 65: 658-759
3.  Jørgensen PL (1974) Isolation of (Na$^+$/K$^+$-ATPase)-ATPase. Methods Enzymol. 32: 277-290
4.  Melzi M, Bertorello A, Fukuda Y, Muldin I, Sereni F, Aperia A.(1989) Na,K-ATPase activity in renal tubule cells from Milan hypertensive rats. Am. J. Hypertension 2: 563-566
5.  Parenti P, Villa M, Anozet GM, Ferrandi M, Ferrari P (1991) Increased Na pump activity in the kidney cortex of the Milan hypertensive rat strain. FEBS 290: 200-204
6.  Tsuruya Y., Ikeda U, Kawakami K, Nagano K, Kamitani T, Oguchi A, Ebata H, Shimada K, Medford RM (1991) Augmented Na,K-ATPase gene expression in spontaneously hypertensive rat hearts. Clin. Exper. Hypertension. A13: 1213-1222

851

# Na$^+$/K$^+$-pump $\alpha$3 Isoform in Rat Vascular Smooth Muscle and its Physiological Role

M. Juhaszova, R.W. Mercer*, D.N. Weiss, D.J. Podbersky, J. Heidrich, M.P. Blaustein

Dept. Physiol., U. MD Med. Sch., Baltimore, MD 21201 & *Dept. Cell Biol. and Physiol., Wash. U. Med. Sch., St. Louis, MO 63110.

## Introduction

The Na$^+$/K$^+$-ATPase is a multimeric ($\alpha$, ß) transmembrane complex expressed in arterial smooth muscle (ASM), that plays a major role in smooth muscle homeostasis by regulating the transmembrane Na$^+$ gradient. The catalytic $\alpha$ subunit contains the binding site for cardiotonic steroids such as ouabain. Ouabain blocks the Na$^+$/K$^+$-pump and raises the cytosolic Na$^+$ concentration; in turn, the cytosolic Ca$^{2+}$ concentration is increased via Na$^+$/Ca$^{2+}$ exchange and Ca$^{2+}$ is then pumped into the sarcoplasmic reticulum (SR) to buffer the cytosolic Ca$^{2+}$ (2). At least three $\alpha$-subunit isoforms, $\alpha$1, $\alpha$2 and $\alpha$3, have been characterized, with, respectively, low (mM), high ($\mu$M) and very high (nM), affinities for ouabain. Although the $\alpha$1 subunit is widely distributed in all tissues, the distribution of the $\alpha$2/$\alpha$3 is less well defined. Recent studies demonstrate that nanomolar plasma levels of ouabain are hypertensinogenic (13), strongly suggesting that a very high affinity $\alpha$ isoform must be present in ASM.

The present study was designed to determine whether rat ASM contains Na$^+$/K$^+$-pumps with a very high affinity to ouabain. Immunochemical analysis was used to determine the presence of different $\alpha$-isoforms of the Na$^+$/K$^+$-pump in mesenteric artery (MA) cells. Antibodies raised against $\alpha$1, $\alpha$2 and $\alpha$3 isoforms were tested. PCR analyses were employed to define the $\alpha$ isoform mRNA composition of the MA. In addition we tested the effects of different ouabain concentrations on caffeine-evoked contractions (CECs) to determine whether rat arteries are sensitive to low nanomolar concentrations of ouabain.

## Na$^+$/K$^+$ pump $\alpha$-isoform composition of rat mesenteric artery

Previous studies have established a developmental and tissue specific expression pattern for $\alpha$ isoforms of Na$^+$/K$^+$-ATPase (reviewed in 6,9 and 10). In the present study we used immunological and molecular genetic techniques to identify the appearance of different $\alpha$ isoforms of Na$^+$/K$^+$-pump in cultured MA cells.

Figure 1 shows immunoblots of microsomal membrane proteins from rat heart, brain and kidney and from the cultured mesenteric artery myocytes. The blots were probed with antibodies raised against $\alpha$1, $\alpha$2 and $\alpha$3 isoforms. A polyclonal antibody, specific for $\alpha$1 isozyme of Na$^+$/K$^+$-ATPase, was raised against the peptide sequence DKYEPAAVSEHGD derived from the N-terminus. Peptide-directed polyclonal antiserum, specific for $\alpha$3 isozyme, was raised against the N-terminal sequence GDKKDDKSSPKKS. The monoclonal antibody McB2 raised against rat brain, specific for the $\alpha$2 isoform, was generously provided by Dr. K.J. Sweadner. Relatively large amounts of α1 isoform were detected in all four rat tissues tested: heart, brain, kidney and artery (Fig. 1). The brain also exhibited prominent bands that crossreacted with both α2 and α3 antibodies. The α2 band from the heart was less prominent, that from the kidney was barely detectable, and there was no detectable α2 isoform in MA. A small amount of α3 crossreactivity was observed in the heart and kidney; the α3 antiserum also crossreacted with a band from the MA, although the labeling was light.

The presence of all three isoforms in rat brain and heart is well established (10), and there is increasing evidence for the presence of α3 as well as α2 isoforms in the kidney (5,8). Our PCR and Western blot data suggest that rat artery myocytes also possess an "α3-like" isoform of the Na$^+$/K$^+$-ATPase α subunit. However, the α3 band from brain migrates more slowly that the α3 band from other tissues. The different mobility raises the possibility that the α3 isoform from the brain may differ from the α3 subunit in the other tissues, but we cannnot rule out the possibility that the α1 isoform in these tissues crossreacts non-specifically with the α3 antiserum.

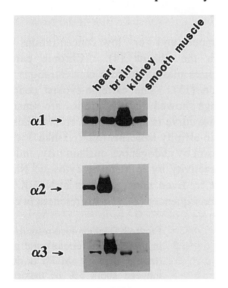

Figure 1: Western blot analysis of rat heart, brain, kidney and mesenteric artery cell microsomal membrane proteins (100 μg/lane). The immunoblots were probed with antibodies raised against the three different Na$^+$ pump α subunit isoforms. These experiments revealed the presence of a relatively large amount of α1 isoform in mesenteric artery membranes and a relatively small amount of α3 isoform; no α2 isoform was detected in the mesenteric artery membranes.

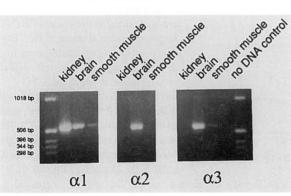

Figure 2: PCR analysis of gene transcription of the Na,K-ATPase α1, α2 and α3 isoforms in vascular smooth muscle. PCR reactions using isoform specific primers were analyzed (see text).

853

PCR analysis of gene transcription of the Na$^+$/K$^+$-ATPase $\alpha$1, $\alpha$2 and $\alpha$3 isoforms in different rat tissue is shown in Figure 2. mRNAs from rat kidney, brain and MA cells were reverse transcribed using random hexamers. Isoform specific 3' primers were: $\alpha$1: 5'-GCTTAGGCTCCGATGCGTTTGGGT; $\alpha$2: 5'-GGCTCTGGGGGCTGTCTTCCCT; $\alpha$3: 5'-ATCGGTTGTCATTGGGGTCCTCAGT. The 5' primer for all reactions consisted of 5'-CTGGCTGGAGGCTG TCATCTTCTTCAT. $\alpha$1 message was detected in each tissue tested: the relative intensities of the message (Fig. 2) parallel the relative intensities of the immunoreactivity (Fig. 1). $\alpha$2 isoform message was detected only in the brain, and $\alpha$3 message only in brain and smooth muscle.

Taken together, the Western blot and PCR data indicate that rat MA cells possess a large amount of the low affinity $\alpha$1 isoform and a modest quantity of $\alpha$3 (or "$\alpha$3-like") isoform of the Na$^+$/K$^+$-ATPase $\alpha$ subunit.

### Low dose ouabain effect on arterial contraction

Ouabain is an adrenocortical hormone, and very low concentrations of this cardiotonic steroid normally circulate in the plasma (4). Chronic parenteral administration of ouabain to normal rats raises plasma ouabain concentrations to low nanomolar levels and induces hypertension (13). The caffeine-evoked contraction experiments on small mesenteric artery rings showed that rat arteries are sensitive to nanomolar concentrations. The dose-response curve is biphasic; this corresponds to the presence of both, low and very high ouabain affinity isoforms ($\alpha$1/"$\alpha$3-like") (Fig.3). Physiologic regulation of the high affinity form by endogenous ouabain may, indirectly, modulate SR Ca$^{2+}$ content and vascular reactivity by influencing cytosolic Na$^+$ and, thus, Na$^+$/Ca$^{2+}$ exchange and cytosolic Ca$^{2+}$. Even minor changes in Na$^+$/K$^+$-pump function may have profound physiological consequences (2). The differences in cardiac

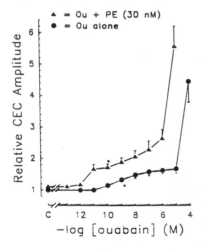

Figure 3: Ouabain dose-response curves in the absence (circles) and presence (triangles) of 30 nM PE (a dose that induced about 3% of maximal PE-evoked tension). The CEC amplitudes are normalized to the control CEC (C) amplitude in the absence of PE. Each data point corresponds to the mean from at least 8 rings. * = lowest ouabain concentration at which the CEC amplitude with ouabain was significantly different (p < 0.05) from the amplitude of the respective control CEC (C). The two curves are significantly different (ANOVA, p < 0.001). Mean control (C) CEC in the absence of PE was 61 ± 8

mg (N=23). Ouabain (10$^{-9}$ M) significantly augmented the CECs, and PE increased the sensitivity to 10$^{-10}$ M ouabain. This shows that ouabain increases SR Ca$^{2+}$.(from Ref. 12, with permission)

854

glycoside sensitivities of $\alpha$ isoforms are well known. The amino acid sequence of the H1-H2 (first extracellular) loop, especially amino acids at the border of this region, may be responsible for ouabain sensitivity (7). Sequences of the H1-H2 regions of $\alpha 2$ and $\alpha 3$ isoforms show high conservation of isoform-specific features. Takeyasu et al. (11) suggested that H1-H2 region is a prime candidate for structural characteristics that might provide isoform-specific susceptibility to regulation by endogenous digitalis-like inhibitors of the $Na^+/K^+$-pump.

We conclude that in rat mesenteric artery the modulation of high affinity "$\alpha 3$" isoform of the $Na^+/K^+$-ATPase $\alpha$ subunit with nanomolar ouabain is responsible for the physiological, pathophysiological response of the cells. Similarly, recent studies (3) also suggest that the very high affinity "$\alpha 3$" (1) isoform is responsible for the sensitizing effects of ouabain to glutamate neurotoxicity in rat neurons.

*Acknowledgements*: NIH grant HL-36573 to R.W M. and HL-45211 to M.P.B.

**References:**

1. Blanco G, Berberian G, and Beaugé L (1990) Detection of a highly ouabain sensitive isoform of rat brainstem Na,K-ATPase. Biochim Biophys Acta 1027:1-7
2. Blaustein MP (1993) The pathophysiological effects of endogenous ouabain: control of stored $Ca^{2+}$ and cell responsiveness. Am J Physiol (Cell Physiol 33):264:C1367-1387
3. Brines ML, Robbins RJ (1992) Inhibition of $\alpha 2/\alpha 3$ sodium pump isoforms potentiates glutamate neurotoxicity. Brain Res 591:94-102
4. Hamlyn JM, Blaustein MP, Bova S, DuCharme DW, Harris DW, Mandel F, Mathews WR, Ludens JH (1991) Identification and characterization of ouabain-like compound from human plasma. Proc Natl Acad Sci USA 88:6259-6263
5. Hansen O (1992) Heterogeneity of Na,K-ATPase from kidney. Acta Physiol Scand 146:229-234
6. Orlowski J and Lingrel JB (1988) Tissue-specific and developmental regulation of rat Na,K-ATPase catalytic $\alpha$ isoform and ß subunit mRNAs. J Biol Chem 263:19436-19442
7. Price EM and Lingrel JB (1988) Structure-function relationships in the Na,K-ATPase $\alpha$ subunit: site-directed mutagenesis of glutamine-111 to arginine and asparagine-122 to aspartic acid generates a ouabain-resistant enzyme. Biochemistry 27:8400-8408
8. Sverdlov ED Akopyanz NS, Petrukhin KE, Broude NE, Monastyrskaya GS, Modyanov NN (1988) $Na^+/K^+$-ATPase: tissue-specific expression of genes coding for $\alpha$-subunit in diverse human tissues. FEBS Lett 239:65-68
9. Sweadner KJ (1989) Isozymes of the Na,K-ATPase. Biochim Biophys Acta 988:185-220,
10. Sweadner KJ (1991) Overview: Subunit diversity in the Na,K-ATPase. In: Kaplan JH, De Weer P (eds), The Sodium Pump: Structure, Mechanism, and Regulation. The Rockefeller University Press, New York, pp 63-76
11. Takeyasu K, Lemas V, and Fambrough DM. (1990) Stability of $Na^+-K^+$-ATPase $\alpha$-subunit isoforms in evolution. Am J Physiol (Cell Physiol 28): 259:C619-C630
12. Weiss DN, Podbersky DJ, Heidrich J and Blaustein MP. Nanomolar ouabain augments caffeine-evoked contractions in rat arteries. Am J Physiol (Cell Physiol) in press
13. Yuan CM, Manunta P, Hamlyn JM, Chen SW, Bohen E, Yeun J, Haddy FJ, Pamnani MB (1993) Chronic ouabain administration produces hypertension in rats. Hypertension 22:178-187

# Hormonal Effects on the Expression of $Na^+/K^+$-ATPase Beta Subunit During Pregnancy

A. Turi, N. Mullner, I. Szanto and Z. Marcsek*

1st Institute of Chemistry and Biochemistry, *Department of Cell Biology Semmelweis Medical School, Budapest, Hungary

## Introduction

The $Na^+/K^+$-ATPase in several tissues is not homogeneous but contains catalytic subunits varying in structure as well as in function (6). The ratio of these isoenzymes is altered during development and influenced also by hormonal levels (5,3). In our previous work we demonstrated the presence of two different kind of isozymes in rat myometrium and one of them showed an increased Ca sensitivity (8).

Changes have been detected in the $Na^+/K^+$-ATPase enzyme activity and in its mRNA content of myometrial samples taken at different periods of time during pregnancy. The highest increase in the enzyme activity as well as alpha1 and beta mRNA synthesis were observed on the 17th day of pregnancy (9).

Oestradiol and progesterone are the two dominant hormones which result most of the myometrial changes during pregnancy, so in our present work we investigated how they influence mRNA and protein expression of the $Na^+/K^+$-ATPase beta subunit, because its mRNA synthesis proved to be the most responsive in our earlier experiments. Furthermore, we studied if there are any other factors which could be responsible for this phenomenon.

## Methods

The myometrial strips of young Wistar rats were isolated in 50mM Tris-EDTA buffer (pH=7.2). All myometrial pools were divided into two parts. One part of the muscle was used to measure the enzyme activity and for Western analysis and from the other part total RNA was isolated. Hormonal treatment: The ovariectomized young Wistar rats (150 g) two weeks after the surgery were treated by Akrofollin (oestradiolum propionicum) 33 microgram and 3.3 mg and by Glanducorpin (progesterone) 25 microgram and 2.5 mg per kg body weight or together by both hormones intramuscularly for one day or eight days. Na+/K+-ATPase activity was measured by the modified method of Everts at al. (2,9). Northern blot hybridization: Total cellular RNA was isolated by Chirgwin's method (1). Hybridization and evaluation of autoradiographs was performed as it was previously described (9,5). Western blot analysis: 60 microgram protein per samples were separated on a 5-15 % SDS poliacrylamide gradient gel (4) and transfered to nitrocellulose filter (7). The visualization of antibody binding was performed with Amersham ECL chemiluminescent kit and the signals on the photonegatives were scanned with densitometer.

# Results and discussion

Until recently the factors and the modulation of the uterus contractility during pregnancy has been an unsolved problem. We examine the requirement of $Na^+/K^+$-ATPase in this process because this enzyme plays crucial role in the maintenance of the membrane potential and it also influences the intracellular Ca level.

Previously we observed that the transcription of $Na^+/K^+$-ATPase subunits (alpha1,3 and beta) were enhanced characteristically in the pregnant rat myometrium (9). Especially the amount of beta mRNA increased at the 17th day of pregnancy. Western analysis however showed only moderate changes in case of beta 1 subunit. Enzyme activity measured with detergent treated microsomes gave significant rise at the same period, too.

In the following set of experiments we tried to answer the question, whether the detected changes in the beta subunit expression during pregnancy are under hormonal control and which hormones play role in this process.

Oestradiol treatment of ovariectomized animals increased the amount of the beta subunit mRNA level, the progesterone proved to be less effective (especially at high concentration) while the combined effect of the two hormones was similar to the oestradiol alone (Fig.1).

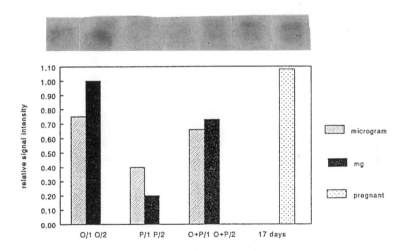

Figure 1: Effect of oestradiol and progesterone on the expression of beta subunit mRNA. (O:oestradiol; P:progesterone; OP: both; /1: 33 microgram (O) or 25 microgram (P) administered once; /1*: the same but administered for 8 days; /2: 3.3 mg (O) or 2.5 mg (P) administered once) control: 17 days pregnant uterus

Oestradiol increased the abundance of beta1 subunit on protein level in a dose dependent manner (Fig. 2.). The progesterone was in this case also less effective, but interestingly according to the simultaneous administration progesterone seems to compensate the oestradiol induced enhancement.

The ATPase activity measurement supported the results of Western blot analysis (Table 1.).

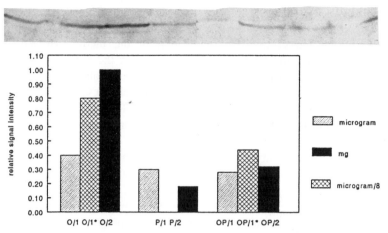

**Figure 2:** Effect of oestradiol and progesterone on the beta 1 subunit abundance (O:oestradiol; P:progesterone; OP: both; /1: 33 microgram (O) or 25 microgram (P) administered once; /1*: the same but administered for 8 days; /2: 3.3 mg (O) or 2.5 mg (P) administered once)

**Table 1: Effect of oestradiol and progesterone on the Na$^+$/K$^+$-ATPase activity**

|     | Oestradiol | Progesterone | O+P |
| --- | --- | --- | --- |
| /1 | 3.98+/-0.57 | 2.15+/-0.22 | 2.82+/-0.24 |
| /1* | 4.95+/-0.35 | 2.38+/-0.64 | 2.92+/-0.53 |
| /2 | 5.31+/-0.43 | 1.94+/-0.33 | 3.11+/-0.23 |

(units: micromole P/mg protein/hour /1: 33 microgram (O) or 25 microgram (P) administered once; /1*: the same but administered for 8 days; /2: 3.3 mg (O) or 2.5 mg (P) administered once)

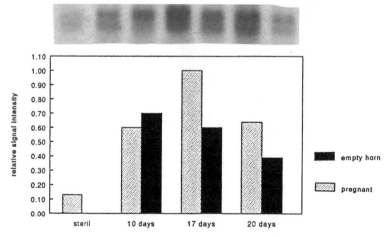

**Figure 3:** Beta subunit mRNA expression in pregnant and nonpregnant uterus horns. Pregnant and empty (no foetus) horns were prepared from the same animal at different stages of pregnancy.

858

Our results proved that the mRNA expression and protein level of beta subunit depend on hormonal conditions. Beside the administrations of ovarian hormones we analysed the possible effect of intrauterin factors originated from the foetus or the placenta. Comparing the samples derived from the pregnant and nonpregnant uterus horns of the same partially ovariectomised animals also showed differences in the expression of beta subunit as well as in transcriptional and posttranscriptonal level. In the empty horns we found them lower in both cases than in the pregnant ones (Fig.3) . This difference was more pronounced at the end of the pregnancy.

The presented results strongly indicate the complex coordination of $Na^+/K^+$-ATPase capacity of myometrium. In the case of beta subunit its abundance in the membrane seems to be determined by both the oestrogen and progesterone level. Furthermore, in the last trimester of pregnancy intrauterin factors also influence the expression of $Na^+/K^+$-ATPase beta subunit.

Acknowledgment: We thank Dr J. B. Lingrel's group, University of Cincinnati OH the $Na^+/K^+$-ATPase beta subunit cDNA probe.Special thanks for the financial help of the Deutsche Forschungsgemeinschaft/Bonn making possible our participation on this conference. This work was supported by Hungarian Ministry.of Social Welfare (ETT-10162) and National Scientific Research Found. (OTKA-T-5328, and V3 2622)

# References

1. Chirgwin JM, Prsybyla AE, MacDonald RJ and Rutter WJ (1979) Isolation of biologically active ribonucleic acid from sources enriched in ribonuclease. Biochemistry 18. 5294-5299

2. Everts ME, Andersen JP, Clausen T and Hansen O (1989) Quantitative determination of Ca++-dependent ATPase from sarcoplasmic reticulum in muscle biopsies. Biochem. J. 260, 443-448

3. Ikeda U, Hyman R, Smith TW and Medford RM (1991) Aldosteron-mediated regulation of Na+/K+-ATPase gene expression in adult and neonatal cardiocytes. J.Biol.Chem. 266, 12058-12966

4. Laemmli EK (1970) Cleavage of structural proteins during the assembly of the head of bacteriophag T4. Nature 227. 680-685

5. Orlowski J, and Lingrel JB (1988) Tissue specific and developmental regulation of rat Na+/K+-ATPase catalytic alpha and beta subunit mRNAs. J.Biol.Chem. 263, 10436-10442

6. Shull GE, Greeb J, and Lingrel JB (1986)   Molecular cloning of three distinct forms of   the Na+/K+-ATPase alpha subunit from rat brain. Biochemistry 25, 8125-8132

7. Towbin H, Staehelin T and Gordon J. (1979) Electrophoretic transfer of proteins from polyacrylamide gels to nitrocellulose sheets: Procedure and some applications. Proc. Natl. Acad. Sci. 76, 4350-4354

8. Turi A, Somogyi J and Mullner N. (1991) The effect of micromolar Ca2+ on the activities of the different Na+/K+-ATPase isozymes in the rat myometrium. Biochem. Biophys. Res. Commun. 174, 969-974

9. Turi A, Marcsek Z, Mullner N, Kucsera M and   Bori Z (1992) The activity of Na+/K+-ATPase and abundance of its mRNA are regulated in rat myometrium during pregnancy. Biochem. Biophys. Res. Commun. 188, 1191-1197

# Does Human Skeletal Muscle Na$^+$/K$^+$-ATPase Concentration Correlate With Capacity for Muscle Performance?

K. Kjeldsen, [*]T.A. Schmidt, [#]K.Ø. Christensen, [o]B. Saltin, [+]M.J McKenna

Departments of Medicine B and [*]Cardiothoracic Surgery RT, Rigshospitalet, University of Copenhagen, Blegdamsvej 9, 2100 Copenhagen, Denmark; [#]Surgical Department E, Århus Municipal Hospital, University of Århus, Nørrebrogade 44, 8000 Århus C, Denmark; [o]Physiology III, Karolinska Institute, P. O. Box 5626, Stockholm, Sweden; [+]Department of Physical Education and Recreation, Victoria University of Technology, P. O. Box 14428 MMC, Melbourne, Victoria, 3000, Australia.

## Introduction

Following denervation, tenotomy or immobilization rodent skeletal muscle Na$^+$/K$^+$-ATPase concentration has been found to be reduced by as much as 30% (2,8,14,19). In dogs moderate training has been reported to induce an increase of 165% in the activity of Na$^+$/K$^+$-ATPase prepared from muscle homogenates (13). Measurements of the total concentration of $^3$H-ouabain binding sites in rat demonstrated that intensive swim training caused an increase of 46% which was reversible upon deconditioning (11).

Based on these animal studies we have during the last years assessed the possibility of inducing change in skeletal muscle Na$^+$/K$^+$-ATPase concentration by physical conditioning. In local anaesthesia muscle biopsies were taken from the vastus lateralis muscle. Na$^+$/K$^+$-ATPase concentration was determined using vanadate facilitated $^3$H-ouabain binding to intact samples (4). These studies were of special interest since rise in interstitial K concentration has been associated with muscular fatigue (18). Thus, changes in capacity for active K removal from the interstitial space in skeletal muscle may be related to capacity for muscle performance.

Since furthermore skeletal muscle K balance is of importance for extrarenal K homeostasis (7,17), studies of plasma K changes during exercise were carried out.

All human studies were approved by the local Ethical Committees according the Helsinki Declaration II.

## Moderate training

Fifteen healthy conscripts were enroled for muscle biopsy and venous plasma K determination during exercise before and after 10 weeks of physical training at the Danish Military Service (9,10). Although these subjects had subjective as well as objective evidence for moderate increase in physical conditioning no change was observed in vastus lateralis muscle Na$^+$/K$^+$-ATPase concentration (p > 0.50). Evaluation of venous plasma K concentration during a bicycle exercise test loading gradually to exhaustion revealed, however, a decreased rise in plasma K. Thus peak plasma K concentration was significantly reduced by 0.5 mmol/l ($p < 0.05$) and throughout the exercise test plasma K was generally educed by 0.2 - 0.5 mmol/l (10). On the basis of these evaluations it was concluded that moderate training in human subjects does not significantly influence muscle $Na^+/K^+$-ATPase concentration whereas it may improve extrarenal K handling probably due to training induced increased rise in plasma cathecholamines during exercise (5,6) causing increased K uptake in resting muscle fibres (1).

Conversely Klitgaard and Clausen (12) have reported a 30 - 40% higher ($p < 0.05$) $^3$H-ouabain binding site concentration in muscle biopsies of active 70 year old men as compared to age matched sedentary men. However, the cross sectional nature of this latter study may have influenced the difference observed between sedentary and active elderly men.

## Sprint training

In this study (15) 6 untrained subject underwent intensive cycle sprint training having muscle biopsies taken before and after this period. This showed a significant increase of 16% ($p < 0.05$) in $Na^+/K^+$-ATPase concentration. It should be noted that in 3 matched subjects left untrained during the same period no change in $Na^+/K^+$-ATPase concentration was observed. Furthermore plasma K was evaluated in blood samples taken from a superficial forearm vein in association with sprint bouts. This showed that the increase in plasma K resulting from each exercise bout was significantly reduced by 19% ($p < 0.05$) after training. The increased muscle $^3$H-ouabain binding site concentration and reduced rise in K during exercise is consistent with improved skeletal muscle and extrarenal K regulation after sprint training.

## Hemiplegia

In 7 patients aged 35 - 64 years who had developed unilateral hemiplegia due to cerebral events, $Na^+/K^+$-ATPase concentration in muscle biopsies of the paretic leg was compared with that of the normal leg of the same patient. Biopsies were taken 1 - 7 months after the cerebral event. All patients had moderately reduced force in the affected leg as evaluated by OB Combiträner. Whereas the study showed no significant reduction in $Na^+/K^+$-ATPase concentration a tendency to a 9% decrease was observed ($p > 0.40$). Moreover a tendency to the largest decrease being observed after the longest duration of hemiplegia was found. The concentration of $Na^+/K^+$-ATPase found in the unaffected leg was in good agreement with that usually found in healthy subjects.

## Osteoporosis/polyarthrosis

In this study 7 patients with osteoporosis and more than 1 spinal fracture and 10 patients with polyarthrosis affecting more than one hip were compared with 5 healthy but otherwise matched subjects. Here a significant 29% reduction ($p < 0.02$) was observed in $Na^+/K^+$-ATPase in vastus lateralis muscle of osteoporotic patients compared with controls. In patients with polyarthrosis a tendency to a decrease of 17% (p > 0.05) was seen.

Taken together with the above mentioned study on hemiplegia the study of patients with osteoporosis and polyarthrosis indicate that decreased capacity for muscle performance may in human subjects be associated with moderate downregulation of skeletal muscle $Na^+/K^+$-ATPase concentration. However nas mentioned in the introduction care should be taken when comparing cross sectionally as in the study of osteoporosis and polyarthrosis. Other factors may be involved in patients with decreased muscular function. In this respect it is of special interest that in heart failure patient skeletal muscle $Na^+/K^+$-ATPase concentration evaluated in vastus lateralis muscle biopsies has been found to be reduced by 25% (16). It may also be noted that K-depletion even without concomitant hypokalaemia in patients on diuretics has been shown to be associated with a significant 18% reduction in skeletal muscle $Na^+/K^+$-ATPase concentration (3). In the relation to extrarenal K homeostasis with a muscle disease it is of interest that an excessive rise in plasm K concentration after exercise has been observed in muscular dystrophia (20).

## Conclusion

Taken together the present results and evidence from the literature are in favour of some relationship between skeletal muscle $Na^+/K^+$-ATPase concentration and capacity for muscle performance. Training seems to modulate plasma K raises during exercise. Moderate alteration of physical capacity may in human subject probably not induce substantial regulation of skeletal muscle $Na^+/K^+$-ATPase concentration whereas major changes may cause moderate regulation. The regulations may bear importance for local as well as overall extrarenal K homeostasis.

## Acknowledgements

Grete Simonsen is thanked for skilled technical assistance and Stig Hausø for valuable discussions. The study was in part supported by grants from the Danish Heart Foundation, the Danish Sports Research Foundation, NOVO's Foundation, Nordic Insulin Foundation and the Danish Medical Research Foundation.

## References

1.    Clausen T, Flatman JA (1977) The effect of catecholamines on Na-K transport and membrane potential in rat soleus muscle. J Physiol Lond 270:383-414
2.    Clausen T, Sellin LC, Thesleff S (1981) Quantitative changes in ouabain binding after denervation and during reinnervation of mouse skeletal muscle. Acta Physiol Scand 111:373-375
3.    Dørup I, Skajaa K, Clausen T, Kjeldsen K (1988) Reduced concentrations of potassium, magnesium, and sodium-potassium pumps in human skeletal muscle during treatment with diuretics. Br Med J Clin Res 296:455-458
4.    Hansen O, Clausen T (1988) Quantitative determination of $Na^+$-$K^+$-ATPase and other sarcolemmal components in muscle cells. Am J Physiol 254:C1-C7

5.  Kjaer M (1989) Epinephrine and some other hormonal responses to exercise in man: with special reference to physical training. Int J Sports Med 10:2-15

6.  Kjaer M, Farrell PA, Christensen NJ, Galbo H (1986) Increased epinephrine response and inaccurate glucoregulation in exercising athletes. J Appl Physiol 61:1693-1700

7.  Kjeldsen K (1991) Muscle Na,K-pump dysfunction may expose the heart to dangerous K levels during exercise. Can J Sport Sci 16:33-39

8.  Kjeldsen K, Bjerregaard P, Richter EA, Thomsen PEB, Nørgaard A (1988) $Na^+,K^+$-ATPase concentration in rodent and human heart and skeletal muscle: apparent relation to muscle performance. Cardiovasc Res 22:95-100

9.  Kjeldsen K, Nørgaard A, Hau C (1990) Human skeletal muscle Na, K-ATPase concentration quantified by $^3$H-ouabain binding to intact biopsies before and after moderate physical conditioning. Int J Sports Med 11:304-307

10.  Kjeldsen K, Nørgaard A, Hau C (1990) Exercise-induced hyperkalaemia can be reduced in human subjects by moderate training without change in skeletal muscle Na,K-ATPase concentration. Eur J Clin Invest 20:642-647

11.  Kjeldsen K, Richter EA, Galbo H, Lortie G, Clausen T (1986) Training increases the concentration of [$^3$H]ouabain-binding sites in rat skeletal muscle. Biochim Biophys Acta 860:708-712

12.  Klitgaard H, Clausen T (1989) Increased total concentration of Na-K pumps in vastus lateralis muscle of old trained human subjects. J Appl Physiol 67:2491-2494

13.  Knochel JP, Blachley JD, Johnson JH, Carter NW (1985) Muscle cell electrical hyperpolarization and reduced exercise hyperkalemia in physically conditioned dogs. J Clin Invest 75:740-745

14.  Leivseth G, Clausen T, Everts ME, Bjørdal E (1992) Effects of reduced joint mobility and training on Na,K-ATPase and Ca-ATPase in skeletal muscle. Muscle and Nerve 15:843-849

15.  McKenna MJ, Schmidt TA, Hargreaves M, Cameron L, Skinner SL, Kjeldsen K (1993) Sprint training increases human skeletal muscle Na,K-ATPase concentration and improves K regulation. J Appl Physiol 75(1):173-180

16.  Nørgaard A, Bjerregaard P, Baandrup U, Kjeldsen K, Reske-Nielsen E, Thomsen PEB (1990) The concentration of the Na,K-pump in skeletal and heart muscle in congestive heart failure. Int J Cardiol 26:185-190

17.  Nørgaard A, Kjeldsen K (1991) Interrelation of hypokalemia and potassium depletion and its implications: a re-evaluation based on studies of the skeletal muscle sodium, potassium-pump. Clin Sci 81:449-455

18.  Sjøgaard G (1986) Water and electrolyte fluxes during exercise and their relation to muscle fatigue. Acta Physiol Scand Suppl 556:129-136

19.  Ward KM, Manning W, Wareham AC (1987) Effects of denervation and immobilisation during development upon [$^3$H]ouabain binding by slow- and fast-twitch muscle of the rat. J Neurol Sci 78:213-224

20.  Wevers RA, Joosten MG, van de Biezenbos JB, Theewes GM, Veerkamp JH (1990) Excessive plasma $K^+$ increase after ischemic exercise in myotonic muscular dystrophy. Muscle Nerve 13:27-32

# Regulation of the $Na^+/K^+$-pump proteins in skeletal muscle:  Acute effects of insulin on α2 distribution and muscle fibre-type specific expression of β subunits

A.Klip, L. Lavoie, A. Marette, H. Hundal, C. Ackerley and J.L. Carpentier

Division of Cell Biology, The Hospital for Sick Children, Toronto, Canada M5G 1X8, and Departement de Morphologie, Universite de Geneve, Suisse.

## Introduction

Insulin is a major anabolic hormone that plays a pivotal role in $K^+$ homeostasis. A decrease in plasma $K^+$ concentration is mentioned in the first reports of insulin effects on the whole organism. The hormone enhances active $K^+$ uptake by extra-renal tissues, primarily muscle and liver (1). This response can be observed rapidly in isolated cell preparations such as excised muscle (2,3) and isolated hepatocytes (4). In isolated skeletal muscle, adipocytes and cardiocytes, addition of insulin results in a rapid hyperpolarization (5). This action has been mainly ascribed to a rapid stimulation of the $Na^+/K^+$-pump (6,7).  In addition to incrementing $K^+$ uptake and storage, the stimulation of the pump by insulin results in $Na^+$ extrusion that is necessary to maintain low cytosolic levels of this cation, which would otherwise increase due to the simultaneous stimulation by the hormone of $Na^+$-coupled amino acid uptake and $Na^+/H^+$ countertransport.

The mechanism underlying the insulin-dependent stimulation of the $Na^+/K^+$-pump has remained elusive, in part due to conflicting results obtained from studies in diverse biological preparations, as summarized below.

(i)  In isolated rat hepatocytes (4) and in BC3H-1 cells in culture (8), a cell line of brain origin that expresses some smooth muscle specific proteins, stimulation of the $Na^+/K^+$-pump by insulin results from a rise in cytosolic $Na^+$ secondary to stimulation of $Na^+/H^+$ exchange by the hormone (4,8,9). Both cell types express only the α1 isoform of α subunits, suggesting that this protein is regulated by insulin only indirectly. We hypothesize that this type of regulation occurs in cells where tight control of intracellular $Na^+$ is not essential, i.e. cells where small fluctuations in $Na^+$ are tolerated. A different mechanism(s) responsible for insulin stimulation of the pump operates in cells expressing the α2 isoform.

(ii) Kinetic evidence shows that in rat adipocytes (10), brain synaptosomes (11) and skeletal muscle (10), insulin increases pump activity of the component of high-sensitivity to ouabain, suggesting that α2 is the major regulated isoform. Even in these tissues, the molecular basis of insulin stimulation of α2 activity differs. In rat fat cells, insulin elevates the Vmax and the affinity for $Na^+$ of the α2 subunit (10,12-15).

This is thought to occur through removal of an inhibitory factor in response to the hormone (15,16) but no details are available concerning the identity of this factor. However, stimulation by insulin did not change the number of pump subunits in the fat cell surface (10,12).

(iii)  In soleus muscle, the mechanism of insulin action is drastically different from that in BC3H-1 smooth myocytes.  In particular, amiloride has no effect on the time course or magnitude of the insulin-stimulation of ouabain-sensitive $^{86}Rb^+$ uptake in this tissue (17,18).  Hence, in mature skeletal muscle, stimulation of the $Na^+/K^+$-ATPase by insulin is not secondary to stimulation of $Na^+/H^+$ exchange. A mechanism of stimulation of pump activity in amphibian skeletal muscle was proposed by Grinstein and Erlij (2,19)  consisting of unmasking latent $Na^+/K^+$-pump units, based on the following: (i) in muscles with the basal-state pumps pre-blocked by extracellular ouabain, insulin effectively stimulated $^{22}Na^+$ efflux; (ii) insulin increased $^3H$-ouabain binding to the surface of the muscle at all concentrations of the ligand; (iii) increased binding persisted after blocking the basal pumps with non-radioactive ouabain. These effects were not prevented by cycloheximide, leading to conclude that insulin unmasks or brings previously inactive pumps to the plasma membrane. These observations were further extended by Omatsu-Kanbe and Kitasato who found higher $Na^+/K^+$-ATPase activity (20) and ouabain binding capacity (21) in plasma membranes isolated from insulin-treated frog muscle relative to controls. An intracellular membrane fraction isolated concomitantly showed reduced $Na^+/K^+$-ATPase activity and ouabain binding relative to control counterparts (21). These results suggested that active $Na^+/K^+$-pumps exist intracellularly and migrate to the plasma membrane with insulin treatment in amphibian muscle.  Neither the pump isoforms involved nor whether pump recruitment takes place in mammalian muscle was defined by those studies.

## Results

1. **Insulin-dependent translocation of $\alpha2$ and $\beta1$ subunits.** We have recently shown that rodent skeletal muscles of mixed fibre type express $\alpha1$, $\alpha2$, $\beta1$ and $\beta2$ but not $\alpha3$ or $\beta3$ isoforms of the $Na^+/K^+$-pump.  Subcellular fractionation studies coupled to Western blotting using isoform-specific antibodies demonstrated that only the $\alpha1$ subunit is restricted to the cell surface. The $\alpha2$ isoform is present both in the isolated plasma membranes and in a fraction of intracellular origin low in sarcoplasmic reticulum markers (intracellular pool). The $\beta$ subunits were found in the plasma membrane, in the intracellular pool, and in a membrane fraction rich in sarcoplasmic reticulum. After a short insulin treatment *in vivo* (30 min), the content of $\alpha2$ and $\beta1$ (but not of $\alpha1$ or $\beta2$) increased in the plasma membrane fraction. Accompanying reductions in $\alpha2$ were recorded in the intracellular pool, but reductions in $\beta1$ occured mostly in the sarcoplasmic reticulum-rich fraction (22).  These results suggest that insulin causes a selective translocation of $\alpha2$ and $\beta1$ subunits to the plasma membrane (22). The increase in $\alpha2$ at the cell surface was corroborated by

immunoelectron microscopy of ultracryosections of lightly fixed muscles through immunogold labelling using both monoclonal and polyclonal antibodies (23). The results in Table 1 show that both antibodies, using different fixation and labelling protocols, recorded statistically significant increases in immunolabelled α2 proteins in the muscle cell surface in response to insulin.

TABLE 1: Quantification of gold-labelled α2 subunit of the $Na^+/K^+$-ATPase in plasma membranes of skeletal muscles from control and insulin-treated rats.

| | Protocol 1 | | | Protocol 2 | | |
|---|---|---|---|---|---|---|
| | (gold part./μm) | # cells | Total μm | (gold part./μm) | # cells | Total μm |
| Control | 1.80 ± 0.53 | 24 | 220 | 0.89 ± 0.09 | 89 | 536 |
| Insulin | 6.64 ± 2.45 | 42 | 603 | 1.31 ± 0.11 | 94 | 591 |
| Ins/basal | 3.69 | | | 1.47 | | |
| P value | $P < 0.001$ | | | $P < 0.005$ | | |

Results are expressed as gold particles per linear μm of membrane. Protocol 1 used monclonal antibody McB2; Protocol 2 used polyclonal anti-α2 (UBI). The two protocols also vary in the type of fixative, detection system, gold particle size and muscle sampled. Two pairs of control and insulinized rats were used, each producing three experiments of ultracryosectioning and immunolabelling.

## 2. Muscle-fibre-type specific expression of β isoforms.

Skeletal muscle is a heterogeneous tissue constituted by different fibre (i.e.cellular)-types. This encompasses metabolic, contractile and dimensional differences. Using muscles rich in type I (slow-twitch oxidative) fibres (e.g. soleus), or in type IIb (fast-twitch glycolytic) fibres (e.g. white gastrocnemius) we investigated their complements of α and β subunits as mRNA and protein. The former was measured by Northern blots of total RNA extracts and the latter by Western blots in total membranes and in purified plasma membranes and intracellular pool fraction. Both muscle-type groups had similar levels of α1 and α2 isoform transcripts and proteins. However, we found that slow- oxidative muscle expresses primarily the β1 isoform, whereas the fast-glycolytic muscle expresses primarily the β2 isoform (24), at both the mRNA (Table 2) and the protein levels (24). These results suggest that expression of β subunits is muscle fibre-type specific, and this may have specific consequences on the regulation of pump activity by hormones and contractile activity in each muscle type. Future investigations will reveal whether insulin can stimulate pump activity and/or cause pump subunit translocation in fast-twitch glycolytic muscles lacking the β1 subunit isoform.

TABLE 2. Relative proportions of β and α mRNAs in rat skeletal muscles.

|      | Soleus          | Red gastrocnemius | White gastrocnemius |
|------|-----------------|-------------------|---------------------|
| β1   | $1.00 \pm 0.17$ | $0.69 \pm 0.09$   | $0.13 \pm 0.04$     |
| β2   | $0.04 \pm 0.01$ | $0.47 \pm 0.03$   | $1.00 \pm 0.50$     |
| α1   | $1.00 \pm 0.16$ | $0.95 \pm 0.08$   | $0.70 \pm 0.11$     |
| α2   | $1.00 \pm 0.09$ | $0.92 \pm 0.07$   | $0.69 \pm 0.01$     |

Total RNA was extracted from individual muscles and analyzed by Northern blots using isoform-specific cDNA probes. RNA from 2-3 independent extractions was run simultaneously. The abundance of each transcript is given in relative units, assigning a value of 1.00 to the muscle with highest content of each transcript. Comparisons can be made for each isoform among muscles, but not among isoforms.

## References

1. Bia M, DeFronzo RA.  Am J Physiol 240: F257-68, 1981
2. Erlij D, Grinstein S.  J Physiol 259: 13-31, 1976
3. Clausen T, Kohn PG.  J Physiol 265: 19-42, 1977
4. Fehlmann M, Freychet P.  J Biol Chem 256: 7449-53, 1981
5. Czech MP.  Annu Rev Biochem 46: 359-84, 1977
6. Moore RD.  J Physiol 232: 23-42, 1973
7. Gavryck WA, Moore RD, Thompson RC.  J Physiol 252: 43-58, 1975
8. Rosic NK, Standaert ML, Pollet RJ.  J Biol Chem 260: 6206-12, 1985
9. Gelehrter TD, Shreve PD, Dilworth VM.  Diabetes 33: 428-34, 1984
10. Lytton J, Lin JC, Guidotti G.  J Biol Chem 260: 1177-84, 1985
11. Brodsky JL.  Am J Physiol (Cell Physiol) 258: C812-7, 1990
12. Resh MD.  J Biol Chem 257: 11946-52, 1982
13. Lytton J.  J Biol Chem 260: 10075-80, 1985
14. Resh MD, Nemenoff RA, Guidotti G.  J Biol Chem 255: 10938-45, 1980
15. McGill DL, Guidotti G.  J Biol Chem 266: 15824-31, 1991
16. McGill DL.  J Biol Chem 266: 15817-23, 1991
17. Weil E, Sasson S, Gutman Y.  Am J Physiol 261: C224-30, 1991
18. Clausen T, Flatman JA.  Am J Physiol 252: 338: 163-81, 1987
19. Grinstein S, Erlij D.  Nature (Lond) 251: 57-8, 1974
20. Omatsu-Kanbe M, Kitasato H.  Biochem J 246: 583-8, 1987
21. Omatsu-Kanbe M, Kitasato H.  Biochem J 272: 727-33, 1990
22. Hundal H, Marette A, Mitsumoto Y, Ramlal T, Klip A.  J Biol Chem 267: 5040-3, 1992
23. Marette A, Krischer J, Lavoie L, Ackerley C, Carpentier J-L., Klip A.  Am J Physiol, in press, 1993
24. Hundal H, Marette A, Ramlal T, Liu Z, Klip A.  FEBS Letts 328: 253-8, 1993

# Insulin as Regulator of Skeletal Muscle Na$^+$/K$^+$-ATPase Concentration in Rats With Experimental Diabetes and Patients With Diabetes Mellitus

H. Vestergaard, [+]T.A. Schmidt, [*]P.A. Farrell, [o]S. Hasselbalch, [#]K. Kjeldsen

Steno Diabetes Centre, Niels Steensens Vej 2, 2820 Gentofte, and [+]Departments of Cardiothoracic Surgery RT, [o]Neurology N and [#]Medicine B, Rigshospitalet, University of Copenhagen, Blegdamsvej 9, 2100 Copenhagen, Denmark; [*]Laboratory for Human Performance Research, Pennsylvania State University, PA 16802, U.S.A.

## Introduction

Using vanadate facilitated $^3$H-ouabain binding to intact samples it has earlier been demonstrated that untreated streptozotocin (STZ) induced diabetes in rats leads to a significant reduction in skeletal muscle Na$^+$/K$^+$-ATPase concentration (8). Aspects regarding Na$^+$/K$^+$-ATPase regulation in humans with diabetes have primarily been investigated using erythrocyte membranes. Both in IDDM (13,16) and NIDDM (10,11) a Na$^+$/K$^+$-ATPase decrease of around 30% has previously been reported in erythrocytes. The attention should, however, be brought to the evidence that studies on erythrocyte membranes regarding Na$^+$/K$^+$-ATPase regulation need not mirror regulatory aspects of Na$^+$/K$^+$-ATPase in other tissues, and should generally not be expected to furnish data of general physiological impact (7,12,14). In e.g. K-depletion Na$^+$/K$^+$-ATPase concentration is reportedly upregulated in erythrocytes but downregulated in muscles. While skeletal muscle in adult human subjects contain 75% of the total body K, erythrocytes contain only 7%. Since furthermore skeletal muscle K continuously equilibrates substantially with blood, erythrocytes, thus, play only a minor role in plasma K regulation compared to skeletal muscle.

Applying vanadate facilitated $^3$H-ouabain binding the present study was carried out to evaluate the mechanism involved in downregulation of skeletal muscle Na$^+$/K$^+$-ATPase concentration as well as to study a putative regulatory change in human skeletal muscle Na$^+$/K$^+$-ATPase with diabetes.

## Results and Discussion

The previous observation of a decrease of Na$^+$/K$^+$-ATPase concentration in skeletal muscle in rats with experimentally induced diabetes (8) was confirmed (Fig.). Thus 8 weeks after induction of diabetes with STZ (n = 9) as well as 10 weeks following partial pancreatectomy (PPX) (n = 15) soleus muscle $^3$H-ouabain binding site concentration was significantly reduced by 23 and 19% respectively (p < 0.05). In accord with earlier studies (9), semistarvation (n = 9) also caused a reduction of soleus muscle Na$^+$/K$^+$-ATPase concentration. Its magnitude (20%; p < 0.05) was similar to that obtained as a result of diabetes. Insulin treatment of STZ induced diabetes caused a significant increase in soleus muscle $^3$H-ouabain binding site concentration of 23% (p < 0.05) above control value.

868

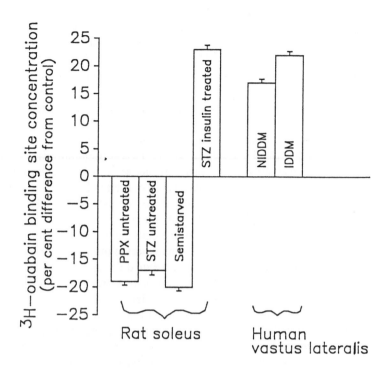

**Fig.** Relative change in ³H-ouabain binding site concentrations. Columns are means, bars denote SEM.

The human study was conducted in accordance with a protocol approved by the local Ethical Committee as being in accordance with the Helsinki Declaration II. Skeletal muscle samples were obtained by muscle biopsy. Compared with normal control subjects (n = 8) ³H-ouabain binding site concentration in vastus lateralis was 17 and 22% (p < 0.05) greater in patients with regulated non insulin (NIDDM) (n = 24) and insulin dependent diabetes mellitus (IDDM) (n = 7), respectively (Fig.). A significant, positive, linear correlation between muscle Na⁺/K⁺-ATPase and plasma insulin concentration was observed in patients with NIDDM and controls.

The present data on rat skeletal muscle seems to be in good accord with the previous study concerning the effects of untreated diabetes on skeletal muscle Na⁺/K⁺-ATPase concentration (8). Increased skeletal muscle Na⁺/K⁺-ATPase concentration above control level as a result of insulin treatment was in the present study observed both in rats and in humans. A tendency to such an upregulation was also seen in the earlier studies on skeletal muscle from STZ rats given insulin (8). From the present study evidence seems to have emerged indicating a linkage between plasma insulin and skeletal muscle Na⁺/K⁺-ATPase concentration. Such a linkage may be associated with the anabolic properties of insulin. Thus hypoinsulinaemia (2) has been reported to be associated with atrophy and decreased tetanic tension output in skeletal muscles from rats with STZ induced diabetes. In addition mounting evidence seems to suggest that insulin is of importance for protein synthesis (1,6). Furthermore, in insulin resistant states growth promoting properties of insulin need not be blunted (4). Hence, also in NIDDM the increased plasma insulin concentration may have been of importance for the greater skeletal muscle $Na^+/K^+$-ATPase concentration. Comparisons were in the present study drawn to skeletal muscle $^3H$-ouabain binding results obtained from animals exposed to semistarvation: In soleus the relative changes associated with untreated diabetes did not discriminate themselves from the results obtained after semistarvation. Hence, it may be argued that the changes seen in skeletal muscle $Na^+/K^+$-ATPase concentration associated with untreated diabetes bear similarities with malnutrition. Here it is noteworthy that semistarvation is associated with reduced plasma insulin concentration (5).

The present study may have implications of importance for extrarenal K handling. During muscle activity K leaks from the muscle cells due to depolarization and thus, interstitial K rises. K is in part transported back to the cells by the Na,K-pump. It has previously been shown that insulin exerts an acute stimulation of the active ion transport of existing Na,K-pumps in skeletal muscles by increasing their activity (3). Thus, in unregulated diabetes concomitant low plasma insulin and skeletal muscle $Na^+/K^+$-ATPase concentration might cause significantly increased interstitial K concentration in active muscles especially during and immediately following physical exertion, while the reverse might be expected in regulated diabetes with increased skeletal muscle $Na^+/K^+$-ATPase concentration. Optimum function of skeletal muscle $Na^+/K^+$-ATPase seems to be of particular importance, since dysregulation of interstitial K concentration may be a cause of muscle fatigue (15). Furthermore when extended to major muscle groups dysregulation of skeletal muscle $Na^+/K^+$-ATPase concentration as well as variations in plasma insulin level may impair extrarenal K regulation. This may be of special importance because K homeostasis is recognized to be severely disturbed in dysregulated diabetes.

## Acknowledgements

We thank Grete Simonsen, Karin Stahr, Annemette Foman and Bente Mottlau for skilled technical assistance; Gitte Moos Knudsen, Johannes Jakobsen and Stig Haunsø for valuable discussions and Lene Theil Skovgaard for expert help with the statistical analysis of the data. Novo Nordisk is thanked for a gift of long acting insulin (Ultralente[R]). This study was in part supported by the Danish Heart Foundation, Nycomed DAK, NOVO's Foundation, Leo Pharmaceutical's Foundation, The Danish Sports Research Foundation, the Nordic Insulin Foundation and the Danish Division of Bristol-Myers-Squibb.

## References

1.  Balon TW, Zorzano A, Treadway JL, Goodman MN, Ruderman NB (1990) Effect of insulin on protein synthesis and degradation in skeletal muscle after exercise. Am J

Physiol 258:E92-E97

2.  Cameron NE, Cotter MA, Robertson S (1990) Changes in skeletal muscle contractile properties in streptozocin-induced diabetic rats and role of polyol pathway and hypoinsulinemia. Diabetes 39:460-465

3.  Clausen T (1986) Regulation of active $Na^+$-$K^+$ transport in skeletal muscle. Physiol Rev 66:542-580

4.  DeFronzo RA, Ferrannini E (1991) Insulin resistance. A multifaceted syndrome responsible for NIDDM, obesity, hypertension, dyslipidemia, and atherosclerotic cardiovascular disease. Diabetes Care 14:173-194

5.  Hunsicker KD:, Mullen BJ, Martin RJ (1992) Effect of starvation or restriction of self selection of macronutrients in rats. Physiol Behav 51(2):325-330

6.  Kent JD, Kimball SR, Jefferson LS (1991) Effect of diabetes and insulin treatment of diabetic rats on total RNA, poly(A)+ RNA, and mRNA in skeletal muscle. Am J Physiol 260:C409-C416

7.  Kjeldsen K (1987) Regulation of the concentration of $^3$H-ouabain binding sites in mammalian skeletal muscle - effects of age, K-depletion, thyroid status and hypertension. Dan Med Bull 34:15-46

8.  Kjeldsen K, Braendgaard H, Sidenius P, Larsen JS, Nørgaard A (1987) Diabetes decreases $Na^+$-$K^+$ pump concentration in skeletal muscles, heart ventricular muscle, and peripheral nerves of rat. Diabetes 36:842-848

9.  Kjeldsen K, Everts ME, Clausen T (1986) Effects of semi-starvation and potassium deficiency on the concentration of [$^3$H]ouabain-binding sites and sodium and potassium contents in rat skeletal muscle. Br J Nutr 56:519-532

10. Mimura M, Makino H, Kanatsuka A, Yoshida S (1992) Reduction of erythrocyte ($Na^+$-$K^+$) ATPase activities in non-insulin-dependent diabetic patients with hyperkalemia. Metabolism 41:426-430

11. Noda K, Umeda F, Hashimoto T, Yamashita T, Nawata H (1990) Erythrocytes from diabetics with neuropathy have fewer sodium pumps. Diabetes Res Clin Pract 8:177-181

12. Nørgaard A, Kjeldsen K, Clausen T (1981) Potassium depletion decreases the number of $^3$H-ouabain binding sites and the active Na-K transport in skeletal muscle. Nature 293:739-741

13. Jourdheuil DR, Mourayre Y, Vague P, Boyer J, Vague IJ (1987) In vivo insulin effect on ATPase activities in erythrocyte membrane from insulin-dependent diabetics. Diabetes 36:991-995

14. Schmidt TA, Holm-Nielsen P, Kjeldsen K (1991) No upregulation of digitalis glycoside receptor (Na,K-ATPase) concentration in human heart left ventricle samples obtained at necropsy after long term digitalisation. Cardiovasc Res 25:684-691

15. Sjøgaard G (1986) Water and electrolyte fluxes during exercise and their relation to muscle fatigue. Acta Physiol Scand Suppl 556:129-136

16. Testa I, Rabini RA, Fumelli P, Bertoli E, Mazzanti L (1988) Abnormal membrane fluidity and acetylcholinesterase activity in erythrocytes from insulin-dependent diabetic patients. J Clin Endocrinol Metab 67:1129-1133

# The maximum capacity for active sodium-potassium transport in the slow and fast skeletal muscle of obese (ob/ob) and lean mice

T. Kovács, T. Bányász, I. Kalapos, and J. Somogyi

Department of Physiology, University Medical School, H-4012 Debrecen, P. O. Box 22, and Department. of Chemistry and Biochemistry I. Semmelweis University Medical School, H-1444 Budapest, Puskin street. 9, Hungary.

**Introduction.**

Mature, genetically obese mice (C57BL/6J ob/ob) display obesity, hyperinsulinemia and resistance to insulin. These findings confirm the hypothesis that the obese mouse can be used as an animal model for the study of type II diabetes (8). Previously it has been shown, that particular fractions obtained from the hind limb muscles of obese mice contain fewer [3H]-ouabain binding sites than those prepared from lean litter mates (6). In contrast, others have demonstrated that the number of [$^3$H]-ouabain binding sites showed no significant difference in intact soleus and extensor digitorum longus muscles obtained from 9-20 week-old obese and lean mice (2). The state of resistance to the action of insulin, however, develops in muscles (e.g. soleus, diaphragm) only during the late phase of obesity (4). The aim of the present study was to compare the vanadate-facilitated binding of [$^3$H]-ouabain and the maximum rate of ouabain-suppressible K$^+$ uptake in samples of extensor digitorum longus muscle and soleus muscle from mature obese mice of C57BL/6J-ob/ob strain and their lean litter-mates.

**Methods**

Experiments were carried out on isolated slow-twitch soleus muscle and fasts-twitch extensor digitorum muscle from 10-12 months old obese mice and their homozygote and heterozygote lean litter-mates. After mild food deprivation the obese mice had the same fasting blood glucose (7.9 ± 1.3 mmol/l) as their lean mates (7.5 ± 0.3 mmol/l), but much higher insulin level (73.3 ± 12.9 mU/l) than did their lean controls (7.8 ± 1.9 mU/l). [$^3$H]-ouabain binding in vitro to muscle preparation was performed by the vanadate-facilitated method as described by Norgard et al (7) and Kjeldsen et al (5). The specific binding of [$^3$H]-ouabain was plotted as a function of specifically bound [$^3$H]-ouabain relative to the amount of free ouabain in the incubating medium. The measurement of the maximum capacity of $^{86}$Rb-uptake in isolated muscle preparations at various concentrations of extracellular K$^+$ was determined by the method of Clausen et al (3).

**Results and discussion**

[$^3$H]-ouabain binding was measured using a final ouabain concentration of $10^{-8}$ - $10^{-6}$ mol/l. Nonspecific binding of [$^3$H]-ouabain was measured in the presence of excess unlabelled ouabain (1 mmol/l). The Scatchard-type plot gave no evidence for more than one population of binding sites. The specific site densities were significantly lower in obese muscles compared to those of the muscles from lean litter-mates (Table 1). The reduction of the number of [$^3$H]-ouabain binding site in slow-twitch soleus muscle was larger than in the fast-twitch extensor digitorum longus muscle.

About 45 per cent of the total Rb-influx in the isolated muscle preparations of lean mice was oubain-suppressible at 30$^o$C in standard incubation medium containing 5.5 mM K$^+$. It seems reasonable to assume that these components reflect the transport across the sarcolemma via the Na$^+$-K$^+$ pump. When the external K$^+$ concentration was increased the ratio of ouabain-suppressible Rb uptake was progressively reduced. Comparing the ouabain-suppressible uptake between the muscles from lean and obese mice over the concentration range of 5.5-120 mM K$^+$, it was found that $^{86}$Rb uptake was higher in both EDL and SOL muscles from lean mice as compared to the muscles from obese animals. The ouabain-suppressible component of $^{86}$Rb uptake could be estimated by applying the Eadie-Hofstee plot. The maximum rates of ouabain-suppressible component of $^{86}$Rb uptake were significantly lower in extensor digitorum longus and soleus muscles from obese mice those of their lean litter-mates (Table 1). These data are in good agreement with results of the ouabain binding site concentrations, and demonstrate that the maximum ouabain-suppressible [86]Rb-uptake varies in proportion to the concentration of [$^3$H]-ouabain sites in muscle preparations obtained either from lean or obese mice.

Table 1. Comparison of maximum ouabain-suppressible [86]Rb uptake (nmol/g wet wt/min); [$^3$H]-ouabain binding site density (pmol/g wet wt) as well as numbers of $K^+$ ions transported through pumping site per min (min$^{-1}$) in extensor digitorum longus (EDL) and soleus (SOL) muscles from lean and obese mice. (Values are mean ± S.E.M., and n)

|  | LEAN | | OBESE | |
|  | EDL | SOL | EDL | SOL |
| --- | --- | --- | --- | --- |
| Ouabain suppressible [86]Rb uptake | 1623.1 ± 58.0 (n=34) | 1605.9 ± 93.1 (n=33) | 1206.9 ±164.1 (n=24) | 1083.0 ±17.1 (n=23) |
| Density of [$^3$H]-ouabain binding site | 512.2 ±49.1 (n=24) | 511.7 ±17.5 (n=24) | 401.9 ±22.3 (n=31) | 381.8 ±43.8) (n=32) |
| Number of $K^+$ ions transported per [$^3$H]-ouabain binding site/min | 3168 | 3138 | 3002 | 2842 |

It should be noted that the number of $K^+$ ions transported per ouabain binding site per minute was around 40 per cent of the theoretical maximum, presumably due to the incomplete $Na^+$ loading during pre incubation.

The present study shows that obese mice of C57BL/6J (ob/ob) strain, in the later phase of obesity, display insulin resistance manifested in vivo by normal blood glucose level in the face of excessive insulin secretion and by an abnormal glucose tolerance test. In this state of obesity significant differences were found in [$^3$H]-ouabain binding and maximum capacity for sodium pumping between obese and lean litter mates. This confirms the important regulatory role of insulin on $Na^+/K^+$-ATPase activity in skeletal muscle. Our finding agree with previous data and confirm that both types I.

and type II. diabetes mellitus are associated to a reduced rate of active $Na^+$-$K^+$ transport in skeletal and cardiac muscle (1).

*Acknowledgements.*

The skilled technical assistance of Katalin Láber and Katalin Horváth is gratefully acknowledged. The study was supported by grant #1466 from the National Research Fund (OTKA).

## References

1.  Bányász T, Kovács T, Kalapos I, Somogyi J (19888) Alteration in sarcolemmal Na,KüATPase activitz in heart ventricular muscle during diabetes. Eur Heart J 9. 70

2.  Clausen T, Hansen O (1982) The $Na^+$-$K^+$-pump, energy metabolism, and obesity. Biochim Biophys Res Comm 104:357-362.

3.  Clausen T, Everts ME, Kjeldsen K (1987) Quantification of the maximum capacity for active sodium-potassium transport in rat skeletal muscle. J. Physiol. 388:163-187

4.  Grundleger ML, Godbole VY, Thenen SW (1980) Age-dependent development of insulin resistance of soleus muscle in genetically obese (ob/ob) mice. Am J Phsyiol 239:E363-E371

5.  Kjeldsen K, Norgard A, Hansen O, Clausen T (1985) Significance of skeletal muscle digitalis receptors for H-ouabain distribution in the guinea pig. J Pharm Exp Ther 234:720-727

6.  Lin MH, Vander Tuig JG, Romsos DR, Akera T, Leveille Ga (1980) Heat production and $Na^+$-$K^+$-ATPase enzyme units in lean and obese (ob/ob) mice. Am J Physiol 238:E193-E199

7.  Norgard A, Kjeldsen K, Hansen O, Clausen T (1983) A simple and rapid method for the determination of the number of H-ouabain binding site in biopsies of skeletal muscle. Biochim Biophys Res Comm 111:319-325

8.  Surwit RS, Kuhn CM, Cochrane C, McCubbin JA, Feinglos MN (1988) Diet-induced type II diabetes in CB57BL/6J mice. Diabetes 37:1163-1167

# Alterations in Na$^+$/K$^+$ATPase Isoforms of Different Tissues in Streptozotocin Induced Diabetes Mellitus

J. Somogyi, Ágota Vér, Ildikó Szántó, K. Kalff, P. Csermely, I. Kalapos, T. Bányász and T. Kovács

Semmelweis University School of Medicine, Department of Biochemistry I. Budapest, and University of Medical School, Department of Physiology, Debrecen Hungary

## Introduction

The Na$^+$/K$^+$–ATPase is a membrane embedded protein primarely responsible for the active transport of sodium and potassium in mammalian cells. There are three isoforms of its catalytic alpha subunit and at least two of the beta chain. The three alpha isoforms differ from each other in several biochemical and physiological characteristics, and show a specific tissue distribution (5,11). Evidence is slowly accumulating that alpha isoforms may be differentially regulated by several physiological and pathological stimuli (10). Some data show for example that isoform ratios change during development in brain and heart, and certain hormones alter both the isoforms ratios and their abundance (4,8,10).

We have demonstrated earlier that Na$^+$/K$^+$–pump sites were considerably reduced in cardiac tissue of streptozotocin induced diabetic rats similarly to other reports (1,5). However, no data are available to show whether changes in the Na$^+$/K$^+$–ATPase activity during diabetes influenced all of the isoforms or only a specific isoform changing the isoform pattern of the tissues. The aim of this study was to investigate the ATPase activity, the P$_i$ facilitated ouabain binding, the isoenzyme composition and the expression of $\alpha$ and $\beta$ mRNA isoforms of Na$^+$/K$^+$–ATPase in different tissues prepared from normal and streptozotocin induced diabetic rats.

## Methods

One group of non-diabetic male rats was used as controls (C). 50 mg/kg i.p. Streptozotocin (STZ) was given to an other group and diabetes was verified 24 h later by measuring hyperglycaemia and glucosuria. A group of the diabetic animals was killed after two weeks (D$_2$) another group after 4 weeks (D$_4$) of STZ administration. In insulin replacement studies insulin was administered to diabetic rats in an individual dose to normalise blood glucose level 2 and 4 weeks after STZ injection for 2 weeks before the time of the assay (D$_2$R, D$_4$R). Kidney, brain, heart, liver and skeletal muscle (m. quadriceps) tissues were used for the assays. Crude microsomal fractions were prepared by differential centrifugation (12). ($^3$H)ouabain binding was measured as previously described (9). Solubilized microsomal proteins were analysed on 10 % SDS polyacrylamide gels (7) and transferred to nitrocellulose sheets for Western blot analysis (11). These sheets were incubated with the antibodies of

Na$^+$/K$^+$–ATPase subunits isoforms (UBI, USA), and a peroxidase conjugated second antibody. The reaction was detected with ECL Western blotting detection kit (Amersham), and the film negatives were analysed by laser densitometry. Total cellular RNA was purified by the guanidium isothiocyanate method (2). RNA electrophoresis, blotting and hybridisation were performed (13). Autoradiographs were analysed by laser densitometry, and each peak of Na$^+$/K$^+$–ATPase gene transcript was related to the area of the ethidiumbromide stained 18S rRNA band. cDNA probes specific for $\alpha_1$ and $\beta$ isoforms were a generous gift of Dr J.B. Lingrel (Univ. Cincinnati).

## Results and Discussion

Streptozotocin treatment caused a reduction of ($^3$H)ouabain binding in all tissue investigated (not shown) except the medullary region of kidney. The decrease in ouabain binding capacity proved to be especially significant in heart, skeletal muscle and brain. The reduction in ouabain binding could be mostly reversed by insulin treatment. Similarly to an earlier report (6) of kidney Na$^+$/K$^+$–ATPase activity, we have also found a significant enhancement in oubain binding capacity after STZ treatment in kidney medulla. The elevation could be also partially reversed by insulin administration. Measuring the ouabain dependence of the Na$^+$/K$^+$–ATPase activity in brain microsomes we have found differences in ouabain sensitivity of Na$^+$/K$^+$–ATPase prepared from control, STZ treated and STZ treated insulin injected rats. The question arises from our and earlier data whether the changes in ATPase activity and ouabain binding capacity in diabetes are due to alterations of biochemical properties of the enzyme or derived from the quantitative changes of the transport molecule or from alterations in isoform pattern of the tissues. To evaluate the expression of Na$^+$/K$^+$–ATPase polypeptide, antisera specific for $\alpha_1$, $\alpha_2$, $\alpha_3$, $\beta_1$ and $\beta_2$ subunits were used to probe Western blots of microsomes prepared from control, STZ treated diabetic and STZ treated insulin injected rats' tissues. Although the $\alpha$ subunit isoform pattern altered in STZ induced diabetes, the tissue specificity of the isoforms did not change, i.e. no extra isoform was detected in diabetic tissues compared to the control ones. STZ treatment resulted in a significant decrease in the amount of one of the $\alpha$ subunits depending on the tissue, either $\alpha_1$ or $\alpha_2$. Insulin treatment did not restore the original pattern completely. While STZ treatment did not influence the amount of $\alpha_3$, the insulin administration to diabetic rats caused a significant enhancement of this isoform in brain. Among the $\beta$ subunits only the amount of $\beta_1$ changed in STZ treated and STZ treated insulin reversed animals. Summarising the results of Western blots we can conclude that the reduction of Na$^+$/K$^+$–ATPase activity and ouabain binding capacity of the tissues in diabetes at least partly derived from the reduction of the amount of transport protein. In contrary to the other tissues in kidney medulla the amount of Na$^+$/K$^+$–ATPase detected by Western blot technique increased after STZ treatment. In Fig.1. the expression of kidney Na$^+$/K$^+$–ATPase $\alpha_1$ and $\beta_1$ subunits and their change after STZ treatment and insulin administration are shown.

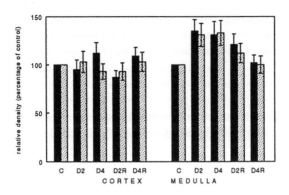

Fig.1. Densitometric analysis of Western blot of $\alpha_1$ (filled bars) and $\beta_1$ (dotted bars) subunits of Na$^+$/K$^+$–ATPase from kidney cortex and medulla of C, D$_2$, D$_4$, D$_2$R, D$_4$R rats. N=5

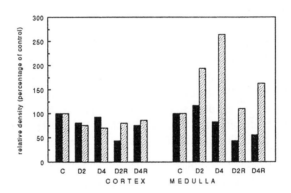

Fig. 2. Densitometric analysis of Northern blot of Na$^+$/K$^+$–ATPase $\alpha_1$ (filled bars) and $\beta$ (dotted bars) subunit mRNAs prepared from kidney cortex and medulla of C, D$_2$, D$_4$, D$_2$R, D$_4$R rats. A representative of 3 experiments. More than 40 percent change was considered significant.

As shown in Fig. 2. we found differences in the isoform specific mRNA transcription between cortex and medulla. In diabetes we could only detect a significant change in medulla in case of the $\beta$ isoform transcript, wich showed a 200 and 260 percent increase in 2 and 4 week diabetic state, respectively. There was no change in $\alpha$ subunit mRNA abundance. Insulin administration caused an approximately 50 percent decrease in $\beta$ mRNA abundance in medulla, while it had no such effect in cortex. The $\alpha_1$ mRNA transcription was reduced by approximately 50 percent in both cortex and medulla after insulin administration. From our data we conclude that the activity changes of $Na^+/K^+$–ATPase in experimental diabetes and in case of insulin administration is a result of the change of protein amount which correlates with the mRNA levels especially with that of the β subunit.

Acknowledgements

We thank E.Végh and O.Herman for excellent technical assistance. This work was supported by grants from Hungarian Scientific Council OTKA 1089, 1466 and from the AB AEGON Hungary

References

1.  Bányász T, Kovács T, Somogyi J (1988) Alteration in sarcolemmal $Na^+/K^+$–ATPase activity in heart ventricular muscle during diabetes. Eur Heart J 9: Suppl pp 70

2.  Chirgwin JM., Przybyla AE, MacDonald RJ, Rutter WJ (1979) Isolation of biologically active ribonucleic acid from sources enriched in ribonuclease. Biochemistry 18:5294-5299

3.  Geering K (1992) Subunit assembly and posttranslational proccessing of $Na^+$-punp. Acta Physiol Scand 146:177-181

4.  Ikeda U Hyman R, Smith T and Medford R M (1991) Aldosteron-mediated regulation of $Na^+/K^+$-ATPase gene expression in adult and neonatal cardiocytes. J Biol Chem 266:12058-12066

5.  Jewell EA, Shamraj OI, Lingrel JB (1992) Isoforms of the α subunit of Na,K-ATPase and their significance. Acta Physiol Scand 146:161-169

6.  Ku DD, Roberts RB, Sellers BM, Meezan E (1987) Regression of renal hypertrophy and elevated renal $Na^+/K^+$-ATPase activity after insulin treatment in streptozotocin-diabetic rats. Endocrinology 120:2166-2173

7.  Laemmli,U.K. (1970) Cleavage of structural proteins during the assembly of the head of bacteriophage $T_4$. Nature 227:680-686

8.  McGill DL, Guidotti G (1991) Insulin stimulates both the α1 and the α2 isoform of the rat adipocyte $Na^+/K^+$-ATPase. J Biol Chem 266:15824-15831

9.  Norgaard A, Kjeldsen K, HansenO, Clausen T (1983) A simple rapid method for the determination of the number of $^3$H-ouabain binding sites in biopsies of skeletal muscle. Biochem Biophys Res Commun 111:319-325

10. Orlowski J Lingrel JB (1988) Tissue-specific and developmental regulation of rat $Na^+/K^+$-ATPase catalytic α isoform and β subunit mRNAs. J Biol Chem 263:10436-10442

11. Shyjan A W, Levenson R (1989) Antisera specific for alpha1, alpha2, alpha3 and beta subunits of $Na^+/K^+$-ATPase : Differencial expression of alpha and beta subunits in rat tissue membrane. Biochemistry 28:4531-4535

12. Sweadler KJ (1978) Purification from brain of an intrinsic membrane protein fraction enriched in $Na^+/K^+$-ATPase. Biochem Biophys Acta 508:486-499

13. Ausubel FM, Brent R, Ekingston R, MooreDD, Seidman JG, Smith JA, Struhl K (1987) Analysis of RNA by Northern hybridization. In Ausubel et al eds. Current protocols in molecular biology Wiley I and sons, Cischester, New York, 4.9.1.- 4.9.8

# Properties of the Oxidized $Na^+/K^+$-ATPase

Y. Wang, A. Askari, W.-H. Huang

Department of Pharmacology, Medical College of Ohio, P.O. Box 10008, Toledo, Ohio
43699-0008 USA

## Introduction

A number of studies in recent years have indicated that $Na^+/K^+$-ATPase is inhibited by
$H_2O_2$ and various oxygen radicals, and that partial inhibition of the sodium pump by the
increased generation of these reactive oxygen species during ischemia-reperfusion is
involved in post-ischemic injury to the heart (2-4 and references therein). Here we report
on the characterization of some properties of the purified canine kidney enzyme that is
oxidatively inhibited by $H_2O_2$.

## Results and Discussion

Time-course of irreversible inactivation of the enzyme by $H_2O_2$ (not shown) suggested
the existence of multiple sites and/or the states of the enzyme with different oxidant
sensitivities. By preincubating the enzyme for a fixed time period with different $H_2O_2$
concentrations (Fig. 1), it was easily possible to show two "populations" with vastly
different $H_2O_2$-sensitivities, and to prepare partially oxidized enzyme that had 40-60% of
the original activity and was relatively resistant to further oxidation by $H_2O_2$.
Comparison of the properties of this "half-oxidized" enzyme with those of the native
enzyme showed the following:

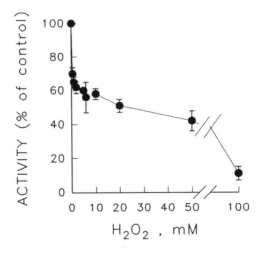

Figure 1. Effects of varying concentrations of $H_2O_2$ on $Na^+/K^+$-ATPase activity. Preincubation
was for 30 min. Enzyme preparation and other procedures as described before (2,3).

Apparent $K_m$ of ATP for the $Na^+/K^+$-ATPase activity of the oxidized enzyme was significantly lower than that of the native enzyme (Fig. 2). Note that the ATP concentration range used was not wide enough to allow the detection of the well known negative cooperativity of the ATP-velocity curve. The data of Fig. 2 indicate that it is the apparent affinity of ATP at the low-affinity site that is increased in the oxidized enzyme.

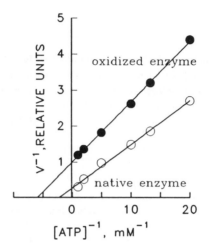

Figure 2. Effects of varying concentrations of ATP on $Na^+/K^+$-ATPase activities of the native and the oxidized enzymes.

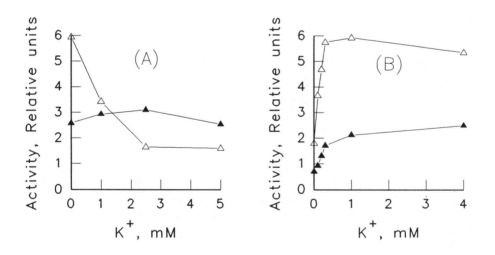

Figure 3. Effects of varying concentrations of $K^+$ on the $Na^+$-ATPase activities of the native (Δ) and the oxidized (▲) enzymes at 1 μM ATP (A) and 50 μM ATP (B). $Na^+$ concentration was 100 mM.

In experiments of Fig. 3 effects of $K^+$ on $Na^+$-ATPase activity of the same two preparations used in experiments of Fig. 2 were examined. When 1 $\mu$M ATP was used, $K^+$ inhibited the $Na^+$-ATPase of the native enzyme as expected, while it had little or no effect on the $Na^+$-ATPase of the oxidized enzyme (Fig. 3A). At 50 $\mu$M ATP, $K^+$ increased activity in both preparations (Fig. 3B). The combined data of Figs. 2 and 3 suggest that a major effect of oxidation is at the $K^+$ occlusion-deocclusion steps of the reaction cycle (1).

In experiments of Table 1 several properties of the native and the "half-oxidized" preparations were compared. Occluded $Rb^+$ (measured at 0.2 mM $Rb^+$) was indeed lower in the oxidized enzyme. (Studies on the kinetics of $Rb^+$ occlusion and deocclusion in the oxidized enzyme are in progress.) Interestingly, $Na^+$-dependent ADP-ATP exchange activities, and the maximal levels of the phosphoenzyme (EP), were the same in both preparations (Table 1).

Table 1. Comparison of some properties of the native and the "half-oxidized" enzymes.

|  | Native Activity[a] | Oxidized activity (% of native)[a] |
|---|---|---|
| $Na^+/K^+$-ATPase ($\mu$mol Pi/mg/h) | $960 \pm 50$ | $60 \pm 3\%$ |
| $Na^+$-ATPase ($\mu$mol Pi/mg/h) | $3.7 \pm 0.4$ | $53 \pm 7\%$ |
| $Rb^+$ occlusion[b] (nmol/mg) | $2.5 \pm 0.2$ | $56 \pm 8\%$ |
| $Na^+$-dependent ADP-ATP exchange (nmol/mg/h) | $94 \pm 15$ | $112 \pm 11\%$ |
| $Na^+$-dependent EP from ATP (nmol/mg) | $2.67 \pm 0.18$ | $92 \pm 6\%$ |

a. Values are Mean$\pm$SEM (average of 4 experiments).
b. Measured as described before (1) at 0.2 mM $Rb^+$.

Properties of the EP of the "half-oxidized" enzyme differed from those of the native EP. This difference was more easily evident, however, when a fully oxidized enzyme was used. Based on the data of Fig. 1, a preparation whose ATPase activity was inhibited by more than 95% was obtained, and its EP was compared with the native EP at 0°C. As the data of Fig. 4 show, the oxidized EP was more stable than the native EP; and the oxidized EP was equally sensitive to $K^+$ and ADP, while the native EP was $K^+$-sensitive and ADP-insensitive.

The implications of these findings are in two areas: First, if the partially oxidized pump of the intact cell does indeed have a lower $K_m$ for ATP as the data of Fig. 2 suggest, this may be an excellent defense mechanism which would allow the post-ischemic tissue to cope with the shortage of ATP. Second, since it is possible to cause partial oxidation of the enzyme inadvertently in the normal course of purification, solubilization, reconstitution, and storage of the enzyme, the data of Fig. 4 warn us that contamination with oxidized enzyme may complicate the interpretation of studies on the reaction mechanism of the enzyme. The dangers of such contamination are increased when preparations are used that contain significant amounts of $\alpha_2$ and $\alpha_3$ isoforms of the catalytic subunit, because these isoforms are more susceptible to oxidation than $\alpha_1$ isoforms (3).

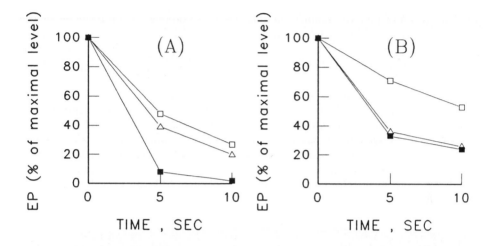

Figure 4. Effects of $K^+$ and ADP on EP decomposition in the native (A) and the oxidized (B) enzymes. Phosphorylation was initiated at 0°C by the addition of 50 $\mu$M ATP to reaction mixtures containing 60 $\mu$g/ml enzyme protein, 100 mM $Na^+$, 1 mM $Mg^{++}$. After 30 sec, CDTA ($\square$), CDTA + $K^+$ ($\blacksquare$), or CDTA + ADP ($\triangle$) were added before reactions were terminated 5 sec or 10 sec later. CDTA, 20 mM; $K^+$, 10 mM; ADP, 2 mM.

Acknowledgement

This work was supported by NIH grant HL-45554 awarded by National Heart, Lung, and Blood Institute, United States Public Health Service/DHHS.

References

1.  Hasenauer J, Huang W.-H, Askari A (1993) Allosteric regulation of the access channels to the $Rb^+$ occlusion sites of $(Na^+ + K^+)$-ATPase. J Biol Chem 268:3289-3297
2.  Huang W.-H, Wang Y, Askari A (1992) $(Na^+ + K^+)$-ATPase: Inactivation and degradation induced by oxygen radicals. Int J Biochem 24:621-626
3.  Huang W.-H, Wang Y, Askari A, Zolotarjova N, Ganjeizadeh M (1993) Different sensitivities of the $Na^+/K^+$-ATPase isoforms to oxidants. Biochim Biophys Acta in press
4.  Xie Z, Wang Y, Askari A, Huang W.-H, Klaunig JE, Askari A (1990) Studies on the specificity of the effects of oxygen metabolites on cardiac sodium pump. J Mol Cell Cardiol 22:911-920

# Modification of Na+/K+-ATPase in a Model of Epilepsy in the Rat

W.R. Anderson, W.L. Stahl, A.A. Maki, and T.J. Eakin,
VA Medical Center and University of Washington School of Medicine, Seattle, WA

Introduction

The Na+/K+-ATPase (Na+ pump) is a transmembrane protein crucial for maintenance of ion homeostasis within cells(5). This enzyme establishes electrochemical gradients that are used by cells for a variety of functions, including synaptic activity within neurons. Due to this involvement in synaptic activity the Na+ pump has been extensively studied in connection with epilepsy research. Results have not, however, provided a clear understanding of how the Na+ pump might be affected by, or contribute to, epileptiform activity. Utilizing a kainic acid model of temporal lobe epilepsy we have further investigated this question. With bilateral, intracerebroventricular injections of kainic acid we have examined pump alterations at 1, 2, and 3 weeks after kainate injection in hippocampal neurons from kainate-lesioned rats showing seizure activity. Our goal was to identify alterations in the Na+ pump, in either protein isozymes levels or subunit isoform mRNA levels, in tissue associated with the development of an epileptic focus.

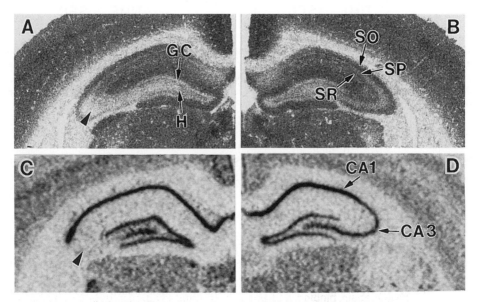

Figure 1. Na+ pump in rat hippocampus. High affinity [3H]ouabain binding in rat brain shown in A (kainate-lesioned) and B (control). Binding carried out in 70 nM [3H]ouabain (1,4) and exposed to hyperfilm for 4 days. Expression of α3 mRNA in rat brain shown in C (2 weeks after kainate lesioning) and D (control). In situ hybridization (2,6) carried out on brain tissue sections with an [35S]-labeled, 30mer oligonucleotide probe, dipped in photographic emulsion, and exposed for 6 days. Arrowheads indicate lesion sites. GC, granule cells; H, hilus; SO, stratum oriens; SP, stratum pyramidalae; SR, stratum radiatum.

## Results and Discussion

Ouabain is one member of a family of cardiac glycosides which bind specifically to the $Na^+$ pump(5) Pump isozymes bind ouabain with varying affinities which are dictated by their subunit composition, with certain $\alpha$ subunit isoforms associated with high affinity binding and others associated with lower affinity binding. We differentiated and quantitated pump isozyme levels on the basis of their [$^3$H]ouabain binding characteristics using previously developed quantitative autoradiographic methods(1,4,6) With [$^{35}$S]-tailed oligonucleotide probes(2) we also made relative determinations of mRNA levels in control and kainate-treated tissue for the 3 different $\alpha$ and 2 different $\beta$ subunit isoforms.

Subsequent to kainate lesioning the hippocampus undergoes extensive restructuring(3). Reactive gliosis, axon sprouting, and synaptic reorganization are all present in the weeks following kainic acid injection. Tissue was examined at 1, 2, and 3 weeks after kainate injection to identify any effects not apparent at a single time point and to monitor development of any kainate effects on hippocampal $Na^+/K^+$-ATPase.

Table 1.  High Affinity [$^3$H]Ouabain Binding levels in rat brain

$B_{max}$ (fmol/mm$^2$)

| Region | Control | Kainate-lesioned |
|---|---|---|
| Dentate Gyrus | | |
|    Hilus | $175.8 \pm 27.6$ | $133.9 \pm 9.4$ |
|    Granule cells | $82.5 \pm 11.6$ | $58.8 \pm 5.3$ |
| CA3, Str. lucidum | $105.0 \pm 9.2$ | $30.9 \pm 5.3^*$ |
| CA1, Str. pyramidalae | $102.5 \pm 11.4$ | $68.4 \pm 10.1^*$ |
| CA1, Str. oriens | $140.9 \pm 44.1$ | $111.4 \pm 12.5$ |
| CA1, Str. radiatum | $174.2 \pm 15.1$ | $129.5 \pm 9.3^*$ |

*Significant difference from control at 95% confidence level or higher.
[$^3$H]Ouabain binding and analysis of autoradiograms were carried out as previously described (1) using 12 μm thick, coronal tissue sections. 5-200 nM [$^3$H]Ouabain (5-200nM) was used for autoradiography of high affinity ouabain binding sites. The data were obtained from 9 rats with recovery times of 1-3 weeks after bilateral injection of kainic acid and from 9 saline-injected control rats.

Results are summarized and presented in Tables 1 and 2. The CA3 region of hippocampus is quite sensitive to intraventricular kainic acid injection and shows extensive loss of neurons in this model (Fig. 1). All components of ouabain binding in CA3 subregions were severely decreased, with losses approaching 70% in some areas. These results are expected as the loss of neurons in this area includes the loss of their associated pump molecules.

The hilar region and granular cell layer of the dentate gyrus showed no cell loss (data not shown) and no statistical differences between control and kainate-treated tissue in ouabain binding levels. This indicates no change in the levels of pump isozymes in this region and is consistent with its undamaged appearance after kainic acid injection.

The CA1 region of hippocampus is also well-preserved morphologically in this model but showed significant kainate effects. Stratum pyramidale and stratum radiatum both showed substantial decreases of 25-33% in high affinity ouabain binding . No changes were evident in low affinity ouabain binding (data not shown). Examination of tissue at 1, 2, and 3 weeks after kainate injection showed that the lesion development and deficit in high affinity ouabain binding was virtually complete at 1 week post injection and continued through week 3.

Table 2. Quantitation of relative mRNA levels versus recovery time for $Na^+/K^+$-ATPase $\alpha 3$ isoform in CA1 pyramidal cell layer.

Grain Counts ( $\pm$S.E.M.)

| Week | Control | Kainate-lesioned |
|------|---------|------------------|
| 1 | $37.1 \pm 2.3$ (n=127) | $43.9 \pm 1.6$ (n=218) |
| 2 | $60.1 \pm 4.0$ (n=109) | $40.3 \pm 2.0$* (n=110) |
| 3 | $23.5 \pm 1.4$ (n=83) | $26.0 \pm 1.2$ (n=163) |

*Significant difference from control at 99% confidence level or higher.
Grain counts in CA1 pyramidal cells from hippocampus. The probe used was specific for the $\alpha 3$ subunit isoform. All slides dipped in photographic emulsion and developed after 9 days. Values are mean grains per cell standardized to unit area. Values for n indicate number of cells sampled. Data taken from 4 kainate-lesioned and 4 control animals at each time point.

Relative mRNA levels between pump subunit isoforms in CA1 pyramidal cells were compared in control and kainate-treated tissue in an attempt to account for the loss of high affinity ouabain binding seen in CA1 (Table 1). CA1 pyramidal cells showed decreases in $\alpha 3$ mRNA levels of 26-50% at 2 weeks after kainic acid injection (Table 2). No changes were seen in other $\alpha$ or $\beta$ subunit isoform mRNA levels. Analysis of

886

kainate-treated tissue showed no differences from control tissue at either 1 or 3 week recovery periods (data not shown).

The relationship between the deficit in high affinity ouabain binding and the decreased α3 mRNA levels is not immediately apparent. The α3 subunit is associated with high affinity ouabain binding in rat hippocampus and lower mRNA levels might result in synthesis of fewer high affinity pump isozymes. This may in turn lead to the observed decrease in high affinity ouabain binding.

The time courses, however, between the two kainate effects are inconsistent. The ouabain binding deficit in CA1 subregions is fully developed when examined 1 week after kainate injection, while the decrease in α3 subunit mRNA levels is not seen until 2 weeks after kainate injection. If the decrease in pump isozyme levels were the direct result of a decrease in mRNA levels, the lower mRNA levels should be present before the protein levels fall. This was not seen. This inconsistency appears to indicate that the two kainate effects may not be directly related.

The factors underlying the transient decrease in α3 mRNA levels in CA1 pyramidal cells are also unclear. In response to the kainate lesion the hippocampus is, however, undergoing extensive restructuring during the time periods examined. While strictly speculative it is possible that the decrease in mRNA levels required 2 weeks to reach a measurable level but then was compensated for at 3 weeks by some of the ongoing responses to injury.

In conclusion the high affinity component of ouabain binding in hippocampal CA1 subregions was substantially decreased at all post-lesion times studied. mRNA levels for one subunit isoform, α3, were significantly decreased in CA1 pyramidal cells at 2 weeks after injection of kainic acid. Mechanisms to account for the observed results are unclear but the reduced pump in CA1 neurons may lead to modified ion homeostasis and the hyperexcitability seen in this model of epilepsy.

Acknowledgements

This research was supported in part by NIH grant NS 20482 and by the Medical Research Service of the Department of Veterans Affairs.

References

1. Antonelli MC, Baskin DG, Garland M, Stahl WL (1989) Localization and characterization of binding sites with high affinity for [$^3$H]ouabain in cerebral cortex of rabbit brain using quantitative autoradiography. J Neurochem 52:193-200
2. Filuk PE, Miller MA, Dorsa DM, Stahl WL (1989) Localization of messenger RNA encoding isoforms of the catalytic subunit of the Na,K-ATPase in rat brain by in situ hybridization histochemistry. Neurosci Res Commun 5:155-162.
3. Franck JE, Kunkel DD, Baskin DG, Schwartzkroin PA (1988) Inhibition in kainate-lesioned hyperexcitable hippocampi: physiologic, autoradiographic, and immunocytochemical observations. J Neurosci 8:1991-2002.
4. Maki AA, Baskin DG, Stahl WL (1992) [$^3$H]-Ouabain binding sites in rat brain: distribution and properties assessed by quantitative autoradiography. J Histochem Cytochem 40:771-779.
5. Stahl WL (1986) The Na,K-ATPase of nervous tissue. Neurochem Int 8:449-476.
6. Stahl WL, Baskin DG (1990) Histochemistry of ATPases. J Histochem Cytochem 38:1099-1122

# Tumor development and changes in $Na^+/K^+$-pump parameters

M.Zilmer, A.Kengsepp, K.Zilmer, C.Kairane, T.Talpsep[1]

Department of Biochemistry, [1]Institute of Chemical Physics,
Tartu University, Jakobi Str. EE2400, Tartu, Estonia

## Introduction

A definite elevation of intracellular $Na^+$ and $K^+$ ratio ($Na^i/K^i$) is associated with intense mitosis and growth stimulation. The $Na^i/K^i$ is increased also in the case of carcinogenesis. For maintenance suitable $Na^i/K^i$, in any case, the crucial role has $Na^+/K^+$-ATPase. Therefore, the examination of parameters (functionality) of this enzyme, isolated from different sources (tumors, oncoprotein transformed cells, carcinogen-affected pretumorated tissue), is important relevant to role of the $Na^+/K^+$-pump in development of tumor.

## Parameters of the $Na^+,K^+$-ATPase from different sources

Potassium bromate (a new renal carcinogen) induces after a single dose oxidative DNA damage (8-OH-deoxyguanosine) in rat kidney (2). The maximum level of 8-OH-dG ( at 24-48 h after single administration of potassium bromate) is followed by its elimination (3). In this work we compare (Table 1) the parameters of the $Na^+,K^+$-ATPase from rat normal kidney with our data on kidney after a single i.p. administration of $KBrO_3$ (practically nontumorated object), on human brain (HB), human brain tumor (glioma,astrocytoma; HBT) and PVO enzyme (5,6). PVO are the fibroblasts-like C127 cells transformed by oncoprotein E5 (5).

Our main results (Table 1) are as follows: 1) Unlike to normal objects ( HB, C127), the enzyme from HBT and PVO did not reveal cooperative $Na^+$ binding and its physiological efficiency ($V/K_{0.5}$) for $Na^+$ was diminished. At the same time no significant changes were established for $K^+$. 2) A new break-area on Arrhenius plot appears for enzyme preparation in the case of HBT and PVO. 3) Unlike to HBT and PVO the physiological efficiency of enzyme for $Na^+$ increased after $KBrO_3$-treatment (the affinity for sodium is increased and a definite transition to Na-form occurs - $Na^+/K^+$ is altered; see Tabel 1). 4) Potassium bromate treatment increases significantly lipid peroxidation (LP) and the LP degree differs in normal objects and in tumorated objects.

Our data allow to suggest that, evidently, the $Na^+/K^+$-pump undergoes functional changes underlaying development (growth) tumor via maintenance of suitable $Na^i/K^i$. May be there occur the definite local conformational changes (LCC) which alter the enzyme functionality as with respect to thermoinactivation, tryptophan residues quenching (with $Na^+$), accessible number of SH-groups, protection effect of ATP at different temperatures against the SH-groups blocation the HB and HBT enzymes differ (6). The unlike TBARS content and a new break-area on Arrhenius plot (Table 1) refer

888

to fact, that also altered lipid-protein interactions underlay enzyme LCC. On the one hand, the latter is supported by data that the LP degree in normal objects and tumorated objects significantly differs and potassium bromate increases LP in kidney (Table 1). On the other hand, the several free radicals (derived also from lipids) are second messengers and mutagenes (4).

## Table 1. The Na$^+$/K$^+$-ATPase from different sources

| | Pathological objects | | | Normal objects | | |
|---|---|---|---|---|---|---|
| | Kidney[g] | HBT | PVO | Kidney | HB | C127 |
| Activity[f] | 30 | 27 | 1 | 28 | 60 | 1.1 |
| Sodium[b]: | | | | | | |
| $K_{0.5}$(mM) | 8.4[a] | 7.5 | 4.7 | 14.3 | 9.7 | 4.9 |
| | ±0.7 | ±0.8 | ±0.5 | ±1.6 | ±1.2 | ±0.6 |
| $n_H$ | 1.3[a] | 1.0[a] | 0.9[a] | 2.0 | 1.7 | 1.2 |
| | ±0.03 | ±0.04 | ±0.03 | ±0.06 | ±0.05 | ±0.04 |
| $V_M/K_{0.5}$ | 0.44[a] | 0.40[a] | 0.09[a] | 0.29 | 0.50 | 0.14 |
| Potassium[b]: | | | | | | |
| $K_{0.5}$(mM) | 1.6 | 3.4 | 1.7 | 1.5 | 2.9 | 2.0 |
| | ±0.1 | ±0.4 | ±0.3 | ±0.1 | ±0.3 | ±0.4 |
| $n_H$ | 1.6 | 1.3 | 1.2 | 1.7 | 1.3 | 1.3 |
| | ±0.04 | ±0.05 | ±0.04 | ±0.04 | ±0.05 | ±0.05 |
| $V_M/K_{0.5}$ | 2.62 | 1.24 | 0.33 | 2.67 | 1.25 | 0.29 |
| Na$^+$/K$^{+c}$ | 100/50 | 130/20 | 130/20 | 130/20 | 130/20 | 130/20 |
| TBARS[d] | 203[a] | 0.9[a] | 35[a] | 116 | 21 | 120 |
| | ±19 | ±0.1 | ±4 | ±14 | ±0.2 | ±10 |
| Breakarea[e] | yes | yes | yes | no | no | no |

[a]p<0.05 vs normal objects; [b]parameters were calculated via Hill's plot (6); [c]from Na$^+$/K$^+$ ratio (mM) curve; [d]TBA-reactive substances: nmol/g wet tissue(kidney); nmol/mg protein (HB,HBT); nmol/10$^9$ cells (C127,PVO); [e]at 26-30$^0$ (via Arrhenius plot); [f]µmol P$_i$/mg/h; [g]KBrO$_3$ (80mg/kg,after 48h)

It is very interesting, that some parameters of sodium pump alter already after a single administration of $KBrO_3$ as carcinogen. Evidently, such changes of "Na-parameters" (Table 1) - increased affinity, a definite transition to Na-form ($Na^+/K^+$ is alterated), increased physiological efficiency for sodium in the case of practically nontumorated tissue are protective (adaptative). This view is supported by several facts: 1) only prolonged (2 yrs) influence of $KBrO_3$ induces renal tumor (2), but its single effect disappears; 2) the $Na^+/K^+$-pump's kinetics changes occur in premalignant mucosa months before gross tumors develop (1); 3) the content of GSH increases ($p<0.05$) in kidney after $KBrO_3$ effect (our unpublished data).

### References

1. Davies RJ, Sandle GJ, Thompson SM (1991) Inhibition of the Na/K-ATPase pump during induc tion of experimental colon cancer. Cancer Biochem Biophys 12: 81-94
2. Kurokawa Y, Maekawa A, Takahashi M, Hayashi Y (1990) Toxicity and carcinogenity of potassium bromate - a new renal carcinogen. Environmental Health Perspectives 87: 309-355
3. Sai K, Takagi A, Umemura T, Hasegawa R, Kurokawa (1991) Relation of 8-hydro-xydeoxyguanosine formation in rat kidney to lipid peroxidation, glutathione level and relative organ weight after a single administration of potassium bromate. Jpn J Cancer Res 82:165-169
4. Schreck R,Bayerle P(1991) A role of oxygen radicals as second messengers. Trends Cell Biology 1:39-42
5. Talpsepp T, Soosaar A, Ustav M, Kask R, Zilmer M (1992) The Na/K-ATPase activity change in BPV-1 transformed cells. 11th Int. Papillomavirus Workshop, University of Edinburgh, p.183
6. Zilmer M, Tähepõld L, Salum T, Sillard R, Kask R (1991) The role of conformational changes in functioning of brain Na,K-ATPase. Cytology 33: 55-60

# Distinct Interaction of Mercury and Silver with Isolated Na$^+$/K$^+$-ATPase

M. S. Hussain, C. Burrus, B. M. Anner

The Laboratory of Experimental Cell Therapeutics, Geneva University Medical Centre, CH-1211 Geneva 4, Switzerland

## Introduction.

The Na$^+$/K$^+$-ATPase (NKA; EC 3.6.1.37) or sodium pump is a vital ubiquitous membrane system regulating the ionic balance across the cell membrane and carrying the receptor for cardioactive steroids; the binding of these compounds to the receptor results in the inhibition of all ATPase and ion pump activities (7). We have shown in our previous studies that mercury binds to NKA so tightly that the universal chelating agent for metals, for instance, EDTA could not prevent mercury inhibition (2). In the present work we show how silver inhibits isolated NKA and how this effect can be modulated by free cysteine. The possible mechanism of on-off mechanism of silver inhibition of NKA is discussed in view of a possibly significant role of metal binding sites of NKA.

### Distinct interaction of NKA with mercury and silver

NKA was purified from the outer medulla of sheep kidneys by a dodecyl sulphate extraction microprocedure adapted from Dzhandzhugazyan and Jørgensen (4). Isolated NKA was incubated with various concentrations of silver alone or in presence of 1 mM cysteine. The NKA activity was measured by the enzyme-linked assay at 37°C as described in legend to Fig. 1. The upper line of the Fig. 1A shows enzyme activity measured continuously before and after the addition of silver. The sudden break in line reveals a rapid and potent inhibition of NKA without a latency period. This suggest that inhibition of NKA activity by silver is strikingly different from other metals. Total inhibition of NKA was obtained at 100 nM with an IC$_{50}$ value of 10 nM (5)

The lower line of Fig. 1A illustrate the enzyme activity measured before and after addition of mercury. The slope of the curve after mercury addition suggest a delayed inhibition. In contrast to silver, mercury showed a latency period prior to inhibition of the enzyme. However, the precise mechanism for delayed mercury inhibition is not yet known. The control experiments were carried out to make sure that silver was not inhibiting the pyruvate-kinase and lactate-dehydrogenase enzyme actvities of the linked-enzyme system used to measure the activity (Fig. 1B).

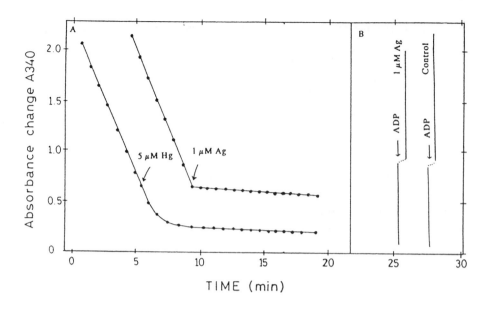

**Figure 1. A.** Inhibition of NKA activity by mercury and silver. One $\mu$g of NKA protein was added to the enzyme solution containing (in mM) 0.3 NADH, 2.5 phosphoenol pyruvate, 30 imidazole, 1 Tris-EDTA, 2.5 ATP, 5 MgCl$_2$, 100 NaCl and 10 KCl, as well as 8 $\mu$l pyruvate-kinase/lactate-dehydrogenase suspension, pH 7.2. the rate of NADH oxidation was recorded at 340 nm in the automated enzyme kinetic accessory of a Philips Unicam SP 1800 spectrophotometer. Experimental points show absorbency change measured at 1-min intervals At 5 min mercury was added and at 10 min silver to the other cuvette **B.** Control experiments showing that silver does not effect the pyruvate-kinase/lactate-dehydrogenase linked-enzyme system used to asses NKA activity by adding 25 $\mu$M ADP to the solution. Mercury effect on the linked-enzyme system has been ruled out previously (3).

Addition of cysteine (1 mM) to NKA before the metals protects from silver and mercury inhibition. Fig. 2 illustrates the protection from silver and mercury inhibition by addition of cysteine to the enzyme. On the other hand if cysteine is added to NKA after addition of silver or mercury the reversibility pattern of NKA was found to be totally different: for instance, addition of 1 mM cysteine reactivated the silver-blocked NKA activity fully whereas mercury remained partially bound (3). In our earlier publications we described in details the mercury interaction with NKA and its possible molecular mechanism and side of action (1,6).

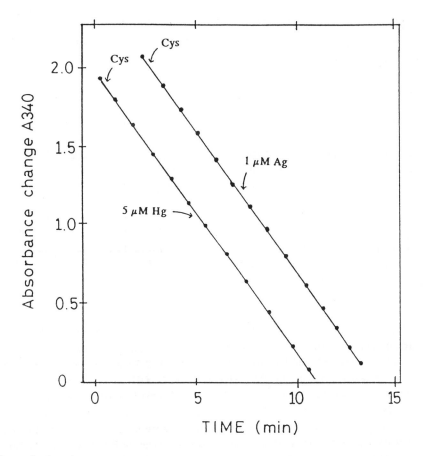

**Figure 2**. Cysteine protection from silver and mercury inhibition; 1 mM cysteine was added before adding silver or mercury to NKA; enzyme activity was measured as described in legend to Fig. 1.

In summary, our results show that silver inhibits NKA by an on-off mechanism whereas inactivation by mercury involves a latency period. Inhibition of NKA by silver or mercury is protected by cysteine pre-addition, e.g., the metals are nor transferred from free cysteine to critical cysteines of the NKA protein. Free cysteine is then able to extract all bound silver form the NKA, restoring full ATPase activity (5) , whereas extraction of mercury from NKA by cysteine is only partial, indicating that mercury has become occluded by the protein, presumably hold by co-operatively acting covalent and co-ordinative cysteine bonds. Taken together these results indicate that silver has a different mechanism of binding to NKA cysteines when compared to mercury. We suppose that silver acts directly on a SH-group critical for active transport whereas mercury action either requires interaction with several SH-groups, some of them buried, or induction of a slow progressive conformational change underlying ATPase inhibition.

*Acknowledgements*

We are grateful to Mrs M. Moosmayer for help with enzyme purification and inhibition studies and to Mrs D. Lacotte for repeating and reproducing experiments. The work was supported by the Swiss National Science Foundation (grant No. 31-25666.88) and the Swiss Public Health Service (grant No. 31-30317.90).

# References

1.  Anner BM, Moosmayer M (1992) Mercury inhibits $Na^+/K^+$-ATPase primarily at the cytoplasmic side. Am J Physiol 262 (Renal Fluid Electrolyte Physiol 31) F843-F848

2.  Anner BM, Moosmayer M, Imesch E (1990) Chelation of mercury by ouabain-sensitive and ouabain-resistant renal $Na^+/K^+$-ATPase. Biochem Biophys Res Commun 167: 1115-1121

3.  Anner BM, Moosmayer M, Imesch E (1992) Mercury blocks $Na^+/K^+$-ATPase by a ligand-dependent and reversible mechanism. Am J Physiol 262 (Renal Fluid Electrolyte Physiol 31) F-830-F8361

4.  Dzhandzhugazyan KN and Jørgensen PL (1985) Asymmetric orientation of amino groups in the alpha-subunit and the beta-subunit of $Na^+/K^+$-ATPase in tight right-side-out vesicles of basolateral membranes from outer medulla. Biochim Biophys Acta 817: 165-173

5.  Hussain S, Anner RM, Anner BM (1992) Cysteine protects $Na^+/K^+$-ATPase and isolated human lymphocytes from silver toxicity. Biochem Biophys Res Commun 189: 1444-1449

6.  Imesch E, Moosmayer M, Anner BM (1992) Mercury weakens membrane anchoring of $Na^+/K^+$-ATPase. Am J Physiol 262 (Renal Fluid Electrolyte Physiol 31) F-837-F842

7.  Schwartz A, Lindenmayer GE, Allen JC (1975) The Sodium-potassium adenosine triphophatase: pharmacological and biochemical aspects. Pharmacol Rev 27: 3-134

894

# Author index

## E

Eakin, T. J.  884
Eakle, K. A.  11, 206
Efendiyev, R. E.  370
Efthymiadis, A.  482
Ellis-Davies, G. C. R.  321
Elsner, S.  482, 553
Emerick, M.  254
Engelmann, B.  694
Erdmann, E.  791
Esmann, M.  605
Ezzaher, A.  828

## F

Faller, L. D.  593
Fambrough, D. M.  254
Farley, R. A  11, 206
Farrell, P. A.  868
Fedorova, O. V.  787
Fedosova, N. U.  561, 565
Fendler, K.  495
Fink, D. J.  226
Fisone, G.  662
Foersom, V.  70
Forbush III, B.  433
Fortes P. A. G.  649
Frank, U.  625
Friehs, S.  214
Froehlich, J. P.  441, 495
Fryckstedt, J.  662
Fukui T.  641
Furukawa, T.  541

## G

Gadsby D. C.  472
Ganjeizadeh, M.  264, 350
Garcia, E.  783
Garrahan, P. J.  425, 429
Gati, I.  759
Gatto, C.  609
Geering, K.  61, 200, 682
Georg, H.  135
Gerbi, A.  820, 824, 828
Gersch, B.  621
Gevondyan, N. M.  374
Ghione, S.  755
Glitsch, H. G.  533, 537
Gloor, S.  203
Goldshleger, R.  309, 397
González-Martínez, L. M.  218
Götz, E.  625
Graves, S. W.  771
Green, N. M.  110, 120
Greengard, P.  662
Grell, E.  569, 617, 621, 625, 629
Grindstaff, K. K.  74

Grupp, I. L.  718
Grupp, G.  718
Guidotti, G.  670
Guitierrez, C.  783

## H

Habermann, E.  385
Hagiwara, K.  641
Hahnen, J.  377, 421
Hamer, E.  332
Hamlyn, J. M.  722
Hamrick, M.  254
Hansen, O.  139
Hara, Y.  537
Harris, T. M.  743
Hasselbalch, S.  868
Haupert Jr., G. T.  732, 848
Hayashi, H.  66
Hayashi, Y.  453, 710
Hazon, N.  57, 246
Heidrich, J.  852
Hemmings Jr., H. C.  662
Hensley, C. B.  170
Hermans, A. N.  537
Hildebrandt, G.  779
Hilgemann, D. W.  507
Hirakawa, K.  541
Hiraoka, M.  541
Hobbs, A. S.  441
Hoffmann, O.  779
Hollenberg, N. K.  771
Holmgren, M.  545
Holtbäck, U.  662
Homareda, H.  362, 710
Horisberger, J.-D.  549
Houda Bouanani, N. el  828
Hoving, S.  309
Huang, W.-H.  417, 690, 880
Hundal, H.  864
Hussain, M. S.  891
Hwang, B.  254

## I

Ikawa, Y.  541
Ikeda, K.  1, 33
Inagami, T.  743, 763
Inoue, S.  581
Inoue, K.  698
Inoue, N.  710
Ishii, T  86, 264
Ito, Y.  541
Izumi, F.  210

## J

Jack-Hays, M.  690
Jaisser, F.  61, 549, 682

896

897

899

# Subject index

901

903

## DATE DUE